ANALYSIS OF BIOLOGICAL DEVELOPMENT

Klaus Kalthoff
The University of Texas at Austin

McGRAW-HILL, INC.
New York   St. Louis   San Francisco   Auckland   Bogotá   Caracas
Lisbon   London   Madrid   Mexico City   Milan   Montreal   New Delhi
San Juan   Singapore   Sydney   Tokyo   Toronto

Cover Illustration

Homeobox genes control development in animals as different as fruit flies and mice. These genes, represented here by colored boxes, occur in clusters on certain chromosomes; their hallmark is the homeobox, a stretch of DNA that has been conserved in evolution for more than 500 million years. Each homeobox gene is expressed in a certain domain within the epidermis of the fly and the central nervous system of the mouse. The location of a homeobox gene on a chromosome corresponds to where it is expressed in the body: the farther downstream it is located on the chromosome the more anterior it is expressed in the body. By regulating the activity of many other genes, homeobox genes control the morphological characteristics that develop in their expression domains. Thus, the antero-posterior body patterns of flies and mice are specified by nearly identical sets of regulatory genes.

ANALYSIS OF BIOLOGICAL DEVELOPMENT

1 2 3 4 5 6 7 8 9 0 VNH VNH 9 0 9 8 7 6 5

ISBN 0-07-033308-4

This book was set in Palatino by York Graphic Services, Inc.
The editors were Kathi M. Prancan, Judith Kromm, and Holly Gordon;
the art editor was Janet Young;
the design was done by Keithley and Associates, Inc.;
text illustrations were done by Linda McVay and Fine Line Illustrations, Inc.;
cover illustration by Patricia Wynne;
the production supervisor was Richard A. Ausburn.
The photo editor was Safra Nimrod;
the photo researchers were Ondine Cleaver, Flora Love, and Mira Schachne.
Von Hoffmann Press, Inc., was printer and binder.

Library of Congress Cataloging-in-Publication Data

Kalthoff, Klaus.
    Analysis of biological development / Klaus Kalthoff.
        p.   cm.
    Includes bibliographical references and index.
    ISBN 0-07-033308-4
    1. Developmental biology.   I. Title.
QH491.K35   1996
574.3—dc20   95-896

To My Teachers, Colleagues, and Students

ABOUT THE AUTHOR

Klaus Kalthoff is Professor of Zoology at The University of Texas at Austin. Born and raised in Germany, he studied biology at the universities of Erlangen, Hamburg, and Freiburg. He received his Ph.D. degree (Dr. rer. nat.) in 1971 from the University of Freiburg, where he completed his dissertation on pattern formation in insect embryos under the direction of Klaus Sander. Subsequent work done by Kalthoff and coworkers, also at Freiburg, showed that eggs of midges contain ribonucleoprotein particles acting as anterior determinants: they are localized near the anterior pole of the egg and are necessary for forming anterior body parts. Kalthoff also discovered that damage from ultraviolet light to certain insect eggs is reversible in a catalyzed reaction that depends on light of longer wavelength.

Kalthoff was awarded the Prize of the Scientific Society of Freiburg in 1975. He moved to Austin in 1978, where he and his coworkers characterized the RNA moiety of the anterior determinants in *Chironomus samoensis* as small, polyadenylated, and cytoplasmic. He is author or coauthor of numerous articles in scientific books and journals, including *Development, Developmental Biology, Nature, Photochemistry and Photobiology,* and *Proceedings of the National Academy of Sciences of the U.S.A.* Kalthoff teaches courses in developmental biology and human biology.

CONTENTS IN BRIEF

CONTENTS

Developmental biology is both a classical and a modern field. Beginning with Wilhelm Roux and Hans Driesch late in the nineteenth century, a new generation of biologists ventured into the causal analysis of development. With hand-made glass needles and hair loops, they removed and transplanted parts of frog and sea urchin embryos to see what the parts would do in isolation and how they would interact with cells that were not their normal neighbors. By defining a new set of terms based on operational criteria, these scientists established experimental embryology as a new discipline. In 1924, Hans Spemann and Hilde Mangold published their famous "organizer" experiment, which showed that a transplanted piece of dorsal blastopore lip could induce surrounding host tissue to form a secondary embryo. This powerful demonstration epitomizes the first golden age of developmental biology.

Developmental biologists with a bent toward genetics soon found an alternative way of analyzing development. Studying mutant strains of various creatures, they began to uncover the logic of the genetic networks that control development. Those working with the fruit fly *Drosophila* learned that mutations in certain genes had dramatic effects on the overall body pattern, such as the replacement of certain body parts with parts that are normally formed elsewhere. This type of analysis was made more powerful by the introduction of mutagenesis screens, which can identify virtually all genes involved in the control of a developmental process. In the 1980s, genetic analysis was boosted again by the arrival of DNA cloning; now, the proteins encoded by patterning genes could be characterized in molecular terms. The prospect of knowing not only the mutant phenotypes of these genes, but also the biochemical properties of their products, energized hundreds of researchers. Today, the development of the *Drosophila* embryo is almost completely understood in terms of a network of gene activities that unfolds in a spatial order. The spectacular success of this work is being emulated by researchers working with other organisms. Developmental biology has entered another golden age.

Living in this second golden age is exhilarating. New and important discoveries are made every week. Just keeping up with the ever-growing literature has become a major effort, and so an undergraduate course in developmental biology is now a challenge for students and instructors alike. On the one hand, the classical foundations of development need to be preserved. On the other hand, the current excitement in developmental biology comes from bringing new and powerful tools to bear on some of the long-standing problems in the field. My answer to the challenge is this textbook, which is written for advanced undergraduates and beginning graduate students. I hope it will help them and

their teachers to appreciate the rich heritage left by the classical development biologists and to savor the power and elegance of the new work.

The organization of the book reflects the union of classical and modern ways of analysis in contemporary developmental biology. Part One emphasizes classical methods of analysis and covers the series of embryonic stages from gametogenesis to histogenesis. Interspersed are basic conceptual topics such as nuclear totipotency, cell determination, cytoplasmic localization, induction, and morphogenesis. Part Two introduces the genetic and molecular analysis of development, beginning with a chapter on the use of mutants, DNA cloning, and transgenic organisms. Subsequent chapters explain the concept of differential gene expression with the goal of understanding how the genomic information is used to build a three-dimensional organism that unfolds in time. A chapter on paragenetic information provides a counterpoint to the emphasis on genetic information. Part Three freely combines classical and modern types of analysis and should be the most enjoyable portion of the book. It illustrates how the application of new research tools has led to a better understanding of long-standing issues in development. This part features a set of chapters on pattern formation, one of the central topics in developmental biology. Cell differentiation, sex determination, hormonal control, growth, and the roles of cell adhesion and extra cellular materials in morphogenesis round out what should be a fair representation of contemporary developmental biology.

Facing a plethora of old and new results, I found it important to bring out about a dozen general principles. For instance, many steps in development rely on synergistic mechanisms that complement or reinforce each other. Spemann referred to this as the principle of double insurance; I introduce this principle in the context of fertilization and then take it up again in chapters on induction and genetic control. Other principles, including stepwise approximation and default programs, are treated in a similar fashion.

The textbook in developmental biology that I used as a student was Alfred Kühn's *Entwicklungsphysiologie*. I have adopted Kühn's habit of discussing key experiments in some detail. When introducing transgenic organisms, I describe the germ line transformation of *rosy* mutant fruit flies with the wild-type transgene. This description makes it necessary to explain the method of Southern blotting. Experiment descriptions are marked with colored bars, and explanations of methods are boxed. I hope that readers will find these sections especially worthwhile. Including experiment descriptions while keeping chapters to a manageable length has made it necessary to present materials selectively. In most chapters, one or two subtopics are discussed in more depth than the others, and these subtopics are marked with color in the list of contents for the chapter.

On various occasions throughout the text, I point out problems whose solutions remain elusive. I hope this will convey the spirit of science as an endeavor in which answers bear new questions and in which one generation passes the torch on to the next.

Acknowledgments

I feel grateful to some special people who provided inspiration and support for writing this text. My mentor, Klaus Sander, long ago introduced me to the history and culture of developmental biology. When I came to The University of Texas at Austin, I met a congenial group of colleagues including Gary Freeman, Antone Jacobson, Bill Jeffery, the late Stephen Meier, and Matt Winkler. In countless journal club sessions and lunch conversations, they broadened my outlook on developmental biology and its practitioners. My wife, Karin, and sons, Christian, Ulrich, and Philipp, gave me the freedom and peace of mind to spend long hours in front of journals and computers screens.

Carrying this text the long way from first draft to bound book has required the help of many people, to whom I express my appreciation. First and foremost, I thank Kathi Prancan, sponsoring editor at McGraw-Hill, for skillfully guiding this complex project to completion, and for going several extra miles in doing so. Judith Kromm read my first draft and nudged me from writing review articles to writing textbook chapters. Richard K. Mickey expertly copyedited the final manuscript. Janet Young supervised the artwork, and Linda McVay prepared the hand-drawn illustrations. Safra Nimrod, Mira Schachne, Ondine Cleaver, and Flora Love helped to assemble the photographs that speak better than words. Many colleagues were very diligent and courteous in providing their prints or negatives. Keithley and Associates designed the cover and text. Holly Gordon assembled all parts of the manuscript with great attention to detail. Lewis Patterson, Rachel Savage, Karen Nordby, Kelli Baxter, and Hemant Makan helped with proofreading, indexing, and clerical work and kept me going.

The following colleagues and students read draft chapters and provided valuable criticisms.

Robert C. Angerer (University of Rochester)
Karl Aufderheide (Texas A&M University, College Station)
Bruce Babiarz (Rutgers—The State University of New Jersey)
Edward Berzin (Maimonides Medical Center, Brooklyn, New York)
Antonie W. Blackler (Cornell University)
Hans Bode (The University of California at Irvine)
Danny Brower (University of Arizona at Tucson)
Rudolf Brun (Texas Christian University)
Susan Bryant (The University of California at Irvine)

James Bull (The University of Texas at Austin)
David Capco (Arizona State University)
Mark Dansker (The University of Texas at Austin)
Thomas Drysdale (The University of Texas at Austin)
Susanne Dyby (U.S. Department of Agriculture, Gainesville, Florida)
Richard P. Elinson (University of Toronto)
David Epel (Stanford University)
Charles Ettensohn (Carnegie-Mellon University)
Joseph Frankel (University of Iowa)
Gary Freeman (The University of Texas at Austin)
Paul B. Green (Stanford University)
Jeffrey Hardin (University of Wisconsin at Madison)
Laurie Iten (Purdue University)
Herbert Jäckle (Max Planck Institute for Biophysical Chemistry, Göttingen, Germany)
Marcelo Jacobs-Lorena (Case Western Reserve University)
Antone G. Jacobson (The University of Texas at Austin)
Laurinda Jaffe (University of Connecticut at Farmington)
Ray Keller (University of California at Berkeley)
Kenneth J. Kemphues (Cornell University)
Michael Kessel (Max Planck Institute for Biophysical Chemistry, Göttingen, Germany)
Chris Kintner (Salk Institute, San Diego)
Mark Kirkpatrick (The University of Texas at Austin)
David Knecht (University of Connecticut at Storrs)
Wallace LeStourgeon (Vanderbilt University)
James Mauseth (The University of Texas at Austin)
Douglas Melton (Harvard University)
John Morrill (New College)
Ken Muneoka (Tulane University)
Gerald L. Murison (Florida International University)
Raymond Neubauer (The University of Texas at Austin)
Richard Nuccitelli (The University of California at Davis)
Gail R. Patt (Boston University)
JoAnn Render (University of Illinois)
Stanley Roux (The University of Texas at Austin)
Klaus Sander (University of Freiburg, Germany)
Amy Sater (University of Houston)
Helmut W. Sauer (Texas A&M University, College Station)
Anne M. Schneiderman (Cornell University)
Hazel Sive (Whitehead Institute, Cambridge, MA)
David L. Stocum (Purdue University)
William S. Talbot (University of Oregon)
William H. Telfer (University of Pennsylvania)
William Terrell (The University of Texas at Austin)
Akif Uzman (University of Houston)
Peter Vize (The University of Texas at Austin)
Stanley Wang (The University of Texas at Austin)
Detlef Weigel (California Institute of Technology)
Gregory A. Wray (State University of New York at Stony Brook)
Paul Wright (Western Carolina University)

I thank all these individuals for their most valuable contributions.

Klaus Kalthoff

Figure I.1 Scanning electron micrograph of a snail embryo (*Ilyanassa obsoleta*) at the 8-cell stage. (The tiny cell at the center is a polar body.) The four smaller cells, or micromeres, are rotated clockwise relative to their larger sister cells, or macromeres. One macromere is larger than the other three macromeres because it has incorporated a special mass of cytoplasm. The asymmetrical localization of this cytoplasm endows the largest macromere and its descendants with specific developmental capabilities.

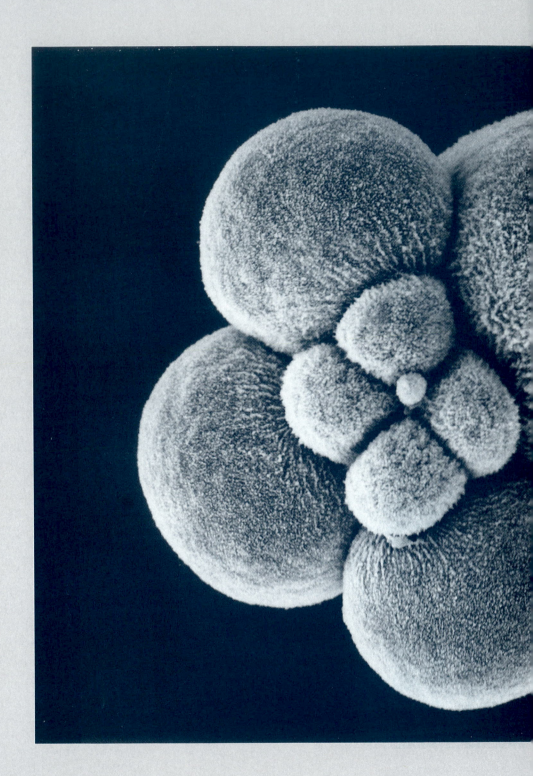

FROM GAMETOGENESIS TO HISTOGENESIS

In Part One of this text, we will follow the development of several organisms through the early stages of life, a period called embryogenesis. Beginning with the production of egg and sperm, we will examine fertilization, the subsequent cleavage of the large egg cell into smaller embryonic cells, the dramatic movements that generate multiple cell layers and form organ rudiments, and finally the differentiation of functioning tissues. Developmental biologists traditionally focus on embryogenesis because it is during this stage that the basic body plan emerges and that later events in development are programmed to a large extent. Therefore, we will concentrate on this period first and explore some aspects of postembryonic development later in Part Three.

As we examine the embryonic stages, we will draw examples from most of the animal species that have been developmental biologists' favorite research subjects, including frogs, chickens, mice, and humans among the vertebrates, and sea urchins, snails, and flies among the invertebrates. Some recent additions to this list, such as the plant *Arabidopsis thaliana* and the roundworm *Caenorhabditis elegans*, will be discussed in detail in Part Three.

Interwoven throughout this part of the text will be discussions of important principles, theories, and research strategies. For example, the principles of cell determination by cytoplasmic localization and by induction will be discussed in several places. The theory of genomic equivalence—that all cells in an organism have a full complement of the organism's genetic information—will be presented in the context of how different types of cells emerge from a single fertilized egg cell. Analysis of embryonic movements in terms of cellular behavior will emerge as a major research strategy.

At the end of Part One, readers should be familiar with the stages of embryonic development, the standard organisms used in developmental research, and the basic lines of reasoning and experimentation that have defined developmental biology as a classical discipline.

OVERVIEW
OF DEVELOPMENT

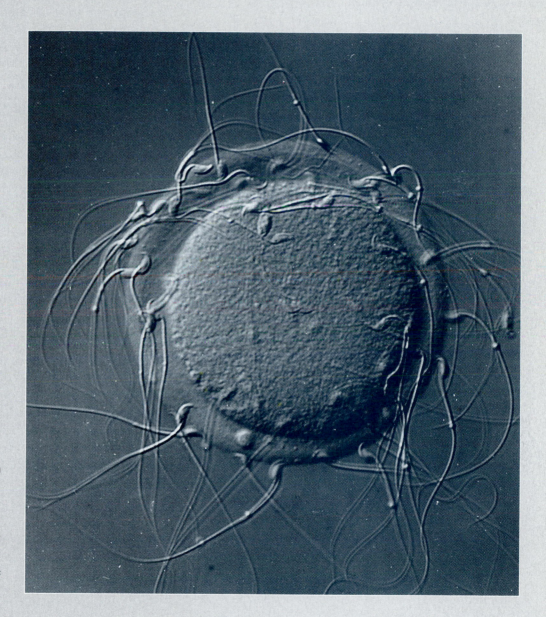

Figure 1.1 Photograph of a mouse egg being fertilized in a test tube. The egg is surrounded by a thick, transparent coat, called the zona pellucida. Many sperm adhere to the zona but none has yet fertilized the egg. Under natural conditions, fewer sperm reach the site of fertilization in the oviduct.

In this opening chapter, we will explore a few central topics in developmental biology. One is the **principle of epigenesis**,* stating that most organisms begin their lives with a relatively simple organization and proceed stepwise to more complex arrangements. Epigenesis unfolds as part of each organism's *life cycle*. Most individuals begin life as a fertilized egg and proceed through ever more complex embryonic and juvenile stages to adulthood. Adults, in turn, produce eggs and sperm, thus beginning the life cycle anew in their offspring. Related to epigenesis is another central topic in developmental biology: *pattern formation*. Each organism attains a certain form and organization by which it can be recognized as, say, a vertebrate embryo and then as a mammalian fetus and finally a mouse. How these characteristic patterns originate from their simpler precursors is a fascinating story that is still being written.

While these topics have always been on the minds of developmental biologists, the methods of investigating them have changed over time. The classical methods of isolating and transplanting embryonic parts have been augmented by genetic and molecular techniques. In our first look at the role of genes in development, we will explore how the abnormalities generated by mutated genes provide clues to the internal "logic" of development. The combination of genetic and other methods allows re-

---

* Technical terms are printed in **boldface italics** at the point where they are fully defined. This is usually where the term first appears. Elsewhere, plain *italics* are used on occasion to remind readers that this is a defined term. If one does not remember its meaning, one can look it up in the index at the end of the book. Plain italics are also used for emphasis and species names, but the context should make clear when italics signal a technical term.

searchers to study how the information present in a fertilized egg is used to build a three-dimensional organism that undergoes epigenesis.

Organismic development is also analyzed in terms of cellular behaviors, which in turn are often explained by molecular events. These types of analysis are now being complemented by new approaches, which aim to reveal how molecular processes are in turn affected by organismic events.

The Principle of Epigenesis

As an organism develops from a fertilized egg, its visible complexity increases. This is not to say that the organization of an egg is simple. An egg has all the organelles and the sophisticated regulatory mechanisms of a living cell, and no cell has yet been synthesized by a human. The egg is also a specialized cell, designed for its particular role in development. It is surrounded by protective envelopes but is prepared to admit a sperm for fertilization (Fig. 1.1). It contains not only nutrients, but also structural information that defines the future body axes, such as the anteroposterior axis. With time, however, the fertilized egg develops into something noticeably more complex. The development of humans and other mammals illustrates this process.

After fertilization, a mammalian egg divides into two cells, then into four, and so on (Fig. 1.2). A few days later, it has become an embryo that comprises about 60 cells of at least two types, which are arranged in a characteristic configuration known as the **blastocyst** (Gk. *blastos*, "germ"; *kystos*, "bag"). About 5 weeks after fertilization, a human embryo already shows the basic characteristics of a vertebrate, including a head with nose and eyes, the rudiments of a backbone and segmental trunk musculature, and paddle-shaped fore- and hindlimbs (Fig. 1.3). By the third month of development, it has become a fetus; it has the external features of a small adult. The limbs grow longer, while fingers and toes separate (Fig. 1.4). During the fourth month, the fetus begins to move in the uterus, indicating that its muscles and nervous system have begun to function. At this stage, the human body consists of about 200 different cell types arranged in intricate patterns. After several more months of growth and cell differentiation, the new individual is born.

The ability of a single cell, the fertilized egg, to develop into a complex adult organism is one of the greatest marvels of nature. The term used to describe this phenomenon is *epigenesis* (Gk. *epi*, "upon"; *genesis*, "pro-

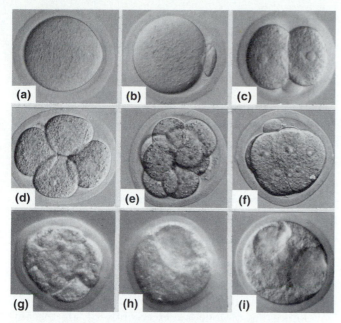

Figure 1.2 Mouse development prior to implantation. **(a)** Oocyte in metaphase II of meiosis. **(b)** Fertilized egg. **(c)** 2-cell stage embryo. **(d)** 4-cell stage. **(e)** Early 8-cell stage before compaction. **(f)** Late 8-cell stage after compaction. **(g)** 16-cell stage. **(h)** 32-cell stage; beginning of cavitation. The two cavities present in **(i)** will merge later.

duction"). The principle of epigenesis applies not only to physical development but also to patterns of gene activity that will be described later (Chapters 21 through 24).

The principle of epigenesis was first recognized by Aristotle (384–322 B.C.), who observed the development of chickens and other animals whose embryos are large enough to see with the naked eye. Despite widespread acceptance of Aristotle as an authority, the principle of epigenesis was rejected by subsequent generations of scholars. As late as the seventeenth century, nearly all biologists favored the theory of *preformation*, according to which each individual is fully formed within a germ cell and merely increases in size during development. The preformationists were divided into two camps, the "ovists" and the "spermists," who bitterly opposed each other. According to the ovists, the ovaries of a female contained miniature versions of her offspring, which in turn would hold miniatures of the following generation in their ovaries, and so forth. The seminal fluid contributed by the male was thought to merely activate the growth process. In contrast, the spermists proposed that sperm contained tiny but completely formed offspring, and that the role of the egg was simply to nourish the sperm. Distinguished scientists claimed to have seen miniature human figures

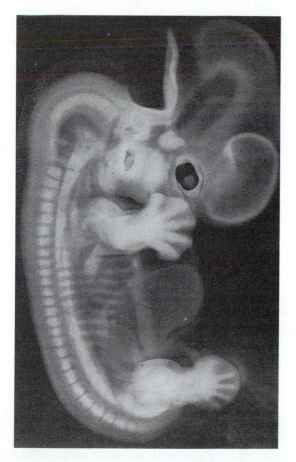

Figure 1.3 Photograph of a 5-week human embryo. Note the paddle-shaped arm and leg buds.

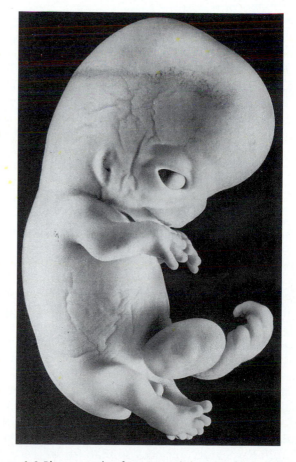

Figure 1.4 Photograph of an 8-week human fetus. Note the separation of fingers and toes.

Figure 1.5 Miniature human ("homunculus") preformed in the sperm, as depicted by Niklaas Hartsoeker in 1694.

curled up in the heads of sperm (Fig. 1.5). We must remember that the microscopes of the time were primitive and did not show any cellular detail, so that it was tempting to fill the observational void with preconceived notions.

With better microscopes and clearer reasoning, the champions of epigenesis came to prevail during the eighteenth century. The German anatomist Kaspar Friedrich Wolff (1733–1794) pointed out that if adults were preformed in the egg or the sperm, so that development meant only growing in size, then embryos would have to look like small adults. However, Wolff's observations showed that, for instance, developing chickens do not look like adult chickens at all. Instead, he found *temporary* adaptations for nourishment, respiration, and waste removal that were very prominent in the embryo but were replaced later in development. As more scientists discovered differences between embryos, juvenile stages, and adults, preformationism faded away.

Epigenesis is now recognized as the guiding principle of development, but the underlying cellular and molecular mechanisms are only partly understood. What is known is that the nuclei of both egg and sperm contain genetic instructions encoded in the nucleotide sequence of their DNA. The genes of a species provide most of the information needed for an individual's growth and development throughout life. In addition, the egg cytoplasm contains the building blocks for the developing embryo and, very important, some localized components that are distributed in a coarse but regular order. These latter components are segregated into different embryonic cells so that, for instance, the cytoplasm of anterior cells differs from that of posterior cells. The different cytoplasmic environments activate different combinations of genes in the nuclei of different groups of cells, so that the embryonic cells form a three-dimensional pattern according to the gene activities they exhibit. The ensuing interactions between different embryonic cells then generate an ever more complex network of cells and organs. It is mostly the anatomical complexity of the ovary in adult females that ensures that the egg will contain all the building blocks and the spatial order of cytoplasmic components needed for the development of an embryo.

Developmental Periods and Stages in the Life Cycle

The developmental sequence that begins with a fertilized egg, proceeds to adulthood, and then repeats itself through reproduction is known as the *life cycle.* Developmental biologists conveniently divide the animal life cycle into three major *periods:* embryogenesis, postembryonic development, and adulthood. Each of these periods is further subdivided into *stages.* To illustrate an animal life cycle, we will use the South African clawed frog, *Xenopus laevis,* which is a favorite research subject of developmental biologists today (Fig. 1.6).

The term *embryo* is generally used to describe the developing individual from fertilization through the formation of differentiated tissues. This period of development, called *embryogenesis,* is subdivided into the stages of fertilization, cleavage, gastrulation, organogenesis, and histogenesis. *Fertilization* is the union of egg and sperm. The egg is an extremely large cell, loaded with nutrients to support development until the new organism can obtain nutrients from outside. Most eggs have at least one polarity axis, known as the *animal-vegetal axis.* The **animal pole** is the pole closest to the egg nucleus, while the opposite pole is called the **vegetal pole.** In the case of *Xenopus,* the animal hemisphere is also pigmented, whereas the vegetal hemisphere is whitish with large amounts of yolk. The sperm is a cell that is highly specialized for its function of finding and fertilizing an egg. It contains a nucleus with very tightly packed chromosomes and a flagellum to propel itself. After fertilization by a sperm, an egg is called a **zygote.** Fertilization triggers or accelerates many metabolic reactions, notably the synthesis of DNA and protein, which are required in large amounts for the subsequent stages of development.

During *cleavage,* all zygotes undergo a series of mitotic divisions that result in the formation of progressively smaller cells called **blastomeres** (Gk. *blastos,* "germ"; *meros,* "part"). By the end of cleavage, the *Xenopus* embryo is a hollow sphere called a **blastula** (pl., *blastulae*). The ensuing stage of **gastrulation** in all embryos involves a complex series of **morphogenetic**

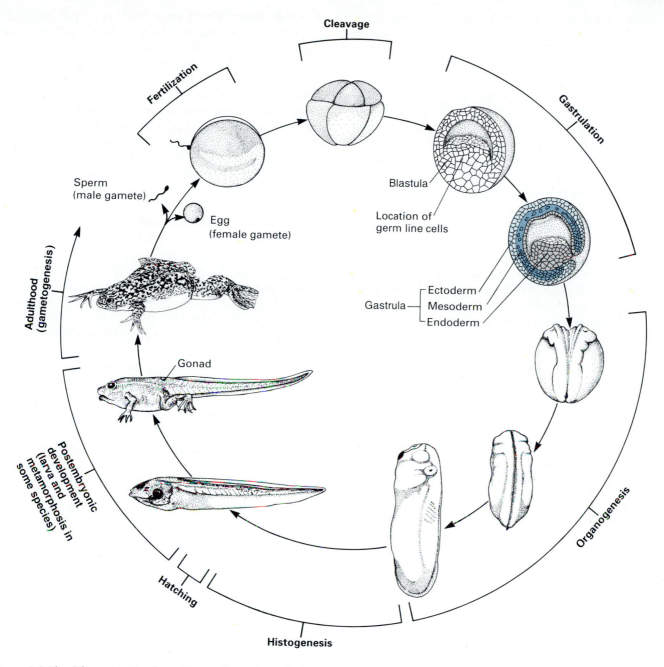

Figure 1.6 The life cycle of a frog. The embryonic period is subdivided into five stages: fertilization, cleavage, gastrulation, organogenesis, and histogenesis. In most species, a larval (tadpole) period begins with hatching from the eggshell and ends with metamorphosis. The adult period begins with the production of eggs or sperm (gametogenesis) and ends with death.

movements (Gk. *morphe,* "form"; *genesis,* "production"), in which cells rearrange and migrate while entire sheets of cells spread, bend, and fold. An embryo in the process of gastrulation is called a ***gastrula*** (pl., *gastrulae*). At the end of gastrulation the embryo consists of three concentric layers of cells called ***germ layers.*** During the next stage of embryogenesis, the germ layers undergo further morphogenetic movements and interact with one another to form the rudiments of the larval organs. At the end of this process, which is known as ***organo-*** *genesis,* the embryo shows the basic body plan of the animal group to which it belongs. Finally, the embryo goes through an extended stage of ***histogenesis,*** during which cells—which by now have reached their final location—acquire their functional specialization.

The ***postembryonic period*** of development lasts from the end of embryogenesis to the beginning of adulthood. During this period, some species, including humans, already look like miniature adults. In contrast, *Xenopus* and many other animals go through an intermediate period in which the organism is known as a *larva* (pl., *larvae*). The larval period generally begins when the individual hatches from the eggshell. A typical larva is mobile, feeds, and gains substantially in body mass. However, the larva differs in form and lifestyle from the adult and often has a different name; for example, the *tadpole* is the larva of the frog. The tadpole has some of the organs found in the adult frog, such as eyes, brain, and heart, but also stage-specific organs that will disappear later, including gills and a finned tail. Conversely, some adult organs, including legs and lungs, are still missing from the tadpole. The transition from the larval to the adult period entails a dramatic sequence of events known as *metamorphosis.*

The *adult period* of the life cycle begins when the individual becomes sexually mature and produces eggs or sperm. The lineage of cells that eventually form eggs or sperm is known as the *germ line.* Only germ line cells have the potential to generate an *indefinite lineage* of daughter cells, as long as each generation produces offspring. All other cells of the body are collectively called *somatic cells;* they perish when the adult dies.

In the following sections, we will discuss some general aspects of animal life cycles. We will emphasize the role each developmental stage plays in relation to other stages, and how the organism at different stages fulfills certain of its basic functions such as feeding, mating, or dispersal of the species. We will also consider which role some life cycle stages may have played during evolution.

Most Animals Are Multicellular and Reproduce Sexually

Two standard features of most animal life cycles are multicellularity and sexual reproduction. Only the simplest animals, the *protozoa,* are unicellular. They generally multiply asexually by cell division. In addition, they exchange nuclei or form sex cells that fuse, thus generating new individuals with new combinations of genes. Most animals develop into adults with specialized organs made up of very large numbers of cells. Such animals are called *metazoa.* The advantages metazoa have over protozoa include the division of labor between different cells and organs and the ability to maintain a favorable internal environment. These features have allowed metazoa to populate a wider range of ecological niches than protozoa could.

Some of the more primitive metazoa, such as polyps, can multiply asexually by budding. However, most metazoa procreate sexually through reproductive cells called *gametes* (Gk. *gamein,* "to marry"). Gametes always exist in two types, usually in the familiar forms of eggs and sperm. Gametes arise from the germ line, the only cell line that is potentially immortal. Building large bodies that die after they have perpetuated their germ

line may seem to be a wasteful way of propagating a species. Indeed, many animals spend an enormous amount of energy on producing gametes and raising their offspring. Furthermore, attracting or finding mates is often done at the risk of life and limb. For instance, male frogs advertise themselves to females by mating calls, and yet the same calls alert predators to their whereabouts. If sexual reproduction comes at such extravagant cost, why is it so prevalent?

The adaptive value of sexual reproduction has long been a matter of intense debate (Fisher, 1930; Michod and Levin, 1987). Most theories emphasize that sexual reproduction generates more genetic diversity than asexual reproduction and thereby facilitates adaptation and evolution. Darwin (1868) observed that sexually produced offspring show greater differences among themselves and from their parents than offspring generated asexually. The causes underlying these differences are genetic. Asexual offspring, such as the stolons of a plant or the buds of a polyp, are genetically identical, or nearly identical, to the parent organism. By contrast, gametes inherit a randomly assorted half of the parental genes, and during fertilization two such samples from the male and female parents can combine in almost unlimited ways. Natural selection can then act on a correspondingly large variety of offspring. Presumably the surviving variants will be those that can move into previously unoccupied ecological niches, and those that are best suited to survival in environments—including predators and parasites—that change over time. One could conclude, then, that it is genetic diversity, including rare combinations of "good" or adaptive genes, that has made sexual reproduction so prevalent.

An apparent drawback of sexual reproduction is that the most favorable genetic variants are not necessarily stable; an adaptive trait that relies on the simultaneous activity of two or more genes may disappear if the genes are separated in subsequent generations. Evolutionary biologists have therefore looked for other advantages of sexual reproduction. A more subtle, long-term advantage may be quicker elimination of "bad," or nonadaptive, genes. Mutability is an inherent feature of genetic information, and the vast majority of mutations are deleterious. If such mutations occur frequently in each generation, then sexual reproduction obliterates them faster than asexual reproduction (Kondrashov, 1988). The need to eliminate bad genes may outweigh the disadvantage of losing favorable combinations of good genes. Another advantage of sexual reproduction is that it segregates mutant genes from their nonmutated counterparts (see Chapter 3) and facilitates the expression of the mutant genes in a population.

Many of the competing theories on the advantages of sexual reproduction are not mutually exclusive, and each of them may have some validity (Maynard-Smith, 1989). For whatever combination of reasons, sexual reproduction is central to the life cycles of most animals.

Embryonic Development Begins with Fertilization and Ends with Histogenesis

Development is a stepwise process that often seems indirect. The *zygote* contains the raw materials and information needed to form a whole new organism, but at this early stage of life it in no way resembles the complex organism it will ultimately become. Only after many intermediate steps do its cells achieve the high degree of specialization in structure and function that characterize muscle cells, neurons, and other types of cells. Moreover, cells undergo extensive movements before reaching their final destinations. All the while, the body pattern typical of the organism's species is gradually taking shape.

The embryonic period of development begins with fertilization and a single large cell, the fertilized egg. During cleavage the embryo is subdivided into smaller and smaller blastomeres. For instance, a *Xenopus* egg is about 1.2 mm in diameter, whereas most cells in a tadpole are less than 20 μm across. In this respect, cleavage differs from cell division at later stages, when cells grow back to their normal size after each division. Cell growth requires the uptake of external nutrients, which are not available to most embryos. The small embryos of mammals are exceptions to this rule. While most eggs support embryonic development until the new or-

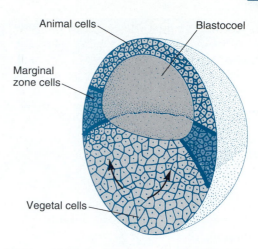

Figure 1.8 Frog blastula showing animal cells with pigmented surface; large, yolky vegetal cells; and an intervening marginal zone. Signals from vegetal blastomeres (arrows) prepare the marginal zone cells for mesoderm formation, which occurs later during gastrulation.

ganism can feed and breathe, mammalian eggs give rise only to the *blastocyst* stage, which is reached after only six rounds of cell division. At this stage, the mammalian embryo implants itself into the uterus and obtains further nourishment from maternal blood. Consequently, mammalian embryos begin to grow in size after implantation.

The blastomeres generated during cleavage contain different amounts and types of egg cytoplasm. These differences provide the basis for subsequent cellular interactions that cause further epigenesis. For instance, blastomeres in the *animal* half of a frog embryo are small and have a darkly pigmented surface, whereas those in the *vegetal* half are larger, are unpigmented, and contain more yolk (Fig. 1.7). In addition to such visible distinctions, animal and vegetal blastomeres also acquire different messenger RNAs and proteins. These differences enable the animal and vegetal blastomeres to interact with each other, generating a *marginal zone* between them (Fig. 1.8). At the end of cleavage, the embryos of frogs and many other animals consist of a hollow blastula, with embryonic cells surrounding a fluid-filled cavity called the *blastocoel* (Gk. *blastos*, "germ"; *koiloma*, "cavity"). At the end of the blastula stage, the embryo is prepared for the ensuing morphogenetic movements, which first generate the *germ layers* and then the *organ rudiments*.

Because gastrulation is quite complex in frogs, we will use the sea urchin as a simpler case to illustrate the gastrulation process in general. Early in gastrulation, the sea urchin embryo shows a depression near the vegetal pole, much like a soft tennis ball pushed in with a finger (Fig. 1.9i). As this depression deepens and elongates, it creates an internal germ layer, or *endoderm*. The cells remaining on the outside of the gastrula constitute the outer germ layer, or *ectoderm*. In the space

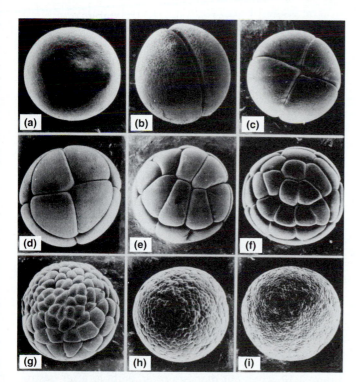

Figure 1.7 Early development of a newt (*Pleurodeles waltlii*) embryo as seen in scanning electron micrographs. **(a)** Fertilized egg. **(b)** 2-cell stage. **(c)** 4-cell stage. **(d)** 8-cell stage. **(e)** 16-cell stage. **(f)** 32-cell stage. **(g)** Early blastula. **(h)** Midblastula. **(i)** Late blastula.

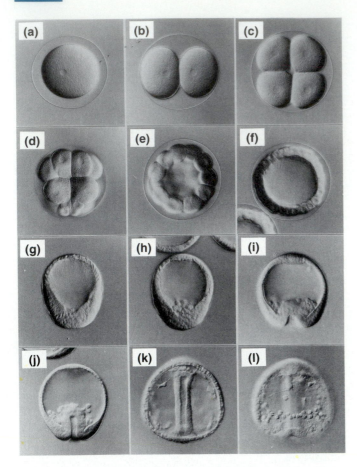

Figure 1.9 Development of a sea urchin embryo (*Lytechinus variegatus*) as seen in photomicrographs of live specimens. **(a)–(f)** The process of cleavage until the blastula stage, when a monolayer of cells surrounds the blastocoel cavity. **(g)** Beginning of gastrulation. The vegetal plate forms as a tall epithelium, visible at the bottom of the picture. **(h)** The ingression of primary mesenchyme cells, which leave the vegetal plate and move into the blastocoel. **(i, j)** Inward buckling of the vegetal plate to generate the rudiment of the gut, the archenteron. **(k, l)** Formation of secondary mesenchyme cells at the tip of the elongating archenteron. The archenteron represents the inner germ layer, or endoderm. The external epithelium is the outer germ layer or ectoderm. The primary and secondary mesenchyme cells form the intermediate germ layer or mesoderm. These three germ layers are formed in all animal embryos, although the movements that generate the germ layers vary considerably among different groups of animals.

between the ectoderm and the endoderm, an intermediate germ layer, or *mesoderm,* is formed. The mesodermal cells separate from the vegetal base of the sea urchin gastrula and from the tip of the elongating endoderm.

How the three germ layers are established varies considerably among groups of animals. Nevertheless, the resulting configuration of three germ layers and the derivatives formed later by each layer are remarkably consistent among animals of different phyla. The major derivatives of the ectoderm are the epidermis and the

nervous system. The endoderm forms the inner lining of the digestive tract and its appendages. The mesoderm gives rise to the skin layers underneath the epidermis and to the muscles of the gut. It also forms a wide range of other organs, which include, depending on the species, bone or other skeletal structures, muscle, heart, blood vessels and blood cells, kidneys, and reproductive organs.

The germ layers differ in microscopic structure and cellular arrangement. Cells that form a boundary to the outside world are joined together tightly in a coherent sheet called an *epithelium*. Epithelia may consist of single or multiple layers of cells. Both the epidermis and the endoderm face the outside world and form epithelia. (The lumen of the gut is connected to the outside world via the mouth and the anus.) The mesoderm, although not in contact with the outside world, may form epithelia as well. But the mesoderm also occurs in loose arrangements of more mobile cells with large amounts of extracellular material between them. Such an arrangement of cells is called a *mesenchyme.* For instance, the mesoderm of the sea urchin originates as a mesenchyme. During development, the arrangement of mesodermal cells may change from mesenchymal to epithelial and vice versa. A simplified representation of an embryo after gastrulation would show ectodermal and endodermal epithelia, and mesenchyme in between (Fig. 1.10). By this time, the embryo is prepared for the development of the organ rudiments.

The stage of *organogenesis* is characterized by further cellular or epithelial movements and interactions. Many of the organs arising during this stage consist of mesenchymal and epithelial components, which are often derived from different germ layers. Proper organ development requires extensive interactions between epithelial and mesenchymal components. The pancreas, for example, is formed by interactions between pancreatic mesenchyme and endodermal epithelium, whereas tooth development requires cooperation be-

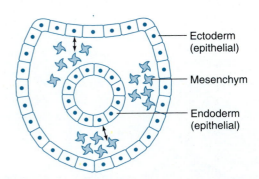

Figure 1.10 Schematic diagram of an embryo after gastrulation. The germ layers facing the outside world (ectoderm and endoderm) are epithelial, whereas the intermediate germ layer (mesoderm) is mesenchymal. The arrows symbolize the interactions between the germ layers that are required for organogenesis.

tween oral mesenchyme and ectodermal epithelium.

To illustrate the process of organogenesis, we will turn from sea urchins back to the more familiar vertebrates. One of the most spectacular events in vertebrate organogenesis is the formation of the central nervous system, a process known as *neurulation*. Ectodermal cells on the dorsal surface of the embryo—the embryo is called a *neurula* during neurulation—elongate perpendicularly to the surface, forming a raised plate called the *neural plate*. The margins of the neural plate bend together and eventually fuse along the dorsal midline (Fig. 1.11). The neural plate thus closes into the *neural tube*, which ultimately gives rise to the brain and spinal cord. The adjacent ectodermal cells join to form the overlying epidermis.

At the end of organogenesis, the embryo has become a complex animal, having the rudiments of most organs in place and typically exemplifying the basic body plan of its class or phylum.

Organogenesis blends into the subsequent stage of *histogenesis* (Gk. *histos*, "tissue"), during which the embryo generates tissues consisting of different mature cell types. Throughout their lifetimes, all cells perform basic "housekeeping" functions, such as energy metabolism and protein synthesis, which are necessary for their own survival. As tissues mature, most cells also acquire a special function, together with certain structural characteristics that facilitate performing that function. For instance, red blood cells, which carry hemoglobin, have a shape that provides a relatively large surface for gas exchange but at the same time allows them to squeeze through narrow capillaries. The acquisition of specialized functional and structural characteristics by cells is called *cell differentiation*.

Postembryonic Development Can Be Direct or Indirect

In most animal species, the *postembryonic period* begins when the developing animal hatches from its eggshell and begins to take up nutrients from outside. Newly hatched tadpoles, for instance, begin to feed on plant materials and small animals. In mammals and other species that nourish their embryos through a placenta, however, the beginning of the postembryonic period is better defined as the period following organogenesis. The developing human, for example, is called an *embryo* during the first eight weeks of gestation, and a *fetus* from the ninth week until birth (see Figs. 1.3 and 1.4). In frogs and mammals alike, postembryonic development is characterized by rapid growth.

During the postembryonic stage some animals look like miniature adults, except that some parts of the body are further advanced in growth than others and the gonads and other sexual characteristics have not yet matured. Such animals undergo *direct development* and are called *juveniles* (or more specific names) during their

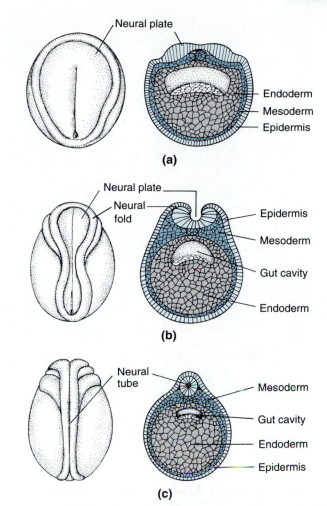

Figure 1.11 Frog embryos during neurulation. Left column, dorsal views; right column, transverse sections. The dorsal ectodermal region forms the neural plate, which closes into the neural tube, which later gives rise to the brain and spinal cord. The remainder of the ectoderm mainly forms the outer cell layer, the epidermis. The endoderm forms the primitive gut, the archenteron. The mesodermal cells in between form bones, muscles, and the bulk of many internal organs. **(a)** Early neurula. **(b)** Middle neurula. **(c)** Late neurula.

postembryonic period. Most vertebrates, including mammals, develop this way. In the human fetus and infant, for instance, the body parts are those of an adult, but the head is disproportionately large compared with the trunk, arms, and legs, which grow rapidly after birth (Fig. 1.12).

In other groups of animals, the postembryonic stages bear little resemblance to adults. Familiar examples include the tadpoles of frogs, the caterpillars of butterflies, and the maggots of flies. Such animals undergo *indirect development* and are called *larvae* during their postembryonic period. The transition between the larval and adult periods, known as *metamorphosis*, is more or less dramatic depending on how radically the larval and adult stages differ from each other.

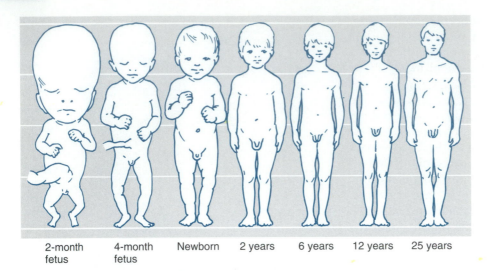

Figure 1.12 Changing proportions of the human body during fetal development and after birth. Note the disproportionately large head in the embryo at 2 months, and the more rapid growth of trunk, arms, and legs thereafter.

| 2-month fetus | 4-month fetus | Newborn | 2 years | 6 years | 12 years | 25 years |

Different insect orders show different degrees of change during metamorphosis. When insects grow or change in body form, they molt the rigid external cuticle and replace it with a larger or different one. In many primitive insects, including grasshoppers and true bugs, the juvenile stages differ from the adults only in their undeveloped reproductive organs and wings (Fig. 1.13). The wings are not visible when such insects hatch from the egg; they grow slowly from molt to molt. Otherwise, the juvenile and adult stages of primitive insects look much alike and have similar lifestyles. In contrast, larvae of more evolved insect orders, including flies, wasps, beetles, and butterflies, differ radically from adults. They molt several times as larvae, growing enormously in the process (Fig. 1.14). Under hormonal control, the larva enters an immobile stage in which it is called a *pupa* (pl., *pupae*); during this stage it is enclosed in a cocoon or cuticular case. During the pupal stage most larval tissues are broken down and used—along with nests of cells set aside early in development—to build the adult insect. The adult that emerges is very different from the larva in its form and lifestyle (Fig. 1.15). This very dramatic transformation is called *complete metamorphosis*; the gradual change from juvenile to adult observed in primitive insects is called *incomplete metamorphosis.*

Complete metamorphosis is a time-consuming procedure and is presumably quite demanding in terms of genetic control. Yet the insect groups with complete metamorphosis account for the largest numbers of species, which means that they have come to occupy the greatest variety of ecological niches. What is the adaptive value of having a larval period in the life cycle? One advantage may be that larvae and adults can exploit alternative food resources, so that the growth of a population is less restricted. Caterpillars eat leaves, but adult butterflies feed on pollen and nectar. Another advantage may result from associating basic activities in life, such as growth and reproduction, with different

periods in the life cycle. Living in the soil is a safe way for a grub to grow in size, but staying there would make it hard to find a mate; subsequent metamorphosis into a free-living beetle seems to provide the best of both worlds. Other basic requirements for organisms are dispersal and protection, which tend to be mutually exclusive. Thus, it seems adaptive to design only one stage of the life cycle for dispersal. In insects, this is the flying adult stage; in sea urchins, dispersal is achieved during the larval stage, when the animals float in the ocean. Finally, metamorphosis enables many animals to synchronize their life cycles with the seasons: many of the insects that undergo complete metamorphosis survive the winter in their pupal cuticles or in cocoons.

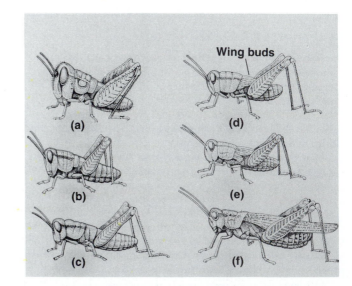

Figure 1.13 Incomplete metamorphosis in primitive insects. **(a–e)** Juvenile (also called nymphal) stages of a grasshopper. **(f)** Adult. Stages are drawn at different scales. Note the gradual appearance of the wings and the overproportionate growth of the adult abdomen containing the reproductive organs.

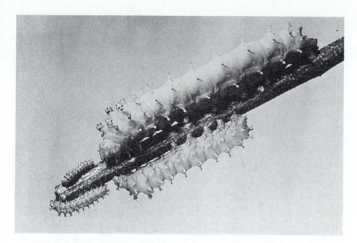

Figure 1.14 Photograph showing silkworm larvae after hatching from the egg, and after one, two, and four additional molts. Feeding on leaves, the larva grows about 5000-fold in weight.

Adulthood Begins with the Onset of Reproduction

The *adult period* begins with the production of mature gametes and ends with death. Mating and parenting are the key issues during adult life. Access to mates may involve competition over territory and the establishment of dominance hierarchies; in some species, females choose males displaying certain appealing characteristics. After mating, adult animals spend varying amounts of time and energy caring for their offspring. Some species simply produce as many offspring as possible and abandon them. Others invest tremendous effort into caring for a few young. Most fall somewhere in between.

Maintaining a living body requires energy, which living things take up either directly from sunlight or as chemical energy from food. However, individual organisms cannot sustain themselves indefinitely. Over time, energy-harvesting and maintenance mechanisms accumulate defects, which eventually overpower the organism's capabilities for repair and adjustment. As a result, all individuals that do not die from accidents or predation eventually undergo a process of deterioration called *senescence.* The length of the adult life span before senescence differs widely in accord with the reproductive strategy of a species. Some insects do not even feed as adults; they mate, produce a large clutch of offspring, and die within a day. At the other end of the scale, large mammals may survive several decades as adults while raising only a few young.

Pattern Formation and Embryonic Fields

Having considered life cycles, we will now turn to another, equally fundamental aspect of development called *pattern formation.* This is the orderly spatial development of different elements. For instance, eyespots are found on the wings of many butterflies; they apparently serve to deflect predators from the main part of the body (Fig. 1.16). Eyespots are formed by the coordinated differentiation of cells with different pigments. Cells that reflect blue light form a spot in the center, and cells with black and light-brown pigments form concentric rings around it. Thus, pattern formation requires that cells behave in different ways. In addition, different cell behaviors must be orchestrated in space and time for the correct pattern to emerge. In the case of the butterfly wing, if the differently pigmented cells were arranged in a salt-and-pepper mixture instead of concentric rings, the protective effect of the eyespots would be completely lost.

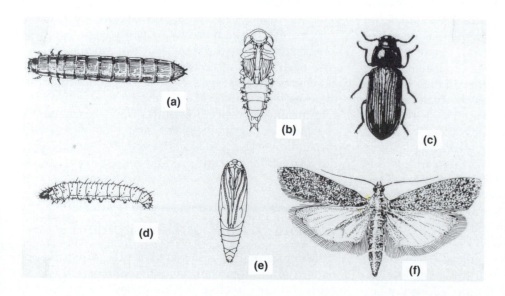

Figure 1.15 Complete metamorphosis in evolved insects. Top row, beetle (*Tenebrio molitor*); bottom row, moth (*Anagasta kühniella*). **(a, d)** Larvae. **(b, e)** Pupae. **(c, f)** Adults.

Figure 1.16 Photograph of *Dynamine melitta*, a butterfly common in Central America. Its wings show eyespots, which apparently deflect predators from the central part of the body. The eyespots and other features of the wing pattern are generated by the coordinated synthesis of different pigments.

Patterns are formed at all stages of development, and on different scales, from the arrangement of microscopic organelles in a cell to the position of macroscopic organs in the body plan. Cells differentiate in distinct patterns when they form tissues. Tissues are arranged in patterns in organs. Organs, in turn, are arranged according to a basic body plan. For example, a vertebrate's heart is situated ventrally, a vertebral column develops dorsally, and a central nervous system forms within the vertebral column and the skull.

A fundamental characteristic of biological patterns is that the same pattern can be formed in widely varying sizes. To illustrate with an analogy, let us consider the simple one-dimensional pattern of letters seen in the world "tone." This pattern is characterized by the number of letters in the word, the individual letters that constitute it, and their sequence. The word "note," although consisting of the same letters, has a different pattern and a different meaning. However, changing the overall space occupied by the letters, as in tone or tone would not change the pattern. Organismal patterns have just this property of *size invariance.* Early embryos of many species can be split, yet each half will give rise to a small but complete and normally proportioned embryo, which may attain normal size by growing more than usual during postembryonic stages. In fact, human identical twins originate in this fashion. Conversely, two embryos stuck together can join and form one normally proportioned embryo (see Chapter 6).

Groups of cells that cooperate to form a particular organ or an entire organism are called *embryonic fields.* The establishment of fields and the formation of patterns are basic phenomena in development, but they are not entirely understood. Not only can cells differentiate in numerous ways, but their differentiation can also be orchestrated so that a well-proportioned organ-

ism results. Moreover cells can communicate with one another to form complete, well-proportioned embryos with widely differing numbers of cells participating.

Two models have been proposed to explain pattern formation in embryonic fields. According to the *gradient model,* certain cells in a field act as a source of a diffusible signal substance, called a *morphogen.* The morphogen is a *long-range* chemical signal transmitted throughout the entire field. However, its concentration decreases with distance from the source (Fig. 1.17a). Depending on the local morphogen concentration, different genes are activated or repressed in the cells of the field. In essence, the morphogen concentration profile is transformed into a pattern of different gene activities, which in turn initiate various pathways of cell differentiation. For instance, the eyespots in butterfly wings can be explained by assuming that the cells in the center of the eye produce a morphogen and that the cells of the adult wing assume different colors depending on the local morphogen concentration to which they were exposed at an earlier stage of development. Cells reflecting blue light would originate where the highest range of morphogen concentration has been, cells making black pigment would arise in the next lower concentration range, and cells with light-brown pigment in the following range.

An alternative model of pattern formation, the *sequential induction model,* postulates a series of *short-range* interactions between neighboring cells. In the case of the eyespot in the butterfly wing, cells that ultimately reflect blue light might communicate with their neighbors by a short-range signal that causes them to make black pigment in the adult (Fig. 1.17b). These cells, in turn, would send their neighbors another short-range signal that stimulates them to make light-brown pigment, and so on. As we will see in Chapters 20 to 22, both morphogen gradients and sequential induction are involved in the formation of at least some embryonic patterns.

All models of pattern formation imply that cells have the capacity to activate different sets of genes. In our example of the butterfly wing, the synthesis of different pigments by different cells is ascribed to the presence of different sets of enzymes (Nijhout, 1980). Each enzyme is encoded by a gene, so that the formation of the eyespot results from the orchestrated activation of certain genes in concentric rings of cells. In this context, as well as in many others, the regulation of gene activities in time and space has emerged as a central topic in developmental biology.

Genetic Control of Development

All the information necessary for generating the body pattern of an organism, for differentiating all types of cells, and for directing the life cycle through its stages

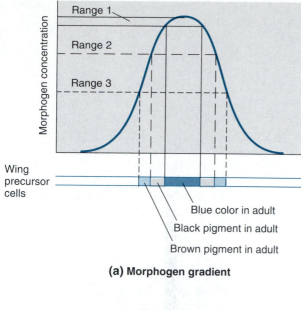

Range 1
Range 2
Range 3

Wing precursor cells

Blue color in adult

Black pigment in adult

Brown pigment in adult

(a) Morphogen gradient

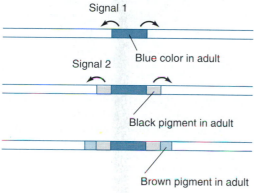

Signal 1

Signal 2   Blue color in adult

Black pigment in adult

Brown pigment in adult

(b) Sequential induction

Figure 1.17 Two models of pattern formation applied to eyespot formation in the butterfly wing. For simplicity, the eyespot is drawn in only one dimension. **(a)** According to the gradient model, the cells forming the center of the eyespot produce a diffusible chemical signal that serves as a long-range morphogen. Depending on its local concentration, the same substance stimulates cells to develop in such a way that in the adult they reflect blue light, make black pigment, or make light-brown pigment. **(b)** According to the sequential induction model, cells in the center of the eyespot, which in the adult will reflect blue light, send out a short-range signal to their neighbors. In response, the neighboring cells, which in the adult synthesize black pigment, send a different signal to their neighbors, and so on.

is initially contained in a single cell. This one cell, the zygote, carries instructions for building and maintaining the entire organism from its first cell division to the end of its life. These instructions direct the physical form and, to a large extent, the behavior of an organism. Farmers, animal breeders, and biologists have long observed that behavioral as well as morphological traits are heritable. Today we know that most of the information for heritable traits is encoded in deoxyribonucleic acid (DNA). The functional units of this information are

called *genes,* and one entire complement of DNA in an organism is called its *genome.* The genetic makeup of an organism—in particular, with respect to the normal or altered state of certain genes—is called its *genotype.* The visible characteristics of an organism—especially with regard to the normal or mutant state of certain heritable traits—constitute its *phenotype.*

Genetics research has traditionally been concerned with mapping genes on chromosomes and with studying their phenotypic effects by morphological and biochemical methods. The raw materials for such studies are *mutants,* that is, variant stocks showing abnormal heritable phenotypes. Most abnormal phenotypes are caused by one or more *mutations,* or heritable changes, in the genomic DNA. Alternative forms of one gene that arise by mutation are called *alleles.* The allele of each gene that predominates in natural populations is called the *wild-type allele;* it is symbolized by a superscript plus following the name of the gene. For instance, *shiverer*[+] refers to the wild-type allele of a particular mouse gene, whereas *shiverer* refers to any other allele of the same gene.

The name of a gene is often misleading because it refers to a mutant phenotype rather than the normal phenotype. For instance, the *white*[+] gene of the fruit fly *Drosophila melanogaster* is named after the white eye color of mutant flies that are defective in the *white* gene. The normal function of the *white*[+] gene is to encode a transport protein that is involved in synthesizing the red eye pigment of normal flies. In the absence of this transport protein, the eyes of the fly lack the normal pigment and appear white. This system of naming genes after a defect caused by improper or lacking gene function, and not according to the normal gene function, can be confusing to the beginner. With the benefit of hindsight, it would have been more descriptive to name the *white*[+] gene *transport*[+] instead. However, scientists name a gene as soon as they isolate a mutant stock, which is normally long before they unravel the normal function of the wild-type allele.

Genes affect the morphology and behavior of cells in an organism at *all* developmental stages of the life cycle. In fact, genes control the production and properties of the very molecules on which development is based (Fig. 1.18). Because genes are transcribed into ribonucleic acid (RNA) and translated into polypeptides, the nucleotide sequence of the genomic DNA strictly determines the sequence of amino acids in polypeptides. Intermolecular forces between the amino acids of a polypeptide determine, to a large extent, its three-dimensional configuration and its assembly into composite proteins and incorporation into cellular organelles. The properties of enzymes and structural proteins determine much of the morphology and behavior of cells, and thus the development of the entire organism.

Initially a zygote's development is guided by its mother's genes (Fig. 1.19). RNA and proteins encoded by *maternal effect genes*—or *maternal genes,* for short—

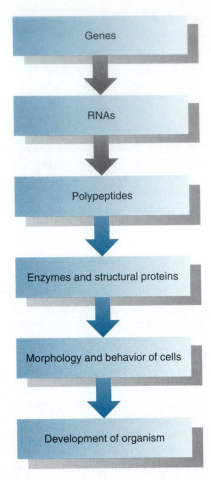

Figure 1.18 Genetic control of development. The black arrows represent transcription and translation under the strict control of an individual's genes. The colored arrows represent a strong but not exclusive genetic control of subsequent processes.

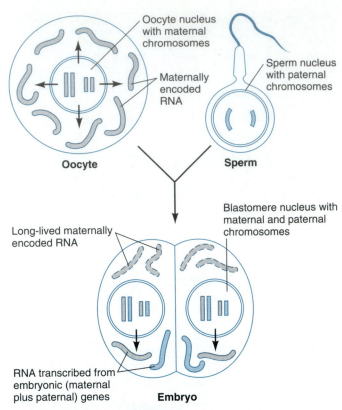

Figure 1.19 Two types of gene expression during embryonic development. Long-lived RNA and protein encoded by maternal genes typically support embryonic development through the cleavage period. In many animal species the transcription of zygotic (maternal *and* paternal) genes begins near the end of cleavage.

are synthesized and deposited in the egg during oogenesis. The presence of these maternally provided products sustains the developing organism while its own DNA is replicating at a very rapid rate and is therefore only minimally transcribed. Typically, the embryo's dependence on maternal gene products lasts through the first cleavage divisions. At this point the embryonic genes, or **zygotic genes**, which are inherited from *both* father and mother, take over the direction of development.

The role of maternal gene products in an embryo's development is twofold. Some of these products, such as DNA polymerase and tubulin messenger RNA (mRNA), are simply stockpiled to promote rapid development and are used by all embryonic cells alike. Other maternal gene products are distributed unevenly in the egg, so that unlike quantities of them end up in different blastomeres. If such localized gene products are altered by mutations in the encoding maternal genes, the developing embryos may show severely disturbed body patterns. For instance, *Drosophila* females with a

defect in both copies of the *bicaudal⁺* gene have offspring that show a mirror-image duplication of the posterior abdomen while head and thorax are missing (Fig. 1.20). This striking phenotype develops independently of the *bicaudal* allele inherited from the offspring's male parent. Genetic and molecular analysis have revealed that the bicaudal protein is translated from maternal mRNA in the ovary. In embryos derived from mothers without a functional *bicaudal⁺* allele, embryonic genes required for head and thorax formation are not activated, and the anterior parts of the body pattern are not formed.

Mutations in zygotic genes can also have profound effects on the embryonic body pattern. For example, *Drosophila* adults with mutations in the *Ultrabithorax⁺* gene have four wings instead of two (Fig. 1.21). The third thoracic segment (T3) of these flies looks like another copy of the second thoracic segment (T2). Normally, T2 carries a pair of wings while T3 carries a pair of balancer organs known as *halteres*. In the *Ultrabithorax* mutant, T3 has been transformed into another T2, thus creating a four-winged fly. Genetic analysis has shown that the *Ultrabithorax⁺* gene promotes the morphology characteristic of T3 and that this gene is normally active in T3 but not in T2. The mutant

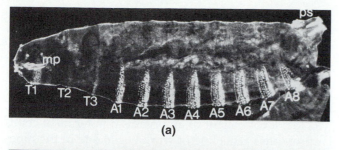

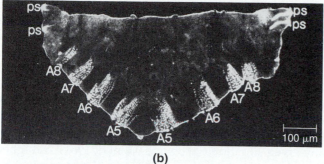

(a)

(b)

Figure 1.20 Abnormal body pattern caused by a mutation in a maternal effect gene. The photomicrographs show *Drosophila* larvae derived from a normal (wild-type) mother **(a)** and a *bicaudal* mutant mother **(b)**. Each micrograph shows the cuticle (exoskeleton) after all inner parts were dissolved. The cuticles are photographed with dark-field illumination in which all light-scattering objects appear bright against a dark background. The normal larva shows mouthparts (mp), three thoracic segments (T1–T3), and eight abdominal segments (A1–A8). The end of the abdomen is marked by respiratory openings called posterior spiracles (ps). The *bicaudal* larva shows a mirror-image duplication of the posterior abdominal segments; it is missing all other segments.

Ultrabithorax phenotype results from a failure to properly activate the gene in T3. In the absence of the normal *Ultrabithorax*+ product, T3 assumes the morphological characteristics of T2.

Other mutant phenotypes result from the activation of genes in regions where they are normally inactive. For instance, the *Antennapedia*+ gene is normally expressed in the thorax but not in the head. Faulty expression of this gene in the head results in the replacement of antennae with legs (see Color Plate 5). Thus, a mutant phenotype may result either from failure of a gene to be activated in its normal expression domain, as in the case of *Ultrabithorax*, or from inappropriate activation of a gene in an area where it should remain silent, as in the case of *Antennapedia*.

Many genes have unique functions in development and are active only during certain developmental stages and in certain groups of cells. Mutants that are abnormal in the expression of these genes have proved to be a rich source of information for the analysis of development. The phenotypes of such mutants—especially those that can be detected early in life—provide valu-

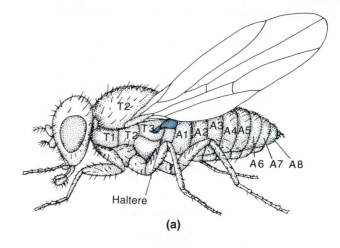

(a)

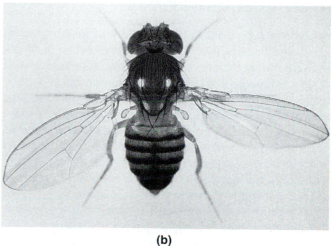

(b)

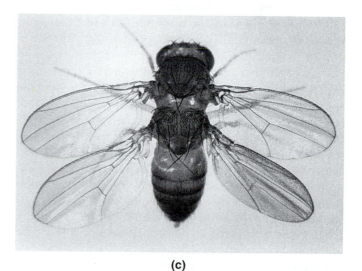

(c)

Figure 1.21 Abnormal body pattern caused by the faulty expression of a zygotic gene in *Drosophila*. **(a)** Schematic drawing of a normal fruit fly. Notice how the second and third thoracic segments (T2, T3) differ in size and appendages. T2 carries a pair of wings; T3 carries a pair of smaller balancer organs called halteres. **(b)** Photograph of a normal fly. **(c)** Four-winged fly caused by three mutations in regulatory regions of the *Ultrabithorax* gene. In the mutant phenotype, T3 (including halteres) is replaced by a duplication of T2 (including wings).

able clues to the *biological role(s)* of these genes in development.

The mutational analysis of a gene often facilitates its *molecular cloning* by recombinant DNA techniques (see Chapter 14). The nucleotide sequence of a cloned gene can be determined, and this information often provides clues to the *biochemical function* of the protein encoded by a gene. Knowing the biological and biochemical functions of a certain gene product has often yielded a clear understanding of developmental processes that were once difficult to investigate.

Reductionist and Synthetic Analyses of Development

Analysis is a separation of a whole into its component parts, and the study of how the features of the parts determine the properties of the whole and vice versa. Most types of analysis tend to be *reductionist,* because they try to explain a complex system in terms of its elements, which are at a lower level of organization. Explaining the properties of an organism in terms of the properties of its cells is an example of a reductionist analysis. At another level of reductionism would be the attempt to explain the properties of cells in terms of the properties of their component molecules.

A potential problem with reductionist analysis is that in separating a system's parts, properties may be lost or destroyed. The properties of cells or molecules studied *in vitro,* which means in culture dishes or test tubes, may differ from the properties *in vivo,* that is, in the intact organism. The advantage of in vitro studies is that investigators can administer nutrients, radioactive marker molecules, and other reagents in well-controlled experiments. However, culture conditions also deprive cells of their contacts with other cells and their normal environment of extracellular materials. As a result, cells may display only some of their normal functions, or altogether abnormal behavior. Even worse, in the course of prolonged culture there is a selection process that favors mutant cells that happen to be well adapted to the culture conditions but have less and less in common with normal cells in vivo. A similar problem occurs when a molecule serving a crucial function is unstable and eludes standard procedures of extraction and purification. Under these circumstances, a different, less active molecule may be mistaken for the biologically active molecule.

To avoid the pitfalls of reductionist analysis, investigators must be aware of the limitations of their methods and of misleading artifacts that can result from conditions in vitro. One safeguard is the application of two independent methods, such as electron microscopy and biochemistry, to clarify the same point. Another safeguard is to use data obtained in vitro for deriving hypotheses that can be tested in vivo. In addition, investigators must pay particular attention to phenomena observed in vivo that cannot be explained by reductionist theories.

Contemporary developmental biology makes use of the two levels of reductionist analysis that were already mentioned. At the first level, scientists attempt to analyze the development of an organism in terms of cellular properties. This line of work began late in the nineteenth century when it became clear that all animals and plants consist of cells (E. B. Wilson, 1925). Since then, understanding morphogenetic movements such as gastrulation or neurulation in terms of cellular properties has been a major goal (Trinkaus, 1984). A second level of reductionist analysis derives from recent progress in genetics and molecular biology. The first model of gene regulation explained how bacteria transcribe only those genes whose products they need in any given situation (Jacob and Monod, 1961). In higher organisms, the control of gene expression proved to be much more complex. Nevertheless, the concept of differential gene expression was soon applied to development (E. H. Davidson, 1986). The core of the theory is that all the cells of an organism have the same complement of genetic information and that cells acquire different structures and functions by *expressing* different sets of genes. The objective now is to identify the key genes and to learn how their expression is orchestrated to bring about the properties of cells and, ultimately, the epigenesis of a complex organism.

The genetic and molecular analysis of development has become a new frontier in science. In three organisms, the fruit fly *Drosophila melanogaster,* the nematode *Caenorhabditis elegans,* and the wall cress *Arabidopsis thaliana,* the genome has been saturated with mutations, and many of the genes involved in certain developmental processes have been identified. An effort is also under way to map and sequence the entire human genome. Other efforts are aimed at developing molecular techniques that will allow investigators to study the expression of certain genes without using mutants. In Part Two of this book, we will introduce as much molecular biology as is needed to understand these techniques, and Part Three will include some of the major success stories from this type of work.

As cellular and molecular events are studied in intact embryos, it becomes clear that at least some events do depend on signal exchanges within the intact organism. For example, certain cells that behave as cancer cells in adult animals and in tissue culture are "tamed" and develop normally when they are properly integrated into an embryo. Somehow, the cell movements and shape changes involved in morphological processes feed back at the molecular signals that control gene expression. Therefore, the molecular events that control cellular behavior and hence the development of the whole embryo are controlled in turn by signals gener-

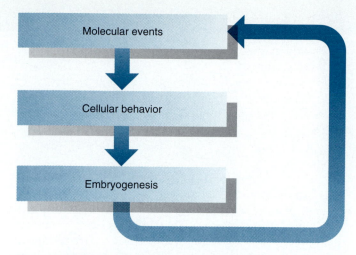

Figure 1.22 Synthetic analysis of development takes into account that molecular events, cellular behavior, and the development of the entire embryo control each other in a cyclic fashion.

ated by the entire embryo (Fig. 1.22). Here, the reductionist analysis comes full circle and turns into *synthetic analysis,* which considers the interdependence of molecular, cellular, and organismic events.

SUGGESTED READINGS

Davidson, E. H. 1986. *Gene Activity in Early Development.* 3d ed. New York: Academic Press.

Trinkaus, J. P. 1984. *Cells into Organs: The Forces That Shape the Embryo.* 2d ed. Englewood Cliffs, N.J.: Prentice-Hall.

Wilson, E. B. 1925. *The Cell in Development and Heredity.* 3d ed. New York: Macmillan.

THE ROLE
OF CELLS IN
DEVELOPMENT

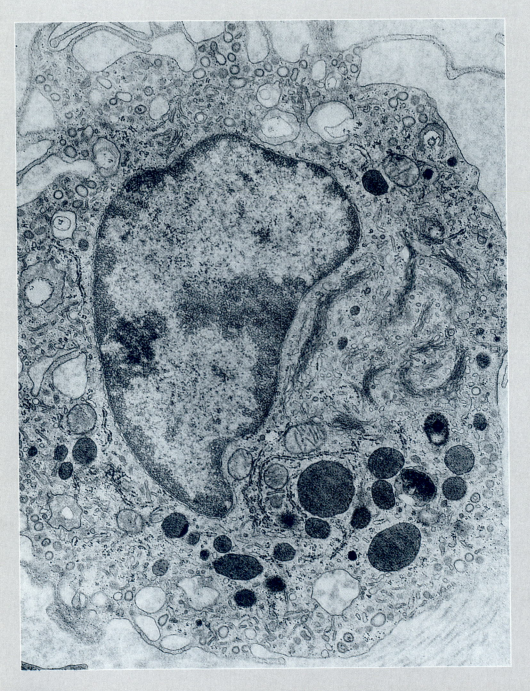

Figure 2.1 Transmission electron micrograph of a macrophage. It contains a fair representation of the many different components of an animal cell. The large organelle in the center—the cell nucleus—contains the chromatin, a meshwork comprised mostly of DNA and protein. The nucleus is surrounded by the nuclear envelope, a double membrane with openings that allow RNA to exit and regulatory proteins to enter. The cytoplasm outside the nucleus contains various smaller organelles. The oval organelles with ribs, such as the one visible in an indentation of the nuclear envelope, are mitochondria, the power plants of the cell. The dark bodies are lysosomes, membranous bags filled with enzymes that digest materials taken up from outside the cell. The plasma membrane surrounding the cell often forms fingerlike projections called microvilli.

Cells are the smallest units of matter having all the attributes of living things. They have a distinct form and maintain a high degree of internal organization (Fig. 2.1). They take up energy and exchange materials with the environment. Most cells have a full complement of genetic information, and they reproduce by dividing into daughter cells. Moreover, cells communicate with each other by signals and receptors, and they have an elaborate system of internal messengers to process the information received from outside.

Cells are also the smallest units that can spontaneously aggregate into organisms. If a sponge is forced through a fine-meshed cloth, the surviving cells can reaggregate into a living sponge (H. V. Wilson, 1907). Such reconstitution of an organism has never been observed with units smaller than cells, such as organelles or molecules.

In animals more complex than sponges, the ability to reconstitute a whole organism from disaggregated cells is limited to the embryonic stage. Isolated blastomeres from sea urchin embryos at the 16-cell stage can reassemble into almost normal-looking larvae (Freeman, 1988). Viable mice can be formed by combining separated blastomeres from three different embryos; such aggregates, when reimplanted into foster mothers, develop into mice of normal size and behavior (Markert and Petters, 1978). Cells isolated at later stages of development no longer assemble into viable organisms. However, they retain the ability to form aggregates that resemble embryos or embryonic parts.

To developmental biologists, cells are the natural units from which organisms are made. Many processes in development have been analyzed in terms of cellular behavior. Growth, for example, can be studied in terms of cell division, and gastrulation can be studied in terms of cell movements and changes in shape. To prepare for this type of analysis, we will review some of the basic properties of cells in this chapter. Our review will be limited to *eukaryotic cells* (Gk. *eu*, "good," "normal"), which make up the organisms studied by most developmental biologists. Eukaryotic cells have a distinct nucleus and cytoplasmic organelles such as mitochondria and an endoplasmic reticulum. We will focus on those cellular properties that are especially relevant to development, including the cell cycle, the cytoskeleton, the cell membrane, cell movement, intercellular junctions, and cell signaling.

The Principle of Cellular Continuity

Today every student of biology learns that all organisms consist of cells. This well-known fact was first proposed as a theory in 1838 by Matthias Schleiden and a year later by Theodor Schwann. During the decades that followed, investigators found that cells arise *only* through the division of existing cells. This was an important discovery, because earlier scientists thought that cells might also form spontaneously from noncellular materials. However, this has never been found to occur under the conditions that now prevail on earth, although cells probably evolved from simpler aggregates of organic matter under different environmental conditions during the early history of our planet. The early insights into the continuity of cells were summed up succinctly by Rudolf Virchow in 1858 in a famous aphorism, *omnis cellula e cellula,* meaning that all cells arise from cells.

Another important advance in cell theory was the discovery that eggs and sperm are specialized cells, and that they arise from less conspicuous cells in the ovary or testis. In 1841, Rudolf Albert Koelliker showed that spermatozoa are special cells originating in the testis and are not, as others had suggested, parasitic animals in the seminal fluid. Likewise, the egg was recognized as a single cell by Karl Gegenbauer in 1861. A few years later, in 1875, Oskar Hertwig observed two nuclei in fertilized sea urchin eggs, one derived from the sperm and the other from the egg.

Together, these discoveries established the **principle of cellular continuity:** that all organisms have evolved through an uninterrupted series of cell divisions, punctuated by gamete formation and fusion. This means that the cells in our bodies today are the temporary ends of an *unbroken chain of cells,* extending back through our parents, grandparents, and earlier ancestors to Cro-Magnons, nonhuman primates, primitive mammals, reptiles, amphibians, fish, and invertebrates, and—ultimately—to a primordial unicellular organism that lived billions of years ago.

Modern biology began when the cell theory was combined with observations on chromosomal behavior. In 1883, Eduard van Beneden noted that the nuclei of conjugating gametes contribute equal numbers of chromosomes at fertilization. Chromosomal behavior during gamete formation and fertilization provided a physical basis for Mendel's laws and other observations about inheritance. At the beginning of the twentieth century it became clear that heredity is based on the continuity of cell division, and on the faithful replication and distribution of chromosomes. The analysis of cell organelles other than the nucleus, such as *centrosomes,* soon followed, and in 1896 the discoveries of cell biologists were summarized in Edmund B. Wilson's groundbreaking book *The Cell in Development and Heredity.*

The Cell Cycle and Its Control

For an organism to develop and reproduce, its cells must increase in number. They do so in a cyclic process of growth and division known as the *cell cycle* (Fig. 2.2). A seemingly quiescent growth period known as *interphase* alternates with the dramatic process of cell division, called *M phase.* The major events that occur during M phase are mitosis and cytokinesis. *Mitosis* (Gk.

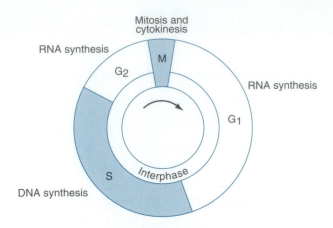

Figure 2.2 Schematic representation of the cell cycle. The relative lengths of the four phases vary among cell types, with G_1 exhibiting the greatest variation.

mitos, "thread," alluding to the appearance of chromosomes during this phase, during which they are visible under the microscope) designates the phase in which the cell nucleus divides. *Cytokinesis* (Gk. *kytos,* "cell"; *kinesis,* "movement") refers to the division of the cell cytoplasm. During interphase, most chromosomes are too thin to be visible under the light microscope. This is the working phase for the cell nucleus, the period when DNA and RNA are synthesized. The replication of DNA is limited to an interval within the interphase called the *S phase* ("S" for synthesis of DNA). The synthesis of RNA occurs during the interval preceding S phase, known as the *G_1 phase* ("G" refers to a gap in DNA synthesis), and during the interval following S phase, known as the *G_2 phase.* Thus, a typical interphase is made up of successive G_1, S, and G_2 phases. Generally, a cell grows during interphase to the predivision size of its parent.

Chromosomes Are Duplicated during S Phase

Nuclear DNA, the carrier of genetic information, is assembled in **chromatin,** which contains histones and other chromosomal proteins in addition to DNA. The natural segments of chromatin are called **chromosomes;** their number and form are characteristic of each animal species. Before S phase, each chromosome contains one long DNA molecule. During S phase, the chromosomal DNA is faithfully replicated to preserve the genetic information; in fact, cells have several enzymatic mechanisms for editing out errors that occur during replication. (The few errors that escape correction are mutations, which are inherited thereafter.) Following DNA replication, the chromatin structure of each of the twin DNA strands is restored by the addition of new histones and other proteins. Thereafter, each chromosome consists of twin **chromatids** held together at a region called the **centromere.** DNA synthesis and chromatid separation are delayed at the centromere; this region plays a key role in mitosis.

Chromosome Duplicates Are Split between Daughter Cells during Mitosis

The events that occur during mitosis are of truly fundamental importance. Each chromosome splits exactly into two chromatids, and an elaborate choreography of chromosome movements ensures error-free allocation of chromatids to the daughter cells. Mitosis is divided into five stages, known as *prophase, prometaphase, metaphase, anaphase,* and *telophase.* A stage-by-stage account of the mitotic events accompanies Figure 2.3.

The cell nucleus changes dramatically during the cell cycle. The *nuclear envelope,* which surrounds the nucleus during interphase, breaks up into small vesicles during prometaphase. At the end of mitosis, the fragments reunite with the separated chromatids to form new envelopes around each of the two daughter nuclei. Chromatids undergo an immense amount of coiling during prophase so that the long, thin structures do not become hopelessly entangled when they segregate. The coiled chromatids are visible as short threads under the light microscope (Fig. 2.3). The two joined chromatids of each chromosome form a distinct V or X shape, depending on whether the centromere is at the tip or near the middle of the chromosome. Two disc-shaped structures called *kinetochores,* facing in opposite directions and each aligned with one chromatid, form next to the centromere.

The key organelle during mitosis is the *mitotic spindle,* which is visible from prometaphase to anaphase (Mitchison, 1989; McIntosh and McDonald, 1989; Fuller and Wilson, 1992). The mitotic spindle extends between two poles, which are marked by *centrosomes* (not to be confused with centromeres). In most animal cells, centrosomes consist of a pair of cylindrical bodies called *centrioles,* which are positioned at right angles to each other and surrounded by a dense matrix (Glover et al., 1993). The plane midway between the spindle poles and perpendicular to the axis connecting them is called the *equatorial plane.* The most conspicuous elements of the spindle—the spindle fibers—consist of *microtubules.* The spindle fibers can be subdivided into three classes. *Astral fibers* radiate outward from the two centrosomes. *Pole-to-pole* fibers project from one centrosome toward the equatorial plane, where they interact with their counterparts from the opposite centrosome. *Kinetochore fibers* project from one centrosome and interact with one kinetochore of a chromosome.

Because the two kinetochores of a chromosome face in opposite directions, each interacts with one pole of the spindle. The chromosome thus becomes the object of a tug-of-war as each kinetochore is pulled by the kinetochore fibers to which it is attached. The force with which a chromosome is pulled to each pole seems to be proportional to the length of the kinetochore fibers. As

a result, all chromosomes move to the equatorial plane, where collectively they form the *metaphase plate.* Each chromosome is oriented with one of its chromatids facing one spindle pole and its sister chromatid facing the opposite spindle pole. This orientation ensures that each chromosome, when it divides, will contribute one of its chromatids to each of the daughter nuclei.

When the centromere splits after metaphase, the two chromatids become independent chromosomes and move to opposite spindle poles. The chromosomal movement during anaphase is driven by two forces. Each chromosome is pulled toward a spindle pole by kinetochore movement along the kinetochore fibers. In addition, the pole-to-pole microtubules elongate and interact to push the two spindle poles apart.

The Cell Cytoplasm Is Divided during Cytokinesis

Mitosis is followed by cytokinesis, when the cytoplasm of the mother cell is divided between the daughter cells. Cytokinesis begins during anaphase with the formation of a furrow in the plasma membrane. This furrow is often formed by the contraction of a circular bundle of filaments known as the *contractile ring* (Fig. 2.4). The contractile ring consists of actin and myosin filaments and assembles in the plane that was occupied by the metaphase plate during mitosis. The filaments of the contractile ring are anchored to proteins embedded in the cell plasma membrane, so that constriction of the ring causes the membrane to furrow. The force exerted by the contractile ring is most likely generated by sliding of the actin and myosin filaments past one another, a mechanism similar to the action of these proteins in muscle contraction. As the ring constricts, the two daughter cells are pinched apart. In some types of cells cytokinesis is incomplete, so that a temporary or permanent cytoplasmic bridge persists between the daughter cells.

The mechanism of cytokinesis in plants differs from that in animals. Instead of being pinched apart by the contraction of actin and myosin filaments, plant cells undergo cytokinesis by building a new cell wall. New cell walls are built from precursor materials contained in membrane-bound vesicles derived from the *Golgi apparatus.* These vesicles fuse to form a disclike structure called the *cell plate* while the vesicle contents assemble into pectin, hemicellulose, and other components of the primary cell wall. The orientation of the cell plate, like that of the metaphase plate, is perpendicular to the mitotic spindle axis.

The Cyclic Activity of a Protein Complex Controls the Cell Cycle

The cells in multicellular organisms divide at different rates. Embryonic cells may divide every 10 minutes, epidermal cells go through a complete cell cycle within

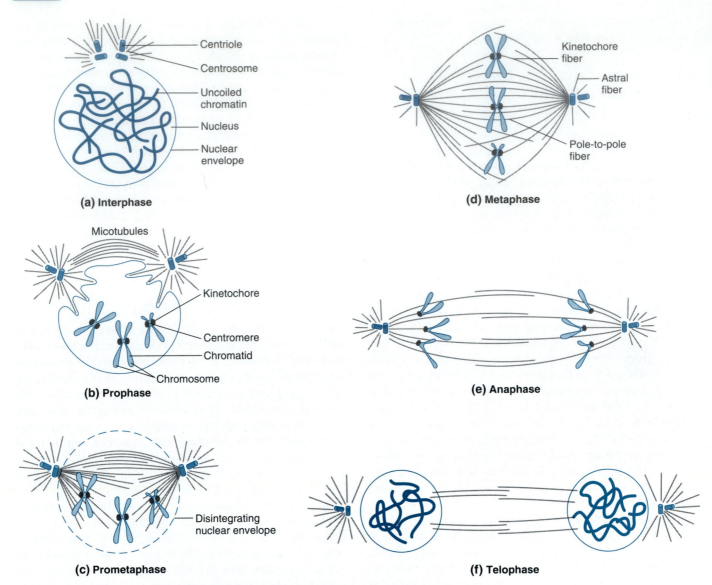

(a) Interphase

- Centriole
- Centrosome
- Uncoiled chromatin
- Nucleus
- Nuclear envelope

(d) Metaphase

- Kinetochore fiber
- Astral fiber
- Pole-to-pole fiber

(b) Prophase

- Micotubules
- Kinetochore
- Centromere
- Chromatid
- Chromosome

(e) Anaphase

(c) Prometaphase

- Disintegrating nuclear envelope

(f) Telophase

Figure 2.3 Stages of the cell cycle. **(a)** Interphase. The nuclear envelope is intact, and chromosomes are *not* seen as distinct structures. The genomic DNA and an extranuclear organelle, the centrosome, are both replicated. Each centrosome consists of two cylindrical bodies, called centrioles, and a surrounding matrix. **(b)** Prophase. Chromosomes condense into threads that are visible under the light microscope. Each chromosome consists of two chromatids, held together at the centromere. At the centromere, each chromatid is associated with a kinetochore. The centrosomes outside the nucleus move apart and begin to organize the spindle fibers. **(c)** Prometaphase. The centrosomes have moved to nearly opposite sides of the nucleus. The nuclear envelope disintegrates into small, membranous vesicles. This allows the spindle fibers to interact with the kinetochores. **(d)** Metaphase. The mitotic spindle is established, with the centrosomes as opposite poles. The spindle fibers can be grouped into three classes: kinetochore fibers project between a kinetochore and a centrosome; pole-to-pole fibers project between the two centrosomes; and astral fibers radiate out from the centrosomes. The chromosomes are aligned in the metaphase plate midway between the centrosomes and perpendicular to the spindle axis. The two kinetochores of each chromosome are pulled to opposite centrosomes, but the two chromatids are still held together at the centromere. **(e)** Anaphase. The chromatids separate, each becoming an independent chromosome. Each chromatid moves toward the centrosome to which it is connected by kinetochore fibers. The kinetochore fibers shorten while the pole-to-pole fibers elongate and interact to push the spindle poles apart. **(f)** Telophase. Kinetochore fibers disappear. A new nuclear envelope forms around each group of chromosomes. Chromosomes uncoil.

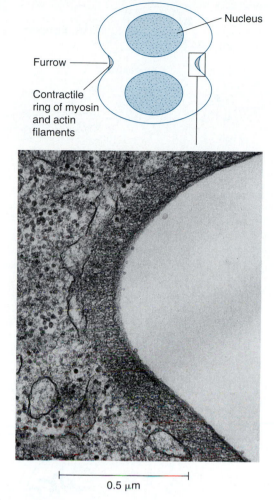

Figure 2.4 The contractile ring of a cell during cytokinesis: schematic drawing and transmission electron micrograph. (From H. W. Beams and R. G. Kessel, Cytokinesis: A comparative study of cytoplasmic division in animal cells, *Am. Sci.* **64**:279–290, 1976.)

clin-dependent kinase, *cdk1*, was characterized in fission yeast as the product of the *cdc2 gene*, but very similar proteins seem to occur in all eukaryotic cells (Nurse, 1990). Yeast have only one cyclin-dependent kinase and one cyclin; mammalian cells have at least five cyclin-dependent kinases (cdk1 through cdk5) and as many cyclins (cyclin A through cyclin E). Different combinations of cyclin-dependent kinases and cyclins regulate the progress of cells through different phases of the cell cycle (Sherr, 1993).

The transition from interphase to mitosis is induced by a protein complex known as **M-phase promoting factor (MPF)**, which consists of cdk1 and cyclin B (Fig. 2.5). (Historically, MPF was first described in the context of *meiosis*—see Chapter 3—as a *maturation promoting factor*. Only later was it found that MPF stimulates mitotic cell divisions as well.) To form active MPF, cdk1 must be modified by phosphorylation of a particular threonine residue, Thr-161, and dephosphorylation of a tyrosine residue, Tyr-15. In addition, cdk1 must associate with *cyclin B*, which is also activated by phosphorylation. The kinase activity of MPF is thought to cause, directly or indirectly, the events that occur during prophase, including chromosome condensation, the breakdown of the nuclear envelope, and the assembly of the mitotic or meiotic spindle. Right after MPF has reached its peak activity at metaphase, cyclin B is

a day, and most adult neurons have stopped dividing altogether. The differences in cell cycle times are due mainly to variations in the length of the G_1 phase. Most cells that stop dividing are arrested in a quiescent G_1 phase, also called a G_0 phase.

The progress of cells through the cell cycle is regulated by certain enzymes known as *cyclin-dependent kinases*. These enzymes carry out *kinase* activity, which means that they help to add phosphate groups to (phosphorylate) other proteins. This function is of great importance in cells because phosphorylation changes the three-dimensional structure and thus the biological activity of proteins. There are many types of kinases with different properties. The activity of cyclin-dependent kinases requires association with a cyclin. The *cyclins* are a family of proteins so named because the abundance of most of them changes during the cell cycle. For instance, cyclin B accumulates during interphase and is abruptly degraded after metaphase. The first cy-

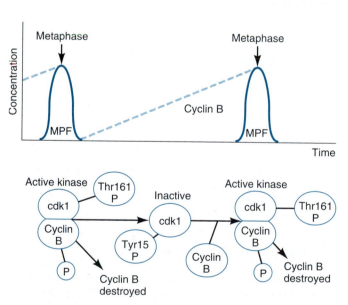

Figure 2.5 Schematic diagram illustrating the cyclic generation and destruction of M-phase promoting factor (MPF) during the cell cycle. At the beginning of each M phase, cdk1 protein is modified and becomes associated with another protein, cyclin B. Specifically, cdk1 is phosphorylated at a particular threonine residue, Thr-161, and dephosphorylated at a tyrosine residue, Tyr-15. Cyclin B is synthesized throughout the cell cycle and rapidly destroyed by specific proteases after each metaphase. During its association with cdk1, cyclin B is also phosphorylated. After proper modifications and association, the cdk1–cyclin B complex is an active kinase.

rapidly degraded by a *protease,* that is, an enzyme that breaks down proteins, and the modification steps that activate cdk1 protein are reversed. Thus MPF becomes inactive and the cell proceeds to the next interphase.

Other combinations of cyclin-dependent kinases and cyclins regulate other steps in the cell cycle. In mammalian cells, for instance, the abundance of cyclin E peaks at the transition from G_1 phase to S phase, and it appears that cdk2 combined with cyclin E controls the ability of adult mammalian cells to enter S phase.

Cyclic MPF activation by cdk1 modification and association with cyclin B is a basic cell cycle control mechanism that has been extremely well conserved in evolution. The cdk1 proteins from yeast and humans are so similar that they can functionally replace each other, and cyclins from clams and sea urchins function in frog oocytes (Draetta et al., 1987). However, the control of the cell cycle is modified in various ways at different stages of development, and some of these modifications seem to vary among groups of animals (Whitaker and Patel, 1990; Murray, 1992). For instance, the entry of embryonic cells into M phase is regulated by the enzyme that dephosphorylates Tyr-15 in cdk1. Mature cells have other checkpoints: unreplicated DNA blocks the cell cycle before entry into mitosis, and absence of a mitotic spindle blocks entry into metaphase.

Cell Shape and the Cytoskeleton

Cells Change Their External Shape As Well As Their Internal Order

Cells have the capacity to either stabilize or change their overall shape and their internal organization. Morphogenetic processes such as gastrulation and neurulation, which shape the whole embryo, are partly the result of coordinated changes in cell shape. For instance, amphibian ectodermal cells assume very different shapes depending on whether they develop into epidermis or into neural plate (see Fig. 1.11). Prospective epidermal cells become *squamous,* which means flat or scalelike; whereas prospective neural plate cells become *columnar,* that is, tall like a column (Fig. 2.6a). As the neural plate gives rise to the neural tube, the cells change shape further through constriction of their *apical* surfaces, that is, the surfaces that originally face outward. This *apical constriction* contributes to the bending of the neural plate into a tube.

In addition to a distinct outward shape, cells also maintain a high degree of inner order. Eggs, for instance, distribute certain cytoplasmic components unevenly, enriching some areas with specific mRNAs, proteins, pigment granules, or other organelles. Eggs and other cells can generate such asymmetrical arrangements in their cytoplasm by oriented transport. The fact that cells maintain these asymmetries against the force

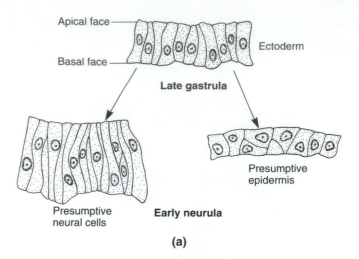

(a)

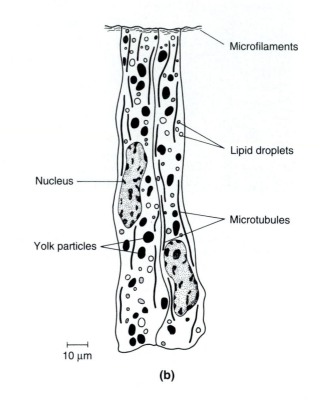

(b)

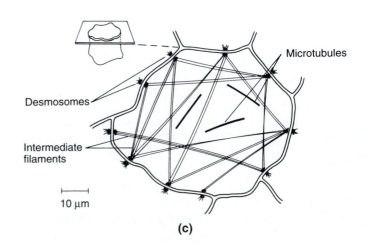

(c)

TABLE 2.1

Cytoskeletal Components

	Microtubules	Microfilaments	Intermediate Filaments
Fiber diameter (nm)	24	7	8–10
Polypeptide subunit	Tubulin	Actin	Keratins; vimentin; desmin; neurofilaments; lamins
Unpolymerized form	Globular, dimer	Globular, monomer	Helical and globular domains, dimer
Inhibitors	Colchicine; nocodazole; low temperature; high pressure	Cytochalasins; regulatory proteins (e.g., profilin, gelsolin)	Specific antibodies
Promoters	Taxol; D₂O	Phalloidin	——

of diffusion suggests that there must be "scaffolds" to hold cellular components in place.

How do cells change or maintain their shape? How do they conduct oriented transport and maintain asymmetries within the cytoplasm? Of particular importance for these activities is a network of cellular filaments known as the **cytoskeleton.** Improved methods of light and electron microscopy have revealed three types of cytoskeletal filaments that crisscross the cytoplasm (Table 2.1). Two types are present in all eukaryotic cells: *microtubules* and *microfilaments,* which have characteristic diameters of 24 nm and 7 nm, respectively (Fig. 2.6b). A third, more heterogeneous class of cytoskeletal fibers are intermediate in diameter (typically 8 to 10 nm) and are therefore called *intermediate filaments* (Fig. 2.6c). We will discuss each type of filament along with their roles in different cell activities.

Microtubules Maintain Cell Shape and Mediate Intracellular Transport

In electron micrographs, *microtubules* look like hollow rods about 24 nm in diameter and up to 500 μm in length. Microtubules self-assemble from subunits, each of which is a dimer consisting of two different globular polypeptides called *tubulins* (Fig. 2.7). The tubulins are a family of very similar polypeptides encoded by a corresponding family of genes. The most common type of dimer consists of one polypeptide from a subfamily

called α-tubulins (alpha tubulins) and another polypeptide from a subfamily called β-tubulins (beta tubulins); dimers composed of other tubulins are found in special cells, such as spermatozoa.

The assembly of microtubules is reversible and reveals an inherent polarity, which is based on the polarity of the tubulin dimers. At one end of the microtubule, designated the (+) end, assembly outpaces disassembly while the converse is true at the other end, called the (−) end. In many microtubules the (−) ends are anchored in a nucleation center, such as the centrosome, for protection against disassembly.

The equilibrium between free tubulin and microtubules can be shifted through normal cellular regulation and by experimental manipulation. In cultured cells under normal growth conditions, about half the total amount of tubulin is contained in microtubules and the other half is present in tubulin dimers and monomers. The assembly of microtubules is modulated by the availability of guanosine triphosphate (GTP) and by *microtubule-associated proteins (MAPs).* Some MAPs are species-specific, while others may be characteristic of certain tissues or developmental stages. Low temperature or high pressure shifts the equilibrium toward free tubulin. The drugs *colchicine* and *nocodazole* bind to tubulin dimers and prevent their assembly into microtubules. This leads to the disappearance of microtubules, because their spontaneous disassembly is not blocked at the same time. A different type of drug, *taxol,* stabilizes microtubules, as does heavy water (D₂O).

Microtubules stabilize the tall, columnar shape that characterizes, for example, the cells in the neural plate of vertebrates (Fig. 2.6b). These cells lose their distinctive shape and shorten when treated with colchicine, low temperature, high pressure, or other agents that destroy microtubules. After being restored to normal conditions, such treated cells rebuild their microtubules and regain their columnar shape. Similar experiments show that microtubules provide the critical support for cell extensions such as cilia and flagella. As described earlier, microtubules are also the principal components of mitotic spindles.

Figure 2.6 Shape changes in the ectoderm cells of a newt embryo during neurulation. **(a)** At the end of gastrulation, the ectoderm is an epithelium of short columnar cells. As neurulation proceeds, the presumptive neural cells form the neural plate of tall columnar cells. Meanwhile, the presumptive epidermal cells become much flatter. **(b)** Two neural plate cells. Numerous microtubules are aligned parallel to the long axis of the cells. Microfilaments are arranged in a bundle encircling the apical end of the cell like purse strings. **(c)** Flattening epidermis cell. Beneath the apical face, bundles of intermediate filaments span the cytoplasm between areas of cell contact known as *desmosomes.* The few microtubules in these cells seem randomly oriented.

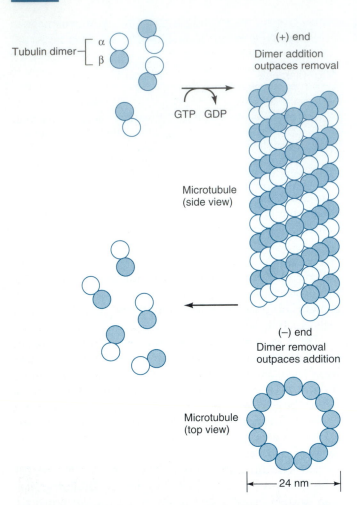

Tubulin dimer — α β

(+) end

Dimer addition outpaces removal

GTP GDP

Microtubule (side view)

(−) end

Dimer removal outpaces addition

Microtubule (top view)

|← 24 nm →|

Figure 2.7 Microtubule assembly. The building blocks of microtubules are dimers consisting of two different tubulin polypeptides, usually α-tubulin and β-tubulin. Dimers are constantly added to microtubules and removed again. Dimer addition outpaces removal at one end of the microtubule, called the (+) end, whereas the converse is true at the other, called the (−) end. The addition of dimers requires guanosine triphosphate (GTP), which is hydrolyzed to guanosine diphosphate (GDP) in the process.

In addition to their role in maintaining cell shape, microtubules are involved in the organization and transport of cell organelles and molecules within the cell. For instance, in the axons of neurons, microtubules convey an extensive traffic of materials in both directions between the cell body and the axonal terminus (Allen, 1987). The transport of certain cell organelles along microtubules is mediated by *motor proteins* such as *dynein* and *kinesin*. Each of these motor proteins has a receptor domain linking it with the organelle to be transported and two binding domains by which it binds reversibly to microtubules. Using energy obtained by hydrolyzing adenosine triphosphate (ATP), these binding domains move along the microtubules, dynein toward the (−) end of the microtubule and kinesin toward the (+) end. Motor proteins also generate the chromosomal movements in the mitotic spindle (Fuller and Wilson, 1992).

Microfilaments Stabilize the Cell Surface and Generate Contracting Forces

A second class of cytoskeletal elements, the *microfilaments,* appear under the electron microscope as straight fibers about 7 nm in diameter. Microfilaments self-assemble as double-helical fibers from a globular polypeptide called *actin* (Fig. 2.8). Actin, which may constitute as much as 10% of the total cellular protein, is one of the best-conserved proteins in evolution. Like tubulins, actins are a group of proteins encoded by a family of genes. Found in most metazoa, these proteins vary slightly in structure, presumably because they have been adapted for different functions. Some types of actin are very abundant in muscle fibers, where they are part of the contractile system. Other types of actin form the microfilaments of nonmuscle cells.

The assembly of microfilaments is similar to that of microtubules in several respects. In vivo as well as in vitro, microfilaments self-assemble from actin in the presence of ATP, calcium or magnesium ions, and proper nucleation centers. Also like microtubules, microfilaments are polarized, with assembly outpacing disassembly at the (+) end.

The equilibrium between assembled microfilaments and free actin can be modified by various drugs and regulatory proteins. A family of drugs called *cytochalasins* bind to the (+) ends of microfilaments, preventing further elongation. *Phalloidin,* a poison made by the mushroom *Amanita phalloides,* binds tightly to microfilaments and inhibits their disassembly. At proper concentrations, these drugs interfere specifically with microfilaments and not with microtubules or intermediate filaments. Thus, cytochalasin blocks cellular locomotion and cytokinesis without affecting overall cell shape or mitosis, and so this drug can be used to create multinucleate cells.

A variety of proteins control the assembly and length of microfilaments (Fig. 2.8). A small protein called *profilin* binds reversibly to actin molecules and inhibits their incorporation into microfilaments, at least in vitro. Profilin and other proteins also function as *capping proteins,* which bind specifically to the ends of microfilaments, thus controlling their further elongation or disassembly. *Severing proteins* can sever and cap microfilaments. For instance, *gelsolin* binds to the side of a microfilament, breaking it at the point of attachment and capping the (+) end created by the breakage. Thus, the combined effect of severing and capping is to break down microfilaments. Most likely, this mechanism plays a major role in cell motility, as we will see. Other proteins, such as α-*actinin,* cross-link microfilaments with one another or with different cytoskeletal components.

The various interactions between actin and other proteins reflect the diverse roles of microfilaments in

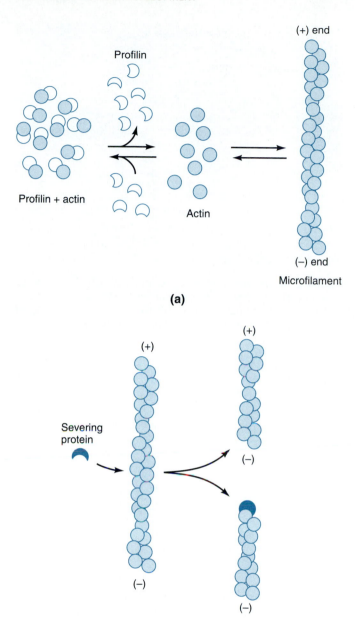

(a)

(b)

Figure 2.8 Assembly of microfilaments from actin in the presence of regulatory proteins. **(a)** Microfilaments assemble from actin. The end of the microfilament where assembly outpaces disassembly is designated the (+) end. Microfilament assembly is inhibited by the presence of profilin, which binds to actin. **(b)** Severing proteins, such as gelsolin, bind laterally to microfilaments so that they break. Severing proteins also cap the (+) end of one or both fragments, thus preventing further elongation.

the cell. A major function of microfilaments is to interact with motor proteins, such as myosin, and generate contracting forces. This has been shown most clearly in muscle cells, which are specialized for this function. Almost certainly the same type of interaction occurs in nonmuscle cells during cytokinesis when the microfilaments that make up the contractile ring constrict the bridge between the dividing cells. This conclusion is

supported by the inhibitory effects of both cytochalasin and myosin antibodies on the progress of cytokinesis. Microfilaments also form a contractile ring beneath the *apical* surface of *epithelial* cells, where they contribute to coordinated changes in cell shape.

As part of the cytoskeleton, microfilaments give dynamic mechanical support to the cell and its extensions. In the surface cytoplasm beneath the plasma membrane, microfilaments form a meshwork that resists deformation and supports the plasma membrane, which by itself is fluid and has no mechanical strength. The reinforcement of the surface cytoplasm with microfilaments is especially prominent in large cells like eggs, which have a more resilient surface cytoplasm, known as the *cortex* (Lat., "rind"). Other roles performed by microfilaments in cell locomotion and in the localization of cytoplasmic components will be discussed later in this chapter and in Chapter 8.

Microfilaments can assemble rapidly from actin monomers. One example is the assembly of microfilaments in the acrosomal process, a projection formed at the tip of many invertebrate spermatozoa when they approach an egg (see Chapter 4). In sea cucumber sperm, the acrosomal process is extruded at a speed of 10 μm per second, penetrating the egg envelopes like a harpoon and bringing the sperm and egg plasma membranes into contact (Fig. 2.9).

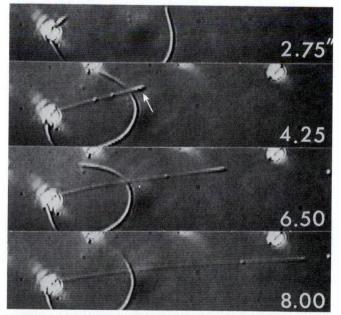

Figure 2.9 Acrosomal process formation in a sea cucumber sperm. The light micrographs were taken at intervals indicated in seconds to the lower right. The sperm head is to the left. The arc to the right of the sperm head is a portion of the sperm tail. The acrosomal process (arrow) is formed when the sperm contacts the jelly layer surrounding the egg. The contact releases actin that is bound to profilin and stored in the tip of the sperm. The free actin polymerizes rapidly into microfilaments, which cause the acrosomal process to extend.

Intermediate Filaments Vary among Different Cell Types

The third class of cytoskeletal elements is a diverse group of fibers collectively called *intermediate filaments.* Under the microscope, they are visible as arrays of straight or curved fibers typically 8 to 10 nm in diameter. There are several types of intermediate filaments, composed of different polypeptides including keratin, desmin, vimentin, glial fibrillary acidic protein, and others (see Table 2.1). The protein composition of intermediate filaments varies more from species to species than that of tubulin and actin. Although intermediate filaments consist of repetitive subunits, they do not seem to undergo the rapid assembly and disassembly that characterize microtubules and microfilaments. Specific inhibitors or promoters of intermediate filaments have not been found; however, their assembly can be disrupted by injecting antibodies against their constituent proteins.

The functions of intermediate filaments are poorly understood. The filaments' variation in distribution and molecular nature suggests that different intermediate filaments may have different functions. One of their roles seems to be stabilizing epithelial sheets. Epithelial cells are traversed by intermediate filaments, which connect at certain cellular junctions (see Fig. 2.6c). The transcellular network of fibers generated by this arrangement gives the epithelium tensile strength.

Other intermediate filaments, composed mostly of *lamin* polypeptides, are found inside the cell nuclei, where they connect parts of the chromatin to the inside of the nuclear envelope. In contrast to the cytoplasmic intermediate filaments, which are fairly stable, the nuclear filaments are broken down and reassembled as part of the cell cycle.

Transport across the Cell Membrane

Cells exchange various materials with their environment. Nutrients are taken up mostly as solutes, although some cells also engulf and digest other cells or cell fragments. These processes are vital to the survival of cells but also serve specific functions in development. For instance, the large amounts of yolk in eggs are accumulated in the oocyte through selective uptake of proteins from the maternal blood.

Cells may release substances of their own making, including enzymes, hormones, and neurotransmitters, into their surroundings. Cells also create their own physical environment by synthesizing and depositing materials collectively known as the *extracellular matrix.* Generally, the extracellular matrix in animal tissues consists of collagen and other fibers embedded

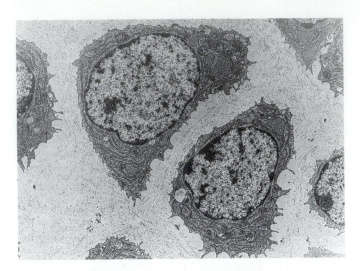

Figure 2.10 Transmission electron micrograph of cartilage, showing the massive extracellular material synthesized by the chondrocytes (cartilage cells). Note the well-developed endoplasmic reticulum in the chondrocytes.

into protein and carbohydrate complexes. In cartilage and other connective tissues, the extracellular material has more mass than the cells that created it (Fig. 2.10). The shape and mechanical properties of connective tissues are largely determined by the nature of their extracellular components.

In the course of development, cells remodel their environment by breaking down extracellular materials and depositing new materials in their place. But the interaction works both ways: the chemical composition of the extracellular matrix largely determines the behavior of cells (see Chapter 26).

The Plasma Membrane Consists of a Phospholipid Bilayer with Embedded Proteins

The key to understanding the transport of different materials into and out of cells is the molecular architecture of the cellular *membrane,* which consists mainly of a *lipid bilayer.* The most common membrane lipids are phospholipids, polar molecules with hydrophilic (water-soluble) "heads" and hydrophobic (water-repellent) "tails." In an aqueous medium, phospholipids self-assemble into a double layer of molecules, with the hydrophilic heads turned outside and two layers of hydrophobic tails facing each other on the inside (Fig. 2.11). The hydrophobic tails thus form an oily barrier to water-soluble molecules, preventing their free diffusion across cell membranes. Cellular membranes are like soap bubbles: they are effective barriers to diffusion, but they are fluid and have minimal mechanical strength. The relative stiffness of the cell surface is due not to its membrane but to a superficial layer of cytoplasm called the *cortex,* which is reinforced with micro-

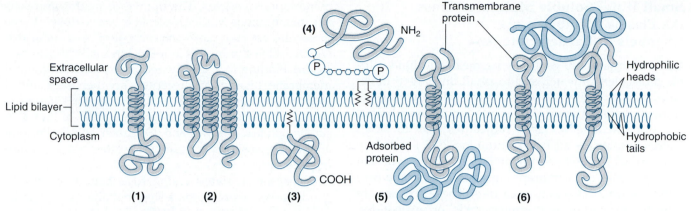

Figure 2.11 Model of a cell membrane. Phospholipid molecules form a bilayer with their hydrophobic tails squeezed together, and with the hydrophilic heads of one layer facing outside the cell and those of the other layer facing the cytoplasm. Some proteins are held in the lipid bilayer with a single hydrophobic domain **(1)** or with multiple domains **(2)**. Other proteins are anchored in the lipid bilayer by a lipid covalently linked to the protein, either directly **(3)** or via a complex sugar **(4)**. Many proteins are held to the membrane only by noncovalent adsorption to other membrane proteins **(5, 6)**.

filaments as described earlier. The membrane surrounding the entire cell is known as the *plasma membrane.* Similar membranes enclose intracellular organelles such as the mitochondria and the endoplasmic reticulum.

Cellular membranes can fuse with one another to form new configurations. This property facilitates the breakdown and reconstitution of membranes during many developmental processes, including mitosis, endocytosis and exocytosis (described in this chapter), and fertilization (see Chapter 4). In some of these events, the ease with which membranes fuse is regulated by special membrane proteins and by the composition of the lipid bilayer, especially the phospholipid:cholesterol ratio.

Plasma membranes contain a variety of proteins; some serve as receptors for signals from outside the cell, while others serve as carriers or channels for small molecules to cross the plasma membrane.

According to the widely accepted *fluid mosaic model* of membrane structure, membrane proteins are anchored in the lipid bilayer yet they are free to float laterally within it. Some membrane proteins have hydrophobic domains that are enclosed by the tails of the membrane lipids and hydrophilic domains that extend into the cytoplasm or out into the extracellular environment (Fig. 2.11). Other proteins are held in the membrane by covalent linkage to lipids or by weak interactions with other membrane proteins.

Many membrane proteins (as was already noted) are connected by linker proteins to cytoskeletal elements, especially microfilaments (Luna and Hitt, 1992). This association prevents the membrane proteins from floating laterally in the plasma membrane and allows cells to concentrate certain membrane proteins in spe-

cific areas of the plasma membrane. Membrane proteins that are connected to the cytoskeleton include proteins that in turn bind to extracellular matrix molecules (Fig. 2.12). This connection extends through the fluid plasma membrane and establishes a firm link between the cytoskeleton and the extracellular matrix. This link gives cells traction, allowing them to move on the extracellular matrix. Traction cannot be achieved in cells whose membrane proteins float freely in the plasma membrane.

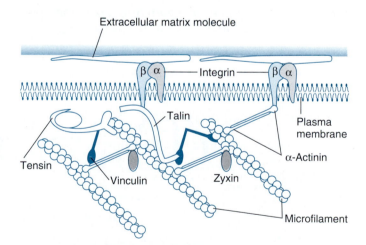

Figure 2.12 Model for the association of microfilaments with membrane proteins that in turn bind to extracellular matrix molecules. The membrane proteins, called *integrins*, are dimers consisting of two polypeptides (α and β). They bind to specific molecules outside the cell and to certain linker proteins—such as talin and α-actinin—inside the cell. These proteins and others—including tensin, vinculin, and zyxin—establish a firm link between microfilaments and membrane proteins.

Small Water-Soluble Molecules Cross the Plasma Membrane via Channels or Carrier Proteins

Cell membranes are generally permeable to lipid-soluble molecules, to water, and to small uncharged molecules such as carbon dioxide or glycerol. However, because of the oily nature of their interior, cell membranes are impermeable to large molecules and to small charged molecules such as sodium ion (Na^+) and chloride ion (Cl^-). Only specific molecules of this type can cross the plasma membrane, through membrane proteins that act as channels or carriers. These proteins are selective for certain ions, amino acids, or nucleotides, which may pass by means of diffusion or active transport.

There are two major classes of membrane transport proteins: channel proteins and carrier proteins (Fig. 2.13). *Channel proteins* are charged, hydrophilic molecules forming water-filled pores that extend across the lipid bilayer; when these pores are open they allow ions of appropriate size and charge to pass through. Such channels are opened and closed in response to changes in electrical potential across the membrane, to mechanical stimuli, and to other signals. In contrast, *carrier proteins* bind the molecule to be transported and undergo an energy-dependent conformational change in order to move the molecule and release it on the other side of the membrane.

Large Materials Enter Cells by Endocytosis and Leave by Exocytosis

Objects too large to be transported by membrane proteins are taken up into the cell or released from it in membranous vesicles. The uptake of matter into cells via membranous vesicles is called *endocytosis.* Large molecules are taken in by *receptor-mediated endocytosis,* in which the molecule to be internalized acts as a *ligand,* binding to a specific receptor protein in the cell membrane. Ligands and receptors are then internalized in small membranous vesicles. This process has been studied in particular detail for the uptake of low-density lipoprotein (Brown and Goldstein, 1984). Presumably, endocytosis is very similar for other receptor-binding molecules.

Receptor-mediated endocytosis begins when a ligand binds to its receptor, a step that allows the uptake of only one type of molecule (Fig. 2.14). A cell may have different types of receptor proteins, but each of them usually accepts only one ligand. After many receptors have been loaded with ligands, they are collected in a depression in the plasma membrane. This depression, called a *coated pit,* is coated on the cytoplasmic side with *clathrin,* a protein that facilitates the concentration of loaded receptors (Fig. 2.15). The pit is then pinched off from the plasma membrane as a *coated vesicle,* which fuses with another organelle called an *endosome.* In the vesicle resulting from this fusion, the ligands dissociate from their receptors, and the two become segregated into different vesicles. The vesicle with the receptor proteins is recycled to the plasma membrane by exocytosis. The remaining vesicle, with the ligands still inside, fuses with another membranous vesicle called a *lysosome.* In the fused vesicle, lysosomal enzymes degrade the ligands into small components, which eventually cross the vesicular membrane into the cytoplasm.

Protozoa and certain white blood cells can also engulf bacteria or other large particles and ingest them in

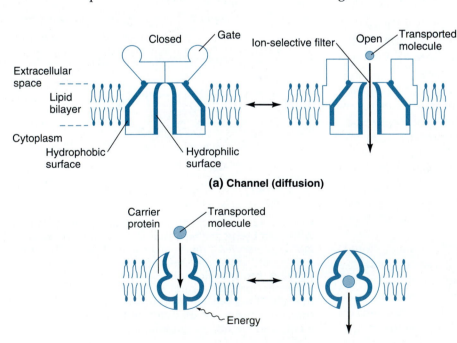

(a) Channel (diffusion)

(b) Carrier (active transport)

Figure 2.13 Membrane transport proteins act as channels for diffusion or as carriers for active transport. Hydrophobic amino acid domains of these proteins are embedded in the lipid bilayer. Hydrophilic amino acid domains are positioned on the inside, where they allow ions and other water-soluble molecules to pass. **(a)** Channel protein for diffusion. **(b)** Carrier protein for active transport.

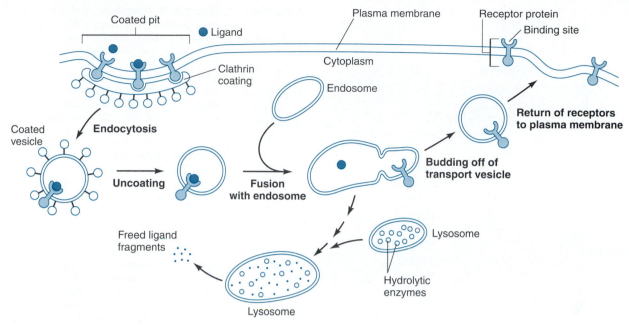

Figure 2.14 Receptor-mediated endocytosis. Large molecules bind as ligands to matching receptor proteins floating in the cell plasma membrane. Receptors with ligands are collected in a pit, which is coated with clathrin on the cytoplasmic side. The coated pit balloons into the cytoplasm and pinches off as a coated vesicle. Upon fusion of the vesicle with an endosome, the ligands dissociate from their receptors and are segregated into a different vesicle, which in turn fuses with a lysosome. Digestive lysosomal enzymes break down the ligand until the fragments can cross the vesicular membrane and enter the cell cytoplasm. The membrane vesicle containing the receptors is recycled by fusion with the plasma membrane.

a droplet of extracellular fluid. This process is specifically called *phagocytosis* (Gk., indicating the eating state of the cell).

The release of matter from cells via membranous vesicles is called *exocytosis.* Cells use exocytosis to recycle receptor proteins, to release water, to expel the remnants of phagocytosed particles, to secrete hormones and digestive enzymes, to increase the size of the cellular membrane, and to build up extracellular

materials. Proteins to be exported from the cell are synthesized on the surfaces of the *endoplasmic reticulum (ER),* an intracellular labyrinth of membranous sacs (Fig. 2.16). The newly synthesized proteins leave the ER in membranous *transport vesicles,* which shuttle the proteins to the *Golgi apparatus,* a stack of flat, membranous vesicles where the proteins are modified and sorted for different destinations. Proteins leave the Golgi apparatus in *secretory vesicles,* some of which move to

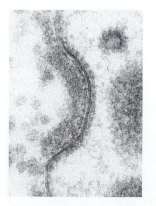 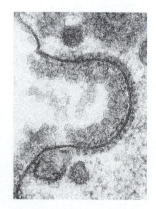 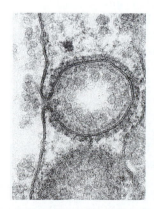 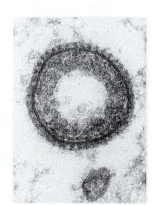

Figure 2.15 Transmission electron micrographs showing formation of a coated vesicle from a coated pit at the surface of a hen's oocyte. The oocyte cytoplasm is to the right and the extracellular space to the left. The coated pits and vesicles take up lipoproteins synthesized in the hen's liver and stored as yolk in the oocyte.

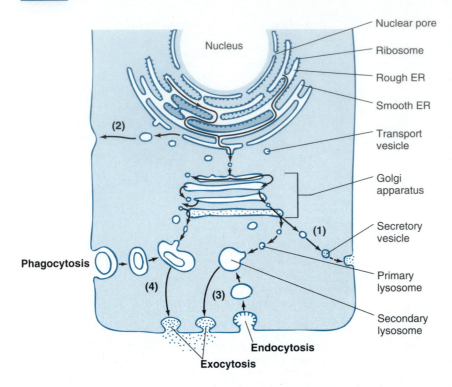

Nuclear pore
Ribosome
Rough ER
Smooth ER
Transport vesicle
Golgi apparatus
Secretory vesicle
Primary lysosome
Secondary lysosome

Nucleus
(2)
(1)
Phagocytosis
(4)
(3)
Endocytosis
Exocytosis

Figure 2.16 Functions of the endoplasmic reticulum (ER), Golgi apparatus, and secretory vesicles. In the first pathway **(1)**, synthesis of a protein begins in the rough ER (which is studded with ribosomes) and ends in the extracellular space, where the protein may serve, for instance, as an extra-cellular matrix component. Another pathway **(2)**, beginning in the smooth ER (which contains no ribosomes), is used simply to add more lipid bilayer to the plasma membrane. A third pathway **(3)**, again beginning in the rough ER, is used to store digestive enzymes in membranous vesicles called lysosomes, which may serve in receptor-mediated endocytosis. Finally **(4)**, lysosomes may serve in the phagocytosis of larger pieces of foreign matter.

the cell surface. When the vesicular membranes fuse with the plasma membrane, the contents of the secretory vesicles are shed outside the cell. The same route is used to assemble membrane vesicles in the ER and add them to the plasma membrane to increase its surface.

Instead of being exported from the cell in secretory granules, certain digestive enzymes leave the Golgi apparatus in *lysosomes*. Lysosomes fuse inside the cell with endocytotic and phagocytotic vesicles and digest the latter's contents before the remnants are expelled.

Cellular Movement

Another important characteristic of cells is motility. This is dramatically demonstrated by certain white blood cells in adult vertebrates, which, when attracted by signals of inflammation, can actually squeeze out of blood vessels and into infected tissue (Fig. 2.17). Similar feats are performed by primordial germ cells, which typically move a long way from their site of origin to the gonad rudiment (see Chapter 3). Mesenchymal cells move by crawling upon extracellular material. Even epithelial cells can move relative to their neighbors, thus causing the epithelium to elongate in one direction and shrink in another.

Cell locomotion has been studied extensively using cultured mammalian *fibroblasts*, cells that adhere in vivo to collagen fibrils and remain fairly stationary except during wound healing. In vitro, they adhere to and move around on the bottom of the culture dish. The *leading edge* of a fibroblast in motion advances forward while the *trailing edge* drags behind. Fibroblasts move by expanding at the leading edge and contracting at the trailing edge (Fig. 2.18). The leading edge continually forms flat extensions of cortical cytoplasm called *lamellipodia* (sing., *lamellipodium*; Lat. *lamella*, "little plate,"

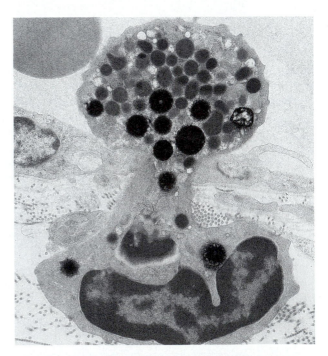

Figure 2.17 Transmission electron micrograph of a white blood cell passing through the wall of a blood capillary.

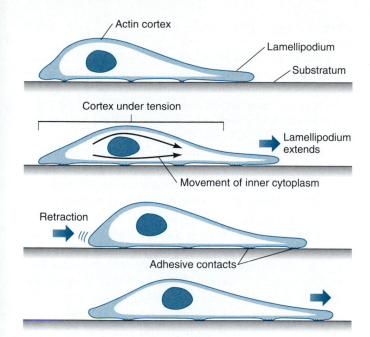

Figure 2.18 Movement of a fibroblast in culture. The cell extends a lamellipodium at the leading edge, makes new points of contact with the substratum, and severs older points of contact as the trailing edge contracts.

and Gk. *podos*, "foot"), some of which adhere to the substratum while previous points of adhesion near the trailing edge are retracted.

Another favorite subject for cell movement studies is the neuron—in particular, the *growth cone* at the tip of the growing axon (Fig. 2.19). The growth cone is rich in microfilaments, while the axon itself contains mostly microtubules for reinforcement and transport. The growth cone forms lamellipodia and also extends spike-like processes called *filopodia* (Lat. *filum*, "thread"). Filopodia are constantly extending and retreating, as if exploring their environment, and they are able to transmit environmental information to their parent growth cone (Davenport et al., 1993). They can grow as much as 50 μm in length, and they contain bundles of microfilaments. If a filopodium adheres to a particular site, the adjacent part of the lamellipodium may also adhere and pull the growth cone in that direction. It appears that this behavior enables the growth cone to reach its proper target cell.

Exactly how cells extend lamellipodia and filopodia, and how cells pull themselves forward, is still under investigation. Microfilaments attached to membrane proteins (Fig. 2.12) must play a major role, because *cytochalasin* stops cell movement. Indeed, new microfilaments are formed at the leading edge as a cell moves (Theriot and Mitchison, 1991). It also seems significant that during lamellipodium formation the cortical cytoplasm undergoes cycles of relaxation and stiffening (Oster, 1984; Bray and White, 1988). The cytoplasm becomes more fluid when a lamellipodium originally

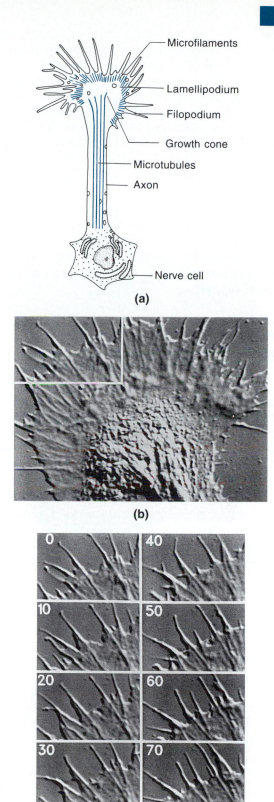

Figure 2.19 Growth cone of a neuron. **(a)** Schematic drawing of a neuron with a growing axon. **(b)** Video-enhanced image of a growth cone. Spikelike filopodia, reinforced with microfilament bundles, extend from the crescent-shaped lamellipodium. The lamellipodium excludes the granular organelles and microtubules present in the axon and the adjacent part of the growth cone. **(c)** A series of images taken every 10 s of the boxed region in part b. Note the constant protrusion and retraction of filopodia.

forms and more gel-like after the lamellipodium has made contact with the substratum. The relaxation may be caused by microfilament-severing proteins such as gelsolin (Fig. 2.8). The stiffening may be accomplished by cross-linking proteins such as α-actinin. The activity of microfilament-severing and cross-linking proteins depends on the intracellular pH and on the concentration of free calcium ions. It is therefore likely that these parameters play important roles in regulating cell movement.

Another area of active research is how cells establish the polarity of their movement. Cells extend their lamellipodia in a preferred direction, often in response to chemical signals. Presumably, such signals act by locally modulating the pH or the calcium ion concentration inside the cell.

Epithelia and Mesenchyme; Cell Junctions

As mentioned in Chapter 1, embryonic tissues are broadly classified as either epithelial or mesenchymal (Fig. 2.20). An *epithelium* is a sheet of contiguous cells that rest on a layer of *extracellular matrix* called a **basal lamina.** Epithelial cells are always polarized: the cell surface facing the basal lamina is the **basal surface,** and the opposite cell surface facing the outside world or certain body cavities is the **apical surface.** Typically, the Golgi apparatus is closer to the apical surface. The apical surfaces of many epithelia have small, fingerlike projections, called *microvilli,* which increase the surface area available for endocytosis and exocytosis. In contrast, the cells of a *mesenchyme* have only sparse contacts with their neighbors or are surrounded on all sides by extracellular material. If they are migratory they

show a leading edge and a trailing edge as discussed earlier.

Epithelial and mesenchymal cells are connected by different types of junctions. The common types of intercellular junctions are *desmosomes* and *tight junctions,* which are formed between epithelial cells, and *gap junctions,* which occur between epithelial cells as well as between mesenchymal cells (Fig. 2.21).

Desmosomes can be thought of as rivets that hold cells firmly together. At the so-called desmosomal core, the adjacent plasma membranes are connected by a thin layer of glycoprotein cement and fibrous material (Fig. 2.21c). The fibrous material seems to be connected to plaques of electron-dense material located on the cytoplasmic face of the plasma membrane on each side of the core. These plaques in turn serve as anchors for keratin and other cytoskeletal filaments that traverse the inside of the cell to end in another desmosome. The desmosomes thus link the intermediate filaments of adjacent cells to form a transcellular network that provides strength and elasticity to an epithelium.

Tight junctions connect epithelial cells (Fig. 2.21b). Strings of transmembrane proteins run continuously around the circumference of each cell in an epithelium. A tight junction is a zipperlike connection between a string of proteins in one cell membrane and the corresponding proteins in the membrane of an adjacent cell. The tight junction forms effective seals beneath the apical surfaces of the epithelial cells: there is no intercellular space left open for large molecules to diffuse between the epithelial cells. Instead, molecules crossing from outside to inside, or vice versa, must be gated through the plasma membranes. The barrier created by tight junctions between epithelial cells permits an organism to control its internal environment.

Tight junctions also prevent membrane proteins from floating between apical and lateral cell surfaces.

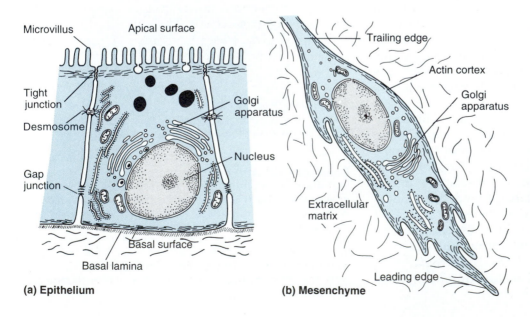

Microvillus Apical surface

Tight junction

Desmosome

Gap junction

Golgi apparatus

Nucleus

Basal surface

Basal lamina

(a) Epithelium

Trailing edge

Actin cortex

Golgi apparatus

Extracellular matrix

Leading edge

(b) Mesenchyme

Figure 2.20 Differences between epithelial cells and mesenchymal cells. **(a)** Epithelial cells rest on a layer of extracellular matrix known as basal lamina. They are polarized, with the basal cell surface adjacent to the basal lamina and the apical cell surface facing the outside world or a body cavity. Epithelial cells are closely connected to their neighbors with tight junctions and desmosomes. **(b)** Mesenchymal cells are only loosely connected to other cells or are surrounded entirely by extracellular matrix. When in motion, mesenchymal cells show a leading edge and a trailing edge.

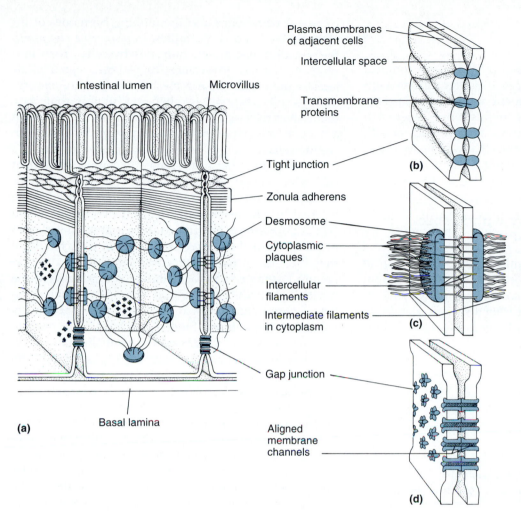

Figure 2.21 Three types of intercellular junctions. **(a)** Schematic diagram showing the inner epithelium of the small intestine in a mammal; the microvilli on top are projections of the apical cell surface. The basal cell surface rests on the basal lamina, a sheet of extracellular material. **(b)** A tight junction is composed of adjacent rows of transmembrane proteins, which hold the membranes together like zippers. Tight junctions effectively seal the tissue inside from the external environment. **(c)** A desmosome consists of a pair of cytoplasmic plaques, just inside the respective plasma membranes of a pair of adjacent cells. The core between the cell membranes is filled with fibrous material. Desmosomes are connected across the cytoplasm by intermediate filaments. Together, these structures form a transcellular network that gives the epithelium strength and elasticity. **(d)** A gap junction is an array of cylindrical channels consisting of membrane proteins. Channels of adjacent plasma membranes are aligned to transmit small molecules.

Thus, cells can maintain different sets of membrane proteins on these surfaces so that they can interact differently with the external environment and with the organism's own intercellular space. Embryos make use of this ability early in development. For instance, in mouse embryos at the 8-cell stage, tight junctions form between the outer cells. These cells later help implant the embryo in the maternal uterus, while the inner cells, which give rise to the embryo proper, maintain a different chemical environment (see Chapter 5).

Right beneath their tight junctions, epithelia often show a *zonula adherens,* or adhesion belt (Fig. 2.21a). Here the membranes are held together by cell adhesion molecules called *cadherins* (see Chapter 25). These molecules traverse the cell membrane and are linked to rings of microfilaments in the cell cytoplasm. This linkage seems critical for holding cells together during apical constrictions.

The third type of intercellular junction is the *gap junction,* which is an array of miniature channels between cells (Fig. 2.21d). Each channel consists of two connecting cylinders, formed by proteins embedded in the abutting cell membranes. Cells joined by gap junctions are electrically coupled and can exchange molecules of up to about 1200 daltons. Thus, gap junctions

serve as conduits for chemical signals traveling between cells. Significantly, cells can modulate the degree of coupling with their neighbors. Even slight increases in the intracellular concentration of free calcium ions or slight changes in pH can trigger a decrease in gap junction permeability. These effects are thought to be mediated by changes in the conformation of the protein units that constitute the two cylinders of a gap junction.

In most early embryos, at least some of the blastomeres are connected by gap junctions. This was revealed by an experiment in which hydrophilic dye molecules were injected into one cell and the dye was observed diffusing into the neighboring cells. Gap junctions often appear at a stage when, according to experimental evidence, the blastomeres begin to communicate with each other. When antibodies to gap junction proteins are injected into specific blastomeres of frog or mouse embryos, the injected cells continue to divide but do not undergo normal development (S. Lee et al., 1987). Developmental defects arise among the descendants of the injected cell (Warner et al., 1984), indicating that some function of the gap junctions—presumably, the passage of signal molecules between embryonic blastomeres—is necessary for normal development.

Cellular Signaling

Throughout the life cycle of an organism, cells communicate by chemical signals. Eggs respond to sperm, vegetal cells modify the development of their animal neighbor cells, and organs grow in response to growth hormones. Among the plethora of signaling mechanisms that have evolved, the ones that will be discussed in this section are especially common and relevant to development.

Intercellular Signals Vary with Respect to Distance and Speed of Action

Animal cells respond to chemical signals coming from distant glands, from neighboring cells, and even from themselves (Fig. 2.22). In *endocrine signaling,* cells respond to a hormone released from a specific gland and distributed by blood or other body fluids. The sex hor-

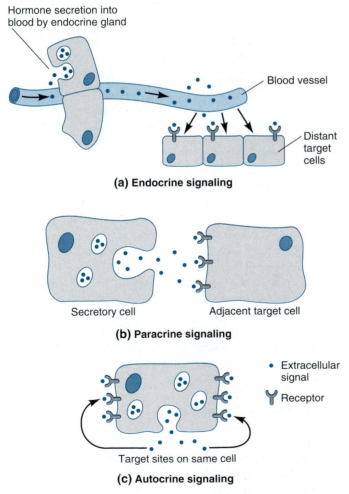

Figure 2.22 General schemes for cell-to-cell signaling by chemicals. The chemical signals may travel over longer or shorter distances, ranging from meters in endocrine signaling (a) to micrometers in paracrine (b) and autocrine signaling (c).

mones of vertebrates and the molting hormones of insects are well-known examples. In *paracrine signaling,* the signaling cell affects only nearby cells. In an amphibian blastula, for instance, vegetal cells signal to adjacent animal cells to form mesoderm. In *autocrine signaling,* cells respond to chemical signals that they themselves release. Several growth factors serve as both autocrine and paracrine signals by stimulating the secreting cells as well as neighboring cells to divide.

Most of the chemical signals exchanged between cells, especially in vertebrates, can be classified as either steroids or peptides. These two groups act through characteristic mechanisms that differ in their speed of action (Fig. 2.23). Steroids, as a rule, diffuse through the plasma membranes of their target cells. Inside the cytoplasm they bind to *cytoplasmic* receptor proteins, and the resulting hormone-receptor complex accumulates in the nucleus. Here, the complex interacts with a small number of target genes whose transcription is specifically enhanced or inhibited. Since RNA transcripts need to be synthesized and processed in the nucleus and functional mRNAs have to be released into the cytoplasm before they can be translated into new proteins, the response to a steroid hormone generally takes hours or even days.

In contrast, *peptides* do not cross plasma membranes, because they are water-soluble rather than lipid-soluble. Instead they act as *ligands,* binding specifically to *membrane* receptor proteins on the surface of their target cells. The loaded receptors then cause the activation of certain proteins that are *already present* in the cell cytoplasm. Therefore, a target cell can respond to a peptide much faster than to a steroid.

The remainder of this section will focus on the diverse and often indirect cellular responses to peptide signals.

Signal Receptors in Plasma Membranes Trigger Different Responses

Different types of membrane receptors trigger different cellular responses. In most cases, the final result is a change in the three-dimensional conformation of the target proteins and thus their biological activity. However, this common result is achieved through a variety of intermediate steps (Fig. 2.24). Some receptors have cytoplasmic domains that act directly as protein kinases, transferring phosphate groups to other proteins and thus changing their conformation and activity. However, most membrane receptors respond more indirectly. Some are ion channels that open in response to the binding of the ligand. The altered ion concentration inside the cytoplasm may, directly or indirectly, change the activity of certain proteins.

Most membrane receptors cooperate with other membrane proteins that act as signal transducers. A large family of common signal transducers are called *G*

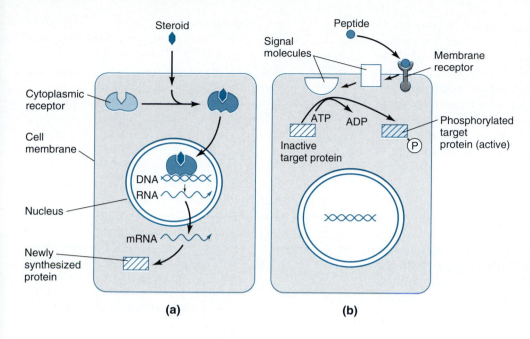

Figure 2.23 Cellular responses to steroid and peptide signals. **(a)** A steroid crosses the plasma membrane and binds to a cytoplasmic receptor protein in the target cell. The loaded receptor accumulates in the nucleus, where it interacts with specific genes to activate or inhibit their transcription into RNA. Transcripts must be processed and released as mRNA into the cytoplasm before new proteins can be synthesized. **(b)** A peptide binds to a receptor protein in the plasma membrane of the target cell. The loaded receptor interacts with other signal molecules to activate cellular proteins that are already present in the cytoplasm. The activation is often mediated through transfer of a phosphate group (P) from adenosine triphosphate (ATP) to the inactive target protein.

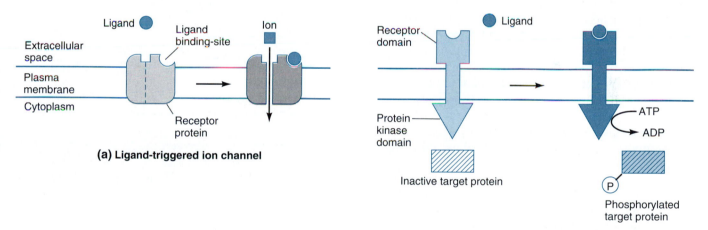

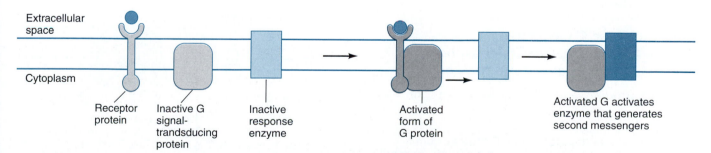

Figure 2.24 Different types of membrane receptor proteins. **(a)** Ligand-triggered ion channel. Ligand binding causes a conformational change in the receptor, which opens an ion channel across the membrane. **(b)** Ligand-triggered protein kinase. Ligand binding activates the cytoplasmic domain of the receptor protein, which acts as a protein kinase, phosphorylating target proteins. **(c)** Ligand-triggered generation of a second messenger. Ligand binding activates a signal-transducing G protein, which in turn activates an enzyme to generate a second messenger molecule inside the target cell.

proteins, because they bind guanosine triphosphate (GTP). More specifically, the term is used for signal-transducing membrane proteins that consist of three polypeptides (M. I. Simon et al., 1991; Fig. 2.25). G proteins activate or inhibit a third group of membrane proteins that act as enzymes or ion channels. These proteins in turn generate small *intracellular* signal molecules known as *second messengers.* How are second messengers produced and what is their role in intracellular signaling?

Adenylate Cyclase Generates cAMP as a Second Messenger

One of the most common second messengers, *cyclic adenosine monophosphate (cAMP)*, is made with the help of the enzyme adenylate cyclase. This enzyme floats in the plasma membrane and is activated (or inhibited) by various receptors via G proteins (Fig. 2.25). This mechanism introduces several important features into the signaling process. First, the extracellular signal is greatly amplified. Since the receptor and the G proteins diffuse freely in the plasma membrane, a single loaded receptor can stimulate many G protein and adenylate cyclase molecules, and each molecule of adenylate cyclase will in turn synthesize many cAMP molecules. Second, different signals can be integrated when their different membrane receptors and associated G proteins act on the same set of adenylate cyclase molecules. Third, the G protein functions only as long as a GTP molecule is bound to it. Since the GTP is hydrolyzed in the course of the G protein's function, cells can dampen an overly strong response by reducing the supply of GTP.

The cAMP molecules generated in response to an external signal perform their second messenger function in different ways. One common pathway is through cAMP-dependent protein kinases, which phosphorylate other proteins, thus regulating their biological activity (Fig. 2.26).

Phospholipase C Generates Diacylglycerol, Inositol Trisphosphate, and Calcium Ions as Second Messengers

Receptor proteins and G proteins may act on other

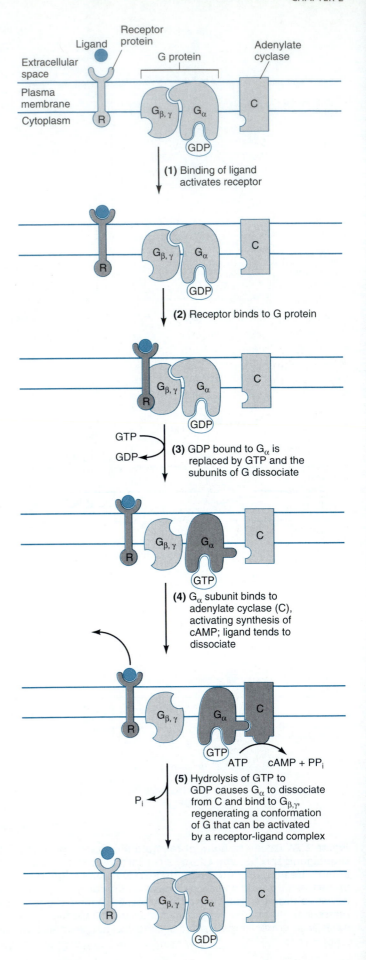

Figure 2.25 Signal transduction by G protein from a loaded receptor to a membrane-bound enzyme, exemplified here as an adenylate cyclase. The G protein consists of three peptides, G_α, G_β, and G_γ. Upon ligand binding, the loaded receptor interacts with the G protein. This allows the α subunit to dissociate from the $\beta\gamma$ subunit and to exchange its bound guanosine diphosphate (GDP) for guanosine triphosphate (GTP). In this short-lived active form, the α subunit activates the membrane-bound enzyme. However, the α unit also hydrolyzes its bound GTP to GDP, whereupon it dissociates from the enzyme and returns to the $\beta\gamma$ subunit.

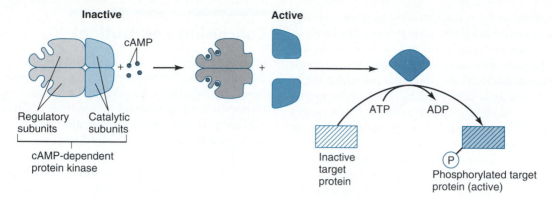

Figure 2.26 Activation of protein kinase by cyclic adenosine monophosphate (cAMP). Binding of cAMP to the regulatory subunits causes the catalytic subunits to dissociate. Only in the dissociated state are the catalytic subunits active, that is, able to phosphorylate target proteins.

Phosphatidylinositol bisphosphate (PIP$_2$) is one of several inositol phospholipids found in the cytoplasmic layer of the plasma membrane. *Phospholipase C* is a lipid-cleaving enzyme floating in the same membrane layer. If stimulated by a receptor via a G protein, phospholipase C cleaves PIP$_2$ into two components: *diacylglycerol (DAG)* and *inositol trisphosphate (IP$_3$)*; (Fig. 2.27). The DAG product remains in the plasma membrane, while the water-soluble IP$_3$ product diffuses into the cell cytoplasm. Here IP$_3$ interacts with the endo-

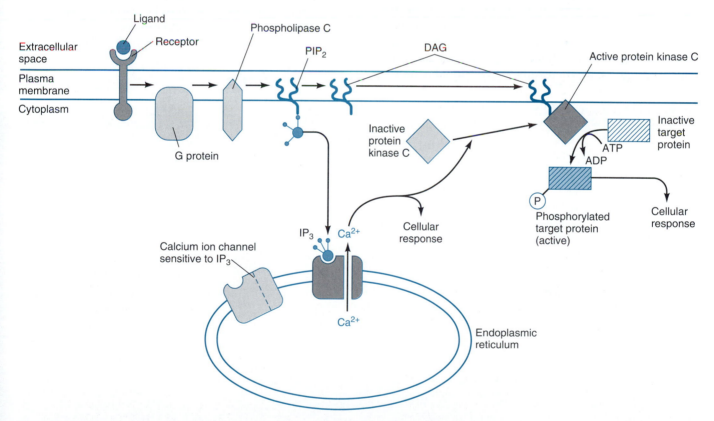

Figure 2.27 Second messengers in the phosphatidylinositol bisphosphate (PIP$_2$) pathway. Ligand binding activates G protein, which in turn activates phospholipase C. This enzyme cleaves PIP$_2$ into diacylglycerol (DAG) and inositol trisphosphate (IP$_3$). The latter diffuses through the cytoplasm to the endoplasmic reticulum, where it releases stored calcium ion (Ca^{2+}). The released calcium ion allows inactive cytoplasmic protein kinase C to insert into the plasma membrane, where protein kinase C is activated by DAG. The active protein kinase C then phosphorylates target proteins, which mediate cellular responses. The released calcium ion may trigger additional cellular responses.

plasmic reticulum, which responds by releasing stored calcium ions. The liberated ions cooperate with DAG in activating another enzyme, *protein kinase C.* In its inactive state, protein kinase C is a cytoplasmic protein. In the presence of calcium ions, it inserts into the plasma membrane, where it is activated by DAG. The active protein kinase C then phosphorylates various target proteins for different cellular responses including changes in gene activity.

In addition to helping with the activation of protein kinase C, calcium has many more second messenger functions. Most of these are carried out by calcium-dependent proteins, such as *calmodulin.* In its active form, with four calcium ions bound to it, calmodulin activates a wide variety of other proteins including protein kinases.

In summary, cells interact by various chemical signals to coordinate development and maintain bodily functions. Most signals act through specific receptor proteins. Steroid receptors act slowly by regulating the transcription of specific genes. Signal peptides are bound by receptor proteins located in the plasma membrane. These receptors act quickly, although often indirectly, by changing the biological activity of cytoplasmic proteins that are already present. The membrane receptor proteins act through various small molecules, known as second messengers, which amplify, integrate, and modulate the received signals. Most second messengers cause the phosphorylation of certain target proteins, thus changing their three-dimensional conformation and their biological activity.

Conclusion and Outlook

Cells in general have many basic abilities in common: the ability to divide, to maintain or change shape and inner organization, to take up or release materials, to move, to form junctions, to become polarized, and to exchange signals with one another. Recent discoveries have greatly improved our understanding of the cell organelles and molecules involved in these activities. Yet many seemingly simple questions—such as how a cell extends a lamellipodium—are still to be answered. In the meantime, current knowledge allows us to analyze developmental processes in terms of cellular behavior—an approach that we will employ repeatedly in Part One of this book.

SUGGESTED READINGS

Alberts, B., D. Bray, J. Lewis, M. Raff, K. Roberts, and D. Watson. 1994. *Molecular Biology of the Cell.* 3d ed. New York: Garland.

Darnell, J., H. Lodish, and D. Baltimore. 1990. *Molecular Cell Biology.* 2d ed. New York: Scientific American Books.

Prescott, D. M. 1988. *Cells.* Boston: Jones and Bartlett.

GAMETOGENESIS

Figure 3.1 Frontispiece of William Harvey's book on the generation of animals (Latin edition, 1651; English edition, 1653). The picture shows Zeus releasing many kinds of animals from an egg. The writing on the egg held by Zeus says, "All living beings come from an egg."

Gulielmus Harveus
de
Generatione Animalium.

This chapter is about *gametogenesis,* the formation of *gametes* (eggs and sperm). Gametes are highly specialized cells that differ greatly from *somatic cells.* Typical somatic cells are *diploid:* they carry two sets of chromosomes, one set derived from the male parent and the other from the female parent. A gamete is *haploid:* the number of chromosomes in its nucleus is half the number in a diploid cell. The reduction in number of chromosomes occurs during gametogenesis as gamete precursor cells undergo two specialized cell divisions called *meiosis.* At the time of fertilization, when an egg and a sperm unite, their chromosomes combine so that the resulting *zygote* is diploid.

Although male and female gametes are both haploid, they differ from each other in all other respects. The male gamete, called a *sperm* or *spermatozoon* (pl., *spermatozoa*), is highly specialized for its sole function of finding and fertilizing an egg. The typical spermatozoon consists of a nucleus with a set of highly condensed chromosomes, a flagellum to propel itself, and mitochondria to provide chemical energy for flagellar movement. At its tip, the spermatozoon has certain recognition molecules that bind to an egg from an individual of the same species.

Moreover, enzymes stored behind the tip of the sperm help it penetrate the egg envelope.

The female gamete, called the *egg* or *ovum* (pl., *ova*), is quite different from the male gamete in its specialization. Besides providing genetic information, its function is to support the development of the new individual until it can obtain nourishment from another source. Accordingly, most eggs are large, having as much volume as thousands of somatic cells. A typical egg is filled with high-energy nutrients, such as glycoproteins, carbohydrates, and lipids. These components, collectively called the *yolk,* are broken down during embryogenesis into small molecules, which are used by embryonic cells to synthesize new macromolecules. Eggs also contain prefabricated building materials that can be rapidly assembled after fertilization. For instance, eggs store histones and tubulin, which are assembled during cleavage into chromosomes and microtubules, respectively. In most animals, the egg also provides spatial information through the uneven distribution of its cytoplasmic components, which later determines one or two of the embryo's body axes.

We will begin this chapter with a few historical comments on the discovery of the mammalian egg. Then we will examine the concept of the *germ line* as a lineage of potentially immortal cells, which is segregated from *somatic cells* early in development and eventually produces eggs or sperm. A discussion of *meiosis* will focus on the different timing of this process in males and females. Sperm development will be covered briefly, with emphasis on the final phase when the typical sperm morphology is generated. Oogenesis will be described in more detail than spermatogenesis; of particular interest are the processes by which the egg is supplied with RNA molecules and storage proteins. We will also discuss the process of egg maturation, when the egg is prepared for fertilization. Finally, we will explore some of the properties of egg envelopes, which protect the fragile egg cell but at the same time allow it to breathe.

The Discovery of the Mammalian Egg

Humans have always been keenly interested in their own procreation as well as in the propagation of cultivated plants and animals. In most cultures, it has long been known that both males and females of a species contribute to the generation of offspring. It has also been recognized that semen is the male contribution to reproduction in both humans and animals. Until relatively recently, the female contribution was less clear.

Aristotle, although he had observed chicken embryos developing inside their eggshells, had no idea that the chicken egg was a single cell. Much less could he imagine a mammalian egg, which is invisible to the naked eye. He believed that mammalian development began from coagulated menstrual flow, a widely held idea in his time. Aristotle also suggested that the male's semen did not contribute any material substance to the offspring. Instead he postulated that the semen contained a form-giving principle that caused the epigenesis of an embryo from the female-derived amorphous material. This mixture of physical and metaphysical concepts prevailed until an interest in biology, including human anatomy and reproduction, was rekindled during the Renaissance.

A major contribution to the eventual recognition of the mammalian egg came from William Harvey (1578–1657), who is also known for his discovery of blood circulation. As a physician to King Charles I of England, Harvey had access to the royal deer herd. He used this privilege for mating experiments and subsequent dissections, which convinced him that mammalian embryos did *not* originate from coagulated menstrual flow. Although he could not *see* mammalian eggs, he *postulated* that every animal, including the human, originates from an egg (Fig. 3.1).

The Dutch biologist Regnier de Graaf (1641–1673) described the ovarian follicles that were later named after him, but he mistook the entire follicle for the egg. Eventually, improved light microscopes made it possible to identify the mammalian egg as a tiny vesicle within the Graafian follicle. The German embryologist Karl Ernst von Baer (1792–1876) made this discovery in 1827. However, the cellular nature of the mammalian egg was not generally recognized until late in the nineteenth century.

The Germ Line Concept and the Dual Origin of Gonads

With the establishment of the cell theory, it became clear that all organisms consist of cells, that gametes are specialized cells, and that chromosomes are the carriers of genetic information. The far-reaching implications of these discoveries for heredity, development, and evolution were recognized by August Weismann (1834–1914). He developed the concept of the *germ line,* which is defined as the lineage of cells in an organism from which gametes arise (Fig. 3.2). The germ line can readily be identified in many animals, because a small number of *primordial germ cells* become segregated early during embryonic development. The primordial germ cells eventually give rise to gametes; the somatic cells produce all other parts of the body. Germ line cells therefore have a chance to live on in future generations, whereas all somatic cells die with the individual. Thus the germ line cells can be viewed as a continuous lineage that extends from generation to generation. From this perspective, the somatic cells and the bodies they constitute are temporary caretakers entrusted with perpetuating the germ line.

In vertebrates and some invertebrates, primordial germ cells are distinguished from somatic cells by their size, shape, division schedule, motility, and stainability for certain marker proteins. However, the germ line is less distinct in other invertebrates, in which dedicated germ cell precursors appear only late in development or cannot be identified at all (Nieuwkoop and Sutasurya, 1979, 1981). Even in animals that do have a clearly recognizable germ line, the distinction between germ line and somatic cells is not irreversible. Under experimental conditions, somatic cells can form functional

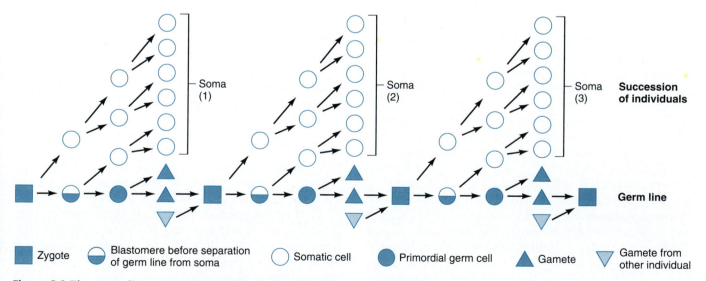

Figure 3.2 The germ line concept according to August Weismann (1834–1914).

gametes and germ line cells can give rise to mature somatic cells. Johannes Günther (1971) eliminated the primordial germ cells from embryos of the wasp *Pimpla turionellae* and found that these embryos nevertheless developed into fertile adults. Subsequently, Gerhard Fleischmann (1975) observed that the lost primordial germ cells were replaced in this wasp by mesodermal cells, which do not normally give rise to gametes. Conversely, primordial germ cells of *Xenopus,* when transplanted within the embryo, may form differentiated muscle, skin, and gut epithelial cells (Wylie et al., 1985). These examples indicate that the initial separation of germ line and somatic cells is tentative and can be overridden by later interactions between cells.

Historically, the germ line concept has played an important role in defining those cells that in normal development pass on genetic information to the next generation. Most of Weismann's contemporaries thought that *all* parts of the parental bodies contributed to the characteristics of their offspring. This view was compatible with the Lamarckian idea that acquired traits are heritable. In contrast, the germ line concept conflicts with Lamarckism, because it holds that only germ line cells, not somatic cells, transfer genetic information to gametes. Thus, the germ line concept had a profound influence on theories of heredity and evolution. Today, the germ line concept is still relevant in the context of gene therapy. Replacement of defective genes with normal ones is feasible in mature somatic cells, in fertilized eggs, or in embryonic cells before the separation of soma and germ line. In humans, gene therapy has been administered via *somatic cells* to combat certain cancers and other genetic disorders (A. D. Miller, 1992). The benefit of this treatment is limited to the treated individual. In contrast, if gene therapy could successfully be applied to eggs, early embryonic cells, or the *germ line,* cured individuals would pass the intact genes on to their offspring. These types of gene therapy are currently not suitable for use in human medicine but are carried out with plants and animals (see Chapter 14).

In most animals, the primordial germ cells associate with somatic cells to form the gonads (ovaries or testes), in which the gametes develop. Typically, the somatic part of the gonad is derived from the *mesoderm,* the intermediate germ layer that forms during gastrulation. The primordial germ cells usually reach the mesodermal gonad rudiment after considerable migration. For instance, in a 3-week-old human embryo, the primordial germ cells are located in an appendage of the embryonic gut known as the *yolk sac* (Fig. 3.3). From there, the cells migrate around the hindgut, up the mesentery suspending the hindgut, and into the mesodermal gonad rudiment, which is next to the embryonic kidney (Eddy et al., 1981). In reptiles and birds, the primordial germ cells travel through the bloodstream and when they reach the vicinity of the mesodermal gonad rudi-

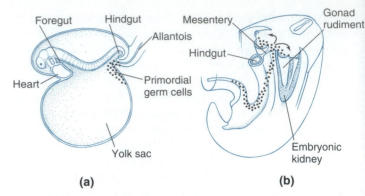

Figure 3.3 The migration of primordial germ cells in the human embryo. **(a)** Schematic drawing of a 3-week-old embryo in median section. **(b)** Posterior half of a 6-week-old embryo in ventrolateral aspect, with abdominal walls cut away. The primordial germ cells appear among the endodermal cells in the caudal wall of the yolk sac. They migrate by amoeboid movement around the wall of the hindgut and up the dorsal mesentery. Eventually, they invade the mesodermal gonad rudiment, located on top of the embryonic kidney.

ment they squeeze out of the blood vessel and into the gonad.

The primordial germ cells arrive at the mesodermal gonad rudiment during the sexually indifferent stage of development, which in humans lasts through the sixth week of gestation. At this time, somatic gonadal cells initiate the development of either a testis or an ovary (see Chapter 27). The primordial germ cells proliferate within the gonad via mitosis and then usually enter a quiescent period. Thereafter, they are distributed in specific parts of the gonad, and the male and female germ cells acquire different cytological characteristics; at this stage, the germ cells are called gonia (Gk. *gone,* "seed")—*spermatogonia* in the male and ***oogonia*** in the female.

Meiosis

During gametogenesis, germ cells divide to reduce the number of their chromosomes by half, from a *diploid* to a *haploid* set, in two successive cell divisions known as meiotic divisions I and II, or ***meiosis*** (Gk. *meioun,* "to diminish"). In diploid cells, each maternal chromosome has a paternal *homologue* (Gk. *homos,* "same"; *logos,* "word" or "sense"). Homologous chromosomes show the same morphology and contain sets of genes that are *allelic,* if not identical. (Only the ***sex chromosomes,*** designated X and Y in mammals, are for the most part nonhomologous.) Thus, diploid cells have two identical or allelic copies of most genes. The hallmark of meiosis is the separation of homologous chromosome pairs and the formation of haploid daughter cells containing only one chromosome from each pair.

Homologous Chromosomes Are Separated during Meiosis

Meiosis is preceded by a round of DNA replication that results in two identical chromatids for each chromosome. Because meiosis entails two successive divisions without intervening DNA replication, the entire process yields four haploid cells. In the first meiotic division, pairs of homologous chromosomes are segregated between two daughter cells. The second meiotic division separates the two chromatids of each chromosome and creates four haploid cells. The names of the stages of meiosis are the same as the ones used for mitosis—prophase, metaphase and so on (see Chapter 2).

During meiotic prophase I, individual chromosomes become visible under the microscope as they coil and shorten. At this stage, the process resembles mitotic prophase. Subsequently, two fundamental differences between mitosis and meiosis become apparent. First, in meiosis, the twin chromatids of each chromosome are held together along their entire length by protein axes (Fig. 3.4a). Second, homologous chromosomes come together in pairs, which are connected by a fibrous element. This coupling, which is called *synapsis*, is characteristic of meiotic prophase I and is *not* observed in mitosis.

The configuration of four chromatids that results from synapsis is known as the *synaptonemal complex*. The formation of the synaptonemal complex facilitates the genetically important process of *crossing over*, in which homologous chromosomes exchange parts of their chromatids (Fig. 3.4a). Crossing over begins with the formation of a *recombination nodule,* which is believed to contain the enzymes necessary for cutting and splicing chromosomal DNA (Prescott, 1988). The chromatids are severed at corresponding positions and rejoined crosswise. Crossover events are not visible until later, when the synaptonemal complex disintegrates and the homologous chromosomes drift apart (Fig. 3.4b). At this time, the two chromatids of each chromosome are clearly recognizable and are seen to be crossed at points called *chiasmata* (sing., *chiasma*). Crossing over leaves most chromatids with combined segments of maternal and paternal homologues.

During metaphase I, each pair of homologous chromosomes is a loosely tangled bundle of four chromatids. Within each chromosome, the two chromatids are still held together at the centromere, which stays intact throughout meiosis I. At this time spindles form, as they do during mitosis. However, during meiosis I, the homologous chromosomes are paired in such a way that *all* kinetochore fibers from one chromosome extend to *one* spindle pole, and *all* kinetochore fibers from the

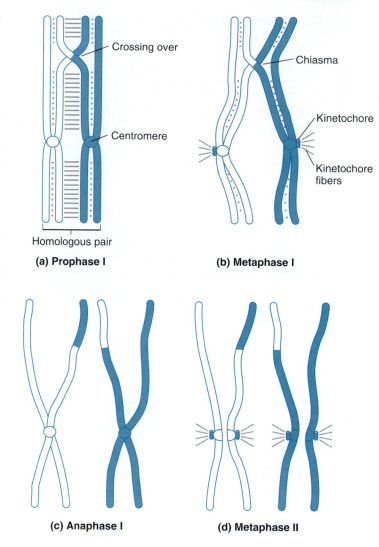

(a) Prophase I **(b) Metaphase I**

(c) Anaphase I **(d) Metaphase II**

Figure 3.4 Meiosis. Schematic diagrams focus on the behavior at various stages of one pair of homologous chromosomes, containing corresponding sets of identical or allelic genes. **(a)** Prophase I. A germ cell entering meiosis is diploid, containing two homologues of each chromosome, one from the male parent and one from the female parent. Each homologue consists of twin chromatids connected by axial proteins (dots). Each pair of homologues is held together by a fibrous central element (short lines), forming the synaptonemal complex. Crossover events lead to the exchange of chromatid segments. **(b)** Metaphase I. Each centromere sends kinetochore fibers to only one spindle pole, while the centromere of the homologous chromosome sends kinetochore fibers to the opposite pole. As a result, the homologous chromosomes are segregated into two daughter cells, while the twin chromatids of each chromosome remain joined. The crossover points, called chiasmata, become visible when the central element breaks up and the homologues move apart. **(c)** Anaphase I. The axial connections between twin chromatids disappear. **(d)** Metaphase II. Each centromere sends kinetochore fibers to both spindle poles, and the twin chromatids of each chromosome are separated. This phase of meiosis resembles the anaphase stage of mitosis.

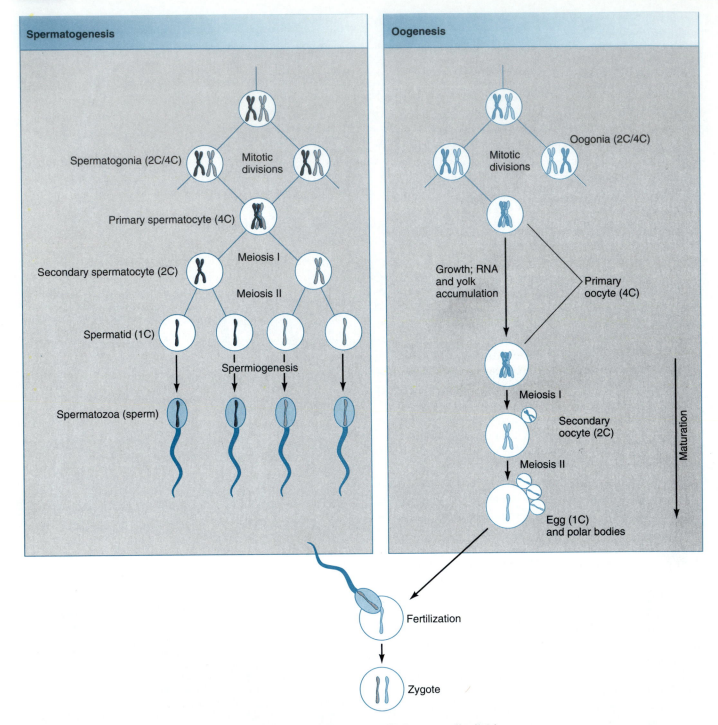

Spermatogenesis

Spermatogonia (2C/4C)

Mitotic divisions

Primary spermatocyte (4C)

Meiosis I

Secondary spermatocyte (2C)

Meiosis II

Spermatid (1C)

Spermiogenesis

Spermatozoa (sperm)

Oogenesis

Oogonia (2C/4C)

Mitotic divisions

Growth; RNA and yolk accumulation

Primary oocyte (4C)

Meiosis I

Secondary oocyte (2C)

Meiosis II

Egg (1C) and polar bodies

Maturation

Fertilization

Zygote

Figure 3.5 Comparison of spermatogenesis and oogenesis. Primordial germ cells divide mitotically, producing spermatogonia in males and oogonia in females. These cells are diploid, containing two or four genomes (2C or 4C), depending on their stage in the mitotic cycle. Before the gonia enter meiosis, their DNA replicates. They are then called primary spermatocytes or oocytes. After the first meiotic division, they contain two genomes (2C) and are called secondary spermatocytes or oocytes. After the second meiotic division, they are haploid (1C) spermatids or eggs. Note that the two rounds of meiosis produce four haploid spermatids, each of which develops into a spermatozoon, but only one egg results from meiosis of a diploid oogonium; the small polar bodies attached to the egg have no known function.

homologous chromosome are oriented toward the *opposite* spindle pole (Fig. 3.4b).

During anaphase I, the homologous chromosomes are pulled apart at the centromeres, and the chromatid tangles at the chiasmata are resolved (Fig. 3.4c). During telophase I, two daughter nuclei are formed. Each nucleus contains a complete set of chromosomes, with each chromosome now consisting of two nonidentical chromatids. The chromatids are separated during the second meiotic division.

During meiosis II, each centromere now sends kinetochore fibers to *both* spindle poles, as in mitosis (Fig. 3.4d). The final step in meiosis II is the formation of four haploid cells. These cells are genetically nonequivalent because homologous genes are allelic but not always identical.

Different terms are used to describe the germ cells at different points in meiosis (Fig. 3.5). When a gonial cell is about to undergo its first meiotic division, it is called a *primary oocyte* or *primary spermatocyte*. Following meiosis I, the germ cells are called *secondary oocytes* or *secondary spermatocytes*. After the completion of meiosis, the male germ cells are referred to as *spermatids*. The female germ cells are called *eggs* after

their release from the ovary, an event called *ovulation*. Ovulation and fertilization occur at different stages of meiosis, depending on the species.

The Timing of Meiosis Is Different in Males and Females

The timing of meiosis is markedly different in females and males. In many cases, the first meiotic division of oogenesis starts even before the female reaches sexual maturity. An extended meiotic prophase I allows animals that produce large eggs to synthesize and accumulate molecular building materials for the embryo. Mammals, which produce unusually small eggs, nevertheless follow the same pattern. In the human female, meiosis begins before birth and is not completed until an oocyte matures during one of the menstrual cycles between puberty and menopause (Fig. 3.6b). Some oocytes are therefore arrested in meiotic prophase I for as long as 50 years.

In the male germ line, meiosis begins later but finishes in much less time. In the human male, spermatogonia enter meiosis in waves from the onset of puberty throughout life (Fig. 3.6a). Each meiotic division is com-

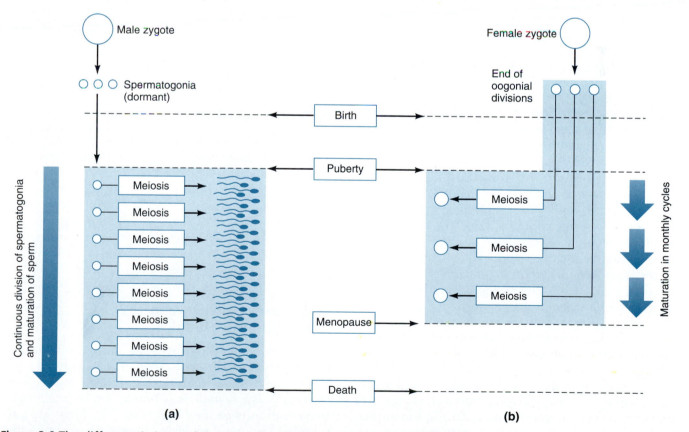

Figure 3.6 The different timing and duration of meiosis in human males and females. **(a)** In males, meioses are initiated continuously from puberty on. Each meiosis is completed within a few weeks. **(b)** In females, all meioses begin prior to birth, but normally only one oocyte completes meiosis during each menstrual cycle between puberty and menopause.

pleted within a few weeks. Immediately thereafter, the spermatids mature into spermatozoa (see the section "Spermatogenesis"). A man produces sperm in this fashion until the end of his life. An ejaculation releases 100 to 500 million sperm; unejaculated sperm are resorbed.

Meiosis Promotes Genetic Variation and Helps to Establish Homozygous Mutant Alleles

Two events during meiosis are especially important for their contribution to genetic variation. First, meiotic anaphase I produces nuclei that may contain any combination of maternally and paternally derived chromosomes. During this phase, a given nucleus with, say, the maternal homologue of one chromosome (Ch1) may receive the maternal or paternal homologue of another chromosome (Ch2) and so forth. This phenomenon, known as *independent assortment*, makes it possible for one human, with 23 homologous chromosome pairs, to generate 2^{23} (about 8 million) *types* of gametes with different combinations of chromosomes.

The second event that promotes genetic variation is crossing over, which results in new combinations of maternally and paternally derived genes, as described earlier. The high degree of organization required for this hazardous process suggests that crossing over has adaptive value.

The contributions of independent assortment and crossing over to genetic variation culminate at fertilization, when an egg with its random assortment of maternal genes unites with a sperm containing a different assortment of paternal genes. Because the activities of genes affect one another, a nearly limitless number of overall gene activity patterns are possible.

Independent assortment, crossing over, and the combination of maternal and paternal genes at fertilization are all associated with sexual reproduction and do not occur during asexual reproduction. This link and the prevalence of sexual reproduction despite its high costs and the hazards of cutting and splicing genomic DNA suggest that genetic variation is indeed beneficial. However, sustaining a genetically diverse population, though important, is not the only benefit derived from sexual reproduction. Another advantage that sexually reproducing organisms have is the opportunity to perpetuate genetic mutations that have adaptive value for their species.

A comparison of the genetics of asexual and sexual reproduction illustrates the competitive advantage that meiosis confers on sexually reproducing organisms. Remember that diploid cells have two copies of each gene. If they are identical, the individual is said to be *homozygous* for this gene. If the two copies are allelic but not identical, the individual is *heterozygous* for the gene. When a mutation occurs, the possibility that a matching mutation will spontaneously arise in a homologous gene is remote. Therefore, mutant alleles in diploid cells of asexually reproducing organisms virtually always exist in heterozygous combinations with wild-type alleles. Since the wild-type alleles generally determine the phenotype, a mutation would be ineffective in such an organism and could not produce a selectable advantage.

In contrast, the *segregation* of homologous chromosomes during meiosis provides a mechanism by which favorable mutations can be quickly established in a homozygous configuration (Kirkpatrick and Jenkins, 1989). For simplicity, let us assume that a diploid germ line cell is homozygous for a gene G. A mutation affecting one of the two homologues might generate a more adaptive allele, G'. The genetic constitution of the cell would then be G/G', but its phenotype would not change. During meiosis, however, the G allele would be segregated from G'. Subsequent matings between offspring that inherited the G' allele would eventually establish the G' allele in a homozygous configuration. In this case, any phenotypic advantage conferred by the G'/G' genotype would enhance selection in favor of the G' allele.

In summary, meiosis promotes genetic variation and provides a mechanism for establishing mutant alleles in homozygous condition. Both effects generate a broader basis for natural selection, which appears to be the major advantage conferred by sexual reproduction.

Spermatogenesis

The entire process of sperm formation, beginning with diploid cells called *spermatogonia* and resulting in mature *spermatozoa*, is referred to as *spermatogenesis*. Spermatogenesis has been studied most extensively in the mammalian testis (Fawcett, 1975; Eddy, 1988). Mammalian spermatogenesis will therefore be presented first and then contrasted with spermatogenesis in other animals. (The role of the testis as a hormone-producing gland will be discussed in Chapter 28.)

Male Germ Cells Develop in Seminiferous Tubules

The mammalian testis is divided into portions called *lobules*, which contain one or more tiny convoluted tubes called seminiferous tubules (Lat. *semen*, "seed"; *ferre*, "to bear"). The *seminiferous tubules* are the main functional subunits of the testis (Fig. 3.7). They are surrounded by a *basal lamina* and a sheath of connective tissue cells. Inside the basal lamina, the developing germ cells are arranged in a distinct order. Cells in the earliest stages—i.e., the spermatogonia—are located at the periphery; those in more advanced stages are suc-

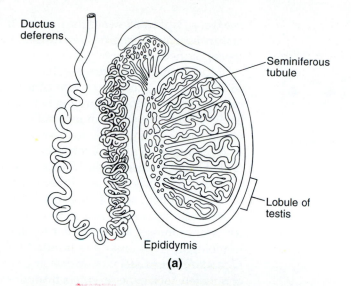

Ductus deferens

Seminiferous tubule

Lobule of testis

Epididymis

(a)

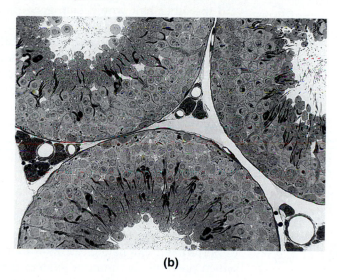

(b)

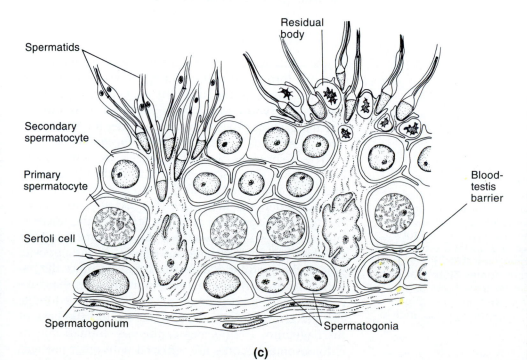

Spermatids

Residual body

Secondary spermatocyte

Primary spermatocyte

Blood-testis barrier

Sertoli cell

Spermatogonium

Spermatogonia

(c)

Figure 3.7 Seminiferous tubules in the mammalian testis. **(a)** Schematic diagram of a testis. Each testis is divided into lobules containing seminiferous tubules. These tubules are connected via a network of collecting ducts and the epididymis to the ductus deferens, which joins the urethra. **(b)** Light micrograph showing partial cross sections of tubules and clusters of interstitial cells in the spaces between them. **(c)** Drawing of a section of a seminiferous tubule, showing Sertoli cells and germ cells at different stages of spermatogenesis.

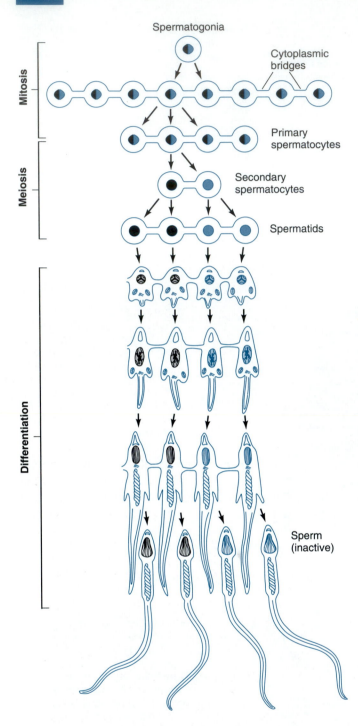

Figure 3.8 The development of mammalian sperm as clones of interconnected cells. The final mitotic divisions of the primordial germ cells are followed by incomplete cytokinesis, resulting in clones of spermatogonia and primary spermatocytes that are connected by cytoplasmic bridges. The subsequent meiotic divisions are also followed by incomplete cytokinesis. The nuclei are colored to show the presence of an X chromosome (black), a Y chromosome (blue), or both sex chromosomes. Note that cytoplasmic bridges between spermatids allow them to exchange gene products of maternal and paternal origin. During the subsequent process of spermiogenesis, individual sperm are separated.

cessively displaced toward the interior. In the final stage, mature spermatozoa are given off into the lumen (open space) of the tubule.

Close examination reveals that the maturing germ cells are connected to one another by cytoplasmic bridges and develop in strict synchrony (Fig. 3.8). The bridges, which result from incomplete cytokinesis after mitotic as well as meiotic divisions, allow the passage of large molecules and even organelles between connected cells. Because of the bridges, the development of all the sperm cells can be directed by gene products arising from both sets of parental chromosomes. This is particularly important in the case of genes located on the sex chromosomes, which for the most part are nonhomologous. Without cytoplasmic bridges, half of the secondary spermatocytes and their descendants would completely lack gene products from the Y chromosome, and the other half would lack gene products from the X chromosome.

In addition to germ line cells at different stages of maturation, the seminiferous tubules contain prominent somatic cells known as *Sertoli cells*. These large, columnar cells span the tubule wall from the basal lamina to the lumen. Sertoli cells provide developmental signals and structural support to spermatocytes and spermatids, which develop in deep recesses (pockets) of the Sertoli cells. By forming extensions connected by tight junctions, the Sertoli cells also create a separate basal compartment in which the spermatogonia multiply (Fig. 3.7). The tight seal allows different environmental conditions to be maintained for germ line cells in different stages of development. In addition, the sheet of Sertoli cell extensions forms a blood-testis barrier, which prevents the immune system from attacking the maturing germ cells (Tindall et al., 1985).

Spermiogenesis

At the conclusion of meiosis, each primary spermatocyte has formed four haploid spermatids, which are inconspicuous, round cells. In most animal species, spermatids mature into uniquely shaped spermatozoa. The maturation process that transforms the spermatid into a spermatozoon is called *spermiogenesis*.

The major steps of spermiogenesis are summarized in Figure 3.9. Vesicles pinched off from the Golgi apparatus of the spermatid coalesce into a membranous organelle called the *acrosome* (Gk. *akron*, "tip"; *soma*, "body"). The acrosome forms a cap over the nucleus, and the nucleus rotates so that the acrosome points to the sheath of the seminiferous tubule. At the opposite pole of the spermatid, a centriole pair becomes the basis for the growth of an array of microtubules that supports a *flagellum* extending into the lumen of the tubule. The cell's mitochondria aggregate around the base of the flagellum. The nucleus condenses dramatically, as chromosomal histones are replaced with other proteins

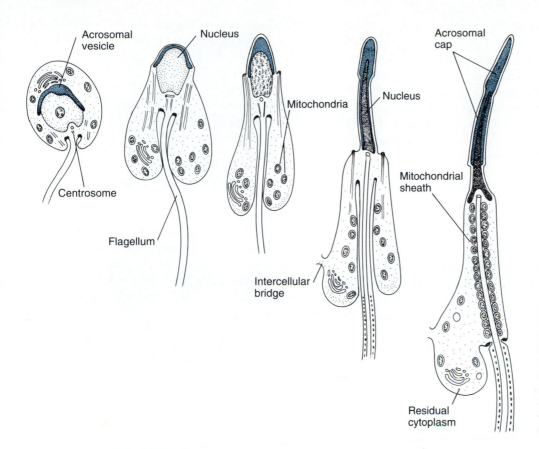

Labels (left to right): Acrosomal vesicle, Nucleus, Acrosomal cap, Nucleus, Centrosome, Mitochondria, Mitochondrial sheath, Flagellum, Intercellular bridge, Residual cytoplasm

Figure 3.9 Successive stages in guinea pig spermiogenesis. The dramatic morphological changes include the formation of an acrosomal cap, condensation of the nucleus, outgrowth of a flagellum, and arrangement of a sheath of mitochondria around the base of the flagellum.

called *protamines*. The bulk of the cytoplasm is displaced, starting in the nuclear region, and is ultimately pinched off. Since spermatids mature in clones connected by cytoplasmic bridges, the spermatozoa leave behind a large residual body of cytoplasm when they enter the lumen of the seminiferous tubule. While still in the testis, spermatozoa assemble different sets of proteins in the plasma membrane regions covering the anterior head, the posterior head, the anterior tail, and the posterior tail. The differentiation of these domains continues as the spermatozoa leave the testis and traverse the epididymis (Phelps et al., 1990; Cowan and Myles, 1993).

The mature mammalian sperm is highly specialized for its function—the delivery of the male genome to the egg (Fig. 3.10). The nucleus and acrosome form the *head* of the sperm, which is joined by a neck to the remainder of the sperm, called the *tail*. The nucleus, of course, carries the genetic information of the male parent. The acrosome contains enzymes for penetrating the cells and envelopes surrounding the egg (see Chapter 4). The proximal part of the tail, called the *midpiece*, contains the mitochondria and serves as the power plant for the sperm. The rest of the tail is a propulsive organ, the *flagellum*. Like any other cell, the spermatozoon is surrounded by a plasma membrane, but its membrane contains special receptors that allow the sperm to respond to chemoattractants and to interact with the egg.

Most nonmammalian spermatozoa generally resemble their mammalian counterparts, but on closer in-

spection show some differences. The sperm of marine and freshwater invertebrates, which are shed into the water, tend to have a less elaborate tail structure and a shorter midpiece. Many invertebrate sperm have a cup-

Figure 3.10 Scanning electron micrograph of human spermatozoa.

shaped depression in the anterior face of the nucleus, which contains globular *actin*. At fertilization, the actin polymerizes and supports the formation of a fingerlike process that plays a role in sperm-egg adhesion (see Chapter 4). The sperm of some animal species have no resemblance at all to typical sperm. For instance, the sperm of some crayfish and crabs are star-shaped, and the sperm of roundworms are large cells that contain much cytoplasm and move like amoebas.

Oogenesis

Eggs differ from sperm in many respects. The basic functions of the egg are to contribute its own genome, to incorporate the genome of one sperm, and to provide molecular building materials, structural information, and protective envelopes for the developing embryo. These functions are anticipated in the provisions made during the egg's formation in the ovary, a process called *oogenesis*. Generally, it takes considerably more resources to make an egg than to make a sperm (Browder, 1985). As we have already seen, meiosis in the female germ line starts much earlier in the life of the organism and takes much longer to complete. In further contrast to the male germ line, the meiotic divisions of an oogonium produce only one large gamete, the egg; the tiny haploid sister cells of the egg, known as *polar bodies*, have no apparent function in later development (Figs. 3.5 and 3.11).

Eggs are storehouses filled with maternally provided building blocks—mostly RNAs and proteins—for the developing embryo. These materials are incorporated into the oocyte during meiotic arrest. After this growth phase, the oocyte is endowed with protective envelopes and prepared for its transition from the ovary to another environment. These processes will be described in the following sections.

Oocytes Are Supplied with Large Amounts of RNA

During its arrest in meiotic prophase I, the oocyte nucleus in many species increases significantly in volume and becomes very active in RNA synthesis. This stage is well suited for RNA synthesis because four chromatids, and thus four sets of genes, are available for transcription. The large nucleus of the primary oocyte is known as the *germinal vesicle*. In frogs and many other animals, the chromosomes in the germinal vesicle assume a characteristic configuration that appears when the synaptonemal complex dissolves and the homologues are held together only by chiasmata. The two chromatids of each chromosome stay aligned with each other but become very elongated. Condensed chromosomal regions alternate with uncoiled chromatid loops extending symmetrically from the main axis of the chromosome. Because the shape of these chromosomes re-

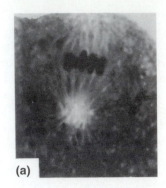

(a)

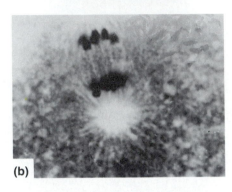

(b)

(c)

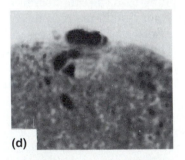

(d)

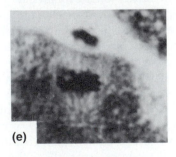

(e)

Figure 3.11 Meiosis and polar body formation in the giant oyster *Ostrea gigas*. **(a)** Metaphase I. **(b)** Anaphase I. **(c)** Telophase I and formation of the first polar body. **(d)** Prometaphase II. **(e)** Metaphase II preceding formation of the second polar body.

minded early researchers of the brushes used to clean the glass cylinders of kerosene lamps, they called them *lampbrush chromosomes* (Fig. 3.12). Their conformation seems to support the active transcription of many genes simultaneously.

In keeping with the vast size of the oocyte relative to somatic cells, most cytoplasmic organelles are much more abundant in oocytes. For instance, a fully grown *Xenopus* oocyte contains 200,000 times as many ribosomes as an average somatic cell (Laskey, 1974). Transcribing the necessary amount of ribosomal RNA (rRNA) is a daunting task. Even somatic cells have several hundred copies of rRNA genes per haploid genome to satisfy their needs for rRNA (Lewin, 1980). These genes are repeated in tandem, forming a distinct region called a *nucleolus organizer*, which is normally present in two homologous chromosomes per diploid set. When a nucleolus organizer is actively transcribed, it forms a densely staining organelle called a *nucleolus*. Although two nucleoli can provide most somatic cells with enough ribosomes, making all the ribosomes for an oocyte in this way would take several years. Instead, oocytes of amphibians and other animals accelerate the synthesis of large rRNAs by *selective amplification* of the nucleolus organizer (D. D. Brown and I. B. Dawid, 1968). As the genes in the amplified organizers become active in rRNA synthesis, they form *extrachromosomal nucleoli*—about 3000 per *Xenopus* oocyte (Fig. 3.13). The

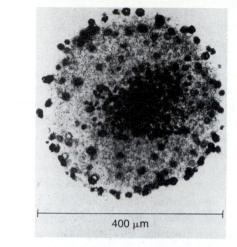

(a)

400 μm

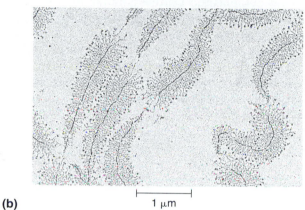

(b)

1 μm

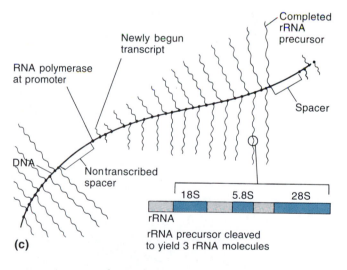

(c)

RNA polymerase at promoter

Newly begun transcript

Completed rRNA precursor

Spacer

DNA

Nontranscribed spacer

18S   5.8S   28S

rRNA

rRNA precursor cleaved to yield 3 rRNA molecules

Figure 3.13 Selective amplification of genes for large ribosomal RNAs in frog oocytes. **(a)** Light micrograph of an isolated nucleus (germinal vesicle) of a frog oocyte. Each dark dot represents a stained nucleolus. A normal diploid cell has only two nucleoli located on a pair of homologous chromosomes, but a *Xenopus* oocyte has about 3000 extra nucleoli that are not attached to chromosomes. **(b)** Electron micrograph of a portion of one nucleolus. The genomic DNA in each nucleolus contains several hundred copies of a gene from which a large ribosomal RNA precursor molecule is transcribed. The RNA molecules become longer as transcription proceeds, giving each transcribed gene the appearance of a Christmas tree. **(c)** Interpretive drawing of part b. The completed RNA transcripts are cleaved to yield three ribosomal RNA molecules having sedimentation values of 5.8S, 18S, and 28S.

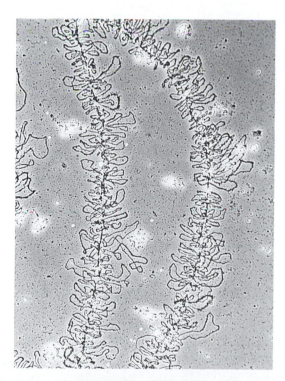

Figure 3.12 Lampbrush chromosomes in a newt (*Notophthalmus*) oocyte. This light micrograph shows loops of chromatin extending from a chromosomal axis. Closer analysis reveals that each chromosome consists of two chromatids with aligned sections that form the chromosome axis and symmetrical loops extending from the axis.

extra nucleoli along with many repetitive rRNA genes per nucleolus ensure that large oocytes accumulate enough ribosomes in a short period of time.

Another strategy for providing the oocyte with large amounts of all types of RNA is seen in certain insects. Primitive insects such as grasshoppers and cockroaches show *panoistic oogenesis* (Gk. *pan*, "all"; *oon*, "egg"), meaning that all their oogonia develop into oocytes. More evolved insects, including flies, bees, butterflies, and beetles, undergo *meroistic oogenesis* (Gk. *meros*, "part"). Only some of their oogonia develop into oocytes while the others become *nurse cells* (Figs. 3.14 and 3.15). Oocytes and nurse cells are sister cells, and both belong to the germ line. Together, they form clones of cells connected by cytoplasmic bridges similar to the clones formed by mammalian spermatogonia (Fig. 3.8). Instead of forming gametes, the nurse cells assume a helper role for those oocytes to which they are connected (Telfer, 1975). The nurse cells become *polyploid* by undergoing multiple rounds of chromosome replication without nuclear division or cytokinesis. The nurse cell nuclei in a

Drosophila ovary, for instance, contain about 1000 copies of each chromosome, so the nurse cells can rapidly synthesize large supplies of RNA for the oocyte. The transfer of at least some mRNAs from the nurse cells to the oocyte depends on microtubules that extend through the cytoplasmic bridges (Spradling, 1993).

The formation of the nurse cell–oocyte complex in *Drosophila* has been reconstructed from ovaries dissected at different stages of development (R. C. King, 1970). Each ovary consists of about 16 spindle-shaped tubes called *ovarioles* (Fig. 3.14). Oogonial stem cells located at the very tip of the ovariole produce cells that would normally be called oogonia. They are referred to as *cystoblasts*, however, because only some of their descendants form oocytes. Each cystoblast divides four times to produce a cluster of 16 interconnected cells called *cystocytes*. Because of the successive order of incomplete divisions, only two of the cystocytes have four cytoplasmic bridges to other cystocytes (Fig. 3.15). These two cystocytes enter meiosis, but only one persists and becomes an oocyte; all of the remaining 15 cystocytes

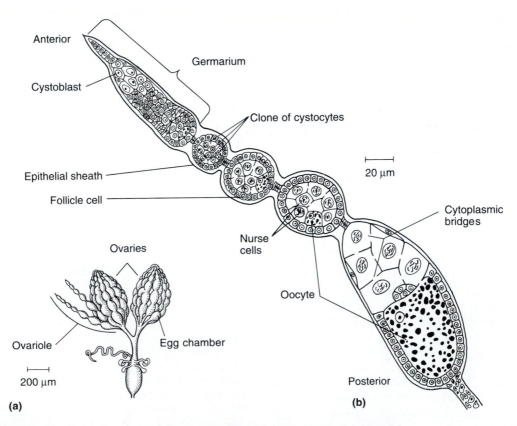

Figure 3.14 Oocyte development in *Drosophila melanogaster*. **(a)** Schematic drawing of the internal reproductive organs of an adult female. Each of the two ovaries consists of about 16 spindle-shaped ovarioles. **(b)** A single ovariole with several egg chambers. The most fully developed egg chamber, at the base of the ovariole, contains a large oocyte undergoing vitellogenesis, several large nurse cells, and an epithelium of follicle cells surrounding the nurse cell–oocyte complex. In the mid-range of the ovariole, the egg chambers are in earlier stages of development. The germarium near the tip of the ovariole contains developing germ cells (future oocytes and nurse cells) and mesodermal cells that form the follicle cells.

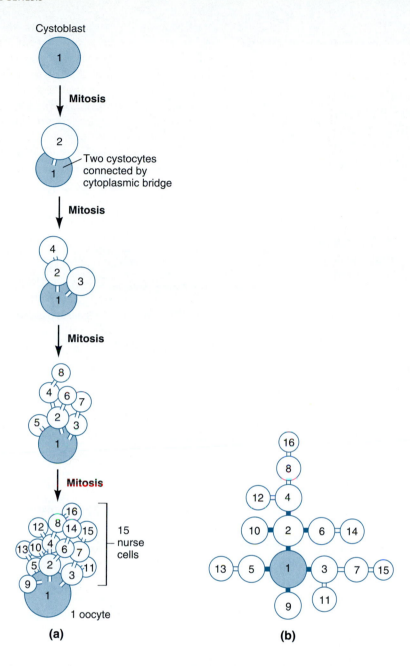

Cystoblast

Mitosis

Two cystocytes connected by cytoplasmic bridge

Mitosis

Mitosis

Mitosis

15 nurse cells

1 oocyte

(a)

(b)

Figure 3.15 Development of the nurse cell–oocyte complex in the *Drosophila* ovary. The cystoblasts correspond to the oogonia of other species. **(a)** A cystoblast undergoes four mitotic divisions with incomplete cytokinesis, resulting in 16 interconnected cystocytes. One of two cystocytes with four cytoplasmic bridges becomes the oocyte. The other 15 cystocytes develop as nurse cells. Instead of undergoing meiosis, the nurse cells become highly polyploid and synthesize large quantities of macromolecules and cell organelles, which are transferred to the oocyte through the cytoplasmic bridges. **(b)** Spread-out view of the nurse cell–oocyte complex showing all cytoplasmic bridges.

become polyploid nurse cells. The whole cluster of 16 germ line cells is surrounded by somatic cells that develop into **follicle cells**, which mediate the uptake of yolk proteins and synthesize eggshell proteins. The entire arrangement of one oocyte, the connected 15 nurse cells, and the surrounding follicle cells is called an **egg chamber**. Similar egg chambers are found in many groups of insects, including those orders that comprise the greatest numbers of species.

The transfer of RNA from nurse cells to the oocyte in fly egg chambers was demonstrated by Bier (1963) using the technique of *autoradiography* (described in

Methods 3.1). To trace the movement of newly synthesized RNA, Bier injected female houseflies with a radioactive RNA precursor, [³H]cytidine or [³H]uridine. Because these nucleotides are turned over rapidly during cell metabolism, they are available for RNA synthesis for only a short time. However, most of the RNA synthesized during this period is much longer-lived. This type of labeling, called **pulse labeling**, allows investigators to monitor the fate of a cohort of long-lived macromolecules that were synthesized during a short labeling pulse.

The flies were allowed to survive for different intervals after injection before their ovaries were removed and prepared for autoradiography. In ovaries removed 1 h after injection, much labeled RNA was located in the nuclei of the nurse cells and the follicle cells, but *not*

Autoradiography

Tracing the synthesis and transport of molecules requires labeling the molecules of interest. Biological molecules can be traced effectively if they are synthesized in the presence of a radioactive precursor—a procedure termed *radiolabeling*. For instance, proteins are radiolabeled by adding radioactive amino acids, such as [^{3}H]leucine or [^{35}S]methionine, to the medium of a cell culture or by injecting them into an experimental animal. Cells take up the labeled amino acids through the plasma membrane and incorporate them, along with their own unlabeled amino acids, into *newly synthesized* proteins. (Of course, proteins synthesized before the addition of radioactive precursors will not be labeled.)

A sensitive method of tracing radiolabeled molecules is known as *autoradiography* (Lat. *radius*, "ray"; Gk. *autos*, "self"; *graphein*, "to write"). This method is based on the same principle as photography: both electromagnetic and particulate radiation, like electrons from radioactive decay, will bring forth silver grains in a photographic emulsion. Thus, the presence of radiolabeled molecules can be recorded on photographic film. Exposure to the radiolabel must be carried out in the dark or under a safety light. When the film is developed, areas of radioactivity show up as black spots.

Autoradiography can be used to reveal size, location, and other properties of radiolabeled molecules. For example, the size of radiolabeled proteins can be determined by extracting *all* proteins from the cells of interest and subjecting them to gel electrophoresis. The resulting gel will contain bands of proteins separated according to size. The gel is placed on a sheet of photographic film and kept in the dark for a suitable exposure time (Fig. 3.16). After development, dark bands on the film reveal the location of the radiolabeled proteins in the gel. The location of an unknown protein in a gel relative to marker proteins of known size is indicative of the unknown protein's size.

All proteins, radioactive or not, can also be detected by stains, provided that enough proteins are present to bind visible amounts of stain. If a protein band is stainable but emits no radioactive signal, then the protein was synthesized before the radiolabeled precursor became available to the cells. If a protein band gives off a radioactive signal but is not stainable, then the protein is newly synthesized but has not accumulated in stainable amounts.

To identify the location of newly synthesized radiolabeled proteins in a tissue, the appropriate tissue is fixed and embedded in wax or plastic so that it can be cut into thin sections (Fig. 3.17). The sec-

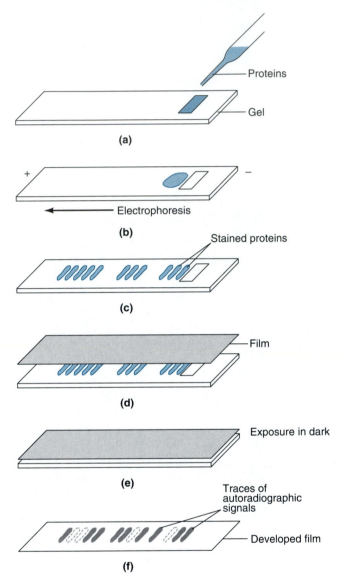

Figure 3.16 Autoradiography to determine the size of newly synthesized proteins. Proteins are synthesized in vivo in the presence of a radioactive amino acid. **(a)** All proteins (labeled and unlabeled) are extracted and separated by gel electrophoresis. **(b)** Sufficiently abundant proteins are visible as stained bands. **(c, d)** To detect newly synthesized proteins, a photographic film is placed on the gel and exposed in the dark. **(e)** The film is developed. **(f)** Dark bands on the processed film reveal the position of radiolabeled proteins in the gel. These bands may or may not coincide with stains produced by abundant proteins.

tions are placed on microscope slides and rinsed to remove any unincorporated amino acids—but not the labeled proteins—from the tissue. The rinsed sections are then covered with a layer of photographic

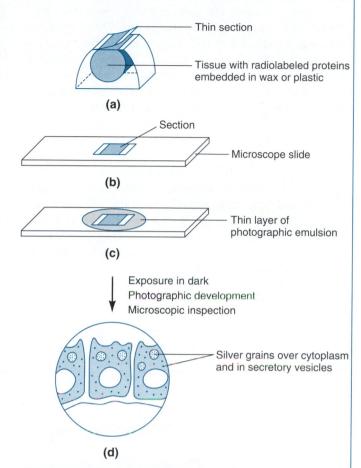

(a)

(b)

(c)

Exposure in dark
Photographic development
Microscopic inspection

Silver grains over cytoplasm and in secretory vesicles

(d)

Figure 3.17 Autoradiography to determine the locations where radiolabeled proteins accumulate in intact tissue. Proteins are synthesized in vivo in the presence of a radioactive amino acid. **(a)** Fixed and embedded tissue is sectioned. **(b)** Section is placed on a microscope slide and rinsed to remove unincorporated amino acids. **(c)** Photographic emulsion is pipetted onto the section and allowed to spread as a thin layer. **(d)** After exposure in the dark, the emulsion, still adhering to the tissue, is developed, and the whole preparation is viewed under the microscope. In transmitted light, dark grains in the emulsion indicate where radiolabeled proteins have accumulated in the tissue.

emulsion in the dark. After appropriate exposure, the emulsion is developed, and the sections are viewed under the microscope. Areas of the tissue that are marked with silver grains from the photographic emulsion are the sites where the radiolabeled proteins accumulated.

Autoradiography is suitable also for determining the size and location of RNA, DNA, and other biomolecules that have been radiolabeled with appropriate precursors.

in the oocyte nucleus (Fig. 3.18). In the ovaries removed 5 h after injection, labeled RNA was not present in nuclei. Instead, the nurse cell cytoplasm—especially in those nurse cells adjacent to the oocyte—contained labeled RNA. Most intriguing was the observation of labeled material in the oocyte cytoplasm next to the cytoplasmic connections to the nurse cells. These observations show that nurse cell nuclei synthesize large amounts of RNA that is transferred first into the nurse cell cytoplasm and then into the oocyte cytoplasm.

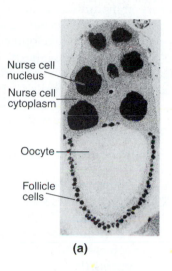

Nurse cell nucleus

Nurse cell cytoplasm

Oocyte

Follicle cells

(a)

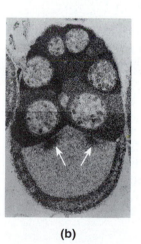

(b)

Figure 3.18 RNA synthesis in ovarian egg chambers of the housefly, *Musca domestica*. Newly synthesized RNA is radiolabeled with [3H]cytidine injected into the abdomen. **(a)** Autoradiograph of an egg chamber fixed 1 h after the labeled precursor was injected. The nuclei of nurse cells and follicle cells are heavily labeled, indicating that they have synthesized large amounts of RNA. Some label is also seen in the cytoplasm of these cells. The oocyte is virtually free of label. **(b)** Somewhat younger egg chamber fixed 5 h after injection. The nurse cell nuclei are only weakly labeled, indicating that most of the radioactive cytidine has been used up. The labeled RNA synthesized previously has been transported to the nurse cell cytoplasm and passes into the oocyte through two cytoplasmic bridges (arrows).

Yolk Proteins for Oocytes Are Synthesized in the Liver or Fat Body

Animal eggs contain proteins, lipids, and glycogen to nourish the developing embryo. These materials, which are collectively called *yolk*, accumulate in the egg cytoplasm. Water-soluble yolk components and some insoluble ones are stored inside membranous vesicles; water-insoluble yolk components, such as lipids, also exist as free droplets. The amounts of yolk are minimal in mammalian eggs, which have to sustain embryogenesis only to a very immature stage. In contrast, the large eggs of birds, reptiles, and sharks have copious amounts of yolk to support embryonic development to advanced stages. In the course of embryogenesis, yolk is broken down into small molecules, such as amino acids, which are then used by embryonic cells in their synthetic processes. By the end of embryonic development, the yolk is used up, or its remnants are enclosed in the larval gut.

Like RNA, yolk accumulates in the oocyte during meiotic arrest and contributes to the oocyte's enormous growth. In most animals, the process of yolk synthesis is similar to RNA synthesis in higher insects in that the oocyte has other cells working for it. In particular, this principle is observed in the synthesis of yolk proteins called *vitellins*, which are formed from precursor proteins called *vitellogenins* (Lat. *vitellus*, "yolk"; Gr. *gennan*, "to produce"). Vitellogenin synthesis occurs outside the oocyte. In vertebrates it takes place in the liver, and in most insects the major production site is the fat body (the insect equivalent of the liver). Newly formed vitellogenins are released into the bloodstream and taken up into the oocyte by *receptor-mediated endocytosis* (see Chapter 2). Within the oocyte, the vitellogenins are modified to reduce their water solubility and, hence, increase their stability. Finally, they are packed as vitellins in a crystal-like arrangement inside membranous yolk bodies. The entire process from vitellogenin synthesis to the sequestration of vitellins in the oocyte is termed *vitellogenesis*. Vitellogenesis has been studied in particular detail in insects and amphibians.

William Telfer (1965) pioneered the study of insect vitellogenesis. Working with large moths, he compared the blood proteins of males and females and discovered a female-specific protein. He also found that the concentration of this protein decreased when yolk formation began in the ovaries, suggesting that the protein may be destroyed or sequestered during vitellogenesis. Since fluid collected from eggs contained the protein in concentrations 20 times higher than that in the blood of females, Telfer devised a test to show whether the protein was taken up from the blood. Blood was transfused

from one moth species, *Hyalophora cecropia,* to another, *Antherea polyphemus.* Following transfusion, female-specific protein from *Hyalophora* accumulated in the recipient's oocytes until its concentration was 20 times greater in the eggs than in the blood. Other proteins that were transfused in the same manner did not accumulate in the recipient's eggs. Telfer's data showed that certain female-specific blood proteins are vitelogenins—the proteins that are selectively taken up and stored by oocytes. Later experiments showed that the selectivity of the uptake is mediated by vitellogenin-specific receptors involved in endocytosis (Hagedorn and Kunkel, 1979).

The amount of vitellins taken up by the oocyte during endocytosis is very impressive. This process is aided by folds and fingerlike extensions of the plasma membrane called *microvilli,* which increase the surface area available for endocytosis severalfold (Fig. 3.19). Even so, endocytosis would use up the entire plasma membrane of a *Hyalophora* oocyte in less than a minute if the membrane were not recycled (Telfer et al., 1982). In addition, in some species of insects, including *Drosophila,* the follicle cells not only synthesize a significant share of vitellogenin, but also facilitate its uptake by the oocyte (Telfer et al., 1982). Whereas most of the time the

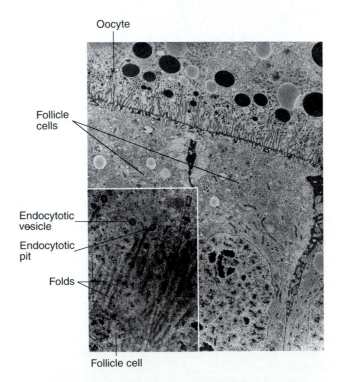

Figure 3.19 Vitellogenesis in a moth egg chamber. The oocyte surface is enlarged by deep folds with spaces between them that contain electron-dense material (vitellogenin). The same material is also present between follicle cells. Endocytotic pits form mostly at the bases of these spaces (arrows in inset). These pits are pinched off as endocytotic vesicles, which initially have a dense coat of a cytoplasmic protein, clathrin.

follicle cells form a tightly sealed epithelium, passageways open up between them during vitellogenesis. The follicle cells themselves create these passageways by shrinking or by secreting extracellular material that keeps the cell membranes apart and allows blood proteins to pass.

The hormonal control of vitellogenesis is very similar in frogs and insects (Fig. 3.20). In both groups of animals, reproduction tends to be seasonal, regulated by environmental cues that are integrated in the brain. In amphibians, the brain stimulates the hypothalamus to secrete *gonadotropin-releasing hormones*. They cause the pituitary gland to release *gonadotropins*. These hormones in turn stimulate the ovarian follicle cells to produce another hormone, *estrogen*, which causes the liver to synthesize vitellogenins. In many insects the hormonal control of vitellogenesis follows a similar pattern. The *corpora allata*, a pair of endocrine glands attached to the insect brain, are analogous in function to the pituitary gland and release the so-called *juvenile hormone*. In dipterans, the juvenile hormone stimulates the ovarian follicle cells to produce a steroid hormone, *ecdysone*. This hormone, like estrogen in amphibians, stimulates both the fat body and the follicle cells themselves to produce vitellogenin. In other insects, juvenile hormone stimulates the fat body directly to synthesize vitellogenin.

In insects that feed on the blood of vertebrates, vitellogenesis is linked with their irregular feeding regimen. Some of these species, including bedbugs and lice, were actually the first insects in which vitellogenesis was studied. It was found that soon after a blood meal, breakdown products of the ingested blood appear in

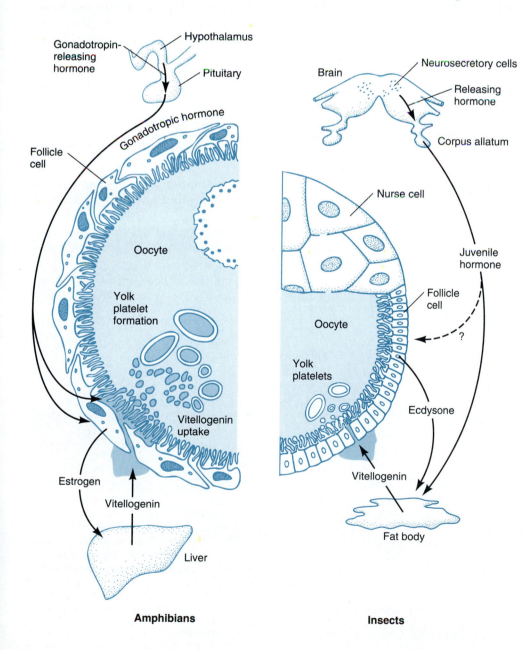

Amphibians

Insects

Figure 3.20 Comparison of the hormonal control of vitellogenesis in amphibians and insects. In both groups of animals, a part of the brain (hypothalamus or neurosecretory cells) stimulates a gland attached to the brain (pituitary gland or corpus allatum). In amphibians, the pituitary releases gonadotropic hormone, which stimulates the ovarian follicle cells to secrete estrogen, which in turn causes the liver to synthesize vitellogenin. In some insects, the corpora allata release juvenile hormone, which stimulates the follicle cells to secrete ecdysone, which in turn causes the fat body to synthesize vitellogenin. In other insects, the juvenile hormone acts directly on the fat body.

the body fluids of the parasites and then in their oocytes (Wigglesworth, 1943).

The Ovarian Anatomy Imposes Polarity on the Oocyte

Providing the oocyte with RNA and protein is a major task, but it is not all there is to oogenesis. The oocytes of most animals also receive signals from surrounding cells in the ovary that cause the cytoplasm to be asymmetrically distributed about an axis. In many species, the oocyte acquires an *animal-vegetal polarity*. The animal pole, defined by its proximity to the asymmetrically located oocyte nucleus, determines where cleavage furrows will begin, and the animal hemisphere typically forms the ectoderm of the embryo. In other organisms, oocytes acquire an anteroposterior polarity that foretells the future head-to-tail axis of the embryo. The insect oocyte illustrates this relationship.

In the egg chambers of more highly evolved insects, the oocyte is associated with a cluster of nurse cells, as described earlier in this chapter. The nurse cells are invariably located on one side of the oocyte, facing the *germarium* at the tip of the ovariole (Fig. 3.14). This side becomes the *anterior* pole of the egg, where the head of the embryo will be formed. The chain of events leading from the arrangement of the nurse cell–oocyte complex in the egg chamber to the anteroposterior body pattern of the embryo has been elucidated in part by genetic analysis. Certain *Drosophila* mutants with nurse cells on both ends of the oocyte form embryos with two heads (see Chapter 8). It is not known yet how mutated genes alter the positioning of the oocyte and nurse cells in the egg chamber. However, it is clear how the proximity of the nurse cells determines which will be the anterior pole of the oocyte. The *bicoid*$^+$ gene of *Drosophila* is transcribed in the nurse cells, and the resulting mRNA is transferred to the oocyte along with other RNAs. Upon entry into the oocyte, the bicoid mRNA is trapped at the pole near the nurse cells, presumably by attachment to cytoskeletal components (see Fig. 8.9). When the trapped mRNA is translated, bicoid protein forms a *morphogen* gradient, with a maximum concentration at this pole. The highest concentration ranges of this morphogen activate those embryonic genes that control the development of head and thorax (see Chapter 21).

Maturation Processes Prepare the Oocyte for Ovulation and Fertilization

A fully grown oocyte contains the maternal genetic information, the molecular building blocks, and the polarity axes necessary to begin the formation of an embryo. Yet, the oocytes of most animals at this stage are not ready for fertilization. Meiosis must still be completed to generate the haploid female *pronucleus.* Before meiosis can resume, the chromatin has to be condensed. The permeability of the plasma membrane to small molecules and ions must be adjusted so that the cell can leave the ovary and function in other environments. The surface of the oocyte has to be prepared to respond to external stimuli such as contact with sperm.

The processes that prepare the grown oocyte for fertilization are collectively called *oocyte maturation*. In most animals, maturation begins with the end of meiotic arrest in prophase I. Early maturation events include chromosome condensation and the disintegration of the nuclear envelope, a process known as *germinal vesicle breakdown*. Maturation also entails the release of the oocyte from the ovary, an event referred to as *ovulation*. After ovulation, the female gamete is called an *egg*.

Eggs of different animal species are ovulated and fertilized at different stages of meiosis (Fig. 3.21). At

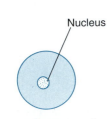
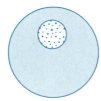

Young primary oocyte	Fully grown primary oocyte	Metaphase I	Metaphase II	Female pronucleus
The annulate worms *Dinophilus* and *Saccocirrus* The polychaete worm *Histriobdella* The flatworm *Otomesostoma* The onychophoran *Peripatopsis*	The roundworm *Ascaris* The mesozoan *Dicyema* The sponge *Grantia* The polychaete worm *Myzostoma* The clam worm *Nereis* The clam *Spisula* The echiuroid worm *Thalassema*	The nemertine worm *Cerebratulus* The polychaete worm *Chaetopterus* The mollusk *Dentalium* The cone worm *Pectinaria* Many insects Starfish and ascidians	The lancelet *Branchiostoma* Amphibians Most mammals	Cnidarians (e.g., sea anemones) Sea urchins

Figure 3.21 Different stages of egg maturation at the time of sperm entry in different animals.

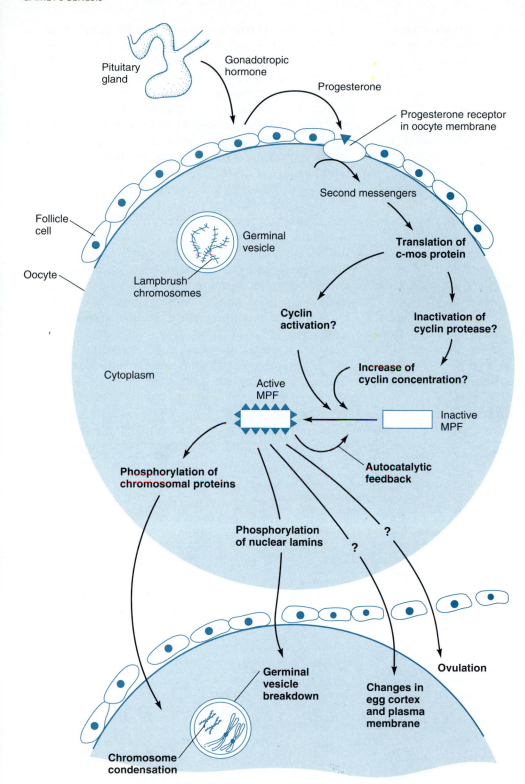

Figure 3.22 A model of oocyte maturation in amphibians. The binding of progesterone to the oocyte plasma membrane triggers second messenger events, which cause the translation of c-mos mRNA stored in the oocyte cytoplasm. The c-mos protein, being a protein kinase, presumably phosphorylates and thereby activates cyclin protein stored in the cytoplasm. Alternatively or in addition, c-mos protein may inactivate specific proteases that would otherwise degrade cyclin. Because cyclin is a component of M-phase promoting factor (MPF), the increased availability of active cyclin boosts MPF activity, which subsequently increases further by autocatalytic feedback. Through its protein kinase activity, MPF causes the phosphorylation of several other proteins, which triggers multiple maturation processes including germinal vesicle breakdown, chromosome condensation, and possibly cortical changes and ovulation.

one extreme, ovulation occurs with the germinal vesicles still intact, and the egg does not resume meiosis until it has been fertilized. This pattern is shown by various worms, including *Ascaris* (a roundworm) and *Nereis* (a polychaete worm). At the other extreme, the eggs of some animals have completed both meiotic divisions by the time of fertilization. This pattern is shown by sea urchins and coelenterates. The eggs of most animals fall somewhere between these extremes. We will

focus on the pattern seen in most vertebrates, including humans. In this case, maturation begins with the release of the oocyte from its arrest at prophase I. Ovulation follows the completion of meiosis I. A second arrest, at metaphase II, marks the end of the maturation process: the egg is now ready for fertilization.

The control of oocyte maturation has been studied most extensively in frogs, in particular, *Rana pipiens* and *Xenopus laevis* (Fig. 3.22). Like vitellogenesis, oocyte maturation is controlled by hormonal interactions of the hypothalamus, pituitary gland, and follicle cells (Wasserman and Smith, 1978). Stimulated by releasing hormones from the hypothalamus, the pituitary gland produces gonadotropic hormones, which in turn prompt the follicle cells to secrete *progesterone.* Progesterone acts on the oocyte surface, apparently by binding to a membrane receptor protein (Blondeau and Baulieu, 1984). This is unusual because progesterone and other steroid hormones generally bind to receptors in the cytoplasm (see Chapter 15). Oocyte maturation occurs within a few hours after hormonal stimulation.

To analyze the effect of progesterone on oocyte maturation, Yoshio Masui and Clement Markert (1971) transplanted cytoplasm from progesterone-treated oocytes at various stages of maturation into fully grown but immature oocytes. The researchers found that the cytoplasm from the maturing oocytes induced maturation in the immature oocytes as well. They concluded that the progesterone-treated oocytes contained a *maturation promoting factor.* Subsequent work showed that this factor is identical to *M-phase promoting factor (MPF),* which plays a central role in controlling the mitotic cycle of somatic cells (see Chapter 2). Thus, the release of the oocyte from meiotic arrest in prophase I is equivalent to the transition from G_2 phase to mitosis in a somatic cell.

How does progesterone promote MPF activity at the beginning of oocyte maturation? The binding of the hormone to its membrane receptor triggers a *second messenger* pathway involving cAMP, for which only grown oocytes are competent (Taylor and Smith, 1987; L. D. Smith, 1989). This pathway seems to initiate several responses that cooperate in triggering oocyte maturation. It appears that at least one response to progesterone requires protein synthesis, because a translation inhibitor, *cycloheximide,* blocks oocyte maturation if administered right after progesterone treatment.

One of the required proteins is encoded by a gene that causes cancer if deregulated and is therefore known as the *c-mos proto-oncogene* (see Chapter 29). The c-mos mRNA is present throughout oogenesis and early embryogenesis, but the c-mos protein is detectable only during oocyte maturation and is destroyed after fertilization (Fig. 3.23). During its brief presence, this protein plays a major role in releasing the oocyte from its dormancy (Sagata et al., 1988, 1989; Watanabe et al., 1989; Yew et al., 1992). Blocking the translation of c-mos mRNA by injecting *antisense oligonucleotides* into immature oocytes prevents germinal vesicle breakdown, whereas injection of c-mos protein activates MPF and triggers oocyte maturation. Thus, c-mos protein synthesis is both necessary and sufficient for completing first meiosis. Similar experiments have shown that c-

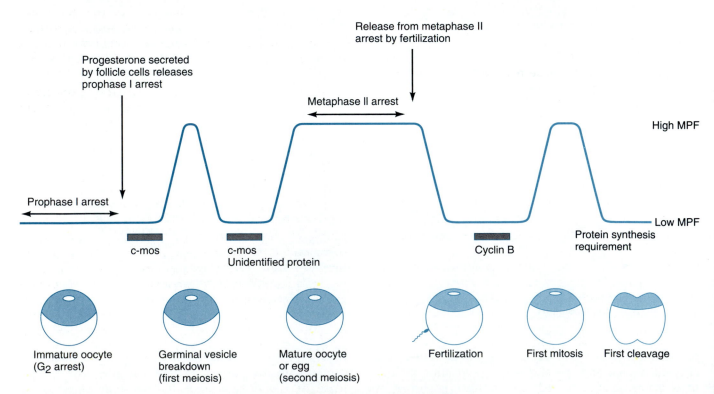

Figure 3.23 Model of *Xenopus* oocyte development during maturation and first cleavage. The curve represents the level of M-phase promoting factor (MPF) at each stage. The two vertical arrows indicate the stages when the oocyte's progress through the cell cycle depends on external signals. The solid bars below the MPF curve show the periods in which protein synthesis is required for entry into the next M phase (meiosis or mitosis). Some of the necessary proteins are identified beneath the bars.

mos protein and another, as yet unidentified protein must be synthesized after first meiosis in order for the oocyte to initiate second meiosis (Minshull, 1993). How the c-mos protein affects MPF activity is still under investigation. Being a kinase, c-mos protein might phosphorylate and thereby activate cyclin proteins stored in the oocyte. Alternatively, or in addition, c-mos protein may inactivate cyclin proteases that would otherwise keep the level of cyclin low.

Once a certain amount of MPF activity is present, oocyte maturation becomes independent of protein synthesis. The active MPF acts as a *protein kinase,* an enzyme that phosphorylates other proteins. MPF increases and maintains its own activity by means of a positive feedback loop, presumably by phosphorylating its own cyclin component. In addition, MPF acts on other target proteins, bringing about the multiple cellular changes that characterize oocyte maturation (Fig. 3.22). Most prominent among these changes are germinal vesicle breakdown, possibly caused by phosphorylation of nuclear envelope–associated proteins called *lamins,* and chromosome condensation caused by phosphorylation of chromosomal proteins. Other target proteins of MPF are thought to trigger ovulation and the changes in egg membrane and cortical cytoplasm that occur during maturation.

Having been released from arrest at prophase I, the oocyte proceeds with meiosis, only to become arrested again in metaphase II (Fig. 3.23). This second arrest is caused by a **cytostatic factor** that was discovered, like MPF, by Masui and Markert (1971) in cytoplasmic transplantation experiments. The researchers injected cytoplasm from unfertilized frog eggs into blastomeres of cleaving embryos and found that the injected blastomeres were arrested in their next metaphase. No such inhibition occurred when cytoplasm from fertilized eggs was transplanted. More recent work has shown that *continued high* MPF activity keeps cells arrested in metaphase. In order for a cell in either mitosis or meiosis to proceed beyond metaphase, MPF activity must drop to a low level. At the beginning of oocyte maturation, MPF activity is raised by the c-mos protein, as previously described. After metaphase II, c-mos protein behaves as a cytostatic factor: c-mos mRNA injected into *Xenopus* blastomeres arrests them in metaphase. Conversely, cytostatic factor activity falls off in cytoplasm from unfertilized eggs that have been treated with c-mos antibody (Sagata et al., 1989). These results raise the question why c-mos protein does not stop maturation at metaphase I. One possible answer is that c-mos protein needs to interact with the other, as yet unidentified protein to generate active cytostatic factor (Minshull, 1993).

The arrest of the egg in metaphase II is normally broken by fertilization. One of the responses to the entry of a sperm is a rise in the concentration of free calcium ions (Ca^{2+}) in the egg's cytoplasm (see Chapter 4).

Ca^{2+}-dependent protein activity is thought to destroy both MPF and the c-mos protein, thus allowing the fertilized egg to exit metaphase II and to begin the mitotic cycles of cleavage (see Fig. 4.20).

In the mammalian ovary, the oocytes are closely associated with somatic cells, in structures known as *follicles.* An immature follicle consists of one oocyte surrounded by several layers of somatic cells called **granulosa cells** (Fig. 3.24). The oocyte synthesizes glycoproteins that are deposited around it as a translucent layer known as the **zona pellucida** (Lat. *zona,* "girdle"; *pellucida,* "translucent"). The zona is traversed by *microvilli* that form gap junctions between the oocyte and the granulosa cells (Fig. 3.25). A mature follicle is often called a *Graafian* follicle, after its discoverer, Regnier de Graaf. It contains a fluid-filled space known as the *antrum,* which is large compared with the tiny oocyte. The antrum is located near the periphery of the ovary. Toward the end of the maturation process, the follicular antrum bursts open and the oocyte, still surrounded by granulosa cells, is ovulated.

The timing of oocyte maturation and ovulation in mammals varies. In some species, ovulation is prompted by seasonal cues. In others, including mink and rabbits, ovulation is triggered by the act of mating. In primates, including humans, ovulation comes about as part of a monthly reproductive cycle. The underlying control mechanisms are based on hormones, including gonadotropin-releasing hormones, produced by the hypothalamus; follicle-stimulating hormone and luteinizing hormone, produced by the pituitary gland; and estrogen, produced by the granulosa cells. The synthesis of these hormones and their receptors is regulated in such a way that only one or a few oocytes mature at a time.

Eggs Are Protected by Different Types of Envelopes

The egg and the early embryo are vulnerable stages in the life cycle of many animals. A predator could hardly find a more nutritious meal. Desiccation and mechanical stress are also hazardous to eggs deposited in the open air. For protection, eggs have different types of envelopes. The plasma membrane of the egg cell is covered by a glycoprotein layer, called the *zona pellucida* in mammals and generally referred to as the **vitelline envelope**, which plays an important role in fertilization (see Chapter 4). Most eggs have additional protective coats, some of which also serve as nutrients for the embryos or hatchlings. The egg coats are produced either before ovulation by the oocyte or follicle cells, or after ovulation by other cells. A jellylike coat surrounds many eggs spawned in water, such as sea urchin or frog eggs. In contrast, many types of eggs deposited on land, such as those of reptiles and birds, have hard shells. For instance, the yolky chicken egg is surrounded initially by

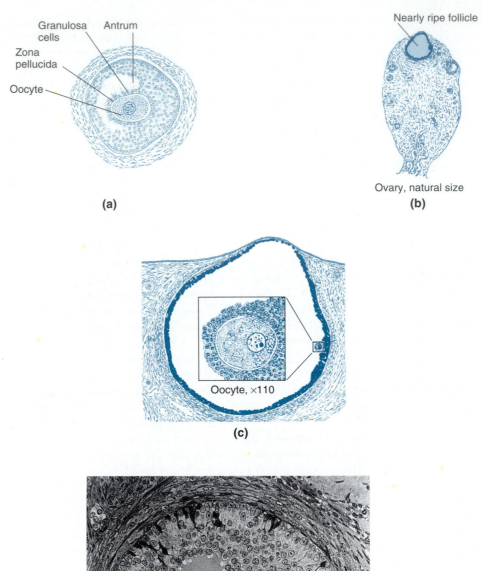

Granulosa cells

Antrum

Zona pellucida

Oocyte

(a)

Nearly ripe follicle

Ovary, natural size

(b)

Oocyte, ×110

(c)

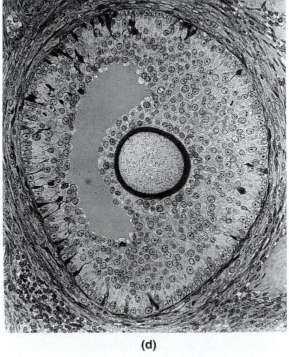

(d)

Figure 3.24 Development of mammalian oocytes in the ovary. **(a)** Human follicle during antrum formation. **(b)** Human ovary shown natural size. **(c)** Mature human follicle; inset shows oocyte. **(d)** Photomicrograph of a follicle from a monkey ovary during the beginning of antrum formation. The oocyte is surrounded by the zona pellucida (dark) and by granulosa (follicle) cells.

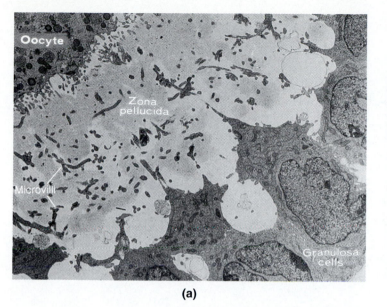

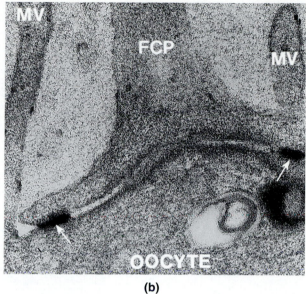

(a)   **(b)**

Figure 3.25 Mammalian oocytes and follicle cells. **(a)** Transmission electron micrograph showing part of an ovulated oocyte (upper left) and some follicle cells, now called granulosa cells (lower right), with the zona pellucida in between. Microvilli extend from both the oocyte and the granulosa cells into the amorphous substance of the zona pellucida. The microvilli look disconnected because they run into and out of the plane of section, but they actually form connections between the oocyte and the granulosa cells. **(b)** A granulosa cell at its point of contact with the oocyte. The dense patches (arrows) are gap junctions stained with lanthanum. FCP = follicle cell process; MV = oocyte microvillus.

a fragile vitelline envelope (Fig. 3.26). As the egg moves down the oviduct, it is invested with several layers of "egg white," consisting of ovalbumin and related proteins. Then two layers of matted keratin fibers, called the *shell membranes,* are added. (These are the membranes found inside the hard outer shell of a hard-boiled egg.) The outermost shell, which is added after fertilization, consists mostly of calcium carbonate with collagen-filled pores, an arrangement that regulates both water retention and respiration.

A very elaborate type of egg envelope, called a **chorion,** surrounds the insect egg. It is laid down by the follicle cells, after they have completed their contribution to vitellogenesis and formed the vitelline envelope. Insects that lay their eggs in water tend to produce thin, transparent chorions, whereas those that deposit eggs in the open air make thick chorions. A thick chorion must allow respiration and at the same time limit the loss of water. This is no easy task, since the water molecule is smaller than the oxygen molecule. The problem

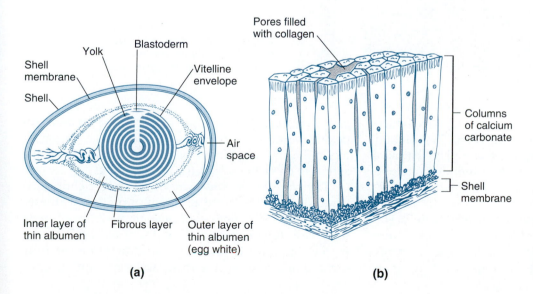

(a)   **(b)**

Figure 3.26 Chicken egg coats. **(a)** Drawing of a longitudinal section of a chicken egg. All egg coats outside the vitelline envelope are laid down after ovulation. **(b)** Drawing of a small segment of a bird's egg shell. The outermost shell consists mainly of calcium carbonate cylinders interspersed with collagen-filled pores to facilitate gas exchange.

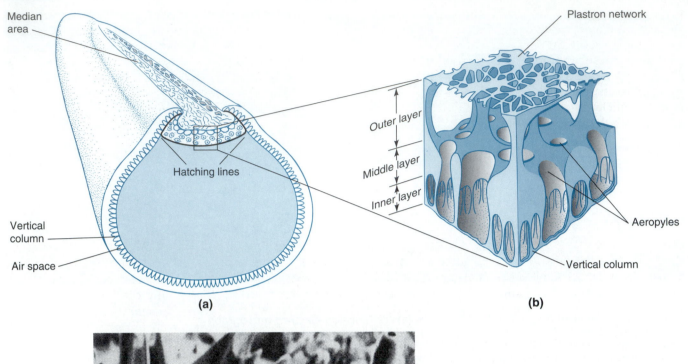

Median area

Hatching lines

Vertical column

Air space

(a)

Plastron network

Outer layer

Middle layer

Inner layer

Aeropyles

Vertical column

(b)

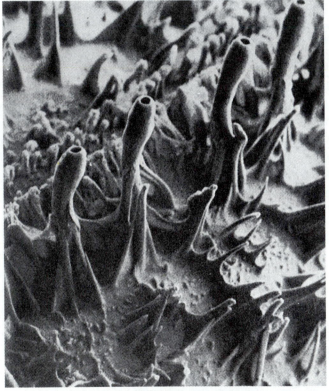

(c)

Figure 3.27 Insect egg chorion. **(a)** Drawing of the chorion of the blowfly *Calliphora erythrocephala.* The plastron (physical gill) structure is formed along a median strip lying between two hatching lines. **(b)** A section of the plastron is enlarged to show an inner layer of thin, vertical columns, an outer layer with larger columns and spaces, a layer of chimneylike connections in between, and a superficial network on the outside. **(c)** Scanning electron micrograph of the chorion of a bug (*Piezodorus lituratus*) showing tall aeropyles that rise from its surface like chimneys.

is compounded by the small size of insect eggs, which makes for a large surface-to-volume ratio. The solution found in most chorions is an inner air space that surrounds the entire egg close to the vitelline envelope (Fig. 3.27). The inner air space is prevented from collapsing by chorionic columns and is vented to the outside by channels called *aeropyles* (Gk. *pyle,* "gate"). The aeropyles can be short ducts or long, chimneylike structures, which presumably create ventilating air currents (Hinton, 1970). Many insect chorions also feature an area that functions as a physical gill, or *plastron,* repelling water to maintain an air-filled space. An opening in the chorion, known as the *micropyle,* allows the sperm to enter the egg at the anterior pole.

The complex architecture of the insect chorion is based on a sophisticated program of protein synthesis. For instance, silk moth chorions are made up of more than 50 different proteins (Kafatos et al., 1977). These proteins are synthesized and laid down in a carefully orchestrated process. If released at the appropriate time, the proteins diffuse to their destinations. Subsequently, covalent bonds form between different proteins to anchor them in their final patterns.

Vitellogenesis and chorion formation are remarkable processes, in terms of both quantity and complexity. For insects, both activities are carried out by the ovarian follicle cells. These cells switch rapidly from helping with the synthesis of bulk vitellogenins to making large amounts of different chorion proteins. The genes for chorion proteins are *selectively* amplified in ovarian follicle cells after they have completed vitellogenesis but before they begin chorion synthesis (see Chapter 7). In this respect, insect chorion genes in amphibian follicle cells behave like the ribosomal RNA genes in amphibian oocytes.

SUMMARY

Animals of most species carry a germ line, defined as the lineage of cells that gives rise to gametes, i.e., eggs or sperm. Primordial germ cells segregate from somatic cells early in development and migrate to the mesodermal gonad rudiments during embryogenesis. In the gonads, the primordial germ cells give rise to the gametes, through a process known as gametogenesis. It entails meiosis—i.e., two cell divisions during which homologous chromosomes separate and the dividing diploid cells give rise to haploid daughter cells.

In males, primordial germ cells, which develop within the testis, are called spermatogonia. Germ cells are present in this form throughout life. At the stages in which they undergo meiotic division, the male germ cells are called primary and secondary spermatocytes. The haploid cells resulting from meiosis are called spermatids. Each primary spermatocyte ultimately gives rise to four spermatids. In most animal species, spermatids undergo a dramatic change in morphology called spermiogenesis, during which the nucleus becomes highly condensed while an acrosome is formed at the tip and a flagellum at the other end. After this transition, the male germ cells are called spermatozoa or sperm. In mammals, the mitotic divisions of committed spermatogonia and the ensuing meiotic divisions are followed by incomplete cytokinesis, so that spermatids form large clones of interconnected cells.

In females, primordial germ cells are found in the developing ovaries and are called oogonia. During the meiotic divisions, female germ cells are known as primary and secondary oocytes. The meiotic divisions begin in the ovaries at an early stage—in mammals, before birth. Primary oocytes are arrested for an extended period of time in meiotic prophase I. During this phase, they accumulate large amounts of RNA, functional proteins, and storage proteins. Other cells often help to synthesize and store these materials. Many oocytes also acquire one or two axes of polarity during this time. The meiotic divisions in female germ cells are characterized by unequal cytokinesis, generating one large egg cell and three tiny sister cells that do not develop further. The final phase of oocyte development is called maturation. It is initiated by gonadotropic hormones that act on somatic cells in the ovary, which then trigger a cascade of events in the oocyte that allow it to complete meiosis. Oocyte maturation includes release from the ovary, an event called ovulation. While sperm are specialized to deliver their nucleus to an egg, eggs provide not only genetic information but also the building blocks and one or two polarity axes for the development of the embryo. Thus, most eggs are very large cells surrounded by various envelopes that protect them against mechanical stress and loss of water.

SUGGESTED READINGS

Browder, L. W., ed. 1985. *Developmental Biology.* Vol. 1, *Oogenesis.* New York: Plenum.

Eddy, E. M. 1988. The spermatozoon. In E. Knovil et al., eds., *The Physiology of Reproduction,* 27–68. New York: Raven Press.

FERTILIZATION

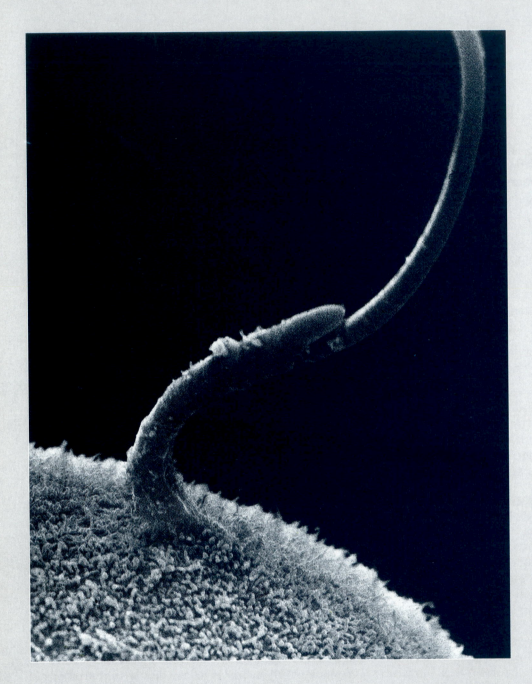

Figure 4.1 Fertilization in the rat. This scanning electron micrograph was made after removal of the egg envelope, or zona pellucida. The sperm's head has been engulfed by the egg; the remainder of the sperm is still outside. The egg surface is covered with fingerlike projections called microvilli, which enlarge the egg surface and provide a reservoir of plasma membrane to be used during cleavage.

Most higher organisms reproduce sexually by forming two types of haploid gametes, eggs and sperm. Both are highly specialized and short-lived unless they unite to form a newly developing organism, through a process called *fertilization*. Fertilization begins with the approach of the sperm to the egg and ends with the formation of the diploid *zygote*.

Gametes are often produced in vast quantities, especially by aquatic organisms such as fish and sea urchins. During each breeding season, a sea urchin female spawns 400 million eggs, while a male releases 100 billion sperm (Epel, 1977). Producing so many gametes seems necessary because the gametes quickly disperse in the ocean. Even mammals, which fertilize internally, produce an excess of sperm, of which only a small fraction arrive in the upper oviduct, where fertilization occurs (Fig. 4.1).

Sperm that encounter an egg face a series of obstacles, because the envelopes that protect the egg are also formidable barriers to sperm entry. Several solutions to this dilemma have evolved. Some of the hardest egg coats, such as the calciferous shells of bird eggs, are laid down only after sperm have been admitted to the egg. The eggs of other animals, including fish and insects, have tiny holes called *micropyles* in their shells through which sperm can enter. In most species, however, the sperm must penetrate one or more egg envelopes before they reach the egg cell proper (Fig. 4.2). Of particular importance is the *vitelline envelope*, a coat of proteins laid down by the oocyte before ovulation. Beneath the vitelline envelope is the egg plasma membrane, which fuses with the plasma membrane of the fertilizing sperm. In mammals, the vitelline envelope is a thick, transparent layer of glycoproteins called the *zona pellucida*. Surrounding the zona are the granulosa cells, which are derived from the ovarian follicle (see Chapter 3). Thus, mammalian sperm must burrow between the granulosa cells and then penetrate the zona before reaching the egg plasma membrane.

A critical step in fertilization, called *sperm-egg adhesion*, is the binding of sperm to the egg surface, typically, to the vitelline envelope. The underlying molecular mechanisms are species-specific: sperm adhere more stably to eggs from the same species than to eggs from another species. This species specificity is particularly important for aquatic organisms, whose eggs may be exposed to sperm from different species.

After adhesion to an egg, sperm penetrate the vitelline envelope using enzymes produced in the acrosome. When the first sperm reaches the egg cell, it establishes *plasma membrane contact*. This important event triggers a chain of events, called *egg activation*, causing resumption of the egg's cell cycle, which has been arrested since oocyte maturation. In many species, egg activation includes measures to protect the fertilized egg against the entry of additional sperm, which would disrupt mitotic divisions during cleavage. Within seconds after plasma membrane contact, egg and sperm proceed with *plasma membrane fusion*. The two gametes form a cytoplasmic bridge, which widens to permit *sperm entry* into the egg. The two haploid nuclei contributed by the egg and sperm unite so that both sets of chromosomes are passed on together during cleavage divisions.

Most of our knowledge about fertilization is derived from observations and experiments in vitro, which are easily carried out with animals that shed their gametes into water. Therefore sea urchins, marine worms, fish, and amphibians have traditionally been the organisms chosen for research on fertilization. More recent studies have focused on mammalian fertilization, with the goal of developing better infertility treatments, new methods of contracep-

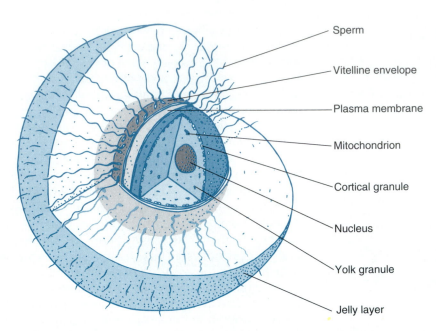

Sperm

Vitelline envelope

Plasma membrane

Mitochondrion

Cortical granule

Nucleus

Yolk granule

Jelly layer

Figure 4.2 Sea urchin egg, drawn to show the egg coats that a sperm must penetrate before reaching the egg cell proper. The sperm first encounters a thick jelly layer, which it penetrates with digestive enzymes from its acrosome. Next the sperm binds to the vitelline envelope. After it crosses the vitelline envelope, the plasma membranes of sperm and egg make contact and fuse. Just beneath the egg plasma membrane are thousands of membranous vesicles called cortical granules, which undergo exocytosis seconds after fertilization and shed their contents into the perivitelline space between the egg and the vitelline envelope. The egg cytoplasm contains one nucleus and numerous mitochondria, yolk granules, and other organelles. The diameter of the egg cell is about 75 μm.

tion, and other medical applications. Since research on fertilization in mammals is now beginning to rival the depth and sophistication of work on aquatic organisms, this chapter will focus on fertilization in sea urchins and mice.

Interactions before Sperm-Egg Adhesion

Many mechanisms have evolved in animals to ensure that gametes meet at just the right time and place and under the right conditions to achieve fertilization. Thus, fertilization follows a series of well-choreographed interactions between males and females, and between sperm and eggs. Parental mating behavior brings eggs and sperm close together before the sperm begin to swim with their flagella. In some organisms, chemical attractants produced by the egg itself or by ovarian cells guide the sperm in their final approach to the egg. In mammals and other animals as well, sperm must undergo a biochemical conditioning process before they are capable of fertilizing an egg.

Some Sperm Must Undergo Capacitation Before They Can Fertilize Eggs

Early attempts to fertilize mammalian eggs in vitro were unsuccessful because experimenters used freshly ejaculated sperm, which are unable to penetrate the *zona pellucida*. Sperm normally acquire this capability in the course of a few hours inside the female genital tract, through a process called *capacitation* (C. R. Austin,

1952). Capacitation also takes place in vitro when sperm are incubated with fluid from uterus or oviduct. Prior to capacitation, mammalian sperm are in a state of low activity, apparently saving their energy and responsiveness until they are most likely to encounter an egg. A capacitated sperm is metabolically more active and beats its flagellum rapidly (Gwatkin, 1977). Similar behavior has also been observed in the sperm of some invertebrates (Wikramanayake et al., 1992).

The molecular mechanisms of capacitation are not known. They may involve changes in the composition of the sperm plasma membrane, such as lowering of the cholesterol:phospholipid ratio (B. K. Davis, 1981). Other models postulate that capacitation results from changes in the *cytoskeleton* at the sperm tip or removal from the sperm surface of "coating factors" that mask egg-binding sites.

Some Sperm Are Attracted to Eggs by Chemical Signals

Although parental mating behavior brings egg and sperm together, most sperm complete their journey to the egg by swimming. In mammals, ejaculated sperm must travel from the vagina to the site of fertilization in the upper oviduct. Muscular contractions of the uterus and oviducts contribute to the movement of the sperm. Nevertheless, of the vast number of sperm (100 million to 500 million in humans) deposited in the vagina, only a few hundred reach the fertilization site. This extreme reduction in number is very intriguing and does not seem to be fully explained as a means of selecting against abnormal sperm.

Sperm movement is guided by chemical attractants produced by the egg or by ovarian cells, at least in some species. Oriented movement in response to an external chemical signal is called *chemotaxis.* Chemotactic behavior in sperm has been observed in various groups of animals, such as hydrozoans, mollusks, echinoderms, urochordates, and mammals, including humans (R. Miller, 1985; Eisenbach and Ralt, 1992). In the hydrozoan *Campanularia,* eggs develop in an ovarylike structure called a *gonangium,* which has a funnel-shaped opening through which sperm must enter to reach the eggs. Richard Miller (1966) showed that the funnel produces an attractant that activates sperm and causes them to turn toward the funnel (Fig. 4.3). This attractant is species-specific; thus, funnel extract from *C. calceofera* attracts sperm from the same species but not from a related species, *C. flexuosa.* In another hydrozoan, *Orthopyxis caliculata,* the attractant is released by the eggs instead of the ovary (R. Miller, 1978). Its release is timed precisely: the sperm of this hydrozoan are attracted by eggs that have completed their second meiotic division, not by less mature eggs or oocytes.

Several sperm attractants have been isolated from sea urchins. One, called *resact,* is found in the jelly layer surrounding the egg of *Arbacia punctulata* (G. E. Ward et al., 1985). Resact is a small peptide consisting of 14 amino acids. At very low concentrations, resact attracts sperm in a species-specific way. However, when sperm are pretreated with high concentrations of resact they lose their chemotactic ability. This result suggests that the sperm surface may have specific receptors that become saturated at a certain resact concentration. In fact, such receptors have been identified in the plasma membrane of *Arbacia* sperm (Bentley et al., 1986).

Preliminary evidence for sperm chemoattractants in mammals came from reports that mammalian sperm accumulate *below* the fertilization site and remain relatively motionless for hours until ovulation has occurred. Then the sperm become motile again, and a small number of them find their way to the fertilization site. The experiments described below indicate that human sperm are chemotactic in response to a substance present in the follicular fluid. However, the chemical nature of the putative attractant is not yet known, nor is it clear whether the signal is released by the egg itself or by other ovarian cells.

▼

For a recent study of chemotaxis in human sperm, Dina Ralt and her colleagues (1991) obtained eggs and follicular fluid from the ovaries of women undergoing in vitro fertilization. To test follicular fluid for its ability to chemically attract sperm, the researchers used vials that each had two wells separated by a filter with small pores (8 μm in diameter). They placed human sperm in the lower wells and fluid samples from indi-

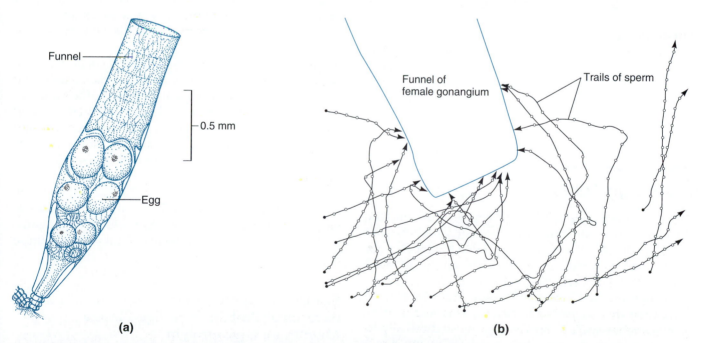

(a)

(b)

Figure 4.3 Chemotaxis in the hydrozoan *Campanularia flexuosa.* **(a)** Drawing of a female reproductive structure, the gonangium, from a colony of animals. The gonangium has a large funnel leading to the eggs within. **(b)** Plot of 21 sperm trails in the vicinity of the gonangium. The solid circles mark the start of each trail, and the open circles give the position of each sperm at 0.45-s intervals. Most trails lead to the funnel. The increasing distances between the circles indicate that the sperm accelerate as they draw closer to the funnel.

vidual follicles in the upper wells. After 15 min incubation time, they measured how many sperm had crossed each membrane and accumulated in the upper well. In about half the cases, the fluid samples caused the accumulation of sperm in the upper wells. Remarkably, the ability of a fluid sample to attract sperm correlated almost perfectly with the fertilizability of the egg obtained from the same follicle. No significant sperm accumulation was observed in control experiments in which the upper well of the vials contained culture medium or blood serum or in which follicular fluid had been added to both wells.

In order to determine whether the follicular fluid changed the swimming pattern of sperm, the investigators traced the paths of individual sperm after injecting follicular fluid into a vial containing sperm. Indeed, some sperm oriented themselves toward the added fluid while others seemed to ignore it. This observation is consistent with the fact that only a small fraction of the sperm that are ejaculated into the vagina arrive at the site of fertilization in the oviduct.

Although the evidence for a human sperm chemoattractant is credible, it does not tell us why only a fraction of all sperm appear to be receptive to the attractant at any particular time. One possible explanation for differences in receptivity, proposed by Michael Eisenbach and Dina Ralt (1992), is that a population of ejaculated sperm is heterogeneous: some sperm are able to fertilize an egg, but others are immature or overmature. The putative chemoattractant in follicular fluid seems to selectively attract sperm that are at the peak of their ability to fertilize. According to this hypothesis, the egg encounters only the selected sperm upon ovulation; the immature and overmature sperm are excluded from the competition.

Sperm-Egg Adhesion

Before a sperm can fertilize an egg, it must burrow through jelly coats and other protective layers surrounding the egg. This activity is guided by signals from the egg and its surrounding covers or cells (C. R. Ward and Kopf, 1993). A critical step in this process is *sperm-egg adhesion*, the binding of the sperm to the *vitelline envelope* or the egg *plasma membrane*. This step is more or less species-specific: sperm bind most effectively to eggs of the same species. This species specificity is mediated by molecules on the surfaces of the egg and sperm that match each other like a lock and a key. This section describes some of the experiments by which the molecules involved in sperm-egg adhesion have been characterized in sea urchins and in mammals.

Sperm Undergo the Acrosome Reaction As They Adhere to Egg Envelopes

Enzymes stored in the acrosomal vesicle at the tip of its head enable the sperm to penetrate the egg coats. Upon contact with the egg coats, the acrosomal vesicle releases its contents by exocytosis, in a process called the *acrosome reaction*. It is triggered by an event in the sperm plasma membrane and mediated by intracellular signals, including a rise in calcium ion (Ca^{2+}) concentration and an increase in pH, much like the egg activation process to be discussed later in this chapter.

In sea urchins, the acrosome reaction occurs when the sperm contacts certain glycoproteins in the jelly coat of the egg (S. H. Keller and Vacquier, 1994). The acrosomal membrane and the overlying portion of the sperm plasma membrane fuse, releasing the contents of the acrosomal vesicle near the tip of the sperm (Fig. 4.4). Acrosomal enzymes digest, or lyse, a hole through the egg jelly so that the sperm gains access to the vitelline envelope (J. C. Dan, 1967). Simultaneously, an extension called the *acrosomal process* forms at the tip of the sperm. The acrosomal process is covered by what was originally the acrosomal membrane and has now become part of the sperm plasma membrane. The extension of the acrosomal process is a rapid reaction driven by the polymerization of actin stored behind the acrosomal vesicle in a cup-shaped depression next to the nucleus (see Figs. 2.9 and 4.4).

Mammalian sperm have less subacrosomal material than sea urchin sperm have and do not form acrosomal processes (Fig. 4.5). During the acrosome reaction in mammals, the acrosomal membrane and the sperm plasma membrane fuse at many points, thus breaking up the two membranes into many small vesicles. Because of the technical problems associated with studying mammalian fertilization in vivo, the timing of the acrosomal reaction is not known with certainty; it seems to vary among species. In rabbits, the acrosome reaction seems to be triggered by the *granulosa cells* adhering to the outside of the zona pellucida (see Fig 3.24). In mice, evidence obtained in vitro indicates that the sperm must first adhere to the *zona pellucida* before the acrosome reaction is elicited by a zona protein designated **ZP3** (Wassarman, 1987). As we will see, the same protein is also involved in the initial adhesion of mouse sperm to the zona.

Sea Urchin Sperm Adhere with an Acrosomal Protein to an Egg Plasma Membrane Glycoprotein

Sea urchin sperm that have penetrated the egg jelly encounter the egg's vitelline envelope. If the egg and sperm are from the same species, they adhere to each other stably. Each spermatozoon adheres to the vitelline envelope by its acrosomal process, which is coated with

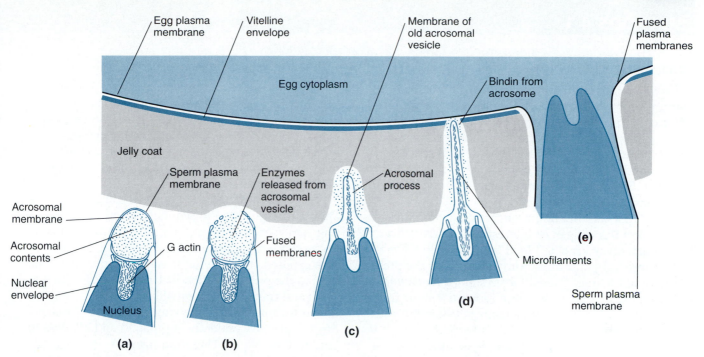

Figure 4.4 Acrosome reaction in sea urchin sperm. **(a)** Drawing of the head of an unreacted sperm, showing the nucleus surrounded by the nuclear envelope, the acrosomal material surrounded by the acrosomal membrane, and a subacrosomal space filled with G actin (short lines). The entire sperm is surrounded by the plasma membrane. **(b)** When the sperm contacts the jelly coat of the egg, the acrosomal vesicle (acrosome) undergoes exocytosis. Proteins (small dots) released from the acrosomal vesicle dissolve the jelly coat. **(c)** The acrosomal process extends and remains covered with acrosomal material, which adheres to the vitelline envelope. **(d)** Sperm-egg plasma membrane contact after lysis of the vitelline membrane. Bindin is the substance that enables the sperm to adhere to the egg. **(e)** Plasma membrane fusion and beginning of sperm entry into the egg.

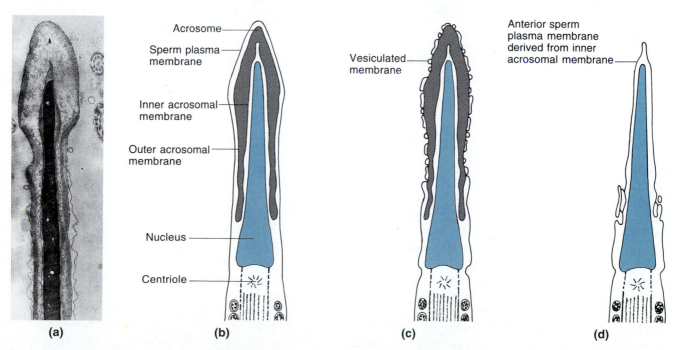

Figure 4.5 Acrosome reaction in mammals. **(a)** Transmission electron micrograph of a hamster sperm before the acrosome reaction. A = acrosome. **(b)** Drawing of sperm head before acrosome reaction, with sperm plasma membrane and acrosomal vesicle intact. **(c)** During the acrosome reaction, the sperm plasma membrane and the outer acrosomal membrane fuse at many places, forming numerous vesicles. **(d)** The anterior sperm plasma membrane, visible after the acrosome reaction, is derived from part of the inner acrosomal membrane.

Immunostaining

The *immunostaining* procedure relies on blood serum proteins known as immunoglobulins or *antibodies.* Antibodies bind to specific *antigens,* that is, foreign molecules that elicit an immune response. In particular, antibodies from one animal species are bound as antigens by antibodies from other species. Scientists use immunostaining to locate antigens at their normal sites of occurrence.

The first step in an immunostaining procedure is to inject the antigen of interest into a laboratory mammal, often a rabbit. Blood is taken from the rabbit at intervals, and the serum is tested for the presence of antibodies against the antigen. *Preimmune serum* prepared from the same rabbit *before* the first antigen injection is kept as a control. When the antibody concentration in the rabbit's blood is sufficiently high, a larger blood sample is taken, and the immunoglobulin fraction (the serum proteins containing the antibodies of interest) is prepared. This preparation is the *primary antibody,* which binds selectively to the antigen of interest.

One way of making the primary antibody visible is to conjugate it to a fluorescent molecule or some other tag that can be detected under the microscope. For more consistent results, researchers often use a *secondary antibody,* prepared in a different mammalian species and directed against the primary antibody as an antigen. Frequently, the secondary antibody is made in a goat or a swine and is directed against antigenic sites found in all rabbit im-

munoglobulins. The secondary antibody is conjugated to a fluorescent dye, such as *fluorescein* or *rhodamine.* The secondary antibody may also be conjugated to a stable enzyme, such as peroxidase, which generates a dark precipitate from colorless reagents, such as diaminobenzidine and hydrogen peroxide.

For immunostaining, the primary and secondary antibodies are added in sequence to the antigen (Fig. 4.6). The specimen containing the antigen is preserved with a fixative that leaves the antigenic sites exposed and fairly undisturbed so that they will bind to the primary antibody. Unless the specimen is small and transparent, it must be embedded and sectioned for microscopic viewing. On each section, a drop of primary antibody is pipetted and allowed to bind to the antigen. Unbound primary antibody is washed off before the secondary antibody is added and allowed to bind to the primary antibody. Unbound secondary antibody is washed off, so that only specifically bound antibody remains on the section. Depending on the conjugate attached to the secondary antibody, further reagents are added to make the conjugate visible.

When viewed under the microscope, the fluorescent stain or visible precipitate shows where the secondary antibody is located. Assuming that secondary antibody is bound *only* to primary antibody, and that primary antibody is bound *only* to the antigen of interest, the conjugate reveals where the antigen of in-

---

material originally contained in the acrosomal vesicle. This observation led investigators to hypothesize that sea urchin sperm and eggs may be attached to each other by a component from the acrosome.

The critical acrosomal component was first prepared from the sperm of the sea urchin *Strongylocentrotus purpuratus* (Vacquier and Moy, 1977; Vacquier, 1980). The sperm were suspended in a solution that dissolved the plasma membrane and acrosomal membrane but preserved the acrosomal contents as a more or less intact granule. The granules were separated from other sperm components by several rounds of centrifugation and filtration before they were dissolved in a strong detergent and analyzed by gel electrophoresis. The researchers observed one major band, a polypeptide with an apparent molecular weight of 30,500, which they termed *bindin.* Two experiments indicated that bindin is the adhesive material by which sea urchin sperm adhere to eggs. The first experiment used a very versatile procedure known as *immunostaining* (Methods 4.1). The

second experiment involved a standard test for cell aggregation.

In their analysis of sperm-egg adhesion in sea urchins, Moy and Vacquier (1979) revealed the location of bindin by immunostaining, using a primary antibody against sea urchin bindin. This method showed a thick coat of bindin covering the acrosomal process of acrosome-reacted sperm (Fig. 4.7). Control sperm, which had not undergone the acrosome reaction, were not stained by the procedure. In newly fertilized eggs, bindin was found at the site where acrosomal processes adhered. Thus bindin is present exactly when and where it should be as a mediator of sperm-egg adhesion.

By *cell aggregation experiments,* Charles Glabe and William Lennarz (1979) showed that bindin agglutinates eggs in a species-specific manner. They mixed equal numbers of dejellied *Strongylocentrotus purpuratus* and *Arbacia punctulata* eggs with bindin from either species in culture dishes. (Bindin is poorly soluble in water without detergent and forms particles in seawa-

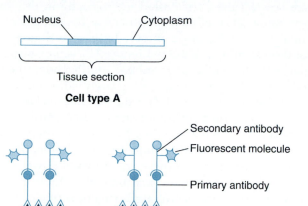

Figure 4.6 Immunostaining is used to reveal the position of a known antigen in sections of fixed tissue. In this hypothetical example, the question is whether the antigen of interest is present in cell type A or B, and whether the antigen is located in the nucleus or in the cytoplasm. Primary antibody directed against the antigen (black triangles) is allowed to interact with the sectioned tissue. Secondary antibody directed against the primary antibody is covalently linked with a fluorescent molecule or some other conjugate. The distribution of the conjugate over the section reveals that the antigen is present here in the cytoplasm of cell type B.

terest is located in the sectioned specimen. These assumptions need to be confirmed with appropriate controls using specimens without antigen or preimmune serum instead of primary antibody.

ter.) The dishes were kept on a rotary shaker for a few minutes and then inspected under the microscope. Each bindin preparation agglutinated mostly eggs from its own species (Fig. 4.8). In order to test whether the bindin particles were directly responsible for holding the eggs together, the researchers labeled bindin with a fluorescent dye, fluorescein. Agglutinated eggs showed fluorescent particles precisely at the spots where they were stuck together. Thus bindins adhere specifically to eggs from the same species.

Bindins prepared from different sea urchin species are similar but not identical in size and amino acid composition, in accord with their similar yet species-specific functions (Glabe and Clark, 1991). Bindins interact with specific molecules on the egg surface (Foltz and Lennarz, 1993). In *S. purpuratus*, digestion of the entire egg plasma membrane and vitelline envelope with proteases has led to the identification of a glycoprotein that adheres to bindin of the same species but not of other species. The molecular characteristics of this glycoprotein indicate that it is embedded in the plasma membrane of the egg (Foltz et al., 1993). The receptor domain, which interacts with bindin, presumably extends into the vitelline envelope.

A Glycoprotein in the Zona Pellucida of Mouse Eggs Adheres to Sperm and Triggers the Acrosome Reaction

Sperm-egg adhesion in mice, as in sea urchins, is mediated by specific molecules on the surfaces of egg and sperm. Analogous to the sperm-binding glycoprotein in the sea urchin plasma membrane is a sperm-binding glycoprotein in the *zona pellucida* of the mouse (Wassar-

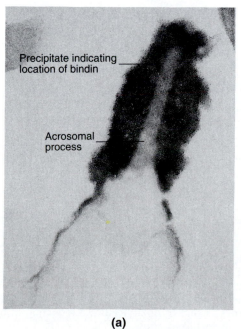

(a)

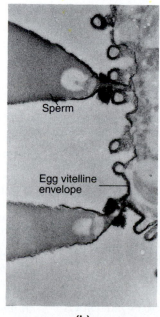

(b)

Figure 4.7 Localization of bindin on the acrosomal process of sea urchin sperm after the acrosomal reaction (a) and after binding to the egg's vitelline envelope (b). The bindin location was revealed by immunostaining thin sections for viewing under the electron microscope. Primary rabbit antibodies were prepared against sea urchin bindin, and secondary swine antibodies against rabbit IgG were conjugated with peroxidase. When furnished with the appropriate substrates, the enzyme produced an electron-dense precipitate, which revealed the location of bindin.

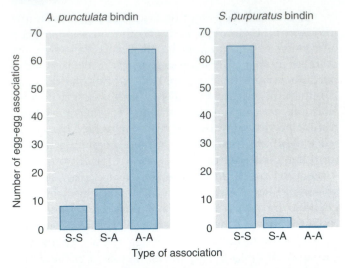

Figure 4.8 Species-specific agglutination of sea urchin eggs by bindin. A suspension containing equal numbers of eggs of *Strongylocentrotus purpuratus* (S) and *Arbacia punctulata* (A) was mixed with small amounts of bindin particles from one of the species and gently shaken for 2 to 5 min. The resulting egg-egg associations were scored on the basis of different egg pigmentations as S-S, A-S, or A-A. Each bindin agglutinated mostly eggs from its own species.

man, 1987, 1990). The zona is synthesized by the growing oocyte as a relatively thick but porous structure. Its major constituents are three glycoproteins (designated ZP1, ZP2, and ZP3), which assemble into long, interconnected filaments. The zonae of other mammals, including humans, also consist of just a few glycoproteins.

The sperm-binding glycoprotein in mice was identified by means of a *bioassay*, that is, a test used to screen various components for their ability to elicit a biological response. In this instance, Jeffrey Bleil and Paul Wassarman (1980) tested zona pellucida glycoproteins for their ability to compete with intact eggs for fertilizing sperm.

▼

To establish their bioassay, the investigators fertilized mouse eggs in vitro under conditions in which the sperm were in limited supply (Fig. 4.9). The percentage of fertilized eggs was determined as the percentage of eggs that began to cleave. This percentage decreased dramatically when the sperm were preincubated with zona pellucida glycoproteins from *unfertilized eggs*. In contrast, the same glycoproteins from *two-cell embryos* did not have this inhibitory effect. The investigators concluded that zona glycoproteins from unfertilized eggs were competing with intact unfertilized eggs for binding sites on the sperm. In two-cell embryos, the zona glycoproteins seemed to be modified so that sperm could no longer bind to them.

Once this *competitive binding* assay for sperm-binding compounds was established, each zona glycoprotein was purified and tested individually in the same way. Only ZP3 inhibited sperm-egg adhesion. The other zona proteins did not compete with unfertilized eggs for the sperm (Fig. 4.10). Evidently, ZP3 is the only zona glycoprotein that binds to mouse sperm during the initial sperm-egg adhesion.

Molecular analysis of ZP3 showed that its backbone is a single polypeptide of about 400 amino acids (Wassarman, 1990). Extending from the polypeptide are many *oligosaccharides*, that is, short chains of connected sugars. Some of these are *O-linked oligosaccharides*, which means that they are linked to an oxygen atom on either one of two amino acids: serine or threonine. To reveal which of the ZP3 components function in sperm adhesion, ZP3 was pretreated with various agents to destroy specific parts of the molecule. The pretreated ZP3 was then tested again as a competitor in

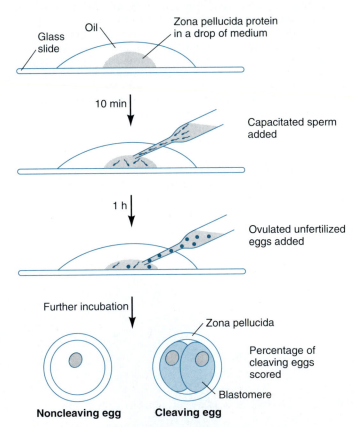

Figure 4.9 Competitive binding assay to identify the mouse sperm receptor in the zona pellucida of the egg. A drop of medium with various zona pellucida proteins is kept on a glass slide under oil to prevent desiccation. Capacitated sperm are added to the drop from a glass micropipette. After 1 h of incubation, unfertilized ovulated eggs are added. After further incubation, the eggs are washed, stripped of adhering sperm, and prepared for light microscopy. The percentage of cleaving eggs is scored. The number of sperm is kept low so that any reduction in the sperm's fertilizing ability causes a reduction in the percentage of fertilized eggs.

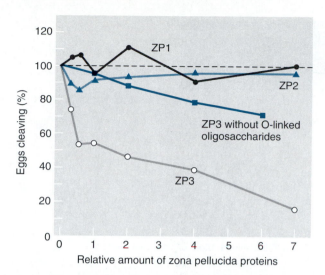

Figure 4.10 Competition of mouse ZP3 with intact egg zona pellucida for sperm binding sites. Sperm were preincubated with purified proteins (ZP1, ZP2, ZP3) from mouse zonae pellucidae before unfertilized test eggs were added. The eggs were scored later for fertilization. Sperm preincubated with ZP1 or ZP2 fertilized the test eggs, but sperm preincubated with ZP3 showed a reduced ability to fertilize. The inhibitory action of ZP3 depended largely on its O-linked oligosaccharides and was lost after fertilization. (Apparent fertilization rates greater than 100% are an artifact of the method used for data processing.)

the bioassay. Removal of the oligosaccharides destroyed the competitive function of ZP3, whereas cutting the polypeptide enzymatically into short pieces did not have this effect (Fig. 4.10). Specifically, removing the O-linked oligosaccharides with a mild alkali left ZP3 an unfit competitor (Florman and Wassarman, 1985). Conversely, the O-linked oligosaccharides alone competed as well as intact ZP3. Finally, Bleil and Wassarman (1988) showed that a terminal galactose residue is critical to the sperm-binding function of the O-linked oligosaccharides.

Three additional lines of evidence confirmed that specific portions of the ZP3 glycoprotein are necessary and sufficient for the initial sperm-egg adhesion in mice (Bleil and Wassarman, 1986). First, purified and radiolabeled ZP3 from *unfertilized eggs* binds only to the head of the sperm, not to its midpiece or tail (Fig. 4.11). This is consistent with the observation that sperm bind by the head to the egg zona pellucida. Second, purified and labeled ZP3 from the *embryonic* zona pellucida does not bind to sperm, confirming the earlier observation that zona pellucida proteins from fertilized eggs are inactive in the competitive binding assay. Third, unfertilized eggs and ZP3 bind to sperm that have an intact acrosome, not to sperm that have undergone the acrosome reaction and have lost much of the plasma membrane that covers the head of an unreacted sperm.

In addition to being the initial sperm-binding molecule on the zona pellucida, ZP3 also triggers the acrosome reaction in mice. However, this function depends on the polypeptide backbone of ZP3 as well as the O-linked oligosaccharides that alone are sufficient for sperm binding. Presumably, several oligosaccharides must bind to the sperm in a certain configuration to cause the changes in the sperm plasma membrane that initiate the acrosomal reaction (Leyton and Saling, 1989).

During the acrosome reaction, the sperm sheds much of the plasma membrane that covers its head and mediates its initial adhesion to the zona pellucida. Nevertheless, the sperm remains bound and proceeds to penetrate the zona. The continued sperm-egg binding seems to rely on the ZP2 protein and on some component of the sperm acrosomal membrane that is exposed after the acrosome reaction (Bleil et al., 1988).

Several mouse sperm membrane proteins that bind selectively to ZP3 have been identified (C. R. Ward and Kopf, 1993). One such protein is an enzyme, *N-acetylglucosamine galactosyltransferase (GalTase)*, which is located at the tip of acrosome-intact sperm (Fig. 4.12). The biochemical function of GalTase is to transfer galactose to an oligosaccharide chain ending with N-acetylglucosamine. However, since galactose is not present in the oviduct, the enzyme is not able to complete its function and so remains stuck to the terminal N-acetylglucosamines of the O-linked oligosaccharides of ZP3.

The role of a GalTase present on acrosome-intact sperm as an egg adhesion molecule in mice is supported by several sets of data (Shur, 1989; D. J. Miller et al., 1992). First, mouse sperm GalTase transfers labeled galactose to ZP3 from unfertilized eggs, but not to ZP2 or ZP1. Second, removing or blocking N-acetylglucosamine from ZP3 inhibits the sperm GalTase's ability to bind. Third, GalTase on acrosome-reacted sperm does not recognize zona proteins from unfertilized eggs. Fourth, GalTase on acrosome-intact sperm does not recognize

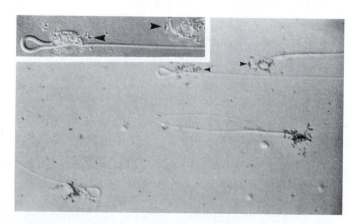

Figure 4.11 Specific binding of mouse sperm to a zona pellucida protein (ZP3) from unfertilized eggs. The autoradiograph shows capacitated mouse sperm after incubation with radiolabeled ZP3. The silver grains (arrowheads) are located specifically over the sperm heads.

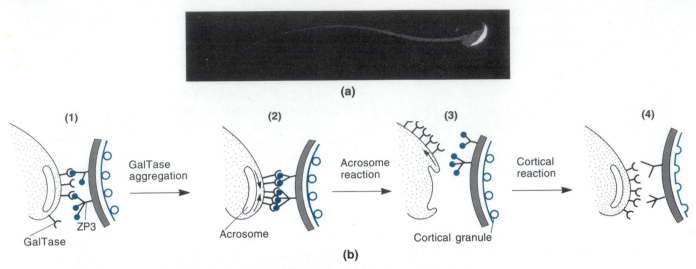

Figure 4.12 The role of *N*-acetylglucosamine galactosyltransferase (GalTase) in mouse sperm-egg adhesion. **(a)** Fluorescence photomicrograph of sperm immunostained with antibody against GalTase. The area of plasma membrane above the acrosome is specifically labeled. **(b)** Model for the interaction between sperm GalTase and mouse egg zona pellucida protein 3 (ZP3) during sperm-egg adhesion. Initial sperm-egg recognition is mediated by sperm GalTase binding to terminal *N*-acetylglucosamine residues (black dots) on ZP3 oligosaccharides **(1)**. Because the second substrate of GalTase, activated galactose, is not normally present in the oviduct, GalTase remains bound to *N*-acetylglucosamine. The multivalent ZP3 binds to two or three GalTase moieties, thereby aggregating the GalTase at the tip of the acrosome **(2)** and initiating the acrosome reaction. During the acrosome reaction, the plasma membrane overlying the acrosome is released, and GalTase moves to the side of the sperm head **(3)**. After the acrosome reaction, the fertilizing sperm binds to the egg in an interaction that does not seem to involve ZP3 and GalTase. The fertilizing sperm penetrates the zona pellucida and contacts the underlying egg plasma membrane (not shown). The resulting egg activation involves the exocytosis of egg cortical granules that contain several enzymes, possibly including *N*-acetylglucosaminidase **(4)**. This enzyme removes terminal *N*-acetylglucosamine residues from the sperm-receptor oligosaccharides, thereby preventing other sperm from binding to the zona pellucida.

zona proteins from fertilized eggs. Fifth, GalTase is also known as a cell adhesion molecule between other cells (see Chapter 25).

Antibodies to Sperm-Egg Adhesion Proteins Act as Contraceptives

In vitro studies on mammalian fertilization may lead to the development of a contraceptive vaccine. Since sperm-egg adhesion depends on a few specific molecules, it may be possible to block this process using antibodies that do not interfere with other vital functions. Such antibodies might be elicited by injecting suitable antigens, such as the zona proteins or sperm proteins that mediate sperm adhesion.

Immunization with zona proteins from hamsters makes female mice infertile, but after the antibody concentration decreases, the mice are capable of bearing normal young again (Gwatkin, 1977). Similarly, female mice immunized with mouse ZP3-produced antibodies that coated the zonae pellucidae of their developing oocytes. These mice were also infertile as long as the antibody concentration remained high (Millar et al., 1989).

Egg-binding sperm proteins are another target for contraceptive antibodies. Clinical data indicate that human infertility may be caused by antisperm antibodies in either the male or the female partner. (Normally, such an autoimmune reaction in males is prevented by the blood-testis barrier described in Chapter 3.) In guinea pigs immunized with guinea pig sperm protein, the vaccination was fully contraceptive in both males and females, and the effects were reversible (Primakoff et al., 1988). However, the transfer of this technique to humans requires better characterization of the human sperm-egg adhesion proteins.

Gamete Fusion

After sperm-egg adhesion, sperm penetrate the vitelline envelope with lytic enzymes stored in the acrosomal vesicle (Colwin and Colwin, 1960). The sperm plasma membrane then comes in contact with the egg plasma membrane. This *plasma membrane contact* is a key event with far-reaching consequences, which will be discussed in the next section, on egg activation.

After plasma membrane contact has occurred, egg and sperm proceed with **plasma membrane fusion,** the event that most narrowly defines fertilization. In sea urchins, membrane fusion always occurs at the tip of the acrosomal process and often involves an egg microvillus (Fig. 4.13). In mice and other mammals, the sperm contacts the egg sideways; the part of the sperm membrane that fuses with the egg membrane is behind the former location of the acrosome. In many species, the part of the egg membrane that fuses with the sperm is restricted to the area near the egg nucleus (Freeman and Miller, 1982).

The molecular components involved in plasma membrane fusion are not yet known, but a few candidates are being investigated. One of these is a protein, designated PH-30, that was isolated from guinea pig sperm. Several characteristics make it a likely agent for plasma membrane fusion (Myles, 1993). First, antibodies against PH-30 inhibit fertilization. Second, PH-30 is spread over the entire head of immature sperm, but becomes concentrated in the posterior head region as the sperm mature. It remains in this location after the acrosome reaction, so that it is in the appropriate place for sperm-egg contact. Finally, PH-30 has the molecular characteristics of the fusion proteins that viruses use to

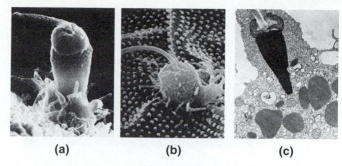

(a)   (b)   (c)

Figure 4.14 Sperm incorporation into sea urchin egg. **(a, b)** Scanning electron micrographs showing the internalization of the sperm head into the egg. **(c)** Transmission electron micrograph showing the sperm head engulfed by a cytoplasmic mound known as the fertilization cone.

penetrate the plasma membranes of their host cells (Blobel et al., 1992). Sea urchin bindin, the same protein that functions in sperm-egg adhesion, may also be involved in membrane fusion: it causes phospholipid bilayers to fuse (Glabe, 1985) and has an amino acid sequence similar to that of viral fusion proteins.

At the site of membrane fusion, the fertilized egg forms a protrusion called the **fertilization cone.** The cone engulfs the sperm as it sinks into the egg (Fig. 4.14). This process involves a movement of egg cytoplasm into the region surrounding the sperm nucleus (Longo, 1989). The fertilization cone often grows for several minutes after sperm entry and then regresses. The formation of the cone is associated with the polymerization of cortical actin into microfilaments; *cytochalasin,* a drug that interferes with microfilament polymerization, inhibits cone formation. It is thought that the fertilization cone and microfilaments facilitate sperm entry, but the mechanisms involved are still unclear.

Under the influence of the egg cytoplasm, the sperm nucleus undergoes certain changes that prepare it to participate in forming the *zygote* nucleus. The envelope of the sperm nucleus disintegrates into small vesicles, thus exposing the sperm chromatin to the egg cytoplasm (G. R. Green and Poccia, 1985). The sperm nucleus then swells and its envelope reconstitutes, whereupon it is called the **male pronucleus.** The egg nucleus has already completed meiosis and is now called the **female pronucleus.** (In most species, however, meiosis is not completed until after fertilization. See Chapter 3.) Both pronuclei move toward each other and to the center of the egg in movements that are mediated by microtubules (Fig. 4.15).

The process by which a diploid nucleus forms in the zygote varies among groups of animals. In sea urchins and their relatives, the envelopes of the pronuclei fuse in a process called *syngamy* and form a common nuclear envelope around the maternal and paternal chromosomes. However, other animals, including mammals,

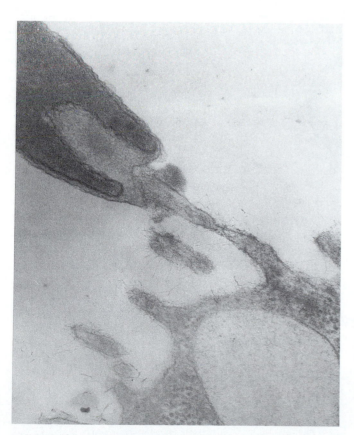

Figure 4.13 Plasma membrane fusion during sea urchin fertilization. The electron micrograph shows a cytoplasmic connection between the acrosomal process of the sperm (top left) and a microvillus from the egg.

Male
pronucleus

Female
pronucleus

Figure 4.15 Fusion of pronuclei in the fertilized egg of the sea urchin *Clypeaster japonicus*. This series of photographs, which covers a time span of 800 s, shows the male pronucleus surrounded by an aster of microtubules. The pronuclei move toward each other and fuse.

skip this step. Instead, the male and female pronuclei retain separate nuclear envelopes until just prior to mitosis. Then, their nuclear envelopes break up, and the maternal and paternal chromosomes align within the same mitotic spindle (Fig. 4.16). Diploid nuclei therefore do not originate until after the first mitosis in these animals.

The fusion of a male and a female gamete concludes the process of fertilization by forming a single cell, the *zygote*. Both gametes contribute corresponding sets of chromosomes to the zygote. Nearly all of the zygote cytoplasm and its organelles are, of course, derived from the egg. The materials stored in the egg cytoplasm will sustain the developing embryo until it is able to obtain food from other sources. In addition to storing the building blocks for the embryo, the egg often provides one or two polarity axes in the form of unevenly distributed cytoplasmic components. These axes form a rudimentary coordinate system for the development of different cells within the embryo. In some cases, one axis is oriented by the site of sperm entry.

In most species, the *basal body* located at the base of the sperm's flagellum replicates and forms the *centrosomes* that organize the first mitotic spindle in the zygote. The flagellum itself and the mitochondria of the sperm usually disintegrate, although in some species the sperm mitochondria are conserved (Giles et al., 1980; Zouros et al., 1992). The plasma membrane of the zygote consists mostly of the egg membrane, with a small patch contributed by the sperm membrane at the entry site.

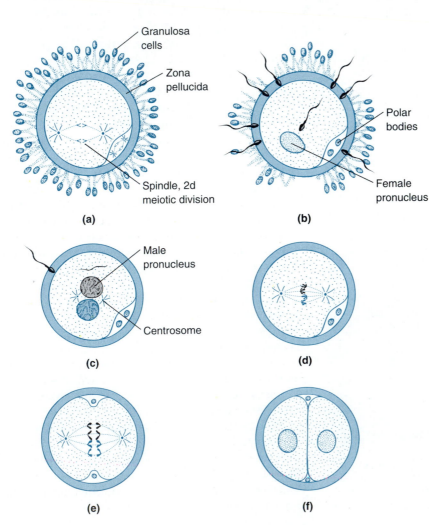

✓ only sperm head enters

✓ 3rd pol. body?

Figure 4.16 Fertilization of the human egg. **(a)** Ovulated egg arrested in metaphase II. **(b)** One sperm has entered the egg, which has completed meiosis II and formed a female pronucleus. **(c)** The sperm nucleus decondenses and forms the male pronucleus. **(d)** The nuclear envelopes of both pronuclei disintegrate, and their chromosomes are incorporated into a common mitotic spindle. **(e)** Anaphase of first embryonic cleavage. **(f)** Embryo at the 2-cell stage.

TABLE 4.1

Events in Sea Urchin Fertilization

Event	Approximate Time after Plasma Membrane Contact[a]
Fertilization potential appears (fast block to polyspermy)	Before 3 s
Plasma membrane fusion	Before 15 s[b]
Inositol trisphosphate and diacyclglycerol produced	Before 15 s[b]
Intracellular calcium release	40–120 s
Cortical reaction (slow block to polyspermy)	40–100 s
Sperm entry	1–2 min[b]
Activation of NAD kinase	1–2 min
Increase in O_2 consumption	1–3 min
Na^+/H^+ exchange, increase in pH	1–5 min
Sperm chromatin decondensation	2–10 min[c]
Sperm nucleus migration to egg center	2–10 min[c]
Activation of protein synthesis	After 5 min
Fusion of pronuclei	20 min[d]
Initiation of DNA synthesis	20–40 min
First mitotic prophase	60–80 min
First cleavage	85–95 min

a. The times listed are estimates based on data from *Lytechinus pictus* kept at 16–18°C (Whitaker and Steinhardt, 1985), except as noted.
b. Laurinda A. Jaffe (personal communication).
c. Based on data from *Clypeaster japonicus* (Hamaguchi and Hiramoto, 1980).
d. Based on data from *Strongylocentrotus purpuratus* (Epel, 1977).

Egg Activation

Membrane contact between sperm and the egg triggers *egg activation,* a programmed series of events that stimulates the quiescent egg to begin development (Table 4.1). Activation entails resuming the cell cycle, which has been in meiotic arrest (see Fig. 3.23). In all animal eggs studied so far, the early activation events include *calcium ion (Ca^{2+}) transients,* that is, relatively short rises or oscillations in Ca^{2+} concentration. Such transients can be monitored with certain fluorescent dyes that are sensitive to Ca^{2+} concentration. The egg activation steps that follow the Ca^{2+} transients lead to an acceleration of the egg's metabolism before cleavage. Another aspect of egg activation in many species is the prevention of multiple fertilizations, which we will discuss separately.

Egg Activation May Be Triggered by a Receptor Protein in the Egg Plasma Membrane

How does fertilization trigger egg activation? The likely locations for the trigger are the sperm cytoplasm or the egg plasma membrane (Whitaker and Swann, 1993). According to the first hypothesis, a component intro- duced with the sperm cytoplasm sets off the egg activation process. For instance, a protein factor isolated from hamster sperm, when injected into eggs, mimics the Ca^{2+} oscillations and subsequent activation responses normally seen after fertilization. However, this protein factor has not yet been characterized.

The second hypothesis is based on observations that *inositol trisphosphate (IP$_3$), diacylglycerol (DAG),* and free Ca^{2+} all increase in concentration early in the activation process (Table 4.1). These are the same second messengers that mediate cellular responses to hormones and neurotransmitters received through membrane-bound receptors associated with *G proteins* (see Chapter 2). These findings led to the hypothesis that membrane contact between sperm and egg activates a *sperm receptor* in the egg plasma membrane, just as other membrane receptors are activated by chemical ligands (Fig. 4.17). (These hypothetical sperm receptors in the egg plasma membrane are thought to be different from molecules like ZP3, which mediate sperm adhesion to the zona pellucida or other vitelline envelopes.) The sperm receptor in the plasma membrane presumably activates, via G proteins, a *phospholipase* that cleaves membrane-bound *phosphatidylinositol bisphosphate (PIP$_2$)* into IP$_3$ and DAG, signal molecules that in turn activate protein kinase C and release Ca^{2+} from cytoplasmic storage sites.

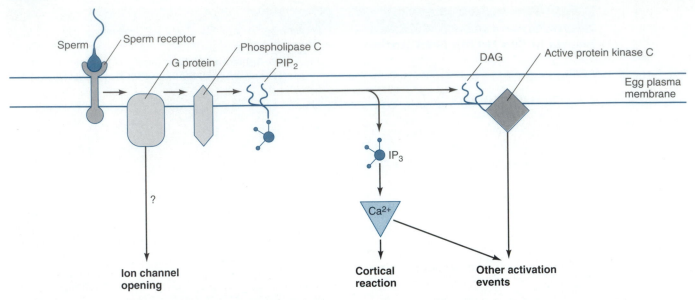

Figure 4.17 Proposed role of a sperm receptor and G protein in egg activation. A hypothetical sperm-receptor protein in the egg plasma membrane is activated by contact with the sperm. The active receptor stimulates a G protein, which activates phospholipase C to cleave phosphatidylinositol bisphosphate (PIP_2) into inositol trisphosphate (IP_3) and diacylglycerol (DAG) as second messengers. The IP_3 releases Ca^{2+} stored in the egg's endoplasmic reticulum. The released Ca^{2+} regulates the cortical reaction and, along with DAG-activated protein kinase C, other egg activation events. G protein may also be involved in the opening of ion channels that generate the fertilization potential.

The sperm receptor model is compatible with results from several experiments. For example, injection of IP_3 into unfertilized sea urchin or frog eggs starts the activation process (Whitaker and Irvine, 1984; Larabell and Nuccitelli, 1992).

▼

To test the involvement of G proteins in egg activation, Douglas Kline and his colleagues (1988) reasoned that if egg activation were mediated by G proteins, then another membrane receptor associated with the same type of G proteins might be able to cause egg activation. Accordingly, they designed an experiment that allowed them to stimulate G proteins in *Xenopus* eggs by exposing the egg plasma membrane to a neurotransmitter (Fig. 4.18).

Because frog eggs do not have receptors for neurotransmitters, the investigators injected oocytes with mRNA isolated from rat brain. Since oocytes were known to translate exogenous mRNA, the researchers expected them to synthesize all the neurotransmitter receptors encoded by the injected brain mRNA. Indeed, two days after mRNA injection, when the oocytes were briefly rinsed with a medium containing a neurotransmitter (acetylcholine or serotonin), they responded with membrane depolarization just as a neuron does. These results showed that the oocytes had synthesized

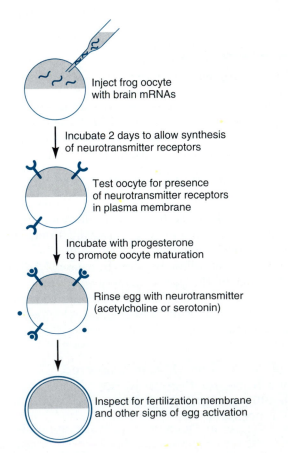

Inject frog oocyte with brain mRNAs

Incubate 2 days to allow synthesis of neurotransmitter receptors

Test oocyte for presence of neurotransmitter receptors in plasma membrane

Incubate with progesterone to promote oocyte maturation

Rinse egg with neurotransmitter (acetylcholine or serotonin)

Inspect for fertilization membrane and other signs of egg activation

Figure 4.18 Design of an experiment to test whether G protein is involved in frog egg activation.

the appropriate neurotransmitter receptors and incorporated them into their plasma membranes as expected. Next the researchers induced the injected oocytes to mature into eggs by incubating them with progesterone. Now they had mature eggs with neurotransmitter receptors in their plasma membranes. These eggs were either rinsed with acetylcholine or with serotonin or fertilized normally for control purposes. Under each regimen, the eggs showed all external signs of normal egg activation, including the formation of the hardened envelope that protects the fertilized egg against additional fertilizations.

Similar results were obtained in corresponding experiments with mouse eggs. In this case, experimental stimulation of a G protein–coupled receptor resulted in complete egg activation and cleavage to the 2-cell stage (G. D. Moore et al., 1993). Also, stimulation of protein kinase C at low Ca^{2+} concentrations triggered some egg activation responses in hamster eggs (Gallicano et al., 1993). Thus, the same second messenger pathway that elicits cellular responses to hormones and neurotransmitters is *sufficient* to cause egg activation.

To test whether this second messenger pathway is *necessary* for egg activation, Shunichi Miyazaki (1988) inhibited G protein activity in hamster eggs by injecting a modified guanosine diphosphate that blocks the binding site of G protein for regular guanosine diphosphate (see Fig. 2.25). This treatment interfered with the Ca^{2+} response of the egg to fertilization. Likewise, injection of antibody to the IP_3 receptor that mediates the release of Ca^{2+} from cytoplasmic storage sites completely blocked the sperm-induced Ca^{2+} waves and oscillations (Miyazaki et al., 1992). However, similar experiments with sea urchin eggs gave a different result: injection of *heparin,* another inhibitor of the IP_3 receptor that releases Ca^{2+}, did not block the Ca^{2+} transient at fertilization (Crossley et al., 1991).

The mechanisms by which fertilization triggers egg activation may vary among animal groups. For mammals, it seems likely that sperm contact activates a receptor protein in the egg plasma membrane that starts an intracellular signal chain using the same G proteins and second messengers involved in processing chemical signals that bind to membrane receptors. On the other hand, the same signal chain may be silent during the normal activation of sea urchin eggs, in which Ca^{2+} seems to be released independently of IP_3.

A Temporary Rise in Ca^{2+} Concentration Is Necessary and Sufficient for Egg Activation

The Ca^{2+} transients that are naturally triggered by fertilization are an early and regular feature of egg activa-

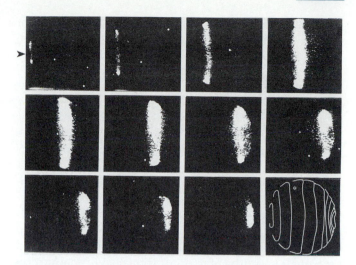

Figure 4.19 Ca^{2+} wave in a fertilized fish egg (*Oryzias latipes*). The egg has been injected with aequorin, a protein that emits light when it binds to Ca^{2+} ions. The photographs were taken at 10-s intervals to show the wave of Ca^{2+} that begins near the sperm entry point (arrowhead at the top left panel) and passes across the egg. The bottom right panel outlines the leading edges of the light front seen in the other 11 panels.

tion. In species with large eggs, the increased concentration of free Ca^{2+} spreads like a wave, beginning at the site of sperm entry (Fig. 4.19). The wavelike propagation seems to follow the release of Ca^{2+} from intracellular stores as a result of stimulation either by IP_3 or by Ca^{2+} itself.

To test whether Ca^{2+} transients are *necessary* for all subsequent activation events to occur, researchers have *removed* Ca^{2+} from *fertilized* eggs. This can be achieved by injecting eggs either with drugs that prevent the release of Ca^{2+} from its intracellular stores or with molecules that bind to Ca^{2+}, rendering it metabolically inactive. Either procedure prevents all subsequent egg activation events. It follows that Ca^{2+} is necessary for these events to occur.

To test whether Ca^{2+} transients are *sufficient* for later activation events to occur, investigators have *added* Ca^{2+} to *unfertilized* eggs. This can be done by injecting Ca^{2+} into unfertilized eggs, or by keeping unfertilized eggs in a Ca^{2+}-containing medium and making them permeable to Ca^{2+} with drugs called *ionophores*. Either treatment triggers subsequent activation steps, showing that Ca^{2+} is sufficient for these steps to occur. These and other results underline the importance of Ca^{2+} as a second messenger in newly fertilized eggs. Together with other second messengers, Ca^{2+} is a key regulator of egg activation in many species.

Egg Activation Causes the Cell Cycle to Resume

Egg activation stimulates many events that are part of the cell cycle (Table 4.1). DNA synthesis, of course, is a prerequisite for subsequent mitosis. The acceleration of protein synthesis, which is also part of egg activation, entails in particular an improved capacity for synthesizing *cyclins*. Because Ca^{2+} can activate many proteins, it may regulate several cell cycle events independently. Alternatively, Ca^{2+} may control one or two master switches to restart the entire cell cycle. The exact mechanism may well depend on the stage of meiosis at the time of sperm entry, which is quite variable among different animal groups (see Fig. 3.21).

The unfertilized eggs of amphibians and most mammals are arrested during the second metaphase of meiosis. The arrest is caused by a sustained high level of *M-phase promoting factor (MPF)*. The high MPF level in turn depends on a high concentration of *c-mos protein,* and presumably another factor, during oocyte maturation (see Chapter 3). This situation changes upon fertilization, when a burst of Ca^{2+} is released in response to fertilization or parthenogenetic activation (Fig. 4.20). Ca^{2+} activates *calmodulin,* a Ca^{2+}-dependent protein, which in turn activates a calmodulin-dependent kinase, desig-

nated *CaM K_{II}* (Lorca et al., 1993). The active CaM K_{II} has a dual effect: First, it triggers the proteolysis of cyclin and thus reduces the high level of MPF that has kept the egg arrested in metaphase II. Second, CaM K_{II} causes the degradation of c-mos, the *cytostatic factor* that has maintained the MPF level high in the unfertilized egg. The molecular mechanisms by which CaM K_{II} governs MPF and c-mos inactivation are still being investigated.

Blocks to Polyspermy

Eggs of many species have thousands of adhesion sites on the vitelline envelope, presumably to ensure that fertilization can occur wherever a lone sperm may land on the egg. However, the excess of adhesion sites is potentially troublesome: it could lead to fertilization of a single egg by more than one sperm. Such a condition, known as *polyspermy,* would be disastrous in many species. The fertilization of a sea urchin egg by two sperm, for example, results in a zygote with three haploid sets of chromosomes. Also, the two centrosomes introduced by the two sperm set up a mitotic spindle with four poles. The ensuing cleavage produces blastomeres with irregular numbers of chromosomes (see Fig. 14.4). Such embryos usually die early (Boveri, 1907).

Multiple mechanisms have evolved to prevent either polyspermy itself or its consequences. Many species, including some insects, most salamanders, reptiles, and birds, are naturally polyspermic, but extra sperm are somehow inactivated in the eggs of these animals. In other species, polyspermy itself is prevented, typically by two independent mechanisms. One mechanism, known as the *fast block to polyspermy,* is rapid but temporary. The other mechanism, called the *slow block to polyspermy,* takes time to get under way but is permanent.

The Fertilization Potential Serves as a Fast Block to Polyspermy

When sea urchin eggs are fertilized in the laboratory with excess sperm, they rarely become polyspermic. As soon as one sperm contacts the egg, conditions change to prevent the plasma membrane from fusing with additional sperm. Some inhibitory signal must spread around the entire egg surface from the site of first plasma membrane contact. This mechanism is effective even when sperm concentrations are high enough for one fusion per second. Given an egg diameter of 0.1 mm, diffusion of a chemical signal would be much too slow for a response time of less than 1 s. Only an electrical signal can travel fast enough.

By inserting an electrode into an egg and placing another electrode on the outside, one can measure an

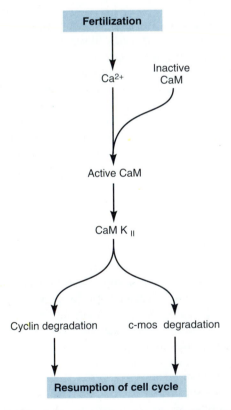

Figure 4.20 Proposed scheme for the resumption of the cell cycle upon fertilization or parthenogenetic egg activation. The Ca^{2+} transient after fertilization activates calmodulin (CaM), a Ca^{2+}-dependent protein. The active calmodulin in turn activates a calmodulin-dependent kinase, designated CaM K_{II}. The active CaM K_{II} reactivates the cell cycle of the amphibian egg by degrading cyclin, and thus the accumulated M-phase promoting factor (MPF), which arrests the egg in meiotic metaphase. In addition, CaM K_{II} destroys c-mos, a protein that keeps the MPF level high.

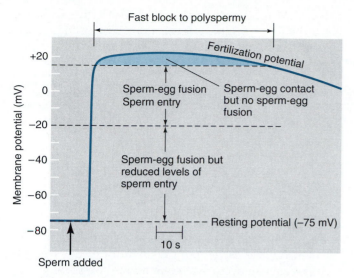

Figure 4.21 Resting potential and fertilization potential in the egg of the sea urchin *Strongylocentrotus purpuratus*. The resting potential is negative inside the cytoplasm (−75 mV). The first sperm-egg contact triggers a fertilization potential, which rises to near +20 mV for about 1 min. Whereas the resting potential permits membrane fusion, sperm entry requires a shift to a membrane potential between −20 and +17 mV. A potential above +17 mV allows the first sperm to complete its entry, but any additional sperm make only membrane contact and do not proceed with sperm-egg fusion; thus the fertilization potential acts as a fast block to polyspermy. This block lasts for about 1 min, after which the membrane potential returns to intermediate values that allow all fertilization events to occur.

electric potential across the plasma membrane. This is known as the ***resting potential***, which is generated—as it is in neurons—by concentration differences in potassium (K⁺) and sodium (Na⁺) ions inside and outside the egg. In unfertilized sea urchin eggs, the resting potential is about −75 mV (Nuccitelli and Grey, 1984). Upon contact with a sperm, the membrane potential changes temporarily to a ***fertilization potential*** of near +20 mV (Fig. 4.21). The membrane potential remains positive for about 1 min and then gradually returns to the resting potential.

The fertilization potential is caused by increased membrane permeability to certain ions. This change in permeability is an immediate response to sperm-egg contact, possibly mediated directly by the G protein associated with the sperm receptor (Fig. 4.17). The nature of the ion channels that undergo permeability changes varies among animal groups (L. A. Jaffe and Gould, 1985; Elinson, 1986). In sea urchin eggs, the fertilization potential is caused by an increase in the membrane's permeability to sodium ions (Na⁺), temporarily allowing these positively charged ions to enter the egg. In frog eggs, the fertilization potential results from an increase in the membrane's permeability to chlorine ions (Cl⁻), allowing these negatively charged ions to leave the egg.

The fertilization potential serves as a fast block to polyspermy in sea urchin eggs. This was shown by Laurinda A. Jaffe (1976), who used an electric current to maintain the membrane potential of sea urchin eggs at different levels and then exposed these eggs to sperm. At a constant membrane potential of +20 mV or more, sperm adhered, but membrane fusion did not occur. Conversely, at a constant potential of −30 mV, eggs fused with more than one sperm. The molecular basis for the observed voltage dependence of membrane fusion is still under investigation.

In addition to controlling membrane fusion, the membrane potential also affects fertilization cone formation and sperm entry, at least in sea urchins (Lynn et al., 1988). If the egg's membrane potential is held between −70 and −20 mV, fertilization cones begin to form. However, the cones do not develop fully, and sperm entry usually fails. At membrane potentials between −20 and +17 mV, membrane contact is followed by membrane fusion, full fertilization cone formation, and sperm entry (Fig. 4.21). At membrane potentials above +17 mV, sperm contact elicits no further response from the egg.

Evidently, a sperm can initiate fertilization at −70 mV, since this is the resting potential, but for sperm entry to proceed, the membrane potential must shift to a more positive level. Only during the fleeting moment while the membrane potential rises from −20 to +17 mV is a sperm admitted to the egg. Once membrane fusion has occurred, entry of the *first* sperm can continue at +20 mV, but *additional* sperm cannot enter because they fail to fuse with the egg. However, the electrical block to polyspermy is only temporary, because the intermediate potential that allows fertilization is restored after about one minute.

The Cortical Reaction Causes a Slow Block to Polyspermy

The Ca²⁺ transients triggered by fertilization are correlated with another egg activation event, the ***cortical reaction***, also known as the ***zona reaction*** in mammals. The cortical reaction is the exocytosis of ***cortical granules***, which are membrane-bound vesicles located beneath the egg plasma membrane. An unfertilized mouse egg has about 4000 cortical granules, while a sea urchin egg has 15,000 or so. During egg activation, the cortical granules undergo *exocytosis*, releasing their contents into the ***perivitelline space*** between the plasma membrane and the vitelline envelope (Fig. 4.22). The cortical reaction begins less than a minute after sperm-egg contact and is completed within a few minutes, depending on egg size.

In sea urchin eggs, the components released during the cortical reaction trigger three major events. First, enzymes modify vitelline envelope components so that sperm no longer adhere to the egg surface. Second, pro-

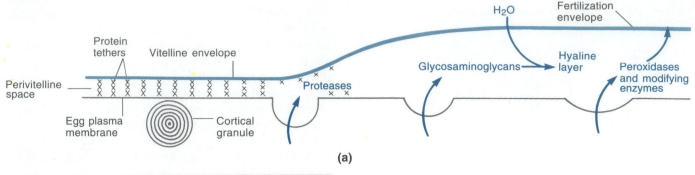

(a)

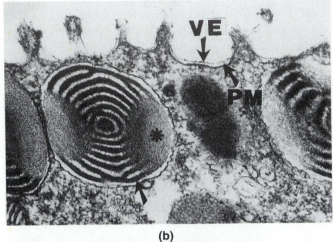

(b)

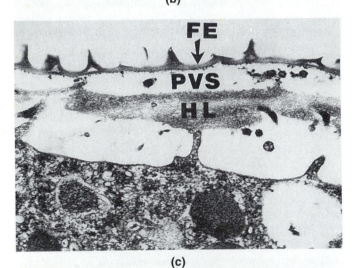

(c)

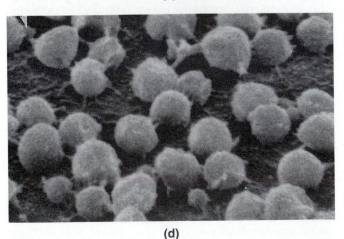

(d)

Figure 4.22 Cortical reaction in the sea urchin egg. **(a)** As part of egg activation, the cortical granules beneath the plasma membrane undergo *exocytosis*, releasing into the perivitelline space several components that support three major processes. First, proteases cleave the proteins that tether the vitelline envelope to the plasma membrane, while glycosaminoglycans attract water into the perivitelline space and form a hyaline layer that lifts the vitelline envelope off the egg plasma membrane. Second, peroxidases harden the vitelline envelope by cross-linking adjacent proteins. The hardened envelope is then called the fertilization envelope. Third, enzymes modify vitelline envelope components so that sperm no longer adhere to the egg surface. **(b)** Transmission electron micrograph showing the cortex of an unfertilized sea urchin egg (*Strongylocentrotus purpuratus*). The vitelline envelope (VL) is closely applied to the plasma membrane (PM) and follows the contours of the microvilli. The cortical granules contain lamellar and amorphous (asterisk) components. **(c)** Ten minutes after fertilization, the fertilization envelope (FE) is separated from the egg by a hyaline layer (HL), which builds up in the perivitelline space (PVS). **(d)** Scanning electron micrograph showing the inner aspect of the plasma membrane of an unfertilized egg. The cortical granules (round bodies) are still intact.

teases cleave the proteins that tether the vitelline envelope to the plasma membrane, and complex polysaccharides known as *glycosaminoglycans* form a *hyaline layer,* which attracts water into the perivitelline space, so that the vitelline envelope is lifted off the egg plasma membrane. Third, *peroxidases* harden the vitelline envelope by cross-linking adjacent proteins. The hardened envelope is then called the *fertilization envelope.*

The cortical reaction causes a slow block to polyspermy. Its protective action is necessary because the egg underneath the fertilization envelope is still fertilizable: if such an egg is stripped of its fertilization envelope and hyaline layer, then additional sperm will enter. However, the fertilization envelope permanently inhibits the penetration of more sperm and even causes previously adhering sperm to fall off (Fig. 4.23).

The *zona reaction* in mammalian eggs is similar to the cortical reaction in sea urchin eggs and has very similar results. The enzymes released from the cortical granules make the zona impermeable to sperm and modify the ZP3 glycoprotein. When isolated from fertilized eggs, ZP3 no longer binds to sperm or elicits the acrosome reaction. These changes in behavior of ZP3 after fertilization are not associated with a detectable change in molecular weight of the glycoprotein. Removal of the O-linked oligosaccharides would not cause a measurable change in molecular weight but would be an effective means of preventing further adhesion of sperm (Wassarman, 1987).

The Principle of Synergistic Mechanisms

Many steps in development rely on two or more mechanisms that complement or reinforce each other. We call this the *principle of synergistic mechanisms.* The eminent embryologist Hans Spemann (1938) referred to it as the "principle of double assurance" or the "synergetic principle of development." To illustrate this principle he described a certain detail of frog metamorphosis. The gills on each side of a tadpole are covered with a skin fold called the *operculum.* Before metamorphosis, the forelimb grows under the operculum. When the forelimb is completed, it must break through the operculum to get out into the open. Where the elbow presses against the operculum from inside, the opercular skin becomes thinner and thinner until it is finally perforated, so that a hole is formed through which the arm is set free. The process gives the impression that the hole in the skin is formed by the pressure and friction from the elbow. Surprisingly, the same hole, *although smaller,* appears in the proper place after the rudiment of the forelimb has been removed at an early stage. Conversely, the limb can also force its way out on its own strength. This is seen when the limb rudiment is

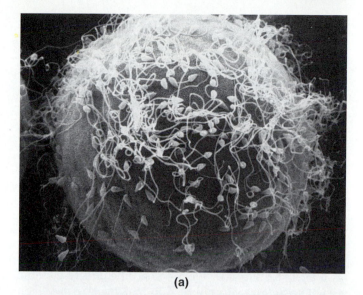

(a)

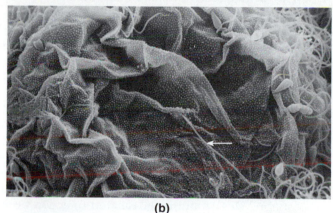

(b)

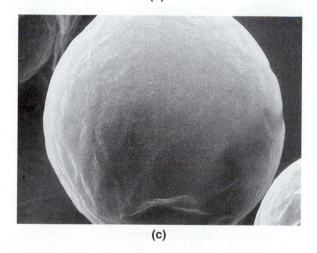

(c)

Figure 4.23 Removal of excess sperm after the cortical reaction in the egg of the sea urchin *Strongylocentrotus purpuratus.* **(a)** Scanning electron micrograph of an egg 15 s after the addition of sperm. **(b)** Thirty seconds after the addition of sperm. The arrow marks the tail of the fertilizing sperm. The progress of the cortical reaction is indicated by the zone of sperm detachment surrounding the fertilizing sperm. The wrinkles in the fertilization envelope are an artifact of fixation. **(c)** Three minutes after the addition of sperm. The fertilization envelope has hardened and no longer wrinkles during fixation. All sperm have detached.

implanted elsewhere under the skin. These experiments show clearly that the limb and the operculum act *independently* and yet cooperate in forming the hole.

The degree of overlap between synergistic mechanisms in development differs from case to case. In the context of fertilization, we saw that many animal eggs have two blocks to polyspermy: a fast block relying on an electric membrane potential, and a slow block relying on the formation of the fertilization envelope. The latter mechanism in turn entails three major effects, namely, the release of sperm-binding molecules from the vitelline envelope; the formation of the hyaline layer, which lifts the vitelline envelope off the egg plasma membrane; and the hardening of the vitelline envelope, which makes it impermeable to additional sperm.

The fast block to polyspermy holds up about as long as it takes for the slow block to become effective (Table 4.1). This is an example of minimal overlap between different mechanisms that serve the same biological function. In contrast, the three inhibitory effects that constitute the slow block appear to be redundant. In theory, there should be no need for the vitelline envelope to harden or to lift off the plasma membrane when the sperm-binding molecules have been modified or shed. There are two possible explanations for the fact that these seemingly redundant mechanisms still exist. First, each mechanism might serve as a fail-safe mechanism to back up the other two. If each mechanism by itself eliminated only 9 out of 10 extra sperm, then the three mechanisms together would eliminate 999 of 1000 extra sperm. This would be adequate protection even for an egg that is exposed to an excess of sperm. Another explanation for the persistence of seemingly redundant mechanisms for one biological function is that they may have additional functions. For instance, the hardened fertilization envelope also provides mechanical protection, and the hyaline layer also forms a firm coat around the blastula cells, which prevents these cells from leaving the embryo. Thus, what seem to be redundant mechanisms for one function may actually be part of a larger *network* of mechanisms serving several functions.

Synergistic mechanisms may be crucial for the organism, but they can also be a challenge to researchers. If a given biological function is supported by multiple mechanisms, it is very difficult to analyze any one of them. One way of testing whether a certain mechanism is *necessary* for a given function is to block the mechanism (by mutation, drugs, or other means) and to observe whether the function fails. This test may not reveal a necessary mechanism among a group of synergistic mechanisms, because each blocked mechanism is backed up, at least to an extent, by others that are still working. In this situation, investigators have to interfere with several mechanisms simultaneously before the biological function of interest fails and yields to analysis.

Parthenogenesis

Many animals can reproduce by *parthenogenesis* (Gk. *parthenos*, "virgin"; *genesis*, origin"), which is the development of viable offspring from unfertilized eggs. In aphids and other insects, parthenogenesis alternates with sexual reproduction, a reproductive strategy known as *facultative parthenogenesis*. Females reproduce parthenogenetically during the summer when food is abundant. At the end of the season, a generation of males and females develops and reproduces sexually. The fertilized eggs from this generation overwinter and develop into females during the next spring. In other insects, including the honeybee, parthenogenesis is coupled with sex determination. Fertilized eggs give rise to diploid females, while unfertilized eggs develop into haploid males called *drones.* Some species, mostly invertebrates but also some lizards, reproduce by *obligatory parthenogenesis.* Their populations consist entirely of females.

Parthenogenetic species are faced with three problems. First, they need to compensate for the reduction in number of chromosomes that normally occurs during meiosis. They can do this through several mechanisms, such as skipping one meiotic division or fusing two haploid nuclei after meiosis. The second problem in parthenogenetic species is egg activation. The silver salamander, *Ambystoma platineum*, solves this problem by enlisting the help of males from another species. There are no *A. platineum* males, but the females mate with males of a closely related salamander, *A. jeffersonianum*. The sperm from these males only activate the eggs of *A. platineum* and do not contribute their genome to the offspring (Uzzell, 1964). Other parthenogenetic species use different stimuli to activate their eggs. In parasitic wasps, unfertilized eggs are activated by friction or distortion when they pass through the narrow ovipositor (Went and Krause, 1973). The third problem in parthenogenesis is the lack of a pair of centrioles that are normally introduced by a sperm and serve to organize the mitotic spindle. However, centrioles may also be contributed by the egg or formed from smaller components in the egg cytoplasm or replaced functionally by other microtubule organizing centers.

In the laboratory, a variety of physical and chemical treatments have been used to activate eggs. One of the most commonly used methods is simply to prick the eggs with a needle, preferably one that has been dipped in blood or other tissues. This treatment may trigger an early event in the activation cascade, such as the Ca^{2+} transient, which then triggers the subsequent events. Impurities on the pricking needle may also serve as nucleating centers for spindle formation. For development to advanced stages, the diploid chromosome number must usually be restored. However, adults have been

TABLE 4.2

Pronuclear Transplantation Experiments with Fertilized Mouse Eggs

Class of Reconstructed Zygotes	Operation	Number of Successful Transplants	Number of Progeny Surviving
Bimaternal		339	0
Bipaternal		328	0
Control		348	18

Source: McGrath and Solter (1984) and Gilbert (1991). Used by permission.

Natural parthenogenesis is unknown in mammals. Mouse eggs activated in a series of experiments did not develop beyond day 11, halfway through their gestation. By transplanting pronuclei between fertilized mouse eggs, one can construct zygotes with two female pronuclei and no male pronucleus (Table 4.2). Such zygotes develop into bimaternal embryos, which cease development at about the same time as a parthenogenetic embryo. Zygotes with two male pronuclei and no female pronucleus develop into bipaternal embryos; they also die midway through gestation but show different defects from the bimaternal embryos. Control zygotes with one male and one female pronucleus, generated by the same transplantation technique, develop normally (McGrath and Solter, 1984). These results indicate that in mammals both a female pronucleus and a male pronucleus are needed for embryonic development. Certain mammalian genes can be activated only in the female pronucleus, whereas other genes must be introduced through the male germ line in order to be expressed (Surani et al., 1986; see also Chapter 15).

obtained, with varying success, through artificial activation of eggs from sea urchins, starfish, silk moths, fishes, and frogs (Beatty, 1967).

SUMMARY

Fertilization is the union of two haploid gametes—one egg and one sperm—to form a diploid zygote from which a new individual develops. In addition to properly timed parental mating behavior, chemical attractants and capacitation mechanisms are found in some animals to ensure that the egg and the sperm are ready for fertilization when they encounter each other. In most species, sperm penetrate several egg coats before reaching the egg plasma membrane. Of particular importance is the vitelline envelope—in mammals called the zona pellucida—which is a coat of proteins laid down by the oocyte before ovulation. The interaction of the sperm with the vitelline envelope or another egg coat triggers the acrosome reaction, in which the sperm releases the contents of its acrosomal vesicle. Enzymes from the acrosome help the sperm to dissolve its way to the egg cell proper. Sperm adhere to the egg surface in a species-specific way, by means of matching molecules on the sperm head and the vitelline envelope or the plasma membrane of an egg from the same species.

The contact between the plasma membranes of egg and sperm triggers a cascade of egg responses collectively called egg activation. It includes plasma membrane fusion, sperm entry, transient rises in calcium ion (Ca^{2+}) concentration and pH, acceleration of many metabolic processes, and resumption of the cell cycle. In many species, egg activation also entails changes at the egg surface that prevent polyspermy. The eggs of such species typically use a fast but temporary electrical block before a slower, permanent block sets in. The slow block results from the cortical reaction, in which cortical granules located beneath the egg plasma membrane release their contents by exocytosis. The cortical reaction alters the vitelline envelope in several ways that make it impermeable to sperm.

Some animals can reproduce by parthenogenesis, which is the development of viable offspring from unfertilized eggs. These species use several mechanisms to avoid or compensate for the reduction in number of chromosomes that occurs during meiosis and to trigger egg activation.

SUGGESTED READINGS

Schatten, H., and G. Schatten, eds. 1989. *The Cell Biology of Fertilization.* San Diego: Academic Press.

Schatten, H., and G. Schatten, eds. 1989. *The Molecular Biology of Fertilization.* San Diego: Academic Press.

Wassarman, P. M. 1990. Profile of a mammalian sperm receptor. *Development* 108:1–17.

Whitaker, M., and K. Swann. 1993. Lighting the fuse at fertilization. *Development* 117:1–12.

CLEAVAGE

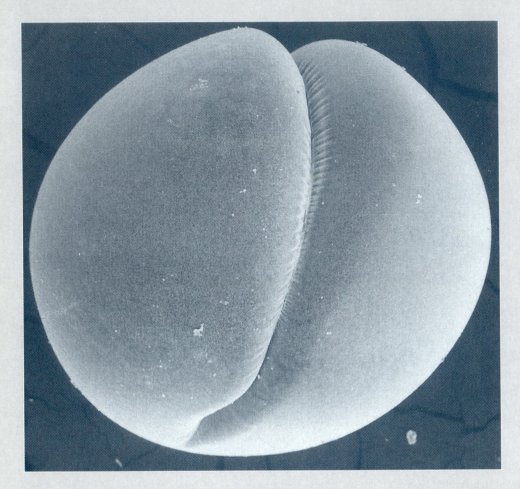

Figure 5.1 Scanning electron micrograph of a frog egg during its first cleavage. The first cleavage furrow forms at the end of the first mitosis. The furrow originates at the animal pole (facing up in this picture), where the asters of the mitotic spindle are closest to the egg surface. The furrow extends toward the vegetal pole, pinching the egg until it is divided into two blastomeres. The mechanical force to cut through the egg is generated by a contractile ring of actin and myosin filaments. These filaments assemble in the cytoplasm at the base of the furrow and must be connected by linker proteins to the plasma membrane. The small folds visible near the base of the cleavage furrow reveal the stress that contraction of the ring imposes on the plasma membrane. (From H. W. Beams and R. G. Kessel, Cytokinesis: A comparative study of cytoplasmic division in animal cells, *Am. Sci.* **64**:279–290, 1976.)

In all metazoa, fertilization is followed by a series of rapid mitotic divisions. The zygote divides first into two blastomeres, which in turn divide to give four blastomeres, and so forth, in exponential progression (Fig. 5.1). This embryonic stage, known as cleavage, has multiple functions in the metazoan life cycle. First, cleavage generates a large number of cells that can move relative to each other and undergo gastrulation and organogenesis. (By analogy, if you wanted to build a house from one large boulder, you would first have to cut the boulder into manageable stones.) Second, the mitotic divisions that take place during cleavage generate many copies of the zygotic genome. This allows cells to express different subsets of the complete genetic information and embark on different pathways of development. Finally, cleavage increases the value of an important cellular parameter, the *nucleocytoplasmic ratio,* which is defined as the ratio of nuclear volume to cytoplasmic volume, where the nuclear volume is taken as an indicator of the number of genomes present. Increasing the nucleocytoplasmic ratio seems to be imperative for the timely replacement of RNA and proteins, which, like most other molecules in a cell, have short life spans; some proteins last only for minutes. Thus, just to sustain itself a cell must constantly synthesize RNA and proteins. The smaller the nucleocytoplasmic ratio, the more difficult it becomes for a cell to transcribe enough RNA in order to replace the protein losses from a large cytoplasm.

Animal eggs are inordinately large cells with small nucleocytoplasmic ratios. Large eggs seem advantageous to many animals because they sustain the development of advanced larvae that can effectively move and feed. However, the extremely unfavorable nucleocytoplasmic ratio of large eggs renders the uncleaved egg highly perishable. Therefore it would seem adaptive to increase the nucleocytoplasmic ratio and pass through this perilous stage of development as quickly as possible (Bier, 1964). Accordingly, cleavage divisions differ from later cell divisions in two respects. First, blastomeres do not grow between cleavage divisions as more mature cells do (Fig. 5.2). Second, the cleavage divisions proceed much more rapidly than later cell divisions; typical cleavage cycles are completed in less than an hour, whereas later cell cycles last several hours to many days.

Another advantage of rapid cleavage might be to reduce the risk of predation during an immobile and defenseless phase of the life cycle. In accord with this hypothesis, cleavage proceeds at a very leisurely pace in mammalian embryos, which are well protected in the maternal uterus. Mammalian eggs are also comparatively small, so the need to increase the nucleocytoplasmic ratio is less urgent.

While increasing the nucleocytoplasmic ratio is a universal aspect of cleavage, animal forms differ in their *cleavage patterns,* that is, in positioning and orientation of mitotic spindles. The cleavage pattern determines the relative sizes of blastomeres and their configuration. This is of great importance because certain cytoplasmic components are distributed unevenly in eggs. The cleavage process segregates these components into different blastomeres and thus ensures the distinctive development of each blastomere.

In the first half of this chapter, we will study the cleavage patterns of several representative animals, including

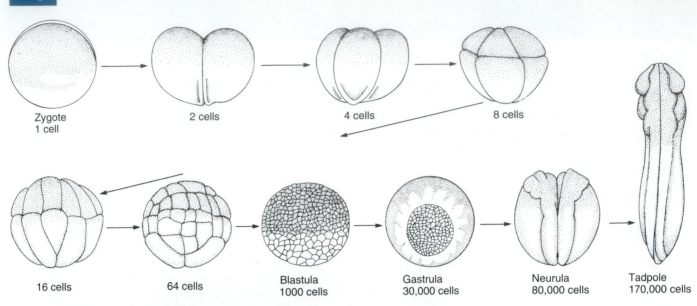

Figure 5.2 Development of *Xenopus laevis* from the fertilized egg to the tail bud stage. During this period, the total volume of the embryo remains nearly constant while the number of cells increases by several orders of magnitude.

sea urchins, frogs, chickens, mammals, and insects. Some animals have very strictly standardized cleavage patterns that each embryo faithfully replicates. The embryos of other species cleave less precisely, and any deviations from the norm are corrected at a later time. While the former group relies on utmost precision, the latter has adopted a strategy of *stepwise approximation.*

In the second half of this chapter, we will discuss mechanical aspects of cleavage. What mechanisms in embryos control the positioning and orientation of mitotic spindles? How is the cell cycle regulated during cleavage and after the transition to a mature cell cycle? Linking these questions is the concept of a "cleavage clock": the positioning and orientation of mitotic spindles seem to be controlled by progressive changes in the cytoplasm that can be disengaged from the mitotic cycle.

Yolk Distribution and Cleavage Pattern

The amount and distribution of yolk in the egg affect the cleavage pattern of the developing embryo. Eggs that have a small amount of evenly distributed yolk in the cytoplasm are called *isolecithal* (Gk. *isos,* "equal"; *lekithos,* "yolk"). Most echinoderms, mollusks, ascidians, and mammals produce this type of egg. (In this context, it is customary to speak of the cleavage of "eggs," although technically they are zygotes or embryos.) Eggs with a large amount of yolk present mostly in the vegetal hemisphere are called *telolecithal* (Gk. *telos,* "end"). Amphibians have moderately telolecithal eggs, while most fish, reptiles, and birds have extremely

telolecithal eggs. Eggs in which the yolk is concentrated in the central cytoplasm are called *centrolecithal.* The eggs of virtually all insects and many other arthropods belong to this type.

The amount and distribution of yolk in an egg have a major impact on its cleavage pattern (Fig. 5.3 and Table 5.1). Isolecithal and moderately telolecithal eggs undergo a type of cleavage known as *holoblastic cleavage* (Gk. *holos,* "whole"; *blastos,* "germ"). This term indicates that the entire egg is cleaved during cytokinesis. Eggs that are characterized by partial cytokinesis are said to undergo *meroblastic cleavage* (Gk. *meros,* "part"). There are two major types of meroblastic cleavage. The first, *discoidal cleavage,* is restricted to a small disc of yolk-free cytoplasm at the animal pole; a large amount of yolk-rich cytoplasm at the vegetal pole remains uncleaved. Discoidal cleavage is characteristic of extremely telolecithal eggs. In the second type of meroblastic cleavage, called *superficial cleavage,* cytokinesis is limited to a surface layer of clear cytoplasm while the yolk-rich inner cytoplasm remains uncleaved. Superficial cleavage is characteristic of centrolecithal eggs.

All large eggs go through meroblastic cleavage, presumably because large masses of yolk interfere with cytokinesis. However, meroblastic cleavage is not limited to large eggs. Virtually all insect eggs, large or small, cleave superficially. Thus, the cleavage pattern of an embryo seems to be part of an overall developmental program that is characteristic of its phylogenetic group. In some groups of animals a particular symmetry or blastomere arrangement gives early embryos a unique appearance. Several typical patterns will be discussed in the following section.

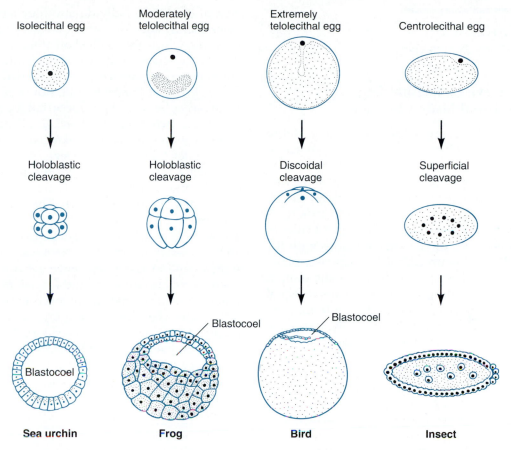

Figure 5.3 Typical cleavage patterns of isolecithal, moderately telolecithal, extremely telolecithal, and centrolecithal eggs.

TABLE 5.1

Cleavage Patterns in Relation to Yolk Distribution in Eggs*

Type of Egg (Based on Yolk Distribution)	Type of Cleavage	Representative Pattern of Cleavage	Blastula	Animal Groups
Isolecithal (little yolk, evenly distributed)	Holoblastic (egg cell completely cleaved)	Radial Bilateral Spiral Rotational	Spherical single layer of blastomeres surrounding large, fluid-filled blastocoel	Echinoderms, ascidians, mollusks, annelids, mammals
Moderately telolecithal (moderate amount of yolk, mostly in vegetal hemisphere)	Holoblastic	Radial	Sphere of multiple cell layers enclosing an eccentric, fluid-filled blastocoel	Amphibians, some fishes
Extremely telolecithal (large amount of yolk, except for blastodisc at animal pole)	Meroblastic (egg cell incompletely cleaved)	Discoidal (blastomeres form disc on uncleaved yolk)	Cellular disc consisting of epiblast and hypoblast, with flat space in between	Most fishes, birds, reptiles
Centrolecithal (yolk concentrated in center of egg)	Meroblastic	Superficial (blastomeres form at the egg surface)	Ovoid consisting of single cell layer (blastoderm) surrounding central yolk	Insects and other arthropods

*See also Figure 5.3.
Source: After B. M. Carlson (1988).

Cleavage Patterns of Representative Animals

Sea Urchins Have Isolecithal Eggs and Undergo Holoblastic Cleavage

Most sea urchin eggs are isolecithal and undergo holoblastic cleavage (Hörstadius, 1973). As discussed in Chapter 4, fertilization in sea urchins leads to a union of pronuclei. Cleavage begins with assembly of the first mitotic spindle, which is oriented perpendicular to the animal-vegetal axis (Fig. 5.4). After mitosis, cytokinesis occurs in the *cleavage plane*, which is perpendicular to the axis of the mitotic spindle and coincides with the plane marked by the preceding metaphase plate. The first cleavage plane passes through the animal and vegetal poles, creating two blastomeres of equal size. This type of cleavage is called *meridional*, because the cleavage furrow runs through the poles like a meridian on a globe. For the second mitosis, spindles form simultaneously in each of the two blastomeres. Their axes are still perpendicular to the animal-vegetal axis, and they are also perpendicular to the first mitotic spindle. Thus the second cleavage, which results in four blastomeres of equal size, is again meridional. For the third mitosis, the spindles are parallel to the animal-vegetal axis. The subsequent cleavage is called *equatorial*, because together the four cleavage furrows circle the embryo like the equator on a globe. This cleavage separates the embryo into four animal and four vegetal blastomeres.

The fourth cleavage in sea urchins shows a unique pattern (Summers et al., 1993). The animal blastomeres cleave in a slightly oblique orientation, generating two tiers of blastomeres that are somewhat offset (Fig. 5.4). The spindles in the vegetal blastomeres are displaced and tilted toward the vegetal pole. At cytokinesis, the vegetal blastomeres divide into four large cells, called *macromeres*, and four much smaller cells, called *micromeres*, which are pinched off at the vegetal pole from their larger sister cells. The eight animal blastomeres are intermediate in size and are known as *mesomeres*. The cleavage of the animal blastomeres features *equal cytokinesis*, because it produces daughter cells of nearly equal size, whereas the cleavage of the vegetal blastomeres shows *unequal cytokinesis*, since it gives rise to daughter cells of unequal size.

During the fifth cleavage, the mesomeres divide equally and meridionally, giving rise to two tiers of eight animal blastomeres each. The macromeres divide similarly to produce a tier of eight vegetal blastomeres. The micromeres divide unequally to produce four large and four small micromeres. At the sixth cleavage, mesomeres and macromeres cleave equatorially. The large micromeres also divide, but the small blastomeres skip this cleavage, so that the embryo now consists of 60 cells. From the 32-cell stage on, the blastomeres are arranged in a blastula configuration, that is, as a sphere surrounding a fluid-filled *blastocoel*.

The cleavage pattern of sea urchin eggs is referred to as *radial cleavage*, because the blastomeres are arranged in radial symmetry around the animal-vegetal axis.

Amphibians Have Moderately Telolecithal Eggs but Still Cleave Holoblastically

Cleavage in most amphibian embryos is both holoblastic and radially symmetric, as in sea urchins. However, amphibian eggs contain much more yolk than sea urchin eggs, especially in the vegetal hemisphere, and are therefore classified as moderately telolecithal. The first cleavage furrow is meridional. It begins to form at the animal pole and progresses toward the vegetal pole,

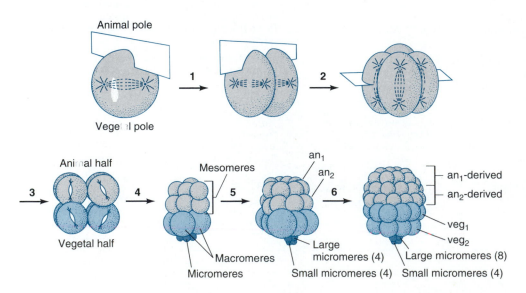

Figure 5.4 Cleavage of the sea urchin egg. The first two cleavages are meridional, passing through the animal-vegetal axis. The third cleavage is equatorial and perpendicular to the animal-vegetal axis. The fourth cleavage features oblique spindle orientations. The animal blastomeres cleave nearly equally, whereas the vegetal blastomeres cleave unequally. The fifth cleavage produces two animal tiers of mesomeres and one vegetal tier of macromeres (each comprising eight cells), a tier of four large micromeres, and a tier of four small micromeres.

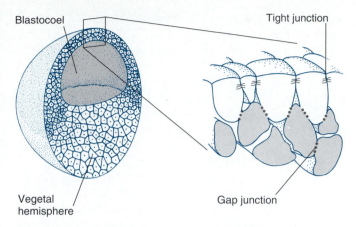

Blastocoel

Tight junction

Vegetal
hemisphere

Gap junction

Figure 5.5 Frog blastula. The wall of the blastula is several cell layers thick. There are more layers—and the blastomeres are larger—in the vegetal hemisphere, so that the blastocoel is displaced toward the animal pole. The outermost cells are connected by tight junctions, which create a seal isolating the interior of the embryo from the external medium. Gap junctions provide channels of communication between the blastomeres.

where it meets with resistance from the yolk-rich cytoplasm (Fig. 5.1). In the axolotl, the furrow elongates at a rate of about 1 mm/min in the animal hemisphere but slows down to about 0.02 mm/min as it nears the vegetal pole (Hara, 1977). While the first furrow is still knifing through the vegetal hemisphere, the second furrow has already started at the animal pole, with its plane running perpendicular to the first.

The third cleavage of the amphibian embryo is equatorial. However, in contrast to the corresponding cleavage in the sea urchin, the mitotic spindles of the amphibian egg are displaced toward the animal pole, resulting in closer proximity of the third cleavage plane to the animal pole (Fig. 5.2). This unequal cleavage separates four small animal blastomeres from four larger vegetal blastomeres. The fourth cleavage is meridional and the fifth equatorial. Because holoblastically cleaving embryos during the 16-cell and the 32-cell stages outwardly resemble mulberries, these stages are called *morula stages* (Lat. *morum*, "mulberry").

In amphibian embryos at the 128-cell stage, the blastocoel cavity becomes apparent. The embryo is now considered a blastula. Through further cell divisions, the blastula develops into a multilayered epithelium, in which the outermost cells become *polarized*. This means that their outer and inner cell faces acquire different structures and functions. Near the outer surface of the epithelium, *tight junctions* seal off the inside of the embryo from the external environment (Fig. 5.5; see also Fig. 2.21). In addition, all adjacent cells in the blastula form *gap junctions*, channels of communication through which small molecules can pass. Both tight junctions and gap junctions also form in other embryos during cleavage.

Snails Have Isolecithal Eggs and Follow a Spiral Cleavage Pattern

Spiral cleavage is seen in several groups of worms and in most mollusks, including snails. This cleavage pattern can be considered a type of radial cleavage in which certain traits that are weakly expressed in other embryos with radial cleavage are so exaggerated that a unique pattern results (Freeman, 1983). The first two cleavage planes run parallel to the animal-vegetal axis of the zygote, dividing it into four blastomeres (Fig. 5.6). In contrast to sea urchins, the first four blastomeres

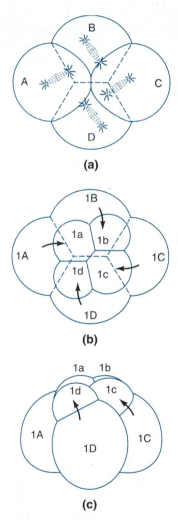

Figure 5.6 Spiral cleavage pattern in mollusks. **(a)** 4-cell stage seen from the animal pole. The first four blastomeres, called macromeres because of their large size, are designated clockwise A, B, C, and D. Blastomeres B and D are located opposite each other and touch at the vegetal pole. (In some species, one of these blastomeres is longer than the other and is then designated the D blastomere.) Blastomeres A and C, also opposite each other, touch at the animal pole. Note the spiral arrangement of the mitotic spindles in preparation for the third cleavage. **(b)** 8-cell stage seen from the animal pole. Arrows connect each micromere with the macromere from which it has budded off. **(c)** Lateral view of the 8-cell stage.

in snails form a tight arrangement resembling a tetrahedron. This tendency to minimize the outer surface area is also observed during later cleavages and makes the embryo look like a cluster of soap bubbles.

During subsequent cleavages, the four large blastomeres (macromeres) bud off quartets of smaller blastomeres (micromeres) that accumulate at the animal pole. However, these micromeres are not aligned with their sister macromeres. The first micromere quartet, given off during the third cleavage, is rotated clockwise when viewed from the animal pole of the embryo. The rotated pattern, which results from the oblique orientation of the mitotic spindles relative to the animal-vegetal axis, gave the spiral cleavage pattern its name. During the fourth cleavage, the spindle orientation changes by about 90° so that the macromeres give off a second micromere quartet that is rotated counterclockwise. At the same time, the first micromere quartet divides in a similar fashion. During the following cleavage, the orientation of the spindles again shifts by 90°, and so forth.

The Ascidian Cleavage Pattern Is Bilaterally Symmetrical

Ascidians, such as the sea squirts, are relatives of the vertebrates, with which they share certain features during the embryonic and larval stages. In addition to animal-vegetal polarity, ascidian eggs have an anteroposterior polarity caused by the asymmetrical distribution of several cytoplasmic components. The resulting bilateral symmetry is preserved throughout the cleavage of the zygote. The first cleavage plane passes through the animal-vegetal axis and divides the asymmetrically distributed cytoplasm evenly between the first two blastomeres (Fig. 5.7). This cleavage plane corresponds to the future *median plane,* which separates the right and left halves of the embryo. The second cleavage is parallel to the animal-vegetal axis but slightly displaced toward the posterior. It separates two large anterior blastomeres from two smaller posterior blastomeres which inherit cytoplasm of a different composition. The different cytoplasmic compositions are critical to the differential development of anterior and posterior blastomeres (see Chapter 8). At this point the cleavage pattern has only one plane of symmetry, the median plane, and is therefore described as *bilaterally symmetrical.* During subsequent cleavages, differences in cell size and shape enhance the bilateral symmetry of the embryo.

Mammalian Eggs Show Rotational Cleavage

The mammalian egg cleaves during its journey down the oviduct, which takes several days in humans (Fig. 5.8). With cell cycles of about 12 hours, cleavage of

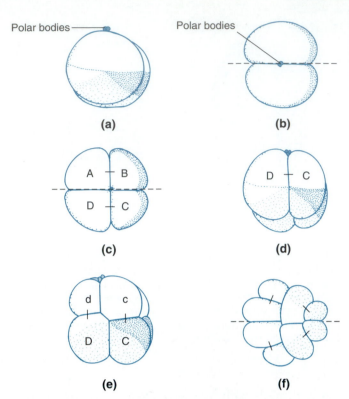

Figure 5.7 Bilaterally symmetrical cleavage pattern in ascidians. **(a)** 2-cell stage in lateral view, anterior to the left, animal pole marked by polar bodies. **(b)** Same stage viewed from animal pole. **(c, d)** 4-cell stage in animal pole and lateral views, respectively. **(e)** 8-cell stage in lateral view. **(f)** 16-cell stage seen from vegetal pole. Note that the embryo has only one plane of symmetry.

mammalian eggs is extremely slow. Mammalian cleavage is also unusual in that it is asynchronous. In other words, all blastomeres do not divide at the same time. An embryo therefore does not always proceed regularly from 2 to 4 to 8 blastomeres, and so on; in between, it may consist of an odd number of cells.

The first cleavage in mammals is meridional, passing through the animal-vegetal axis as in other eggs that undergo holoblastic cleavage (Fig. 5.9). However, the second cleavage is described as a *rotational cleavage,* because the two blastomeres divide in different planes: one blastomere divides meridionally, while the other divides equatorially. This results in a crosswise arrangement of cleavage furrows and blastomeres that is characteristic of most mammals, including humans. However, in a given strain of laboratory rabbits, only about half of the embryos may show rotational cleavage. In the rest, the second cleavage occurs more or less simultaneously in both blastomeres and with the same meridional orientation. Both types of embryos develop similarly thereafter, indicating that the orientation of the second cleavage is not critical, at least in rabbits.

At the 8-cell stage, the mammalian embryo undergoes a striking change known as *compaction.* At first, the blastomeres form a loose arrangement, touching

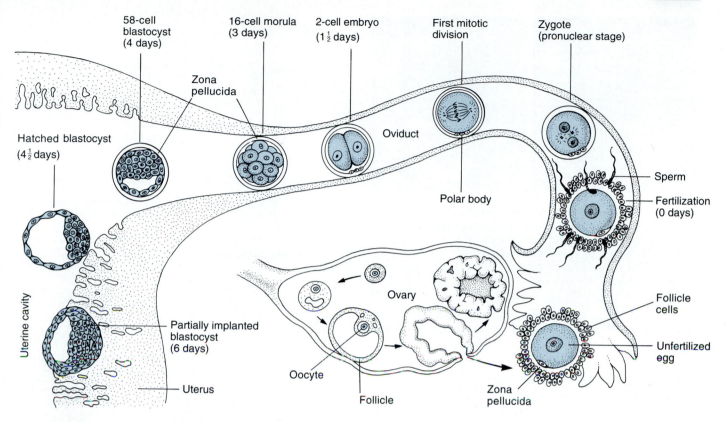

Figure 5.8 First week of human development. Fertilization occurs in the upper third of the oviduct. The zygote undergoes cleavage as it travels down the oviduct. About 4.5 days after fertilization, the embryo hatches from the zona pellucida and becomes implanted in the inner layer of the uterus.

their neighbors only at small areas of their surfaces. After compaction, the blastomeres adhere tightly, maximizing the areas of contact and forming a solid ball of cells (Fig. 5.10). During compaction, each of the eight blastomeres undergoes a *polarization* process like the one that occurs in the outermost cells of amphibian blastulae (Fig. 5.5). Each blastomere has an external surface that faces the outside environment as the apical face of an epithelial cell does. The other surfaces are internal and adjacent to other blastomeres. During compaction, *tight junctions* develop beneath the external surfaces, and *gap junctions* form between the internal

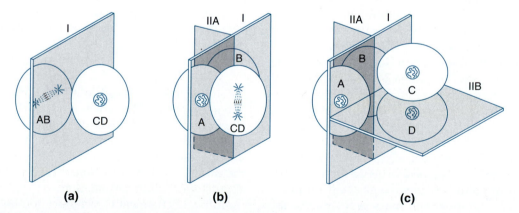

Figure 5.9 Rotational cleavage in mammals. **(a)** The first cleavage is on a meridional cleavage plane (I) passing through the animal-vegetal axis. The resulting two blastomeres are designated AB and CD. **(b)** The AB blastomere divides first, again on a meridional cleavage plane (IIA). The CD blastomere, which lags behind, elongates parallel to the furrow between A and B. Its mitotic spindle is oriented parallel to the long axis of the cell. **(c)** The CD blastomere divides on a cleavage plane (IIB) that is perpendicular to both I and IIA.

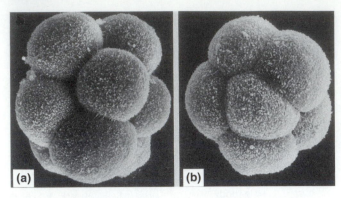

Figure 5.10 Scanning electron micrographs of mouse embryos at the 8-cell stage before compaction **(a)** and after compaction **(b)**. The rough appearance of the blastomere surfaces is caused by numerous microvilli.

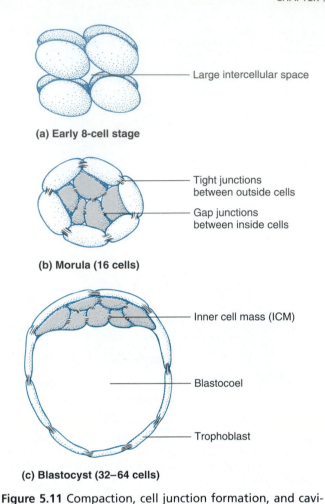

(a) Early 8-cell stage — Large intercellular space

(b) Morula (16 cells) — Tight junctions between outside cells — Gap junctions between inside cells

(c) Blastocyst (32–64 cells) — Inner cell mass (ICM) — Blastocoel — Trophoblast

Figure 5.11 Compaction, cell junction formation, and cavitation in the mammalian embryo. **(a)** 8-cell stage before compaction. **(b)** At the morula stage, the embryo consists of 9 to 14 outside cells and 2 to 7 inside cells. The outside cells are connected by tight junctions to form a seal between the inside of the embryo and the external environment. The outside cells take up fluid and nutrients from the environment by endocytosis and give off fluid to the inside by exocytosis, causing cavitation (i.e., formation of the blastocoel). **(c)** The blastocyst consists of the inner cell mass (ICM), which gives rise to the embryo proper, and the trophoblast, which plays a role in hatching the embryo from the zona pellucida and in implantation in the uterus.

surfaces. Polarization and the formation of tight junctions allow the blastomeres to create an inner embryonic environment that is different from the outside environment. Similar processes occur in other types of embryos, but compaction is an especially dramatic event in mammals.

The morula stage in mammals begins when the embryo consists of 16 blastomeres. In humans this occurs 3 to 4 days after fertilization, when the embryo passes from the oviduct into the uterus. A 16-cell morula consists of 9 to 14 polarized *outside blastomeres* surrounding 2 to 7 nonpolarized *inside blastomeres* (Fig. 5.11). As both groups of cells continue dividing, the outside blastomeres begin to pump fluid from the uterus into the embryo. The cells take up uterine fluid through their external membranes by means of endocytosis. This fluid is given off again, in a somewhat altered composition, by exocytosis to the interior of the embryo. This process, which results in the formation of fluid-filled cavities inside the embryo, is known as *cavitation*. These cavities merge and form the blastocoel.

Between the 32- and 64-cell stages, the mammalian embryo acquires the *blastocyst* configuration, which is the hallmark of early mammalian development. At the blastocyst stage the embryo consists of two groups of cells (Fig. 5.11). The outer layer of cells is called the *trophoblast* while the inner group of cells, known as the *inner cell mass,* forms a small mound eccentrically attached to the inside of the trophoblast. The trophoblast (Gk. *trophe,* "nourishment") participates in forming the placenta, that is, the disc-shaped structure in the uterus through which the fetus obtains oxygen and nutrients from its mother. The inner cell mass gives rise to the embryo proper.

As the mammalian embryo moves down the oviduct, the zona pellucida prevents premature implantation there. Only after floating in the uterus for a day or two does the blastocyst hatch from the zona. The act of hatching involves the local digestion of the zona

by an enzyme produced in a patch of trophoblast cells situated opposite the inner cell mass (Fig. 5.12). This location minimizes the risk of collateral enzymatic damage to the embryo (Perona and Wassarman, 1986). Having escaped from the zona, the blastocyst begins the process of *implantation,* which in humans occurs about 1 week after fertilization.

Intricate chemical communication between the embryo and the mother allows implantation to occur and enables the embryo to reside in the uterus and grow in its protective environment. When trophoblast cells come into contact with uterine cells, they proliferate and form two layers. The inner layer, called the *cytotrophoblast,* remains cellular. The outer layer is called the *syncy-*

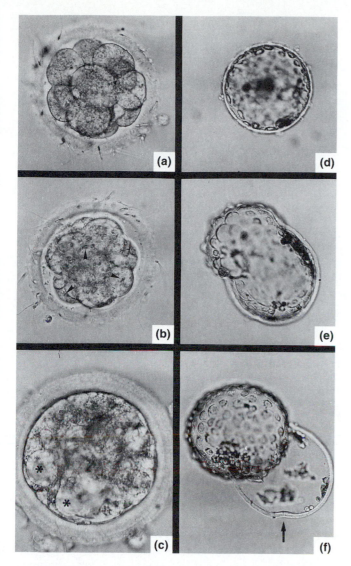

Figure 5.12 Hatching of a rhesus monkey blastocyst from the zona pellucida. **(a)** 16-cell embryo. **(b)** Morula after compaction. **(c)** Blastocyst with unfused blastocoel cavities (asterisks). **(d)** Fully formed blastocyst still inside the zona pellucida. **(e)** Blastocyst begins to hatch from zona. **(f)** Fully hatched blastocyst.

tiotrophoblast, because cells fuse to form a multinuclear mass of cytoplasm, or *syncytium*. Syncytiotrophoblast cells produce another enzyme that digests a small hole in the uterus, which the embryo penetrates. The syncytiotrophoblast erodes small maternal blood vessels, and the emanating maternal blood nourishes the embryo by diffusion. This is the beginning of placenta formation, which becomes more elaborate with the development of the heart, blood vessels, and umbilical cord (see Chapter 13). Meanwhile, trophoblast cells produce a hormone, known as *chorionic gonadotropin*, which, in humans and other primates, interrupts the menstrual cycle so that the uterus will not discharge the embryo. Other signals from the trophoblast cells regulate the maternal immune response so that the embryo will not be rejected like a tissue graft.

Birds, Reptiles, and Many Fishes Have Telolecithal Eggs and Undergo Discoidal Cleavage

As explained earlier, the amount and distribution of yolk in animal eggs are correlated with their cleavage patterns. In contrast to the holoblastic cleavage patterns discussed so far, in which the entire egg cell is cleaved during cytokinesis, meroblastic cleavage patterns are the result of incomplete cytokinesis, which leaves the yolk-rich portion uncleaved. There are two types of meroblastic cleavage, discoidal and superficial.

Discoidal cleavage is typical of fish, reptiles, and birds. The cleavage pattern of the zebra fish *Brachydanio rerio* is not far removed from the holoblastic pattern of amphibians (Fig. 5.13). In the unfertilized zebra fish egg, the cytoplasm is distributed as a thin layer around the central yolk mass. Upon fertilization, cytoplasm streams to the animal pole, forming a mound called the *blastodisc.* The first cleavage cuts the blastodisc in half, with the cleavage plane oriented perpendicular to the surface. The cleavage furrow begins to form at the animal pole, as in amphibian eggs, but instead of cutting all the way through the egg, it stops at the yolk. The second cleavage is perpendicular to the first and to the egg surface. Again cytokinesis is incomplete, and the resulting four blastomeres remain continuous with the yolk below and with the cytoplasmic layer at their outer margins. The ensuing cleavages continue making shallow cuts in the blastodisc in a fairly regular pattern. Complete cells, surrounded on all sides by plasma membranes, are first formed when cleavage planes are parallel to the surface, near the center of the blastodisc. Continued cleavage builds up the embryo atop an uncleaved mass of yolk.

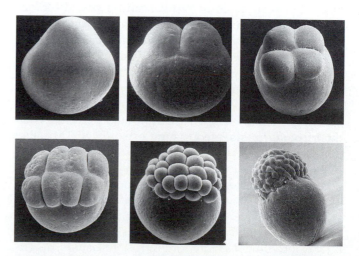

Figure 5.13 Discoidal cleavage in the zebra fish as seen in scanning electron micrographs. The cleavage furrows begin at the animal pole, but only the blastodisc cleaves, not the yolk-rich vegetal portion of the egg. (From H. W. Beams and R. G. Kessel, Cytokinesis: A comparative study of cytoplasmic division in animal cells, *Am. Sci.* **64**:279–290, 1976.)

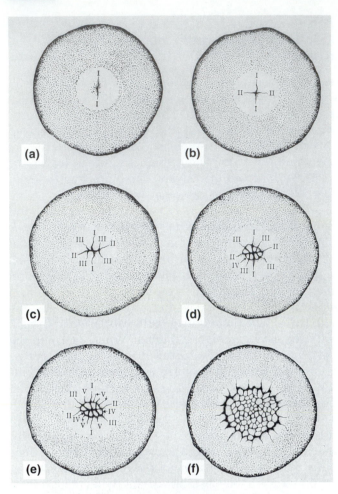

Figure 5.14 Cleavage in a pigeon's egg viewed from the animal pole. The drawings represent the blastodisc atop the uncleaved portion of the egg. Egg white and eggshells have been omitted. Roman numerals indicate the order in which the cleavage furrows appear.

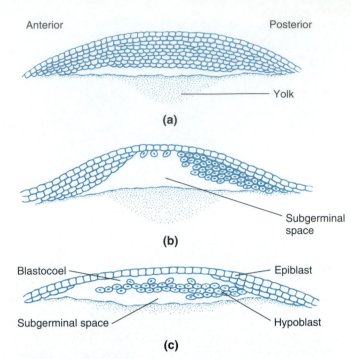

Figure 5.15 Bird development. The drawings show median sections. **(a)** Blastoderm stage. **(b, c)** Hypoblast formation. The separation of the hypoblast cells from the epiblast begins at the future posterior pole of the embryo.

In the large eggs of reptiles and birds, discoidal cleavage occurs in a similar fashion, except that the egg cell is much larger in relation to the blastodisc (Fig. 5.14). When cleavage has progressed to the point at which the cleavage planes become irregular and the number of blastomeres considerable, the term *blastoderm* is applied to the entire group of cells. (The fertilized "egg" laid by a hen actually contains an embryo at the blastoderm stage consisting of some 60,000 cells.) Between the blastoderm and the uncleaved yolk, a fluid-filled space, called the *subgerminal space*, is formed.

The next developmental step is the formation of two blastoderm layers, a lower layer, called the *hypoblast,* and an upper layer, the *epiblast* (Fig. 5.15). The separation of the hypoblast and the epiblast is brought about by the detachment of interior groups of blastoderm cells from the overlying cells and by the formation of a solid lower shelf of cells near the posterior pole of the future embryo (Eyal-Giladi, 1984). The hypoblast is smaller than the epiblast, and so the two layers touch inside the margin of the subgerminal space. The cavity between the epiblast and the hypoblast, which forms at the expense of the subgerminal space, is the *blastocoel.* The epiblast gives rise to the embryo proper, while the hypoblast forms the extraembryonic endoderm that later surrounds the yolk (see Chapters 10 and 13).

Insects Have Centrolecithal Eggs and Undergo Superficial Cleavage

In most insect eggs, cleavage is limited to a superficial layer of yolk-free cytoplasm called *periplasm*. The yolk-rich central cytoplasm, known as *endoplasm*, remains uncleaved. For insect embryos, the term "cleavage" is something of a misnomer, because cytokinesis is delayed until many rounds of mitosis have occurred. The division of the zygote nucleus starts deep in the endoplasm (Fig. 5.16). The multiplying daughter nuclei, with surrounding jackets of cytoplasm, move gradually toward the periplasm. Here they undergo further rounds of mitosis but are still not surrounded by plasma membranes. A few nuclei stay behind and become *vitellophages* (Gk., "yolk eaters"), which regulate the breakdown of yolk components.

Cleavage in *Drosophila* exemplifies the superficial cleavage pattern of insects (Fullilove and Jacobson, 1971; F. R. Turner and A. P. Mahowald, 1976; Foe and Alberts, 1983). The zygote nucleus and its progeny undergo eight rounds of mitosis in the yolk-rich endoplasm before a few nuclei reach the periplasm at the posterior

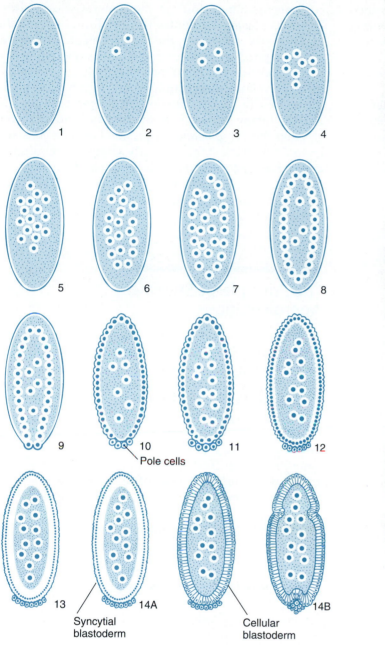

Figure 5.16 Superficial cleavage of the *Drosophila* embryo. In each diagram, the anterior pole is at the top. The number at the bottom of each diagram indicates the nuclear cycle. A cycle begins with the start of interphase and ends with the conclusion of M phase. Cycle 1 extends from fertilization through the first interphase and the first mitosis. All embryos are shown during interphase. The pole cells are pinched off during cycle 10. The blastoderm is cellularized during cycle 14, which also marks the onset of gastrulation (14B).

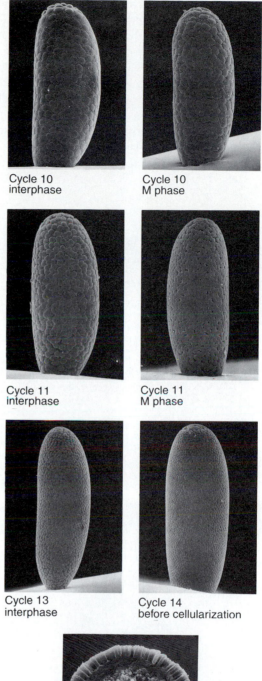

Cycle 10
interphase

Cycle 10
M phase

Cycle 11
interphase

Cycle 11
M phase

Cycle 13
interphase

Cycle 14
before cellularization

Cycle 14
during cellularization

Figure 5.17 *Drosophila* embryos during preblastoderm and blastoderm stages shown in scanning electron micrographs with the posterior pole up. The bulges in the surface of the embryos are caps of cytoplasm, stiffened by microtubules and microfilaments, that lie over the nuclei near the surface. Notice that the bulges become increasingly smaller and more numerous with each cycle.

pole. Here they become enclosed by plasma membranes to form the primordial germ cells, which are known as ***pole cells.*** After another round of mitosis, most nuclei arrive in the periplasm, each lodging beneath a small cytoplasmic mound that protrudes into the perivitelline space (Fig. 5.17). This stage is referred to as the ***syncy-***

tial blastoderm stage, since the nuclei are still contained in a common cytoplasmic layer at the surface. During the ensuing rounds of mitosis, the rapid pace of the cell cycle is gradually slowed down. The pole cells divide with regular cytokinesis, although with a slower cell cycle than the somatic nuclei.

After 13 rounds of mitosis, about 5000 nuclei are crowded into the periplasm of a *Drosophila* embryo, forming a dense hexagonal array. Before the somatic nuclei are enclosed by plasma membranes, the thickness of the yolk-free periplasm increases as the yolk

components move closer to the center. Cellularization begins as plasma membrane furrows form between the nuclei. At the same time, the nuclei elongate as they become enclosed by cages of microtubules (Fig. 5.18). The membranes deepen simultaneously between all nuclei in the periplasm, thus cutting a honeycomb pattern into the entire egg surface. Having cut deeper than the nuclei, the furrows broaden at their bases, gradually constricting the cytoplasmic connection between cells and endoplasm. This stage of nearly complete cellularization is referred to as the *cellular blastoderm.* It corresponds to the blastula stage of other embryos, although there is no fluid-filled blastocoel: the center of the egg is still filled with yolk-rich endoplasm.

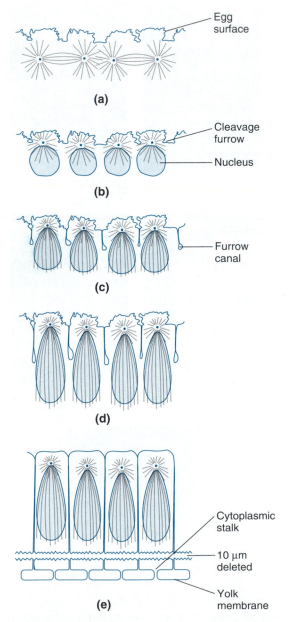

Figure 5.18 Cellularization of the blastoderm in *Drosophila.* **(a)** 13th mitosis (late cycle 13). **(b–e)** Cellularization during cycle 14. Cell membranes develop from furrows that surround each nucleus by folding in from the egg plasma membrane. Cytoplasmic stalks between blastoderm cells and uncleaved central yolk persist until they are severed at the onset of gastrulation.

The Principle of Stepwise Approximation

As noted earlier in this chapter, cleavage is very precise in some groups of animals: all embryos of a species then cleave in exactly the same way, generating a constant number of cells in a stereotypical arrangement. Ascidian embryos (Fig. 5.7) fit into this category, as do roundworm embryos, which will be described in Chapter 24. In most of the animals discussed so far, however, cleavage is a somewhat variable process. The embryos of such animals illustrate the *principle of stepwise approximation:* instead of being under perfect control at each step of development, they tolerate a margin of variation at first; any imbalances are corrected at a later time. To appreciate this important strategy, let us look again at cleavage in mammalian embryos.

▼

The allocation of cells to the *inner cell mass (ICM)* and the *trophoblast* of mouse embryos has been studied in detail. From direct observation it appears that much or all of the trophoblast originates from the outside cells at the morula stage (16 cells). Because only outside cells of embryos are capable of *endocytosis*, Tom Fleming (1987) was able to label the outside cells selectively by "feeding" them tiny fluorescent latex beads. Using this technique, he found that the ratio of outside to inside cells in different embryos ranged from 9:7 to 14:2. When the labeled embryos were allowed to develop to the blastocyst stage (32 to 64 cells), the trophoblast proved to be derived mostly from morula outside cells. The origin of the ICM cells in the blastocysts was more varied. On average, about 75% of the ICM cells were unlabeled, that is, descended from cells that had been inside cells at the morula stage. The remainder of the ICM cells were found to be labeled with latex beads and therefore must have been derived from outside morula cells.

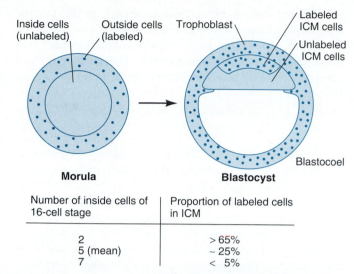

Number of inside cells of 16-cell stage	Proportion of labeled cells in ICM
2	> 65%
5 (mean)	~ 25%
7	< 5%

Figure 5.19 Stepwise formation of the inner cell mass (ICM) in mammalian embryos. Most of the ICM cells are derived from those cells that are in an inside position at the morula stage. Thus, after selectively labeling cells on the outside of a morula, most ICM cells of the developing blastocyst are unlabeled. However, in embryos that have few inside morula cells, additional ICM cells are generated by differential cleavage of outside morula cells.

The contribution of labeled outside morula cells to the ICM was highest in embryos that had the fewest inside cells at the morula stage (Fig. 5.19). In blastocysts derived from morulae with only two inside cells, more than 65% of the ICM cells were labeled. However, in blastocysts that developed from morulae with seven inside cells, fewer than 5% of the ICM cells were labeled. Thus, the outside morula cells "filled up" the ICM of embryos that did not have enough inside cells to begin with. Each additional ICM cell seems to be generated by differential cleavage of an outside morula cell into one trophoblast cell and one ICM cell. This was indicated by the position of the labeled ICM cells, which were always found near the trophoblast, not near the blastocyst cavity.

These results show that the ICM, which gives rise to the embryo proper, originates by stepwise approximation. The first step occurs during the fourth cleavage, when some blastomeres cleave parallel to their outer surface, thus giving off the first inside cells of the morula. Depending on the number of inside cells generated during this first step, a variable number of outside cells produce additional ICM cells by further cleavages parallel to their outer surface. In other words, the mouse embryo forms its ICM through a strategy of *stepwise approximation* rather than through a precise initial allocation of cells. This principle is widely observed in animal development.

Spatial Control of Cleavage: Nuclear Position and Mitotic Spindle Orientation

In the remainder of this chapter, we will consider the cellular mechanisms that bring about cleavage. This section will cover the spatial control of cleavage, in particular, the positioning of the nuclei within the blastomeres and the orientation of mitotic spindles. The next section will deal with the timing of cleavage divisions.

The Mitotic Spindle Axis Determines the Orientation of the Cleavage Plane

Embryonic cleavage, like the division of adult cells, involves mitosis and cytokinesis. The key organelle in mitosis is the *mitotic spindle*, which is organized by a pair of *centrosomes* (see Chapter 2). The key organelle in cytokinesis is the **cleavage furrow**. In small eggs that undergo holoblastic cleavage, the cleavage furrow constricts like a tightening belt around the entire cell; it is then called a *contractile ring*. In eggs that undergo meroblastic cleavage, cleavage furrows begin as infoldings of the egg plasma membrane at the animal pole and advance like cutting knives until they reach the yolk mass, which remains uncleaved.

Mitosis is usually followed by cytokinesis. Exceptions to this rule include embryos with superficial cleavage, in which many rounds of mitosis occur without cytokinesis (Fig. 5.16). A similar situation can be created in the laboratory with a procedure called **cleavage arrest**, which is often carried out by inhibiting cytokinesis with *cytochalasin*. For instance, ascidian embryos treated in this way proceed with their cell cycles and even with cellular differentiation in the absence of cytokinesis (Whittaker, 1973). Conversely, cytokinesis has been observed in sea urchin eggs from which the nuclei have been removed so that they cannot undergo mitosis (Harvey, 1956).

In most cases, mitosis not only sets the timing of cytokinesis but also determines the orientation of the cleavage plane (Strome, 1993). Invariably, cytokinesis occurs in a plane perpendicular to the axis of the mitotic spindle and equidistant between the two spindle poles (Fig. 5.20). This indicates that either the spindle orientation itself determines the cleavage plane or both are determined by a third event preceding mitosis. The latter alternative was ruled out by several investigators performing similar experiments on a variety of embryos. Squeezing blastomeres between two glass slides forced the mitotic spindles out of their normal orientation and into an orientation parallel to the plates. The plane of the cytokinesis that followed was always perpendicular to the new orientation of the spindle. Thus, the spindle orientation itself must control the orientation of cytokinesis.

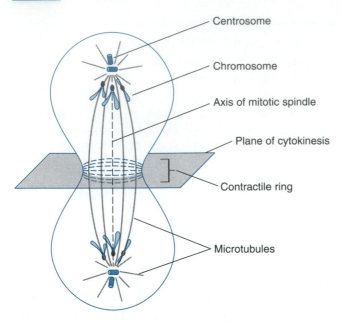

Centrosome

Chromosome

Axis of mitotic spindle

Plane of cytokinesis

Contractile ring

Microtubules

Figure 5.20 Cell division. The plane of cytokinesis develops perpendicular to the axis of the mitotic spindle.

The role of the mitotic apparatus in cleavage plane orientation was analyzed by Raymond Rappaport (1974). His experiments grew out of comparative observations on cleavage in isolecithal and telolecithal eggs. When the mitotic spindle is centered, as it typically is in an isolecithal egg, the cleavage furrow forms simultaneously all around the circumference. If the mitotic spindle is displaced toward the animal pole, as it is in moderately telolecithal eggs, the furrow first appears at the animal pole and then cuts toward the vegetal pole (Fig. 5.1). If the mitotic spindle is very eccentric, as is the case in extremely telolecithal eggs, the furrow forms only near the mitotic apparatus (Figs. 5.13 and 5.14). Thus it seems that proximity between the egg cortex and one or more components of the mitotic apparatus is required for furrowing to occur.

▼

Which parts of the mitotic apparatus are sufficient for the development of a normal-looking furrow? To answer this question, Rappaport pushed a small glass sphere through a sand dollar zygote, thus forcing it into a doughnut shape (Fig. 5.21). As a result, the mitotic apparatus was displaced to one side, and the first cleavage produced a horseshoe-shaped cell with two nuclei. Next, a mitotic apparatus, complete with spindle fibers and chromosomes between asters at opposite spindle poles, was set up in each "leg" of the horseshoe. In the horseshoe bend, two asters belonging to different spindles were facing each other but had no chromosomes between them. Under these conditions, *three* cleavage furrows formed, generating four nucleated cells. Two of these furrows were positioned normally, that is, along

the corresponding metaphase plates where the mitotic spindles had been. The third cleavage furrow appeared in the horseshoe bend although no chromosomes had been in this location. Nevertheless, this furrow looked normal and appeared at the same time as the other two.

Rappaport showed that interactions between two asters and the egg cortex provide whatever is necessary to form a cleavage furrow. Indeed, in insect embryos, the majority of the furrows that form simultaneously around the superficial nuclei arise between asters that belong to different spindles (Fig. 5.18a). What signals are generated between asters and egg cortex and how these signals direct the formation of cleavage furrows is still unclear.

Actin and Myosin Form the Contractile Ring in Cytokinesis

Cytokinesis must involve mechanical elements that generate the force to advance cleavage furrows. Under the electron microscope, a thin, dense layer is visible immediately under the plasma membrane of cleavage furrows (see Fig. 2.4). Schroeder (1972) showed that this layer contains many filaments oriented parallel to the plasma membrane and parallel to the cleavage plane. These filaments thus form part of the *contractile ring*. Immunostaining (see Methods 4.1) revealed that the contractile ring contains both actin and myosin (Fig. 5.22).

The contractile ring, which constricts the cell like a purse string or a belt, must be attached in some way to the cytoplasmic face of the plasma membrane. The force exerted by the contractile ring is strong enough to bend fine glass needles inserted into cleaving sea urchin embryos (Rappaport, 1967). Most likely, this force is generated by a musclelike sliding of actin and myosin filaments in the contractile ring. This has been demonstrated by inhibiting cytokinesis with agents that are known to interfere with actin and myosin filaments, such as cytochalasin or antibodies against myosin. After cleavage, the contractile ring disappears as quickly as it assembled. In fact, the filaments seem to be dismantled even before cytokinesis is complete, because the contractile ring does not become thicker as it constricts.

The contractile ring exemplifies how quickly cytoskeletal components can be assembled for a particular purpose and then disassembled until needed again, perhaps for a different function.

Specific Sites in the Egg Cortex Attract Nuclei and Anchor Spindle Poles

The cleavage pattern of an embryo is defined by the position and orientation of the cleavage furrows during cytokinesis. In some cases, cytokinesis is unequal, giv-

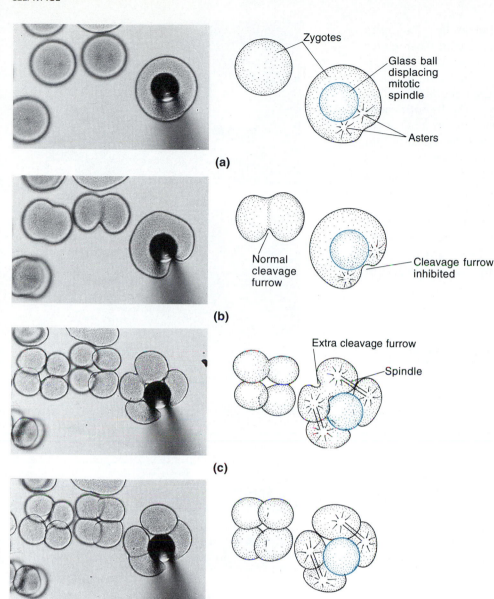

(a)

(b)

(c)

(d)

Zygotes

Glass ball displacing mitotic spindle

Asters

Normal cleavage furrow

Cleavage furrow inhibited

Extra cleavage furrow

Spindle

Figure 5.21 Formation of a new cleavage furrow between juxtaposed asters. The diagrams to the right are interpretations of the photographs shown to the left. **(a)** A glass ball forced into a sand dollar zygote creates a doughnut-shaped cell with an eccentric mitotic spindle. **(b)** The cleavage furrow stops at the glass ball, thus forming a horseshoe-shaped cell with two nuclei. **(c, d)** At the next cleavage, an extra furrow forms between adjacent asters, although no mitotic spindle was there before.

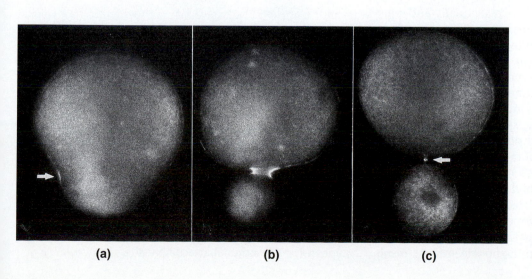

(a)   **(b)**   **(c)**

Figure 5.22 Localization of contractile rings in sea urchin blastomeres during fourth cleavage. Photographs show the unequal cleavage of an isolated vegetal blastomere into a macromere (top) and a micromere. The blastomeres were immunostained for myosin. In an optical section **(a)**, the contractile ring appears to be crescent-shaped (arrow). However, focusing through the entire cell reveals that the whole ring is stained. Stainability for myosin begins at telophase **(a)** and lasts until the end of cytokinesis **(b, c)**.

ing rise to daughter cells that differ in size. Examples include the fourth cleavage of the vegetal blastomeres in sea urchins, which generates macromeres and micromeres (Fig. 5.4), as well as the meiotic divisions in the female germ line, which produce one large oocyte and three small polar bodies (Fig. 3.11). In both cases, the eccentric position of the nucleus before meiosis or mitosis determines the unequal size of the daughter cells after cytokinesis. Some cleavages are associated with obliquely oriented cleavage planes, as seen during the fourth cleavage in sea urchins (Fig. 5.4) and in spiral cleavage (Fig. 5.6). These cleavage plane orientations are dictated by the orientation of the mitotic spindle, as discussed previously. Thus, the critical parameters of a cleavage pattern are the *positions* of the nuclei before mitosis and the *orientations* of the mitotic spindles. How are these parameters controlled?

▼

In sea urchins, the fourth cleavage divides the vegetal blastomeres unequally into macromeres and micromeres (Fig. 5.22). During the interphase preceding the fourth mitosis, the nuclei of the vegetal blastomeres migrate to the inner vegetal corners of their respective cells, where the micromeres will be pinched off. No comparable nuclear movement is observed in the animal blastomeres, which divide equally. Each vegetal nucleus moves quickly and directly to its destination as if it were attracted to the site. To establish this unique site, at least one component of the egg must be localized within the vegetal hemisphere before fertilization. This was demonstrated by Hörstadius (1973), who cut unfertilized sea urchin eggs into animal and vegetal halves and fertilized both halves. Both halves cleaved, but only the vegetal halves formed micromeres. Another step in establishing the attractive vegetal site occurs between the second and third cleavages. When embryos are treated with a detergent solution during this period, the nuclei in the vegetal blastomeres remain centrally located, and these blastomeres cleave equally and equatorially like the animal blastomeres. The same treatment at other times has no effect on micromere development (Tanaka, 1976; K. Dan, 1979). Translocation of the vegetal nuclei is also prevented by microtubule inhibitors, but not by microfilament inhibitors (Lutz and Inoué, 1982). As the vegetal nuclei enter mitosis, the vegetal poles of their mitotic spindles must somehow be anchored near the vegetal pole.

The eccentric anchorage of meiotic spindles has been studied in maturing oocytes of a polychaete worm, *Chaetopterus*. Before polar bodies form, the germinal vesicle moves from a central position to the animal pole, where a meiotic spindle is formed and one spindle pole is anchored to the egg cortex. This anchorage keeps the spindle axis perpendicular to the oocyte surface, result-

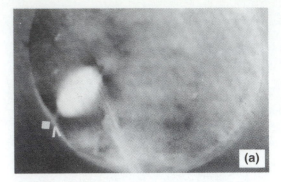

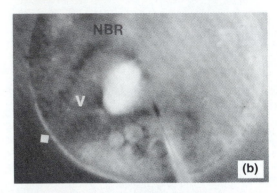

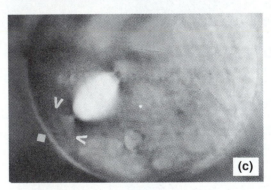

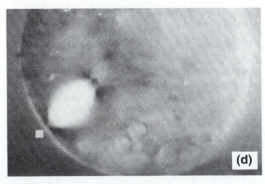

Figure 5.23 Detachment and return of a displaced meiotic spindle. Videomicrographs show a *Chaetopterus* oocyte under polarized light, which makes the meiotic spindle appear bright. The arrowheads mark astral fibers radiating out from the outer centrosome. **(a)** As the meiotic spindle is pulled away from the animal pole by a fine glass needle, a small dimple forms on the cell surface (white square), revealing a mechanical link between the spindle and the oocyte cortex. **(b)** If the glass needle pulls the spindle further to the interior, the spindle detaches from the cortex and the dimple disappears. **(c, d)** After withdrawal of the glass needle, the spindle returns spontaneously to its original attachment site.

ing in cleavage parallel to the surface. (If the spindle were parallel to the surface, the oocyte would cleave meridionally as during the first zygotic cleavage.) Using a fine glass needle mounted on a micromanipulator, Lutz and coworkers (1988) dislodged the spindle from its normal position (Fig. 5.23). If the spindle was pulled toward the cell interior, the cell surface adjacent to the outer spindle pole dimpled inward. As the spindle was pulled further inward, the dimple suddenly receded, indicating the rupture of a mechanical link between the egg cortex and the outer spindle pole. When released from the glass needle, the spindle spontaneously returned to its original attachment site. If removed too far from the cortex, however, the spindle remained stationary until pushed closer to its attachment site. Spindles that were inverted after detachment reattached to the former cortical site but with the pole that had formerly faced the cytoplasm. Spindle poles pushed against other cortical regions, however, did not attach.

The experiments described above demonstrate the existence of localized cortical attachment sites in oocytes, eggs, and blastomeres. These sites can attract and hold either pole of a meiotic or mitotic spindle. The attachment sites are generated at the appropriate times and places during development. The nature of these

sites and the programs for their generation remain to be elucidated.

Maternal Gene Products and Spatial Constraints Orient the Mitotic Spindle

As noted, the *orientation of the mitotic spindle* also plays a role in determining cleavage pattern. Spindle orientation is controlled by cytoplasmic factors encoded by genes expressed during oogenesis. In small blastomeres, mechanical constraints also orient mitotic spindles parallel to the longest cell axis.

▼

The role of cytoplasmic factors in the orientation of mitotic spindles is known from genetic experiments and cytoplasmic transplantations. The freshwater snail *Lymnaea peregra* has a shell that normally shows *dextral coiling*, meaning that if one looks down on the top of the shell, its coil winds toward the opening in a clockwise or right-handed (dextral) spiral. However, in some broods of *Lymnaea*, the shell winds in the opposite direction, showing *sinistral coiling*. Crampton (1894) observed that embryos from dextral and sinistral broods differ in the orientation of their mitotic spindles during the second cleavage (Fig. 5.24). All subsequent cleav-

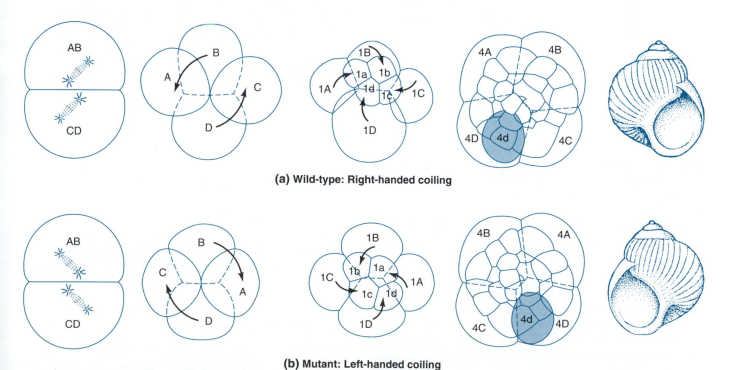

(a) Wild-type: Right-handed coiling

(b) Mutant: Left-handed coiling

Figure 5.24 Genetic control of embryonic cleavage pattern and shell coiling in the snail *Lymnaea peregra*. The wild-type shell shows a right-handed (dextral) coil **(a)**. A mutant allele generates a left-handed (sinistral) coil **(b)**. The direction of shell coiling depends on the position of the 4d blastomere, which gives rise to the shell gland. The position of the 4d blastomere can be traced back to the orientation of the previous cleavages, which mirror each other in dextral and sinistral embryos. The dextral cleavage depends on the activity of a single gene during oogenesis.

ages in sinistral embryos are mirror images of those in dextral embryos. Both shell coiling and the direction of embryonic cleavage are controlled by one gene. The *wild-type allele* of this gene, D, which produces the dextral phenotype, is dominant over the mutant allele, d, which results in the sinistral phenotype. Most important, the sinistral form is a maternal effect mutant, as defined in Chapter 1. Thus, eggs derived from d/d individuals give rise to sinistral embryos, regardless of the allele introduced by the sperm, whereas all eggs from D/d individuals become dextral embryos even if fertilized by d sperm. Therefore, the orientation of the mitotic spindles during cleavage must be controlled by components laid down in the egg cytoplasm during oogenesis.

To investigate this phenomenon further, Gary Freeman and Judith Lundelius (1982) transplanted cytoplasm between dextral and sinistral zygotes. Cytoplasm from dextral donors made recipients from sinistral broods cleave dextrally, whereas the converse transplantation had no effect. These results, together with the dominance of the genetic allele producing the dextral phenotype, indicate that the dextral pattern requires a certain gene product in the egg cytoplasm. In the absence of this product, the zygote switches to a default program, which results in sinistral cleavage. The nature of the critical product is still unknown.

Mechanical constraints on spindle orientation have been observed in small eggs and blastomeres from many species. Investigators have spun eggs in centrifuges and squeezed blastomeres between glass slides to alter cell shape. Invariably, the mitotic spindles formed in such cells are oriented with their axes parallel to the longest cell dimension.

▼

Ethel Harvey (1956) observed that the spindle orientation in sea urchin zygotes can be changed by centrifugation. At low centrifugal force (3000g), yolk and other cytoplasmic components became stratified, but the zygote retained its spherical shape and cleaved normally (Fig. 5.25a and b). At higher centrifugal force (10,000g), the stratified yolk components, because of their different buoyant densities, pulled the zygote into a dumbbell-like shape. If the zygote cleaved while still elongated, the cleavage plane was perpendicular to its long diameter, regardless of the animal-vegetal axis (Fig. 5.25c and d). The shape of the zygote had constrained the spindle so that it aligned with its axis parallel to the longest cell diameter. The same behavior can also be observed in normal embryos. Whenever a small blastomere is not spherical, the mitotic spindle tends to align its axis parallel to the longest cell diameter. For in-

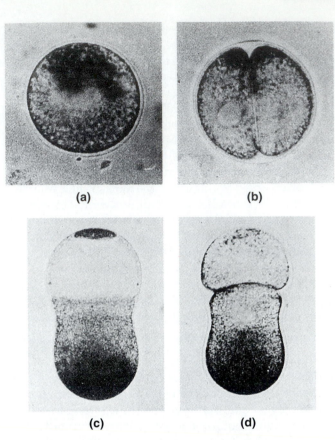

Figure 5.25 Cleavage in centrifuged sea urchin eggs. **(a)** Photograph of a sea urchin zygote centrifuged at 3000g, with the direction of the centrifugal force randomly oriented relative to the egg's original animal-vegetal axis. **(b)** Centrifuged zygote after cleavage, with the cleavage plane parallel to the direction of the centrifugal force. **(c)** Zygote centrifuged at 10,000g. **(d)** Centrifuged zygote after cleavage, with the cleavage plane perpendicular to the direction of the centrifugal force.

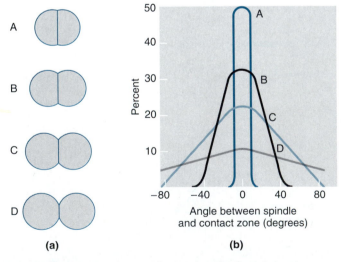

Figure 5.26 Dependence of mitotic spindle orientation on the extent of cell contact between snail blastomeres. **(a)** 2-cell embryos A through D show different degrees of contact reduction between blastomeres. **(b)** Graph showing distributions of the angle between the plane of blastomere contact and the spindle axes during second mitosis.

stance, during third mitosis in sea urchin and frog embryos, the longest diameter of the blastomeres is parallel to the animal-vegetal axis, and so is the orientation of the mitotic spindles (Fig. 5.4).

The shape of blastomeres, and thus the orientation of spindles within them, is affected by the area of contact between the blastomeres. Meshcheryakov (1978) systematically altered the extent of contact between snail blastomeres during cleavage by adjusting the concentration of calcium ions (Ca²⁺) in the culture medium. When there was extensive contact between the blastomeres at the 2-cell stage, their shape was hemispherical, and all spindles were aligned parallel to the contact area (embryo A in Fig. 5.26). As the area of contact was reduced and the blastomeres became more spherical, the orientation of the mitotic spindles began to vary until eventually it was almost random (embryos B, C, and D in Fig. 5.26).

The available data indicate that the cleavage pattern of embryos is determined by the positioning of nuclei during mitosis, which determines the relative sizes of daughter cells, and by the orientation of mitotic spindles, which determines the planes of cytokinesis. Specific cortical sites in eggs and blastomeres can attract and hold nuclei in eccentric positions. Mitotic spindle orientation is influenced by maternally encoded cytoplasmic factors in the egg. Presumably these factors position the centrosomes that organize the mitotic spindle (see Chapter 2). However, the resulting spindle orientations can be overridden in small cells by mechanical constraints that orient the spindle axis parallel to the longest cell dimension.

The Timing of Cleavage Divisions

The fast pace of embryonic cleavage is possible because the egg contains all necessary proteins, or at least the maternal mRNAs from which these proteins can be rapidly translated. These reserves allow blastomeres to speed through several cycles of mitosis without having to transcribe the genes, or even translate the mRNAs, for required proteins such as DNA polymerase, histones, and tubulin. As development proceeds, maternal supplies are depleted or actively destroyed, and embryos come to rely on their own RNA synthesis. When a surplus of molecular building blocks is no longer guaranteed, certain "checkpoints" are built into the cell cycle. The cycle does not proceed beyond these checkpoints unless their DNA is completely replicated, the cell has a minimum size, and other requirements are met. When the cleavage stage ends, with the onset of gastrulation, cell division continues at a much slower

pace and in regional patterns. As the nucleocytoplasmic ratio reaches a certain level, cells also begin to grow between mitoses. The necessary nutrients are mobilized from yolk reservoirs or are obtained from outside.

The Cell Cycle Slows Down during the Mid-Blastula Transition (MBT)

Unlike most mature eukaryotic cells, which have division cycles lasting from several hours to many days, an early *Drosophila* embryo undergoes a new round of mitosis every 9 min. Several features allow for a very rapid mitotic cycle in early embryos. For instance, the S phase is very short because there are an exceptionally large number of initiation sites for DNA replication (Alberts et al., 1989). In addition, cleaving blastomeres skip entire phases of the cell cycle (Fig. 5.27). In mature cells, the cell cycle consists of mitosis (M phase), a postmitotic gap phase (G_1 phase), DNA synthesis (S phase), and a premitotic gap phase (G_2 phase). In contrast, the blastomeres of most early embryos have no G_1 phase; their DNA begins to replicate during the last stage (telophase) of mitosis. Sea urchin blastomeres have a short G_2 phase, during which a small amount of RNA is synthesized. In other species, such as *Drosophila* and *Xenopus*, early blastomeres skip the G_2 phase as well and synthesize no measurable amounts of RNA.

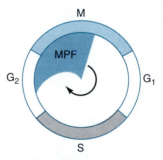

(a) Late embryonic and mature cell cycle

(b) Early embryonic cell cycle

Figure 5.27 Cell cycle in early embryonic and mature cells. **(a)** In mature cells, the cell cycle consists of mitosis (M), postmitotic gap phase (G_1), DNA synthesis (S), and premitotic gap phase (G_2). Synthesis of RNA is limited to the gap phases. **(b)** In early embryonic cells, there is no G_1 phase, because S phase begins at the end of M phase. S phase is accelerated, and G_2 phase is short or nonexistent. Both embryonic and mature cell cycles are controlled by M-phase promoting factor (MPF). MPF activity increases gradually during mitotic prophase and metaphase and decreases rapidly thereafter.

Toward the end of the cleavage stage, the cell cycle slows down as gap phases are added or elongated. Around the same time, many embryos begin to synthesize substantial amounts of RNA. In *Xenopus*, for instance, 12 rapid synchronous cleavages are followed by a period of slower, asynchronous cleavages. At the same time, blastomeres become motile and active in RNA synthesis. Together, these changes make up the *mid-blastula transition* (Newport and Kirschner, 1982a). In other species, however, the changes that occur simultaneously in *Xenopus* are more dissociated and spread out over several mitotic cycles. In *Drosophila*, for example, 10 rapid synchronous cell cycles followed by three somewhat slower and nearly synchronous cell cycles precede much longer, asynchronous cycles (Fig. 5.28). During the 11th cell cycle, nuclear RNA synthesis becomes detectable (B. A. Edgar and G. Schubiger, 1986). During the 14th cycle, the G_2 phase is extended, and large amounts of RNA are transcribed. Other animals, including mammals and sea urchins, undergo RNA synthesis throughout cleavage and thus have no mid-blastula transition in the originally defined sense.

The following section focuses on cell cycle length and the nucleocytoplasmic ratio, the aspects of the mid-blastula transition that are most directly related to the cleavage process.

Stage- and Region-Specific Gene Activities Modulate the Basic Cell Cycle

As cell divisions slow down and become part of region-specific morphogenetic processes, the cell cycle becomes more and more restricted. As discussed in Chapter 2, the cell cycle is regulated by a biochemical oscillator in the cytoplasm that generates *M-phase promoting factor (MPF)*. This oscillator involves the periodic synthesis of *cyclins* and the concomitant activation of *cyclin-dependent kinases* (see Fig. 2.5). Cyclins are synthesized during interphase and mitotic prophase and are degraded by specific proteases between metaphase and anaphase. Cyclins form MPF by associating with cyclin-dependent kinases, which are activated before each M phase by the phosphorylation of certain amino acid residues and dephosphorylation of others. Thus MPF activity increases during mitotic prophase, peaks during metaphase, and is low at all other times (Fig. 5.27). The control of the cell cycle by this basic oscillator appears to be universal, from yeasts to mammals (Nurse, 1990).

During early cleavage, the cell cycle runs freely, controlled only by the basic oscillator while all accessory components needed to keep the oscillator running are in abundant supply. At other stages of development, the basic cell cycle control is altered through modulation of the synthesis, activity, or breakdown of one of

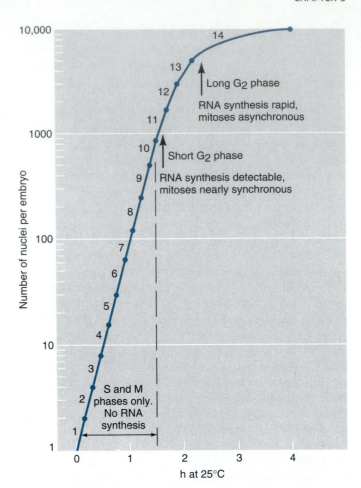

Figure 5.28 Nuclear divisions during early *Drosophila* embryogenesis. The numbers on the curve indicate the cell cycles. The first 10 cell cycles are synchronous and last 9 min each; little or no nuclear RNA is synthesized. The cell cycle during this period consists only of M and S phases. As the cell cycle slows down somewhat after cycle 10, a short G_2 phase is added, and nuclear RNA synthesis becomes detectable; mitoses are still nearly synchronous. After cell cycle 13, the G_2 phase becomes much longer, the rate of RNA synthesis increases dramatically, and cell divisions become asynchronous.

the two MPF components, cyclin or cyclin-dependent kinase. We have already discussed one example of such modulation in the context of frog oocyte maturation: between prophase I and metaphase II, c-mos protein translated from stored maternal mRNA seems to activate stored cyclin and to inactivate a cyclin protease (see Fig. 3.22).

Cell cycle modifications that occur toward the end of the cleavage stage rely on other mechanisms. The mid-blastula transition (MBT) in amphibians begins when the nucleocytoplasmic ratio reaches a certain level. In haploid embryos, the MBT is delayed by one cycle, whereas in polyploid embryos, the MBT is accelerated. Experimental elimination of three-quarters of the egg volume by constriction induces the MBT two cell cycles early. Injection of DNA also hastens the MBT, and the amount of DNA needed to trigger the MBT is equal to the amount of nuclear DNA present after 12 mitoses (Newport and Kirschner, 1982b). A signal generated by the nucleocytoplasmic ratio appears to slow down the basic cell cycle oscillator. This is evident because addition of a protein synthesis inhibitor, which prevents the synthesis of cyclin and thus stops the cell cycle, is followed by the onset of RNA synthesis and cell mobility (Kimelman et al., 1987). These experiments show that amphibian blastomeres are capable of both movement and RNA synthesis before MBT; these activities are suppressed prior to MBT by the rapid cell cycles and begin spontaneously once the cell cycle slows down.

A similar control seems to operate in *Drosophila* embryos, where the final part of MBT occurs at the blastoderm stage during the 14th mitotic cycle. Haploid embryos, which start at half the normal nucleocytoplasmic ratio, undergo an extra mitotic cycle before MBT. Conversely, increasing the ratio by ligation (tying off some of the cytoplasm) can result in the omission of a mitotic cycle (B. A. Edgar et al., 1986). As in frog embryos, RNA synthesis can be triggered precociously by inhibiting protein synthesis and stopping the cell cycle (B. A. Edgar and G. Schubiger, 1986).

Genetic and molecular analysis have provided additional insight into the variations of cycle control in *Drosophila* embryos (B. A. Edgar et al., 1994). During the first seven cell cycles, cdk1 and cyclin are continuously present and show little variation in abundance or activity, suggesting that these cycles do not require the typical MPF oscillation. During cycles 8 to 13 a cyclic degradation of cyclins leads to increasing oscillations of MPF activity. Mutants deficient in cyclin mRNAs show cell cycle delays, indicating that cyclin accumulation becomes rate-limiting during this period. During cell cycles 14 to 16, the protein encoded by the *string+* gene is necessary for initiating M phase (B. A. Edgar and P. H. O'Farrell, 1989, 1990). By modifying the supply of string protein, the *Drosophila* embryo now switches from rapid overall cleavage to regional patterns of mitoses at much longer intervals.

Throughout the first 13 cell cycles in *Drosophila,* the string protein is translated from an ample supply of maternal mRNA, and the mitoses that are part of these cycles occur rapidly and synchronously throughout the embryo. During the 14th cell cycle, the maternal string mRNA is actively degraded, so that all subsequent mitoses depend on transcription of embryonic *string+* genes. These latter mitoses occur in distinct regional patterns, and each of these division patterns is forecast precisely by the transcription of the embryonic *string+* gene during G_2 phase (Fig. 5.29). The string protein acts on the basic cell cycle oscillator by activating one MPF component, the cyclin-dependent kinase cdk1

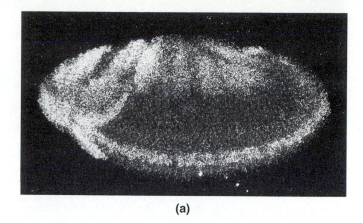

(a)

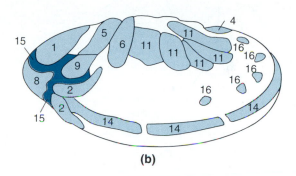

(b)

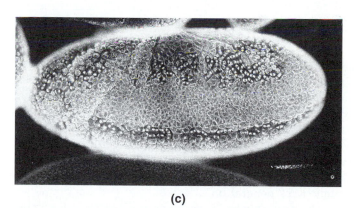

(c)

Figure 5.29 Correlation of mitotic patterns with *string+* gene expression in the *Drosophila* embryo. **(a)** Embryo late in cycle 14 (early gastrula). The white grains represent a labeled probe that binds specifically to string mRNA (see Methods 15.1 for technique). The larger gray dots are nuclei stained with a fluorescent dye. **(b)** Tracing of the embryo shown in part a, showing the correlation of *string+* expression (shaded) with domains of cells undergoing mitosis (outlined and numbered areas). The numbers were assigned by Victoria Foe (1989) to each mitotic domain and indicate the sequence in which mitosis is initiated in each domain. This embryo has initiated mitosis in domains 1 through 6, where *string+* expression is most intense. **(c)** Embryo photographed at a slightly later stage after immunostaining with antitubulin antibodies (see Methods 4.1). This procedure imparts a bright stain to mitotic spindles. Embryos prepared in this fashion were used to map the mitotic domains shown in part b.

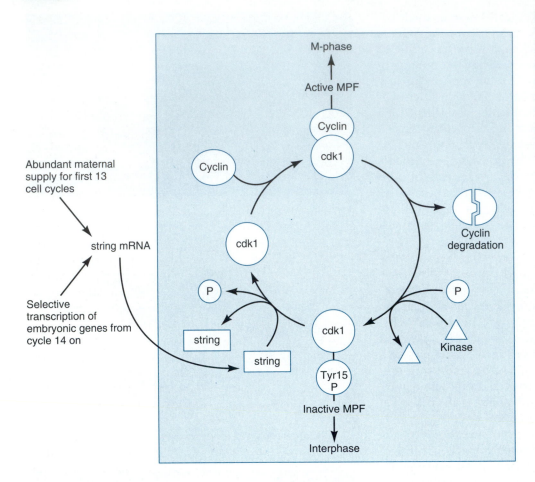

Figure 5.30 Control of the basic cell cycle oscillator by the string protein in *Drosophila*. In the basic oscillator (colored panel), a cyclin-dependent kinase (cdk1) combines with periodically synthesized cyclin protein to form active M-phase promoting factor (MPF). When cyclin is degraded after metaphase, MPF becomes inactive and allows the rest of M phase and the following interphase to proceed. The tyrosine-15 residue of the dissociated cdk1 is phosphorylated by a kinase. (Other modifications of cdk1 and cyclin also occur but are not shown.) The tyrosine-15 phosphorylation is reversed by the string protein, a phosphatase, before the cdk1 protein can form active MPF again. This basic oscillator runs freely during cleavage while a large supply of maternal string mRNA is available. After 13 mitotic cycles, the maternal string mRNA is destroyed, at which point mitosis becomes dependent on the selective transcription of embryonic *string*+ genes.

(Fig. 5.30). Specifically, string protein acts as a phosphatase, removing an inhibitory phosphate group from the tyrosine-15 residue of cdk1 (Dunphy and Kumagai, 1991; Gautier et al., 1991). Proteins encoded by genes similar to *string*+ have also been found in *Caenorhabditis elegans*, frogs, and mammals, including humans (Sadhu et al., 1990). These data indicate that the activation of cdk1 by a stringlike protein is a common feature of cyclic MPF generation.

The critical roles in the *Drosophila* MBT of both the nucleocytoplasmic ratio and the maternal-zygotic transition in string protein supply raise the question of how these phenomena might be related. Experimental manipulations that delay the increase of the nucleocytoplasmic ratio also delay the maternal-zygotic transition in *string*+ expression (Yasuda et al., 1992). The simplest interpretation of these results is that the nucleocytoplasmic ratio somehow regulates the breakdown of maternal string mRNA and the beginning of embryonic *string*+ expression.

The nucleocytoplasmic ratio serves to control cell division not only in amphibians and *Drosophila*, but throughout the animal kingdom. For example, this strategy has also been found in sea urchins and hydrozoa (Kühn, 1971). In sea urchins, the nucleocytoplasmic ratio varies among the macromeres and micromeres that are generated by the fourth cleavage. The macromeres, which have the smallest nucleocytoplasmic ratio, undergo 10 additional cleavage divisions, whereas the micromeres, which have the greatest nucleocytoplasmic ratio, undergo only six additional divisions (Masuda and Sato, 1984).

The general significance of the nucleocytoplasmic ratio and the apparent universality of the basic cell cycle oscillator invite speculation that there may be a single mechanism by which the nucleocytoplasmic ratio is "read" and transmitted to the cell cycle oscillator. Many researchers have proposed a *titration model* to explain how cells might sense the nucleocytoplasmic ratio (Fig. 5.31). According to this model, a signal substance in the egg cytoplasm is bound (titrated) by DNA or some other nuclear component that increases exponentially during cleavage. When all of the signal substance is bound, the cell cycle is altered. The nature of these hypothetical components remains to be elucidated.

A "Cleavage Clock" Can Be Uncoupled from the Cell Cycle

As discussed earlier in this chapter, the embryonic cleavage pattern depends on the generation of cortical sites that attract nuclei and hold them in eccentric locations

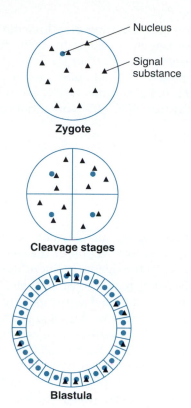

Figure 5.31 Titration model for the change in cell cycle control that occurs at the end of cleavage. A hypothetical cytoplasmic signal substance (black triangles) is bound (titrated) by genomic DNA or another nuclear component (colored circles). As the DNA or nuclear component increases exponentially during cleavage, all of the signal substance will eventually be bound. In the absence of free signal substance, a change in cell cycle control occurs.

and on the positioning of centrosomes that organize mitotic spindles. These events seem to be regulated by a pacemaker, or *cleavage clock,* which is started at fertilization and can be dissociated from the MPF oscillator that controls the mitotic cycle.

The idea of a cleavage clock is supported by the research of Sven Hörstadius (1973), who found that shaking sea urchin embryos in dilute seawater delayed or suppressed cleavage. This treatment blocked mitosis as well as cytokinesis. Upon being returned to normal seawater, the embryos resumed cleavage. Surprisingly, nuclear positioning and spindle orientation in the experimental embryos proceeded on the same schedule as in control embryos kept in normal seawater. For instance, after the first cleavage had been inhibited in dilute seawater, the subsequent cleavages in normal seawater were patterned like the second, third, and fourth cleavages in control embryos (Fig. 5.32). In other words, the delayed zygote behaved as if it were a normal 2-cell blastomere. If the first and second cleavages were suppressed, the cleavages that resumed in normal seawater were patterned like the third and fourth cleavages in normal embryos. If cleavages were delayed rather than suppressed in dilute seawater, the spindle orientations that were observed after return to normal seawater were intermediate between normal cleavage patterns. Hörstadius concluded that progressive changes in the cytoplasm orient the mitotic spindles and that these cytoplasmic changes proceed at set times after fertilization whether cleavages occur or not.

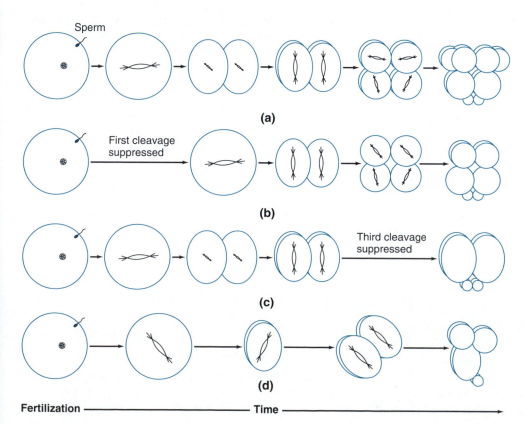

Figure 5.32 Cleavage pattern of sea urchin embryos in which one or two cycles of mitosis and cytokinesis have been suppressed. **(a)** Normal cleavage. **(b)** First cleavage suppressed. **(c)** Third cleavage suppressed. **(d)** All cleavages delayed. Note the oblique spindle orientations.

In similar experiments with a snail, *Ilyanassa obso-leta,* cleavage was inhibited reversibly with *nocodazole,* an antitubulin drug (Cather et al., 1986). Again the cleavage pattern changed progressively, regardless of whether the embryos were actually cleaving or not. Very little is known about the nature of the hypothetical cleavage clock that is thought to control the position and orientation of mitotic spindles. Like the MPF oscillator, the cleavage clock seems to reside in the cytoplasm, because cytoplasmic transplantations can change the cleavage pattern of snail embryos from sinistral to dextral, but it is not clear how the cleavage clock is synchronized with the MPF oscillator in normal development.

SUMMARY

The first divisions of the zygote are called cleavage divisions, and the resulting cells are called blastomeres. Unlike dividing mature cells, blastomeres do not grow back to their original size between cleavages. Correspondingly, cleavage divisions are usually fast, with cell cycles that have no G_1 phase and short S and G_2 phases.

The cleavage patterns of different animal groups are correlated with the amount and distribution of yolk in their eggs. The cleavages are termed holoblastic or meroblastic, depending on whether the egg cells are cleaved totally or partially. Holoblastic cleavage patterns include rotational cleavage, seen in mammals, and spiral cleavage, seen in mollusks and other groups of animals. Meroblastic cleavage is called discoidal when it is limited to a disc of cytoplasm at the animal pole and superficial when it leads to a layer of cells surrounding a central mass of yolk.

Embryonic cleavage, like mature cell division, involves mitosis and cytokinesis. Invariably, the plane of cytokinesis is perpendicular to the spindle axis of the preceding mitosis. The positioning and orientation of mitotic spindles are therefore key events that determine the cleavage pattern. In cases of highly unequal cleavage, one spindle pole is anchored in a particular region of the cell cortex, thus maintaining an eccentric spindle position. The spindle orientation is affected by spatial constraints and by cytoplasmic factors that are controlled by the maternal genotype.

The timing of cleavage is controlled by the same biochemical oscillator that governs meiosis and the division of mature somatic cells. This basic oscillator relies on the cyclic synthesis, activation, and breakdown of two proteins that together form M-phase promoting factor (MPF). In early embryos, this basic oscillator is free-running, because abundant maternal supplies of the accessory components required to maintain the oscillator are present. At the end of cleavage, the maternal supplies are degraded, and the cell cycle becomes dependent on the transcription of embryonic genes. This so-called mid-blastula transition is a distinct event in some species. The timing of this transition is controlled by the nucleocytoplasmic ratio. The positioning of nuclei and orientation of mitotic spindles progress as if they were driven by an autonomous pacemaker that can be dissociated from the MPF oscillator.

SUGGESTED READINGS

Beams, H. W., and R. G. Kessel. 1976. Cytokinesis: A comparative study of cytoplasmic division in animal cells. *Am. Sci.* **64**:279–290.

Edgar, B. A., F. Sprenger, R. J. Duronio, P. Leopold, and P. H. O'Farrell. 1994. Distinct molecular mechanisms regulate cell cycle timing at successive stages of *Drosophila* embryogenesis. *Genes Dev.* **8**:440–452.

Freeman, G. 1983. The role of egg organization in the generation of cleavage patterns. In W. R. Jeffery and R. A. Raff, eds., *Time, Space, and Pattern in Embryonic Development,* 171–196. New York: Alan R. Liss.

CELL FATE, POTENCY, AND DETERMINATION

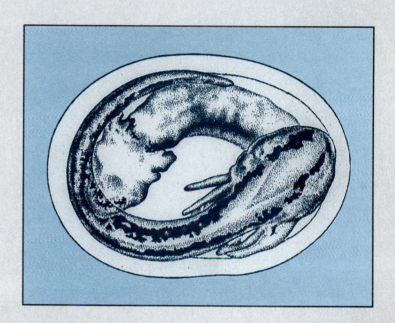

Figure 6.1 Twin embryos of the newt *Triton taeniatus* obtained after the first two blastomeres were separated by ligation with a hair loop.

As a new individual develops from a zygote, embryonic cells diverge from one another in size, shape, rate of mitosis, and other properties. This process of **cell diversification** is progressive and occurs at different stages depending on the species. In frogs, for example, blastomeres of different size and pigmentation are generated during the third cleavage, whereas in *Drosophila,* all somatic cells look alike until the 14th cell cycle (see Figs. 5.5 and 5.17). Later in development, each organism forms various types of mature cells, such as neurons, muscle cells, or gland cells. When these cells have attained their final structure and specialized function, they are called *differentiated cells. Cell differentiation,* the process of forming differentiated cells, will be discussed in Chapter 19; this chapter is concerned with aspects of cell diversification that precede differentiation.

The way in which cells diversify during cleavage varies among groups of animals (E. H. Davidson, 1990). Some groups are characterized by **invariant cleavage,** or stereotypical cleavage patterns that are followed exactly by each embryo. Roundworms, mollusks, sea urchins, and ascidians show invariant cleavage. Other groups of animals, including vertebrates and insects, show **variable cleavage** patterns. They follow the *principle of stepwise approxima-*

tion, which eliminates the need for accuracy by allowing for later correction. We have seen this principle at work in mouse embryos, where the ratio of inside to outside cells at the 16-cell stage can vary greatly (see Chapter 5).

The invariant and variable cleavage types are correlated with different mechanisms of cell diversification. Invariant cleavage depends strongly on the unequal distribution of different kinds of cytoplasm to establish differences among blastomeres (see Chapter 8). Variable cleavage depends to a greater extent on cellular interactions to set and reset the developmental programs of cells. The existence of variable cleavage implies that embryonic cells have more capabilities than they actually use. Only if this is true can they modify their development according to the signals they receive from other cells.

A dramatic example of the ability of embryonic cells to adjust to different circumstances is the development of twins. This was first observed by Hans Driesch (1892), who found, much to his surprise, that each blastomere from a 2-cell or 4-cell sea urchin embryo could form a complete larva by itself. Other investigators, including Hans Spemann (1938), obtained similar results with amphibian embryos (Fig. 6.1). In fact, the monozygotic ("identical") twins of humans arise by spontaneous division of the embryo at the 2-cell stage or later. Apparently, embryonic cells communicate with each other; they sense whether their normal neighbors are present and adjust their development if not. Twinning is characteristic of species whose embryos show variable cleavage and is not observed in forms with invariant cleavage. Nevertheless, most embryonic cells respond to some signals from neighboring cells, even in embryos that undergo invariant cleavage.

In both types of embryos, cells are eventually committed to certain developmental pathways, from which they rarely depart. Describing and analyzing this process of gradual commitment is one of the central goals of developmental biology. The basic concepts used in this analysis are *fate, potency,* and *determination.* Each of these terms is borrowed from human psychology but is also defined by objective criteria as a concept in developmental biology. The terms are applied to single cells as well as groups of cells or entire embryonic regions.

The *fate* of a cell is what becomes of it in the course of normal development. Cell fate can be identified by continuously monitoring the location of a cell and its descendants in an embryo. More reliable ways of mapping a cell's fate involve the use of labels that can be introduced into cells and are then passed on to their descendants but not to other cells. To establish the *potency* and *determination* of cells, scientists use **operational criteria,** which are experimental procedures designed to reveal specific features. The usefulness of an operational definition depends on how well the operation can be reproduced by different people using different techniques. As we will see, the operations that define cell potency and determination depend on many details that experimenters must be aware of.

In this chapter, we will define cell fate, potency, and determination and illustrate these concepts using insect, amphibian, and mammalian embryos as examples. Next we will discuss some properties of the determined state, in particular, its stability. Finally, we will return to the strategies of invariant and variable cleavage, redefining them in terms of determination as *mosaic* versus *regulative* development. It will become apparent that both strategies are at work in all embryos, although to varying extents depending upon the species.

Fate Mapping

The *fate* of a cell is the sum of all structures that its descendants will form at a later stage of development. A *fate map* is a diagram of an organism at an early stage of development, indicating the fate of each cell or region at a later stage. Fate maps are an essential tool in most embryological experiments. Only if we know the normal fates of cells can we establish how their behavior is altered by grafting, isolation, or other experimental treatments. Thus, developmental biologists have established fate maps for many of the species they study.

As an example, Figure 6.2 shows the fate map of a 32-cell amphibian embryo, indicating the structures each blastomere will have formed at the neurula stage (Dale and Slack, 1987a). The map indicates which blastomeres will contribute to endoderm, mesoderm, and the two primary derivatives of the ectoderm: epidermis and neural plate. In other words, the fate map delineates the precursor cells of each neurula region. Groups

of precursor cells, especially those that form a morphologically distinct region, are also called *primordia* (sing., *primordium*) or *rudiments.* Thus, the cells that will form the neural plate are called the *neural plate primordium,* or *neural plate rudiment,* or *prospective neural plate.* Fate maps change over time because cells multiply and move relative to each other. For this reason, fate maps usually indicate fates at a closely following stage. A series of fate maps at consecutive stages shows the progression of different cells or regions through longer periods of development.

There are different ways of establishing fate maps. Some are constructed from observations on live embryos under the microscope, augmented with histological sections and time-lapse motion pictures. For the mapping of large or opaque embryos, it is often necessary to label specific cells. A label for this purpose must be easy to apply to the desired cell or small embryonic region, should not spread to neighboring cells, has to be readily detectable at later stages, and should not disturb normal development.

One commonly used labeling method is the injection of fluorescent dyes conjugated to large, metabolically inert carrier molecules that will not cross cell membranes or pass through gap junctions (Weisblat et al., 1980; Strehlow and Gilbert, 1993). When such molecules are microinjected into a cell, they stain the cell and all its descendants very distinctly (Fig. 6.3). Alternatively, dyes may be injected into a zygote to label all cells of the developing embryo; at a later stage, individual cells from the labeled embryo are transplanted to an unlabeled host at the same stage to replace corresponding unlabeled cells. In effect, this operation generates an embryo with one or a few specific cells stained. The fate map shown in Figure 6.2 was generated with this technique. Some species can be infected with compatible viruses engineered so that their DNA encodes an enzyme that produces a visible stain; the viruses are also engineered not to leave the infected cells but to be passed on to daughter cells during mitosis.

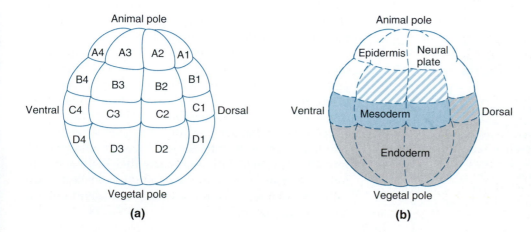

(a)

(b)

Figure 6.2 Fate map of the *Xenopus* embryo. **(a)** Lateral view of the embryo at the 32-cell stage. Each blastomere is identified by a letter and a number. **(b)** Fate map indicating which blastomeres will give rise to endoderm (gray), mesoderm (color), and the major ectodermal derivatives: neural plate and epidermis (white). Blastomere B2 contributes to both epidermis and mesoderm, B3 to neural plate and mesoderm, and C1 to endoderm and mesoderm.

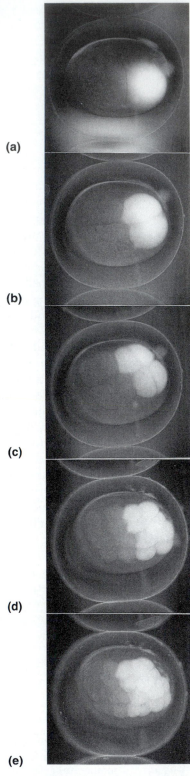

(a)

(b)

(c)

(d)

(e)

Figure 6.3 Injection of a fluorescent macromolecule as a label for fate mapping. This photograph shows a zebra fish embryo injected at the 2-cell stage with a fluorescent dye conjugated to dextran. **(a)** Embryo immediately after injection. The uninjected cell is seen in faint outline to the left. **(b–e)** Same embryo at the 4-, 8-, 16-, and 32-cell stages, respectively. The pictures were taken at intervals of about 30 min. At each stage, only half of the cells are labeled, indicating that the dye is passed on clonally but does not leak into other cells.

The accuracy of fate maps is subject to several biological limitations, even if the labeling technique itself is nearly ideal. First, in embryos with variable cleavage, the size and position of primordia may vary from one embryo to the next, so cells in the same topographical position contribute to different primordia in different embryos. Second, there may be some mixing between the cells of adjacent primordia from the time of cell labeling until the time when cell fates become fixed. For instance, cell B3 in the *Xenopus* fate map shown in Figure 6.2 forms epidermis but also contributes to mesoderm (Dale and Slack, 1987a). Similarly, blastomere B2 gives rise to varying amounts of neural plate and mesoderm, while blastomere C1 may form endoderm or mesoderm or both.

The magnitude of statistical error in fate maps depends on developmental stage and varies widely between species with invariant and variable cleavage. Variability of blastomere fates is characteristic of species with variable cleavage, such as *Xenopus*. An example of highly invariant cleavage is the roundworm *Caenorhabditis elegans*, fate maps for which are very precise at every stage of development (see Chapter 24).

Clonal Analysis

Clonal analysis is a special form of fate mapping. A *cell clone* consists of all surviving descendants of a single cell, which is called the *founder cell* of the clone. To make a clone visible, it is necessary to label its founder cell in such a way that the label is passed on to all descendants of the founder cell but not to cells outside the clone. The label may be a fluorescent dye or a visible homozygous mutation generated by X-ray–induced somatic crossover (see Methods 6.1). The later in development a founder cell is labeled, the smaller the labeled clone will be.

Even clones generated randomly by somatic crossover provide much useful information. The outline of a clone indicates the degree of cell mixing and any preferred orientation of cell division or locomotion during development. In addition, the decreasing size of clones labeled at successive stages can be used to chart the growth kinetics of an organ. Clone sizes can also be used to estimate the number of cells in an organ primordium at a certain stage. If a founder cell labeled at a particular stage produces clones covering a fraction $\frac{1}{n}$ of the organ, then the organ primordium presumably consisted of $\frac{1}{2}n$ cells at the time of labeling. (The factor $\frac{1}{2}$ accounts for the fact that after crossover, a cell divides into two daughters, only one of which is homozygously mutant.) This calculation assumes that there is no differential growth, although this may occur both naturally and as a result of the X-irradiation used to induce somatic crossover.

Labeling Cells by Somatic Crossover

As we have seen, the founder cell of a clone can be labeled by microinjecting a dye. Another way of labeling founder cells is by X-ray–induced *somatic crossover.* This technique is frequently used to label clones of cells in *Drosophila* and other species in which suitable mutants are available. The schematic diagram in Figure 6.4 represents a fly heterozygous for the mutant allele *multiple wing hairs (mwh)*. Homozygous *(mwh/mwh)* wing cells exhibit several irregular wing hairs as opposed to the one straight hair present on wild-type (+/+) and heterozygous (+/mwh) cells. In cells between S phase and mitotic anaphase, X-rays may induce crossovers between chromatids of homologous chromosomes. If such a cell is heterozygous for a recessive allele such as +/mwh, then there is a 50% chance that the next cell division will produce one wild-type (+/+) daughter cell and one homozygous mutant *(mwh/mwh)* daughter cell. The latter will become the founder cell of a clone of *mwh/mwh* cells having multiple wing hairs (Fig. 6.5).

The greatest advantage of the somatic crossover procedure is that the genetic label is passed on undiluted during cell divisions. Some disadvantages are that somatic crossovers are rare events and that X-rays cannot be focused on small target areas. Therefore, investigators must apply whole-body irradiation to thousands of fly larvae and screen the individuals as adults to find a small number of specimens with randomly positioned labeled clones.

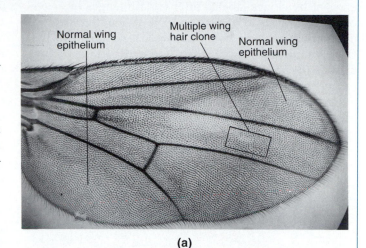

(a)

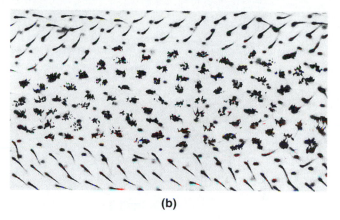

(b)

Figure 6.5 *Drosophila* wing with a *multiple wing hair* clone. **(a)** Photomicrograph in which clone is boxed. **(b)** Clone shown at higher magnification.

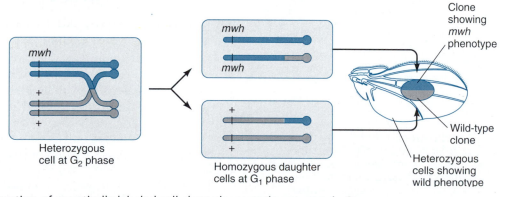

Figure 6.4 Generation of genetically labeled cell clones by somatic crossover in *Drosophila*. To induce the crossover, individuals heterozygous for a suitable mutant allele such as *multiple wing hair (mwh)* are X-irradiated. In a cell at G_2 phase, a crossover may occur between chromatids of homologous chromosomes. Depending on the orientation of the kinetochores in the mitotic spindle, there is a 50% chance that the chromatids will be separated so that one daughter cell becomes homozygous for the wild-type allele mwh^+ while its sister cell becomes homozygous for the mutant allele. This latter cell becomes the founder cell of a clone expressing the *mwh* mutant phenotype. Its sister cell produces a clone of homozygous wild-type cells, which are phenotypically indistinguishable from the heterozygous "background" cells.

Clonal analysis has led to the unexpected discovery of restriction lines in insect epidermis that are not crossed by clones labeled after certain stages in development (Garcia-Bellido et al., 1979). When investigators analyzed *Drosophila* clones labeled at the larval stage, they found that *none* of them crossed certain boundaries in the adult epidermis. For instance, clones did not extend from the underside to the top of the wing. And surprisingly, labeled clones in *all* individuals "respected" the *same* straight line running down the middle of the wing although this line did not coincide with any visible anatomical structure (Fig. 6.6). In fact, the boundary line was completely invisible unless demarcated by one of the labeled clones. In this case, the edge of the clone along the boundary line was straight, whereas everywhere else the edge of the clone was ragged. The boundary line was especially well demarcated by labeled clones that had been engineered by a genetic trick to grow faster than the surrounding cells. Despite their enormous size, all these clones stopped and formed straight edges at the same boundary lines. Areas of insect epidermis that are defined by such clonal restriction lines are called **compartments.**

A compartment is never filled entirely by one clone but rather is made up of several clones, which together can be called a **polyclone** (Crick and Lawrence, 1975). Polyclones have several regulatory properties that clones lack. Labeled clones generated in different individuals, when superimposed on the same drawing, do *not* fit each other like a jigsaw puzzle; they overlap and differ in size and shape (Fig. 6.6a). However, polyclones *always* occupy a compartment that has the same distinct shape and position within all individuals of a species. Polyclones are also units of size and shape regulation. Even if one member of the polyclone is given a genetic growth advantage so that it outgrows all the other clones, the compartment is still normally sized and shaped (Fig. 6.6b).

Moreover, polyclones seem to be units of gene regulation. In *Drosophila*, there is evidence that each compartment is characterized by a unique combination of active and inactive *selector genes*, which characterize each compartment like a binary zip code (Garcia-Bellido, 1975). Differences in selector gene activity seem to keep the boundaries between compartments straight (see Chapter 21).

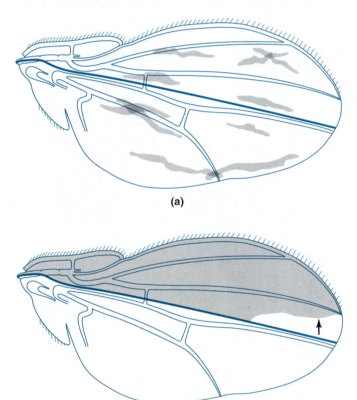

(a)

(b)

Figure 6.6 Clones and compartments in the wing of *Drosophila*. **(a)** Composite drawing in which clones (gray) generated in the wings of *different* flies are superimposed. Each clone was founded by a homozygous mutant cell generated randomly by X-ray–induced somatic crossover. Note that the clones have jagged edges and that some of them overlap. Although many more clones were generated than shown here, none of them transgressed the compartment boundaries located at the wing margin and along an invisible line running down the middle of the wing (bold line). **(b)** The boundaries of the anterior wing compartment are demarcated especially well by a large clone that was genetically engineered to grow faster than all other cells in the fly. Note that the wing is still normally sized and shaped. Also note that the large clone has smooth edges where it runs up against compartment boundaries, but a jagged edge where it abuts other cells within the same compartment (arrow).

Potency of Embryonic Cells

An embryonic cell during early development usually has a larger range of capabilities than its actual fate, especially in species with variable cleavage. This wider range of abilities is covered by the term "potency." The *potency* of a cell is the total of all the structures a cell or its descendants can form if placed in the appropriate environment (Slack, 1991). This is an operational criterion that can be satisfied only approximately, because in practice only a limited selection of environments can be tested.

The two operations used to define potency are isolation and heterotopic transplantation (Gk. *heteros,* "other"; *topos,* "place"). An *isolation experiment* tests the potency of a cell or region when it has been removed from the influence of other parts of the organism. Cells are excised from an embryo and kept in vitro in a medium that nourishes them but supposedly does not enhance or inhibit the development of any particular structure. *Heterotopic transplantation* tests the potency of a cell or region when it is influenced by cells other than its normal neighbors. Cells are excised from a *donor* and reimplanted into different regions of a *re-*

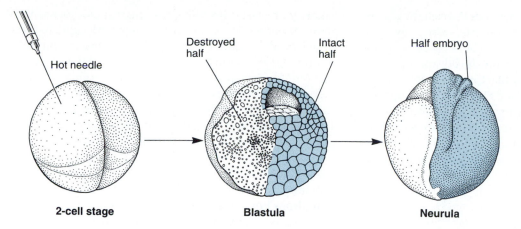

2-cell stage   **Blastula**   **Neurula**

Hot needle

Destroyed half

Intact half

Half embryo

Figure 6.7 Partial development after one blastomere of a 2-cell frog embryo was killed with a hot needle. In this experiment, carried out by Wilhelm Roux in 1888, the surviving blastomere developed into a half embryo.

cipient, or *host,* organism. In practice, the distinction between isolation and transplantation is not always clear: tissues are sometimes transplanted into body cavities of a host, where they are nourished by the host's body fluid but do not necessarily receive any instructive signals from the host environment.

As an illustration, if cells from the prospective neural plate of an amphibian embryo are isolated prior to gastrulation, they form epidermis. If they are transplanted elsewhere in the same embryo or another embryo, the graft forms nose sensory epithelium, eye lens, inner ear, or various mesodermal structures, depending on the site of implantation. Thus, on the basis of just a few tests, it can be shown that the potency of prospective neural plate cells prior to gastrulation includes epidermis, nose, lens, ear, and various mesodermal structures as well as neural plate. Note that the potency of a region always includes its fate. If cells or regions can form more structures than their fate, they are called *pluripotent,* whereas if they can give rise to a complete individual, they are called *totipotent.*

On the basis of a large number of isolation and grafting experiments, one could draw a *potency map* of an embryo. A potency map would look like a fate map except that each area would have multiple labels. Thus, the area labeled simply "neural plate" on the fate map would be labeled "neural plate, epidermis, nose, eye, ear structures, or mesoderm" on the potency map.

A problem with the operational definition of potency is that the outcome of isolation and grafting experiments depends very much on technical detail. This is best illustrated by some of the earliest embryological studies. In 1888, Wilhelm Roux published the results of experiments in which he had killed certain blastomeres of frog embryos with a hot needle. After one blastomere of a 2-cell embryo was killed, the other blastomere frequently developed according to its fate: it produced a lateral half embryo (Fig. 6.7). Other investigators completely separated amphibian blastomeres from each other by ligating embryos with a hair loop along the first cleavage plane. The separated blastomeres often developed into twin embryos that were half-sized but normally proportioned (Fig. 6.8). Thus, the potency of a

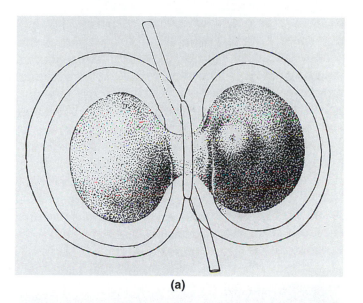

(a)

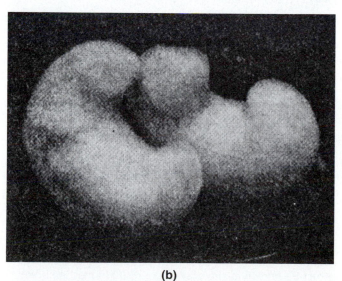

(b)

Figure 6.8 Development of twin embryos after complete separation of newt blastomeres. The blastomeres were separated by ligation of a 2-cell embryo along the first cleavage furrow with a hair loop **(a)**. In many cases, both blastomeres developed into half-sized but normally proportioned embryos **(b)**. This experiment was first attempted by Oskar Hertwig in 1893 and later carried out successfully by others.

blastomere at the 2-cell stage depended dramatically on the method of isolation. A blastomere isolated by ligation produced a complete embryo, whereas a blastomere isolated by killing its neighbor produced a lateral half embryo. Apparently, the dead blastomere in Roux's experiment was still sending signals to the surviving blastomere, thus limiting its potency to a lateral half.

The outcome of the preceding isolation experiments also depended on the orientation of the first cleavage. The results reviewed so far were obtained only in the

majority of cases, when the first cleavage occurred in a plane bisecting the embryo into left and right halves. However, if the first cleavage bisected the zygote into ventral and dorsal halves, then only the isolated dorsal blastomere formed a complete embryo, while the isolated ventral blastomere produced a ball of cells without overt differentiation (Fig. 6.9). This result indicates that the orientation of a cleavage plane may have a major effect on the potency of isolated blastomeres.

Since the first investigators did not always recognize how much the outcome of isolation experiments depended on the technique used, their different results sometimes led to spirited controversies with philosophical overtones. Roux believed that embryos were like machines in which each part had a strictly limited function, and he found this idea confirmed by the development of a half embryo from the blastomere that was left after its sister was killed at the 2-cell stage. Another pioneer of developmental biology, Hans Driesch (1909), adopted a radically different view. When he observed the development of whole sea urchin larvae from isolated blastomeres, he was so struck by this amazing result that he later postulated the existence of a goal-directed force (entelechy) in living organisms. With time, the effects of different isolation and transplantation techniques have become better understood, and some of the reasoning pitfalls can be avoided today, as long as researchers remember that operational criteria, such as the outcome of an isolation experiment, may depend very much on experimental detail.

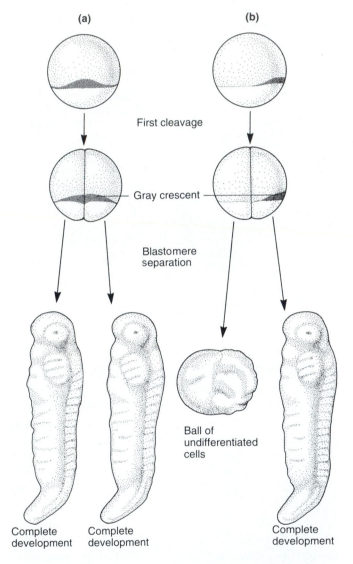

Figure 6.9 Dependence of twinning in amphibian embryos on the orientation of the first cleavage plane. Blastomeres were separated at the 2-cell stage as shown in Figure 6.8. The dorsal side of the eggs used was marked by a cortical structure known as the gray crescent. **(a)** In most embryos, the first cleavage plane bisected the gray crescent, creating a right blastomere and a left blastomere. In these cases, blastomere separation was followed by complete development of both blastomeres. **(b)** Separation parallel to the gray crescent was less frequent and was followed by complete development of the dorsal blastomere containing the gray crescent. The ventral blastomere formed only a round piece of tissue without overt differentiation.

Determination of Embryonic Cells

In most animal species, the potency of an early blastomere will exceed its fate, as the previous section demonstrated. However, as development progresses, the range of structures that a cell may form becomes more limited. When the potency of a cell is restricted to its fate, the cell is said to be *determined.* If the fate of a cell is to form heart, and if after various isolation and transplantation experiments it always forms heart, then this cell is said to be determined for heart development. The process, therefore, by which the fate and the potency of a cell become identical is called *determination.*

Cell Determination Is Discovered through Operational Criteria

To decide whether a cell or region is determined, one has to compare its fate with its potency (Slack, 1991). Often, the fate is already known from a fate map. Then, only the potency needs to be found, by isolation and heterotopic transplantation. These two procedures therefore serve as *operational criteria for embryonic determination.* According to these criteria, an embryonic cell or region is said to be determined when it develops according to fate upon isolation as well as heterotopic transplantation.

Since finding the determined state of cells or regions involves assessing their potency, the difficulties with potency tests also affect the assessment of determination. For instance, the behavior of cells upon transplantation may depend on how many of them are transplanted (Gurdon, 1988). Single cells or small groups tend to blend in with their new environment, whereas larger groups of cells are more likely to hold to their original fate. This *community effect* may reflect a need for intercellular signaling to activate and maintain the expression of certain genes. Because single cells are deprived of such signals, they may be unable to develop according to fate upon isolation or transplantation. Therefore, single cells or small groups may show an apparent lack of determination in isolation and transplantation tests while larger groups of the same cells seem to be determined.

In addition to isolation and transplantation tests, there is a third operational criterion for determination, which is based on clonal analysis. If a labeled clone overlaps two different structures, such as leg and wing in a fly (Fig. 6.10), then the potency of the founder cell (at the time of labeling) was obviously not limited to either leg or wing—that is, the founder cell was *not* determined to form either structure. Note that the *lack* of clonal restriction is used here as a *negative* criterion for determination. It is not possible to use clonal restriction as a positive criterion for assessing cell determination. Since clonal analysis reveals the fate of the founder cells but may underestimate their potency, this procedure cannot positively show that a cell was determined. In short, the absence of clonal restriction also means no determination, whereas clonal restriction shows *only* a restriction in fate.

The three operational criteria for assessing cell determination, and the problems with these criteria, are summarized in Table 6.1. The application of these criteria will be illustrated as we discuss determination in insects, amphibians, and mammals.

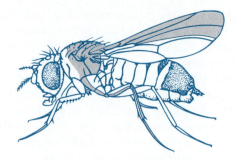

Figure 6.10 Clonal analysis as a test for nondetermination of the clonal founder cell. The diagram shows the outline of a *Drosophila* adult fly with a labeled clone (gray shading) extending over most of the anterior compartment of the second thoracic segment, including wing and second leg. The clone was generated by X-ray–induced somatic crossover at the cellular blastoderm stage. The clone cells were homozygous for a cuticular marker gene and for a gene that confers a growth advantage. The fact that the clone contributes to both leg and wing shows that the potency of the clone founder cell was not limited to either leg or wing development.

Drosophila Blastoderm Cells Are Determined for a Single Segment but Not for Dorsal or Ventral Structures within One Segment

Cell determination in insect embryos has been tested by clonal analysis and heterotopic transplantation. The following experiments with *Drosophila* embryos were designed to reveal whether cells at the blastoderm stage were already determined to form specific segments or parts thereof.

X-ray–induced somatic crossover (Fig. 6.4) was applied to *Drosophila* embryos at the cellular blastoderm stage (Wieschaus and Gehring, 1976). All labeled clones were restricted to a single segment. However, within a segment some clones overlapped dorsal and ventral structures such as wings and second legs. The possibility that the overlap was an appearance caused by two separately induced clones was excluded by the low frequency of clones in either wings or legs. Among 664

TABLE 6.1	
Operational Criteria for Cell Determination	
Operation	**Potential Problems**
Heterotopic transplantation	Dependence of cell potency on number of transplanted cells and contact between them
	Limited number of transplantation sites that can be tested
Isolation	Dependence of cell potency on number of isolated cells and contact between them
	Additional dependence of cell potency on method of isolation and culture medium
Clonal analysis	Only negative conclusion possible

flies irradiated at the blastoderm stage and scored in the adult thoracic region, only 13 clones were found in second legs and 25 in wings. The probability of finding *one* fly with two independently generated clones in wing and second leg on the same side of the body was therefore $(13 \times 25)/(664 \times 2) = 0.24$. Actually, the investigators obtained *seven* flies with clones in both wings and second legs on the same side, indicating that in virtually all of these cases a single labeled clone had split during development. In an extension of these experiments, marked clones endowed with a genetic growth advantage were found to almost fill entire anterior or posterior *compartments* of the mesothorax, including wing and leg, on one side of the body (Fig. 6.10). The results show that blastoderm cells of *Drosophila* are not yet determined to form only dorsal or only ventral structures within a segment.

The restriction of labeled clones to a segment does *not* reveal, as discussed in the previous section, whether the founder cells were restricted in their potency to a single segment. To find out whether they were, Simcox and Sang (1983) transplanted groups of about six blastoderm cells between *Drosophila* embryos at the cellular blastoderm stage (Fig. 6.11). Both donors and recipients

were raised to adulthood. While the donors were genetically wild-type, the recipients carried several mutant alleles that gave the adults phenotypes such as yellow cuticles and forked bristles. Wild-type donor tissue could therefore be distinguished from mutant host tissue anywhere in the adult epidermis. The recipients were scanned for patches of donor tissue, while the donors were examined for defects in those areas that, according to known fate maps, might have been damaged by the removal of cells. Whenever matched pairs of donors and recipients survived, the donors were indeed lacking those structures that were formed by the transplants in the recipients. After heterotopic transplantation, the donor cells still expressed fates corresponding to their origin. For instance, when a foreleg primordium was transplanted to the abdominal region of a host at the blastoderm stage, the donor lacked the leg as an adult while the recipient had an extra leg showing the genetic labels of the donor. These results indicated that the potencies of blastoderm cells are restricted to one segment. Similar experiments at the preblastoderm stage, before cellularization, showed that transplanted nuclei could still alter their fates according to their new positions (Kauffman, 1980).

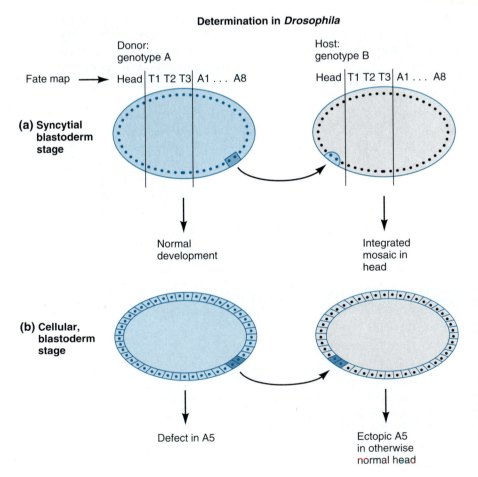

Determination in *Drosophila*

Donor: genotype A

Host: genotype B

Fate map → Head | T1 T2 T3 | A1 . . . A8

Head | T1 T2 T3 | A1 . . . A8

(a) Syncytial blastoderm stage

Normal development

Integrated mosaic in head

(b) Cellular, blastoderm stage

Defect in A5

Ectopic A5 in otherwise normal head

Figure 6.11 Transplantation as a test for cell determination in *Drosophila*. Genetic labels were used to distinguish structures in the adult formed by the donors' cells and those formed from the hosts' cells. Fate maps indicate the approximate positions of nuclei and cells that form head, thoracic segments (T1 to T3), and abdominal segments (A1 to A8). **(a)** At the *syncytial blastoderm stage*, nuclei and surrounding cytoplasm were grafted from a donor region fated to form A5 to the prospective head region of the host. The graft contributed to a normal head consisting of both donor and host cells (integrated mosaic), indicating that the transplanted nuclei had not been determined yet. **(b)** In a similar experiment, cells were transplanted at the *cellular blastoderm stage*. After metamorphosis, the host was defective in A5, while the donor had a head with ectopic A5 structures carrying the genetic markers of the donor. Thus, the potency of the donor cells was limited to the segment corresponding to their original fate.

Together, the foregoing experiments show that *Drosophila* blastoderm cells are determined for a single segment but not for dorsal or ventral structures within one segment. The restriction of potency to one segment occurs around the time of cellularization.

Prospective Neural Plate Cells of Amphibians Are Determined after Gastrulation

As mentioned earlier, the neural plate rudiment of an amphibian morula is *fated but not yet determined* to become neural plate. The same is still true at the early gastrula stage, as shown by heterotopic transplantations carried out by Spemann (1938). To understand this experiment, we need to look briefly at the processes of *gastrulation* and *neurulation* in amphibians, which will be described more fully in Chapters 10 and 11.

At the *blastula* stage, the embryo is a hollow sphere surrounding a fluid-filled cavity called the *blastocoel* (Fig 6.12). The vegetal half of the blastula consists of cells that are larger and contain more yolk than the cells in the animal half. Gastrulation begins at a depression called the **blastopore**, which is located below the egg equator near the area that was marked by the *gray crescent* in the zygote. Here the vegetal cells buckle into the blastula, forming a curved groove that later extends into a circle. When the blastopore has just formed, the embryo is said to be at the **early gastrula** stage. During gastrulation, the animal and equatorial cells flatten and expand and then move inside over the blastopore rim. When the gastrulation process is complete, the embryo is called a **late gastrula.** At this stage, the embryo consists of an outer layer, called *ectoderm,* and an inner layer that gives rise to *mesoderm* and *endoderm*. The endoderm cells surround the embryonic gut, or *archenteron.* The major derivatives of the ectoderm are the *epidermis* and the *neural plate*. The neural plate originates from the dorsal ectoderm between the blastopore and the animal pole. The neural plate first forms as an elevated area on the dorsal side of the embryo. Subsequently, the neural plate closes into a tube that is internalized and gives rise to brain and spinal cord.

To test the prospective neural plate region for its determined state, Spemann transplanted prospective neural plate from an *early* donor gastrula to a prospective epidermis region of an early host gastrula (Fig. 6.13a). The transplant developed in accord with its new surroundings and formed part of the host's epidermis. Similarly, prospective epidermis transplanted into the neural plate primordium contributed to the host's neural plate. In order to secure a reliable and lasting distinction between the transplant and the host, Spemann used two different newt species, the lightly pigmented *Triton cristatus* and the darkly pigmented

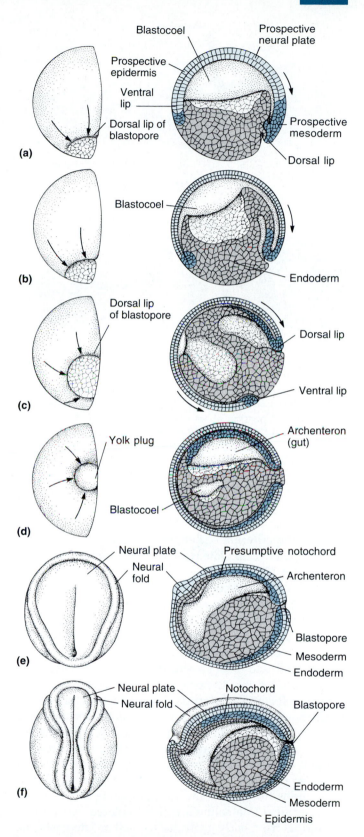

Figure 6.12 Gastrulation and neurulation in an amphibian embryo. **(a–d)** Gastrula stages. **(e, f)** Neurula stages. Left column shows cutaway left halves in surface view from a posterodorsal **(a–d)** or dorsal **(e, f)** angle. Right column shows corresponding right halves viewed from the left side.

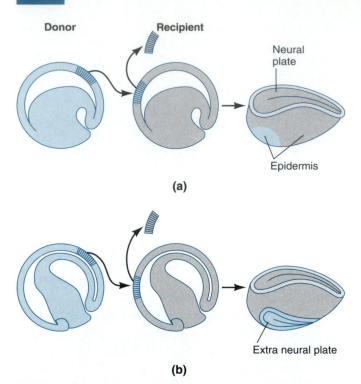

Donor · Recipient

Neural plate

Epidermis

(a)

Extra neural plate

(b)

Figure 6.13 Neural plate determination in the amphibian embryo. **(a)** The prospective neural plate, if transplanted to the ventral portion at an *early gastrula stage*, forms epidermis. **(b)** If the same experiment is carried out at a *late gastrula stage*, the transplant forms an additional neural plate. These results show that the prospective neural plate region is determined during gastrulation.

Triton taeniatus (Fig. 6.14). The results showed clearly that, by the criterion of heterotopic transplantation, the prospective neural plate was not determined at the early gastrula stage.

An entirely different result was obtained when the same experiment was carried out at the *late* gastrula stage. This time prospective neural plate developed into brain or spinal cord *wherever* it was placed in the embryo (Fig. 6.13b). Likewise, prospective epidermis had lost its ability to form structures of the nervous system. If transplanted into the prospective neural plate, it nevertheless formed epidermis. Spemann concluded that both the prospective epidermis and the prospective neural plate had been determined during gastrulation.

The same conclusions were reached by Holtfreter, a student of Spemann's, from a corresponding series of isolation experiments. Holtfreter (1938) excised different regions from newt and frog gastrulae and cultured them in vitro in a saline solution. Isolated prospective epidermis from early gastrulae as well as late gastrulae developed into epidermis and gland cells. The same structures were derived from prospective neural plate isolated *before* gastrulation. *After* gastrulation, isolated prospective neural plate formed neural structures, including cells normally found in brain or eye. These re-

sults confirmed the conclusion reached from Spemann's transplantation experiments that the prospective neural plate region is determined during gastrulation.

Mouse Embryonic Cells Are Determined at the Blastocyst Stage

Isolation and transplantation experiments indicate that mammalian blastomeres are totipotent up to the 8-cell stage. To follow the development of isolated blastomeres, Susan Kelly (1977) dissociated mouse embryos at the 4-cell or 8-cell stage by gently sucking them into and out of a micropipette. Because single mouse blastomeres do not have enough mass to develop into complete blastocysts, she combined isolated "donor" blastomeres with "carrier" blastomeres of a different genotype (Fig. 6.15). This method generates *chimeras,* that is, organisms composed of cells with two or more distinct genotypes. The chimeric mouse embryos were kept in vitro to form blastocysts before they were implanted into foster mothers and raised to midpregnancy or to term. Because of different genetic labels in the donor and the carrier, it was possible to decide which organs of the developing mice had been formed by the isolated donor blastomeres. In several cases three blastomeres obtained from the same 4-cell embryo contributed to live-born mice, and donor characters were found in their skin, germ cells, and other organs. Together, the results strongly suggest that *each* blastomere was totipotent up to the 4-cell stage and could contribute substantially to the embryo or trophoblast or both. Similar results were obtained with embryos containing a single 8-cell–stage blastomere.

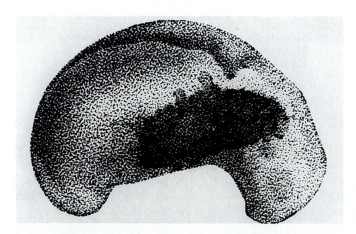

Figure 6.14 *Triton cristatus* embryo (lightly pigmented) with a graft from a *T. taeniatus* embryo (darkly pigmented). The graft, taken from the prospective neural plate of an early gastrula and grafted to prospective lateral epidermis of the host as shown in Figure 6.13a, contributed to the epidermis of the host.

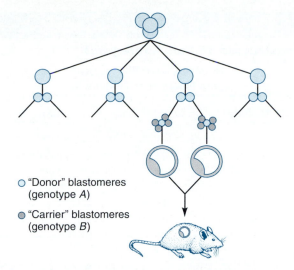

○ "Donor" blastomeres
(genotype *A*)

● "Carrier" blastomeres
(genotype *B*)

Figure 6.15 Test of isolated mouse blastomeres for totipotency. "Donor" embryos of genotype *A* were dissociated into single blastomeres at the 4-cell stage. Each blastomere cleaved again to reach the equivalent of the 8-cell stage. The pairs of 8-cell blastomeres were separated, and each was combined with four 8-cell–stage "carrier" blastomeres of genotype *B*. The composite embryos were cultured in vitro until the blastocyst stage, when they were implanted into the uterus of a foster mother. The developing offspring were of genotype *A*, genotype *B*, or composites of both genotypes.

At the 16-cell stage, some mouse blastomeres are still pluripotent, if not totipotent. At this stage, normal embryos consist of polarized cells with tight junctions on the outside and nonpolarized cells with gap junctions on the inside. The *fate* of the inside cells is to form the *inner cell mass*, while most of the outside cells contribute to the *trophoblast* (see Chapter 5). To investigate the *potencies* of inside and outside cells at the morula stage, Carol Ziomek and Martin Johnson (1982) combined heterotopic transplantation with clonal analysis (Fig. 6.16). They dissociated 16-cell embryos into single cells and classified the cells as outside or inside cells on the basis of morphological criteria. After staining with a fluorescent dye, single stained cells were aggregated with 15 age-matched, unstained host blastomeres to reconstitute the equivalent of a morula. This reconstitution was done in two steps. First, six blastomeres were incubated for 1 h to form the core of the aggregate; then 10 more cells were added to form the shell of the aggregate. The aggregates were kept in vitro for 24 h, by which time they had formed blastocysts.

The results of this experiment are summarized in Table 6.2. When a stained outside cell was placed back in its "native" shell position, it contributed to the trophoblast in 13 of 14 cases. Similar results were obtained in 16 of 18 cases when an outside cell was placed in an "alien" position in the core of the aggregate. Thus, most of the outside cells developed according to fate regard-

less of their position in the host. In the exceptional cases, outside cells gave rise to contiguous clones that extended from the trophoblast to the inner cell mass. When a stained inside cell was placed back in its native core position, it contributed to the inner cell mass in 16 of 19 cases; stained parts appeared in the trophoblast in the other 3 cases. However, when an inside cell was placed in an alien shell position, the results were quite varied.

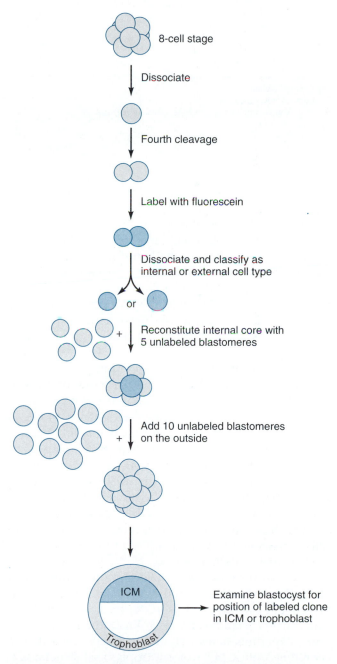

8-cell stage

Dissociate

Fourth cleavage

Label with fluorescein

Dissociate and classify as internal or external cell type

or

Reconstitute internal core with 5 unlabeled blastomeres

Add 10 unlabeled blastomeres on the outside

ICM

Trophoblast

Examine blastocyst for position of labeled clone in ICM or trophoblast

Figure 6.16 Experimental procedure used to test whether isolated 16-cell mouse blastomeres are determined to form inner cell mass or trophoblast. In this diagram, it is assumed that one of the labeled cells is placed inside the core. In a similar experiment, one of the labeled cells was placed in the outer shell.

TABLE 6.2

Fate of Isolated Labeled Mouse $\frac{1}{16}$ Blastomeres in Reconstructed 16-Cell Aggregates

Type of Aggregate and Position of Labeled Cell	Number of Blastocysts Analyzed	Number (%) of Blastocysts with Fluorescent Progeny in:		
		Tr* Only	ICM† Only	ICM + Tr
Labeled external $\frac{1}{16}$ cell				
Native, outside position	14	13 (93)	0	1 (7)
Alien, inside position	18	16 (89)	0	2 (11)
Labeled internal $\frac{1}{16}$ cell				
Native inside position	19	3 (17)	16 (84)	0
Alien, outside position	13	7 (54)	4 (31)	2 (15)

*Tr = trophoblast.
†ICM = inner cell mass.
Source: Ziomek and Johnson (1982). Used by permission.

Only in 4 of 13 cases did the donor cells develop according to fate, contributing to inner cell mass only. In the majority of cases, the transplanted inside cells contributed to trophoblast only or to both trophoblast and inner cell mass. These cells then developed in accord with their new positions in their hosts and not according to their fates in the donor embryos.

The experiments reviewed so far indicate that neither the outside nor the inside blastomeres of the mouse morula are strictly determined. Each type reveals a wider potency after heterotopic transplantation. Likewise, the observation of overlapping clones between trophoblast and inner cell mass indicates a genuine pluripotency of at least some blastomeres at the 16-cell stage. This pluripotency does not seem to be an artifact caused by heterotopic transplantation, because some of the stained cells gave rise to overlapping clones even if they were returned to their native positions. By comparison, the outside morula cells show a stronger bias toward developing in accord with their original fate than the inside morula cells. However, even this stronger bias can be overridden if the number of inside cells is too small to form a sufficient inner cell mass in the blastocyst (see Chapter 5).

The conclusions reached from the foregoing transplantation experiments and clonal analysis were confirmed and extended to the blastocyst stage by isolation experiments (Handyside, 1978; Rossant and Lis, 1979). These experiments focused on the ability of isolated *inner cell masses (ICMs)* to form trophoblast structures or even complete embryos. Together, the results of transplantation and isolation experiments with mouse embryos indicate that blastomeres are totipotent through the 8-cell stage and perhaps longer. Inside and outside cells from morula stages are biased but not fully determined to form ICM and trophoblast, respectively. Even

at the early blastocyst stage, entire ICMs are pluripotent, if not totipotent. At the late blastocyst stage, however, both ICM and trophoblast cells seem to be determined. Thus, mammalian cells show a very prolonged process of gradual determination.

Properties of the Determined State

In this section, we will explore some of the mechanisms by which cells are determined, and the connections between cell determination and other embryonic processes including pattern formation and cell division. Finally, we will discuss a set of experiments that illuminates the stability of the determined state by pushing it to its limits.

Determination Is a Stepwise Process of Instruction and Commitment

The operational criteria described in this chapter may create the impression that determination is essentially a negative event. It seems to amount to a loss of potency, or a commitment to follow one path of development at the exclusion of others. However, it is important to realize that selecting the path of fate from among all other possibilities often entails an element of instruction. For instance, the prospective neural plate region in the amphibian blastula and early gastrula will form epidermis upon isolation. Thus, forming epidermis is the *default pathway* of this region, that is, the pathway taken in the absence of further signals. However, in order to develop in accord with its fate, the prospective neural plate needs additional signals, which it receives in its normal location in the embryo.

There are two types of signals by which embryonic cells are determined to develop toward their fate. In most animals, certain components in the egg cytoplasm are distributed unevenly and partitioned into distinct blastomeres during cleavage. Such components are called localized cytoplasmic determinants; they cause blastomeres to be determined for certain cell lineages or major body regions. For example, the posterior pole plasm of many insect eggs contains localized germ cell determinants. Cells that contain this type of cytoplasm are determined to form primordial germ cells (see Chapter 8).

Another type of determining signal is received when different blastomeres interact with one another, creating new types of cells in the boundary regions. This type of determination by the close proximity of other cells is known as *embryonic induction* (see Chapters 9 and 11). During morphogenetic movements in gastrulation and organogenesis, cells acquire several new neighbors and impart inductive signals to one another. As an example, prospective neural plate cells in amphibian gastrulae are induced in part by mesodermal tissue that moves into the blastocoel and contacts the prospective neural plate from underneath (Fig. 6.13).

In addition, mutual activations and inhibitions among cells of the same tissue modulate their inductive and responsive capabilities. Thus, the final body pattern results from a *hierarchy of determinative events* in which each step is based on the previous steps.

As an example of a hierarchy of determinative events, let us again consider an amphibian embryo. As a result of localized cytoplasmic determinants, animal cells become different from vegetal cells during cleavage. Next, the animal blastomeres closest to their vegetal neighbors are induced to form an intervening *marginal zone*. The three blastula regions resulting from this induction already differ in their potencies as indicated by isolation experiments (Fig. 6.17). Isolated regions from the animal cap develop into spheres of epidermis. Isolates from the marginal zone form *notochord* (an embryonic skeletal element), muscle, and mesenchyme. Vegetal isolates develop into an undifferentiated, gut-like tissue. Upon gastrulation the animal cells will form the outer germ layer called *ectoderm*. Within the ectoderm, the next determinative step is the induction of the neural plate by adjacent mesoderm, leaving the remaining animal cells in a still general state of determination for epidermis. Within the epidermal area, further determinative steps will lead to distinctions between epidermis proper and epidermal derivatives such as glands and sensory epithelia for nose and ear. A similar series of determinative steps are followed in the neural plate and in the other two germ layers.

Figure 6.18 is a simplified diagram showing some of the major determinative steps during amphibian development. In this diagram, each branching point is thought to be a determinative step, when one set of cells receives an instructive signal while other cells do

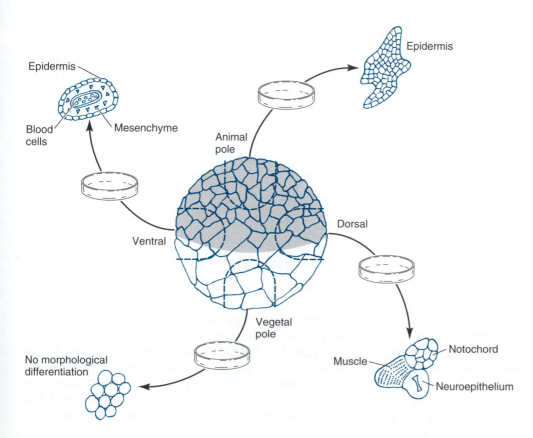

Figure 6.17 Potencies of isolated amphibian blastula regions in vitro.

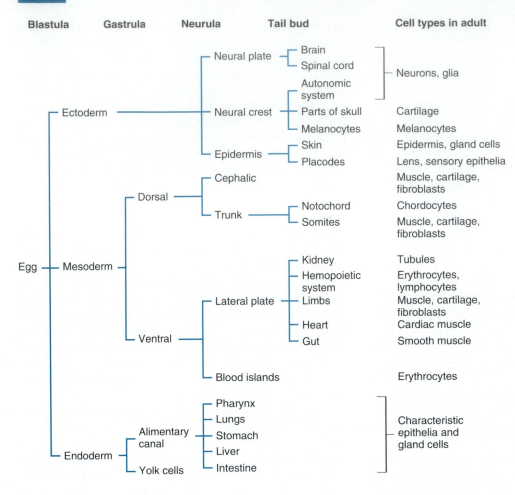

| Blastula | Gastrula | Neurula | Tail bud | Cell types in adult |

Figure 6.18 Series of determinative steps in amphibian development. The organism develops as a result of a hierarchy of determinative events, beginning with localized cytoplasmic determinants in the egg and continuing with inductive interactions. At the tail bud stage, the embryo shows the typical vertebrate body plan and has formed the major organ rudiments. Several further developmental decisions will in most cases be taken before the cells differentiate into the terminal cell types listed on the right-hand side. Note that some cell types, such as cartilage and gland cells, arise from more than one lineage.

not. An extended series of determinative events eventually generates differentiated cells such as neurons, pigment cells, cartilage, muscle fibers, and blood cells.

Cell Determination Occurs as Part of Embryonic Pattern Formation

In the larva and the adult, differentiated cells are arranged in *spatial patterns* that enable the proper function of each organ and the entire organism. The visible patterns of adult organisms—for instance, those on the wings of butterflies or in the plumage of birds—are often most elaborate and appealing to the human eye. We therefore tend to associate body patterns with differentiated cells. However, it is important to realize that organized body patterns unfold from the earliest stages of development and *precede* the appearance of differentiated cell types. A coarse distribution of cytoplasmic determinants in the egg sets up the initial pattern of animal and vegetal blastomeres, and, in some instances, the difference between dorsal and ventral cells. Subsequent inductive interactions generate patterns of tissue and organ *rudiments,* which in turn set the stage for further inductive interactions.

Cells are usually determined long before they reach their mature differentiated states. In other words, cell determination is part of the diversification process that precedes cell differentiation. It must be kept in mind, though, that cell determination is assessed by operational criteria, and that morphologically and biochemically diversified cells are not necessarily determined. For example, mammalian embryos at the morula stage consist of outside and inside cells. Of these two cell types, the outside cells can be distinguished by their larger size, the microvilli on their apical faces, and their ability to pump fluid from outside to inside. Yet the transplantation experiments described in this chapter show that outside cells can still acquire the fate of inside cells and vice versa.

The Determined State Is Stable and Is Passed On during Mitosis

The determined state is passed on from each cell to its descendants, at least under conditions of normal development. The mitotic progeny of cells determined to form cartilage, for example, are likewise determined to form cartilage. This stability is remarkable, and its underlying molecular mechanisms are still being investigated. In order to test the endurance of determination mechanisms, determined cells have been subjected to conditions of excessive proliferation. The most exten-

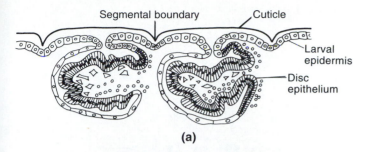

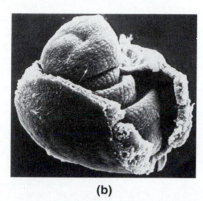

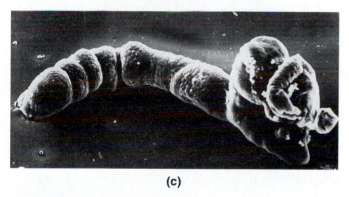

Figure 6.19 Eversion of imaginal discs during insect metamorphosis. **(a)** Drawing of a section of the wing imaginal discs of an ant. Note the continuity of the disc epithelium with the larval epidermis. **(b, c)** Scanning electron micrographs showing a leg disc before and after eversion.

inside of the larval epidermis but have no role in larval life except growth: an initial group of 20 to 50 cells increases to tens of thousands of cells. Under the influence of the hormones that control metamorphosis, the folded discs evert and form a cuticle with hairs, bristles, and other markings. For instance, a pair of wing discs forms the wings and the dorsal parts of the second thoracic segment to which the wings are attached (Fig. 6.20). Most imaginal discs are paired. The only unpaired disc is the genital imaginal disc, which gives rise to the external genital and anal structures of the adult.

To test the determined state of imaginal discs, Ernst Hadorn (1963) transplanted discs from donor larvae to the epidermis or body cavity of host larvae. When the host larva went through metamorphosis, its hormones acted on the implanted disc, and an extra adult structure formed in the metamorphosing fly. When this extra structure was removed from its host and inspected under the microscope, its morphology always corresponded to the fate of the transplant in the donor, independently of its location in the host. Thus, according to the criterion of heterotopic transplantation, each imaginal disc is determined to form the imaginal structure for which it is fated. In fact, by transplanting disc fragments, Hadorn found that each part of an imaginal disc was already determined to form a particular portion of the appropriate imaginal structure. Thus, certain regions of a male genital disc formed the penis, others the sperm pump, and still others the hindgut. Finally, Hadorn and his coworkers separated and reaggregated cells from imaginal discs with different genetic labels. They found that even individual cells were determined to develop according to their fate.

sive test of this kind was carried out on those cells in *Drosophila* larvae that later form the epidermis of the adult.

The life cycle of *Drosophila* as well as other highly evolved insects includes a dramatic *metamorphosis*, during which the larva first enters an immobile and transient stage, in which the organism is called a *pupa,* and then reaches the adult stage, in which it is known as an *imago* (see Chapter 28). During this type of metamorphosis, most larval organs are resorbed while adult structures are built from groups of progenitor cells in the larva. The entire epidermis of the adult head and thorax, including legs, wings, and other appendages, is formed from pockets of larval epithelia called *imaginal discs* (Fig. 6.19). The imaginal discs are attached to the

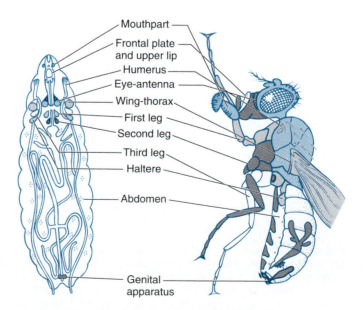

Figure 6.20 Imaginal discs of *Drosophila*. The diagram shows the position of the discs in the larva and their adult derivatives.

Mouthpart
Frontal plate and upper lip
Humerus
Eye-antenna
Wing-thorax
First leg
Second leg
Third leg
Haltere
Abdomen
Genital apparatus

In order to test how extra rounds of mitosis would affect the stability of the determined state, Hadorn developed the serial transplantation technique shown in Figure 6.21. Here, an imaginal disc is first implanted into the body cavity of an *adult* host. Under these conditions, the disc bypasses the hormones present in fully grown larvae and pupae that serve to trigger metamorphosis; the adult fly's body fluid merely serves as a convenient culture medium in which the disc tissue can proliferate. Since the culture period is limited by the life span of the host, the implant must be transferred to a new adult host every few weeks. During the transfers, the implants can be cut and the fragments reimplanted into two or more new adult hosts. At the same time, the determined state of a disc may be checked by transferring part of it back into a *larva*, where the test implant is challenged to metamorphose along with its larval host.

Using the serial transplantation technique, Hadorn (1968) studied the determined state of imaginal discs that were allowed to proliferate for several years. One extended series began with the implantation of a genital disc. During the first few transfer generations, the test implants formed only genital and anal structures, indicating that they had maintained their original determination (Fig. 6.22). Genital and anal structures were obtained for a total of 55 transfer generations. During this time, the imaginal disc tissue proliferated and passed on its determination for genital and anal structures to at least some of its descendants. However, from the eighth transfer generation on, head and leg structures appeared in addition to the genital and anal struc-

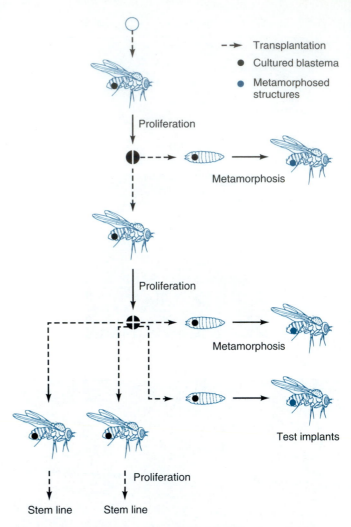

Figure 6.21 Serial transplantation technique for culturing imaginal discs in vivo.

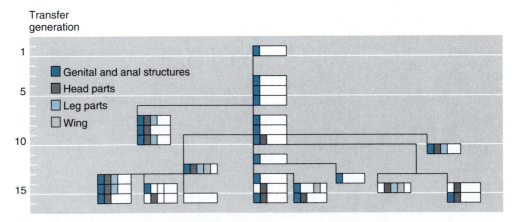

Figure 6.22 Transdetermination in imaginal discs of *Drosophila*. This chart summarizes the results of serial transplantations using the technique illustrated in Figure 6.21. The original transplant was taken from the genital disc, whose fate is to form sex organs and anal structures. The transplanted cells and their descendants formed only genital and anal structures during seven transfer generations (indicated by horizontal lines and the scale to the left). However, some cells transdetermined to head parts and leg structures by the eighth transfer, and soon thereafter to wing. The transdetermined states were passed on stably until another transdetermination occurred.

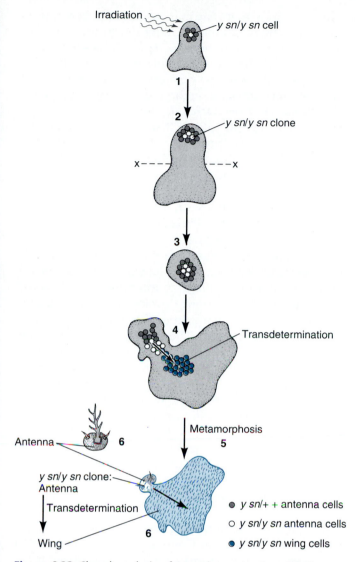

Figure 6.23 Clonal analysis of transdetermination. **(1)** X-ray treatment of larvae heterozygous for the recessive mutant alleles *yellow (y)* and *singed (sn)* and induction of a homozygous *(y sn/y sn)* clone in the eye-antenna disc. **(2)** Dissection of the antennal disc part from the mature larva. **(3)** Transfer to an adult host. **(4)** Proliferation of the disc fragment and growth of the *y sn/y sn* clone, in which transdetermination occurs (arrow, color). **(5)** Transplantation into a larval host with which the cultured disc tissue undergoes metamorphosis. **(6)** Examination of metamorphosed structures, which include *y sn/ + +* antennal parts and a clone of *y sn/y sn* cells. The clone consists of antennal cells and wing cells, indicating that the founder cell of the clone was determined for antenna before it transdetermined to wing.

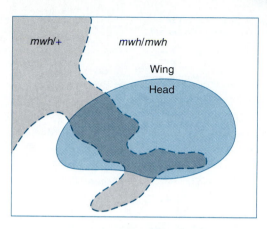

Figure 6.24 Simultaneous transdetermination in more than one cell. In a serial transplantation experiment as illustrated in Figure 6.21, transdetermination occurred from wing to head (colored area). The area of transdetermination overlapped with an X-ray–induced clone homozygous for *multiple wing hair (mwh/mwh)* in an *mwh/ +* background (gray area). Consequently, the transdetermination must have occurred simultaneously in at least one *mwh/mwh* cell and one *mwh/ +* cell.

state to which many implants had switched by the 100th transfer.

How can we be sure that transdetermination really represents a switch from one determined state to another, such as from antenna to wing? Could this be a false appearance generated by the *first* determination of previously unrecognized and *undetermined* reserve cells? To address this question, Gehring (1968) induced *somatic crossover* in antennal discs before they were subjected to serial transplantation. In some of the adult tissue showing transdetermination to wing, he observed single clones extending from antennal to wing structures (Fig. 6.23). This showed that the founder cell of the labeled clone was indeed determined for antenna, because it passed this determination on to daughter cells before transdetermination to wing took place.

To examine whether transdetermination occurs in single cells or in groups of cells, Gehring (1968) analyzed cases in which transdetermination had occurred in imaginal discs with genetically labeled clones. In some of these cases, the transdetermined area comprised labeled and unlabeled domains (Fig. 6.24). Therefore, the transdetermination event must have happened simultaneously in two or more cells inside and outside the labeled clone. Thus, transdetermination may occur in groups of cells rather than in individual cells.

tures. The imaginal disc tissue in which these switches had occurred continued to form head and leg structures during subsequent test implantations. It behaved as if it were now determined to form head or leg. Each switch to a new determined state was termed a ***transdetermination.*** Further transdeterminations, those to wing and other thoracic structures, occurred during transfers 13 and 19, respectively. The thorax represented a ground

Hadorn's experiments showed that the original determination of imaginal disc cells is stable and passed on faithfully as the cells divide. Only after excessive proliferation in culture did the determined state change in distinct steps, called *transdeterminations.*

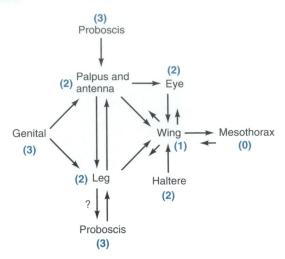

Figure 6.25 Flow diagram of transdetermination steps in *Drosophila*. The lengths of the arrows represent the relative frequencies of transdetermination. The numbers represent the minimum number of transdetermination steps needed before the disc can reach the determined state of mesothorax, which seems to be the ground state.

In contrast to genetic mutations, transdeterminations occur frequently and in small groups of cells simultaneously. Moreover, transdeterminations do not occur randomly but follow certain rules. Some transdeterminations can occur in one step, while others require two or more steps (Fig. 6.25). Some of these steps are reversible, while others are almost irreversible. While mutations are changes in the genetic information itself, transdeterminations are changes in the expression patterns of genetic information. Nevertheless, these expression patterns are also passed on during mitotic divisions.

The molecular mechanisms underlying determination and transdetermination are still being investigated. We will discuss in Chapter 15 how interactions between genes and certain regulatory proteins might generate *bistable control circuits* that can flip-flop between two states depending on the concentration of the regulatory proteins. It may be that the transdeterminations occur when such regulatory proteins are diluted as a result of many extra mitoses.

Regulation in Development

As mentioned in the introduction to this chapter, embryos have different ways of orchestrating the development of each blastomere so that a complete and harmonious organism results. The two strategies called *invariant cleavage* and *variable cleavage* can now be redefined in terms of determination as mosaic and regulative development. An embryo or embryonic region is said to undergo *mosaic development* when its potency

map is identical to its fate map, or, in other words, when all its cells are determined. Under the criterion of isolation, an embryo or embryonic region shows mosaic development when it can be cut into pieces and each piece develops in the same way as it would in normal development. In contrast, an embryo or embryonic region is said to undergo *regulative development* when the potencies of its cells are greater than their fates, or, in other words, when its cells are not yet determined. The ability of a cell or of an embryonic region to deviate from its fate and to respond to varying circumstances is called *regulation.*

These terms have been used to characterize, for example, nematode and ascidian embryos as "mosaic embryos" while amphibian and sea urchin embryos were classified as "regulative embryos." However, this dichotomy is misleading. As explained in this chapter, determination proceeds in steps, and most organisms will eventually undergo mosaic development. The question is, *How early* does the transition from regulative to mosaic development occur? The answer depends not only on the type of animal but also on the embryonic region under consideration. In a given embryo, development in some regions may be mosaic while in others it is still regulative. Most regions of an amphibian early gastrula, for example, are still capable of regulation, while the *dorsal blastopore lip,* located just above the blastopore, is already determined (Figs. 6.12 and 11.14).

A common type of regulative development is *twinning,* that is, the development of two complete individuals from an embryo that has spontaneously divided into two parts or has been split by an experimenter. Many embryos have this capability at the 2-cell stage, and some retain it until they consist of hundreds or thousands of cells. Human twins arising through the division of one embryo are known as identical or *monozygotic twins.* (Fraternal, or *dizygotic,* twins result from two separately fertilized eggs.) Monozygotic twins can originate through division of the embryo at several stages of development (Fig. 6.26). The earliest possible separation is at the 2-cell stage, resulting in two complete blastocysts that develop and implant separately. In this case, each twin will develop its own placenta and amniotic cavity. In most cases of human twinning, however, the inner cell mass of the embryo seems to divide during the early blastocyst stage. The resulting embryos have separate amniotic cavities but a common placenta. In rare cases the separation occurs at later embryonic stages, resulting in two partners with a common amniotic cavity and placenta. Incomplete separation at a late embryonic stage leads to the formation of conjoined twins that, depending on the nature of the bridge between them, can sometimes be separated surgically.

The converse of twinning is the experimental *fusion* of two or more embryos resulting in the development of one large embryo. Such fusion has been carried out

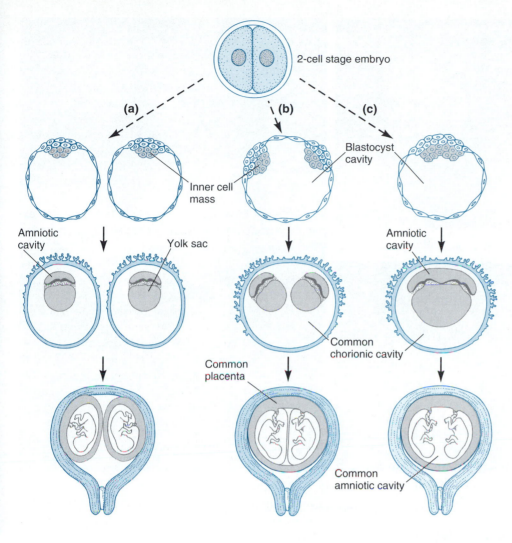

2-cell stage embryo

(a) (b) (c)

Blastocyst cavity

Inner cell mass

Amniotic cavity

Amniotic cavity

Yolk sac

Common placenta

Common chorionic cavity

Common amniotic cavity

Figure 6.26 Monozygotic twin development in humans. **(a)** After splitting at the 2-cell stage, each embryo develops its own placenta, chorionic cavity, and amniotic cavity. **(b)** After complete splitting of the inner cell mass, the two embryos have a common placenta and chorionic cavity, but separate amniotic cavities. **(c)** After incomplete or late splitting, the embryos share a common placenta, chorionic cavity, and amniotic cavity.

extensively with mouse embryos of different genetic constitution (Mintz, 1965). The embryos stick to each other readily and form an unusually large blastocyst, which then implants and develops into a *chimeric* individual of normal size (Fig. 6.27).

Chimeric mice were initially used to analyze the role of cell mixing and cell position in development, and to estimate the minimum number of cells needed for founding certain cell lineages. Later, fusion experiments were extended to combine embryonic and nonembryonic cells. For instance, investigators showed that teratocarcinoma cells injected into blastocysts contributed to apparently healthy mice, forming functional tissues including blood, liver, skin, and germ cells (Mintz and Illmensee, 1975; Brinster, 1976). These results show that when cancer cells are transplanted back to a normal embryonic environment, their malignant properties may be lost or suppressed. Another application of the chimeric mouse technique is the addition of genetically engineered mouse cells to the *inner cell mass* of a recipient embryo (see Chapter 14). If an added cell contributes to the germ line of the chimera, one can breed a new strain of mice that have the genotype of the genetically engineered cell.

Twinning and fusion experiments dramatically demonstrate regulative development. But the ability of embryonic cells in certain species to switch from generating an entire embryo to generating half of one, or vice versa, and the development of chimeric mouse embryos are not the only examples of this phenomenon. In embryos of many animal species, when cells are removed, their neighbors often produce additional descendants that take over the role of the lost cell. Conversely, the addition of extra cells is compensated by fewer mitoses of the neighbors, which thus give up some of their normal fates to the extra cells. Such adjustments allow regulative embryos to develop according to the *principle of stepwise approximation*, relying on the ability to compensate errors rather than on utmost accuracy.

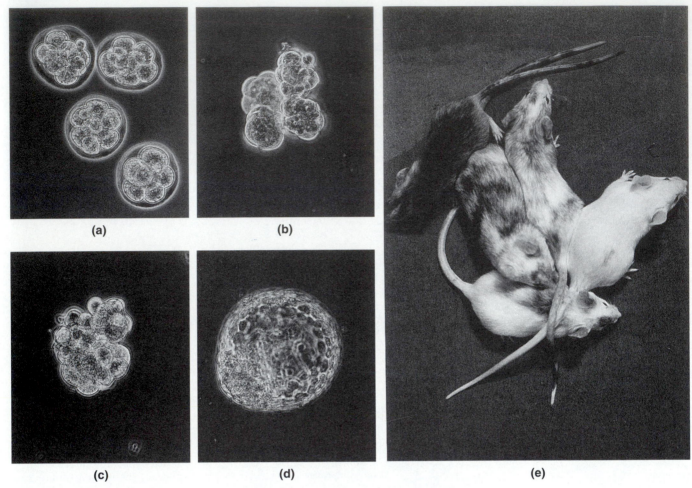

(a) (b)

(c) (d) (e)

Figure 6.27 Regulation in mammalian embryogenesis. **(a–d)** Photographs showing the aggregation of four mouse embryos to form a single, large blastocyst. The embryos are removed from the zona pellucida at the 8-cell stage and placed into contact. They form an aggregated blastocyst, which is implanted into the uterus of a foster mother and brought to term. **(e)** Mouse chimeras made by combining embryos from strains with different pigmentation.

SUMMARY

It is important to distinguish the fate of embryonic cells from their potency and determination. A cell's fate is what becomes of it in the course of normal development. A fate map is a diagram of an embryo indicating which structures each region will form at a later stage. Two common methods for establishing fate maps are time-lapse cinematography and the injection of dyes that do not impair development. A special form of fate mapping is clonal analysis. A cell clone consists of all surviving descendants of a single cell, the founder cell of the clone. A clone can be labeled by injecting a single cell with a dye that is not transferred to neighboring cells but is passed on to daughter cells. In organisms with suitable mutants, clones can also be labeled by X-ray–induced somatic crossover generating a homozygously mutant cell in a heterozygous background. In the adult epidermis of *Drosophila,* labeled clones do not cross certain boundary lines that define compartments. Each compartment is filled by several clones, which collectively form a polyclone. Polyclones have the same size and place in each individual, whereas individual clones do not.

Embryonic cells can often form a range of structures that is wider than their fate. The total of all structures that a cell can form, given the appropriate environment, is the potency of the cell. The potency of cells is tested by isolation in a culture medium and by transplantation to different positions in a host. The outcome of such experiments, however, depends very much on experimental parameters, including the way cells are isolated and the number of cells that are isolated or transplanted together.

Embryonic cells undergo a stepwise process, called determination, in which their potency becomes limited to their fate. To assess the determined state of a cell or of an embryonic region, its fate must be compared with its potency as revealed by isolation or heterotopic transplantation. The nondetermination of cells can be tested by clonal analysis, because each clone reveals a sample of those structures that its founder cell was still able to form.

The process of cell determination entails a loss of potency as well as the receipt of specific signals from localized cytoplasmic components or from other cells. The determination of each cell therefore depends on its position in the embryo. There are indications that determination occurs in groups of cells rather than in single cells. The determined state of cells is normally stable and is passed on to daughter cells during mitotic divisions. Under conditions of excessive proliferation, however, the determined state of cells may change—an event that presumably reflects alterations in control circuits of gene activity. Cells differ in the stage at which they become determined. Within a given organism, some cells become determined earlier than others. In certain groups of animals, such as roundworms and ascidians, determination tends to occur early. These animals are said to show mosaic development as opposed to regulative development, which is characterized by late determination as observed in sea urchins and vertebrates. One characteristic of regulative embryos is the capacity for twinning, that is, the formation of small but complete individuals from isolated blastomeres.

SUGGESTED READINGS

Davidson, E. H. 1990. How embryos work: A comparative view of diverse models of cell fate specification. *Development* **108:**365–389.

Gehring, W. 1968. The stability of the determined state in cultures of imaginal discs in *Drosophila.* In H. Ursprung, ed., *The Stability of the Differentiated State,* 136–154. New York: Springer-Verlag.

Slack, J. M. W. 1991. *From Egg to Embryo: Determinative Events in Early Development.* 2d ed. Cambridge: Cambridge University Press.

Spemann, H. 1938. *Embryonic Development and Induction.* New Haven: Yale University Press. Reprinted 1967. New York: Hafner Publishing Co.

GENOMIC EQUIVALENCE AND THE CYTOPLASMIC ENVIRONMENT

Figure 7.1 Carrot plant grown in an Erlenmeyer flask from a single differentiated root cell. Plant cells from which the rigid cell walls have been removed can be grown in culture. If treated with the appropriate nutrients, growth hormones, and light conditions, individual cells may give rise to complete plants. This ability demonstrates that differentiated plant cells retain a full complement of genetic information.

Embryonic cells acquire distinct features as they undergo *differentiation* to become muscle fibers, skin cells, nerve cells, and other kinds of cells. Even before cells become overtly different, they are usually *determined* to follow certain pathways of differentiation at the exclusion of others. What is the molecular basis of cell determination and cell differentiation? Cells become different through the synthesis of different proteins, which can ultimately be traced to differential gene activity.

Do all cells retain a full complement of genetic information and use it selectively, or is cell differentiation based on the selective loss and/or multiplication of certain genes? To answer this question, researchers have compared chromosomes as well as specific segments of nuclear DNA from different tissues. As a biological test for completeness of the genetic information present in differentiated cells, researchers have challenged whole cells or their nuclei to promote the development of complete new organisms (Fig. 7.1). Some of their experiments have met with great difficulties, and in the absence of clear-cut results, additional evidence has had to be obtained and carefully weighed in order to arrive at a reasonable conclusion. Most researchers in the life sciences today accept as a general rule that all cells of an organism have complete and equivalent sets of genetic information but use the information selectively. This raises the question, What determines which genes a given cell will express? Toward the end of this chapter, we will see that *gene expression* in a cell nucleus is controlled by its cytoplasmic environment.

Theories of Cell Differentiation

As cells undergo determination and differentiation, diverse sets of genes are activated to synthesize specific combinations of proteins. The mixture of protein products differs qualitatively or quantitatively from one cell type to another. Since the late nineteenth century three hypotheses have been put forward to account for these differences. All are supported to some extent by observations of genetic activity in certain animals. Only one, however, seems to apply over a broad spectrum of cell types in most organisms. This section describes the three hypotheses, and subsequent sections present some of the evidence on which the current thinking is based.

According to one hypothesis, first proposed by August Weismann (1834–1914) and Wilhelm Roux (1850–1924), cells become different by the *differential loss of genetic material*. Weismann thought that each cell would inherit only the particular fraction of the entire genetic material that it required to develop in accord with its fate (Fig. 7.2). For example, he proposed, the genetic information present in the arm would differ from that in the leg, and the information in the upper arm would differ from that in the lower arm, and so forth. This view seemed to be supported by observations on the behavior of chromosomes during cleavage in the embryos of a few animal species. For instance, in the nematode *Ascaris megalocephala,* early cleavage divisions are associated with the loss of chromosome frag-

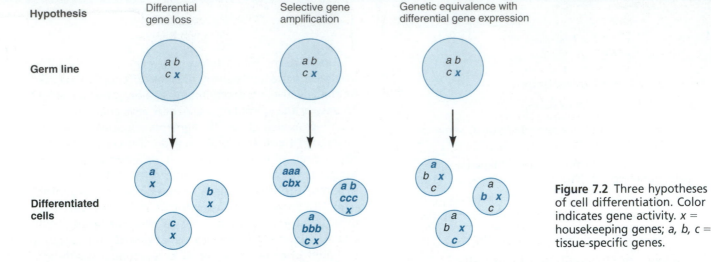

Figure 7.2 Three hypotheses of cell differentiation. Color indicates gene activity. x = housekeeping genes; a, b, c = tissue-specific genes.

ments, a phenomenon termed ***chromatin diminution*** (Fig. 7.3). Chromatin diminution occurs in all somatic cells of *Ascaris;* only the germ line cells retain the full complement of chromatin (Boveri, 1887).

Other studies have found that, rather than eliminating parts of their chromatin from somatic cell nuclei, a few types of cells undergo *selective gene amplification;* that is, they replicate specifically those segments of nu-

clear DNA that they will utilize most intensely (Fig. 7.2). Frog oocytes, for example, have thousands of extra nucleoli, which contain the genes for large ribosomal RNAs (see Fig. 3.13). The selective amplification of these genes makes it possible to supply the oocyte with enough ribosomes corresponding to its huge mass of cytoplasm.

A third theory of cell differentiation is based on the

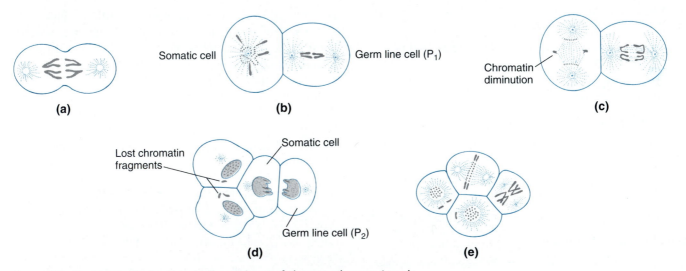

Figure 7.3 Chromatin diminution in the embryo of the roundworm *Ascaris megalocephala.* **(a)** The first cleavage is normal, dividing the zygote into a somatic cell and a germ line cell designated P$_1$ (for propagation). **(b, c)** During the second mitosis, the somatic cell shows chromatin diminution while the P$_1$ cell undergoes normal mitosis. In the somatic cell, the large composite chromosome breaks up into fragments. Only the inner fragments are connected to mitotic spindle fibers; the outer fragments remain outside the spindle. These outer fragments are not integrated into the daughter nuclei during telophase and are eventually digested in the cytoplasm. **(d, e)** During the second cytokinesis, P$_1$ divides into another somatic cell and a new germ line cell designated P$_2$. After a rearrangement of the new somatic cell and P$_2$, the third mitosis begins. Now the newly formed somatic cell undergoes chromatin diminution while P$_2$ divides normally.

theory of *genomic equivalence* and the *principle of differential gene expression*. Genetic equivalence means that all cells in an organism have complete and equivalent sets of genetic information, without losses, alterations, or selective amplifications. Differential gene activity means that cells utilize different subsets of their entire genetic information (Fig. 7.2). According to these concepts, cells use their genetic information the way different people read the same newspaper. Most readers scan the front page news and the weather forecast. In addition, some read the sections on national politics and sports, while others prefer local news and the comics. Similarly, most life scientists today assume that most cells in most organisms have the same genetic information but use it differently. As we will see, the evidence for genetic equivalence in plants is direct and positive. For animals, the data are less compelling, but they have allowed scientists to build a good case.

Observations on Chromosomes

The events that take place during the cell cycle seem to be designed to ensure that chromosomes are faithfully replicated and delivered in complete sets to both daughter cells during cell division (see Chapter 2). Consequently, there should be no loss or gain of chromosomes, and all cells in an organism should have the same complement of genetic information. This expectation is generally met when chromosomes from different tissues of the same organism are viewed under the microscope.

Different Cells from the Same Individual Show the Same Sets of Chromosomes

Cells prepared from different organs of the same individual usually show the same numbers of chromosomes, with the same sizes, shapes, and staining patterns (Tijo and Puck, 1958). In some cell types, however, the number of *entire chromosome sets* is reduced or increased. While most eukaryotic cells are *diploid* (have two sets of chromosomes), gametes are *haploid* (having one chromosome set) and some specialized somatic cells are *polyploid* (having more than two chromosome sets). Cells become polyploid by replicating their chromosomes and separating the chromatids without undergoing mitosis or cytokinesis. In this way, cells may vastly increase their number of chromosomes. Polyploidy is generally found in large, postmitotic cells that synthesize a few proteins in large quantities. However, since all chromosomes are multiplied by the same factor, there is no *selective* amplification or loss of genes in polyploid cells.

Most chromosomes are small, and inspecting them under the microscope would reveal losses or duplications only of large segments that contain hundreds of genes. Better resolution is obtained with chromosomes of a special kind in which the loss or duplication of individual genes can be detected visually. This type of chromosome will be discussed next.

Polytene Chromosomes of Insects Show the Same Banding Pattern in Different Tissues

A special form of polyploidy is polyteny. *Polytene chromosomes* originate through several rounds of chromosome replication without subsequent mitosis. The chromatids stick together and remain in register so that a polytene chromosome looks much like a cable composed of many wires (Fig. 7.4). In addition, the pairs of homologous chromosomes align side by side within the same polytene chromosome. Thus, in place of each pair of chromosomes in a normal cell, polytene cells have one polytene chromosome consisting of hundreds or even thousands of chromatids. Polytene chromosomes occur in trophoblast cells of mammalian embryos as well as in specialized insect cells. The best-known examples are found in the salivary gland cells of *Drosophila* larvae. In these cells, each chromosome has gone through 10 rounds of replication, resulting in 1024 ($= 2^{10}$) chromatids. Since, further, the paternal and maternal homologues are paired within a polytene chromosome, a salivary gland chromosome consists of 2048 chromatids. Because chromosomes can convert from the polytene to the normal polyploid state and vice versa, the structure of a polytene chromosome seems to be similar to that of a normal chromosome (Ribbert, 1979).

Polytene chromosomes are easily seen under a light microscope. If stained for DNA, they show a distinct *banding pattern*, in which dense bands alternate with weakly stained interbands (Fig. 7.4). The bands are regions of very dense DNA coiling; their higher concentrations of DNA account for their more intense staining. In salivary gland chromosomes of the genus *Drosophila*, each species has a distinct banding pattern, whereas different individuals of a given species have the same banding pattern. In *Drosophila melanogaster*, detailed drawings of salivary gland chromosomes have been used to map mutations that are associated with small deletions, duplications, or inversions of chromosome sections. The overall number of bands in these chromosomes is about 5000, corresponding approximately to the number of genes identified by genetic techniques (Judd and Young, 1973). However, certain bands contain more than one gene and some interbands are transcribed as well.

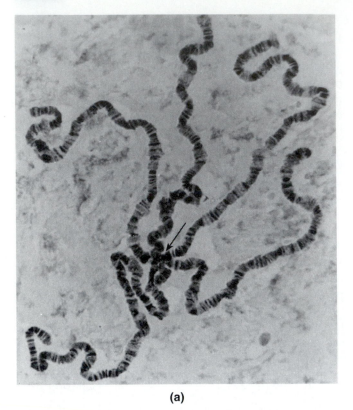

(a)

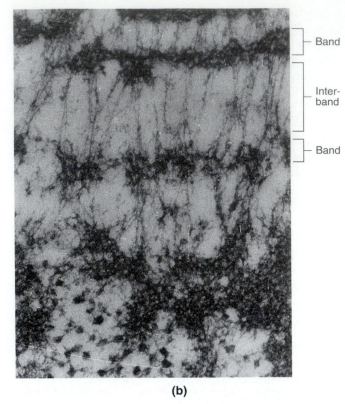

Band

Inter-
band

Band

(b)

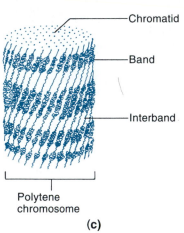

Chromatid

Band

Interband

Polytene
chromosome

(c)

Figure 7.4 Polytene chromosomes from a *Drosophila* salivary gland cell. **(a)** Light micrograph of a complete chromosome set. The chromosomes were squashed between a microscope slide and a cover slip for better viewing. The centromeres of the chromosomes aggregate and form one chromocenter (arrow). The five long arms extending from the chromocenter are the X chromosome and the left and right arms of chromosomes 2 and 3. The short arm is chromosome 4. Each polytene chromosome represents a pair of homologous chromosomes, each of which has been replicated 10 times to yield $2^{10} = 1024$ chromatids. Because the polytene chromosomes are in interphase, they are less coiled and therefore much longer than normal metaphase chromosomes. Each chromosome shows bands of high DNA concentration between interbands of lower DNA concentration. **(b)** Transmission electron micrograph of a thin longitudinal section of a polytene chromosome, showing bands of highly coiled chromatin alternating with straight chromatin strands in the interbands. **(c)** Schematic diagram showing how a polytene chromosome is composed of many chromatids.

Since the loss or duplication of single bands is visible in polytene chromosomes, these chromosomes, as noted previously, allow researchers to study the problem of genetic equivalence at much higher resolution. Wolfgang Beermann (1952a, b) investigated the banding patterns of polytene chromosomes from different larval tissues of the midge *Chironomus tentans* (Fig. 7.5). He found that these chromosomes looked identical in different nuclei from the same tissue of a given individual. Chromosomes from different tissues or developmental stages differed in diameter, reflecting varying numbers of DNA replication rounds. However, the *banding pattern* of chromosomes from different tissues looked remarkably similar. Thus, whatever the molecular mechanism underlying the banding pattern may be, it obviously produces very similar results in different tis-

sues, suggesting that the chromosomal DNA sequences are also very similar or even identical between tissues.

Although the banding pattern of polytene chromosomes is generally constant for a given species, these chromosomes show variable patterns of local decondensations called *puffs*. Puffs appear and disappear in stage-dependent and tissue-specific patterns (Fig. 7.5; see also Fig. 28.15). Molecular studies indicate that the puffed areas of polytene chromosomes are areas of strong RNA accumulation, probably caused by intense RNA synthesis (see Chapter 15). The variability of the puffing pattern therefore reflects stage-dependent and tissue-specific gene activity. Thus, inspection of polytene chromosomes confirms not only the theory of genetic equivalence but also the principle of differential gene expression.

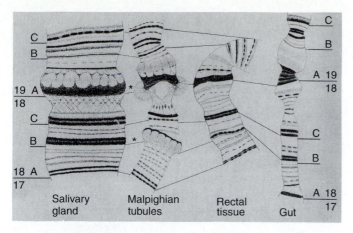

Figure 7.5 Drawings of polytene chromosome segments from different organs of a midge, *Chironomus tentans*. The degree of polyteny, indicated by the thickness of the chromosomes, varies among tissues. The puffed regions (asterisks) are also variable. However, the banding pattern of the chromosomes is remarkably constant among the tissues.

Chromosome Elimination Is Associated with the Dichotomy between Germ Line and Somatic Cells

Although chromosomes generally occur in the same number and morphology across all tissues of an individual, there are some cases in which germ line cells and somatic cells of the same organism differ in genetic constitution (Beams and Kessel, 1974). One well-known example is the chromatin diminution in the somatic cells of *Ascaris megalocephala* mentioned earlier. Several insect species undergo a similar process called *chromosome elimination*, in which whole chromosomes are eliminated rather than chromosomal fragments (Fig. 7.6). It occurs spontaneously during certain mitotic divisions, except in primordial germ cells. The protection

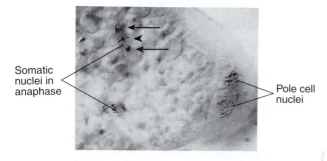

Figure 7.6 Light micrograph showing chromosome elimination in an insect embryo, *Smittia* sp. When somatic nuclei undergo mitosis in the yolk cytoplasm, some chromosomes (arrows) move normally to the spindle poles during anaphase. However, other chromosomes (arrowhead) are left behind in the metaphase plate, where they later disintegrate. Only the nuclei of pole cells (primordial germ cells) are protected from chromosome elimination and retain a full complement of chromosomes.

of the primordial germ cells from chromosome elimination depends on a cytoplasmic component, presumably RNA, that is sensitive to ultraviolet light and localized near the posterior pole of the egg (Von Brunn and Kalthoff, 1983).

Chromosome elimination is relatively rare. The chromosomes that are retained only in germ cells contain genes required for gametogenesis and are obviously dispensable for all somatic functions. It also appears that the same chromosomes are eliminated from all somatic cells. Thus, chromosome elimination is correlated with the dichotomy between germ line and somatic cells; differences in number or identity of chromosomes have not been found among somatic cells.

Molecular Data on Genomic Equivalence

With the techniques of molecular biology, it is possible to clone any segment of DNA and to quantify its abundance in any type of cell or tissue (see Chapter 14). These techniques can reveal deletions or additions of DNA segments that are too small to detect by microscopic inspection of polytene chromosomes. Applications of these techniques have confirmed the rule of genetic equivalence, at least for most cell types. Normally, genes that are abundantly expressed in certain cell types are not amplified. For instance, there are no extra copies of globin genes in red blood cells and no extra copies of ovalbumin genes in chicken oviduct cells (P. R. Harrison et al., 1974; Sullivan et al., 1973). Likewise, randomly chosen *Drosophila* genes are present in different tissues in the same copy number, and in DNA segments of identical length (M. Levine et al., 1981).

Notable exceptions include the genes encoding *ribosomal RNA*, which are selectively amplified in the oocytes of frogs and other animals, as described earlier. Also, the *Drosophila* genes encoding the proteins from which the egg envelope or *chorion* is made are multiplied by normal polyploidy and by selective amplification (Spradling and Mahowald, 1980; Spradling, 1981). The selective amplification occurs when the genomic DNA containing the chorion genes undergoes several rounds of extra replication. The selective amplification of chorion genes is limited to the ovarian follicle cells, which synthesize chorion proteins, and takes place immediately before chorion synthesis. Selective amplification is also common among genes encoding drug-detoxifying enzymes. This has been seen in cultured cells exposed to various drugs and in insect populations that have become resistant to pesticides (Mouches et al., 1990). Apparently, eukaryotic cells have the potential to amplify specific genes. However, such amplifications seem to be maintained only in the presence of strong selection pressure.

A different kind of deviation from the rule of genetic equivalence is observed in maturing B lymphocytes, the blood cells that produce antibodies against foreign molecules and cells. Mammals can produce many more different antibody molecules than they have genes. Nobumichi Hozumi and Susumu Tonegawa (1976), who received the 1987 Nobel Prize in physiology or medicine, discovered that each lymphocyte clone undergoes a different combination of DNA excisions and rearrangements, thus generating the enormous variety of antibody-encoding genes. This process entails irreversible losses and alterations of genomic DNA. However, this way of generating specialized cells appears to be an exception rather than the rule.

Taken together, the available data indicate that selective DNA amplifications, losses, and rearrangements are rare occurrences. However, these data do not exclude the possibility of permanent changes in chromosomal DNA or its associated proteins that would leave genes present but nonfunctional. This issue is discussed in the following section.

Totipotency of Differentiated Plant Cells

A very sensitive way of testing somatic cells for completeness and intactness of their genetic information is to challenge them to regenerate a whole new organism. If an isolated cell could accomplish this in an experiment, then it would be *totipotent*, and it would follow that no essential parts of its genetic information were missing.

Indeed, botanists are able to grow whole new plants from several types of differentiated plant cells. Related techniques were developed in experiments originally designed to study the metabolism of plant cells stimulated to rapid growth (Steward, 1970). Small pieces of carrot root phloem (vascular tissue) were cultured in rotating flasks in a medium containing coconut milk, which normally serves to nourish the developing coconut plant. Under these conditions the carrot tissue grew in unorganized masses called **calluses.** Studying cell behavior in these cultures, Steward and coworkers observed other groups of cells called **embryoids,** which resembled normal stages in the embryonic development of carrots. Presumably, the embryoids had arisen from carrot cells that had rubbed off the calluses. Under suitable culture conditions the embryoids grew into complete carrot plants (Fig. 7.1).

In a similar experiment, Vimla Vasil and A. C. Hildebrandt (1965) generated whole tobacco plants from stem pith cells. These experimenters isolated *single* cells sloughed off from calluses and kept them in microcultures. The cells grew into new calluses, which were placed on a semisolid culture medium with plant growth factors. The calluses formed plantlets with roots, shoots, and leaves, which, after transfer to soil, matured into flowering plants. This spectacular result, obtained under well-controlled conditions, shows that single cells from tobacco stem pith retain all the genetic information to build a whole new plant. In addition, the result shows that such cells can revert to rapid growth and to sending and receiving the signals that are necessary for forming a complete and well-proportioned organism.

Currently, the regeneration of whole plants from single cells in the laboratory often starts with the preparation of **protoplasts** (Shepard, 1982). These are naked cells, from which the cell walls have been removed enzymatically (Fig. 7.7). Protoplasts, which grow and divide in culture, have several properties that make them useful in research aimed at crop improvement. First, they can be used to grow a virtually unlimited number of genetically identical plants. Second, protoplasts from different plant strains or even different species can be fused to create hybrid protoplasts. Plants grown from such hybrids combine traits of both parent plants. This method of making hybrids is faster than traditional genetic crossing, and it may be the only way of obtaining crosses between unrelated plants. Third, protoplasts can be genetically manipulated with cloned DNA that confers traits such as resistance to cold or pesticides. After these genetic manipulations, it is often necessary to screen thousands of hybrids or transformants before a favorable variety is found. Many of these screens can be done more quickly and economically with protoplasts or small calluses in the laboratory than with plants in the field.

In summary, many regeneration experiments have clearly shown that differentiated plant cells are still totipotent. This result does not exclude the possibilities of selective gene amplification or DNA rearrangements, but it proves that all essential genes are still present and functional.

Totipotency of Nuclei from Embryonic Animal Cells

Attempts to grow complete *animals* from isolated animal cells have met with very limited success. In many species, including sea urchins, frogs, and mammals, it is possible to isolate blastomeres at the 2-cell or 4-cell stage and raise them to normally proportioned larvae or even fertile adults. However, blastomeres isolated from later embryonic stages do not normally have this ability. Somehow, animal cells seem to lose their totipotency during cleavage. One possible explanation, proposed by Weismann, is that animal cells lose part of their genetic information in the course of development. Alternatively, it may be that animal cells preserve a full complement of genetic information in their nuclei but

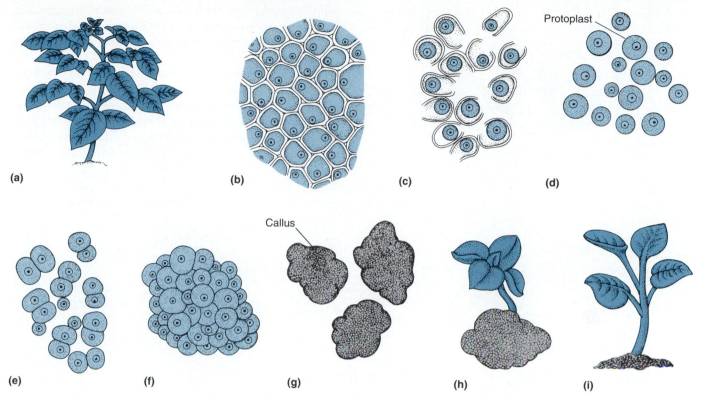

Figure 7.7 Regeneration of a complete potato plant from a single leaf cell. **(a–c)** A leaf from a young potato plant is placed in an enzyme solution to remove the cell walls. **(d, e)** The isolated protoplasts are transferred to a culture medium, in which they divide and grow. **(f, g)** Each protoplast gives rise to a clump of undifferentiated cells, called a callus. **(h)** Callus cells differentiate to form a primordial shoot. **(i)** The shoot forms a plantlet that can be transferred into soil.

the properties of animal cytoplasm are too restrictive to allow proper expression of all the genetic information that is still present. To test the latter possibility, experimenters have sought to bring *nuclei* from advanced cells back into the cytoplasmic environment of an egg. If a transplanted nucleus promoted the development of a complete organism, then it must have retained a full complement of genetic information. This would rule out Weismann's hypothesis that somatic cells retain only those genes that are required for their normal functions.

Newt Blastomeres Develop Normally after Delayed Nucleation

Spemann, at the University of Freiburg in Germany, where Weismann had worked before him, sought to test Weismann's hypothesis by a simple but ingenious experiment. With a baby hair tied into a loop, he ligated a newt zygote shortly after fertilization. The ligature was tight enough to confine the zygote nucleus to one egg half but sufficiently loose to maintain a cytoplasmic bridge between the nucleate and the anucleate egg halves (Fig. 7.8). While the nucleate half cleaved at a normal pace, the anucleate half remained undivided

until eventually a nucleus slipped through the ligature. At this point, Spemann pulled the hair loop tight to completely separate the two embryonic halves. The newly nucleated half now contained a nucleus that was already a product of several rounds of mitosis. According to Weismann's hypothesis, this nucleus should have lost some portion of its genetic information. However, both halves developed into complete tadpoles, provided that the ligature had been in the median plane of the egg. The only difference between the twin embryos was that one lagged behind in its development because of the delayed nucleation. The result showed clearly that newt nuclei are still totipotent after several rounds of mitosis. Spemann (1938) suggested that nuclei from more advanced cells should be tested in similar ways, but he did not have the technical means to do so at the time.

Nuclei from Embryonic Cells Are Still Totipotent

A straightforward way of testing the nucleus of any cell for totipotency is to transplant it into an egg of the same species. The recipient egg's own nucleus must be removed so that any subsequent nuclear activity can be

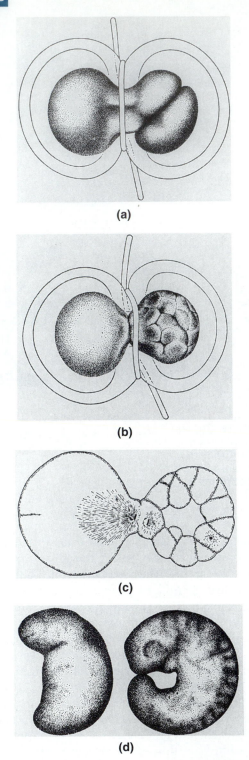

(a)

(b)

(c)

(d)

ascribed with certainty to the transplanted nucleus. The recipient egg must also be artificially activated in the absence of the activation trigger normally provided by the sperm.

A technique that meets these requirements was developed by Robert Briggs and Thomas J. King (1952). These investigators chose to work on the leopard frog, *Rana pipiens*. The recipient eggs were activated by pricking with a glass needle, and the egg nucleus was pushed out through the puncture (Fig. 7.9). Nuclei were taken

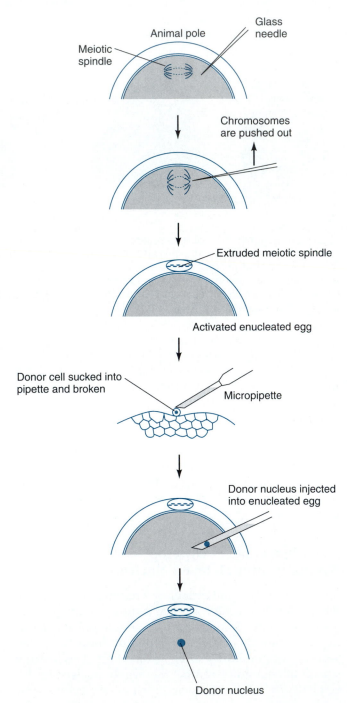

Figure 7.8 Delayed nucleation of a newt blastomere. **(a)** A fertilized egg was ligated, with the nucleus retained in one half. The ligature was incomplete, so that the halves were connected by a cytoplasmic bridge. Because the bridge was initially too narrow for a nucleus to pass through, cleavage was restricted to the nucleated half. **(b)** Several cleavages later, a nucleus slipped through the ligature. Now the other half egg was nucleated by a descendant of the zygote nucleus that had resulted from several mitoses. **(c)** Histological section of a stage similar to part b. **(d)** The ligature was closed completely, so that the two egg halves were fully separated. Each developed into a normally proportioned embryo, although the late-nucleated half lagged behind.

Figure 7.9 Technique used by Briggs and King (1952) for transplanting nuclei from blastula cells into enucleated eggs of the leopard frog, *Rana pipiens*. The relative size of the meiotic spindle is exaggerated.

from isolated embryonic donor cells. These cells were ruptured by being sucked into a glass pipette with a bore smaller than the cell but large enough to accommodate the nucleus. The aspirated nuclei and surrounding cytoplasm were then injected into the enucleated host eggs.

The first experiments were carried out with nuclei from donor cells at the blastula stage, and the results were very encouraging: about 60% of all transferred nuclei promoted the development of their host eggs into swimming tadpoles. Some of the tadpoles even grew into fertile adults, proving that nuclei from frog blastomeres at the blastula stage were truly totipotent. Moreover, the technique employed in this experiment was so versatile that cells from later stages and from any tissue could now be used as donors of nuclei. Some further experiments employing this technique will be described later in this chapter.

Nuclei Can Be Transferred by Cell Fusion

The nuclear transplantation technique pioneered by Briggs and King has been applied to several animal species. The application of this technique to mammals is of practical interest, because if successful it would allow the cloning of domestic animals. However, microinjection of nuclei into mammalian eggs has proven difficult. A glass pipette large enough to accommodate a cell nucleus causes much trauma when inserted into a mammalian egg, the diameter of which is only one-tenth that of a frog egg. To avoid this problem, Willadsen (1986) fused enucleated sheep eggs with whole blastomeres from 8-cell embryos. More than 30% of the fused cells developed into blastocysts, and three of four blastocysts transferred into foster mothers were born as full-term lambs.

Nuclear transplantations in *Drosophila* have yielded results similar to those in frogs. Transplanted nuclei from donors at the early gastrula stage still promoted the development of fertile adults (Illmensee, 1973). Taken together, the transplantation experiments described so far indicate that the cell nuclei of animal embryos up to the blastula or the early gastrula stage remain totipotent.

Pluripotency of Nuclei from Differentiated Animal Cells

The totipotency of nuclei from mature or differentiated animal cells has been more difficult to ascertain than expected. This section describes the modest results obtained as well as the difficulties encountered.

Nuclei from Older Donor Cells Show a Decreasing Ability to Promote the Development of a New Organism

The nuclear transplantation experiments that worked so well with blastula cells were extended to older cells. However, nuclei from *Rana* embryos beyond the gastrulation stage showed an ever-decreasing ability to promote normal development in host eggs (Fig. 7.10). Nuclei from donors at the tail bud stage were no longer capable of supporting the development of new tadpoles (McKinnell, 1978). Similar results were obtained with other amphibian species (Briggs, 1977; DiBerardino et al., 1984).

The results from these experiments showed that nuclei from mature animal cells lose their ability to promote the development of an entire organism. However,

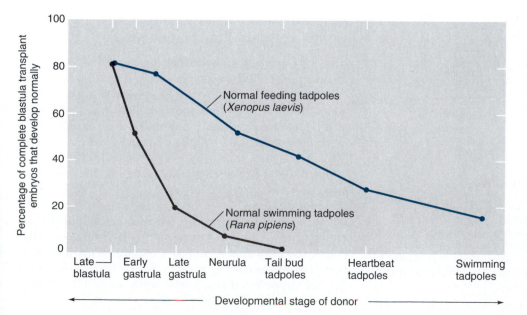

Figure 7.10 Success rate of nuclear transplantation experiments. The abscissa shows the developmental stage of the donors of the nuclei. Among the transplants, only those that formed a complete blastula were counted. The percentage of complete blastulae developing into normal tadpoles was plotted on the ordinate. The success rate decreased steeply for *Rana pipiens*, in which nuclei from donors older than the tail bud stage did not support normal development. With *Xenopus laevis*, there was still a decline with age, but a major fraction of nuclei taken from swimming tadpole cells supported the development of complete feeding tadpoles.

it is uncertain *which* nuclear capacity diminishes with age. One possibility is that parts of the genetic information have been lost, as proposed by Weismann, or rendered unusable. If this were the *only* reason for the diminishing potency of animal nuclei, then the nuclei of germ cells, which are known to have a full complement of genetic information, should perform much better in transplantation tests than somatic cell nuclei. However, transplanted nuclei from spermatogonia gave the same low yields of viable embryos (DiBerardino and Hoffner, 1971). It must be concluded that nuclei from maturing animal cells undergo *another* change that causes, or at least contributes to, their diminished function after transplantation into an egg.

Nuclei from Mature Cells Are Unprepared for the Fast Mitotic Cycles of Early Embryos

To promote the development of a new organism after transplantation into an egg, a nucleus must participate in the rapid mitotic cycles that characterize embryonic cleavage. Failure to do so might result in incomplete DNA replication and abnormal mitoses. It is quite possible, then, that transferred nuclei lose parts of their genomic DNA when they are rushed into a fast series of mitoses for which they are no longer prepared.

This interpretation is supported, although not proven, by the following lines of evidence (Gurdon, 1968; DiBerardino et al., 1984). Both the S phase and the entire cell cycle are very brief during cleavage (see Chapter 5). In normally fertilized frog eggs, the S phase of the mitotic cycle lasts only a few minutes (Fig. 7.11).

In contrast, when nuclei from mature cells are transplanted to eggs, they replicate their DNA much more slowly. Nevertheless, the cytoplasmic pacemaker that controls the embryonic cleavage cycle proceeds at its normal rate. Unlike mature cells, embryonic cells have no control mechanism that would inhibit mitosis as long as DNA replication is incomplete. And indeed, nuclear-transplant embryos frequently show chromosome breakages, which are apparently caused by precocious mitoses. Thus, abnormal development of these embryos most likely does result from the loss of genetic information, but this loss is an experimental artifact and not an original attribute of nuclei in mature cells. Presumably, the genomic DNA is associated with different molecules in mature nuclei and in early embryos, and the different associations either enhance or inhibit DNA replication.

The need of transplanted nuclei to adjust to rapid cell division is also indicated by the following observations. Nuclei from erythroblasts, which still divide, were found to promote the development of nuclear-transplant embryos to tadpoles while nuclei from erythrocytes, which no longer divide, supported only the development of gastrulae (Brun, 1978). But in another group of experiments, frog erythrocyte nuclei that were preincubated in oocyte cytoplasm before transfer into enucleated eggs promoted the development of swimming tadpoles, while control transplants without oocyte preincubation again yielded gastrulae at best (DiBerardino et al., 1984). Also, the success rate of nuclear transplantations is improved markedly by serial transfers. This strategy salvages poorly developing embryos by using their most normal-looking cells as donors for nuclei to be transplanted into yet a new set of enucleated eggs. Apparently, this procedure selects for nuclei that have survived the first mitotic cycles intact and are then better adjusted to proceed normally through additional cycles.

At Least Some Differentiated Cells Contain Highly Pluripotent Nuclei

A series of nuclear transplantation experiments similar to those described so far was started by John Gurdon (1962). He used a different frog species, *Xenopus laevis*, and a different technique from the one favored by Briggs and his coworkers. Gurdon employed ultraviolet light instead of surgery to inactivate the host egg nuclei (Fig. 7.12). To be sure that any developing individuals were supported by the transplanted nucleus and not by a surviving egg nucleus, he used a nuclear marker. The donor individuals were from a mutant *Xenopus* strain with one nucleolus per diploid nucleus, while the recipient eggs had the normal number of nucleoli, two per diploid nucleus. Like other researchers before him, Gurdon observed a steady decrease in the developmental capacity of the nuclei with advancing donor

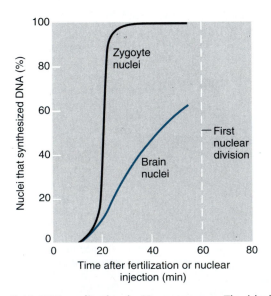

Figure 7.11 DNA replication in *Xenopus* eggs. The black curve represents a plot of DNA synthesis in normal zygote nuclei. The colored curve traces DNA synthesis in brain nuclei transferred into enucleated eggs. Note that many of the transfer nuclei had not completed DNA synthesis at the time of first mitosis.

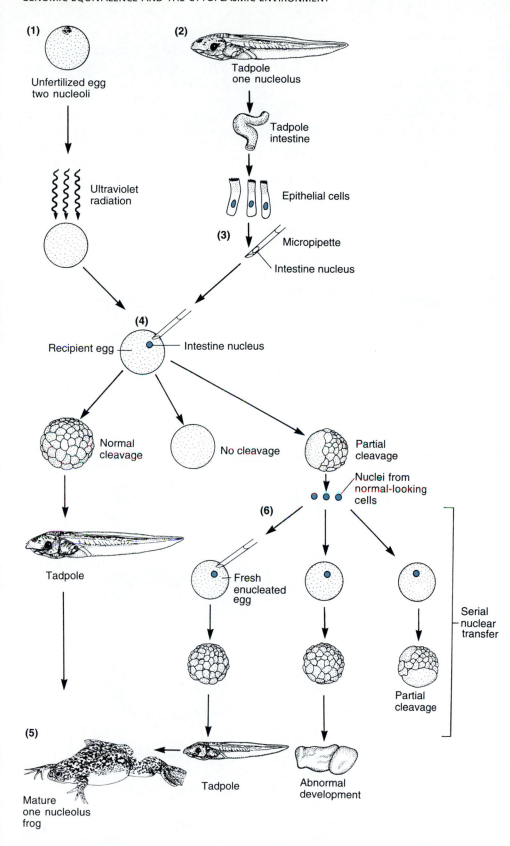

(1) Unfertilized egg two nucleoli

Ultraviolet radiation

(2) Tadpole one nucleolus

Tadpole intestine

Epithelial cells

(3) Micropipette

Intestine nucleus

(4) Recipient egg — Intestine nucleus

Normal cleavage

No cleavage

Partial cleavage

Nuclei from normal-looking cells

(6) Fresh enucleated egg

Serial nuclear transfer

Tadpole

Partial cleavage

(5) Mature one nucleolus frog

Tadpole

Abnormal development

Figure 7.12 Nuclear transplantation technique used by Gurdon (1962). An egg of the African clawed frog, *Xenopus laevis*, is prepared, by destruction of its own nucleus with ultraviolet light, to receive a nuclear transplant **(1)**. A feeding tadpole carrying a genetic marker (one nucleolus instead of the normal two nucleoli per diploid nucleus) serves as a donor **(2)**. A single cell from the intestinal epithelium is drawn into a fine glass pipette **(3)**. The cell wall breaks, leaving the nucleus free. The nucleus is injected into the prepared egg **(4)**. Host eggs cleaving normally are allowed to develop **(5)**. Host eggs showing partial cleavage are salvaged for serial nuclear transfer; their most normal-looking daughter cells are used again to donate nuclei, which are transferred into fresh enucleated eggs **(6)**. The presence of the genetic marker in developing tadpoles or adults indicates that their development has been promoted by a transplanted nucleus.

age, but the decline occurred less rapidly in his experiments (Fig. 7.10). In particular, nuclei from tadpole intestinal cells, in 10 of 726 transfers, promoted the development of feeding tadpoles of normal appearance. The yield of successful cases was boosted to 7% by serial transfers, and some of the tadpoles metamorphosed to adult frogs (Fig. 7.13). Gurdon concluded that many of the transplanted intestinal nuclei were highly pluripotent while those that gave rise to complete adults were totipotent.

(a) **(b)**

Figure 7.13 Normal frog and nuclear transplant frog *(Xenopus laevis)*. **(a)** Normal frog raised in the laboratory from an egg that had been fertilized in the usual way by a sperm. **(b)** Transplant frog raised from an egg cell from which the nucleus had been removed and into which the nucleus of an intestine cell had been transplanted. The frog is normal in all respects, indicating that the transplanted nucleus had a full complement of genetic information.

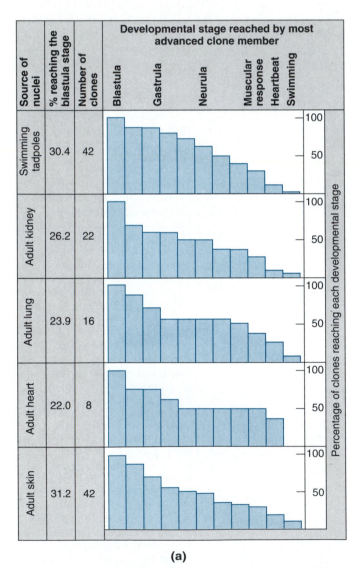

(a)

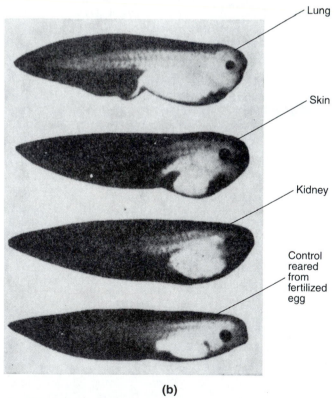

(b)

Figure 7.14 Results of transplanting nuclei from cultured *Xenopus* cells into eggs. **(a)** Nuclei were obtained from cell cultures prepared from tadpoles and four adult tissues. For each cell type, the percentage of nuclear transfers forming a partial or complete blastula is recorded in the second column. Several nuclei from each blastula were transplanted into fresh enucleated eggs to obtain the number of clones indicated in the third column. The resulting embryos were monitored, and the developmental stage reached by the most advanced embryo of each clone was tallied. **(b)** Swimming tadpoles obtained by serial nuclear transfer from lung, skin, and kidney cells.

These conclusions were criticized on the grounds that the totipotent nuclei might have accidentally been taken from migrating primordial germ cells or other undetermined cells. To counter this criticism, Laskey and Gurdon (1970) transplanted nuclei from cultured cells derived from adult *Xenopus* kidney, lung, heart, and skin. Again, the donors were of the single-nucleolus strain. Of the cultured skin cells, 99.9% contained large amounts of keratin and were therefore considered to be terminally differentiated (Reeves and Laskey, 1975). About 30% of the eggs injected with nuclei from such cells cleaved, but none of the recipients developed beyond neurulation. Serially transferred nuclei, however, did support development to advanced stages. Nuclei from skin, kidney, and lung cells gave rise to swimming tadpoles with functional muscles and nerves, a heartbeat and blood circulation, eyes with lenses, and other types of differentiated cells (Fig. 7.14). Most of them showed one nucleolus per diploid nucleus, indicating that they were derived from donor cell nuclei. Although these tadpoles could not be distinguished from control tadpoles raised normally, most of them died before reaching the feeding stage. Similar results were obtained using as donor cells lymphocytes from adult *Xenopus* (Wabl et al., 1975) or erythrocytes from adult *Rana* (DiBerardino and Hoffner, 1983).

Taken together, the transplantations of frog nuclei from differentiated cells to enucleated eggs show that at least small fractions of such cells still have highly pluripotent nuclei. The low yields of successful transplantations, and the failure of most transplants to undergo complete development, are open to different interpretations. However, these failures can plausibly be ascribed to a mismatch between the slow DNA replication in the transplanted nuclei and the fast mitotic oscillator in the egg cytoplasm.

Cells Change Their Differentiated State during Regeneration

Pluripotency of mature cells is also shown by *transdifferentiation,* the ability of a cell to convert to a different cell type. A striking example is the regeneration of a surgically removed eye lens from iris cells in the salamander (Fig. 7.15). This process involves a series of events in the iris cells including shape changes, loss of pigment, DNA replication and mitosis, ribosomal RNA synthesis, shaping of a new lens, and synthesis of lens-specific proteins (T. Yamada, 1967). These events are *not* a recapitulation of the normal steps of lens development: normal lens formation is induced in ectodermal cells that would otherwise form epidermis, whereas the iris—along with the retina—arises from an outpocketing of the brain rudiment (see Chapter 12). Rather, lens regeneration exemplifies the ability of cells to transdifferentiate from one cell type into another.

Other examples of transdifferentiation can be observed during the regeneration of lizard tails or amphibian limbs. When a limb of a salamander or newt is amputated, epidermal cells first spread over the cut surface and form an *apical epidermal cap.* Underneath the cap, all tissues undergo a dramatic process of *dedifferentiation,* in which the extracellular materials of bone, cartilage, and connective tissues dissolve and release the embedded cells. Multinucleated muscle fibers fragment into mononucleated cells. Cells from all these sources become histologically indistinguishable and form a cone of mesenchymal cells called a *regeneration blastema.* After a period of further proliferation, the blastema cells redifferentiate and reconstruct the missing limb. Are the new bones formed only by former bone cells, the new muscles only by former muscle fibers, and so on, or can blastema cells redifferentiate into cell types that they have not formed before? To study this question, limbs were X-irradiated before amputation to prevent the resident cells from dividing, and unirradiated cartilage was transplanted so that it would serve as the only available resource for limb regeneration (Fig. 7.16). The limbs regenerated under these conditions were stunted and poorly shaped, but the blastema cells derived from cartilage cells were able to form cartilage, dermis, and muscle, provided that the cartilage transplant came from a young donor (Wallace et al., 1974). Likewise, a muscle transplanted in a corresponding experiment regenerated all limb tissues including cartilage (Namenwirth, 1974). These and other data indicate that differentiated cells, if challenged by amputation, can dedifferentiate, proliferate, and then redifferentiate into other cell types.

Cells Express Different Genes at Different Times

The cases of transdifferentiation observed during regeneration are triggered by traumatic loss of body parts. However, some cells are known to undergo radical changes in structure and function as part of their *normal* development. This type of change is called *sequential polymorphism* (Gk. *poly-,* "many"; *morphe,* "form") and can be observed in several tissues (Kafatos, 1976). The ovarian follicle cells of insects are one example (see Chapter 3). During vitellogenesis, the follicle cells form an epithelium of tall, columnar cells. Channels between the follicle cells give blood proteins access to the oocyte surface. The follicle cells actively create these passageways by synthesizing extracellular material through which the blood proteins can diffuse. Depending on the species, the follicle cells may also synthesize vitellogenin. At the end of vitellogenesis, both the shape and the protein synthesis of the follicle cells change dramatically (Fig. 7.17). The cells assume a flat shape, with no openings between them, and they begin to synthesize large amounts of chorion proteins while the synthesis of other proteins declines sharply. Interspersed between the two differentiated states of the follicle cells is a period of "retooling," during which the follicle cells

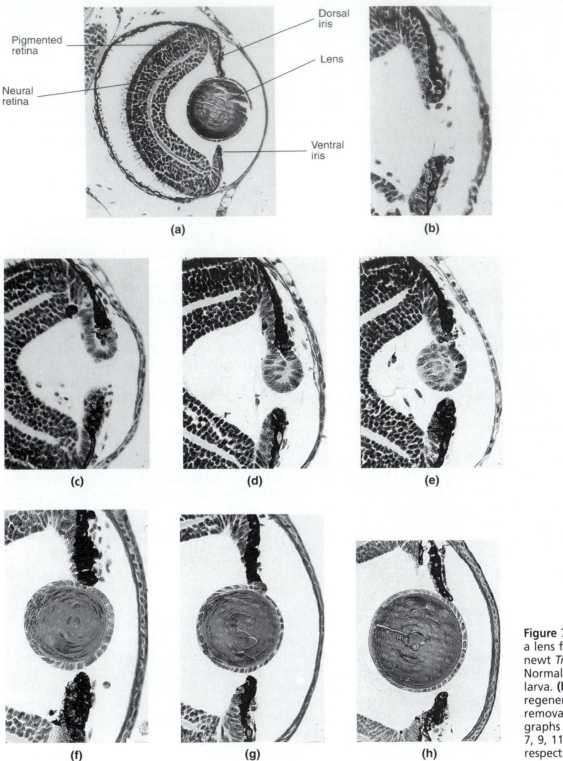

Pigmented retina

Neural retina

Dorsal iris

Lens

Ventral iris

(a)

(b)

(c)

(d)

(e)

(f)

(g)

(h)

Figure 7.15 Regeneration of a lens from the iris in the newt *Triturus viridescens*. **(a)** Normal eye in an unoperated larva. **(b–h)** Stages of regeneration after surgical removal of the lens. Photographs show specimens at 5, 7, 9, 11, 16, 20, and 30 days, respectively, after surgery.

go through several rounds of DNA replication. After this period, at least in *Drosophila*, the chromosomal regions containing the genes for chorion proteins are also selectively amplified, as described earlier in this chapter.

In the context of both regeneration and sequential polymorphism, it becomes clear that cells have the ability to differentiate, undergo a process of dedifferentiation, and then differentiate again in a different way. As will be detailed in Parts Two and Three of this book,

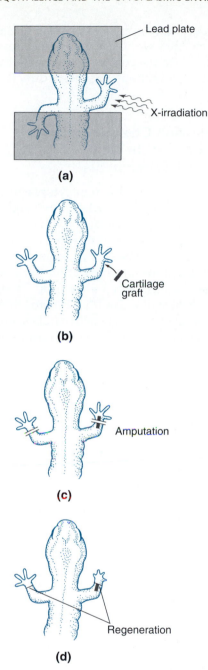

(a)

(b)

(c)

(d)

Figure 7.16 Experimental scheme to study the transdifferentiation of cells during limb regeneration in the axolotl *Ambystoma mexicanum*. **(a)** X-irradiation. The shaded areas were shielded by lead plates. **(b)** Transplantation of cartilage from an unirradiated donor. The cartilage was cleaned of other tissues before transplantation. **(c)** Amputation of the irradiated limb with cartilage transplant (right) and of the unirradiated control limb (left). **(d)** Regeneration of a deformed limb (right) and a complete limb (left).

cells become different by expressing different sets of genes. In terms of gene activity, then, regeneration and sequential polymorphism involve repressing one set of genes and activating a different set. For example, the iris cells regenerating a lens stop synthesizing pigment

and eventually make large amounts of *crystallins*, the proteins characteristic of eye lenses. Clearly, the genes for lens crystallins must be present but not expressed while the progenitor cells of a regenerated lens are still functioning as iris cells. And presumably, the genes for the enzymes necessary to make the iris pigments are still available but silent in the regenerated lens cells. The same argument can be made for the sequential polymorphism of follicle cells. The genes for chorion proteins must be present but dormant during vitellogenesis, and the genes expressed during vitellogenesis are probably retained but inactive during chorion formation.

Taken together, the evidence from many observations suggests that, as a rule, nuclei retain a complete or an almost complete set of genetic information, and that cells can express different sets of genes during their lives. Consequently, genetic equivalence has become a widely accepted theory. The known cases of chromosome loss, selective gene amplification, and DNA rearrangements are considered exceptions to the rule. The

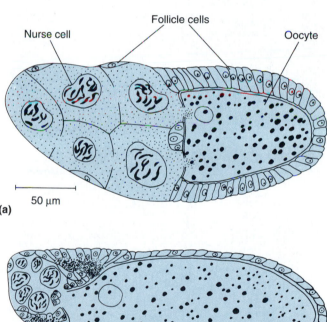

(a)

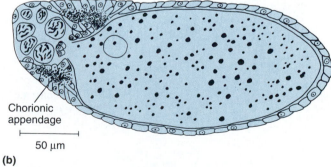

(b)

Figure 7.17 Sequential polymorphism of insect ovarian follicle cells. **(a)** Egg chamber of *Drosophila melanogaster* during vitellogenesis. Follicle cells surrounding the oocyte are columnar and aid in vitellogenin synthesis and uptake. Follicle cells surrounding nurse cells are squamous. **(b)** Egg chamber at a later stage. Follicle cells surrounding the oocyte are now squamous and synthesize chorion proteins. Follicle cells surrounding the collapsed nurse cells are now forming the chorionic appendages.

most important implication of genetic equivalence is that cells become different by using selected subsets of their total genetic information. Consequently, one would expect a multitude of regulatory signals to select the combination of gene activities appropriate for each cell type. Indeed, there are whole networks of genes whose function is to activate subordinate genes at the right time and in the right groups of cells (see Chapters 15 and 21 through 24).

Control of Nuclear Activities by the Cytoplasmic Environment

Having accepted genetic equivalence as a reasonable basis for further analysis, we want to know how various signals cause cells to express different portions of their genetic information. This is one of the focal ques-

tions in modern developmental biology. In the remainder of this chapter, we will explore two experimental situations in which the control of gene activities by the cytoplasmic environment has been directly demonstrated.

Gene Expression Changes upon Transfer of Nuclei to New Cytoplasmic Environments

To analyze the effects of different cytoplasmic environments on nuclear activity, Gurdon (1968) injected nuclei from one type of cell into the cytoplasm of another type of cell. He monitored three basic types of nuclear activity, namely, DNA synthesis, RNA synthesis, and chromosome condensation in preparation for meiosis (Fig. 7.18). The test nuclei came either from *Xenopus* adult brain cells, which synthesize RNA but no DNA, or from early blastula cells, which are active in DNA replication but synthesize virtually no RNA. As host cells, Gurdon

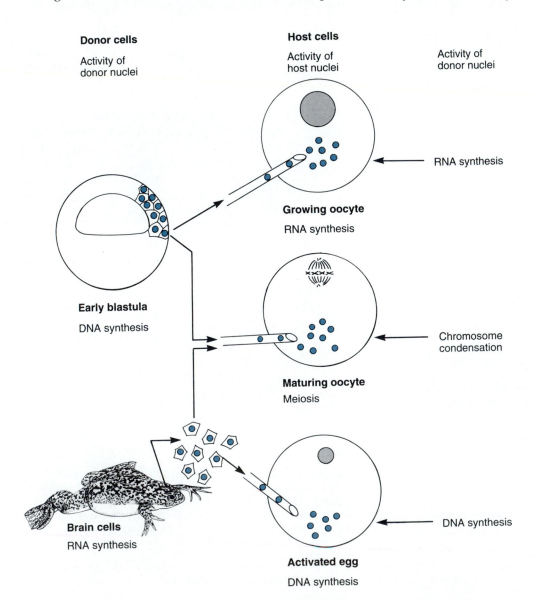

Figure 7.18 Transplantation experiments with nuclei from *Xenopus* showing that nuclear activities are controlled by the cytoplasmic environment.

used growing oocytes, which synthesize RNA but no DNA; or maturing oocytes, in which the nucleus forms a meiotic spindle; or activated eggs, the nuclei of which synthesize DNA but no RNA. In every combination, the transplanted nucleus adjusted its synthesis of macromolecules to the new cytoplasmic environment. Blastula nuclei injected into growing oocytes stopped synthesizing DNA and entered a phase of continuous RNA synthesis. (This was shown by coinjecting radiolabeled precursors of either DNA or RNA along with the blastula nuclei. The injected nuclei incorporated the RNA precursor but not the DNA precursor.) Conversely, brain cell nuclei injected into activated eggs stopped RNA synthesis and began DNA synthesis. When the same nuclei were injected into maturing oocytes, they synthesized neither DNA nor RNA but underwent chromosome condensation and formed a meiotic spindle.

Taking this experiment a step further, De Robertis and Gurdon (1977) injected nuclei from kidney cells into growing oocytes and assayed for the reprogramming of the nuclei in terms of individual gene products. Would the transplanted nuclei continue to express the same set of genes in oocytes as they had done in kidney cells, or would they stop expressing those genes? Would the transplanted kidney cell nuclei begin to express genes that were normally active in oocytes only? The experimenters identified individual polypeptides (proteins), which are the products of single genes; the proteins were labeled with radioactive amino acids and separated by *two-dimensional gel electrophoresis.* This technique involves two rounds of electrophoresis through polyacrylamide gel, during which proteins are separated according to isoelectric point in one dimension and according to molecular weight in the other dimension. The resulting patterns of radioactive spots in the gel were made visible by *autoradiography* (see Methods 3.1).

The authors first analyzed and compared the proteins synthesized normally in *Xenopus* oocytes and kidney cells (Fig. 7.19). They found that most proteins were common to both kinds of cells; these are the "housekeeping" proteins that cells need to make their basic metabolic enzymes, for instance, or to assemble their cytoskeletons and their ribosomes. However, a few proteins occurred in kidney cells but not in oocytes or vice versa. These were termed *oocyte-specific proteins* or *kidney-specific proteins,* respectively.

Next, for their nuclear transplantation experiments, De Robertis and Gurdon injected nuclei from kidney cells of the frog *Xenopus laevis* into oocytes of the newt *Pleurodeles waltlii* (Fig. 7.20). This allowed them to distinguish proteins encoded by the transplanted nuclei from proteins encoded by stable mRNA in the host oocyte, since the proteins of the two species run differently on two-dimensional gels. After transplantation of the kidney cell nuclei, the host oocytes were cultured

for several days, during which time mRNA was synthesized and accumulated in the cytoplasm. Then radioactive amino acids were added to the culture medium to label any polypeptides synthesized afterward. After a labeling period of 6 h, oocyte proteins were analyzed by two-dimensional gel electrophoresis. Among the radiolabeled proteins, at least seven were recognized as encoded by the *Xenopus* genome. They included three housekeeping proteins and four oocyte-specific proteins but no kidney-specific proteins. This meant that the oocyte environment had reprogrammed the kidney nuclei to express housekeeping genes and some oocyte-specific genes, but had inhibited the expression of kidney-specific genes. Thus, whatever programming signals are present in *Pleurodeles* oocyte cytoplasm were at least partially effective in reprogramming *Xenopus* nuclei, even though newts and frogs have evolved independently for more than 200 million years.

Cell Fusion Exposes Nuclei to New Cytoplasmic Signals

A similar reprogramming of nuclei by their cytoplasmic environment was observed in cell fusion experiments. Cells can fuse to form a combined cell with two or more nuclei (Fig. 7.21). This happens during normal development of skeletal muscle and other tissues. Cells that do not normally fuse can be stimulated to do so in vitro by exposing them to inactivated viruses or polyethylene glycol, either of which alters the plasma membranes so that they tend to fuse with one another. These procedures have been employed to fuse cells obtained from different tissues of different species, such as rabbit macrophages with chicken erythrocytes. A hybrid cell with two or more nuclei from different species is called a *heterokaryon* (Gk. *heteros,* "other"; *karyon,* "nucleus"). Heterokaryons are capable of synthesizing DNA, RNA, and proteins and show rapid mixing of the cytoplasmic components and membrane proteins from their constituent cells.

Although heterokaryons may live for several weeks under favorable conditions, their continued survival depends on the formation of daughter cells containing a single hybrid nucleus. In binucleate heterokaryons, simultaneous mitosis of both nuclei often leads to the formation of a single spindle, and subsequent cytokinesis produces two *hybrid cells,* each with one nucleus containing the chromosomes of both species (Fig. 7.21). For instance, a heterokaryon derived from one diploid human cell (containing 2×23 chromosomes) and one diploid mouse cell (containing 2×20 chromosomes) may divide into two hybrid cells, each containing a nucleus with 23 human plus 20 mouse chromosomes.

Heterokaryons and hybrid cells have been used in a variety of studies. For the study of nucleocytoplasmic interactions, a particularly useful heterokaryon was

Xenopus kidney cells

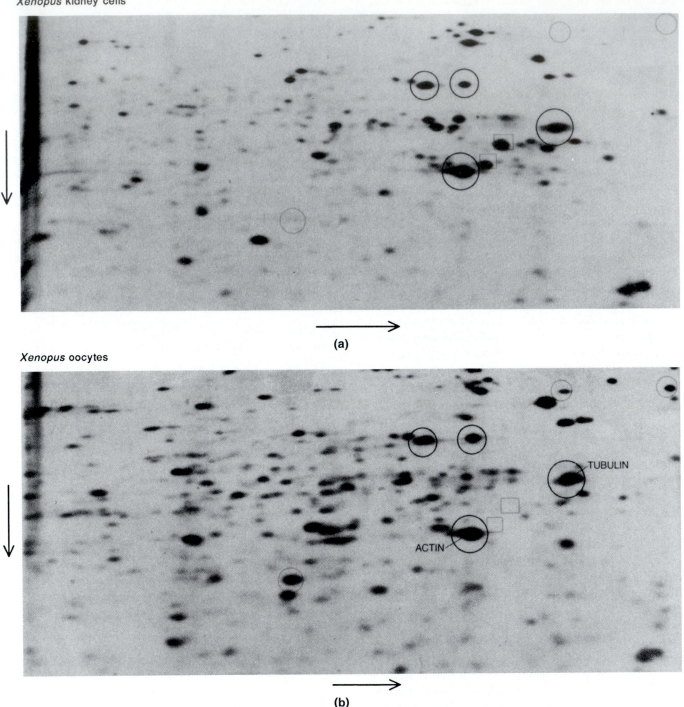

(a)

Xenopus oocytes

TUBULIN

ACTIN

(b)

Figure 7.19 Comparison of *Xenopus* proteins synthesized by cultured kidney cells **(a)** and oocytes **(b)**. Proteins were radioactively labeled in vivo before they were extracted and separated by two-dimensional gel electrophoresis. The arrows indicate the directions of protein movement during separation. The proteins are seen as gray spots on an X-ray film exposed to the gel. Most proteins, including actin and tubulin, are made by both cell types. However, some proteins are oocyte-specific (empty circles in A) while others are kidney-specific (empty squares in B).

generated from chicken erythrocytes and human tumor cells known as HeLa cells (Harris, 1974). These two cell types display very different levels of DNA and RNA synthesis. HeLa cells, as actively proliferating cells, synthesize both DNA and RNA at high levels. By contrast, chicken erythrocytes are terminally differentiated, non-dividing cells that retain a nucleus but synthesize no DNA and very little RNA. Erythrocytes also react in a peculiar way with the inactivated Sendai virus used for cell fusion: their plasma membranes become so leaky that their entire cytoplasm flows out. The depleted cells, which consist of little more than a cell nucleus in a leaky

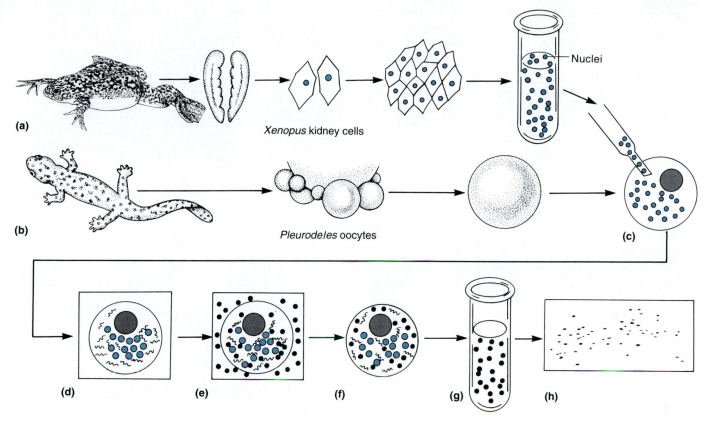

Figure 7.20 Nuclear transplantation experiment showing that oocyte cytoplasm can reprogram nuclei from kidney cells to express oocyte-specific proteins. **(a)** Kidney cells from the frog *Xenopus laevis* were used to prepare nuclei (color). **(b)** Oocytes were removed from the newt *Pleurodeles waltlii*. **(c)** The oocytes were injected with the kidney nuclei and cultured for three days. **(d)** Messenger RNAs (wavy lines) accumulated in the cytoplasm. **(e)** Radiolabeled amino acids (black dots) were added to the medium. **(f–h)** Proteins synthesized during the following 6 h were extracted and separated by two-dimensional gel electrophoresis. Analysis of X-ray film exposed to the gel revealed proteins characteristic of *Xenopus* oocytes but none of the kidney-specific proteins.

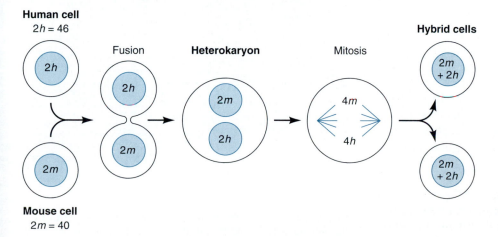

Figure 7.21 Somatic cells from different species, such as human and mouse, can be fused by membrane-altering agents, including Sendai virus or polyethylene glycol. In the resulting heterokaryons, the nuclei of the two original cells stay separated. Simultaneous mitosis of the nuclei often generates two hybrid cells, each containing the chromosomes from both original cells.

membrane, are known as erythrocyte "ghosts." When such ghosts fuse with HeLa cells, the heterokaryons contain the two different nuclei surrounded by cytoplasm that is derived almost exclusively from HeLa cells. Therefore, any changes in the erythrocyte nuclei can be ascribed to signals from HeLa cell cytoplasm.

Upon fusion of erythrocyte ghosts with HeLa cells, Harris observed a dramatic reactivation of the previously dormant erythrocyte nuclei (Harris, 1974). It began with a 20- to 30-fold increase in nuclear volume, part of which resulted from the uptake of cytoplasmic proteins. As the nuclear volume increased, so did the rates of DNA and RNA synthesis (Fig. 7.22). This was shown by culturing the heterokaryons in the presence of radiolabeled thymidine or uridine and subsequent autoradiography (see Methods 3.1). The results showed that an almost completely inactive nucleus can resume DNA and RNA synthesis in response to cytoplasmic factors from a cell type with an active nucleus. In addition, the data indicated that the activating components must have been conserved remarkably well during evolution, because chicken nuclei respond to signals from human cytoplasm, although birds and mammals diverged about 200 million years ago.

In a similar experiment, J. E. Brown and M. C. Weiss (1975) showed that chromosomes from mouse fibroblasts were able to direct the synthesis of liver-specific enzymes. By fusing rat liver tumor cells with mouse fibroblasts, the investigators constructed hybrid cells containing rat chromosomes as well as mouse chromosomes. The hybrid cells divided and secreted various proteins into the culture medium. Among these were liver-specific proteins such as albumin, aldolase, and tyrosine aminotransferase. Most important, the synthesis of these proteins was directed by genes from *both* species, rat and mouse. This was shown by electrophoretic mobility and other molecular characteristics that distinguished the mouse proteins from their rat counterparts. Since the mouse fibroblasts, prior to fusion, had not expressed their genes for the liver-specific enzymes, these genes must have been activated by factors in the hybrid cell cytoplasm. These results show again that at least some of these gene-activating factors have been conserved remarkably well between species.

Conclusion and Outlook

One of the basic questions in development is whether or not differentiated cells are still genetically equivalent. Observations on chromosome number and morphology indicate that most cells inherit complete and unaltered sets of chromosomes. Cases of chromosome loss or selective gene amplification seem to be exceptions to the rule. Molecular investigations also indicate that most cells do not undergo loss, alteration, or selective amplification of their genomic DNA as part of normal differentiation.

As a biological test for genetic equivalence, single cells or nuclei have been examined for totipotency, that is, their ability to promote the development of a whole organism. Experiments with several types of differentiated plant cells indicate that they are clearly totipotent. Nuclei from various differentiated animal cells are also at least highly pluripotent. In addition, regeneration and sequential polymorphism indicate that cells can differentiate, revert to a dedifferentiated state, and then differentiate in another way.

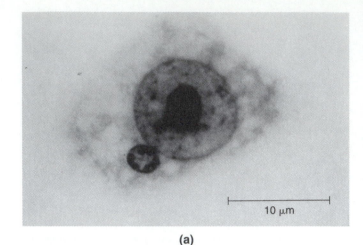

10 μm

(a)

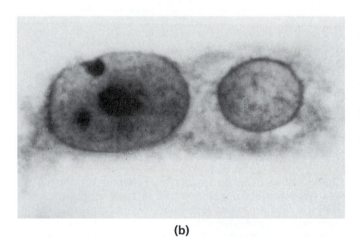

(b)

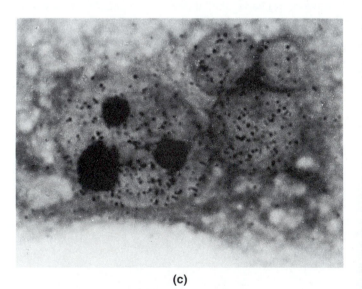

(c)

Figure 7.22 Photomicrographs showing heterokaryons of human HeLa cells and hen erythrocytes. **(a)** Soon after fusion, the erythrocyte nucleus (lower left) was much smaller than the HeLa cell nucleus, and its chromatin was highly condensed. **(b)** 24 h after fusion, the erythrocyte nucleus (right) had enlarged, and the chromatin was dispersed. **(c)** Autoradiograph of a heterokaryon incubated with [³H]uridine. The dark dots (silver grains) indicated that all nuclei, the HeLa cell nucleus (left) as well as the three erythrocyte nuclei, were synthesizing RNA.

Together, these results indicate that, as a rule, all cells in an organism have complete and equivalent sets of genetic information. Although the evidence for genetic equivalence of mature animal cells is not unequivocal, most biologists have adopted genetic equivalence as a well-founded theory. It implies that cells express different parts of their genetic information as they specialize for various roles in the organism. This principle of differential gene expression will be discussed more fully in subsequent chapters.

Gene activities are controlled by the cytoplasmic environment, as indicated by nuclear transplantation and cell fusion experiments. New cytoplasmic signals can silence previously active genes and activate previously silent genes. This type of control operates continuously during embryonic development, generating an ever-increasing variety of cells with different gene activities. Most fertilized eggs have cytoplasmic asymmetries that define one or two polarity axes. Certain cytoplasmic components are partitioned into different blastomeres as the zygote cleaves, creating diverse cytoplasmic environments that initiate a primary pattern of embryonic cells that have distinct programs for gene expression. Later during development, interactions between blastomeres release new signals, which then determine additional groups of cells to activate new sets of genes.

SUGGESTED READINGS

Beermann, W. 1952. Chromosomes and genes. In W. Beermann, ed., *Developmental Studies on Giant Chromosomes*, 1–33. New York: Springer-Verlag.

De Robertis, E. M., and J. B. Gurdon. 1979. Gene transplantation and the analysis of development. *Sci. Am.*, December, 74–82.

DiBerardino, M. A., N. J. Hoffner, and L. D. Etkin. 1984. Activation of dormant genes in specialized cells. *Science* **224**:946–952.

Harris, H. 1974. *Nucleus and Cytoplasm.* 3d ed. Oxford: Clarendon Press.

Vasil, V., and A. C. Hildebrandt. 1965. Differentiation of tobacco plants from single isolated cells in microcultures. *Science* **150**:889–892.

LOCALIZED CYTOPLASMIC DETERMINANTS

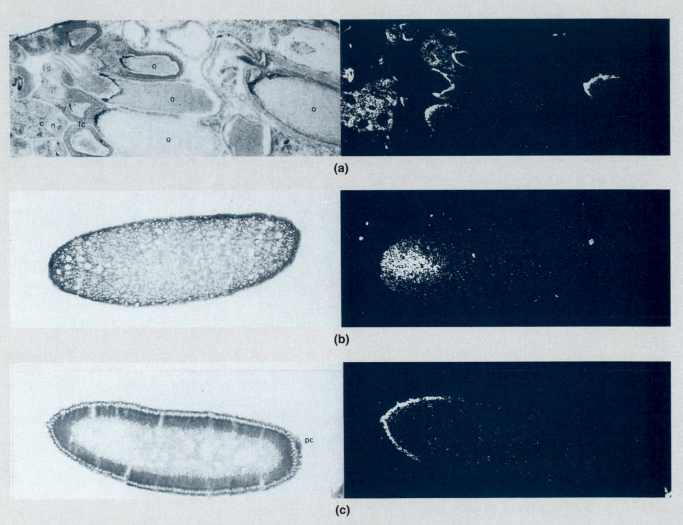

Figure 8.1 Localization of bicoid mRNA in *Drosophila* oocytes and embryos. All photographs show histological sections treated with a cloned DNA probe that binds specifically to bicoid mRNA present in the sectioned tissue. Radioactive label in the probe is made visible by autoradiography, generating the black grains seen to the left under bright-field illumination, and the white grains seen in the same specimens to the right under dark-field illumination. All sections are oriented with the anterior pole to the left. Bicoid mRNA accumulates in the nurse cell cytoplasm and along the anterior margin of the oocytes **(a)**, in a deep anterior cone of cytoplasm during cleavage **(b)**, and in the anterior periplasm at the syncytial blastoderm stage **(c)**. c = nurse cell cytoplasm; fc = follicle cell; n = nurse cell nucleus; o = oocyte; pc = pole cells.

Since the beginning of the twentieth century, embryologists have proposed that certain components in the egg cytoplasm are segregated into different blastomeres, in which they determine different pathways of development (E. B. Wilson, 1925; E. H. Davidson, 1986). The importance of this concept has been widely recognized, although the molecular nature of localized components remained enigmatic for a long time. With new genetic and experimental tools, investigators have now characterized several cytoplasmic determinants with respect to their molecular nature, their species specificity, and other properties. These advances have rekindled interest in localized determinants, so that today they are again a favorite topic of developmental biologists.

The concept of cytoplasmic localization is based on both genetic and experimental evidence. Geneticists noticed that certain *maternal effect mutants* have *regional abnormalities*; for example, offspring of *bicoid* mutant *Drosophila* females lack anterior body parts (see Fig. 14.2). According to the simplest interpretation, such phenotypes are due to an absence or a defective condition of RNAs or proteins that would normally be encoded by maternal genes and localized in specific areas of egg cytoplasm during oogenesis. Indeed, the molecular cloning of such genes led to the identification of bicoid mRNA (Fig. 8.1; Color Plate 1) and several localized maternal gene products.

In this chapter, we will first discuss cytoplasmic localization as a basic principle of embryonic development. A few examples will show that the process of localization may occur during oogenesis, but also as part of egg activation or cleavage. Next we will explore two types of bioassay that can be used to prove the existence of localized cytoplasmic determinants and to characterize their molecular nature, their species specificity, and other properties. Finally, localized cytoplasmic determinants will be used to introduce the principle of *default programs* in development.

The Principle of Cytoplasmic Localization

The *principle of cytoplasmic localization* is simple and applies to most types of animals. Eggs have unevenly distributed cytoplasmic components, which are partitioned among different blastomeres as the egg cleaves. The dissimilar cytoplasmic environments elicit diverse gene activities in the blastomeres, thus initiating different pathways of development. Localized cytoplasmic components that affect the fate of the blastomeres containing them are called *localized cytoplasmic determinants*—or *localized determinants, cytoplasmic determinants,* or *localizations,* for short. Some cytoplasmic determinants are associated with visible markers, such as pigmented yolk particles or other organelles, that directly reveal their localization. Other cytoplasmic determinants are devoid of such convenient markers. Conversely, some cytoplasmic components are localized but have no effect on cell determination. For instance, the dark pigment in the cortex of amphibian eggs is limited to the animal hemisphere, but albino mutants lacking this localized pigment develop normally otherwise. To prove that a cytoplasmic component is a localized determinant, one has to demonstrate that the material of interest is unevenly distributed and that it affects the fate of the cells containing it.

Experimental evidence for localized cytoplasmic determinants started with the simple observation that, in most eggs, yolk and other components are distributed unevenly along the animal-vegetal axis. When an egg is cut *along* this axis and each half fertilized separately, both halves often develop into complete larvae (Fig. 8.2). When the same

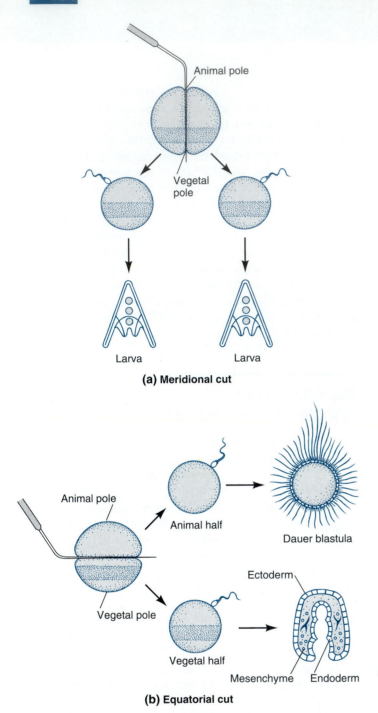

Figure 8.2 Development of sea urchin eggs cut in halves and fertilized. **(a)** After a meridional cut (parallel to the animal-vegetal axis), each half develops into a half-sized but normally proportioned larva. **(b)** After an equatorial cut (perpendicular to the animal-vegetal axis), the animal half develops into a dauer blastula (a ciliated blastula that does not gastrulate). The vegetal half forms a disproportionate embryo.

operation is carried out except with the egg cut *perpendicular* to the animal-vegetal axis, both halves tend to develop abnormally. These results show that animal as well as vegetal cytoplasm contains localized components and that both sets of components are necessary for normal development.

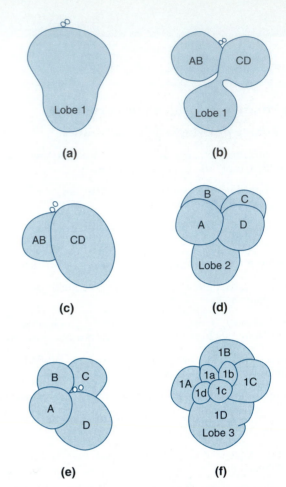

Figure 8.3 Polar lobe formation during early cleavages in the snail *Ilyanassa obsoleta.* **(a–c)** The first lobe forms at the vegetal pole of the zygote before the first cleavage and is thereafter resorbed into one blastomere designated CD. **(d, e)** The second lobe forms at the vegetal pole of the CD blastomere before the second cleavage and is thereafter resorbed into one blastomere designated D. **(f)** The third polar lobe is formed incompletely and resorbed before D divides.

② Dramatic cases of cytoplasmic localization are observed in many animals with spiral cleavage, including mollusks and annelids. The first embryonic cleavages involve the cyclic appearance of a cytoplasmic protrusion at the vegetal pole known as the ***polar lobe*** (Fig. 8.3). The lobe formed during the first mitotic cleavage is referred to as the *first polar lobe,* although in some species, earlier polar lobes are observed during meiotic divisions. As the first two blastomeres become separated by cytokinesis, the polar lobe constricts until it is connected by only a thin cytoplasmic strand to the rest of the embryo (Fig. 8.4). This stage, when the polar lobe looks like a third blastomere, is called the ***trefoil stage.*** After completion of the first cytokinesis, the polar lobe is resorbed into one of the two blastomeres, which is then designated the CD blastomere. Similarly, the second polar lobe is formed by the CD blastomere and resorbed into the D blastomere after the second cleavage.

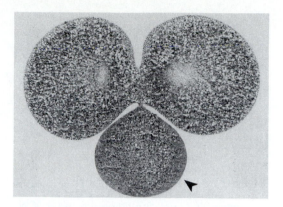

Figure 8.4 Photomicrograph of a mollusk embryo *(Dentalium)* at the trefoil stage. The arrowhead points to the polar lobe, which is connected by a thin cytoplasmic bridge to the dividing zygote.

As a result, the D blastomere becomes larger than blastomeres A, B, and C.

As cleavage continues, the A, B, C, and D blastomeres give off successive quartets of small blastomeres called *micromeres*. The first quartet originates when blastomere A divides into macromere 1A and micromere 1a, blastomere B divides into macromere 1B and micromere 1b, and so forth (Fig. 8.3). Similarly, the second quartet is given off when blastomere 1A divides into macromere 2A and micromere 2a, blastomere 1B divides into macromere 2B and micromere 2b, and so on. Of particular interest within the fourth quartet is the 4d micromere (Fig. 8.5). It divides into two cells, designated *mesentoblasts*, each of which gives rise to a yolk-rich endoblast cell and a clear primary mesoblast cell. The endoblast and mesoblast pairs impose a bilateral symmetry as development progresses. Proper formation of these cells is critical to the formation of a normal larva.

To explore the significance of polar lobe formation, investigators removed the lobe at different cleavage stages. The most detailed investigations were carried out with embryos of the scaphopod *Dentalium* (E. B. Wilson, 1904) and the mud snail *Ilyanassa obsoleta* (Clement, 1962). Cutting off the polar lobe at the trefoil stage had dramatic consequences for subsequent development. The larger size and special cleavage schedule of the D blastomere were no longer observed, no mesoblast cells were formed, and the operated embryos cleaved in a radially symmetrical way. The developing larvae were severely disorganized and almost devoid of inner structures (Fig. 8.6). The same type of abnormal development was observed in larvae raised from AB blastomere pairs and other combinations lacking the D blastomere. In contrast, CD blastomere pairs and other combinations containing D developed normally. Similar results were obtained with other mollusks (Verdonk and Cather, 1983).

The general conclusion from these experiments was that the polar lobe contains cytoplasmic components,

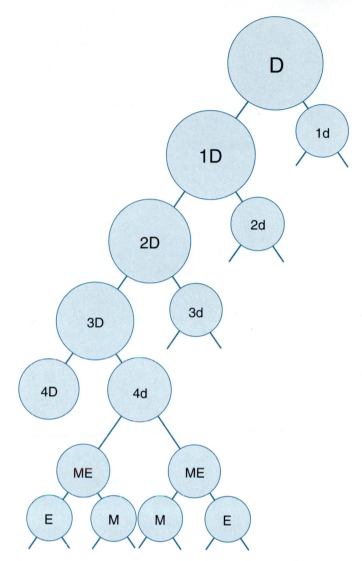

Figure 8.5 Diagram of the D cell lineage in the snail *Ilyanassa obsoleta.* The D macromere gives off successive micromeres, 1d, 2d, and 3d. At the following division, the 3D macromere divides into a 4D and a 4d blastomere, which are similar in size. Much of the polar lobe material originally allocated to the D macromere is passed on to 4d. Subsequently, 4d divides into mesentoblasts (ME), each of which divides into endodermal (E) and mesodermal (M) founder cells.

most of which are funneled selectively into the D blastomere and are, directly or indirectly, required for forming the missing structures. It appears that the polar lobe material is set aside at each cleavage to ensure its selective allocation to the D blastomere and its descendants.

While polar lobes are found in many species with spiral cleavage, embryos of other such species are lobeless. In these, the first two cleavages generate four macromeres of equal size. The entire embryo remains radially symmetrical with regard to the animal-vegetal axis until the interval between the third quartet formation and the fourth. At this time the vegetal macromeres

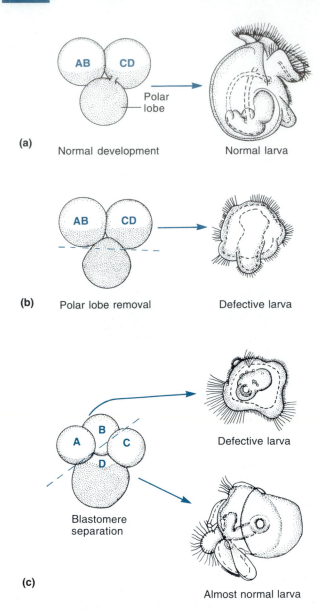

(a) Normal development → Normal larva

(b) Polar lobe removal → Defective larva

(c) Blastomere separation → Defective larva / Almost normal larva

Figure 8.6 The effect of polar lobe removal on the development of *Ilyanassa*. **(a)** Development of a normal larva. **(b)** A highly defective larva is formed after polar lobe removal at the trefoil stage. **(c)** Blastomere separation at the 4-cell stage. The CD pair gives rise to an almost normal larva, while the AB pair forms a larva with internal structures severely disorganized or missing.

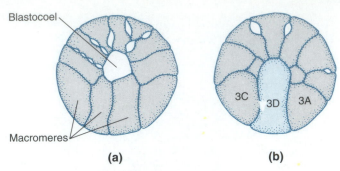

(a) **(b)**

Figure 8.7 D-blastomere induction in an equally cleaving snail, *Lymnaea palustris*. **(a)** After the third micromere quartet has been given off, the embryo consists of 24 cells surrounding a fluid-filled blastocoel. The section drawn passes through three macromeres and eight micromeres. **(b)** Before the fourth micromere quartet is formed, macromeres and micromeres grow together, thus filling the blastocoel. The macromere contacting the greatest number of micromeres (five in this section) becomes the 3D cell. (The 3B cell is outside the section.)

elongate and contact the animal micromeres as they fill the blastocoel (Fig. 8.7). The macromere that forms contacts with the greatest number of micromeres becomes the 3D blastomere, which then gives rise to the 4d micromere and subsequently to the ME cells (van den Biggelaar and Guerrier, 1983).

To test whether the contacts with cells actually cause the macromere to acquire the 3D fate, Mark Martindale and his colleagues (1985) did the following experiment. Shortly before expected formation of the animal-vegetal cell contacts, the embryos were treated with the microfilament inhibitor *cytochalasin B,* which prevented the elongation of the macromeres. When the drug was

washed out, the macromeres elongated but their contact with micromeres apparently came too late: the embryos remained radialized, and typical 4d cell descendants were not formed. From these and other results the researchers concluded that the animal-vegetal interaction is necessary for forming the typical 4d cell lineage in lobeless mollusks.

An interaction between two embryonic cells that changes the fate of one partner is called an *induction.* In this interaction the cells whose fate changes are called the *responding cells,* while their partners are called *inducing cells.* Induction implies that the responding and inducing cells are different prior to their interaction. Their already-existing differences may result from an earlier induction or from cytoplasmic localization. Generally, cytoplasmic localization tends to generate a coarse pattern of a few differently determined cells early during development. However, these coarse-pattern localizations provide the basis for subsequent inductions, which greatly increase the number of different cell states.

Embryonic induction is as basic to development as cytoplasmic localization. More cases illustrating the principle of induction will be discussed in Chapters 9, 11, and 24. For the remainder of this chapter, the focus will be on cytoplasmic localization.

The Timing of Cytoplasmic Localization

In most animals, cytoplasmic localization occurs mainly during oogenesis. However, there are also cases in which localization takes place as part of egg activation or during cleavage.

Localization during Oogenesis Is Based on Asymmetries in the Germ Line Cells or in the Ovary

As oocytes develop in the maternal ovary, they often acquire asymmetries that foreshadow the animal-vegetal polarity of the egg. Such asymmetries may be intrinsic to the oocyte itself, perhaps initiated by cytoskeletal elements that set up directional transport systems. Alternatively, oocytes may be surrounded by nurse cells or follicle cells in an asymmetrical configuration that imposes an asymmetry. The following examples will illustrate these two situations.

Fully grown amphibian eggs show animal-vegetal polarity, which is obvious in the dark pigmentation of the egg cortex in the animal hemisphere. The same animal-vegetal polarity is also apparent in the distribution of several components in the deep cytoplasm. At the vegetal pole, for example, many amphibian eggs contain dense granular material that stains intensely for RNA. The polar granules are later incorporated into primordial germ cells and are thought to be germ cell determinants. Amphibian oocytes and eggs also contain other localized RNAs and proteins that are not associated with visible markers (Jäckle and Eagleson, 1980; Dreyer et al., 1982; Rebagliati et al., 1985). A few proteins are strictly localized near the vegetal pole. Others are distributed in a gradient fashion along the animal-vegetal axis.

The molecular nature of some of the localized components in *Xenopus* embryos has been characterized. An mRNA localized in the vegetal pole region, designated Vg1, encodes a protein that belongs to a family of *growth factors,* which are involved in embryonic induction and in the control of cell division (see Chapters 9 and 29). An mRNA localized in the animal hemisphere, called An3, encodes an RNA helicase, an enzyme that unwinds hybridized RNA duplex structures (Weeks et al., 1985). This activity may be important during development for the release of information stored in other maternal mRNAs. Because of their localization and molecular characteristics, Vg1 and An3 mRNAs are likely to act as cytoplasmic determinants, and indeed, there is strong evidence that Vg1 protein is involved in establishing the dorsoventral body axis.

The mechanism of Vg1 mRNA localization has been clarified in several experiments. To monitor the distribution of Vg1 mRNA at different stages of oogenesis, researchers have used the method of *in situ hybridization.* This widely used technique is based on the fact that single-stranded DNA and RNA molecules form double-stranded hybrids if their nucleotide sequences are complementary to each other. For any known RNA sequence, one can synthesize a complementary strand of labeled DNA or RNA (see Methods 15.1). Such a labeled strand binds specifically to its counterpart wherever the two meet, even in a histological section.

Using this method, Douglas Melton (1987) studied the localization of Vg1 mRNA during frog oogenesis. Vg1 is uniformly distributed in young oocytes, but it becomes localized during vitellogenesis to form a cap at the vegetal pole (Fig. 8.8). During the localization period, the overall amount of Vg1 remains approximately constant, suggesting that its localization at the vegetal pole results from translocation and not from conservation at the vegetal pole and a breakdown everywhere else. Even external Vg1 mRNA, synthesized in vitro and microinjected near the equator of vitellogenic oocytes, will "home" to its normal location at the vegetal pole. Other mRNAs, including histone and globin mRNA, do not show this remarkable ability to localize. These results show that the oocyte has a mechanism for translocating specific RNA sequences and anchoring them in certain areas.

The transport and eventual anchorage of the Vg1 mRNA depend on cytoskeletal components within the oocyte. Microtubules are required to transport Vg1 mRNA to the vegetal hemisphere, and microfilaments are necessary to anchor the mRNA in the cortical cytoplasm around the vegetal pole (Yisraeli et al., 1990). In addition to these general cytoskeletal components, the transport mechanism must also involve proteins that interact specifically with Vg1 mRNA and cytoskeletal components. While these proteins remain to be characterized, the matching recognition sequence in Vg1 mRNA has already been identified in the nontranslated mRNA *trailer* region (Mowry and Melton, 1992). Upon removal of this sequence, Vg1 mRNA is no longer localized. Conversely, if this sequence from Vg1 mRNA is added to an unrelated mRNA such as globin mRNA,

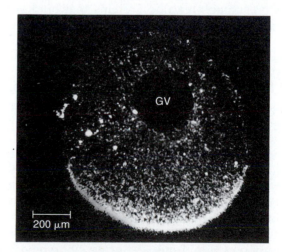

Figure 8.8 Localization of Vg1 mRNA near the vegetal pole of a *Xenopus* oocyte, as shown by in situ hybridization (see Methods 15.1). Histological sections of the fixed oocyte were incubated with a radiolabeled RNA probe that formed stable hybrids with Vg1 mRNA. The radioactivity generates silver grains in a coat of film emulsion spread over the section. The silver grains appear as white dots under the dark-field illumination used for photography here. Control sections hybridized with labeled RNA probes binding to other mRNAs showed an even coverage with silver grains. GV = germinal vesicle.

then the globin-Vg1 hybrid mRNA is directed to the vegetal pole just as the Vg1 mRNA is.

The polarized transport systems that localize RNAs and proteins in the oocyte must themselves be generated at an earlier stage of oogenesis. Indeed, the animal-vegetal polarity can be traced back to previtellogenic stages, when a cloud of mitochondria and dense granular material is located asymmetrically between the germinal vesicle and the vegetal pole. This early polarity seems to arise independently of the oocyte's position in the ovary.

A different localization system has been observed for bicoid mRNA in the *Drosophila* egg. Here, the constellation of oocytes and other ovarian cells clearly affects cytoplasmic localization during oogenesis. The bicoid mRNA appears to be trapped upon entry into the oocyte. As described in Chapter 3, *Drosophila* oocytes are connected to nurse cells by cytoplasmic bridges. In ovaries of wild-type individuals, all nurse cells are located as a group next to one egg pole, and the part of the oocyte that faces the nurse cells becomes the anterior pole of the egg (Fig. 8.9a). Thus, the large amounts of RNA that are synthesized in the nurse cells and pass through the cytoplasmic bridges enter the oocyte from the future anterior pole. Unlike most RNAs, bicoid mRNA accumulates near the anterior pole of the oocyte after passing the cytoplasmic bridges between nurse cells and the oocyte (Fig. 8.9b). This localization mechanism depends on a large secondary structure formed by the trailer region of bicoid mRNA (Macdonald and Struhl, 1988; Macdonald, 1990).

The localization of bicoid mRNA also depends on the activity of other genes (Berleth et al., 1988). These genes presumably encode proteins that bind specifically to the secondary structure of the bicoid mRNA trailer and link it to anchor proteins, which in turn may be connected to the cytoskeleton of the oocyte. Indeed, the mRNA for an adducinlike protein that seems to promote the assembly of membrane-associated cytoskeletons is also localized at the anterior pole of *Drosophila* oocytes and eggs (Ding et al., 1993). However, it is not necessary to assume that other molecules involved in the anterior localization of bicoid mRNA have to be localized as well. The polarized influx of bicoid mRNA into the oocyte, in conjunction with nonlocalized oocyte molecules that attach bicoid mRNA to cytoskeletal components, might be sufficient to trap bicoid mRNA wherever it enters the oocyte. Subsequent changes in the localization of bicoid mRNA during embryonic development, as indicated by in situ hybridization data shown in Figure 8.1, may be caused by changes in the cytoskeletal system, to which the bicoid mRNA is attached.

The importance of the polar arrangement of nurse cells and the oocyte in the ovary—nurse cells located at one pole of the oocyte, the one that will become the anterior pole—is confirmed by the phenotype of the *dicephalic* mutant of *Drosophila*. Females lacking the nor-

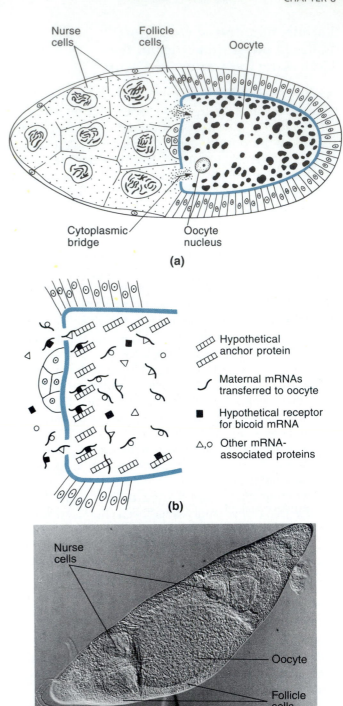

Figure 8.9 Apparent dependence of bicoid mRNA localization in *Drosophila* egg chambers on the arrangement of oocytes and nurse cells. **(a)** Wild-type ovarian egg chamber. Cytoplasm flowing from the nurse cells into the oocyte (arrows) contains maternal mRNAs, including bicoid mRNA. **(b)** Model of possible interactions between bicoid mRNA, a specific receptor protein for bicoid mRNA, and an anchor protein that links the receptor to a cytoskeletal component. These interactions would trap the bicoid mRNA at the point of entry into the oocyte. **(c)** Photomicrograph of an egg chamber from a *dicephalic* mutant female. Two clusters of nurse cells adjacent to opposite poles of the oocyte supply the *dicephalic* oocyte with nurse cell cytoplasm. Developing eggs from *dicephalic* females give rise to embryos with mirror-image duplications of cephalic and thoracic structures.

mal dicephalic function have egg chambers with two clusters of nurse cells located on opposite sides of the oocyte (Fig. 8.9c). Embryos developing from such oocytes show cephalic and thoracic structures on both the anterior and posterior ends (Lohs-Schardin, 1982). Apparently, the bipolar influx of nurse cell cytoplasm into the oocyte causes the localization of bicoid mRNA at two poles of the oocyte, and both of them later give rise to anterior embryonic structures.

The localization of other mRNAs in *Drosophila* oocytes cannot be ascribed to a simple trapping mechanism. Several mRNAs are localized to the posterior pole of the oocyte, the pole that is most distant to the nurse cells. The proteins encoded by these mRNAs are involved in germ cell determination and in establishing the posterior pole of the embryonic body pattern (see Chapter 21). The mRNAs that are localized posteriorly are also synthesized in the nurse cells, from where they are transported across the oocyte to their anchorage site near the posterior pole. Such targeting mechanisms are currently being investigated in several laboratories (Ding and Lipshitz, 1993). Presumably they will be based, in a more complex way than trapping, on the basic polarity of the ovarian egg chamber.

Cytoplasmic Components of Ascidian Eggs Are Segregated upon Fertilization

Ascidians are considered to be related to vertebrates, because some ascidian larvae have a structure called a *notochord*, an embryonic skeletal element that is later replaced by the backbone. These ascidian larvae look like tadpoles and swim with long, muscular tails. Ascidians have been studied by developmental biologists for more than a century, partly because of various pigmented inclusions in their egg cytoplasm. Early investigators described as many as five differently colored types of cytoplasm in ascidian eggs. Most important, Conklin (1905) reported that differently pigmented cytoplasms are segregated in different embryonic cells, which then proceed to form distinct tissues or organs. This observation suggested that the colored cytoplasms might be localized determinants. Aside from the function of these determinants, the process of their localization has become of interest in itself, because it consists of a dramatic set of movements.

Adult ascidians shed their eggs as primary oocytes still arrested in meiotic prophase I. These oocytes are radially symmetrical relative to the animal-vegetal axis, and several organelles in the egg *cortex* are distributed with an animal-vegetal polarity (Sardet et al., 1992). Oocyte maturation is triggered by sperm entry, which may occur anywhere on the egg surface. In newly fertilized eggs of *Styela partita*, a layer of cytoplasm containing yellow pigment granules lies just beneath the plasma membrane (Fig. 8.10a). This yellow cytoplasm is called *myoplasm* (Gk. *mys,* "muscle"), because it is

enclosed by cells that later form most of the tail muscle. A different mass of clear cytoplasm, derived from the contents of the germinal vesicle, is located in the animal half of the egg. This material is called *ectoplasm*, because the cells inheriting it will form mostly ectodermal structures. The remainder of the egg is filled with gray, yolk-rich cytoplasm that will be segregated in endodermal cells, which form the gut.

Shortly after fertilization, the contents of *Styela* eggs become dramatically rearranged, with the result that different types of cytoplasm become distinctly localized (Fig. 8.10b–d). The rearrangements, which are referred to as *ooplasmic segregation,* can be divided into two phases (Jeffery, 1984). The first phase occurs while the egg is still completing its meiotic divisions. The myoplasm streams down the egg periphery toward the vegetal pole, where it accumulates as a yellow cap. In the wake of the myoplasm, islets of ectoplasm also flow to the vegetal pole and form a clear layer above the yellow cap. As both myoplasm and ectoplasm are streaming into the vegetal half of the egg, the gray ectoplasm is displaced toward the animal pole. At the end of the first phase, the gray cytoplasm, ectoplasm, and myoplasm are stratified perpendicular to the *animal-vegetal axis,* with the myoplasm occupying the vegetalmost position.

The second phase of ooplasmic segregation begins after the meiotic divisions have been completed. The sperm nucleus has then become the male pronucleus and is located near the vegetal pole, presumably because the sperm entered there or because the sperm nucleus was swept there along with the myoplasm. The male pronucleus forms an aster of microtubules and migrates along the periphery of the egg in the direction of the animal pole. Both myoplasm and ectoplasm move along with the male pronucleus. At a position below the egg equator, the myoplasm is left behind and forms a *yellow crescent,* which marks the future posterior end of the embryo. After the yellow crescent has formed, the ectoplasm returns to the animal pole along with the male pronucleus. Here the ectoplasm will form a specialized region around the fused pronuclei. At the same time, a major portion of the gray cytoplasm returns to the vegetal hemisphere. A fourth cytoplasmic region, the *chordoplasm,* forms opposite the yellow crescent and is later included into the blastomeres that give rise to an embryonic skeletal rod, the *notochord.*

At the close of the ooplasmic segregation, the ascidian egg has two axes of polarity. One is the *animal-vegetal axis,* with the vegetal pole predicting the site of *blastopore* formation. In addition to the animal-vegetal axis, the ascidian egg has acquired an *anteroposterior axis,* with the posterior pole marked by the yellow crescent.

The yellow crescent of myoplasm has received much attention because its localization is so dramatic, and because it is segregated in a cell lineage that will form nearly all of the larval tail muscle (Whittaker, 1979;

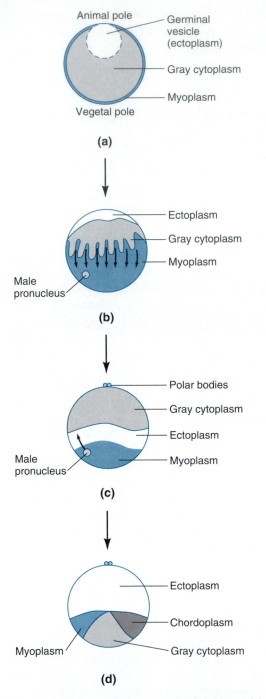

Figure 8.10 Ooplasmic segregation in *Styela partita*. **(a)** Unfertilized egg: the entire cortical cytoplasm contains yellow pigment granules (myoplasm). **(b, c)** Fertilized egg during the first phase of ooplasmic segregation. Upon fertilization, myoplasm streams down the egg periphery and accumulates as a yellow cap at the vegetal pole. In the wake of the myoplasm, clear cytoplasm (ectoplasm) also flows to the vegetal pole and forms a layer on top of the myoplasm while gray yolky cytoplasm is displaced to the animal hemisphere. **(d)** Second phase of ooplasmic segregation. The male pronucleus, which presumably has been swept to the vegetal pole by the myoplasm, migrates along the egg periphery toward the animal pole. Both myoplasm and ectoplasm move with the male pronucleus. While the ectoplasm returns to the animal hemisphere, the myoplasm is left behind to form the yellow crescent below the egg equator. An area of dark-gray chordoplasm forms opposite the yellow crescent.

Meedel et al., 1987; Nishida, 1987). Some component in the myoplasm acts as a morphogenetic determinant for muscle development. The yellow pigment in the myoplasm is probably not the determinant itself but rather an associated marker. Nevertheless, the visibility of the pigment granules makes it much easier to analyze the mechanism for its localization.

To analyze the mechanism of yellow crescent formation, William Jeffery and Stephen Meier (1983) used biochemical methods and scanning electron microscopy. They concluded that the yellow pigment granules are attached, presumably by intermediate filaments, to a *plasma membrane lamina (PML)*. The PML forms a thin meshwork right underneath the plasma membrane and includes microfilaments as major components. In unfertilized eggs, the PML is present in the entire cortex. When the yellow granules recede to the vegetal pole after fertilization, so does the PML (Fig. 8.11). The authors propose that contraction of the PML provides the driving force for the movement of the myoplasm to the vegetal pole. They also suggest that PML contraction drives the gray cytoplasm into its temporary position near the animal pole. This would explain the bulge of cytoplasm at the animal pole that widens considerably before the spherical shape of the egg is restored (Fig. 8.12).

The role of cytoskeletal elements in ooplasmic segregation was also tested by application of appropriate inhibitors. The first phase of segregation, but not the second, was inhibited by cytochalasin B (Sawada and Schatten, 1988). This indicates again that microfilaments must function during the first phase as concluded from the observations of Jeffery and Meier. Conversely, the microtubule inhibitors colcemid and nocodazole inhibited the second phase of ooplasmic segregation but not the first. Sawada and Schatten concluded that the aster of microtubules originating near the male pronucleus is necessary for the second segregation phase. They proposed that during the first phase, microfilament action sweeps the sperm nucleus and the myoplasm together near the vegetal pole. Then, as the microtubules of the sperm aster assemble, the growth and movement of the aster toward the animal pole might serve to bring the myoplasm into its final crescent position.

In summary, ascidian eggs acquire their animal-vegetal polarity during oogenesis and their anteroposterior polarity upon fertilization. The entry of the sperm triggers a dramatic reorganization of cytoplasmic components, which are segregated along the two egg axes. The segregation process depends on microfilaments in the egg cortex and on microtubules in the aster surrounding the male pronucleus. The segregated cyto-

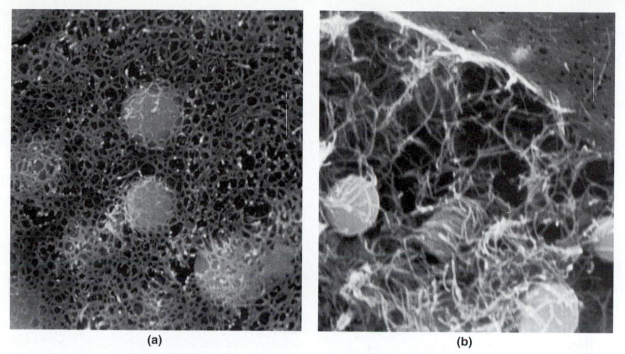

(a)   (b)

Figure 8.11 Association of pigment granules with a plasma membrane lamina in the eggs of an ascidian, *Styela plicata*. The scanning electron micrographs show eggs that were extracted with a detergent to make cytoskeletal structures more visible. **(a)** Plasma membrane lamina with underlying pigment granules at the conclusion of the first phase of ooplasmic segregation. **(b)** The plasma membrane lamina (top) and deep filamentous lattice (below) viewed from the side. Note the association of the lattice filaments with both the plasma membrane lamina and pigment granules (lower left).

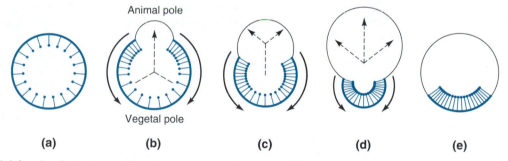

(a)   (b)   (c)   (d)   (e)

Figure 8.12 Model for the first phase of ooplasmic segregation in ascidians. In each diagram, the thick, colored part of the egg boundary represents the part of the plasma membrane with plasma membrane lamina underneath, and the thin, black part of the egg boundary represents the part of the plasma membrane without the plasma membrane lamina. The structures attached to the inside of the plasma membrane lamina, including the deep filamentous lattice and pigment granules, are represented by radial lines and dots. **(a)** Unfertilized egg. **(b–d)** First phase of ooplasmic segregation. Direction of ooplasmic movement: arrows with solid lines, myoplasm; arrows with broken lines, gray cytoplasm. **(e)** Zygote that has completed meiosis and the first phase of segregation.

plasmic components are enclosed by different blastomeres, whose fate they seem to determine.

Cytoplasmic Determinants in Ctenophores Are Localized during Cleavage

Cytoplasmic localizations that occur during oogenesis or fertilization are generally stable, presumably because the localized components somehow become anchored to the *cytoskeleton*. In some species with localized determinants, investigators have shown that eggs develop normally even if the orientation of cleavage furrows is variable or is manipulated experimentally (Henry et al., 1990). In other species, cytoplasmic determinants are localized while cleavage is under way, and the orientation of the cleavage furrows is critical to the localization process.

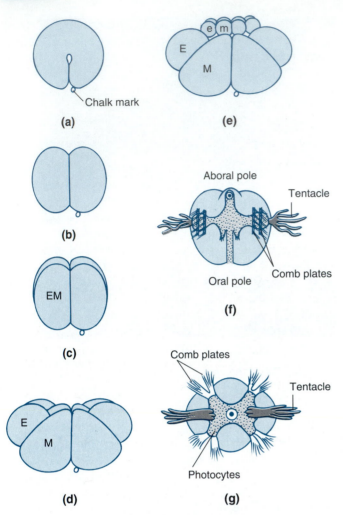

Figure 8.13 Normal development of the ctenophore *Mnemiopsis leidyi.* **(a)** First cleavage. A chalk mark placed on the site where the first cleavage furrow originates is traced to the oral pole, where the mouth of the larva is formed later. The opposite pole is the aboral pole. **(b–e)** 2-, 4-, 8-, and 16-cell stages of embryogenesis, viewed from the side. **(f)** Larva viewed from the side, with tentacles sticking out. **(g)** Larva viewed from the aboral pole, showing eight rows of comb plates. The photocytes are located in the radial canals beneath the comb plates. Comb plates and photocytes are derived from the E and M blastomeres, respectively.

A clear example of localization during cleavage is seen in the embryo of the ctenophore *Mnemiopsis leidyi.* As a larva, this aquatic animal has several distinct cell types including ciliated comb plate cells for swimming and photocytes for producing light. During the third embryonic cleavage, each blastomere forms an inner M macromere and an outer E macromere (Fig. 8.13). This is the first division during development in which cytoplasmic determinants become segregated in different cell lineages. Isolated E macromeres continue to cleave in culture and eventually form comb plate cells but no photocytes. Conversely, isolated M macromeres form photocytes but no comb plate cells. The following ex-

periments indicate that the localization of cytoplasmic determinants in *Mnemiopsis* depends on mitotic spindle orientation. Only if the appropriate orientation is attained during a certain time interval do comb plate cell determinants and photocyte determinants segregate normally.

Using chalk particles as markers, Gary Freeman (1976, 1979) mapped the prospective E and M blastomeres to certain blastomere regions at the 2-cell and 4-cell stages. Then he isolated the corresponding blastomere regions by pinching their plasma membranes together with fine glass tools. Nucleated blastomere fragments created in this fashion were cultivated in isolation and monitored for their *potency* to generate comb plate cilia or photocytes. According to this analysis, determinants for comb plate cells or photocytes are not localized at the 2-cell stage. At the 4-cell stage, the comb plate determinants are localized exclusively in the prospective E macromere region, while the photocyte determinants are still found in both regions.

To analyze the effects of cleavage on the localization process in *Mnemiopsis*, Freeman inhibited *cytokinesis* reversibly with cytochalasin B or 2,4-dinitrophenol (Fig. 8.14). When the second cytokinesis was suppressed, then the subsequent cleavage belonged to one of two types. In most embryos, the orientation of the mitotic spindles during the subsequent cleavage was the same as in a normal third cleavage (Fig. 8.14c, upper right). In these cases, the spindle orientation was appropriate to the third cleavage cycle, and the determinants for comb plate cells and photocytes segregated normally. However, in some embryos the first cleavage after release of the cytokinetic block reassembled a normal *second* cleavage (Fig. 8.14c, lower right). In these cases, the spindle orientation was inappropriate to the cycle counted by the hypothetical cleavage clock, and the determinants for comb plate cells and photocytes did not segregate normally.

In another experiment, Freeman forced the mitotic spindle orientation out of synchrony with the cleavage clock by compressing embryos during second cleavage (Fig. 8.14d). The resulting blastomere configuration was the same as shown in Figure 8.14c, upper right, except that it was generated one cycle early. Under these circumstances, the cytoplasmic determinants were not segregated normally.

The results of Freeman (1976, 1979) suggest that the localization mechanism depends on a cleavage "clock" as described earlier for sea urchins (see Chapter 5). This hypothetical clock is started by the first cleavage and counts off mitotic cycles regardless of whether cytokine-

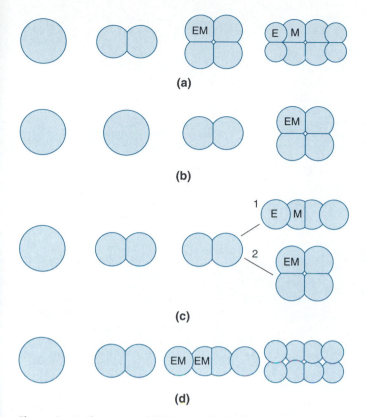

Figure 8.14 Cleavage inhibition and spindle reorientation experiments in *Mnemiopsis* embryos, viewed from the aboral pole. **(a)** Normal development. **(b)** First cytokinesis reversibly inhibited. **(c)** After reversible inhibition of the second cytokinesis, two results were observed. The resulting cleavage pattern resembled the one normally produced by the third cleavage (upper pattern) or the second cleavage (lower pattern). **(d)** Second cleavage occurred under compression, producing a cleavage pattern similar to C1 but with a different effect on localization. The letters E and M represent the potency to form comb plates and photocytes, respectively.

sis occurs or is inhibited. The cytoplasmic determinants acquire their normal localization only if the mitotic spindles are oriented appropriately to the cleavage cycle.

Bioassays for Localized Cytoplasmic Determinants

Two methods allow investigators to decide unequivocally whether a cytoplasmic region contains a localized determinant. One method, called *rescue*, restores a defective embryo to normal development; the defect in the test embryo may result from either experimental manipulation or from mutation in a maternally expressed gene, and the rescuing material—in this context—consists of transplanted cytoplasm or cytoplasmic components. The other method, known as *heterotopic*

transplantation, involves transplanting the cytoplasm of interest to an abnormal site. Both rescue and heterotopic transplantation can be used as *bioassays* to characterize localized determinants with regard to their molecular nature, time of localization, species specificity, and other properties.

Rescue Experiments Restore Defective Embryos to Normal Development

In many insect embryos, the primordial germ cells form at the posterior pole and are known as *pole cells*. They are easy to distinguish from somatic cells because they form earlier, are larger than the *blastoderm* cells, and are round rather than columnar in shape (Fig. 8.15). The cytoplasm that is enclosed in pole cells is free of yolk and contains fibrous granules, called **polar granules,** that stain for RNA. In transmission electron micrographs, polar granules appear dark and are often associated with *polysomes,* i.e., aggregates of mRNA and ribosomes that synthesize protein. To explore the function of polar granules, investigators have removed or damaged them by heat or irradiation.

When the posterior pole of a *Drosophila* embryo is irradiated with ultraviolet light (UV), no pole cells form, and the developing adults have gonads without gametes. This demonstrates that some components of the posterior cytoplasm are *necessary* for pole cell formation. To find out whether the activity of these components is *transplantable,* Masukichi Okada and his colleagues (1974) injected sterilized embryos with posterior pole plasm from unirradiated donor embryos (Fig. 8.16). Indeed, many of the injected embryos were restored to pole cell formation and normal gonad development. So far, this experiment does not show that the rescuing components, those necessary for pole cell formation, are *localized.* These components could have been nonlocalized organelles, such as ribosomes or mitochondria, that are necessary for the survival of any cell.

In order to prove that the rescuing components are localized, Okada and coworkers performed the following control. They took cytoplasm from the anterior pole and injected it in the same fashion into the posterior pole region of UV-irradiated embryos. No rescue was observed: the recipients were unable to form pole cells. Together, the experiment and the control show that components required for pole cell formation are present in the posterior but not in the anterior pole of *Drosophila* eggs. Because these components are localized and change the fate of cells from somatic to pole cells, they act as pole cell determinants. Since the fate of pole cells is to form gametes, these determinants are usually called **germ cell determinants.**

Rescue bioassays can also be used in conjunction with maternal effect mutants. In these studies, the recipient eggs are derived from mutant females instead of being inactivated experimentally.

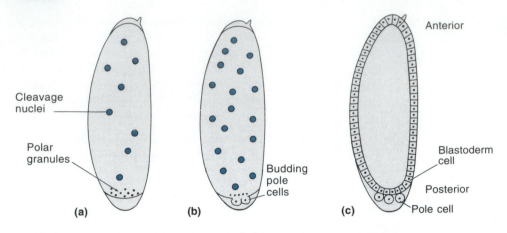

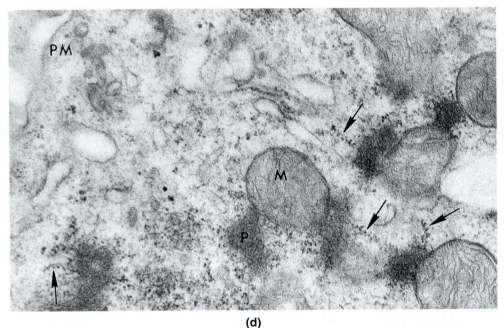

Figure 8.15 Polar granules and pole cell formation in the *Drosophila* embryo. **(a)** Schematic outline of an embryo during early cleavage, indicating the position of polar granules near the posterior pole. **(b)** As the pole cells form, they include the cytoplasm containing polar granules. **(c)** Embryo at the blastoderm stage. Pole cells differ in size and shape from blastoderm cells. **(d)** Transmission electron micrograph showing polar plasm of a *Drosophila* embryo 15 min after fertilization. The polar granules (P) are attached to mitochondria (M) and have helical structures resembling polysomes (arrows). PM = egg plasma membrane.

For instance, eggs from *Drosophila* females that are mutant for the *oskar* gene have no polar granules, fail to form pole cells, and give rise to sterile adults. However, Ruth Lehmann and Christiane Nüsslein-Volhard (1986) were able to restore such eggs to normal development and fertility by injecting cytoplasm from wild-type donor eggs. To determine the location of the rescuing activity in the donor embryo, they tested cytoplasm taken from various positions along the anteroposterior axis. The activity was localized near the posterior pole and dependent on the expression of the wild-type *oskar*$^+$ allele in the females that produced the donor eggs. Eggs from *oskar* mutant females failed to provide rescuing activity. Jointly, these results show that rescuing activity is derived from maternal *oskar*$^+$ genes and becomes localized near the posterior pole of the egg.

Similarly, eggs from *Drosophila* females homozygous for *bicoid* develop into larvae lacking head and thorax (see Fig. 14.2). However, such eggs are restored to normal development by injecting anterior cytoplasm from wild-type donor eggs (Frohnhöfer and Nüsslein-Volhard, 1986). These experiments, which combine the rescue technique with the power of genetic analysis, will be discussed more fully in Chapter 21.

Heterotopic Transplantation Reveals Cytoplasmic Determinants by Their Action in Abnormal Locations

Another way of demonstrating the involvement of localized cytoplasmic determinants in the development of an embryo is by *heterotopic transplantation*. In this procedure, a presumed determinant is taken from its normal region in the donor and transplanted to a different region in the recipient. Using this method, Karl Illmensee and Anthony Mahowald (1974) showed that posterior pole plasm from *Drosophila* eggs caused *ectopic* pole cell formation (Gk. *ektopos*, "out of place"). The first step in their experiment was the microinjection of posterior pole plasm from a donor embryo during

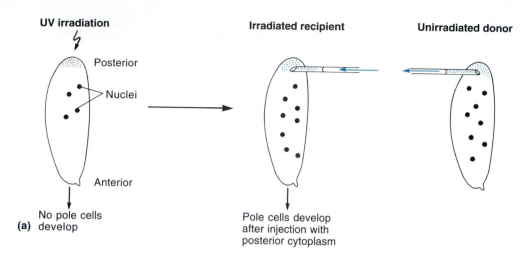

UV irradiation

Posterior

Nuclei

Anterior

(a) No pole cells develop

Irradiated recipient

Unirradiated donor

Pole cells develop after injection with posterior cytoplasm

Pole cells

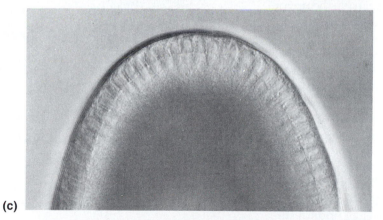

(b)

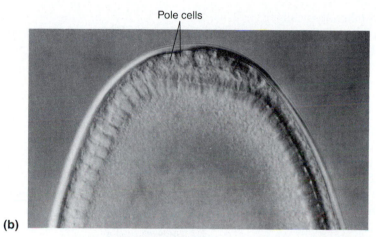

(c)

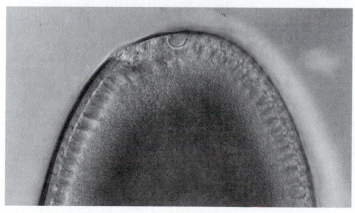

(d)

Figure 8.16 Rescue of *Drosophila* embryos sterilized by UV irradiation by subsequent injection with cytoplasm from an unirradiated donor. **(a)** Experimental technique. The cytoplasm was transferred with a fine, beveled glass needle. Only posterior cytoplasm restored the ability of the irradiated recipient to form pole cells. Light micrographs showing **(b)** normal embryo with pole cells, **(c)** UV-irradiated embryo without pole cells, and **(d)** rescued embryo.

early cleavage to the anterior pole of a recipient at the same stage of development (Fig. 8.17). As a result, the recipient formed cells at the anterior pole that resembled pole cells in their size, shape, and ultrastructure. Specifically, the modified anterior cells incorporated the polar granules transferred with the posterior pole plasm and formed electron-dense nuclear bodies that are characteristic of pole cells.

In order to test whether these modified anterior cells would be able to function as primordial germ cells, the researchers transplanted them to the posterior pole of a new host embryo, from where the cells could reach the host's gonads. The gonads then became populated by transferred cells as well as the host's own pole cells. To distinguish between the gametes produced by the two different sets of pole cells, the investigators used recessive genetic markers. (Such markers are phenotypically expressed only if *all* copies of the gene in a cell are mutated.) The transplanted cells carried the nuclei of the first recipient, which was homozygous for one set of genetic markers. In contrast, the host embryos were homozygous for another set of genetic markers. When adult flies developing from the host embryos were mated with partners of the same genotype, most of the offspring showed the host phenotype. However, 4 of 92 matings also produced offspring showing the *wild* phenotype, which therefore had to be heterozygous for any of the recessive genetic markers used. Such offspring could arise only with the participation of gametes derived from the transplanted cells. The results showed that the posterior pole plasm transferred from the donor had caused the ectopic formation of functional primordial germ cells in the recipient embryo.

An experimental technique equivalent to heterotopic transplantation entails forcing mitotic spindles of early embryos into abnormal orientations. This may lead to abnormal partitioning so that localized cytoplasmic components end up in blastomeres that would not normally inherit them. By observing the fate of these blastomeres, one can establish whether the abnormally partitioned material acts as a determinant. Such an experiment was carried out with embryos of the ascidian *Styela plicata*. As described earlier, a crescent of yellow *myoplasm* becomes localized in *Styela* eggs upon fertilization. The yellow myoplasm is enclosed by certain blastomeres that subsequently give rise to most of the tail musculature.

In order to test whether yellow myoplasm actually determines muscle cell development, J. Richard Whittaker (1980) performed the following experiment. By compressing *Styela plicata* embryos between glass plates during third cleavage, he changed the orientation of the cleavage plane from equatorial to meridional. This resulted in a faulty segregation of the yellow myoplasm (Fig. 8.18). Instead of being allocated to the two dorsal vegetal blastomeres only, the myoplasm was now inherited by four out of eight blastomeres. The subsequent development of these blastomeres was difficult

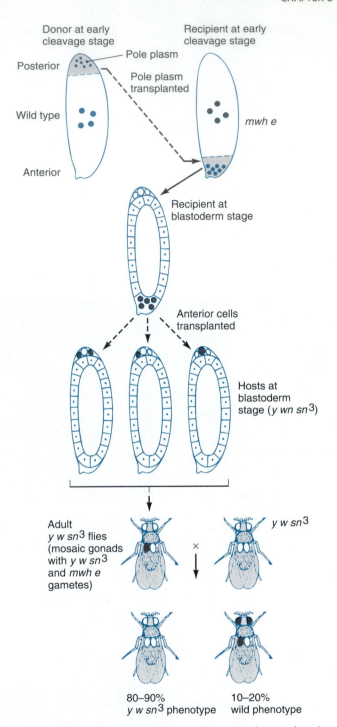

Figure 8.17 Heterotopic transplantation experiment showing the localization of germ cell determinants at the posterior pole of *Drosophila* eggs. Posterior pole plasm from wild-type donors was transplanted to the anterior pole of recipients at the same early cleavage stage. As a result, cells with the appearance of pole cells developed at the injection site. These cells (shaded black) were transplanted to the posterior pole of host embryos, from which they could reach the host's gonad. To distinguish any gametes formed by the transplanted cells from those formed by the host's own pole cells, the transplanted cells were marked by the mutations *multiple wing hair (mwh)* and *ebony (e)*, whereas the host carried a different set of mutant alleles, *yellow (y)*, *white (w)*, and *singed (sn³)*. Hosts grown to adulthood and outcrossed with *y w sn³* partners produced some phenotypically wild offspring, indicating that some of their gametes were derived from the transplanted pole cells.

to monitor, because the development of the compressed embryos was too disorganized. However, Whittaker still obtained strong evidence that the four blastomeres containing yellow myoplasm were programmed for muscle development.

Having released the embryos from the compression between the glass plates, Whittaker incubated them in a medium containing *cytochalasin B*. Treatment with this antimicrofilament drug suppressed cytokinesis and kept the embryos in a state of cleavage arrest. Control embryos that had passed through their third cleavage without compression were kept in cleavage arrest in the same way. At the stage when normal embryos began to form tail muscle, the cleavage-arrested embryos were subjected to a treatment that stains acetylcholinesterase, an enzyme indicative of muscle cell differentiation. As expected, the embryos that had passed through their third cleavage normally showed two cells stained for acetylcholinesterase: the two cells that inherit yellow myoplasm and proceed to form tail muscle. In contrast, four of eight blastomeres in the compressed embryos stained positively for acetylcholinesterase. This result indicated that the allocation of yellow myoplasm to additional blastomeres was sufficient to switch their fate to muscle cells.

Rescue and Heterotopic Transplantation Can Be Used as Bioassays

Heterotopic transplantations, as well as rescue experiments, can be used as bioassays to further characterize the determinant of interest. The following examples will illustrate how such bioassays have been used to define the exact stage of localization for a determinant, its species specificity, or its molecular nature.

The heterotopic transplantation of posterior pole plasm in *Drosophila* was used to define the *developmental stage* at which the germ cell determinants become localized at the posterior pole. To this end, the transplanted cytoplasm was taken from the posterior poles of eggs and oocytes at progressively earlier stages (Illmensee et al., 1976). In each case, the cytoplasm was injected near the anterior pole of a genetically marked host embryo. The ectopic pole cells that developed were examined for their morphological characteristics and for their ability to give rise to gametes, as described previously. Functional ectopic pole cells were obtained with posterior cytoplasm from oocytes as early as the stage at which vitellogenesis is complete and the nurse cells have just transferred their contents to the oocyte. In donor oocytes at earlier stages, polar granules had accumulated in the posterior cytoplasm, but this cytoplasm did not cause ectopic pole cell formation. Thus, it appears that the polar granules visible in younger oocytes are not yet associated with germ cell determinants.

In a similar set of experiments, Mahowald and coworkers (1976) explored whether localized germ cell

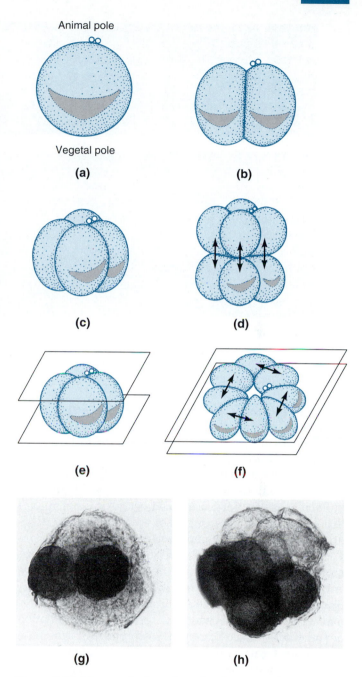

Figure 8.18 Abnormal allocation of yellow myoplasm after compression of *Styela plicata* embryos during third cleavage. **(a–d)** Normal allocation into B4.1 blastomeres. **(e, f)** Mitotic spindles during third cleavage were reoriented by compression between glass plates. As a result, four blastomeres instead of two inherited yellow myoplasm. **(g)** A normal embryo was subjected to cleavage arrest at the 8-cell stage and later stained for acetylcholinesterase. Two blastomeres stained. **(h)** When a compressed embryo was treated the same way, four blastomeres stained for acetylcholinesterase.

determinants from one species were compatible with the responding system of another species. To this end, these researchers transplanted posterior pole plasm from *Drosophila immigrans* donor embryos to the anterior pole region of *Drosophila melanogaster* recipients. As a result, the recipients developed hybrid pole cells con-

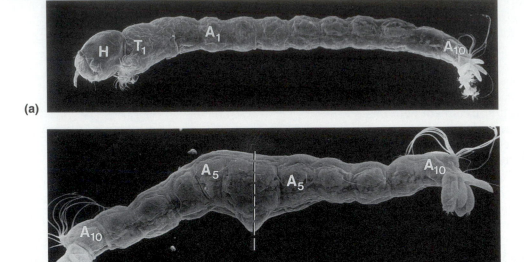

(a)

(b)

Figure 8.19 (a) Scanning electron micrographs of normal larva of *Chironomus samoensis* and **(b)** double-abdomen larva obtained after anterior UV irradiation of the embryo during cleavage. H = head; T1 = first thoracic segment; A1, A5, A10 = first, fifth, and last abdominal segments.

taining a *D. melanogaster* nucleus and *D. immigrans* cytoplasm. When these cells were transplanted back to the posterior pole of a genetically marked host, they gave rise to functional gametes. The results show that germ cell determinants have been conserved well enough during the process of evolution to be recognized by cells of a closely related species.

Rescue experiments with eggs of a midge, *Chironomus samoensis*, have been used as a bioassay to identify the molecular nature of a localized anterior determinant that specifies the anterior part of the body pattern. Eggs of *Chironomus* and related species form strikingly abnormal body patterns in response to various experimental manipulations. In the **double-abdomen pattern,** anterior segments including head and thorax are replaced by a mirror-image duplication of abdominal segments (Fig. 8.19). This pattern is formed after irradiation of the anterior pole region of wild-type embryos with ultraviolet light (Yajima, 1964), or as a mutant pheno-

type (Percy et al., 1986). Embryos programmed for double-abdomen development are restored to normal development by the microinjection of cytoplasm from unirradiated wild-type donors (Fig. 8.20). The rescuing activity is present in anterior, but not posterior, egg cytoplasm of *Chironomus* and closely related species (Kalthoff and Elbetieha, 1986). The same rescuing activity was found in total egg RNA, and in RNA containing a poly(A) segment, a characteristic of mRNA.

To find what size of RNA has anterior determinant activity in *Chironomus* eggs, total egg RNA was sedimented through a sucrose gradient, and fractions containing different size classes of RNA were collected. Each size class was tested separately in the rescue bioassay (Elbetieha and Kalthoff, 1988). The strongest rescue activity was associated with the smallest size fraction, containing RNA molecules between 250 and 600 nucleotides in length (Fig. 8.21). Control experiments with corresponding size fractions from *Xenopus* eggs and

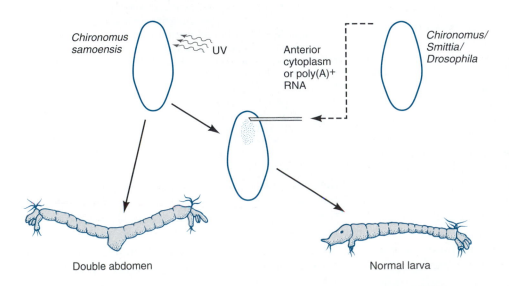

Double abdomen

Normal larva

Figure 8.20 Rescue bioassay for anterior determinants in eggs of *Chironomus samoensis*. Eggs are programmed for double-abdomen development by anterior UV irradiation. Embryos can be restored to normal development by injecting anterior cytoplasm or poly(A)-containing RNA from unirradiated donor eggs. Anterior determinant activity is measured as an increase in the percentage of normal embryos among the surviving embryos.

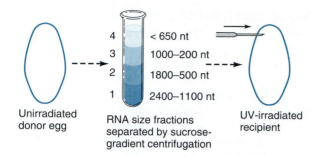

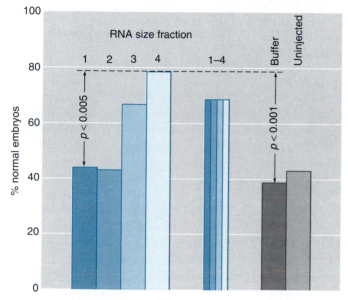

Figure 8.21 Association of anterior determinant activity in *Chironomus* eggs with a small RNA size class. RNA extracted from unirradiated eggs was divided into size classes by centrifugation on a sucrose gradient. The larger the RNA, the faster it sedimented to the bottom of the tube. Size fractions were tested for anterior determinant activity in UV-irradiated recipients as shown in Figure 8.20. The greatest activity was found in the smallest size class, enriched for RNA smaller than 650 nucleotides in length. UV-irradiated eggs that were uninjected or injected with salt solution served as controls. *p* values in double-headed arrows indicate that the probabilities for obtaining the observed differences by chance were very low.

with UV-irradiated RNA from *Chironomus* eggs showed no rescuing activity. The data indicate that *Chironomus* eggs contain an unusually small mRNA that is localized anteriorly and serves as a cytoplasmic determinant for the anterior body pattern.

Properties of Localized Cytoplasmic Determinants

Localized Cytoplasmic Determinants May Be Activating or Inhibitory

Most localized determinants exert function by *activation*, because their presence causes the development of a particular structure that would otherwise not be formed. In other words, an activating determinant endows a cell or an embryonic region with a capability that otherwise would not be within its potential. However, some cytoplasmic determinants act through *inhibition*, by preventing cells from forming certain embryonic parts. Those cells may still be able to form more than one structure, but the inhibitory determinant restricts their potential without adding a new capability.

Both activating and inhibitory determinants are present in polar lobes of the polychaete worm *Sabellaria cementarium*. Polychaete larvae have an apical tuft of long cilia and an equatorial band of shorter cilia (Fig. 8.22). Jo Ann Render (1983) showed that no apical tuft will form if the polar lobe that protrudes before the first cleavage is removed. This suggested that the first polar lobe contains cytoplasmic components required for apical tuft formation. In another set of experiments, the first cleavage was allowed to occur normally before the polar lobe preceding the second cleavage was removed. When blastomeres were isolated thereafter, C and D blastomeres each formed an apical tuft while A and B blastomeres did not. When the second cleavage was equalized by addition of a detergent to the culture medium, so that both C and D blastomeres received polar lobe material, no apical tuft was formed. These results indicated that the second polar lobe contains an inhibitor of apical tuft formation. When the first two cleavages were allowed to occur normally, any combination of cells containing the C blastomere but not the D blastomere, such as ABC, formed apical tufts. Conversely, cell combinations containing the D blastomere but not the C blastomere, such as ABD, did not form apical tufts. Apparently, the inhibitor present in

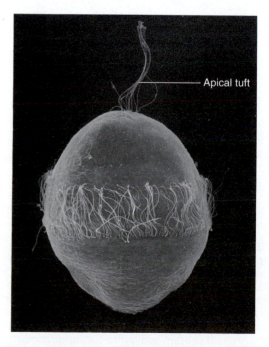

Figure 8.22 Scanning electron micrograph of *Sabellaria* larva showing apical tuft.

the second polar body is shunted selectively into the D blastomere during second cleavage.

Together, the results show that the first polar lobe contains both activating and inhibitory determinants of apical tuft formation (Fig. 8.23). Both determinants are shunted into the CD blastomere after the first cleavage, but only the activating determinants remain there; the inhibitory determinants move back into the second polar lobe. These determinants are shunted selectively into the D blastomere after completion of the second cleavage. As a result, the C blastomere inherits only activating determinants, while the D blastomere inherits activating as well as inhibitory determinants. The inhibitory ones prevail in the D blastomere.

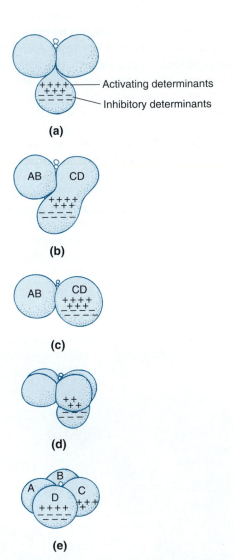

Figure 8.23 Hypothetical distribution of activating (+) and inhibitory (−) determinants for apical tuft formation in the polar lobes of *Sabellaria*. **(a)** First polar lobe at trefoil stage. **(b, c)** Fusion of the polar lobe with the CD blastomere. **(d)** Second polar lobe containing only the inhibitory determinants for tuft formation. **(e)** 4-cell stage after fusion of second polar body with D blastomere. Both C and D blastomeres now contain activating tuft determinants, whereas the D blastomere also contains inhibitory determinants.

Many Localized Cytoplasmic Determinants Are Maternal mRNAs

The molecular nature of cytoplasmic determinants, and the ways they control differential gene activity, have been of particular interest. Presumably, a variety of mechanisms are involved, many of which may still be unknown. Proteins would seem to be most suitable as localized determinants, because of their roles in intracellular signaling and their ability to control the expression of specific genes. Other likely candidates are the mRNAs that are translated into such regulatory proteins. Because mRNAs are generally more stable than proteins, they can be transcribed from the maternal genome during oogenesis and stored until they become functional during early embryogenesis. Indeed, about 20 localized mRNAs have been found in the oocytes and embryos of *Drosophila* and *Xenopus* alone (Ding and Lipshitz, 1993). Several of these mRNAs are active in rescue and heterotopic transplantation bioassays, as described in this chapter and in Chapters 9 and 21.

The mRNAs that serve as localized cytoplasmic determinants, like other mRNAs, are translated into proteins. The lifespan and/or diffusion of the protein products tends to be limited so that the concentration of these proteins is greatest in the vicinity of their mRNA templates. Some of the protein products bind to nuclear DNA and control the activity of embryonic genes. Others interact with different mRNAs, affecting their translation and stability. Still other proteins translated from localized mRNAs are signal molecules that act locally in the cell of origin or bind to receptors on the surface of other cells.

Several cytoplasmic determinants are associated with distinct organelles, such as the yellow pigment granules in the myoplasm of ascidians and the polar granules in the posterior pole plasm of *Drosophila* and other insects. In some cases, the association of a determinant with a visible marker may be fortuitous: the determinant would function as well without the marker. In other cases, a visible organelle may aid, at least temporarily, in the localization of a determinant.

Polar granules are visible in germ line cells throughout the life cycle of *Drosophila* (Mahowald, 1971). The association of polar granules with germ cell determinants, as measured by rescue or heterotopic transplantation, lasts for a limited period of time from late oogenesis until pole cell formation. During this period, polar granules are associated with helical structures resembling polysomes, that is, mRNA in the process of translation (Fig. 8.15e). Taken together, these observations suggest that polar granules serve as temporary anchor sites for mRNAs acting as germ cell determinants. (The same polar granules are also repositories for determinants specifying the posterior body pattern. See Chapter 21.) Organelles resembling the polar granules in *Drosophila* are also associated with germ cell determinants in amphibians and other organisms (Wakahara, 1990).

In summary, protein products of maternal mRNAs are the most likely components to act as localized cytoplasmic determinants, although other types of molecules may still be found. During the localization process, these determinants are probably anchored to cytoskeletal components, which may function as transport vehicles as well as anchor sites. Other cytoplasmic organelles, such as fibrous granules or yolk components, may also serve as temporary repositories and visible markers for localized determinants.

Cytoplasmic Determinants Control Certain Cell Lineages or Entire Body Regions

As we have seen, localized cytoplasmic determinants may control the fate of just one cell lineage, such as the germ line or the muscle cell lineage. Thus, inactivation of germ cell determinants in eggs eliminates the germ line but usually leaves all other cell types unaffected; the affected eggs develop into animals that are normal in their anatomy and behavior except for their empty gonads. Other localizations, such as the anterior determinants in *Chironomus* eggs, control the development of entire body regions. Instead of determining specific cell lineages or tissues, these localizations specify parts of the overall body pattern.

These different roles have implications for the ways in which determinants exert their regulatory functions. A determinant of one cell lineage or tissue type may affect one or more tissue-specific master genes that in turn control a set of subordinate genes. A determinant of an entire body region is more likely to interfere with a hierarchy of patterning genes that specify a harmonious array of entire body parts. We will return to this topic in Chapters 20 through 23.

In conclusion, many localized cytoplasmic determinants are maternally encoded mRNAs and proteins, which regulate embryonic gene activity. Most determinants seem to have activating functions, although some are inhibitory or mutually antagonistic. Determinants' mode of action on target genes may differ, depending on whether they control a specific tissue or an entire body region.

The Principle of Default Programs

The determination of embryonic cells by cytoplasmic localizations illustrates the *principle of default programs*, which can be observed in several contexts of development. Many developmental pathways require that cells receive a specific signal from a cytoplasmic localization or by induction. However, the absence of such a signal does not typically result in chaos or death. Instead, cells fall back on a program of development that does not depend on receiving the additional signal. Such programs, known as *default programs*, are very common in development. To illustrate this principle, the *Chironomus* embryos introduced earlier will again serve as an example.

In the eggs of *Chironomus* and related species, maternal mRNAs localized near the anterior pole are necessary for head and thorax formation. In the absence of these anterior determinants, eggs develop into *double-abdomen* embryos (Fig. 8.19). Double abdomens look the same, whether the anterior determinants have been removed by centrifugation or inactivated by means of enzymes or ultraviolet light. Thus, formation of a double-abdomen embryo is a default program followed by the egg in the absence of anterior determinants.

To test whether forming the anterior half of a double abdomen requires any interactions with the posterior half of the egg, Walter Ritter (1976) performed a combined UV irradiation and ligation experiment (Fig. 8.24). Transverse ligation separated an anterior embry-

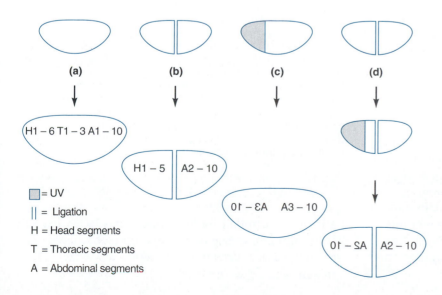

Figure 8.24 Combined ligation and UV irradiation experiment with *Smittia* embryos. **(a)** The normal segment pattern comprises the head, 3 thoracic segments, and 10 abdominal segments. **(b)** Upon transverse ligation, the anterior egg fragment develops an incomplete head, and the posterior fragment forms a partial abdomen. **(c)** UV irradiation of the anterior pole region causes the formation of double-abdomen embryos, similar to the one shown in Figure 8.19b. **(d)** After ligation and subsequent UV irradiation of the anterior pole region, the anterior fragment forms an abdomen with reversed polarity, indicating that all components necessary for abdomen formation are present in the anterior egg half.

onic fragment, which produced a head, from a posterior fragment producing a set of abdominal segments. However, when the anterior pole region was UV-irradiated after ligation, the anterior fragment formed, instead of a head, another set of abdominal segments with reversed polarity. Therefore, everything required for abdomen development must be present in both anterior and posterior halves of the egg. In other words, formation of an abdomen is the default program of the anterior egg half in the absence of anterior determinants. This result raises the question whether the posterior egg half, if provided with anterior determinants, will form head and thorax; as we will see, it does.

In addition to normal embryos and double-abdomen specimens, the eggs of *Chironomus* and related species can form two additional types of abnormal body pattern, called double-cephalon embryos and inverted embryos (Fig. 8.25). *Double-cephalon* embryos have mirror-image duplications of the head but are missing thorax and abdomen. *Inverted embryos* look normal, but the abdomen develops at the anterior end while the head originates posteriorly. Consequently, the pole cells formed at the posterior pole end up in the head. Both types of abnormal body pattern are obtained after centrifugation of eggs, an operation that apparently causes a variable displacement of anterior determinants (Rau and Kalthoff, 1980; Kalthoff et al., 1982; Yajima, 1983). In accord with this interpretation, anterior UV irradiation after centrifugation enhances the yield of double-abdomen and inverted embryos, presumably by inactivating anterior determinants that have remained anterior in the egg. Conversely, posterior UV irradiation of centrifuged eggs enhances the yield of double-abdomen and normal embryos, apparently by inactivating anterior determinants that have moved toward the posterior during centrifugation.

The various data obtained with *Chironomus* and related species indicate that both anterior and posterior egg halves can give rise to either anterior or posterior body parts. In the presence of anterior determinants, each egg half will form the anterior part of the body pattern. If these determinants are weakened by mutation or experimental means, then each egg half falls back independently on the default program of forming the posterior part of the body pattern.

The principle of default programs is also observed in other cases of cytoplasmic localization. In the absence of germ cell determinants, somatic cells are formed. For instance, UV irradiation of *Drosophila* eggs on the posterior pole prevents the formation of pole cells. Instead, the posterior pole is covered with the same type of blastoderm cells that are formed elsewhere. Moreover, the principle of default programs applies to embryonic induction. In fact, the definition of induction implies that the responding cells will follow a default pathway in the absence of the inducing signal. There are also default programs in the developmental

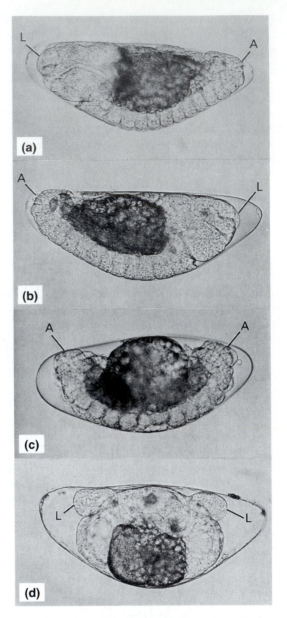

Figure 8.25 Four basic body patterns observed in embryos of *Smittia* sp. and related midges. In these photomicrographs, the anterior pole as marked by the micropyle is to the left, the ventral side facing down. **(a)** In the normal body pattern, the head is marked by the labrum (L) while the abdominal end is tipped with anal papillae (A). **(b)** Inverted embryos develop with their anteroposterior axis completely reversed, except for the pole cells, which end up in the head. **(c)** Double-abdomen embryos show a mirror-image duplication of abdominal segments in the absence of head and thorax. **(d)** In double-cephalon embryos, cephalic segments are duplicated while thorax and abdomen are missing. All four body patterns were obtained after centrifugation during cleavage, a treatment that apparently displaces or inactivates localized anterior determinants.

pathways controlled by homeotic genes (see Chapters 21 through 23), in sex determination (see Chapter 27), and in hormonal control (see Chapter 28).

SUMMARY

Most eggs have unevenly distributed components, called localized cytoplasmic determinants, that affect the fates of the blastomeres to which they are allocated. Such determinants become localized during oogenesis, upon fertilization, or during cleavage. As they are partitioned among different blastomeres, they create different environments for their nuclei and cause differential gene expression. Some determinants control the development of particular cell lineages or tissue types, while others direct the formation of entire body regions encompassing many different cell types. Most determinants have an activating effect, although some are inhibitory.

Two experimental designs are generally used to identify localized cytoplasmic determinants. In one of these, known as the rescue strategy, a mutant or experimentally manipulated embryo is restored to normal development by transplantation of cytoplasm from a normal donor. If the rescuing activity is restricted to a specific area in the donor, then it follows that one or more components in the transferred cytoplasm are localized and necessary for the development of the parts that would otherwise be missing or abnormal.

Another way of demonstrating the involvement of a localized cytoplasmic determinant in embryonic development is by heterotopic transplantation. In this procedure, cytoplasm containing a presumed determinant is transplanted from its normal location in the donor to a different region in the recipient. The same structure that would have formed at the donor site where the cytoplasm was removed may then be formed in the recipient at the site of transplantation. In this case, it can be concluded that one or more components of the transplanted cytoplasm are sufficient to cause the formation of the structure of interest. Both the rescue strategy and heterotopic transplantation can be used as bioassays to characterize the molecular nature and other properties of cytoplasmic determinants.

Localized cytoplasmic determinants may consist of different components. Maternal mRNAs and their protein products are the most likely molecules to act as localized cytoplasmic determinants, although other active molecules may also be found. During the localization process, determinants are probably anchored to cytoskeletal components, which may function as transport vehicles as well as anchor sites. Other cytoplasmic organelles, such as fibrous granules or yolk components, may serve as temporary repositories and visible markers for localized determinants. In their active phase, protein determinants may regulate gene expression or interact with other cytoplasmic localizations.

SUGGESTED READINGS

Davidson, E. H. 1986. *Gene Activity in Early Development,* 3d ed., Chap. 6. New York: Academic Press.

Ding, D., and H. D. Lipshitz. 1993. Localized mRNAs and their functions. *Bioassays* **15**:651–658.

Freeman, G. 1979. The multiple roles which cell division can play in the localization of developmental potential. In S. Subtelny and I. R. Konigsberg, eds., *37th Symposium of the Society for Developmental Biology: Determinants of Spatial Organization,* 53–76. New York: Academic Press.

Verdonk, N. H., and J. N. Cather. 1983. Morphogenetic determination and differentiation. In K. M. Wilbur, ed., *The Mollusca,* Vol. 3, 215–252. New York: Academic Press.

AXIS FORMATION AND MESODERM INDUCTION

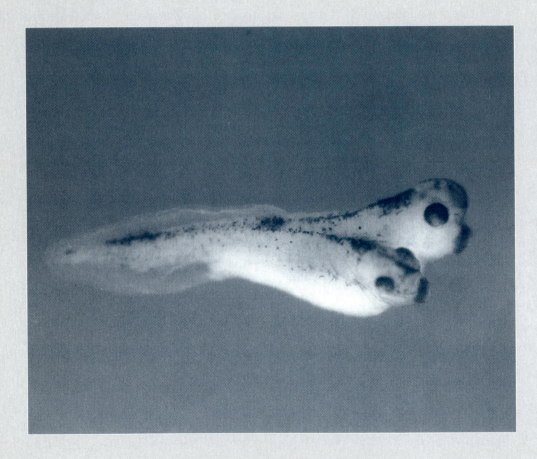

Figure 9.1 Duplication of dorsal organs in *Xenopus* embryos after ectopic injection of Vg1 mRNA, which encodes a peptide of the transforming growth factor-β family. Injection of modified Vg1 mRNA into ventral blastomeres during cleavage causes the formation of a second embryonic axis including a head. This result indicates that Vg1 protein plays a central role in establishing the dorsal side of the embryo.

In many cases, body axes are determined by *cytoplasmic localizations* in the egg or early embryo. As discussed in Chapter 8, certain anterior localizations in insect embryos determine the anterior *pole* of an anteroposterior body axis. Similarly, rearrangements of cytoplasm in amphibian eggs bias one half of the egg to develop as the future dorsal side. Some of these early polarities are *labile:* they may be overruled by subsequent events. In extreme cases, polarities can even be oriented by environmental factors such as light or gravity. As development proceeds, though, the polarities stabilize and become integral parts of the emerging body pattern. The stabilization process often involves *inductive interactions* by which cells direct the development of their neighbors; mimicking such inductive signals in wrong places in the embryo severely disturbs axis development (Fig. 9.1). In addition to making polarity axes stable, inductive interactions increase the complexity of the developing body pattern. From a simple axis with two poles, inductions generate *axial patterns* with several elements in a distinct order.

In this chapter, we will explore axis formation mainly in two organisms: a brown alga of the genus *Fucus* and the frog *Xenopus laevis. Fucus* has only one polar axis, which is induced by light or other environmental factors. In *Xenopus,* three body axes—animal-vegetal, dorsoventral, and left-right—form in a prescribed sequence. After surveying the development of all three body axes, our discussion will focus on the dorsoventral axis, which originates with the rearrangements of cytoplasm after fertilization and is fixed prior to the first cleavage. The further establishment of the dorsoventral axis is linked to the development of the intermediate germ layer, the *mesoderm,* which arises from inductive interactions between animal and vegetal blastomeres.

Because induction plays a key role in stabilizing and elaborating embryonic body axes, the *principle of induction* will be introduced in this chapter, along with the criteria by which investigators prove the occurrence of inductive interactions between cells. The chapter will conclude with a survey of current efforts to elucidate the molecular mechanisms underlying mesoderm induction.

Most animals have three *body axes* oriented at right angles to one another: *anteroposterior, dorsoventral,* and *left-right.* Some animals, and most plants, have only one axis. For instance, adult sea urchins have one axis, the *oral-aboral* axis, which connects the oral pole (the mouth) with the aboral pole (near the anus). An axis is defined by two opposite poles, and various structures are aligned along each axis in a specific order. For example, the order of bony elements along the anteroposterior axis of a mammal is the skull, the cervical (neck) vertebrae, the shoulder girdle with forelimbs, and so forth.

Body Axes and Planes

Most metazoa have three *body axes:* an *anteroposterior axis,* a *dorsoventral axis* (Lat. *dorsum,* "back"; *venter,* "belly"), and a *left-right axis* (Fig. 9.2a). An additional axis, used mostly for designating limb parts, is the *proximodistal axis. Proximal* (Lat. *proximus,* "next," "closest") describes the portion close to the point of attachment, to the center of the body, or to another point of

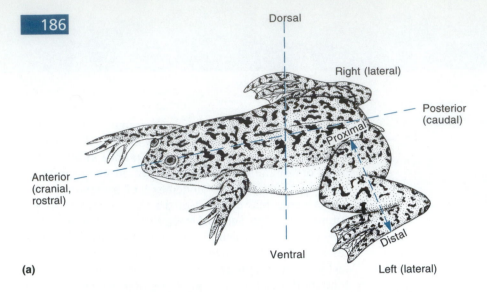

(a)

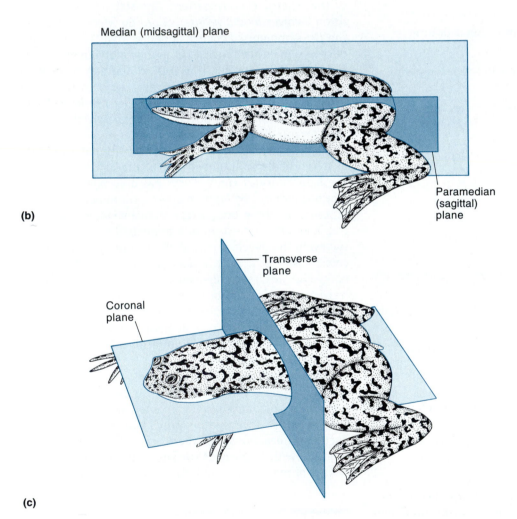

(b)

(c)

Figure 9.2 Body axes and planes in a vertebrate. See text.

reference. *Distal* (Lat. *distare*, "stand apart") refers to a part that is away (distant) from the point of reference. We also describe views of the body with regard to the poles of axes. For instance, a dorsal aspect is a view from the dorsal pole, and a lateral aspect is a view from the right or left side.

Sections of the body are named according to specific planes (Fig. 9.2b and c). The *median* (or *midsagittal*) *plane* divides the left and right sides of the body and is a plane of symmetry for most organs. A *paramedian* (or *sagittal*) *plane* is any plane parallel to the median plane but displaced to the left or right side. The median plane and all paramedian planes are perpendicular to the left-right axis. Any plane perpendicular to the anteroposterior axis is a *transverse plane*, or, colloquially, a *cross section*. Any plane perpendicular to the dorsoventral axis is a *coronal plane*, also known as a *frontal plane* in human anatomy.

Generation of Polarity in *Fucus* Eggs

The genus *Fucus* comprises large brown algae that are commonly seen in intertidal zones of North America and Europe. These algae have been favorite subjects for developmental studies because their organization is relatively simple (Fig. 9.3). They are also easy to keep and manipulate in the laboratory. In particular, large numbers of zygotes can be treated so that they develop in synchrony and with their polarity axes oriented the same way.

The egg of *Fucus* is a perfectly spherical cell with no apparent polarity. After fertilization, the egg forms a fertilization membrane, which is later modified into a cell wall like the one that generally surrounds plant cells. At about 12 h after fertilization (at 15°C), the zygote bulges at one pole and becomes pear-shaped, in a

process called *germination.* About one day after fertilization, the zygote undergoes its first cell division, separating a small, pointed cell that includes the germination bulge from a larger, round cell. The large, round cell will form the leafy bulk of the plant called the *thallus.* The small, pointed cell gives rise to the *rhizoid,* which anchors the plant to a rock or similar substrate. The emerging axis between a thallus pole and a rhizoid pole is the *Fucus* plant's only polar axis.

How is the rhizoid-thallus axis of *Fucus* established? In the absence of any orienting clues from the environment, the rhizoid pole forms at the site of sperm entry: the entering sperm somehow breaks the spherical symmetry of the egg, and the resulting bias prevails until the rhizoid pole becomes fixed many hours later. However, the sperm's polarizing effect is easily overruled by various environmental clues. The best investigated of these signals, and presumably the natural orienting clue, is the direction of the incident light. The thallus pole forms on the side of the egg facing the light, and the rhizoid pole forms on the opposite side. This behavior seems highly adaptive because it maximizes the chance for the embryo to become anchored instead of being swept ashore or out to sea.

Under experimental conditions, the *Fucus* zygote can be polarized by light between 4 and 10 h after fer-

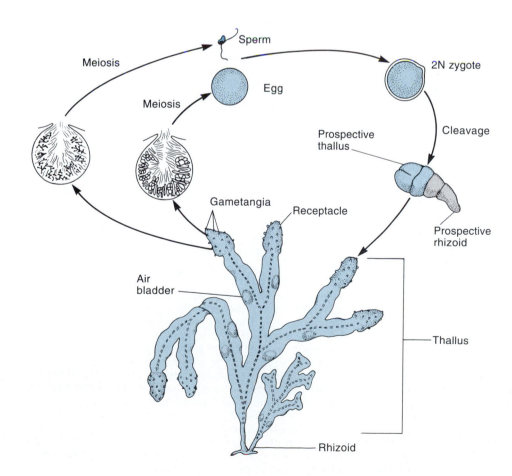

Figure 9.3 Life cycle of the brown alga *Fucus.* The adult plant consists of a rhizoid, which attaches the plant to a rock, and a large thallus, which is the photosynthetic organ. The receptacles at the tips of the thallus branches have openings leading to sex organs called gametangia, which produce eggs or sperm. Cleavage of the zygote is asymmetrical, separating the prospective thallus cell from the prospective rhizoid.

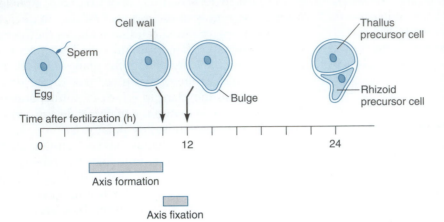

Figure 9.4 Axis formation and axis fixation in *Fucus*. Between 4 and 10 h after fertilization, the zygote appears spherically symmetrical but can be reversibly polarized by light and other environmental factors. Between 10 and 12 h, the polarity of the zygote is fixed according to the last polarizing signal received. At about 12 h, a bulge appears at the future rhizoid pole. At about 24 h, the first cell division separates the rhizoid precursor cell from the thallus precursor cell.

tilization (Fig. 9.4). During this period, the light-induced polarity remains as labile as the sperm-induced polarity: another light pulse from a different direction will induce a new polarity axis. However, between 10 and 12 h of development, the axis is fixed so that new orienting signals can no longer change it. We therefore distinguish a period of **axis formation,** when the thallus-rhizoid axis is set up in a preliminary way, from a period of **axis fixation,** when the axis is irreversibly established.

The process of axis formation in *Fucus* zygotes was first analyzed by E. J. Lund (1922), who observed that zygotes grown in an electric field germinate toward the anode. Several decades later, Lionel F. Jaffe (1966) found that in the absence of an external field, the germinating *Fucus* zygote polarizes electrically by itself, with the rhizoid end becoming negative. Similar observations were made on the zygote of a related genus of algae, *Pelvetia*.

▼

To measure the small voltages across the germinating *Fucus* zygote, Jaffe placed about 200 fertilized eggs in a glass capillary and induced axis formation by shining light from one end of the tube (Fig. 9.5). This effectively connected the zygotes in series, thallus to rhizoid, so that the small electric potentials across the individual cells added up to a measurable potential across the entire tube. No voltage developed until 12 h after fertilization, when the zygotes began to germinate. However, as germination proceeded, there was a parallel rise in voltage across the tube; the end toward which the rhizoids formed became increasingly negative. No such voltage was observed in a control tube with zygotes that were illuminated from all sides and germinated randomly in all directions.

This experiment was extended later with a vibrating electrode that can detect very small electric currents. Using this instrument, Richard Nuccitelli and Jaffe

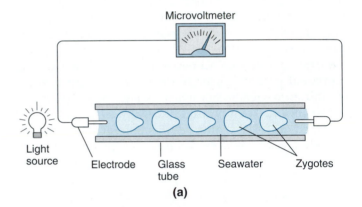

(a)

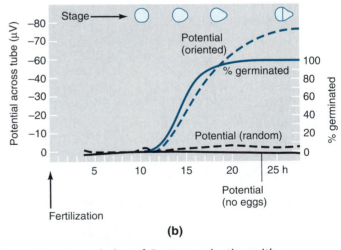

(b)

Figure 9.5 Association of *Fucus* germination with a spontaneously generated electric potential. **(a)** Experimental design. Fertilized *Fucus* eggs were loaded into a glass tube with seawater and illuminated from one end of the tube, so that each zygote formed a rhizoid oriented away from the light source. The electric potential along the tube was monitored with a microvoltmeter. In control experiments, glass tubes were filled with seawater only, or the fertilized eggs were illuminated from all sides so that they germinated randomly in all directions. **(b)** Electric potential across the glass tube during the first day after fertilization. As the zygotes began to germinate, the rhizoid end of the tube became increasingly negative. No such polarization was observed in the control tubes.

(1974) detected a small inward current at the prospective rhizoid site more than 1 h before germination and often as early as 5 or 6 h after fertilization. In these cases, when the direction of the unilateral light was changed, the inward current shifted to the new prospective rhizoid position (Nuccitelli, 1978). Thus, the electric current through the zygote begins during axis formation and continues through axis fixation.

Further experiments were directed at identifying the carrier of the inward-directed current. Initial observations suggested that the current may result from the flux of calcium ions (Ca^{2+}), which are present at much higher concentration in a seawater than in living cells. Experiments with a calcium ionophore, a lipid-soluble agent that makes plasma membranes permeable for Ca^{2+}, indicated that the rhizoid pole forms near the site of Ca^{2+} influx, which is then the site of highest Ca^{2+} concentration inside the cell. Further measurements showed that Ca^{2+} indeed enters germinating zygotes at the prospective rhizoid pole and leaves everywhere else (Robinson and Jaffe, 1975). However, subsequent experiments with artificial seawater demonstrated that, in the absence of external Ca^{2+}, both *Fucus* and *Pelvetia* zygotes nevertheless polarize (Kropf, 1992). Under these circumstances, the zygotes may use internal Ca^{2+} stores or rely on Ca^{2+}-independent mechanisms for polarization, in accord with the *principle of synergistic mechanisms*.

Taken together, the experiments of Lund and of Jaffe and coworkers show that *Fucus* and *Pelvetia* zygotes drive an electric current through themselves as part of the axis formation process. However, the carrier(s) of the current and the current's role in axis formation still need to be clarified.

Axis formation under the influence of unilateral light depends on microfilaments (Quatrano, 1990; Kropf, 1992). When *cytochalasin B*, a drug known to interfere with microfilament formation, is added to seawater with *Fucus* or *Pelvetia* zygotes during the axis formation period, photopolarization is inhibited. If the drug is subsequently removed, the rhizoids grow out in random directions. A molecular probe that binds selectively to microfilaments stains *Fucus* zygotes symmetrically before axis fixation, but following axis fixation, staining occurs selectively near the future rhizoid pole (Fig. 9.6). The localization of the microfilaments coincides with the axis fixation process (Kropf et al., 1989).

In another series of experiments, researchers removed the cell wall surrounding the zygote and allowed it to form again. Axis fixation occurred only when a cell wall was present (Kropf et al., 1988). The researchers concluded that axis fixation involves the assembly of molecular bridges across the plasma membrane between cytoplasmic microfilaments and extra-

cellular material. The resulting molecular complex may anchor other membrane proteins to the future rhizoid pole and may provide cytoskeletal tracks for the transport of vesicles and other organelles. In addition, the complex may serve to anchor one pole of the first mitotic spindle near the rhizoid pole, thereby positioning the spindle so that the zygote divides unequally into the large prospective thallus cell and the small prospective rhizoid cell.

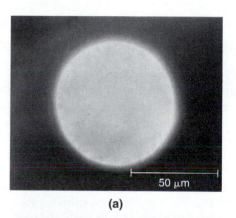

(a)

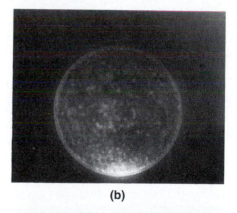

(b)

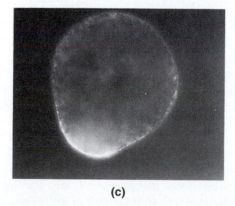

(c)

Figure 9.6 Accumulation of microfilaments (F actin) near the future rhizoid pole of a *Fucus* zygote. **(a)** 7 h after fertilization. F actin is evenly distributed in the cortical cytoplasm. **(b)** 11 h after fertilization. Cortical F actin accumulates near the future rhizoid pole. **(c)** 16 h after fertilization. As the zygote becomes pear-shaped, F actin remains most abundant in the future rhizoid region.

In summary, studies on *Fucus* and *Pelvetia* have shown that the thallus-rhizoid axis is first formed in a preliminary way, oriented either by the site of sperm entry or by environmental clues. After a labile period, the axis is fixed by a stable molecular assembly at the future rhizoid pole. A similar two-step process directs the development of at least one body axis of the *Xenopus* embryo, which will be discussed in the following sections.

Axis Determination in *Xenopus* Eggs

The development of *Xenopus* involves the formation and fixation of three embryonic axes. The *anteroposterior* axis develops during gastrulation from the *animal-vegetal* axis, which originates during oogenesis. The *dorsoventral* axis is formed after fertilization and fixed before the first cleavage. The *left-right* asymmetry is not overt until the heart and other internal organs become asymmetrical during embryogenesis. However, molecular clues to left-right asymmetry have been detected during the blastula and early gastrula stages.

The Animal-Vegetal Axis Originates by Oriented Transport during Oogenesis

The anteroposterior axis of amphibians develops approximately parallel to the animal-vegetal axis of the egg. The *animal pole* is defined as the pole next to the oocyte nucleus. The animal hemisphere typically contains a larger amount of clear cytoplasm, some of which is released when the germinal vesicle disintegrates during meiosis. The *vegetal pole* is not distinguished by a unique structure; it is simply the pole opposite the animal pole.

The animal hemisphere of amphibian eggs is easy to recognize because its cortical cytoplasm is darkly pigmented. There is no such pigment in the vegetal hemisphere, which therefore has a whitish appearance derived from densely packed yolk platelets. Fertilized amphibian eggs rotate within the fertilization envelope so that the dark side faces up and the light side faces down. This orientation seems to offer the best camouflage against predators from the air above and from the water below.

The animal-vegetal polarity develops during oogenesis. As discussed in Chapter 8, the localization of Vg1 mRNA in the vegetal cortex relies on oriented transport. The same holds for the accumulation of yolk protein in the vegetal hemisphere. A precursor yolk protein, *vitellogenin,* is synthesized in the maternal liver, transported by blood, and taken up into the oocyte by *receptor-mediated endocytosis* (see Chapter 3). Within the oocyte, vitellogenin collects in transitional yolk bod-

ies before being processed into *vitellin* and stored in crystal-like *yolk platelets*. During this process, there is an overall transport of yolk protein toward the vegetal hemisphere, which eventually contains about 70% of the total vitellin in the oocyte. The same transport sets up a gradient in yolk platelet size. The largest platelets, having a diameter of 10 to 15 μm in *Xenopus,* are all located in the vegetal part of the egg. The animal region contains smaller platelets, about 2 to 4 μm in diameter; and intermediate sizes are found between the animal and vegetal hemispheres. Thus, in the *Xenopus* oocyte, oriented transport plays a major role in the development of animal-vegetal polarity. The polarity of the transport mechanism appears to be fixed long before the egg is laid.

The animal-vegetal polarity in amphibian eggs determines the spatial organization of the germ layer rudiments, which develop during the blastula stage. Most of the animal half of the blastula forms ectoderm, and most of the vegetal half forms endoderm, while an intermediate zone gives rise to mesoderm (Fig. 9.7). This zone is called the *marginal zone,* because it forms the margin on the upper (animal) side of the blastopore during gastrulation (see Chapter 10). During gastrulation, the marginal zone and endoderm turn inward while the ectoderm expands to cover the entire embryo (see Fig. 6.12). The marginal zone plays the leading role during gastrulation and in the subsequent induction of the brain and spinal cord, as will be discussed in Chapters 10 and 11. The vegetal portion of the marginal zone turns inside first during gastrulation and forms anterior mesoderm. The animal portion of the marginal zone turns inside later and forms posterior mesoderm.

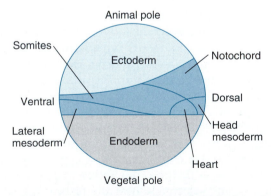

Figure 9.7 Fate map of the *Xenopus* embryo before the onset of gastrulation. The ectoderm arises from animal cells, while the endoderm originates from vegetal cells. An intervening zone, called the marginal zone, gives rise to endoderm (superficial cells) and mesoderm (deep cells). The map also indicates the positions of major mesodermal derivatives: notochord, head mesoderm, heart, somites (which give rise to muscle, among other structures), and lateral mesoderm. During gastrulation, mesoderm and endoderm move inside the embryo through a circular opening, the blastopore (see Chapter 10).

In the overlying ectoderm, large dorsal animal portions form the brain and sense organs of the head while smaller dorsal vegetal portions form the spinal cord. Through this sequence of events, the original animal-vegetal polarity of the egg translates into specific cell behaviors during gastrulation and neurulation, which in turn determine the further development of the anteroposterior axis and the other two body axes as well.

The Dorsoventral Axis Is Determined by Cytoplasmic Rearrangements Following Fertilization

The second polarity axis of the amphibian embryo, the *dorsoventral axis,* is oriented by the sperm entry, which may occur anywhere in the animal hemisphere. The entering sperm causes a temporary concentration of pigment in the egg cortex, to which experimenters can apply a bit of dye to permanently mark the **sperm entry point.** Sperm entry is followed by the appearance of a **gray crescent** at the egg surface near the equator. The crescent, which usually forms opposite the sperm entry point, marks the future dorsal aspect of the embryo. Its formation reflects global rearrangements of egg cytoplasm in which the thin outer shell of cytoplasm, called the **cortex,** rotates relative to the massive inner core of the egg, called the **endoplasm.** This movement is termed **cortical rotation** (Fig. 9.8).

In some amphibians, including *Xenopus laevis,* the cortical rotation is hard to observe because the dark pigment of the animal hemisphere is contained mostly in the subcortical endoplasm and the cortex is nearly transparent. In other amphibians, including *Rana pipiens,* the cortex contains more of the animal pigment. The gray crescent forms where this cortical pigment rotates toward the animal pole and bares some of the underlying endoplasm. The tips of the crescent define the axis of cortical rotation, which is perpendicular to the animal-vegetal axis; and the crescent is centered over the meridian of greatest cortical displacement toward the animal pole (Fig. 9.8). Thus, the gray crescent—if visible—predicts exactly where the dorsal part of the embryo will form.

To make cortical rotation better visible in *Xenopus,* Jean-Paul Vincent and his colleagues (1986) applied two types of dye marks to fertilized eggs. One dye was used to stain the cortex, a layer only 2 to 5 µm thick, and the fertilization envelope. Another dye was applied to the endoplasm. Both sets of dye marks retained their size and, more or less, their spacing until the first cleavage division. About halfway through the first cell cycle, the cortical shell rotated relative to the endoplasm by 30° about a horizontal axis (Fig. 9.9). Invariably, the dorsal portion of the embryo developed along the meridian of greatest cortical displacement toward the animal pole. In most cases, this meridian was located opposite the sperm entry point.

The physical basis of cortical rotation is not fully understood. Normally, the cortex rotates while the core is stationary, because gravity keeps the heavy vegetal part

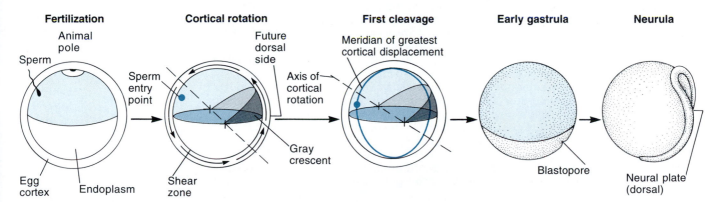

Figure 9.8 Relationship between fertilization, cortical rotation, gray crescent formation, and establishment of the dorsoventral axis in amphibians; animal pole at the top, sperm entry point oriented to the left. The cortex of the egg is shown as a shell surrounding the deeper endoplasm. (The thickness of the cortex is exaggerated in the drawing; in reality, the cortex accounts for less than 1% of the egg radius.) The cortical rotation (arrows) pivots on an axis (passing through crosses) that is perpendicular to the animal-vegetal axis. A shear zone marks the boundary between the rotating cortex and the endoplasm. In many amphibians, a gray crescent forms near the egg equator and usually opposite the sperm entry point. The gray color of the crescent is ascribed to a decrease in the amount of pigment, some of which moves away from the area with the rotating cortex. The gray crescent marks the future dorsal side of the embryo, where the blastopore will originate and the neural plate will form later. The meridian that bisects the gray crescent is the line of greatest cortical displacement toward the animal pole. This meridian is located within the future median plane separating the right and left halves of the body.

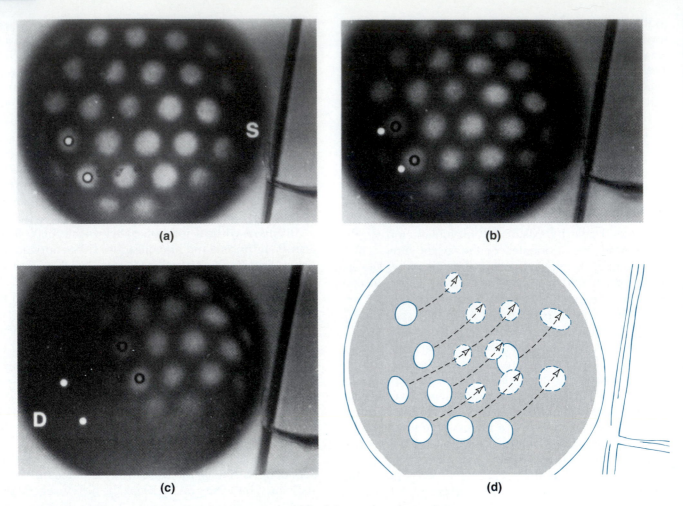

Figure 9.9 Evidence for cortical rotation in *Xenopus* eggs. **(a–c)** Successive photomicrographs of the vegetal half of a newly fertilized egg in the process of cortical rotation. The egg was embedded in gelatin and the perivitelline space dehydrated so that the egg cortex was immobilized. Under these conditions, the yolky endoplasm rotated inside the cortical shell. The endoplasm was stained with a dye that produced the large bright spots on the seemingly dark egg when viewed under a fluorescence microscope. The original positions of the stained endoplasmic spots were marked on the egg surface. On the photographs, the original positions of two stained spots are marked by sharp white dots in black rings. As the embryo proceeded through its first cell cycle, the pattern of stained endoplasmic spots rotated relative to the marks on the surface. The black rings in the photographs mark the moving subcortical spots, while the sharp white dots preserve their original positions. The endoplasmic spots move toward the future ventral side of the embryo, which in most cases is also the side of sperm entry (S). In part c, D marks the embryo's future dorsal side. **(d)** Tracing of the subcortical spots in parts a and c. Arrows show the movement of the spots over time. See also Figure 9.10.

of the egg core facing down. If eggs are held in a fixed position, the cortex remains stationary and the endoplasm rotates away from its gravitational equilibrium. The rotation must therefore be an autonomous and energy-consuming process. The fact that eggs activated not by fertilization but by needle puncture undergo cortical rotation at the appropriate time but with unpredictable orientation indicates that the required energy comes from the egg and that the sperm merely *orients* the rotation.

Normal cortical rotation depends on a parallel array of *microtubules* in the shear zone between the cortex and the endoplasm. Agents that depolymerize microtubules, such as *colchicine, nocodazole,* cold shock, hydrostatic pressure, and UV irradiation, inhibit the cortical rotation (Malacinski et al., 1977). The shear-zone microtubules are present only during cortical rotation; they point with their (+) ends in the direction of the cortical rotation (Elinson and Rowning, 1988; Houliston and Elinson, 1991). Aligned in the direction of the shear, the microtubules seem to act as tracks for the rotation. Microtubules of the *sperm aster,* which extend through the entire egg cytoplasm, conceivably bias the orientation of the shear-zone microtubules. The bias may direct the beginning of cortical rotation, and the ongoing rotation may in turn reinforce the initial bias in microtubule orientation (Gerhart et al., 1989). There is evidence that microtubule-associated *motor proteins* moving toward the (+) end of the microtubules may pull the entire cortex with them (Houliston, 1994).

The cytoplasmic rearrangement revealed by the cortical rotation is crucial to the dorsoventral organization of the embryo. The area of cortical displacement toward the animal pole—the gray crescent, if visible—marks the future dorsal side of the embryo, where gastrulation will begin and the neural plate will form. Eggs in which cortical rotation has been inhibited by UV irradiation or other means are **radially ventralized.** Such eggs cleave and gastrulate in radial symmetry around the animal-vegetal axis, but neurulation does not occur, and the distinctive dorsal organs of the vertebrate body do not form (Scharf and Gerhart, 1983). This result demonstrates that some event associated with cortical rotation is necessary for the development of dorsal organs.

Radially ventralized eggs can be *rescued* if the normal cortical rotation is replaced with a similar but gravity-driven rotation (Fig. 9.10). For this operation, the fluid between the fertilization envelope and the plasma membrane is removed osmotically, so that the egg sticks to the inside of the fertilization envelope. The egg can then be held in any position if it is placed in a mold of wax. If the animal-vegetal axis is held at an angle, the endoplasm, or *core,* of the egg undergoes a rotation that is driven by gravity: the yolk-laden vegetal cytoplasm sinks down, while the lighter animal core cytoplasm moves up until gravitational equilibrium is restored. Under these circumstances, the direction of the gravity-driven rotation, and *not* the point of sperm entry, determines the dorsoventral axis. The future dorsal side of the embryo originates along the meridian of maximal core displacement toward the vegetal pole. When viewed as a relative movement between core and cortex, maximum core displacement toward the vegetal pole is equivalent to maximum cortical displacement

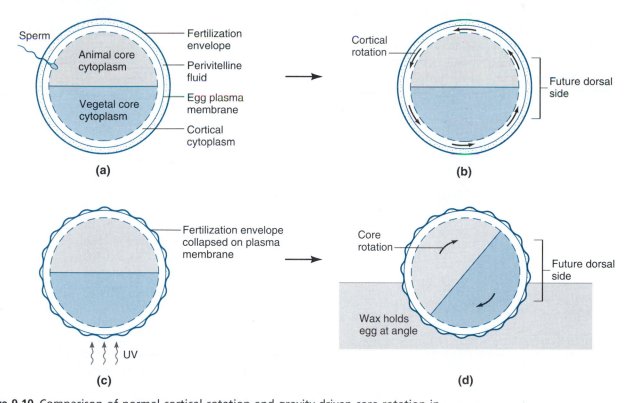

Figure 9.10 Comparison of normal cortical rotation and gravity-driven core rotation in *Xenopus* eggs. **(a)** Normally, the egg plasma membrane is separated from the fertilization envelope by perivitelline fluid. **(b)** The cortical cytoplasm therefore rotates around the core cytoplasm, which remains stationary, with the heavy vegetal core cytoplasm pointing down and the lighter animal core cytoplasm pointing up. **(c)** The perivitelline fluid can be removed osmotically by placing the egg into a hygroscopic (water-attracting) medium. This treatment can be applied to normal eggs and to eggs UV-irradiated on the vegetal pole. The latter do not form dorsal organs unless they are treated as shown in part d. **(d)** An egg without perivitelline fluid can then be held in a wax mold so that its animal-vegetal axis is at an angle to the direction of gravity. The core cytoplasm will then rotate back toward gravitational equilibrium. The dorsal side of a UV-irradiated embryo originates along the meridian where the core cytoplasm undergoes the maximal downward displacement. Note that this is the same relative movement as the upward displacement of dorsal cortical cytoplasm toward the animal pole.

toward the animal pole (Fig. 9.10). Thus, it is consistent that either movement determines the dorsal surface of the embryo.

Gravity-driven core rotation can not only restore the dorsoventral polarity of UV-irradiated eggs, but also override the polarizing effect of the normal cortical rotation in nonirradiated eggs (Kirschner et al., 1980; Gerhart et al., 1981). When eggs are immobilized after gray crescent formation and kept so that the gray crescent faces upward, all embryos begin blastopore formation where the gray crescent has been, just as untreated embryos do (Fig. 9.11). However, in eggs oriented with the gray crescent facing down for 15 to 60 min, the blastopore develops *opposite* the former gray crescent position. These results show that the dorsoventral polarity of *Xenopus* eggs, as indicated by the gray crescent, can be overruled by later rearrangements under the influence of gravity. This means that the gray crescent by itself is neither necessary nor sufficient for blastopore formation. Rather, some other event, which is normally associated with cortical rotation but can be mimicked by gravity-driven core rotation, must determine the dorsal side of the embryo.

In summary, an event associated with cytoplasmic rearrangement in the fertilized egg is critical for establishing the dorsoventral axis of the amphibian embryo. Like the thallus-rhizoid axis of *Fucus*, the dorsoventral axis of *Xenopus* is labile at first. Once fixed during the first cell cycle, the axis determines the dorsal side, where blastopore formation will begin and where the dorsal organs of the embryo will form.

Left-Right Asymmetry Is Defined with Regard to the Anteroposterior and Dorsoventral Axes

Most animals are *bilaterally symmetrical* relative to the median plane, which is defined by the anteroposterior and dorsoventral axes (refer to Fig. 9.2). The left side is nearly a mirror image of the right side, at least externally. Even if this symmetry were perfect, there would still be a *mediolateral polarity*, because medial organs, such as ribs, differ from lateral organs such as the hands. Thus, the two lateral halves of the body share the same anteroposterior and dorsoventral axes, but their mediolateral axes point in opposite directions. As anyone can easily test, the two hands of one person can be aligned at most with two, but never with all three, of their polarity axes. Any two objects with this property differ in their **handedness,** and we define them as left and right in analogy to the two sides of the body. Every three-dimensional object with three polarity axes—molecules, gloves, cars—exists or could be made in a right-handed and a left-handed version. Thus, even if the two bilateral halves of an organism are perfect mirror images of one another, their two mediolateral axes can be defined as one *left-right axis*, an axis that is perpendicular to the median plane and has a defined polarity.

In many so-called bilaterally symmetrical organisms, some of the inner organs have a **left-right asymmetry.** In humans, the stomach curves to one side in the abdominal cavity while most of the liver, the entire spleen, and the appendix of the colon are on the other side. A typical sea urchin larva is bilaterally symmetrical, but the rudiment from which the adult is formed arises on the left side only. Most left-right asymmetries are constant in their handedness: the human stomach normally curves to the left while liver, spleen, and appendix are to the right. Such a **handed asymmetry** differs from a **random asymmetry,** in which asymmetrical organs would be randomly positioned one way or another.

Cell determination in an organism with a handed asymmetry requires additional information that is not required in the case of perfect bilateral symmetry. The nature of this information, and how it arises in its necessary relation to the anteroposterior and dorsoventral polarities, has intrigued researchers for a long time. Since only a few studies on this subject have been carried out with *Xenopus*, we will discuss experiments with other species as well. Studies on sea urchin embryos have shown that an experimentally induced inversion of the dorsoventral axis is associated with an inversion of the left-right asymmetry, so that the handedness of the developing larva is the same as in a normal larva.

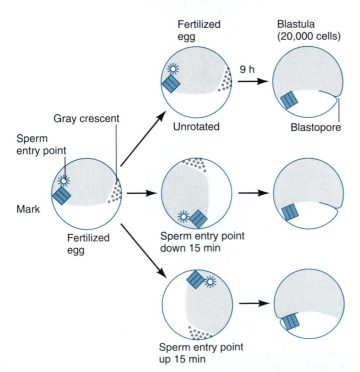

Figure 9.11 Rotation experiments with fertilized *Xenopus* eggs. The sperm entry point was marked with dye (color). In unrotated controls, the blastopore formed at the site of the gray crescent opposite the sperm entry point. The same result was obtained with eggs rotated 90° with the sperm entry point facing down. However, after 90° rotation with the sperm entry point facing up, the blastopore formed opposite the original site of the gray crescent.

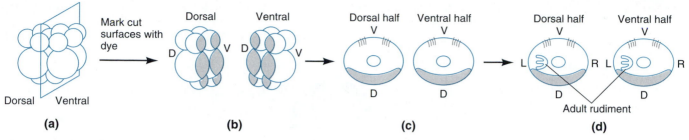

Figure 9.12 Specification of left-right asymmetry with regard to the anteroposterior and dorsoventral axes in sea urchin *(Lytechinus variegatus)* embryos. **(a)** Embryos at the 16-cell stage were cut into ventral and dorsal halves, and the cut surfaces were marked with dye. **(b)** Thus, the dye was applied to the ventral surface of the dorsal half and to the dorsal surface of the ventral half. **(c)** Both halves developed into normally proportioned larvae, shown here in a schematic view from the animal pole. In most cases, however, the dorsal half became inverted in dorsoventral polarity. The dye, originally marking the ventral surface, was associated with dorsal structures instead. Such an inversion did not occur in the ventral half, and both halves maintained their animal-vegetal polarity. **(d)** When the larvae had developed far enough to see the rudiment of the adult sea urchin, the adult rudiment in both halves was on the left side, as it is in the normal embryo. It follows that the left-right axis must have been inverted along with the dorsoventral axis in the dorsal half.

Fate mapping experiments with sea urchin embryos have shown that the first cleavage plane divides the embryo into left and right halves, and that the second cleavage plane separates the dorsal half from the ventral half. Thus, by staining one blastomere at the 2-cell stage, and recognizing the animal-vegetal axis from morphological criteria, one can infer the plane that separates dorsal and ventral halves at subsequent stages. When the dorsal and ventral halves of an embryo are separated at the 16-cell stage, both halves develop into normally proportioned larvae. Moreover, fate-mapping experiments show that the ventral half retains both its animal-vegetal axis and its dorsoventral axis. In contrast, the dorsal half maintains its animal-vegetal axis while frequently reversing its dorsoventral axis (Fig. 9.12). Stain applied to the cut surface (that is, the original ventral surface) of the dorsal half usually appears on the dorsal side of the developing larva (Hörstadius, 1973). Nevertheless, the rudiment of the adult sea urchin always forms on the left side of both halves as it does in the normal embryo (McCain and McClay, 1994). This means that the dorsal half changes in both its dorsoventral axis and its left-right axis (in most cases) or in neither of the two axes (in a few cases).

The experiments with sea urchins indicate that the embryo orients its left-right axis with regard to the anteroposterior and dorsoventral axes, so that the larva develops the normal handed asymmetry. Similar observations have been made with embryos of the roundworm *Caenorhabditis elegans* (Priess and Thomson, 1987; Wood, 1991). How this dependent specification of the left-right axis occurs is not understood. According to a model proposed by Nigel Brown and Lewis Wolpert (1990), the dependent specification occurs through a handed molecule, which becomes oriented with respect to the anteroposterior and dorsoventral axes of the embryo and then determines the left-right axis of the embryo.

Other studies on the specification of handed asymmetry have exploited an abnormal condition known as *situs inversus*, in which the internal organs, such as heart and intestine, are arranged asymmetrically but their handedness is inverted: organs normally positioned on the left are on the right instead, and vice versa. Mutations and experimental treatments causing situs inversus have therefore been used to study the origin of handed asymmetry. Typically in such studies, no more than 50% of the homozygous mutants or experimentally altered embryos developed situs inversus (Oppenheimer, 1974; N. A. Brown and L. Wolpert, 1990). Instead of inverting the normal handedness, the mutations and experiments seemed to *randomize* the asymmetry. For instance, the *situs inversus viscerum* mutation in mice causes situs inversus in about 50% of all homozygous individuals (Layton, 1976). Similar randomizations of asymmetry were observed after experimental interference.

▼

The molecular basis of left-right asymmetry in *Xenopus* was studied by H. J. Yost (1992). In blastulae and early gastrulae, he disturbed the layer of extracellular material (ECM) deposited on the basal surface of the animal cells that form the roof of the *blastocoel*. Transplanting pieces of ectoderm *locally* randomized the asymmetry of the heart and/or parts of the viscera (Fig. 9.13). Specifically, the organs that were randomized had been in direct contact, during gastrulation, with ECM that had been disturbed by the transplantation. Yost

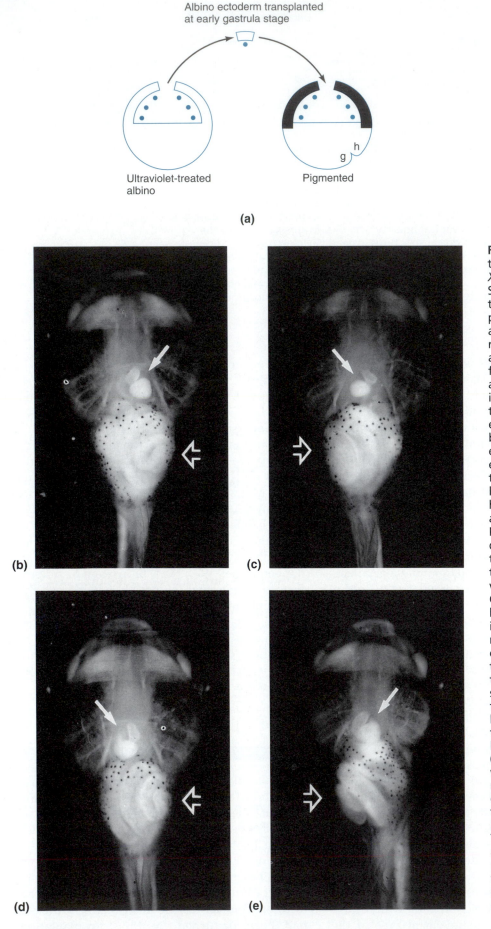

Figure 9.13 Perturbation of the left-right asymmetry in *Xenopus* gastrulae. **(a)** Schematic illustration of the transplantation procedure. A patch of ectoderm was excised and replaced with a corresponding patch from an albino donor. The positions of future heart mesoderm (h) and gut endoderm (g) are indicated. The dots represent the fibronectin-containing extracellular matrix produced by the basal surface of the ectoderm. The albino epidermis was fate-mapped at the tail bud stage, and the left-right orientation of the heart and viscera was scored at the tadpole stage. **(b–e)** Photographs of tadpoles that developed from the transplantation experiments. Each tadpole is shown in ventral view, anterior at the top. Closed arrow points to the heart, open arrow to the intestinal coil. **(b)** Normal left-right asymmetry. The intestine coils counterclockwise, with the shield of the coil to the tadpole's left. **(c)** Complete situs inversus. In this specimen, the transplanted ectoderm had become directly apposed to the heart, upper gut, and lower gut at the tail bud stage. **(d)** Tadpole in which the heart was inverted but the gut was normal. The transplanted ectoderm had been apposed to the heart and upper gut, but not the lower gut. **(e)** Tadpole in which the heart and upper gut were positioned normally but the lower gut was inverted. The transplanted ectoderm had been apposed only to the lower gut.

achieved *global* randomization of left-right asymmetries by injecting certain peptides or enzymes that interfere with the deposition of *fibronectin,* a major ECM component. These results indicate that information about the handedness of inner organs is contained in the extracellular matrix on the inside of the *blastocoel* roof and that this information can be transmitted locally and independently to cardiac and visceral primordia.

To account for the frequent cases in which random asymmetry seemed to result rather than inverse handedness, Brown and Wolpert (1990) proposed that handedness is controlled by an intrinsically random process that is biased toward one side by an additional signal, such as the product of the mouse *situs inversus viscerum*[+] gene or some extracellular matrix component in the frog blastula. However, this model is difficult to reconcile with the discovery of another mouse mutant, in which 100% of all homozygotes develop situs inversus.

▼

Takahiko Yokoyama and colleagues (1993) unintentionally generated a new mouse mutant with situs inversus. In the course of a study on the synthesis of melanin (black skin pigment), they injected fertilized eggs from a mouse albino strain with a modified gene for tyrosinase, the first enzyme in the pathway of melanin synthesis. In several of the injected eggs, the tyrosinase "transgene" was incorporated into a mouse chromosome and passed on during cell divisions as if it were one of the host's own genes (endogenous genes). Each of the transgenic mice developing from the injected eggs was bred into a separate strain. (The technique of making transgenic mice will be discussed more fully in Chapter 14.)

In one of the strains, the transgene had "landed" in a particular spot on chromosome 4. In this strain, offspring *homozygous* for the transgenic chromosome—that is, carrying the transgene in both copies of chromosome 4—did not survive for more than 7 days. (The homozygotes could be distinguished from their *hemizygous* and nontransgenic littermates by their pigmentation.) When the researchers autopsied the dead animals, they found that *all* of them had situs inversus combined with an enlarged spleen and a severe kidney abnormality. As the earliest sign of left-right inversion, they observed that homozygous mutant embryos underwent a counterclockwise turn inside the amniotic cavity, whereas hemizygous and nontransgenic embryos made a clockwise turn.

Offspring homozygous for the tyrosinase transgene survived well in 10 other strains, each of which carried the transgene in a different chromosomal position. Therefore, the researchers concluded that the situs in-

versus and associated abnormalities did not result from the activity of the transgene but stemmed from the interruption of the endogenous gene in which the transgene had been inserted. They named the endogenous gene, which had been mutated fortuitously by the insertion of the transgene, *inversion of embryonic turning.* Because the mutant phenotype (counterclockwise turning, situs inversus, etc.) appeared only if both *inversion* genes were interrupted, the researchers concluded that the mutant phenotype resulted from a *loss* of the normal *inversion*[+] function rather than from an abnormal *gain* of function. On the basis of this conclusion, they proposed that the situs inversus in mice represents a *default program,* which is carried out in the absence of *inversion*[+] gene activity. This means that the function of the normal, uninterrupted *inversion*[+] gene is to reverse the handedness of the default program.

It seems difficult at this time to accommodate all available data on handed asymmetry in a simple, plausible model. The mouse *situs inversus viscerum*[+] and *inversion of embryonic turning*[+] genes have been mapped (Brueckner et al., 1989) and may be cloned in the future. The nucleotide sequences of the cloned genes will allow researchers to predict the amino acid sequence of the encoded protein, which in turn may provide clues to the proteins' biological functions.

It is clear that even in one species, *Xenopus,* a variety of mechanisms contribute to the development of the three body axes. Oriented transport of molecules and organelles plays a critical role in establishing the animal-vegetal axis, and possibly the left-right asymmetry. Cortical rotation as part of a major cytoplasmic rearrangement is the hallmark of dorsoventral axis formation. The three axes also differ in the freedom of orientation that is possible when each axis is fixed. The direction of the animal-vegetal axis during oogenesis is unrestricted because no other axis exists at the time. The dorsoventral axis, formed during the first embryonic cell cycle, can be fixed in any orientation perpendicular to the animal-vegetal axis. The right-left axis, formed last, is completely determined by its perpendicular alignment relative to the other two axes and may change only in polarity, as in situs inversus.

The remainder of this chapter will focus on the further development of the dorsoventral axis.

The Role of Vegetal Cytoplasm in Establishing Dorsoventral Polarity

The cytoplasmic rearrangements that occur in amphibian eggs after fertilization have attracted much attention. The gray crescent, which forms as part of these rearrangements, marks the site of the future *dorsal blastopore lip,* where gastrulation begins, as will be described in Chapter 10. The dorsal blastopore lip forms the dorsal mesoderm and also induces neighboring cells to cooperate in the formation of the entire set of dorsal vertebrate organs including the brain, spinal cord, axial skeleton, and dorsal aorta. The organizing capability of the dorsal blastopore lip came to light in a landmark experiment conducted by Hans Spemann and Hilde Mangold (1924; see Chapter 11). Hence, the dorsal blastopore lip is also known as ***Spemann's organizer.*** Because the dorsal blastopore lip develops at the site previously occupied by the gray crescent, it was tempting to speculate that the gray crescent acts as a cytoplasmic determinant for Spemann's organizer. An experiment designed to test this notion seemed to confirm it until a reinvestigation showed that Spemann's organizer must originate through a different and more complex cascade of events (Gerhart et al., 1981). Several generations of investigators have contributed toward elucidating this cascade, which is still incompletely understood.

Instead of tracing the discoveries of the known key events historically, we will discuss them in the sequence in which they occur in the embryo, proceeding from the cortical rearrangements after fertilization toward the events of later stages.

Cortical Rotation Is Associated with Deep Cytoplasmic Movements

As described earlier, the cytoplasmic rearrangements normally associated with cortical rotation are necessary for the formation of the dorsal blastopore lip and subsequent gastrulation. Eggs in which cortical rotation has been inhibited gastrulate with radial symmetry and do not form dorsal organs. Several observations indicate that the proper formation of the dorsal blastopore may depend on the *deep cytoplasmic movements* that are associated with cortical rotation. For instance, the dorsal blastopore lip always appears where cells containing large yolk platelets abut cells with small yolk platelets, suggesting that the intervening cytoplasm with intermediate platelets has been displaced (Pasteels, 1964).

To analyze the deep cytoplasmic movements in *Xenopus* eggs, Michael Danilchik and J. M. Denegre (1991) *pulse-labeled* layers of yolk platelets by injecting females with a fluorescent dye that binds to vitellogenin. Eggs subsequently produced by such females were allowed to develop for different periods of time before they were fixed and sectioned for microscopy. This study revealed complex movements of deep cytoplasm. During the first cell cycle following fertilization, these movements produced a *swirl of cytoplasmic layers* on the future dorsal side of the embryo (Fig. 9.14). This swirl consisted of alternating labeled and unlabeled layers of cytoplasm that had not been in contact previously. The

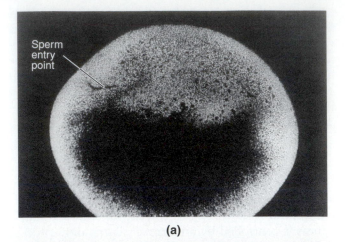

(a)

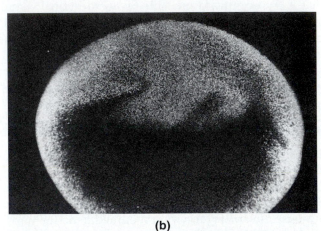

(b)

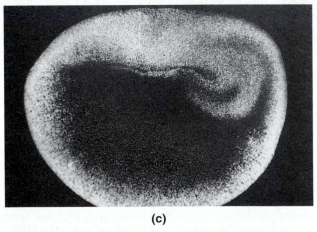

(c)

Figure 9.14 Rearrangement of cytoplasm in *Xenopus* embryos during the first cell cycle. *Xenopus* females were injected prior to spawning with a fluorescent dye that binds to vitellogenin. Yolk platelets that formed after the injection appear white in these photomicrographs. Eggs were fertilized, and embryos were fixed for histological sectioning at various times during the first cell cycle. All sections are oriented with the animal pole up and the sperm entry point to the left. Thus, the future dorsal side is to the right. **(a)** 30 min after fertilization. Labeled cytoplasm in the central animal hemisphere has shifted slightly in the dorsal direction. **(b)** 45 min after fertilization. Central animal cytoplasm has shifted further dorsally. **(c)** 90 min after fertilization (first mitosis). Shifted animal cytoplasm has generated a swirl, consisting of alternating labeled and unlabeled cytoplasm, in the dorsal region.

swirl originated simultaneously with cortical rotation and occurred in fertilized eggs as well as in eggs activated by electric current.

These observations indicate that cortical rotation and the deep cytoplasmic movements are different aspects of the cytoplasmic rearrangements that occur midway through the first cell cycle. Some aspects of these rearrangements must establish the dorsoventral axis, but the molecular mechanisms by which cytoplasmic mixing might determine cells and set up an embryonic axis are not yet understood. Conceivably, the process combines two previously segregated partners of a chemical reaction. For instance, a *kinase* or *protease* may be brought together with its substrate protein, which may then be phosphorylated or cleaved so that it can snap into its biologically active conformation. Such an activated protein in turn could, for example, control the translation of a maternal mRNA or modify the assembly of a cytoskeletal component.

Although the most dramatic cytoplasmic mixing occurs on the future dorsal side, it is not clear whether the mixing affects primarily the animal or the vegetal blastomeres. Experiments described in the following section have proved that a critical cytoplasmic determinant is first localized near the vegetal pole and then shifts dorsally. This shift seems to impart special properties on dorsal vegetal cells that are not present in ventral vegetal cells. However, further observations to be discussed later in this chapter indicate that the animal cells are biased along the same axis as early as the 8-cell stage.

A Cytoplasmic Component That Promotes Dorsal Organ Formation Is Localized near the Vegetal Egg Pole and Shifts Dorsally after Fertilization

The vegetal pole of amphibian oocytes seems to contain two components—both of them UV-sensitive—that are necessary for establishing the dorsoventral axis. One component is most sensitive to UV after fertilization and may consist of guanosine triphosphate associated with the microtubules that form tracks for the cortical rotation. When this component is destroyed by UV irradiation, the resulting inhibition of cortical rotation can be offset by gravity-driven core rotation, as discussed earlier. The other component at the vegetal pole is most UV-sensitive in oocytes, and damage to this component cannot be offset by gravity-driven core rotation (Holwill et al., 1987; Elinson and Pasceri, 1989).

▼

To examine the second UV-sensitive component more closely, M. Yuge and colleagues (1990) established a *bioassay* for cytoplasmic components that promote the formation of the set of **dorsal organs** that characterize

vertebrate embryos, including brain, spinal cord, notochord, and axial musculature. Initially, the researchers took cytoplasm from the *dorsal* vegetal blastomeres at the 16-cell stage and injected it into the *ventral* vegetal cells of a recipient at the same stage (Fig. 9.15). The transplantation caused the formation of a secondary set of dorsal organs in 42% of the surviving recipients. In contrast, injection of cytoplasm from ventral vegetal cells into the same recipient area never caused secondary dorsal organ formation, and most recipients developed into normal tadpoles.

This bioassay was subsequently used by two research groups to test cytoplasm from various regions of eggs and 16-cell embryos for their dorsal organ–promoting activity (Fujisue et al., 1993; Holowacz and Elinson, 1993). Cortical (superficial) cytoplasm taken from the *vegetal* pole of eggs before cortical rotation, when injected into ventral vegetal cells of a host at the 16-cell stage, caused the formation of a secondary set of dorsal organs. In contrast, cortical cytoplasm from other

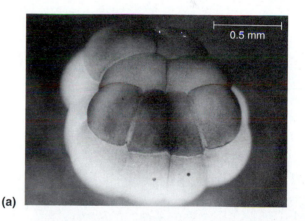

(a)

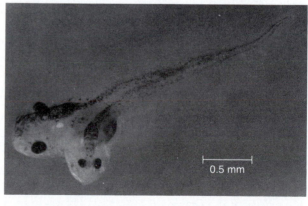

(b)

Figure 9.15 Bioassay for dorsalizing activity of transplanted cytoplasm. **(a)** A *Xenopus* embryo at the 16-cell stage is stripped of its fertilization envelope and oriented with the animal pole up. The cytoplasm to be tested is injected into the equatorial area of the two ventral vegetal blastomeres (injection sites marked with dots). **(b)** If the transplanted cytoplasm has dorsalizing activity, the host embryo forms an extra set of dorsal organs (including head with brain and sense organs, spinal cord, vertebrae, and dorsal trunk musculature). Dorsalizing activity is present near the vegetal pole of fertilized eggs before it shifts dorsally as shown in Figure 9.17.

egg regions or deep vegetal cytoplasm did not induce dorsal organ formation.

The dorsal organ–promoting activity present in the vegetal cortex of eggs disappeared from the vegetal pole area during the latter half of the first cell cycle, that is, before the first cleavage. During the same time interval, cytoplasm from the *dorsal subequatorial* region acquired dorsal organ–promoting activity, which had not been present there during the first half of the first cell cycle. The activity then remained present in the dorsal subequatorial region through the 16-cell stage.

When donor eggs were UV-irradiated after fertilization to block cortical rotation, the cytoplasmic activity that promoted dorsal organ formation remained at the vegetal pole and did not shift to a dorsal subequatorial position. This result shows that the normal shift of the activity depends on cortical rotation or some associated event. In contrast, donor eggs that were UV-irradiated during the oocyte stage completely lost their activity, indicating a greater UV sensitivity of the active component at the oocyte stage.

Taken together, cytoplasmic transplantation experiments with *Xenopus* eggs indicate that an activity promoting dorsal organ formation is initially present around the vegetal pole and that it shifts to a dorsal subequatorial position as part of the cytoplasmic rearrangements associated with cortical rotation. This activity is required in the dorsal subequatorial region for dorsal organ formation to occur.

Dorsal Vegetal Blastomeres at the 32-Cell Stage Establish the Dorsal Side of the Embryo

The cytoplasmic activity that promotes dorsal organ formation is normally allocated to the dorsal vegetal blastomeres and can be transplanted with these blastomeres at the 32-cell and 64-cell stages. This was observed by Robert Gimlich and John Gerhart (1984), who transplanted various blastomeres from normal donor embryos to recipients in which the dorsoventral axis had been abolished by UV irradiation (Fig. 9.16). As described earlier, UV irradiation of the vegetal hemisphere

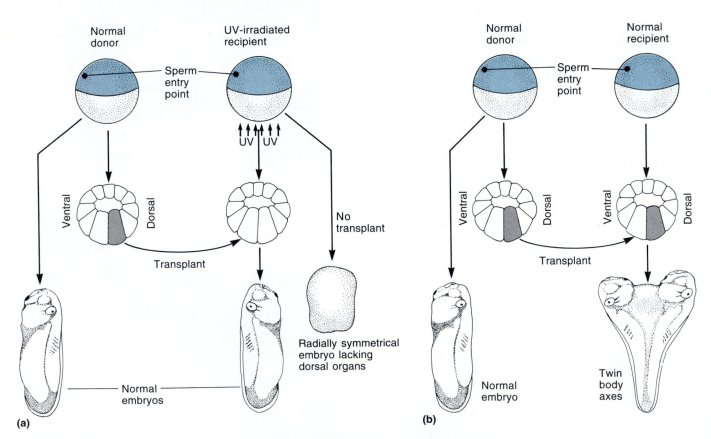

Figure 9.16 Transplantation of dorsal vegetal blastomeres in *Xenopus* embryos. **(a)** Rescue experiment. The dorsoventral axis of recipients was abolished by UV irradiation of the vegetal hemisphere of fertilized eggs. Eggs so treated give rise to radially symmetrical embryos lacking all dorsal organs such as the brain, eyes, ears, spinal cord, and notochord. The radially ventralized embryos were restored to normal development by transplanted dorsal vegetal blastomeres from normal donors. **(b)** Heterotopic transplantation. Transplantation of dorsal vegetal blastomeres to the ventral side of normal recipients resulted in the development of embryos with two sets of dorsal organs.

of uncleaved eggs prevents the cytoplasmic rearrangement associated with cortical rotation. The resulting embryos are radially symmetrical around the animal-vegetal axis and lack dorsal organs. The investigators used such radially ventralized embryos as recipients for blastomeres from normal donors. The question was which, if any, of the grafted cells would restore dorsoventral polarity to the recipients.

▼

Gimlich and Gerhart, who performed their experiments before the work of Yuge, Fujisue, Holowacz, and their colleagues, were guided by earlier observations of Pieter Nieuwkoop (1969b), who had rotated the animal hemisphere of amphibian blastulae relative to the vegetal hemisphere and observed that the *vegetal* hemisphere determined the dorsoventral polarity of the reconstituted embryos. Gimlich and Gerhart therefore expected that vegetal blastomeres would be critical to the establishment of dorsoventral polarity. To test this hypothesis, they transplanted vegetal cells from normal to radially ventralized embryos at the 32- or 64-cell stage. Within the vegetal tier of cells, they considered the quadrant centered over the meridian of the sperm entry point to be the ventralmost quadrant (blastomeres D4 and D4' in Fig. 9.17); the opposite quadrant they designated the dorsalmost quadrant, and the quadrants in between they called lateral quadrants. In each experiment, the researchers removed the cells of an entire quadrant from the vegetal tier of a radially ventralized recipient and replaced them with the corresponding cells of a normal donor. Because of variations in the cleavage pattern, the actual number of blastomeres transplanted varied from 1 to 3. The grafts healed into place within 1 h, and the recipients were allowed to develop until normal control embryos reached the tadpole stage. Then the recipients were scored for completeness of their dorsal organs.

The results of the transplantations depended on the quadrant from which the grafts were taken. With grafted vegetal blastomeres from the *dorsalmost* quadrant, the recipients were substantially rescued. Many of these embryos had a notochord, a brain and spinal cord, and the rudiments of eyes and ears. Only a very few of the UV-irradiated control embryos, which received no grafts, had any of these dorsal structures. In contrast, similar-sized grafts of lateral and ventralmost vegetal cells had no rescuing effect. The recipients of these cells showed the same average level of UV-induced radialization as the UV-irradiated controls without grafts. The investigators concluded that within the vegetal tier of blastomeres the ability to establish dorsoventral polarity is restricted to the dorsalmost quadrant.

On the basis of existing fate maps, the researchers expected that dorsal vegetal blastomeres at the 32- and

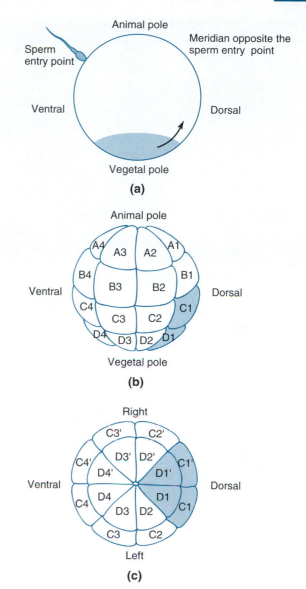

Figure 9.17 Dorsalizing activity (color) in the *Xenopus* embryo, revealed by the transplantation of cytoplasm or blastomeres as shown in Figures 9.15 and 9.16. Dorsalizing activity first occurs in vegetal cytoplasm and then shifts to a dorsal subequatorial region, where it is enclosed in the dorsal vegetal blastomeres. At the 32-cell stage, the dorsalizing activity is restricted to the dorsal quadrant of the vegetal blastomeres (D1 and D1') and marginal blastomeres (C1 and C1'). **(a)** Lateral view of the egg. **(b)** Lateral view of a 32-cell embryo. **(c)** Vegetal view of a 32-cell embryo.

64-cell stages would contribute exclusively to the ventral gut region. They confirmed this prediction by labeling vegetal blastomeres in normal embryos with a fluorescent dye and tracing the progeny of the labeled cells in normal development and after transplantation into UV-irradiated recipients. These experiments also revealed an interesting detail: at the early gastrula stage, the progeny of labeled *dorsalmost* vegetal blastomeres were directly adjacent to Spemann's organizer (Fig. 9.18).

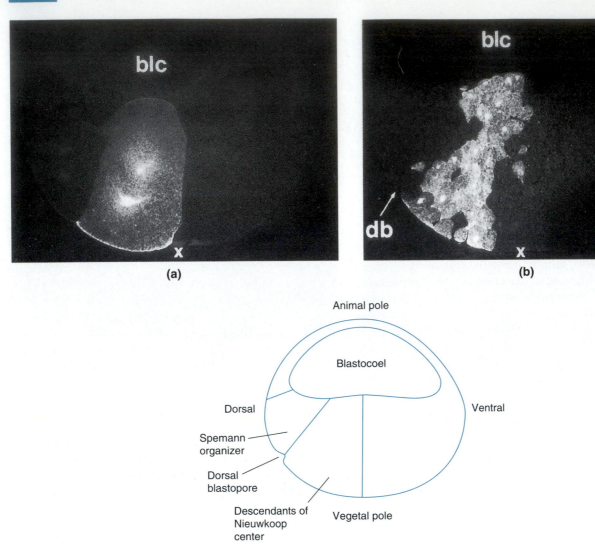

Figure 9.18 Fate of dorsal vegetal cells in normal *Xenopus* embryos. A dorsal vegetal blastomere (corresponding to D1 in Figure 9.17) was injected with fluorescent dye. **(a)** Photomicrograph of a histological section of an embryo fixed shortly after injection. The labeled blastomeres extend from the egg surface near the vegetal pole (marked x) to the floor of the blastocoel (blc). **(b)** Section of a corresponding embryo fixed at the early gastrula stage. **(c)** Schematic drawing based on part b. At the early blastula stage, the area filled by descendants of the labeled blastomeres is called the Nieuwkoop center. The cells adjacent to the Nieuwkoop center (not fluorescent) form the dorsal blastopore lip (db), also known as Spemann's organizer, at the gastrula stage.

In an extension of these experiments, Gimlich (1986) transplanted marginal blastomeres (from the tiers labeled B and C in Fig. 9.17). *Dorsal* marginal cells, but not their lateral or ventral counterparts, caused the development of dorsal organs in the irradiated hosts. The transplanted cells themselves formed much of the notochord and some of the adjacent mesodermal structures. These are the same structures they would have formed in the normal donor. In addition, the grafts induced the neighboring host cells to contribute to the mesoderm and to form most of the brain and spinal cord. The frequency with which the *marginal* grafts caused complete sets of dorsal organs to form increased when they were made at later and later cleavage stages.

In contrast, this capacity decreased in *vegetal* (tier D) blastomeres late during cleavage. Thus, the ability to induce dorsal organs was originally present in the dorsal vegetal blastomeres and later in dorsal marginal blastomeres.

In summary, an activity that promotes the formation of dorsal embryonic organs occurs in the vegetal cytoplasm of *Xenopus* eggs. This activity shifts to a dorsal subequatorial position late during the first cell cycle and is subsequently allocated to dorsal vegetal blastomeres. At the 32-cell stage, the dorsal vegetal blastomeres are sufficient to cause the formation of brain, spinal cord, and other dorsal organs that characterize the vertebrate embryo. The dorsal vegetal blastomeres and their descendants are often called the **Nieuwkoop center** in honor of Peter Nieuwkoop who discovered their special role in the development of dorsoventral polarity in amphibians. However, the Nieuwkoop center itself does not form any of the dorsal organs. Instead, its marginal neighbor cells, the precursors of *Spemann's organizer,* contribute much of the dorsal mesoderm. In short, the Nieuwkoop center causes the adjacent marginal cells to form Spemann's organizer, which in turn instructs its neighbors to cooperate in the development of a complete set of dorsal organs (see Chapter 11). This kind of cellular interaction, in which one group of cells determines the fate of other cells, is known as *induction.* It represents an important general principle, to which we will turn our attention next.

The Principle of Induction

Embryonic cells in many species are pluripotent: their *potency* for producing various structures encompasses more than their actual *fate.* The developmental process by which the potency of a cell becomes limited to its fate is called *determination.* Determination entails loss of potency as well as receipt of specific instructions. The earliest instructive signals in most embryos are *localized cytoplasmic determinants,* which are sequestered selectively in certain blastomeres as described in Chapter 8. Such localizations generate a coarse pattern of differently determined cells. This pattern is complemented and refined by further determinative steps based on cellular interactions.

As already defined in Chapter 8, an interaction between cells that changes the determined state of at least one partner is called *induction* (Jacobson and Sater, 1988). The cells that undergo a change in their determined state are called *responding cells,* and the cells that cause this change are called *inducing cells.* Most important, inductive interactions ensure that cells acquire their fates in a coordinated way.

Induction was already recognized as a basic principle by the earliest experimental embryologists. At the beginning of the twentieth century, the concept of induction was buttressed with rigorous *operational criteria* by the work of Hans Spemann in Germany and Warren Lewis in America (Hamburger, 1988). One way of demonstrating an inductive interaction is the *rescue experiment,* as illustrated by the transplantation of dorsal vegetal blastomeres into radially ventralized *Xenopus* embryos (Fig. 9.16a). This experiment showed that dorsal vegetal blastomeres are necessary to form the dorsal set of organs, although the transplanted cells *do not themselves* give rise to these organs. Another way of proving an inductive interaction is *heterotopic transplan-*

tation, exemplified by the grafting of dorsal vegetal blastomeres into a ventral region of normal *Xenopus* embryos (Fig. 9.16b). As a result of the grafting, cells that normally form ventral structures give rise to dorsal organs instead. From either experiment, one must conclude that the transplant *induces* other cells to form dorsal organs.

The operational criteria for embryonic induction are the same as those used to test the action of cytoplasmic determinants, except that cells or tissues are transplanted instead of cytoplasm. However, this is not the only parallel between cytoplasmic localization and induction. Both are also based on the *principle of default programs.* The absence of a particular localized determinant or inductive signal does not result in chaos or cell death; instead, cells have an alternative pathway of development to fall back on if they do not receive the particular signal. For instance, animal cells of a frog blastula that receive no mesoderm-inducing signal give rise to ectoderm. Also, like cytoplasmic determinants, inductive signals may activate or suppress certain cell fates. For instance, the formation of the lens of the eye is promoted by inductive signals from pharyngeal endoderm, heart mesoderm, and the prospective retina, but inhibited by inductive signals from neural crest.

Inductive interactions occur during certain sensitive phases of development. The ability of the responding tissue to react to an inducer by changing its determined state is called its **competence.** The period of competence typically begins some time before the normal inductive interaction occurs and ends thereafter. Likewise, the ability of an inducing tissue to affect a responding tissue is limited to a window of time before and after the normal inductive interaction. The periods of inductive ability and responsive competence can be defined by combining an inducer of a given age with younger or older responsive tissues, and vice versa.

Induction is a pervasive principle in embryonic development. It may occur between individual cells early during cleavage, as seen, for example, in snails (see Fig. 8.7) and roundworms (see Chapter 24). Many inductive interactions take place during organogenesis, when different germ layers cooperate in the formation of the organ rudiments (see Chapter 11).

For the remainder of this chapter, we will focus on the induction of the mesodermal germ layer, which becomes the principal carrier of dorsoventral polarity in vertebrates.

Mesoderm Induction

The *fate map* of a *Xenopus* blastula (Fig. 9.7) shows the position of the prospective germ layers and their major derivatives: prospective *ectoderm* of the animal hemisphere will form epidermis and neural tissue. Prospective *endoderm* cells in the vegetal hemisphere will form the embryonic gut. The intervening marginal cells will form the *mesoderm,* which gives rise to a wide variety of structures, including the notochord, head mesoderm, heart, somites, lateral plate, and blood cells. A fate map established at the 32-cell stage presents a very similar picture (see Fig. 6.2). However, a major difference between the two stages emerges if the *potency* of isolated regions is mapped rather than their fate. Marginal groups of cells *isolated* at the 32-cell stage do not develop according to fate: instead they form endodermal or ectodermal derivatives. Corresponding groups of cells isolated at the 128-cell stage form small mesodermal structures, and the proportion of mesoderm increases as the cells are isolated later, until at the blastula stage isolated marginal cells form mesodermal structures in accord with their fate. These observations show that mesodermal cells are determined progressively before the blastula stage.

In fact, the mesodermal state arises by *induction* between vegetal cells and animal cells during the blastula stage. Moreover, as we will see, the inductive interactions in the dorsal portion of the embryo are different from those in the ventral and lateral portions. Subsequent interactions between dorsal mesodermal cells and their ventral neighbors establish determined states for dorsal, lateral, and ventral mesodermal derivatives.

During the Blastula Stage, Vegetal Cells Induce Their Animal Neighbors to Form Mesoderm

The induction of mesodermal cells by inductive interactions between vegetal and animal blastomeres was demonstrated by Pieter Nieuwkoop (1969a) with embryos of the axolotl *Ambystoma mexicanum,* and by Sri Sudarwati and Pieter Nieuwkoop (1971) with *Xenopus* embryos. The investigators isolated and combined different parts in the mid to late blastula stages, kept them in tissue culture, and analyzed the resulting structures by microscopy. Figure 9.19 summarizes their findings. Isolated core pieces of *vegetal* cells persisted as large, yolk-laden cells similar to those forming the floor of the embryonic gut during normal development. Isolated *animal* sections, so-called *animal caps,* contracted into spheres of ciliated epidermis. Isolated *marginal zone* elongated considerably in culture and formed mostly mesodermal structures, including a notochord, muscle precursors, embryonic kidney tubules, and blood cells.

The most revealing result was obtained when isolated vegetal cores were combined with isolated animal caps. This experiment was equivalent to heterotopic transplantation, because these two regions are not normally in contact. Under such conditions, mesodermal structures were formed, including a notochord, muscle tissue, and embryonic kidney. While neither animal caps nor vegetal cores by themselves gave rise to meso-

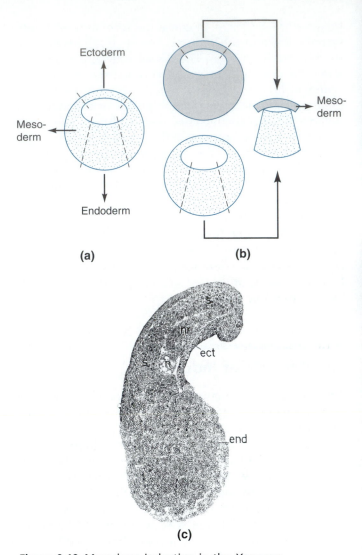

Figure 9.19 Mesoderm induction in the *Xenopus* blastula. **(a)** Fates of isolated blastula areas in tissue culture: caps of animal cells give rise to ectodermal tissue (ciliated epidermis), while cores of vegetal cells form endodermal tissue (large, yolk-laden cells). The marginal cells, constituting a circular band between the animal cap and the vegetal core, form mesodermal structures including the notochord, muscle, mesenchyme, and blood cells. **(b)** Experiment demonstrating mesoderm induction: if animal caps from a fluorescently labeled donor are cultured in contact with unlabeled vegetal cores, the animal cells near the vegetal core are found to form mesodermal structures. **(c)** Histological section showing induced mesoderm, including notochord (n) and muscle tissue (s), between endodermal (end) and ectodermal (ect and nr) structures.

derm, the combined tissues did. In other words, mesoderm formed as the result of an inductive interaction between animal and vegetal cells. This conclusion also explained the increasing formation of mesodermal structures by marginal zones isolated at successive stages as described earlier.

In order to determine whether the induced mesodermal structures derived from animal cells or vegetal cells, researchers have used various labels or combined tissues from different species with different pigments. Leslie Dale and his colleagues (1985) labeled animal caps with a fluorescent dye and combined them with unlabeled vegetal cores. They found that all clearly identifiable mesodermal structures were labeled and therefore derived from animal caps. The same experiments indicated that approximately half of the tissue derived from induced animal caps was mesodermal while the rest was ectodermal. Moreover, animal caps from early gastrulae had lost their competence to respond to the inductive stimulus.

Several investigators have carried out similar experiments using molecular markers instead of histological analysis as a measure of mesoderm induction. For instance, John Gurdon and his coworkers (1985) monitored the synthesis of the mRNA for cardiac actin, an abundant protein in embryonic muscle. They found that this mRNA was synthesized not in isolated animal caps or vegetal cores, but in combinations of animal caps with vegetal cores and in isolated marginal zones. Thus, the pattern of actin mRNA synthesis parallels the formation of the histological structures characteristic of mesoderm.

In summary, the vegetal cells of *Xenopus* blastulae induce their animal neighbor cells to form mesoderm. The response to the inductive signal is limited to the marginal region of the animal hemisphere, that is, to those animal cells that are close to vegetal cells. The cells of the animal cap, which are separated from vegetal cells by many other animal cells and the blastocoel,

maintain their determination to form ectodermal derivatives. Presumably, the range of the inductive signals is limited by slow transport or rapid breakdown. Such properties would allow different region-specific signals to spread side by side. We will see that this is indeed the case.

Mesoderm Is Induced with a Rudimentary Dorsoventral Pattern

During gastrulation, marginal cells move toward the blastopore, where they turn inside and form a layer of mesodermal cells underneath the ectoderm, which stays outside (see Fig. 6.12). The beginning of marginal cell movements in the dorsal area of the embryo suggests that dorsal marginal cells differ from lateral and ventral marginal cells. There is a corresponding regional difference in the inductive capability of vegetal blastomeres at the 32-cell and 64-cell stages. Only transplanted *dorsal* vegetal cells cause the formation of dorsal organs in the embryo (Fig. 9.16). Very likely, then, the types of mesoderm induced by dorsal vegetal cells differ from those induced by lateral and ventral vegetal cells.

To test the mesoderm-inducing capacities of different vegetal blastomeres, Dale and Jonathan Slack (1987b) combined single vegetal blastomeres from 32-cell embryos with animal caps as shown in Figure 9.20. The animal cells were labeled with fluorescent dye to distinguish their descendants from those of the vegetal cells, which were not labeled. The researchers found that dor-

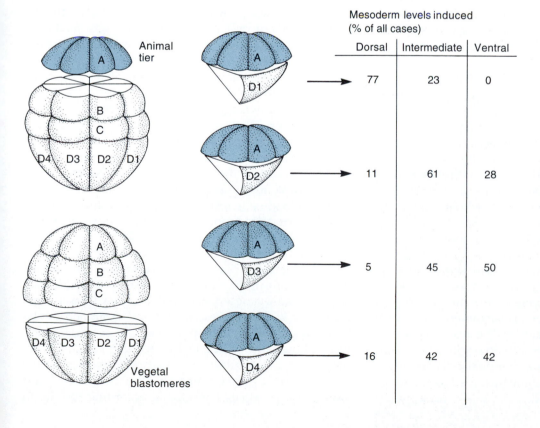

Mesoderm levels induced (% of all cases)

	Dorsal	Intermediate	Ventral
D1	77	23	0
D2	11	61	28
D3	5	45	50
D4	16	42	42

Figure 9.20 Regional specificity of mesoderm induction in *Xenopus*. Animal blastomere tiers (A) were combined with single vegetal blastomeres (D1 to D4) from 32-cell embryos. The animal tiers (color) were labeled with fluorescent dye to distinguish their descendants from those of the vegetal blastomeres. After culture in vitro, the resulting tissues were fixed and sectioned. The mesodermal structures formed by the labeled animal tiers were classified as dorsal mesoderm, intermediate mesoderm, or ventral mesoderm. Dorsal vegetal blastomeres induced primarily dorsal mesoderm, while lateral and ventral vegetal blastomeres induced intermediate and ventral mesoderm in similar proportions.

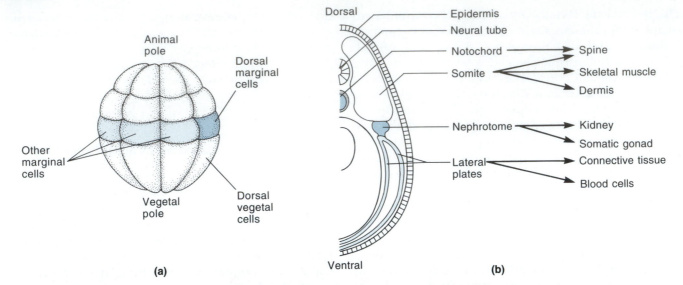

Figure 9.21 Development of the dorsoventral mesoderm pattern in the *Xenopus* embryo. **(a)** Side view at the 32-cell stage. Induction and isolation experiments reveal only two mesodermal states, one in the dorsal marginal cells and the other in all other marginal cells (Fig. 9.20). **(b)** Transverse section at an advanced embryonic stage. The mesoderm contains a complex dorsoventral pattern of different elements, including notochord, somites, nephrotomes, and lateral plates.

sal vegetal blastomeres induced the formation of dorsal mesodermal derivatives, in particular, notochord and muscle (Fig. 9.21). In contrast, ventral vegetal blastomeres induced the formation of ventral mesodermal derivatives, in particular, blood cells, mesenchyme, and mesothelium (epithelial mesoderm), and also some lateral mesodermal structures, including kidney and muscle. Lateral vegetal blastomeres induced a similar spectrum of mesodermal structures, although muscle was formed more frequently and blood cells less often than with ventral inducers.

The experiments just described show that *dorsal* vegetal blastomeres differ from their *lateral and ventral* neighbors in the mesodermal structures they induce. These results corroborate those obtained in the rescue and heterotopic transplantation experiments with vegetal blastomeres, in which only dorsal vegetal blastomeres induced the formation of an entire set of dorsal organs (Fig. 9.16). Together, the studies confirm that the dorsal vegetal blastomeres (Nieuwkoop center) induce dorsal marginal cells to form Spemann's organizer, which in turn induces the formation of an entire set of dorsal embryonic structures.

The mesodermal induction process shows how the dorsoventral polarity arising from cytoplasmic movements during the first cell cycle is perpetuated through different capabilities of vegetal blastomeres to induce different mesodermal characters. The resulting dorsoventral pattern within the mesoderm begins with a simple difference between dorsal marginal cells and other marginal cells. How this rudimentary pattern develops further will be explored next.

Dorsal Marginal Cells Interact with Other Marginal Cells to Generate the Full Mesodermal Pattern

In the experiment shown in Figure 9.20, the mesodermal characters induced by ventral vegetal blastomeres did not differ significantly from those induced by lateral vegetal blastomeres. The data do not reflect the detailed pattern of mesodermal structures observed later during organogenesis, when we can distinguish the notochord, flanked left and right (from dorsal to ventral) by somites, nephrotomes, and lateral plates (Fig. 9.21; see also Chapter 13). The *notochord* is a precursor of part of the spine. *Somites* give rise to other parts of the spine, to skeletal muscle, and to the deeper parts of the skin. *Nephrotomes*, also known as *intermediate mesoderm*, generate the embryonic kidneys. *Lateral plates* form the smooth muscle of internal organs, the connective tissues of trunk and limbs, and the circulatory system, including blood cells. How does this complex mesodermal pattern arise from the simpler pattern of dorsal marginal cells and other marginal cells?

To analyze the further development of mesoderm in *Xenopus*, Dale and Slack (1987b) compared the *fate* and *potency* of prospective mesoderm cells at early blastula stages. To study mesodermal cell fate, they labeled marginal cells with a fluorescent dye and examined the labeled tissues later at a tadpole stage. To explore the potency of the same cells, they isolated them and let them develop in tissue culture. Comparing the results of the two procedures, they found that isolated dorsal and ventral marginal cells developed according to fate. In

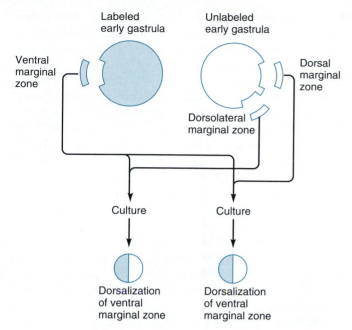

Figure 9.22 Dorsalization of the ventral marginal zone by dorsal or dorsolateral zone of the early gastrula. A piece of ventral marginal zone from a gastrula labeled with fluorescent dye was combined with an unlabeled piece of dorsal or dorsolateral marginal zone. The mesodermal structures formed by each zone were analyzed after culture. The ventral zones developed more dorsal structures than they would have in isolation or in the intact embryo. The development of their dorsal or dorsolateral partners was not changed by the coculture.

contrast, isolated *lateral* marginal cells developed more ventral structures than expected from the fate map. In particular, the lateral isolates produced substantial numbers of blood cells, which according to the fate map were to be expected only of ventral marginal cells. The investigators concluded that in the intact embryo the lateral marginal cells receive a signal that diverts them from ventral to lateral fates.

On the basis of earlier data, the researchers postulated that lateral marginal cells in the intact embryo receive an inductive signal from their dorsal neighbors. To test their hypothesis, they combined labeled ventral marginal parts with unlabeled dorsal or dorsolateral marginal parts (Fig. 9.22). After culture in combination, most of the ventral marginal parts were dorsalized, forming large amounts of muscle instead of blood. Only in a few cases, where the unlabeled dorsolateral part itself had formed ventral structures, did its labeled counterpart do the same. Conversely, the unlabeled dorsal and dorsolateral parts were not ventralized by the labeled ventral parts. Thus, signals received from the dorsal marginal cells changed the development of their lateral and ventral neighbors toward more dorsal structures.

As an overall interpretation of their experiments, Dale and Slack (1987b) proposed a ***three-signal model of mesoderm induction*** (Fig. 9.23). According to this model, mesoderm induction in *Xenopus* depends on three inductive interactions: (1) dorsal vegetal blastomeres induce their marginal neighbors to form dorsal mesoderm; (2) at the same time, ventral and lateral vegetal blastomeres induce their marginal neighbors to form ventral mesoderm; and (3) subsequently, the dorsal marginal cells induce their neighbors to form the intervening lateral elements between the dorsal and ventral mesoderm, thus generating the full spectrum of mesodermal structures along the dorsoventral axis. This model is almost certainly an oversimplification. For instance, it does not accommodate data indicating a ventralizing effect that ventral marginal cells may have on their lateral neighbors (Sive, 1993). Also, it is not clear how the operationally defined signals of this model relate to specific molecules that affect mesoderm induction (discussed in the next section). Nevertheless, the three-signal model has become widely accepted as a first approximation.

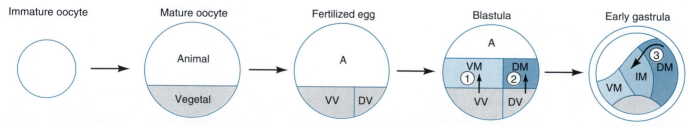

Figure 9.23 Three-signal model of mesoderm induction in *Xenopus* embryos. During oogenesis, cytoplasmic transport establishes the difference between animal and vegetal cytoplasm. Rearrangement of cytoplasm after fertilization generates two types of vegetal cytoplasm—dorsal vegetal (DV) and ventral vegetal (VV)—which are allocated to different cells. During the blastula stage, dorsal vegetal and ventral vegetal cells induce adjacent marginal cells to form dorsal mesoderm (DM) and ventral mesoderm (VM), respectively (signals 1 and 2). In the early gastrula, signal 3 from the dorsal marginal cells induces the adjacent marginal cells to form intermediate mesoderm (IM).

Molecular Mechanisms of Mesoderm Induction

The relative ease with which *Xenopus* blastomeres can be combined and cultured in vitro has made mesoderm induction a favorite system for analysis of the molecular mechanisms involved in embryonic induction. Such experiments have shown that signaling proteins known as *growth factors* can act as mesoderm-inducing signals. Other studies suggest that second messengers, such as *inositol trisphosphate* (IP_3), may mediate the cellular responses to these signals.

Several Growth Factors Have Mesoderm-Inducing Activity

To explore which kind of signal transfer is required for mesoderm induction, Horst Grunz and Lothar Tacke (1986) modified the experiment shown in Fig. 9.19 by placing a filter with very fine pores (diameter 0.4 μm) between vegetal and animal blastula cells. They reasoned that the filter would interfere with induction if direct cell contact was required. However, mesoderm was induced across the filter. The reacting ectoderm formed mostly ventral mesodermal structures, but a few animal caps also formed dorsal mesoderm and neural structures. Electron-microscopic inspection of the filters failed to reveal any cell outgrowths in the filter pores, indicating that transmission of diffusible molecules is sufficient for mesoderm induction to occur.

To further analyze the nature of the diffusible signals that induce mesoderm formation, animal caps from blastulae were cultured in media containing various cell extracts or molecular fractions. In response, blastula ectoderm was often induced to form mesoderm, as indicated by strong elongation of the animal caps, by histological features characteristic of mesodermal tissues, and by the synthesis of mesoderm-specific mRNAs or proteins. These observations revealed the mesoderm-inducing capacity of several *growth factors,* secreted peptides otherwise known for their role in controlling cell division and cell differentiation (Kimelman et al., 1992; Sive, 1993; J. C. Smith, 1993). In particular, several peptides of the *fibroblast growth factor (FGF)* family and the *transforming growth factor-ß (TGF-ß)* family can act as mesoderm-inducing signals. Also involved are secreted glycoproteins of the Wnt family, named after its prototype, Wnt-1, a signal molecule involved in growth control and tumorigenesis (Nusse and Varmus, 1992). The same glycoproteins, if isolated from *Xenopus,* are designated *Xwnt proteins.*

Are the peptides that show mesoderm-inducing activity in vitro the natural signals in the living embryo? A candidate for the natural signal must fulfill the following minimum requirements. First, the molecule must be present in the embryo at the required concentration, in the predicted region, and at the appropriate time. Second, it must be active in rescue and/or heterotopic transplantation experiments with whole embryos. Third, blocking the action of the molecule in vivo must interfere with mesoderm induction.

Basic fibroblast growth factor (bFGF) meets several criteria of a natural signal that would be necessary for ventral mesoderm formation. Receptors for FGF are present in the early embryo (Gillespie et al., 1989). Suppressing the function of FGF receptor eliminates molecular markers of muscle development in vitro and interferes with gastrulation and subsequent trunk mesoderm development in vivo (Amaya et al., 1991, 1993). Maternal mRNA for XeFGF (for *Xenopus* embryonic FGF) is present in the fertilized egg, although zygotic expression occurs only late during the blastula stage (Isaacs et al., 1992).

Several members of the TGF-ß family have been considered as possible dorsal mesoderm inducers. Perhaps the strongest candidate so far is known as the Vg1 protein (Thomsen and Melton, 1993; Dale et al., 1993). One of its attractive features is the localization of its source, Vg1 mRNA. As described in Chapter 8, maternal Vg1 mRNA is transported and anchored to the vegetal pole region during oogenesis. Since embryonic synthesis of Vg1 mRNA is also limited to the vegetal half of the blastula, Vg1 protein occurs primarily in the vegetal region, from which the mesoderm-inducing signals emanate (Dale et al., 1989; Tannahill and Melton, 1989). Like other TGF-ß peptides, Vg1 protein is synthesized as a large, biologically inactive precursor peptide. However, only a small fraction of the precursor peptide is processed into the active Vg1 protein in *Xenopus* eggs. The inefficient processing of Vg1 precursor explains why Vg1 mRNA injected into eggs has not been found to have mesoderm-inducing or dorsal organ–promoting activity. However, an experimentally modified Vg1 mRNA, which encodes a precursor peptide that is readily processed into active Vg1 protein, does elicit biological activity: injection of this mRNA construct into radially ventralized embryos restores normal development, and injection into normal embryos induces the formation of extra dorsal organs (Fig. 9.1).

These results raise the intriguing possibility that the activity of Vg1 protein may be controlled at some of the processing steps that occur after the precursor peptide has been synthesized. Conceivably, an enzyme or other cofactor required for a limiting processing step might be activated during the first cell cycle. Such a cofactor may be activated mostly in the dorsal portion of the embryo in the course of the cytoplasmic rearrangements that occur during the first cell cycle. Thus, the vegetal localization of Vg1 mRNA, and the dorsal activity of the hypothetical cofactor, together could generate active Vg1 protein exactly where mesoderm-inducing activity is first observed: in the dorsal vegetal blastomeres also known as the *Nieuwkoop center.*

Another member of the TGF-β family, *activin,* was discovered as the active component in the first soluble mesoderm inducer observed in vitro. J. C. Smith (1987) found that a particular line of cultured *Xenopus* cells released a mesoderm-inducing factor into the medium. This factor induced isolated animal caps to differentiate into muscle and notochord, while control caps in medium without the factor formed epidermis. The active component in this factor, and in similar factors released by other cell lines, was identified as activin, previously known as a mammalian hormone involved in controlling pituitary cell activities and blood cell development (J. C. Smith et al., 1990; Thomsen et al., 1990). Activin is present in unfertilized *Xenopus* eggs and blastulae (Asashima et al., 1991). Cloned activin from *Xenopus* induces dorsal mesoderm formation in explanted animal caps. Also, suppressing the function of activin receptor inhibits mesoderm induction and formation of dorsal organs in *Xenopus* embryos (Hemmati-Brivanlou and Melton, 1992). Moreover, injection of synthetic activin mRNA into ventral blastomeres at the 32-cell stage causes the formation of a second axis, including muscle.

Other data show that activin acts in cooperation with other factors to induce dorsal mesoderm formation. Sergei Sokol and Douglas Melton (1991) demonstrated that exposure to activin stimulates only dorsal, but not ventral, halves of animal caps to form dorsal mesoderm. The same investigators found that animal caps from embryos radially ventralized by UV irradiation did not form dorsal mesoderm in response to activin. However, activin does induce mesoderm formation in similar animal caps isolated from embryos that have been radially ventralized and injected with mRNA for another signal protein, *Xwnt-8* (Sokol and Melton, 1992). In similar experiments, Jan Christian and her colleagues (1992) showed that Xwnt-8 cooperates with bFGF in inducing dorsal mesoderm whereas neither peptide by itself has this effect. However, Xwnt-8 seems to be a mimic rather than a natural dorsal signal: it is synthesized too late during the blastoderm stage, and Xwnt-8 mRNA accumulates mainly ventrally.

Another factor that lacks mesoderm-inducing activity alone but acts as a modifier is the *noggin protein,* which does not belong to any of the known growth factor families. Maternal noggin protein is present in the egg, and embryonic synthesis is limited to prospective dorsal mesoderm. When applied to ventral marginal zone tissue, which by itself would form blood and mesenchyme, noggin protein causes the formation of muscle (W. C. Smith et al., 1993). These results make noggin a strong candidate for being the third signal in the three-signal model shown in Figure 9.23.

Although the true identity of the mesoderm-inducing signals is still uncertain, work with some of the known candidates and mimics has provided valuable insight. The dorsoventral bias in animal caps, revealed by the different response of dorsal and ventral cap halves to activin, indicates that some of the early steps in establishing the dorsoventral axis may have lasting effects. For instance, the rearrangements of cytoplasm during the first cleavage cycle may polarize not only the vegetal but also the animal blastomeres. The polarization of animal blastomeres appears to be weak, because it can be overridden experimentally, but it may nevertheless generate a bias that modulates the response to later signals. Another lesson from experiments with inducer candidates and mimics is that the three-signal model is probably a simplification of a much more complex natural process. Any signal in the sense of the model may turn out to consist of two or more signal molecules that cooperate or modify one another's effects.

The apparent need for the responding cells to integrate the response to several signal molecules received over time raises the question of which mechanisms might be available for this integrating function. Signal peptides activate receptors, which in turn act through kinases or through G proteins that generate *second messengers* inside the responding cell (see Chapter 2). We will explore the possible role of a particular second messenger next.

Inositol Trisphosphate Levels Specify Dorsoventral Polarity

Early embryologists experimented with various chemicals to study their effects on development. Their studies revealed that lithium is a strong *teratogen* (Gk. *terat-,* "monster") for many organisms, including frogs (T. H. Morgan, 1903). More recently, Kenneth Kao and his colleagues (1986) dramatically reprogrammed the dorsoventral polarity of *Xenopus* embryos at the 32-cell stage by injecting lithium chloride (LiCl) into specific blastomeres. Injecting LiCl solution into ventral *vegetal* blastomeres produced embryos in which an extra set of dorsal anterior structures, including notochord, neural tube, and eyes, formed on the ventral side (Fig. 9.24). No such duplication occurred after injection of sodium chloride and various other salt solutions. These experiments were extended by William Busa and Robert Gimlich (1989), who obtained the same duplications by injecting smaller amounts of LiCl into a pair of ventral *marginal* cells. The results indicated that lithium ion (Li$^+$) reprograms ventral marginal cells to develop like dorsal marginal cells.

Lithium ion blocks enzymes of the phosphoinositide (PI) cycle, which produce *myo-inositol,* a precursor for the regeneration of *inositol trisphosphate (IP$_3$)* and *diacylglycerol (DAG)* (Berridge et al., 1989). As explained in Chapter 2, IP$_3$ and DAG are second messengers that mediate the cellular response to signals acting on cell surface receptors. Figure 9.25 shows a simplified scheme of Li$^+$ action on the PI cycle.

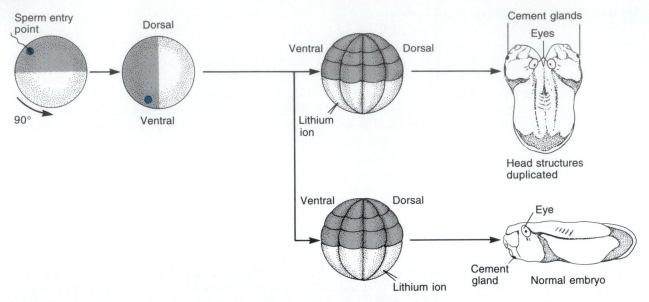

Figure 9.24 Duplication of anterior dorsal structures after injection of lithium chloride (LiCl) into ventral vegetal blastomeres of *Xenopus* embryos. The sperm entry point of fertilized eggs was marked with dye, and the eggs were temporarily rotated to ensure that the dorsal midline developed precisely along the meridian opposite the marked sperm entry point. At the 32-cell stage, 0.3 *M* LiCl was injected into a ventral vegetal blastomere. Most of the developing tadpoles had duplicate anterior dorsal structures including notochord, neural tube, eyes, and cement gland (the anteriormost organ by which the tadpole attaches to a substratum). No such duplications occurred after injection of LiCl into dorsal vegetal blastomeres.

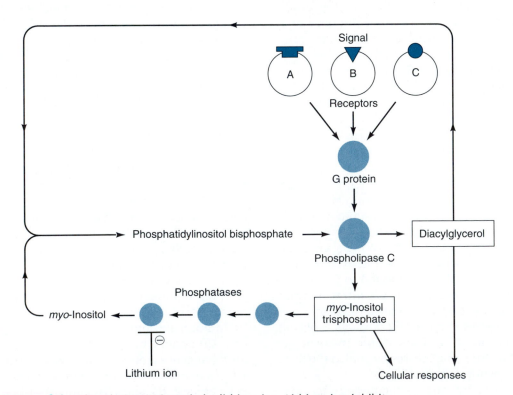

Figure 9.25 Inhibition of the phosphoinositide cycle by lithium ion. Lithium ion inhibits the last enzyme in a series of phosphatases that regenerate *myo*-inositol. The inhibition leads to a depletion of two second messengers, diacylglycerol and *myo*-inositol trisphosphate.

To test whether the teratogenic action of Li⁺ is caused by its interference with *myo*-inositol production, Busa and Gimlich (1989) coinjected Li⁺ and *myo*-inositol. Indeed, *myo*-inositol counteracted the effect of Li⁺: whereas Li⁺ alone caused anterior dorsal duplications, coinjection of Li⁺ and *myo*-inositol was followed by normal development. *epi*-Inositol, a biologically inactive isomer of *myo*-inositol, did not have this protective effect. Further experiments showed that *myo*-inositol does not reduce the intracellular concentration of Li⁺. Therefore, *myo*-inositol must counteract the effect of Li⁺ instead of removing it from the injected cells. Moreover, fate-mapping experiments showed that the cells that were double-injected with Li⁺ and *myo*-inositol developed according to their fate in normal embryos, ruling out the explanation that the double-injected cells simply died and allowed the surviving cells to form a normal-looking embryo by *regulation*. Follow-up studies confirmed that normal mesoderm induction is correlated with an increase in IP₃ concentration, and that Li⁺ causes a decrease in the amounts of both *myo*-inositol and IP₃ in an embryo (Maslanski et al., 1992).

Taken together, the results of experiments with Li⁺ suggest that its dorsalizing effect on *Xenopus* embryos stems from inhibition of the PI cycle, which causes a depletion of IP₃ in ventral marginal blastomeres. This hypothesis implies that in normal embryos the rate of the PI cycle, and the resulting concentration of IP₃, are high in the ventral marginal blastomeres and low in dorsal marginal blastomeres. Thus, one would expect that an increase in the IP₃ concentration on the dorsal side of the embryo will ventralize organ development there. The following experiment shows that this is the case.

Katherine Ault and her colleagues (1994) injected the mRNA for mammalian serotonin receptor into either the two dorsal or the two ventral blastomeres of *Xenopus* embryos at the 4-cell stage and observed that the blastomeres synthesized the receptor protein and integrated it into their plasma membranes. In mammalian cells, the serotonin receptor acts through G protein and releases IP₃ as second messenger (Fig. 9.25; see also Chapter 2). Incubation of the injected *Xenopus* embryos in serotonin is therefore likely to increase the IP₃ concentration in the injected blastomeres and thus to ventralize embryonic development. Indeed, ventral injection of serotonin receptor mRNA had little or no effect on subsequent development, whereas dorsally injected embryos gastrulated abnormally and formed tadpoles that were grossly deficient in dorsoanterior structures. Moreover, animal caps from such embryos, when cultured in the presence of activin, did not synthesize a molecular marker for dorsoanterior mesoderm.

The experiment described above confirms the view that IP₃ and presumably other second messengers integrate the signals conveyed by multiple peptide signals and serve to elicit a unified cellular response. It seems to be a general characteristic of induction processes that more than one inducing signal is involved (see Chapters 11 and 12). Therefore, the concept that second messengers act to integrate several inductive signals is of great interest beyond the subject of mesoderm induction.

SUMMARY

Most organisms acquire one or more body axes in the course of development. The brown alga *Fucus* has perfectly spherical eggs in which one axis develops between the future thallus and the future rhizoid. This axis may be determined by sperm entry or by environmental factors, such as light. A period of axis formation, when the thallus-rhizoid axis is set up in a preliminary way, is followed by a period of axis fixation, when the axis is irreversibly established. Axis formation by light is accompanied by electric polarization, with the rhizoid pole becoming negative. Axis fixation involves the assembly of molecular bridges across the plasma membrane at the rhizoid pole between microfilaments on the inside and extracellular material on the outside.

Metazoa are, for the most part, bilaterally symmetrical and have three body axes: anteroposterior, dorsoventral, and left-right. In *Xenopus*, the anteroposterior axis develops from the animal-vegetal axis, which originates during oogenesis. The dorsoventral axis is formed after fertilization and fixed before the first cleavage. Molecular clues to left-right asymmetry have been detected during the blastula and early gastrula stages.

The animal-vegetal polarity in amphibian eggs determines the spatial organization of the germ layer rudiments during the blastula stage. Most of the animal half of the blastula forms ectoderm, and most of the vegetal half forms endoderm, while a marginal zone in between gives rise to mesoderm. Specific behaviors of the germ layers during gastrulation transform the original animal-vegetal polarity of the egg into the anteroposterior body pattern of the postgastrula embryo.

The dorsoventral axis in the amphibian embryo forms after fertilization. The point of sperm entry orients a rotation of the egg cortex relative to the endoplasm. The meridian of greatest cortical displacement, opposite the sperm entry point, becomes the dorsal midline of the embryo. The cortical rotation is balanced by rearrangements within the endoplasm, which generate a dorsoventral asymmetry. After a labile period during the first cleavage cycle, this asymmetry is fixed primarily in the vegetal tier of blastomeres.

Originating from cytoplasmic localization, the dorsoventral polarity develops further by embryonic induction. Induction, a pervasive principle of development, is defined as an interaction between cells or tissues in which one partner (the inducing partner) changes the determined state of a responding partner. The development of the dorsoventral polarity in *Xenopus* embryos depends on mesoderm induction.

A current model of mesoderm induction in *Xenopus* involves three inductive interactions. Dorsal vegetal blastomeres induce their marginal neighbors to form dorsal mesoderm. At the same time, ventral and lateral vegetal blastomeres induce their marginal neighbors to form ventral mesoderm. Subsequently, the dorsal marginal cells induce their neighbors to form the intervening elements between the dorsal and ventral mesoderm, thus generating the full spectrum of mesodermal structures along the dorsoventral axis.

The molecular nature of the inducing signals is under active investigation. It appears that a given signal, in terms of the three-signal model, may be conveyed by two or more types of molecule. To integrate their response to multiple molecular signals, cells may use second messengers, such as inositol trisphosphate.

SUGGESTED READINGS

Gerhart, J., M. Danilchik, T. Doniach, S. Roberts, B. Rowning, and R. Stewart. 1989. Cortical rotation of the *Xenopus* egg: Consequences for the anteroposterior pattern of embryonic dorsal development. *Development* 1989 Supplement:37–51.

Kropf, D. L. 1992. Establishment and expression of cellular polarity in fucoid zygotes. *Microbiol. Rev.* **56**:316–336.

Smith, J. C. 1993. Mesoderm-inducing factors in early vertebrate development. *EMBO J.* **12**:4463–4470.

GASTRULATION

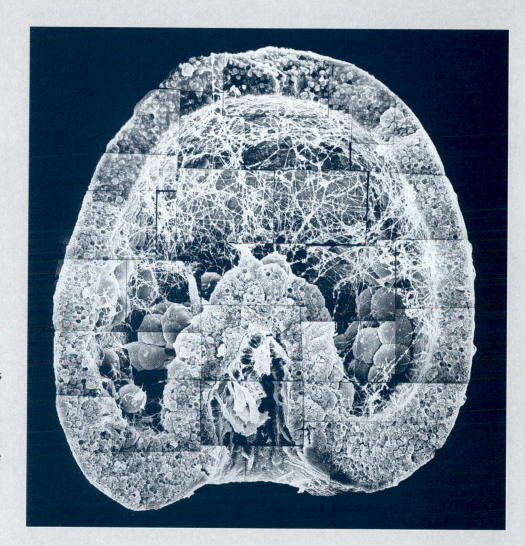

Figure 10.1 Scanning electron micrograph showing the formation of three germ layers in the sea urchin gastrula. The endoderm originates from a depression at the vegetal pole (bottom), which deepens and elongates. Mesodermal cells separate from the vegetal pole area and move into the embryonic cavity, the blastocoel. The ectoderm remains on the outside of the gastrula.

The Analysis of Morphogenesis

Cell Movements and Other Cell Behaviors Involved in Morphogenesis

Gastrulation in Sea Urchins

Gastrulation in Amphibians

Different Gastrula Areas Show Distinct Cellular Behaviors

Bottle Cells Generate the Initial Depression of the Blastopore

Deep Marginal Zone Cells Are Necessary for Involution

Deep Zone Cells and Involuting Marginal Zone Cells Migrate on the Inside of the Blastocoel Roof

Convergent Extension Is Especially Strong in the Dorsal Marginal Zone

The Animal Cap and the Noninvoluting Marginal Zone Undergo Epiboly

Specific Gene Products Control Cell Movements during Gastrulation

Gastrulation in Birds

Gastrulation in Humans

During cleavage, the fertilized egg is divided into hundreds or thousands of cells, which typically form a fluid-filled sphere, the blastula. To become a functioning organism, the embryo must undergo rearrangement of the blastula cells according to the body plan characteristic of its species (Fig. 10.1). The first phase of this process is called *gastrulation* (Gk. *gaster,* "stomach"), because, among other things, it generates the rudiment of a stomach. Gastrulation transforms the spherical blastula into a more complex configuration of three *germ layers* (Fig. 10.2). The outer layer, which is exposed to the external environment, is the *ectoderm.* It gives rise, for the most part, to the epidermis and the nervous system of the developing organism. The inner layer of cells is the *endoderm,* which encloses the primitive gut, or **archenteron** (Gk. *arche,* "origin"; *enteron,* "gut"). The numerous structures in between develop from an intermediate layer called the *mesoderm.* The ectoderm and endoderm are *epithelia,* that is, sheets of closely packed cells connected by tight junctions and resting on a basal lamina (see Fig. 2.20). The mesoderm is sometimes epithelial, but in other cases it forms a *mesenchyme,* that is, a loose arrangement of cells containing more extracellular material. An embryo in

the process of gastrulation is called a *gastrula* (pl., *gastrulae*).

A common feature of early gastrulae is the *blastopore,* an indentation or groove through which cells move inside to form the endoderm and mesoderm. The fate of the blastopore is used as a basis for classifying animals into two groups. Those in which the blastopore develops into the mouth are known as **protostomes** (Gk. *proto-,* "first"; *stoma,* "mouth"). Animals in which the blastopore becomes the anus, and mouth formation is secondary, are called **deuterostomes** (Gk. *deutero-,* "second"). Deuterostomes include vertebrates and echinoderms (sea urchins and their relatives), whereas the protostomes include most other animals. Among the deuterostomes, as well as among the protostomes, there are further differences in regard to the cellular movements that establish the germ layers.

To analyze gastrulation, we will identify a few types of elementary movements that cells carry out individually, in small groups, or in the form of epithelial sheets. We will then see how the actual gastrulation processes in representative animals can be understood as combinations of those elementary movements. The elementary movements in turn will be analyzed in terms of cellular behaviors such as migration or changes in shape. This approach was adopted by the early embryologists and was boosted later with the advent of electron microscopy and improved methods of cell culture (Trinkaus, 1984). It provides a better understanding not only of gastrulation but also of other morphogenetic events to be described in Chapters 11 through 13. In this chapter the emphasis will be on gastrulation in the frog *Xenopus laevis,* which has been analyzed in a particularly incisive series of experiments.

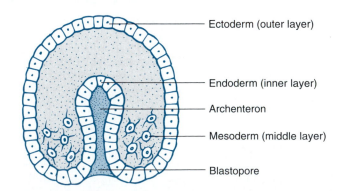

Figure 10.2 Schematic diagram of a gastrula. Formation of the primitive gut (archenteron) has begun at the blastopore. The embryo consists of three germ layers. The inner and outer layers, ectoderm and endoderm, are epithelia. The intermediate layer, mesoderm, consists of mesenchyme, that is, single cells surrounded by extracellular matrix.

The Analysis of Morphogenesis

Gastrulation marks the onset of cellular movements by which the embryo's organ rudiments are generated as part of an overall body plan. These movements are collectively called *morphogenetic movements* or **morphogenesis** (Gk. *morphe*, "form"; *genesis*, "creation").

During morphogenesis, individual cells migrate, often over large distances, while entire epithelia spread, fold, bend, and roll dramatically. These movements bring about gastrulation and the subsequent phase of development, known as *organogenesis*, when cells from the three germ layers cooperate to form the organ rudiments.

Morphogenetic movements are striking in their complexity and order. When seen with time-lapse photography, the behavior of cells during morphogenesis gives the impression of a miniature ballet. Each "performance" creates a new organism of gradually increasing complexity. Once the organ rudiments have formed, the embryo has acquired its **basic body plan;** the appearance of an embryo at this stage is often characteristic of its phylogenetic group. For instance, any insect embryo at this stage begins to look unmistakably like an insect, and any vertebrate embryo at this stage can be recognized as such. The generation of the basic body plan by morphogenetic movements is found in all developing organisms and is always stunning to observe.

As we examine morphogenetic events more closely, we will see how cells divide, wedge themselves between other cells, and even detach from epithelia and move into body cavities. Note that the same behaviors later in life can prove fatal if performed by cancer cells. Thus, the harnessing of these cellular behaviors for the purposes of morphogenesis and the subsequent limitation of them represent a significant regulatory achievement.

The analysis of morphogenetic movements presents researchers with formidable challenges. How can complex movements involving the entire embryo be broken down into smaller, more readily analyzable events? Which of the signals exchanged between cells are most important: the *physical* forces of cells pulling and pushing one another, or the *chemical* signals that diffuse through gap junctions or act on plasma membrane receptors? What kind of control hierarchy should one assume for an initial hypothesis? Does gene expression control cell shape, or does cell shape control gene activity? Most researchers today assume that cells are first programmed to express certain master genes, which in turn coordinate other genes with more limited functions. The latter class includes, for example, genes encoding the proteins that are involved in microfilament assembly or in other processes that promote or inhibit cell movements and shape changes. However, other researchers propose that all cells in an embryo initially produce the same combinations of proteins until their physical interactions generate groups of cells with different mechanical states, which then secondarily cause these cells to express specific combinations of genes (R. Gordon and G. W. Brodland, 1987).

Clearly, the overall hypothesis from which one starts will circumscribe the kind of data obtained. It is the approach taken by the majority of investigators today that we will explore in the following section.

Cell Movements and Other Cell Behaviors Involved in Morphogenesis

The patterns of gastrulation and organogenesis vary from one group of organisms to another, but they all involve combinations of only a few types of elementary cellular movements, which are defined below, and diagrammed in Figure 10.3. Two of these movements, ingression and migration, are carried out by individual cells or by cells in small groups, whereas the other movements are performed by entire epithelia or cellular layers. The specific term "epithelium" is used for cellular layers that rest on a basal lamina and form tight junctions at the apical surface as described in Chapter 2. The more generic term "cell layer" is used for both epithelia and sheets of cells that are not clearly epithelial but show more mesenchymal characteristics, at least at certain stages of development. Transitions between epithelial, layered, and mesenchymal cell configurations occur frequently in embryos.

Migration: Movement of individual cells over a substratum of other cells or extracellular material.

Ingression: The transition of individual cells or small groups of cells from an epithelium into an embryonic cavity.

Invagination: Local inward buckling of an epithelium, much like the indentation produced by poking a soft tennis ball.

Involution: Inward movement of a cell layer around a point or an edge.

Convergent extension: Elongation of a cell layer in one dimension and shortening in another dimension.

Epiboly: The spreading movement of a superficial cell layer to envelop a yolk mass or deeper cell layer.

Delamination: Splitting of one layer of cells into two parallel layers.

Passive movements: Movements in which cells are simply pushed or dragged along by other cells.

Different combinations of these movements yield a wide variety of morphogenetic processes in different animals. Underlying these movements are a small

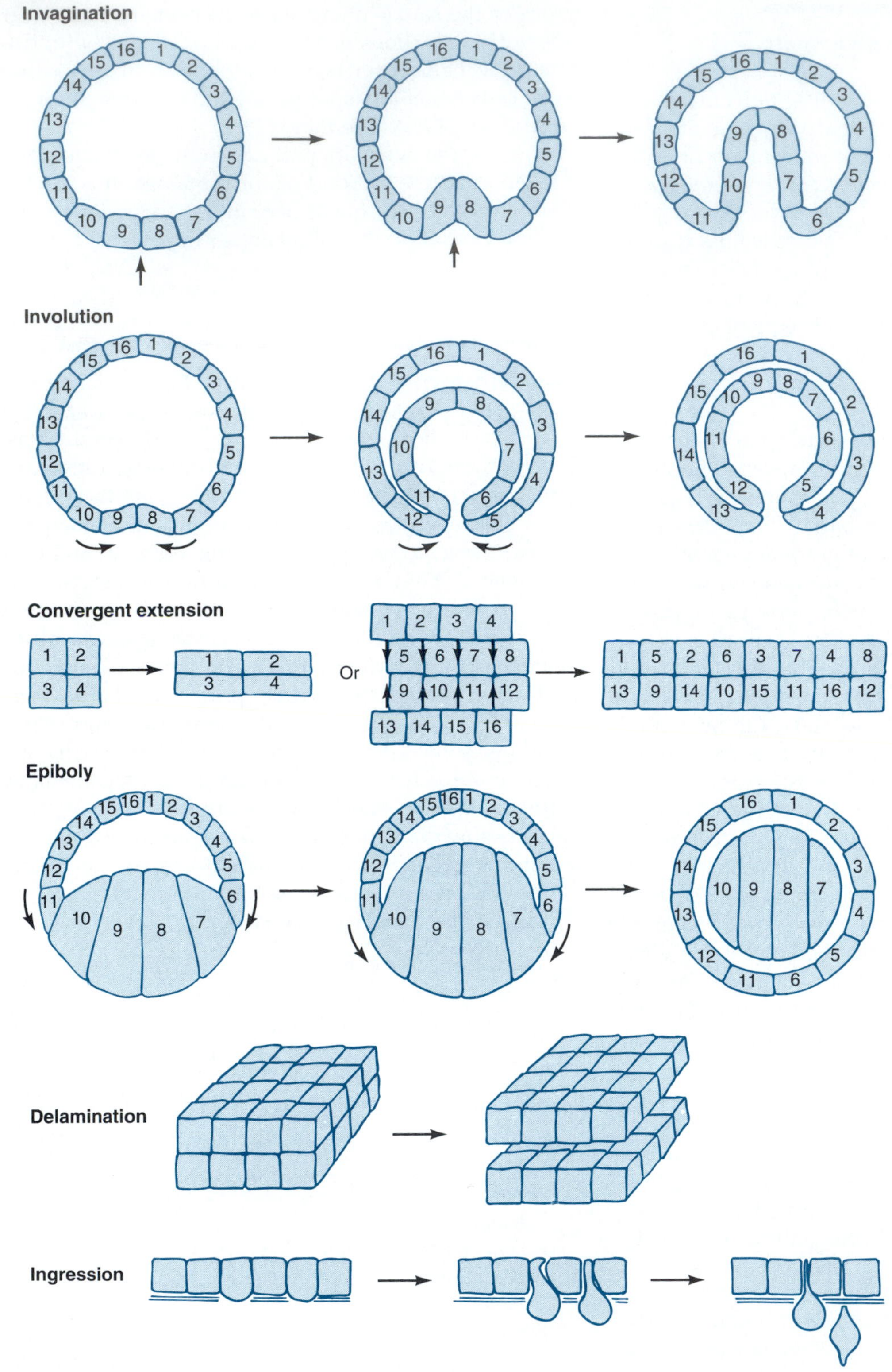

Invagination

Involution

Convergent extension

Or

Epiboly

Delamination

Ingression

Figure 10.3 Schematic diagrams of typical morphogenetic movements.

repertoire of other cellular behaviors that include the following:

Cell division: Oriented cell divisions combined with cell growth can change the size and shape of a cell layer.

Cell shape changes: Coordinated changes in cell shape can cause cell layers to buckle, to roll, or to simply extend in one dimension while shrinking in another.

Changes in adhesiveness: Cells adhere to one another, and to extracellular material, by means of certain membrane proteins. Changes in adhesiveness contribute to cell migration, ingression, delamination, and epiboly.

Intercalation: A cell can wedge itself between two of its neighbors. Coordinated intercalations may cause convergent extension or epiboly.

Cell death: Cells can die in a programmed way. Such programmed cell death may carve out, or "sculpt," certain organs such as the digits of hands and feet.

The cellular behaviors just listed are thought to be regulated by hierarchies of gene activity. For instance, a hierarchy of genes that researchers are now unraveling in the roundworm *Caenorhabditis elegans* controls cell death (see Chapter 24). Similarly, the synthesis of plasma membrane proteins known as *cell adhesion molecules* and *substrate adhesion molecules* governs cell adhesion. These molecules are thought to play key roles in morphoregulatory cycles, in which gene activities control cellular behaviors, which then regulate morphogenetic movements, which in turn feed back on gene expression (see Fig. 1.22 and Chapters 25 and 26).

Gastrulation in Sea Urchins

Sea urchins have long been favorite organisms for the study of gastrulation, because their transparency allows observation of the whole process in a living animal (Fig. 10.4; see also McClay et al., 1992). The sea urchin blastula is a sphere of about 1000 cells surrounding a *blastocoel* filled with a gelatinous fluid. These cells are arranged as a single-layered epithelium with a *basal lamina* on the inside, and with cilia and a *hyaline layer* on the outside. After the cilia have formed, the blastula hatches from the fertilization envelope. The arrangement of cells in the free-swimming blastula, and the fates of these cells after gastrulation, reflect the tiers of blastomeres present at the 60-cell stage (Ettensohn, 1992). A patch of blastomeres with long and coarse cilia that make up the **apical tuft** are derived from the most animal tier of blastomeres. These blastomeres and all their animal and equatorial neighbors will form the larval ectoderm after gastrulation. A girdle of vegetal cells, derived from the vegetal 2 tier of blastomeres at the 60-cell stage, will form the endoderm and part of the mesoderm. The descendants of the large and small micromeres, still located at the vegetal pole of the blastula, will form other portions of the mesoderm. The entire vegetal area of the blastula flattens at the beginning of gastrulation to form the **vegetal plate.**

Gastrulation begins in earnest with the *ingression* of a group of cells near the center of the vegetal plate. First these cells, which are descendants of the large micromeres, begin to pulsate and bulge on their basal sur-

faces while their apical surfaces constrict (Fig. 10.5). Then they undergo a *change in adhesion:* they lose their affinity for neighboring vegetal cells, and they withdraw their cilia and microvilli from the hyaline layer that covers the outside of the embryo (see Chapter 25). At the same time, they develop an affinity for the basal lamina on the inside of the blastula (Fink and McClay, 1985; Anstrom et al., 1987). As a result of these changes, the large-micromere descendants detach from the vegetal plate and ingress into the blastocoel as **primary mesenchyme cells.** They migrate on the basal lamina by extending long, thin processes called *filopodia*. When the tip of a filopodium finds a spot to which it can adhere, it contracts, pulling the entire cell after it. After a period of migration, the primary mesenchyme cells settle into a ringlike pattern on the vegetal side of the blastocoel. Here they fuse to form a *syncytium*, a multinucleate body of cytoplasm that remains anchored to the outer epithelium by filopodia. The syncytium secretes the skeleton of spicules supporting the *pluteus larva*, which is characteristic of sea urchins.

As the primary mesenchyme cells begin to migrate, the remaining cells of the vegetal plate start to form the *archenteron* (Fig. 10.4e). Archenteron formation in sea urchins occurs in three distinct phases. During the first phase, the blastopore forms as the vegetal plate *invaginates*, buckles inward, and extends about one-third into the blastocoel. At this point the embryo looks much like a soft tennis ball pushed in with a finger (Fig. 10.6). During the second phase, the archenteron *elongates by convergent extension* until its tip almost reaches the animal pole. During the third and final phase, the archenteron tip *is pulled by filopodia* toward the ectodermal cells that will form the mouth.

The forces driving invagination are generated within the vegetal plate itself and are independent of the rest of the embryo. When A. R. Moore and A. S. Burt (1939) isolated vegetal halves or thirds of starfish embryos at the beginning of gastrulation, the vegetal parts invaginated on their own, and the rim of each invaginating plate rolled up and closed over the archenteron, forming a gastrula-shaped vesicle (Fig. 10.7). Also, when a radial cut was made from the periphery to the center of an isolated vegetal plate, the cut edges sprang apart, indicating that the vegetal plate was under tension. These results indicated that the forces generated within the vegetal part of the embryo were sufficient to drive the first phase of archenteron formation.

The cellular and molecular mechanisms generating these forces remain to be established (Ettensohn, 1984a). One of the current models is built on the observation that the *hyaline layer* covering the vegetal plate cells consists of several laminae of extracellular material. If an inner lamina were to attract more water and thus swell more strongly than the outer laminae, then the entire hyaline layer and the adjacent vegetal plate would

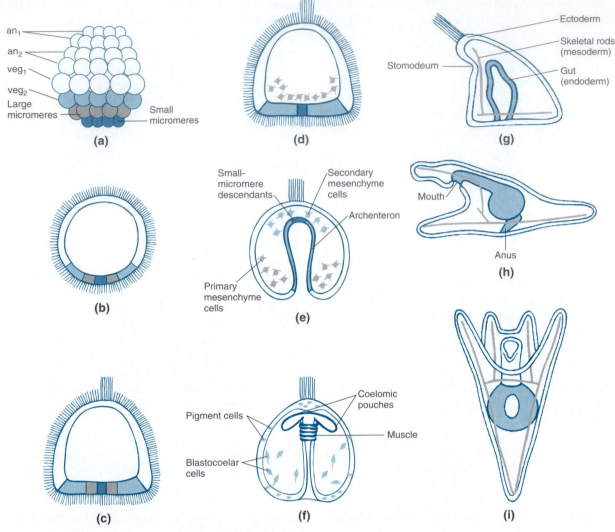

Figure 10.4 Sea urchin gastrulation. **(a)** Early blastula consisting of about 60 cells. The top two tiers, each comprising eight blastomeres, are the animal 1 (an_1) tiers, and the next two are the animal 2 (an_2) tiers. Following are one vegetal 1 (veg_1) tier, one vegetal 2 (veg_2) tier, eight large micromeres, and four small micromeres. **(b, c)** Mid and late blastula. The descendants of the vegetal 2 blastomeres, large micromeres, and small micromeres form a vegetal plate of columnar cells. **(d)** Early gastrula, showing ingression of primary mesenchyme cells, the descendants of the large micromeres. **(e)** Mid-gastrula with archenteron and secondary mesenchyme cells, both derived from the vegetal 2 tier, and the small micromeres, the descendants of which remain at the archenteron tip. **(f)** Late gastrula, showing derivatives of the secondary mesenchyme cells and the small-micromere descendants; the latter contribute to the coelomic pouches that extend from the archenteron tip. Primary mesenchyme derivatives are omitted for clarity. **(g)** Prism-stage larva showing skeletal rods derived from primary mesenchyme cells. The archenteron bends toward an ectodermal area, which forms a depression, the stomodeum. The secondary mesenchyme cell derivatives have been omitted for clarity from this and the following diagrams. **(h, i)** Pluteus larva viewed from the side and from the vegetal pole. The archenteron tip and the stomodeum have joined to form the mouth, while the blastopore develops into the anus.

buckle like the bimetallic strip in a thermostat (Fig. 10.8). In accord with this model, the vegetal plate cells secrete a hygroscopic (water-attracting) material prior to invagination, and treatments that disturb this process also interfere with invagination (Lane et al., 1993).

The second phase of archenteron formation is a dramatic case of convergent extension: the archenteron elongates from a short, squat cylinder to a long and slender one (Fig. 10.9). This approximately threefold increase in overall length cannot be explained by cell division and growth, since only 10 to 20% of the cells divide during this time. Involution does not seem to

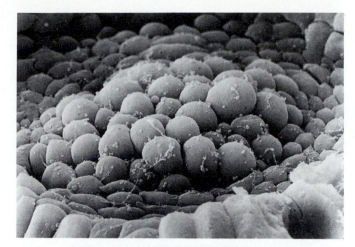

Figure 10.5 Primary mesenchyme formation in the sea urchin embryo. In this scanning electron micrograph, the vegetal plate is viewed from inside the blastocoel. Cells near the center of the vegetal plate are losing the epithelial connections with their neighbors and are beginning to ingress into the blastocoel.

Figure 10.6 Scanning electron micrograph of a sea urchin (*Lytechinus variegatus*) early gastrula after invagination of the vegetal plate to form the blastopore. The cilia on the outside of the embryo have been removed with 2× concentrated seawater.

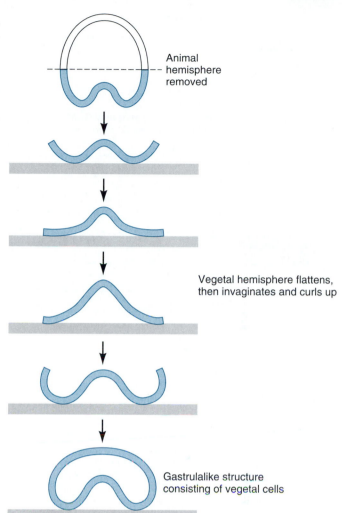

Animal hemisphere removed

Vegetal hemisphere flattens, then invaginates and curls up

Gastrulalike structure consisting of vegetal cells

Figure 10.7 Analysis of gastrulation in starfish. After experimental removal of the animal hemisphere, the vegetal hemisphere invaginates and closes into a gastrulalike vesicle.

play a major role either, since few cells turn inside at the blastopore during the elongation. Two coordinated cell behaviors could account for the convergent extension of the archenteron: cell *shape changes* and cell *intercalation* (Fig. 10.10).

▼

Measurements by Charles Ettensohn (1985b) and Jeffrey Hardin and Louis Cheng (1986) on two sea urchin species revealed that both types of cellular behavior occur during the second phase of archenteron formation. Archenteron cells grew longer and wider, while their

depth decreased by about 40%, indicating that a coordinated change in cell shape plays a role in the elongation of the archenteron. At the same time, the *number* of cells along the length of the archenteron increased while the circumferential number of cells decreased from 16 to 10 in one species (*Strongylocentrotus purpuratus*) and from 18 to 7 in another species (*Lytechinus pictus*). This decrease could not have resulted from cell death, since there were no observations of dying cells and the overall mass of the archenteron did not change. This means that convergent extension of the archenteron is associated with cell rearrangement. The same cellular behavior was observed in embryos treated experimentally so that the archenteron extended to the outside of the embryo rather than invaginating to the inside. This observation showed that the cell rearrangement was driven by forces generated within the archenteron itself.

Tracing cells by videomicroscopy revealed that cells *intercalated,* that is, wedged between their neighbors

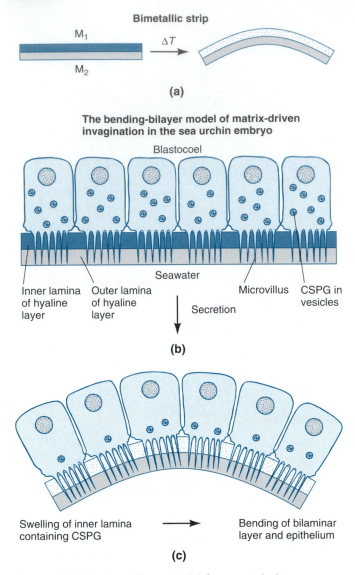

Bimetallic strip

The bending-bilayer model of matrix-driven invagination in the sea urchin embryo

Blastocoel

Inner lamina of hyaline layer | Outer lamina of hyaline layer | Seawater | Microvillus | CSPG in vesicles

Secretion

(b)

Swelling of inner lamina containing CSPG ⟶ Bending of bilaminar layer and epithelium

(c)

Figure 10.8 Bending-bilayer model for vegetal plate invagination in sea urchin embryos by the secretion of a hygroscopic (water-attracting) molecule into the hyaline layer that covers the vegetal plate. **(a)** The model is analogous to a bimetallic strip that bends in response to temperature changes (ΔT) because one metal (M_1) has a greater thermal expansion coefficient than the other metal (M_2). **(b)** Vegetal plate before invagination. The hyaline layer covering the apical surface of the vegetal plate cells is shown for simplicity as consisting of one inner and one outer lamina. The cells contain secretory vesicles with chondroitin sulfate proteoglycan (CSPG, colored stipples on white background). **(c)** Secretion of CSPG makes the inner lamina more hygroscopic, so that it expands more strongly than the outer lamina. Thus, the bilayered hyaline layer and the adjacent epithelium buckle.

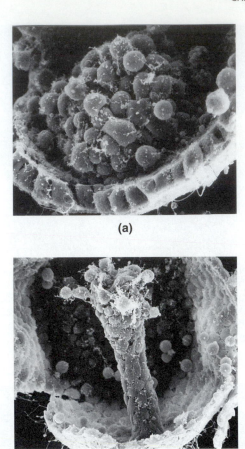

Figure 10.9 Archenteron elongation in the sea urchin *Lytechinus variegatus*. The scanning electron micrographs show the interior of the embryo after removal of the animal ectoderm. **(a)** Mid-gastrula. **(b)** Late gastrula.

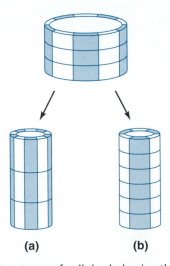

(a) **(b)**

Figure 10.10 Two types of cellular behavior that could explain the convergent extension movement observed during the second phase of archenteron formation in sea urchins. In these schematic diagrams, the archenteron is represented as a cylinder. **(a)** Change of cell shape. Each cell may elongate while shrinking in width and possibly also in depth. **(b)** Cell intercalation may increase the number of cells longitudinally while reducing the number of cells in the cylinder's circumference.

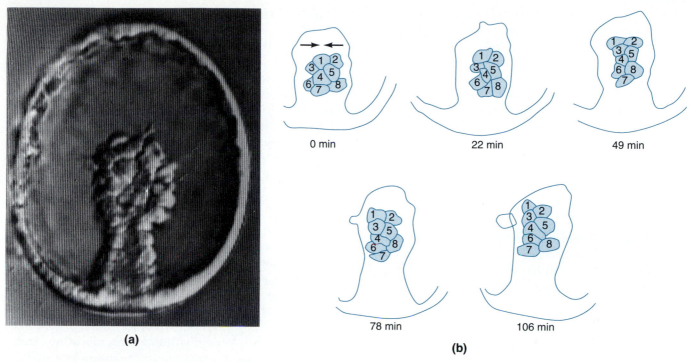

Figure 10.11 Cell rearrangements during gastrulation in the sea urchin *Eucidaris tribuloides*. **(a)** Mid-gastrula photographed from a video image. **(b)** Tracings made on the video monitor at the times indicated. Note that cell intercalations occurred in a circumferential direction (arrows), with cell 3 wedging between cells 1 and 4 and cell 6 wedging between 4 and 7. During the period of observation, the array of cells narrowed from about three cell diameters to two, while elongating from about three cell diameters to more than four. The length of the archenteron increased correspondingly.

(Hardin, 1989). The intercalations were oriented in such a way that the *number* of cells in the longitudinal dimension of the archenteron increased (Fig. 10.11). The intercalating cells seem to displace their neighbors actively, beginning with the formation of small *lamellipodia* near the basal surface. The signals that orient the cell intercalations so that they work in one direction are not yet known.

While the elongation of the archenteron is still under way, another set of mesenchymal cells become motile at the archenteron tip (Fig. 10.9b). These cells, called **secondary mesenchyme cells**, are derived from the vegetal 2 tier of blastomeres formed at the 60-cell stage (Fig. 10.4a and e).

While still attached to the archenteron tip, the secondary mesenchyme cells initiate the third and final phase of archenteron elongation by extending thin filopodia through the blastocoel. The filopodia probe the inside of the blastocoel wall, attach at suitable sites, and then contract, further extending the archenteron and orienting the tip toward its destination (Hardin and McClay, 1990). If secondary mesenchyme cells are killed with a laser beam, the elongation of the archen-

teron remains incomplete (Hardin, 1988). Thus, the shape change (elongation) of the archenteron cells is, at least in part, caused by the contraction of secondary mesenchyme cells. The tip of the extending archenteron is met by the *stomodeum*, a depression in the ectoderm. Fusion of the stomodeum with the archenteron tip generates the mouth of the pluteus larva, while the blastopore becomes the anus. Thus, sea urchin gastrulation follows the pattern typical of deuterostomes.

As the archenteron tip approaches the stomodeum, the secondary mesenchyme cells form two clusters of cells on either side of the foregut. These cells, along with the small-micromere descendants that have remained at the archenteron tip, form pouches of mesodermal epithelium, which displace the blastocoel and form a secondary body cavity called the **coelom.** Other descendants of the secondary mesenchyme cells form pigment cells, muscles, and blastocoelar cells (Fig. 10.4f). While in most sea urchins there is little or no overlap in the *fates* of primary mesenchyme cells and secondary mesenchyme cells, the *potency* of secondary mesenchyme cells includes the major fate of primary mesenchyme cells, that is, skeleton formation. Charles Ettensohn and David McClay (1988) observed that secondary mesenchyme cells substitute for primary mesenchyme cells in skeleton formation if the number of primary mesenchyme cells is experimentally reduced. They found that after the surgical removal of different numbers of primary mesenchyme cells, a fraction of secondary mesenchyme cells converted to the primary mesenchyme cells' fate, that is, participated in skeleton formation. In fact, some sea urchins have only one burst of mesenchyme formation (Schroeder, 1981).

In summary, sea urchin gastrulation combines several different types of morphogenetic movement. The mesoderm is formed by the *ingression* of one or two waves of mesenchyme cells. The endoderm arises through the *invagination* of the vegetal plate followed by the *convergent extension* of the archenteron. The convergent extension movement is based on both cell intercalations and coordinated shape changes. The latter are caused in part by constrictions of secondary mesenchyme cells that pull the archenteron tip toward the prospective mouth region of the ectoderm.

Gastrulation in Amphibians

Amphibian gastrulation has been analyzed in terms of region-specific behaviors of cellular layers and individual cells. Johannes Holtfreter (1943, 1944) and his contemporaries pursued this type of analysis more than 50 years ago, but theories about gastrulation have changed substantially over the past decade, so that today the topic is both classical and new. Amphibian embryos are especially suitable for gastrulation studies, because they are large enough for transplantation experiments. Since much recent work has been done with embryos of the frog *Xenopus laevis*, our discussion will focus on this species.

The outcome of gastrulation is the same in amphibians as in sea urchins: the endoderm moves inside the blastula, the ectoderm remains outside, and the mesoderm fills the spaces in between. However, gastrulation movements in amphibians are more complex than those in sea urchins, partly because the solid mass of cells in the vegetal half of the egg is an impediment to invagination. Also, the amphibian blastula is several cell layers thick, and in some areas the outer cell layers and the inner ones behave differently. Patient study of the following description of gastrulation in *Xenopus* will make it possible to appreciate the elegant experiments described throughout the rest of this section.

Different Gastrula Areas Show Distinct Cellular Behaviors

Gastrulation in *Xenopus* begins when the embryo consists of about 20,000 cells. Along the animal-vegetal axis, John Gerhart and Ray Keller (1986) have defined six regions that are characterized by particular cellular behaviors. These regions are mapped in Figure 10.12,

which also serves as a reference for much of the subsequent discussion. (Color Plate 2 shows a multicolored version of the same map.) Each region of the early gastrula is initially symmetrical about the animal-vegetal axis. However, most cellular behaviors begin first, and are strongest, on the dorsal side of the embryo. The proper placement, timing, and graded intensity of region-specific cell activities are all essential conditions for successful gastrulation. The six regions and their characteristic movements are as follows:

1. The *animal cap*, about three cell layers deep, is derived from the pigmented animal hemisphere of the egg. The animal cap cells engage in *epiboly*, expanding until they cover about half of the late gastrula surface (Fig. 10.12a–c).

2. Adjacent to the animal cap is the *noninvoluting marginal zone*, a wide girdle of cells about four to five layers deep. It is so named because it has certain cell behaviors in common with the adjacent *involuting marginal zone.* The noninvoluting marginal zone and, even more so, the involuting marginal zone extend along the dorsal midline in the animal-vegetal direction while shrinking in other dimensions. This *convergent extension* movement begins midway during gastrulation and causes major distortions in the fate maps of gastrulae at successive stages of development (Fig. 10.12a–c, g–l). In addition, the noninvoluting marginal zone, like the animal cap, undergoes *epiboly*.

3. Adjacent to the noninvoluting marginal zone on the vegetal side is the *involuting marginal zone.* This zone forms the upper margin of the blastopore, which *involutes* during gastrulation. In contrast, the noninvoluting marginal zone does not involute but fills the surface area vacated by the involuting marginal zone. Therefore, the boundary between the involuting marginal zone and the noninvoluting marginal zone, called the *limit of involution*, reaches the edge of the blastopore during gastrulation (Fig. 10.12a and b). Within the involuting marginal zone, we distinguish between the *superficial involuting marginal zone* and the *deep involuting marginal zone.* The superficial involuting marginal zone forms the *archenteron roof*, which becomes part of the endoderm (Fig. 10.12e, j–m), while the deep involuting marginal zone gives rise to much of the mesoderm (Fig. 10.12e, g–i). The term "marginal zone" was used in previous chapters as a short form for the involuting marginal zone, or, more specifically, for the deep involuting marginal zone.

4. The *deep zone* is a ring of cells between the deep involuting marginal zone and the large vegetal cells (Fig. 10.12d). The deep zone cells *migrate* up the wall and roof of the blastocoel (inside the deep involuting marginal zone, noninvoluting marginal zone, and animal cap), thus leading the involuting marginal zone (Fig. 10.12g–l). At the same time, the deep

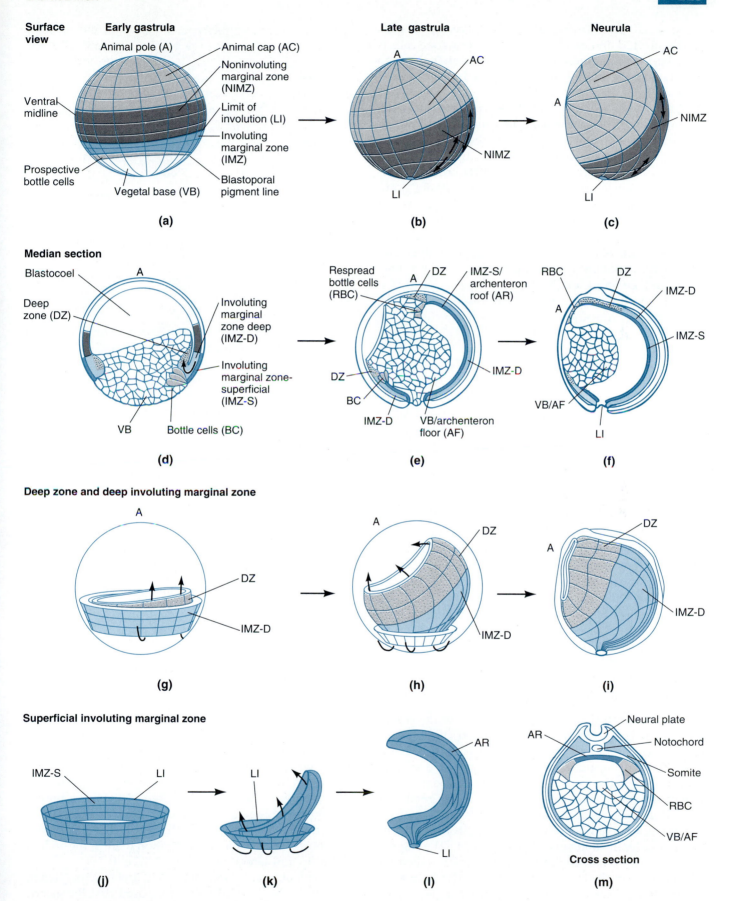

Figure 10.12 Gastrulation in *Xenopus laevis;* dorsal side to the right, animal pole up. Arrows indicate the direction of movement. (See also Color Plate 2.)

zone cells seem to drag along the yolky vegetal cells (Fig. 10.12d and e). The fate of the deep zone cells is to become head mesoderm and heart, while the deep involuting marginal zone cells form the trunk mesoderm.

5. At the vegetal border of the involuting marginal zone, a ring of superficial cells contract at their apical surface while expanding at the base, thus acquiring a bottle shape (Fig. 10.12a, d, and e); from this they were given the name **bottle cells.** The *apical contractions* concentrate cortical pigment granules, so that the bottle cells collectively generate a thin, dark ring called the *blastoporal pigment line.* The apical contractions begin at the dorsal midline, where the blastopore first becomes visible, and progress bilaterally over a period of 2 h until they reach the ventral midline.

6. The *vegetal base* is formed by large yolky cells, extending from the bottom of the blastocoel to the outside of the vegetal hemisphere (Fig. 10.12d). During gastrulation the vegetal base is *tilted to the inside* and *displaced ventrally;* in its new location it becomes the *archenteron floor.* Note that the vegetal base is still contiguous with the respread bottle cells forming the archenteron sides and with the superficial involut-ing marginal zone forming the archenteron roof (Fig. 10.12m). As the vegetal base is tilted inside, the involuting marginal zone undergoes involution, the noninvoluting marginal zone and animal cap spread by epiboly, the limit of involution moves toward the vegetal pole, and the blastopore closes.

In the course of these complex movements, the ectoderm is formed by the animal cap and the noninvoluting marginal zone. The endoderm (archenteron) is composed of the vegetal base (the floor), the respread bottle cells (the sides), and the superficial involuting marginal zone (the roof). The mesoderm is derived from the deep zone in the head region and from the deep involuting marginal zone in the trunk region.

Bottle Cells Generate the Initial Depression of the Blastopore

Amphibian gastrulation begins when the *bottle cells* acquire their characteristic shape by constricting at their apical surfaces and expanding at their bases. The bottle shape becomes especially prominent in the dorsal region of the embryo, where this activity begins, and where the bottle cells sink deep into the vegetal base (Fig. 10.13). These events initiate the formation of the *blastopore,* a curved depression that begins dorsally and expands laterally and ventrally (Fig. 10.14). The ridge of cells on the animal side of the blastopore is called the **blastopore lip.** It, too, originates dorsally before extending laterally and ventrally to form a circle.

To test whether bottle cells assume their characteristic shape independent of other cells, Jeff Hardin and

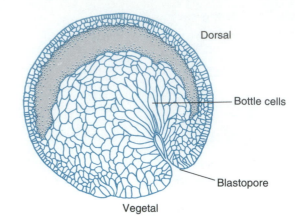

Figure 10.13 Drawing of early amphibian gastrula in median section. Bottle cells initiate the formation of the blastopore.

Ray Keller (1988) monitored their behavior in cultured explants and in surgically altered embryos. They found that isolated bottle cells constrict at the apical surface but assume the characteristic bottle shape only when in contact with other cells surrounding them. In particular, bottle cells in vivo show apical constriction—primarily along the animal-vegetal dimension of their apical surface—and a concurrent apical-basal elongation, whereas explanted bottle cells show a uniform apical constriction and remain fairly rotund. These results indicate that the cells of the blastoporal pigment line—normally, the bottle cells—have a built-in tendency toward apical constriction but depend on mechanical interaction with surrounding cells to form the normal bottle shape.

The bottle cells on the dorsal side remain in the deepest part of the blastopore and become the tip of the advancing archenteron (Fig. 10.15). It has therefore been thought that the bottle cells continue to migrate toward the interior of the gastrula and that they pull the rest of the archenteron with them. To test this hypothesis, R. E. Keller (1981) surgically removed the bottle cells. In some experiments, only the dorsal bottle cells were extirpated, while in others the lateral and ventral bottle cells were removed as well. The embryos healed within 30 min of the operation, and virtually all of them gastrulated and developed normally except that the archenteron was truncated at the cephalic end. The results show that the bottle cells are *not necessary* for continued gastrulation in *Xenopus.* Their role seems to be limited to generating the initial depression of the blastopore. In accord with this conclusion, the bottle cells respread in the course of archenteron elongation to form the archenteron tip and lateral walls (Fig. 10.12e, f).

As an alternative explanation of archenteron elongation, R. E. Keller (1981, 1986) proposed that the gastrulation process is driven to a large extent by *involution* and *convergent extension* of the deep involuting marginal

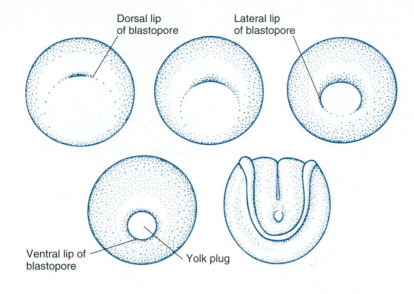

Figure 10.14 Drawings showing successive stages of blastopore formation in a vegetal view. Note that the blastopore surrounds the vegetal base as it first extends laterally and then closes into a circle. The part of the vegetal base that protrudes from behind into the blastopore is called the yolk plug. After the blastopore is closed, the dorsal surface of the embryo flattens and neurulation begins.

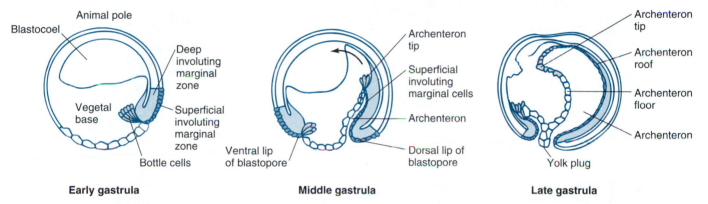

Early gastrula   **Middle gastrula**   **Late gastrula**

Figure 10.15 Archenteron formation in the *Xenopus* gastrula: median sections, animal pole up, dorsal side to the right. The bottle cells form the tip of the archenteron and later its sides (above and below the plane of section and therefore not visible here; see Fig. 10.12). The archenteron floor is formed by the vegetal base. The archenteron roof is formed by the superficial layer of the involuting marginal zone. Note that the archenteron roof moves along with the deep involuting marginal zone, which forms trunk mesoderm.

zone. According to this hypothesis, the main forces driving archenteron elongation would be generated in prospective mesodermal cells, and the adjacent endodermal cells would simply be carried along. We will look at evidence supporting this hypothesis next.

Deep Marginal Zone Cells Are Necessary for Involution

In contrast to sea urchins, in which *invagination* is a major step in archenteron formation, amphibians gastrulate in large part by *involution*. As illustrated in Figure 10.12, the involuting marginal zone turns inward over the blastopore lip, and then back on itself. Because this movement is slow, it is difficult to observe without time-lapse cinematography. However, a convincing demonstration of involution was given by Walther Vogt (1929), using dyes that stained embryonic cells but did not interfere with normal development. Small agar pieces

soaked with such dyes were pressed against newt embryos that were immobilized in wax (Fig. 10.16). When enough stain had been taken up by the embryos, they were released from their molds and observed at successive stages. During gastrulation, the stain marks disappeared into the blastopore. When the embryos were fixed and sectioned, the dye marks were found—in the original sequence—along the archenteron. Some of the marks had elongated considerably, indicating that *convergent extension* had also occurred in those areas. From these observations, it was clear that cells *involuted* around the entire circumference of the blastopore, but most extensively on the dorsal side (Fig. 10.17). In addition, the blastopore lip enclosed the vegetal base by *epiboly*, constricting the blastopore to a small opening at the vegetal pole. The small protrusion of the vegetal base into the constricted blastopore is called the **yolk plug,** and the circular furrow between the yolk plug and the blastopore lip is called the **blastopore groove.**

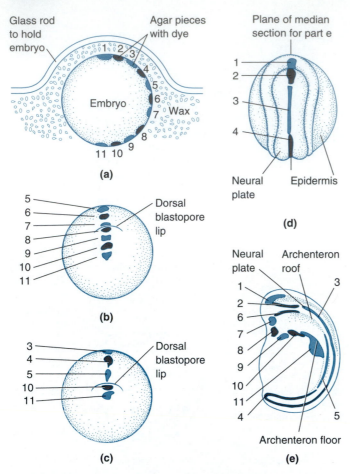

Figure 10.16 Involution and epiboly during gastrulation of the newt *Triturus*. **(a)** Vogt's method of placing dye marks on the surface of the embryo. **(b)** Early gastrula, surface view from vegetal pole. **(c)** Mid-gastrula, same view. Dye marks 7 and 6 have turned inside by involution. Dye marks 8 and 9 were covered as the dorsal blastopore lip moved toward the vegetal pole as a result of epiboly. **(d)** Embryo after neural plate formation, seen from a dorsal-anterior angle. Note elongation of marks 3 and 4. **(e)** Median section of embryo at tail bud stage showing the archenteron floor and roof. Note elongation of marks 3, 4, and 5.

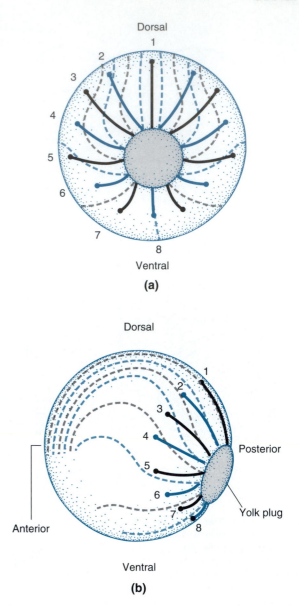

Figure 10.17 Trajectories of involuting marginal zone cells during amphibian gastrulation. **(a)** Posterior view. **(b)** Lateral view.

To determine whether involution requires any special cellular capabilities, R. E. Keller (1981) replaced involuting marginal zone layers with other gastrula tissues. For instance, he excised the deep layer, the superficial layer, or both from the dorsal marginal zone and replaced them with the corresponding layers of cells from the animal cap (Fig. 10.18). The grafted patches were thus challenged to involute. Those animal patches replacing both the *superficial layer* and the *deep layer* of the involuting marginal zone balked at the blastopore lip and did not move in, while the host tissue on both sides of the graft involuted properly. In contrast, animal patches replacing *only the superficial layer* of the marginal zone involuted and became part of the archenteron roof. The same results were obtained with grafts taken from anywhere above the normal limit of involu-

tion. Control embryos in which patches of deep and superficial layers of marginal zone had been excised and reinserted in the same place showed only minor defects. These results indicate that cells in the deep layer of the involuting marginal zone have unique properties that are necessary for involution.

Deep Zone Cells and Involuting Marginal Zone Cells Migrate on the Inside of the Blastocoel Roof

During gastrulation, the deep zone cells migrate from their original position near the equator to the animal pole (Fig. 10.12d–i). The deep involuting marginal zone cells follow suit as they turn inside at the blastopore lip. At this point, the shape and behavior of the involuting

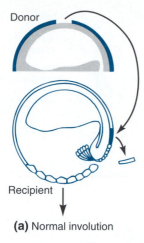

Donor

Recipient

(a) Normal involution

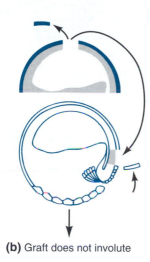

(b) Graft does not involute

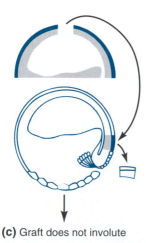

(c) Graft does not involute

Figure 10.18 Analyzing the ability of different areas of *Xenopus* early gastrulae to undergo involution; animal pole up, dorsal side to right. Each drawing shows the animal cap of the donor and the entire recipient. **(a)** Replacing the superficial layer alone did not interfere with gastrulation. **(b)** Replacing the deep layer alone stopped the involution of the graft area. **(c)** Replacing both layers had the same result as replacing the deep layer alone.

cells change (R. E. Keller, 1986). This becomes evident when deep marginal cells, before and after involution, are placed on the inside of a cultured blastocoel roof. Preinvolution cells quickly and actively insert themselves between the roof cells, while postinvolution cells spread laterally on the inside of the roof. Microcinematography of cultured, open gastrulae shows waves of involuting mesodermal cells advancing across the inside of the blastocoel roof.

What role does cell migration along the blastocoel roof play in the overall process of gastrulation? In newts, migration of both deep zone cells and involuting marginal zone cells seems necessary for the internalization of the marginal zone (see Chapter 26). In frog embryos, however, the marginal zone is internalized even in the absence of migration. This was shown by Holtfreter (1933), who completely removed the blastocoel roof from the early gastrula stages of *Hyla* (Fig. 10.19). The

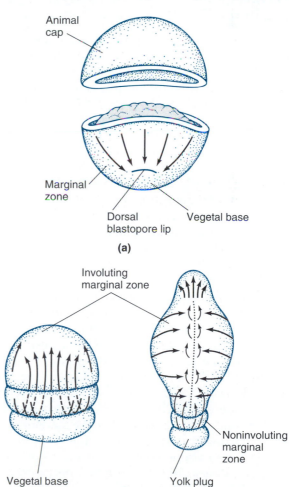

Animal cap

Marginal zone

Dorsal blastopore lip

Vegetal base

(a)

Involuting marginal zone

Vegetal base

Yolk plug

Noninvoluting marginal zone

(b) **(c)**

Figure 10.19 Gastrulation movements of *Hyla* embryos after removal of the blastocoel roof; embryos shown in dorsal view, animal pole up. **(a)** The animal hemisphere was removed completely. **(b)** The isolated vegetal hemisphere underwent invagination and involution. The epiboly movement surrounding the vegetal base was disturbed. **(c)** The involuted marginal zone extended in the anteroposterior direction while shrinking laterally.

result was stunning. Invagination and involution proceeded almost normally, although the deep zone and the involuting marginal zone had no blastocoel roof to migrate on. In addition, the blastopore constricted, and *convergent extension* of the involuted mesoderm occurred in a completely autonomous way. R. E. Keller and S. Jansa (1992) repeated Holtfreter's experiment with *Xenopus* and obtained similar results. It appears that in anurans, cell migration does not need to contribute to the forces that move the marginal zone inside. However, migration seems necessary for the deep zone mesoderm to spread out properly for future head organization. Furthermore, even subtle disturbances of the blastocoel roof, on which the mesodermal cells migrate, interfere with the right-left asymmetry of both heart and gut as described in Chapter 9.

Convergent Extension Is Especially Strong in the Dorsal Marginal Zone

As gastrulation proceeds, amphibian fate maps undergo major distortions, especially in the dorsal area. The greatest shape changes occur during the second half of gastrulation, when the involuting and noninvoluting marginal zones undergo *convergent extension*. Both extend dramatically in the animal-vegetal direction while shrinking in circumference and depth. These movements drive the dorsal blastopore lip toward the vegetal pole. Thus, the marginal zone behaves like a constriction ring that helps to narrow the blastopore and to conclude the involution process. Because the convergent extension movement is most forceful on the dorsal side, the blastopore actually shifts ventrally as it closes.

The autonomy of the convergent extension process was demonstrated by isolation experiments (R. E. Keller et al., 1985). Dorsal sectors from two early gastrulae, each consisting of one superficial cell layer and several deep layers, were excised, sandwiched with their inner sides together, and kept in tissue culture (Fig. 10.20). The explants comprised parts of the animal cap, noninvoluting marginal zone, involuting marginal zone, and deep zone (Fig. 10.12). After the edges of the sandwiches had healed together, no major changes occurred until the time at which untreated control embryos reached the mid-gastrula stage. Then the marginal zones began to narrow and lengthen while the animal cap and deep zone became spherical or knoblike. At the postgastrula stage, cell differentiation became visible. The animal cap cells formed fluid-filled epithelial vesicles. The noninvoluting marginal zone gave rise to an elongated mass of neural cells. The involuting marginal zone differentiated into a notochord flanked by somitic mesoderm. The deep zone developed into a ball of mesenchymal cells. Thus, as far as recognizable, each zone developed according to fate.

More important, the marginal zone parts underwent convergent extension in isolation, showing that no

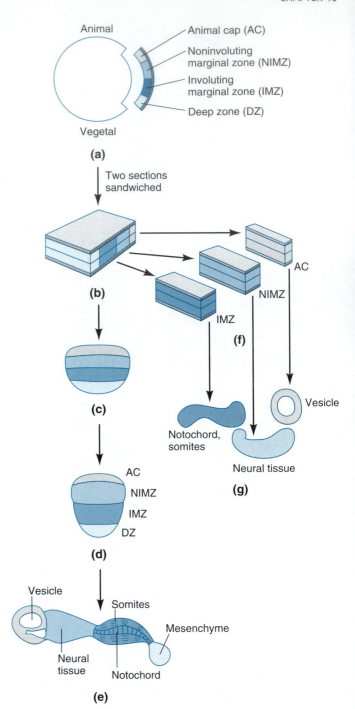

Figure 10.20 Isolation of a dorsal sector from an early *Xenopus* gastrula. **(a)** Embryo shown in lateral view, dorsal side to the right. The isolated sector included portions of the animal cap, noninvoluting marginal zone, involuting marginal zone, and deep zone. **(b)** Two such sectors were sandwiched with their inner surfaces together. **(c–e)** Upon culture in vitro, the "sandwich" extended markedly in the anteroposterior direction while shrinking laterally. In addition, each region differentiated according to its fate in the normal embryo, forming a fluid-filled vesicle, neural tissue, and notochord as well as somites and mesenchyme. **(f, g)** The same differentiations occurred when the regions were isolated after sandwiching and cultured separately.

forces or signals from other parts of the embryo were required. Similar explants of lateral or ventral sectors showed progressively less capacity for convergent extension. These results demonstrate that convergent extension is an inherent property of the marginal zone, especially in the dorsal area, and plays a dominant role in the gastrulation process.

How do the marginal cells bring about convergent extension? To answer this question, Ray E. Keller and coworkers (1985) transplanted pieces of involuting marginal zone that were labeled with a fluorescent dye to an unlabeled early gastrula (Fig. 10.21). Each labeled transplant healed in quickly as a continuous patch. After the recipients had developed past the gastrula stage, they were fixed, sectioned, and studied by fluorescence microscopy. The behavior of the grafts depended strongly on where they had been implanted. In the prospective *notochord* or *somite* region, the labeled graft cells intercalated extensively with unlabeled recipient cells, so the labeled cells spread over nearly the full length of the dorsal axis. These cells contributed to notochord

and somites. In contrast, grafts forming *head* or *ventral mesoderm* showed little or no cell intercalation (R. E. Keller and P. Tibbetts, 1989). These results indicate that the prospective notochord and somite areas undergo convergent extension by means of intercalation.

By excising a dorsal piece of involuting marginal zone, keeping it in tissue culture, and recording the cellular movements by videomicroscopy, R. E. Keller and coworkers (1985) were able to observe the process of cell intercalation directly (Fig. 10.22a). They found that

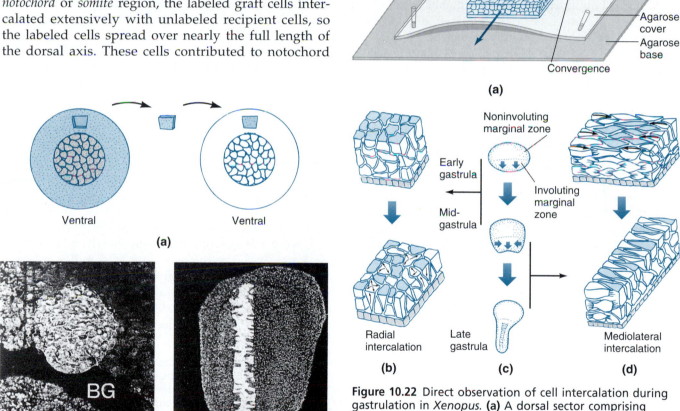

(a)

Ventral   Ventral

(a)

BG

(b)   **(c)**

Figure 10.21 Convergent extension in *Xenopus* gastrulae. **(a)** A fluorescently labeled donor (color) and an unlabeled recipient are shown in vegetal view, dorsal side up. A piece of dorsal involuting marginal zone, comprising both deep and superficial layers, was grafted. **(b)** Photomicrograph of a histological section showing the fluorescent graft after healing in. The graft was a compact patch, separated by the blastoporal groove (BG) from the yolk plug below. **(c)** In a specimen fixed and sectioned after neurulation, the graft had contributed to notochord and somites. The labeled cells had intercalated with unlabeled recipient cells and extended over almost the length of the embryo.

Radial intercalation   Late gastrula   Mediolateral intercalation

(b)   **(c)**   **(d)**

Figure 10.22 Direct observation of cell intercalation during gastrulation in *Xenopus*. **(a)** A dorsal sector comprising both involuting marginal zone and noninvoluting marginal zone is cultured between two sheets of agarose. The inner, deepest cell layer is on top while the epithelial layer that normally forms the outer surface of the embryo is at the bottom. As in the intact embryo, the cultured tissue extends in the animal-vegetal direction (solid arrows) while converging in the mediolateral direction (dashed arrows). **(b)** Radial intercalation during early and mid-gastrulation. The thin arrows indicate radial cell intercalation (perpendicular to the surface), which causes an increase in surface area and a corresponding thinning of the tissue. **(c)** Overall changes in surface area and shape of the cultured tissue. Convergent extension is stronger in the involuting marginal zone. **(d)** Mediolateral intercalation during mid- and late gastrula stages. The thin arrows show mediolateral cell intercalation (parallel to the surface), which causes convergent extension.

cell intercalations occur in two directions. Early gastrulae show *radial intercalation,* meaning that cells intercalate *perpendicular* to the surface of the embryo (Fig. 10.22b). This behavior reduces the number of deep cell layers while increasing their surface area. For reasons not yet understood, the surface area increases only in the animal-vegetal direction, rather than in all directions as is observed in epiboly (discussed later in this section). Radial intercalation is followed by *mediolateral intercalation:* now the deep cells intercalate *parallel* to the surface of the embryo, moving mostly from lateral to medial (Fig. 10.22d). This activity causes extension of the involuting marginal zone in the animal-vegetal direction and a proportional convergence in the mediolateral direction.

Radial and mediolateral cell intercalation, and the resulting convergence and extension of cellular layers, occur not only in the involuting marginal zone but also in the dorsal region of the noninvoluting marginal zone, the part that gives rise to hindbrain and spinal cord (R. Keller et al., 1992b). Transplantations of fluorescent-labeled cell patches as shown in Figure 10.21 confirm that both types of intercalation occur not only in cultured explants but also in whole embryos.

How cells might receive the orienting clues for directional intercalation is being investigated by computer modeling (Weliky et al., 1991) and by detailed observation of the intercalating cells (R. Keller et al., 1992a). Deep involuting marginal zone cells change their intercalation behavior during gastrulation. In early gastrulae, these cells divide once and show rapidly forming and randomly oriented protrusions (Fig. 10.23a and b). This activity ceases at the transition to the midgastrula stage, and subsequent protrusions are directed medially and laterally. At the same time, cells become spindle-shaped, or bipolar, and align in the mediolateral direction (Fig. 10.23c). They then intercalate to form a longer and narrower array.

The extending effect of mediolateral intercalation is enhanced by certain boundaries that run parallel to the anteroposterior axis and are not transgressed by intercalating cells (Fig. 10.23d and e). In the deep involuting marginal zone, such boundaries originate between the *axial mesoderm,* which extends along the dorsal midline and will form a skeletal element called the *notochord,* and the *paraxial mesoderm,* which lies on both sides of the axial mesoderm and will form segmental units known as *somites.* When the boundary between axial and paraxial mesoderm arises late during gastrulation, any bipolar cells that enter the boundary zone by intercalation first show rapid protrusive activity, then spread along the boundary, and finally cease protruding. Thus, they become monopolar and are stabilized in positions at the boundary while still exerting traction on neighboring cells that are not yet at the boundary. Over time, more and more internal cells are pulled to the boundary and trapped there, with convergent extension resulting.

The same cellular behavior has been proposed by Jacobson et al. (1986), on the basis of a general model of cell motility and on cell shapes observed in fixed and sectioned neural plate tissue.

The Animal Cap and the Noninvoluting Marginal Zone Undergo Epiboly

While the involuting marginal zone and the vegetal base turn inside, the animal cap and—to a smaller extent—the noninvoluting marginal zone undergo the spreading movement called *epiboly.* Microcinematography reveals that the superficial cells in these regions divide, with their mitotic spindles oriented parallel to the surface. They also become thinner as gastrulation proceeds. At the same time, major cellular rearrangements occur in the deep layer underneath (R. E. Keller, 1978). Scanning electron micrographs, repro-

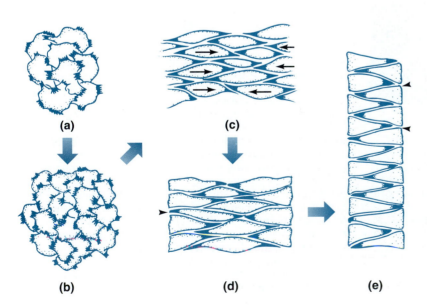

(a) (b) (c) (d) (e)

Figure 10.23 Mediolateral cell intercalation during *Xenopus* gastrulation. The drawings summarize video recordings of cultured horizontal slices of deep involuting marginal zone. **(a, b)** At the early gastrula stage, cells divide once and show rapid protrusive activity in all directions. **(c)** At mid-gastrulation, cells become bipolar with spindle-shaped, stable protrusions at their medial and lateral surfaces. The protrusions exert traction on neighboring cells, and cells intercalate from lateral to medial (black arrows). **(d, e)** Late during gastrulation, boundaries appear between axial mesoderm, which gives rise to notochord, and paraxial mesoderm (not shown), which forms somites. These boundaries inhibit protrusive activity (arrowheads), leaving the cells trapped at the boundary with only one intercalating pole that faces away from the boundary and continues to intercalate. As more and more cells are captured at the boundary, the adjacent area extends and converges.

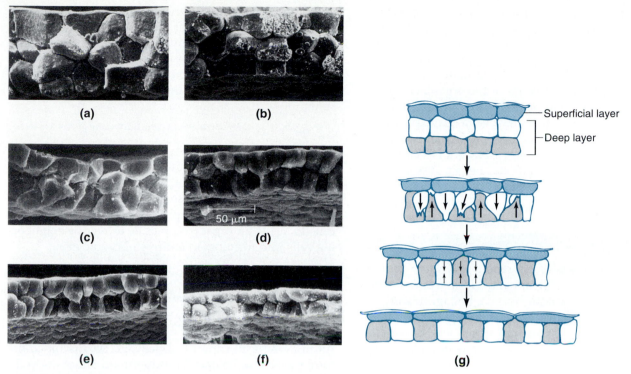

Figure 10.24 Cellular basis of epiboly in the animal cap of the *Xenopus* gastrula. **(a–f)** Scanning electron micrographs of animal cap cells at successive stages of gastrulation; superficial cell layer at the top. **(g)** Interpretative diagram showing the superficial layer (color) and the deep layer (gray and white). The superficial cells flatten, and the deep cells undergo radial intercalation (small arrows).

duced in Figure 10.24, show that the number of deep cell layers in the animal cap decrease from about two at the onset of gastrulation to one at the end of gastrulation (R. E. Keller, 1980). During the transition, the deep cells elongate temporarily and extend lamellipodia from one to another in a radial direction, that is, perpendicular to the surface of the embryo. After the radial intercalation is complete, the deep cells become flatter again. Thus, the *epiboly* of the animal cap and the noninvoluting marginal zone is associated with *cell division* and *cell flattening* in the superficial layer, and with *radial intercalation* in the deep layer. Whether these cellular behaviors actually drive the epiboly movement or whether they occur in response to forces generated elsewhere is still uncertain.

Specific Gene Products Control Cell Movements during Gastrulation

As discussed in previous parts of this section, amphibian gastrulation is composed of several cellular movements that complement one another. Interaction of bottle cells with the vegetal base produces a small *invagination* at the dorsal blastopore. The *involution* of a part of the marginal zone extends the archenteron and places the mesoderm between the archenteron (endoderm) and the ectoderm, which remains outside.

Migration brings deep zone and involuting marginal zone cells to the animal pole. *Convergent extension* causes anteroposterior elongation of the embryo on the dorsal side and constricts the blastopore. As the vegetal parts of the gastrula turn inside, they are enveloped by *epiboly* of the animal portions of the gastrula. How are all these movements orchestrated so that they produce a well-formed gastrula that will proceed with organogenesis to generate the basic body plan?

There are at least two general hypotheses that link gastrulation movements to the earlier processes of cell determination by localized cytoplasmic determinants and inductive interactions. First, one could assume that the cellular behaviors observed in each gastrula region (e.g., animal cap, noninvoluting marginal zone) are controlled locally and individually by the cytoplasmic localizations and inductive events that occur in that region. Thus, localized mRNAs in animal cap cells might cause the descendants of these cells to undergo epiboly, while the specific combination of inductive signals received in the dorsal marginal zone might promote convergent extension, and so forth. According to the second hypothesis, one region of the early gastrula could be determined to carry out its own cellular movements *and* to organize the movements of the other regions. Both hypotheses may be valid to an extent, according to the *principle of synergistic mechanisms*.

The most suitable candidate for an organizer role would be the dorsal blastopore lip, where gastrulation begins. Indeed, transplantation of this region to the ventral side of a host embryo leads to the formation of a second invagination site and eventually a second embryo (see Chapter 11). Because of the organizer properties of the dorsal blastopore lip, researchers have focused on several genes that are specifically expressed in this region and encode proteins that regulate the activity of other genes. Messenger RNA transcribed from one of these genes, *goosecoid*$^+$, has been tested for its effects on gastrulation movements.

▼

To explore the effects of goosecoid mRNA in various cell lineages, Christof Niehrs and his colleagues (1993) injected this mRNA along with a lineage tracer into *Xenopus* blastomeres at the 32-cell stage. Control embryos were injected with lineage tracer only, or with lineage tracer and a modified goosecoid mRNA that encoded a nonfunctional protein. When goosecoid mRNA was injected into *ventral* marginal blastomeres (labeled C4 in Fig. 6.2), the descendants of the injected cells acquired many properties of the normal dorsal blastopore lip; they formed notochord and other dorsal structures, and they recruited neighboring cells to contribute to the formation of a secondary embryonic axis. Moreover, the progeny of cells injected with goosecoid mRNA involuted earlier, and moved farther anterior, than control-injected cells. Thus, the ectopic goosecoid function promoted cell movements and cell fates normally displayed by dorsal mesoderm. When goosecoid mRNA was injected into *dorsal* marginal blastomeres (labeled C1 in Fig. 6.2), the descendants of the injected cells migrated farther anterior, as deep zone cells normally do, instead

of undergoing mediolateral intercalation, as normal dorsal involuting marginal zone cells do. The injected cells also traded their normal fate (making trunk mesoderm) for the fate of deep zone cells (forming head mesoderm). Thus, overexpression of *goosecoid*$^+$ changed both the movements and the fate of C1 progeny correspondingly.

The results of Niehrs and coworkers (1993) demonstrate that the expression of *goosecoid*$^+$ has major effects on the cell movements observed during gastrulation in *Xenopus.* Moreover, the expression of this gene enables cells to act as an organizer, that is, to recruit neighboring cells into forming an entire embryonic axis.

Gastrulation in Birds

As an example of superficial cleavage, we discussed the formation of the chicken blastoderm on top of the yolk-filled, uncleaved portion of the egg (see Chapter 5). The bird blastoderm consists of an upper layer, the *epiblast*, and a lower layer, the *hypoblast*, with a shallow blastocoel in between (Fig. 10.25). The central and peripheral portions of the blastoderm have different optical properties. The central portion, which is separated from the yolk by the *subgerminal space*, is the **area pellucida** (Lat. *pellucidus*, "transparent"). The peripheral portion, which is in direct contact with the yolk, is called the **area opaca** (Lat. *opacus*, "dark"). Similar blastulae are formed by reptiles, many fishes, and other animals that produce large, yolky eggs that undergo discoidal cleavage.

Although the blastulae of birds and amphibians are remotely similar, the bird hypoblast does *not* correspond

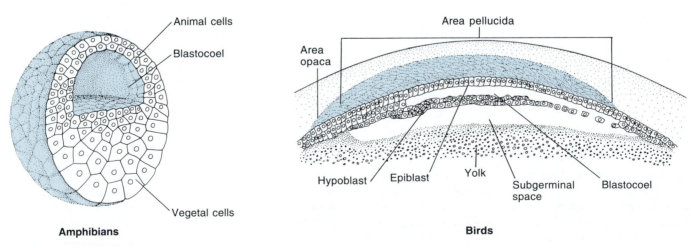

Figure 10.25 Blastula stages in amphibians and birds. In birds, the epiblast gives rise to all three germ layers.

to the vegetal hemisphere of the amphibian blastula. *All three* germ layers of the bird embryo are derived from the epiblast, so that a fate map of the bird embryo prior to gastrulation is essentially a map of the epiblast. The hypoblast gives rise to extraembryonic endoderm that later surrounds the yolk.

Gastrulation in birds is initiated by extensive cell movements within the epiblast. Cells first converge to the posterior midline and then move anteriorly, generating a fountainlike flow (Fig. 10.26). Like the convergent extension movements observed during amphibian gastrulation, this process may be driven by cell intercalation. As a result of these movements, the cells that were spread over the posterior half of the epiblast are rearranged into a solid median ridge, the *primitive streak*, while the entire blastoderm assumes the outline of a pear. As the primitive streak elongates, a furrow called the *primitive groove* forms between two *primitive ridges* along the dorsal midline. The primitive ridges terminate anteriorly in a thickening known as *Hensen's node*. The node contains a funnel-shaped depression, called the *primitive pit*, which marks the anterior end of the primitive groove.

The primitive groove and primitive pit are the sites of major gastrulation events in birds. These are (1) the *involution* of epiblast cells over the edges of the primitive groove and primitive pit and (2) the subsequent *ingression* of the involuted cells into the blastocoel (Fig. 10.27). The cells that constitute the edges of the primitive groove and pit change constantly, as do those on the amphibian blastopore lip described earlier (compare Fig. 10.26d with Fig. 10.17). The primitive groove and pit in birds are therefore the functional equivalents of the amphibian blastopore. As the epiblast cells enter the primitive groove, they undergo major changes. Their apical ends constrict while the basal ends expand, so that the cells take on a bottle shape, as seen in the amphibian blastula (compare Fig. 10.28 with Fig. 10.13). However, instead of moving in as a contiguous cell layer, avian epiblast cells break their junctions and ingress as single cells. Once inside the blastocoel, they flatten and migrate, an activity that is facilitated by the synthesis of certain extracellular matrix molecules. Because of its loose configuration, the bird mesoderm is first classified as a *mesenchyme*. Later, however, the loose and wandering mesenchymal cells will form segmental sacs called *somites* and other contiguous structures (see Chapter 13).

Fate maps of avian embryos have been established by various methods, among them the transplantation of radiolabeled tissue patches to corresponding areas of unlabeled embryos (Rosenquist, 1966; Nicolet, 1971; Vakaet, 1984; Mittenthal and Jacobson, 1990). The maps indicate the position of the germ layers and some of their subdivisions during primitive streak formation (Fig. 10.29). The ectodermal primordia are located ante-

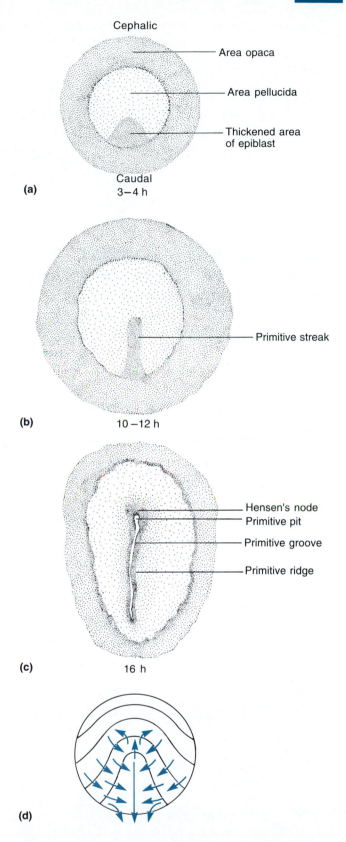

Figure 10.26 Primitive streak formation in the chicken embryo. All drawings show dorsal views of the epiblast. Most of the large uncleaved yolk has been omitted. **(a)** 3–4 h incubation. **(b)** 10–12 h incubation. **(c)** 16 h incubation. **(d)** Interpretative diagram of cell movements.

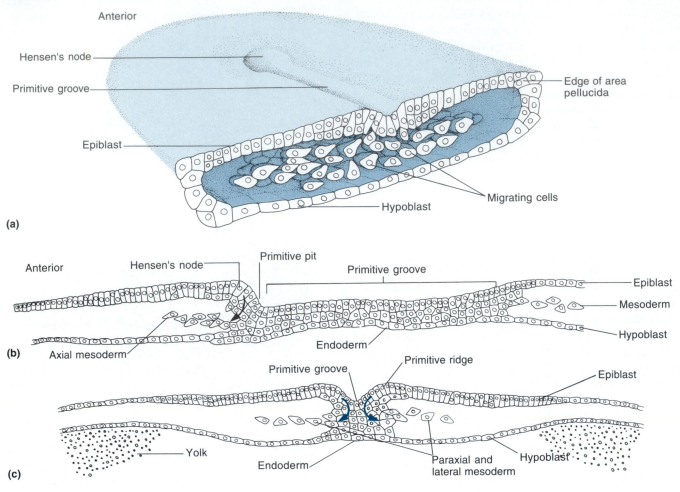

(a)

(b)

(c)

Figure 10.27 Ingression of endodermal and mesodermal cells through the primitive groove in bird embryos. **(a)** Three-dimensional view. The endodermal cells ingress first and displace the hypoblast. **(b)** Median section. Cells ingressing through the primitive pit form axial mesoderm including notochord. **(c)** Transverse section. Cells ingressing through the primitive groove form paraxial and lateral mesoderm.

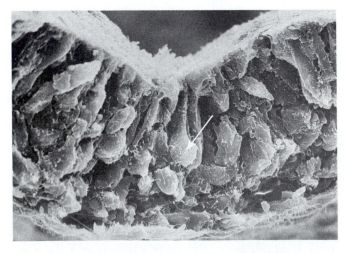

Figure 10.28 Primitive streak of a gastrulating bird embryo. The scanning electron micrograph shows a transversely fractured embryo. Cells entering the primitive groove become bottle-shaped (arrow) as they prepare to ingress into the blastocoel.

riorly in the epiblast and do not ingress. The endodermal and mesodermal primordia are arranged in concentric strips within the posterior half of the epiblast. Among the first cells to ingress through the primitive groove are the endodermal cells. They join the hypoblast cells and gradually displace them toward the margin.

A fate map of Hensen's node in particular was established by staining the membranes of cell groups with a water-insoluble dye and by injecting individual cells with a fluorescent dye (Selleck and Stern, 1991). Cells from the anterior sector of Hensen's node form the *axial mesoderm* along the dorsal midline, while cells from lateral portions of Hensen's node contribute to the *paraxial mesoderm* on both sides of the axial mesoderm. Other contributions to the paraxial mesoderm come from cells that involute from the primitive ridges posterior to Hensen's node. These cells also establish the mesodermal primordia that come to lie more laterally, the *intermediate mesoderm*, the *lateral plates*, and the *extraembryonic mesoderm*.

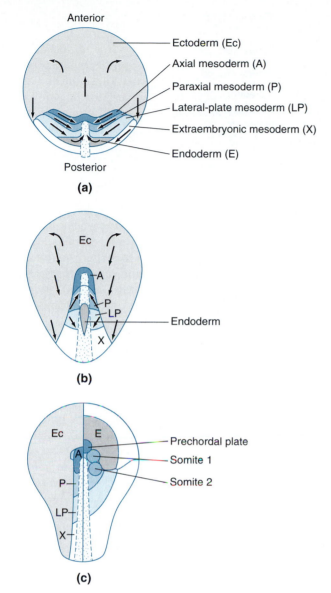

Figure 10.29 Fate maps of the chicken embryo during successive stages of development from early primitive streak formation **(a)** to maximum primitive streak length **(c)**. All maps show dorsal views of the epiblast. The dotted area represents the primitive streak. The epiblast is removed on the right side **(c)** to reveal the underlying endoderm and mesoderm.

while the lateral plates form a wide range of mesodermal structures. Because of the gradual regression of the primitive streak, the anterior of the bird embryo is ahead of the posterior in its development. Thus, the neural tube is already closing in the anterior while the posterior is still gastrulating.

After gastrulation, the remaining epiblast consists only of ectoderm, the hypoblast has been replaced by endoderm, and the mesenchymal cells spread between them. Thus, the result of the combined gastrulation movements is again an embryo consisting of three germ layers. However, in contrast to sea urchins and amphibians, in which the germ layers originate as concentric spheres or tubes, the germ layers of birds begin more like a stack of three discs resting on a large mass of uncleaved yolk.

Once the germ layers have been established, they spread beyond the embryo proper to form *extraembryonic membranes* that surround the entire egg yolk. The ectodermal cells spread by *epiboly*, maintaining their epithelial junctions at the apical surface and advancing as a circular front from the margin of the embryo to the vegetal pole. The leading cells adhere to the vitelline membrane on the inside of the eggshell. At the same time, hypoblast cells spread along the surface of the yolk mass while mesodermal cells expand in between. The extraembryonic membranes serve to break down the yolk into smaller molecules that can be used by the embryonic cells. These are delivered to the embryo through blood vessels that form in the extraembryonic mesoderm and join the embryonic circulation. As the embryo grows, it lifts itself off the yolk, beginning with the formation of the *subcephalic space* under the head (Fig. 10.30). This process involves the closing of the ventral side of the body, which has been open to the uncleaved yolk. As this ventral closure proceeds, first from the anterior and then from the posterior, the remaining yolk is enclosed in a ventral appendage of the gut, the *yolk sac*. Other extraembryonic membranes protect the embryo against dehydration and mechanical stress and also play a role in the storage and exchange of waste products (see Chapter 13).

While the ingression of mesodermal cells is still under way, the primitive streak shortens, gradually shifting Hensen's node to a more posterior position. As the node retreats, it leaves in its wake axial mesoderm cells ingressing from the anterior sector of the primitive pit (Fig. 10.30). The cells that are the first to ingress give rise to the *prechordal plate*; subsequently ingressing cells form the anterior and then the posterior parts of the *notochord*. The paraxial mesoderm cells that ingress from lateral sectors of the primitive pit and from the primitive groove give rise to the segmental *somites*. The intermediate mesoderm forms *nephrotomes* (primitive kidney units)

Gastrulation in Humans

Gastrulation has not been studied as extensively in mammals as in other animals. This is because mammalian embryos are very difficult to maintain in culture beyond the *blastocyst* stage, when they normally implant themselves into the uterus. After the blastocyst stage, mammalian embryos have different ways of establishing a **bilaminar germ disc**, consisting of two cellular layers that correspond to the epiblast and hypoblast of chicken embryos. Our description of this period will focus on the development of primates, in-

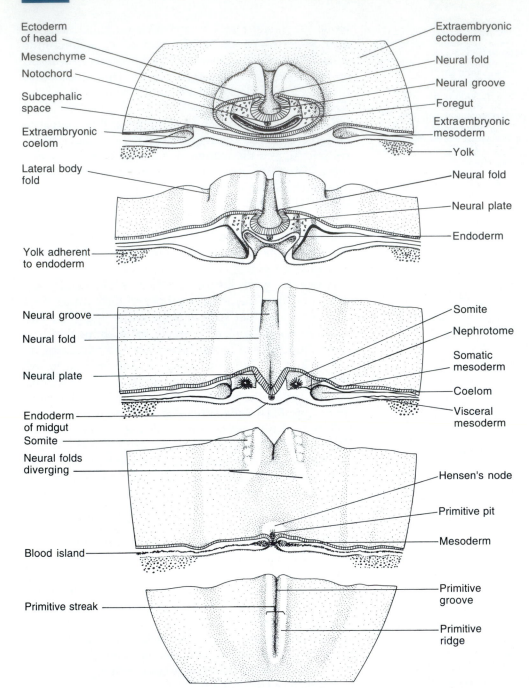

Ectoderm of head
Mesenchyme
Notochord
Subcephalic space
Extraembryonic coelom

Extraembryonic ectoderm
Neural fold
Neural groove
Foregut
Extraembryonic mesoderm
Yolk

Lateral body fold

Neural fold
Neural plate
Endoderm

Yolk adherent to endoderm

Neural groove
Neural fold
Neural plate
Endoderm of midgut
Somite
Neural folds diverging

Somite
Nephrotome
Somatic mesoderm
Coelom
Visceral mesoderm

Hensen's node
Primitive pit
Mesoderm

Blood island

Primitive streak

Primitive groove
Primitive ridge

Figure 10.30 Stereogram of a 24-h chicken embryo. Note how much the anterior end of the embryo (top) is ahead of the posterior in development.

cluding humans. (Rodents and some other mammals develop differently during this period). The gastrulation events in mammals following the bilaminar germ disc stage seem to be similar to corresponding stages in birds.

At the blastocyst stage, the mammalian embryo consists of an outer *trophoblast* surrounding a blastocoel and an eccentric inner cell mass (Fig. 10.31). The subsequent events leading to the establishment of embryonic germ layers and extraembryonic tissues are summarized in Figure 10.32. A thin layer called the *hypoblast* delaminates from the inner cell mass of the blastocyst. As in bird embryos, the hypoblast does *not* contribute

to the embryo proper. Rather, hypoblast cells spread out to surround the blastocoel, thereafter called the *primitive yolk sac,* or *yolk sac* for short. After delamination of the hypoblast, the remainder of the inner cell mass is referred to as the *epiblast*. Subsequently, the epiblast delaminates another cell layer, the *amniotic ectoderm.* The remainder of the epiblast is now referred to as the *embryonic epiblast,* and the fluid-filled space between the amniotic ectoderm and the embryonic epiblast is the *amniotic cavity* (Fig. 10.33). Together, the embryonic epiblast and the underlying hypoblast form the *bilaminar germ disc*. This stage is comparable to the bird blastula before primitive streak formation.

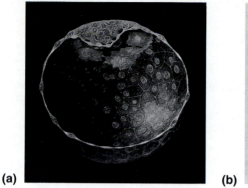

(a)

(b)

Figure 10.31 Blastocyst stage in primates. **(a)** Graphic reconstruction of a monkey blastocyst on day 9 after fertilization. The light areas of the inner cell mass represent hypoblast cells. **(b)** Photomicrograph of a sectioned human blastocyst about 5 days after fertilization.

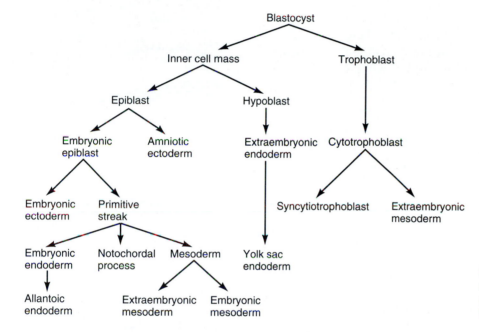

Figure 10.32 Flow diagram showing the origin of the germ layers and their derivatives in primates.

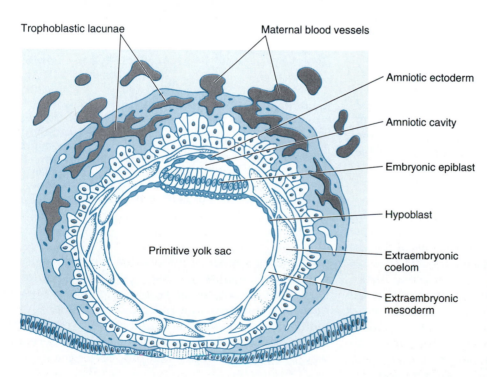

Figure 10.33 Drawing of a human blastocyst of approximately 12 days. The trophoblastic lacunae at the embryonic pole connect with maternal blood vessels. The embryo proper forms the bilaminar germ disc consisting of the epiblast and the hypoblast. Extraembryonic mesoderm proliferates and fills the space between the trophoblast on the outside and the embryo with yolk sac and amnion on the inside.

The bilaminar disc stage shown in Figure 10.33 is reached toward the end of the second week of human gestation. At this point, the disc is still very small (0.1 to 0.2 mm). The upper layer, the epiblast, will give rise to the embryo proper. The other structures will form extraembryonic tissues. The yolk sac will become an appendage to the gut, as in bird embryos. The amniotic cavity will expand and surround the entire embryo as a fluid-filled cushion.

While the bilaminar germ disc is forming from the inner cell mass, the other portion of the blastocyst—the trophoblast—undergoes major changes as well. Its syncytial portion, the *syncytiotrophoblast*, forms internal spaces called **lacunae** and, at the same time, opens maternal blood vessels in the uterus. This is particularly evident at the embryonic pole of the blastocyst, where maternal blood begins to fill the lacunae and to nourish the embryo by diffusion (see Fig. 13.40). In the cellular layer of the trophoblast, the *cytotrophoblast*, a new population of cells appears on the side facing the yolk sac. These cells form a loose meshwork that makes up part of the *extraembryonic mesoderm*. Large cavities that open up in this layer become confluent and fill with fluid. The fluid-filled space, referred to as the **extraembryonic coelom**, surrounds the entire yolk sac, the amniotic cavity, and the bilaminar germ disc in between.

The formation of the *primitive streak* in mammals begins, as in birds, in the posterior part of the embryonic epiblast (Langman, 1981). The cell movements causing primitive streak formation in mammals seem to be very similar to those observed in birds (compare Fig. 10.34 with Fig. 10.27). The embryonic endoderm and mesoderm, as well as part of the extraembryonic mesoderm, originate by *involution* and *ingression* through the primitive streak. Epiblast cells ingressing through the primitive pit move anteriorly to form the prechordal plate and the notochord. Cells ingressing through the primitive groove spread laterally to form other mesodermal structures as well as the endoderm. At the end of gastrulation, the cells remaining in the epiblast have become the ectoderm, the hypoblast has been replaced with endodermal cells, and the embryonic mesoderm has spread in between. This stage, also known as the **trilaminar germ disc**, is reached toward the end of the third week of human gestation.

Although the embryos of placental mammals are very small and cleave holoblastically, their gastrulation movements are remarkably similar to those observed in birds. The bilaminar germ disc of placental mammals behaves as if it were resting on a mass of yolk, although there is none. This is surprising from a functional point of view, because the small embryos of placental mammals could easily gastrulate by invagination or involution. Instead, gastrulation in placental mammals recapitulates a pattern established by their reptilian ancestors. Reptiles are also the ancestors of birds, and many reptiles gastrulate the same way as birds. When the placenta evolved in mammals, their eggs became

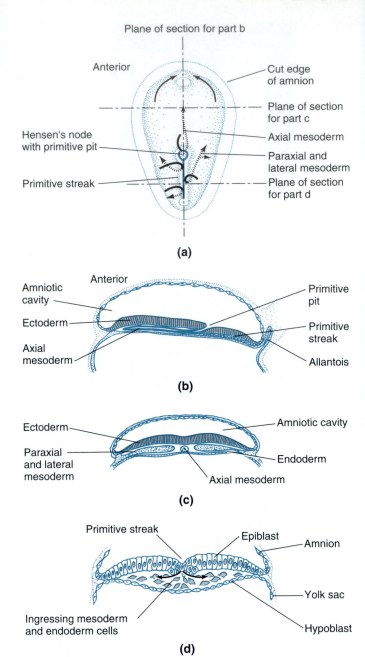

Figure 10.34 Gastrulation in the 16-day human embryo. **(a)** Surface view of the epiblast, with the amnion cut away. Endodermal and mesodermal cells converge toward the primitive streak (solid lines), ingress through the primitive pit and the primitive groove, and spread between the epiblast and the hypoblast (broken arrows). **(b)** Median section. **(c)** Transverse section anterior to Hensen's node. **(d)** Transverse section posterior to Hensen's node.

small but their mode of gastrulation *did not* change correspondingly. Neither did they revert to the pattern of their more distant ancestors, the amphibians, nor did they jump forward to a radically new pattern, even though doing so would seem functionally more direct and economical. This is an example of the conservative way in which developmental processes have evolved (see Chapter 13).

SUMMARY

During the period of gastrulation, the embryo is transformed from a simple sphere or disc to a more complex geometrical configuration of three germ layers. The outer layer, or ectoderm, gives rise to the epidermis and the nervous system. The inner layer, or endoderm, forms the inner lining of the intestine and its appendages. The numerous structures in between are derived from the middle layer, or mesoderm. Although these germ layers are formed universally in the animal kingdom, the cellular movements that establish them differ from one animal group to another.

Gastrulation, as well as the subsequent period of organogenesis, involves extensive morphogenetic movements. Through these movements the embryo molds itself into the three-dimensional form that characterizes its phylogenetic group.

The morphogenetic movements observed in different animals are combinations of the following elementary movements: migration, ingression, invagination, involution, convergent extension, epiboly, and delamination. While migration and ingression are carried out by individual cells or small cell groups, the other movements are performed by entire cellular layers including epithelia. The elementary movements in turn are based on a few types of cellular behavior: division, shape changes, changes in adhesiveness, intercalation, and programmed cell death. Many morphogenetic processes have been analyzed effectively in terms of these elementary movements and cellular behaviors.

Gastrulation in sea urchins involves the ingression of two waves of mesenchymal cells and the formation of the primitive gut, the archenteron. After an initial invagination phase, the archenteron undergoes convergent extension resulting from a combination of cell intercalation and shape changes. Finally, filopodia pull the archenteron tip toward the ectodermal site where the mouth will be formed.

During amphibian gastrulation, different areas of the blastula cooperate in carrying out several types of morphogenetic movements. First, bottle cells in the vegetal hemisphere form the blastopore, which begins as a small invagination on the dorsal side and later extends laterally and ventrally. Blastopore formation is followed by involution of the marginal zone, which moves toward the blastopore lip and turns inside. The involuting cells migrate along the inside of the blastocoel, pulling or at least guiding the cells that involute after them. The involution process generates the archenteron and places the mesoderm between the archenteron (endoderm) and the ectoderm, which remains outside. Convergent extension of the marginal zone completes the involution process, constricts the blastopore, and causes anteroposterior elongation of the embryo on the dorsal side. As the vegetal parts of the gastrula turn inside, they are enveloped by the animal portions of the gastrula in a process of epiboly.

In birds, gastrulation begins with the formation of the primitive streak along the posterior midline of the embryonic epiblast. A dorsal depression in the primitive streak, the primitive groove, is analogous in function to the blastopore in amphibians. Epiblast cells converge toward the primitive streak and ingress through the primitive groove. The endodermal cells turn inside first and displace the hypoblast. The mesodermal cells ingress thereafter and spread as a mesenchyme between the endoderm and the remainder of the epiblast, which remains outside to form the ectoderm. Once the germ layers are established, they spread by epiboly until the entire yolk is surrounded. Many reptiles gastrulate in the same way as birds.

Although placental mammals have small eggs and undergo holoblastic cleavage, their mode of gastrulation is patterned after their reptilian ancestors. In humans and other primates, the inner cell mass splits into the hypoblast, which forms the yolk sac, and the epiblast, which gives rise to the embryo proper. As observed in birds, gastrulation begins with primitive streak formation in the epiblast and ends with a trilaminar germ disc consisting of the endoderm adjacent to the yolk sac, the ectoderm facing the amniotic cavity, and the mesoderm in between.

SUGGESTED READINGS

Ettensohn, C. A., and E. P. Ingersoll. 1992. Morphogenesis of the sea urchin embryo. In E. F. Rossomando and S. Alexander, eds., *Morphogenesis*, 189–262. New York: Marcel Dekker.

Gerhart, J., and R. Keller. 1986. Region-specific cell activities in amphibian gastrulation. *Ann. Rev. Cell Biol.* **2**:201–229.

Keller, R. E., M. Danilchik, R. Gimlich, and J. Shih. 1985. The function and mechanism of convergent extension during gastrulation of *Xenopus laevis*. In J. Slack, ed., *J. Embryol. Exp. Morph.* Vol. 89 Supplement: *Early Amphibian Development*, 185–209.

Vakaet, L. 1984. Early development of birds. In N. M. Le Douarin and A. McLaren, eds., *Chimeras in Developmental Biology*, 71–88. London: Academic Press.

NEURULATION AND NEURAL INDUCTION

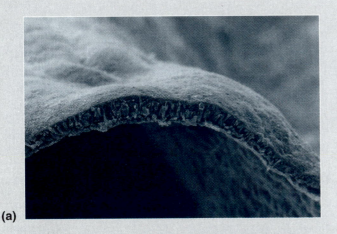

(a)

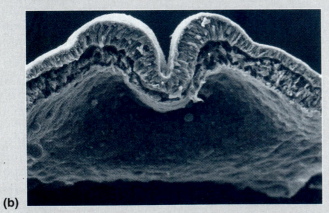

(b)

(c)

Figure 11.1 Neurulation in the chicken embryo. The scanning electron micrographs show transversely fractured embryos at successive stages of development. **(a)** Fracture made anterior to Hensen's node and the primitive streak, through which endodermal and mesodermal cells have ingressed. The ectodermal cells on both sides of the dorsal midline (top) become columnar and form a neural plate. **(b)** The neural plate cells have formed a neural groove. **(c)** The boundaries between the neural plate and the adjacent epidermal ectoderm have formed raised neural folds. The folds will bend together to close the neural plate into the neural tube, which will give rise to brain and spinal cord.

With the completion of gastrulation, the embryo has taken a major developmental step. The germ layers are now arranged according to their ultimate positions in the body. During the subsequent period of organ formation, or *organogenesis,* the germ layers interact to form the organ rudiments. In vertebrates, the most striking part of organogenesis is *neurulation,* the beginning of brain and spinal cord formation. Both organs originate from the same dorsal rudiment, the *neural plate,* which closes to form the *neural tube* (Fig. 11.1). Neurulation, like many other events in organogenesis, begins right after gastrulation. The end of organogenesis is less well defined, be-

cause organogenesis blends into a long period of tissue and cell differentiation during which organ rudiments are transformed into functioning organs. Also, some organs, such as the heart and kidney of vertebrates, begin to function in an embryonic state while they still undergo major changes toward their adult form. In the human, organogenesis is considered to be essentially completed after 6 to 8 weeks of development (Fig. 11.2).

As organ rudiments take shape in the appropriate positions, a *basic body plan* emerges that is characteristic not only of a particular species but also of the entire phylogenetic group to which it belongs. For instance, a 5-week-old human embryo (Fig. 11.2a) has a head with brain, eye, and ear rudiments, and a dorsally segmented trunk with tail and limb buds. Inside the embryo, the rudiments of gut, heart, lung, kidney, and most other organs have formed. The shapes of these organ rudiments and their positions relative to one another are characteristic of mammals and vertebrates in general.

Much of organogenesis occurs deep inside the embryo, where events may be difficult to see and even harder to manipulate experimentally. A major exception is *neurulation* in vertebrates, because the brain and spinal cord are formed on the dorsal surface of the embryo. Because of its accessibility and its central importance, neurulation is the best-studied process in organogenesis. Consequently, this chapter will focus on the morphogenetic movements and the inductive interactions that bring about neurulation. Other examples of organogenesis will be discussed in Chapters 12 and 13.

The shaping of organ rudiments involves many of the same cellular behaviors that drive gastrulation (see Chapter 10). In particular, cell shape changes and convergent extension play strong roles in neurulation.

Many events in organogenesis are controlled by inductive interactions. Neurulation, for example, is induced in the dorsal ectoderm by adjacent mesoderm. This process, called **neural induction,** was revealed in a landmark experiment by Hans Spemann and Hilde Mangold (1924). They transplanted mesoderm from the dorsal marginal zone of newt embryos to the ventral ectoderm of early gastrula hosts. The transplant induced the surrounding host tissue to cooperate in the formation of an entire secondary embryo. This experiment, for which Hans Spemann received the 1935 Nobel Prize in physiology or medicine, is widely seen as the epitome of classical experimental embryology.

Spemann's discovery was followed by a rush of biochemical investigations into the molecular basis of neural induction. Although the quest was originally unsuccessful, it has been taken up again with an armamentarium of new research tools. Modern researchers prefer to work with *Xenopus* rather than with newts, because *Xenopus*

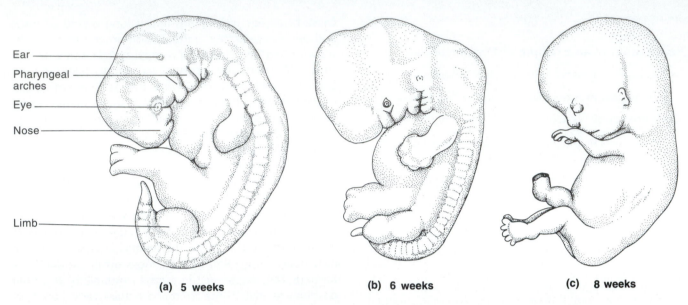

Figure 11.2 Drawings of human embryos during organogenesis, showing the development of nose, eye, ear, and limbs: **(a)** at 5 weeks, **(b)** at 6 weeks, **(c)** at 8 weeks.

eggs can be obtained year-round in the laboratory. In addition to morphological criteria, current studies rely on *molecular markers* such as specific mRNAs or proteins to reveal neural development. Most important, many investigators have been testing neural induction properties of proteins derived from cloned genes. Using *Xenopus* and these new methods, researchers have found that neural induction is a multistep process that involves the action of many genes.

Neurulation as an Example of Organogenesis

Neurulation is a dramatic sequence of morphogenetic events culminating in the formation of the central ner-

vous system. Briefly, an area of dorsal ectoderm is transformed into a plate of tall columnar cells, the *neural plate*. The neural plate subsequently closes into a hollow tube, the *neural tube*, which gives rise to the brain and the spinal cord (Fig. 11.3). An embryo in the process of neurulation is referred to as a *neurula* (pl., *neurulae*), just as a gastrulating embryo is called a *gastrula*.

Neurulation Is of Scientific and Medical Interest

Neurulation is of major scientific interest because the size and accessibility of the neural plate facilitate observations and surgery that would be more difficult to conduct on other developing organs. In addition, a better understanding of neurulation could have significant medical benefits. Defects or delays in the closure of the

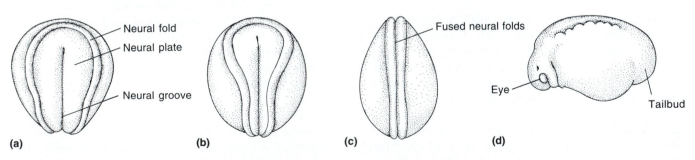

Figure 11.3 Neurulation in the salamander *Ambystoma maculatum*. **(a, b)** Early and late keyhole stages, so named after the outline of the neural plate in dorsal view. **(c)** Closed neural tube, dorsal view. **(d)** Early tail bud stage, lateral view.

neural tube affect the bone, muscle, and skin surrounding the brain and spinal cord. Malformations involving the spinal cord and/or the vertebral arches enclosing the cord are known as **spina bifida** (Lat., "divided spine"). Its mildest form, called *spina bifida occulta* (Lat. *occultus*, "obscure," "concealed"), results from failure of the two arches of a vertebra to fuse dorsally (Fig. 11.4). As many as 10% of all people have this defect, which causes no pain or neurological disorder (K. L. Moore, 1982). The only overt sign of its presence may be a dimple in the skin or a small tuft of hair over the affected area. If more than one or two vertebrae are involved, the *meninges* (membranes surrounding the spinal cord and brain) bulge out dorsally so that a cyst covered with skin forms on the outside. This congenital malformation, known as *spina bifida cystica*, occurs about once in every 1000 births. It is associated with neurological disorders, the severity of which depends on the extent to which nerve tissue protrudes into the cyst.

Failure of the cephalic part of the neural tube to close leads to **anencephaly** (Gk. *a(n)-*, "not"; *enkephalos*, "brain") associated with **acrania** (lack of the vault of the skull, from Gk. *kranion*, "skull") and severe spina bifida. The frequency of anencephaly varies greatly among human populations, ranging from 0.1 to 6.7 per 1000 births (Shulman, 1974). Infants with anencephaly are stillborn or die shortly after birth. Multidisciplinary studies have shown that increasing the amount of folic acid in the diet lowers the frequency of anencephaly and spina bifida in humans. Folic acid is a vitamin that is metabolized into coenzymes that play important roles in several biosynthetic processes.

Neurulation is similar in all vertebrates, but it has been examined most closely in amphibian and chicken embryos. The analysis of neurulation in amphibians is simpler, because their nervous system does not grow during embryogenesis. Measurements of the neural plate and tube show that their volume actually decreases slightly during neurulation (Jacobson, 1978). Furthermore, cell division is slow during amphibian neurulation, and because there is no cell growth during this stage, mitosis cannot be a major factor in amphibian neurulation. Among amphibians, newts have the added advantage that the neural plate cells form a sin-

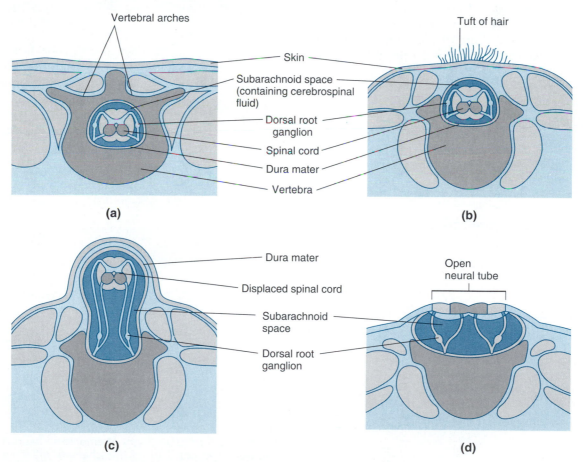

Figure 11.4 Degrees of spina bifida and associated malformations of the spinal cord.
(a) Schematic drawing of the normal human spine in transverse section. **(b)** Spina bifida occulta, a failure of the vertebral arches to close over the spinal cord. This is a common defect of the fifth lumbar and/or first sacral vertebra, causing no clinical symptoms.
(c) Spina bifida cystica with protruding meninges and displaced spinal cord.
(d) Spina bifida with open neural tube.

gle layer, as opposed to the multiple layers found in frogs. Most of the classical work discussed in this chapter has therefore been done with newt embryos.

Neurulation in Amphibians Occurs in Two Phases

In the amphibian embryo, neurulation is divided into two phases. The <u>first phase</u> is the <u>formation of the *neural plate*</u>. This <u>phase concludes with the **keyhole stage**</u>, so named for the peculiar outline of the neural plate at that stage (Fig. 11.3). The <u>main event of the second phase is the closure of the neural plate into the *neural tube*</u>.

The first phase of neurulation in the newt embryo begins with a change in the behavior of the **neural ectoderm** cells, which occupy the dorsal surface of the late gastrula. During gastrulation, these cells move *posteriorly* toward the blastopore along with the rest of the ectoderm. This epibolic movement ceases as soon as gastrulation is completed. Thereafter, neural ectoderm cells begin to move *toward the dorsal midline and the anterior*. Simultaneously, the cells elongate and form a raised plate, the *neural plate,* on the dorsal side of the embryo (Fig. 11.5). A depression called the **neural groove** develops along the midline of the neural plate (Fig. 11.3). At the same time, ridges of cells called **neural**

folds arise along the boundary between the neural plate and the surrounding epidermis. <u>Once the neural folds have emerged, the neural plate extends anteroposteriorly and shrinks laterally, especially in the posterior. As a result, the surface area of the neural plate decreases, and it assumes the characteristic keyhole shape.</u> The anterior part of the keyhole region gives rise to the brain while the posterior part forms the spinal cord.

Unlike the neural ectoderm cells, the remaining ectodermal cells assume a *squamous,* or flat, shape (Fig. 11.5). Because these cells are fated to form epidermis, they are called the **epidermal ectoderm.**

The second phase of neurulation begins shortly after the keyhole stage has ended. The neural plate undergoes a spurt of anteroposterior extension and simultaneously curls up so that the neural folds meet along the dorsal midline, thus closing the neural plate into the *neural tube* (Fig. 11.6). At the same time, the ad-

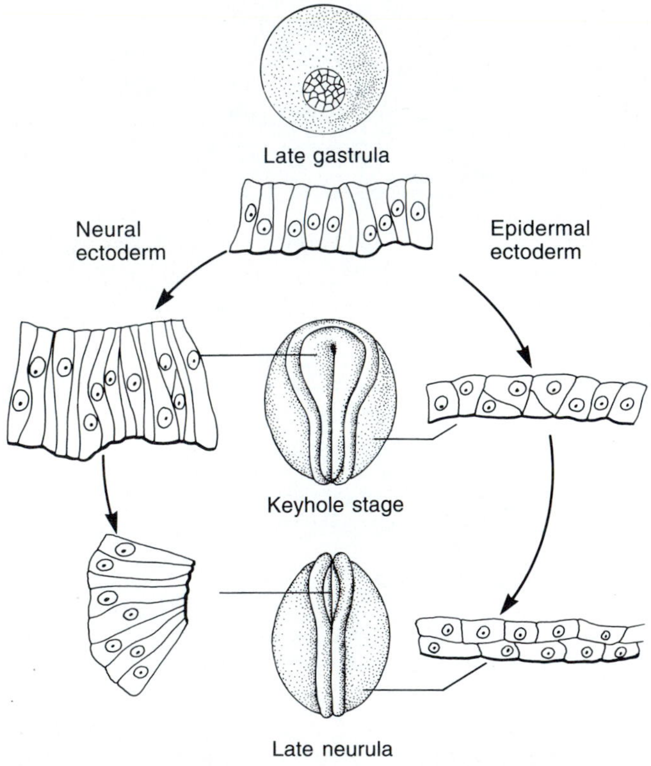

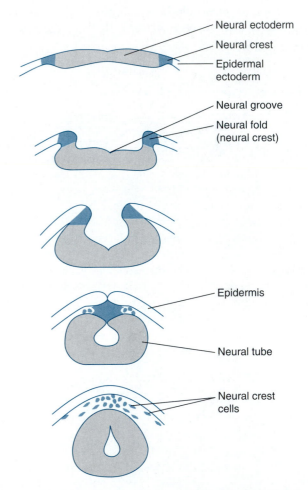

Figure 11.6 Successive stages of amphibian neurulation in transverse sections. Each of the schematic drawings shows the neural ectoderm (gray), the epidermal ectoderm (white), and the neural crest (color). When the neural folds fuse, they close the neural tube under a continuous layer of epidermis. The cells occupying the crests of the neural folds, known as neural crest cells, migrate to different locations and have a wide variety of fates.

Figure 11.5 Changes in shape of ectodermal cells during neurulation in the newt *Taricha torosa.* The neural ectoderm cells become columnar, while the epidermal ectoderm cells become squamous. See also Figure 2.6.

Late gastrula

Neural ectoderm

Epidermal ectoderm

Keyhole stage

Late neurula

jacent sheets of epidermal ectoderm fuse above the neural tube along the dorsal midline. When the neural tube detaches from the future epidermis, the intervening cells located at the crest of the neural folds generate a lineage of their own. These cells, called **neural crest cells**, detach and assume a temporary perch on top of the neural tube. Later, they will migrate to different positions all over the body and give rise to a wide variety of differentiated cells (see Chapter 12).

The region where the neural tube first closes varies from one class of vertebrates to another. In amphibians, the neural tube closes almost simultaneously throughout its length. In birds, the regression of Hensen's node from anterior to posterior gives the anterior end a head start. Consequently, the formation of the neural plate and its closure into a tube begin anteriorly and progress to the posterior (see Fig. 10.30). A similar pattern is observed in mammals. However, the anterior part of the neural plate is bulkier and seems to resist closure into a tube. As a result, it is the middle of the mammalian neural plate that closes into a tube first. The temporarily open ends of the neural tube are called the *anterior neuropore* and the *posterior neuropore*.

Mechanisms of Neurulation

In Chapter 10, we discussed the ability of the embryo to mold itself into a characteristic form through coordinated cellular movements and shape changes. We saw how one morphogenetic event, gastrulation, can be analyzed in terms of a small number of cellular movements and shape changes despite the variation of the process among different animals. In this chapter, we will see how similar types of cellular behavior can also explain neurulation. The appeal of this kind of reductionist analysis is the prospect of understanding a wide variety of complex organismic processes in terms of a small number of cellular behaviors.

Neurulation Is Independent of Surrounding Epidermis but Depends on Axial Mesoderm

The analysis of neurulation has occupied embryologists for more than 100 years (reviewed by R. Gordon, 1985; Schoenwolf and Smith, 1990). In the analysis of any epithelial movement, it is important to assess whether the movement is autonomous, relying only on forces generated by the epithelium itself, or whether it depends on forces created elsewhere. In an early study of neurulation, W. His (1874) proposed that the epidermis lateral to the neural plate expands actively and thereby compresses the neural plate, making it buckle and close into a tube. However, the notion of lateral compression was soon dismissed by Wilhelm Roux (1885), who iso-

lated neural plates from their adjacent epidermis and observed their behavior in tissue culture. He observed that the isolated neural plates closed *even faster than normally*. Although this result might have been an artifact of primitive tissue culture methods, Roux concluded that the forces driving neurulation must reside within the neural plate itself and that the surrounding epidermis resists neurulation rather than helping it.

This conclusion was supported by an experiment conducted by Antone Jacobson and Richard Gordon (1976). When making incisions in the epidermis surrounding the neural plate, these investigators found that the cuts gaped widely regardless of their orientation. They concluded that the epidermis is under considerable tension in every direction and that it cannot possibly be pushing the neural folds together when the neural plate closes into a tube.

Another embryonic tissue that might be involved in neurulation is the mesoderm underlying the neural plate. When neural plates are excised, they separate readily from the underlying mesoderm, except the *axial mesoderm* along the midline. This tissue, the prospective notochord, adheres tightly to the overlying neural plate and presumably was not removed in Roux's isolation experiments. Almost a century later, Jacobson and Gordon (1976) tested whether neurulation would still occur when the notochord was completely removed. Neural plates isolated without prospective notochord at the beginning of neurulation failed to elongate and did not assume the proper keyhole shape. Likewise, isolated notochord did not elongate. However, neural plates with notochords attached did form keyhole-shaped plates. Similar experiments carried out at later stages had somewhat different results. Neural plates isolated without notochord at the keyhole stage elongated considerably, although not as much as control neural plates isolated with notochords (Jacobson, 1985).

Taken together, these results indicate that the neural plate cells must interact with the underlying notochord cells prior to the keyhole stage for neurulation to proceed normally. After the keyhole stage has been reached, the notochord's role is less critical.

Neural Plate Cells Undergo Columnarization

In the course of neurulation, the cells of the neural plate undergo major shape changes while retaining the same volume. During the phase preceding the keyhole stage, the neural plate cells elongate perpendicular to the surface of the embryo while their apical and basal surfaces shrink (Fig. 11.7). This process, called **columnarization**, is commonly observed in epithelial areas at the beginning of morphogenetic movements.

The columnarization of epithelial cells is generally correlated with the alignment of their microtubules. At the beginning of neurulation, neural plate cell micro-

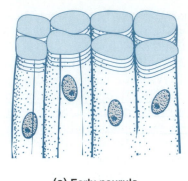

(a) Early neurula

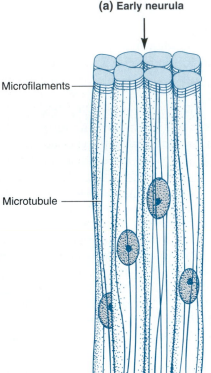

Microfilaments ——

Microtubule ——

(b) Late neurula

Microfilaments   Intercellular
                    junction

(c)

Figure 11.7 Columnarization of neural plate cells in the newt *Taricha torosa*. The same neural plate cells are shown **(a)** immediately following gastrulation and **(b)** after closure of the neural tube. Each cell elongates perpendicular to the surface while maintaining a constant cell volume. Consequently, its apical and basal surfaces shrink. Microtubules are oriented mostly parallel to the direction of elongation. Microfilaments accumulate around the apical circumference in purse string fashion. **(c)** A planar (glancing) section shows that the apical portions of cells are coupled to one another via intercellular junctions. Therefore, the shrinkage of one cell at the apical surface pulls the other cells closer to it, thereby reshaping the epithelium.

tubules are oriented parallel to the axis of cell elongation, whereas previously they were scattered throughout the cells in a seemingly random manner. Microtubules appear to be necessary for elongation, since the process is stopped by colchicine, an inhibitor of microtubule polymerization (Burnside, 1973). It is not clear whether microtubules actually cause the columnarization of cells by orienting their cytoplasmic transport or whether they simply stabilize the elongated cell shape after it has been generated independently.

The degree of columnarization varies among different regions of the neural plate. The longest cells occur in a crescent-shaped anterior area, and the shortest cells, along the posterior midline (Fig. 11.8). When Jacobson (1981) transplanted pieces of neural plate from an area of greater elongation to an area of lesser elongation and vice versa, he found that the transplanted cells elongated according to the region from which they were taken. Johannes Holtfreter (1946) observed that isolated neural plate cells completed their normal elongation in vitro. Therefore, by the criteria of transplantation and isolation, cells in each neural plate area are *determined* to elongate by a certain amount.

The columnarization of individual neural plate cells adds up to an overall reduction in the surface area of the entire neural plate. However, the plate's anteroposterior extension and its transformation to the keyhole shape cannot be explained by columnarization alone.

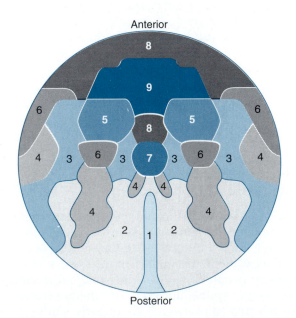

Figure 11.8 Pattern of cell elongation in the neural plate of the newt. The drawing shows a dorsal aspect of the neural plate. The amount by which each cell elongated (perpendicular to the plane of the paper) was mapped and classified on a scale from 1 to 9, with 9 representing the greatest elongation.

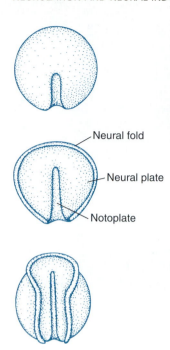

(a)

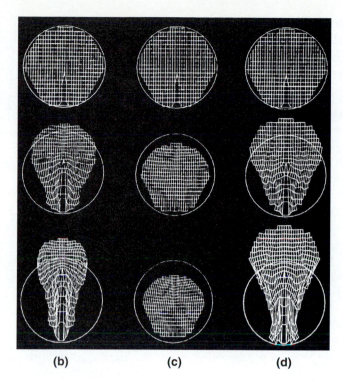

(b)   (c)   (d)

Figure 11.9 Computer modeling of neural plate formation in the newt. **(a)** Sequence of diagrams of the developing neural plate, drawn from time-lapse frames. **(b)** Computer simulation incorporating two forces: columnarization, especially of anterior neural plate cells, and convergent extension, especially of posterior neural plate cells. The resulting transformation of a coordinate grid placed over the neural ectoderm resembled closely the formation of the keyhole-shaped neural plate in the embryo. **(c)** Turning off the convergent extension part of the computer program resulted in a poor simulation. **(d)** Turning off the columnarization part resulted in a fair simulation, but the surface of the anterior neural plate was too large.

Intercalation of Neural Plate Cells Causes Convergent Extension

In addition to columnarization, neural plate cells undergo *convergent extension*. Using time-lapse cinematography, Burnside and Jacobson (1968) surveyed the movements of neural plate cells in the California newt, *Taricha torosa*, from the beginning of neurulation to the keyhole stage (Fig. 11.9a). In this newt, which has a salt-and-pepper pigmentation, they were able to trace the movements of individual cells. In a frame-by-frame analysis, they mapped the movements of cells at the intersections of a superimposed coordinate grid. The pathways turned out to be very consistent from one embryo to another, and the overall cell movements were toward the midline and the anterior. The most conspicuous anteroposterior extension of the grid occurred in its posterior two-thirds along the midline (Fig. 11.9a). This area of the neural plate was termed the *notoplate*, because it originates next to the notochord in the early gastrula.

Analysis of cellular movements in the notoplate revealed *mediolateral cell intercalation*, that is, the same type of cellular behavior that takes place in the *involuting marginal zone* during amphibian gastrulation (see Chapter 10). At the beginning of neurulation, the notoplate cells occupy a semicircular region at the posterior margin of the neural plate (Fig. 11.10). As lateral notoplate cells wedge between medial cells, the notoplate converges laterally and extends anteroposteriorly. At the keyhole stage, the notoplate has been repacked into a narrow strip extending along most of the midline.

Convergent extension also occurs elsewhere in the neural plate, although to a lesser degree.

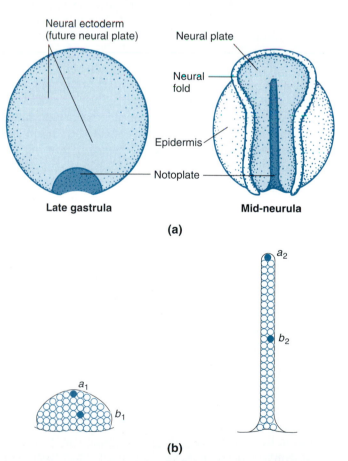

Figure 11.10 Convergent extension of the notoplate (region of the neural plate overlying the notochord) in the newt. **(a)** Schematic drawings, showing dorsal views. **(b)** Cell rearrangement in the notoplate. The two highlighted cells move from a_1 to a_2 and from b_1 to b_2. Thus, different sections of the notoplate extend proportionally.

Computer Simulation Shows That Both Columnarization and Cell Intercalation Contribute to Generating the Keyhole Shape

So far, we have discussed two types of cellular behavior involved in neural plate formation: columnarization and mediolateral intercalation. Their relative importance in bringing about the keyhole shape of the neural plate could be assessed if each could be selectively inhibited. However, since both behaviors are presumably driven by microfilament action, inhibitors like cytochalasin B are not useful for isolating each behavior. Neural plates isolated from their notochords, as described previously, shrink in surface area but do not extend anteroposteriorly, and they fail to assume the keyhole shape. Presumably, the removal of all mesoderm deprives the neural plate cells of the basal support needed for intercalation movements, thus effectively preventing convergent extension. Unfortunately, there is no apparent way of eliminating the columnarization of neural plate cells without affecting their intercalation behavior as well.

In cases like this, computer simulations can be helpful. Jacobson and Gordon (1976) simulated both columnarization and intercalation in a comprehensive computer program, which closely approximated the generation of the keyhole shape in the newt embryo (Fig. 11.9b). By omitting the intercalation part of the program, they obtained a poor simulation that featured a reduction in surface area but no anteroposterior extension, much like the behavior of the neural plate isolated from the underlying mesoderm (Fig. 11.9c). Conversely, turning off the columnarization part of the computer program produced a good simulation of the anteroposterior extension and keyhole formation, but the surface of the simulated neural plate was too large (Fig. 11.9d). The investigators concluded that both mediolateral intercalation and columnarization are necessary to shape the neural plate properly, but that intercalation makes the greater contribution to anteroposterior extension and generation of the keyhole shape.

Neural Tube Closure Is Associated with Apical Constriction, Rapid Anteroposterior Extension, and Cell Crawling

During the second phase of neurulation, when the neural plate closes into a tube, the apical cell surfaces continue the shrinking process that began with columnarization in the first phase (Fig. 11.5). This process of *wedging* gives the neural plate cells a shape that allows the plate to curl into a tube (Fig. 11.11). Beth Burnside (1971) suggested that this shape change may be caused by *apical constriction,* that is, the constriction of a band of microfilaments arranged like a purse string beneath

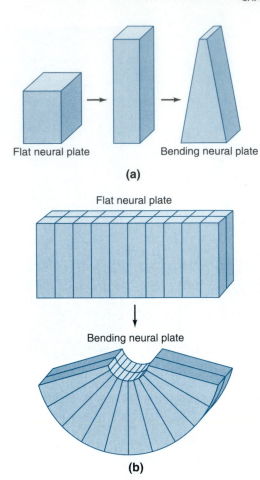

Figure 11.11 Cell wedging during neural tube closure. **(a)** Shape changes shown for a single neural plate cell: columnarization during the first phase of neurulation, and wedging during the second phase. **(b)** Wedging during the second phase of neurulation, shown for two transverse rows of cells.

the apical surface (Fig. 11.7). In accord with this hypothesis, she found that the bundle of microfilaments thickened during apical constriction, suggesting that the constriction might be caused by interdigitation of the microfilaments. However, it is not clear how apical constriction by itself would generate a wedge-shaped rather than a pyramidal or cone-shaped cell. Investigators have therefore looked for additional mechanisms that might be involved in neural tube closure.

Closure of the neural tube coincides with a spurt of rapid anteroposterior extension. To quantify this observation, Jacobson and Gordon (1976) excised neural plates and neural tubes from newts, laid them flat on agar plates, and measured their length under a microscope with a scale built into the eyepiece. Plotting overall length at successive stages, they found that the extension rate changed abruptly and was *10 times faster during neural tube closure* than before and after closure (Fig. 11.12). Similarly, as neurulation in chickens proceeds from anterior to posterior, a wave of rapid extension accompanies tube closure. Most likely, the rapid

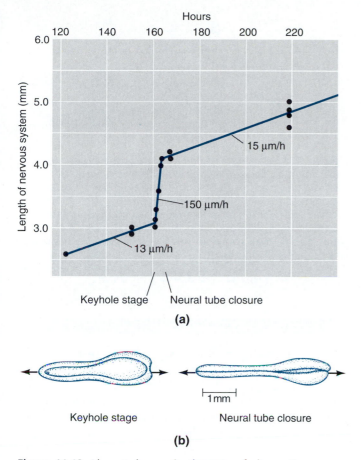

(a)

Keyhole stage   Neural tube closure

(b)

Figure 11.12 Abrupt change in the rate of elongation observed during the closure of the neural tube in the newt *Taricha torosa*. **(a)** Neural plates or tubes were excised and their lengths measured. The *rate* of elongation, indicated by the *slope* of the curve, increased 10-fold between the keyhole stage and neural tube closure, when it returned to the old rate. **(b)** Neural plate and neural tube are drawn to scale to accurately show the degree of extension during the interval.

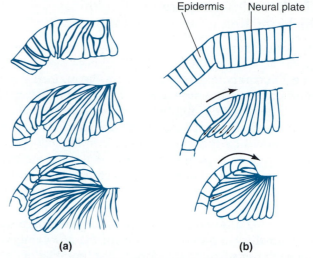

(a)   (b)

Figure 11.13 Cell crawling at the epidermis–neural plate boundary. **(a)** Enlarged tracings show cells from transverse sections of newt neurulae at successive stages. **(b)** Interpretative diagrams highlight major cell movements (arrows). The neural plate cells seem to crawl with their basal ends under the adjacent epidermis cells, thereby attenuating their apical surfaces. The combination of basal crawling and apical constriction appears to generate a movement that lifts up the neural folds and rolls them toward the dorsal midline.

extension of the closing neural plate is driven by cell intercalation (Jacobson, 1981).

At the neural folds, where the columnar neural plate cells are juxtaposed to the squamous epidermal ectoderm cells, it appears that the neural plate cells attempt to crawl under the epidermal cells (Fig. 11.13). Conceivably, this could generate a rolling movement that may contribute to neural fold formation and eventually to neural tube closure. To test this notion experimentally, Jacobson (1994) isolated neural plates including the underlying notochord, with and without margins of adjacent epidermis. If epidermal margins were present, the isolated plates rolled into complete tubes and elongated normally. However, if no epidermis was included, only a U shape was achieved and the neural plates did not elongate as much as normal.

Together, the observations discussed in this section suggest that neural tube closure is based on at least three types of cellular behavior: apical constriction, in-

tercalation, and crawling. Again, the *principle of synergistic mechanisms* (see Chapters 4, 9, 10, 19, 21, 22, and 26) applies. All three cellular behaviors must work together to bring about neural tube closure. If one of them is disturbed, as by eliminating the neural folds to prevent cell crawling, the other activities produce only a partial result.

Neural Induction

The neural tube is an organ rudiment that is always formed in association with an underlying notochord, laterally adjacent somites, and other mesodermal structures (see Figs. 9.21 and 10.30). This set of mostly dorsal organ rudiments in vertebrates is collectively called the *embryonic axis.* (The term "axis" meaning this set of organ rudiments should not be confused with the use of the same term to mean a line of orientation, such as the anteroposterior axis of an embryo. The stereotypical arrangement of different organ rudiments in the embryonic axis suggests that their development is coordinated by inductive interactions. Indeed, the famous *organizer* experiment of Hans Spemann and Hilde Mangold (1924) showed that all axis organs are either formed or induced by the *dorsal blastopore lip* in the early gastrula.

The Dorsal Blastopore Lip Organizes the Formation of an Entire Embryo

The organizer experiment was the culmination of Spemann's long quest to understand how the cells that form the embryonic axis organs are determined. From twinning experiments with newt eggs, Spemann already knew that the ability to form an axis depends on the presence of cytoplasm from the dorsal half of the egg, marked by the gray crescent (see Fig. 6.9). From transplantation experiments before and after gastrulation, he also knew that the neural ectoderm changes its state of determination during gastrulation (see Fig. 6.13).

Pursuing a general plan of mapping the progress of determination in different regions of the newt embryo, Spemann transplanted the upper blastoporal lip of *early* gastrula donors to the prospective flank epidermis of host embryos at the same stage of development (Fig. 11.14). Most of the host embryos developed two neural plates. When three specimens were fixed and sectioned at the tail bud stage, two of them were found to have in their flanks an additional embryonic axis, including a neural tube, notochord, and somites. This experiment was first done in 1916, at a time when fate maps and trajectories of gastrulation movements in amphibians were not yet available. In the absence of such basic information, Spemann thought that the *entire* additional axis had been formed by the transplant—the neural tube from the superficial layer of the blastopore lip, which he presumed was ectodermal, and the mesodermal organs from the deep layer. He did not realize at the time that both layers were still to involute and that neither of them would form ectoderm.

When the experiment was resumed in 1921, Spemann had learned that the dorsal blastopore lip of an *early* gastrula does not include an ectodermal layer. He had also developed a **heterospecific transplantation** technique, in which he used donor and host embryos from different species, so that he could distinguish graft-derived tissues from host tissues on the basis of their different pigmentation. However, Spemann had also become the head of the zoology department at the University of Freiburg in Germany. Being occupied with teaching and administration, he assigned the task of repeating the transplantation of the blastopore lip to one of his graduate students, Hilde Proescholdt.[*] Her experiment became the crowning achievement of Spemann's efforts to understand the determination of axial organs in vertebrate embryos.

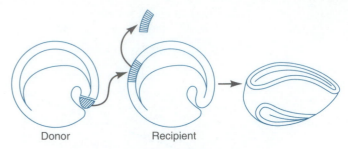

Figure 11.14 Spemann's earlier version of the organizer experiment. A dorsal blastopore lip from a newt early gastrula was transplanted to the ventral ectoderm of a recipient at the same stage of development. The recipients formed two neural plates and later two embryonic axes. However, since the donor and the recipient were of the same species, their contributions to the secondary axis could not be safely distinguished.

Proescholdt transplanted a median piece from the dorsal blastopore lip. In terms of the gastrula regions defined in Chapter 10, the transplant consisted of *involuting marginal zone* and, depending on the developmental stage of the donor, a smaller or greater portion of *noninvoluting marginal zone* (see Fig. 10.12). The transplant was taken from an (unpigmented) *Triturus cristatus* gastrula and implanted in prospective ventral epidermis of a (pigmented) *Triturus taeniatus* gastrula (Fig. 11.15). The graft performed the same morphogenetic movements it would have carried out normally: it involuted and extended into a long strip underneath the host ectoderm. As in Spemann's previous experiments, the host embryo developed an additional embryonic axis at the implantation site. However, because the host tissue was pigmented while the transplant was not, it was now unmistakable that much of the additional axis had been formed by the host.

Five experimental cases were described in detail. In the most completely developed case, the donor was an *advanced* gastrula. In the host, the transplant involuted *almost* completely. Two days after the operation, when the embryo had developed to the neurula stage, a second neural plate complete with neural folds was clearly visible on the flank. Most of the neural plate was pigmented and therefore derived from the host. Only a median strip stretching over the posterior two-thirds of the neural plate was unpigmented and thus derived from the transplant. A day later both the original host embryo and the secondary embryo had advanced to the tail bud stage. The secondary embryo was shorter and lacked the anterior head parts. Fixation and sectioning showed that the secondary embryo had a complete set of axial organs including a neural tube, a notochord, somites, gut, and embryonic kidneys.

The outstanding feature of the secondary embryo was that it was an almost complete and normally pro-

---

[*] The report of the organizer experiment was published under the names Hans Spemann and Hilde Mangold (Spemann and Mangold, 1924), since Proescholdt had married Otto Mangold, a junior colleague of Spemann's. At about the time of publication, Hilde died of severe burns suffered when an alcohol heater in her kitchen exploded (Hamburger, 1988). She could not partake of the excitement and glory that the organizer experiment brought to Spemann's laboratory.

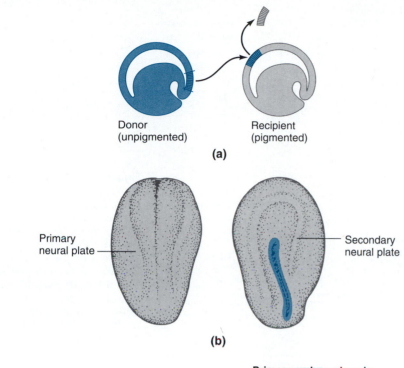

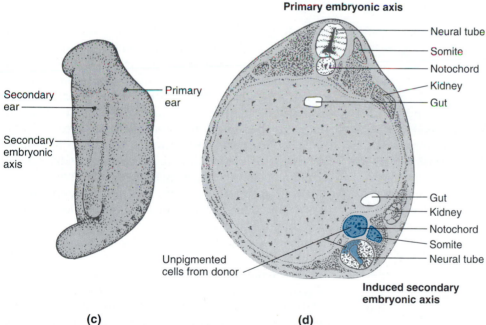

Figure 11.15 The organizer experiment of Spemann and Mangold (1924). **(a)** A median piece taken from just above the dorsal blastopore lip of a gastrula of *Triturus cristatus* (an unpigmented newt species, shown here in color) was transplanted to the ventral ectoderm of a gastrula of *Triturus taeniatus* (a pigmented species). **(b)** At the neurula stage, the recipient formed a secondary neural plate with a streak of unpigmented tissue (color) in the notoplate area. **(c)** The recipient proceeded to form a secondary embryonic axis. **(d)** Transverse section of the recipient shows the structures derived from the unpigmented transplant (color) and the pigmented recipient.

portioned embryo and also a chimera consisting of both host and graft tissues. The neural tube consisted almost entirely of host tissue, except for a ventral strip derived from the transplant. Most likely, this strip was derived from the most dorsal part of the transplant, the part that had not involuted into the host blastopore. It had undergone a sharp elongation characteristic of the notoplate, which is located exactly in this position after gastrulation (Fig. 11.10). The notochord was entirely unpigmented, that is, derived from the transplant. The somites were composed of transplant and host tissue, while the other axial structures were mostly formed by the host.

Three major conclusions can be drawn from the organizer experiment. First, the graft of dorsal blastopore lip *developed according to its fate:* it underwent the normal morphogenetic movements and proceeded to form, for the most part, notochord. Second, the graft *dorsalized the host's mesoderm:* tissue that would normally have formed hypodermis, blood, or other ventral mesodermal structures instead contributed to secondary somites and kidneys. Third, the graft *acted as a neural inducer.* It stimulated the host ectoderm to form a secondary neural plate, which closed into a neural tube running parallel to the graft-derived notochord. In short, the transplanted blastopore lip developed in accord with its own fate and marshaled the host tissues in such a way as to

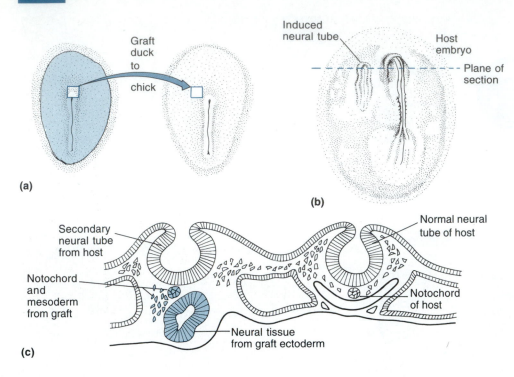

(a)

(b)

(c)

Figure 11.16 Induction of a secondary embryonic axis in birds. **(a)** Hensen's node was grafted from a duck donor to a chicken recipient. **(b)** The recipient developed an accessory neural tube. **(c)** Histological sections revealed that the graft itself formed a notochord and a neural tube underneath. In addition, the graft induced the formation of a neural tube from host tissue.

produce one integrated embryo. To emphasize this remarkable capability of the dorsal blastopore lip, Spemann called it the *organizer.*

Neural induction in birds and mammals appears to occur much as it does in amphibians. As described in Chapter 10, the primitive groove and pit in birds correspond to the blastopore in amphibians, and Hensen's node is the equivalent of the dorsal blastopore lip. If Hensen's node from a duck donor is grafted beneath the epiblast of a chick host, it induces a secondary axis including neural tube, notochord, and somites (Fig. 11.16). Corresponding experiments with mice have yielded similar results (Beddington, 1994). Indeed, similar molecular mechanisms seem to cause mesoderm induction and neural induction in *Xenopus* and the house mouse (Blum et al., 1992).

Neural Induction Shows Regional Specificity

The organizer experiment showed the remarkable capacity of a small transplant to organize almost an entire embryo. However, even the best-developed embryo in Proescholdt's original series had one flaw: it lacked the anterior part of the head. There were at least two possible explanations. First, the ventral host tissue might not have been *competent* to contribute to a head. This was considered unlikely, since Spemann had shown previously that early gastrulae, in which the animal half had been rotated 180° relative to the vegetal half still developed normally, demonstrating that the ventral portion of the animal hemisphere did respond to inducing signals from the dorsal marginal zone. Alter-

natively, the signals from the inducing mesoderm could have an anteroposterior specificity. In Proescholdt's best-developed embryo, the transplanted dorsal blastopore lip had been taken from an *advanced* gastrula. It is possible that the transplant had not included the brain-inducing part of the organizer because this part had already involuted into the donor embryo. To test this hypothesis, Spemann (1931) grafted the dorsal blastopore lip from an *early* gastrula. Under these conditions, the additional embryo had a complete head but no tail (Fig. 11.17). It seemed, therefore, that the early-involuting mesoderm, which moved the farthest anteriorly, tended to induce anterior axial structures whereas the later-involuting mesoderm was more likely to induce trunk and tail.

The regional specificity of the inducing mesoderm was demonstrated with more extensive data, but also at a later stage, by Otto Mangold (1933). He removed the neural plate from an early neurula to expose the underlying tissue, the tissue that had been the *dorsal involuting marginal zone* before it involuted. *After* involution, this tissue is referred to as **chordamesoderm,** although its fate includes not only notochord but also adjacent somites. In Mangold's experiment, the chordamesoderm was divided into four parts along the anteroposterior axis, and the isolated parts were inserted into the blastocoel of early host gastrulae (Fig. 11.18). The secondary structures that protruded from the bellies of the hosts varied depending on which chordamesoderm part had been grafted.

The first chordamesoderm quarter induced mostly structures that lie in front of the brain in the embryo. This result was similar to the one obtained after the

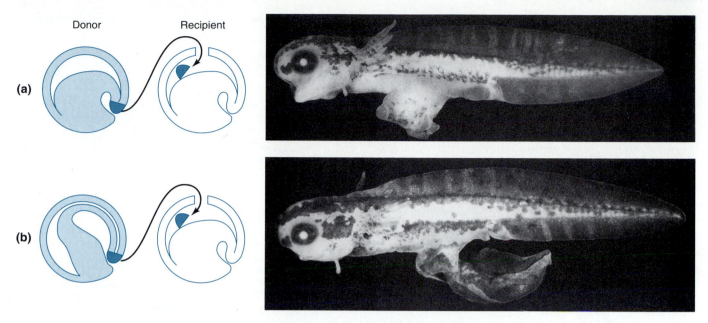

Figure 11.17 Stage dependence of the inductive capacity of the dorsal blastopore.
(a) The dorsal blastopore lip from an *early* gastrula, upon insertion into the blastocoel of a host, induces the formation of a secondary head. **(b)** Inserting a *late* blastopore lip in the same way causes the formation of secondary trunk and tail.

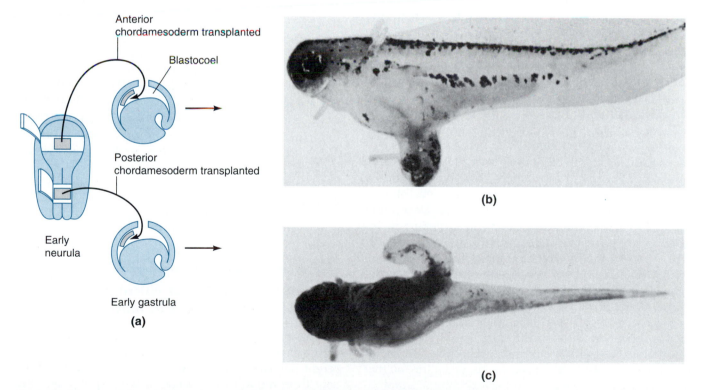

Figure 11.18 Regional specificity of neural induction. **(a)** Design of an experiment carried out by Otto Mangold (1933). He obtained different regions of chordamesoderm (gray shading) after removing the overlying neural plate from early neurulae and inserted the chordamesoderm pieces into blastocoels of early gastrulae. **(b)** If anterior chordamesoderm was inserted, the recipient formed a secondary head with balancers, forebrain, and eyes. **(c)** Posterior chordamesoderm inserted in the same way induced a secondary trunk and tail.

grafting of *early* dorsal blastopore lip, which forms anterior chordamesoderm after involution. The second chordamesoderm quarter induced a head with brain, nose, eye, and ear vesicles. The third quarter induced primarily hindbrain, spinal cord, and musculature, and the fourth quarter induced spinal cord, somites, kidney, and tail. There was considerable overlap among the structures induced by successive areas of chordamesoderm. However, the fourth-quarter graft had an effect similar to the one obtained by grafting *late* dorsal blastopore lip, which forms posterior chordamesoderm after involution. Clearly, blastopore lip removed at different stages of gastrulation, and the corresponding anteroposterior regions of chordamesoderm from late gastrulae, induce different sets of dorsal structures.

Neural Induction Can Be Studied in Vitro As Well As in Vivo

In the original experiments of Spemann and Proescholdt, dorsal blastopore lip was *implanted* into ventral prospective epidermis (Figs. 11.14 and 11.15). These experiments required great manual skill, and the number of successful operations was small. This problem was compounded by the lack of sterile techniques, which caused many embryos to die before they could be analyzed. Subsequent investigators developed alternative methods that were technically less demanding and allowed them to do more experiments with relative ease.

In collaboration with Spemann, Otto Mangold developed the **insertion method,** in which the grafted inducer is inserted into an early gastrula through a slit in the blastocoel roof (Figs. 11.17 and 11.18). During gastrulation, the vegetal base pushes the graft from inside against the ventral ectoderm. This procedure allows little control over the ectodermal area that comes to lie over the inducer, but fortunately, the *competence* of the ectoderm to respond to the inducer showed little regional variation in the newt species used. Depending on the location of the insert during gastrulation, its inductive properties may also be modified by interactions with host endoderm or ventral mesoderm.

As an alternative procedure, Holtfreter (1936) developed the **sandwich method,** in which a flap of ectoderm is wrapped around the inducing tissue and heals to enclose it (Fig. 11.19). This assay is more standardized and eliminates the stimulating or inhibitory effects of other tissues present in intact embryos. Holtfreter also improved the media used for tissue culture so that sandwich preparations and other explants of amphibian embryos could be maintained in vitro for up to 2 weeks. These procedures have since been adopted by many laboratories.

Amphibian tissues are most suitable for in vitro culture because each cell is endowed with its own yolk supply, allowing it to survive in simple inorganic salt

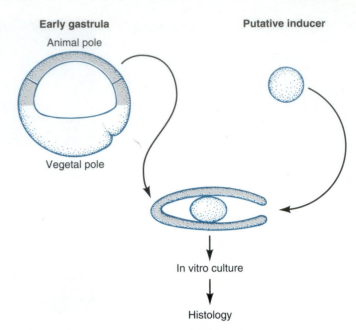

Figure 11.19 Sandwich method of testing the inducing capacities of various preparations by wrapping them into a segment from the animal pole region of an early gastrula.

solutions. The development of sandwiches or other preparations is routinely monitored under a stereomicroscope. If the structures observed need closer inspection, specimens are fixed and sectioned for histological analysis under the compound microscope. Using these standard methods, researchers tested the ability of a wide variety of tissues to induce the formation of axial structures in amphibian epidermis. In many of these experiments, the inducers were *heterologous,* that is, derived from other organisms or even consisting of nonliving materials, such as agar soaked in different fluids. The heterologous inducers were used for convenience, because they were easier to obtain than amphibian blastopore lips, especially in the large quantities required to narrow down the chemical nature of the inducing signal by biochemical fractionation.

The Elusive Molecular Nature of the Neural Inducer in Newts

The regional specificity of the chordamesoderm, as well as the characteristic types of structures induced by different heterologous inducers, showed that the signal sent from the inducing tissue to the responding tissue must have a degree of specificity. Spemann (1927) asked whether the specificity resulted from the *chemical* nature of certain molecules given off by the inducer, or from *mechanical* forces exerted by the inducing tissue on the responding tissue. He reasoned that a mechanical effect could be exerted only by living cells and proceeded to test the inductive capabilities of blastopore lips that had been minced with glass needles or squeezed between glass slides. The crushed cells induced neural tubes in gastrula ectoderm nevertheless. Similar results were obtained by Spemann's coworkers with inducers killed by heat, drying, freezing, or immersion in alcohol (Bautzmann et al., 1932). However, the amount of neural tissue obtained with dead inducers was smaller, and it was not as well formed as that induced by live cells.

The conclusion drawn from the dead organizer experiments was that the neural inducer must be a chemical signal. This inference triggered a flurry of biochemical investigations. Their basic design entailed soaking agar or similar carrier materials in cell extracts and testing their neural inducer activity in a *bioassay* such as the sandwich procedure. An active cell extract could be fractionated using biochemical procedures to identify the active molecule. The original enthusiasm was dampened by the discovery of a plethora of heterologous inducers and by the observations of Holtfreter (1945) that neuralization may be induced by stimuli as unspecific as changes in the ionic composition or pH of the culture medium.

With the benefit of hindsight, it is now clear that neural induction, like many other kinds of induction, is a multistep process. Apparently, the neural ectoderm of newts is primed so well by events preceding its interaction with chordamesoderm that many types of stimuli may give it the final push to form neural tissue. Then why don't more embryonic regions develop into neural tissue? Presumably, inhibitory signals from ventral embryonic parts help limit neural development to the dorsal area of the ectoderm. These complications made it futile to try to determine the nature of the neural induction signal by biochemical procedures. Only recently have the techniques of molecular biology provided new experimental designs that may yield a better understanding of how neural induction works; some of these will be explored in the next section.

Mechanisms of Neural Induction in *Xenopus*

Recent studies of neural induction differ in several respects from their classical forerunners. First, most of the current work is carried out on *Xenopus laevis*, a species that develops faster and is easier to maintain in the laboratory than the newts used by Spemann's group. Second, instead of biochemical fractions, modern investigators prefer to test the inducing activity of mRNAs and proteins derived from cloned genes (see Chapter 14). These preparations are pure and can be obtained in unlimited quantities. As an added advantage, the amino acid sequence of proteins from cloned genes can be de-

termined and compared with sequence data on other proteins in computerized data banks. Such comparisons often provide clues to the biological activity of the protein under investigation. Third, researchers today measure the response to inductive signals not only by morphological criteria but also by the accumulation of *molecular markers*, such as mRNAs or proteins that are synthesized in neural tissue but not in epidermis, or vice versa.

There Are Two Signaling Pathways— Planar and Vertical—for Neural Induction

Spemann considered two signaling pathways for neural induction (Fig. 11.20). First, in the blastula or early gastrula, when the organizer forms the dorsal *blastopore lip*, it might send *planar signals* through the plane of the ectoderm. At this stage, the prospective neural plate occupies a short and wide area that could easily be reached by a planar signal from the dorsal blastopore lip (R. Keller et al., 1992b). Second, after involution, when the organizer has formed the *chordamesoderm*, it might send *vertical signals* into the overlying ectoderm. At this later stage, the prospective neural plate has expanded in the anteroposterior direction and is more accessible to vertical signals from the chordamesoderm.

Originally, Spemann favored the notion of planar signals. However, the results of subsequent experiments by his collaborators Otto Mangold and Johannes Holtfreter were more compatible with vertical signals. Mangold (1933) showed that chordamesoderm by itself was sufficient for neural induction in newts, as described earlier (Fig. 11.18). The dependence of the induced structures on a specific chordamesoderm region argued strongly in favor of two or more vertical signals.

In another experiment, Holtfreter (1933) stripped early axolotl gastrulae of their fertilization envelopes

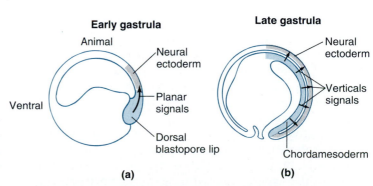

Figure 11.20 Planar and vertical signals in neural induction. **(a)** Early gastrula. The planar signals travel from the dorsal blastopore lip (color) directly into the prospective neural ectoderm (gray). **(b)** Late gastrula. The vertical signals pass from the same inducing tissue, now chordamesoderm, into the overlying neural ectoderm. The relative importance of the two sets of signals varies between species.

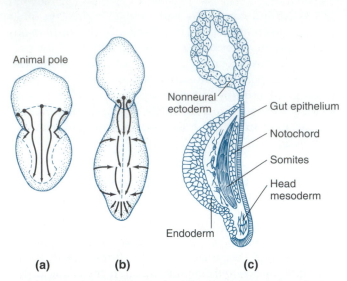

Animal pole

Nonneural
ectoderm

Gut epithelium

Notochord

Somites

Head
mesoderm

Endoderm

(a)   (b)   (c)

Figure 11.21 Exogastrulation in an axolotl embryo. **(a, b)** Marginal zone cells move (arrows) toward the blastopore region (constriction), but instead of involuting, these and the vegetal base cells are pushed outside, where the mesoderm nevertheless undergoes convergent extension. **(c)** Median section through exogastrula, showing an empty ectodermal hull (above) and endodermal plus mesodermal structures (below). Note that the ectoderm forms no brain or spinal cord.

and kept them in a slightly hypertonic salt solution. Under these conditions, the gastrulation movements took a completely abnormal course called *exogastrulation.* The prospective endoderm and mesoderm, instead of turning inside the embryo, turned to the outside and left the ectoderm behind as an empty bag (Fig. 11.21). Under these circumstances, neural induction was still possible via planar signals, while the possibility of vertical signals was excluded. In the exogastrulated embryos, the endodermal and mesodermal derivatives developed more or less normally. In particular, the notochord, somites, and embryonic kidney structures were formed. However, the ectodermal bag, deprived of its underlying mesoderm, failed to develop clearly identifiable neural tissue. Thus, although the inducing tissue developed almost normally, and although planar inducing signals could have passed to the ectoderm, no inductive response was detected.

From the results of these experiments with newt embryos, Spemann and his coworkers concluded that vertical signals were indispensable for neural induction, and that planar signals played a minor role if any. Later investigators reached almost the reverse conclusion from their experiments with *Xenopus* embryos.

Planar Induction Plays a Major Role in *Xenopus* Embryos

Repeating the organizer experiment with *Xenopus* embryos confirmed the results of Spemann and Mangold

(Gimlich and Cook, 1983). However, subsequent experiments using morphological criteria as well as molecular markers have shown that neural induction relies more on planar induction in *Xenopus* than it does in newts and salamanders (Slack and Tannahill, 1992).

In a molecular study of neural induction in *Xenopus*, Chris Kintner and Douglas Melton (1987) found that ectoderm cells express N-CAM, a neural *cell adhesion molecule*, as an early response to neural induction. They showed that in normal embryos, neural plate and neural tube express N-CAM while prospective epidermis does not (see Fig. 25.18). In induction experiments with explanted germ layers, ectoderm did not synthesize N-CAM unless it was in contact with inducing mesoderm. Having established the usefulness of N-CAM as a molecular marker for neural induction, the researchers examined the expression of N-CAM in exogastrulae. Unexpectedly, they found that exogastrulae synthesized nearly as much N-CAM as normal embryos, and that almost all the N-CAM synthesis occurred in the ectodermal portion of the exogastrula. Thus, *Xenopus* ectoderm did express a molecular marker for neural development under conditions that allowed vertical but no planar induction.

In an extension of this study, Ariel Ruiz i Altaba (1992) used antibodies directed against specific neuronal molecules and also examined the histological structures in exogastrula ectoderm more closely. He found that posterior neural tissue was present, although poorly organized, in the ectodermal portions of *Xenopus* exogastrulae. In contrast, he saw no forebrain structures in exogastrulae.

As an alternative test of *Xenopus* ectoderm development under conditions of planar induction only, Ray Keller and Mike Danilchik (1988) isolated patches comprising involuting and noninvoluting marginal zone from the dorsal side of early gastrulae. To keep these patches from curling up, they combined two of them with their deep zones face to face, creating a so-called Keller sandwich. This arrangement allowed only planar signals to travel between the involuting marginal zone (dorsal blastopore lip, organizer) and the noninvoluting marginal zone (the part that would normally form neural plate). Despite these constraints, the morphogenetic movements and cell differentiations in all regions of the sandwich corresponded to those in intact embryos (Fig. 11.22). In particular, the noninvoluting marginal zone underwent convergent extension and formed a neural plate. The neural plate developed into a mass of cells that resembled multipolar neurons and was later shown to stain with a neuron-specific probe.

The researchers extended this experiment under more rigorous conditions by making sandwich prepa-

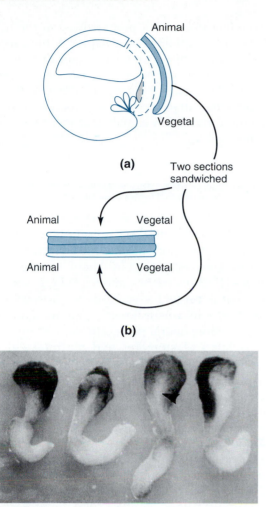

Figure 11.22 Test for neural induction in *Xenopus* by planar signals only. **(a, b)** Two dorsal flaps of tissue comprising involuting marginal zone, noninvoluting marginal zone, and animal cap were cut from two early gastrulae and stuck together with their inner cell layers face to face and with their animal-vegetal axes aligned. **(c)** After a few hours of culture, the involuting marginal zone (below arrowhead) formed notochord and somites while the noninvoluting marginal zone (above arrowhead) formed neural tissue.

rations in which animal caps were brought into planar contact with involuting marginal zone (R. Keller et al., 1992c). By taking either the inducing tissue (involuting marginal zone) or the responding tissue (animal cap) from a donor labeled with a fluorescent dye, they ascertained that no inducer cells invaded the responding tissue. Nevertheless, animal cap cells were stimulated to make the convergent extension movements that are characteristic of neural plate formation.

In the experiments described so far, neural induction was judged by the convergent extension movements and the histological differentiation that are characteristic of developing brain and spinal cord (the central nervous system, or CNS). These criteria did not

reveal the development of specific CNS regions, such as forebrain, midbrain, hindbrain, and spinal cord. On the basis of Otto Mangold's experiments with newts, it had been assumed that the regional specificity of the induced CNS regions relied on vertical signals from the chordamesoderm (Fig. 11.18). However, the new results obtained with *Xenopus* raised the question whether the *anteroposterior pattern of different brain regions* could also be generated by planar induction alone.

To explore this possibility, Tabitha Doniach and her colleagues (1992) used *in situ hybridization* (see Methods 15.1) and *immunostaining* (see Methods 4.1) to detect the mRNA and protein products of certain marker genes that are expressed only in specific CNS regions. These methods made it possible to identify the products of three marker genes (*engrailed-2+*, *Krox-20+*, and *XlHbox6+*) as a distinct series of transverse stripes in the developing midbrain, hindbrain, and spinal cord of intact embryos (Fig. 11.23a). To see whether this pattern would originate in the absence of vertical induction signals, the investigators prepared Keller sandwiches, which allowed only the exchange of planar signals between the dorsal involuting marginal zone and prospective neuroectoderm (Fig. 11.22). In culture, these isolates underwent convergent extension and neural differentiation. When the developed tissues were tested for expression of region-specific marker genes, they indeed showed the same pattern of transverse stripes as the normal embryos (Fig. 11.23b). This stunning result was not caused by vertical signals from migrating meso-

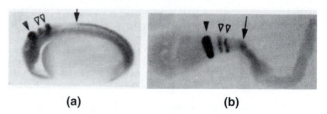

(a)   (b)

Figure 11.23 Expression of neural marker genes in *Xenopus* whole embryos and sandwiches prepared as shown in Figure 11.22. **(a)** Whole embryo, anterior to the left, dorsal side up. The dark areas in the brain and spinal cord are mRNAs labeled by in situ hybridization (see Methods 15.1). The anteriormost band (filled arrowhead) represents mRNA transcribed from the *engrailed-2+* gene, which is expressed selectively at the boundary between midbrain and hindbrain. The doublet of bands (open arrowheads) represents Krox-20 mRNA, which is synthesized in two hindbrain segments. The dark streak (to the right of the arrow) represents XlHbox6 mRNA, which accumulates throughout the spinal cord. **(b)** Tissue developed from a sandwich prepared as shown in Figure 11.22, animal pole to the left. The narrowest region (to the right of the arrow) is prospective spinal cord, the flared region to the left is prospective brain, and the curved region to the right is prospective mesoderm. Note that contact between the neural region and the mesoderm is strictly planar. Nevertheless, the marker mRNAs are synthesized in the same pattern as in the intact embryo.

derm cells or misaligned sandwich layers, as the investigators showed by appropriate controls. They concluded that a pattern closely resembling the anteroposterior pattern of different CNS elements can be induced by planar signals alone.

In a similar experiment, Kathryn Zimmerman and her colleagues (1993) monitored the expression of another neural marker gene, XASH-3+, in normal Xenopus embryos and in Keller sandwiches. They found that the normal expression pattern of the gene in early neurulae includes two longitudinal stripes on both sides of the midline. These stripes are also expressed in cultured Keller sandwiches, indicating that some mediolateral patterning of the neural plate can occur independently of vertical neural induction.

The observations discussed so far indicate that neural induction in amphibians depends on at least two types of signal, planar and vertical. The planar signals are sent within the outer plane of the early gastrula from the dorsal blastopore lip (organizer) to the adjacent dorsal ectoderm. The vertical signals are sent in the late gastrula from the same organizer, then called the chordamesoderm, to the overlying dorsal ectoderm. The classical results indicate that in newt embryos vertical signals play a dominant role whereas planar signals by themselves are insufficient. In contrast, more recent experiments with Xenopus laevis show that in this species planar signals alone cause neural differentiation, including the proper anteroposterior pattern of gene expression in midbrain, hindbrain, and spinal cord. However, forebrain structures seem to require vertical inductive signals as well. These results attest to the specificity of planar neural induction in Xenopus, but they also point out the necessity of vertical inductive signals, in particular for forebrain development.

Why do newts and Xenopus differ in relative importance of planar and vertical neural induction signals? This difference may be related to other differences that also distinguish these two forms. First, Xenopus mesoderm is located deep in the gastrula (see Fig. 10.12), whereas newt mesoderm includes the surface layer. Xenopus development is also very rapid, among the fastest of the amphibians. Species that develop rapidly often evolve in a way that shifts developmental signals to earlier stages; thus, planar neural induction, which occurs before vertical induction, plays a greater role in Xenopus than it does in newts. Indeed, the experiments described next indicate that even earlier events in Xenopus embryos prepare, or bias, the dorsal ectoderm to respond to the subsequent neural induction signals.

Neural Induction Is a Multistep Process

Whether ectoderm develops into neural plate or epidermis depends not only on the proximity of the inducing tissue but also on the preparedness of the responding tissue. (As explained in Chapter 9, the preparedness of a responding tissue to a specific inductive signal is called the competence, or bias, of the responding tissue.) At least in Xenopus, the dorsal ectoderm is more competent than the ventral ectoderm to respond to the neural induction signals that emanate from the dorsal mesoderm.

▼

Differences between dorsal ectoderm, which will form neural plate, and ventral ectoderm, which will form epidermis, can be traced back to the 8-blastomere stage in Xenopus. This was shown by Cheryl London and her colleagues (1988) using as a marker Epi 1, a cell surface antigen that is present specifically in prospective epidermis. At the mid-neurula stage, epidermal cells express Epi 1 while neural plate cells do not. The investigators isolated blastomeres and blastula regions of Xenopus embryos and kept them in culture until control embryos had reached the mid-neurula stage. The descendants of the isolated cells were then fixed for immunostaining (see Methods 4.1) with an antibody against Epi 1. Descendants of ventral animal cells isolated at the 8-blastomere stage expressed the Epi 1 antigen more strongly and more consistently than the descendants of dorsal animal cells.

In subsequent experiments, Robert Savage and Carey Phillips (1989) used the same immunostaining procedure to monitor the inhibitory effect of neural inducers on Epi 1 expression. Ventral ectoderm kept in edge-to-edge contact with dorsal blastopore lip did not express Epi 1. Similarly, chordamesoderm sandwiched between ventral ectoderm layers inhibited Epi 1 expression.

The experiments monitoring Epi 1 expression indicate that ventral animal blastomeres are already biased toward forming epidermis. This early predisposition can be overridden by strong neural inducers such as a grafted dorsal blastopore lip. Conversely, the early bias of dorsal animal cells is not sufficient for them to form neural plate: they still express Epi 1 at low levels, and when isolated at the blastula stage, they form epidermis. However, subsequent neural induction signals completely inhibit Epi 1 expression and determine the cells for neural development. In short, a stepwise inhibition of Epi 1 expression in dorsal ectodermal cells is paralleled by a stepwise determination toward neural plate formation.

In another experiment, C. R. Sharpe and colleagues (1987) compared the inductive effects of chordamesoderm on dorsal and ventral ectoderm (Fig. 11.24). They used two marker mRNAs: one encoding N-CAM and the other transcribed from XlHbox6+, a gene expressed specifically in posterior neural cells of late gastrulae.

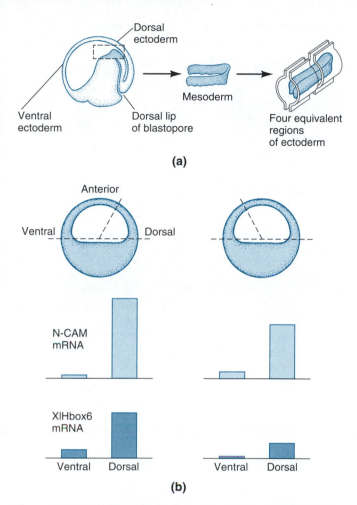

Figure 11.24 Predisposition of dorsal ectoderm to respond to neural induction. **(a)** Experimental design. Mesoderm (color) was taken from the ridge of the involuting tissue at a mid-gastrula stage. The mesoderm was wrapped with four equivalent pieces of ectoderm taken from early gastrulae as shown in part b. **(b)** The ectoderm pieces were cut differently so that the anteriormost ectoderm (top portion) was included either ventrally or dorsally. After in vitro culture, the accumulation of two neural marker mRNAs (N-CAM and XlHbox6) was measured by ribonuclease protection (see Methods 16.1). Marker mRNA amounts are indicated by column height. The dorsal ectoderm responded more strongly than ventral ectoderm to neural induction by mesoderm. Ectoderm incubated without mesoderm did not accumulate measurable amounts of marker mRNAs.

The dorsal ectoderm responded more strongly to inductive signals from chordamesoderm: both markers were present in larger amounts in *dorsal* ectoderm than in ventral ectoderm. Control pieces of dorsal and ventral ectoderm that had been kept without chordamesoderm showed no major accumulation of either marker mRNA. The greater *competence* of the dorsal ectoderm, that is, its better ability to respond to inductive signals from chordamesoderm, must result from signals that were received earlier by dorsal ectoderm but not by ventral ectoderm.

Taken together, the experiments discussed in this chapter and in Chapter 9 support the view that neural induction is a multistep process (Fig. 11.25). The earliest dorsoventral bias is introduced by cytoplasmic movements in the fertilized egg during the first cleavage cycle. This bias is evident from the reduced expression of Epi 1 in dorsal animal blastomeres and from the enhanced response of dorsal animal caps at the blastula stage to mimics of mesoderm inducers (see Chapter 9). During cleavage, only dorsal vegetal blastomeres (the Nieuwkoop center) are able to induce adjacent marginal cells to form dorsal mesoderm. The dorsal mesoderm cells give rise to the dorsal blastopore lip (Spemann's organizer), which sends planar neural induction signals at the early gastrula stage. After the dorsal blastopore lip has formed the chordamesoderm, it augments neural induction by sending vertical signals. The latter seem to be especially important for inducing formation of anterior brain structures; these structures develop from the anterior neuroectoderm, which is located most distant from the source of the planar signal.

In the series of events that lead to neural induction, individual steps can be prevented experimentally with the result of a reduced level of neural development. These observations are in accord with the *principle of synergistic mechanisms*, which pervades many developmental processes (see Chapters 4, 10, 17, and 21). This principle is especially evident in inductive interactions, not only in neural induction, but also in the induction of the nose, eye, ear, and heart (Jacobson, 1966; Jacobson and Sater, 1988). The synergism of several mechanisms in neural induction may also help to explain how the differences between newts and *Xenopus* with regard to planar and vertical induction may have evolved. If a biological function is based on several mechanisms with overlapping effects, one mechanism may increase while another decreases in its relative importance without jeopardizing the function. In this way, fast-developing species like *Xenopus* may have come to rely more strongly on the earlier events in the sequence leading to neural development.

Neural Induction Is Associated with the Region-Specific Expression of Certain Genes

Since neural induction is being viewed as the result of several determinative events, some laboratories have searched for *second messenger* pathways that may be able to integrate multiple neural induction signals. Indeed, the concentration of a common second messenger, *cyclic adenosine monophosphate*, increases substantially during neural induction (Otte et al., 1989). Neural induction is also associated with a shift of *protein kinase C* from the cytoplasm to the plasma membrane of *Xenopus* ectoderm, and pharmacological activation of

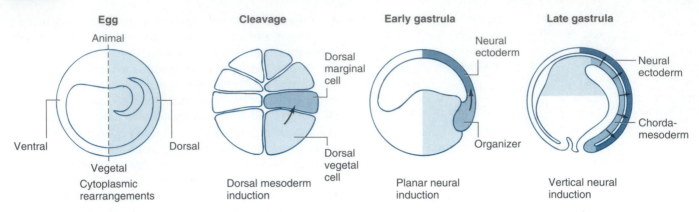

Figure 11.25 Neural induction as a multistep process. See text.

protein kinase C is followed by neural development in competent ectoderm (Otte et al., 1988). Moreover, the intracellular pH increases by approximately 0.1 units during neural induction in explants of *Xenopus* dorsal marginal zone (Sater et al., 1994).

Other researchers have focused on the identification of genes that appear to be critically involved in neural induction. The strategy of identifying critical genes has been used most successfully in the fruit fly *Drosophila melanogaster* and is now being applied to other organisms as well (see Chapters 14 and 21 through 24). In the context of neural induction in amphibians, it is desirable to identify genes that are expressed specifically in the organizer region during gastrulation. One might expect to find several genes with this expression pattern, since the organizer affects the development of both mesoderm and ectoderm and has different qualities along the anteroposterior axis.

The first organizer-specific gene in *Xenopus* was named *goosecoid*⁺ because it shares nucleotide sequences with two *Drosophila* genes, *gooseberry*⁺ and *bicoid*⁺ (Blumberg et al., 1991). At the early gastrula stage, when a

dorsal blastopore lip has formed, the *goosecoid*⁺ gene is transcribed in a patch of marginal cells occupying about a 60° arc centered over the dorsal blastopore (Fig. 11.26a). Specifically, goosecoid mRNA accumulates between the *mid-blastula transition* and the onset of neurulation in the deep layer of the involuting marginal zone, precisely where and when Spemann's organizer is thought to be active in mesoderm dorsalization and neural induction.

▼

To test whether goosecoid mRNA is a reliable marker for organizer activity, Ken Cho and his colleagues (1991) monitored the accumulation of goosecoid mRNA in embryos that had been experimentally dorsalized or ventralized. As expected, goosecoid mRNA circled the entire blastopore lip in embryos that had been dorsalized by lithium treatment during cleavage (Fig. 11.26b; see Chapter 9 for an explanation of the lithium treatment). Conversely, goosecoid mRNA synthesis was suppressed in embryos that had been ventralized by UV irradiation during the first cell cycle (Fig. 11.26c). To learn whether the goosecoid gene product by itself can cause axis formation, the researchers microinjected goosecoid mRNA into the two ventral blastomeres at the 4-cell stage. Some of the treated embryos formed two dorsal blastopore lips and two embryonic axes, although this effect was highly variable. To be sure that the secondary embryo resulted from the biological activity of the injected goosecoid mRNA and not from some side effect of the microinjection procedure, the investigators did control experiments in which they replaced active goosecoid mRNA with an inactive form, from which a critical section of the mRNA had been deleted. Neither this "mutated" goosecoid mRNA nor other mRNAs caused the formation of a secondary embryonic axis. The researchers concluded that the *goosecoid*⁺ gene plays a central role in establishing the properties of Spemann's organizer.

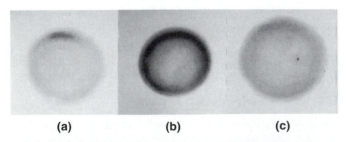

Figure 11.26 Transcription of the *goosecoid*⁺ gene in *Xenopus* early gastrulae. The goosecoid⁺ mRNA is detected by a specific hybridization probe that causes dark staining (see Methods 15.1). All photographs show ventral views. **(a)** Normal embryo. The stained region encompasses a small arc of involuting marginal zone centered dorsally above the blastopore. **(b)** Radially symmetrical accumulation of goosecoid⁺ mRNA in a gastrula dorsalized by lithium treatment during cleavage. **(c)** In a gastrula ventralized by UV irradiation before the first cleavage, no goosecoid⁺ mRNA is synthesized.

William C. Smith and Richard M. Harland (1992) identified another organizer-specific gene product on the basis of its ability to restore normal development to *Xenopus* embryos ventralized by UV irradiation. They named the gene *noggin+*, because injection of its mRNA transcript in high doses causes excessive head development. A small amount of maternal noggin mRNA is present in the oocyte and during cleavage but is not localized. Zygotic transcripts, in contrast, are more abundant and are localized to the organizer region in normal embryos during gastrulation. As with goosecoid mRNA (Fig. 11.26), dorsalized embryos accumulate an entire ring of noggin mRNA around the marginal zone, whereas ventralized embryos lack noggin mRNA.

▼

The noggin protein has the properties of a secreted protein, possibly a signal protein. To test whether this protein mimics the dorsalizing effect of Spemann's organizer on embryonic mesoderm, W. C. Smith and colleagues (1993) isolated ventral marginal zones from gastrulae and kept them in a culture medium containing noggin protein. As an indicator for dorsalization, they chose the synthesis of muscle actin mRNA. This transcript is made abundantly in skeletal muscle, which develops from dorsolateral mesoderm, but not in blood and mesenchyme, which develop from ventral mesoderm. Indeed, the ventral marginal zone accumulated muscle actin mRNA if cultured with noggin protein. The cultured tissue also elongated, another characteristic of dorsal and dorsolateral mesoderm. Neither of these responses was observed in control experiments,

in which ventral marginal zone was cultured with activin (one of the mesoderm inducers discussed in Chapter 9) or just with medium. In another experiment, the investigators injected ventralized embryos with an engineered *noggin* gene designed to be transcribed during gastrulation. Although the engineered gene was transcribed abundantly, the rescue of the ventralized embryos was limited to the formation of posterior and dorsolateral structures. These results were in accord with the hypothesis that noggin protein, like Spemann's organizer, induces the formation of dorsolateral and lateral mesoderm but not dorsal mesoderm itself.

To test the ability of noggin protein to mimic the neural induction effect of Spemann's organizer, Theresa Lamb and her colleagues (1993) kept animal caps of *Xenopus* blastulae in medium with or without noggin protein. They found that noggin protein induced the animal caps to synthesize two proteins that are characteristic of neural tissue, the cell adhesion molecule N-CAM, and the intermediate filament protein XIF3.

Additional genes expressed in the dorsal blastopore lip of *Xenopus* gastrulae include *Xlim-1+* (Taira et al., 1992) and *XGKH1+* (Dirksen and Jamrich, 1992). That such a multitude of genes are active in the organizer region indicates that the genetic control of the organizer function may be complex. The analysis may be further complicated if certain signal pathways must be inhibited for neural induction to occur. Thus, unraveling the molecular mechanisms of Spemann's organizer may still be a distant goal, but several promising lines of analysis have begun.

SUMMARY

After gastrulation, germ layers interact with one another to form the organ rudiments. During this period of organogenesis, a basic body plan emerges that is characteristic of the phylogenetic group to which an organism belongs. The shaping of the organ rudiments involves extensive morphogenetic movements. The process is particularly well studied in neurulation, that is, the formation of the central nervous system in vertebrates. During neurulation, a dorsal layer of ectodermal cells forms the neural plate, which closes into the neural tube and eventually generates the brain and spinal cord. Neurulation has been analyzed in terms of specific cell behaviors including coordinated shape changes, convergent extension, and cell crawling, which cooperate as synergistic mechanisms.

Neurulation and other events in organogenesis involve inductive interactions. The inductive events that promote neurulation are known as neural induction, although they entail the organization of the entire embryonic axis—not only neurulation, but also the organization of mesoderm into dorsal, lateral, and ventral components. The neural inducer was identified by Spemann and Mangold (1924), who transplanted dorsal blastopore lip to the ventral ectoderm of newt embryos during the early gastrula stage and found that a secondary embryonic axis developed. Grafting of chordamesoderm, which arises from the dorsal blastopore lip during gastrulation, had the same effect. The grafted material (called the "organizer") developed in accord with its own fate and marshaled the surrounding host tissue to form a virtually complete and normally proportioned embryonic axis.

Modern work with *Xenopus* embryos has shown that neural induction is a multistep process. It begins with the rearrangement of cytoplasm in the fertilized egg and dorsal mesoderm induction in the blastula, which bias the prospective dorsal ectoderm for neural induction. During the gastrula stage, neural induction depends on both planar and vertical signals. The planar signals are sent from the dorsal blastopore lip within the outer plane of the embryo directly to the adjacent dorsal ectoderm. The vertical signals are sent in the late gastrula from the same inducer, which has then formed the chordamesoderm, to the over-lying dorsal ectoderm. Presumably, the planar signals enhance the competence of the dorsal ectoderm to respond to the vertical signals received later. Several genes that play key roles in neural induction are now being cloned and characterized.

SUGGESTED READINGS

Gordon, R. 1985. A review of the theories of vertebrate neurulation and their relationship to the mechanics of neural tube birth defects. In J. Slack, ed., *J. Embryol. Exp. Morph.* Vol. 89 Supplement: *Early Amphibian Development,* 229–255.

Hamburger, V. 1988. *The Heritage of Experimental Embryology: Hans Spemann and the Organizer.* New York: Oxford University Press.

Slack, J. M. W., and D. Tannahill. 1992. Mechanism of anteroposterior axis specification in vertebrates. *Development* **114:**285–302.

ECTODERMAL

ORGANS

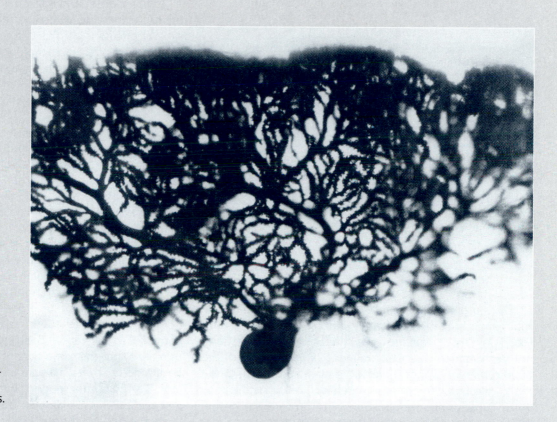

Figure 12.1 Purkinje cell from the cerebellar cortex. The numerous dendritic branches provide an extensive area for thousands of synapses with other neurons.

During organogenesis, the embryo has formed organ rudiments, and the basic body pattern has been established on a small scale. However, the organ rudiments cannot yet function: the eye rudiment cannot see, and the brain rudiment cannot serve as the main coordinating system of the body. To perform these tasks, cells must first acquire functional specialization. Photoreceptors in the retina must assemble certain pigments so that absorption of light triggers the release of neurotransmitters. Neurons in the nervous system must send out different types of processes to form connections with one another (Fig. 12.1). This process of functional specialization is associated with further growth and occupies a relatively long time. In humans, cleavage takes only about 2 weeks, gastrulation another week, and organogenesis 4 more weeks, whereas the subsequent growth and functional specialization occupy the last 7 months of gestation.

The functional specialization of embryonic organs has been studied at different levels. Classical embryologists, using mostly microscopy to monitor the maturation of organ rudiments into functional organs, have coined the term *histogenesis* (Gk. *histos*, "web," or "tissue"; *genesis*, "origin"). This term places the emphasis on tissues, a level of organization between cells and organs. A *tissue* (Lat. *texere*, "to weave") consists of cells and extracellular materials that perform a particular set of functions. Typically, a tissue contains only one or a few different cell types. For instance, *nervous tissue* consists mostly of *neurons* and *glial cells* plus extracellular material, functioning together in transmitting electrical signals. Modern developmental biologists have added the analysis of functional specialization at the cellular and molecular levels, with emphasis on differential gene expression (see Chapter 19).

In this chapter and the one following, we will examine the histogenesis of several vertebrate organs. In doing so, we will discuss examples of the four major tissue types, namely *nervous*, *epithelial*, *muscular*, and *connective*. Because the discussion of organogenesis in Chapter 11 was limited mostly to the central nervous system, the early development of other organs will be included in Chapters 12 and 13 as needed.

The organization of Chapters 12 and 13 will be based on the three germ layers and their major subdivisions. Figure 12.2 is a flowchart relating these embryonic structures to adult tissues and organs. It reflects the important principle that the basic body plan develops through a *hierarchy of determinative events*, although the chart does not show the localizations and inductions involved in generating tissue diversity. The chart is also fairly general: it can be applied to all vertebrates. It is worth noting at the outset that so-called ectodermal and endodermal organs are not formed solely from the corresponding germ layers. In the majority of these organs, only the epithelial portions derive from ectoderm or endoderm, and the mesoderm makes substantial contributions. It is also important to realize that the same type of tissue may arise from different germ layers. Cartilage and muscle arise, for the most part, from mesoderm, but certain cranial cartilages and eye muscles are derived from ectoderm.

In the present chapter, we will review the major derivatives of the ectodermal germ layer, namely the *neural tube*, the *neural crest*, the *ectodermal placodes*, and the *epidermis*. Tracing the development of the neural tube will take some effort because of the sheer complexity of the central nervous system. The origin of the neural crest cells will be considered in some detail because their extensive migrations and pluripotentiality have made them favorite objects for studies on cell adhesion and cell determination.

Neural Tube

The neural tube gives rise to the *central nervous system (CNS)*, a development that is impressive by its sheer numbers. From neural tube closure in the embryo until after birth, the human organism forms an average of

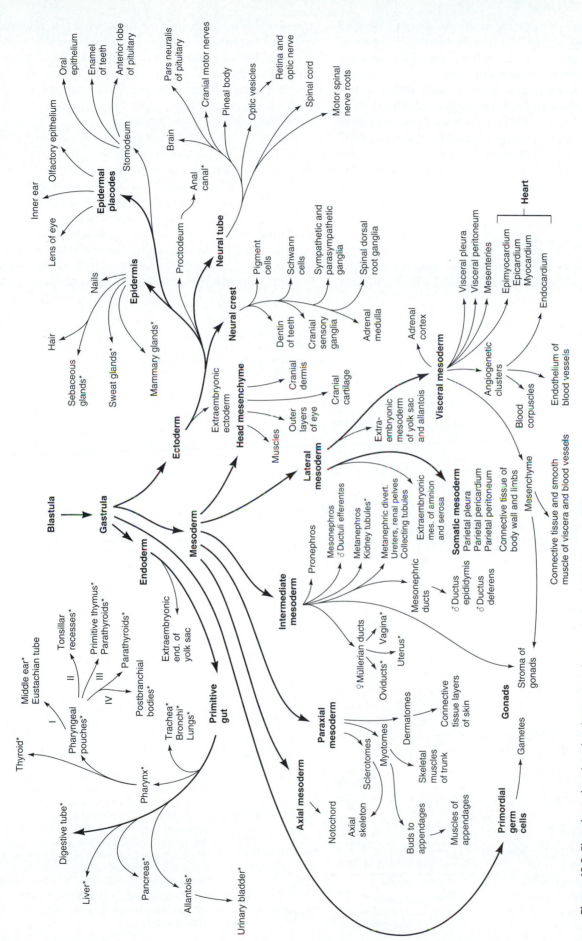

Figure 12.2 Flowchart showing the three germ layers, their subdivisions, and the organs or tissues derived from them. An asterisk indicates that only the epithelial portion of the marked organ is derived as indicated and that the organ has additional investments of mesodermal origin.

250,000 neurons per minute until the central nervous system contains about 100 billion neurons. The morphological complexity of the CNS is equally imposing. However, following its development from the simple organization of the neural tube actually helps us understand and memorize the more intricate structures of later stages (Langman, 1981). First we will examine how the posterior neural tube develops into the *spinal cord.* Then we will see how the anterior neural tube—the future *brain*—follows a similar development before it becomes more complicated.

Nervous Tissue Consists of Neurons and Glial Cells

The wall of the newly closed neural tube consists of a layer of ***neuroepithelial cells*** (Fig. 12.3). On the outside, the neural tube is covered by an ***external limiting membrane*** of extracellular material, which corresponds to the *basal lamina* generally found on the basal surface of an epithelium (see Fig. 2.20). The apical face of the neuroepithelium is directed toward the inner space, or lumen, of the neural tube. Here the cells form a seal of tight junctions as in all epithelia.

Once the neural tube has closed, the neuroepithelial cells begin to behave as *stem cells,* that is, cells that can divide asymmetrically, with one daughter a new stem cell and the other daughter a *committed progenitor cell* that will divide and produce certain types of differentiated cells (Fig. 12.4; see also Chapter 19). Neuroepithelial stem cells produce neuroblasts and glioblasts as committed progenitor cells. *Neuroblasts* are the precursor cells of *neurons* (nerve cells). They migrate toward the external limiting membrane, where they

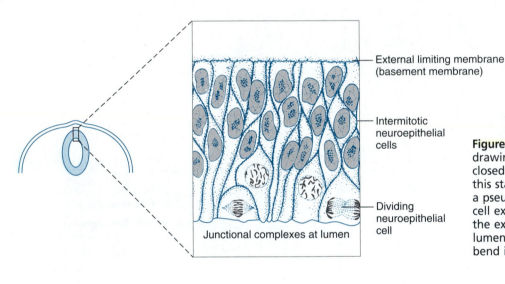

External limiting membrane
(basement membrane)

Intermitotic
neuroepithelial
cells

Dividing
neuroepithelial
cell

Junctional complexes at lumen

Figure 12.3 Neuroepithelium. The drawing shows a portion of the recently closed neural tube in cross section. At this stage, the neuroepithelial cells form a pseudostratified epithelium, with each cell extending the full width between the external limiting membrane and the lumen of the neural tube. (Some cells bend into or out of the plane of section.)

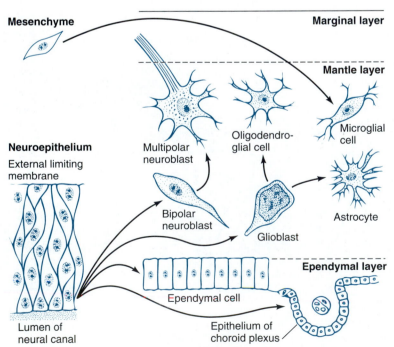

Mesenchyme

Marginal layer

Mantle layer

Oligodendro-
glial cell

Microglial
cell

Neuroepithelium

Multipolar
neuroblast

External limiting
membrane

Bipolar
neuroblast

Astrocyte

Glioblast

Ependymal layer

Ependymal cell

Lumen of
neural canal

Epithelium of
choroid plexus

Figure 12.4 Development of the neuroepithelium. Its major derivatives are neuroblasts, which develop into neurons; and glioblasts, which give rise to two types of glial cells: oligodendrocytes and astrocytes. Oligodendrocytes also derive from neural crest cells. Another type of glial cell, called microglia, is derived from mesenchymal cells. The remaining neuroepithelial cells later form the ependymal cells lining the lumen of the spinal cord.

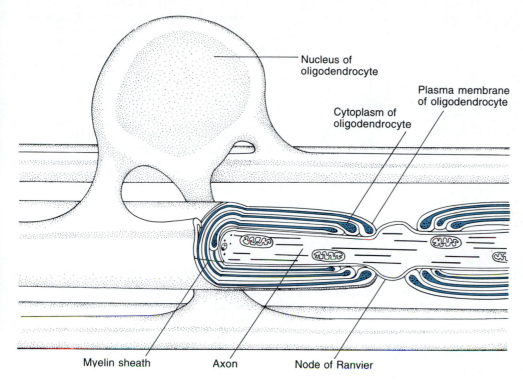

Nucleus of
oligodendrocyte

Cytoplasm of
oligodendrocyte

Plasma membrane
of oligodendrocyte

Myelin sheath   Axon   Node of Ranvier

Figure 12.5 Origin of the myelin sheath surrounding vertebrate axons. Oligodendrocytes in the central nervous system wrap around the axons of neurons, thus surrounding the axon with multiple layers of oligodendrocyte plasma membrane. The compacted layers are called a myelin sheath. Each oligodendrocyte covers only a segment of an axon, leaving uncovered small intervening sections of axon, known as nodes of Ranvier. Similar myelin sheaths are formed by Schwann cells in the peripheral nervous system.

form a progressively thicker layer of nondividing cells, the *mantle layer.* A neuroblast matures by sprouting first two, then several processes called *neurites.* The longest neurite is usually the *axon,* which transmits signals away from the body of the neuron to muscle cells or other neurons. *Dendrites* are usually shorter and transmit signals toward the body of the neuron.

After giving off large numbers of neuroblasts, the neuroepithelial cells produce a second type of committed progenitor cell, the *glioblast.* Glioblasts give rise to *glial cells,* which have supporting, insulating, and nutritive functions. There are about 4 times more glial cells than neurons in the CNS. The two main types of glial cells derived from glioblasts are oligodendrocytes and astrocytes, although oligodendrocytes also derive from neural crest cells. *Oligodendrocytes* wrap around the axons of neurons repeatedly to create an insulating layer of multiple plasma membranes called the *myelin sheath* (Figs. 12.5 and 12.6). *Astrocytes* are associated with the *endothelial cells* of blood capillaries and form the *blood-brain barrier,* which restricts the passage of large molecules from the blood to the central nervous system. Additional glial cells, called *microglia,* are derived from mesenchymal cells. Tissues containing neurons and glial cells are classified as *nervous tissue.*

As the developing neurons send their axons toward the outside, they create another layer, called the *marginal layer,* around the mantle layer (Fig. 12.7). Because of the myelination of the axons, the marginal layer takes on a whitish appearance and is called the *white matter* of the spinal cord. This is in contrast to the *gray matter,* which derives from the mantle layer and contains the

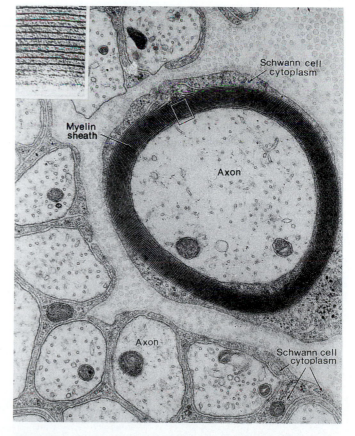

Schwann cell cytoplasm

Myelin sheath

Axon

Axon

Schwann cell cytoplasm

Figure 12.6 Electron micrograph of a small cross section of rat sciatic nerve. A nerve contains many axons. Most are embedded in the cytoplasm of certain glial cells called Schwann cells. Long axons are often surrounded by a myelin sheath, which consists of multiple layers of Schwann cell plasma membranes (inset).

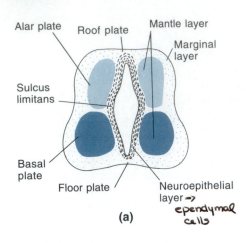

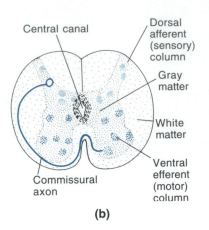

Figure 12.7 Development of the spinal cord in humans. **(a)** At 6 weeks. The neuroepithelium is surrounded by neuroblasts forming the mantle layer, later called the gray matter, which will contain the cell bodies of neurons. The mantle layer is divided, from dorsal to ventral, into roof plate, alar plate, basal plate, and floor plate. Outside the mantle layer is the marginal layer, which consists of sprouting neurites. This layer is later called the white matter; it will contain the myelinated axons of neurons. **(b)** At 9 weeks. Ventral efferent columns arise from basal plates and dorsal afferent columns from alar plates.

cell bodies of neurons as well as unmyelinated axons and dendrites.

As more neuroblasts and glioblasts are added to the mantle layer, they form dorsal and ventral ridges of gray matter along each side of the neuroepithelium. The dorsal ridges are called *alar plates,* later referred to as *dorsal columns* or *afferent columns* (Lat. *afferre,* "to carry toward"). They transmit sensory impulses arriving from the skin, muscles, tendons, and internal organs. The ventral ridges of gray matter are called *basal plates,* later referred to as *ventral columns* or *efferent columns* because they transmit signals to muscles and glands (Lat. *efferre,* "to carry away"). The most dorsal portion of the neuroepithelium between the two alar plates is called the *roof plate.* Correspondingly, the most ventral portion of the neuroepithelium is known as the *floor plate.* One function of the floor plate is to promote and orient the growth of *commissural axons* that originate from neurons in a dorsal column and end on a neuron in the ventral column of the opposite side (Fig. 12.7).

The spinal cord is characterized by a distinct dorsoventral pattern comprising the roof plate, dorsal column, ventral column, and floor plate, each having unique cell types and function. How does this pattern originate? We have already seen that the *chordamesoderm* induces the formation of the *neural plate* (see Chapter 11). Once the neural tube has formed, it stays in close contact with the underlying notochord. Does the notochord continue to send inductive signals and thus establish the dorsoventral pattern of the neural tube?

To explore the inductive effects of the notochord on the neural tube, Marysia Placzek and her colleagues (1990) grafted notochord from chicken embryos to the side of closing neural tubes at a corresponding stage (Fig. 12.8). In response, the adjacent part of the neural tube formed

a wedge-shaped area resembling a floor plate. Both the primary and the secondary floor plate were flanked by efferent columns sending out bundles of axons. The extra floor plate also promoted and guided the outgrowth of commissural fibers in vivo and in vitro. Conversely, removal of the notochord from the neural tube left the ventral neural tube cells unable to guide commissural fibers and resulted in failure to develop efferent columns. Thus, the presence of notochord is both sufficient and necessary for the development of floor plate and efferent columns.

In an extension of this study, Toshiya Yamada and colleagues (1991) used immunostaining with antibodies against antigens in the floor plate, or antigens expressed in both floor plate and efferent columns, or antigens expressed in a wide intermediate band of neural tube cells (shaded in Fig. 12.8). The immunostaining patterns shifted after notochord removal or transplantation, and the shifts corresponded to the morphological changes observed in the neural tube. In particular, the vicinity of a transplanted notochord repressed the synthesis of certain gene-regulatory proteins and intercellular signals that are normally found in the dorsal portion of the notochord (Fig. 12.9; Goulding et al., 1993; Basler et al., 1993).

The availability of molecular markers that normally foretell the development of floor plate and motor columns ventrally, or the formation of neural crest cells dorsally, has facilitated the in vitro analysis of the inductive signals that are required for floor plate and motor column development. In particular, researchers tested whether the inductive effect of the notochord required physical contact with the responding tissue (Placzek et al., 1993; T. Yamada et al., 1993). They found that neural tube cells needed direct contact with either notochord or floor plate to form more floor plate. In contrast, motor neurons developed without physical contact in media that had merely been conditioned by the presence of notochord or floor plate.

Corresponding observations were made in the zebra fish *Brachydanio rerio,* in which the availability of mu-

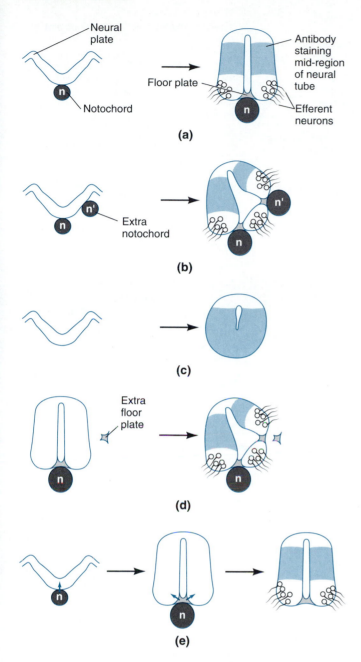

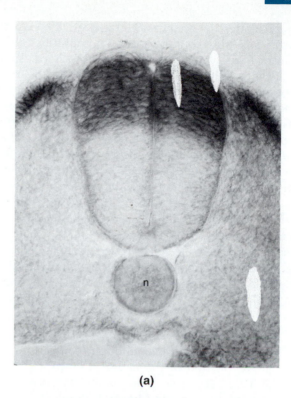

(a)

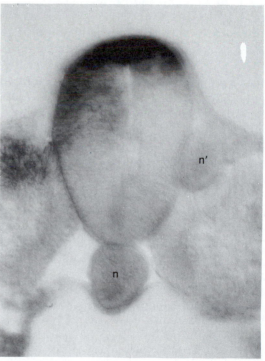

(b)

Figure 12.8 Establishment of the dorsoventral pattern in the spinal cord of the chicken embryo. **(a)** In normal development, the floor plate develops directly above the notochord. Subsequently, columns of efferent neurons develop in the basal plates on both sides of the floor plate. Immunostaining reveals antigens specific to a wide intermediate band within the neural tube (color). **(b)** Grafting an extra notochord alongside the closing neural plate results in the formation of another floor plate flanked by efferent neurons. The immunostainable antigen disappears near the grafted notochord. **(c)** Removing the notochord results in the absence of both floor plate and efferent neurons, and the immunostainable antigen appears in the entire ventral region. **(d)** Grafting a floor plate alongside the closed neural tube has effects much like the operation shown in part b. **(e)** The simplest hypothesis is that the notochord induces adjacent neural tube to form floor plate, and that the prospective floor plate in turn induces the neighboring neural tube region to form more floor plate and efferent neurons.

Figure 12.9 Inhibitory effect of the notochord on the expression of a regulatory gene (Pax-3^+) in the chicken embryo. The photos show cross sections of chicken embryos treated by in situ hybridization (see Methods 15.1) to make Pax-3 mRNA visible as a dark stain. **(a)** Normal embryo. Pax-3 mRNA accumulates in the dorsal part of the neural tube, opposite the notochord (n). **(b)** Embryo that received an extra notochord (n′) transplanted laterally under the open neural plate. A few hours later, when the neural plate had closed to form a tube, the extra notochord inhibited the expression of the Pax-3^+ gene in its vicinity.

tants facilitates the analysis (Hatta et al., 1991). In the *cyclops (cyc-1(b16))* mutant, ventral neural tube cells do not form floor plate in response to notochord. To test whether this failure results from a defect in the inducing tissue (notochord or floor plate) or the responding tissue (neural plate or neural tube), the investigators transplanted labeled precursor cells for each tissue between wild-type and mutant embryos. This type of *mosaic analysis* showed that the *cyclops* mutation interferes with the response of neural cells to an inductive signal from notochord, but not with the response of neural cells to an inductive signal from floor plate: wild-type ectoderm differentiated into floor plate in *cyclops* hosts, whereas wild-type notochord did not induce floor plate formation in *cyclops* hosts. However, transplanted wild-type floor plate cells induced *cyclops* neural cells to make more floor plate and adjacent efferent columns.

These experiments indicate that notochord induces the adjacent portion of the neural tube to form floor plate, and that the first floor plate cells induce their neighbors to generate additional floor plate cells. Moreover, both notochord and floor plate seem to release diffusible signals that induce neural tube cells to form efferent columns and may also cause further refinement of the dorsoventral pattern of the neural tube. One of the signals that induce floor plate formation is encoded by the *hedgehog⁺* gene, which is also involved in patterning vertebrate limbs and *Drosophila* segments (Roelink et al., 1994; J. C. Smith, 1994; see also Chapters 21 and 22).

As the spinal cord develops, both white and gray matter increase in thickness, and the gray matter assumes a butterfly shape in cross section. Mitotic activity in the neuroepithelium decreases. The neuroepithelial cells remaining in the adult spinal cord are called *ependymal cells;* they surround the lumen of the spinal cord, now called the **central canal.** On the outside of the spinal cord, mesenchymal cells form three covering epithelia, the *meninges.*

In summary, the mature spinal cord consists, from the inside out, of the central canal, the surrounding ependymal cells, the gray matter, the white matter, and meninges. The gray matter contains dorsal columns transmitting afferent signals and ventral columns transmitting efferent signals.

The Basic Organization of the Spinal Cord Is Modified in the Brain

The brain develops from the cranial part of the neural tube, and the initial steps in the development of brain and spinal cord are quite similar. However, the brain becomes more complex as the central canal dilates and

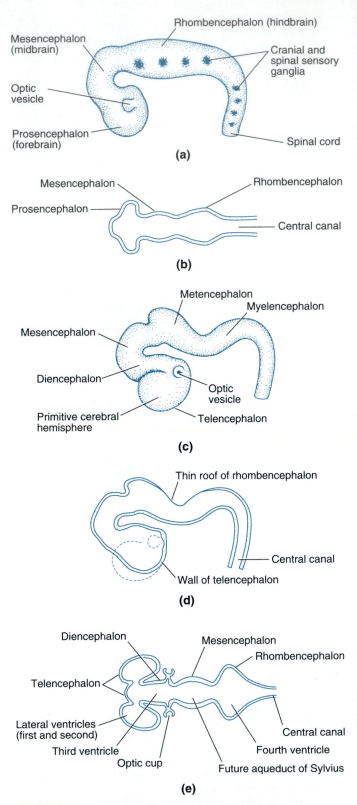

Figure 12.10 Brain development in the human embryo. **(a)** At 4 weeks, lateral view. The cranial sensory and spinal sensory (dorsal root) ganglia are derived from the neural crest (see also Fig. 12.25). **(b)** At 4 weeks, stretched-out brain in frontal section to show dilated portions of the central canal. **(c)** At 5 weeks, lateral view. **(d)** At 5 weeks, median section passing between the two cerebral hemispheres. **(e)** Stretched-out brain in frontal section to show the brain vesicles.

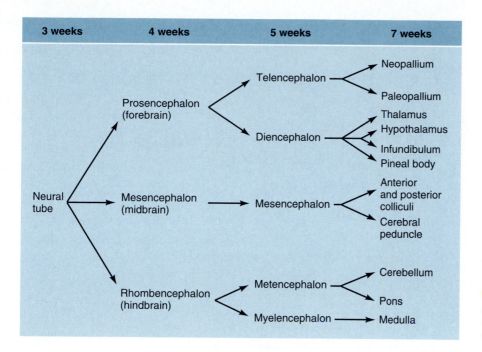

Figure 12.11 Flowchart of the subdivisions of the human brain at subsequent stages of development.

forms fluid-filled spaces, called *ventricles*, which are surrounded by *brain regions* that develop regionally different structures and functions. Also, the relatively simple organization of the spinal cord into a mantle layer (gray matter) on the inside and a marginal layer (white matter) on the outside becomes more refined in the brain. Some clusters of mantle cells move out into the marginal layer and form islands of gray matter called *nuclei* that assume distinct functions. Other mantle cells move all the way to the periphery so that gray matter comes to surround white matter as the multilayered *brain cortex.*

The development of the major adult brain regions is outlined in Figures 12.10 and 12.11. In a 4-week human embryo, three bulges appear in the cephalic end of the neural tube: *prosencephalon, mesencephalon,* and *rhombencephalon.* A week later the prosencephalon has divided into the telencephalon and diencephalon. Anteriormost, the *telencephalon* consists of a median portion and two lateral outpocketings, the future hemispheres of the *cerebrum* (Lat., "brain"). The cerebral hemispheres surround a pair of ventricles (first and second). The adjacent *diencephalon* surrounds the third ventricle and is characterized by the outgrowth of the two *optic vesicles.* The mesencephalon contains a fluid-filled lumen called the *aqueduct of Sylvius,* which connects the third ventricle with the more posteriorly located fourth ventricle. The rhombencephalon subdivides into an anterior and dorsal portion, the *metencephalon,* and a ventral part, the *myelencephalon.* The latter surrounds the fourth ventricle, which extends into the central canal of the spinal cord.

In the 7-week human embryo, the subdivisions of the brain have developed further. We will briefly characterize the brain regions at this stage, beginning caudally.

Medulla. The myelencephalon becomes the *medulla oblongata* (Lat., *medulla,* "innermost part," such as the spinal cord inside the spine; *oblongata,* "extended"), a fairly descriptive term, since this part of the brain retains the greatest similarity to the spinal cord (Fig. 12.12). The medulla controls the reflexes of the neck, throat, and tongue, just as the spinal cord mediates the reflexes of the trunk and appendages. However, the medulla differs from the spinal cord in that its lateral walls are arranged as if they had been rotated around an imaginary axis in the floor plate, a movement similar to opening a book. In the process, the neural canal expands into the fourth ventricle. As in the spinal cord, the basal plates of the medulla oblongata contain efferent neurons, and the alar plates contain afferent interneurons. The roof plate of the fourth ventricle consists of a layer of ependymal cells, which are in direct contact with the meninges. These membranes, along with the blood vessels contained in the meninges, form a *choroid plexus,* which releases a blood filtrate, *cerebrospinal fluid,* into the fourth ventricle. Similar plexuses are also formed in the other ventricles.

Cerebellum. The metencephalon gives rise to the cerebellum and the pons. The *cerebellum* (Lat., "little brain") develops dorsally; it received its name because its cortex is folded and consists of gray matter like that of the cerebrum. The cerebellum arises from extensions of the alar plates that form the *cerebellar plate* anterior to the fourth ventricle (Fig. 12.13). Initially, the cerebellar plate consists of a neuroepithelium, a mantle layer, and

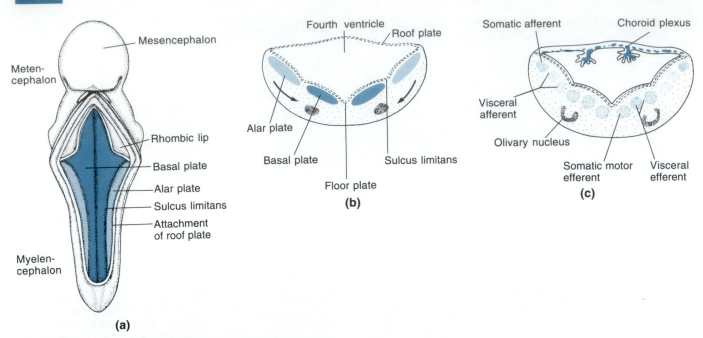

Figure 12.12 Development of the medulla oblongata. **(a)** Dorsal view of the posterior brain in the 6-week-old human embryo. The roof of the fourth ventricle has been cut away to expose the alar plates and basal plates, which are separated by the sulcus limitans. **(b, c)** Transverse sections showing the development of the choroid plexus of the fourth ventricle. Compare the position of the alar and basal plates here and in the spinal cord as shown in Figure 12.7. The olivary nucleus moves from its original position in the alar plate into the ventral white matter, from where it will project sensory input to the cerebellum.

a marginal layer (Fig. 12.14). Cells formed by the neuroepithelium migrate into the marginal layer to form the *external granular layer.* Cells given off later by the neuroepithelium include the *Purkinje cells,* which form one axon plus an enormous number of dendrites (Fig. 12.1). A Purkinje neuron may form thousands of synapses (connections) with other neurons (Purves and Lichtman, 1985). Cells from the granular layer and Purkinje cells eventually form the cortex of the cerebellum, which functions as a coordination center for posture and movement.

Pons. Opposite the cerebellum, on the ventral side of the metencephalon, the *pons* is formed. It serves as a pathway for nerve fibers between the spinal cord and the cerebellum as well as the cerebral hemispheres.

Mesencephalon. The mesencephalon is similar in its morphology to the spinal cord (Fig. 12.15). The marginal layer enlarges ventrally to accommodate nerve fibers connecting the cerebral cortex with the pons and spinal cord. The alar plates in the dorsal mesencephalon initially form two ridges, which become subdivided by

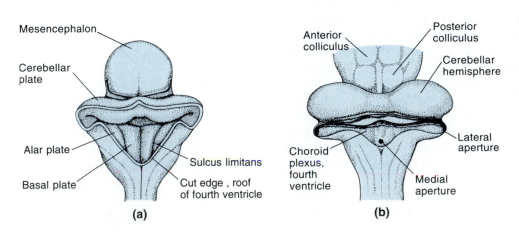

Figure 12.13 Dorsal views of the metencephalon and mesencephalon. **(a)** 8-week-old human embryo. The roof of the fourth ventricle has been cut away to expose the floor of the ventricle. **(b)** 4-month-old human fetus. The choroid plexus covering the fourth ventricle has acquired three apertures through which cerebrospinal fluid drains into the space between the meninges. The cerebellum develops from the cerebellar plate. The anterior and posterior colliculi have formed in the mesencephalon.

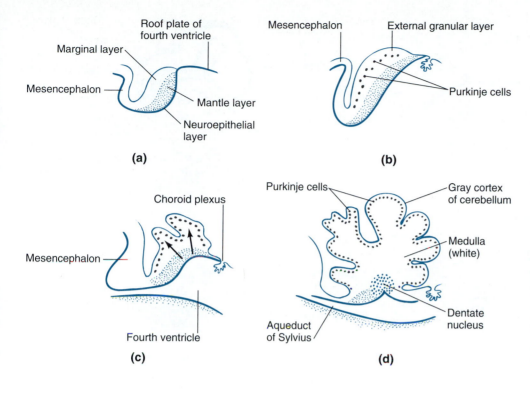

Figure 12.14 Development of the human cerebellum shown in median section at successive stages: **(a)** 8 weeks; **(b)** 12 weeks; **(c)** 13 weeks; **(d)** 15 weeks. The cells of the external granular layer and Purkinje cells are given off successively by the neuroepithelium. Both groups of cells move to the surface of the marginal layer, where they form the cerebellar cortex. The dentate nucleus is one of the deep cerebellar nuclei.

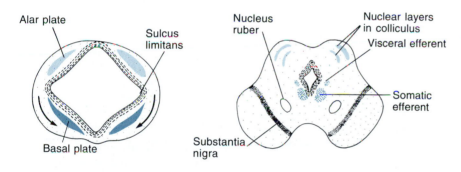

Figure 12.15 Development of the human mesencephalon, shown in transverse section at two stages. The arrows indicate the path followed by alar plate cells to form the nucleus ruber and the substantia nigra in the ventral marginal layer. This movement is similar to that of the cells that form the olivary nucleus in the medulla, as shown in Figure 12.12. Other groups of alar plate cells form stratified nuclear layers in the colliculi.

transverse grooves into four elevations, or *colliculi* (Lat. *colliculus,* "little hill"). In mammals, the colliculi are relatively small relay stations for visual and auditory reflexes, known as anterior colliculi and posterior colliculi, respectively (Fig. 12.13). In nonmammalian vertebrates, the anterior colliculi are called the *optic lobes;* they are much larger and contain the visual projection areas that in mammals are located in the cerebral hemispheres.

Diencephalon. The diencephalon seems to consist of two alar plates, delimiting a narrow third ventricle, while lacking basal plates (Fig. 12.16). The dorsal roof of the third ventricle is again formed by a choroid plexus. Located caudally to the plexus is the *pineal gland,* which through its hormonal secretions generates the circadian rhythm and, in some vertebrates, modulates the yearly reproductive cycle. A lateral groove on the inside of each alar plate separates the dorsal *thalamus* from the ventral *hypothalamus.* The thalamus has been

dubbed the "antechamber of the cerebrum," because it serves as a gateway for sensory fibers passing from the spinal cord and the brain stem to the cerebral hemispheres. In the hypothalamus, several *nuclei* act as regulatory centers for visceral functions, including sleep, digestion, and body temperature, as well as aggression and other emotional behaviors. A prominent landmark at the bottom of the hypothalamus is the *optic chiasma* (Gk. *khiasma,* "cross"), an incomplete crossing of the optic nerves. Right behind the optic chiasma there is a median extension of the hypothalamus, the *infundibulum* (Lat., "funnel"), which forms the posterior lobe of the *pituitary gland,* or *hypophysis.* The anterior lobe of the pituitary gland originates from *Rathke's pouch,* an invagination from the *stomodeum,* the ectodermal epithelium lining the anterior oral cavity (Fig. 12.17).

Telencephalon. The telencephalon, the anterior-most part of the brain, consists mainly of the two cerebral hemispheres, each surrounding a ventricle with a choroid

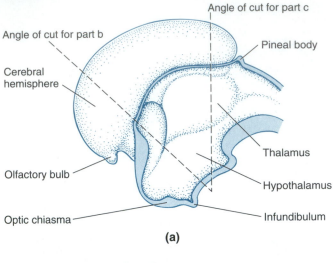

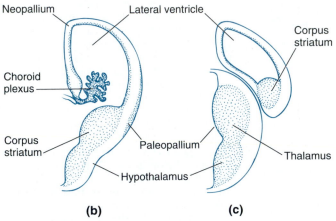

Figure 12.16 Diencephalon and telencephalon of the 8-week-old human embryo. **(a)** Medial view of the right half, from which the left half has been cut away. **(b, c)** Transverse sections at different angles as indicated in part a.

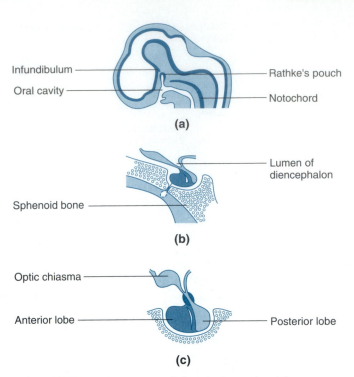

Figure 12.17 Development of the pituitary gland from Rathke's pouch, a median invagination from the stomodeum; and the infundibulum, an extension of the floor of the diencephalon.

plexus (Fig. 12.16). Each of these ventricles opens into the third ventricle through large interventricular openings. To an extent, the development of the cerebral hemispheres reflects their phylogeny. A region that differentiates early in mammals and is relatively large in more primitive vertebrates is known as the *paleopallium* (Gk. *palaios*, "old"; Lat. *pallium*, "mantle"). It is located anteriorly and laterally, extends into the *olfactory bulb*, and is associated with the sense of smell. The phylogenetically newest part of the cerebrum, the *neopallium*, develops relatively late but then grows at a rapid rate. In humans, the neopallium eventually occupies about 90% of the cerebral hemispheres and, together with the cerebellum, covers most other parts of the brain.

In the adult, the right and left hemispheres are connected by several *commissures*, which cross the midline (Fig. 12.18). The first crossing axons to appear form the *anterior commissure*, which connects the olfactory bulb and related brain areas of one hemisphere to those of the other hemisphere. The most important commissure is the *corpus callosum* (Lat., "hard body"). It is formed

gradually after the 10th week of human development and connects the nonolfactory portions of the cerebral hemispheres as they extend over the diencephalon and mesencephalon. Corpus callosum formation begins anteriorly above the *septum pellucidum*, the thin medial layer that separates the two lateral ventricles. From there, the corpus callosum extends posteriorly, arching over the thin roof of the diencephalon.

The Peripheral Nervous System Is of Diverse Origin

The *peripheral nervous system* includes all the nervous tissue outside the central nervous system. In part, the peripheral nervous system consists of the long axons that grow out of neurons located in the central nervous system. In addition, it includes many *ganglia* (sing., *ganglion*, a group of neurons) that originate from the neural crest and the ectodermal placodes (discussed later in this chapter). Most peripheral nerve fibers are myelinated in the same way as the axons in the white matter of the central nervous system (Fig. 12.5). The myelin layers in the peripheral nervous system are built up by *Schwann cells*, which are analogous to the oligodendroglial cells in the central nervous system. A typical peripheral *nerve* contains thousands of axons and/or dendrites, often bundled in groups and surrounded by connective tissue and fat cells derived from mesenchyme.

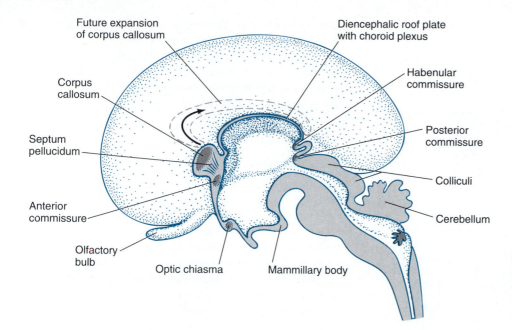

Figure 12.18 Brain of a 4-month human embryo. The brain is cut in the median plane, and the right half is shown from a medial view. The cut regions (gray shading) include several commissures, bundles of axons that connect corresponding portions on the right and left sides. The arrow and broken line indicate the future expansion of the largest commissure, the corpus callosum.

Efferent axons of the spinal cord originate from neurons in the ventral columns and leave the cord in segmental bundles called *ventral roots* (Fig. 12.19). The sensory (afferent) axons entering the spinal cord are bundled in segmental *dorsal roots.* These axons originate from neurons located in the *dorsal root ganglia,* which are derived from neural crest cells. The neurons of the dorsal root ganglia receive input via their long dendrites from sensory cells in the skin and elsewhere.

Dorsal and ventral roots unite to form *spinal nerves,* which leave the spine through canals formed between successive vertebrae. Outside the spine, each spinal nerve gives off a *dorsal ramus* (Lat. *ramus,* "branch"; pl., *rami*), supplying its respective segment of skin and muscles. Farther on, each spinal nerve sends out two *communicating rami* as connections to visceral ganglia, which derive from the neural crest. The remainder of a spinal nerve continues as a large *ventral ramus* to supply ventral regions of the body and—in the case of some spinal nerves—the appendages.

There is a segmental pattern in the adult skin innervation by the sensory components of the spinal nerves that largely parallels the segmental pattern of the spinal nerves themselves (Fig. 12.20). However, the segmenta-

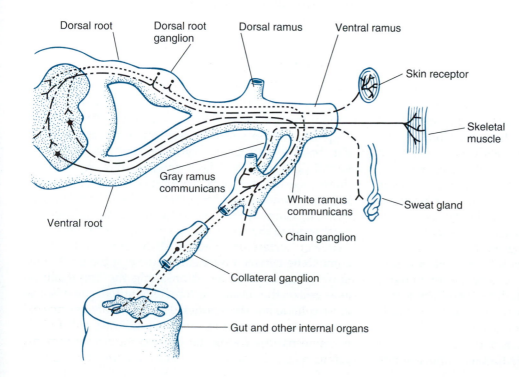

Figure 12.19 Diagram of a spinal nerve with its dorsal (sensory) and ventral (motor) roots. The chain ganglia of the sympathetic nervous system run parallel to the spinal cord outside the vertebral column. They are connected to each spinal nerve by two communicating rami. The nerve fibers are marked as somatic sensory (– – – –), somatic motor (——), visceral sensory (·····), and visceral motor (– – – –).

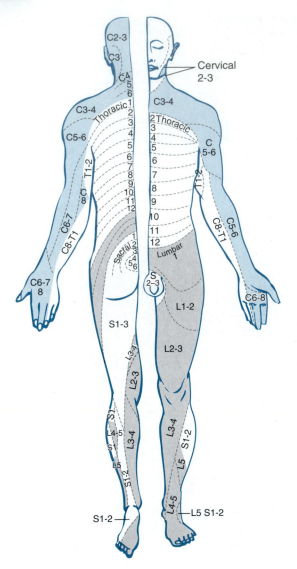

Figure 12.20 Segmental distribution of the skin areas innervated by the sensory components of the spinal nerves. C = cervical, T = thoracic, L = lumbar, S = sacral.

serve mainly—like the communicating rami given off by the spinal nerves—as connections to visceral ganglia. In contrast to the dorsal root ganglia of the spinal nerves, which originate from neural crest cells only, the sensory cranial nerves arise from both the neural crest and ectodermal placodes.

Neural Crest

The cells that originate atop the neural tube, in an area known as the **neural crest,** migrate to their different destinations before assuming their functional characteristics. The *neural crest cells* have been called the chameleons of developmental biology, because they give rise to a bewildering variety of cell types. Their derivatives range from cartilage elements of the head to pigment cells in the epidermis, different types of neurons, and hormone-producing gland cells. The extensive migrations and various fates of neural crest cells make them favorite objects for studying some basic issues in development (B. K. Hall and S. Hörstadius, 1988; Le Douarin et al., 1993; Selleck et al., 1993).

In this section, we will explore the embryonic origin of neural crest cells, their main routes of migration, and their major fates. Turning again to the question of cell determination, we will examine to what extent neural crest cells are "born" with certain restrictions on their potency and how much their fate depends on the environment in which they settle. The role of specific cell adhesion and extracellular matrix molecules in guiding and determining neural crest cells will be explored further in Chapters 25 and 26.

Neural Crest Cells Arise at the Boundary between Neural Plate and Epidermis

Neural crest cells are found only in vertebrates; they originate during the process of neurulation. The designation "neural crest cells" is indicative of their topographical position at the crests of the neural folds (see Fig. 11.6). A neural fold comprises the margins of the neural plate and the adjoining epidermis. As neurulation proceeds, the folds rise higher and bend toward the dorsal midline of the embryo. When the folds fuse, epidermis joins epidermis, and neural plate joins neural plate to form the neural tube. During this process, cells that were originally located at the crest of the folds come to lie between the neural tube and the epidermis. Depending on their position along the craniocaudal axis, and on the class of vertebrates studied, some neural crest cells leave the epithelial folds before the neural tube closes. In other cases, the neural crest cells form a longitudinal band atop the neural tube before they begin to migrate (Fig. 12.21). Before tracing the multiple

tion of the vertebrate body does not originate in the nervous system but is imposed by the segmental arrangement of the *paraxial mesoderm* (see Chapter 13).

Cranial nerves are associated with the brain much in the same way spinal nerves are associated with the spinal cord. However, the cranial nerves do not emerge at regularly spaced intervals, nor do they have similar functions. The union of dorsal and ventral roots characteristic of spinal nerves is not maintained in the brain. Several cranial nerves, including those innervating the eye and tongue muscles, have mostly efferent functions like the ventral roots of the spinal nerves. Other cranial nerves are strictly sensory, including the *olfactory tract* and the *optic nerve,* which are really connections within the telencephalon and the diencephalon, respectively. The *statoacoustic nerve* is an exclusively sensory nerve entering the myelencephalon. Still other cranial nerves

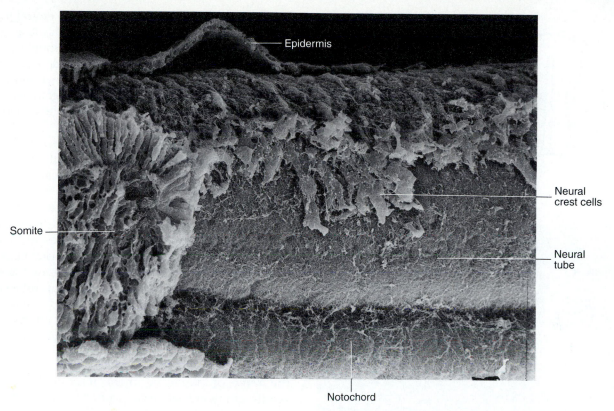

Figure 12.21 Scanning electron micrograph showing a lateral view of a chicken embryo from which much of the epidermis has been removed. The neural crest cells are migrating ventrally from atop the neural tube.

fates of neural crest cells, we will examine their embryonic origin more closely.

Do neural crest cells arise from neural plate cells, or epidermal cells, or both? Also, how do these cells acquire the properties that distinguish them from the neighboring cells of the neural plate and epidermis? Neural crest cells could possibly form where the inductive influence from the *chordamesoderm* is at an intermediate level of strength, too weak to induce neural plate but too strong to permit epidermis formation. Alternatively, the neural crest could arise as a result of the juxtaposition of the neural plate and the adjacent epidermis.

To test the juxtaposition hypothesis, David Moury and Antone Jacobson (1989) juxtaposed pieces of neural plate and epidermis that would not normally participate in neural fold formation (Fig. 12.22a). Working with embryos of a salamander, *Ambystoma mexicanum*, they transplanted an inner piece of neural plate into lateral epidermis. Soon after the graft healed into its new position, folds arose around the edges and curled toward the center of the graft until eventually they closed above it (Fig. 12.22b). Conversely, when epider-

mal pieces were transplanted into neural plate, the entire graft was raised into a ridge. Thus, wherever neural plate and epidermis came in contact with each other, they formed a fold or raised edge.

To see whether or not the interactions between epidermis and neural plate were just part of the wound-healing process, Moury and Jacobson transplanted epidermal pieces into epidermis, and neural plate pieces into neural plate. No fold formation or other disturbances were observed in these control experiments. Also, the type of mesoderm underlying the transplants did not seem to affect the results. Thus, it was the juxtaposition of neural plate and epidermis, rather than wound healing or inductive influences from the mesoderm, that caused neural fold formation.

To follow the behavior of the cells at the interface between neural plate and epidermis, the investigators transplanted neural plate pieces from a pigmented axolotl embryo into the epidermis of an albino host. In some cases, pigmented cells migrated away from the implant, often showing the branched morphology of *melanocytes*, black pigment cells that normally arise from neural crest cells (Fig. 12.22c). Thus, even at a considerable distance from the dorsal midline, *ectopically* generated neural folds contained cells that behaved like a typical neural crest cell derivative. To examine the be-

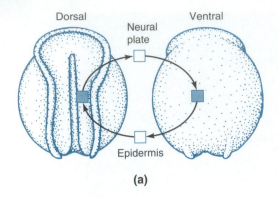

Dorsal   Neural plate   Ventral

Epidermis

(a)

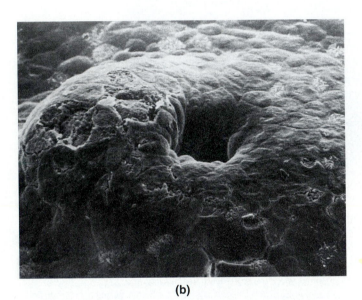

(b)

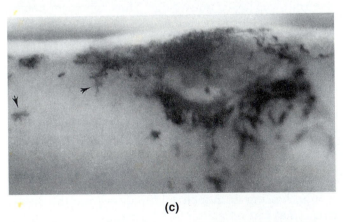

(c)

Figure 12.22 Boundary effect between neural plate and epidermis in the salamander neurula. **(a)** Diagram of transplantation experiments. **(b)** Results of grafting a patch of neural plate into lateral epidermis as seen in a scanning electron micrograph. About 14 h after the transplantation, folds have arisen around the graft and have curled toward its center, leaving an opening that will disappear as the folds fuse over the graft. **(c)** Light micrograph of a pigmented neural plate graft in the epidermis of an albino host. Ten days after the transplantation, dark pigmented cells with the morphology of melanocytes (arrows) migrate away from the graft.

havior of cells from artificially generated neural folds in a more normal environment, Moury and Jacobson (1990) replaced a portion of a *neural fold* of an albino host with a piece of *epidermis* from a pigmented donor. Some of the grafted cells contributed to the spinal and cranial sensory ganglia of the host. These ganglia, as well as the melanocytes observed in the previous experiment, are typical neural crest derivatives.

In related experiments with chicken embryos, Talma Scherson and her colleagues (1993) ablated (removed) the neural crest either alone or in combination with adjoining portions of the neural tube in the midbrain and hindbrain regions. The remaining neural tube cells at the site of ablation were stained with a nondiffusible dye. Labeled cells detached from the neural tube, migrated along the normal pathways of neural crest cells, and appeared to give rise to the normal range of neural crest cell derivatives. The ability of neural tube cells to regulate and form neural crest cells lasted several hours beyond the onset of normal neural crest migration.

The results of these experiments indicate that both epidermis and neural plate can form cells that behave like certain neural crest derivatives. They suggest that in normal development, the neural crest cells may also arise from inductive interactions between juxtaposed neural plate and epidermis.

Neural Crest Cells Have Different Migration Routes and a Wide Range of Fates

Neural crest cells have long held the attention of developmental biologists because of their extensive migration and multiple fates. Investigators have used different techniques to establish fate maps of the neural crest. Weston (1963) and Johnston (1966) introduced a radioactive marker by culturing chicken embryos in the presence of [³H]thymidine and grafting labeled segments of neural tube, including neural crest, in equivalent positions in unlabeled embryos (Fig. 12.23). The hosts were allowed to develop for several days before microscopic sections were prepared for autoradiography (see Methods 3.1). Silver grains generated by radioactivity revealed the position to which the neural crest cells had migrated.

A genetic label, which has the advantage of not being diluted during cell divisions, was introduced by Le Douarin (1969, 1986). She found that embryonic parts can be transplanted between quail and chickens, and that both animals develop and behave normally thereafter. Because quail nuclei have intensely staining *heterochromatin,* they can readily be distinguished from chicken nuclei in histological sections (Fig. 12.24). More

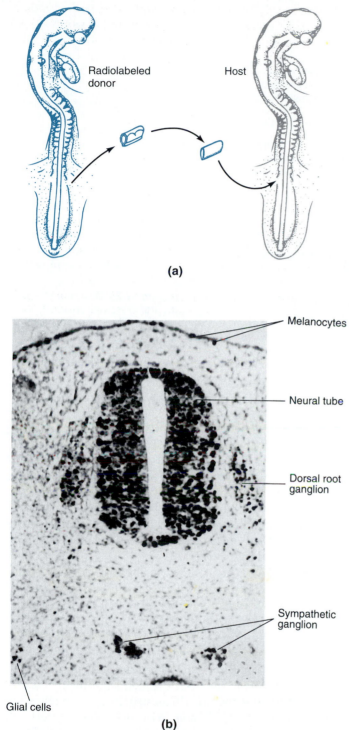

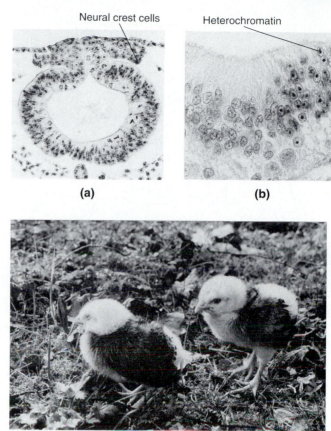

Figure 12.24 Photographs showing the use of quail-chicken chimeras for fate mapping. **(a)** Transverse section of the midbrain of a quail embryo. **(b)** Section of chicken embryo (left part) with a quail graft (right part). The quail cells are marked with deeply staining heterochromatin. **(c)** Chickens that have received grafts of quail neural tube and neural crest display normal behavior. The pigment in their wing feathers is caused by melanocytes derived from quail neural crest cells.

Figure 12.23 Fate-mapping technique for neural crest cells. **(a)** A piece of dorsal axis is removed from a donor labeled with [³H]thymidine. The neural tube and adjacent neural crest are cleaned of other tissues and implanted in a host from which the corresponding piece has been removed. The descendants of the labeled cells are traced in the host by autoradiography. The neural tube cells remain stationary while the neural crest cells migrate extensively. **(b)** Auto-radiograph showing locations of the transplanted neural tube and the descendants of neural crest cells that have formed melanocytes, dorsal root ganglia, sympathetic ganglia, and glial cells.

recently, the use of fluorescent dyes and immunostaining (see Methods 4.1) has further clarified the migration routes of neural crest cells (G. C. Tucker et al., 1984; Bronner-Fraser and Fraser, 1991). The combined use of these techniques has led to the following picture of neural crest development.

In the neural crest of the trunk, cells leave by two major pathways (Fig. 12.25). Cells taking a dorsolateral pathway enter the skin and develop into *melanocytes* (black pigment cells) or *xanthophores* (yellow pigment cells). Cells that take the ventral pathway migrate to a variety of destinations. Some settle close to the neural tube and form the *dorsal root ganglia* (Fig. 12.19). Other neural crest cells form most of the *visceral nervous system,* including the **autonomic nervous system,** which innervates all internal organs. The autonomic nervous system consists of **sympathetic ganglia,** derived from neural crest cells in the cervical, thoracic, and lumbar

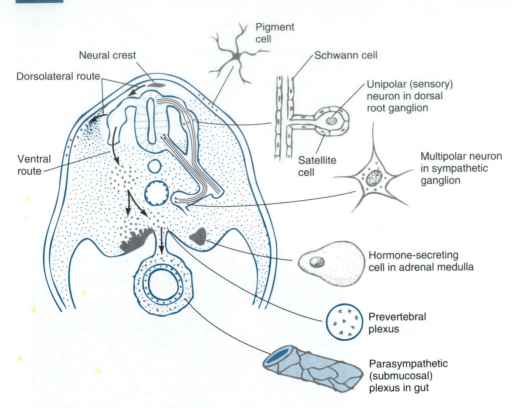

Figure 12.25 Development of neural crest cells in the trunk region. Schematic cross section of a vertebrate embryo showing the main pathways of migration (left side) and the major fates (right side).

regions, and *parasympathetic ganglia,* derived from cephalic and pelvic neural crest cells. Sympathetic ganglia form two chains on the ventral vertebral column, whereas parasympathetic ganglia are located nearer their target organs. Sympathetic and parasympathetic neurons work antagonistically to regulate visceral activities that support either fight or flight responses or rest and recreation, respectively.

Still other neural crest cells of the trunk form the *Schwann cells,* which myelinate the peripheral nerves. Additional neural crest cells participate in forming the *dorsal fin* in amphibians. Finally, cells from the neural crest form the hormone-secreting *adrenal medulla.* Under stress, the adrenal medulla releases epinephrine and norepinephrine into the blood, thus enhancing the activity of the sympathetic neurons, which use the same hormones as transmitters.

In the head, neural crest cells form pigment cells, sensory cranial ganglia, parasympathetic ganglia, and hormone-producing cells, in ways that largely parallel the neural crest of the trunk. In addition, head neural crest cells give rise to bones and other connective tissues in the heads of humans (Fig. 12.26) and chickens (Le Douarin et al., 1993). Moreover, the neural crest cells contribute to the eyes, the teeth, and the blood vessels of the neck region. These structures, as well as the dorsal fins in amphibians, arise from neural crest cell descendants that resemble mesenchymal cells. These cells, because of their ectodermal origin, have been called *ectomesenchymal cells.* They do not fit the general pattern of the germ layers, because bone and blood vessels are generally derived from mesoderm rather than from ectoderm (see Chapter 13).

Neural Crest Cells Are a Heterogeneous Population of Pluripotent Cells

How are neural crest cells determined? Is their potential limited to their fate when they leave the neural

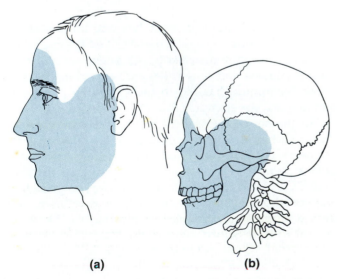

Figure 12.26 Presumed distribution of neural crest cell descendants in the dermis **(a)** and the skeleton **(b)** of the human head.

crest? If not, at which points along their pathways, and in response to which signals, does determination occur? Transplantation experiments indicate that the potential of forming cranial cartilage is limited to the neural crest of the head region (Fig. 12.27). No cartilage is formed when this section of the neural crest is removed or replaced with neural crest from the trunk (B. K. Hall and S. Hörstadius, 1988). However, the potential of head neural crest cells is not limited to their fate. When transplanted to the trunk region, they contribute to dorsal root ganglia, sympathetic ganglia, adrenal medulla, and Schwann cells (Schweizer et al., 1983).

With the exception of cranial cartilage, each region of the neural crest cells seems to be capable of forming almost any neural crest derivative (Fig. 12.27). This conclusion is supported by neural crest transplantations between the posterior rhombencephalon and the thoracic region in chicken embryos. These sections differ in the types of visceral ganglia that they normally give rise to. Rhombencephalic crest cells give rise to parasympathetic neurons, which use acetylcholine as a transmitter, whereas thoracic neural crest cells form sympathetic neurons, which use norepinephrine as a transmitter. However, upon reciprocal transplantation, each type of neural crest cell gives rise to the type of visceral ganglion that corresponds to its *new* location (Le Douarin, 1986). Therefore, the neural crest cells in each region have the potential of forming both types of visceral neurons, although different sets of enzymes are required for the synthesis of the two transmitters. In fact, premigratory neural crest cells from head and trunk contain the enzymes to make *both* acetylcholine *and* norepinephrine (Kahn et al., 1980).

Because the transplantation experiments just described involved *potentially heterogeneous populations* of neural crest cells, the results are open to different interpretations (Fig. 12.28). According to one view, the *pluripotency hypothesis*, each individual *cell* has the potential to form *any* neural crest derivative (except cranial skeleton, unless the cell originates in the head neural crest). This implies that the cells are *not determined* to form any particular neural crest derivative until they receive environmental signals. According to a radically different view, the *selection hypothesis*, any segment of neural crest contains a *mixed population of al-*

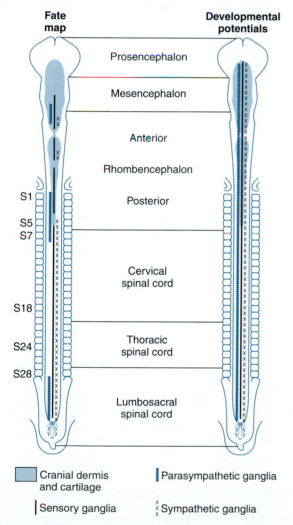

Figure 12.27 Fate map and developmental potency map of the chicken neural crest. Upon transplantation to appropriate sites, neural crest cells from any level of the neural axis give rise to a wide range of different cell types. However, the potential of forming cranial cartilage and other connective tissues is limited to the head area down to the fifth somite (S5).

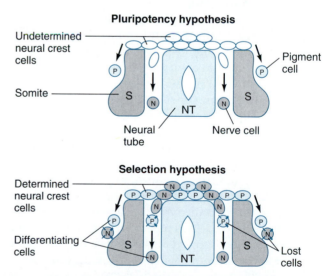

Figure 12.28 Alternative models of neural crest cell determination. The neural crest cell population is depicted by the ovals dorsal to the neural tube (NT). Empty ovals denote undetermined cells, whereas ovals labeled P or N represent cells determined to form pigment cells or nerve cells, respectively. For simplicity, the diagrams show only these two determined states, and only the two major pathways of neural crest cell migration. Circles symbolize the expression of specific neural crest phenotypes at their eventual destination. Crossed circles represent lost cells that do not arrive at a proper destination or die.

ready determined cells, each of which has only one possible fate. This implies that environmental signals received en route enhance the changes of the appropriately determined cells to *arrive* at their destination, and to further *divide* and *differentiate*. These two hypotheses represent two extremes. The following observations, however, suggest that a realistic picture should emerge somewhere in between.

▼

One way of testing whether individual cells are pluripotent is to grow clones from single cells in vitro and to examine whether cells from such clones can differentiate into more than one cell type. Alan Cohen and Irwin Konigsberg (1975) established such cultures from trunk neural crest cells of the Japanese quail. A trunk segment of neural tube with adhering neural crest was kept briefly in tissue culture until the neural crest cells had migrated away from the neural tube. When the neural crest cells adhered to the bottom of the culture dish, the neural tube was removed. Neural crest cells were then dissociated and plated at low density so that individual cells could form new colonies. Three types of clones were obtained in this manner: all pigmented cells, all unpigmented cells, or some pigmented and some unpigmented cells within the same clone. The pigmented cells were melanocytes, and the unpigmented cells were identified as sympathetic neurons or adrenal medulla cells because they contained norepinephrine (Sieber-Blum and Cohen, 1980).

In corresponding experiments on cephalic neural crest cells, Anne Baroffio and her colleagues (1991) found that some neural crest cells gave rise to neurons, glial cells, and cartilage whereas other neural crest cells were more restricted in their potential.

Another technique for studying the potential of single cells by clonal analysis can be used in vivo: here, single cells are labeled with a dye or virus that does not spread to other cells, and the descendants of the labeled cell are traced at a later stage of development. Taking this approach, Marianne Bronner-Fraser and Scott Fraser (1991) microinjected chicken embryo neural crest cells with a fluorescent dye. The labeled cells divided and migrated in an apparently normal fashion, passing on the dye to their daughter cells at each division. After two days of development in vivo, the descendants of the labeled cells were classified according to their location and morphology (Color Plate 3). Clones derived from *premigratory* neural crest cells included as many as four different cell types: dorsal root neurons, sympathetic neurons, Schwann cells, and either adrenal medulla or pigment cells. Many clones derived from *migratory* neural crest cells still contained more than one cell type. Similar results were obtained by Eric Frank and Joshua Sanes (1991), who labeled neural crest cells with a recombinant virus, a technique that allowed them to follow the progeny of the labeled cells for a

longer time, so that the possibility of selective cell death at the target site could be excluded.

The high degree of pluripotency observed in neural crest cells raised the question which kind of signals might eventually determine the fates of these cells in accord with their various locations. Such signals might reside in the abundant *extracellular matrix* that neural crest cells encounter en route to their destinations (Fig. 12.29). Indeed, when neural crest cells are removed from atop the neural tube and cultured in extracellular material taken from the dorsolateral or ventral pathway, the cells develop as pigment cells or neurons, respectively (see Chapter 26). These results, along with data from clonal analyses discussed above, support the pluripotency hypothesis.

Evidence in favor of the selection hypothesis has also been obtained. Hans Epperlein and Jan Löfberg (1984) observed in axolotl embryos that the premigratory neural crest already contained distinct clusters of cells that appeared to be prospective *xanthophores*. Nicole Le Douarin (1986) observed that certain cell culture media enhanced the development of some neural crest subpopulations while inhibiting others. When chicken neural crest cells were kept in a fully defined serum-free medium, a subpopulation of nondividing cells readily differentiated into neurons that did not

Figure 12.29 Neural crest cells migrating in extracellular matrix, which appears as strandy material in this electron micrograph. The extracellular matrix contains at least some of the signals that determine the fates of neural crest cells in accord with their various locations.

synthesize catecholamines (norepinephrine or dopamine) as neurotransmitters. If serum plus chick embryo extract was added to the medium, these neurons disappeared and neurons of a new, catecholamine-containing type formed from dividing cells. These results support the selection hypothesis by showing that the developing neural crest contains at least two subpopulations of neuronal precursors with different behavior and differentiation requirements.

The available evidence, on the whole, indicates that the determination of neural crest cells involves both the selection of subpopulations and the restriction of individual cell potentials (Le Douarin et al., 1993; Selleck et al., 1993). Both processes occur during the migration of neural crest cells to their respective destinations, which takes them through different environments and exposes them to different signals. As discussed in Chapter 6, cells are determined in a stepwise process involving multiple signals. In the case of the neural crest cells, these signals not only affect the potential of individual cells but may also cause selection of appropriately biased subpopulations of cells.

Ectodermal Placodes

Several areas of epidermis in the head region are induced by underlying parts of the brain to form *placodes,* that is, patches of columnar epithelium in a more squamous background (Fig. 12.30). The *ectodermal placodes* of vertebrates emerge in two rows: a lower row of epibranchial placodes, and an upper row of dorsolateral placodes. The epibranchial placodes, together with the neural crest of the head region, form the sensory ganglia of cranial nerves. The dorsolateral placodes also contribute to the cranial sensory ganglia and, in addition, form parts of the ear, eye, and nose. Because of their ability to form sense organs, neurons, and cranial cartilage, and because of their development in the vicinity of the central nervous system, the ectodermal placodes and the neural crest have many properties in common.

The Otic Placode Forms the Inner Ear

The *otic placode,* which will form the inner ear, is the first ectodermal placode to develop. In chicken and amphibian embryos, the otic placode is induced by underlying mesoderm and rhombencephalon. In human embryos, the otic placode appears during the third week on both sides of the rhombencephalon. During the fourth week, the placode invaginates to form the otic pit, which is subsequently pinched off as the *otic vesicle* (Fig. 12.31). On its median surface, the vesicle gives off a group of cells that develop into the *statoacoustic ganglion.*

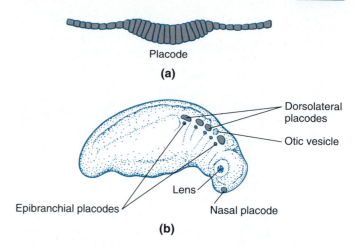

Figure 12.30 (a) Schematic diagram of a placode in an epithelium. **(b)** Ectodermal placodes in the head of a salamander embryo.

Soon the otic vesicle begins to expand, pushing aside the surrounding mesenchyme cells. The expansion is unequal: the vesicle bulges out in some places and constricts in others until it assumes a complicated shape aptly called the *labyrinth* (Fig. 12.32). Some parts of the labyrinth consist of squamous epithelium while others form areas of columnar cells. These develop into sensory epithelia, which receive mechanical stimuli caused by sound, gravity, and body movements and transmit them to the neurons of the statoacoustic ganglion. We will look in greater depth at the part of the labyrinth called the *cochlea* (Gk. *kochlos,* "snail"), which develops in close association with the surrounding tissue (Fig. 12.33).

As the labyrinth develops, it is surrounded by mesenchymal cells, which produce a snugly fitting cartilage capsule. The cartilage later recedes along the cochlea, giving way to two fluid-filled ducts, the *scala vestibuli* and *scala tympani.* The cochlea is then separated from the scala vestibuli by the *vestibular membrane* and from the tympanic duct by the *basilar membrane.* The outside of the cochlea is anchored to the cartilage capsule by the *spiral ligament.* The inside angle of the cochlea is attached to a spiral-shaped cartilage, the *modiolus.* The modiolus also houses the *spiral ganglion,* the part of the statoacoustic ganglion associated with the cochlea.

Sound waves collected by the outer ear are transmitted by tiny bones of the middle ear (see Chapter 13) to the scala vestibuli. The resulting vibrations of the basilar membrane stimulate the sensory cells in the *organ of Corti,* which contains the sensory epithelium of the cochlea. Finally, the spiral ganglion transmits the excitation to the brain. The other sensory epithelia of the labyrinth, receiving stimuli from gravity and body movements, develop and function in similar ways but with less elaborate auxiliary structures.

In most vertebrates, the cartilaginous capsule surrounding the ear is later replaced with solid bone.

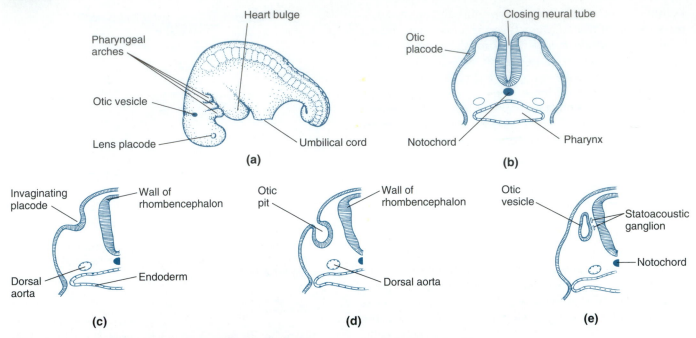

Figure 12.31 Formation of the otic vesicle in the human embryo. **(a)** Side view of a 28-day-old embryo showing the otic vesicle. **(b–e)** Schematic transverse sections through the region of the rhombencephalon showing the formation of the otic vesicle. Note the appearance of the statoacoustic ganglion. **(b)** 22 days; **(c)** 24 days; **(d)** 27 days; **(e)** 31 days.

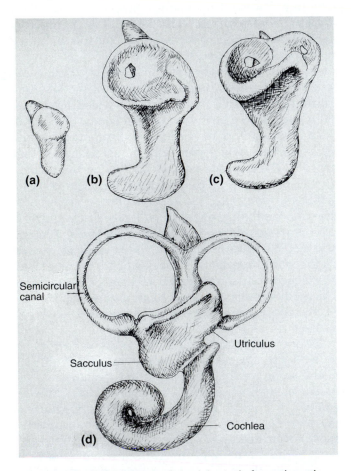

Figure 12.32 Development of the labyrinth from the otic vesicle in the human embryo. The drawings show lateral views of the left labyrinth.

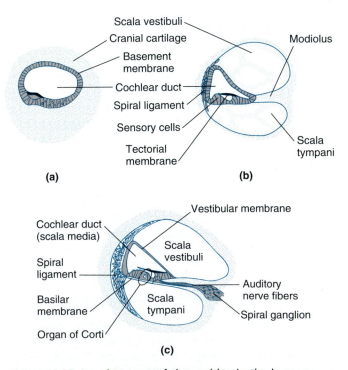

Figure 12.33 Development of the cochlea in the human embryo. **(a)** The cochlea is surrounded by the cartilage of the skull. **(b)** During the 10th week, large vacuoles appear in the cartilage surrounding the cochlea. **(c)** The scala vestibuli and scala tympani have formed, separated from the cochlear duct by the vestibular membrane and the basilar membrane, respectively. Vibrations of the basilar membrane stimulate the sensory cells in the organ of Corti.

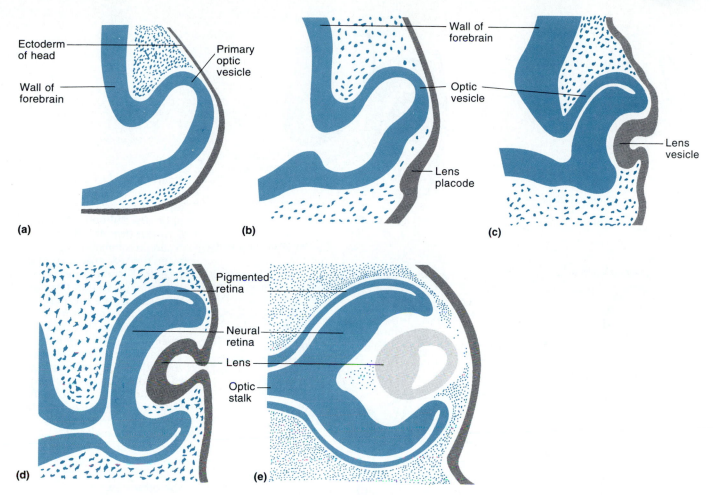

Figure 12.34 Development of the optic cup and lens in transverse sections of human embryos. **(a)** 28 days; **(b)** 29 days; **(c)** 30 days; **(d)** 33 days; **(e)** 39 days.

The Lens Placode Develops Together with the Retina

The *lens placode* is formed in the head ectoderm as the result of a series of inductive interactions with pharyngeal endoderm, heart mesoderm, neural crest cells, and the *optic vesicle,* an extension of the diencephalon (Jacobson, 1966; R. M. Grainger et al., 1988). In the human embryo, the lens placode becomes visible after 4 weeks and invaginates to form the lens during the fifth week (Fig. 12.34). Soon thereafter, the cells of the proximal layer of the vesicle elongate toward the distal layer, gradually filling the lumen of the vesicle. The elongated cells undergo a series of specializations including the synthesis of large amounts of proteins called *lens crystallins* (Piatigorsky, 1981). The differentiated cells, known as *lens fibers,* do not divide. Instead, the lens continues to grow by transforming more outer epithelial cells into inner lens fibers (Fig. 12.35). The transformation begins in a *germinative region,* where the cells divide, and continues in an equatorial *region of cell elongation.* Thus, new lens fibers are continuously lay-

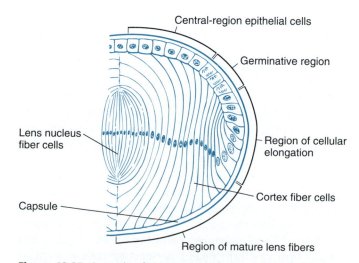

Figure 12.35 Growth of the vertebrate lens. Cells divide in the germinative region of the outer lens epithelium. As cells pass into the region of elongation, they stop dividing and begin to synthesize lens crystallins. The lens fiber cells at the margin of the lens are the youngest. Toward the center (nucleus) of the lens, the lens fibers are progressively older.

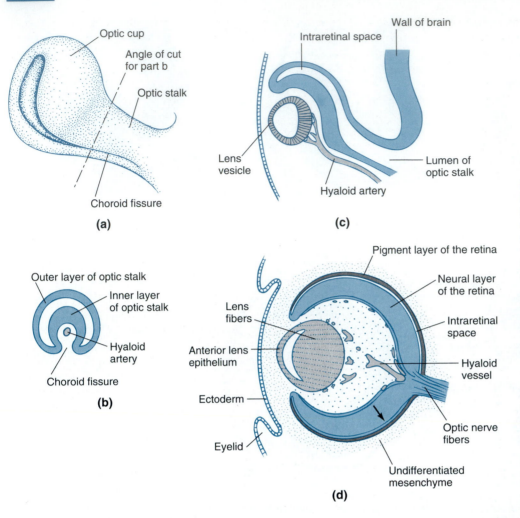

(a)

Optic cup
Angle of cut for part b
Optic stalk
Choroid fissure

(b)

Outer layer of optic stalk
Inner layer of optic stalk
Hyaloid artery
Choroid fissure

(c)

Wall of brain
Intraretinal space
Lens vesicle
Hyaloid artery
Lumen of optic stalk

(d)

Pigment layer of the retina
Neural layer of the retina
Intraretinal space
Hyaloid vessel
Optic nerve fibers
Undifferentiated mesenchyme
Eyelid
Ectoderm
Anterior lens epithelium
Lens fibers

Figure 12.36 Eye development in the human embryo. **(a)** Ventrolateral view of the optic cup and optic stalk at 6 weeks. Note the choroid fissure accommodating the hyaloid artery on the ventral face. **(b)** Section as indicated in part a. **(c)** Section through the plane of the choroid fissure. **(d)** Section from a 7-week embryo. The outer layer of the optic cup has formed the pigment layer of the retina, and the inner layer of the optic cup has formed the neural layer of the retina. The tip of the arrow indicates the position of the light receptors. Nerve fibers emerging at the base of the neural layer (rear end of arrow) converge toward the optic nerve.

ered on top of the preexisting ones. The smallest lens fibers, which constitute the *lens nucleus,* are the first lens fibers formed in the embryo.

While the lens placode invaginates to form the lens vesicle, the *optic vesicle* invaginates to form the *optic cup* (Fig. 12.36). The invagination of the optic cup involves both the outer half and part of the ventral surface, where a groove called the **choroid fissure** forms. This fissure accommodates the **hyaloid artery,** which supplies blood to the interior of the eye. When the choroid fissure closes around the hyaloid artery during the seventh week, the opening of the optic cup acquires the round shape that characterizes the final derivative of this opening, the *pupil.*

The outer layer of the optic cup develops into the **pigmented retina,** while the inner lining of the optic cup forms the **neural retina.** The cells of the neural retina divide much like the neuroepithelium in other areas of the brain. The innermost cells facing the pigmented retina form the light-receptive sensory cells known as **rods** and **cones** (Fig. 12.37). Adjacent to this layer of photoreceptors is the mantle layer consisting of two layers of neurons. The axons of these neurons converge toward the *optic stalk,* which connects the optic cup with the diencephalon. Once the optic stalk is filled with axons, it is referred to as the **optic nerve.**

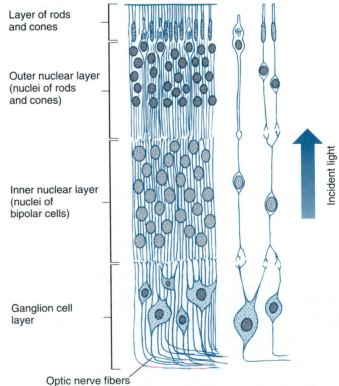

Layer of rods and cones
Outer nuclear layer (nuclei of rods and cones)
Inner nuclear layer (nuclei of bipolar cells)
Ganglion cell layer
Optic nerve fibers
Incident light

Figure 12.37 Schematic representation of the neural retina in a 25-week human fetus.

The sequence of cell layers in the neural retina is contrary to what one might expect from a functional point of view. Light entering the eye has to pass through several layers of axons and neurons before reaching the rods and cones. Also, the light-sensitive tips of these photoreceptors are oriented away from the light instead of facing it. Functionally paradoxical as it may appear, this arrangement is in keeping with its developmental origin. The apical surface of the neural plate, from which the retina ultimately derives, faces the outside world (Fig. 12.38). A simple sensory cell in the epidermis would have the same orientation, with its apical (sensi-tive) surface facing out. But when the neural plate closes to form the neural tube, the apical surface comes to face inside toward the neural canal or the ventricles of the brain. Because the optic cup is an extension of the brain, its apical surface now faces inside, as do the rods and cones. Later, when the neural retina divides to form several cell layers, it gives off neuroblasts away from the lumen, as the neuroepithelium does elsewhere in the brain and spinal cord (Fig. 12.7).

Nasal Placodes Form Olfactory Sensory Epithelia

Development of the *nasal placodes,* located at the anterior tip of the embryo, is induced by the underlying endoderm and telencephalon. In humans, the nasal placodes appear at the end of the fourth week. During the fifth week, two fast-growing ridges, a *lateral nasal swelling* and a *medial nasal swelling,* surround each placode, which then forms the floor of a depression called the *nasal pit* (Fig. 12.39). During the following 2 weeks, the *maxillary swellings,* which will form most of the upper jaw, grow and push the nasal pits medially until the medial nasal swellings fuse. Thus, the upper lip is formed from parts of the maxillary and medial nasal swellings. The fusion of the latter extends also to a deeper level of the face, where they form part of the upper jaw and the *primary palate* (Langman, 1981).

Meanwhile, the nasal pits have deepened until only a thin *oronasal membrane* separates each pit from the oral cavity (Fig. 12.40). Ruptures in the oronasal membranes then create openings, the *primitive choanae,* between the oral cavity and the nasal pits, which are then called *nasal chambers.* After the formation of the *secondary palate,* and with further elongation of the nasal chambers, the *secondary choanae* connect the nasal chambers with the *pharynx.* The original epithelium of the nasal pit is now lining the "roof" of the nasal cavity, where it forms the *olfactory epithelium.* Its sensory cells send their axons directly into *olfactory bulbs,* stemlike extensions of the telencephalon.

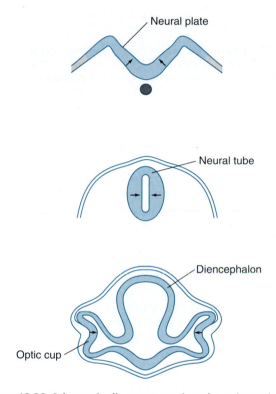

Figure 12.38 Schematic diagrams tracing the orientation of the sensory cells in the neural retina to the apical-basal polarity of the neural plate. The arrow tips indicate the apical (sensory) surface of the cells.

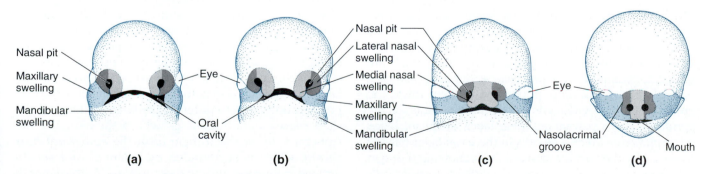

Figure 12.39 Nose development in the human embryo as seen in frontal aspects of the face. The nasal placodes are surrounded by the nasal swellings to form the nasal pits. The medial nasal swelling is separated by the nasolacrimal groove from the maxillary swelling, which forms the upper jaw. **(a)** 5 weeks; **(b)** 6 weeks; **(c)** 7 weeks; **(d)** 10 weeks.

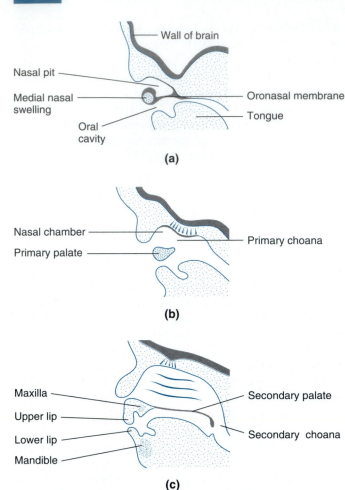

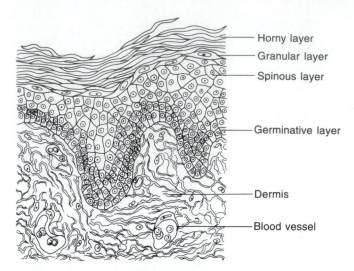

Figure 12.41 Section through the skin of the human shoulder. Epidermal cells originate by mitosis in the germinative or the spinous layer and are pushed via the granular and horny layers to the surface as they mature.

Figure 12.40 Nose development in the human embryo as seen in paramedian sections through one of the two nasal pits. **(a)** At 6 weeks. The nasal pit is separated from the oral cavity by the oronasal membrane. **(b)** At 7 weeks. The oronasal membrane has broken down, so that the nasal chamber is connected with the oral cavity by an opening, the primary choana. **(c)** At 9 weeks. After the development of the secondary palate, the nasal cavity is connected to the pharynx by a secondary choana.

Epidermis

The largest derivative of the ectoderm is the *epidermis*, which forms the outer layer of the skin. The epidermis represents a type of tissue known as *epithelium*. It is characterized by the close apposition of cells, which are sealed together by *tight junctions* at the apical surface (see Chapter 2). Initially only one cell layer thick, the epidermis soon divides into a temporary outer layer, the *periderm*, and a permanent inner layer, the *basal layer*, or *germinative layer*. The cells in the basal layer behave as *stem cells;* they divide actively, thus renewing the stem cell population and creating *committed progenitor cells* that differentiate into epidermal cells. The committed progenitor cells are pushed into the upper layers as they

mature (Fig. 12.41); those that lose contact with the underlying dermis form the *spinous layer.* As part of their differentiation, the epidermal cells synthesize massive amounts of *keratin,* a protein that accumulates in granules. Cells at this stage cease to divide; they form the *granular layer* of the epidermis. When the cells are filled with keratin, they die and form a tough outer layer called the *horny layer.* Throughout life, the outermost cells of the horny layer are sloughed off and replaced by deeper cells, which ultimately originate from the dividing cells in the basal and spinous layers. In adult human skin, a cell born in the basal layer takes about 7 weeks to travel to the surface (Halprin, 1972).

The epidermis is supported by a mesenchymal component of the skin, the *dermis.* The dermis induces the overlying epidermis to form various derivatives and appendages, which include—depending on the body region and the class of vertebrates—scales, feathers, hair, and different glands. As examples, we will briefly consider the development of hair and mammary glands, both characteristic of mammals.

The formation of a hair begins with a proliferation of epidermal cells, a *hair follicle,* penetrating into the underlying dermis (Fig. 12.42). At the base of the follicle, mesenchymal cells form a *hair papilla,* in which blood vessels and nerve endings develop. Soon the core cells of the hair follicle are keratinized and form the *hair shaft,* which is eventually pushed outside. Further proliferation of the epidermal cells at the base of the shaft causes a continuous growth of the hair. The peripheral cells of the follicle form the *epidermal hair sheath,* which is surrounded by a *dermal root sheath* derived from surrounding mesenchyme. A little smooth muscle, also formed by mesenchyme, usually attaches to the dermal root sheath. Contraction of this muscle

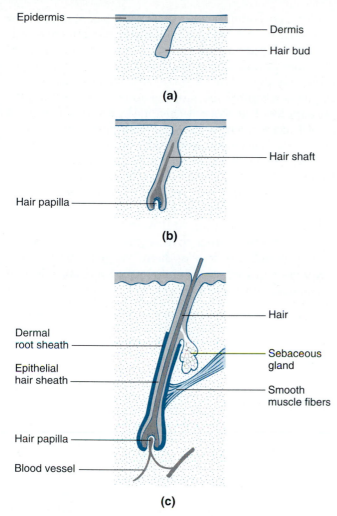

Figure 12.42 Hair development in the human: (a) 4 months; (b) 6 months; (c) newborn.

tilts the follicle perpendicular to the skin surface so that the hair "stands on end." The epidermal hair sheath usually buds out to form a *sebaceous gland,* which gives off an oily substance into the hair follicle, from where it reaches the surface of the skin.

The development of feathers begins in a similar fashion. However, their epidermal sheaths show intricate patterns of local proliferation and programmed cell death, which generate the more complicated structures of feathers.

Epidermal gland development, like hair and feather development, begins with epidermal proliferation resulting from inductive interaction with the underlying mesenchyme. The mammalian epidermis forms several types of glands, including sebaceous glands, sweat glands, scent glands, and mammary glands. The latter develop from a pair of bandlike epidermal thickenings, the *mammary ridges* (Fig. 12.43). Depending on the species, one or more segments of the ridge persist on each side. In each of these locations, solid epithelial cords sprouting lateral buds penetrate the underlying mesenchyme. At birth the buds have developed into

small glands, and the cords have formed hollow *lactiferous ducts* (Lat. *lac,* "milk"; *ferre,* "to bear"). The lactiferous ducts at first open into a small pit, which is later transformed into a nipple by proliferation of the underlying mesenchyme.

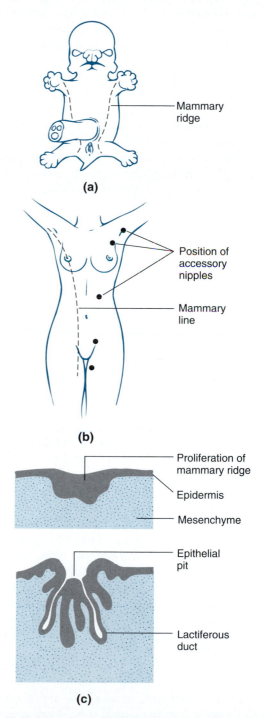

Figure 12.43 Mammary gland development. (a) Generalized mammalian embryo. Dashed lines indicate the normal position of the mammary ridge (milk line) on each side. (b) Accessory nipples or breasts in the adult human result from an abnormal development of the embryonic mammary ridge outside the thoracic region. (c) Sections of the developing mammary gland in the human during the third and the eighth month.

In a 7-week-old human embryo, the mammary ridge extends from the armpit to the groin, as it does in other mammalian embryos. In normal development, only a small portion of the mammary ridge persists in the midthoracic region, giving rise to one pair of breasts in women or nipples in men. Rarely, other segments of the mammary ridge fail to degenerate, so that accessory nipples or breasts are formed. This condition makes one wonder why the human mammary ridge is not limited to the midthoracic region to begin with.

What seems to be an error-prone detour can be explained against the background of phylogeny. The primitive mammals from which primates, including humans, evolved were small creatures, much like tree shrews, that presumably nursed litters of several young. An extended mammary ridge that gave rise to multiple mammary glands would appear to be adaptive for these species. For humans, and other primates that nurse only one or two young at a time, two breasts are sufficient and may, in fact, be more adaptive. Nevertheless, human development recapitulates the embryonic stage of the more primitive mammals. It is noteworthy that *normally* only the *embryonic* stage with extended mammary ridges is recapitulated, not the adult stage with multiple nipples. The *occasional* reappearance of a primitive *adult* feature, such as multiple mammary glands, in an evolved species is an abnormality known as **atavism.**

Presumably, it took fewer steps of mutation and selection to keep the extended mammary ridge and then let most of it degenerate than to develop a new way of inducing just one mammary gland in the thorax. This principle of preserving and modifying old developmental patterns, rather than replacing them, is pervasive in vertebrate development. We will encounter it frequently as we turn to endodermal and mesodermal derivatives in Chapter 13.

SUMMARY

The development of organ rudiments is followed by an extended period of histogenesis, during which tissues mature to their functional state. Surveys of histogenesis often use the germ layers as an organizing principle, starting from the major derivatives of each layer and tracing their development in turn. The major derivatives of the ectoderm are the neural tube, the neural crest, ectodermal placodes, and the epidermis.

A newly closed neural tube consists of a layer of neuroepithelial cells, which give rise to large numbers of neuroblasts and glioblasts. These develop into neurons and glial cells, the cell types that characterize nervous tissue. The neural tube gives rise to the central nervous system. Its anterior portion forms the brain, and its posterior portion forms the spinal cord. The brain is divided into five major regions, called (from cranial to caudal) telencephalon, diencephalon, mesencephalon, metencephalon, and myelencephalon. Axons grow out from motor neurons located in the central nervous system and leave via spinal or cranial nerves to become part of the peripheral nervous system. In addition, the peripheral nervous system comprises sensory and visceral ganglia derived from neural crest cells and ectodermal placodes.

The neural crest cells, which are unique to vertebrates, are located at the crests of the neural folds during neurulation. During neural tube closure, the neural crest cells come to lie on top of the neural tube, from where they migrate along different pathways. Some neural crest cells move dorsolaterally to enter the skin and form pigment cells. Other neural crest cells migrate ventrally and form dorsal root ganglia, visceral ganglia, Schwann cells, and adrenal medulla. In the head region, neural crest cells also form ectomesenchyme contributing to cartilage, other connective tissues, muscle, teeth, and blood vessels. The determination of neural crest cells for these various fates appears to be a stepwise process involving both the restriction of cellular pluripotency and the selection of appropriate subpopulations.

Formation of ectodermal placodes is induced in head epidermis by interactions with underlying tissues. Most of these placodes contribute to cranial sensory ganglia together with the cranial neural crest. Three ectodermal placodes form the lens of the eye and the sensory epithelia of the nose and ear. The remainder of the ectoderm gives rise to the epidermis. Together with the underlying mesenchyme, it forms the skin. The epidermis is a stratified epithelium consisting of several layers. The cell divisions occurring in the basal layers push older cells outward into the upper layers, where they fill with keratin, die, and are eventually sloughed off. Interactions between the epidermis and the underlying mesenchyme induce the formation of various epidermal derivatives, including hair, feathers, scales, and different types of glands.

SUGGESTED READINGS

Carlson, B. M. 1988. *Patten's Foundations of Embryology.* 5th ed. New York: McGraw-Hill.

Hall, B. K., and S. Hörstadius. 1988. *The Neural Crest.* London: Oxford University Press.

Langman, J. 1981. *Medical Embryology.* 4th ed. Baltimore: Williams and Wilkins.

Selleck, M. A. J., T. Y. Scherson, and M. Bronner-Fraser. 1993. Origins of neural crest cell diversity. *Dev. Biol.* **159**:1–11.

ENDODERMAL
AND MESODERMAL
ORGANS

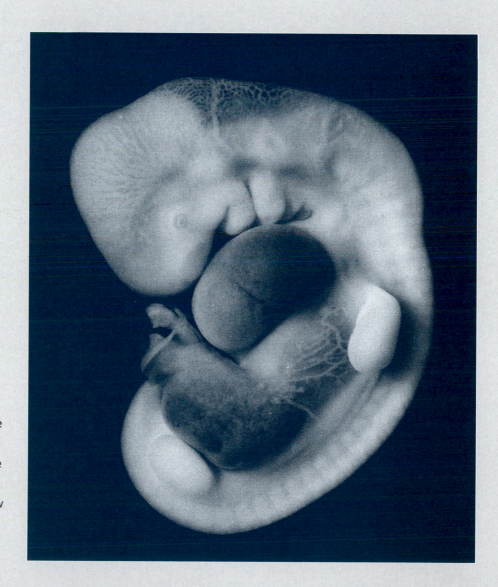

Figure 13.1 Scanning electron micrograph of a human embryo at 31 days. This stage of development represents the phylotypic stage characteristic of all vertebrates. The deep grooves in the neck region are pharyngeal clefts, which in fishes and amphibians would form gill slits. The bulge below the pharyngeal clefts contains the heart rudiment. The lateral protrusions are the limb buds.

Continuing the discussion of histogenesis begun in Chapter 12, we will now consider the derivatives of the endoderm and mesoderm. The major tissue types to be considered in this chapter are connective tissue and muscle tissue. The chapter will also describe the extraembryonic membranes, on which embryos rely for their nutrition, respiration, and waste removal.

The material presented here will provide opportunities to discuss two general topics in development. The first is the occurrence of a *phylotypic stage* in development, when all species of a phylum look very similar (Fig. 13.1). This stage is generally reached after organogenesis, when most organ rudiments are present in a basic body plan. The second general topic to be discussed in this chapter is the *principle of reciprocal interaction.* Many organs originate from two embryonic rudiments—usually one epithelium and one mesenchyme—derived from different germ layers. As the organs develop, the rudiments exchange signals with each other, apparently to ensure that the partners develop harmoniously.

Endodermal Derivatives

The general fate of the embryonic endoderm is to form the inner epithelium of the gut and its appendages. The connective tissues and muscles of the gut, which contribute much more mass, are of mesodermal origin.

The configuration of the endoderm in the vertebrate gastrula varies, as described in Chapter 10. In amphibians, the endoderm begins as the innermost sphere of the gastrula, surrounding the *archenteron.* In most other vertebrate embryos, the endoderm originally forms a disc adjacent to the *yolk* or to a cavity called the *yolk sac.* As these embryos develop, their flanks bend together ventrally, closing the endoderm to create the *digestive tube* (Fig. 13.2). This process is referred to as *lateral folding.* It is enhanced by *craniocaudal flexion* of the embryo, driven mainly by the rapid extension of the neural plate (Fig. 13.3). In the course of these movements, the originally wide opening between the gut and the yolk sac is reduced to a narrow passage called the *vitelline duct.*

The cranial and caudal ends of the digestive tube are closed temporarily by the *buccopharyngeal membrane* and the *cloacal membrane,* respectively. The rudiments of certain appendages to the gut are convenient landmarks for scientists to subdivide the tube into sections (Fig. 13.4). A small ventral outpocketing formed near the cranial end of the digestive tube is the rudiment of the *trachea,* or windpipe, which will convey air to and from the lungs; the section of the digestive tube between the buccopharyngeal membrane and the tracheal rudiment is called the *pharynx.* The next section, the *foregut,* extends from the trachea to the rudiments of the *liver* and the *pancreas.* From there to the cloacal membrane extend the *midgut and the hindgut,* which are not clearly distinguished in the embryo.

The Embryonic Pharynx Contains a Series of Arches

The pharynx and adjoining mesodermal structures are crucial to the development of the entire neck region. The outstanding features are the *pharyngeal arches,* found in all vertebrates and their chordate ancestors.

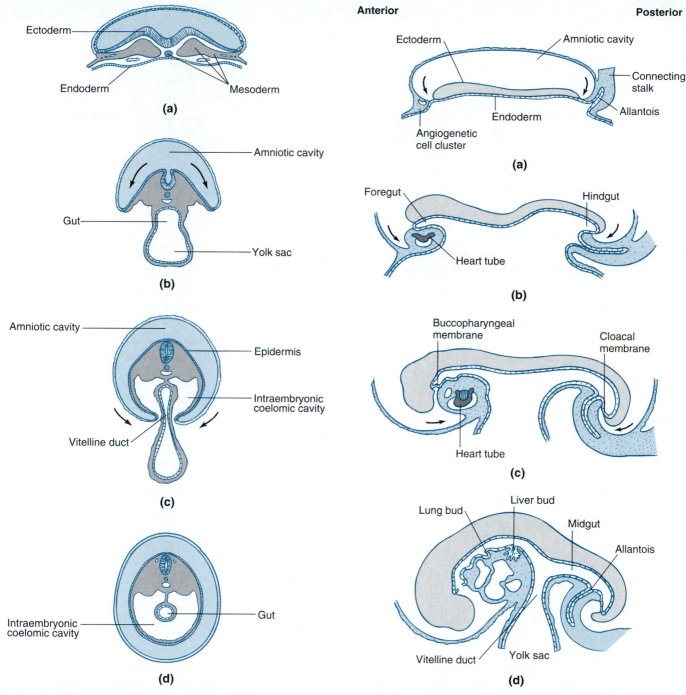

Figure 13.2 Lateral folding in the human embryo. **(a)** At 19 days. **(b)** At 20 days. **(c)** At 21 days. Section through the midgut region shows the vitelline duct connecting the gut with the yolk sac. Lateral folding (arrows) closes the body cavity. **(d)** At 21 days. Section through the hindgut region shows closed abdominal wall.

Figure 13.3 Craniocaudal flexion in the human embryo: **(a)** 18 days; **(b)** 21 days; **(c)** 24 days; **(d)** 30 days. Angiogenetic clusters give rise to blood cells, blood vessels, and heart. The embryonic gut is closed anteriorly by the buccopharyngeal membrane and posteriorly by the cloacal membrane. The yolk sac and the allantois are ventral extensions of the gut.

These arches originate as bars of mesenchyme sculpted from the sides of the neck by pairs of *pharyngeal pouches,* which bulge out from the pharyngeal endoderm and displace the surrounding mesenchyme (Figs. 13.1 and 13.5). Where a pharyngeal pouch approaches the overlying epidermis, it induces the formation of a *pharyngeal cleft.*

In primitive vertebrates and their chordate relatives, the pouches and clefts fuse to form slits through which water can stream from the pharynx to the outside. The

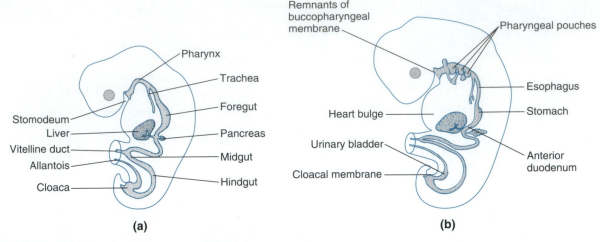

Figure 13.4 Endoderm of the human embryo: **(a)** 4 weeks; **(b)** 5 weeks. Note the development of the pharyngeal pouches, a characteristic of all vertebrate embryos.

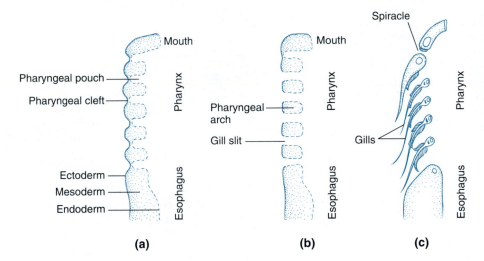

Figure 13.5 Schematic drawings showing the left half of the pharyngeal region in frontal section. **(a, b)** Generalized vertebrate embryo at successive stages. The pharyngeal pouches formed by the endoderm induce the formation of pharyngeal clefts in the ectoderm. Clefts and pouches fuse to form gill slits separated by pharyngeal arches. **(c)** Adult shark. The first gill slit has developed into a circular opening known as the spiracle, and gills have developed on the posterior pharyngeal arches.

vertebrate ancestors probably used this arrangement to strain food particles from a continuous flow of water moving through the pharynx (Torrey and Feduccia, 1991). Later, the pharyngeal arches became the basis for the formation of gills and jaws in primitive fish. In land-dwelling vertebrates, the respiratory function of the gills has been taken over by lungs, and the pharyngeal pouches and clefts have been put to other uses.

All vertebrate embryos form several pairs of pharyngeal arches, each containing precartilage cells of neural crest origin, premuscle mesenchyme, a blood vessel, and a cranial nerve. The first pharyngeal arch forms the rudiments of the upper and lower jaw, known as the *palatoquadrate cartilage* and the *mandibular cartilage* (Fig. 13.6; Table 13.1). These primary jaw elements become part of the adult jaws of most vertebrates. The second pharyngeal arch contains two embryonic cartilages, the *hyomandibular* and the *hyal.* In fishes, the hyomandibular cartilage develops into a stout bone that connects the jaw to the braincase.

During mammalian development, the primary jaws are replaced with a second generation of jaw bones, the *dentary* for the lower jaw, and the *maxillary bone* and the *squamosal* (a portion of the temporal bone) for the upper jaw (Fig. 13.7). However, the vestiges of the palatoquadrate and mandibular cartilages persist as two tiny middle ear bones, the *incus* (Lat., "anvil") and *malleus* (Lat., "hammer"), respectively. The malleus attaches to the eardrum and transfers its vibrations to the incus. The incus, in turn, is linked to a third middle ear ossicle, the *stapes* (Lat., "stirrup"), which is homologous to the hyomandibular bone of fishes. Together, the three middle ear ossicles of mammals act as levers to increase the pressure of the sound waves as they travel from the eardrum to the inner ear. The other cartilage of the second pharyngeal arch, the *hyal,* forms part of the mammalian *hyoid bone* (Gk. *hyoides,* "U-shaped"), which supports the base of the tongue. The following pharyngeal arches also contribute to the hyoid bone and to the voice box, or *larynx.*

TABLE 13.1

Cartilage Elements in Pharyngeal Arches and Their Derivatives

Pharyngeal Arch Number	Embryonic Cartilage Elements	Adult Cartilages or Bones		
		Fishes	Amphibians, Reptiles, Birds	Mammals
I	Palatoquadrate Mandibular	Quadrate Articular	Quadrate Articular	Incus Malleus Meckel's cartilage
II	Hyomandibular Hyal	Hyomandibular Hyal	Columella Part of hyoid	Stapes, styloid process Part of hyoid
III	(Several elements)	Gill bar	Part of hyoid	Part of hyoid
IV–VI	(Several elements)	Gill bars	Laryngeal and tracheal skeleton	Laryngeal and tracheal skeleton

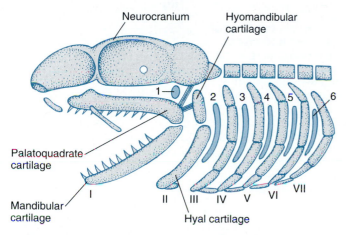

Figure 13.6 Schematic drawing of a developing shark skull in lateral view. In the neurocranium, which contains the brain, the nasal region is anteriormost (to the left); it is followed (in order) by the eye region (with an opening for the optic nerve), the otic region, and the occipital region. The cartilage elements of the pharyngeal arches (Roman numerals) are separated by gill slits (Arabic numerals), the first of which has formed the spiracle. The major cartilage elements of the first arch, the palatoquadrate and the mandibular, form the upper and lower jaw elements, respectively. The cartilage elements of the second pharyngeal arch are the hyomandibular and the hyal. Note that the palatoquadrate is only loosely attached to the neurocranium. A more stable connection is formed in bony fish by the hyomandibular.

The pharyngeal pouches and clefts still develop in mammalian embryos as they do in primitive vertebrates, but the pharyngeal endoderm and the adjoining epidermis are not perforated to form openings. The pouches and clefts, no longer needed for gill slits, assume a variety of other functions (Fig. 13.8). The first pharyngeal pouch, located between the first and second pharyngeal arches, gives rise to the middle ear, which contains the sound-transmitting ossicles already described. The original connection of the pouch to the pharynx persists as the *pharyngotympanic tube*, or *Eustachian tube*, which equalizes the pressure between the middle ear and the outside. The first pharyngeal cleft gives rise to the outer ear. Where the first pouch meets the first cleft, the *tympanic membrane* (eardrum) is formed. The endoderm of the second pouch proliferates, penetrating the surrounding mesenchyme and forming the *palatine tonsil*. Part of the pouch remains and is still found as a groove in the adult tonsil. The third pharyngeal pouch forms the *thymus gland* and the lower portion of the *parathyroid gland*. Both glands lose their connection with the pharynx and move caudally and medially to their adult positions. Similarly, the fourth pharyngeal pouch gives rise to the upper portion of the parathyroid gland, which also associates with the thyroid gland. The second to fourth pharyngeal clefts disappear during later development.

Embryos Pass through a Phylotypic Stage after Organogenesis

The manner in which different structures derive from the pharyngeal arches illustrates the indirect course by which many mammalian organs develop. The German embryologist C. B. Reichert, who discovered the origin of the middle ear ossicles in mammals in 1837, could hardly believe what he saw. Why would these tiny ear bones develop from cartilages that look like the rudiments of jaws? Why do mammalian embryos have deep grooves in the neck that look like the gill slits of fish? Answers to these questions had to await the study of the fossil record on vertebrate evolution (Romer, 1976; Colbert, 1980; S. J. Gould, 1977, 1990; McKinney and McNamara, 1991). We will briefly summarize the findings in this area and their implications for development before we return to our discussion of endodermal derivatives.

Fishes have an inner ear with which they perceive gravity and their own body movements. The same organ can provide a sense of hearing, provided that there

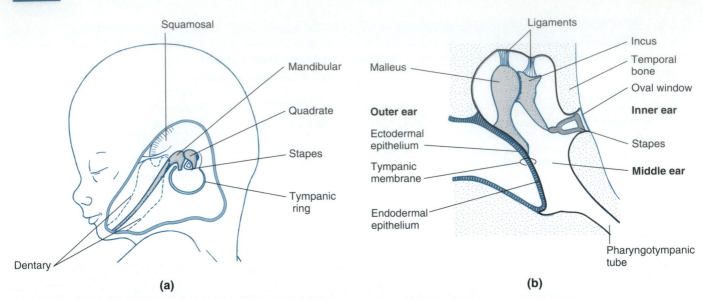

(a) **(b)**

Figure 13.7 Development of the jaws and the middle ear ossicles in mammals. **(a)** Detail of a human fetus at 10 weeks. Two jaw joints are visible: the primary jaw joint is formed by the mandibular cartilage and the quadrate cartilage; the latter is the proximal part of the palatoquadrate, which originally supports the upper jaw. The secondary jaw joint is formed by the squamosal and the dentary. While the secondary jaw joint becomes the definitive one, vestiges of the primary jaw elements are transformed into middle ear ossicles. The malleus, which descends from the proximal part of the mandibular cartilage, will attach to the eardrum. The incus, which develops from the quadrate, will articulate with the malleus as it would in the primary jaw joint. The stapes, derived from the hyomandibular cartilage, will connect the incus with the inner ear. The squamosal and tympanic ring, along with other rudiments, will form the temporal bone of the adult skull. **(b)** Schematic drawing of the middle ear in the human adult. The malleus transmits vibrations from the eardrum via the incus to the stapes, the foot plate of which is in contact with the scala vestibuli of the cochlea (see Fig. 12.33).

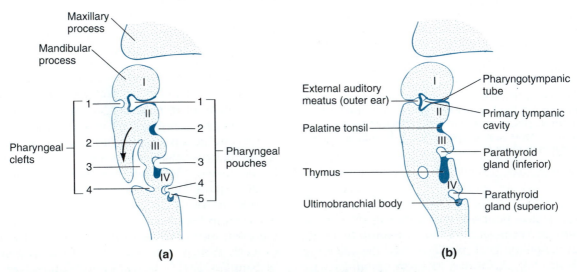

(a) **(b)**

Figure 13.8 Fates of the pharyngeal pouches and clefts in the human embryo, drawn in frontal sections: **(a)** 5 weeks; **(b)** 6 weeks. Roman numerals designate pharyngeal arches; Arabic numerals designate pharyngeal pouches and clefts. The first cleft gives rise to the outer ear, while the other clefts are overgrown. The first pouch develops into the middle ear and the pharyngotympanic tube. The other pouches give rise to palatine tonsils and several endocrine glands.

is a device to transmit sound waves from the environment. One device of this kind evolved from a bone element located in the vicinity of the inner ear. The *hyomandibular bone* of fish, derived from the dorsal part of the second pharyngeal arch, is a stout bone that anchors the jaws to the braincase. The same hyomandibular bone of some ancestral fishes apparently also played a role in hearing, as indicated by the insertion of the bone into the ear region of the braincase (Clack, 1989). As land-dwelling vertebrates evolved, their upper jaws were attached directly to the braincase, so that the hyomandibular bone was no longer necessary as a brace. This bone then became more slender, in accord with its secondary function in hearing. Amphibians, reptiles, and birds still have a single middle ear ossicle, the **columella,** which is derived from the hyomandibular cartilage.

A similar shift of functions occurred during the evolution of the mammalian jawbones. The **primary jaw joint,** still found in all vertebrates except mammals, is formed by the **quadrate** and **articulare,** two bones that develop from the *palatoquadrate cartilage* and the *mandibular cartilage* of the first pharyngeal arch. During mammalian evolution, a **secondary jaw joint** between the *dentary bone* (lower jaw) and the *squamosal,* a part of the temporal bone, was added to the primary jaw joint. Mammal-like reptiles that lived during the Jurassic period (about 180 million years ago) actually used both the primary and the secondary jaw joints as adults. But the existence of a secondary jaw joint eventually relieved the bones of the primary jaw joint of their original function and allowed them to evolve into two additional middle ear ossicles, the *malleus* and the *incus.*

It is a striking phenomenon that the evolution of jawbones into middle ear ossicles is not only shown in the fossil record but is also reenacted in the development of each mammal today. Every mammal, including the human, retraces in its normal fetal development the evolution of the middle ear ossicles from jawbones (Fig. 13.7). This process exemplifies a phenomenon—termed **recapitulation**—that is very pervasive in vertebrate development. It appears that nature often prefers the modification of what it has over the construction of novelties.

It is important to realize that advanced vertebrates recapitulate the *embryonic rather than the adult* stages of their phylogenetic ancestors. In normal human development, it is the embryonic mammary ridge, not the adult condition of multiple mammary glands, that is recapitulated (see Fig. 12.43). Similarly, the development of the middle ear ossicles recapitulates the embryonic stages, not the adult stages, of the primary jaw joint and the hyomandibular bone. A *false* impression that adult conditions are recapitulated might stem from the fact that in primitive vertebrates, adult structures are similar to embryonic structures, whereas in advanced ver-

tebrates, the adult stage is farther removed from the embryonic stage. For instance, the gills of an adult fish remain similar to the pharyngeal arches of a fish embryo, but the neck of an adult mammal is very different from its embryonic rudiment.

Comparative studies on vertebrate development were first carried out in a systematic way by Karl Ernst von Baer, who discovered the notochord, the mammalian egg, and the human egg (see Chapter 3). In his treatise on animal development, von Baer (1828) pointed out that the *general* features of a large group of animals appear earlier in development than the *special* features. For instance, at certain embryonic stages all vertebrates are very similar, especially during the stages shortly after organogenesis, when the different classes of vertebrates are virtually indistinguishable (Fig. 13.9). At early fetal stages, fish and amphibians look different from mammals, reptiles, and birds. Only at later fetal stages can different orders within the class of mammals be distinguished. Curiously, the stages *preceding* organogenesis, such as cleavage and gastrulation, also differ considerably among the vertebrate classes (see Chapters 5 and 10). Thus, it is not the earliest stage but the stage after organogenesis that is most characteristic of the entire vertebrate (sub)phylum and is therefore called the **phylotypic stage** (Sander, 1983). Insects and other arthropods also have a phylotypic stage after organogenesis, the *segmented germ band* (see Fig. 21.46).

The phylotypic stage has been especially resistant to change, although the nature of the evolutionary constraints is still speculative. Conceivably, the morphogenetic movements preceding the phylotypic stage are so complex that any changes caused by mutations have disastrous consequences.

The morphogenetic movements that occur *after* the phylotypic stage have been amenable to gradual changes during evolution. The tendency of these changes to recapitulate phylogenetically old patterns suggests that the changes may be caused by gene activities that *modify* existing regulatory networks rather than creating entirely new ones. This view is supported by the occurrence of *atavisms,* that is, occasional reappearances of phylogenetically older morphological traits in adults. Examples of atavisms include multiple nipples or breasts in humans (see Fig. 12.43), abnormal forms of a divided uterus in humans that are reminiscent of the uteri of more primitive mammals, and occasional supernumerary hoofs in horses and mules, which resemble the three-toed feet of ancestral horses that lived 15 million years ago. Such atavisms demonstrate vividly that phylogenetically old patterns of development are still present in evolved species but are normally suppressed or modified. The genes that control the suppression or modification are best known in insects. The *Ultrabithorax*[+] gene of *Drosophila,* for example, modifies the development of the third thoracic

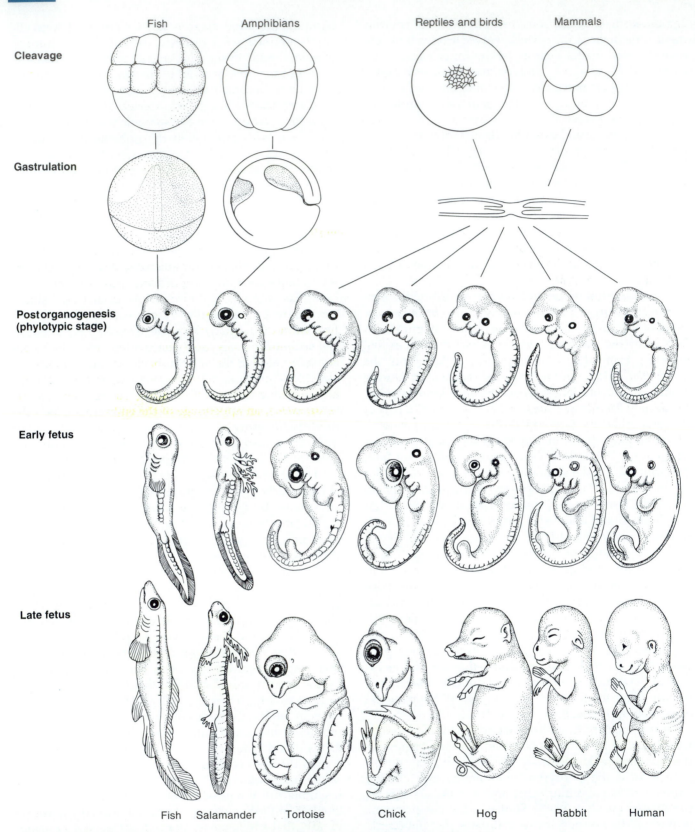

Figure 13.9 The phylotypic stage. At early stages of development, vertebrates differ with regard to cleavage and gastrulation patterns. The stage after organogenesis—the phylotypic stage—is the one most strictly conserved during evolution, during which embryos from different vertebrate classes are virtually indistinguishable. During subsequent development, fetuses become recognizable first as members of their class, then their order, etc., and finally their species.

segment so that it will differ from the second thoracic segment. Failure of the *Ultrabithorax*⁺ gene to function properly leads to the development of a four-winged fly (see Fig. 1.21), a phenotype reminiscent of fossil ancestors of modern flies.

The Endoderm Lines the Inside of the Intestine and Its Appendages

Soon after growing out from the foregut, the tracheal rudiment forms two lateral buds. In the human, these buds divide again into three branches on the right side and two on the left (Fig. 13.10). These branches develop into the main *bronchi*, in advance of the development of the lobes of the lung. The bronchi divide many times into more and finer *bronchioli*. Eventually, the epithelial cells in the terminal sacs of the bronchioli change from a cuboidal to a *squamous* shape, forming the grapelike *alveoli*. The alveoli become intimately associated with blood capillaries so that a very large and thin interface is formed between the air in the alveoli and the blood in the capillaries. Only the inner epithelium of the trachea, bronchi, and lungs is derived from endoderm. The other tissues, including the cartilage reinforcements of the tra-

chea and bronchi, the blood vessels supplying the lungs, and the connective tissue in which these structures are embedded, are of mesodermal origin.

The foregut caudal to the tracheal rudiment develops into the *esophagus, stomach,* and the anterior part of the *duodenum* (Fig. 13.4). The esophagus is a straight connection between the pharynx and the stomach. The stomach, already discernible in the 4-week human embryo, is the most muscular and expandable portion of the digestive tract. The remainder of the embryonic foregut, between the stomach and the liver, is the rudiment of the anterior duodenum. The adjoining midgut, beginning with the posterior duodenum, forms the *small intestine*, in which the epithelium is specialized so as to maximize the area available for taking up nutrients. Mesodermal layers surrounding the epithelium provide the blood vessels and smooth muscle layers involved in digestion. The embryonic midgut also forms a small part of the *colon,* the remainder of which is derived from the hindgut.

The hindgut gives rise to the greater part of the *colon,* the *rectum,* and the *anal canal* of the adult. In addition, the endoderm of the hindgut forms the inner lining of the bladder and urethra. The latter structures arise from the *allantois,* an appendage of the embryonic hindgut,

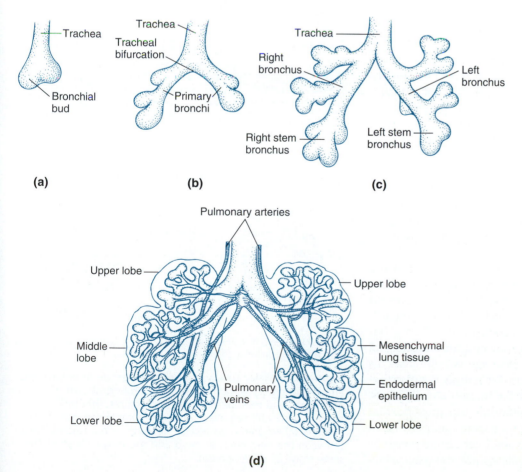

Figure 13.10 Development of the human lung. The diagrams show ventral aspects. The branching endodermal epithelium is surrounded by mesodermal tissues including blood vessels. **(a)** 28 days; **(b)** 33 days; **(c)** 39 days; **(d)** 50 days.

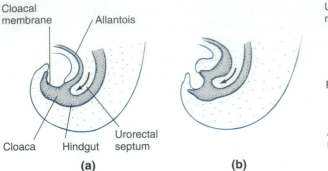

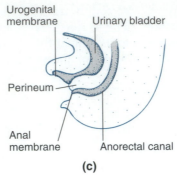

Cloacal membrane — Allantois

Cloaca — Hindgut — Urorectal septum

(a)

(b)

Urogenital membrane — Urinary bladder

Perineum

Anal membrane — Anorectal canal

(c)

Figure 13.11 Development of the cloacal region in the human embryo: **(a)** 5 weeks; **(b)** 7 weeks; **(c)** 8 weeks. Arrows indicate the movement of the urorectal septum, which forms the perineum and divides the cloacal membrane into the anal membrane and the urogenital membrane.

and the embryonic *cloaca* (Fig. 13.11). In most vertebrates, the cloaca serves as a common opening for the gut, the bladder, and the reproductive system. In mammals, the cloaca is partitioned by the development of a transverse mesenchymal ridge, the *perineum,* which divides the *cloacal membrane* into the *urogenital membrane* and the *anal membrane.* The anal membrane is displaced to the interior by an ectodermal depression, the *proctodeum,* before it ruptures to open a passage between the rectum and the outside.

Axial and Paraxial Mesoderm

The two germ layers we have looked at so far in this text, the ectoderm and the endoderm, give rise mostly to *epithelia,* including the epidermis of the skin and the inner epithelium of the gut. The intervening germ layer, the mesoderm, forms both epithelia and large amounts of *mesenchyme.* Mesenchyme consists of isolated cells surrounded by large amounts of extracellular material, as opposed to epithelia, which are sheets of tightly joined cells (see Fig. 2.20). Mesenchymes are typically derived from the mesodermal germ layer, but may also arise from ectoderm, as discussed in Chapter 12 in connection with the *neural crest cells* and some *ecto–dermal placodes.*

The overall shape of the mesoderm in the gastrula varies with the class of vertebrates. In typical amphibians, the germ layers are arranged like three nested tubes. In vertebrates with superficial cleavage, including reptiles and birds, the mesoderm arises as a flat layer between an ectodermal roof and an endodermal floor. In mammals, although cleavage is holoblastic as it is in amphibians, mesoderm formation is nevertheless patterned after the mammals' immediate ancestors, the reptiles. Regardless of whether the mesodermal germ layer is tubelike or flat, it becomes subdivided in a pattern that is characteristic of all vertebrate embryos (Fig. 13.12); its major subdivisions are the *axial meso-*

derm, the *paraxial mesoderm,* the *intermediate mesoderm,* and the *lateral plates.*

Axial Mesoderm Forms the Prechordal Plate and the Notochord

The *axial mesoderm* is located along the dorsal midline. In amphibians, it assumes its position mostly by convergent extension (see Fig. 10.21c). In mammals, birds,

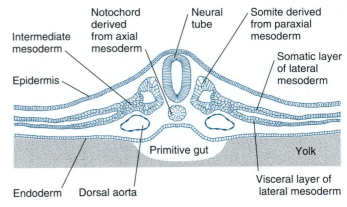

Intermediate mesoderm — Epidermis

Notochord derived from axial mesoderm — Neural tube — Somite derived from paraxial mesoderm

Somatic layer of lateral mesoderm

Primitive gut — Yolk

Endoderm — Dorsal aorta

Visceral layer of lateral mesoderm

(a) Section of normal chick embryo

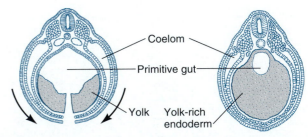

Coelom

Primitive gut

Yolk — Yolk-rich endoderm

(b) Chick embryo removed from yolk, edges pulled together

(c) Section of frog embryo

Figure 13.12 Major subdivisions of the mesoderm in bird and amphibian embryos. **(a)** Normal chick embryo showing mesoderm and mesoderm-derived structures. **(b)** Chick embryo with yolk removed for easier comparison with frog embryo. **(c)** Frog embryo section. Note the close resemblance between the bent chick embryo and the frog embryo at the same stage of development.

and some reptiles, the axial mesoderm ingresses through *Hensen's node* as the *primitive streak* regresses (see Figs. 10.29 and 10.30). The axial mesoderm gives rise to the **prechordal plate** in the anterior head and to the **notochord** in the posterior head, neck, trunk, and tail (Fig. 13.13). The prechordal plate is a source of mesenchyme which, together with *ectomesenchyme* derived from neural crest cells, forms cranial cartilage. The notochord (Lat. *notum*, "back"; Gk. *chorde*, "cord") is a dorsal rod of cartilagelike connective tissue. The notochord persists in the adult stages of small marine creatures making up the subphyla Urochordata and Cephalochordata, which, together with the vertebrates, constitute the phylum Chordata. In most vertebrates, much of the notochord is replaced by the vertebral column. In mammals, including humans, small portions of notochord persist as a pulpy, elastic tissue inside the **intervertebral discs** located between vertebrae.

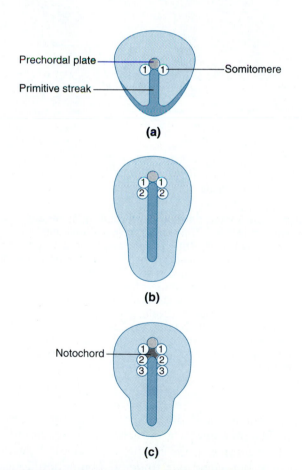

(a)

(b)

(c)

Figure 13.13 Development of the axial mesoderm (prechordal plate and notochord) and paraxial mesoderm (somitomeres). The diagrams show bird embryos during **(a)** primitive streak elongation, **(b)** maximal elongation of the streak, and **(c)** streak regression. The epiblast has been removed to reveal the underlying mesoderm; the position of the primitive streak is represented by the darkly colored area. The prechordal plate is the most anterior axial mesoderm. The somitomeres are numbered in the sequence in which they form.

Paraxial Mesoderm Forms Segmental Plates and Somites

The *paraxial mesoderm* (Gk. *para*, "beside") originates on both sides of the axial mesoderm. About the same time the neural tube closes, the paraxial mesoderm subdivides into segments called *somites*. If the mesoderm is formed with Hensen's node regressing from anterior to posterior, the somites appear in the same sequence, and the number of visible segments is used to identify developmental stages. Before the somites become distinct under the light microscope, the paraxial mesoderm forms two ridges of mesenchymal cells that are not overtly segmented, the *segmental plates*. In chickens, the segmental plates reach from the last-formed somite to Hensen's node (Fig. 13.14). As the node regresses, the segmental plate extends posteriorly and gives rise to overt somites anteriorly.

Close inspection of chicken segmental plates with the scanning electron microscope reveals that the segmental plates already consist of whorls of mesenchyme cells, termed *somitomeres* (Meier, 1979). The size, position, and morphology of the somitomeres indicate that they are precursors of somites. The most mature somitomeres, closest to the emerging somites, are visually distinct from one another by cellular orientations and the progressive buildup of fibrous extracellular material. The least mature somitomeres, barely discernible in stereoscopic pairs of scanning electron micrographs, appear posteriorly on both sides of Hensen's node. As the node regresses, leaving in its wake the axial mesoderm, the left and right rows of somitomeres remain separated by the developing notochord.

▼

Experimental evidence that somitomeres are the precursors of somites was obtained by Packard and Meier (1983). They isolated segmental plates from both chickens and quail and split them into right and left halves. One half was fixed immediately for scanning electron microscopy to determine its somitomere number, while the other half was kept in tissue culture for 5 to 8 h before fixation. The cranial end of the cultured half always formed several somites. The total number of somites and somitomeres in the cultured half matched the number of somitomeres in the half fixed immediately, indicating that for each somitomere lost, a somite had been gained.

Somitomeres develop into somites by changing from mesenchymal to epithelial organization. Each somite originates as an epithelial sac of cells surrounding a central cavity (Fig. 13.15). The apical surface, connected by tight junctions, is oriented toward the cavity, while a

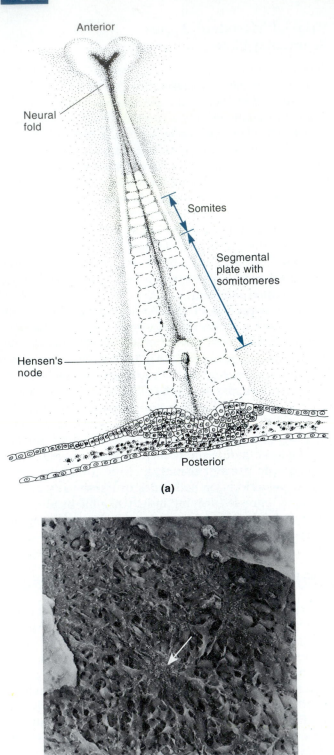

(a)

(b)

Figure 13.14 Somite development in the chicken embryo. **(a)** Survey diagram showing the anterior end of a chicken embryo sectioned posterior to Hensen's node. The dashed outlines represent somitomeres in the segmental plates. The somites anterior to the segmental plate are discernible under the light microscope. Note how the anterior end of the embryo is ahead of the posterior with respect to neurulation and somite formation. **(b)** Scanning electron micrograph of a somitomere in the chicken embryo. The "hub" in the center (arrow) is formed by a cluster of cell extensions (microvilli).

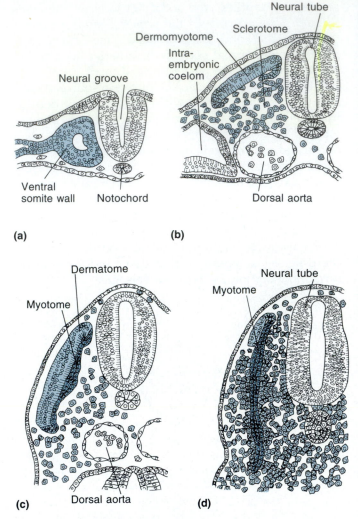

Figure 13.15 Somite development in the human embryo at successive stages. **(a)** The somite forms an epithelial sac around a small cavity. **(b)** The ventral and medial cells of the somite lose their epithelial connections and migrate toward the notochord. These cells, which form vertebrae, are called the sclerotome; the remainder of the somite is called the dermamyotome. **(c)** The medial portion of the dermamyotome gives rise to a new layer of cells, the myotome, which form skeletal muscle and limb buds. **(d)** The remaining cells of the somite, called dermatome, lose their epithelial arrangement and spread out under the epidermis to form the dermis.

basal lamina is laid down around the outer surface. Thereafter, different parts of each somite give rise to different structures. Cells in the ventromedial portion, known as the *sclerotome,* undergo a burst of mitotic activity and become mesenchymal again. The sclerotome cells migrate away from the remainder of the somite and surround the notochord and the neural tube. Secreting large amounts of extracellular material, they form cartilage elements, which are later transformed into vertebrae. The remainder of the somite forms a two-layered structure, the *dermamyotome.* Its dorsolateral layer is called the *dermatome,* since it gives rise to the *dermis,* the tough part of the skin underlying the epidermis. Like the sclerotome, the dermatome cells become mesenchymal before migrating to the surface of the body. The remaining part of the dermamyotome, the *myotome,* proceeds to form skeletal muscle.

Because the segmental organization is blurred in the adult head, it was of particular interest to find that cranial somitomeres are present in representatives of all vertebrate classes during development. The first pair of somitomeres develop at the level of the prosencephalon (Jacobson, 1988). The subsequently formed somitomeres are associated with more posterior brain regions (Fig. 13.16). In most vertebrates, the first seven pairs of somitomeres do not form somites, but give rise to corresponding derivatives including skeletal elements, muscle, and some connective tissues, including the meninges enclosing the brain (Noden, 1983). The eighth pair of somitomeres, located just caudal to the *otic placode,* are the first to develop into a somite delineated by visible clefts. Somites 8 to 11 contribute to the posterior skull. This pattern is characteristic of mice, quail, snapping turtles, and a teleost fish. Amphibians and sharks have a similar arrangement except that they have a single somitomere for each pair the other vertebrates have.

Since somitomeres have been identified in mammals, reptiles, amphibians, and teleosts, it is believed that somitomeres are the first visible signs of segmentation in all vertebrates (Jacobson, 1988). Later expressions of segmentation such as nerves or blood vessels are determined by the pattern of somitomeres and somites (C. L. Anderson and S. Meier, 1981). The total number of somites in a vertebrate depends upon the species.

Connective Tissue Contains Large Amounts of Extracellular Matrix

Several types of vertebrate tissue, including cartilage, bone, tendons, and adipose tissue, are collectively called *connective tissue.* These tissues are characterized by large amounts of *extracellular matrix,* the specific composition of which determines the mechanical properties of the tissue. The extracellular matrix consists of two types of molecules, which are linked together covalently in a meshwork (see Chapter 26). Molecules of one type, called *proteoglycans,* attract water and form a gelatinous mass. The other molecules are proteins that assemble into fibrils reinforcing the proteoglycans and limiting their expansive properties.

All extracellular matrix components are synthesized locally by connective tissue cells. The most common type of connective tissue cell is the *fibroblast* (Fig. 13.17). Fibroblasts are migratory cells that are prevalent in embryonic mesenchyme and play an important part in wound healing. They are also convertible into other types of connective tissue cells, such as *osteoblasts* (bone-forming cells), *chondrocytes* (cartilage cells), and *adipocytes* (fat cells). In addition, fibroblasts can form smooth muscle cells, as we will see later on. For the remainder of this section, we will consider the formation of two abundant connective tissues: cartilage and bone.

Cartilage tissue originates from mesenchyme in several regions of the embryo, including the skull, vertebrae, and limbs. The chondrocytes secrete large amounts of extracellular matrix, which changes in composition as the cartilage matures. Cartilage grows on the inside when chondrocytes divide and secrete more matrix, and grows on the outside by converting adjacent fibroblasts into chondrocytes. As the amount of extracellular material increases, the chondrocytes become more widely separated (Fig. 13.18). Typical cartilage contains no blood vessels; the chondrocytes are sustained instead by diffusion of materials through the extracellular matrix.

The formation of bone tissue involves different types of cells. The osteoblasts, which develop from fibroblasts, as noted previously, lay down the extracellular bone matrix and eventually become entrapped in it as *osteocytes.* The extracellular matrix produced by osteoblasts has a tendency to calcify instead of taking up water as cartilage does. Another part of bone development is bone removal, a function carried out by multi-

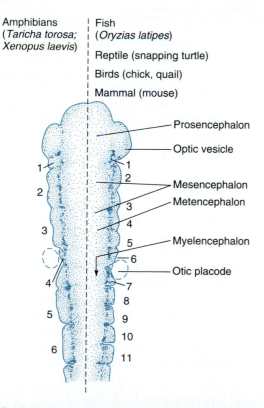

Amphibians
(*Taricha torosa;*
Xenopus laevis)

Fish
(*Oryzias latipes*)

Reptile (snapping turtle)

Birds (chick, quail)

Mammal (mouse)

— Prosencephalon

— Optic vesicle

— Mesencephalon

— Metencephalon

— Myelencephalon

— Otic placode

Figure 13.16 Somitomeres (numbered) in the heads of vertebrates. In all forms examined, the first somitomere is next to the prosencephalon. Caudal to the first somitomere, for each somitomere in amphibians, other vertebrates have two. The somitomeres behind the otic placode form somites separated by clefts. The more anterior somitomeres contribute to head mesenchyme that in turn forms connective tissues.

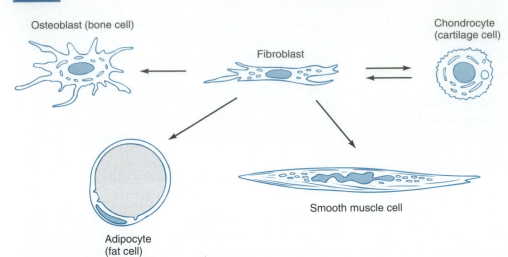

Osteoblast (bone cell)

Fibroblast

Chondrocyte (cartilage cell)

Smooth muscle cell

Adipocyte (fat cell)

Figure 13.17 Cells that are derived from fibroblasts in vertebrates. The arrows indicate the interconversions observed. For simplicity, only one type of fibroblast is shown, although there may actually be several types, perhaps with restricted potentials.

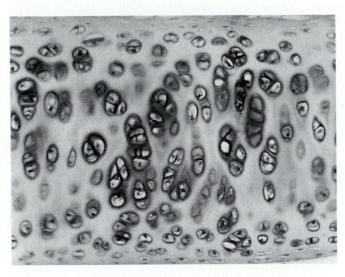

Figure 13.18 Cartilage from growing human bone shown in a light micrograph of a fixed and stained section. The cartilage cells, or chondrocytes, are embedded in abundant extracellular material. The round or oval cavities in the extracellular material are called lacunae. A living chondrocyte completely fills its lacuna, but the fixation process causes chondrocytes to shrink.

nucleate cells called *osteoclasts.* They belong to the family of blood cells, as do the bone marrow cells.

The growth and maintenance of bone requires a delicate balance between the deposition of new matrix and the degradation of previously deposited material. Typically, a healthy mammalian organism destroys and regenerates 5 to 10% of its bone every year. This constant remodeling makes it possible for bone to respond to external pressure or tension with the formation of properly oriented beams or ridges of extracellular material known as *trabeculae* (Lat. *trabecula,* "little beam"). If bone degradation outpaces the deposition of new bone material, bones become fragile, a condition known as *osteoporosis.*

Bone tissue originates in two ways. Some bones, such as the flat bones of the skull, are formed directly from mesenchyme in a process called *membranous ossification* (Fig. 13.19). It begins with the assembly of type I collagen, a fibrous extracellular matrix component (see Chapter 26). Osteoblasts line up along the fibrils and synthesize a matrix that hardens from deposition of calcium hydroxyapatite. The resulting *trabeculae* eventually interconnect to form a meshwork of bone, with marrow filling the intervening spaces.

For most bones, however, chondrocytes first form a cartilage model, which is subsequently replaced in a process known as *endochondrial ossification.* This mode of bone formation is observed especially in the vertebrae and in the long bones of limbs. During endochondrial ossification, some of the cartilage matrix becomes mineralized from the deposition of calcium crystals. The chondrocytes swell and die, leaving large cavities. Osteoclasts and cells that form blood vessels invade the cavities and erode the residual cartilage, while osteoblasts deposit additional bone matrix. In the long bones of newborn vertebrates, cartilage is formed at the ends of the bones, called *epiphyses,* while mineralization occurs in the central part of the bone, the *diaphysis* (Fig. 13.20). In juvenile vertebrates, the terminal ends of the epiphyses also mineralize while subterminal portions, the *epiphyseal discs,* remain cartilaginous until longitudinal growth ceases. In adults, only thin layers of cartilage remain at those surface areas of bones that form joints.

Skeletal Muscle Fibers Arise through Cell Fusion

Muscle cells, like many other differentiated cells, have taken a general cellular feature to a high degree of specialization. A contractile system involving the proteins actin and myosin is present in most animal cells, and families of actin and myosin genes encode differ-

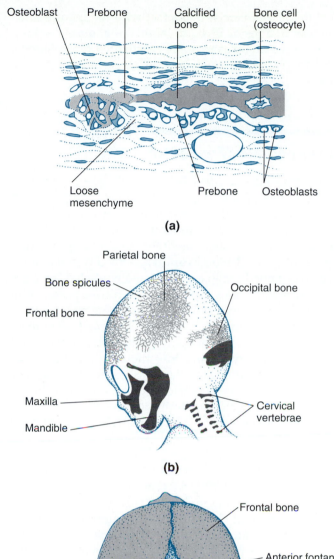

(a)

(b)

(c)

Figure 13.19 Membranous ossification in the human skull. **(a)** Mesenchymal fibroblasts are converted into osteoblasts, which will lay down prebone extracellular matrix. This material turns into bone as it calcifies, trapping some of the osteoblasts as osteocytes. **(b)** Bones of the skull in a 3-month-old human embryo. The webbed areas represent bone spicules that will grow together to form the flat bones of the skull. (The bones shown in black originate, at least in part, through endochondral ossification.) **(c)** Skull of a newborn seen from above. Note the open areas (fontanelles) where the flat bones are incompletely closed.

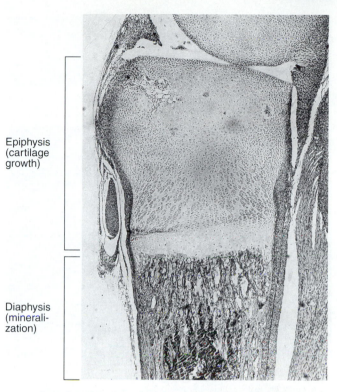

Figure 13.20 Bone formation by endochondrial ossification in the newborn rat. The light micrograph shows the proximal (upper) end of the tibia (shin bone) in longitudinal section. In the cartilaginous ends of the bone (epiphyses), chondrocytes proliferate and grow in size. In the middle part of the bone (diaphysis), extracellular matrix laid down by the chondrocytes hardens by deposition of calcium crystals (mineralization).

ent versions of these proteins that are optimized for different functions. Muscle cells have developed especially large and organized filament systems composed of certain actins and myosins along with other molecules that control muscle contraction. *Skeletal* muscle tissue develops from the *myotome* portions of the somites (Fig. 13.15). Other types of muscle tissue, such as *smooth muscle* and *cardiac muscle,* derive from lateral plates.

The cells in the myotomes that form skeletal muscle are called **myoblasts.** After a period of proliferation, myoblasts fuse with one another to form multinucleate **myotubes** (Fig. 13.21). Myotubes develop into **muscle fibers** by recruiting additional myoblasts and by assembling a characteristic arrangement of actin and myosin fibrils. Muscle fibers are the building blocks of skeletal muscle; they are about 0.1 mm in diameter and can be half a meter long in humans. Muscle fibers are contractile, have a characteristic striated appearance, and are surrounded by a *basal lamina.*

The process of skeletal muscle fiber formation, called **myogenesis,** has been studied in vitro and in vivo. In vitro, myoblasts migrate and form chains before they fuse (Nameroff and Munar, 1976). Chain formation is sufficient to stop further mitoses of myoblasts. Actual

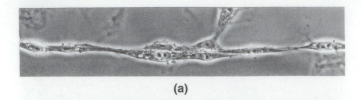

(a)

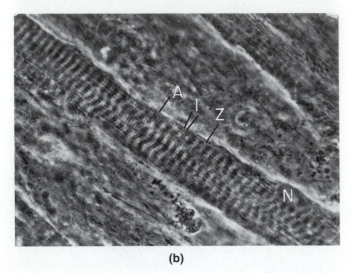

(b)

Figure 13.21 Myogenesis of embryonic cells in tissue culture. **(a)** After 2 days of culture, myoblasts line up to fuse and form myotubes. **(b)** After 12 days of culture, the cytoplasm of myofibrils shows the characteristic cross striation (A, I, and Z bands) of skeletal muscle fibers. The position of the nucleus (N) at the periphery is also typical of advanced differentiation.

cell fusion seems to require some recognition event between myoblasts, since they do not fuse with other cell types. In vivo, the differentiation of trunk muscles depends on signals that emanate from the notochord and the neural tube. Separation of the neural tube–notochord complex from the somites by a surgical slit on one side of the embryo is followed by massive cell death in the somites and failure of most myotome-derived muscles to develop. These effects are observed only on the operated side; the nonoperated side develops normally (Rong et al., 1992). Thus, a signal originating from the axial organs must be necessary for the survival of most somitic cells. Surprisingly, limb muscles and some body-wall muscles, although derived from somites, develop in the absence of contact with the axial organs. In parallel to this distinction, the regulatory gene activities that control myogenesis differ between trunk and limb muscles (see Chapter 19).

The adult number of muscle fibers is attained early in development, before birth in mammals. The subsequent increase in muscle mass is achieved mostly through the enlargement of existing muscle fibers. They grow in length by recruiting more myoblasts at their ends. Growth in girth occurs in response to training and depends mainly on an increase in the number of actin and myosin fibrils per muscle fiber. However, a few myoblasts persist in contact with the mature muscle fiber and inside its basal lamina. If a muscle is damaged, these myoblasts can be reactivated to proliferate, and their descendants can fuse to form new muscle fibers.

Intermediate Mesoderm

The *intermediate mesoderm* is located between the somites and the lateral plates (Fig. 13.12). It gives rise to the kidneys and some of the reproductive structures. The kidneys of all vertebrates are composed of similar functional units called *nephrons* (Fig. 13.22). Each nephron contains a coiled blood capillary, called a *glomerulus,* which is embedded in a cup-shaped epithelial structure known as *Bowman's capsule.* Blood filtrate passes from the glomerulus into the lumen of Bowman's capsule. The capsule is drained by a *nephric tubule* (called *renal tubule* in adults), where many components of the blood filtrate are absorbed back into the surrounding tissue. The remaining fluid, the urine, is collected in *nephric ducts.* There is one nephric duct on each side of the body, and the two ducts drain into the *cloaca.*

Embryologists distinguish three types of vertebrate kidney (Fig. 13.23), which are referred to as the *pronephros* (Gk. *pro,* "before"; *nephros,* "kidney"), the *mesonephros* (Gk. *mesos,* "middle"), and the *metanephros* (Gk. *meta,* "after"). All three types of kidney arise from the *nephrogenic mesenchyme,* a dorsolateral cord of mesenchyme that derives from the intermediate mesoderm and extends from the head to the cloaca. The more advanced vertebrates repeat in their development the

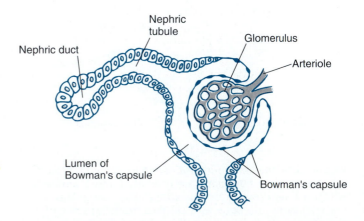

Figure 13.22 Schematic drawing of an embryonic nephron in transverse section. Blood from a cluster of blood capillaries, the glomerulus, is filtered into the lumen of Bowman's capsule. Each capsule opens into a nephric tubule, where many components are recovered from the blood filtrate. The remaining fluid, the urine from all tubules, is collected in two nephric ducts, which empty into the cloaca.

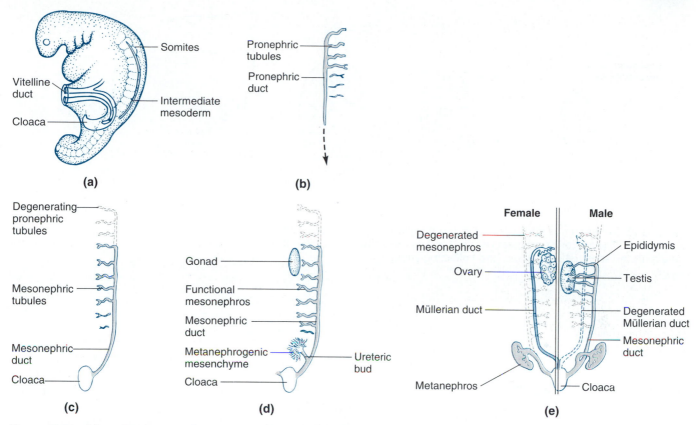

Figure 13.23 Kidney development in mammals. (a) Schematic diagram to show the location of the intermediate mesoderm. (b) Stage of pronephros development, with pronephric duct extending toward the cloaca. (c) Establishment of the mesonephros. (d) Early stage of metanephros development. (e) Differentiation of the male and female urogenital systems. In the male, the mesonephric duct is converted to the epididymis of the reproductive system. In the female, the mesonephros degenerates, and the Müllerian duct forms much of the reproductive tract. The metanephros develops similarly in both sexes.

embryonic kidney forms of their ancestors, again illustrating recapitulation (Torrey and Feduccia, 1991).

The most primitive type of vertebrate kidney, the *pronephros,* is located cranially in the body (Fig. 13.23). Segmental units of the pronephros, known as *nephrotomes,* give rise to nephrons, which filter and drain body fluid via *pronephric tubules* into a longitudinal canal, the *pronephric duct.* The pronephros and the pronephric duct originate from adjacent portions of the intermediate mesoderm; the pronephric duct grows posteriorly toward the cloaca, apparently being guided by cues from the extracellular matrix deposited by other cells (Fig. 13.24). Although the pronephros is formed in all vertebrate embryos, it persists as an adult excretory organ only in a few fishes.

In most vertebrates, the pronephros degenerates early in development and is replaced by the *mesonephros,* a new set of nephrons that form in the central section of the intermediate mesoderm. The *mesonephric tubules* hook up to the pronephric duct, which is then called the *mesonephric duct* or *Wolffian duct.* In the hu-

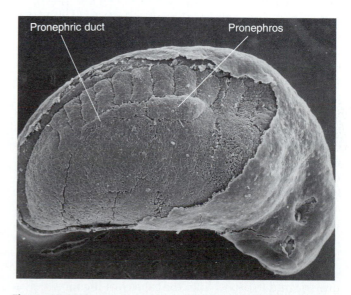

Figure 13.24 Pronephros of a salamander embryo. For this scanning electron micrograph, the epidermis has been removed to expose the underlying mesoderm. Intermediate mesoderm gives rise to the pronephros and the pronephric duct.

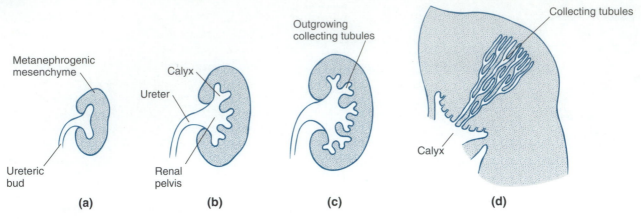

Figure 13.25 Metanephros development in the human embryo: **(a–c)** 5–7 weeks; **(d)** newborn. The ureteric bud gives rise to the structures that drain urine from the kidney: ureter, renal pelvis, calyces, and collecting tubules.

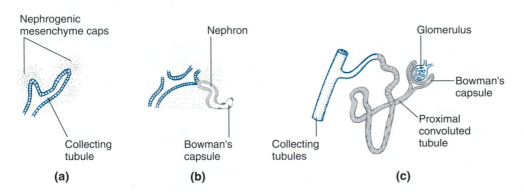

Figure 13.26 Nephron development in the metanephros. The tips of the collecting tubules induce caps of nephrogenic mesenchyme to form nephrons. The nephrons establish open communication with the collecting ducts, enabling urine to flow from the glomerulus to the renal pelvis.

man embryo, the mesonephros forms an elongated organ on the inside of the dorsal body wall. The mesodermal portion of the gonad, the *genital ridge,* develops on the medial border of the mesonephros. The mesonephros is the principal excretory organ of almost all vertebrate embryos and remains as the adult kidney in most fishes and amphibians.

A third excretory organ, the *metanephros,* develops in reptiles, birds, and mammals. The metanephros arises from two embryonic rudiments. One rudiment is a caudal portion of the nephrogenic mesenchyme, and the other rudiment is the *ureteric bud,* a small outgrowth of the mesonephric duct near the cloaca (Fig. 13.23). The ureteric bud gives rise to the *ureter,* the *renal pelvis* and its branches (*calyces*), and *collecting tubules* (Fig. 13.25). As the bud penetrates the nephrogenic mesenchyme, it induces the mesenchyme to form nephrons (Fig. 13.26). Conversely, the mesenchyme is necessary for the ureteric bud to branch into the calyces of the renal pelvis and into the numerous collecting ducts. Similar forms of reciprocal tissue interaction have been observed in other organs that develop from more than one rudiment.

As the metanephros develops, the mesonephros degenerates, except that the Wolffian duct gives rise to parts of the male reproductive system. The *Müllerian duct,* originally formed parallel to the Wolffian duct in both sexes, degenerates in males but gives rise to parts of the female reproductive system (see Chapter 28).

Lateral Plates

The lateral mesoderm occupies a lateral or ventral position in the body (Fig. 13.12). Very early in development, it splits into two layers collectively known as the *lateral plates.* The *somatic layer* develops in association with the ectoderm, while the *visceral layer* develops in association with the gut and other internal organs. The space between the two layers is called the embryonic *coelom.* Thus, the lateral plates form the lubricated membranes that line the body cavities derived from the coelom. Another major derivative of the visceral layer is the cardiovascular system. In addition, the visceral layer gives rise to smooth muscle and connective tissues surrounding the blood vessels, intestines, and other internal organs. The somatic layer, on the other hand, contributes the connective tissues of the body wall and the limbs.

The Lateral Plates
Surround the Coelomic Cavities

The subdivision of the coelom is a complicated process and is only summarized here. Initially, the coelom is divided into two lateral compartments separated by a median partition, the *primary mesentery,* which consists of the visceral layers from each side. As it is formed, the primary mesentery attaches the embryonic gut to the dorsal and ventral midlines. However, the ventral part of the mesentery disappears soon thereafter; what is left

is called the *dorsal mesentery* (Fig. 13.27a). Next, a ventral anterior portion of the coelom is closed off as a small *pericardial cavity* surrounding the heart (Figs. 13.27b–d). The membranes of the pericardial cavity, together called the *pericardium,* consist of a visceral layer applied to the heart and a somatic layer originally applied to the body wall. Later, the somatic layer is displaced from the body wall, so that the heart comes to lie in a separate pericardial sac.

In land-dwelling vertebrates, the lungs grow out from the foregut and around the heart. The part of the coelomic cavity surrounding each lung is called a *pleural cavity* (Gk. *pleura,* "side"), again with the visceral layer applied to the lung and the somatic layer lining the body wall (Fig. 13.28). The pleural cavities are separated from each other and from the pericardial cavity. In mammals, these cavities are also walled off by the *diaphragm* from

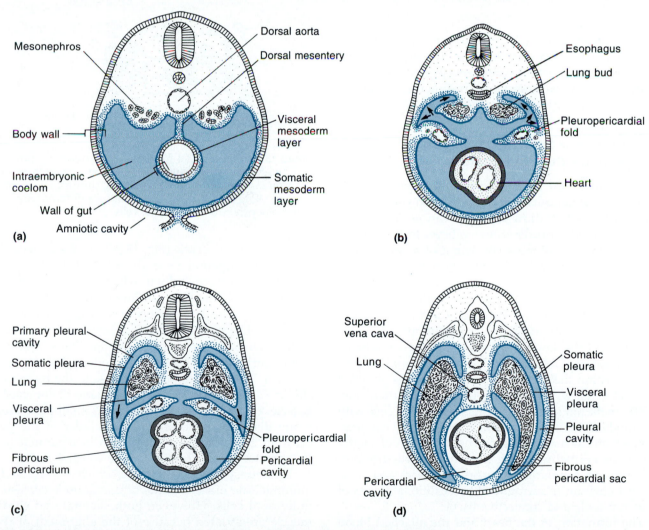

Figure 13.27 Coelomic cavities in the human embryo. **(a)** Transverse section through the abdominal region at the end of the fourth week. The ventral part of the primary mesentery has disappeared, leaving the dorsal mesentery. **(b)** Transverse section through the heart region at the end of the fifth week. The pleuropericardial folds begin to close off the pericardial cavity from the pleural cavity. **(c, d)** During subsequent stages, the growing lungs displace the pleuropericardial folds, and a separate pericardial sac is formed.

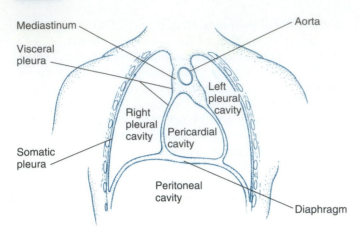

Figure 13.28 Coelomic cavities in the human adult. The abdominal (peritoneal) cavity is separated by the diaphragm from the thoracic cavities. In the thorax, two pleural cavities containing the lungs are separated from each other by the mediastinum, a mass of connective tissue that contains major blood vessels and the pericardial cavity with the heart.

the *peritoneal cavity* (abdominal cavity), which contains the stomach, the remainder of the digestive tract, and other internal organs. The dome-shaped diaphragm arises from several embryonic rudiments, including skeletal muscle derived from myotomes.

Cardiovascular System Development Recapitulates the Phylotypic Stage

The *cardiovascular system,* consisting of heart, blood vessels, and blood cells, begins to function very early in development when diffusion alone can no longer satisfy the metabolic needs of the embryo. In the human embryo, a heart rudiment consisting of a single tube, known as the *heart tube,* begins to beat around the 23d day of development.

All components of the cardiovascular system develop from the visceral layer of the lateral plates. Mesenchymal cells in this layer proliferate and form clusters of cells known as *angiogenetic*—or *angiogenic*—*clusters* (Gk. *angeion,* "vessel"; *gennan,* "to produce"). Within these clusters, the central cells give rise to blood cells, while the peripheral cells flatten (Fig. 13.29). As the angiogenetic clusters sprout and fuse with one another, the flat peripheral cells form the *endothelium,* which has the histological characteristics of a tight, squamous epithelium. This elementary structure persists in blood capillaries, whereas in arteries and veins the endothelium is surrounded by additional layers of smooth muscle and connective tissue.

The formation of the heart and the attached blood vessels in the ventral neck region occurs in the same fashion. The inner lining of the heart originates from paired endothelial tubes on the right and left sides, which in the heart region are called *endocardial tubes,* but which are continuous with the endothelium of the

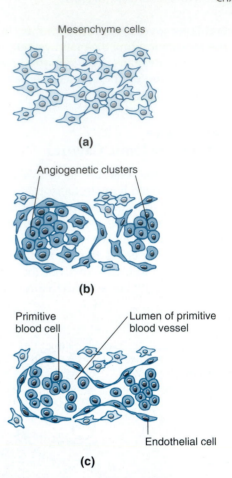

Figure 13.29 Blood vessel formation in vertebrates: **(a)** undifferentiated mesenchymal cells in the visceral layer of the lateral mesoderm; **(b)** angiogenetic clusters; **(c)** primitive capillary. Note that the mesenchymal cells differentiate into either blood cells or endothelial cells.

adjoining blood vessels (Fig. 13.30). As the endocardial tubes grow together ventrally beneath the gut, they form a united *endocardium,* which is the rudiment of the inner heart. The endocardium is surrounded by two layers of visceral mesoderm: *cardiac jelly* and *myocardium.* The *cardiac jelly* is a layer of extracellular matrix rich in *hyaluronic acid,* which facilitates cell movements during the subsequent remodeling of the endocardium. The *myocardium* gives rise to the heart muscle, or *cardiac muscle.* On the outside of the cardiac muscle, other mesoderm cells form a slippery membrane that represents the visceral layer of the pericardium. The visceral pericardium is separated from the somatic pericardium by the pericardial cavity.

In contrast to skeletal muscle, which is composed of multinucleate muscle fibers, cardiac muscle consists of individual cells. However, both skeletal and cardiac muscle are striated because of the alignment of their actin and myosin fibrils.

The embryonic heart tube is subdivided into four longitudinal sections, named—from posterior to anterior, and in the direction of the blood flow—*sinus venosus, atrium, ventricle,* and *truncus arteriosus* (Fig.

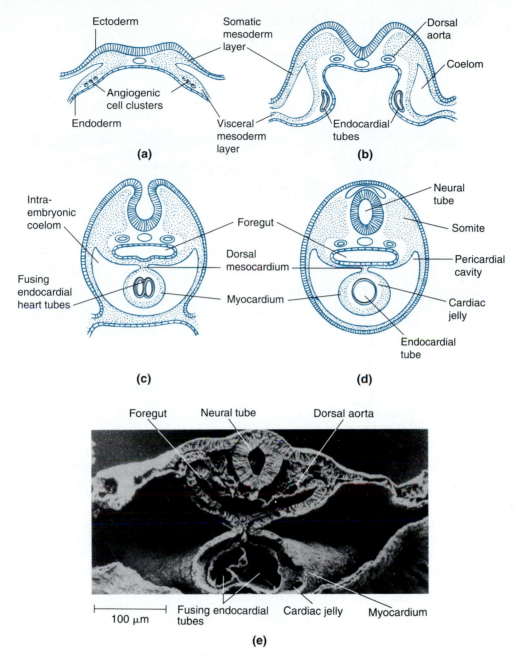

(a)

(b)

(c)

(d)

(e)

Figure 13.30 Formation of the endocardium by fusion of paired endocardial tubes and subsequent investment with cardiac jelly and myocardium. **(a–d)** Schematic cross sections of human embryos aged 17 days, 18 days, 21 days, and 22 days, respectively. **(e)** Scanning electron micrograph of a transversely fractured chicken embryo at a stage comparable to the 21-day human embryo.

13.31). This primitive arrangement is fairly well conserved in adult sharks and bony fishes (Fig. 13.32). In these animals, blood from the truncus arteriosus enters the *ventral aorta.* The ventral aorta gives off paired arteries called *aortic arches,* which are located inside the pharyngeal arches described earlier in this chapter (Fig. 13.5). Here the deoxygenated blood is pumped with maximum pressure through the narrow capillaries of the gills. From the gills the blood is collected in paired *dorsal aortae,* which unite caudally and supply the large arteries of the body with oxygenated blood.

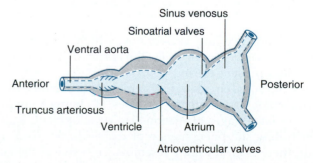

Figure 13.31 Schematic diagram of the primitive vertebrate heart.

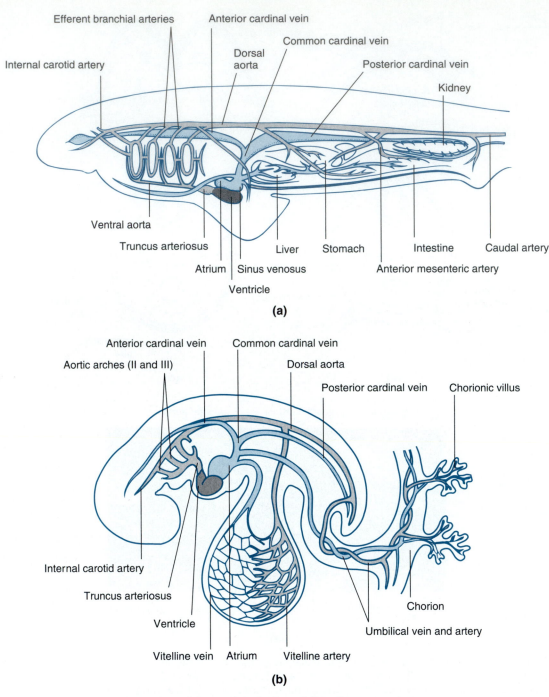

Figure 13.32 Heart and main blood vessels in **(a)** an adult shark and **(b)** a human embryo at the end of the fourth week. Only the blood vessels on the left side are shown.

In land-dwelling vertebrates, which do not need blood vessels to supply gills, the aortic arches are nevertheless present during the *phylotypic stage*. Whereas this embryonic circulatory system is modified little during further development of fishes, it changes greatly during later stages of mammalian development, again illustrating the phenomenon of *recapitulation*. As a result, there is an uncanny resemblance between the circulatory systems of a human embryo and an adult shark (Fig. 13.32). However, as often occurs in recapitulation, development traces evolution in a rather sloppy way. In humans the six aortic arches are never present simultaneously. The two most cranial arches are modified early, and the fifth arch is often underdeveloped or missing altogether.

Most of the aortic arches in mammals acquire secondary fates (Fig. 13.33). The first two aortic arches, located in pharyngeal arches I and II, form minor arteries

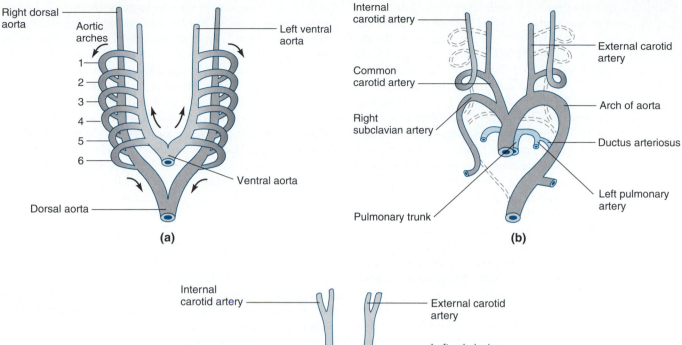

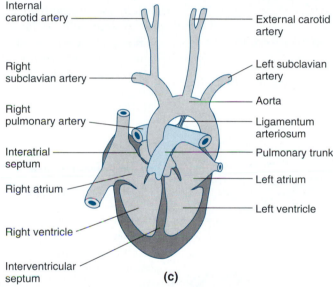

Figure 13.33 Fate of the aortic arches in mammalian development. **(a)** Early embryonic pattern. Arrows indicate the blood flow from the ventral aorta through the aortic arches to the dorsal aortae. **(b)** Degeneration of arches 1, 2, and 5 and transformation of arches 3, 4, and 6. Note the ductus arteriosus as a vestige of the sixth aortic arch on the left side. **(c)** Heart and major arteries in the adult. An interatrial septum has separated the right atrium from the left atrium. Similarly, the ventricle has been partitioned by an interventricular septum. The truncus arteriosus has been divided into the aorta, which emerges from the left ventricle; and a pulmonary trunk, which emerges from the right ventricle. A vestige of the ductus arteriosus persists as the ligamentum arteriosum.

in the head. The third aortic arch forms the large *carotid arteries.* The fourth arch forms part of the *aorta* on the left side and a segment of the *right subclavian artery* on the right. The fate of the sixth arch is particularly interesting, because it gives off a branch toward the developing lung on each side. This branch, together with the proximal part of the sixth aortic arch, gives rise to the *pulmonary artery* on each side. The distal part of the sixth arch on the left side forms the *ductus arteriosus* (not to be confused with the *truncus arteriosus*). The

ductus arteriosus, which connects the left pulmonary artery with the aorta, shunts blood past the lung before birth and constricts soon thereafter into a ligament, the *ligamentum arteriosum* (Fig. 13.33c).

Also recapitulated during mammalian development is the evolution of the *pulmonary circulation,* which drives blood through the lungs and is separate from the *systemic circulation* supplying the rest of the body. The embryonic heart tube, which undergoes little further modification in fishes, undergoes a dramatic series of additional morphogenetic movements in mammals. As part of these movements, an *atrial septum* develops between the right half of the atrium, which receives blood from the body, and the left half, which receives blood from the lungs (Fig. 13.33c). Similarly, the ventricle is divided into right and left halves by the growth of a *ventricular septum*. At the same time, the atrioventricular canal between the old atrium and ventricle is remodeled so that the new atria and ventricles on each side have separate connections. Moreover, the truncus arteriosus splits into the aorta and the pulmonary trunk. The *aorta* connects with the left ventricle and pumps blood into the body, while the *pulmonary trunk* carries blood from the right ventricle to the pulmonary arteries. The split of the truncus arteriosus is accomplished by ridges on the inside, which grow together in a spiral pattern foreshadowing the way in which the pulmonary trunk and the aorta later twist around each other.

Smooth Muscle Consists of Single Cells

A third type of muscle tissue, which under the microscope does not have the striated appearance of skeletal and cardiac muscle, is known as *smooth muscle.* It consists of single cells that are derived from mesoderm or ectomesenchyme. Smooth muscle tissue derived from the visceral layer of lateral mesoderm is found in the digestive tract and its appendages, such as glands and respiratory passages. Smooth muscle in blood vessels controls vessel diameter and thus the blood pressure and the rate of blood flow. However, smooth muscle also occurs in the urinary and genital ducts, notably in the oviducts and uterus. Mammalian skin contains minute smooth muscle cells that can erect the hairs. In the eye, smooth muscle is responsible for lens accommodation and regulating the diameter of the pupil.

A smooth muscle cell is long and spindle-shaped, with a single nucleus positioned about halfway along its length (Fig. 13.34). When smooth muscle cells form bundles or sheets, the cells are offset so that the thick central portion of one cell is juxtaposed to the thin end of another. Smooth muscle represents a primitive type of muscle cell, in its similarity to nonmuscle cells. Smooth muscle is not striated, because its contractile filaments are not arranged in the strictly ordered pattern observed in skeletal and cardiac muscle. Instead, the fibers are attached obliquely to the plasma membrane at disclike junctions that connect smooth muscle cells.

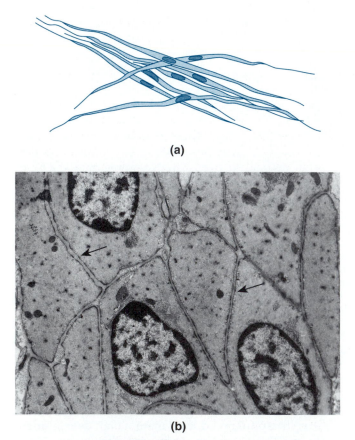

(a)

(b)

Figure 13.34 Smooth muscle cells. **(a)** Isolated smooth muscle cells from the stomach wall of a cat. **(b)** Transmission electron micrograph of smooth muscle in transverse section. The dense areas (arrows) around the inside of the plasma membrane are sites of insertion of contractile filaments.

Limb Formation

Vertebrate limbs develop relatively late during embryogenesis, when most other organs have already been established. In the human embryo, paddle-shaped *limb buds* emerge during the fifth week, and digits are formed thereafter (Fig. 13.35). Limb buds have a core of mesenchyme cells and a covering layer of ectoderm. The ectodermal layer usually forms the *apical ectodermal ridge (AER),* a terminal thickening running in the anteroposterior direction at the end of the limb bud. The mesenchymal core originates in part from the somatic layer of the *lateral plate* and in part from *myotomes.*

To investigate the contributions from these two sources, B. Christ and colleagues (1977) removed a strip of newly formed somites from chicken embryos and replaced them with a corresponding strip of quail somites. (Quail cells and their descendants can be distinguished from chicken cells by their brightly staining het–erochromatin; see Chapter 12.) After incubation, the investigators found that the somitic cells formed muscle while somatic lateral plate mesoderm gave rise to cartilage and other connective tissues of the limb.

The limb bud is gradually shaped into a limb by differential growth, by *programmed cell death,* and by histogenesis. Cell divisions accompanied by cell growth during interphase generate the mass of the limb and its

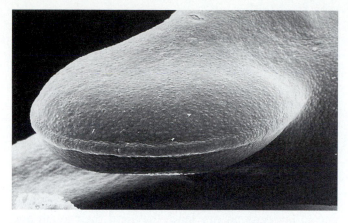

Figure 13.35 Limb bud of a hamster embryo as seen with the scanning electron microscope. Note the apical ectodermal ridge at the distal rim. Figure 13.1 shows an entire human embryo at a similar stage of development.

The Principle of Reciprocal Interaction

The analysis of limb morphogenesis reveals the *principle of reciprocal interaction* during development. Many interactions between tissues are not one-way communications but are more of a dialogue. Developmental cues are exchanged in both directions to balance and maintain the activities of all parts involved so that a harmoniously proportioned organ results.

Reciprocal interactions between the AER and the mesenchymal core of limb buds have been analyzed especially well in chicken embryos (see Hinchliffe and Johnson, 1980). First, limb bud mesenchyme induces the formation of the AER in the overlying ectoderm. Reuss and Saunders (1965) demonstrated this effect by transplanting wing bud mesenchyme that had been stripped of its ectodermal cover into the flank of a chicken embryo. Regenerating flank ectoderm soon healed over the wound. In many cases, an AER was induced over the graft, which then developed into an extra wing. The specific character of a developing limb is determined by the mesenchyme: leg mesenchyme combined with wing ectoderm forms a leg, whereas wing mesenchyme combined with leg ectoderm forms a wing. The inductive capacity of the mesenchyme, and the competence of the ectoderm to respond to it, disappear after the stage when the AER is established in normal development.

Once established, the AER becomes necessary for the further proliferation of the limb bud mesenchyme. If the AER is removed, the mesenchyme ceases to proliferate, and the developing limb is truncated. The earlier the removal of the AER, the more the distal limb structures that fail to develop (Fig. 13.37). When two AERs are combined with a single mesenchymal core, the mesenchyme grows out under each AER. However, apart from promoting limb outgrowth, the action of the

approximate shape. The limb bud's contour is refined by waves of cell death sweeping along the edges. Cell death also causes erosion of the tissues between the digits, thus sculpting the fingers and toes. (In ducks and other web-footed vertebrates, interdigital cell death is limited to the distal margin of the limb.) While the external shape of the limb is still being established, mesenchyme condensations near the center of the limb form the cartilage models that foreshadow the development of bones (Fig. 13.36). Other cells positioned more to the outside of the limb bud form myogenic condensations, the future muscles of the developing limb.

The genetic control of programmed cell death has been investigated most extensively in roundworms and will be discussed in Chapter 24. Models that try to account for the spatial patterning in which growth, cell death, and different types of histogenesis occur will be presented in Chapters 20 and 22. In the following section, we will examine some of the cellular interactions between the AER and the underlying mesenchyme.

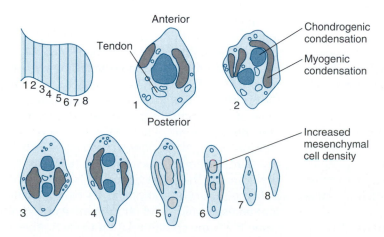

Figure 13.36 Cell condensations foreshadowing the formation of cartilage and muscle in limb buds. Outline of a chicken hindlimb (top left) indicating the position of the accompanying sections (1–8). Chondrogenic condensations (color) form cartilage, myogenic condensations (dark gray) form muscle, areas of increased mesenchymal cell density (light gray) give rise to either. Blood vessels are depicted in outline only.

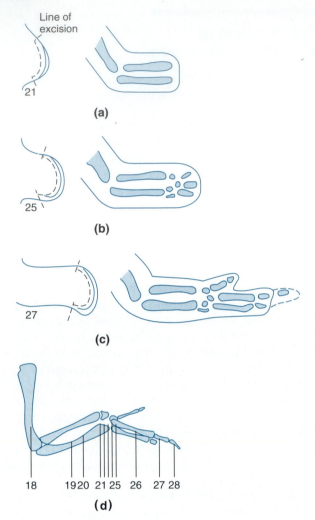

Figure 13.37 Effect of excising the apical ectodermal ridge (AER) from the chicken wing bud. **(a–c)** Excision at later stages of development is followed by truncation at more distal levels. **(d)** Summary diagram showing the level of truncation after AER excision at stages numbered 18 (earliest) to 28 (latest).

AER seems permissive rather than instructive. AERs can be exchanged between wing and leg buds without affecting the developing wing or leg pattern of the mesenchyme (Zwilling, 1955). Similarly, if the anteroposterior axis of the AER is reversed relative to the underlying mesenchyme, the developing limb pattern is not affected (Zwilling, 1956).

Conversely, the persistence of the AER depends on a maintenance factor from the mesenchyme. If limb mesenchyme is removed from a limb bud and replaced with nonlimb mesoderm, such as flank lateral plate or somites, the AER degenerates within 2 days. However, the AER survives if only a small piece of limb mesenchyme is added.

The principle of reciprocal interactions is also encountered in the development of the metanephros from the ureteric bud and the nephrogenic mesenchyme, as discussed earlier, and in the organogenesis and histogenesis of many other organs that develop from mesenchymal and epithelial components.

Extraembryonic Membranes

To fish and amphibian embryos, the water in which they develop affords easily accessible food, protection against desiccation and mechanical shock, and a vast repository for metabolic waste. Since reptiles first began to lay eggs on dry land, they and their avian and mammalian descendants have had to compensate for the benefits lost by leaving the water. Birds and most reptiles develop inside hard eggshells, and most mammals inside the maternal uterus. In these three classes of vertebrates, membranes outside the embryo proper—collectively called *extraembryonic membranes*—provide for nourishment, protection, respiration, and excretion. One of these membranes, directly surrounding the embryo, is known as the *amnion*. Because they have this feature in common, reptiles, birds, and mammals have become known by the collective term *amniota.*

The amniota have four principal extraembryonic membranes (Fig. 13.38). Two of them, the *amnion* and *chorion,* originate from raised folds of a double membrane called *somatopleure* because it consists of an ectodermal layer and the adjacent somatic layer of lateral plate mesoderm. The two other extraembryonic membranes, the *yolk sac* and the *allantois,* develop essentially as parts of the gut. Accordingly, these membranes consist of an endodermal layer with the adjacent visceral layer of lateral plate mesoderm, together called *splanchnopleure.* The terms "somatopleure" and "splanchnopleure" are also used for the corresponding double membranes within the embryo.

The Amnion and Chorion Are Formed by Layers of Ectoderm and Mesoderm

The amnion and chorion of reptiles and birds originate together when folds of the somatopleure arise on all sides of the embryo and fuse dorsally above (Fig. 13.38c and d). The embryo is then covered by two layers of somatopleure. The inner layer, with the ectoderm facing the embryo, is the *amnion.* The outer layer, with the ectoderm facing the eggshell or uterus, is called the *chorion* or *serosa.* The space between amnion and chorion is the *extraembryonic coelom,* which is continuous with the embryonic coelom. The cavity enclosed by the amnion, the *amniotic cavity,* is filled with fluid so that the embryo is afloat in its own private pool. In mammals, amnion and chorion arise differently but assume the same spatial relation to each other and to the embryo.

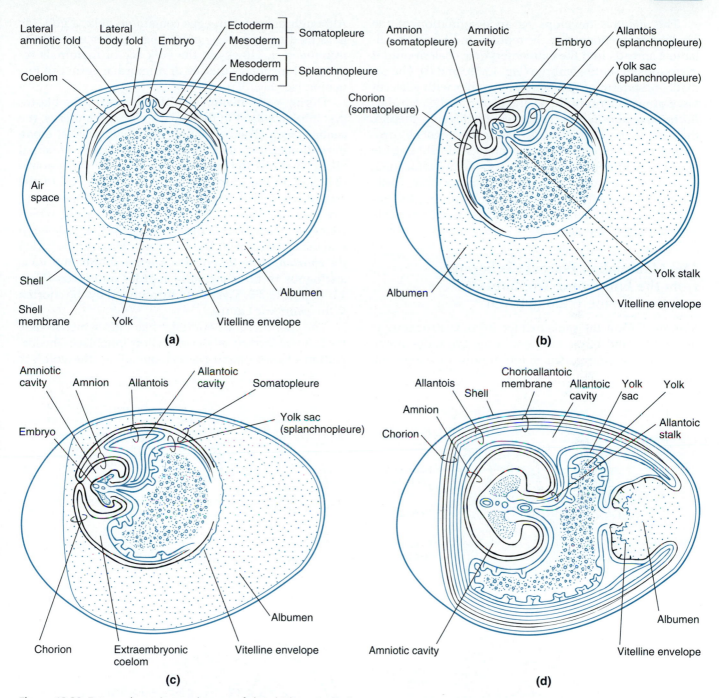

Figure 13.38 Extraembryonic membranes of the chicken. Each diagram represents a transverse section through the embryo. After **(a)** 2 days of incubation; **(b)** 3 days; **(c)** 5 days; **(d)** 14 days.

The Yolk Sac and Allantois Are Formed by Layers of Endoderm and Mesoderm

In reptile and bird embryos, both splanchnopleure and somatopleure grow out in all directions over the uncleaved egg yolk. Where these membranes extend beyond the region of embryo formation, they are considered extraembryonic. When the splanchnopleure has overgrown the egg yolk, it is called the *yolk sac*. The yolk sac turns into a large appendage of the midgut when the hindgut and foregut close to form tubes (Figs. 13.2 and 13.3). Since the embryonic and the extraembryonic portions of the splanchnopleure are continuous, and because the formation of angiogenetic clusters occurs simultaneously in both portions, the developing blood vessels carry the digested yolk components directly to the embryo. A yolk sac is found in all vertebrates except amphibians, whose embryonic cells have their own yolk supply.

The *allantois* develops as an evagination of the hindgut splanchnopleure. In reptiles and birds, the allantois spreads into the extraembryonic coelom until it surrounds the entire embryo (Fig. 13.38c and d). The somatic mesoderm of the chorion, together with adjacent visceral mesoderm of the allantois, forms the ***chorioallantoic membrane.*** As blood vessels develop therein, the blood circulates directly under the eggshell, providing for an effective gas exchange with the outside world. In addition to its function as an embryonic lung, the allantois serves as a repository for metabolic waste products, especially uric acid. In some groups of mammals, the allantois retains some of these functions while contributing to the placenta.

The Mammalian Placenta Is Formed from the Embryonic Trophoblast and the Uterine Endometrium

Mammals form the same extraembryonic structures as do reptiles and birds, but certain modifications occur, apparently as adaptations to intrauterine development.

Although mammalian eggs contain no yolk, a yolk sac is formed nevertheless, providing another example of *recapitulation.* However, the yolk sac of mammals retains its function as the site of primordial germ cell formation (see Fig. 3.3).

During early mammalian development, the blastocyst implants itself in the inner lining of the uterus, the ***endometrium.*** The cells of the *syncytiotrophoblast* erode the endometrium, including its blood vessels. Maternal blood then fills the *lacunae* created by the syncytiotrophoblast. In the human, this occurs at the end of the second week when the embryo is a bilaminar germ disc (see Figs. 10.33 and 13.39). The ***connecting stalk,*** which connects the embryo to the *cytotrophoblast,* will develop into the *umbilical cord.* The cytotrophoblast is now called the ***chorion,*** since its position relative to the embryo is analogous to the chorion in reptiles and birds (Figs. 13.38c and 13.39). The cavity surrounded by the chorion is the ***chorionic cavity.***

The chorion sends out fingerlike *villi,* which invade the lacunae formed by the syncytiotrophoblast. The formation of *angiogenetic clusters* spreads to the yolk sac

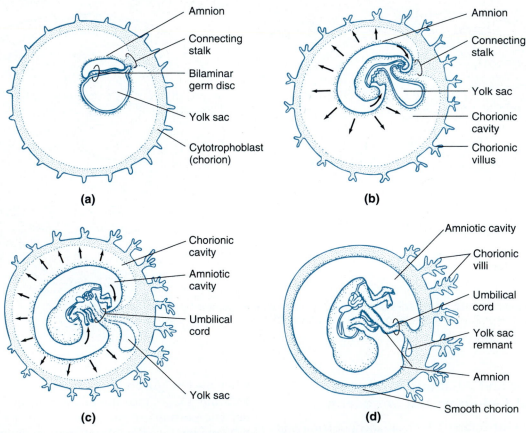

Figure 13.39 Extraembryonic membranes in human development: **(a)** at 3 weeks; **(b)** at 4 weeks; **(c)** at 10 weeks; **(d)** at 20 weeks. The connecting stalk develops into the umbilical cord. The amniotic cavity expands (arrows) until it completely fills the chorionic cavity and envelops the umbilical cord plus the remnant of the yolk sac. The chorionic villi near the umbilical cord branch and form the embryonic portion of the placenta. The other villi disappear.

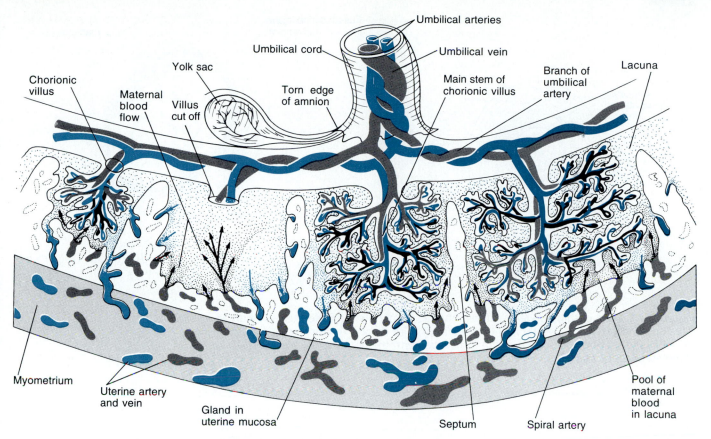

Figure 13.40 The human placenta. The embryonic tissues are labeled along the top of the drawing and the maternal tissues along the bottom. The chorionic villi are drawn as progressively further developed from left to right across the illustration. Note that the chorionic villi float in maternal blood; one villus has been removed in the drawing to show the maternal blood flow from uterine arteries into the lacunae and back into uterine veins.

and the allantois, which reach down the connecting stalk and into the chorionic villi. From the fourth week on, *umbilical arteries and veins* in the stalk connect the embryonic cardiovascular system with the chorionic villi (Fig. 13.40). The villi, in turn, are surrounded by maternal blood, which gives off nutrients and oxygen to the embryonic blood and takes up waste products from it. However, all molecules transferred between embryonic and maternal blood need to diffuse through the tissues of the villi.

As pregnancy advances, the majority of the chorionic villi disappear (Fig. 13.39c and d). The villi near the connecting stalk grow and form branches, anchor-ing themselves firmly in the endometrium. The villous portion of the fetal chorion and the corresponding part of the uterine endometrium make up the organ known as the *placenta* (Fig. 13.40). The placenta connects the fetus with the uterus, mediating the exchange of nutrients, oxygen, hormones, and waste products through-out pregnancy. As the fetus and the amniotic cavity enlarge, the amnion fuses with the inside of the chorion, so the chorionic cavity disappears. At the same time, the growing fetus and the amniotic cavity distend the covering area of the endometrium, thus displacing most of the uterine cavity (Fig. 13.41).

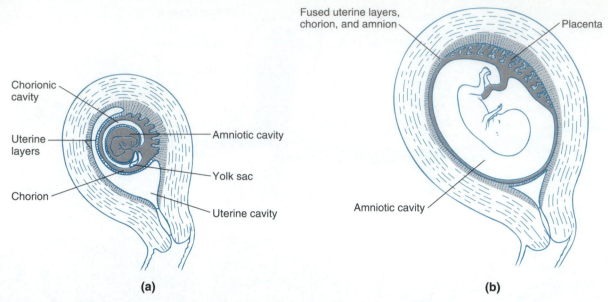

Fused uterine layers,
chorion, and amnion

Placenta

Chorionic
cavity

Amniotic cavity

Uterine
layers

Yolk sac

Chorion

Uterine cavity

Amniotic cavity

(a)

(b)

Figure 13.41 Relationship between extraembryonic membranes and the uterus in human development: **(a)** end of second month; **(b)** end of third month. As the amniotic cavity expands, the uterine layer covering the fetus fuses with the opposite uterine wall, almost completely obliterating the uterine cavity.

SUMMARY

The embryonic endoderm forms the inner linings of the gut and its appendages, such as lungs, liver, and pancreatic gland. Of particular interest in vertebrates are the pharyngeal pouches, which bulge out from the endoderm in the neck region and induce the formation of matching pharyngeal clefts in the overlying ectoderm. Where the pharyngeal pouches and clefts fuse, they create slits connecting the pharynx with the outside world. The tissue between the slits forms pharyngeal arches, which are the sites of gill formation in fishes and amphibian larvae. In more advanced vertebrates, pharyngeal clefts and pouches give rise to other structures, including parts of the ear and various glands. These events reflect the conservation of a phylotypic stage, which is reached after organogenesis. The primitive species of a phylum remain close to the phylotypic stage in their adult organization, whereas the advanced species of a phylum usually undergo major changes in their development from the phylotypic stage to later stages.

The intermediate germ layer, the mesoderm, originates partly in epithelial form like most of the ectoderm and endoderm. In addition, the mesoderm forms large amounts of mesenchyme, an embryonic tissue that is characterized by a loose aggregation of cells surrounded by extracellular matrix. The connective tissues of adult vertebrates, including bone, cartilage, tendons, and adipose tissue, are generally derived from mesenchyme.

The mesoderm of vertebrate embryos consists of axial mesoderm, paraxial mesoderm, intermediate mesoderm, and lateral plates. The axial mesoderm, located along the dorsal midline, gives rise to the prechordal plate and the notochord. The paraxial mesoderm forms segmental plates, which become subdivided first into somitomeres and then into somites. The somites give rise to vertebrae, dermis, skeletal muscle, and limb buds. The intermediate mesoderm forms the kidneys and some reproductive structures. The lateral plates consist of a somatic layer and a visceral layer. Together, they form the inner linings of the coelom and its derivatives, the pericardium, the pleural cavities, and the peritoneal cavity. The visceral layer also produces smooth muscle and the entire cardiovascular system, whereas the somatic layer contributes the body wall and the limb buds.

Limb buds have a core of mesenchyme and a covering layer of ectoderm, on which there is often an apical ectodermal ridge. The mesenchymal core and the apical ectodermal ridge exchange developmental cues in both directions, illustrating the principle of reciprocal interaction. Limb buds are gradually shaped into limbs by growth, programmed cell death, and histogenesis.

In addition to the embryo proper, developing vertebrates form several extraembryonic membranes on which they depend for their protection, nutrition, respiration, and waste removal. A yolk sac is formed by all vertebrates except most amphibians. The other extraembryonic membranes, known as amnion, chorion, and allantois, characterize reptiles, birds, and mammals. The chorion and allantois participate in the formation of the placenta in mammals.

SUGGESTED READINGS

Gould, S. J. 1977. *Ontogeny and Phylogeny.* Cambridge, Mass.: Harvard University Press/Belknap Press.

Hinchliffe, J. R., and D. R. Johnson. 1980. *The Development of the Vertebrate Limb: An Approach through Experiment, Genetics, and Evolution.* Oxford: Clarendon Press.

Larsen, W. J. 1993. *Human Embryology.* New York: Churchill Livingstone.

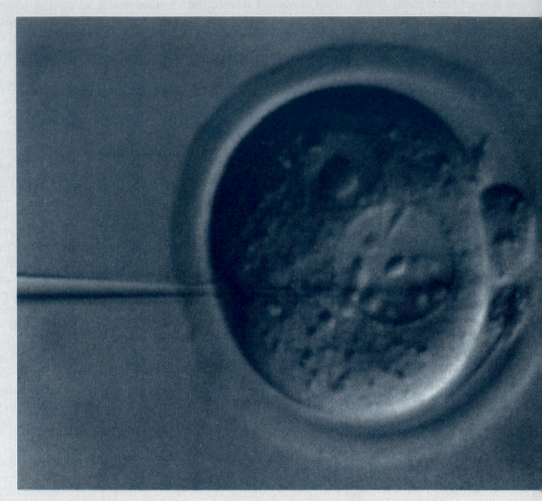

Figure II.1 Genetic transformation of animals. This photomicrograph shows the microinjection of cloned DNA into one pronucleus of a fertilized mouse egg. The fine-tipped glass pipette to the left contains the DNA; the large blunt pipette to the right holds the egg. The injected DNA is integrated into the host DNA and then passed on to daughter cells and offspring as part of the mouse genome. Transgenic organisms are most valuable for the genetic analysis of development and for many applications in medicine and agriculture.

CONTROL OF GENE EXPRESSION IN DEVELOPMENT

In Part One of this text, we analyzed morphogenetic processes in terms of *cellular* behavior. Part Two will introduce the *genetic* and *molecular* analysis of development. We will examine how the genetic information encoded in linear DNA molecules is used to build a three-dimensional organism that unfolds in time. The advent of molecular biology in the 1960s ushered in a new era of breathtaking discoveries in genetics and in cellular and developmental biology. Work at the new frontier was boosted again in the 1970s by a set of modern techniques using *cloned DNA.* This term refers to the amplification of DNA in vitro or in bacterial host cells, so that any gene of interest can be produced in large amounts for detailed analysis.

In Chapter 14, we will explore the use of *mutants,* organisms with altered genes, in the analysis of development. Many mutants display abnormal traits that provide clues to the normal functions of the affected genes. For instance, the additional pair of wings (see Fig.

1.21) seen in *Ultrabithorax* mutants of *Drosophila* indicates that the normal function of this gene is to somehow cause the formation of balancer organs (halteres) instead of wings.

Investigators also make increasing use of *transgenic cells* and *transgenic organisms*, which result from insertion of an engineered gene into the nucleus of a host (Fig. II.1). The engineered gene, called a *transgene*, may be from the same species as the host organism or from a different species. If integrated into the genome of an egg or a germ line cell, a transgene is passed on to offspring like an ordinary mutation. The power of this technique was demonstrated in a spectacular way when mice transformed with a rat growth hormone gene grew to twice the normal size, apparently as a result of excessive growth hormone synthesis (Palmiter et al., 1982). Today, genetic transformation has found a wide range of applications in basic research as well as in medicine, agriculture, and pharmacology.

Chapters 15 through 17 examine the flow of information from genes to proteins in specific developmental events chosen as examples. The first step in gene expression, *transcription*, is the synthesis of RNA from a DNA template. The transcripts of most genes are *messenger RNAs (mRNAs)*, which in turn serve as templates during protein synthesis, a step that is known as *translation*. In eukaryotes, there are many levels of gene regulation, reflecting the eukaryotic cell's advanced organization (Fig. II.2). The nuclear envelope separates transcription in the nucleus from translation in the cytoplasm. Nuclear pre-mRNA undergoes several processing steps before mRNA is released into the cytoplasm, where most mRNAs are immediately translated into proteins while some are stored as inactive messenger ribonucleoprotein (mRNP) particles to be translated later. Proteins undergo several modifications before they

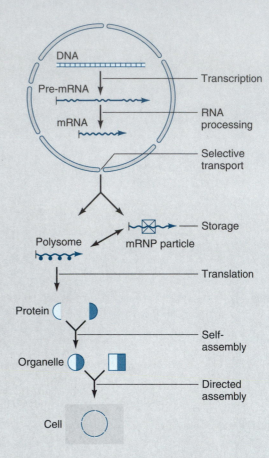

Figure II.2 Gene expression in eukaryotic cells. The genetic information encoded in nuclear DNA is transcribed into pre-mRNA, which is processed into mRNA. Certain mRNAs are transported selectively from the nucleus into the cytoplasm. Cytoplasmic mRNA may be stored in messenger ribonucleoprotein (mRNP) particles or recruited into polysomes. The polysomes translate mRNA into protein. Proteins and other molecules self-assemble into cellular organelles. Organelles are directed by existing cellular structures to assemble into additional cellular structures.

associate with other molecules to form organelles such as ribosomes.

Many molecular complexes and even some cell organelles snap together in a process of self-assembly, driven only by forces between the participating molecules. Other cell organelles, as we will see in Chapter 18, do not have this capability. These organelles can be assembled only using preexisting organelles of the same type as templates. An assembly process that requires structural templates in addition to molecular building blocks is called *directed assembly*. As an example of directed assembly, in Chapter 18 we will study a unicellular organism, *Paramecium*. This protist has an intricately structured cell cortex, with cilia for swimming and sting organelles for defense. Occasional injuries may cause cortical irregularities that are not associated with any genetic changes. Nevertheless, these abnormalities are passed on for hundreds of generations as the animal multiplies by cell division. Apparently, some of the cortical structures, be they normal or abnormal, are duplicated by directed assembly.

Structural information that is passed on during cell division but not encoded in DNA is called *parage-netic information*. Animal cells inherit both genetic and paragenetic information.

During the heady years of beginning molecular biology, a leading researcher expressed his hope that within a few years one would be able to "compute a mouse." He thought that the application of genetic and molecular methods would lead to a complete understanding of development. We are still far from reaching this goal. So far, nobody has even "computed" a cell. However, geneticists and molecular biologists have made great forays, and their success stories attest to the power as well as the limitations of their tools.

THE USE OF MUTANTS, DNA CLONING, AND TRANSGENIC ORGANISMS IN THE ANALYSIS OF DEVELOPMENT

Figure 14.1 Gigantic mouse (left) raised from an egg injected with a cloned gene for rat growth hormone. The mouse grew to about twice the weight of his uninjected littermate (right).

The Historical Separation of Genetics from Developmental Biology

Modern Genetic Analysis of Development

Drosophila Mutants Reveal an Unexpected "Logic" in Embryonic Development

Caenorhabditis elegans Mutants Uncover Gene Activities Controlling Cell Lineages

Genetic Analysis of the Mouse *Mus musculus* Uses Embryonic Stem Cells

DNA Cloning

Transfection and Genetic Transformation of Cells

Germ Line Transformation

The Genetic Transformation of *Drosophila* Utilizes Transposable Genetic Elements

Mammals Are Transformed by Injection of Transgenes Directly into an Egg Pronucleus

DNA Insertion Can Be Used for Mutagenesis and Promoter Trapping

Significance of Transgenic Organisms

Transgenic Organisms Help Researchers Analyze Gene-Regulatory Regions

Transgenic Organisms Serve as Models in Medical and Applied Research

One way of analyzing a complex system is to modify parts of it and then observe its performance. This **method of controlled modification** is especially instructive if the modifications affect single components instead of causing more widespread damage. In terms of an everyday analogy, one could learn what the alternator in an automobile does by removing it—or by cutting the drive belt—and watching the electrical system fail after the battery has been drained. Dropping a boulder on the engine would cause general damage and would be much less informative.

Developmental biologists have used the strategy of controlled modification extensively, transplanting or removing cells or cytoplasms and then analyzing the resulting defects. Biochemists have applied this method to study the functions of specific molecules by deleting, modifying, or adding them in appropriate bioassays. However, there are disadvantages to these applications of controlled modification: the procedures often cause much damage to the organism or entail the culture of cells under conditions that cannot completely simulate their situations in vivo. Geneticists try to avoid these problems by going straight

to the genes, each of which controls the synthesis of one or a few specific proteins in the organism.

The tools for genetic analysis are *mutants,* stocks of individuals in which one or more genes are altered. Some mutants show easily detectable abnormalities that can provide clues to the normal function of the affected gene. For instance, *Drosophila* females in which both copies of the *bicoid* gene are mutated give rise to strikingly abnormal embryos (Frohnhöfer and Nüsslein-Volhard, 1986). These embryos have no head or thorax; instead, they have an extended abdomen, with a duplication of the last abdominal element, the *telson,* attached to the anterior end of the abdomen in reverse polarity (Fig. 14.2). It was shown

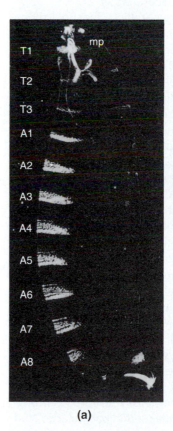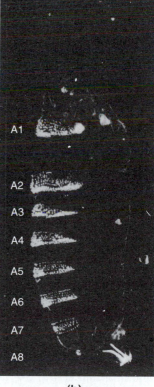

(a)   (b)

Figure 14.2 *Drosophila* larvae derived from a normal (wild-type) mother **(a)** and a *bicoid* mutant mother **(b)**. The photographs show cuticle (skin) preparations of fixed and cleared specimens under dark-field illumination. The bright bands are denticle belts, which the larvae use for traction during crawling. The belts are located near the anterior margins of thoracic (T1–T3) and abdominal (A1–A8) segments. The head contains mouth parts (mp) that are tucked inside the thorax. The tip of the abdomen is marked by anal plates and tracheal openings. The embryo from the *bicoid* mother is lacking head and thorax and shows a partial duplication of the abdominal tip at the anterior end.

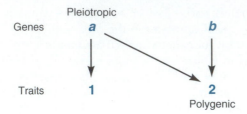

Figure 14.3 Relationships between genes (*a*, *b*) and observable traits (1, 2). Gene *a*, affecting more than one trait, is pleiotropic. Trait 2, affected by more than one gene, is polygenic.

that the normal *bicoid*[+] gene belongs to a group of *maternal effect genes* whose products establish the antero-posterior polarity in the egg (see Chapter 21).

Although mutations are generally a finer tool than a surgical scalpel, even the smallest mutation often affects several observable traits, especially if the gene is expressed in different tissues (Fig. 14.3). This property of genes is known as **pleiotropy** (Gk. *pleion*, "more"; *tropos*, "turn," "direction"). Conversely, many phenotypic traits are **polygenic,** meaning that they are influenced by several genes. Therefore, it is often necessary to identify several genes before the network of gene activities underlying a particular developmental process can be unraveled.

We will begin this chapter with a historical note on the strangely unproductive relationship geneticists and developmental biologists had before molecular biology brought them together. Next, we will explore how mutants are used today in the genetic analysis of development of four organisms: the fruit fly *Drosophila melanogaster;* the house mouse, *Mus musculus;* the roundworm *Caeno–rhabditis elegans;* and the wall cress *Arabidopsis thaliana.* For each species, particularly interesting mutants have been isolated and used to investigate developmental events. Subsequently, we will see how investigators analyze the functions of certain genes even in the absence of suitable mutants. In such cases, the functional equivalents of mutants are generated by cloning the gene of interest and inserting it into the genome of a cultured cell or an entire organism. This strategy is used in particular to study the regulatory sequences that confer a specific expression pattern to each gene.

The Historical Separation of Genetics from Developmental Biology

The science of genetics dates from 1865, when Gregor Mendel made public the results of his breeding experiments with peas. From these results, he inferred the laws of genetic segregation and independent assortment. However, Mendel's abstract laws were not fully accepted until after 1900, when several investigators identified *chromosomes* as the physical carriers of heritable traits.

A critical experiment to test the role of chromosomes in early development was performed by Theodor Boveri (1902). Boveri added excess amounts of sperm to sea urchin eggs to obtain eggs fertilized by two sperm, called *dispermic eggs* (Fig. 14.4). These eggs had two centrosomes, one introduced by each sperm, and three haploid nuclei, two from the sperm and one from the egg. Each chromosome replicated its DNA and formed two chromatids. The two centrosomes also replicated and set up four mitotic spindles, which competed for the three sets of chromosomes. During mitotic anaphase,

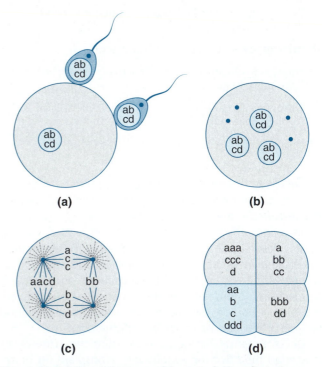

Figure 14.4 Dispermic sea urchin egg during first cleavage. **(a)** Egg being fertilized by two sperm. Each sperm introduces a pronucleus and a centrosome. Each of the male pronuclei, as well as the female pronucleus in the egg, contains a haploid set of chromosomes, designated a, b, c, and d. **(b, c)** Both centrosomes replicate in the fertilized egg and set up four mitotic spindles, which attract random assortments of chromosomes. **(d)** The first cleavage generates four blastomeres simultaneously. In the case shown, only one blastomere (color) inherits at least one copy of each chromosome.

each spindle pole attracted various combinations of chromatids from multiple spindles. Cytokinesis then generated four blastomeres simultaneously.

Boveri separated the blastomeres from one another and observed their development. From previous experiments it was known that quarter blastomeres derived from a normal egg that had been fertilized by one sperm would usually develop into a complete larva. However, only a few isolated blastomeres from dispermic eggs developed into complete larvae. Most important, these blastomeres stopped developing at different stages and with different signs of abnormality. These results could not be explained by the sheer *number* of chromosomes present in each blastomere. It was known that a haploid but *complete* set of chromosomes was sufficient to support normal embryonic development. Most blastomeres had more than a haploid number of chromosomes, but apparently only a few had a complete set. Boveri concluded that *each chromosome was different* and had a specific effect on development.

Geneticists were originally concerned with the *transmission* of heritable traits from one generation to the next. However, their attention soon turned to the study of *gene expression*, that is, how genes bring about their effects in a given individual. Early in the twentieth century, the English physician Sir Archibald E. Garrod made the connection between genes and the new concept of enzymes as catalysts of metabolic processes. He studied rare human diseases such as alkaptonuria and albinism. These conditions are apparent at birth and often affect siblings in spite of their rarity among the entire population. Alkaptonuria results from the presence of alkapton (homogentisic acid) in the urine. Upon exposure to air, the urine darkens and stains the diapers of alkaptonuric babies black. Garrod (1909) reasoned that an enzyme for normal metabolism of alkapton might be absent in alkaptonuric children. He saw that the parents of such children were often first cousins and suspected that the disease was hereditary. The implication was that a mutant *recessive* allele inherited from both parents caused the absence of the enzyme.

Even though the function of genes in adult metabolism was soon apparent, both geneticists and embryologists were reluctant to pursue the idea that genes direct development (Horder et al., 1985). The leading geneticist at the beginning of the twentieth century, Thomas H. Morgan, was also an accomplished embryologist. Although he recognized the basic relationship between genes and development, he and his coworkers published only one analysis of a mutant undergoing abnormal embryonic development, a paper on the sinistrally coiling mutant of the snail *Lymnaea peregra* (Sturtevant, 1923; see also Fig. 5.24). Similarly, preeminent embryologist Hans Spemann, who was a student of Boveri and admired him greatly, had no use for genes in his own experimental designs or interpretations.

Only a few insightful pioneers recognized the value of genetics to the analysis of development. The German geneticist Valentin Haecker (1912) was one of them. Using a method of analysis that he termed *phenogenetics*, Haecker compared wild-type and mutant individuals at different stages of development. He identified the apparent deviation point when the two started to develop differently. Other researchers, including Ernst Hadorn (1948), focused on the stages of developmental arrest in mutants carrying **lethal factors**, mutated genes that cause early death. These studies showed clearly that the activity of certain genes was critical at certain stages of development and in specific organs. In addition, the importance of *homeotic genes*, which are critical for the normal development of overall body pattern, was recognized early (Goldschmidt, 1938) and investigated systematically (Lewis, 1978). In general, however, most researchers in developmental biology remained uninterested in genetics, and vice versa. Until fairly recently, the gap between genes and their manifestation in development seemed too wide to bridge.

The relationship between embryology and genetics began to change with the advent of molecular biology. DNA was revealed as the carrier molecule of genetic information (Avery et al., 1944), and its double-helical structure was clarified (Watson and Crick, 1953). DNA transcription into messenger RNA (mRNA) and other RNAs, and the translation of mRNA into protein, were recognized a few years later. The first model of gene regulation (Jacob and Monod, 1961) provided biologists with a general concept of how genes could be turned on and off in living cells. The model proposed that the transcribed portion of a gene was associated with **regulatory DNA sequences**, to which **regulatory proteins** would bind to control the transcription of the gene. Since the regulatory proteins themselves were gene products, their genes in effect controlled the activity of the regulated genes. Thus, development came to be seen as a "program" encoded in DNA and enacted through networks of gene regulation.

Modern Genetic Analysis of Development

The primary tools for the genetic analysis of development are mutants in which key developmental steps are inhibited or altered. Some general strategies for the generation and maintenance of mutants apply to most organisms. These strategies, along with some general concepts and technical terms, are surveyed in Methods 14.1. More comprehensive discussions can be found in genetics textbooks.

Mutants useful for the analysis of development have been collected for a number of animals. So far, three animal species have attracted sufficient interest to generate hundreds or thousands of mutant strains and an extensive literature. These species are the fruit fly *Drosophila melanogaster;* the roundworm *Caenorhabditis elegans* (commonly abbreviated as *C. elegans*); and the house mouse, *Mus musculus* (Wilkins, 1993; Glover and Hames, 1989). Smaller numbers of mutants have been collected for the axolotl *Ambystoma mexicanum* and the flour beetle *Tribolium castaneum.* Currently, a strong push is being made for collecting a large number of mutants and transgenic lines in the zebra fish *Brachydanio rerio* (Mullins et al., 1994; Solnica-Krezel et al., 1994; Postlethwait et al., 1994), in the hope of making a vertebrate more accessible to genetic analysis than has been possible in the axolotl or the mouse. Each of these animals has its own advantages for developmental analysis. We will look at the three species that are most commonly used today.

Drosophila Mutants Reveal an Unexpected "Logic" in Embryonic Development

The career of *Drosophila melanogaster* as a laboratory animal began in Thomas H. Morgan's "fly room" at Columbia University, where he and a group of enthusiastic coworkers began genetic studies on *Drosophila* around 1910. Fruit flies are easy to keep in the laboratory, and they mate as single pairs. Female fruit flies have two large **X chromosomes,** whereas males have one X chromosome and a smaller **Y chromosome.** Most important, the flies survive well with both chemically and X-ray–induced mutations. Many mutations affect external markers (such as eye color and shape, cuticle color, or bristle shape) that are easy to detect under the dissecting microscope. Our knowledge about mutagenesis, the linear order of genes on a chromosome, genetic sex determination, and other genetic concepts arose mostly from the work of Morgan and his associates. In recognition of these achievements, Morgan was awarded the 1933 Nobel Prize in physiology or medicine.

In previous chapters, we examined *Drosophila* in connection with several topics, including oogenesis (Chapter 3), cleavage (Chapter 5), imaginal discs (Chapter 6), and polytene chromosomes (Chapter 7). The characteristic banding pattern of polytene chromosomes in *Drosophila* makes it possible to identify even small chromosomal deletions, duplications, and inversions. Therefore, the *cytogenetic map* (Fig. 14.5) of *Drosophila* is much more detailed than that of any other organism. With about 5000 identifiable chromosome bands and an estimated 10,000 genes, many genes have been mapped to unique chromosome bands. The estimated number of genes in *Drosophila* is lower than that of mammals but higher than that of simpler organisms

METHODS 14.1

The Generation and Maintenance of Mutants

The term *gene* originally denoted a functional unit of heredity. Today, the word usually means a segment of DNA that is necessary to encode and regulate the synthesis of a specific RNA or protein product. Likewise, the *genome* of an organism denotes either the totality of its functional genetic units or one complete copy of its nuclear and extranuclear DNA. The *genome size* of a species is the length of one genome measured in DNA base pairs (bp).

Most genes are arranged in a linear order on chromosomal DNA. In diploid organisms, which carry two sets of chromosomes, the corresponding maternal and paternal genes and chromosomes are *homologues* of each other. Chromosomes that differ between males and females of the same species are called *sex chromosomes;* they are often symbolized by the letters X and Y. All other chromosomes are collectively called **autosomes.** Genes located on the same chromosome form a **linkage group.** Genes located on sex chromosomes are called **sex-linked** (or **X-linked** or **Y-linked**) genes.

Alternative forms of a gene are referred to as *alleles.* For instance, different mutant versions of a gene, as well as the two homologues of a gene in a diploid organism, are allelic. The allele of a gene most commonly found in natural populations is called the **wild-type allele;** it is marked by a superscript plus sign following the name or abbreviated name of the gene. Thus, *Ultrabithorax$^+$* refers to the wild-type allele of the gene, while *Ultrabithorax* designates any mutant allele. A plus sign (*not* superscripted) is also used by itself to designate the wild-type allele in descriptions of genetic crosses when the context explains which gene is referred to. The position on a chromosome occupied by a particular set of homologous alleles is called the **locus** for those alleles.

A diploid organism carrying the same allele on two homologous chromosomes is *homozygous* for the allele. An organism carrying different alleles of a gene is *heterozygous* for the gene. Sometimes, a diploid organism has only one copy of a gene because it is sex-linked or because the homologous gene has been lost. The organism is then described as **hemizygous** for the allele that is present. If two alleles *A* and *a* of a gene produce different phenotypes and the phenotype of the heterozygote *A/a* is the same as the phenotype of the homozygote *A/A*, then *A* is called the **dominant** allele and *a* the **recessive** allele. (Note that capital and lowercase letters are sometimes used for dominant and recessive alleles, respectively.) Most mutant alleles are recessive to the wild type.

The *genotype* of an organism, in principle, refers to *all* the genes of an individual, regardless of whether they are expressed. Likewise, the *phenotype* of an individual encompasses *all* its observable characteristics. The common phenotype in a natural population is the **wild phenotype.** In practice, the genotypes and phenotypes of organisms are usually compared selectively for the gene(s) under consideration.

Of particular advantage to developmental biologists are **temperature-sensitive (ts) mutants,** mutants that survive and develop the wild phenotype when raised at a **permissive temperature** but express the mutant phenotype when raised at a **restrictive temperature.** A restrictive temperature is usually hotter than the organism's normal habitat. Most ts mutants have a single amino acid substitution in the protein encoded by the mutated gene, a change that renders the protein unstable at restrictive temperatures. Stocks of ts mutants can be maintained conveniently at permissive temperatures, even if the mutant phenotype includes lethality. To study the effects of temperature-sensitive mutations on development, ts mutants are simply subjected to a restrictive temperature. Temperature shifts at regular intervals make it possible to identify the **temperature-sensitive period (TSP),** the period during which mutant individuals must be kept at a permissive temperature in order to survive. The TSP reveals the stage at which the wild-type gene function is required for normal development. The TSP is also used in experiments to eliminate individuals with certain genotypes from a study population.

For the analysis of early embryonic development, *maternal effect genes* are of particular interest. These genes reveal themselves in genetic crosses when the offspring's phenotype is controlled solely by the maternal genotype and unaffected by the paternal genotype. Maternal effect genes are necessary for normal *oogenesis,* and females that are mutant in one of these genes produce defective eggs. The defects may be overt, as in abnormally shaped eggs, or more subtle, as in missing or abnormally distributed cytoplasmic components. The latter type of defect is critical between fertilization and *midblastula transition,* that is, during the period when embryonic development is supported by maternal RNA and proteins rather than by embryonic gene products. Previously discussed maternal effect genes include the gene that controls the orientation of the mitotic spindles dur-

ing cleavage of the snail *Lymnaea peregra* (see Fig. 5.24) and the *bicoid*[+] gene of *Drosophila* (see Color Plate 1 and Fig. 14.2). More maternal effect genes that establish the body axes and determine the primordial germ cells in *Drosophila* will be discussed in Chapter 21.

Some mutants have been found by chance in natural or laboratory populations. They can be simply inbred to generate a mutant strain. However, most currently used mutants have been isolated by means of **mutagenesis screens.** These screens are of great importance to the genetic analysis of biological processes, because they help researchers identify *virtually all* genes involved in the process of interest. In most mutagenesis screens, large numbers of individuals are treated with a chemical mutagen, *ethylmethane sulfonate (EMS),* which causes single-base substitutions in DNA, referred to as **point mutations.** Larger chromosomal rearrangements can be produced with *X-rays,* which break the phosphodiester "backbone" of DNA. This procedure results in chromosomal **deletions** (missing segments), **translocations** (segments moved from one chromosome to another), or **inversions** (segments cut out and reinserted with reversed orientation). Still other mutagenesis screens rely on mobile DNA elements, which will be discussed later in this chapter.

By crossing mutagenized individuals with suitable partners, it is possible to recover stocks, or *lines,* of individuals that are homozygous for one mutagenized chromosome. These lines are then examined for abnormal phenotypes that might be useful to the investigator. Someone interested in spermatogenesis, for example, would screen for lines producing infertile males.

If two mutants have similar phenotypes, it is important to determine whether their mutations are in the same gene or in different genes—in other words, whether the mutated genes in these lines are *allelic.* This is done by a **complementation test.** First, each line is crossed with wild-type individuals to determine whether the mutant allele is recessive to the wild-type allele; most mutant alleles are recessive. Next, the recessive and possibly allelic lines are crossed with each other. The resulting offspring carries both mutant alleles, one on each homologue of the affected chromosome. If the offspring shows the wild phenotype, then each mutated gene from one mutant line is apparently complemented by the wild-type allele of the same gene from the other mu-

Continued on next page

tant line. Thus, the two lines are probably mutated in different, or nonallelic, genes. However, if a cross between two mutant lines produces offspring that does *not* show the wild phenotype, then the mutated genes fail to complement each other and are probably allelic.

Saturation mutagenesis screens are designed to cover the genome with mutations by generating at least one mutant line for *nearly every* mutable gene. Under these conditions, most genes are altered several times, so that multiple alleles of the same gene are obtained. This is useful because each allele may cause unique changes in gene expression.

From each group of allelic mutants, at least one is mapped to its chromosomal position. There are two types of genetic map (Fig. 14.5). The *cytogenetic map,* also called a *physical map,* shows the actual position of a gene on its chromosome. The physical map position of a gene is revealed by examining the chromosomes of mutants that have chromosomal abnormalities large enough to be visible under the microscope. The other type of map, called a *genetic map* or *recombination map,* is based on a statistical evaluation. During meiosis, genes located on the same chromosome are often separated by *crossing over* (see Fig. 3.4). The probability of crossing over

can be calculated from the frequencies of recombinant phenotypes scored in the offspring of appropriate crosses. These data can be used to construct a genetic map because the probability of crossing over between two genes increases with their distance from each other on the chromosome.

Finally, each mutant allele is classified according to the way in which the mutation interferes with the function of the gene. Most mutant alleles cause some loss of normal gene function. Mutations that completely abolish a gene function produce **null alleles.** Their phenotypes are equivalent to the phenotypes of deletions lacking the gene altogether. Null alleles are often marked by a superscript minus sign. For instance, *hairy*[−] is the null allele of the *hairy* gene. Mutations that weaken the function of a gene produce **loss-of-function alleles.** A loss-of-function allele is usually recessive to its wild-type allele except when one copy of the wild-type allele is insufficient to produce the wild phenotype; genes of this kind are called **haploinsufficient,** and their loss-of-function alleles can be dominant. Another type of mutation, which causes a gene to be activated in the wrong place or at the wrong time, generates a **gain-of-function allele.** This type of allele also tends to be dominant.

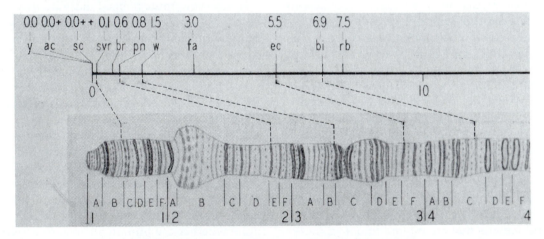

Figure 14.5 Genetic map and cytogenetic map of a portion of the *Drosophila* X chromosome. The genetic map at the top was established by calculating the frequencies of recombinant gametes formed. The letters are abbreviations of gene names; for instance, *y* stands for *yellow*[+]. The position of each gene on the map is indicated in centimorgans (cM). One cM equals 1% recombination. The cytogenetic map at the bottom, drawn by Bridges, a coworker of Morgan, represents the same section of a polytene X chromosome. Such drawings can be used to locate the duplications, deletions, or inversions that are associated with mutant phenotypes. The correlations allow researchers to establish a cytogenetic map. Note that the linear order of loci is the same on the genetic and the cytogenetic map. The apparent crowding of loci at the end of the genetic map is caused by the reduced crossover frequency at the ends of chromosomes.

TABLE 14.1

Genetic Characteristics of Representative Organisms

	Escherichia coli	Caenorhabditis elegans	Drosophila melanogaster	House Mouse	Human
Haploid number of chromosomes	1	6	4	20	23
Genome size (bp)	4.2×10^6	8×10^7	1.5×10^8	2.7×10^9	3.1×10^9
Nonrepetitive DNA (% of total DNA)	100%	83%	76%	60%	70%
Estimated number of genes	3000	5000	10,000	50,000	> 50,000
Generation time (from egg to fertile adult)	(30 min)	3 days	3 weeks	2 months	15 yr

(Table 14.1). About 4000 *Drosophila* genes have been identified through mutation and mapped, and many of these have been individually cloned (Ashburner and Gelbart, 1991). The genome size of *Drosophila* is 1.5×10^8 base pairs (bp), small enough to lend itself to thorough analysis with recombinant DNA techniques.

The detailed cytogenetic map of *Drosophila* and the large number of X-ray–induced mutants have allowed geneticists to do a fair amount of "chromosome engineering." Particularly useful are **balancer chromosomes**, which carry dominant marker mutations, recessive lethal factors, and multiple inversions, which suppress recombination (Fig. 14.6). These chromosomes are used to maintain any *homologous* chromosome carrying an-

other recessive lethal factor. In a strain with such a pair of chromosomes, both types of homozygotes die while the heterozygotes survive because each lethal factor is covered by the wild-type allele on the other homologue. This kind of strain is called a **balanced stock**.

The genetic properties of *Drosophila* have made it possible to carry out *saturation mutagenesis screens*. Using such screens, developmental biologists can estimate the *total* number of genes involved in processes such as oogenesis, cleavage, or establishing the embryonic body pattern. At the same time, a large number of useful mutants are obtained. Their phenotypes are often striking and reveal what types of "decisions" cells make in the course of development and which embryonic areas are affected by these decisions. The following example illustrates this kind of analysis.

In 1980, Christiane Nüsslein-Volhard and Eric Wieschaus published the first part of a genetic analysis that has snowballed into a concerted effort involving hundreds of investigators (see Chapter 21). At the outset, the researchers asked which types of maternally expressed genes are active in setting up the axes of polarity in the *Drosophila* embryo, how the products of these genes control their embryonic target genes, and how embryonic genes interact to form the body pattern seen in the hatched larva.

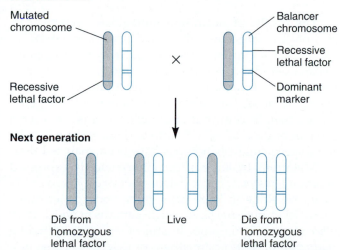

Balanced stock

Next generation

Figure 14.6 Use of a balancer chromosome to maintain a homologous mutated chromosome carrying one or more lethal factors. Since the balancer chromosome carries a lethal factor itself, both homozygotes die while the heterozygotes survive. In the heterozygotes, the genes on the mutagenized chromosome are permanently linked to each other because inversions in the balancer chromosome prevent crossing over.

Nüsslein-Volhard and her coworkers (1984) were particularly interested in genes that could be mutated to cause abnormal embryonic body patterns. They expected that such mutant larvae would be unable to hatch from the *chorion* (egg envelope), because larvae that are not fully vigorous tend to fail at this step. Therefore, they set up saturation screens for embryonically expressed genes that could be mutated to lethality. Figure 14.7 shows the crossing scheme employed to identify such genes on the second chromosome. (Similar screens were carried out for mutants on the first and third chromosomes.) Males homozygous for a second chromosome carrying a set of recessive marker genes were fed *EMS* so that they would produce sperm with mutagenized chromosomes. The overall goal was to recover *balanced stocks*, each with a unique mutagenized second chromosome present in each individual of that stock. To this end, the mutagenized males were mated with females having a pair of specially engineered second chromosomes. One was a *balancer chromosome*, while the other carried a dominant lethal but *temperature-sensitive* mutation. From the offspring (F₁), *individual* males were mated again with females of the mother's genotype. *All* their offspring (F₂) therefore carried the same mutagenized second chromosome derived from the founder male. Making use of the temperature-sensitive mutation and the balancer chromosome introduced by the females, the researchers killed all offspring except

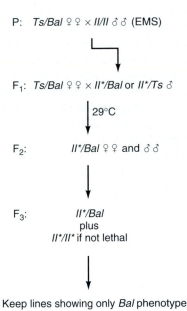

P: *Ts/Bal* ♀♀ × *II/II* ♂♂ (EMS)

F₁: *Ts/Bal* ♀♀ × *II*/Bal* or *II*/Ts* ♂

29°C

F₂: *II*/Bal* ♀♀ and ♂♂

F₃: *II*/Bal*
plus
II*/II* if not lethal

Keep lines showing only *Bal* phenotype

Figure 14.7 Crossing scheme used by Nüsslein-Volhard et al. (1984) to establish mutant *Drosophila* lines, each with a particular lethal factor. Males homozygous for a chromosome II carrying a set of recessive marker genes (*II*) are mutagenized with ethylmethane sulfonate (EMS). They are mated (P generation) with females having specially engineered chromosomes II, one being a balancer chromosome (*Bal*) and the other carrying a dominant lethal but temperature-sensitive mutant allele (*Ts*). In the offspring, each individual carries a particular mutagenized (*II**) second chromosome. *Individual* males are mated again with females of the same type (F₁ generation). Their offspring are kept at the restrictive temperature to kill all individuals with a *Ts* chromosome. The surviving individuals (F₂ generation) are heterozygous *II*/Bal*. Their offspring (F₃ generation) are *II*/Bal* (showing the *Bal* phenotype) or *II*/II** (not showing the *Bal* phenotype). Lines showing only the *Bal* phenotype as adults are kept, since their *II** chromosome must carry a lethal factor.

those that carried the mutagenized chromosome over the balancer chromosome. These F₂ individuals were crossed among each other to test the lethality of those of their F₃ offspring that were homozygous for the mutagenized chromosome.

From about 10,000 F₁ crosses set up, 4580 lines were recovered as balanced stocks—each with a unique second chromosome carrying a homozygous lethal mutation.

The 4580 lines mutagenized to lethality were tested further to identify those lines having *embryonic* lethality. In these lines, about 25% of all eggs laid were expected not to hatch because the embryos were homozygous for a mutation interfering with embryogenesis. Eggs were inspected microscopically from the 2843 lines selected under this criterion. A total of 272 lines produced embryos that were clearly distinguishable from wild-type embryos by their cuticle wholemounts (Fig. 14.2). *Complementation tests* among these lines showed that they represented only 61 genes. Of these, 48 genes had been mutagenized several times, yielding multiple alleles. Only 13 genes had been altered only once. At this point, any additional mutant lines would have produced mostly additional alleles of those genes already mutagenized. In other words, the screen was close to saturation. The investigators concluded that the 61 genes identified represented the majority of the genes on the second chromosome that could be mutated to an embryonic lethal phenotype with a recognizable alteration in the larval cuticle. Similar estimates were obtained for the other chromosomes, except for the fourth, which is very small.

The results of saturation mutagenesis screens carried out by Nüsslein-Volhard and coworkers (1984) indicate that about 200 embryonically expressed genes make a detectable contribution to the larval body pattern. This is a relatively small fraction, representing about 2% of all *Drosophila* genes. Some of these genes make dramatic contributions, while the effects of others are barely detectable. Of course, those producing the most spectacular phenotypes have been analyzed first.

Mutations in some of the embryonically expressed genes in *Drosophila* cause deletions of *whole groups of segments,* resulting in large gaps in the body pattern (Fig. 14.8). Other mutants have cuticle patterns in which *every other segment* is missing. In still others, corresponding parts of *each segment* are absent or defective. These mutant phenotypes suggest that the embryonic body pattern is built stepwise, beginning with large blocks of cells corresponding to several segments, then in intermediate units corresponding to two segments, and finally in small units that have a segmental periodicity. Before these mutant phenotypes were discovered, few people would have thought that insect embryos might

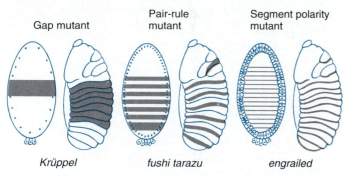

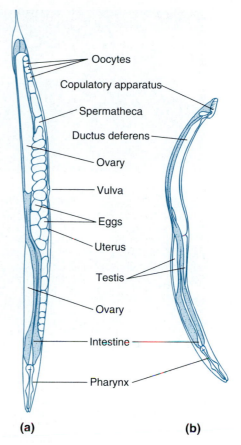

Figure 14.8 *Drosophila* mutants with distinct embryonic pattern abnormalities. The diagrams show one example from each of three classes of mutants. *Krüppel (Kr)* is a gap mutant having a major gap in the larval segment pattern with several segments missing. *fushi tarazu (ftz)* is a pair-rule mutant having a loss of tissue in every other segment. *engrailed (en)* is a segment polarity mutant with a defect in the same part of every segment. For each mutant, the left-hand diagram shows the domains (gray stripes) where and the stage of development at which the gene is first expressed in the wild-type embryo. The right-hand diagram indicates which body parts are defective in the mutant larvae (gray areas). Note that the *Krüppel+* gene is expressed during late cleavage, *fushi tarazu+* during the syncytial blastoderm stage, and *engrailed+* during the cellular blastoderm stage.

Figure 14.9 *Caenorhabditis elegans* hermaphrodite **(a)** and male **(b)**. The spermatheca in the hermaphrodite is a repository for sperm produced at an earlier stage of development. If no male is available to fertilize its eggs, the hermaphrodite uses its own sperm for this purpose.

"think" temporarily in blocks of two segments, a unit size that later is no longer apparent. Many of the patterning genes in *Drosophila* have been investigated in detail, revealing a fascinating hierarchy of genetic control (see Chapter 21).

Caenorhabditis elegans Mutants Uncover Gene Activities Controlling Cell Lineages

Caenorhabditis elegans is a tiny, free-living nematode, measuring just over 1 mm in length and 70 μm in diameter (Fig. 14.9). Adults of *C. elegans* are mostly self-fertilizing **hermaphrodites**, with gonads producing both sperm and oocytes. Males develop infrequently but hermaphrodites, when fertilizing their eggs, prefer sperm obtained from males over their own. In laboratory cultures, the worms can be handled with the ease of bacteria. *C. elegans* completes its entire life cycle in just 3 days, and 100,000 individuals can live in a Petri dish. Because the worm is transparent, investigators can watch its development live under the microscope.

The tiny worm rose to scientific fame through the determined efforts of Sidney Brenner and his coworkers in Cambridge, England. They wanted to study the development of the nervous system in a simple animal that had a small repertoire of standardized behaviors. Nematodes seemed best suited to this project, because

they are simple but "real" animals: they have skin, muscles, a gut, and a nervous system, and when touched with a brush, they react.

Most important, each nematode species has a fixed number of cells, which always develop in the same lineages. In *C. elegans,* hermaphrodites have exactly 959 somatic cells and about 2000 gametes, while males have exactly 1031 somatic cells and about 1000 sperm. The anatomy of each developmental stage has been reconstructed, cell by cell, from serial sections viewed under the electron microscope (Sulston et al., 1983). Thus, *C. elegans* became the first animal with a complete "parts list" (13 pages long!) and detailed cell lineage maps from egg to adult (Wood, 1988a).

The constancy of *C. elegans* cell lineages has made it possible to isolate mutants with abnormal lineages (Fig. 14.10). These mutants provide unique opportunities for scientists to study the network of gene activities that control the normal lineage. The exact knowledge of all cell lineages in *C. elegans* has also allowed investigators to analyze inductive interactions between cells, the timing of cell divisions, and programmed cell death with unparalleled precision (see Chapter 24).

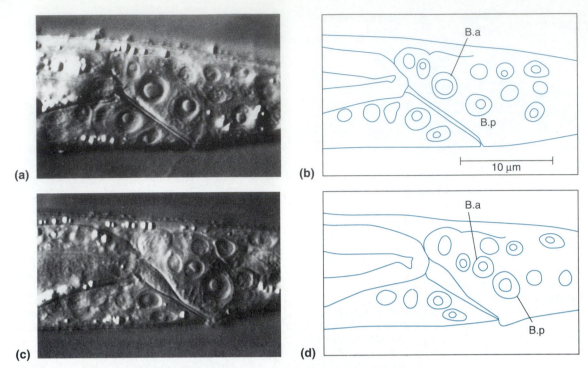

(a) **(b)** **(c)** **(d)**

Figure 14.10 Photomicrographs and tracings of male *C. elegans* larvae. The exact knowledge of every cell lineage in this species allows investigators to study the effects of mutations on individual cells. The example shown here illustrates one effect of a *mab-9* allele (*mab* stands for *males abnormal*) on the B cell lineage, which contributes to the copulatory organ of the adult male. Loss of function in *mab-9* transforms the B cell lineage so that it resembles the lineage of another cell. **(a, b)** Wild type. One characteristic of the wild-type B cell lineage is the unequal division of B into a larger anterior daughter cell (Ba) with a large nucleus and a smaller posterior daughter cell (Bp) with a small nucleus. **(c, d)** *mab-9* loss-of-function allele. The size differential between Ba and Bp disappears or reverses, and the cell lineage undergoes several other changes not shown here.

C. elegans is very amenable to genetic and molecular analysis (Brenner, 1974; Kemphues, 1989). The propagation of *C. elegans* by self-fertilization makes the isolation of homozygotes for recessive mutations very easy. A mutagenized hermaphrodite is simply allowed to reproduce for two generations to obtain homozygous mutants. Diploid cells have five pairs of *autosomes* plus one X chromosome in males and two X chromosomes in hermaphrodites (Table 14.1). The total number of essential genes is estimated to be about 5000, of which about 800 have been mapped and many of these individually cloned (Glover and Hames, 1989). The genome size of *C. elegans* (8×10^7 bp) is one of the smallest in the animal kingdom. A strong effort is under way to sequence the entire genome.

Genetic Analysis of the Mouse *Mus musculus* Uses Embryonic Stem Cells

The mouse has been used for genetic experiments since the beginning of the twentieth century (Wilkins, 1993; I. J. Jackson, 1989). Several human genetic disorders have close parallels in the mouse. The development of humans is also similar to that of mice, except for the period between the blastocyst stage and gastrulation. Therefore, the mouse is often considered a model for human genetics and development. Unfortunately, extensive *mutagenesis screens* are impractical with mice because of their relatively small litter size, long generation time, and large genome size (Table 14.1). However, the ability to produce chimeric mice, along with advanced techniques for cloning DNA, has enhanced prospects for generating additional mutants in the future.

Chimeras can be generated in a variety of ways. One way, described in Chapter 6, is to isolate blastomeres from genetically different embryos during cleavage stages and combine them. The combined cells are allowed to develop in vitro until the blastocyst stage, when they are implanted into a foster mother. Many of the implanted blastocysts develop and are born normally.

Chimeras have also been generated by injecting blastocysts with different types of cultured cells (Fig. 14.11). Particularly useful for this purpose are *embryonic stem cells (ES cells)*, which are derived from normal mouse

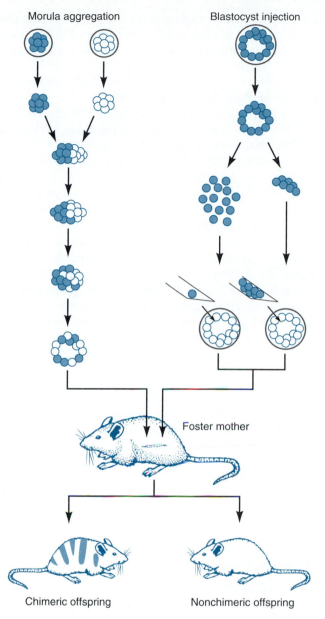

Figure 14.11 Two methods of generating chimeric mice. In the morula aggregation method shown to the left, 8-cell embryos are removed from the zona pellucida and stuck together. In the blastocyst injection method shown to the right, isolated cells or groups of cells are injected into early blastocysts.

inner cell mass cultured in vitro. ES cells lend themselves to mutagenesis and genetic transformation in vitro. When injected into blastocysts, transformed ES cells give rise to all types of tissues including the germ line (Bradley, 1987). This method allows investigators to introduce an engineered gene with a known mutation into ES cells and then to breed mice that have the same mutation in *all* their cells.

Chimeric mice containing transformed ES cells have been used in an experimental strategy known as *gene targeting* or *gene knockout*. The following example shows the application of this technique to the analysis of a *homeotic gene,* that is, a gene that determines the morphological character of specific body segments.

Mutations in homeotic genes typically cause the replacement of a segment with the likeness of another segment. For instance, the four-winged fly mutant discussed earlier, in which the third thoracic segment is replaced by another second thoracic segment (see Fig. 1.21), is attributed to a loss of function in the *Ultrabithorax* gene. Homeotic genes were first identified in insects, especially in *Drosophila,* and most of these genes have been cloned. Since normal mutagenesis has not yielded any mouse mutants with homeotic phenotypes, researchers have used purely molecular procedures to clone mouse genes with nucleotide sequences similar to those of *Drosophila* homeotic genes. In order to establish whether the cloned mouse genes have a homeotic function, that is, determine the morphological character of certain mouse segments, the cloned genes had to be altered in vitro and then reintroduced into living mice.

The mouse gene *Hoxb-4*$^+$ is similar in sequence to the *Drosophila* gene *Deformed*$^+$, which controls the morphological development of certain head segments. Does *Hoxb-4*$^+$ have a corresponding function in mice? To answer this question, Ramiro Ramirez-Solis and his colleagues (1993) modified a cloned *Hoxb-4*$^+$ gene by inserting another gene, *neo*^{R} (Fig. 14.12). The *neo*^{R} gene is a *selectable marker* that confers upon cells the ability to grow in the presence of neomycin, a drug that kills other cells. The cloned *Hoxb-4* gene with the *neo*^{R} insertion was introduced into mouse ES cells. DNA strands with identical nucleotide sequences tend to pair up in cells, and such configurations of paired DNA segments facilitate *homologous recombination,* that is, crossing-over events between homologous sequences. Thus, a significant fraction of ES cells swapped the cloned exogenous DNA containing the *neo*^{R} insert for the corresponding segment in a resident copy of the *Hoxb-4*$^+$ gene.

ES cells that underwent this replacement gained the ability to grow in neomycin-containing medium. At the same time, one of the two *Hoxb-4*$^+$ genes in these cells was changed into a null allele (*Hoxb-4*$^-$), because the insertion resulted in a nonfunctional gene product. To select for cells with the *neo*^{R} containing insert, the researchers cultured them in the presence of neomycin, and performed additional tests to ensure that the inserts had indeed been swapped for corresponding segments of the resident *Hoxb-4*$^+$ genes. Positive-testing ES cells were injected into mouse blastocysts, which were brought to term in foster mothers. Chimeric males were screened for the presence of the

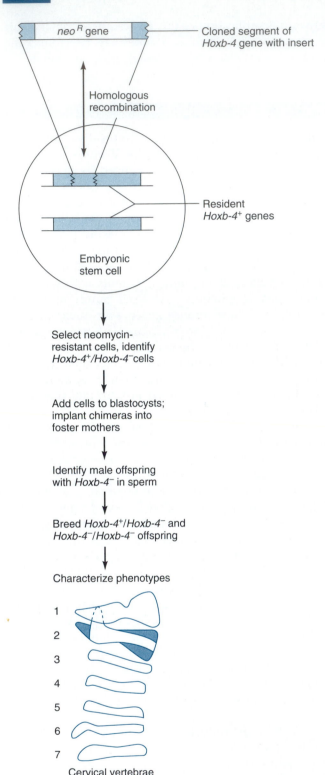

Figure 14.12 Gene knockout technique in mice. The upper diagram shows the experimental procedure for replacing the wild-type allele of the *Hoxb-4+* gene with a mutant (*Hoxb-4−*) allele that has been inactivated by the insertion of the *neoR* gene. The diagram at the bottom shows the outlines of the cervical vertebrae in the wild-type (open) and in the homozygous *Hoxb-4−/Hoxb-4−* mutant (shaded). The seven cervical vertebrae are numbered, starting at the head. Dorsal is to the right. In the mutant, the second cervical vertebra has been transformed into the likeness of a first cervical vertebra.

replacement gene in their sperm, and males having the replacement gene were used to breed mice that were either heterozygous (*Hoxb-4+/Hoxb-4−*) or homozygous (*Hoxb-4−/Hoxb-4−*) in all their cells.

The heterozygotes developed normally and could therefore be used to breed homozygotes. The homozygotes, which were lacking both copies of the normal *Hoxb-4+* allele, had various abnormalities including a striking homeotic transformation in the neck region: their second cervical vertebra (C2) was transformed into the likeness of another first vertebra (C1), as shown in Figure 14.12. In wild-type mice, these two vertebrae differ clearly from each other and from other cervical vertebrae: C1 has a ventral process, called the *tubercle*, which is not present in C2. Also, C1 has a neural arch that is wider than its counterpart in C2. With regard to these criteria, the *Hoxb-4−/Hoxb-4−* homozygotes showed a transformation of the C2 segment into the likeness of a C1 segment. This result indicates that the *Hoxb-4+* gene in mice has both the molecular and the functional characteristics of a homeotic gene.

The method of gene targeting is a great asset for the genetic analysis of mouse development. This procedure allows investigators to modify any cloned gene at will and then test the biological effect of the modification in developing mice. Of course, this method requires that the gene of interest first be cloned and modified. These latter procedures will be introduced in the following section.

DNA Cloning

DNA cloning techniques allow investigators to produce billions of copies of any gene for analysis and modification (Methods 14.2). Since their invention in the 1970s, these techniques have revolutionized many areas of science, medicine, agriculture, and law. For developmental biologists, DNA cloning has provided the means for studying differential gene expression at the molecular level. Cell differentiation and pattern formation can now be studied in terms of specific DNA sequences and regulatory proteins.

Any cloned gene lends itself to detailed analysis. *DNA sequencing* is a method of determining the precise nucleotide sequence of a DNA segment. From the known nucleotide sequence and the genetic code, it is possible to predict the amino acid sequence of the protein encoded by the gene. The amino acid sequence often provides clues to the function of the protein in vivo. The RNA and protein products of cloned genes can also by synthesized in vitro. In addition, it is possible to prepare specific molecular probes that allow investigators

Recombinant DNA Techniques

One way of cloning genes uses *recombinant DNA*, that is, DNA spliced together from different sources. Spliceable pieces are generated by cutting DNA with bacterial enzymes called *restriction endonucleases.* These enzymes cut double-stranded DNA at distinct

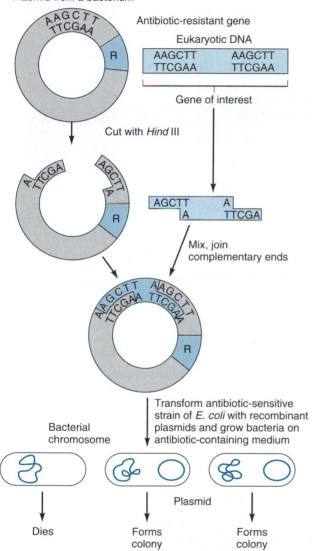

Plasmid from a bacterium

Figure 14.13 Cloning genes using recombinant DNA. A eukaryotic DNA fragment containing the gene of interest is cut with a restriction enzyme such as *Hind* III and joined with a plasmid cleaved by the same enzyme. The recombinant DNA is used to infect host bacteria. The plasmid used as a vector for the gene of interest also confers resistance (R) to an antibiotic on the host bacteria. The host bacteria are plated on an antibiotic-containing medium so that only bacteria containing the R gene grow. Individual colonies are tested for the presence of the gene of interest. A positive colony is grown to obtain a large clone of recombinant DNA.

restriction sites, which are specific sequences of 4 to 8 nucleotide pairs. This technique allows investigators to cut precisely reproducible *restriction fragments* from genomic DNA. Fragments cut by the same restriction endonuclease can be ligated together even if they come from different organisms. Thus, any gene of interest can be cloned by insertion into a carrier, or *vector,* which is usually a modified plasmid or bacteriophage DNA (Fig. 14.13). (A plasmid is an extra-chromosomal circle of bacterial DNA that replicates independently.) The vector will then multiply in host bacteria, creating billions of copies of itself along with the inserted gene.

Recombinant DNA can be used to prepare "libraries" of plasmid or bacteriophage clones, each containing a specific DNA fragment from the organism of interest. A *genomic DNA library* represents the complete genome of an organism; it is prepared by cutting the entire genomic DNA into fragments and cloning them in suitable vectors. Another type of library, called a *cDNA library,* represents all the mRNA sequences present in a cell type of interest; it is prepared by reverse-transcribing mRNA into *complementary DNA (cDNA)* and cloning the cDNA molecules in suitable vectors (Fig. 14.14).

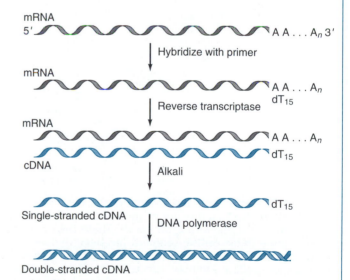

Figure 14.14 Synthesis of complementary DNA (cDNA) from messenger RNA (mRNA). Most mRNAs end with a series of adenosine residues (AA . . . A_n) called a poly-A tail. The poly-A tail hybridizes to a primer consisting of about 15 thymidine residues (dT_{15}). Reverse transcriptase, an enzyme encoded by certain viruses, transcribes the mRNA into cDNA, starting from the primer. The cDNA is isolated by alkaline digestion of the mRNA. Finally, the single-stranded cDNA is made double-stranded with DNA polymerase.

Continued on next page

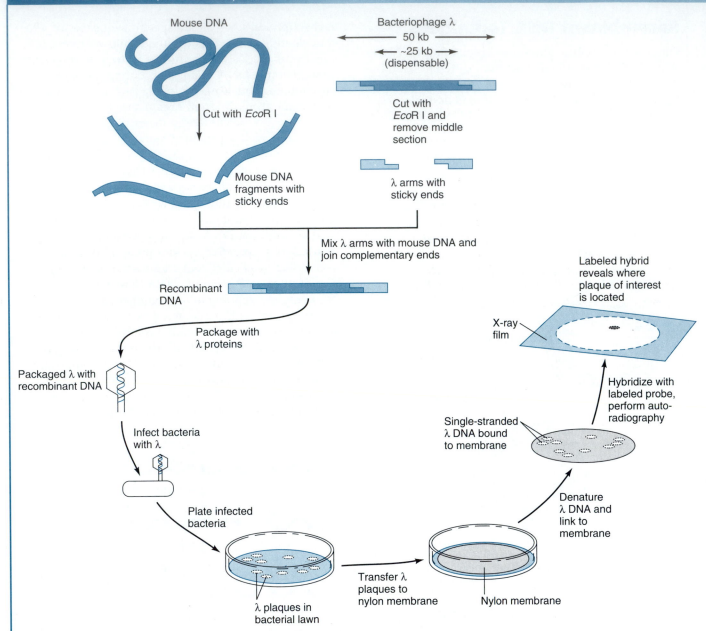

Figure 14.15 Preparation and screening of a genomic or cDNA library prepared in bacteriophage (phage) DNA. A mixture of genomic DNA fragments or cDNAs is cut with *Eco*R I restriction enzyme or fitted with *Eco*R I ends. These DNA inserts are spliced into phage DNA, from which a dispensable central segment has been removed. The recombinant DNA is packaged in phage proteins, and the reconstituted phages are added to suitable host bacteria. These bacteria, some of which are now infected with phage, are spread on agar plates. As the bacteria form a lawn on the agar, each phage inside a bacterium multiplies, lyses the host cell, and reinfects surrounding cells. This leads to the formation of a plaque of lysed bacteria with millions of copies of cloned phage DNA. Each clone contains one specific DNA insert that has been replicated along with the vector DNA. To identify the clone that contains the DNA of interest, a replica of the agar plate is prepared by placing a nylon membrane on it. From each plaque, the membrane picks up some of the phage DNA, which is then made single-stranded and linked to the membrane. Next, the membrane is incubated with radiolabeled DNA or RNA probes representing the gene or cDNA of interest. The probe hybridizes selectively to any complementary DNA sequences attached to the membrane. The position of labeled spots on the membrane, which is detected by autoradiography (see Methods 3.1), reveals which of the plaques contain the insert of interest. These plaques are picked and amplified for testing and analysis.

Both genomic and cDNA libraries can be screened with specific probes to identify, among hundreds of thousands of clones, those containing the specific gene or cDNA of interest. Most screening methods are based on *nucleic acid hybridization,* the formation of double-stranded hybrids from single-stranded DNA or RNA molecules with complementary sequences. The protocols vary according to the vector and host cell chosen. Figure 14.15 illustrates a commonly used procedure to screen a library prepared in a bacteriophage vector.

Pulling a specific DNA sequence from a library requires a suitable probe for screening. For instance, globin genes can be identified by screening a genomic library with labeled mRNA from red blood cells, which contain abundant globin mRNA, and then with a control mRNA from another tissue where globin genes are not transcribed. Clones hybridizing with the first probe but not the second are likely to contain globin genes. Likewise, cDNAs representing RNA sequences that are localized during oogenesis can be identified by screening an egg cDNA library with RNA probes extracted from different egg regions. Michael Rebagliati and his colleagues (1985), for example, have cloned maternal mRNAs localized near the animal or vegetal pole of *Xenopus* eggs (Fig. 14.16). Yet another strategy relies on the observation that many DNA sequences have been conserved so well during evolution that corresponding genes from different species form stable DNA hybrids. Several cloned *Drosophila* genes have been used to identify genes with similar nucleotide sequences from DNA libraries of other organisms, such as the mouse *Hoxb-4*[+] gene examined in the previous section.

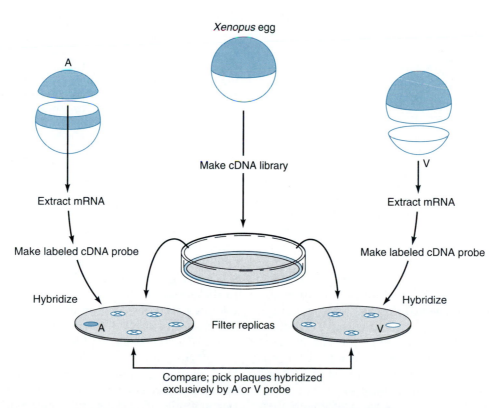

Figure 14.16 Differential screening procedure used by Rebagliati and coworkers (1985) to isolate mRNAs localized to either the animal or the vegetal pole of *Xenopus* eggs. Frozen eggs were cut to isolate animal (A) or vegetal (V) caps. From each, mRNA was extracted and transcribed into labeled single-stranded cDNA probes. These probes were used to screen a cDNA library made from total egg mRNA. Clones hybridizing only with the animal probe represented mRNAs localized to the animal pole. The converse criterion was used for mRNAs localized to the vegetal pole.

to trace these gene products in living cells or in histological sections. Moreover, cloned genes can be engineered by removing or adding certain DNA segments, and by changing single base pairs at will. Again based on the strategy of controlled modification, the testing of engineered genes is a powerful method of learning how genes work.

Transfection and Genetic Transformation of Cells

Recombinant DNA techniques allow researchers to modify various parts of a gene, in particular the regulatory regions that control when and where the gene is transcribed. By introducing the modified gene into a cell, investigators can study how the activity of the modified gene affects the activity of other genes and the differentiation of the host cell. For example, the $MyoD^+$ gene, which is specifically expressed in skeletal muscle, has been cloned and modified so that it will be expressed in any type of cell. If the modified *MyoD* gene is introduced into *fibroblasts*, which normally give rise to connective tissue, these cells instead form skeletal muscle! This stunning observation has allowed investigators to analyze how the MyoD protein activates other muscle-specific genes (see Chapter 19). Thus, the reinsertion of cloned genes into cells provides researchers with direct access to the molecular mechanisms that control cell differentiation and other developmental processes.

Many types of cells, from bacterial to mammalian, take up foreign DNA by *endocytosis,* or can be forced to take it up by other means. A gene introduced into a host cell this way is called a ***transgene.*** If a transgene is complete with its regulatory regions, it is expressed according to the host cell's control signals. Transgenes may persist in the host cell for a limited time without being integrated into the genome; this method of generating a temporarily transgenic cell is called ***transfection*** (Fig. 14.17). Under certain conditions, however, transfected cells integrate the transgene into their own genome, where it is replicated and passed on during cell divisions along with the host cell's own genes. This process of generating a stable genetic change by introducing an exogenous gene into the genome of a host cell or organism is called ***genetic transformation.*** It was first demonstrated in bacteria, in the same experiments that also established the role of DNA as the carrier of genetic information (Griffith, 1928; Avery et al., 1944). Researchers later extended this procedure to eukaryotic cells by showing that DNA from mouse tumor cells transformed fibroblasts into tumor cells.

Transformed cells often behave as the equivalents of *null, loss-of-function,* and *gain-of-function* mutants. The

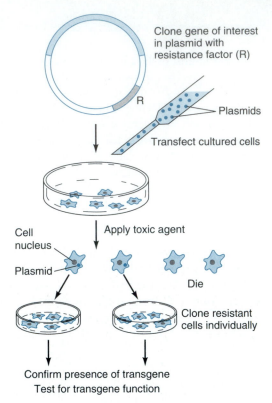

Figure 14.17 Cell transfection. The gene of interest (color) is cloned in a plasmid vector. The vector contains a gene conferring resistance against a toxic agent, such as neomycin. The cloned DNA is added to cultured cells, some of which take up the DNA by endocytosis. Only these cells survive in the presence of the toxic agent. These cells are cloned individually, and the presence of the transgene is confirmed by Southern blotting (see Methods 14.3). Transfected cells are then examined to determine whether the transgene functions as expected.

generation of a null allele in mouse embryonic stem cells by insertion of another gene was discussed earlier. To mimic a loss-of-function mutant, one can insert a "flipped" transgene in such a way that the normally untranscribed DNA strand serves as a template for RNA synthesis (Fig. 14.18a). The resulting RNA, called ***antisense RNA,*** has a nucleotide sequence complementary to the normal mRNA. It will therefore hybridize with the mRNA synthesized by the resident gene in the cell. The resulting double-stranded RNA is often untranslatable and is degraded rapidly in cells. Another method for generating the equivalent of a loss-of-function mutant, known as ***dominant interference,*** is to insert a transgene that encodes a defective polypeptide in excess quantity. The defective polypeptide then outcompetes its normal counterpart in interactions with other molecules, so the normal gene function is sharply reduced (Fig. 14.18b).

To generate the equivalent of a gain-of-function mutant, researchers can engineer a cloned gene so that it is expressed continuously or in response to external signals such as elevated temperature. Cells transformed

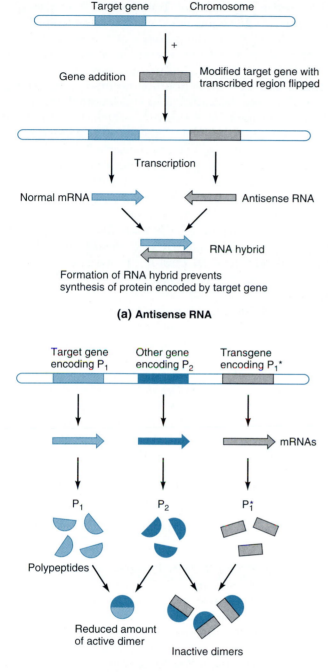

Figure 14.18 Two strategies for mimicking a loss-of-function mutation of a known target gene. **(a)** Antisense RNA strategy. Cells are transfected with a modified version of the target gene in which the transcribed region has been flipped so that transcribed RNA is complementary in sequence to the mRNA transcribed from the normally oriented target gene. This antisense RNA, if produced in excess quantity, inactivates most of the normal mRNA by forming a double-stranded hybrid that is not translated into protein. **(b)** Dominant interference strategy. The target gene encodes a polypeptide (P_1) that must combine with another polypeptide (P_2) to form a functional protein dimer. Cells are transfected with a mutant allele of the target gene that encodes P_1^*, an inactive version of P_1. An excess of P_1^* will bind up most of P_2 in inactive dimers, thus diminishing the function of P_1.

with such a gene can express it under conditions when it would normally be silent.

Once a *cell* has been transformed successfully and responds as expected to experimental manipulation, this cell can be used to generate *whole organisms* with the same transformed genome. This result can be achieved by positioning transformed cells to serve as primordial germ cells in chimeric embryos, as described earlier (Figs. 14.11 and 14.12). Since 1980, it has also been possible to introduce cloned genes directly into the germ lines of flies and mammals. These procedures will be described in the following sections.

Germ Line Transformation

Germ line transformation is a procedure in which an exogenous gene is stably integrated into the genome of *germ line* cells. This strategy for producing transgenic organisms is especially useful for genes that have complex expression patterns or participate in cell interactions that cannot be mimicked in vitro. Many genes of interest to developmental biologists fall into this category. Consequently, germ line transformation has been instrumental in providing a better understanding of gene activity in development. So far, germ line transformations have been performed mostly with *Drosophila, C. elegans,* and the mouse, although this procedure can be carried out, in principle, with any organism. We will first explore how this technique is used in *Drosophila,* where gene insertions are aided by a naturally occurring mobile genetic element. Then we will examine germ line transformation in the mouse, the favorite animal for medical applications. Finally, we will see how germ line transformation can be used to interrupt and isolate previously unknown genes.

The Genetic Transformation of *Drosophila* Utilizes Transposable Genetic Elements

Germ line transformation in *Drosophila* makes use of naturally occurring *transposable DNA elements (transposons)* that jump randomly into and out of genomic DNA. The existence of transposons was first proposed on the basis of genetic experiments with maize by Barbara McClintock (1952). For this discovery, she received the 1983 Nobel Prize in physiology or medicine. Transposons are common in the genomes of mammals, including humans, where they amount to approximately 10% of the genome size (Deininger and Daniels, 1986).

The best-investigated transposons are the *P elements* in *Drosophila.* Molecular analysis shows that P elements encode an enzyme, *transposase,* which is necessary for cutting the transposon out of its place in the genome

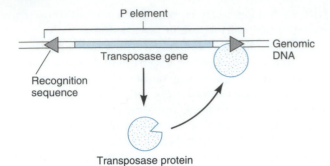

Figure 14.19 Structure of the *Drosophila* P element. A full-length P element has specific recognition sequences flanking a gene that encodes the enzyme transposase. The enzyme binds to the recognition sequences in the process of cutting out the P element from genomic DNA and reinserting it elsewhere. Full-length P elements are about 2.9 kilobases (kb) long. Short P elements have the recognition sequences but an incomplete transposase gene.

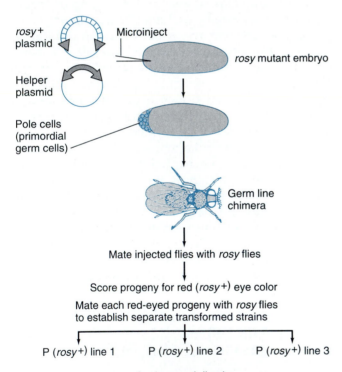

Figure 14.20 Strategy for administering germ line therapy to a *Drosophila* mutant in the *rosy* gene. The wild-type allele (*rosy*⁺) of the gene encodes xanthine dehydrogenase, an enzyme required for synthesizing the wild-type red eye pigment. Injecting the cloned wild-type allele as a transgene into the germ line results in offspring that are permanently cured of the genetic deficiency. Because the *rosy*⁺ transgene is embedded in a P element, the transgenic lines of flies are designated P(*rosy*⁺).

and inserting it elsewhere. Transposase recognizes specific nucleotide sequences on both ends of the P element (Fig. 14.19). In the presence of transposase, the P element is excised from its resident position in germ line DNA and inserted into a new position. This can cause a mutation if the insertion occurs in a functional part of a gene. In addition to full-length P elements, there are *incomplete P elements,* which do not encode active transposase. These elements do not move spontaneously but are excised and reinserted if transposase activity is provided by another complete P element. In fact, any piece of DNA flanked by P-element recognition sequences is treated by transposase as an incomplete P element. P elements are therefore convenient vehicles for inserting foreign DNA into *Drosophila* genomic DNA. This method, known as *P-element transformation,* has been used successfully to insert a wide variety of transgenes into the germ line cells of *Drosophila* embryos.

The first P-element transformation in *Drosophila* had the overall design of a germ line therapy (Fig. 14.20). It was carried out on a mutant homozygous for a *null* allele of the *rosy* gene. This gene encodes xanthine dehydrogenase (XDH), an enzyme required to synthesize the red eye pigment of *Drosophila*. Gerald Rubin and Allan Spradling (1982) inserted the corresponding wild-type allele (*rosy*⁺) into a short P element and injected the construct into the primordial germ cells of early embryos. Some of the surviving flies produced offspring that had the P element along with the inserted *rosy*⁺ allele stably integrated into their genomes. Therefore, these flies and their progeny were permanently cured of their genetic defect.

In preparation for their transformation experiment, Rubin and Spradling constructed a circular P-element vector from a plasmid and a piece of *Drosophila* genomic DNA containing an incomplete P element (Fig. 14.21). The plasmid portion allowed the entire DNA molecule to be cloned in bacterial host cells, while the P element provided the recognition sequences for the transposase. Inserted between the recognition sequences was a fragment of *Drosophila* genomic DNA containing the *rosy*⁺ gene. This insert was 8.1 kilobases (kb) long (1 kb = 1000 base pairs). The entire construct was designated *pry 1*.

The cloned *pry 1* DNA was microinjected into the posterior pole region of *rosy* host embryos. The injection was carried out before *pole cell* formation, so that the injected DNA would be included in the pole cell cytoplasm. Since neither the embryos nor the *pry 1* construct provided transposase activity, the investigators mixed the *pry 1* DNA with a small amount of a so-called helper plasmid containing a full-length P element. Upon entering the egg cytoplasm, the helper plasmid was to

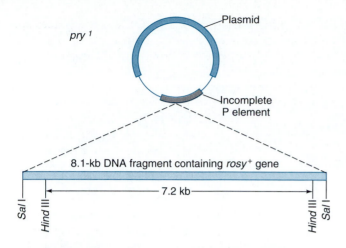

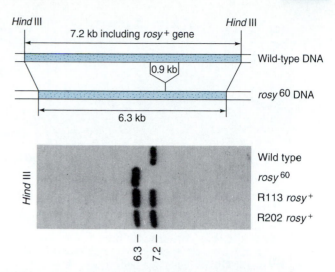

Figure 14.21 Construction of the vector *pry 1*, used by Rubin and Spradling (1982) to genetically transform the *Drosophila* mutant *rosy*. An 8.1-kb insert (straight bar) of *Drosophila* genomic DNA containing the wild-type allele *rosy*+ was isolated with *Sal* I restriction enzyme and cloned. The fragment was inserted into a P-element vector consisting of an incomplete P element (gray bar), small adjacent pieces of *Drosophila* DNA (lines), and a plasmid for replication. The DNA fragment between the *Hind* III restriction sites in the insert was 7.2 kb long. The corresponding fragment from the *rosy* mutant was only 6.3 kb long.

Figure 14.22 Analysis of genomic DNA from *rosy* transformant lines by Southern blotting. *Drosophila* genomic DNA fragments cut with *Hind* III were hybridized with a *rosy* probe. The probe hybridized to a 7.2-kb fragment of wild-type DNA, as expected from the known size of the *Hind* III fragment containing the *rosy*+ gene. The *rosy* mutant, *rosy*⁶⁰, had a 0.9-kb deletion, which reduced the length of the corresponding *Hind* III fragment to 6.3 kb. This was exactly the length of the genomic fragment hybridized by the *rosy* probe in the Southern blot of *rosy*⁶⁰ DNA. In the DNA from transformants designated R113 and R202, two fragments were hybridized by the probe: one fragment from the resident *rosy*⁶⁰ gene, and the other from the *rosy*+ transgene.

produce a burst of transposase activity that would excise the P element from *pry 1* and insert it into the pole cell genomic DNA.

Of 1111 injected embryos, 82 developed into fertile adults. Some or all of these were expected to contain the transgene in their germ line. To identify these germ line chimeras, the investigators mated each adult individually with *rosy* mutant partners. Their progeny were screened for red (wild-type) eye color indicating the presence of *rosy*+ allele in the genome. About 25% of the matings produced offspring with red eyes. Each of these offspring was maintained as a **transformant line**, a separate strain of transgenic flies. Within each line, the red eye color was passed on as a stable, dominant marker.

To confirm that the red eye color in their transformation lines was due to the integration of the *rosy*+ transgene into the host germ line, Rubin and Spradling (1982) used the *Southern blotting* technique (Methods 14.3). Genomic DNA from wild-type flies, from the *rosy* mutant strain that supplied the host embryos, and from each line of transformants was hybridized with a labeled probe representing the *rosy*+ gene (Fig. 14.22). Wild-type DNA showed one labeled *Hind* III fragment corresponding to the known size of the *rosy*+ gene. In DNA from the mutant *rosy* allele, the labeled probe hybridized with a shorter *Hind* III fragment because the mutation had caused a deletion. Both bands were labeled in the transformant lines, since they contained the resident mutant *rosy* allele and the *rosy*+ transgene.

As controls, the investigators probed genomic DNA from some of the lines derived from injected embryos that had no red-eyed progeny. Their Southern blots showed the same labeling pattern as the uninjected *rosy* mutant. Thus, the inheritance of the red eye color correlated perfectly with the integration of the transgene in genomic DNA.

To survey the chromosomal sites at which the transgene had been integrated, the investigators applied a labeled *rosy*+ DNA probe to *polytene chromosomes* prepared from their transformants (Fig. 14.23). In each case, the probe hybridized to the location of the resident *rosy* gene on the *cytogenetic map*. Also, each transformant line had one or two additional sites of transgene integration. The compiled results showed that the transgenes had been inserted at a wide variety of sites in each chromosome.

In a subsequent study, Spradling and Rubin (1983) examined whether the activity of a *rosy*+ transgene depended on its site of integration into the host genome. They measured the specific activity of the rosy enzyme, xanthine dehydrogenase (XDH), in each of their transformant lines of flies. In most lines, the activity level was between 30% and 130% of the wild-type activity. For

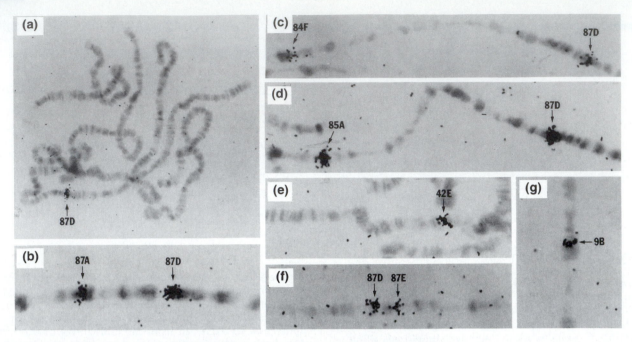

Figure 14.23 Chromosomal sites of *rosy*⁺ transgene insertion located by in situ hybridization (see Methods 15.1) to polytene chromosomes. **(a)** In the *rosy*⁶⁰ host strain, the cytogenetic map location of the *rosy*⁺ gene was 87D. **(b–g)** Additional chromosomal bands were labeled in transformants, indicating sites of transgene insertion.

each transformant line, however, the activity level was stable from one generation to the next. When the investigators examined XDH activity in different tissues, they found high XDH levels in fat body and Malpighian tubules in both wild-type flies and transformants, and this activity distribution in transformants was found regardless of the chromosomal position of the transgene. Thus, the transgene was expressed with the normal tissue specificity and at nearly normal levels regardless of the transgene integration site. These results indicated that the transgene contained not only the transcribed region of the *rosy*⁺ gene but also its regulatory regions necessary for tissue-specific expression.

In later experiments, investigators used transgenes that conferred phenotypic effects more difficult to score than eye color. To facilitate screening the surviving hosts for transformants, researchers constructed new P-element vectors with visible or selectable marker genes. These markers were included between the recognition sequences of the P elements so that they were excised from the vector and reinserted along with the gene of interest. The *rosy*⁺ gene is a convenient *visible* marker. Even more convenient are *selectable* markers such as the bacterial *neo*^R gene, which renders *Drosophila* larvae resistant to the drug G418 (Steller and Pirrotta, 1985). When kept on food containing G418, only those larvae that express the *neo*^R gene survive. These individuals can be expected to carry both the *neo*^R marker and the gene of interest as transgenes.

The results of many genetic transformation experiments in *Drosophila* show that P elements can be used effectively to insert exogenous genes into the germ line. The transgenes are stably integrated into the genome and passed on to future generations. Most transgenes, if their regulatory sequences are present, show normal expression patterns and provide a nearly normal level of gene activity that varies somewhat with the site of integration.

Mammals Are Transformed by Injection of Transgenes Directly into an Egg Pronucleus

Transgenic mammals are generated much as are transgenic fruit flies, but with three differences. First, the cloned transgene is injected directly into a *pronucleus* of the fertilized mammalian egg. This procedure is facilitated by the leisurely pace of events after fertilization in mammals. Second, the transgene is injected without the aid of a natural transposon. This method is therefore applicable to all mammals, whereas the P-element transformation is limited to *Drosophila melanogaster* and a few closely related species. Third, mammalian transgenes are injected as linear DNA fragments without a cloning vector, since plasmid or bacteriophage DNA inserted into the host genome may have undesirable effects. In mammals, tandem arrays of up to 100 copies of the transgene are often integrated into the same site of the host genome. For the most part, these multiple in-

Southern Blotting and Northern Blotting

Frequently, it is necessary to test whether a mixture of DNA fragments contains a particular sequence. The technique shown in Figure 14.24 is named *Southern blotting,* after its inventor, Edward Southern (1975). First, the DNA is cut into fragments with an appropriate *restriction enzyme.* The restriction fragments are then separated according to size by gel electrophoresis. Next, the fragments in the gel are made single-stranded by alkali treatment, and the denatured fragments are transferred by suction or an electric field to a nylon membrane placed on the gel. The DNA fragments are thus blotted onto the membrane *in the same order* (according to size) into which they had been sorted inside the gel. Further treatment permanently links the DNA fragments to the nylon membrane, so that the fragments will not wash off when the membrane is immersed in various fluids for analysis.

To detect a particular DNA sequence on the membrane, the membrane is incubated with a fluid containing a labeled RNA or single-stranded DNA probe. (This step in the Southern blotting procedure is also used in screening recombinant DNA libraries; see Figure 14.15.) Wherever a DNA fragment complementary to the probe is located on the membrane, a labeled hybrid will form. The position of the labeled hybrid reveals the size of the DNA fragment containing the sequence of interest. Moreover, the intensity of the label provides an estimate of the quantity of the fragment of interest relative to known controls.

An application of the Southern blotting technique is shown in Figure 14.22.

A corresponding technique, called *Northern blotting,* identifies specific RNA sequences in an RNA mixture. In this procedure, the first step is to separate the RNA molecules according to size by gel electrophoresis. After hybridization with a complementary probe, the position and amount of labeled hybrid indicate the size and relative amount of the RNA of interest.

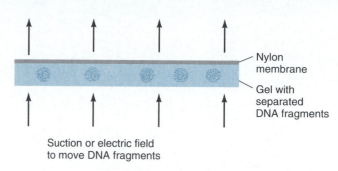

Suction or electric field
to move DNA fragments

Figure 14.24 The Southern blotting technique. DNA separated according to size by gel electrophoresis is transferred from the gel to a nylon membrane, to which it becomes permanently linked. The membrane can then be hybridized with labeled probes to detect sequences of interest.

---

tegrations are stable in successive generations. Figure 14.25 shows how a transgene is inserted into a mouse egg.

Injected mammalian eggs are allowed to develop in vitro until the blastocyst stage, when surviving embryos are implanted into foster mothers. Their offspring are tested directly for transgene integration by *Southern blotting.* Many of the offspring that test positive have the transgene in *all* tissues, not just in some of the germ line cells as in *Drosophila.* Apparently, foreign DNA injected into a mammalian pronucleus is usually integrated either before the first genomic DNA replication or not at all. The site of transgene integration into the host chromosomes seems to be random, as in *Drosophila.* Most transgenes in mammals are expressed at a lower-than-normal level, although exact quantification is difficult because of the unknown number of transgene copies inserted. Some chromosomal sites seem to silence inserted transgenes so that they are not expressed.

A particularly striking demonstration of transgene activity in mice came from an experiment conducted by Richard Palmiter and his colleagues (1982). They worked with the cloned mouse gene for *metallothionein (MT),* a protein synthesized mostly in liver and kidney that removes ingested heavy metals such as cadmium or zinc. The investigators isolated the regulatory region, which controls the expression of the MT gene. An important feature of this control region is that its activity is inducible by the presence of heavy metals. The investigators also cloned the rat gene for *growth hormone (GH).* Normally produced in the pituitary gland, GH enhances the growth of bones and muscle and reduces fat deposition. Using recombinant DNA techniques, the investigators spliced the mouse MT regulatory region to the transcribed region of the rat GH gene, thus creating a composite gene, or *fusion gene* (Fig. 14.26). They expected the fusion gene to be expressed primarily in liver and kidney, because the regulatory region of a gene, and not its transcribed region, controls its expression (see Chapter 15).

The fusion gene was amplified in a plasmid vector, excised from the vector, and injected into pronuclei of

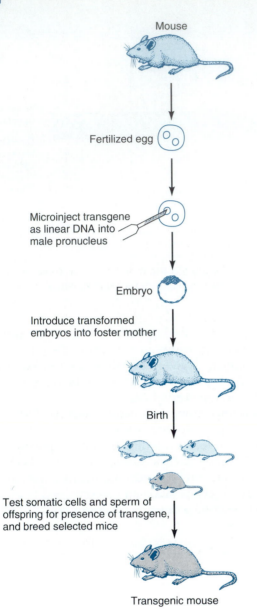

Figure 14.25 Technique for generating transgenic mice by microinjection of DNA into eggs. Fertilized eggs are removed from the oviduct of the donor mother. The cloned foreign gene is injected directly into a pronucleus (see also Fig. II.1). Developing blastocysts are implanted into the oviducts of foster mothers. Offspring are screened for transgenes in somatic cells and sperm by Southern blotting. Transgenic mice are bred to obtain individuals that are either heterozygous or homozygous for the transgene.

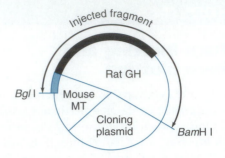

Figure 14.26 Mouse metallothionein (MT)-rat growth hormone (GH) fusion gene created by Palmiter et al. (1982). The regulatory region of the MT gene (color) was spliced to the transcribed region of the GH gene (black). The fusion gene was cloned in a plasmid vector. The DNA fragment to be injected was cut out with *Bgl* I and *Bam*H I restriction enzymes.

nontransgenic littermates that served as controls (Figs. 14.1 and 14.27).

It remained to be tested whether the unusual growth of the transgenic mice was indeed caused by growth hormone encoded by the transgene or resulted from unexpected activation of the resident mouse GH gene. As proof, the investigators showed that the liver RNA in the transgenic mice contained a 74-nucleotide segment present in transgene mRNA but not in the resident growth hormone mRNA or metallothionein mRNA.

The growth rate of the transgenic mice was not always proportional to the number of transgene copies in their genomes as revealed by the Southern blot; it is possible that not all the transgene copies integrated into the host genome were biologically active. Another unexpected result was obtained later when the transgene expression patterns were examined (Swanson et al., 1985). In addition to the tissues normally expressing the MT gene, the investigators also found novel expression areas of the MT-GH fusion gene. It was expressed, for instance, in the hypothalamus and in distinct regions of the cerebral cortex, where neither MT nor GH is normally produced. These results were not readily explainable and suggested that there might be unexpected interactions between transgenes and neighboring resident genes in mammals.

Palmiter's experiment confirmed and extended results of an earlier study of Jon W. Gordon and his colleagues (1980), who had shown that a viral gene injected into an egg pronucleus integrated stably into the mouse genome. Since then, mice and other mammals have been the subject of germ line transformations with a large variety of transgenes, many of them taken from species other than the host. Once integrated into the genome of the host's germ line, the transgenes are passed on to offspring in a fairly stable manner. The expression pattern of the transgenes is usually normal, but low rates and abnormal sites of transcription are observed more frequently than in *Drosophila*.

mouse eggs. From 170 developing embryos implanted into foster mothers, 21 mice were born. Small biopsies of tail tissue were used to prepare genomic DNA for Southern blotting. Upon probing with a fragment of the rat GH gene, the DNA from seven mice showed the hybridization pattern expected after integration of the transgene. When these seven mice were placed on a diet containing zinc sulfate (ZnSO$_4$) to induce the MT promoter activity, they underwent extraordinary growth. One of them reached twice the weight of the

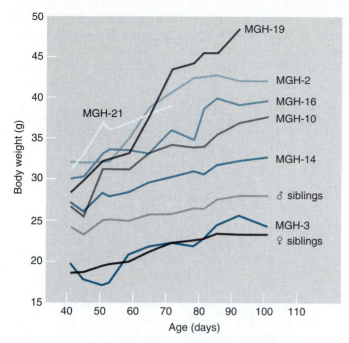

Figure 14.27 Growth of mice transformed with a mouse metallothionein (MT)-rat growth hormone (GH) fusion gene. At 33 days, the mice were weaned, and their drinking water was supplemented with zinc sulfate to stimulate transgene expression. The body weight of each transgenic mouse (designated MGH-2, MGH-3, etc.) was measured individually. For comparison, the average weight of the mice's nontransgenic siblings was also recorded.

DNA Insertion Can Be Used for Mutagenesis and Promoter Trapping

The site in a host genome where a transgene inserts itself appears to be random. If insertion occurs within a functional part of a gene, the activity of the host gene will be disrupted. The generation of mutants by random insertion of foreign DNA into a host genome is called *insertional mutagenesis.* It may occur as an unwanted side effect in a genetic transformation experiment. However, insertional mutagenesis also provides a *method* of generating new mutant strains. This method has been used most successfully so far in *Drosophila,* but it has also been applied to other organisms.

Mutants generated by insertion of engineered segments of DNA greatly facilitate the molecular analysis of the affected gene. If the inserted DNA contains a selectable marker, only transformed individuals survive on an appropriate medium. If the insert contains a unique sequence that occurs nowhere else in the host genome, this sequence can be used as a tag to identify the mutated gene. A labeled probe hybridizing with the tag will identify the mutated gene in a genomic library prepared from the mutant strain. With this method, it is but one step from the genetic to the molecular analysis of a gene.

For *Drosophila,* Lynn Cooley and her colleagues (1988) have worked out a scheme in which two strains are crossed so that a modified P element is mobilized in the offspring and inserts itself elsewhere in the genome. The offspring are then crossed and screened as if they were offspring obtained from a parent treated with a chemical mutagen (Fig. 14.7). Instead of mutant lines with a chemically induced DNA alteration, this scheme yields lines with P-element insertions. From a collection of such lines, investigators select those of particular interest for further analysis. This procedure bypasses the tedious microinjection of cloned transgenes and makes insertional mutagenesis so convenient that it can be used for *saturation mutagenesis screens.*

Two variants of the insertional mutagenesis method are known as **promoter trapping** and **enhancer detection.** Promoters and enhancers are regulatory gene elements that will be described more fully in Chapter 15. These elements are not transcribed themselves; instead, they control the region- and stage-specific transcription of other genetic elements in their vicinity. If the transcribed region of a transgene is inserted between the regulatory elements and the transcribed region of a resident gene, then the effect is that of a cuckoo laying an egg in another bird's nest. The resident regulatory elements "adopt" the transgene and transcribe it along with the resident gene. If the transgene causes the formation of a readily detectable product—say, a blue stain—then the staining pattern in an embryo will shed light on the normal transcription pattern of the neighboring resident gene. Thus, for example, researchers can detect genes involved in neurulation by randomly inserting stain-producing transgenes and selecting those transgenic lines that show selective staining in the neural plate. An application of this technique is shown in Figure 22.1.

Significance of Transgenic Organisms

Transgenic Organisms Help Researchers Analyze Gene-Regulatory Regions

Transgenic organisms have a wide range of research applications (Wakimoto and Karpen, 1988; J. W. Gordon, 1988). One important use is to confirm the identity of cloned genes. Many cloning strategies leave some uncertainty whether the cloned gene does in fact encode the protein of interest. In these cases, proof comes from genetically transforming the entire organism with the cloned gene and monitoring the gene's function in vivo. This test is especially convincing if the transgene restores the normal phenotype for the gene of interest in a *loss-of-function* mutant. Loss-of-function mutants are usually available for *Drosophila.*

Investigators studying species for which mutants are not available can test the functions of cloned genes by using molecular means to generate the equivalents of loss-of-function and gain-of-function mutants. Gain-of-function alleles can be produced by placing the transgene of interest under the control of a promoter that is inducible by heat, metal, or other environmental factors. Loss-of-function alleles can be constructed with transgenes producing *antisense RNA* or defective polypeptides, as described earlier in this chapter.

The most important application of transgenic organisms in basic research is in the analysis of gene-regulatory regions. As will be described in Chapter 15, genes have regulatory sequences controlling their expression. The location and function of each regulatory sequence can be revealed by removing DNA fragments from cloned genes. Testing the effects of such modifications in transgenic organisms is especially valuable for investigating regulatory sequences that respond to signals not easily mimicked in cell culture, such as region- or tissue-specific signals.

Transgenic Organisms Serve as Models in Medical and Applied Research

In addition to their invaluable role in basic research, transgenic organisms have many applications in medicine, agriculture, and pharmacology. For instance, many types of cancer are caused by genes called *oncogenes*. Most oncogenes are mutant alleles of normal cellular genes that encode hormones, hormone receptors, and intracellular signals (see Chapter 29). In some cases, a single oncogene is sufficient to cause cancer. Such genes, when introduced as transgenes into healthy mice, inevitably lead to the development of a specific cancer (Hanahan, 1989). In other cases, an oncogene causes cancer only in combination with other mutations or environmental factors. While some of these combinations can be tested in cell culture, transgenic animals are unsurpassed in modeling the full range of factors contributing to development of cancer or other diseases in humans.

Mice have proven to be useful models for types of gene therapy that involve the germ line and whose effects are passed on to offspring. For example, mice homozygous for the *shiverer* mutation suffer from tremors and convulsions leading to early death. The mutant gene encodes a defective cell membrane protein that is necessary for forming myelin sheaths around axons in the central nervous system (Fig. 14.28). When *shiverer* mice are provided with a *shiverer*+ transgene, they form improved myelin sheaths and no longer shiver or die prematurely (Readhead et al., 1987). Similarly, researchers have been able to correct other genetic disorders in mice affecting the synthesis of hemoglobin (Constantini et al., 1986) and reproductive functions (Mason et al., 1986).

Gene therapy of *germ line* cells is not at present carried out in humans, because it could benefit only their offspring and because the uncontrollable effects of transgene insertions pose an unacceptable risk. However, ex-

perimental gene therapy of human *somatic cells* has begun for patients with certain genetic diseases that would otherwise be fatal (A. D. Miller, 1992). In this type of therapy, the abnormal somatic cells are removed from

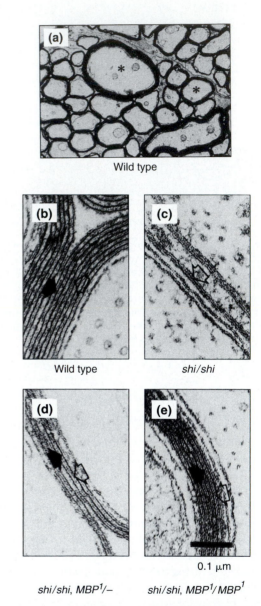

Figure 14.28 Germ line therapy of *shiverer* mutant mice. Electron micrographs show cross sections of optic nerves. **(a, b)** Wild type. Each axon (asterisk) has a myelin sheath consisting of multiple plasma membranes of Schwann cells (see Figs. 12.5 and 12.6). The cytoplasmic faces of plasma membranes compact to form major dense lines (solid arrows); they alternate with weaker lines formed where the outer surfaces of two plasma membranes abut (open arrows). The formation of the dense lines seems to require myelin basic protein (MBP). **(c)** Mouse homozygous for *shiverer (shi)*, a mutant allele of the gene that encodes MBP. A few uncompacted plasma membranes are present, but no dense lines are formed. **(d)** *shi/shi* mouse heterozygous for the *MBP*[1] transgene, which encodes wild-type MBP protein. Some dense lines have formed. **(e)** *shi/shi* mouse homozygous for the *MBP*[1] transgene. Almost normal myelin sheaths have assembled.

the patient and transfected with a cloned wild-type allele of the defective gene. After testing in vitro, successfully transformed cells are reintroduced into the patient's body. Somatic gene therapy is effective only if the transformed cells reach their appropriate destination and last for some time. To work out these problems, experiments on mice and other mammals have been very useful.

Transgenes are also being used to improve agricultural plants and livestock. For example, growth hormone transgenes from cattle were inserted into pigs' eggs in an attempt to produce pigs with more meat and less fat. The transgenic pigs had the desired properties but also suffered from a wide range of malformations and ailments, which then had to be treated with undesirable medications (Pursel et al., 1989). Experiments with transgenic plants have been more successful (C. S. Gasser and R. T. Fraley, 1992). So far, more than 40 agricultural plants have been genetically engineered to improve crops or make them resistant to pests.

SUMMARY

Life scientists frequently use the strategy of controlled modification. Developmental biologists remove or transplant cells or molecular fractions in order to learn from the resulting abnormalities about the normal functions of the manipulated cells or molecules. Mutant alleles of genes are similarly used to infer the functions of the wild-type alleles. Mutant alleles may be characterized as causing a complete loss of function, a partial loss of function, or a gain of function.

In the fruit fly *Drosophila melanogaster,* and in the roundworm *Caenorhabditis elegans,* large mutagenesis screens have been used to identify virtually all genes involved in certain developmental processes, such as the formation of the embryonic body pattern. Saturating chromosomes with mutations is a method that enables researchers to estimate the total number of genes involved in embryonic pattern formation. Some of the mutant phenotypes obtained reveal in a striking way what types of decisions cells make during development. They also show which embryonic areas are affected by these decisions. Saturation mutagenesis is not practical in the mouse, because of its smaller brood size, longer generation time, and larger genome size. However, greater numbers of mutant mouse strains may be generated in the future by new methods of mutagenizing embryonic stem cells. Mutant stem cells, if injected into host blastocysts, may contribute to the germ line of the developing chimeras, thus passing on their mutant genotype to the offspring.

With recombinant DNA techniques, it is often possible to clone and characterize genes using purely molecular criteria. To examine the function of a cloned gene, it is best to reinsert it into a cell or an organism. The insertion of the exogenous gene may be temporary (transfection), or a permanent and heritable integration into the host genome (genetic transformation). In either case, the inserted gene, called a transgene, is expressed along with the resident alleles of the same gene in the cell. With suitably engineered transgenes, the cellular equivalents of both loss-of-function and gain-of-function mutants can be constructed. Transformed cells can also be used to generate the equivalents of mutant organisms if the transformed cells can be added to embryos so that they will contribute to the germ line.

In fruit flies and mammals, it is possible to genetically transform the entire organism by microinjecting cloned transgenes directly into fertilized eggs. Transgenes seem to insert randomly into the host genome. In *Drosophila melanogaster*, the expression pattern of the transgene is usually normal and fairly independent of its genomic insertion site, provided that all regulatory sequences are included. In the house mouse, the insertion site affects transgene expression more strongly.

The insertion of a transgene into a resident gene may cause a mutation. Such insertional mutagenesis can be an unwanted side effect, but is also used as a method for generating additional mutants. If the inserted DNA contains suitably engineered sequences, these can greatly facilitate the cloning of the disrupted gene.

Transgenic organisms have become invaluable for the genetic and molecular analysis of development. They are widely used to confirm the identity of cloned genes, and to delimit the positions and functions of regulatory sequences. In particular, they allow investigators to identify specific gene-regulatory regions that control gene expression in response to localized cytoplasmic determinants, inductive interactions, and other signals. In addition to these uses in basic research, there are many applications in medicine, agriculture, and pharmacology.

SUGGESTED READINGS

Glover, D. M., and B. D. Hames. 1989. *Genes and Embryos.* Oxford: IRL Press.

Gordon, J. W. 1988. Transgenic mice. In G. M. Malacinski, ed., *Developmental Genetics of Higher Organisms: A Primer in Developmental Biology,* 477–498. New York: Macmillan.

Horder, T. J., J. A. Witkowski, and C. C. Wylie, eds. 1985. *A History of Embryology.* Cambridge: Cambridge University Press.

Wakimoto, T., and G. H. Karpen. 1988. Transposable elements and germ-line transformation in *Drosophila.* In G. M. Malacinski, ed., *Developmental Genetics of Higher Organisms: A Primer in Developmental Biology,* 275–303. New York: Macmillan.

Wilkins, A. S. 1993. *Genetic Analysis of Animal Development.* 2d ed. New York: Wiley.

TRANSCRIPTIONAL CONTROL

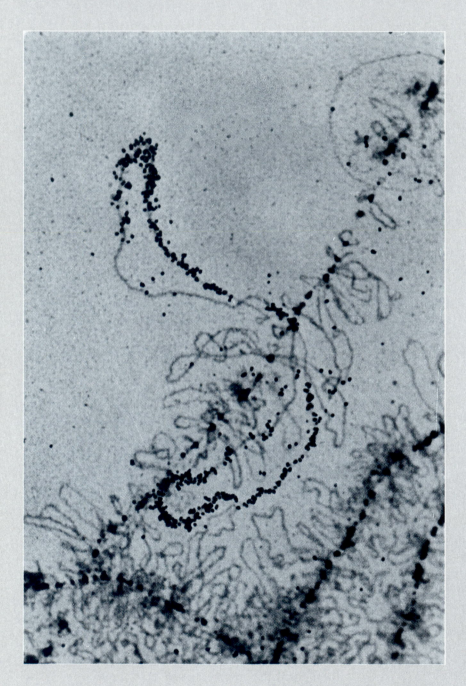

Figure 15.1 Transcription from a lampbrush chromosome in a newt oocyte. Each chromosome consists of an axis with paired loops extending from it. In situ hybridization (see Methods 15.1) with a radiolabeled histone DNA probe identifies a pair of loops with histone genes in the process of transcription. The probe hybridizes to the transcripts while they are still attached to the genes. As the length of the transcript increases, so does the intensity of the label.

The nuclear transplantation experiments and other observations discussed in Chapter 7 support the rule of *genomic equivalence*—that most differentiated cells retain a full and functional complement of the organism's genetic information. Nevertheless, cells become clearly different in their morphology and function as development proceeds. At the molecular level, cells differ in the proteins they synthesize. This is shown unequivocally by comparing newly synthesized proteins from different cells using *two-dimensional electrophoresis* and *autoradiography* (see Fig. 7.19). Analogous experimental methods reveal that differ-

ent types of cells also accumulate distinct combinations of mRNAs. It follows that different cells in an organism generally *have* the same genetic information but *use* or *express* different subsets of genes. In Chapters 15 through 17, we will examine various mechanisms of differential gene expression.

The Principle of Differential Gene Expression

The **principle of differential gene expression** is based on the well-founded theory of genomic equivalence and on the observation that cells accumulate stage-dependent and region-specific mRNAs and proteins. The principle states that cells in an organism develop differently not because they *have* different genetic information, but because they *express* different subsets of the same information. As suggested in Chapter 7, the principle of differential gene expression can be compared to the way different people read the same newspaper. Most readers scan the front page news and the weather forecast. However, some people then turn to the business section, whereas others prefer the sports section, and still others enjoy the comics. Similarly, most cells in an organism express a large set of **common genes,** thought to encode all the proteins that cells need to maintain themselves. In addition, different cells express different **cell-specific genes** to support their particular functions as nerve cells or muscle cells, for example. However, the expression of cell-specific genes is often a matter of *quantitative* differences. Only a small percentage of proteins are expressed *exclusively* in one cell type.

Some cell-specific proteins are synthesized in large amounts by fully differentiated cells. For example, hemoglobin accounts for the bulk of all proteins in red blood cells. Other cell-specific proteins are present in only small quantities but nevertheless have important regulatory functions. Some of these proteins are synthesized early in development, before embryonic cells become visibly different from one another. In a *Drosophila* embryo, for instance, the fushi tarazu and even skipped proteins are each synthesized in seven distinct stripes of blastoderm cells (see Color Plate 4). These genes are part of a hierarchy of locally expressed genes that determine the embryonic body pattern (see Chapter 21).

In the first step of differential gene expression, each cell *transcribes* a selected set of cell-specific genes in addition to the common genes (Fig. 15.1). The selection of genes transcribed in each cell depends on its stage of development, tissue type, position in the body, and physiological conditions (Fig. 15.2). This *selective* use of DNA as a template for RNA synthesis—as a result of any of these factors—is called ***transcriptional control.*** The concept of transcriptional control was first proposed by François Jacob and Jacques Monod (1961), who found that bacteria transcribed certain groups of genes only when the enzymes encoded by those genes were needed for metabolism. Since then it has been shown that transcriptional control is the most prevalent type of gene regulation in bacteria as well as eukaryotes.

In this chapter, we will review how transcriptional control works in development. The selectivity of transcription depends on *regulatory DNA sequences* associated with the transcribed regions of genes. These sequences interact with *regulatory proteins* that enhance or inhibit transcription. Some of these regulatory proteins are translated from localized maternal mRNAs and activate genes in certain areas of developing embryos. Other regulatory proteins are translated from zygotic mRNAs. Some regulatory proteins are themselves regulated by hormones and other signal molecules that cells use to communicate with one another.

In our study of transcriptional control, we must take into account that the genomic DNA of eukaryotes is tightly associated with structural proteins in chromosomes. We will examine how the availability of genomic DNA for transcription depends on local decondensation of the chromatin structure. Then, having familiarized ourselves with the molecular mechanism of transcriptional control, we will relate it to the process of cell determination in development and to the *principle of default programs,* which is based in part on transcriptional control.

Evidence for Transcriptional Control

The steps of gene expression in eukaryotes are RNA synthesis, RNA processing, protein synthesis, and protein modifications. The synthesis of RNA from a DNA template is known as *transcription*. Since transcription is the first step, transcriptional control would seem to be the most economical way to achieve differential gene expression. Indeed, many genes are transcribed only in distinct embryonic regions, in specific organs, or at certain stages of development.

The prevalence of transcriptional control is indicated by several lines of evidence. Comparison of *cDNA libraries* (see Methods 14.2) from different sources shows that certain mRNAs occur specifically in certain organs or at certain stages of development. Some mRNAs are found only in liver, not in kidney or brain. In embryos, several mRNAs are present only at the gastrula stage and not at other stages of development. Another technique for revealing the stage- or region-specific occurrence of specific RNA sequences is called ***in situ hybridization*** (Methods 15.1). It allows investigators to see the distribution of a particular RNA sequence directly in wholemounts or histological sections of entire embryos.

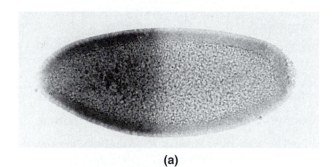

(a)

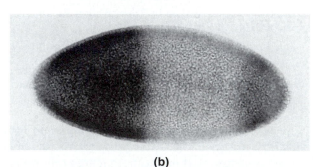

(b)

Figure 15.3 Differential expression of the *hunchback*+ gene in a *Drosophila* embryo; anterior pole to left. **(a)** *hunchback*+ expression is limited to a large anterior domain at the syncytial blastoderm stage. **(b)** An additional posterior domain of *hunchback*+ expression appears about 1 h later at the cellular blastoderm stage. The hunchback mRNA is made visible by in situ hybridization (see Methods 15.1) with a modified DNA probe detected by an antibody. Immunostaining of the hunchback protein (Fig. 15.10) shows a very similar distribution.

Figure 15.2 Expression of the *Cwnt-8C*+ gene in the hindbrain neuroepithelium of a chicken embryo. The accumulated Cwnt-8C mRNA is visualized by in situ hybridization (see Methods 15.1). The gene is transiently expressed in a specific region of the developing hindbrain (arrow).

In Situ Hybridization

In situ hybridization is conceptually simple, although in practice it can be tricky. Like *Southern blotting* analysis and recombinant *DNA library* screening, in situ hybridization is based on *nucleic acid hybridization*. In this case, one hybridization partner is a cellular RNA left in its natural location—hence, the designation "in situ"—after histological fixation. It hybridizes with a labeled RNA or DNA probe of complementary nucleotide sequence. The label associated with the probe then reveals the position of the sequence of interest in the fixed cell or tissue.

In preparation for in situ hybridization, tissues are treated with a fixative that preserves much of the cellular organization. After fixation, opaque embryos or tissues are cut into thin sections, whereas small and transparent embryos are processed whole. Making the RNA of interest accessible to the labeled probe means freeing it at least partially from associated proteins. This is achieved by mild digestion with a proteinase, which is gauged carefully to leave most of the tissue structure intact. After hybridiza-

tion, all probe that is not bound in stable hybrids is washed away. Next, the label associated with the probe is made visible. This is done by *autoradiography* if the label is radioactive (see Methods 3.1). Many modern probes, however, use chemically modified nucleotides that are bound specifically by antibodies. These antibodies, in turn, are associated with a visible marker, such as an enzyme that generates an insoluble colored product. This version of in situ hybridization is very similar to *immunostaining* (see Methods 4.1), except that the first reaction is a nucleic acid hybridization instead of an antigen-antibody binding.

In situ hybridization can also be used to map the locations of genes on chromosomes. The labeled probe is easier to detect if the gene of interest is present in many adjacent copies. This is the case for repetitive or amplified genes. Particularly suitable for in situ hybridization are *polytene chromosomes*, in which hundreds of chromatids are bundled neatly in register. The mapping of transgenes to *Drosophila* polytene chromosomes by in situ hybridization was discussed earlier (see Fig. 14.23).

In situ hybridization has been used to probe the expression patterns of many genes in a variety of organisms. The results show directly and dramatically that many genes are expressed in specific tissues or regions and at specific stages during development. Of particular interest are those expression patterns that appear before embryonic cells become determined or overtly differentiated. These patterns offer us a glimpse into the gene activities that must be part of the determination and differentiation process itself. For example, the *hunchback*+ gene of *Drosophila* is expressed selectively in the anterior half of the embryo *before* the blastoderm becomes cellularized (Fig. 15.3). This is about an hour before the *potency* of cells or nuclei is limited to the structures of individual segments.

A critical reader might object that the assays discussed so far are suggestive of transcriptional control but do not actually prove it. Indeed, both the screening of cDNA libraries and in situ hybridization detect all RNAs *present*, not just RNA that is *newly transcribed*. The RNA distributions observed with these techniques can therefore be biased by posttranscriptional events, including RNA transport and RNA degradation. To test whether transcription itself is being controlled, one must use a "nuclear run-on" procedure that is carried out with isolated

nuclei: newly synthesized RNA is labeled during a brief period of transcription, but the subsequent steps of RNA modification and transport are curtailed. Such an experiment was performed by Eva Derman and her colleagues (1981). They prepared nuclear run-on RNA from mammalian liver, kidney, and brain. The labeled RNA fractions were hybridized with various cloned cDNAs that were unlabeled. Some cDNAs, including actin cDNA and tubulin cDNA, hybridized with nuclear run-on probes from liver, kidney, and brain. This showed that actin and tubulin genes were transcribed in all three of these tissues. Other cDNAs hybridized with only one or two probes, demonstrating that those gene products were made selectively in these tissues. These experiments proved that the expression of some genes was controlled directly at the level of transcription.

DNA Sequences Controlling Transcription

Transcription requires enzymes known as **RNA polymerases**. In eukaryotes, there are three types of RNA polymerase, which synthesize different classes of RNA. We will focus on the activity of RNA polymerase II, which synthesizes mRNA precursors known as *pre-*

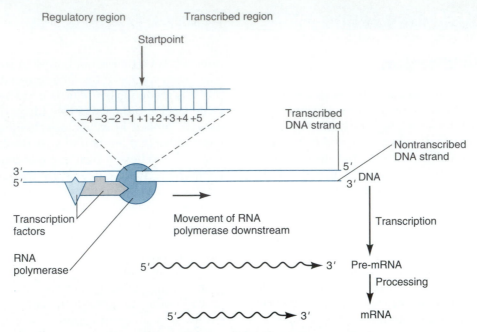

Figure 15.4 Regulatory region and transcribed region of a eukaryotic gene. In the transcribed region, nucleotides are numbered 1, 2, 3, etc., with 1 marking the startpoint of transcription. The nucleotides of the regulatory region are numbered −1, −2, −3, etc., with −1 directly upstream of the startpoint. Regulatory proteins called transcription factors bind to specific DNA sequences within the regulatory region. These proteins interact with RNA polymerase to position it at the startpoint. The movement of the RNA polymerase is downstream, or from 3′ to 5′, on the *transcribed* DNA strand of the gene. Most regulatory DNA sequences are located upstream of the transcription startpoint. This end of the gene is also called the 5′ end of the gene. It corresponds to the 5′ end of the *nontranscribed* DNA strand, which has the same polarity and nucleotide sequence as the RNA transcript. The primary transcript, called pre-mRNA, undergoes several processing steps to yield mRNA. The processing includes the removal of pre-mRNA segments that are not translated into proteins.

mRNA or *heterogeneous nuclear RNA (hnRNA)*. In particular, we will examine how the activity of this enzyme is directed toward specific genes in different cells.

A typical eukaryotic gene is shown in Figure 15.4. It consists of a *transcribed region,* plus several *regulatory DNA sequences.* The first transcribed nucleotide of a gene is called the *startpoint* and is given the number 1. In the course of transcription, the RNA polymerase moves *downstream* from nucleotide 1 to nucleotides 2, 3, 4, and so forth. The transcribed region of the gene consists of *exons,* which will be represented in the final mRNA, and *intervening sequences (introns).* Both are transcribed into pre-mRNA, but the introns are removed during the subsequent processing of pre-mRNA into mRNA (see Chapter 16).

Most regulatory DNA sequences are located upstream of the transcription startpoint. The nucleotides located here are assigned negative numbers, with −1 directly upstream of 1. The function of regulatory DNA sequences is to enhance or inhibit the binding of RNA polymerase II to the transcription startpoint. This regulation is indirect. The regulatory DNA sequences are recognized and bound by proteins called *transcription*

factors, which in turn interact with the RNA polymerase II to help or hinder its binding to the transcription startpoint. The regulatory DNA sequences of a given gene are the same in all cells of an organism, because they are part of the genome. In contrast, combinations of transcription factors acting on a gene may vary from one cell type to another.

In the remainder of this section and in the next section, we will examine first some properties of regulatory DNA sequences and then some features of transcription factors.

Regulatory DNA Sequences Are Studied in Fusion Genes

To analyze the regulatory region of a eukaryotic gene in detail, researchers often replace the transcribed region of the eukaryotic gene with the transcribed region of a bacterial *reporter gene,* a gene that encodes a product that is readily detected and measured. The composite gene consisting of the regulatory region from the eukaryotic gene of interest and the transcribed region of the reporter gene is called a *fusion gene*. If a fusion gene is introduced into a eukaryotic cell, the fusion gene's regulatory region responds to the host cell's signals, and the reporter part of the fusion gene is transcribed accordingly.

A commonly used reporter gene encodes the bacterial enzyme **chloramphenicol acetyltransferase (CAT)**. Its expression in cultured cells is easily measured by the acetylation of labeled chloramphenicol included in the culture medium (Fig. 15.5). Because CAT is not produced in normal eukaryotic cells, its synthesis in a transfected cell accurately reflects the activity of the reporter

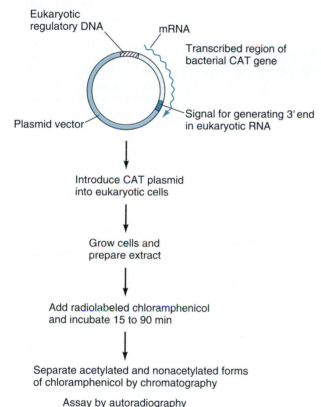

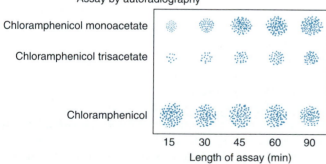

Figure 15.5 Use of a bacterial reporter gene to analyze the regulatory region of a eukaryotic gene. The transcribed region of the chloramphenicol acetyltransferase (CAT) reporter gene is fused with the regulatory region from the eukaryotic gene of interest. The activity of the fusion gene in transfected cells is measured by the ability of a cell extract to convert radiolabeled chloramphenicol to its acetylated forms (monoacetate and trisacetate). The proportion of acetylated chloramphenicol increases with time and with the amount of CAT present.

gene. Another bacterial reporter gene, designated *lacZ*, encodes β-galactosidase. This enzyme produces a blue stain in the presence of a suitable precursor substance. The *lacZ* reporter gene is especially useful for studying region-specific gene expression in whole embryos because the cells expressing the reporter gene stain dark blue against an unstained background (Fig. 15.6).

Fusion genes are frequently used for the techniques known as **deletion mapping** and **point mutagenesis.** The regulatory DNA segment of interest is modified by deletions or single-base substitutions. The effects of these manipulations on the expression level of the fusion gene show which regulatory DNA sequences are functionally important. For instance, deletion mapping has identified different regulatory elements in the *fushi tarazu*$^+$ gene of *Drosophila* (Hiromi and Gehring, 1987). One regulatory element, called the *zebra element*, is required for early expression of the *fushi tarazu*$^+$ gene in stripes of embryonic ectoderm (Fig. 15.6). Another element, the *upstream element*, is needed for expression in early embryonic mesoderm. A third regulatory element, the *neurogenic element*, is necessary for later expression in the central nervous system.

Point mutagenesis, on the other hand, is a way of analyzing regulatory regions by changing each single base pair. Most single substitutions do not change the level of transcription from the reporter gene. However, mutations in certain base pairs cause sharp downturns or increases in the transcription rate. These base pairs are clustered in short, distinct sequences of about 4 to 15 nucleotides called **modules.** Such modules are the critical parts of regulatory DNA regions; most of them are located upstream of the transcription startpoint, but some have been found at the 3' end or in introns.

Promoters and Enhancers Are Regulatory DNA Regions with Different Properties

Two types of regulatory DNA regions—promoters and enhancers—are found in eukaryotic genes (Fig. 15.7). The **promoter** is necessary for transcription. It is located upstream of the transcription startpoint and contains two or more *modules*. Comparison of different promoters has revealed that certain modules are found in most promoters. Such modules, called **consensus sequences,** form the central parts of the recognition regions bound by the gene-regulatory proteins described later in this chapter. The first consensus sequence, named the *TATA box* because it consists only of A/T base pairs, is usually located at −25 to −30. That the TATA box is almost perfectly conserved in life forms from yeast to humans underlines its fundamental importance (Hoffman et al., 1990). Another consensus sequence, the **CAAT box,** is typically located near −80. A third consensus sequence, the **GC box,** contains the sequence GGGCGG. Copies of the GC box may be present at more than one location

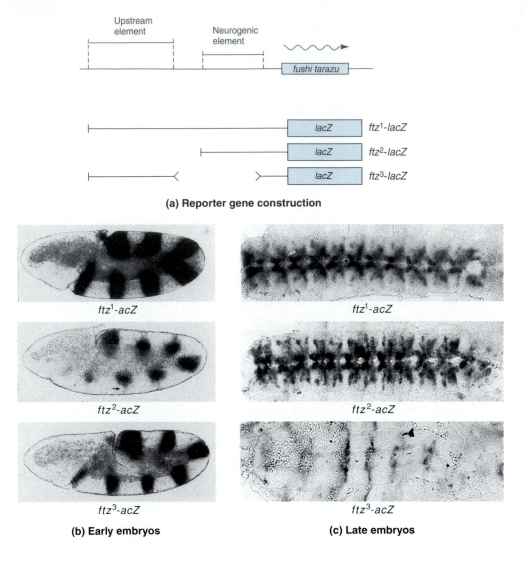

Figure 15.6 Deletion analysis of the regulatory DNA sequences of the *fushi tarazu*⁺ gene in *Drosophila* using the *lacZ* reporter gene. **(a)** Construction of fusion genes with different regulatory regions from the *fushi tarazu*⁺ gene. The *ftz¹-acZ* construct includes a complete *fushi tarazu*⁺ regulatory region and the transcribed region of *lacZ*. In the *ftz²-lacZ* construct, a large upstream element is deleted. The *ftz³-lacZ* construct is missing a different regulatory element, called the neurogenic element. **(b)** Expression pattern of the fusion genes in early embryos. The complete fusion gene (*ftz¹-lacZ*) is expressed in segmentlike areas (dark bands) of both the ectoderm and the mesoderm. Deletion of the neurogenic element (in the *ftz³-lacZ* construct) does not change this pattern. However, deletion of the upstream element (in the *ftz²-lacZ* construct) limits expression to the mesoderm. **(c)** Expression pattern of the fusion gene after development of the central nervous system. Deletion of the upstream element does not change the normal expression pattern. However, deletion of the neurogenic element precludes expression in the central nervous system.

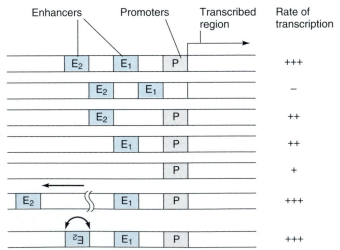

Figure 15.7 Effects of a promoter (P) and two enhancers (E_1, E_2) on the rate of transcription. Promoters are indispensable for transcription and need to be located just upstream from the transcription startpoint. Removing enhancers causes only partial loss of activity. Enhancers can be moved or flipped in orientation without losing their activity. Some enhancers (unlike those shown) have inhibitory effects on transcription.

within a promoter. None of these modules is indispensable for promoter function. Some promoters lack a TATA box while others lack a CAAT box or have no GC box. However, some combination of these is required for transcription to occur and to begin precisely at the startpoint of the gene. In short, every gene needs a promoter. Some genes, such as the *hunchback* gene of *Drosophila*, use two promoters in conjunction with alternate patterns of RNA processing.

In addition to promoters, eukaryotic genes have one or more regulatory sequences known as *enhancers* or *response elements*, which *increase or inhibit* the activity of any promoter in their vicinity. Enhancers are distinguished from promoters in three ways. First, enhancers need promoters to work, while promoters will work alone, albeit slowly. Second, most enhancers are effective even if they are moved thousands of base pairs upstream or downstream from their normal position. Third, most enhancers are effective in reverse orientation whereas promoters are not. Enhancers are modular, like promoters. A typical enhancer may consist of a half dozen modules, each module 6 to 15 bp long. Some modules are unique. Others may occur repeatedly within the same enhancer. A given module may also be found in different enhancers, indicating that groups of genes are regulated in concert. Some enhancer modules are also present in promoters, an observation that blurs the distinction between the two. Conceivably, enhancers evolved from promoters (Yamamoto, 1989).

In accord with the principle of genetic equivalence, the promoter and enhancer sequences of a gene appear to be identical in all cells of a given organism. The stage- and organ-specific transcription of genes must there- fore be mediated by other molecules that interact with promoter and enhancer modules.

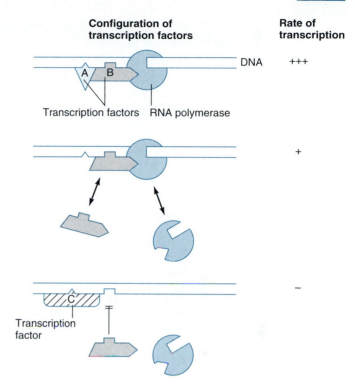

Figure 15.8 Cooperative action of transcription factors. The situation drawn involves three transcription factors (A, B, and C) and RNA polymerase. The binding of A to its DNA module stabilizes the binding of B, which in turn po- sitions RNA polymerase at the transcription startpoint. In the absence of A, the binding of B becomes labile, and so does the positioning of RNA polymerase. The binding of C, which recognizes the same module as A, prevents the binding of B. As a result, transcription is inhibited.

Transcription Factors and Their Role in Development

As mentioned earlier, the binding of RNA polymerase II to the transcription startpoint depends on regulatory proteins that are called *transcription factors*. All tran- scription factors have at least two active domains: a **DNA-binding domain**, which binds to promoter and enhancer modules, and one or more **activation do- mains**, by which transcription factors interact with one another and/or with RNA polymerase II (Fig. 15.8). Multiple transcription factors binding to neighboring DNA modules fit together like a three-dimensional jig- saw puzzle. The resulting superstructure *enhances or in- hibits* the binding of RNA polymerase more or less strongly, depending on what combination of transcrip- tion factors is present.

Transcription factors connect genes in complex regu- latory networks. Frequently, a given enhancer interacts with several transcription factors, each of which has a different effect on RNA polymerase activity. The net out- come then depends on the relative concentration of each family member. Conversely, a given transcription factor may interact with similar regulatory modules in many genes, including its own encoding gene and genes en- coding other transcription factors. Thus, a maze of in- teractions between different transcription factors, their encoding genes, their target modules, and RNA poly- merase form the molecular basis of transcriptional con- trol. As key elements in these interactions, transcription factors are of particular interest for a molecular under- standing of transcriptional control.

General Transcription Factors Bind to All Promoters

Some transcription factors, called **general transcription factors,** bind to the promoters of virtually all genes transcribed by RNA polymerase II. A specific configu- ration of more than twenty general transcription fac- tors, known as the **transcription complex,** is required to

position the right type of RNA polymerase at the tran- scription startpoint and to initiate transcription (Buratowski, 1994). The general transcription factors as- sociated with RNA polymerase II have been named TFIIA, TFIIB, and so forth, in the order of their discov- ery. TFIID consists of a TATA box–binding protein (TBP) and several associated factors. The binding of TBP to the TATA box causes a major change in the DNA con- formation (Kim et al., 1993). This change and the pres- ence of other TFIID components allow the binding of TFIIB, followed by RNA polymerase II and other fac- tors. TFIIB is of particular interest because it is thought to be the target of several other transcription factors that activate the transcription process (Colgan et al., 1993; S. G. E. Roberts et al., 1993).

Similar sets of general transcription factors are pres- ent in *all cells* of an organism. Therefore, these factors, like RNA polymerases, *cannot* account for the stage- dependent or region-specific transcription of certain genes. Such specific regulatory effects depend on other transcription factors that help to assemble and activate the transcription complex. This most interesting class of transcription factors will be discussed next.

TABLE 15.1

General and Specific Transcription Factors

	General Transcription Factors	Specific Transcription Factors
Target genes	All genes	Selected genes
Occurrence in	All cells	Certain cells
Recognition modules located in	Promoters	Some promoters, mostly enhancers
Examples	TBP, TFIIB	bicoid protein, steroid receptors

Specific Transcription Factors Associate with Restricted Sets of Genes and Occur Only in Certain Cells

Specific transcription factors, as their name implies, act selectively on specific target genes. While some transcription factors are neither fully general nor clearly specific, the two groups of transcription factors differ in conceptually important respects (Table 15.1). Whereas general transcription factors bind strictly to promoters, associate with the majority of all genes, and occur in all types of cells, specific transcription factors bind to some promoters but primarily to enhancers, and associate with a restricted number of target genes. Most important, the combination of specific transcription factors present varies from one cell type to another. Thus, it is the combined action of specific transcription factors that selects the array of genes to be transcribed in each cell.

A central issue in understanding transcriptional control is how a relatively small number of specific transcription factors can control the activity of many thousands of genes in a stage-dependent and tissue-specific manner (Tjian and Maniatis, 1994). One part of the answer is that specific transcription factors, like their general counterparts, act in combination, forming multimolecular arrays with numerous combinations of enhancing or inhibitory interactions. Another part of the answer lies in the nucleotide sequences of the enhancer modules, or *DNA response elements,* to which transcription factors bind. Some enhancer modules bind only one transcription factor, but with appropriately spaced modules, several transcription factors can mutually enforce or inhibit their actions. Alternatively, two or more transcription factors may bind simultaneously, or competitively, to the same enhancer module. Some of these interactions will be illustrated in the last two parts of this section, using as examples first the bicoid protein in *Drosophila* and then the steroid receptor family in mammals.

Transcription factors within the same species, but also between distantly related species, can be grouped into families. These families are based on the transcription factors' functional domains, especially their *DNA-binding domains* and their *activation domains*. In addition,

some transcription factors are themselves multimeric proteins, and their constituent polypeptides are held together by *multimerization domains*.

One common DNA-binding domain in transcription factors is the **zinc finger,** a sequence of about 30 amino acids folded around a zinc atom. Typically, two or more zinc fingers are found in a row, presumably extending into neighboring sections of the major DNA groove. Another DNA-binding domain is the **homeodomain,** a sequence of 60 amino acids. First discovered in the products of *homeotic genes* in *Drosophila,* homeodomains have since been found in transcription factors of many organisms including humans (see Chapters 21 and 22). The activation domains of transcription factors have been classified as acidic, glutamine-rich, and proline-rich, although the functional significance of this classification is still under investigation. A common multimerization domain is the **leucine zipper,** a leucine-rich strip of amino acids that zips two polypeptides together into a dimeric transcription factor (S. L. McKnight, 1991).

The functional domains of general and specific transcription factors alike have been exceedingly well conserved in evolution. For example, an enhancer element and a specific transcription factor activating the alcohol dehydrogenase gene are present in both the human liver and the corresponding organ of the fruit fly, the fat body (Abel et al., 1992; Falb and Maniatis, 1992). The similarity of transcription factors from widely divergent species is a discovery of fundamental importance. A similar conservation of functional protein domains has been observed in growth factors and their receptors, cell adhesion molecules, and extracellular matrix molecules (see Chapters 25, 26, and 29). These discoveries indicate that, in widely divergent organisms, similar sets of key molecules take part in regulatory processes in development. Learning of the generality of these processes has been both intellectually rewarding and of practical importance—intellectually rewarding because general mechanisms are viewed as more noteworthy than species-specific ones, and of practical importance because the conserved protein domains reflect genomic DNA sequences that are often similar enough that a cloned segment from one species can be used to identify corresponding DNA segments from other species.

The bicoid Protein Enhances Transcription of the *hunchback*⁺ Gene in the *Drosophila* Embryo

Specific transcription factors, conferring special gene activities on different cells, are very important regulatory elements during embryogenesis. As an example, let us examine the bicoid protein in *Drosophila*. Embryos derived from mothers lacking a functional *bicoid*⁺ gene have no head or thorax; they consist of an extended abdomen carrying at its front end a duplication of terminal abdominal structures (see Fig. 14.2). In wild-type females, ovarian nurse cells synthesize bicoid mRNA and transfer it to the anterior pole of the oocyte. Following fertilization, the bicoid mRNA serves as a localized source of bicoid protein, which diffuses posteriorly, forming a concentration gradient with a maximum near the anterior pole of the embryo (Fig. 15.9; Color Plate 1.)

The bicoid protein contains a *homeodomain*, providing a first indication that this protein may act as a transcription factor. A plausible target gene of bicoid protein is the *hunchback*⁺ gene, which is one of the first genes transcribed in the *Drosophila* embryo. Embryos homozygous for loss-of-function alleles of *hunchback*

lack posterior head parts and the entire thorax. The similarity between the *bicoid* and *hunchback* phenotypes suggests that much of the *bicoid* phenotype may be caused by a failure to activate the embryonic *hunchback*⁺ gene. This hypothesis is confirmed by monitoring the synthesis of hunchback RNA or protein in embryos derived from wild-type versus *bicoid* mutant mothers (Schröder et al., 1988; Tautz, 1988).

In offspring from wild-type mothers, the *hunchback*⁺ gene is expressed in the anterior half of embryos at preblastoderm stages (Figs. 15.3a and 15.10a). After formation of the cellular *blastoderm*, an additional *hunchback*⁺ expression domain appears as a subterminal posterior

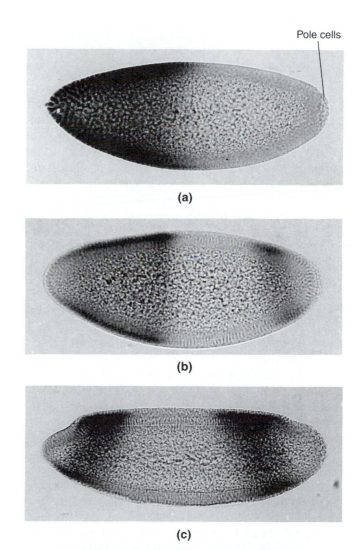

(a)

(b)

(c)

Figure 15.10 Expression of the *hunchback*⁺ gene in *Drosophila* embryos shown by immunostaining. The posterior pole (to the right) is marked by pole cells. **(a, b)** Wild-type embryos. Before blastoderm formation **(a)**, hunchback protein is found only in the anterior half. At the cellular blastoderm stage **(b)**, an additional expression domain appears as a subterminal band near the posterior pole. **(c)** Embryo derived from a *bicoid* mutant mother. The anterior *hunchback*⁺ domain is missing, while the posterior *hunchback*⁺ domain is duplicated anteriorly.

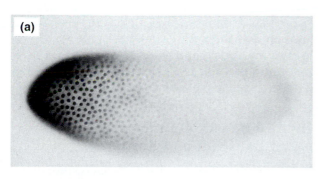

(a)

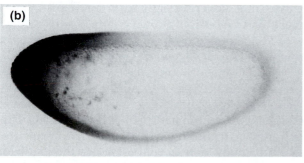

(b)

Figure 15.9 *Drosophila* embryos immunostained for bicoid protein (see Methods 4.1). The photomicrographs show wholemounts with the anterior pole oriented to the left, dorsal side up. **(a)** Embryo at the syncytial blastoderm stage, with focus on the surface. **(b)** Same, but with focus on the median plane. Note the anteroposterior gradient in bicoid protein concentration and the accumulation of the protein in the nuclei. The bicoid protein acts as a transcription factor enhancing the transcription of the embryonic *hunchback*⁺ gene.

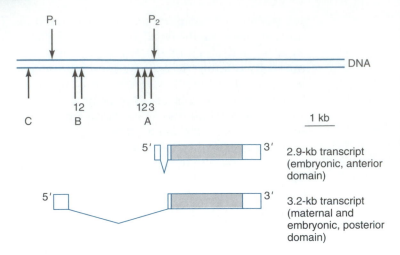

Figure 15.11 Map of the *hunchback⁺* gene showing the binding sites for bicoid protein. The *hunchback⁺* gene is transcribed from two promoters, P_1 and P_2. The mRNA originating from P_2 is 2.9 kb long and is expressed in the anterior half of the embryo. The mRNA originating from P_1 is 3.2 kb long and is expressed near the posterior pole of the embryo. (This mRNA is also transcribed maternally and deposited in the oocyte during oogenesis.) The translated regions (gray) of the two mRNAs are identical. A, B, and C are enhancer regions protected from DNase I in the presence of bicoid protein. Most strongly protected are the modules A_1, A_2, and A_3, which are located next to the P_2 promoter.

band (Figs. 15.3b and 15.10b). In offspring from *bicoid* loss-of-function mothers, the large anterior *hunchback⁺* expression domain is missing while the posterior domain appears twice—in its normal posterior location and as an anterior duplication corresponding to the duplication of the *telson* (Fig. 15.10c). Molecular investigations show that transcripts for the normal posterior and anterior domains begin at two different promoters, designated P_1 and P_2, respectively (Fig. 15.11). Together, these results suggest that the activity of promoter P_2 depends on the presence of bicoid protein as a transcription factor while promoter P_1 is activated without bicoid protein.

To further analyze the effects of maternal bicoid protein on the *hunchback⁺* promoter P_2, Wolfgang Driever and Christiane Nüsslein-Volhard (1989) used a technique known as **DNase footprinting.** For this procedure the investigators cloned a *hunchback⁺* gene segment containing the P_2 promoter and the upstream region, which most likely contained any enhancers. This DNA was allowed to bind to bicoid protein and then was exposed to deoxyribonuclease (DNase I), an enzyme that digests naked DNA but not DNA associated with protein. Those DNA sequences that were bound to bicoid protein were protected from DNase I attack. The most strongly protected DNA modules, termed A_1, A_2, and A_3, were located upstream of the P_2 promoter (Fig. 15.11).

To test the function of modules A_1 through A_3 in vivo, the researchers spliced a DNA segment containing these modules and the P_2 promoter in front of the transcribed region of the *lacZ reporter gene*. The fusion gene was introduced into the germ line of host embryos by *P-element transformation* (Driever et al., 1989a). The expression pattern of the transgene mimicked the expression of the resident *hunchback⁺* gene very closely (Figs. 15.10a and 15.12a). However, after progressive deletion

of modules A_1 to A_3, the expression of the fusion gene became progressively weaker (Fig. 15.12b–d). Further experiments showed that the DNA domain containing A_1 to A_3 behaved as an enhancer: it could be reversed or moved more that 1 kb from the promoter and still remain active.

The results just described show that the maternally encoded bicoid protein acts as a specific transcription factor activating the P_2 promoter of the embryonic *hunchback⁺* gene. In addition, the bicoid protein enhances or inhibits the region-specific expression of several other embryonic genes (see Chapter 21).

Steroid Hormones Activate Specific Transcription Factors

Given the importance of transcriptional control, it is no surprise that several regulatory mechanisms have evolved that tie transcriptional control to environmental or endogenous signals. All species seem to have a set of genes, called *heat shock genes*, that are transcribed in response to heat shock and other types of environmental stress. In vertebrates, toxic substances when present in the blood trigger the transcription of genes that encode detoxifying proteins such as the *metallothioneins*, which remove cadmium and other heavy metals from the blood serum. Researchers often use the regulatory regions of such genes in transgenes designed to be expressed conditionally.

Many transcription factors are synthesized in an inactive precursor form, so that their activation itself becomes a level of regulation. As mentioned earlier, some factors are composed of two polypeptides that form an active protein only upon dimerization. Other transcription factors are activated by phosphorylation and/or dephosphorylation of certain amino acids (T. Hunter and M. Karin, 1992). The enzymes that add or remove phos-

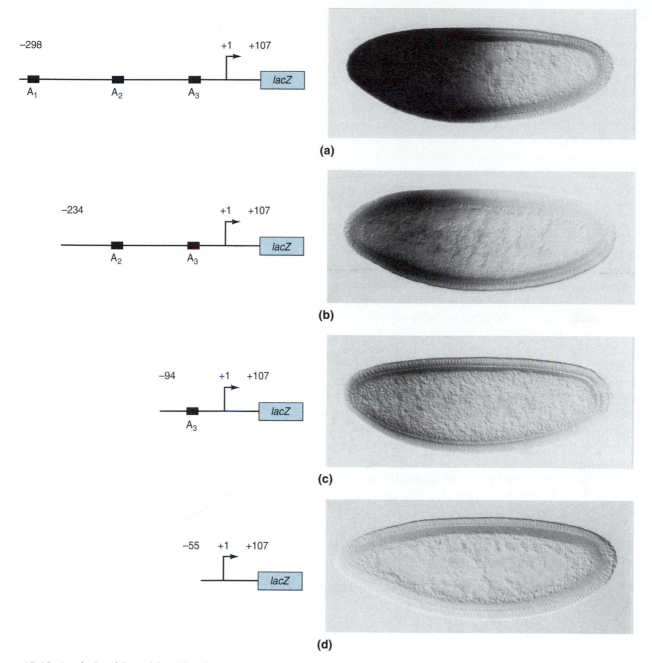

Figure 15.12 Analysis of *hunchback*+ enhancer sequences by fusion genes consisting of the *hunchback* P$_2$ promoter and a variable number of enhancer modules (A$_1$ to A$_3$) spliced to the *lacZ* reporter gene. Each fusion gene is used to create transgenic *Drosophila* lines by P-element transformation. The diagrams show the *hunchback* enhancer sequences present in the respective fusion genes. The photomicrographs show embryos stained for *lacZ* expression. **(a)** When enhancer modules A$_1$ to A$_3$ all are present, the fusion gene is expressed in a pattern similar to that of the resident *hunchback*+ gene in the anterior domain (compare with Fig. 15.3). **(b–d)** As the enhancer modules are progressively removed, *lacZ* expression becomes progressively weaker.

phate groups are known as *protein kinases* and *protein phosphatases,* respectively. The activity of these enzymes may depend upon their own phosphorylation state, so that they often form regulatory cascades that ultimately regulate the activity of a transcription factor (see Chapter 29). Still other transcription factors are acti-

vated by the binding of a *ligand,* that is, a small matching molecule such as a hormone. In relation to the ligand, such transcription factors are also called *receptors.*

Hormones are signal molecules that are carried by the bloodstream to all cells in the body. Yet each hormone elicits a response only in some cells, termed *tar-*

get cells, and only from a few genes, called *target genes.* For instance, the production of egg white proteins in the chicken oviduct is controlled by the sex hormones estrogen and progesterone. In the presence of one of these hormones, certain oviduct cells synthesize ovalbumin and other egg white proteins in large quantities. After 10 days of hormone stimulation, the ovalbumin level in these cells increases dramatically from trace amounts to more than 50% of all newly synthesized proteins. The surge in ovalbumin synthesis is associated with a steep increase in the production of ovalbumin mRNA, indicating transcriptional regulation of the ovalbumin gene (Fig. 15.13). The astounding specificity of the hormone action is due to the molecular mechanisms by which hormones act on their target genes.

There are different chemical classes of hormones. Steroid hormones in vertebrates include the glucocorticoids and mineralocorticoids produced in the adrenal glands and the sex hormones produced in the gonads. Steroid hormones are small lipid-soluble molecules. They cross plasma membranes by simple diffusion and permeate all the cells of an organism. Target cells are distinguished from nontarget cells by the presence of a matching *receptor* in the cell cytoplasm. Each receptor is composed of two polypeptides and associated proteins (Fig. 15.14). Upon hormone binding, the receptor undergoes an activation process involving the release of one or more inhibitory proteins. The hormone-receptor complex then moves into the nucleus and binds to specific enhancer modules known as *steroid response elements.* Such modules are the characteristic feature of

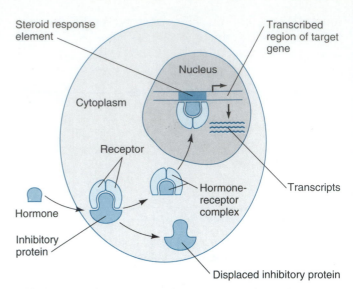

Figure 15.14 General model of gene activation by a steroid hormone. The hormone enters *all* cells in an organism by diffusion. However, only some cells (called the target cells of the hormone) have a cytoplasmic receptor protein to which the hormone can bind. The receptor is a dimeric protein consisting of two identical polypeptides. The binding of the hormone displaces one or more inhibitory proteins associated with the receptor. The hormone-receptor complex then enters the nucleus and binds to certain genes (called the target genes of the hormone) that have enhancers with the matching steroid response elements. The binding usually activates, but sometimes inhibits, transcription of the target gene.

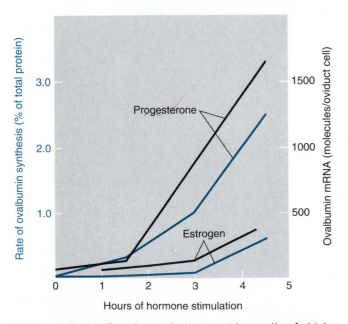

Figure 15.13 Ovalbumin synthesis in oviduct cells of chickens injected with estrogen or progesterone. Both hormones caused sharp increases in the concentration of ovalbumin mRNA (black lines) and the rate of ovalbumin synthesis (colored lines).

the *target genes* that respond to a steroid hormone. For a typical steroid hormone, there are about 50 target genes.

Several *steroid receptors* form a family of similar proteins, all of which act as transcription factors. The best-known family member is the mammalian glucocorticoid receptor (Yamamoto, 1985; R. M. Evans, 1988). Similar receptor proteins exist for other steroids including mineralocorticoids, sex hormones, and vitamin D. In addition, this family includes the receptor for retinoic acid, a molecule that plays a prominent role in vertebrate limb morphogenesis (see Chapter 22). Comparing the amino acid sequences of all these receptors reveals striking similarities. The most highly conserved domain is a central region containing two *zinc fingers* serving as the DNA-binding domain. Also well conserved is a domain near the C terminus of the receptor, containing the hormone-binding and dimerization functions. The similarities within the family of steroid receptors suggest that their genes have evolved by duplication and subsequent modification.

Specific Transcription Factors May Act Individually or in Combination

So far, we have noted that the specificity of steroid hormone action relies on the selective presence of *receptors* in target cells, and on the specific binding of the hormone-receptor complex to target gene *enhancer* modules. However, the effect of the receptor binding may also depend on the presence of other transcription factors that bind to the same enhancer module or to neighboring modules. Enhancer modules have been classified as **simple response elements,** which are bound by single transcription factors, or **composite response elements**, which are bound by two or more transcription factors simultaneously.

As an example of a composite response element, Marc Diamond and his colleagues (1990) studied an enhancer element, designated plfG, that is associated with the mouse *proliferin⁺* gene and responds to the glucocorticoid receptor. When fused with a minimal promoter to the CAT reporter gene and transfected into cultured mouse cells, the plfG element behaved as a composite response element: the effect of bound glucocorticoid receptor depended on another transcription factor, known as AP-1 (Fig. 15.15). AP-1 is composed of two different polypeptides, designated c-fos and c-jun, which can form homodimers (c-fos/c-fos or c-jun/c-jun) or heterodimers (c-jun/c-fos). In the presence of AP-1, the glucocorticoid hormone receptor complex either enhanced or inhibited transcription, depending on the subunit composition of AP-1. If the AP-1 bound at plfG was composed predominantly of c-jun/c-jun homodimers, glucocorticoid enhanced transcription from plfG. If instead AP-1 consisted predominantly of c-jun/c-fos heterodimers, glucocorticoid inhibited transcription. Finally, in the absence of AP-1, glucocorticoid had no effect on transcription from plfG.

In *composite response elements*, different transcription factors act *cooperatively* with one another and with the DNA motifs to which they bind. The glucocorticoid receptor appears to adopt different three-dimensional conformations, depending on the bound response element and on other transcription factors present (Yamamoto et al., 1992). Other composite response elements have overlapping recognition sequences that are bound competitively to different transcription factors with activating or inhibitory functions (see Chapter 21). Still other mechanisms will presumably be discovered as more composite response elements are analyzed. The molecular mechanisms by which transcription factors bound to *simple response elements* interact with the transcription complex at the promoter are also under investigation. Many promoters interact with more than one simple response element, but with each one *independently*. It may be that the DNA upstream of the promoter can bend in different ways so that different enhancers may be positioned next to the promoter.

The interactions of transcription factors with one another and their DNA recognition sequences are of great interest to developmental biologists. They offer a possible explanation for the generation of thousands of different gene expression patterns by various combinations of a comparatively small number of transcription factors. The concentration of these regulatory proteins depends on their rates of synthesis and degradation. Their activity may also be affected in more subtle ways by posttranslational modifications that affect one functional domain while leaving others unchanged. For example, phosphorylation of a certain amino acid residue in one domain of a specific transcription factor may change its interaction with another specific transcription factor but leave its domains for DNA binding and interaction with general transcription factors unaffected. Thus, a relatively small number of transcription factors

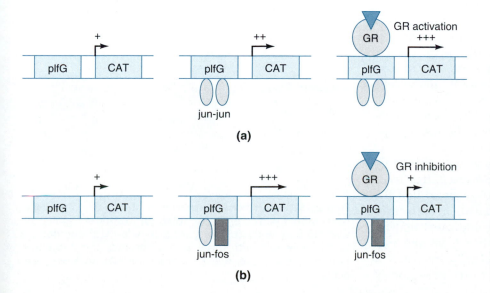

Figure 15.15 Cooperative action of the glucocorticoid receptor (GR) with another transcription factor, AP-1, which is a dimeric protein consisting of c-jun and/or c-fos monomers. GR and AP-1 interact at a complex response element (plfG) normally associated with the mouse *proliferin⁺* gene. In the experiment summarized here, the plfG element was spliced along with a promoter to the CAT reporter gene. **(a)** If AP-1 was a c-jun/c-jun homodimer, simultaneous binding of glucocorticoid receptor (with hormone) caused further activation of the reporter gene. **(b)** If AP-1 was a c-jun/c-fos heterodimer, simultaneous glucocorticoid receptor binding inhibited the reporter gene.

and modifying enzymes may generate the large number of molecular configurations needed to account for differential gene expression.

Chromatin Structure and Transcription

So far in our discussion of transcriptional control, we have assumed that genomic DNA is readily accessible for interactions with RNA polymerase and transcription factors. However, the genomic DNA of eukaryotes is contained in *chromatin,* where the DNA is tightly associated with histones and a variety of nonhistone proteins. These associations, which change during the cell cycle and depend on the developmental stage of a cell, may greatly affect the availability of genes for transcription. As a rule, DNA in condensed chromatin is transcribed less actively than DNA in decondensed chromatin. We will find that this general correlation applies to whole chromosomes, to chromosomal regions, and at the molecular level.

Heterochromatic Chromosome Regions Are Not Transcribed

During the cell cycle, most chromosomes alternate between a highly condensed state and a more diffuse state. The condensed state during M phase allows the chromosomes to untangle so that mitosis can occur. No RNA synthesis is observed during this phase. The diffuse state during interphase allows the chromosomal DNA to be transcribed. However, certain chromosome regions remain highly compacted even during interphase. The chromatin in these regions is called ***heterochromatin.*** Autoradiographs of nuclei incubated with [³H]uridine show that little or no RNA is synthesized in heterochromatin (Fig. 15.16).

Some heterochromatic regions, called *constitutive heterochromatin,* are present in all individuals and cells of a species. Other chromosomal regions, known as *facultative heterochromatin,* are heterochromatic only in one sex or at certain stages of development. A prominent case of facultative heterochromatin can be found in the cells of female mammals, in each of which one of the two X chromosomes is heterochromatic. This heterochromatic X chromosome, called the ***Barr body,*** after its discoverer, is absent from males (Fig. 15.17).

The inactivation of one X chromosome in female mammals is a means of *dosage compensation,* that is, of adjusting the expression of X-linked genes so that the same amounts of gene products are made in both sexes (see Chapter 27). Genetic evidence indicates that females express X-linked genes either from the maternally inherited or from the paternally inherited X chromosome; only some genes in a certain region of

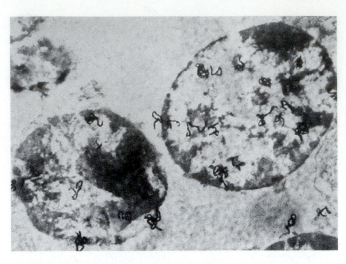

Figure 15.16 Correlation between transcription and decondensed chromatin status. This transmission electron micrograph shows a calf thymus nucleus after incubation with [³H]uridine. The dark wiggly lines are autoradiographic signals (see Methods 3.1) indicating the presence of newly synthesized RNA. Note that these signals appear mostly over the decondensed (light) areas of chromatin.

the X chromosome escape inactivation and are expressed from both X chromosomes. The inactivation of one X chromosome occurs during cleavage and appears to occur randomly in the cells of the *inner cell mass,* which give rise to the embryo proper. From then on, the active or inactive state of each X chromosome is passed on clonally.

Puffs in Polytene Chromosomes Are Actively Transcribed

While nontranscribed chromatin is often heterochromatic, actively transcribed chromatin is generally in a loose, decondensed state. In most cells in interphase it is difficult to distinguish decondensed chromatin regions under the light microscope. However, this is easy with *polytene chromosomes,* which consist of many chromatids

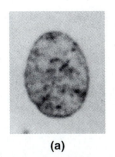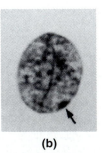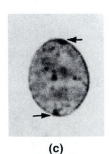

(a)　　　　(b)　　　　(c)

Figure 15.17 Barr bodies in nuclei of human oral epithelial cells stained with cresylecht violet. **(a)** Normal XY male, with no Barr body. **(b)** Normal XX female, with one Barr body. **(c)** Female with three X chromosomes and two Barr bodies. In each case, one X chromosome per nucleus is left active.

aligned neatly in register (see Chapter 7). Polytene chromosomes have areas of decondensation known as *puffs*. Puffs respond to signal molecules such as hormones, suggesting that they might represent sites of gene regulation. Experiments in which cells with polytene chromosomes are pulse-labeled with [³H]uridine confirm this (Pelling, 1959). Autoradiographs prepared from such cells show that the puffed chromosome regions are areas of very active RNA accumulation (Fig. 15.18). These results confirm that decondensed chromosomal regions are transcribed more actively than condensed regions. However, the data do not show whether puffing is a consequence or a prerequisite of transcription.

DNA in Transcribed Chromatin Is Sensitive to DNase I Digestion

The general correlation between chromatin decondensation and transcription is also found at the molecular level. To appreciate investigations in this area, we must first familiarize ourselves with the basic packaging units of chromatin, the *nucleosomes* (Fig. 15.19). Each nucleosome consists of about 160 bp of DNA wrapped nearly twice around an octamer of core histone proteins. Neighboring nucleosomes are connected like beads on a string by stretches of "linker" DNA. This *beaded string configuration* seems to leave the chromosomal DNA fairly accessible to transcription factors and RNA polymerase. More condensed chromatin fibers measure 30 nm in diameter. These fibers are arranged in a helical configuration of six nucleosomes per turn. In this *solenoid configuration,* the nucleosomes are associated with

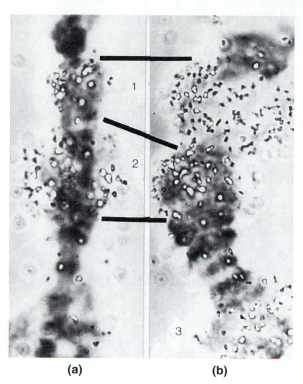

(a)   **(b)**

Figure 15.18 Correlation between puffing and RNA synthesis in polytene chromosomes. **(a)** Salivary gland chromosome from a normal larva of the midge *Chironomus tentans* after injection with a radiolabeled RNA precursor, [³H]uridine. **(b)** Corresponding chromosome from larva also injected with the molting hormone, ecdysone. Note size increase in puff 1. The grains over the puffs were generated through autoradiography (see Methods 3.1) and indicate newly synthesized RNA. The rate of RNA synthesis over puff 1 increases with the puff's size.

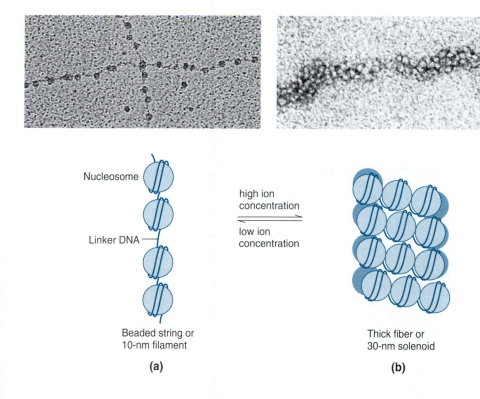

Nucleosome

Linker DNA

high ion concentration

low ion concentration

Beaded string or 10-nm filament

(a)

Thick fiber or 30-nm solenoid

(b)

Figure 15.19 Nucleosome structure of eukaryotic chromatin. Each nucleosome consists of a core of eight histone proteins with about 160 bp of DNA wrapped around it in 1.8 turns. Nucleosomes are connected by stretches of linker DNA, which vary in length but measure about 60 bp on average. **(a)** Electron micrograph and schematic diagram of decondensed chromatin, obtained after spreading at low ion concentration. In this beaded string configuration, the linker DNA is unprotected. **(b)** Chromatin prepared at high ion concentration forms thicker fibers, measuring about 30 nm in diameter. The nucleosomes are packed tightly in a solenoid arrangement, in which the nucleosomes are associated with another histone protein designated H1 (not shown).

a special histone, designated H1. The solenoid arrangement seems to leave chromosomal DNA less accessible to transcription factors and RNA polymerase than the beaded string configuration.

Observations on a wide range of organisms indicate that nucleosome packing is indeed relaxed in transcribed chromatin regions. This correlation can be inferred from tests exposing chromatin to very low concentrations of DNase or other DNA-digesting enzymes. After a mild digestion of chromatin or whole nuclei with DNase, the DNA extracted from the chromatin can be tested for the presence and integrity of any gene for which a labeled hybridization probe is available. The tests show regularly that those genes that are most actively transcribed are also most vulnerable to DNase attack. Transcribed genes must therefore be in a less condensed state than other genes.

▼

A typical experiment of this type, carried out by Jürg Stalder and his colleagues (1980), is summarized in Figure 15.20. The investigators tested the DNase sensitivity of the chicken β_2-globin gene in chromatin from

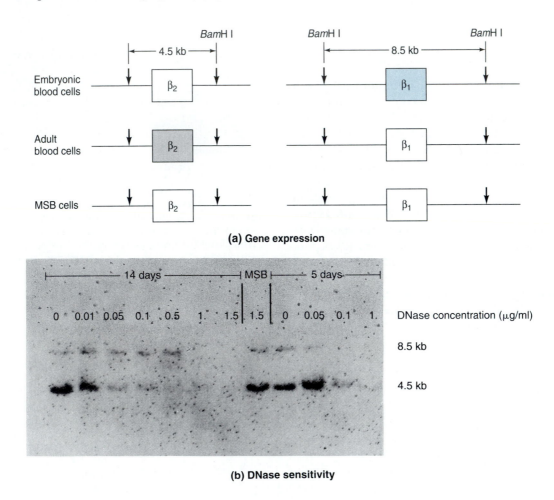

(a) Gene expression

(b) DNase sensitivity

Figure 15.20 Correlation between transcription and DNase sensitivity of chromatin. **(a)** Schematic diagram showing two chicken globin genes in fragments of genomic DNA cut with *Bam*H I restriction enzyme. The β_1-globin gene was transcribed in embryonic (5-day) red blood cells, and the β_2-globin gene was transcribed in adult (14-day) red blood cells. Neither gene was expressed in the cultured chicken cells (MSB) used as controls. **(b)** Nuclei from each type of cell were isolated and exposed to increasing concentrations of DNase (0 to 1.5 μg/ml). Nuclear DNA was then extracted, digested with *Bam*H I restriction enzyme, and analyzed by Southern blotting (see Methods 14.3). A labeled globin probe hybridized with DNA fragments of 8.5 kb (containing the β_1-globin gene) and 4.5 kb (containing the β_2-globin gene). The 8.5-kb fragment began to disappear from adult blood cell nuclei at 1.0 μg/ml DNase and from embryonic blood cell nuclei at 0.05 μg/ml DNase, indicating 20-fold greater sensitivity of the β_1-globin gene in embryonic blood cells. Conversely, the β_2-globin gene contained in the 4.5-kb fragment was only slightly more sensitive in adult blood cells. Neither gene was digested in MSB cells, even at the highest DNase concentration. When the same Southern blot was hybridized with an ovalbumin probe, no DNase effect was detected.

red blood cells and a line of cultured cells designated MSB cells. They expected that this gene would be DNase-sensitive in the red blood cells because it is actively transcribed there. The MSB cells, which do not synthesize globin, were used as a source of chromatin in which the globin gene was expected to be DNase-resistant. The investigators isolated nuclei from each cell type and exposed them to increasing concentrations of DNase. Then they extracted the nuclear DNA and digested it with the restriction enzyme *Bam*H I. The closest *Bam*H I restriction sites flanking the β_2-globin gene were 4.5 kb apart. This restriction fragment hybridized strongly with a labeled globin probe in a *Southern blot* (see Methods 14.3). However, the DNA fragment disappeared from blood cell nuclei after DNase treatment, indicating that the globin gene was digested. This did not occur, even at the highest DNase concentration used, in MSB cell nuclei. When the investigators probed the Southern blot with labeled ovalbumin and other cDNAs, the corresponding genes turned out to be DNase-resistant in both types of cells. They concluded that the DNase sensitivity was a specific trait of the globin gene in red blood cells, not a general trait of the globin gene in any cell or a general property of all genes in red blood cells.

This study of the globin gene was particularly interesting because chickens express at least two β-globin genes, one (β_1) in embryonic and the other (β_2) in adult blood cells. Both genes were almost equally DNase-sensitive in embryonic cells even though these cells did not produce β_2-globin. In contrast, the embryonic gene (β_1) was more resistant in adult cells, where it was no longer expressed. Apparently, the β_2-globin gene in the embryonic blood cells was in a pretranscriptional state of decondensation, as if it were already prepared or marked for future transcription.

The experiments described above demonstrated that globin genes are especially sensitive to DNase in red blood cells but not in other cells. In chicken oviduct cells, the ovalbumin gene is DNase-sensitive while the globin genes are DNase-resistant. In each type of cell, the most actively transcribed genes are also most DNase-sensitive. These results show that transcribed genes are in a decondensed state that leaves them more accessible to DNase, and, by implication, to transcription factors and RNA polymerase.

DNA regions sensitive to DNase are often very large, spanning 100 kb or more. Within such a region, there are much smaller **DNase-hypersensitive sites**. These are degraded by even lower concentrations of DNase. In contrast to *DNase sensitivity*, which reflects a decondensed state in which the chromatin is uncoiled but the *nucleosome* structure is left intact, *DNase hypersensitivity* may result from a disruption of nucleosomes and subsequent displacement of histones (C. C. Adams and J. L. Workman, 1993). The hypersensitive sites are of particular interest because they coincide with promoters and enhancers. However, it is only partially understood how the hypersensitive sites are generated. Certain nuclear proteins—transcription factors and/or other chromatin modifying factors—are able to compete with histones for binding to promoter and enhancer sites (Felsenfeld, 1992). Some of these factors first interact with a defined region of histone H4 at the nucleosome that contains the TATA box (Grunstein, 1992). This interaction seems to cause a partial breakup of the core, so that the TATA box becomes accessible. Experiments with chromatin assembled in vitro indicate that the GAGA transcription factor can bind to nucleosomes and cause nucleosome disruption as well as DNase I hypersensitivity at the TATA box and other promoter modules (Tsukiyama et al., 1994).

The observations described in this section support a two-step model of transcriptional activation (Sippel et al., 1987). In step 1, chromatin regions are decondensed without dissolving the nucleosome structure. This step confers DNase sensitivity to large stretches of DNA and makes them fairly accessible to proteins. This step may occur before transcription actually begins. In step 2, the nucleosome structure is disrupted as transcription factors and RNA polymerase bind to promoters and enhancers. This step confers DNase hypersensitivity to the gene-regulatory regions. When both steps are completed, transcription proceeds.

Transcriptional Control and Cell Determination

Transcriptional control affects cells early and late in development. The control of some abundantly expressed genes, such as the ovalbumin gene in certain oviduct cells, is part of terminal cell *differentiation*. In other cases, transcriptional control is part of the earliest steps of cell *determination*. For instance, the regional expression of the *fushi tarazu$^+$* and *hunchback$^+$* genes in the *Drosophila* embryo begins *before* the cellular blastoderm stage, when cell determination occurs. The timing suggests that the regional expression of certain embryonic genes is part of the cell determination process itself. In this section, we will examine the relationship between transcriptional control and cell determination.

Cell Determination May Be Based on Bistable Control Circuits of Switch Genes

The determined states of cells are passed on during cell division. This is shown most clearly in the *imaginal discs* of *Drosophila* larvae (see Chapter 6). During serial transplantations, the determined states of imaginal disc cells are passed on faithfully through many rounds of mitosis. Only rarely do cells undergo *transdetermination,* that is, a switch to another determined state. In these cases, the transdetermined state of a cell is then inherited by its descendants during subsequent mitoses. The heritability of determined states is basic to development. Only in this way is it possible for determination to occur as a stepwise process while mitotic divisions are still going on.

The molecular basis of cell determination and its heritability are not yet fully understood. In the bacteriophage *lambda,* two mutually exclusive developmental stages are controlled by two genes. Each encodes a protein that inhibits the transcription of the other gene. Because this type of control system has two stable states, it is called a **bistable control circuit.** Stuart Kauffman (1973) showed that an assumption of five bistable control circuits could explain most of the data on transdetermination in *Drosophila.* He also suggested that *homeotic genes,* such as *Antennapedia⁺* and *Ultrabithorax⁺,* might be involved in forming these bistable control circuits. Since the time of Kauffman's study, the genetic and molecular analysis of development has turned up many **switch genes,** which decide, by their activity or nonactivity, between two possible fates of a cell. The *homeotic* genes of *Drosophila* are switch genes. Both gain-of-function and loss-of-function alleles of these genes cause substitutions of body parts with other parts that are normally found elsewhere. Another type of switch gene, exemplified by the *MyoD⁺* gene of vertebrates, can switch fibroblasts from connective tissue to muscle fiber development (see Chapter 19).

The active or inactive states of genes may be passed on by different molecular mechanisms, which may vary between animal groups. In the following, we will examine two proposed models. One is based on bistable control circuits, in which transcription factors act as signals. The other relies on DNA methylation.

Switch Genes Are Linked by Transcription Factors in Bistable Control Circuits

The analysis of pattern formation in *Drosophila* embryos has revealed an extensive network of switch genes, also known as *selector genes,* linked by transcription factors (see Chapter 21). As examples, we will examine the *Ultrabithorax⁺* and *Antennapedia⁺* genes of *Drosophila.* As described earlier, individuals carrying certain *Ultrabithorax* loss-of-function alleles develop as four-winged flies (see Fig. 1.21): the third thoracic segment (T3) develops as a repeat of the second thoracic segment (T2). Therefore, the normal activity of the *Ultrabithorax⁺* gene must be necessary for generating the morphological characteristics of T3 instead of T2.

Similarly, the normal function of the *Antennapedia⁺* (*Antp⁺*) gene is necessary for generating the characteristics of the second thoracic segment (T2). In gain-of-function alleles of *Antennapedia,* the gene is also expressed in the head, causing the development of legs instead of antennae (Color Plate 5).

The following experiments have shown that the *Ultrabithorax⁺* gene encodes a protein that enhances the transcription of its own gene and inhibits the expression of the *Antennapedia⁺* gene. Both interactions have been analyzed in whole embryos and cultured cells.

▼

Mariann Bienz and Gaby Tremml (1988) studied the expression pattern of the *Ultrabithorax⁺* gene in genetically transformed *Drosophila* embryos. The transgene was a *lacZ* reporter gene driven by an *Ultrabithorax⁺* promoter and enhancer region. They found that this fusion gene was not expressed at all in *Ultrabithorax* loss-of-function mutants, which could not synthesize functional Ultrabithorax protein. They concluded that the Ultrabithorax protein acts as a positive transcription factor on its own gene.

Ernst Hafen and his colleagues (1984) investigated the expression pattern of the *Antennapedia⁺* gene using *in situ hybridization* (Fig. 15.21). Antennapedia mRNA accumulated strongly in an area overlapping the first and second thoracic segments. Posterior to this domain, *Antennapedia⁺* was expressed only weakly in wild-type embryos. However, the expression domain of *Antennapedia⁺* was extended dramatically in *Ultrabithorax* mutant embryos, which transcribed the *Antennapedia⁺* gene throughout the thoracic region. These results indicate that the wild-type function of *Ultrabithorax⁺* is involved in restricting the expression of *Antennapedia⁺* to its normal domain.

Results obtained with cultured *Drosophila* cells, which can be manipulated more easily than entire embryos, confirmed the conclusions drawn from the preceding observations. Mark Krasnow and his colleagues (1989) cotransfected cultured cells with two types of plasmids (Fig. 15.22). One plasmid, called a **reporter plasmid,** contained the transcribed region of the bacterial *CAT* gene and the regulatory region of a *Drosophila* gene. The *Drosophila* DNA segment contained the promoter and enhancer sequences from the switch gene of interest. The other plasmid, called an **effector plasmid,** encoded a *Drosophila* regulatory protein binding to the enhancer in the reporter plasmid. Cultured *Drosophila* cells were transfected with both plasmids simultaneously for studying the impact of effector plasmid expression on the activity of the reporter plasmid.

In one experiment, Philip Beachy and his colleagues (1988) used a reporter plasmid containing the regulatory DNA from the *Ultrabithorax⁺* gene. The effector

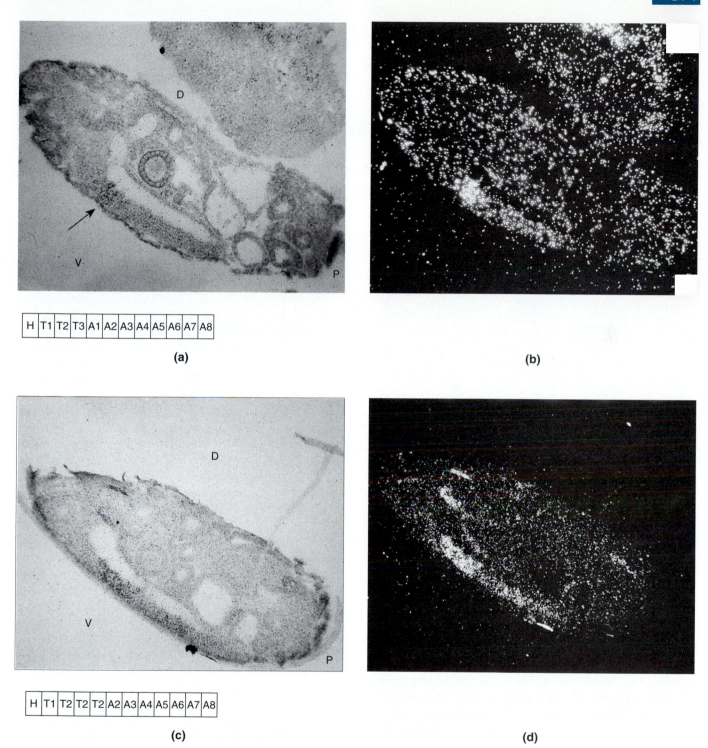

Figure 15.21 Expression of the *Antennapedia*[+] gene in normal and mutant *Drosophila* embryos. The transcription of *Antennapedia*[+] is monitored by in situ hybridization with a radiolabeled cDNA probe (see Methods 15.1). Photomicrographs of paramedian sections are shown in transmitted light (left) and in dark-field illumination to make the autoradiographic signal more visible (white grains, right). The diagrams show the embryonic segment pattern, with H representing the head, T1 to T3 the thoracic segments, and A1 to A8 the abdominal segments. **(a, b)** Normal embryo expressing the *Antennapedia*[+] gene in a region overlapping T2 [indicated by the arrow in part (a)]. **(c, d)** Embryo mutant for *Ultrabithorax* (*Ubx*[105]). This allele transforms segments T3 and A1 toward T2. Correspondingly, the expression domain of *Antennapedia*[+] is extended posteriorly. D = dorsal side of the embryo; P = posterior pole of the embryo; V = ventral side of the embryo.

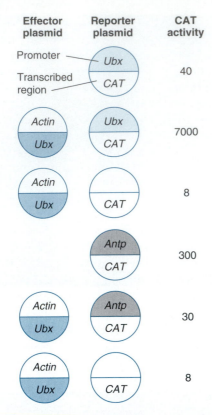

Effector plasmid	Reporter plasmid	CAT activity
	Ubx / *CAT*	40
Actin / *Ubx*	*Ubx* / *CAT*	7000
Actin / *Ubx*	/ *CAT*	8
	Antp / *CAT*	300
Actin / *Ubx*	*Antp* / *CAT*	30
Actin / *Ubx*	/ *CAT*	8

Figure 15.22 Enhancement of *Ultrabithorax* (*Ubx*) gene expression by Ultrabithorax protein, and negative regulation of *Antennapedia* (*Antp*) gene expression by the same protein. Cultured *Drosophila* cells are infected simultaneously with an effector plasmid and a reporter plasmid. The effector plasmid contains the promoter from the *actin*⁺ gene and the transcribed region of the *Ultrabithorax*⁺ gene. The reporter plasmid contains the promoter from either the *Ultrabithorax*⁺ gene or the *Antennapedia*⁺ gene, or no promoter, and the transcribed region of the *CAT*⁺ gene. The CAT activity is determined quantitatively (see Fig. 15.5). While the activity of the *Ultrabithorax*⁺ promoter is enhanced by the Ultrabithorax gene product, the *Antennapedia*⁺ promoter is inhibited by the Ultrabithorax gene product.

Regulatory mechanisms observed in the *Ultrabithorax*⁺ and *Antennapedia*⁺ genes explain some of the general properties of switch genes (Fig. 15.23). Typical switch genes are activated early in development, long

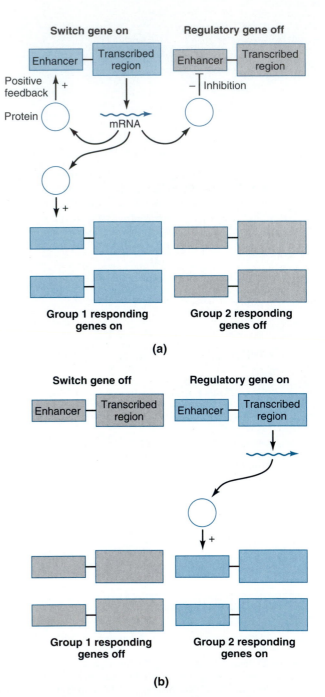

Figure 15.23 Properties of switch genes. **(a)** Through a positive feedback loop, a switch gene can propagate its own activity during cell divisions. At the same time, it exerts positive control over a group of responding genes (group 1) and keeps them on. By means of negative control, the switch gene keeps another regulatory gene and its responding genes (group 2) off. **(b)** If the switch gene itself becomes inactive, the other regulatory gene becomes active. Then group 1 of the responding genes goes off and group 2 comes on.

plasmid contained the transcribed region of the *Ultrabithorax*⁺ gene driven by an actin promoter. Transfection of host cells with the reporter plasmid alone produced only weak CAT activity. However, cotransfection with both the reporter and the effector plasmid caused a 200-fold increase in reporter gene activity. In a similar experiment, an *Antennapedia* reporter gene was inhibited by the expression of the *Ultrabithorax* effector gene. Thus, the presence of the *Ultrabithorax* gene product strongly enhanced the activity of the *Ultrabithorax* promoter while inhibiting the *Antennapedia* promoter. In accord with these observations, the researchers found that at least one protein encoded by the *Ultrabithorax*⁺ gene binds tightly to specific DNA sequences, presumably enhancer modules, near the *Ultrabithorax*⁺ promoter and near one of the two *Antennapedia*⁺ promoters.

before cells reach their state of final differentiation. During the period of switch gene activity, cells still divide. The active or inactive state of a switch gene must therefore be stable and inherited during cell divisions. This can be achieved by a *positive feedback loop.* If the protein product encoded by a switch gene reenters the nucleus and binds to an enhancer module of its own gene, then the activity of the gene is stable. If there is enough protein to maintain the feedback loop after a cell division, then the switch gene remains active in both daughter cells.

In addition to stabilizing their own expression, switch genes *inhibit* the expression of other regulatory genes (Fig. 15.23). On the basis of this mechanism, the *inactivity* of one switch gene would permit the *activation* of other genes. Thus, a given switch gene can have specific effects in both the active and inactive states, in accord with the *principle of default programs* discussed in Chapter 8. Of course, some genes under the negative control of the switch gene might themselves be switch genes, as illustrated by the *Antennapedia*[+] gene in the experiments just described. In this fashion, switch genes are linked by transcription factors in control circuits of activation and inhibition.

DNA Methylation Maintains Patterns of Gene Expression

Another molecular basis for the inheritance of gene activity patterns, known as *DNA methylation,* is observed in mammals (Christy and Scangos, 1986; Riggs, 1990). About 4% of the cytosine nucleotides in mammalian DNA combine with a methyl group to form *5-methyl cytosine.* This step occurs in an enzymatic process *after* DNA replication. The methylation is restricted to cytosine in the dinucleotide sequence CG, which often occurs in clusters called **CG islands.** Any existing CG islands are faithfully restored after each round of DNA replication. The methylating enzyme recognizes the half-methylated CG sequences that exist after DNA replication and methylates the opposite CG sequences (Fig. 15.24).

The effects of methylation on gene expression are mostly inhibitory. Several studies have shown that most genes are methylated when they are inactive and demethylated when they are expressed, but the causal relationship between demethylation and transcriptional activation is not clear. In some cases demethylation precedes transcription, while the reverse has been observed in other cases. Usually, DNA methylation follows the end of transcription. This sequence suggests that methylation may preserve rather than cause the inactive state of a gene. Such a secondary role of DNA methylation would be in line with its phylogenetic limitation: DNA methylation is found in few organisms other than mammals. Thus, DNA methylation may be a way of reinforcing the inactive state of genes in a large genome.

The molecular mechanism by which DNA methylation affects transcription is unknown. Methylation clearly changes the binding of proteins to DNA, as shown by the inability of restriction enzymes to cut

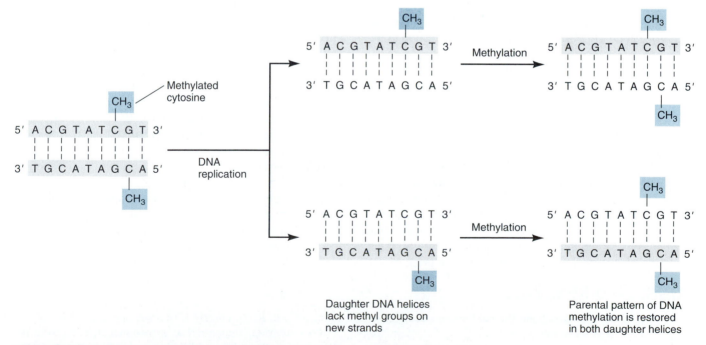

Figure 15.24 Maintenance of DNA methylation patterns. Once a pattern of cytosine methylation is established, it is propagated after each DNA replication by an enzyme, maintenance methylase. The enzyme recognizes those CG sequences in a new DNA strand that are base-paired to a methylated CG sequence in the old strand.

their recognition sites when they are methylated. (Bacterial DNA is protected against the organism's own restriction enzymes by methylation of the corresponding recognition sites.) Therefore, methylation may interfere with the binding of gene-regulatory proteins, mostly preventing but occasionally helping it. Methylation may also affect transcription indirectly by causing changes in chromatin conformation. This is suggested by correlations between methylation and DNase I sensitivity (Groudine and Weintraub, 1981; Groudine et al., 1981). In either case, DNA methylation would establish a gene expression pattern that is passed on during cell division.

DNA methylation may also play a role in a phenomenon termed *gene imprinting.* Mammalian embryos express certain genes only if they are inherited through the sperm. Other genes are expressed only if inherited through the egg. The parent-dependent activation of certain genes seems to be the reason why mammalian eggs need both a male and a female pronucleus for normal development, and do not develop parthenogenetically (see Chapter 4). A striking example of imprinting was observed by Judith Swain and her colleagues (1987) in a transgenic mouse strain containing a tumor-producing gene (c-*myc*). If the transgene was inherited from the male parent, it was expressed in the heart. If inherited from the female parent, the transgene was not expressed at all. This pattern of expression correlated exactly with the parentally imprinted methylation state. Methylation of the transgene was acquired in the gene's passage through the female parent and eliminated during gametogenesis in the male. Other genes were found to be highly methylated during male gametogenesis and extensively undermethylated during female gametogenesis (Sanford et al., 1987). However, it is not clear how the imprinting of genes by methylation is passed on during early embryogenesis when the entire genome becomes highly undermethylated (Brandeis et al., 1993).

SUMMARY

The cells in an organism have thousands of proteins in common but are distinguished by a few hundred cell-specific proteins. Some of the latter, such as globin or ovalbumin, are synthesized in large quantities and reflect the terminal differentiation of the cells producing them. Other cell-specific proteins have gene-regulatory functions and are part of the cell determination process itself.

Transcription, the first step in gene expression, is the DNA-dependent synthesis of RNA by an enzyme, RNA polymerase. The control of transcription involves regulatory DNA sequences. One regulatory region, the promoter, is located immediately upstream of the transcription startpoint. Other regulatory regions, called enhancers, occur in various positions.

Transcription factors are gene-regulatory proteins that bind to specific nucleotide sequences in both promoters and enhancers. General transcription factors occur in all cells of an organism. They bind to the promoter and to each other, forming a transcription complex that positions the RNA polymerase at the transcription start site. Other transcription factors bind to enhancer modules of specific genes to increase or decrease their rates of transcription. These specific transcription factors occur preferentially in certain cells. While most transcription factors are involved in regulatory interactions between genes within the same nucleus, other factors are part of more global interactions.

Some of these factors are translated from localized maternal mRNAs and control the region-specific expression of embryonic genes. Other transcription factors are activated by the binding of a hormone or similar signal molecule.

Chromosome regions containing actively transcribed genes are less condensed than nontranscribed chromosome regions. The decondensation may occur in two steps, one encompassing the entire gene and the second affecting in particular the promoter and enhancer regions.

Specific transcription factors are the molecular basis of differential gene transcription. Some of these factors are encoded by switch genes, which, by their activity or nonactivity, can shift cells between two possible fates. These transcription factors link switch genes in control circuits of autocatalytic activation and mutual repression. Such control circuits have alternate stable states that are passed on during mitosis. In mammals, the methylation of inactive genes is an additional mechanism for stabilizing the inactive state of genes and propagating it during mitosis. This mechanism may account for the phenomenon of gene imprinting in mammalian embryos.

Switch genes exemplify the principle of default programs in development. According to this principle, decisions between two pathways of development are made by the presence or absence of a specific activity, such as the transcription of a gene or the processing of its product.

SUGGESTED READINGS

Evans, R. M. 1988. The steroid and thyroid hormone receptor superfamily. *Science* **240**:889–895.

Felsenfeld, G. 1992. Chromatin as an essential part of the transcriptional mechanism. *Nature* **355**:219–224.

Tjian, R., and T. Maniatis. 1994. Transcriptional activation: A complex puzzle with few easy pieces. *Cell* **77**:5–8.

Yamamoto, K. R., D. Pearce, J. Thomas, and J. N. Miner. 1992. Combinatorial regulation at a mammalian composite response element. In J. L. McKnight and K. R. Yamamoto, eds., *Transcriptional Regulation,* 1169–1192. Plainview, N.Y.: Cold Spring Harbor Laboratory Press.

RNA
PROCESSING

Figure 16.1 Electron micrograph of the nuclear envelope of a newt oocyte. The envelope contains numerous nuclear pore complexes, openings that allow the active transport of macromolecules between nucleus and cytoplasm. Each nuclear pore complex has an eightfold radial symmetry and a central plug, or transporter element. These and other features show that nuclear pore complexes are elaborate gates for the active transport of proteins and ribonucleoprotein particles. The passage of messenger RNP particles through nuclear pore complexes is one of many controlled processing steps that mRNAs undergo between transcription and translation.

Posttranscriptional Modifications of Pre-Messenger RNA

Control of Development by Alternative Splicing

A Cascade of Alternative Splicing Steps Controls Sex Development in *Drosophila*

Alternative Splicing of Calcitonin and Neuropeptide mRNA Is Regulated at the Calcitonin-Specific Splice Acceptor Site

Messenger RNP Transport from Nucleus to Cytoplasm

The Nuclear Architecture Provides a Basis for the Regulated, and Possibly Sequence-Specific, Transport of RNPs to the Cytoplasm

Experiments with Cloned cDNAs Indicate Differential mRNP Transport

Messenger RNA Degradation

The Half-Life of Messenger RNAs in Cells Is Regulated Selectively

The Degradation of mRNAs in Cells Is Controlled by Proteins Binding to Specific RNA Motifs

---

As soon as a gene is transcribed into RNA, the transcript undergoes a series of modifications inside the nucleus. While the transcription process is still going on, the nascent transcript associates with proteins to form *ribonucleoprotein particles (RNP particles, or RNPs).* All subsequent modifications are made not to RNAs, but to RNPs, although we will sometimes call them RNAs in order to focus on the RNA moiety. We will limit our discussion to *messenger RNPs (mRNPs),* that is, RNPs containing mRNA. Most mRNPs are cut and spliced and become extended at both ends before they pass from the nucleus into the cytoplasm. Their passage usually takes the form of active transport through gates in the nuclear envelope known as *nuclear pore complexes* (Fig. 16.1). These processing and gating steps are characteristic of eukaryotes, and all of them have evolved into posttranscriptional levels of controlling gene expression. These controls affect not only the longevity and translatability of an mRNP and the speed with which it is released into the cytoplasm, but also the very genetic information that is passed on.

For several decades it was thought that each gene encoded exactly one protein. This idea, known as the *one gene–one enzyme hypothesis,* originated from the work of George Beadle and Edward Tatum (1941), who showed that yeast cells metabolize their nutrients in a series of discrete biochemical steps. Each step is catalyzed by a specific enzyme, and each enzyme depends on the function of one gene. Since the 1970s, it has become clear that in eukaryotes many genes encode more than one protein.

Most eukaryotic genes are transcribed and processed in the same way movies are made. While many scenes from the original footage of a movie are edited out and never shown, other parts of the total footage are spliced together for the film that is released to movie theaters. Likewise, only certain segments of each eukaryotic gene are represented in the mRNA product that leaves the nucleus (Fig. 16.2). These parts of the gene are called *exons.* Exons are separated by *intervening sequences (introns),* which are edited out during mRNA generation. The primary transcript, called *pre-mRNA,* is a faithful copy of the entire gene and contains all exons and introns. However, in *splicing* steps that occur inside the nucleus, the introns are removed and the exons are ligated together into one mRNA molecule. The exonintron structure, also known as *split-gene organization,* is found in nearly all genes of higher eukaryotes.

Many genes produce a pre-mRNA transcript that is spliced in two or more ways, yielding different mRNAs and different proteins. In terms of the film analogy, the footage produced can be edited in different ways to make more than one movie. The splicing of two or more different mRNAs from the same type of pre-mRNA is called *alternative splicing.* Along with differential transcription, alternative splicing is a most important molecular mechanism by which cells assume different fates during development. We will examine two examples to illustrate this point. One example is sex differentiation in *Drosophila:* whether a fly grows up to be male or female depends on alternative splicings of several gene products. The second example is a rat gene that is spliced in two ways: one type of splicing is carried out in gland cells and yields the mRNA for a hormone, and the other way of splicing the same gene product occurs in neurons and yields the mRNA for a neuropeptide instead.

After splicing and further modifications, mRNPs pass through the *nuclear pore complexes* into the cytoplasm. There the association of mRNA with proteins and possibly other factors controls its longevity and translatability. For instance, in the mammary glands of nursing mammalian females, the mRNA for casein, an abundant milk protein, is associated with certain proteins that protect it against degradation. As a result, a large number of casein mRNA molecules are available for milk production. In nonnursing females, however, casein mRNA is associated with a different set of proteins, targeting it for rapid degradation by RNA-degrading enzymes.

In this chapter, we will first examine the processing of pre-mRNA into mRNA, with an emphasis on alternative splicing. We will consider next the export of mature mRNPs from the nucleus, and finally the controlled degradation of mRNA in the cytoplasm. All of these steps are regulated and provide additional mechanisms for differential gene expression.

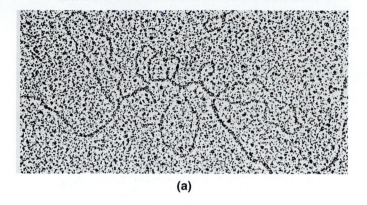

(a)

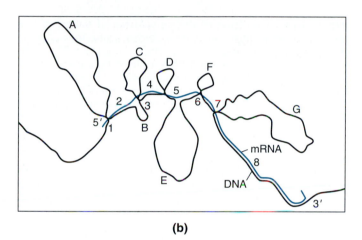

(b)

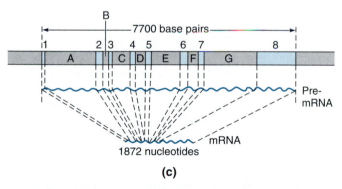

(c)

Figure 16.2 Exon-intron structure of eukaryotic genes. **(a)** Electron micrograph showing the transcribed strand of the chicken ovalbumin gene after hybridization to ovalbumin mRNA (magnification 130,000 ×). **(b)** Interpretative diagram indicating that only small sections of the genomic DNA (black) hybridize with the mRNA (color), the molecule from which the ovalbumin protein is translated. These DNA sections are known as exons. Between the exons, large DNA segments, called introns, are looped out because they have no mRNA sequences to hybridize with. **(c)** Schematic representation of the ovalbumin gene showing the eight exons labeled 1–8 (color) and the seven introns labeled A–G (gray). The primary gene transcript (pre-mRNA) is 7700 bases long and contains both exons and introns. The mRNA processed from this pre-mRNA consists only of exons and is 1872 bases long.

Posttranscriptional Modifications of Pre-Messenger RNA

The major steps in the *posttranscriptional modification* of pre-mRNA are its association with proteins, the "capping" of the 5′ end, the "tailing" of the 3′ end, and the splicing of exons to one another.

Association with Proteins Newly made pre-mRNA in eukaryotes immediately associates with proteins to form a series of closely spaced particles of about 20 nm diameter called *heterogeneous nuclear ribonucleoproteins* or *ribonucleosomes.* Each ribonucleosome consists of about 700 nucleotides (nt) of RNA folded through and around particles consisting of several proteins (Beyer and Osheim, 1990). This arrangement is similar to the packaging of DNA in nucleosomes. The function of ribonucleosomes is not well understood. Conceivably, they keep the nascent RNA single-stranded and accessible to processing proteins by preventing intrastrand hybridization.

Capping of the 5′ End While the pre-mRNA molecule is still being synthesized, its 5′ end is extended by the addition of a methylated guanosine triphosphate (7-methylguanylate), which is attached enzymatically to the pre-mRNA in an unusual 5′-to-5′ phosphotriester linkage (Fig. 16.3). The first two original (transcribed) nucleotides are also methylated, and collectively these modifications of the 5′ end are called the *cap* of the mRNA. The cap protects the growing RNA transcript against degradation by nucleases that attack unprotected 5′ ends. After the processed mRNA has been released into the cytoplasm, the cap also promotes the translation of many mRNAs.

Polyadenylation of the 3′ End The 3′ end of most pre-mRNAs is *not* formed simply by termination of transcription. Rather, the pre-mRNA is cleaved by an enzyme complex that recognizes a specific nucleotide sequence, the *polyadenylation signal,* which in mammals consists of AAUAAA and additional GU-rich sequences farther downstream (Fig. 16.4). The pre-mRNA is cut downstream of the AAUAAA signal, which remains part of the mRNA *trailer* region. The 3′ end of the mRNA is then extended by another enzyme, poly(A) polymerase, which adds more adenosine nucleotides. This activity generates a *poly(A) tail,* which in most vertebrate mRNAs is about 200 nucleotides long. The polyadenylation of mRNAs is a reversible, dynamic process that continues after the mRNA has been transported into the cytoplasm and plays an important role in the regulation of translation (see Chapter 17). For most mRNAs, the poly(A) tail seems also necessary for nucleocytoplasmic transport and for protection against enzymatic degradation.

Aside from its biological functions, the poly(A) tail of mRNAs is a gift of nature to molecular biologists. It

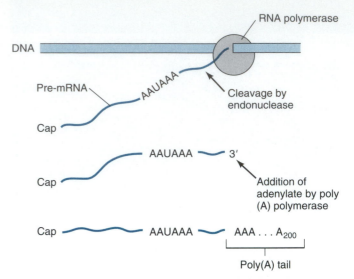

Figure 16.4 Generation of the poly(A) tail of eukaryotic pre-mRNA and mRNA. RNA polymerase transcribes past the site corresponding to the 3′ end of the pre-mRNA. An endonuclease cuts the pre-mRNA, guided by a polyadenylation signal (AAUAAA in mammals) and additional GU-rich nucleotide sequences. A poly(A) polymerase adds about 200 single adenosine nucleotides, which make up the poly(A) tail.

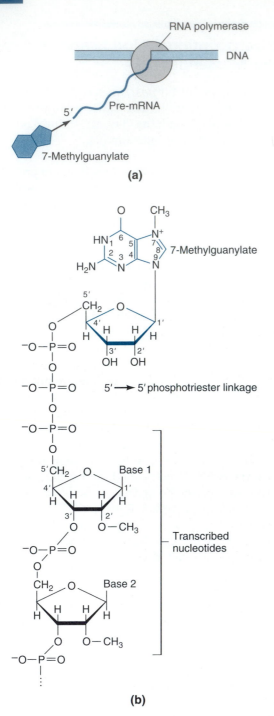

Figure 16.3 Cap structure at the 5′ end of eukaryotic pre-mRNA and mRNA. **(a)** A "capping" enzyme adds 7-methylguanylate to the 5′ end of the pre-mRNA while the rest of the pre-mRNA is still being transcribed. **(b)** The 7-methylguanylate (color) is linked via a 5′-to-5′ phosphotriester linkage to the first transcribed nucleotide of the mRNA. The first two transcribed nucleotides are also methylated.

allows them to isolate the relatively rare mRNAs from the much more abundant ribosomal and transfer RNAs. This is accomplished by hybridization of the poly(A) tails with poly(U) or poly(T) bound to chromatographic columns or papers.

Splicing of Exons Pre-mRNA associates with nuclear ribonucleoprotein particles known as *small nuclear RNP particles (snRNPs).* Each snRNP consists of one or two small RNA molecules and up to 10 different proteins. (The protein composition of snRNPs has been intensively investigated because humans with an autoimmune disorder, *lupus erythematosus,* produce antibodies against their own snRNPs.) There are about a dozen different snRNPs carrying out several vital functions in the processing of pre-mRNA. They recognize specific pre-mRNA nucleotide sequences and bind to them through base pair complementarity. One or more snRNPs are involved in cutting newly synthesized pre-mRNA at the nuclear polyadenylation site. Other snRNPs mediate the removal of introns from the pre-mRNA during its processing into mRNA (Maniatis, 1991; Gall, 1991).

The basic steps of intron removal are summarized in Figure 16.5. Each intron has a *splice donor site* at its 5′ end and a *splice acceptor site* at its 3′ end. These sites have *consensus sequences* that are very similar among the introns of vertebrates and other groups of organisms. These consensus sequences extend mostly over the ends of the intron but extend slightly into the adjoining exons. One type of snRNP binds to the splice donor site, and another binds to a *branchpoint sequence* close to the acceptor site. These two snRNPs, together with other snRNPs and regulatory proteins, assemble into a large, composite RNP particle known as a *spliceosome.* The spliceosome cleaves the pre-mRNA at the donor site and joins the free intron end to the branch-

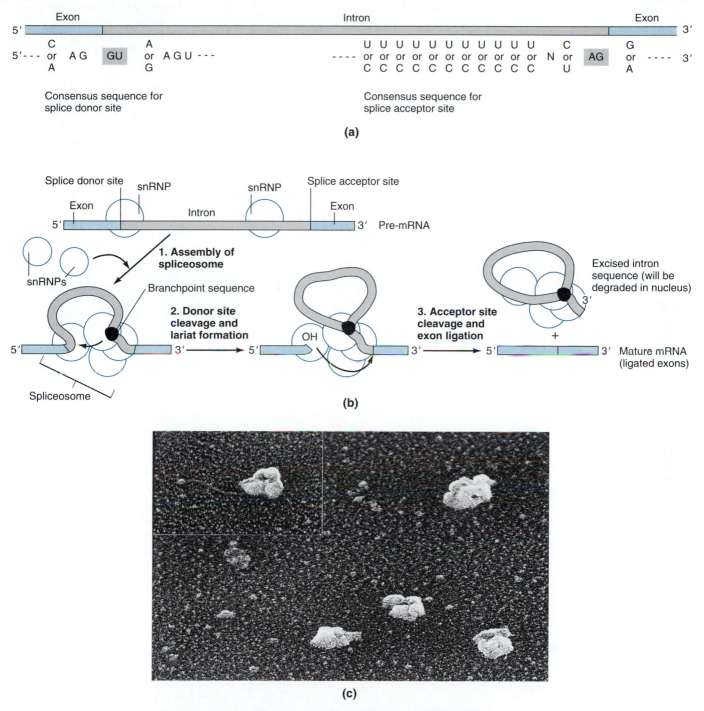

Figure 16.5 Intron removal and exon splicing during pre-mRNA processing. **(a)** Consensus sequences at the splice donor site (5′ end of intron) and splice acceptor site (3′ end of intron). The shaded GU and AG dinucleotides at these sites are nearly invariant. N stands for any of the four RNA nucleotides. **(b)** Three major steps in intron removal and exon splicing. Step 1: spliceosome formation from several snRNPs. Step 2: donor site cleavage and lariat formation. Step 3: acceptor site cleavage and exon ligation. **(c)** Electron micrograph of spliceosomes isolated by gel column chromatography, fixed with glutaraldehyde, mounted onto thin carbon film, and shadow-cast with tungsten.

point sequence. As a result, the intron is bent into the shape of a lariat. Next the intron is cleaved at the acceptor site, and the adjacent exons are simultaneously joined together. The excised intron is degraded. Exon splicing must be done with great precision because any deletion or addition of nucleotides would disrupt the reading frame of nucleotide triplets in the mRNA, which is translated into an amino acid sequence later.

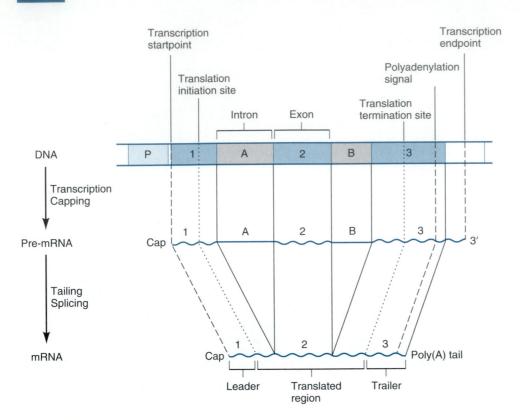

Figure 16.6 Summary of eukaryotic gene transcription and RNA processing. The transcribed region of the gene extends from the transcription startpoint downstream of the promoter (P) to a poorly defined transcription endpoint. The transcript (pre-mRNA) is capped at the 5′ end and cleaved behind the polyadenylation signal. The resulting 3′ end is extended into a poly(A) tail. The introns (A, B) are removed and the exons (1, 2, 3) spliced together. The translated region of the mRNA extends from the translation initiation site to the translation termination site. Between the cap and the translation initiation site there is an untranslated RNA segment called the leader sequence. Likewise, there is an untranslated trailer sequence between the translation termination site and the beginning of the poly(A) tail.

The processing of a pre-mRNA into an mRNA is summarized in Figure 16.6. The processed mRNA consists of a cap, a 5′ untranslated region called the *leader sequence,* a translated region, or *coding region,* a 3′ untranslated region called the *trailer sequence,* and a poly(A) tail. The leader and trailer sequences of mRNAs are involved in several functions, including mRNA localization, regulation of longevity, and translational control (see Chapter 17). The processed mRNA is now ready to be transported into the cytoplasm, where it will be translated into a polypeptide.

Control of Development by Alternative Splicing

Generally, the splicing snRNPs combine the splice donor and acceptor sites of the *same* intron, so that adjacent exons are joined. However, certain proteins can regulate the step of spliceosome formation and cause *alternative splicing.* In such cases, there is a delicately balanced competition between a *regulated splicing pattern,* which depends on one or more regulatory proteins, and a *default splicing pattern,* which occurs in the absence of these regulatory proteins. Frequently, both patterns lead to functional proteins. Even if one of the proteins is nonfunctional, its lack of function often triggers a viable developmental pathway, again illustrating the *principle of default programs* in development.

Three common modes of alternative splicing are illustrated in Figure 16.7. First, a splice *acceptor* site may be subject to *blockage* by an inhibitory protein. The donor site of the intron will then be combined with the acceptor site of the next intron. Second, an intrinsically poor splice *acceptor* site may be skipped in the default splicing pattern, whereas in the regulated pattern it is subject to *enhancement* by an activating protein. Third, pre-mRNAs initiated from different promoters may generate alternative splice *donor* sites that are combined with the same acceptor site.

Alternative splicing patterns usually generate mRNAs with different translated regions, which give rise to different proteins. However, some alternatively spliced mRNAs have *the same* translated regions and differ only in their leader or trailer sequences. Although these mRNAs encode the same protein, the different untranslated regions may contain different regulatory sequences for mRNA localization and longevity. The use of alternative promoters also places the synthesis of one protein under two sets of transcriptional control signals (Fig. 16.7c).

Alternative splicing patterns complicate the definitions of *exon* and *intron,* because what is an exon in one pattern may be part of an intron in another. Generally, though, the context clarifies any ambiguities. The following two examples will show how alternative splicing mediates sexual differentiation in fruit flies and tissue-specific gene expression in mammals.

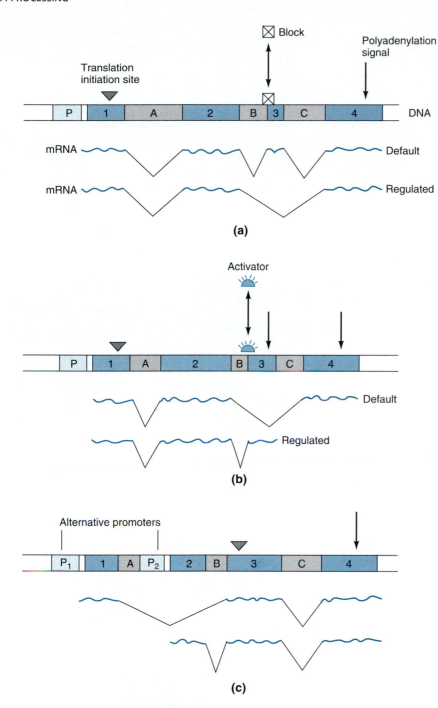

Figure 16.7 Three common modes of alternative splicing. **(a)** Regulated inhibition of a splice acceptor site (a block on the 3' end of intron B, in this diagram) causes the alternative use of a neighboring acceptor site. The resulting two mRNAs in this case are transcribed from the same promoter, with the same transcription and translation initiation sites and the same polyadenylation signal. **(b)** An inherently poor splice acceptor site at the end of intron B is skipped in the default program. However, a regulated enhancement of this acceptor site causes its preferred use over the default acceptor site at the end of intron C. In the case shown, exons 3 and 4 both have polyadenylation signals. **(c)** Alternatively processed pre-mRNAs may arise from the use of different promoters (P_1 and P_2). In the case shown, the alternative mRNAs have the same translated region but different leaders.

A Cascade of Alternative Splicing Steps Controls Sex Development in *Drosophila*

X:A Ratio Most animal species have a genetic mechanism that determines their sex (see Chapter 27). In both fruit flies and mammals, normal females have two X chromosomes and two sets of *autosomes* (abbreviated "A") per diploid cell. Normal males have one X chromosome, one Y chromosome, and two sets of autosomes per diploid cell. However, this similarity between mammals and flies is deceptive. In mammals, the presence or absence of a gene on the Y chromosome decides whether an individual develops as a male or a female. In *Drosophila*, the Y chromosome is of secondary importance: it contains only a few genes necessary for spermatogenesis and is not involved in sex determination.

The primary sex-determining signal in *Drosophila* is the **X:A ratio,** that is, the number of X chromosomes relative to sets of autosomes per diploid cell. Individuals with one X chromosome and two sets of autosomes have an X:A ratio of 0.5 and become males. Individuals with two X chromosomes and two sets of autosomes have an X:A ratio of 1.0 and become females. How *Drosophila* cells sense the X:A ratio is only partially understood. The numerator of the X:A ratio is measured in the product of two X-linked genes, *sisterless-a*⁺ and *sisterless-b*⁺ (J. W. Erickson and T. W. Cline, 1993): the more X chromosomes there are, the more sisterless product is made. How the denominator of the X:A ratio is measured is still being investigated (see Chapter 27).

A hierarchy of regulatory genes transmits the X:A ratio signal to batteries of *realizator* genes, or *effector* genes, that control the sexual differentiation of *Drosophila* and lead to the formation of either eggs or sperm (Fig. 16.8). The low X:A ratio in males also causes the genes located on their single X chromosome to be transcribed at twice the normal rate. This *male hypertranscription* ensures that males and females, although they have dif-

ferent numbers of X chromosomes, still transcribe the same amounts of RNA from X-linked genes. (Male hypertranscription does not start until *after* the *sisterless*⁺ gene activity to measure the X:A ratio is over. Otherwise, male hypertranscription would defeat this measurement.)

We will now trace—starting from the X:A ratio—the development of the somatic sexual characteristics in *Drosophila*—such as reproductive organs, the pigmentation of the abdomen, and the "sex comb" at the male foreleg (Fig. 16.9). The main regulatory genes control-

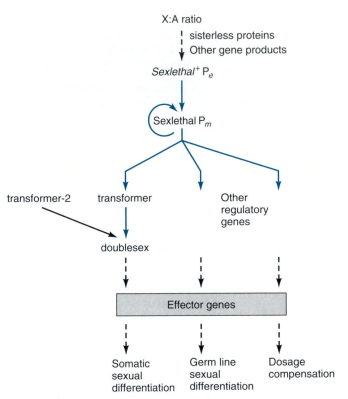

Figure 16.8 Regulatory hierarchy controlling sex development in *Drosophila*. The sex-determining signal is the X:A ratio; it is measured by the amounts of sisterless proteins and other gene products. At an X:A ratio greater than 0.7, the sisterless proteins activate the establishment promoter (P_e) of the *Sexlethal*⁺ gene. The Sexlethal protein controls the female-specific splicing pattern of pre-mRNA transcribed later from the maintenance promoter (P_m) of the same gene. The Sexlethal protein also controls the female-specific splicing pattern of transformer pre-mRNA. The resulting transformer protein, together with the transformer-2 protein, controls the splicing of doublesex pre-mRNA. The resulting two mRNAs, and their protein products, are sex-specific and control, via many effector genes, the somatic sexual differentiation of the fly. In addition, the Sexlethal protein controls other regulatory and effector genes that direct the sexual differentiation of the germ line and the transcription of the X-linked genes as a means of dosage compensation. The colored arrows represent control steps known to be based on RNA splicing.

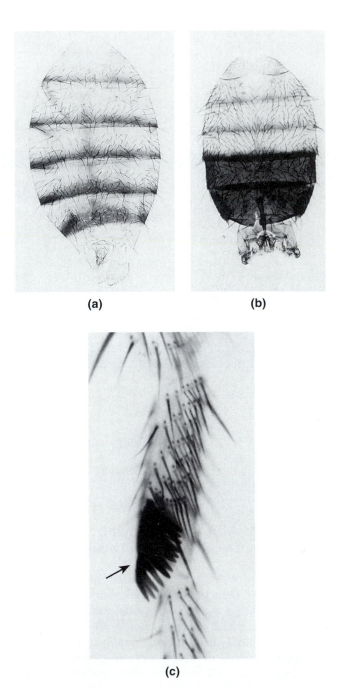

Figure 16.9 Somatic sexual differentiation in *Drosophila*. (a) Female abdomen. (b) Male abdomen. (c) Sex comb (arrow) on the male foreleg.

ling these somatic sexual dimorphisms are the *Sexlethal*+gene, the *transformer*+ gene, and the *doublesex*+ gene (Baker, 1989). These genes form a regulatory hierarchy, in which the protein encoded by each gene controls the splicing of the pre-mRNA transcribed from its subordinate gene (Fig. 16.10). A general characteristic of this hierarchy is that default splicing patterns yield non-sex-specific or male-specific mRNAs, whereas the regulated splicing patterns generate female-specific mRNAs. The following paragraphs outline the functions of these genes and the alternative splicing patterns of their pre-mRNAs.

Alternative Splicing of Sexlethal Pre-mRNA   The gene that is *directly* controlled by the X:A ratio is the *Sexlethal*+ gene, a master gene that in turn controls all aspects of sexual development in *Drosophila* (Fig. 16.8). The control of *Sexlethal*+ expression by the X:A ratio is a two-step process involving two separate promoters of the *Sexlethal*+ gene, called the *establishment promoter (P_e)* and the *maintenance promoter (P_m)*. The first step is the *activation* of P_e by the sisterless proteins, which act as *transcription factors* (Keyes et al., 1992). At an X:A ratio

below 0.7, the amount of sisterless protein made is not sufficient to activate P_e. At an X:A ratio greater than 0.7, there is enough sisterless protein to activate P_e, a step that always leads to the synthesis of active Sexlethal protein. The second step in the control of *Sexlethal*+ expression begins when P_e shuts down and P_m becomes active (in females), or P_e remains inactive and P_m becomes active (in males). In contrast to the pre-mRNA transcribed from P_e, the pre-mRNA transcribed from P_m can be spliced in two different ways, yielding either a *male-specific mRNA* or a *female-specific mRNA* (Bell et al., 1988). The male-specific mRNA includes exon 3, which is absent from the female-specific mRNA (Fig. 16.10). Exon 3 contains a *stop codon,* i.e., a nucleotide triplet terminating translation, so male-specific mRNA yields a truncated Sexlethal protein that is biologically inactive. The female-specific mRNA, because it excludes exon 3, avoids this stop codon and gives rise to a biologically active Sexlethal protein.

Which splicing pattern is initiated depends on the availability of functional Sexlethal protein. In female development, the activation of the P_e promoter by sisterless protein "jumpstarts" the synthesis of Sexlethal

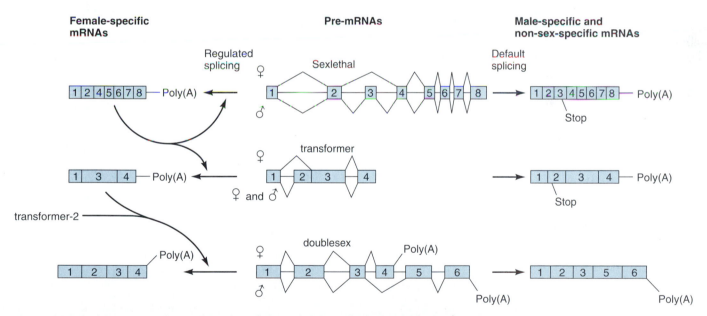

Figure 16.10 Alternative splicing patterns of the transcripts from the *Sexlethal*+, *transformer*+, and *doublesex*+ genes. The respective primary transcripts (pre-mRNAs) of these genes are identical in males and females (central portion of figure). However, they are spliced differently. The default patterns give rise to the mRNAs shown on the right-hand side. These mRNAs are male-specific in the cases of Sexlethal and doublesex. The default splicing of transformer pre-mRNA results in the production of non-sex-specific mRNA in both sexes. The Sexlethal and transformer default mRNAs contain early translation stop codons, giving rise to truncated and inactive proteins. Among the default mRNAs, only the male-specific doublesex mRNA encodes an active protein. The alternative splicing patterns produce female-specific mRNAs shown on the left-hand side. All these mRNAs produce active proteins. The Sexlethal protein promotes the female-specific splicing patterns of both Sexlethal pre-mRNA transcribed from the maintenance promoter and transformer pre-mRNA. The transformer protein, along with the transformer-2 protein, promotes the female-specific splicing pattern of doublesex pre-mRNA.

protein. The Sexlethal protein then directs the female-specific splicing of pre-mRNA transcribed from P$_m$, thus establishing a positive feedback loop for its continued synthesis. Males do not produce enough sisterless protein to activate P$_e$ and therefore produce no "start-up" amount of Sexlethal protein. Upon activation of P$_m$, they initiate the male-specific splicing pattern by default and maintain this pattern because the resulting Sexlethal protein is not functional.

RNA-Binding Domain in Sexlethal Protein Sequence analysis of the Sexlethal protein has revealed an **RNA-binding domain** consisting of about 90 amino acids. This domain has been conserved during evolution in different RNA-binding proteins in life forms ranging from yeasts to humans, indicating that it fulfills critical functions in all organisms (Bandziulis et al., 1989). Generally, proteins with RNA-binding domains interact with specific RNA sequences; their interactions are similar to the regulatory interactions of transcription factors with their DNA recognition motifs in promoters and enhancers. In the case of the Sexlethal protein, the analogy with transcription factors extends even further. Just as several transcription factors act cooperatively in regulating promoter activity, the Sexlethal protein appears to interact with other RNA-binding proteins in blocking the male-specific splice acceptor site of exon 3 in Sexlethal pre-mRNA (J. Wang and L. R. Bell, 1994). In addition to promoting the female-specific splicing pattern of its own pre-mRNA, the Sexlethal protein has a similar effect on the splicing of the pre-mRNA of the *transformer*$^+$ gene.

The *transformer*$^+$ Gene In the genetic hierarchy controlling somatic sexual differentiation, the *transformer*$^+$ gene is directly subordinate to the *Sexlethal*$^+$ gene. The *transformer*$^+$ gene, and the related *transformer-2*$^+$ gene, behave similarly in genetic tests; loss-of-function alleles cause male sexual development, irrespective of the X:A ratio.

Molecular analysis shows that the expression of *transformer*$^+$ is also regulated by alternative splicing. The default pattern is called *non-sex-specific* because it occurs in both sexes (Fig. 16.10). The resulting mRNA includes exon 2, which contains a stop codon truncating the translated protein (Boggs et al., 1987). The regulated splicing pattern is female-specific. It avoids exon 2, and leads to the synthesis of a functional transformer protein. The female-specific pattern depends on functional Sexlethal protein (Nagoshi et al., 1988). This is clearly indicated by the transformer mRNAs produced in males and females carrying different alleles of *Sexlethal*$^+$ (Fig. 16.11). The female-specific splicing pattern is used when the non-sex-specific splice acceptor site is blocked by the Sexlethal protein (Valcárcel et al., 1993). The Sexlethal protein is not very effective at this function; about half of the transformer pre-mRNA in females is wasted because it follows the non-sex-specific splicing pattern. However, enough functional trans-

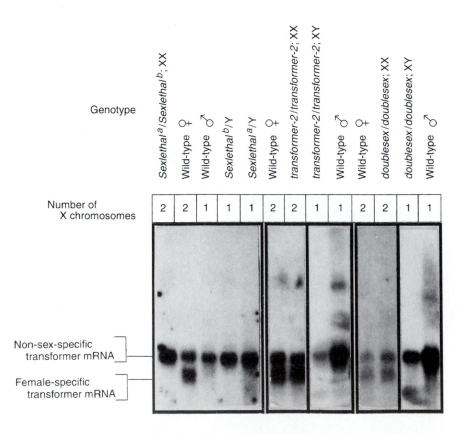

Figure 16.11 Splicing of transformer mRNAs in fruit flies with different mutations. The transformer male-specific and non-sex-specific mRNA were identified by Northern blotting (see Methods 14.3). RNA samples from different fly stocks were separated by gel electrophoresis, blotted, and probed with a labeled cDNA hybridizing with both female-specific and non-sex-specific mRNA. All flies tested produced the non-sex-specific mRNA. The female-specific mRNA was produced only in flies with two X chromosomes (resulting in an X:A ratio of 1.0) and two wild-type alleles of *Sexlethal*$^+$. (*Sexlethal*^{a} and *Sexlethal*^{b} are two different loss-of-function alleles.) Mutant alleles of *transformer-2* or *doublesex* did not affect the splicing pattern of transformer pre-mRNA. The data show that the splicing pattern depends on the expression of *Sexlethal*$^+$ but not of *transformer-2*$^+$ or *doublesex*$^+$.

former protein is apparently generated to ensure female development. The active transformer protein cooperates with the transformer-2 protein, which itself is not sex-specific, in regulating the splicing pattern of the next gene in the genetic hierarchy.

The *doublesex*⁺ Gene   The target gene controlled cooperatively by the transformer and transformer-2 proteins is the *doublesex*⁺ gene. Various mutant alleles of *doublesex* cause the development of intersexes as well as normal females and males. Molecular analysis shows that the *doublesex*⁺ gene encodes two functional mRNAs. One mRNA is female-specific and inhibits male development. Its lack of function causes mutant females to develop as intersexes, which have both male and female genitalia and intermediate secondary sex characteristics. The other doublesex mRNA is male-specific and inhibits female development; its lack of function causes mutant males to develop as intersexes.

The two doublesex mRNAs are generated by alternative splicing combined with the use of different polyadenylation sites (Fig. 16.10). The female-specific mRNA comprises exons 1 through 4. The male-specific mRNA comprises exons 1 through 3, 5, and 6. As in the cases of Sexlethal and transformer pre-mRNA, the male-specific splicing pattern of doublesex pre-mRNA is the default program depending only on the basic cellular splicing machinery. The female-specific splicing pattern requires the presence of the transformer and transformer-2 proteins.

The shift from male- to female-specific splicing of doublesex pre-mRNA might be explained by two models (Fig. 16.12). According to the *blockage model*, the transformer and transformer-2 proteins block the default

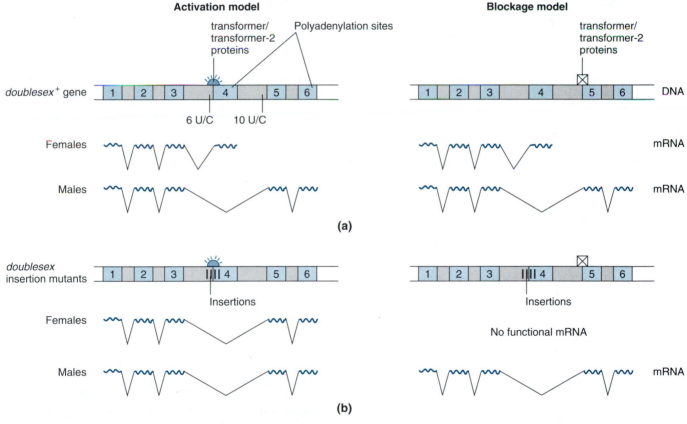

Figure 16.12 Splicing of doublesex mRNAs. The female-specific mRNA comprises exons 1 to 4. The male-specific mRNA comprises exons 1 to 3, 5, and 6. The male-specific splicing pattern is the default pattern, presumably because the acceptor site for exon 5 has a stretch of ten U or C nucleotides as compared with six in the acceptor site for exon 4. **(a)** For wild-type doublesex pre-mRNA, the activation model predicts that the transformer and transformer-2 proteins in females will associate with the acceptor site for exon 4 to make it more attractive to the splicing machinery. According to the blockage model, the transformer and transformer-2 proteins will block the acceptor site for exon 5 in females, forcing the splicing machinery to use the acceptor site for exon 4 instead. **(b)** For *doublesex* mutants with insertions near the splice acceptor site for exon 4, the activation model predicts that both males and females will produce male-specific mRNA. In contrast, the blockage model predicts that only males will produce male-specific mRNA while females will produce no functional mRNA, because they cannot use either splice acceptor site. Experimental data contradict the blockage model.

(i.e., male-specific) splice acceptor site at the end of intron 4; this model has a parallel in the action of Sexlethal protein on the transformer pre-mRNA described previously. According to the *activation model*, the same two proteins make the female-specific acceptor site at the end of intron 3 more attractive to the splicing machinery. The following observations indicate that the activation model is correct.

▼

To decide between the blockage and activation models, Kenneth Burtis and Bruce Baker (1989) compared the splice acceptor sites in the *doublesex*+ gene with other acceptor sequences in *Drosophila*. They found a consensus sequence of (C/U)$_n$NCAG, with (C/U)$_n$ standing for a stretch of *n* pyrimidines (U or C) and N representing any one nucleotide (Fig. 16.5). It seemed that sequences with long pyrimidine stretches generated more competitive acceptor sites than otherwise equal sequences with shorter pyrimidine stretches. By this criterion, the female-specific acceptor site, which has 6 pyrimidines, was deficient relative to the male-specific and other common acceptor sites, which have 9 to 11 pyrimidines. This comparison supported the activation model, according to which the female-specific site was a poor acceptor unless it was enhanced by the transformer and transformer-2 proteins.

The activation model is also strengthened by genetic evidence that contradicts the blockage model. Four *doublesex* mutants with small deletions or insertions near the female-specific splice acceptor site produced only male-specific mRNA *regardless* of their chromosomal sex (Fig. 16.12). This was difficult to reconcile with the blockage model, according to which the females should have been unable to form male-specific doublesex mRNA because their male-specific splice acceptor sites were supposed to be blocked by transformer and transformer-2 proteins.

Finally, the activation model was confirmed by splicing reactions carried out in vitro (Tian and Maniatis, 1992) and in cultured *Drosophila* cells transfected with the *doublesex*+ gene and with cDNAs representing transformer-2 mRNA and the female-specific transformer mRNA (Hoshijima et al., 1991). Both studies showed that the female-specific splicing of doublesex pre-mRNA is positively regulated by the binding of transformer-2 and transformer proteins to the female-specific splice acceptor site.

In summary, development of sexual dimorphism in *Drosophila* depends on alternative splicing of a hierarchy of gene products. At each level of the regulatory cascade, the default splicing pattern leads to male development, whereas the female-specific pattern requires an active protein made at the same level or the next higher

level. The Sexlethal protein regulates the splicing of transformer pre-mRNA by blockage, whereas the transformer and transformer-2 proteins regulate the splicing of doublesex pre-mRNA by activation. Thus, both types of splicing regulation operate in the same cascade.

It is not immediately clear why there should be so many steps in this splicing cascade. Why doesn't the Sexlethal protein control the doublesex pre-mRNA splicing directly? One might speculate that the transformer and transformer-2 gene products evolved as intermediates because they afforded a more precise regulation of *doublesex*+ expression. The splicing of doublesex pre-mRNA must be controlled very effectively to avoid the development of intersexes, which are detrimental to a population. The Sexlethal protein might be insufficiently specialized for this purpose because it regulates several other gene functions at the same time. (Remember that regulation of transformer pre-mRNA splicing by the Sexlethal protein is quite ineffective. Only about half of the transformer pre-mRNA is spliced correctly in females.) In contrast, the transformer and transformer-2 proteins control the splicing of doublesex pre-mRNA very effectively.

Alternative Splicing of Calcitonin and Neuropeptide mRNA Is Regulated at the Calcitonin-Specific Splice Acceptor Site

Alternative splicing of the same pre-mRNA can also generate different mRNAs that have entirely unrelated functions. As an example, we will examine a rat gene that encodes two proteins: *calcitonin* and a *neuropeptide*. The hormone **calcitonin** is produced in the parathyroid glands and acts to increase the level of calcium in the blood. The particular **neuropeptide** encoded by the gene is synthesized in certain neurons of the brain and in the pituitary gland. The two proteins are translated from *different* mRNAs, which are polyadenylated and spliced alternatively from the *same* type of pre-mRNA (Fig. 16.13). The calcitonin/neuropeptide gene and its pre-mRNA transcript contain six exons and two polyadenylation sites. The first three exons (1, 2, and 3) are used in parathyroid cells and neurons alike. Parathyroid cells use the polyadenylation signal poly(A)$_1$, which is located in exon 4, and splice exons 1 through 3 to exon 4. Neurons use the polyadenylation signal poly(A)$_2$ present in exon 6 and splice exons 1 through 3 to exons 5 and 6. Thus, differential polyadenylation and splicing of the same pre-mRNA results in two different mRNAs: calcitonin mRNA in the parathyroid gland and neuropeptide mRNA in certain neurons.

The two mRNAs are translated into the corresponding proteins, each in the appropriate cell type. Both these proteins are initially synthesized as inactive precursors. Functional proteins are generated by proteases that clip off the parts corresponding to exons 1 through 3. Thus, in their final active form, the two proteins encoded by the same gene have no parts in common.

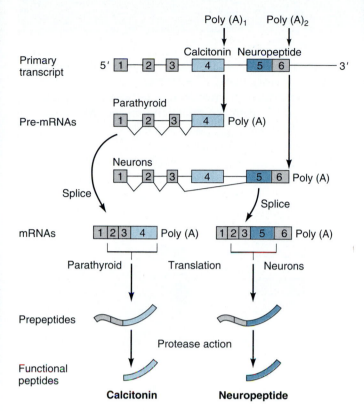

Figure 16.13 Alternative polyadenylation and splicing of the pre-mRNA transcribed from the calcitonin/neuropeptide gene. In parathyroid gland cells, mRNA for calcitonin, a hormone, is formed as polyadenylation takes place at the site in exon 4, and exon 3 is spliced to exon 4. In neurons, mRNA for a neuropeptide is created through polyadenylation at the site in exon 6 and splicing of exon 3 to exons 5 and 6. The proteins translated from both types of mRNA are modified by proteases that remove the domains encoded by exons 1, 2, 3, and 6.

To further analyze the alternative splicing of calcitonin/neuropeptide pre-mRNA, it was necessary to establish two lines of cultured cells that mimicked the respective splicing patterns observed in the two rat tissues. Only with cultured cells was it practical to introduce suitable reporter genes with engineered splice acceptor sites and other alterations. A line of cancerous human epithelial cells, when transfected with the rat calcitonin/neuropeptide transcribed region under a suitable promoter, mimicked the splicing pattern of the rat parathyroid gland. For simplicity, we designate these the C (calcitonin-producing) cells. Another cell line, derived from a mouse teratocarcinoma and containing the same transgene, paralleled the splicing pattern of rat neurons. We designate the cells of this line the N (neuropeptide-producing) cells. To monitor the splicing pattern in each cell line, Ronald Emeson and his colleagues (1990) used the *ribonuclease protection assay* described in Methods 16.1. It showed that

METHODS 16.1

Ribonuclease Protection Assay

The *ribonuclease protection assay* is used to detect small quantities of a particular RNA sequence in a sample containing a mixture of many other RNAs. The assay is based on the fact that the enzyme ribonuclease A degrades single-stranded RNA but not double-stranded RNA (Fig. 16.14). Therefore, the

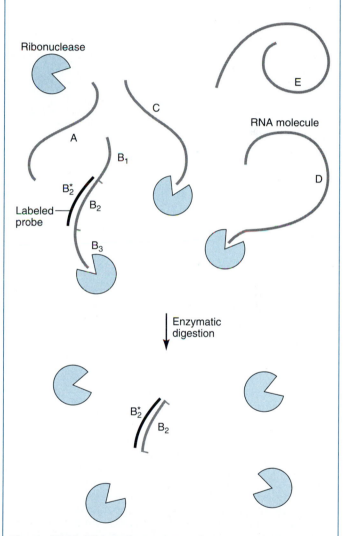

Figure 16.14 Ribonuclease protection assay. A mixture of unlabeled RNA molecules (A through E) is tested for the presence of a particular sequence, B_2. B_2 is part of a larger molecule, $B_1 + B_2 + B_3$. To test for the presence of B_2, a complementary labeled probe, B_2^*, is added to the mixture and allowed to hybridize. Then a small amount of ribonuclease is added. This enzyme digests single-stranded but not double-stranded RNA. After a short period, all RNA is digested except for the B_2/B_2^* hybrid. Upon electrophoresis, the label associated with B_2^* still migrates in one piece. The protection of the labeled probe shows that B_2 was present in the mixture.

Continued on next page

presence of the RNA sequence of interest can be tested by adding the complementary RNA sequence as a labeled probe. If the sequence of interest is present, it will protect the probe from enzymatic degradation, so that the probe can still be identified by gel electrophoresis. If the RNA sequence of interest is not present in the mixture, then the probe is degraded by the enzyme and can no longer be detected.

As an example, Figure 16.15 shows a ribonuclease protection assay using as a labeled probe antisense RNA transcribed in vitro from a template comprising exons 4 and 5 from the rat calcitonin/neuropeptide gene.

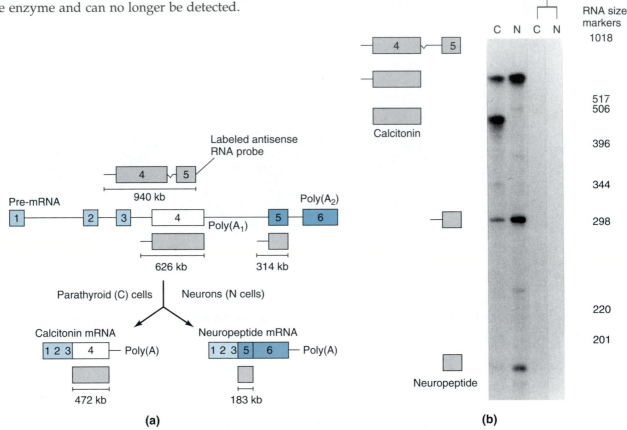

(a)

(b)

Figure 16.15 Differential splicing of calcitonin/neuropeptide pre-mRNA in parathyroid glands and neurons was simulated in cultured cell lines designated C cells and N cells, respectively. The two mature mRNAs and their splicing intermediates were identified by RNase protection (see Fig. 16.14). **(a)** Schematic representation of a labeled antisense RNA probe (gray), the calcitonin/neuropeptide pre-mRNA, and the two mRNAs for calcitonin and neuropeptide. The probe was an antisense RNA transcript from an engineered cDNA consisting of exons 4 and 5 along with their adjoining splice acceptor sites. For the pre-mRNA and each mRNA, the protected probe segments are indicated in gray. **(b)** Results of a ribonuclease protection assay, using RNA from C cells and N cells to protect the probe shown in part a. RNA from untransfected cells protected no piece of the probe, indicating that these cells produced no complementary RNA sequences. RNA from C cells transfected with an actively transcribed calcitonin/neuropeptide transgene included the mature calcitonin mRNA and splicing intermediates of both mRNAs. RNA from N cells containing the same transgene included the mature neuropeptide mRNA and both splicing intermediates.

each cell line produced not only the correct final mRNA but also the expected intermediate splice products (Fig. 16.15).

Once the cell lines were established, the researchers used the assay to test several hypotheses. One possibility was that the alternative use of the two polyadenylation signals, poly(A)$_1$ in exon 4 and poly(A)$_2$ in exon 6, was directing the splicing machinery to the adjacent acceptor sites. Experiments showed that this was *not* the case; pre-mRNA from an engineered transgene in which

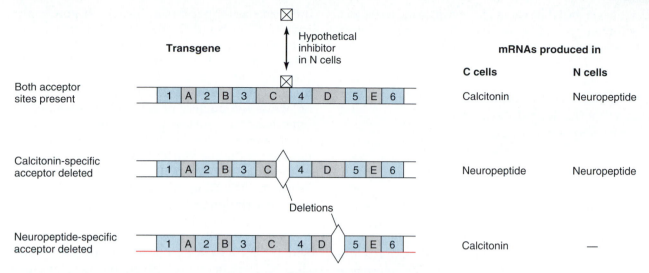

Figure 16.16 Analysis of calcitonin/neuropeptide pre-mRNA splicing by deletion of the alternatively used splice acceptor sites (for exons 4 and 5). With a transgene containing both acceptor sites, C cells produced calcitonin mRNA and N cells produced neuropeptide mRNA. With a transgene lacking the acceptor site for exon 4, both cell types produced neuropeptide mRNA. With a transgene lacking the acceptor site for exon 5, only C cells produced calcitonin mRNA while N cells produced neither mRNA. The results are best explained by assuming that the calcitonin mRNA is produced by default, and that N cells contain an inhibitor that blocks the calcitonin-specific (default) splicing acceptor site for exon 4.

the poly(A)$_1$ signal was replaced with another copy of the poly(A)$_2$ signal was still spliced correctly in both C cells and N cells. Next the investigators tested the hypothesis that the splice acceptor sites for exons 4 and 5 were critical to the alternative splicing. They transfected C cells and N cells with modified calcitonin/neuropeptide genes from which one of the two acceptor sites had been removed (Fig. 16.16). With a transgene lacking the acceptor site for exon 4, C cells no longer produced calcitonin mRNA but spliced neuropeptide mRNA instead. With a transgene lacking the acceptor site for exon 5, N cells no longer produced neuropeptide mRNA and *did not* splice calcitonin mRNA either. The investigators concluded that N cells contained an *inhibitor* that prevented the use of the calcitonin-specific acceptor. No corresponding inhibitor of the neuropeptide-specific acceptor seemed to exist in C cells, because they were able to use this site when the alternative site was deleted. Thus, the calcitonin-specific splicing pattern seems to be used by default.

The experiments just described indicate that the rat calcitonin/neuropeptide gene gives rise to two products by means of alternative splicing. The default splicing pattern is carried out in parathyroid cells and gives rise to calcitonin mRNA. Neurons contain an inhibitor that prevents the use of the calcitonin-specific splice ac-

ceptor site. Consequently, the splicing machinery is redirected to the alternative acceptor site and neuropeptide mRNA is generated instead.

Messenger RNP Transport from Nucleus to Cytoplasm

Once pre-mRNPs have been processed into mRNPs, most of them are released into the cytoplasm. Usually, the release occurs during the interphase of the cell cycle, when the nucleus is partitioned from the cytoplasm by the *nuclear envelope*. Under these conditions, molecules entering or leaving the nucleus must pass through the *nuclear pore complexes* (Fig. 16.1). Traffic through these pore complexes is regulated in both directions (M. Miller et al., 1991; Feldherr, 1992). Certain pore proteins serving as gatekeepers seem to recognize specific attributes of the molecules approaching the pore complexes. Small molecules are generally allowed to pass by simple diffusion. Large molecules, such as proteins and RNPs, are either rejected or actively moved through the gates. The molecular basis of this selective treatment is not understood. However, the available data indicate that nucleocytoplasmic transport is yet another means of differential gene expression.

In this section, we will first survey some observations on the ultrastructure of nuclear pore complexes and some general data on nucleocytoplasmic transport. We will then present some quantitative measurements on the selective transport of specific RNPs.

The Nuclear Architecture Provides a Basis for the Regulated, and Possibly Sequence-Specific, Transport of RNPs to the Cytoplasm

The nucleus is a crowded place where chromatin, RNP particles, and some proteins are present at very high concentrations. Immunostaining indicates that several nuclear molecules are restricted to certain nuclear zones (Gerace and Burke, 1988). Nuclear proteins called *lamins* form a meshwork, known as the **nuclear lamina,** which lines the inside of the nucleus (Fig. 16.17). Internal to this lamina are other proteins that provide attachment sites for chromatin, much of which is therefore concentrated in domains near the inner aspect of the nuclear envelope. Pre-mRNPs originating in these domains are processed in tracks that originate in spaces between the chromatin domains in the interior of the nucleus (Xing et al., 1993). Presumably, these tracks lead to the *nuclear pore complexes* that connect the nucleus with the cytoplasm (Carter et al., 1993).

As a result of this organization, nuclear mRNPs may follow distinct pathways from the site of their synthesis to the site of their export from the nucleus. Such structural constraints could make the processing and transport of mRNPs more efficient than if all nuclear components were freely floating and randomly colliding.

Nuclear pore complexes have an intricate ultrastructure, which has been highly conserved during the evolution of eukaryotes (Figs. 16.1 and 16.17). Ultrastructural studies and experiments with molecules of different sizes suggest that the nuclear pore complexes have two functions. On the one hand, they simply act like the openings of a sieve: they allow small molecules such as nucleotides or amino acids to *diffuse* freely into and out of the nucleus. Larger molecules pass more slowly, and molecules larger than 20 to 40 kd do not cross the nuclear envelope by diffusion. On the other hand, the nuclear pore complexes are part of an active transport system that interacts with carrier proteins to shuttle specific RNP particles and protein molecules into and out of the nucleus (Silver, 1991; Feldherr, 1992; Dingwall and Laskey, 1992). For example, transcription factors, after being synthesized in the cytoplasm, are *targeted* to the nucleus by *active transport* mechanisms that recognize certain *signal sequences* within the transcription factors. Conversely, some nuclear RNP particles are actively released into the cytoplasm after they have undergone a series of modification steps.

Experiments with Cloned cDNAs Indicate Differential mRNP Transport

Several investigators, working with various animal species, have used cloned cDNAs to monitor the processing of pre-mRNPs into mRNPs and the transport of mRNPs from the nucleus into the cytoplasm. Such cDNA probes are reverse-transcribed from cytoplasmic or polysomal mRNAs and cloned in a suitable vector. Because cDNAs represent individual mRNAs and are free from introns and nontranscribed genomic sequences, they are highly specific tracers. Two examples will illustrate this type of analysis.

As discussed in Chapter 4, egg activation entails a rapid increase in the overall rate of protein synthesis. Most of this increase results from the recruitment of

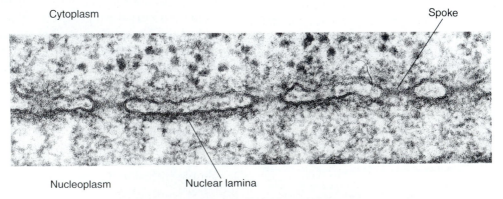

Cytoplasm

Spoke

Nucleoplasm

Nuclear lamina

Figure 16.17 Nuclear lamina and nuclear pore complexes in a frog oocyte. This transmission electron micrograph shows the sectioned nuclear envelope with the outer membrane facing the cytoplasm at the top and the inner membrane facing the nucleoplasm at the bottom of the photograph. The dark, fibrous material adjacent to the inner membrane is the nuclear lamina. The openings in the nuclear envelope are nuclear pore complexes (see also Fig. 16.1). Radially oriented proteins inside the pore complex are known as spokes.

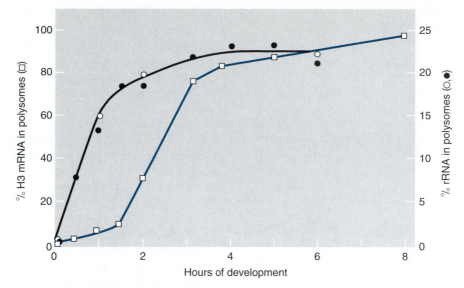

Figure 16.18 Delayed recruitment of histone H3 mRNA into polysomes after fertilization in sea urchin eggs. The fraction of H3 mRNA in polysomes (left ordinate, squares) was measured at intervals after fertilization (abscissa). The amounts of polysomal H3 mRNA and total H3 mRNA were determined by hybridization with a radiolabeled cDNA probe. For comparison, the fraction of ribosomal RNA in polysomes was measured in the same way (right ordinate, open circles) or by optical density (filled circles). This measurement reflected the amount of all mRNAs loaded into polysomes. The comparison showed that the overall assembly of polysomes reached a plateau within 90 min after fertilization. In contrast, histone H3 mRNA was loaded into polysomes mostly between 90 min and 3 h after fertilization. The cause for this specific delay is illustrated in Figure 16.19.

previously untranslated mRNPs into polysomes. When tracing the recruitment of mRNP for the histone H3, Dan Wells and coworkers (1981) observed that it lagged about 90 min behind the recruitment of most other mRNPs (Fig. 16.18). A similar time interval passed between fertilization and the breakdown of the zygote nucleus before first cleavage. Thus, the data were compatible with the hypothesis that the histone mRNP was retained selectively in the nucleus. This conclusion was confirmed by *in situ hybridization* (see Methods 15.1) with probes specific for several histone mRNPs (DeLeon et al., 1983). These mRNPs were found to be sequestered in the nucleus until first mitosis, when they were rapidly released into the cytoplasm (Fig. 16.19). This sequestration also occurs during the subsequent mitotic cycles and is specific to embryonic histone mRNPs. These mRNPs are made only during cleavage, and their passive release during mitosis occurs shortly before each S phase, when histone synthesis is required. Thus, no active transport of histone mRNP from the nucleus into the cytoplasm is necessary.

The following study on the slime mold *Dictyostelium discoideum* shows that some mRNAs are transported quickly from the nucleus into the cytoplasm while others move more slowly. In the course of their life cycle, *Dictyostelium* cells first grow and multiply as individual amoebae feeding on bacteria. When starved, the amoebae aggregate, differentiate into a fruiting body, and

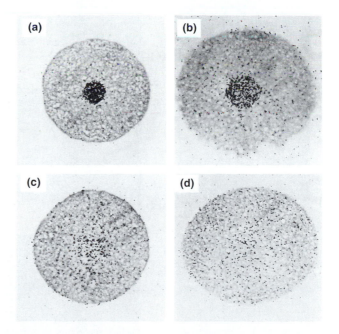

Figure 16.19 In situ hybridization (see Methods 15.1) of a radiolabeled histone mRNA probe with sections of sea urchin eggs fixed at 70 min **(a)**, 80 min **(b, c)**, and 90 min **(d)** after fertilization. The intensely labeled area in the center of the egg is the zygote nucleus, which breaks down about 80 min after fertilization during the first mitosis. These data explain the delay, relative to other mRNAs, before histone mRNA is recruited into polysomes (see Fig. 16.18).

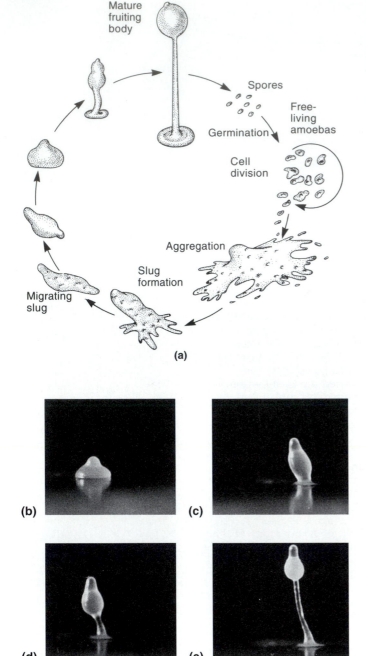

Figure 16.20 Life cycle of the slime mold *Dictyostelium discoideum*. **(a)** Individual amoebalike cells feed on bacteria and divide. When starved, the cells aggregate and form migrating slugs. The slugs form fruiting bodies, which release spores. The spores germinate when living conditions are favorable, and the life cycle begins anew. **(b–g)** Photographs showing successive stages in the development of the fruiting body.

form spores that can survive adverse conditions for a long time (Fig. 16.20). Some mRNAs, called *common mRNAs*, are present in the cytoplasm during all stages of development. Other mRNAs, called *developmentally regulated mRNAs*, are present only at certain stages.

To analyze the mechanisms involved in the regulation of mRNA transport in *Dictyostelium*, Giorgio Mangiarotti and coworkers (1983) traced individual mRNA sequences known from previous investigations to be either common or aggregation-specific. Hybridization of individual cDNA clones with *pulse-labeled* nuclear and cytoplasmic mRNA fractions revealed different patterns of processing and/or nucleocytoplasmic transport (Fig. 16.21). *Common* mRNAs remained in the nucleus for about 45 min before they appeared in the cytoplasm. In contrast, *aggregation-specific* mRNAs did not appear in the cytoplasm until after 90 min. Depending on the particular mRNA, between 20 and 60% of the nuclear transcripts were transported into the cytoplasm. A third type of RNA, expressed predominantly during cell differentiation, was degraded extensively within the nucleus; what remained of it was released into the cytoplasm after a delay of over 2 h.

The foregoing studies show that mRNP processing and/or transport from the nucleus into the cytoplasm can regulate the timing and quantity of gene expression. However, the molecular mechanisms underlying these control steps remain to be elucidated.

Messenger RNA Degradation

Once in the cytoplasm, most mRNPs are incorporated into polysomes to synthesize proteins. (The storage of some mRNPs in the form of inactive RNP particles will be discussed in Chapter 17.) The rates at which different proteins are synthesized in a cell are precisely regulated. The synthesis rate for a protein in a cell is determined to a large extent by the *number* of available mRNA molecules, which ranges from a few to many thousands. This number in turn depends on the rates of mRNA *production* and *degradation*. The importance of the rate of production is obvious and has been discussed previously. However, for quick adjustment of the available mRNA, the rate of degradation is just as critical (Ross, 1989). As an analogy, consider the task of adjusting fluid in a tank to varying levels. To raise the level, one needs a controlled influx. To lower the fluid level, waiting for evaporation or leakage to do the job would be slow. A quicker adjustment requires a controlled efflux. This is precisely what controlled degradation of mRNA does in a cell.

The Half-Life of Messenger RNAs in Cells Is Regulated Selectively

The longevity of mRNA in cells is measured by its *half-life*, which is defined as the time after which an origi-

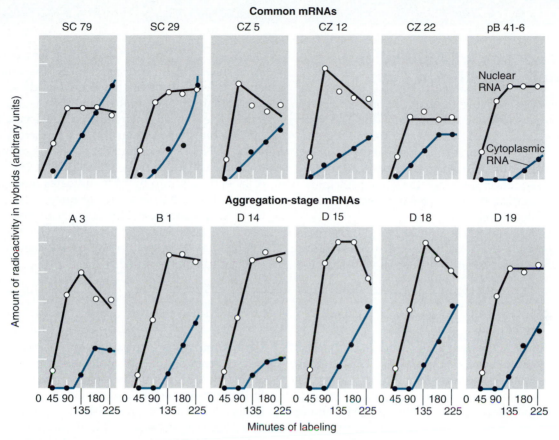

Figure 16.21 Differential processing or transport of mRNAs in the slime mold. In this experiment, all newly synthesized RNA was labeled in vivo, beginning at 14 h of development. During the labeling period, both common and aggregation-specific mRNAs were synthesized. After labeling periods ranging from 45 min to 225 min, cells were harvested to prepare nuclear and cytoplasmic RNA. Each fraction was hybridized separately with 12 cloned cDNAs known to represent either common mRNAs or aggregation-specific mRNAs. The cloned cDNAs were made single-stranded and immobilized on membranes. The amount of label binding to each cDNA indicated how much of the corresponding mRNA was present in the nucleus (open circles) and in the cytoplasm (filled circles). Every sequence was detected in the nucleus within 45 min. The common mRNAs also appeared in the cytoplasm within 90 min, except for one sequence (pB 41-6), which was also unusual in other respects. However, the aggregation-specific mRNAs were not detected in the cytoplasm until after 90 min. The results indicate a sequence specificity in the processing and/or transport of different mRNAs from the nucleus to the cytoplasm.

nal number of intact molecules has been reduced to one-half (Methods 16.2). The half-life of eukaryotic mRNAs in active cells ranges from minutes to days.

The half-life of mRNAs is controlled in cells in adaptive and selective ways. A few examples follow.

Globin mRNA   As bone marrow cells mature into red blood cells, they undergo several steps of differentiation, including mass production of hemoglobin and, in mammals, elimination of the cell nucleus. In the enucleate cells, which are called *reticulocytes*, no additional RNA can be synthesized. Nevertheless, the RNA composition continues to change by differential degradation. This is shown by comparing the proteins synthesized and the mRNAs present at different stages of development. While

young reticulocytes synthesize globins plus several other proteins, mature reticulocytes synthesize globins almost exclusively. Analysis of the mRNAs shows that globin mRNA in rabbit reticulocytes has a half-life about 4 times as long as that of the mRNA for protein I, the next-most abundant reticulocyte protein (Lodish and Small, 1976). Since differential mRNA production cannot occur in this case, the result must be caused by differential degradation. The apparent adaptive value of this type of regulation is that it maximizes the synthesis of globins by reducing the concurrent synthesis of other proteins.

Casein mRNA   The mammary gland is a target tissue for several hormones that interact during pregnancy to cause growth and differentiation of the gland cells.

Pulse Labeling of Molecules and Their Half-Life in Cells

Researchers can measure the half-life of a molecule in a cell by *pulse-labeling* the molecule during a short period of synthesis and then measuring the time until half of the labeled molecules have decayed. The following procedure is used to measure the half-life of an RNA sequence; however, the same principles apply to measuring the half-lives of proteins and other molecules in cells.

To pulse-label an RNA synthesized in cells, the investigator adds a labeled nucleotide to the culture medium for a short period of time. The labeled nucleotide is then "chased" by adding an excess of the same nucleotide in unlabeled form. The unlabeled nucleotide dilutes the labeled nucleotide so that the subsequent uptake of label into RNA is negligibly small. The cells are then allowed to survive for various intervals before their RNA is extracted. An unlabeled cDNA representing the sequence of interest is hybridized with the labeled cell RNA. The amount of label driven into hybrids indicates how many copies of the RNA sequence of interest are present in the RNA preparation. This number decreases exponentially with time. The time after which the label in hybrids has decreased to half of its original value is the half-life (abbreviated $t_{1/2}$).

In nursing females, the hormone *prolactin* stimulates lactation, the actual production of milk. This response can also be observed in cultured mammary gland tissue, using the synthesis of *casein,* an abundant milk protein, as a biochemical marker (Fig. 16.22). Upon pro-lactin stimulation, the availability of casein mRNA increases from 50 to 2000 molecules per cell (Guyette et al., 1979). Most of this increase is caused by an increase in the half-life of the casein mRNA, which is approximately 1 h without prolactin and 28 h with prolactin. This dramatic increase is highly selective because the half-life of *total* mRNA changes little in response to the hormone treatment. Similarly, the presence of estrogen causes a substantial increase in the half-lives of vitellogenin mRNA in liver cells of frogs and birds, and of ovalbumin mRNA in hen oviduct cells (Shapiro et al., 1987).

Oncogene mRNA Many cancer-causing genes, called *oncogenes*, are mutant alleles of normal cellular genes called *proto-oncogenes* (see Chapter 29). For example, the c-*myc* gene is a proto-oncogene encoding a protein involved in the control of cell proliferation. However, certain mutant alleles of this gene cause lymph node cancer.

Comparing the c-*myc* proto-oncogene with its cancer-causing alleles (designated *myc),* Marc Piechaczyk and coworkers (1985) observed that the mRNAs transcribed from *all* alleles had identical translated regions, so that the encoded proteins must have been the same. The mutated sequences were in the *leader* regions of the mRNAs. In correlation with these changes, the mutant mRNAs had half-lives 3 to 5 times longer than wild-type mRNA. The resulting overproduction of the c-*myc* protein appeared to be causing the cancers. Evidently, the half-life of a proto-oncogene mRNA must be strictly controlled.

This example illustrates not only that maintaining the correct half-life times of an mRNA may be critical, but also that the regulation process may depend on sequences in the leader region.

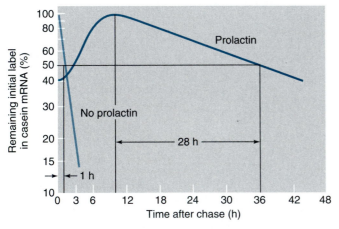

Figure 16.22 Stabilization of the mRNA for the mammalian milk protein casein in the presence of the pituitary hormone prolactin. In this experiment, breast tissue from lactating rats was placed in a medium without prolactin to dilute the endogenous hormone. One sample was then incubated with prolactin and pulse-labeled with labeled uridine (see Methods 16.2). Another sample was also pulse-labeled with uridine, but without prolactin. After 3 h, the labeled uridine was removed, and unlabeled uridine was added to chase the labeled uridine from the tissues. Samples taken at intervals thereafter were analyzed for casein mRNA by hybridization with cloned, unlabeled cDNA. The culture with prolactin accumulated labeled casein mRNA for several hours, presumably because the chase of labeled uridine was ineffective. The eventual decay of mRNA was much slower ($t_{1/2} = 28$ h) in the presence of prolactin than without the hormone ($t_{1/2} = 1$ h).

The Degradation of mRNAs in Cells Is Controlled by Proteins Binding to Specific RNA Motifs

The cellular agents of mRNA degradation are enzymes called *ribonucleases (RNases)*. The destruction of RNA molecules by RNases is regulated by cellular proteins that bind to specific nucleotide sequences or secondary structures of the RNA. These regulatory proteins in some cases mark the bound mRNAs for enzymatic breakdown and in other cases inhibit the RNase action. In either event, the specificity with which regulatory proteins bind to certain RNA motifs explains the selective control of mRNA longevity. The cases investigated so far have revealed several regulatory proteins binding to various regions of their target mRNAs (Sachs, 1993). The importance of the *leader* region in the c-myc mRNA has already been mentioned. In other cases, regulatory proteins bind to the poly(A) tail or to the *trailer* region.

The poly(A) tail at the 3′ end prolongs the half-life of most mRNAs. The tail's stabilizing effect depends on a *poly(A)-binding protein (PABP)* that binds specifically and tightly to poly(A) segments. Presumably, PABP protects the poly(A) tail (and indirectly the entire mRNA) against ribonuclease attack. At the same time, PABP is thought to promote the initiation of translation (R. J. Jackson and N. Standart, 1990; see Chapter 17).

The trailer region affects the half-lives of many mRNAs, as shown by cell transfection experiments with chimeric genes. For instance, globin genes can be engineered to encode mRNAs with 3′ untranslated segments from other genes (Fig. 16.23). If the trailer segment comes from a stable mRNA, the chimeric globin mRNA is stabilized; if the trailer segment comes from an unstable mRNA, the half-life of the chimeric globin mRNA is reduced. Instability is conferred in particular by sequences rich in A and U nucleotides (Shaw and Kamen, 1986). These results suggest that certain sequences of these two bases are recognized by the regulatory proteins that bind to specific mRNAs and increase their susceptibility to ribonuclease.

Some of the regulatory proteins that enhance or inhibit the degradation of mRNAs are themselves activated by a *ligand*, such as prolactin in the case of prolactin mRNA protection described earlier. This situation is similar to the regulation of transcription factors by ligands, as described in Chapter 15 for steroid hormone receptors. As a well-investigated example, we will explore a protein that regulates the half-life of an mRNA in response to the iron concentration in blood cells.

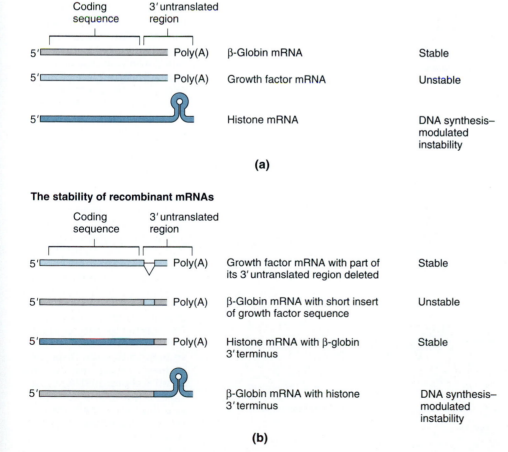

The stability of natural mRNAs

(a)

The stability of recombinant mRNAs

(b)

Figure 16.23 Importance of the trailer region for mRNA stability. **(a)** Three naturally occurring mRNAs with different half-lives. Globin mRNA is stable, with $t_{1/2} = 10$ h. Growth factor mRNA is unstable, with $t_{1/2} = 0.5$ h. Histone mRNA, which is unusual in that its trailer ends with a stem-loop structure instead of a poly(A) tail, has a modulated instability: its half-life is 1 h when cells are synthesizing DNA and 0.2 h during the rest of the cell cycle. **(b)** By cell transfection with engineered cDNA transgenes, it is possible to create recombinant mRNAs. Growth factor mRNA becomes stable when a certain part of its trailer region has been deleted. Grafting this trailer sequence from growth factor mRNA to globin mRNA makes the globin mRNA unstable. Swapping trailer regions between histone mRNA and globin mRNA renders histone mRNA stable and confers modulated instability on globin mRNA.

Young red blood cells take up large amounts of iron from blood serum for hemoglobin synthesis. At the same time, the intracellular concentration of free iron needs to be kept at nontoxic levels. A key element in iron uptake is a receptor protein in the plasma membrane (Fig. 16.24). It selectively binds a blood serum protein, *transferrin*, which in turn serves as an iron carrier. The synthesis of *transferrin receptor* protein is regulated by the amount of available mRNA. The half-life of receptor mRNA is controlled by a *regulatory protein* using iron as a ligand (Casey et al., 1989). At low cellular iron concentration, the regulatory protein dissociates

from its ligand and assumes its active form. It binds to the *iron response element,* a stem loop in the trailer region of transferrin receptor mRNA. The bound protein protects the transferrin receptor mRNA against intracellular ribonuclease and thus increases the amount of receptor being made. Consequently, more iron is imported into the cell. As the cellular iron concentration increases, it binds to the regulatory protein and renders it inactive. This leaves the iron response element unprotected, and the transferrin receptor mRNA becomes susceptible to RNase breakdown. As a result, the iron concentration is kept at nontoxic levels.

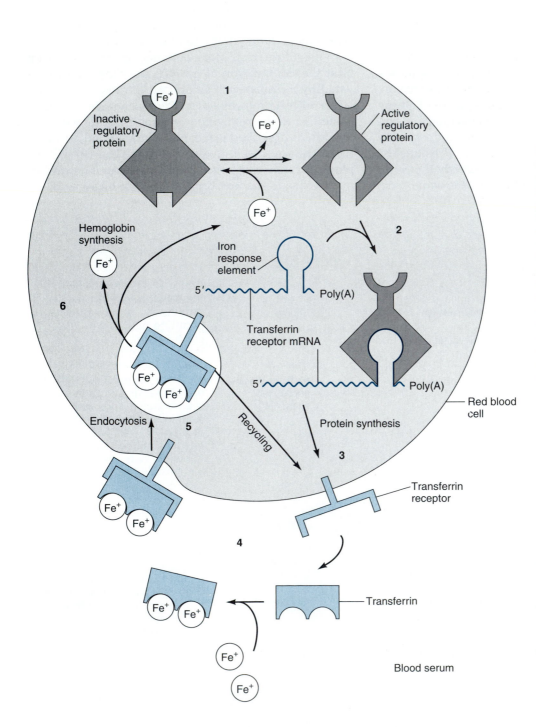

Figure 16.24 The half-life of transferrin receptor mRNA in red blood cells is controlled by a regulatory protein. The protective function of the regulatory protein depends on the absence of iron (Fe^+), which acts as a ligand. At high Fe^+ concentration, the regulatory protein is inactive **(1)**. At low Fe^+ concentration, the regulatory protein binds to a stem-loop structure, called iron response element, in the trailer region of the mRNA **(2)**. The bound protein protects the mRNA against breakdown by intracellular RNases. Conservation of the mRNA enhances the synthesis of transferrin receptor protein, which is inserted into the plasma membrane **(3)**. The receptor binds transferrin, an extracellular protein serving as a carrier of Fe^+ **(4)**. The loaded receptor undergoes endocytosis **(5)**. Transferrin and its receptor are recycled to the plasma membrane. Fe^+ is released into the cytoplasm **(6),** where it is used for hemoglobin synthesis. Any excess Fe^+ inactivates the regulatory protein, rendering it unable to bind to the iron response element. This leaves the transferrin receptor mRNA susceptible to RNase breakdown and thus limits the intracellular Fe^+ concentration to nontoxic levels.

Taken together, these and many other studies show that mRNA longevity is controlled by several destabilizing signals that may be located in various regions of the molecule from its cap to its tail. Some of these signals are nuclease-sensitive sites that can be masked by regulatory proteins, which in turn may be regulated by hormones or other ligands. Other destabilizing signals are recognized by proteins that recruit ribonucleases that will degrade the RNA bound to the recognition protein. Very similar interactions between proteins and their recognition sequences also regulate the *translation* of mRNA, as we will see in the following chapter.

SUMMARY

Newly synthesized nuclear RNA undergoes several modifications before it is exported into the cytoplasm. The processing of pre-mRNA into mRNA includes an association with proteins to form mRNP particles, the addition of a 7-methylguanylate cap at the 5′ end, the polyadenylation of the 3′ end, and the removal of introns. Capping and polyadenylation protect the mRNA against rapid degradation by ribonucleases. For intron removal, pre-mRNA associates with a family of small nuclear RNP particles (snRNPs), which interact to form complexes called spliceosomes. They remove introns from pre-mRNA and splice exons together to form mature mRNA. Most of these processing steps serve as mechanisms for differential gene expression.

During the splicing of two exons, a splice donor site at the 5′ end of the intron is usually combined with a splice acceptor site at the 3′ end of the same intron. However, the blocking of an acceptor site by a regulatory protein may cause the donor site of the intron to be combined with the acceptor site of another intron. The splicing machinery may also deviate from its usual pattern if an intrinsically poor acceptor site is made more attractive by a regulatory protein. Finally, the presence of more than one promoter within the same gene may generate alternative donor sites cooperating with the same acceptor site. These regulatory processes allow cells to generate two or more different mRNAs from one pre-mRNA, an ability called alternative splicing.

Alternative splicing controls gene expression in a variety of organisms at different stages of development. In *Drosophila*, for instance, the development of sexual dimorphism depends on splicing patterns resulting in a regulatory cascade of gene products. At each level of the cascade, a default splicing pattern leads to male development; the female-specific pattern requires an active protein made at the same or a higher regulatory level. Alternative splicing can also generate entirely different proteins in different types of cells. For example, the pre-mRNA transcribed from one rat gene is spliced in two ways, giving rise to either the hormone calcitonin or a neuropeptide.

After pre-mRNPs have been processed into mRNPs, most of the mRNPs are transported into cytoplasm. Nucleocytoplasmic transport provides another opportunity for differential gene expression. This has been shown by monitoring both the processing of individual transcripts and their release into the cytoplasm with cloned cDNA tracers. The ultrastructure of the nuclear envelope, and the restricted movements of macromolecules through nuclear pore complexes, suggest that the nucleocytoplasmic transport of mRNPs is selective and energy-dependent.

The half-life of eukaryotic mRNAs in cytoplasm ranges from minutes to months and is regulated in selective and adaptive ways. The agents of active mRNA degradation are cytoplasmic ribonucleases. These enzymes in turn are regulated by a poly(A)-binding protein and other proteins that recognize certain nucleotide sequences in different regions of the mRNA.

SUGGESTED READINGS

Baker, B. S. 1989. Sex in flies: The splice of life. *Nature* **340**:521–524.

Dingwall, C., and R. Laskey. 1992. The nuclear membrane. *Science* **258**:942–947.

Gall, J. G. 1991. Spliceosomes and snurposomes. *Science* **252**:1499–1500.

Maniatis, T. 1991. Mechanisms of alternative pre-mRNA splicing. *Science* **251**:33–34.

Sachs, A. B. 1993. Messenger RNA degradation in eukaryotes. *Cell* **74**:413–421.

Wickens, M. 1990. In the beginning is the end: Regulation of poly(A) addition and removal during early development. *Trends in Biochemical Science* **15**:320–324.

TRANSLATIONAL CONTROL AND POSTTRANSLATIONAL MODIFICATIONS

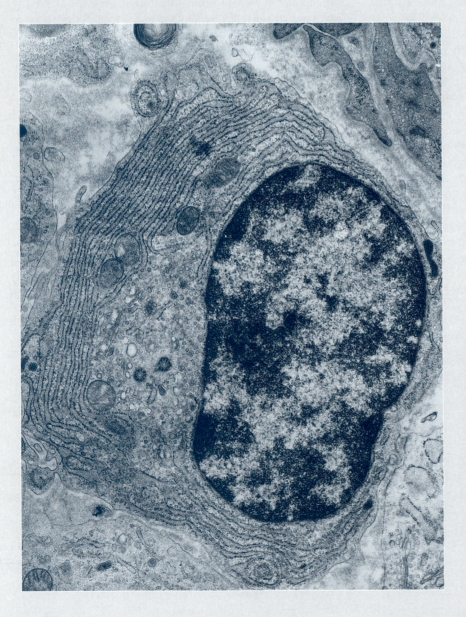

Figure 17.1 This transmission electron micrograph shows a plasma cell from human bone marrow. The large dark organelle to the right is the nucleus. The membranous stacks to the left enclose a system of flat intracellular spaces, known as the rough endoplasmic reticulum (rough ER) because the cytoplasmic side of each membrane is studded with ribosomes, which appear as dark granules. The rough ER is the site of synthesis for proteins that are released from cells. In the case of active plasma cells, these proteins are primarily antibodies.

Cells owe their form and function primarily to proteins. Cytoskeletal proteins such as *tubulins* and *actins* support the cell's shape and its motility. Cell cycle proteins such as *cyclins* control cell divisions. Many proteins are enzymes that serve as the catalysts of cell metabolism. Some of the cell's key regulatory molecules are DNA-binding and RNA-binding proteins. Thus, an organism develops essentially through changes in the protein composition of its cells or regulation of protein activity. Indeed, cells expend much energy synthesizing and modifying proteins (Fig. 17.1).

The overall activity of a given type of protein in a cell depends on how many molecules of the protein are pres-

ent and on the percentage of the molecules that are in their active conformation. In many cases, the number of protein molecules present is controlled at the levels of transcription and RNA processing as discussed in previous chapters. However, additional levels of control operate after mRNAs have been exported to the cytoplasm. *Translational control* mechanisms regulate the translation of mRNAs into polypeptides. *Posttranslational control* mechanisms control modification and assembly of polypeptides to form active proteins.

Translational and posttranslational controls are based on a wider variety of mechanisms than are transcriptional and posttranscriptional control. It appears that many translational and posttranslational controls have evolved to meet the specific needs of certain cell types. We will discuss in particular the "masked messenger RNA" hypothesis. It states that some mRNAs are made temporarily untranslatable by association with "masking" proteins that can be removed quickly and selectively in response to certain signals.

Organisms especially rely on translational and posttranslational control mechanisms when transcriptional control and RNA processing are not operational or would be too slow. Such conditions occur during certain stages of development including egg activation and spermatogenesis.

Animal oocytes are generally provided with large amounts of maternally synthesized RNA that carry them through cleavage. The translation of stored maternal mRNA is carefully regulated. Fully grown oocytes are generally quiescent and preserve the maternal endowment until embryogenesis actually begins. Maturation or fertilization activates the egg and boosts the overall rate of translation. Eggs also have the ability to translate certain maternal mRNAs selectively before oocyte maturation and to translate different mRNAs during cleavage. These changes in protein synthesis occur while the egg nucleus is transcriptionally inactive; even if it were active, the amount of RNA transcribed would be insignificant because of the extremely small *nucleocytoplasmic ratio*. Thus, transcriptional control and RNA processing cannot account for the changes in protein synthesis observed at the time of egg activation.

Spermatogenesis is another developmental process in which translational control is critical. Most sperm undergo dramatic changes in their morphology, which require the synthesis of different proteins at different times. While these changes are taking place, the chromatin in the nucleus is packed very densely, making it difficult or impossible to transcribe. Again, translational control is used to orchestrate protein synthesis during a stage when transcriptional control has been lost.

Toward the end of this chapter, we will survey how newly synthesized proteins undergo posttranslational modifications before they assume their cellular functions.

Some of these modifications direct proteins to their appropriate cellular compartments, such as the endoplasmic reticulum. Other modifications give proteins the three-dimensional conformation and chemical substitutions that are necessary for their biological functions. These modifications also control the power of proteins to assemble spontaneously with other molecules, building complex organelles such as ribosomes and microtubules (see Chapter 18). Posttranslational modifications occur selectively in certain proteins and promote their particular functions and assembly properties.

Finally, we will examine how the longevity of proteins is controlled in cells. As with mRNA, we will see that the breakdown of proteins is an active, adaptive enzymatic process carried out in a sequence-specific way.

Formation of Polysomes and Nontranslated mRNP Particles

During *translation*, the nucleotide sequence of the mRNA directs the formation of a *polypeptide* from amino acids. This process depends on the ability of

transfer RNAs (tRNAs) to bind, or load, a specific amino acid and to recognize a corresponding nucleotide triplet (codon) in an mRNA. Translation also relies on the ability of ribosomes to assemble two loaded tRNAs in adjacent positions.

In each polypeptide, the first amino acid has a free amino group that marks the *N terminus*, and the last amino acid has a free carboxyl group that marks the *C terminus*. In this chapter, the term "polypeptide" will be used for the immediate product of translation, and the term "protein" for processed and assembled polypeptides. In contexts where this distinction is not important, the shorter term "protein" will be used for both proteins and polypeptides.

Most mRNAs Are Immediately Recruited into Polysomes and Translated

In most cell types, mRNAs released from the cell nucleus are immediately translated into polypeptides. The process of *translation* consists of three phases: initiation, elongation, and termination. The most important phase for the control of translation is *initiation,* in which the mRNA and the first loaded tRNA are bound to a ribosome (Fig. 17.2). This process involves the *initiation codon (AUG)* of the mRNA, the small ribosomal subunit, a special initiator tRNA, and several catalytic pro-

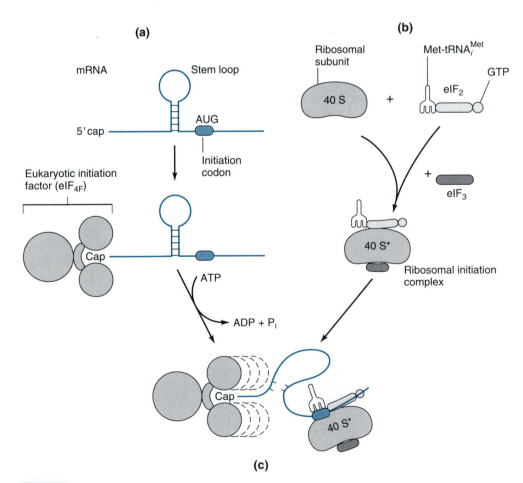

Figure 17.2 Model of initiation of protein synthesis. **(a)** Several proteins collectively known as eukaryotic initiation factor 4F (eIF$_{4F}$) associate with the cap of the mRNA, forming a cap-binding complex. This complex promotes the unwinding of any stem-loop structures upstream of the initiation codon (AUG). **(b)** The small ribosomal subunit (40 S) associates with additional eukaryotic initiation factors and a loaded tRNA to form the initiation complex. The first loaded tRNA (Met-tRNA$_i$^{Met}) is a special initiator tRNA, which is always loaded with the amino acid methionine. **(c)** When stem loops or other secondary structures near the initiation site have been unwound, the initiation complex binds to the initiation codon (AUG).

teins called *eukaryotic initiation factors (eIFs)*. The first initiation step includes the unwinding of any stem loop or other secondary structures in the leader region of the mRNA. These structures otherwise tend to block initiation, especially if they are stabilized by regulatory proteins. The unwound mRNA associates with an *initiation complex* consisting of the small ribosomal subunit, *methionyl-initiator transfer RNA (met-tRNA),* and further initiation factors. The initiation complex travels down the mRNA until the met-tRNA is positioned over the *translation initiation codon,* AUG. Finally, the large ribosomal subunit joins the properly positioned initiation complex, thus completing the initiation step.

During the next translation phase, called *elongation,* the ribosome travels along the mRNA. At each elongation step, the ribosome binds a newly loaded tRNA next to one bound previously (Fig. 17.3a). This arrangement promotes the formation of another peptide bond between the new amino acid and the growing polypeptide. As the ribosome vacates the initiation site, another ribosome can be assembled there. In this fashion, mRNAs associate with several ribosomes to form a polyribosome, or *polysome.*

The last phase of translation, *termination,* occurs when the ribosome encounters one of the three *stop codons,* none of which is recognized by a tRNA. Instead, proteins known as release factors bind to the stop codon and release the polypeptide from the ribosome (Fig. 17.3b). Next, the ribosome dissociates into its two subunits, which can then engage in another initiation event.

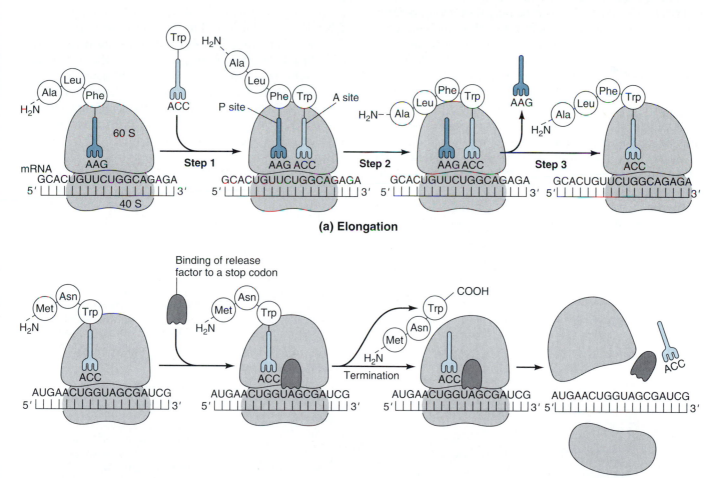

(a) Elongation

(b) Termination

Figure 17.3 Elongation and termination phases of protein synthesis. These steps involve three binding sites of the ribosome: the site for the mRNA, the aminoacyl-tRNA binding site (A site), and the peptidyl-tRNA binding site (P site). **(a)** During elongation, a three-step cycle is repeated over and over. In step 1, a new tRNA loaded with an amino acid binds to the A site of the ribosome. Meanwhile, the growing polypeptide chain is still bound to the P site. In step 2, a new peptide bond is formed between the polypeptide and the newly bound amino acid held by the tRNA in the A site. In step 3, this tRNA moves from the A site to the P site, ejecting the old tRNA. **(b)** Termination occurs when the A site is lined up with any of three stop codons: UAA, UAG, or UGA. Each stop codon binds to a release factor instead of a loaded tRNA. This causes translation to stop, the growing polypeptide to be released, and the ribosomal subunits to dissociate from the mRNA.

The rates of elongation and termination are not known to vary significantly from one mRNA to another. In contrast, the rate of initiation varies considerably among different mRNAs. Most studies on translational regulation have therefore focused on the initiation step.

Some mRNAs Are Stored as Nontranslated mRNP Particles

Some cell types, including gametes and early embryonic cells, can store mRNAs in the form of nontranslated *messenger ribonucleoprotein (mRNP)* particles. In these particles, mRNAs are complexed with proteins and possibly other components that render the mRNPs temporarily nontranslatable. Messenger RNP particles are also called *subribosomal RNP particles* or *postribosomal RNP particles*, because of the way they separate from polysomes and ribosomal subunits when they are centrifuged on sucrose gradients. In the centrifugal field, these fractions separate according to their velocity of sedimentation, with poly-somes sedimenting fastest, followed by ribosomal subunits and a broad band of subribosomal RNP particles (Fig. 17.4).

Analysis of subribosomal RNP particles reveals that they represent temporary storage forms of mRNAs. The *existence* of mRNA molecules in mRNP particles is shown by the presence of poly(A) tails in the RNA moieties extracted from mRNP particles, and by the ability of the RNA moieties to hybridize with specific cDNA probes. The *translatability* of mRNA extracted from sub-

ribosomal mRNP particles has been tested in *cell-free translation systems*. Such systems contain all components required for protein synthesis—including ribosomal subunits, tRNAs, amino acids, and initiation factors—*but not mRNAs*. Cell-free translation systems are prepared by breaking up actively translating cells, such as immature red blood cells or wheat germ cells, and removing their nuclei, their membranous organelles, and their own mRNAs. Any protein synthesis in these systems then depends on the addition of mRNA from another source. The process of translating mRNA in a cell-free system is called *in vitro translation*. When mRNAs are extracted from mRNP particles and added to cell-free translation systems, they stimulate protein synthesis. Therefore, subribosomal RNP particles are known to contain fully functional mRNAs.

Results obtained from gradient centrifugation, cDNA hybridization, and in vitro translation prove that gametes and early embryos contain nontranslated mRNP particles. These cells can control the translation of mRNAs by shifting them from subribosomal mRNP particles, where they are not translated, to polysomes, where they are translated. The shift is apparently caused by the association of mRNAs with different sets of proteins or other regulatory factors. One set of factors, which presumably blocks the initiation site of mRNAs, keeps them in their storage form as untranslated mRNP particles. Another set of associated factors enhances the mRNAs' recruitment into polysomes and subsequent translation.

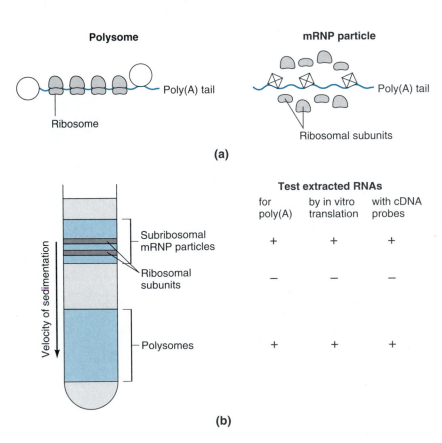

(a)

(b)

Figure 17.4 Analysis of messenger ribonucleoprotein (mRNP) particles. **(a)** Schematic diagram showing cytoplasmic mRNAs associated with two sets of proteins or other factors. One set of factors (open circles) makes mRNAs available for recruitment into polysomes. With other factors (crossed boxes), they form temporarily untranslatable mRNP particles. **(b)** Centrifugation on a sucrose gradient separates polysomes, ribosomal subunits, and subribosomal mRNP particles (so named because some of them sediment behind the ribosomal subunits). Each fraction can be assayed for the presence of mRNA by extracting RNA from it and testing the RNA moiety for the presence of poly(A) tails, translatability in vitro, and hybridization with cloned cDNA probes. On the basis of these criteria, mRNA has been found in both polysomes and subribosomal RNP particles.

Mechanisms of Translational Control

Unlike transcriptional control, translational control is based on a wide variety of molecular mechanisms. Some of these are more suited to regulate the *overall* rate of protein synthesis, while others control—directly or indirectly—the translation of *specific* mRNAs.

Some Translational Control Mechanisms Regulate the Overall Rate of Protein Synthesis

Control mechanisms affecting the overall rate of protein synthesis use signals that do not require the recognition of specific RNA sequences.

Free Calcium Ion Concentration and Intracellular pH   Fertilization in sea urchin eggs triggers the *activation* process, which entails many metabolic changes including an acceleration in the rate of protein synthesis (see Chapter 4). Two key steps in the activation process are an increase in the concentration of free calcium ions (Ca^{2+}) and an elevation of the intracellular pH. These changes are thought to promote the shift of mRNAs from subribosomal RNPs to polysomes, and to accelerate elongation (Winkler, 1988). However, the intermediate steps linking these events are as yet unknown.

Availability of Initiation Factors   Regulating the availability of any of the translation initiation factors is an effective way of controlling protein synthesis. Although strictly sequence-specific initiation factors are not known, some mRNAs have been found to have different affinities for the same initiation factor. Therefore, any change in the availability of an initiation factor affects both the absolute and the relative rates of protein synthesis. Either event may severely upset the control mechanisms of a cell. For instance, overexpression of a specific initiation factor, eIF_{4E}, causes cultured cells to form tumors (Lazaris-Karatzas et al., 1990).

A well-investigated example of translational control due to the availability of an initiation factor is the synthesis of hemoglobin, the most abundant protein in the red blood cells of vertebrates. The hemoglobin molecule consists of four polypeptides, two α-globins and two β-globins. Each of the globins is associated with a prosthetic (helper) group, *heme*, which mediates respiratory exchange by associating with oxygen or carbon dioxide. The heme molecule plays a key role in regulating the overall amount of hemoglobin per cell, and in establishing the correct heme:globin ratio. Heme limits its own synthesis by means of a negative feedback loop: excess heme inhibits δ-aminolevulinate synthetase, the enzyme catalyzing the first step of heme synthesis from glycine and succinyl coenzyme A. In addition, heme stimulates the translation of globin polypeptides from their mRNAs (Fagard and London, 1981; Grace et al., 1984). It accomplishes this task indirectly by keeping the initiation factor eIF_2 available, as illustrated in Figure 17.5.

Polyadenylation and Deadenylation Control the Translation of Specific mRNAs

Translational control mechanisms that regulate the synthesis of *specific* proteins do so by recognizing distinct RNA sequences. Just as some regulatory proteins recognize certain DNA motifs, other regulatory proteins interact with specific RNA sequences. Some of these proteins have an *RNA-binding domain* that is highly conserved in evolution. We have already discussed the involvement of such proteins in RNA splicing and in controlling the longevity of mRNAs (see Chapter 16). Similar proteins bind to mRNAs in ways that, directly or indirectly, regulate translation.

In many developing systems, there is a direct correlation between the length of the poly(A) tail of certain mRNAs and their active translation. In oocytes of the clam *Spisula*, Eric Rosenthal and coworkers (1983) traced individual mRNAs for which cloned cDNAs were available. They found four mRNAs that were inactive in the oocyte but were recruited into polysomes after fertilization. All four of these mRNAs had very short poly(A) tails in the oocyte, which lengthened considerably after fertilization. Conversely, one mRNA was long-tailed

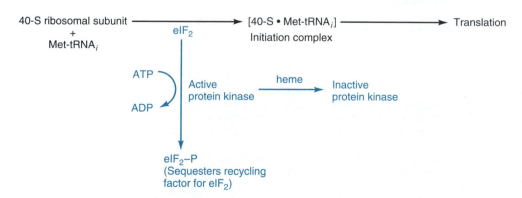

Figure 17.5 Translational control of globin synthesis by its prosthetic group, heme. Heme inhibits a protein kinase that would otherwise phosphorylate and thereby inactivate the initiation factor eIF_2 (see Fig. 17.2). Phosphorylated eIF_2 would bind up a recycling factor for eIF_2 and thus bring translation to a halt.

and actively translated in the oocyte and became completely deadenylated and disappeared from polysomes after fertilization. These results revealed a striking correlation between the polyadenylated state and translational activity, although the correlation did not hold for all mRNAs investigated. These results imply that poly(A) polymerase, the enzyme that extends the poly(A) tail, is not limited to the nucleus, where it is part of the machinery that processes pre-mRNA into mRNA. Instead, enzymes that extend and shorten poly(A) tails of mRNAs must also be present in the cytoplasm.

A correlation between polyadenylation and translational activation is also found in *Xenopus* oocytes. The mRNAs synthesized during oogenesis can be grouped broadly into two classes (R. J. Jackson, 1993). One class is represented by actin and ribosomal protein mRNAs, which have long poly(A) tails and are actively translated during oogenesis. But shortly after the beginning of egg activation, their poly(A) tails are shortened, and their translation ceases. Other mRNAs are stored in the oocyte as nontranslated mRNPs with short poly(A) tails; these mRNAs undergo polyadenylation and transcriptional activation either during egg activation or upon fertilization.

▼

A representative of the second class of mRNAs, termed G10, was investigated by Lynn McGrew and her colleagues (1989). G10 mRNA is recruited for translation during oocyte maturation. Coincident with its translation, the poly(A) tail of this mRNA is elongated from about 90 to 200 adenylate residues, while other mRNAs remain unchanged. To identify the mRNA sequence that is required for this specific polyadenylation and translation, the researchers synthesized radiolabeled G10 mRNAs that were lacking certain segments and injected these defective transcripts into *Xenopus* oocytes. The oocytes were then induced to mature with progesterone, and the injected mRNAs were monitored for size and association with polysomes (Fig. 17.6). This deletion analysis identified a 50-nucleotide sequence including the AAUAAA polyadenylation signal in the *trailer region* of the G10 mRNA, which was necessary for polyadenylation and translation. The same sequence, when fused with globin mRNA, was also sufficient to confer polyadenylation and translation. Further results indicated that the 50-nucleotide sequence directed primarily polyadenylation, and that it was the polyadenylation process rather than the result of a long poly(A) tail that promoted translation.

Subsequent studies revealed that the synthesis of c-mos and cyclins, proteins involved in oocyte maturation, is also regulated by polyadenylation (Sheets et al., 1994). The trailer regions of the corresponding mRNAs control not only when these mRNAs are translated but also at which rate. In each case, the translational activation requires polyadenylation: if polyadenylation is prevented by modifying the polyadenylation signal or adjacent sequences, translation is not activated.

Immediately upstream of the polyadenylation signal, many mRNAs have uridine-rich sequences known

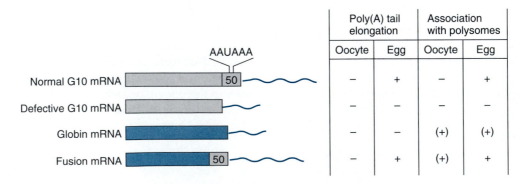

	Poly(A) tail elongation		Association with polysomes	
	Oocyte	Egg	Oocyte	Egg
Normal G10 mRNA	–	+	–	+
Defective G10 mRNA	–	–	–	–
Globin mRNA	–	–	(+)	(+)
Fusion mRNA	–	+	(+)	+

Figure 17.6 Correlation between polyadenylation and translational activation of G10 mRNA in *Xenopus* eggs. During oocyte maturation, normal G10 mRNA undergoes an extension of its poly(A) tail and is recruited into polysomes. Defective G10 mRNA lacking a 50-nucleotide sequence including the polyadenylation signal does not undergo poly(A) extension and is not recruited into polysomes. Normal globin mRNA undergoes no poly(A) tail extension at maturation, and only a small fraction of this mRNA is associated with polysomes in both oocytes and eggs. However, if truncated globin mRNA is fused with the 50-nucleotide sequence from G10 mRNA, then the fusion RNA behaves like G10 mRNA during oocyte maturation: its poly(A) tail is extended and it is strongly recruited into polysomes.

as *adenylation control elements,* or *cytoplasmic poly-adenylation elements.* These elements appear to be bound by proteins that regulate polyadenylation and de-adenylation. The nature of these regulatory proteins and the control of their activity by other signals are still under investigation.

The mechanism that links polyadenylation with translational activation is also currently being studied. As described in Chapter 16, the poly(A) tail is tightly associated with a *poly(A)-binding protein,* which protects mRNAs against premature breakdown. The same protein seems also involved in the last step of translation initiation, the joining of the large ribosomal subunit (R. J. Jackson and N. Standart, 1990). The implied cooperation of the two ends of the mRNA during the initiation step may seem counterintuitive, but it would guarantee that only full-length, translation-competent mRNA complete the initiation step.

"Masking" Proteins Bind to the Leader Regions of Specific mRNAs and Inhibit Their Translation

As mentioned earlier, part of the mRNA in eggs and other cells is stored in the form of temporarily untranslated mRNP particles. Researchers have spent much effort to find out what prevents the mRNAs stored in these particles from being translated. According to the *masked mRNA hypothesis,* the mRNAs contained in mRNP particles are *untranslatable* because their association with certain proteins leaves them inaccessible to initiation factors. Alternatively, the mRNAs stored in mRNP particles may be *untranslated by default,* simply because the cell does not have enough initiation factors to recruit them into polysomes.

To test the masked mRNA hypothesis, Joel Richter and L. Dennis Smith (1984) prepared subribosomal RNP particles from young *Xenopus* oocytes and isolated five major proteins from them (Fig. 17.7). These proteins were mixed with oocyte mRNA to allow the reconstitution of mRNP particles. When these particles were injected into mature oocytes, they were poorly translated compared with naked oocyte mRNA. The mRNP particles reconstituted with oocyte proteins were also less active than control mRNP particles reconstituted with proteins from other frog tissues. The investigators concluded that proteins found in subribosomal mRNP particles from oocytes specifically inhibit the translation of the associated mRNAs.

A logical binding site for a masking protein would seem to be the leader region, where it would impair formation of the initiation complex. In accord with this hypothesis, the leader regions of many mRNAs have stem loops and/or other secondary structures that are known to be bound by regulatory proteins. An example is the

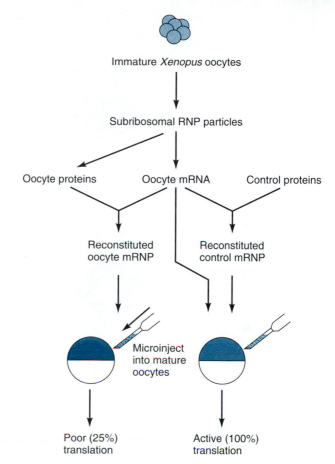

Figure 17.7 Demonstration of "masking" proteins that inhibit mRNA translation. In this experiment, subribosomal RNP particles were prepared from immature *Xenopus* oocytes. From these RNP particles, five oocyte proteins were isolated. These proteins were combined with oocyte mRNA to reconstitute oocyte mRNP particles. When injected into mature *Xenopus* oocytes, these mRNPs translated poorly in comparison with naked oocyte mRNA and with control mRNP reconstituted from oocyte mRNA and control proteins from other *Xenopus* tissues (liver, leg muscle, or heart).

mRNA encoding *ferritin,* a protein that stores iron resorbed from food in intestinal cells. Ferritin mRNA has a stem loop in its *leader* region, which interacts with a regulatory protein (Fig. 17.8). Binding of the protein to the stem loop inhibits translation of ferritin mRNA (Goossen et al., 1990).

The regulatory protein binding to ferritin mRNA is the same *iron-responsive regulatory protein* discussed previously in the context of the controlled degradation of transferrin receptor mRNA (see Fig. 16.24). Also, the stem-loop structures in the leader region of ferritin mRNA and the trailer region of transferrin receptor mRNA are almost identical *iron-response elements.* The interactions between the two iron-responsive elements and their regulatory protein have synergistic effects on the concentration of iron in blood plasma and blood cells. At low iron concentration, the regulatory protein

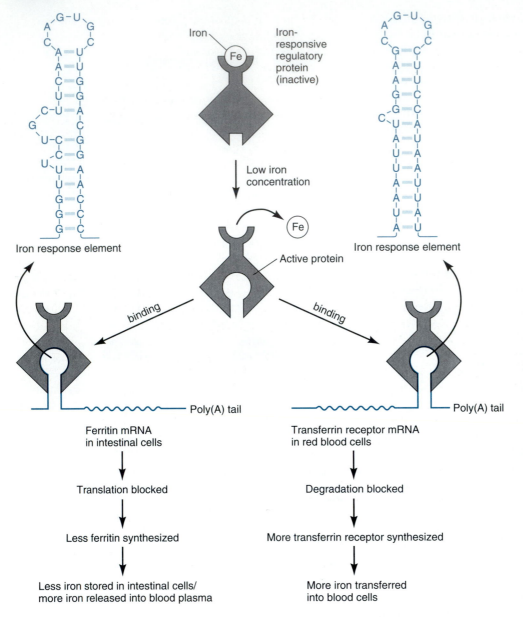

Iron response element

Iron response element

Ferritin mRNA
in intestinal cells

Transferrin receptor mRNA
in red blood cells

Translation blocked

Degradation blocked

Less ferritin synthesized

More transferrin receptor synthesized

Less iron stored in intestinal cells/
more iron released into blood plasma

More iron transferred
into blood cells

Figure 17.8 The translation of ferritin mRNA and the degradation of transferrin receptor mRNA are controlled by the same iron-responsive regulatory protein binding to nearly identical stem-loop structures (i.e., iron response elements) in the two mRNAs. At low iron concentration, the iron-responsive regulatory protein releases bound iron and becomes capable of binding to the mRNA response elements. The response element of ferritin mRNA is located in the leader region; binding of the iron-responsive protein inhibits translation of the mRNA. The response element of transferrin receptor mRNA is located in the trailer region; binding of the iron-responsive protein inhibits degradation of the mRNA. The overall effect is to increase the free iron concentration in the blood plasma and in the blood cells.

binds to both response elements. The resulting inhibition of ferritin mRNA translation means that intestinal epithelial cells synthesize less iron storage protein and therefore release more iron into blood serum. At the same time, red blood cells take up more iron by blocking the degradation of transferrin receptor mRNA.

In summary, there are a wide variety of translational control mechanisms, many of which are also involved in regulating other steps of gene expression. General signals, such as the intracellular pH or the availability of initiation factors, control the overall rate of protein synthesis. Regulatory proteins binding to specific RNA sequences control the translation of individual mRNAs. Some of these proteins bind to recognition sites in mRNA trailer regions and seem to affect translation via polyadenylation. Other regulatory proteins, binding to recognition sites in mRNA leader regions, interfere directly with translation initiation.

Translational Control in Oocytes, Eggs, and Embryos

Translational control is a common way of regulating gene expression in oocytes, eggs, and early embryos. As discussed in Chapter 3, large amounts of RNAs, including mRNAs, are deposited in growing oocytes. These RNAs are stored to support development during the phase of rapid cleavages, when most embryos do not transcribe enough RNA from their own genome. It is critical to conserve these stores during the time before fertilization. Thus, molecular mechanisms that keep oocyte mRNAs in an untranslated state until embryonic development is imminent would be highly adaptive. Observations on eggs from different animal species indicate that such mechanisms are indeed widespread.

In this section, we will discuss evidence that maternally encoded mRNA is translated during early embryonic development. We will see that normal development in the early stages, including protein synthesis, does not require transcription of the embryonic genome. Next we will consider translational control as a *general* mechanism affecting the overall rate of protein synthesis in oocytes and eggs. Then we will examine data showing that translational control in eggs can be *sequence-specific*, that is, that it can inhibit the synthesis of specific proteins while allowing the synthesis of others.

Early Embryos Use mRNA Synthesized during Oogenesis

Interspecies Hybrids The control of early embryonic development by maternally derived cytoplasmic factors was first observed in embryos obtained by crossing different animal species. These hybrid embryos were monitored for the expression of maternal, paternal, or intermediate traits. Working with sea urchin hybrids, the German embryologist Hans Driesch (1898) observed that the number of *primary mesenchyme* cells formed at the beginning of gastrulation was characteristic of the species that contributed the *egg* and was unaffected by the species contributing the sperm (Table 17.1). When reexamining the same hybrids at the pluteus stage, after the skeletal rods had formed, Driesch found characteristics of *both* parental species. These results showed that early embryonic traits were controlled only by maternal factors while traits expressed later were influenced by both parents.

Maternal Effect Mutants The control of early embryonic development by maternal gene products is confirmed by the effects of certain mutations on development. Most embryonic phenotypes are controlled by both the maternal *and* the paternal genotype. However, in *maternal effect mutants*, the embryonic phenotype is controlled *only by the maternal* genotype, and is completely independent of the paternal genotype. Examples of maternal effect genes include the gene controlling the orientation of mitotic spindles in snail embryos (see Chapter 5) and the *bicoid+* gene of *Drosophila* (see Chapters 8, 15, and 21). Genetic analysis of these maternal effect mutants shows that the embryonic phenotype depends on *all alleles* (wild-type or mutant) of the gene present in the *mother*, not just on the *one* allele inherited by each *embryo*. For instance, *all* offspring of *bicoid*/+ mothers develop normally, not just the 50% that inherit the wild-type allele. These observations indicate that maternal effect genes control embryonic phenotypes through maternally encoded *gene products* (RNA and/or protein) that are stored during oogenesis.

Enucleated Eggs The use of mRNA accumulated during oogenesis is also demonstrated by the development of egg fragments without nuclei. Such fragments can be obtained by centrifugation of sea urchin eggs in a sucrose solution (Fig. 17.9). The egg contents then stratify under the influence of the centrifugal force until they are torn in half. One half contains the egg nucleus, and the other half is enucleated. When placed in hypotonic seawater to activate them, *both* halves develop. The enucleated half undergoes several rounds of cytokinesis, forming an irregular blastula that hatches successfully (Harvey, 1940). Activated enucleated eggs synthesize proteins at a rate comparable with their nucleated counterparts (Denny and Tyler, 1964). These results indicate that the cytoplasm of sea urchin eggs contains maternally encoded mRNAs and other components for synthesizing proteins. These components can support early development without the presence of nuclear DNA.

Inhibition of mRNA Synthesis An alternative to physical enucleation is a biochemical experiment in which eggs are incubated with an inhibitor of transcription, such as actinomycin D. Using this strategy,

---

TABLE 17.1

Primary Mesenchyme Cells in Sea Urchin Hybrids

Egg		Sperm	Average Number of Primary Mesenchyme Cells*
Echinus	×	*Echinus*	55 ± 4
Spherechinus	×	*Spherechinus*	33 ± 4
Spherechinus	×	*Echinus*	35 ± 5
Strongylocentrotus	×	*Strongylocentrotus*	49 ± 3
Spherechinus	×	*Strongylocentrotus*	33 ± 3

*The mesenchyme cells of 15, 25, 47, 15, and 22 embryos were counted in the five samples, respectively. The average and *range* of the counts are given.
Source: Davidson (1976), after data of Driesch (1898). Used with permission.

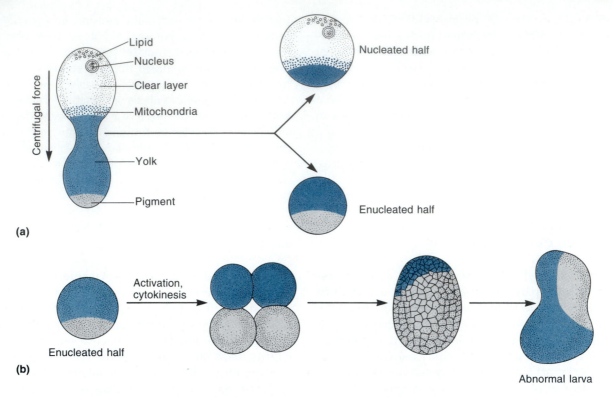

(a)

(b)

Abnormal larva

Figure 17.9 Development of enucleated egg halves. **(a)** When sea urchin eggs are centrifuged in sucrose solution, their components stratify under the influence of the centrifugal force. The half containing the light components of the cytoplasm is pulled to the top, while the other half, containing the heavy cytoplasmic components, is pulled to the bottom. The upper half retains the egg nucleus. **(b)** The enucleated lower half may nevertheless develop to stages resembling an abnormal blastula or larva.

Gross and Cousineau (1964) fertilized sea urchin eggs in water containing enough actinomycin D to shut down almost all RNA synthesis. Nevertheless, the eggs developed and synthesized protein at the same rate as controls treated equally but without the inhibitor (Fig. 17.10). Both samples of eggs developed to the blastula stage and hatched normally. From this stage on, the treated sample ceased to develop, and it did not undergo a second burst of protein synthesis as the controls did. Corresponding experiments with inhibitors of protein synthesis caused immediate arrest of development.

Mid-Blastula Transition Revisited The point at which many embryos begin to replace maternal mRNA with their own occurs during the *mid-blastula transition (MBT)*. Before MBT, embryos undergo very rapid mitoses, which leave little or no opportunity for RNA synthesis (see Chapter 5). Even if some RNA is synthesized, the amount produced by the few nuclei present in the embryo at this stage is generally insufficient to produce the proteins needed to support the relatively large masses of cytoplasm surrounding them. Nevertheless, many proteins are synthesized during this time, some of them in large quantities. The synthesis of these proteins relies initially on maternally supplied RNA. At

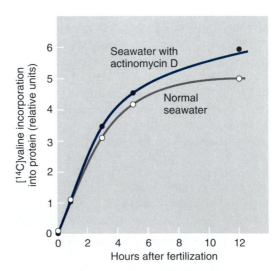

Figure 17.10 Protein synthesis in the absence of RNA synthesis in the sea urchin *Arbacia punctulata*. In this experiment, eggs were fertilized in seawater containing actinomycin D, an inhibitor of RNA synthesis. At the concentration used, actinomycin D almost completely inhibited mRNA synthesis. Control eggs from the same batch were fertilized in normal seawater. The overall amount of protein synthesized in each batch was measured by incorporation of radiolabeled valine. Until the mid-blastula stage, there was no significant difference in the amounts of protein synthesized.

MBT, when an embryo has produced many nuclei and mitosis has slowed down, the embryo can transcribe enough RNA from its own genome, and the maternal RNA supply is degraded. Typically, MBT occurs during late cleavage, but the timing of the transition varies among species, and not all animals have a conspicuous MBT.

Multiple Regulatory Mechanisms Limit Translation in Oocytes

In most animal species, fully grown oocytes are kept in a quiescent state until oocyte maturation or fertilization triggers an *activation* response including a steep increase in the overall rate of protein synthesis. This surge is accompanied by a rapid recruitment of maternal mRNAs from subribosomal mRNP particles into polysomes and by a corresponding increase in the percentage of all ribosomes that are integrated in polysomes.

The molecular mechanisms limiting translation in preactivation oocytes have been a subject of controversy. According to the *masked messenger RNA hypothesis,* most of the mRNA in these oocytes is *nontranslatable* because certain proteins block mRNA sites critical to translation. Alternatively, the rate of protein synthesis may be limited by the availability of an initiation factor or some other component of the translational machinery. It is important to realize that these two hypotheses are not mutually exclusive. Rather, the *principle of synergistic mechanisms* (see Chapter 4) suggests that a feature as vital as conserving the protein-synthesizing capacity of an egg might be secured by more than one molecular mechanism. This expectation is confirmed by the following experiments on frog and sea urchin eggs.

The reconstitution experiments discussed previously show that mRNP particles from frog oocytes contain specific proteins that inhibit translation (Fig. 17.7). As a direct test for masked mRNA in sea urchins, several investigators have compared the translatability of subribosomal RNP particles and purified mRNA from unfertilized eggs in cell-free translation systems. In such an experiment, James L. Grainger and Matthew Winkler (1987) prepared a cell-free translation system from red blood cells that was *known to be limited by the availability of translatable mRNA.* They also used a gentle preparation technique for sea urchin mRNPs that left them in as natural a state as possible. Under these conditions, polysomes and deproteinized mRNA stimulated the cell-free translation system while subribosomal mRNPs were inactive. The researchers concluded that most of the maternal mRNAs present in sea urchin eggs are associated with proteins or other factors that inhibit their translation.

Other results indicate that the translational capacity of both frog oocytes and unfertilized sea urchin eggs is limited not only by the availability of mRNA but also by other factors. Microinjecting *Xenopus* oocytes with additional mRNA from other cells does not increase the oocytes' overall protein synthesis. Some exogenous mRNAs are translated, but only at the expense of the endogenous mRNAs (Laskey et al., 1977). Corresponding microinjections of sea urchin eggs would be tedious because of their small size. (A thousand sea urchin eggs have about the same volume as one frog egg.) However, a cell-free system that was prepared from unfertilized sea urchin eggs and included endogenous mRNPs and polysomes has been studied instead (Winkler et al., 1985). This system translates exogenous rabbit globin mRNA only at the expense of the endogenous sea urchin mRNAs. While exogenous *mRNA* does not stimulate the cell-free sea urchin system, the addition of exogenous *initiation factors* from rabbit reticulocytes does.

Together, the results obtained with frog and sea urchin eggs show that protein synthesis in unfertilized eggs is kept at a low level by both the eggs' limited translational capacity and the masking of maternal mRNA. Both types of block are released as part of the egg activation process. The molecular mechanisms underlying the release are still being investigated.

Specific Messenger RNAs Are Shifted from Subribosomal mRNP Particles to Polysomes during Oocyte Maturation

A fundamental question in the analysis of the translational control is whether or not it can be *sequence-specific.* If so, then a developing organism could adjust both the overall rate of protein synthesis *and* the kinds of proteins made to meet the requirements of different stages. This ability would most likely be adaptive during oocyte maturation and fertilization. These events mark the transition from the ovarian to the embryonic environment, and from growth to cleavage. The proteins required before and after this transition are different, although both sets of proteins must be translated from the same pool of maternal mRNAs. The following studies focus on the role of sequence-specific translational control during this transition.

▼

Convincing evidence for the sequence-specific control of translation in oocytes and early embryos was obtained in experiments with the surf clam *Spisula solidissima.* (In *Spisula,* oocytes are fertilized while still in meiotic prophase I. The events normally referred to as oocyte maturation, beginning with germinal vesicle breakdown, start 10 min after fertilization.) Eric Rosenthal and coworkers (1980) compared proteins synthesized in unfertilized oocytes, in fertilized eggs during maturation, and in embryos after first cleavage. Each stage was incubated for 20 min with radioactive amino acids; this method labels only proteins synthesized during the time that the radioactive precursor is present. Following the extraction and separation of all proteins by electrophoresis, the labeled proteins were identified by autoradiography (see Methods 3.1). At least three proteins (designated X, Y, and Z) were synthesized abundantly in oocytes, but much less so in eggs or embryos (Fig. 17.11). Conversely, at least three other proteins (designated A, B, and C) were synthesized abundantly in eggs and embryos, but only sparsely in oocytes. Proteins A and B were later identified as *cyclins* A and B, two major proteins that control the cell cycle during cleavage. Protein C turned out to be the small subunit of ribonucleotide reductase, a major enzyme required for DNA synthesis during cleavage. Each of these proteins is required in quantity during cleavage but not during oogenesis.

Thus, the pattern of proteins synthesized in *Spisula* eggs changed significantly after fertilization. Notably, there was a marked increase in the synthetic rate of some proteins needed specifically during cleavage.

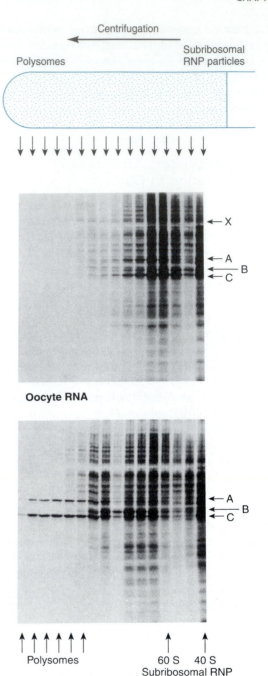

Oocyte RNA

Embryonic RNA

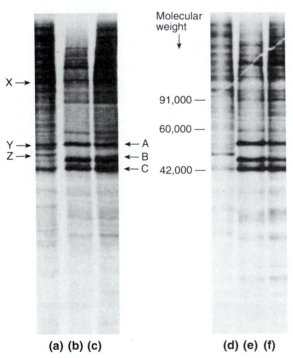

(a) (b) (c)   **(d) (e) (f)**

Figure 17.11 Sequence-specific translational control in the surf clam *Spisula solidissima*. The autoradiographs show proteins synthesized by oocytes, eggs, and embryos labeled for 20 min with [^{35}S]methionine **(a–c)** or [^{3}H]leucine **(d–f)**. **(a, d)** Oocytes; **(b, e)** eggs labeled after germinal vesicle breakdown; **(c, f)** embryos labeled after first cleavage. The locations of molecular weight markers (91,000, 60,000, and 42,000) are indicated. Proteins X, Y, and Z were synthesized specifically in oocytes but not in eggs and embryos. Conversely, proteins A, B, and C (later identified as cyclin A, cyclin B, and ribonucleotide reductase) were synthesized in eggs and embryos but not in oocytes.

Figure 17.12 Sequence-specific shift of mRNAs from subribosomal RNP particles to polysomes in *Spisula solidissima*. In this experiment, polysomes and subribosomal RNP particles were separated by sucrose gradient centrifugation. RNA was extracted from each gradient fraction and translated in vitro. The diagram on top shows the position of the fraction from which the mRNA for each lane was collected. Proteins A through C and X through Z are the same as in Figure 17.11. The mRNAs encoding proteins A and C were present in subribosomal mRNP particles at the oocyte stage and in polysomes at the embryonic stage. The mRNA encoding protein B underwent only a minor shift, and the causes for this different behavior were not clear. The mRNA encoding protein X was present in some of the oocyte polysomal fractions. However, the protein was not synthesized abundantly enough to be detected at the embryo stage. Proteins Y and Z were not detected at either stage.

To determine whether the changes in the pattern of protein synthesis were associated with any changes in mRNA synthesis or degradation, the investigators prepared *total* mRNA from oocytes, eggs, and embryos. Samples from each mRNA preparation were translated in a cell-free translation system. The results were striking: mRNAs from all stages directed the synthesis of *the same* proteins in vitro. In particular, those proteins that had shown major synthetic rate changes *in vivo* were synthesized at the same rate *in vitro* using mRNA from all stages. This outcome showed that oocytes, eggs, and embryos contained *identical* sets of mRNAs. The investigators concluded that some mRNAs formed variable associations with other factors in vivo that rendered them translatable only at specific stages.

According to the masked mRNA hypothesis, the changing patterns of in vivo protein synthesis should be associated with shifts of maternal mRNAs between *polysomes* and *subribosomal RNP particles*. To test whether they were, Rosenthal and coworkers (1980) prepared polysomal and subribosomal RNP fractions from *Spisula* oocytes, eggs, and embryos. From each preparation, RNA was extracted and translated in vitro. As expected, mRNAs encoding the proteins A, B, and C described previously were present in subribosomal RNP particles at the oocyte stage and in polysomes at the embryonic stage (Fig. 17.12). Moreover, the mRNA encoding protein X was present in oocyte polysomal fractions. Because of quantitative limitations, it was not clear whether the mRNAs encoding the oocyte-specific proteins were later degraded or stored in subribosomal mRNP particles.

Similar shifts of mRNAs from subribosomal mRNP particles to polysomes during egg maturation have also been observed in other organisms. In a corresponding study on *Drosophila* oocytes and embryos, J. Mermod and coworkers (1980) focused on proteins that bind to DNA and are therefore presumed to have gene-regulatory functions. To distinguish these proteins, the researchers prepared mRNA from polysomes and subribosomal mRNPs of oocytes and embryos, translated the mRNAs in vitro, and separated the synthesized proteins by two-dimensional gel electrophoresis. Most of the proteins were synthesized similarly in oocytes and embryos. Three mRNAs encoding DNA-binding proteins were present in subribosomal mRNP particles but absent from polysomes at the oocyte stage, and present in both fractions at the embryo stage. Because there was little RNA synthesis between the stages, the investigators concluded that some maternal mRNAs were recruited *selectively* into polysomes after fertilization. This conclusion has since been confirmed for specific transcription factors. For instance, bicoid mRNA is made abundantly in *Drosophila* oocytes, but no synthesis of bicoid protein is detected before egg deposition (Driever and Nüsslein-Volhard, 1988).

The experiments just described show that eggs can control the translation of maternally supplied mRNA *in a sequence-specific manner*. This ability is noteworthy from both a biological and a molecular perspective. Biologically, it is important that eggs can adjust the utilization of at least some maternal mRNAs to meet different stage-specific requirements. This seems especially significant for mRNAs encoding transcription factors and other regulatory proteins. With regard to molecular mechanisms, sequence-specific translational control means that certain mRNAs must be earmarked by specific nucleotide sequences or secondary structures. Biologists are now at work identifying these signals.

Translational Control during Spermatogenesis

During spermatogenesis, male germ line cells first undergo meiosis and then go through a series of dramatic morphological changes. These changes include formation of the acrosome, growth of the flagellum, condensation of nuclear chromatin, and loss of most of the cytoplasm (see Chapter 3). These processes require carefully orchestrated synthesis of many proteins, some of them in bulk quantities. At the same time, meiosis and chromatin condensation impede RNA synthesis, thus excluding transcriptional control as an effective means of regulating protein synthesis. Therefore, spermatogenesis is another biological situation in which *sequence-specific* translational control would be highly adaptive.

Protamine mRNA Is Stored in Subribosomal RNP Particles before Translation

Studies on translational control during spermatogenesis have focused on the synthesis of protamines, abundant proteins that replace histones during chromatin condensation. To analyze protamine synthesis in the rainbow trout, Kostas Iatrou and coworkers (1978) induced spermatogenesis in juvenile males by injecting pituitary extracts until their testes contained all stages of maturing sperm. They dissociated testes into cells and separated different cell types from each other by sedimentation in a density gradient. Cell fractions collected from the gradient were identified histologically as spermatogonia, primary spermatocytes, secondary spermatocytes, and early or late spermatids. To determine when protamines were synthesized, the investigators analyzed the basic proteins from each cell fraction by gel electrophoresis. The results showed that protamines are not synthesized until the spermatid stage. However, the following experiment indicates that protamine mRNA is synthesized much earlier and stored in subribosomal mRNP particles.

To trace the synthesis of protamine mRNA, the investigators prepared polysomes and subribosomal RNP particles from all stages of testis cells they had isolated. RNA extracted from each of these fractions was hybridized with radiolabeled protamine cDNA until half of the probe had been driven into hybrids. The time needed to reach this point was inversely proportional to the concentration of protamine mRNA in each RNA frac-

tion. (The more protamine mRNA there was in an RNA fraction, the sooner it drove the probe into hybridization.) The hybridization curves indicated that, in spermatogonia, little protamine mRNA was found in polysomes or subribosomal RNP particles (Fig. 17.13). In primary spermatocytes, virtually all of the protamine mRNA was present in subribosomal RNP particles. In secondary spermatocytes, most of the protamine mRNA was still in the subribosomal RNP particle fraction. However, by the spermatid stage, about half of the total protamine mRNA had been recruited into polysomes.

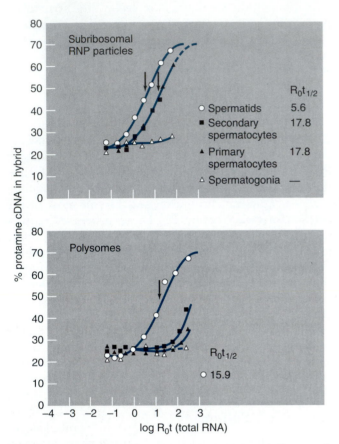

Figure 17.13 Storage of protamine mRNA in subribosomal mRNP particles during spermatogenesis in the rainbow trout. In this experiment, polysomes and subribosomal RNP particles were prepared from spermatogonia, primary spermatocytes, secondary spermatocytes, and spermatids. RNA extracted from each fraction was hybridized with labeled protamine cDNA. The abscissa represents the product of total RNA concentration and hybridization time, abbreviated "R_0t." The ordinate shows the percentage of labeled protamine cDNA driven into hybrids. The $R_0t_{1/2}$ mark indicates the R_0t value at which half of the protamine cDNA tracer was driven into hybrids. A low $R_0t_{1/2}$ value means that the RNA sample contained many copies of protamine mRNA. However, to compare $R_0t_{1/2}$ values from polysomes and subribosomal RNP particles, the values must be corrected for the 1:3 ratio of subribosomal to polysomal RNA per cell. After this correction, the measurements indicated that in spermatids, which synthesize protamine, about half of the total protamine mRNA was present in polysomes and the other half in subribosomal RNP particles. At the earlier stages, 90% or more of the total protamine mRNA was stored in subribosomal RNP particles.

Iatrou and coworkers found that trout protamine mRNA is synthesized in primary spermatocytes and stored as subribosomal RNP particles until translation during the spermatid stage. The time of storage lasts between 15 and 30 days, during which many other mRNAs are translated. It must be concluded that protamine mRNAs contain a specific nucleotide sequence that earmarks them for storage. Comparisons of mouse, human, and bovine protamine mRNAs show a remarkable sequence similarity in the *trailer* region, although this region is not well conserved in the evolution of most other mRNAs. The unusual degree of sequence conservation suggests that the protamine mRNA trailer region has an important function. This has been confirmed by the experiments with transgenic mice and fruit flies described below.

Messenger RNA May Be Earmarked for Storage by Its Leader or Trailer Sequence

The specific nucleotide sequences that earmark mRNAs for storage during spermatogenesis are found in either the mRNAs' trailer regions or their leader regions, depending on species. Scientists have been able to map such sequences by testing engineered mRNAs in transgenic organisms.

To study the role of the trailer region in the translation of mouse protamine mRNA, Robert E. Braun and coworkers (1989) transformed mice with a matched pair of transgenes (Fig. 17.14). Both transgenes contained the regulatory region (promoter and enhancers) and the leader sequence of a mouse protamine (mP) gene as well as the coding (translated) region of a human reporter (hR) gene. The transgenes differed only in the sequences encoding the *trailer* regions of their mRNAs: one encoded a short segment of the mP trailer sequence, while the other encoded the hR trailer region. *Northern blots* and *in situ hybridization* indicated that the transgenic mice transcribed both transgenes with virtually the same stage and tissue specificity as the resident mP gene (for the techniques, see Methods 14.3 and Methods 15.1). In this respect, both transgene mRNAs could be considered valid substitutes for endogenous mouse protamine mRNA. The mouse transcripts first appeared at

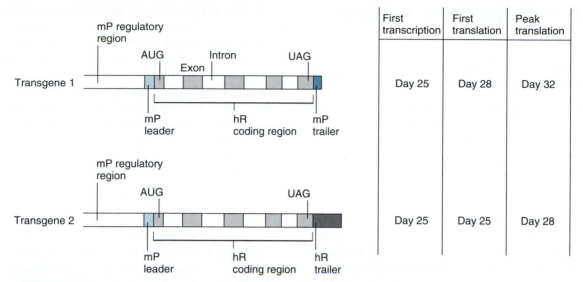

	First transcription	First translation	Peak translation
Transgene 1	Day 25	Day 28	Day 32
Transgene 2	Day 25	Day 25	Day 28

Figure 17.14 Role of the trailer region in the translational control of mouse protamine mRNA. The diagram shows a matching pair of transgenes used to genetically transform mice. Exons are gray or colored; introns are white. Both transgenes contain the coding region of a human reporter (hR) gene, the leader region of a mouse protamine (mP) gene (light color), and 4.1 kb of 5' regulatory region (promoter and enhancers) of the mP gene. The two transgenes differ in their trailer region: one trailer is from the mP gene (156 bp, dark color) and the other from the hR gene (dark gray). Both transgenes are transcribed first on day 25 after birth. However, the mP trailer causes a marked delay in mRNA translation. Protein synthesis from the mRNA with the mP trailer begins on day 28 and peaks on day 32, whereas translation from the control mRNA with the hR trailer begins right after transcription on day 25 and peaks on day 28.

the round-spermatid stage (day 25 after birth). At the elongating-spermatid stage (day 28), however, the transgene mRNA with the hR trailer *disappeared*, while the transgene mRNA with the mP trailer and the endogenous mP mRNA *persisted*. This observation suggested that the mRNA with the hR trailer was less stable than the other two mRNAs, which had mP trailers.

The translation of endogenous mouse protamine mRNA as well as the two transgene mRNAs was monitored by *immunostaining* (see Methods 4.1). In mice containing the transgene with the mP trailer region, hR protein was first detected on day 28. This was the same time when endogenous mP protein was also first observed. However, in mice containing the transgene with the hR trailer region, hR protein already present on day 25. Thus, although both transgenes were under the same transcriptional control, the mRNA with the mP trailer was translated later than its counterpart with the hR trailer. The delayed translation had to be ascribed to the mP trailer sequence. In essence, a short segment of mP trailer sequence was sufficient to enhance the stability of a transgenic mRNA and to confer on it an mP-like translational regulation.

Similar results were obtained in studies on gene expression during *Drosophila* spermatogenesis. In this organism, transcription is very active in primary sperma-

tocytes but ceases before the meiotic divisions. All later protein synthesis relies on premeiotic transcripts. To analyze the control mechanisms involved, Rainer Kuhn and coworkers (1988) transformed fruit flies with a transgene containing the control regions of a gene that was specifically expressed in the male germ line. Transcription of the transgene occurred before meiosis, but translation of the transgenic mRNA was delayed by 3 days until spermatid elongation was completed. In this case, the stability and delayed translation of the transgenic mRNA depended on a short sequence within the *leader* region.

Posttranslational Polypeptide Modifications

After translation, many polypeptides undergo further processing steps before they become functional proteins. First, polypeptides are directed to one of several cellular destinations. Second, many polypeptides are cleaved or receive certain chemical substitutions before attaining their biologically active form. Finally, polypeptides are degraded in a controlled fashion. These processing steps are regulated, at least in part, by certain amino acids, known as *signal sequences,* that are part of the polypeptide.

Polypeptides Are Directed to Different Cellular Destinations

The synthesis of all polypeptides begins in the *cytosol,* i.e., the cytoplasmic matrix that fills the spaces between the membranous organelles of a cell. Many polypeptides, including RNA-binding proteins, cytoskeletal proteins, and many enzymes, stay in the cytosol. Other polypeptides are formed with signal regions that target them to other cellular compartments.

A large class of polypeptides enter the *rough endoplasmic reticulum (ER)* while they are still being synthesized. The rough ER is a closed system of folded membranes connected with the nuclear envelope; the cytoplasmic face of the rough ER membrane is studded with polysomes (Fig. 17.1). From the rough ER, many polypeptides are given off in *transport vesicles* that are directed by a system of signal and docking proteins to the *Golgi apparatus,* a stack of closed membranous sacs. Inside the Golgi apparatus, polypeptides are modified, and many are then released in *secretory vesicles.* These vesicles are similarly targeted to the plasma membrane—where they release their contents via exocytosis—or to other cellular organelles.

Some of the targeting signals that regulate the protein traffic in cells take the form of a *signal sequence,* typically a continuous stretch of 15 to 60 amino acids located at the N terminus. Other targeting signals occur as *signal patches,* three-dimensional arrangements of amino acids that may be located at different positions within the polypeptide but come together when polypeptides fold into their active conformation (explained later in this chapter). The targeting signals interact with different types of receptors that mediate the transport of the polypeptide across a membrane into another cellular compartment.

For instance, polypeptides destined for the rough ER have a signal sequence of about 30 amino acids at their amino terminus. This sequence becomes bound to a *signal recognition particle (SRP),* a small RNP particle floating in the cytoplasm. The SRP stops further translation until it binds to an *SRP receptor,* a protein in the ER membrane (Fig. 17.15). The SRP receptor is associated with a *signal sequence receptor protein* in the ER membrane, with which the signal sequence becomes aligned so that the polypeptide is pointed into the ER lumen. The signal sequence is clipped off, and translation continues while the nascent polypeptide is guided into the lumen of the ER. Most polypeptides targeted to the ER enter it completely; these polypeptides are passed on through the Golgi apparatus and are eventu-

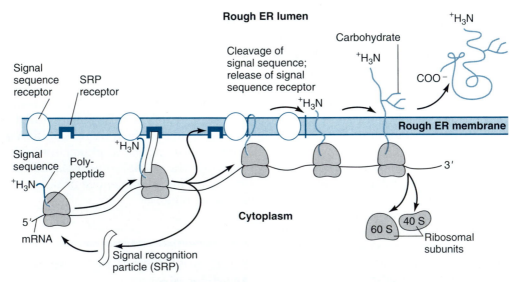

Figure 17.15 Model for the synthesis of a secreted polypeptide on the rough endoplasmic reticulum (ER). Polypeptides to be released into the ER have a signal sequence of about 30 amino acids at their N terminus. When the signal sequence emerges from a ribosome, it combines with a signal recognition particle (SRP), which binds to an SRP receptor protein on the outside of the ER. The signal sequence inserts itself into the ER membrane, presumably by association with a signal sequence receptor protein next to the SRP receptor. During this step, the SRP and its receptor dissociate from the ribosome-peptide complex. Next, the signal sequence is cleaved off and degraded and the signal sequence receptor released. As its elongation continues, the polypeptide is threaded into the ER lumen, where carbohydrates are added. When the C terminus of the polypeptide disappears in the ER lumen, the ribosomal subunits are released.

ally secreted. However, some polypeptides have a *stop transfer sequence* that keeps them embedded in the ER membrane; these polypeptides become plasma membrane proteins when ER vesicles bud off and fuse with the cell plasma membrane.

Polypeptides May Undergo Several Posttranslational Modifications

Many newly synthesized polypeptides are inactive until they have undergone further modifications. These modifications may include formation of disulfide bridges, cleavage of the polypeptide chain between specific amino acids, folding of the linear polypeptide into a three-dimensional configuration, and addition of carbohydrates or other substitutions. Some of these modifications are necessary to stabilize the three-dimensional shape of the polypeptide. Others contribute to the formation of signal patches that direct the polypeptide further along its pathway in the cell. Still other modifications are required to form the active sites for the

polypeptide's biological function. We will briefly survey these modifications and then consider, as an example, how they contribute to the synthesis of collagen.

Disulfide Bridges Disulfide bridges between two cysteine residues (Cys-S-S-Cys) bridges are critical to the stability of many proteins (Fig. 17.16). Disulfide bridges are generated enzymatically in the ER, which provides a more oxidizing environment than the cytosol. Hence, these bridges are commonly found in membrane proteins and in proteins that are excreted from the cell.

Limited Proteolysis Many polypeptides undergo enzymatic cleavages between specific amino acids, a process called *limited proteolysis.* This is known from several hormones and enzymes, which are synthesized as inactive prohormones or proenzymes. For instance, the hormone insulin is synthesized in pancreatic cells as a long polypeptide called *preproinsulin* (Fig. 17.16). Inside the ER, the signal sequence is removed and three disulfide bridges are formed. These disulfide bridges

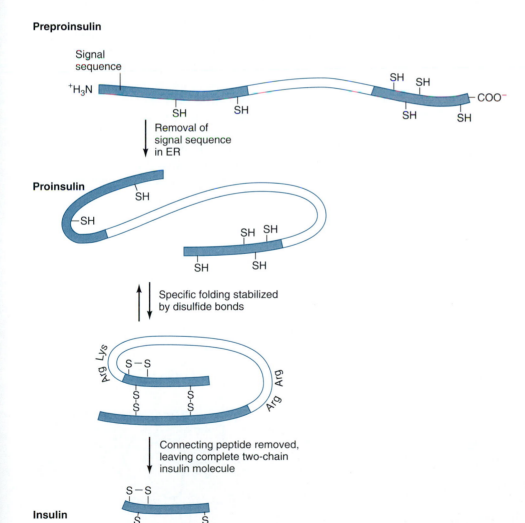

Figure 17.16 The hormone insulin is synthesized as a much larger precursor peptide, preproinsulin. Removal of the signal sequence in the ER generates proinsulin. The proinsulin folds into a specific shape that is determined by the position of the cysteine residues forming disulfide bridges. A central portion of the peptide is then removed by proteases recognizing the flanking arginine and lysine residues. Now the remaining insulin has acquired its biologically active form.

stabilize the shape of the molecule, which is now called *proinsulin*. A central domain is then removed in two proteolytic steps while the two lateral domains are held together by disulfide bridges. The resulting *insulin* is now in its biologically active form.

Folding   All biologically active polypeptides have a characteristic conformation, which can be described at four levels (Fig. 17.17). The *primary structure* of a polypeptide is its linear sequence of amino acids; it is determined directly by the genetic code.

The term *secondary structure* refers to the folding of certain polypeptide sections into regular patterns such as α *helices* and β *pleated sheets* (Fig. 17.18). These patterns are stabilized by hydrogen bonds between C=O and N−H residues. In most polypeptides, α helices and β pleated sheets alternate with a less regular pattern known as a *random coil*.

The *tertiary structure* of a polypeptide is formed by the folding of the helices, sheets, and coils into a three-dimensional shape. This shape is often stabilized by disulfide bridges and hydrophobic interactions. The secondary and tertiary structures dictate which amino acids are located at the surface of the polypeptide and therefore determine its chemical properties.

Finally, a protein may consist of two or more polypeptides arranged in a *quaternary structure*. The formation of tertiary and quaternary structures may bring into close proximity amino acids that are far apart according to the primary structure alone. Such grouping of amino acids may form active sites of a particular shape suitable for interacting with complementary sites on other molecules.

Many polypeptides seem to have just one secondary and one tertiary structure, which are dictated by the primary structure and formed under normal conditions in cells. In particular, some sequences of amino acids are conducive to α helix formation whereas other sequences favor the formation of β pleated sheets. In addition, the positions of cysteine residues limit the disulfide bridges that can be formed. However, some polypeptides exist in multiple conformations that can be converted into one another via unfolding and refolding (Fox et al., 1986).

The ability of some polypeptides to fold in more than one way seems to have prompted the evolution of a class of proteins referred to as *molecular chaperones:* they associate with newly translated polypeptides in ways that allow the formation of proper protein domains while translation is still going on, and direct the final folding of the polypeptide once it is released from the ribosome (Frydman et al., 1994). Molecular chaperones include the class of *heat shock* proteins, which are synthesized by all cells in the presence of heat or under other stressful conditions. They seem to rescue unfolded polypeptides by restoring them to their active conformations and to hasten the degradation of polypeptides that are denatured beyond repair (Craig, 1993).

Oligomeric Proteins   Many proteins consist of two or more *subunits* and are therefore called *oligomers* (Gk. *oligo-*, "few"; *meros*, "part"). Each subunit is a polypeptide encoded by a separate gene. The configuration of an oligomeric protein is also known as its *quaternary structure*, as discussed previously. For example, the quaternary structure shown in Figure 17.17 is that of adult hemoglobin, which is composed of four polypeptides, two α-globins and two β-globins. For another example, many transcription factors consist of two polypeptides, as discussed in Chapter 15.

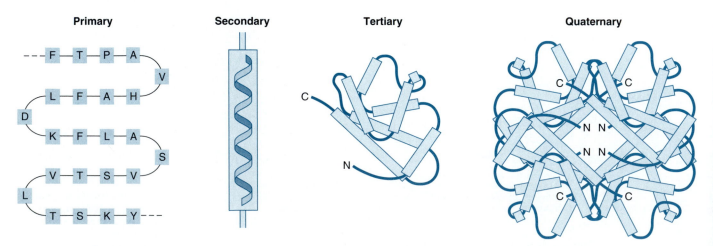

Figure 17.17 Primary, secondary, tertiary, and quaternary structure of proteins. The quaternary structure shown here is made up of four of the tertiary structures.

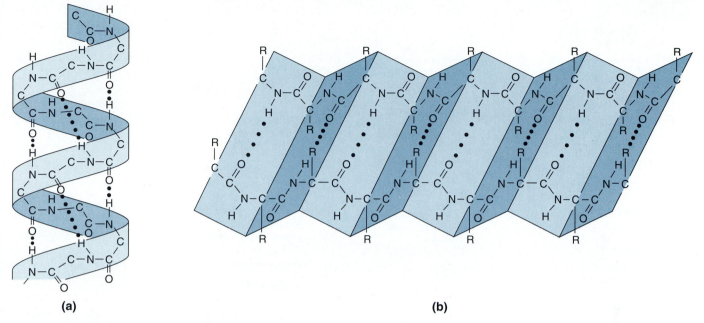

Figure 17.18 Common secondary structures of proteins: α helix and β pleated sheet. **(a)** Ribbonlike representation of the α helix. The amino acid residues (not shown, for clarity) stick out laterally from the helix. Note that each C=O residue forms a hydrogen bond (dotted line) with an N−H residue about 3.5 positions away. **(b)** β pleated sheet consisting of two antiparallel peptide domains. The β pleated sheet is another common secondary structure, forming curved stacks in many proteins. Each amino acid is hydrogen-bonded to a neighboring one. The three-dimensional nature of the covalent bonds causes the sheet to be pleated. The amino acid residues (R) extend above and below the sheet.

The formation of oligomeric proteins from two or more polypeptides occurs spontaneously under the conditions prevailing in cells. The subunits are attracted to each other by hydrogen bonding and hydrophobic interactions. Some associations between subunits are later stabilized by disulfide bonds. Other oligomeric associations are *reversed* as part of an activation process. For instance, the binding of a steroid hormone to its receptor is associated with the release of an inhibitory protein from the receptor (see Fig. 15.14). Similarly, an enzyme known as **cAMP-dependent protein kinase** originates as an inactive oligomer consisting of two regulatory subunits and two catalytic subunits (see Fig. 2.26). At high levels of cAMP (cyclic AMP), the catalytic subunits dissociate from the regulatory units and become active enzymes.

Phosphorylation, Glycosylation, and Other Substitutions The biological functions of many polypeptides in the cytosol are regulated by *protein kinases,* which add phosphate groups to certain amino acids, and by *phosphatases,* which remove phosphate groups. The resulting changes in the tertiary structure of the polypeptide increase or reduce its biological activity. Phosphorylation is reversible and serves to regulate the

activity of enzymes, transcription factors, eukaryotic initiation factors, and many other proteins. Other reversible polypeptide modifications are methylation and acetylation. Polypeptides synthesized in the ER and passed through the Golgi apparatus may undergo further enzymatic modifications. Some enzymes add carbohydrates to certain amino acids, a process called *glycosylation.* Other enzymes add hydroxyl groups, a step known as *hydroxylation.* Still others link fatty acids to polypeptides, thus targeting them for incorporation into membranes.

Collagen Synthesis To illustrate the modification steps that have been summarized, let us consider the formation of *collagen,* the most abundant structural protein in vertebrates. Collagen is synthesized by *fibroblasts,* an abundant cell type in connective tissue. After synthesis inside fibroblasts, collagen undergoes *exocytosis* to become part of the *extracellular matrix.* There are at least 12 types of collagen, each optimized for a different function. The following description fits type I collagen, which forms the fibrous components of tendons and other connective tissues.

As a future component of the extracellular matrix, the *preprocollagen* polypeptide is synthesized on the

Rough ER

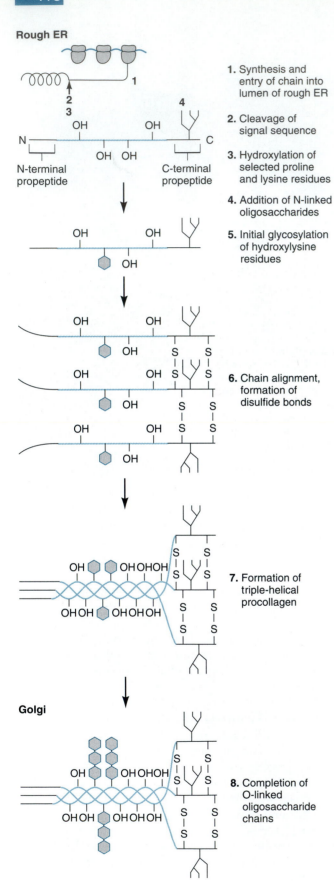

1. Synthesis and entry of chain into lumen of rough ER

2. Cleavage of signal sequence

3. Hydroxylation of selected proline and lysine residues

4. Addition of N-linked oligosaccharides

5. Initial glycosylation of hydroxylysine residues

6. Chain alignment, formation of disulfide bonds

7. Formation of triple-helical procollagen

Golgi

8. Completion of O-linked oligosaccharide chains

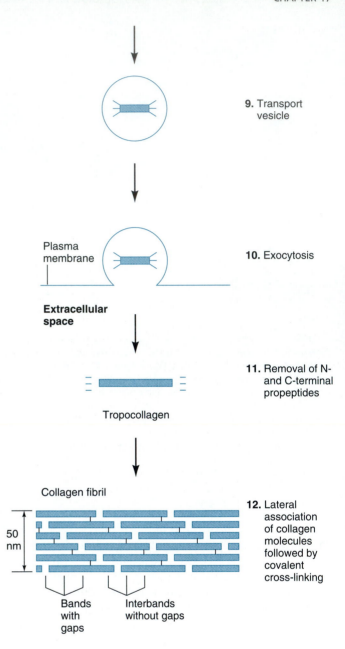

9. Transport vesicle

10. Exocytosis

11. Removal of N- and C-terminal propeptides

12. Lateral association of collagen molecules followed by covalent cross-linking

Figure 17.19 Summary of the major steps in collagen biosynthesis. Modifications of the collagen polypeptide include hydroxylation, glycosylation, disulfide bond formation, and limited proteolysis. These steps occur in a precise sequence in the rough ER, the Golgi complex, and the extracellular space. The modifications allow formation of a stable triple helix and also the staggered lateral association and covalent cross-linking of helices into fibrils of 50 nm diameter. The staggered association of the collagen molecules lines up the gaps between consecutive molecules in transverse bands. Because these bands stain differently from the interbands, which have no gaps, fibrous collagen has a striated appearance in electron micrographs (see Fig. 17.20).

rough ER (Fig. 17.19). After the signal peptide has been clipped off, the remaining *procollagen* polypeptide is hydroxylated by different enzymes recognizing proline and lysine residues in specific contexts of other amino acids. Next, hydroxylysine residues are glycosylated by addition of galactose. In the next step, three procollagen polypeptides are aligned and disulfide bonds form among them. The linked polypeptides then form a triple helix over most of their length, resulting in a rod-shaped *procollagen protein* held together by disulfide bridges and hydrogen bonds. Only the ends of the procollagen protein remain globular. At this point, the protein is transported to the Golgi apparatus, where further glycosylation steps occur. After passage through secretory vesicles and exocytosis, the globular ends are clipped off. The remaining strictly helical protein is about 300 nm long. It is called a *collagen protein*.

The collagen proteins spontaneously form staggered lateral associations called *collagen fibrils* (Fig. 17.20). Type I collagen fibrils average about 50 nm in diameter and can be up to several micrometers in length. The stability of these fibrils is greatly enhanced by the formation of additional disulfide bridges *between* adjacent collagen proteins. The fibroblasts that synthesize collagen also control the density and orientation in which the fibrils assemble.

In addition to providing strength to tendons, bones, dermis, and other connective tissues, collagens play an important role in cellular interactions (see Chapter 26).

The Longevity of Proteins Is Differentially Controlled

Like most molecules in an organism, proteins are subject to **turnover**. This means that they are broken down and resynthesized at short intervals relative to the organism's life span. As discussed previously in the context of mRNA longevity, both the synthesis and breakdown of a gene product are effective means of controlling gene expression. In mammalian cells, the *half-lives* of proteins range from minutes to weeks. Proteins with short half-lives include cyclins, which are broken down rapidly after each M phase in the cell cycle (see Chapter 2). Likewise, bicoid protein and many other regulatory proteins have short half-lives, a property that makes it possible to limit their activity to short periods of time.

Protein degradation is carried out by cellular enzymes called *proteases*. Many proteins are earmarked for degradation by the addition of a small protein, *ubiquitin*. The ubiquitin additions, carried out enzymatically and in a sequence-specific way, make proteins susceptible to protease attack. Once the breakdown of a

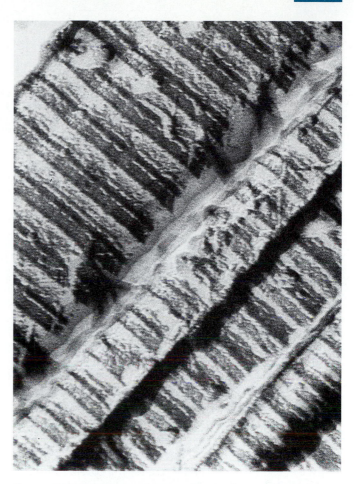

Figure 17.20 Electron micrograph of collagen fibrils from a tendon. The transverse banding reflects the fact that collagen molecules assemble in a staggered pattern but with the gaps between consecutive molecules at distinct levels (see Fig. 17.19).

protein has begun, the digestion continues to the level of single amino acids.

The selectivity of the ubiquitin addition process depends in part on signals provided by the structure of the protein to be broken down. For instance, incomplete or misfolded polypeptides are degraded rapidly. Another signal for rapid degradation seems to be provided by amino acid clusters that are rich in proline, glutamate, serine, and threonine. Such clusters are common in polypeptides with half-lives of less than 2 h but rare in long-lived polypeptides (Rogers et al., 1986).

In summary, cells are able to degrade proteins quickly and selectively in response to appropriate signals. The initiation of the breakdown process depends on the presence of certain amino acids in the polypeptide, much as the selective degradation of mRNAs depends on certain nucleotide sequences in their leader or trailer regions.

SUMMARY

During translation, the nucleotide sequence of mRNA is used to order and join amino acids into a polypeptide. The initiation phase of translation, which involves several proteins called eukaryotic initiation factors, is most important for translational control. In cells, most mRNAs released from the nucleus are immediately recruited into polysomes and translated into polypeptides. However, certain cell types can store mRNA in the form of messenger ribonucleoprotein (mRNP) particles. According to the masked messenger hypothesis, the mRNA present in mRNP particles is temporarily untranslatable because a protein or another factor blocks a site critical to initiation. There are a wide variety of additional translational control mechanisms, including polyadenylation and deadenylation of mRNAs.

Translational control is a prominent way of regulating gene expression in oocytes, eggs, and embryos. Many of the mRNAs stored in the oocyte are left untranslated until embryonic development begins. Upon oocyte maturation or fertilization, these stored mRNAs are shifted from mRNP particles into polysomes. At the same time, the translational machinery accelerates, and both processes together result in an overall increase in protein synthesis.

Also as a result of oocyte maturation, the synthesis of some oocyte-specific proteins decreases while the synthesis of some cleavage-specific proteins increases. These changes, which occur in the absence of transcription, demonstrate that the translation of maternally supplied mRNA is controllable in eggs in a sequence-specific way. Similarly, the translation of protamine mRNA is delayed selectively during spermatogenesis.

After translation, many polypeptides undergo further processing before they become functional proteins. The presence of specific amino acid sequences directs a polypeptide to the cytosol or into the endoplasmic reticulum (ER), and further into various cellular compartments. Many polypeptides are cleaved by limited proteolysis and form disulfide bridges before they assume their active three-dimensional configuration. Several cytosolic polypeptides are phosphorylated to modulate their activity. Polypeptides synthesized in the ER are often glycosylated by the addition of carbohydrates. These processing steps, and finally the controlled degradation of polypeptides, rely on signals provided by certain amino acid sequences or other structural aspects of the polypeptide.

SUGGESTED READINGS

Jackson, R. J., and N. Standart. 1990. Do the poly(A) tail and 3' untranslated region control mRNA translation? *Cell* **62**:15–24.

Winkler, M. 1988. Translational regulation in sea urchin eggs: A complex interaction of biochemical and physiological regulatory mechanisms. *BioEssays* **8**:157–161.

GENETIC AND PARAGENETIC INFORMATION

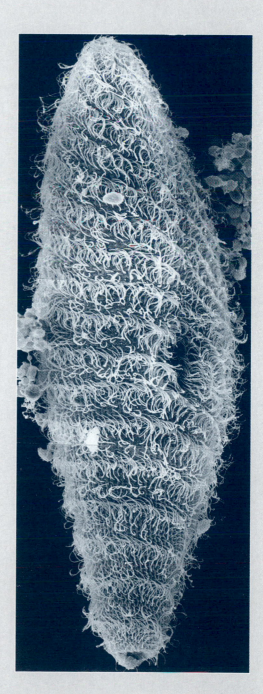

Figure 18.1 Scanning electron micrograph of *Paramecium*, a ciliate protozoon (anterior at the top, ventral to the right). This unicellular animal is about 170 μm long and 50 μm wide. It is covered with thousands of cilia, which propel it through the water by coordinated strokes. Modified cilia also drive bacteria and other small matter into an oral apparatus, the opening of which is visible on the ventral surface. *Paramecium* reproduces by cell division, and the pattern of the mouth, cilia, and other organelles is duplicated before each division. Studies of *Paramecium* and other ciliates have played an essential role in establishing the concept of paragenetic information.

In the three previous chapters, we traced the expression of genetic information from genes to proteins. Yet, we are still far from understanding how a single cell is made. How do polypeptides reach their target sites within a cell? We have learned that signal sequences in polypeptides interact with signal recognition particles (SRPs) and that matching receptors guide the loaded SRPs to their destinations. But how are the receptors laid out so that they will guide the cellular protein traffic properly? This question has been difficult to answer, because no one has ever assembled a living cell in a test tube. No matter how carefully one breaks a cell down into its component molecules, the molecules will not reassemble into a cell. The only way for a cell to be generated is by division of a progenitor cell. This fundamental observation was summarized succinctly in Latin by the German pathologist Rudolf Virchow: *Omnis cellula e cellula* ("every cell comes from a cell"). Coined in 1858, this aphorism still holds true.

The Principle of Genetic and Paragenetic Information

Virchow's aphorism is restated, somewhat less concisely, in the *principle of cellular continuity* (see Chapter 2): all cells of currently living organisms are the temporary ends of *uninterrupted* cell lineages extending back through their ancestors' germ lines to primordial cells billions of years ago. Two types of information have been passed on through these lineages. One is *genetic information,* which is encoded in RNA or DNA. Equally important is the unbroken chain of *structural organization* that is passed on directly, without being encoded in genes, from each cell to its daughter cells. Because this organization is inherited in parallel to the genetic information, it is called *paragenetic information* (Gk. *para,* "near," "beside," "beyond"). Genetic and paragenetic information complement each other functionally but are inherited in different ways.

According to the *principle of genetic and paragenetic information,* cells pass on both types of information to their daughter cells. As a result, organisms pass on both types of information to the next generation, and have done so throughout the course of evolution. Whereas the last four chapters dealt with the role of genes in development, this chapter will address the nature and inheritance of paragenetic information.

To define the role of paragenetic information, we will examine different modes of molecular assembly in cells. In the simplest types, *self-assembly* and *aided assembly,* molecules snap together without paragenetic information. By describing specific examples, we will explore the role of conformational changes and auxiliary proteins in these processes. Of particular interest are *seed* structures, which form nucleation centers for the assembly process. In self-assembly and aided assembly, seeds are not absolutely necessary, but they greatly accelerate the initiation phase.

Directed assembly is the most prevalent mode of molecular assembly in cells. For this process, genetic information alone is not sufficient. Paragenetic information in the form of preexisting seeds or more complex *organizing centers* is also required. These organizing centers not only start the assembly process, but may also determine the type of product to be assembled. Frequently, cells use alternative organizing centers to assemble different products from the same pool of building blocks. Eukaryotic cells, for example, use the same tubulin molecules to assemble different arrays of microtubules in the cytoplasm and in the cores of cilia and flagella. Which of these microtubular arrays are assembled depends on the available organizing centers. By controlling the replication and placement of these organizing centers and by modifying their molecular composition, cells direct the assembly of various microtubular arrays according to cell type and phase of the cell cycle. The role of paragenetic information in directed assembly is most evident in ciliated protozoa (Fig. 18.1). The surface of each of these animals embodies much paragenetic information. This becomes apparent when occasional errors in surface structures are propagated indefinitely during cell division, even though the genetic information is completely unaltered.

In addition to directed assembly, ciliates have independent mechanisms of *global patterning.* They possess a type of paragenetic information that preserves the global features of the cell pattern—including the number and orientation of major organelles—during cell division and *encystment.* Like ciliated protozoa, metazoan cells also pass on features of global organization during cell division.

Self-Assembly

Self-assembly is the simplest way in which molecules combine and form larger structures: the molecules just snap together spontaneously. All the energy and information needed to form the final structure are contributed by its constituent molecules. No other molecules are required as scaffolds, catalysts, or energy sources. Only water and small ions must be present to provide a solution of suitable pH and ionic strength. Self-assembly processes are energetically favorable; that is, they release energy. Some examples of self-assembly have already been discussed in previous chapters. Hemoglobin self-assembles from two α-globin polypeptides, two β-globin polypeptides, and four heme molecules; even larger structures, such as collagen fibrils, self-assemble from collagen proteins. A simple in vitro test can be used for classifying an assembly process as self-assembly: if a structure can be generated from its component molecules in a simple aqueous solution, it is self-assembling. Thus, self-assembly does not require paragenetic information.

Self-Assembly Is under Tight Genetic Control

Self-assembly can generate entire cell organelles. This process has been investigated in detail for bacterial ribosomes (Nomura, 1973). Like eukaryotic ribosomes, bacterial ribosomes consist of a large and a small sub-

unit. Each subunit can be broken down into ribosomal RNA and proteins (Fig. 18.2). When these components are recombined in appropriate salt solutions in a test tube, they spontaneously reconstitute a functional ribosomal subunit. The same self-assembly process observed in vitro may also occur during the normal formation of ribosomal subunits in vivo.

The self-assembly of ribosomes is under tight genetic control. If ribosomal rRNA from one bacterial species is mixed with ribosomal proteins from a *closely related* species, functional particles still form. However, components from *distantly related* bacteria do not self-assemble. Apparently, ribosome self-assembly requires a close fit between certain RNA sequences and some protein domains. The quality of the fit degenerates as mutations accumulate during evolution.

Stringent genetic control is also indicated by the effects of some point mutations that cause abnormal self-assembly. A dramatic example is the heritable human disease known as *sickle cell anemia.* It is caused by a single base pair change that results in the substitution of valine for glutamine in position 6 of the β-globin polypeptide. The mutant β-globin assumes a different *tertiary structure,* which in turn causes the mutant hemoglobin molecules to attach to each other to form long rods. These rods distort red blood cells, forcing them

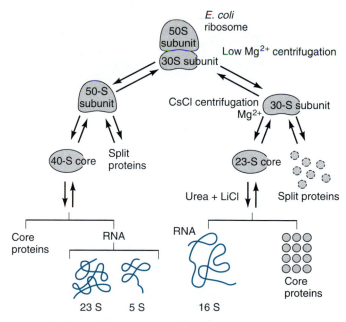

Figure 18.2 Dissociation and self-assembly of bacterial ribosomes. The ribosome of *Escherichia coli* consists of a large (50-S) and a small (30-S) subunit, which can be dissociated by centrifugation at low magnesium concentration. The small subunit is dissociated, by centrifugation in cesium chloride, into a 23-S core and 7 split proteins. Finally, the core is dissociated by denaturing agents into 16-S rRNA and 12 core proteins. Under suitable conditions, each step is reversible and complete ribosomes will self-assemble from their single components. Similar dissociation and reassembly steps are found with the large ribosomal subunit.

into an abnormal sickle shape (Fig. 18.3). The sickle cells are less elastic than normal red cells and therefore tend to clog small blood vessels, causing painful, life-threatening conditions. Thus, a small change in the genetic information can affect not only the structure of a polypeptide but also the self-assembly properties of an oligomeric protein.

The Initiation of Self-Assembly Is Accelerated by Seed Structures

Some products of self-assembly, such as microtubules and microfilaments, consist of many repeating units (see Chapter 2). The assembly of such polymers usually starts from *seed structures* that provide nucleation centers for polymerization. The importance of seed structures has been documented especially well in the self-assembly of a simple virus.

The *tobacco mosaic virus* (*TMV*) infects tobacco plants, causing their leaves to become mottled and wrinkled. This virus is rod-shaped, with a coat of proteins surrounding an RNA core in a helical pattern (Fig. 18.4).

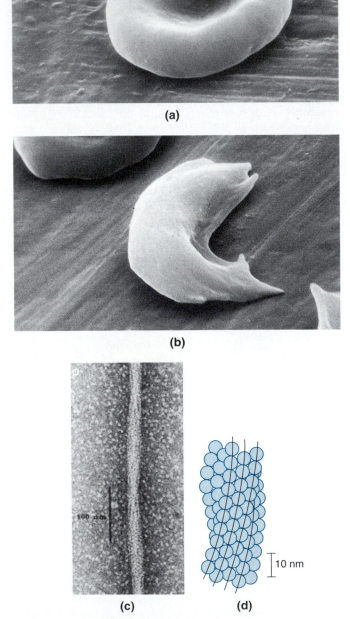

Figure 18.3 Abnormal self-assembly of hemoglobin molecules in humans with sickle cell anemia. **(a)** Scanning electron micrograph of a normal red blood cell. **(b)** Sickled red blood cell. **(c)** Electron micrograph of fibers formed by hemoglobin S (hemoglobin tetramer with β-globin carrying the sickle mutation). **(d)** Model of self-assembled hemoglobin S. Each circle represents a complete hemoglobin S tetramer. Normal hemoglobin tetramers do not assemble in this fashion.

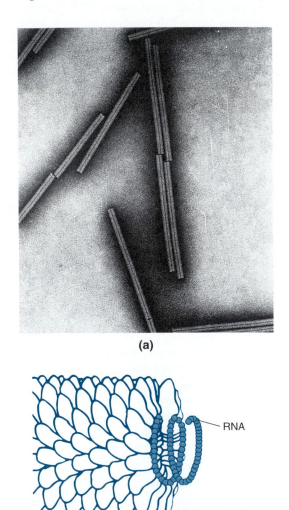

Figure 18.4 Tobacco mosaic virus (TMV). **(a)** Electron micrograph of the rod-shaped virus particles. **(b)** Schematic drawing of a portion of TMV with some of the coat proteins removed to expose the viral RNA. Note the spiral arrangement of the RNA and the coat proteins.

Each rod consists of one 6400-nucleotide RNA molecule and 2130 identical polypeptides that form the cylindrical coat. TMV can easily be dissociated into its components. The resulting mixture of RNA and proteins will spontaneously reassemble into a complete virus, and this reassembly will proceed in vitro under proper conditions of ionic strength and pH. The biological activity of self-assembled TMV can be tested by spreading it on tobacco plant leaves: depending on assembly conditions, at least some of the TMV particles are infective.

Studies of TMV have provided much insight into the self-assembly process (P. J. Butler and A. Klug, 1978). In one of the earliest experiments, when TMV protein and RNA were mixed together in vitro, it was several hours before maximum yields of infective virus were reached (Fig. 18.5). Throughout the self-assembly process, the investigators took samples of the reaction mixture and analyzed them under the electron microscope. Initially, they saw only single coat proteins and aggregates of several coat proteins. Only after a few hours did they find a more advanced intermediate product in the reaction mixture: circular discs consisting of two layers of coat proteins. When such discs were added in quantity to TMV RNA, the results were dramatic: complete virus particles formed within minutes. This outcome suggests that incomplete discs are unstable, and that therefore the formation of a complete disc is a time-consuming step. In contrast, a complete disc is a stable intermediate. Also, after one disc has associated with a specific segment of the viral RNA, this complex acts as a seed structure, to which further discs can be added in rapid succession.

The Conformation of Proteins May Change during Self-Assembly

When molecules fall together during self-assembly, they release energy and become thermodynamically more stable. Participating proteins may change in tertiary or quaternary structure in the process. Each conformational change associated with a self-assembly step may in turn facilitate the addition of the next building block (Fig. 18.6). Such self-propagating waves of conformational changes are thought to play a role in many assembly processes.

To illustrate this phenomenon, we will continue our exploration of TMV self-assembly. As already noted, TMV coat proteins self-assemble into two-layered discs. Such discs interact with a unique initiation region of the TMV RNA (Fig. 18.7). This region is about 50 nucleotides long and has a specific initiation sequence that binds stably to the disc of viral coat proteins. (RNAs lacking this sequence will not support TMV self-assembly.) The binding of the initiation sequence causes a conversion of the disc into two turns of a flat helix that looks like a lock washer. This conformational change traps the

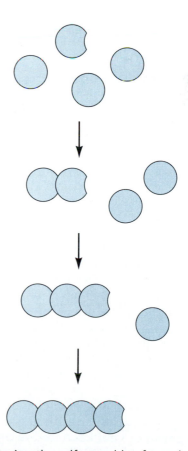

Figure 18.6 During the self-assembly of a molecular complex or an organelle, some proteins undergo a change in tertiary or quaternary structure (represented by the shallow depression the circle). The new conformation of each added protein in turn facilitates the addition of the next protein. This self-propagating conformational change imposes a direction on the self-assembly process.

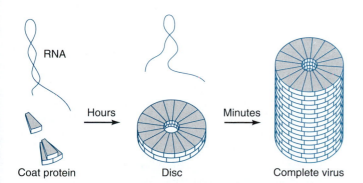

Figure 18.5 Self-assembly of TMV from its component RNA and coat proteins. As an intermediate product, protein discs devoid of RNA are formed. Each disc consists of two circular layers of coat proteins, each layer containing 17 protein units. This number corresponds almost exactly to the number of coat proteins per helical turn in the fully assembled virus. The self-assembly of a disc from single coat proteins takes several hours. However, once a disc has formed and has associated with a specific segment of the viral RNA, the RNA-disc complex serves as a seed structure, facilitating the assembly of the complete virus within minutes.

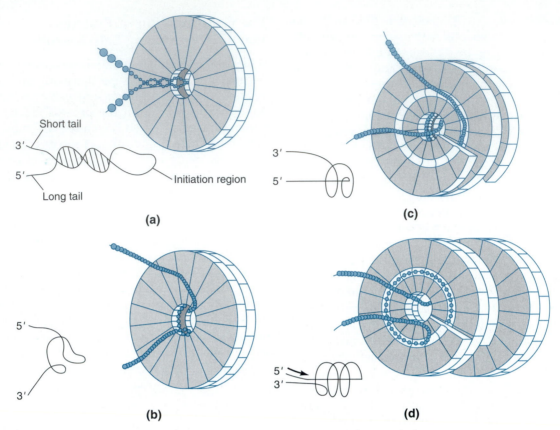

Figure 18.7 Nucleation process in self-assembly of TMV. **(a)** The viral RNA forms a hairpin loop at the initiation region, which is about 50 nucleotides long. This region is located about 1 kb away from the 3′ end (short tail) of the viral RNA, and about 5 kb away from the 5′ end (long tail). The loop inserts into a disc of coat proteins. **(b)** The RNA initiation region intercalates between the two layers of the disc, which form an open cleft on the inside. **(c, d)** The intercalation triggers a change in the conformation of the disc, which now assumes a helical shape like a lock washer. This change also closes the cleft between the protein layers and traps the RNA inside. In addition, the change activates a new ring of RNA binding sites on the outside surfaces of the lock washer. The lock washer–RNA complex serves as a nucleation site for the self-assembly of a complete virus by the rapid addition of further discs.

RNA initiation sequence between the lock washer proteins and generates new binding sites for RNA on the outside of the proteins. The resulting lock washer–RNA complex serves as a nucleation site for the rapid addition of further coat protein discs (Fig. 18.8). As each new disc interacts with the viral RNA, it is converted to the lock washer shape and added on top of the growing helix. The resulting structure is increasingly stable, and further additions are no longer dependent on the specific nucleotide sequence of the viral RNA.

In summary, TMV self-assembly involves the initial formation of protein discs, the association of a disc with a specific segment of viral DNA, a resulting change of the original disc to a lock washer conformation, and a self-propagating wave of disc additions and changes to the lock washer conformation.

Together, these observations show that self-assembly is a powerful process that generates cell organelles and simple viruses just by permitting their parts to snap together. Many cellular structures self-assemble in vitro and presumably originate by self-assembly in vivo. These structures include the oligomeric proteins, collagen fibers, ribosomes, microtubules, and microfilaments already mentioned. In addition, simple cell membranes self-assemble in vitro.

The self-assembly of proteins depends on their *tertiary* and *quaternary* structure. These are determined, to a large extent, by the *primary* and *secondary* structure of their component polypeptides and are therefore under direct genetic control. Self-assembly processes are greatly accelerated in the presence of seed structures and are often associated with conformational changes in their building blocks. These properties help to restrict self-assembly processes to certain areas within cells and to control the direction in which they progress.

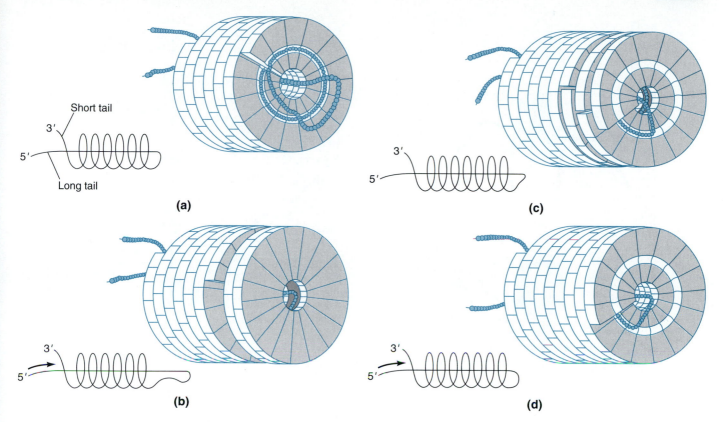

Short tail

3′
5′
Long tail

(a)

3′
5′

(c)

3′
5′

(b)

3′
5′

(d)

Figure 18.8 Elongation of TMV. **(a)** The nucleation complex of Figure 18.7 is now seen from the opposite side. The longer tail (5′ end) of the RNA trapped in the lock washer is doubled back through the central hole. **(b)** The loop extends beyond the surface of the lock washer and interacts with another disc of coat proteins. The loop inserts into the open cleft of the disc as shown in Figure 18.7b. **(c)** The disc is converted into a helical lock washer. **(d)** The new lock washer is stacked on top of the previous one, adding two turns of helix. This process repeats rapidly until the long tail of the RNA is completely coated with viral protein. At the same time, the short RNA tail (3′ end) is also coated with viral protein; the mechanism of this process is not well understood.

Aided Assembly

Notwithstanding its great powers, self-assembly is limited in the complexity of the structures it can produce. The limits of self-assembly are best illustrated by the morphogenesis of another virus, the *T4 bacteriophage* (Wood, 1980). The phage is composed of about 40 different coat proteins surrounding the genomic DNA. If the phage components are mixed together in aqueous solution, phage parts assemble, but not a complete phage. Indeed, mutational analysis has shown that T4 assembly depends on at least six auxiliary proteins that are encoded by the phage genome but are not present in the mature phage. These proteins seem to serve as catalysts or scaffolds that hold some of the coat proteins in place until their assembly is completed. The assembly of T4 therefore does not fit the definition of self-assembly. Rather, this type of assembly is classified as *aided assembly* to reflect its dependence on auxiliary components. Like self-assembly, however, aided assembly does not require paragenetic information.

Bacteriophage Assembly Requires Accessory Proteins and Occurs in a Strict Sequential Order

Detailed studies of T4 assembly have revealed two features of general interest that are also important principles underlying the assembly of cellular organelles. The first is the dependence of assembly on auxiliary proteins, as already noted. The second interesting feature is the *strict sequential order* of phage assembly. Each assembly step depends on completion of the preceding steps. These features are emphasized in the following description of T4 assembly.

The T4 bacteriophage consists of a head, a tail, and six tail fibers (Fig. 18.9). In order to multiply, the virus injects its DNA into a host bacterium. Once inside, the viral DNA commandeers the bacterium's synthetic machinery to replicate the viral DNA and to synthesize viral proteins. About 40 of these proteins are combined with viral DNA to form new bacteriophages. The remaining proteins are used in DNA synthesis and other life cycle functions of the virus. The assembly of new

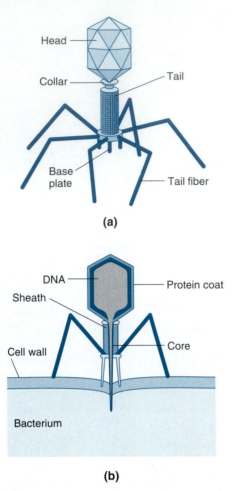

(a)

(b)

Figure 18.9 The T4 bacteriophage. **(a)** View of the phage with head, collar, tail, base plate, and tail fibers. **(b)** Schematic section showing how the phage infects a host bacterium. The phage binds to the bacterial surface, using receptor proteins at the tail fiber ends. The contact triggers a contraction of the tail, and the viral genome is injected into the bacterium. Inside the host, the phage DNA is replicated and transcribed, and the transcripts are translated into proteins.

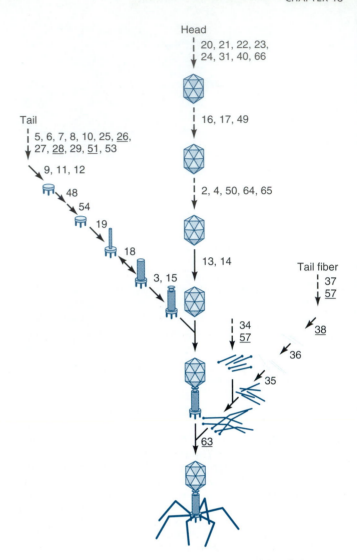

Figure 18.10 Flowchart for the aided assembly of the T4 bacteriophage. Each number represents a gene function revealed by mutant analysis. The underlined numbers refer to accessory proteins that are required for assembly but are not constituents of the final structure. The major components of the phage (head, tail, and tail fibers) are assembled independently before they are mounted together. The components are assembled in a strict sequential order.

phage particles occurs inside the host bacterium, which is eventually lysed by the new phages so that they can move on to infect other hosts.

To study T4 assembly, investigators have isolated a large number of conditionally lethal mutants of the phage. Under permissive conditions, such as low temperature, these mutant phages assemble normally. Under restrictive conditions, such as high temperature, assembly stops. Each mutant is characterized by an abnormal or a missing protein and by a set of partially assembled phage particles inside the host bacteria. This type of analysis has provided a detailed picture of T4 assembly (Fig. 18.10). The major viral components (head, tail, and tail fibers) are assembled independently before they are finally mounted together. A defect that stops the assembly of one of the components does not affect the assembly of the other two.

The number of genes involved in phage assembly exceeds the number of coat proteins, indicating that the assembly of the coat proteins depends on auxiliary proteins. In particular, the tail fibers consist of only four proteins but require at least seven genes for their assembly. Likewise, three proteins necessary for making the base plate of the tail do not become part of the final structure. Moreover, the assembly of the head capsule requires prior formation of an inner scaffold or core that is later destroyed enzymatically. Finally, some of the head proteins are cleaved by limited proteolysis before assembly. The need for enzymes or scaffold proteins constitutes a clear departure from self-assembly and justifies the term *aided assembly*.

Components are assembled in a fixed sequence. For instance, the sheath of the tail is not assembled unless the base plate has been assembled first. The order in which parts are assembled cannot be attributed to the timing of protein synthesis, because all proteins are synthesized simultaneously. Rather, it seems that each assembly step is associated with a conformational change generating the seed structure that accelerates the following step. This type of sequential assembly allows different building blocks to be present simultaneously without making "wrong" connections.

Aided Assembly Is Common in Prokaryotic and Eukaryotic Cells

The aided assembly of proteins into complex structures is a common occurrence in prokaryotic and eukaryotic cells. Indeed, some of the proteins that aid in the assembly of bacteriophage heads are encoded by their host bacteria. These proteins belong to a large group of proteins called *molecular chaperones* (Hemmingsen et al., 1988). Chaperone proteins are also found in mitochondria, in plant chloroplasts, and in the nuclei of various eukaryotic cells. Their main function, as discussed in Chapter 17, is to ensure the proper folding of polypeptides into their secondary and tertiary structures, so that they will assemble correctly into oligomeric proteins or other complex structures. For instance, the assembly of nucleosomes depends on an acidic nuclear protein, *nucleoplasmin*, which itself does not become part of the nucleosomes.

Directed Assembly

In the types of assembly discussed so far, we have assumed that a given building block, such as a T4 tail fiber protein, can serve in the assembly of *only one* structure, in this case a T4 bacteriophage. In many instances, however, the same pool of building blocks can participate in the assembly of *two or more* different structures. Which of the possible structures is assembled depends on what kind of nucleation center is present. A nucleation center may be a simple seed structure, as in self-assembly, or may be a more complex *organizing center* such as an entire centrosome. The type of assembly in which the presence of a particular seed or organizing center directs the assembly of one structure at the expense of another structure is called *directed assembly*. The term is also used if a seed or an organizing center directs the assembly of only one structure but is *required* for the assembly process rather than just *accelerating* it as in self-assembly. (This distinction may be difficult to make in practice, though.) Both types of directed assembly depend on paragenetic information in the form of seeds or organizing centers. Of all molecular assem-

bly types, directed assembly is probably the most prevalent (Grimes and Aufderheide, 1991).

The following examples will illustrate directed assembly at different levels of complexity.

One Bacterial Protein Can Assemble into Two Types of Flagella

Many bacteria swim by means of a *flagellum* attached to a rotating "hook," which is driven by a "motor" that traverses the bacterial plasma membrane. A bacterial flagellum is built much more simply than a eukaryotic flagellum, consisting of a helical filament that assembles from one type of protein, *flagellin.* Flagella harvested from bacterial cells can be dissociated into flagellin monomers, and the assembly of flagella from flagellin monomers can be studied in vitro. These studies show that flagella assemble from flagellin and that the shape of the assembled flagellum depends on the available seed structure, such as a hook protein or a segment of a flagellum.

In a related study, Sho Asakura and coworkers (1966) used different strains of the *Salmonella* bacterium. One strain had wild-type flagella bent in a long, sinusoidal pattern with a 2.5-μm wavelength (Fig. 18.11). A *curly* mutant had flagella with a much shorter wavelength of 1.1 μm. Instead of swimming straight, the *curly* bacteria rotated about themselves. The mutant phenotype was caused by a single amino acid substitution in the flagellin monomer. The substitution restricted the ability of the monomer to assemble into a flagellum: whereas the wild-type flagellin could assemble into either wild-type flagella or *curly*-like flagella, the mutant flagellin could assemble only into *curly* flagella. This was revealed by the following observations.

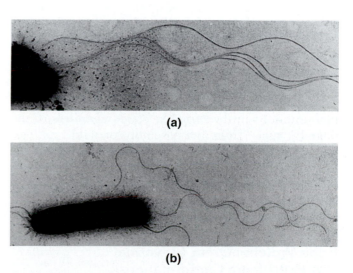

(a)

(b)

Figure 18.11 Electron micrographs of *Salmonella* bacterial cells. **(a)** Wild type. **(b)** *curly* mutant with flagella of shorter wavelength.

TABLE 18.1

Directed Assembly of a Bacterial Flagellum

Seed Structure	Genotype of Flagellin Monomer	Phenotype of Flagellum
Wild-type bacterial "hook"	Wild type	Wild type
Wild-type fragment	Wild type	Wild type
Wild-type fragment	*curly*	Curly
curly 0.2-μm fragment	Wild type	Wild type
curly 1.0-μm fragment	Wild type	Curly
curly fragment	*curly*	Curly
Any fragment	Wild type + *curly*	Curly

When either wild-type or curly flagellin was mixed with different primers, the phenotype of the reassembled flagella was generally determined by the nature of the flagellin monomer (Table 18.1). However, when wild-type flagellin monomers were mixed with moderately long (1.0-μm) *curly* flagellum segments as seeds, the result was surprising: the reassembled flagella had the *curly* phenotype. In contrast, the same wild-type monomers, when mixed with short (0.2-μm) *curly* flagellum segments or wild-type flagellum segments, assembled into wild-type flagella. The investigators also found that a combination of wild-type and curly monomers assembled into *curly*-like flagella with any type of seed. Moreover, wild-type filaments assumed the *curly* phenotype when incubated for a long time. The transformed flagella reverted to the wild-type conformation after treatment with low concentrations of pyrophosphate or adenosine triphosphate.

These results indicate that wild-type flagellin can reversibly change its tertiary structure between a state assembling into wild-type flagella and another state assembling into *curly*-like flagella. Which conformation the protein assumes depends on the chemical composition of the medium and on the *nature of the primer* that nucleates its assembly. With a wild-type primer, each added wild-type flagellin molecule seems to assume the wild-type configuration, thus propagating the assembly of monomers in this state. In contrast, a *curly* primer seems to start a self-propagating wave of monomer additions in the *curly*-like state, regardless of the flagellin type. Thus, the assembly of wild-type *Salmonella* flagellin is a simple case of *directed assembly*, although otherwise it has all the characteristics of a *self-assembly* process. Many cases of directed assembly also depend on accessory proteins and energy sources, characteristics of *aided assembly*.

Prion Proteins Occur in Normal and Pathogenic Conformations

The directed assembly of flagellin has an uncanny parallel in the transmission of fatal brain diseases such as scrapie. Scrapie results in progressive degeneration of

the central nervous system of sheep and goats, causing its victims to scratch and scrape, and eventually leads to debility and death. Similar diseases are bovine spongiform encephalopathy (BSE) in cows, and Creutzfeldt-Jakob disease (CJD) and Gerstmann-Sträussler syndrome (GSS) in humans (Prusiner, 1992). These diseases can be propagated by transferring brain extracts from diseased to normal individuals. However, the diseases can also be contracted by eating the brains of diseased individuals. For instance, an epidemic of BSE in England was traced to the addition of processed sheep carcasses to cow feed.

Although the infective agent for each of these diseases was originally thought to be a virus, no nucleic acid component has been identified (Weissman, 1994). Instead, Stanley Prusiner (1982) isolated from the brains of scrapie-infected hamsters an apparently nonviral agent that was capable of transmitting scrapie, and for which he coined the term **prion** (for *in*fectious *proteina*cous particle). He proposed that mammals produce a normal cellular prion protein (PrP^C) of unknown function, and that in each species a prion represents an altered, scrapie-inducing form (PrP^{Sc}) of that species' normal prion protein. The pathogenic PrP^{Sc} form appears to originate from the harmless PrP^C form through a posttranslational conversion, in which the PrP^{Sc} form (from the same species or another) acts as a template or catalyst. The PrP^C form is a secreted protein that is tethered to the plasma membrane of neurons and, to a lesser extent, of lymphocytes and other cells. The PrP^{Sc} form collects mostly in membranous vesicles inside the cells. The accumulation of these vesicles seems to cause the symptoms of scrapie and related diseases.

Prion proteins have been found in all mammals investigated and in chickens. In each species, the PrP^{Sc} form has a large core that is extremely resistant to acids and proteases, in accord with epidemiological data showing that the agents of scrapie are transmitted when brains from diseased individuals are eaten.

Physicochemical studies indicate that the three-dimensional structure of PrP^{Sc} is different from that of PrP^C (Pan et al., 1993). Whereas the PrP^C form has a higher proportion of α helices, the PrP^{Sc} form has a

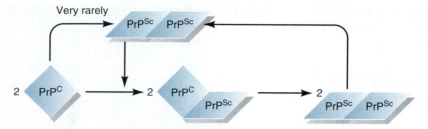

Figure 18.12 Model of the posttranslational modification of prion protein from the normal form (PrP^C) to the scrapie form (PrP^Sc). The modification is thought to be a conformational change that occurs very rarely in wild-type prion protein but more frequently in certain mutant prion proteins. When present, PrP^Sc can associate with PrP^C and distort it to yield more PrP^Sc.

higher content of β pleated sheets. The mechanism of the conversion of PrP^C to PrP^Sc is not fully understood, but the conversion is accelerated by a direct interaction between PrP^C and PrP^Sc, at least in vitro. In a cell-free system that favors the unfolding and refolding of proteins, PrP^C is converted to protease-resistant forms similar to PrP^Sc if preexisting PrP^Sc is added as seed (Kocisco et al., 1994). A similar process presumably occurs in vivo: once PrP^Sc is present, it can associate with PrP^C in such a way that more PrP^Sc is generated (Fig. 18.12). The prion diseases would then originate through a directed change in the conformation of a molecule, as observed in the flagellin protein of *Salmonella* and in other cases of directed assembly.

The conversion of PrP^C to PrP^Sc seems to occur spontaneously, although very rarely, in wild-type prion proteins. However, the probability of spontaneous conversion is greater in prion proteins encoded by certain mutant alleles of prion genes. Genes encoding prion proteins have been cloned in several species, and the prion genes of human families with hereditary CJD or GSS have been screened for mutations. In each family, at least one prion gene was found to be mutated. In 10 families from 9 different countries, the mutations caused the same amino acid substitution: leucine for proline at residue 102 (G. A. Carlson et al., 1991). Other investigators examined the prion genes of individuals with nonhereditary CJD; almost all these patients were found to be homozygous at amino acid residue 129 of the prion protein, a site where the majority of the normal population is polymorphic and heterozygous (M. S. Palmer et al., 1991). These data suggest that the homozygous genotype at residue 129 predisposes its carriers to CJD. Similarly, sheep homozygous for a prion allele that encodes a glutamine instead of an arginine residue in position 171 are more prone to developing scrapie (Westaway et al., 1994).

Transgenic mice in which both copies of the normal prion (PrP^C) gene have been replaced with a disrupted allele develop like normal mice, although in electrophysiological experiments they show impaired synaptic inhibition (Büeler et al., 1992; Collinge et al., 1994). Such mice without functional prion genes were tested

for a response to infection with PrP^Sc (Büeler et al., 1993; Prusiner et al., 1993). The animals remained healthy for more than 500 days after infection (Fig. 18.13). In contrast, normal mice subjected to the same treatment showed neurological dysfunctions within less than 165 days, and heterozygous mice with one normal prion gene developed dysfunctions between 400 and 465 days after infection. More-over, brain homogenates from infected mice without functional prion genes failed to make normal mice sick, whereas brain homogenates from infected normal mice transferred the disease. These data show that the scrapie agent cannot propagate in the absence of normal prion protein.

In summary, the available data indicate that there are two forms of the prion protein, PrP^C and PrP^Sc, which differ in three-dimensional conformation. The PrP^Sc conformation by itself seems sufficient to cause

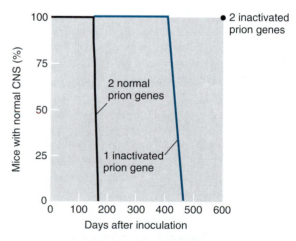

Figure 18.13 Dependence of scrapie on the expression of the prion gene. Scrapie agent from diseased mice was transferred to healthy mice with two normal prion genes, and to healthy transgenic mice in which one prion gene or both prion genes had been inactivated. The normal mice began to show neurological dysfunction within less than 165 days, and the mice with one functional prion gene developed the same symptoms after 400 to 465 days. In contrast, the mice without functional prion genes were all symptom-free after more than 500 days.

scrapie and related neurodegenerative diseases, presumably by directing the conversion of PrP^C into more PrP^Sc.

Tubulin Dimers Assemble into Different Arrays of Microtubules

The *tubulin* polypeptides synthesized in eukaryotic cells form heterodimers, consisting of one α- and one β-tubulin. These dimers assemble into the cylinders known as *microtubules* (see Chapter 2). Depending on cell type and phase in the cell cycle, microtubules form different arrays (Fig. 18.14). In mitotic cells, many short and straight microtubules radiate out from two *centrosomes* at the spindle poles (see Fig. 2.3). During interphase, a loose network of long and wavy microtubules, often forming a tight meshwork around the nucleus, extend

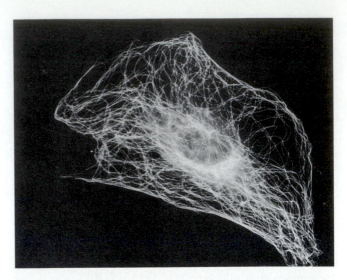

Figure 18.15 Microtubules in a cultured mouse 3T3 cell during interphase. The cells were fixed, and microtubules were made fluorescent by immunostaining (see Methods 4.1). The microtubules radiate out from the centrosome, which lies just outside the nucleus.

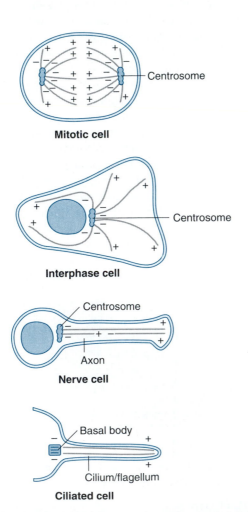

Figure 18.14 Microtubular arrays in various eukaryotic cells at different stages of the cell cycle. Most microtubules originate from a microtubule organizing center (MTOC). The two most common MTOCs are centrosomes and basal bodies. The position and properties of centrosomes change during the cell cycle. A plus sign indicates the growing end of a microtubule. A minus sign indicates the end that is shrinking unless it is anchored in an MTOC.

from one centrosome (Fig. 18.15). In neural plate cells, oriented microtubules support the columnar cell shape (see Fig. 11.7). In many protozoa and epithelial cells, microtubules form the core of *cilia* and *flagella*, as we will see. In neurons, microtubules are part of the cytoskeletal lattice stabilizing the axon and providing a bidirectional transport system.

A particular type of microtubular array, known as the **axoneme**, forms the core of all cilia and flagella in eukaryotic cells. (Eukaryotic flagella and cilia are very similar. Flagella are generally longer, and there are usually one or a few per cell. Here, we refer to *both* structures as cilia.) In all axonemes, microtubules are arranged in a universal pattern of nine *doublets* and two *singlets* (Fig. 18.16). Each singlet consists of 13 *protofilaments* (longitudinal rows of tubulin dimers) and is similar to a cytoplasmic microtubule. A doublet consists of two joined subfibers, designated A and B. In cross section, subfiber A contains the normal circular arrangement of 13 protofilaments. Subfiber B is sickle-shaped in cross section and consists of only 10 or 11 protofilaments. The doublet is associated with *dynein* and other proteins that generate the ciliary beat.

The different arrays of microtubules and axonemes assemble from the same cellular pool of tubulin dimers. Nevertheless, each type of cell displays a selective microtubular array rather than bits of everything. Indeed, the tubulin needed for the assembly of one array is often provided by the disassembly of another array. For instance, during mitosis, the microtubular meshwork characteristic of interphase disappears while the mitotic spindle is built. These observations indicate that cells *direct* the assembly of microtubular arrays to specific sites depending on cell type and developmental stage.

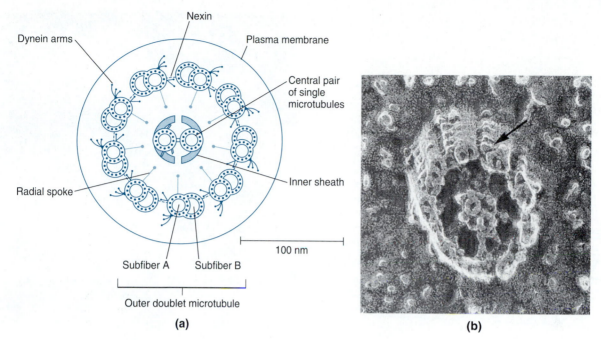

Figure 18.16 Axoneme structure shown in cross section. **(a)** Schematic diagram showing the universal arrangement of two singlet and nine doublet fibers. Each doublet fiber is composed of one circular subfiber (subfiber A) and a sickle-shaped subfiber (subfiber B). Associated with each A subfiber are "arms" of dynein molecules, which interact with the adjacent subfiber to generate the ciliary beat. Holding adjacent microtubule doublets together are proteins called nexins. Radial spoke proteins extend from each circular subfiber to the center of the axoneme. Sheath projections extend from the central singlet microtubules. The entire cilium is surrounded by a protrusion of the cell's plasma membrane. **(b)** Electron micrograph of an axoneme of a flagellate protozoon, *Chlamydomonas*. Arrow points to dynein arm.

Which type of microtubular array is assembled in a given cell at a particular time depends on available organizing structures and on other proteins that associate with microtubules. The organizing structures for microtubule assembly in cells are called ***microtubule organizing centers (MTOCs)***. The assembly of each microtubule begins with the (−) end (see Chapter 2), which is stabilized by the MTOC. Construction of most cytoplasmic microtubules is initiated by MTOCs known as *centrosomes*. In animal cells, centrosomes contain a pair of centrioles, surrounded by a cloud of amorphous material. Other MTOCs known as *basal bodies* initiate the assembly of the axonemes in cilia and flagella. *Microtubule-associated proteins* act to stabilize microtubules against disassembly and to mediate their interactions with other cell components. By controlling the localization and molecular composition of these MTOCs, and by gauging the availability of microtubule-associated proteins, cells can direct the assembly of different microtubular arrays.

Centrioles and Basal Bodies Multiply Locally and by Directed Assembly

Having recognized the importance of *centrosomes* and *basal bodies* as MTOCs, we will now examine how these critical structures are replicated and positioned in cells, and how their composition changes during the cell cycle. Basal bodies will also figure prominently in our subsequent discussion of the directed assembly of larger cortical structures in ciliates.

A ***basal body*** is located at the base of each cilium right beneath the plasma membrane; the basal body anchors the axoneme of the cilium in the cortical cytoplasm of the cell. Each basal body is a small cylinder consisting of nine microtubule triplets (Fig. 18.17). Each triplet consists of three subfibers, designated A, B, and C. Subfibers A and B extend directly into the corresponding subfibers of the axoneme (Fig. 18.16). In ciliate protozoa, new basal bodies originate at right angles to preexisting basal bodies. Later, the daughter basal body tilts up and assumes its definitive position parallel to the parent basal body and perpendicular to the cell surface.

Most animal cells during interphase contain one *centrosome* near the nucleus. It consists of a pair of *centrioles* and some surrounding amorphous material. Each centriole is built like a basal body, except that it is not connected to an axoneme. The two centrioles are oriented at right angles to each other, forming an L-shaped configuration. The centrosomes of higher plants are de-

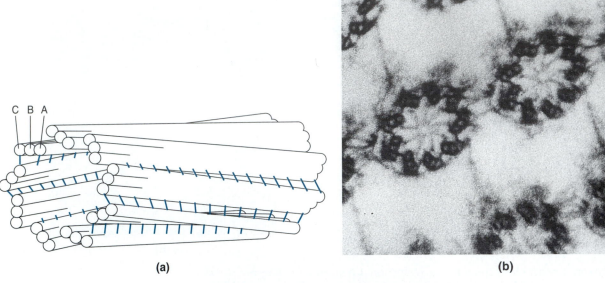

C B A

(a)

(b)

Figure 18.17 Structure of a basal body. **(a)** Schematic diagram. A basal body has an overall cylindrical shape, about 0.4 μm long and 0.2 μm wide. For the most part, it consists of nine microtubule triplets. The inner two subfibers of each triplet (designated A and B) continue into the corresponding subfibers of axoneme doublets (see Fig. 18.16). The outer subfiber terminates in a transition zone between the basal body and the axoneme shaft. The two central singlets of the axoneme also begin in this transition zone. Other proteins (color) link the triplets together. **(b)** Electron micrograph of basal bodies in the cortex of a protozoon. In the proximal portion of a basal body, faint protein spokes radiate from a central core to each of the triplets.

void of centrioles, suggesting that the functional part of centrosomes in all types of cells may be the amorphous material. This material is poorly characterized except that it associates with γ-tubulin as an apparent link to the ($-$) end of microtubules (Stearns and Kirschner, 1994).

The replication of centrioles is coordinated with the cell cycle (Fig. 18.18). During S phase, each centriole nucleates the assembly of a daughter centriole. The daughter centriole is oriented perpendicular to the parent centriole and grows at the end facing away from the parent. During mitotic prophase, the two daughter centrosomes move to opposite poles of the nucleus, where they direct the assembly of the mitotic spindle.

The molecular composition of both centrosomes and microtubule-associated proteins changes with the cell cycle. Each centrosome seems to be able to organize a maximum number of microtubules. This number increases considerably during mitotic prophase. The change may explain the transition from the interphase array of relatively few, long microtubules to the mitotic spindle containing many short microtubules.

The stereotypical arrangements of replicating centrioles and basal bodies indicate that each parent struc-

ture directs the assembly of its daughter. It has been a matter of considerable debate whether basal bodies and centrioles contain DNA or RNA. Nucleic acids may encode proteins that play special roles in the replication of basal bodies and centrioles. Conceivably, the local availability of special proteins and the *spatial order* of the preexisting MTOCs could promote the replication of the MTOCs.

Although basal bodies and centrioles usually originate by duplication of their respective parent structures, they can also arise from less conspicuous precursors. Some epithelial cells form hundreds of cilia on their apical surfaces. The basal bodies of these cilia arise simultaneously from electron-dense "satellites" formed in close association with the centrosome of the nonciliated precursor cell. Likewise, the eggs of many animal species lack conspicuous centrosomes and use the centrosome introduced by the sperm to set up their first mitotic spindle. If such eggs are activated parthenogenetically, some nevertheless form a centrosome, presumably from some precursor structure. Such precursors seem to exist in eggs and normally remain quiescent in the presence of a fully formed centrosome introduced by the sperm.

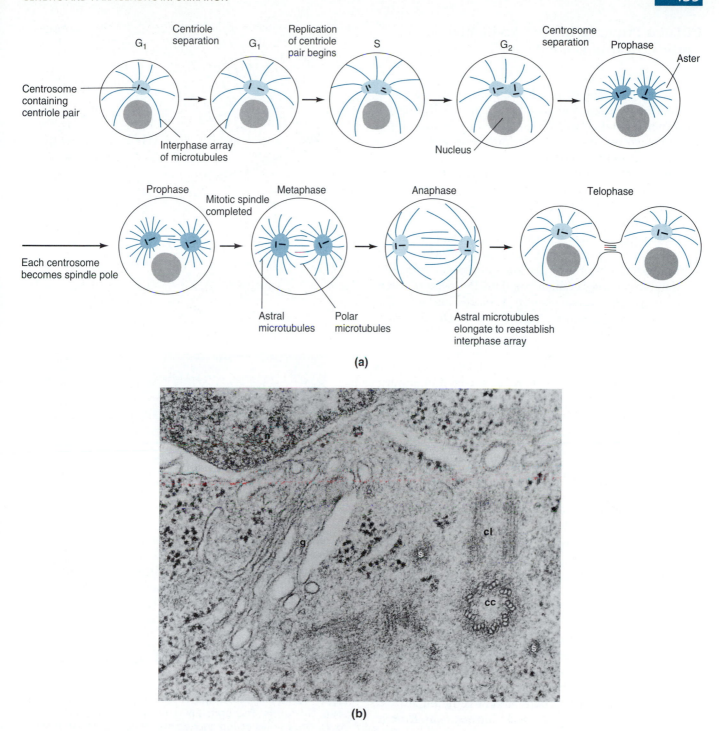

(a)

(b)

Figure 18.18 Centrosome replication during the mitotic cycle. **(a)** Schematic diagram. The centrosome consists of a pair of centrioles (black bars) surrounded by amorphous material (light color). During interphase (G_1, S, G_2), the centrosome anchors a relatively small number of long microtubules. During S phase, each centriole directs the assembly of a daughter centriole oriented at right angles to its parent centriole. During prophase, the two centriole pairs split and move apart, forming two centrosomes. At the same time, the molecular composition of the amorphous material changes (dark color). Each centrosome then nucleates a larger number of shorter microtubules, forming an aster. After the nuclear envelope breaks down, the two asters set up the mitotic spindle. During cytokinesis, each daughter cell inherits one centrosome to organize a new interphase array. **(b)** Transmission electron micrograph showing two pairs of centrioles at G_2 phase. One centriole of each pair appears in longitudinal section (cl) and the other in cross section (cc) because they are oriented at right angles to each other. g = Golgi apparatus; s = satellite.

Ciliated Protozoa Inherit Accidental Cortical Rearrangements

A spectacular instance of directed assembly occurs in the ciliated protozoon *Paramecium aurelia.* To appreciate the experimental results obtained with this unicellular animal, we must first familiarize ourselves with its morphology and reproduction.

Dubbed the "slipper animal" because of its overall shape, *Paramecium* consists of one large cell (Fig. 18.1) and is about 0.1 mm long. The cell has an anteroposterior and a dorsoventral polarity. The anterior end moves ahead during normal swimming, and the cilia at the posterior end are elongated. An *oral apparatus* on the ventral side gathers food for phagocytosis, while two contractile vacuoles, which regulate the osmotic pressure, open on the dorsal surface. The right and left sides of the cell are defined by analogy to bilaterally symmetrical metazoa.

The surface layer, called the *cortex,* is more viscous and structured than the interior cytoplasm. The cortex consists of about 4000 *cortical units* arranged in longitudinal rows. Each unit has at least one cilium anchored in a *basal body.* A **kinetodesmal fiber** extends anteriorly from the basal body, and a **parasomal sac** is located to the right of the kinetodesmal fiber (Fig. 18.19). This arrangement gives each cortical unit an anteroposterior polarity and a right-left polarity. Normally, all cortical units in an individual are oriented the same way. This uniformity is functionally important because it provides for a uniform orientation of the cilia, which in turn is a prerequisite for the coordinated ciliary strokes that propel the animal.

Paramecium reproduces by a transverse division process called *fission* (Fig. 18.20). Each fission is accompanied by both longitudinal growth and the duplication of each cell organelle. Likewise, the number of cortical units doubles during each growth cycle. This involves the generation of new basal bodies, a process that closely resembles the generation of new centrioles (Fig. 18.18). Each new basal body arises *immediately anterior* to an existing basal body and at right angles to it. Once formed, the new basal body moves away from its parental basal body and tilts upward to make contact with the cell surface. Here it is positioned exactly in line with the other basal bodies of the same row. Meanwhile, the other organelles are also duplicated and the cortical unit elongates. The unit is then cross-partitioned, giving rise to two daughter units next to each other in the same row.

Independently of its reproduction by fission, *Paramecium* engages in a sexual process called *conjugation* (Fig. 18.21), during which two animals adhere to each other with their oral apparatuses. Each conjugant forms two haploid nuclei and transfers one of them to its mate. Thereafter, the partners separate.

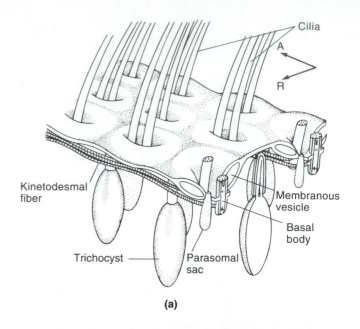

(a)

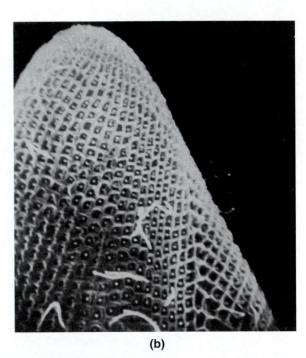

(b)

Figure 18.19 Organization of the cortex of *Paramecium* in repetitive units. The units are arranged in longitudinal rows. Organelles called trichocysts, which function in self-defense, are located between each unit and its neighbors in the same row. Each cortical unit is covered by a hexagonal membranous vesicle. From a depression in each vesicle emerge one or two cilia, each anchored in a basal body. A striated band called a kinetodesmal fiber originates from one basal body and extends anteriorly past the basal bodies of the anteriorly adjacent units. To the right of the kinetodesmal fiber in each unit is a blind sac called the parasomal sac. The arrangement of these structures gives each unit an anteroposterior and a right-left polarity, indicated by the intersecting arrows. (In describing ciliates, the same right-left convention is used as for the human body: in a ventral view, the cell's right is on the observer's left-hand side.) **(a)** Interpretative drawing; **(b)** scanning electron micrograph of the anterior ventral surface.

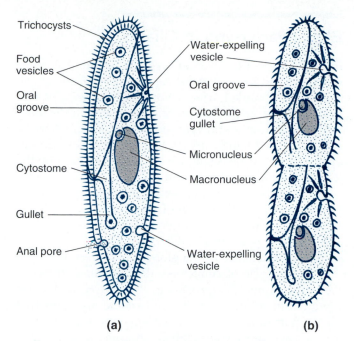

(a)   **(b)**

Figure 18.20 Organization and fission of *Paramecium caudatum*. **(a)** The ventral aspect shows the oral groove, through which cilia drive algae and other food particles into the cytostome (cell mouth), from where the food passes down a gullet. Food vesicles bud off the gullet and circulate through the cell until they are exocytosed at the anal pore. Two water-expelling vesicles regulate the osmotic pressure inside the cell. The genes of the polyploid macronucleus are actively transcribed, whereas the diploid micronucleus undergoes meiosis during conjugation. **(b)** *Paramecium* propagates by transverse fission. Before each fission, all cell organelles are duplicated.

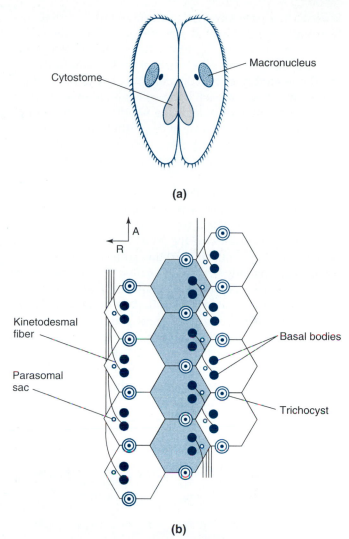

Figure 18.21 Conjugation and "cortical picking" in *Paramecium*. **(a)** Drawing of a conjugating pair before meiosis and exchange of haploid nuclei. When the partners separate after conjugation, one may take a piece of cortex from the other. **(b)** Transplanted row of cortical units observed after conjugation. This diagram shows part of two normally oriented rows of cortical units flanking a single row that is rotated 180° (color). Arrows point to the anterior (A) and to the right (R) of the entire cell.

When two individuals separate after conjugation, they sometimes take portions of their mate's cortex with them. This "cortical picking" amounts to a transplantation of one or two partial rows of cortical units from one animal to another. Because of the orientation in which the animals conjugate and separate, the transplanted row may wind up rotated 180° in the plane of the cell surface (Fig. 18.21b). As a result, the stroke of the cilia is reversed in the transplanted row. This causes an abnormal swimming pattern, which is readily identifiable under the dissecting microscope. Animals having this feature can be isolated and propagated into a clone of cells.

Observation of *Paramecium* individuals with cortical transplants has provided the most striking example of directed assembly. The pioneering study of this type was carried out by Janine Beisson and Tracy M. Sonneborn (1965). They observed that the transplanted patches grew during fission cycles until they extended over the full length of the cell surface. In the transplant, new basal bodies were formed with the correct orientation to their cortical units—i.e., to the *posterior* (with regard to the entire cell) of the preexisting basal bodies. After extension of the transplant to a full row, the progeny kept inherit-

ing complete inverted rows. Occasionally, transplanted rows were lost during fission, as revealed by the recovery of the normal swimming pattern in one daughter cell. When such animals were eliminated from the culture, the original inversion could be maintained for hundreds of generations and remained unaffected by subsequent conjugations with normal animals. The experiment has been repeated, with the same results, by several investigators.

The experimental results indicate that the assembly of new cortical units in *Paramecium* is directed by the "local geography" of preexisting units in the same row. The molecular basis of this directed assembly is not known. It is noteworthy that normal and inverted cortical units are duplicated side by side. Thus, the cellular genome encodes the molecular building blocks for both normal and inverted cortical units without directing the assembly of one or the other. The genome does not even provide a bias, because conjugations between normal and altered mates have no effect on the propagation of inverted cortical units. Rather, there seems to be an unbroken chain of *paragenetic information* that is passed on from one cortical unit to the next unit originating in the same row. This conclusion holds true whether or not the basal bodies in *Paramecium* contain DNA or RNA. The act of cortical picking (which starts the inversion of cortical rows) cannot possibly change the *nucleotide sequence* of the basal body DNA or RNA of dozens of cortical units in the same way. Of course, the cortical picking would invert the *spatial orientation* of any basal body DNA or RNA, as it does with all other molecules in a cortical unit. And basal body DNA or RNA might be part of the local geography that perpetuates itself when cortical units are duplicated before fission. Thus, with or without basal body DNA or RNA, it is the *local orientation of supramolecular assembly* that is passed on from each cortical unit to its daughter units.

Global Patterning in Ciliates

In addition to the orientation of their small cortical units, ciliates have a *global body pattern* of major cell organelles (Frankel, 1989). For instance, *Pleurotricha lanceolata* (Fig. 18.22) has an oral apparatus in the anterior left quadrant of its ventral side and a characteristic pattern of **cirri** (sing., *cirrus*), that is, clusters of hexagonally packed cilia. A single row of cirri is found on the left margin, a double row on the right margin, and a constant pattern of cirri in between. The dorsal surface is covered with a thin lawn of cilia (Fig. 18.22a). These structures confer distinct anteroposterior, dorsoventral, and right-left polarities on the cell.

The maintenance of the global cell pattern throughout the life cycle of ciliates is called **global patterning.** We will examine this process in *Pleurotricha* with three thoughts in mind. First, it will become apparent that global patterning relies on paragenetic information. Second, we will see that global patterning preserves the overall body pattern but not necessarily all the details. Third, we will discuss a critical observation showing that global patterning is independent of the directed assembly of individual organelles.

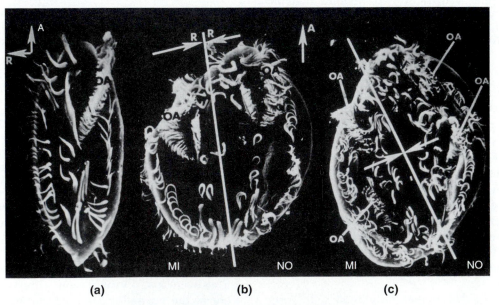

(a) (b) (c)

Figure 18.22 Scanning electron micrographs of *Pleurotricha lanceolata,* a ciliate. **(a)** Ventral view of a normal cell. Arrows indicate the position of the anterior pole and the right-hand side. On the ventral surface are an oral apparatus (OA) and additional arrangements of packed cilia. The most conspicuous part of the oral apparatus is an oblique band of membranelles in the anterior left quadrant. Each membranelle consists of about 30 cilia packed in a rectangular array. In addition, clusters of cilia are arranged in one row to the left, in two rows to the right, and in a specific pattern in between. **(b)** Mirror-image doublet, fused at the right cell margins and showing duplications of the oral apparatus and the left ciliary row on both sides of the symmetry plane (white line). **(c)** Prefission doublet, showing duplication of the oral apparatus and other ciliary pattern elements. MI = mirror-image half of doublet; NO = normal half of doublet.

The Global Pattern in Ciliates Is Inherited during Fission

Each time a ciliate multiplies by fission, the cell's *global* pattern also duplicates. However, the mechanisms underlying the duplication of the global pattern appear to be different from the directed assembly processes discussed so far. For instance, a second oral apparatus is formed *far away* from the original apparatus in the posterior half of the cell, so that after fission both daughter cells have an oral apparatus in the appropriate place (Fig. 18.20). The new oral apparatus, along with other duplicated pattern elements, is positioned with great precision. This suggests that ciliates have some *long-range* control mechanisms that maintain their global pattern during fission. These mechanisms seem to differ from the *close-range* mechanisms of directed assembly discussed so far, in which the newly assembled structures originate next to their seed structures or organization centers.

Using the *method of controlled modification,* investigators have analyzed global patterning by generating abnormal cellular patterns. This they can do by heat-shocking, by preventing cells from separating after conjugation, and by microsurgery (X. Shi et al., 1991). Incidentally, none of these treatments is known to cause any changes in DNA or RNA. The most spectacular of the resulting pattern abnormalities are **mirror-image doublets** (Fig. 18.22b). The doublets are like a cell and its mirror image, joined at the right margins, with both left margins turned laterally, anterior poles pointing in the same direction, and ventral structures facing the same way. The overall arrangement resembles the two hands of a person held side by side with both palms facing in the same direction. The global ciliary pattern of the left half corresponds to that of a normal singlet cell, while the right half takes on a mirror-image pattern. We will refer to the two parts as the *normal half* and the *mirror-image half* of the doublet.

Before doublets divide by fission, the components of the ciliary patterns are duplicated in both halves (Fig. 18.22c). For instance, each half generates a second oral apparatus in its posterior region. Thus, doublets beget doublets. Occasionally doublets revert to singlets by resorption of the right (mirror-image) component. However, when such revertants are removed from a culture, the doublet pattern is stable for hundreds of generations. These observations show that the global cell pattern, whether normal or abnormal, is maintained during fission.

The Global Cell Pattern Is Maintained during Encystment

Pleurotricha and related ciliates can survive periods of starvation by **encystment:** they form a structure called a *cyst,* which is surrounded by a protective coat. When environmental conditions are favorable again, the cells undergo *excystment:* they emerge from the protective coat and resume a normal growth and division cycle. What happens with respect to global patterning during encystment and excystment is striking. All ciliary structures seem to disappear during encystment; no parts of cilia or basal bodies have been found in cysts by electron microscopy (Grimes and Hammersmith, 1980). However, these structures apparently reassemble from smaller organizing centers after excystment, perhaps in a process similar to the assembly of centrosomes in unfertilized eggs as discussed earlier. The disappearance of all ciliary structures during encystment means the loss of an *overt* global body pattern. However, the faithful reappearance of the same body pattern after excystment shows that the pattern is preserved by some other, as yet unknown type of paragenetic information.

Investigators have studied encystment to find out which aspects of *abnormal* body patterns are maintained in the absence of ciliary structures. Under varying environmental conditions, doublets encyst and excyst like normal cells. Surprisingly, and so far inexplicably, doublets always excyst as doublets, thus restoring the *global* body pattern that existed before encystment (Grimes, 1990). In contrast, *local* irregularities, such as some ventral cilia placed on the dorsal side of otherwise normal cells, are *not* reconstituted upon excystment. (Remember that similar irregularities are propagated faithfully during fission.) Apparently, the paragenetic information passed on through encystment is sufficient to preserve the *global* aspects of both normal and abnormal body patterns. Also, the global body pattern is maintained in such a way that the cirri can reconstitute themselves from cilia. However, the paragenetic information maintained through encystment does not suffice to preserve *local abnormalities* in the ciliary pattern.

Global Patterning Is Independent of Local Assembly

The global pattern of *Pleurotricha* mirror-image doublets gives the impression of perfect mirror-image symmetry. For example, the oral apparatus of the normal half curves to the left, while that of the mirror-image half curves to the right (Fig. 18.22b). However, the impression of symmetry vanishes when we study the *local* organization of individual pattern elements. For instance, each oral apparatus consists of many **membranelles,** each of which is typically made up of four rows of basal bodies with cilia (Fig. 18.23). These rows assemble from pairs of basal bodies that have a *handedness,* which they impose on the membranelle. This handedness is the *same,* rather than *symmetrical,* in the two oral apparatuses of the doublet (Grimes et al., 1980). This is significant because it shows that the same local membranelle organization is compatible with *both* the normal global pattern and its mirror image. In this case, therefore, the global pattern and

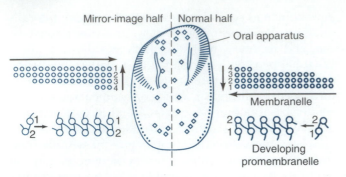

Figure 18.23 Discrepancy between the global pattern and the local assembly of oral apparatus membranelles in *Pleurotricha lanceolata*. The diagram in the center represents the ventral aspect of a mirror-image doublet, with each part having its own oral apparatus. The overall arrangement of the other ciliary structures is also approximately symmetrical with respect to a plane indicated by the dashed line. The top diagrams on each side represent mature membranelles, consisting of four rows of basal bodies. The bottom diagrams show the assembly of the first two rows of a membranelle from dimers of basal bodies. The membranelles in the two halves of the doublet are not mirror images of each other. Instead, the membranelles in the mirror-image half of the doublet are rotated 180° relative to the membranelles in the normal half. It appears that the basal body dimers can be assembled in only one configuration, so that true mirror-image assembly of a membranelle is not possible.

the local assembly of the pattern elements are not coordinated and, presumably, are determined by different types of paragenetic information.

A close look at the local assembly of oral apparatus membranelles will illustrate this important point. Each membranelle of the oral apparatus is composed of four rows of cilia, as we have seen. In a normal animal, and in the normal part of a doublet, the two posterior rows are the longest. The anteriormost row is composed of only three cilia at the anterior right corner of each membranelle. If this local assembly were coordinated with the overall pattern, then the membranelles in the two parts of the doublet would be mirror images of each other. However, this is not the case. In the mirror-image part of the cell, *both* the anteroposterior polarity and the right-left polarity are reversed. It is as though the membranelle had gone through a planar 180° rotation instead of developing as a mirror image. The local membranelle assembly and the overall pattern of the oral apparatus are therefore incongruent. This inconsistency is also reflected in the abnormal function of the oral apparatus in the mirror-image half of the doublet. Instead of moving food particles into the oral apparatus, the membranelles move them out. Consequently, doublets feed only through the oral apparatus of the normal half, and an isolated mirror-image half of a doublet will starve to death.

The progress of oral apparatus formation suggests that membranelles can be assembled only in their nor-

mal configuration. This restriction seems to be imposed by the three-dimensional structure of basal body pairs. The development of true mirror images of membranelles would require right-handed and left-handed versions of basal body pairs. Since these do not exist, the anteroposterior axis has to be inverted along with the left-right axis. What is important for us to observe is that the rotated assembly pattern of the membranelles does *not* impose a rotated pattern on the entire oral apparatus. Therefore, the global pattern of the cell must be determined independently of the local membranelle assembly pattern.

Ciliates, then, employ more than one type of paragenetic information to maintain their body pattern through fission and encystment. Cortical units, membranelles, and other local pattern elements originate through local directed assembly. However, the global cellular pattern, including the shape and position of the oral apparatus, is determined independently.

The physical basis of global patterning in ciliates is not yet known. The observations reviewed in this section suggest that it is not the close-range mechanisms that underly the cases of directed assembly discussed previously. Nor can genetic mutations explain the global pattern changes discussed here, for two reasons. First, the experimental manipulations that generate doublets are not mutagenic. Second, the doublet pattern is unaffected by conjugation with either normal cells or doublets. In corresponding experiments with *Paramecium*, Sonneborn (1963) showed that a doublet phenotype is not associated with nuclear genes or with components of internal cytoplasm that are exchanged during conjugation. It appears that the factors determining the global pattern are associated with the cell cortex. However, the loss of cilia and basal bodies during encystment indicates that the pattern determinants are independent of these overt structures.

Paragenetic Information in Metazoan Cells and Organisms

What do the lessons learned from ciliates tell us about paragenetic information in the cells of metazoans, such as fruit flies or mammals? Clearly, the cell surfaces of ciliates are highly specialized for multiple functions in swimming, self-defense, and nutrition. However, the cells of metazoa also have an internal organization that is critical to their survival and function. Is this spatial organization of a cell passed on paragenetically, or is it lost at each cell division and reestablished through external cues, such as contact with other cells or extracellular materials? Both processes occur to varying degrees in different cells. We will focus here on paragenetic information.

In every cell, newly synthesized polypeptides have to be targeted to the appropriate cellular compartments. As discussed in Chapter 17, polypeptides have specific *signal sequences* or *signal patches,* which interact with matching *receptor proteins.* The interactions either initiate the transfer of the polypeptide across a cellular membrane or keep the polypeptide where it is (Fig. 18.24). For instance, proteins entering the endoplasmic reticulum (ER) have at their amino terminus a signal sequence that associates with a *signal recognition particle* and then with a *receptor protein* in the ER membrane (see Fig. 17.15). In addition, proteins enclosed in transport vesicles are guided to their destinations by proteins on the vesicle surfaces. These mechanisms direct the traffic of vesicles to mitochondria, to the nucleus, and toward other cellular destinations. Without the appropriate guidance of its polypeptide traffic, a cell would not be viable.

Targeting the correct destination depends on each polypeptide's own signal sequences and on matching

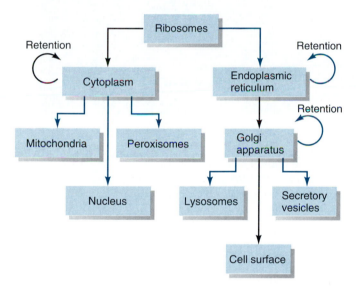

Figure 18.24 Simplified "road map" showing how polypeptides reach their cellular destinations. The signals that guide a polypeptide's movement along its route are contained in its amino acid sequence and in peptides on the outside of transport vesicles. These signals interact with matching receptors on the membranes of the appropriate cellular compartments. The journey of a polypeptide begins with its synthesis on a ribosome. At each station in its journey, a decision is made whether the polypeptide is to be retained or transported further. A signal may be required either for retention or for moving on. (The signal-dependent pathways are shown in color.) In the absence of the required signal, the polypeptide takes a default pathway (black). The signal sequences that determine the route of a polypeptide are encoded by the gene for the polypeptide. The molecular properties of the receptors with which the signal sequences interact are also encoded genetically. However, the distribution of these receptors in the cell is part of the preexisting cell structure. This structure is passed on, independently of DNA replication, during cell division.

sequences in signal recognition particles and receptors, all of which are controlled by the genes encoding these components. However, the *correct distribution* of receptors is not encoded in genes but is part of the preexisting spatial order in the cell. During mitosis, each new cell inherits a cytoplasmic organization that already has in place a complement of organelles with receptors for the signal sequences of new polypeptides. The correct positioning of the receptors presumably depends on the preexisting membrane structure and therefore would be classified as *directed assembly.*

Evidence for the direct propagation of cytoplasmic organization during cell division has also been obtained from observations on cultured 3T3 mouse fibroblasts (Albrecht-Buehler, 1977). During mitosis, these cells produce pairs of sister cells that look either identical or like mirror images of each other. The similarity is especially striking with respect to cell shape and microfilament organization (Fig. 18.25). Moreover, paired daughter cells leave behind mirror-image tracks as they move. Corresponding observations have been made on neuroblastoma cells (Solomon, 1981). These cells display a wide repertoire of morphological characteristics. Mitotic pairs of these cells are very similar with respect to cell shape and number of neurites as well as length, thickness, and branching pattern. These observations show that cytoplasmic structures controlling cell morphology and cytoskeletal organization are passed on during cell division.

Finally, one can argue that spatial attributes of multicellular organisms, such as their anteroposterior and dorsoventral polarity, are passed on as paragenetic information throughout the life cycle. A fertilized egg contains all the genetic and paragenetic information necessary to carry the organism through all the subsequent stages in its life. Genetic information is encoded in the DNA of the nucleus, of the mitochondria, and possibly of other organelles. Paragenetic information is contained in several egg features, including *localized cytoplasmic determinants* (see Chapter 8).

How do animal eggs propagate the paragenetic information encoded in their cytoplasmic organization? Clearly, the egg organization is *not* passed on in its entirety to all blastomeres. Instead, different blastomeres may inherit different cytoplasmic components and may thus be determined to form different parts of the developing embryo. Inductive interactions between different cells, and morphogenetic movements guided by these events, generate the embryonic body pattern, which in turn gives rise to the adult body pattern. The paragenetic information that was originally localized in specific regions of the egg cytoplasm is now manifest in spatial relationships between the adult cells and organs. These relationships include the configuration of oocytes and other cells in the ovary.

How does the adult anatomy regenerate cytoplasmic localization in the next generation of eggs? As de-

(a)

(b)

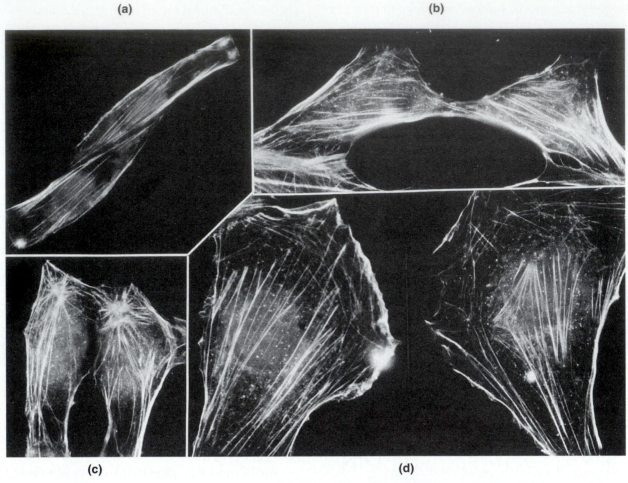

(c)

(d)

Figure 18.25 Microfilament bundles in mouse 3T3 fibroblasts immunostained for actin. Each photograph shows a pair of sister cells after mitosis. Note that the overall cell shape and the distribution of microfilaments are either identical **(a, c)** or mirror-symmetrical **(b, d)**.

scribed in Chapter 8, *Drosophilia* females have egg chambers, in which all nurse cells point toward the distal end of the ovariole and toward the head of the mother. The influx of cytoplasm from the nurse cells causes bicoid mRNA to be localized near the point of entry into the oocyte. The site of bicoid mRNA localization in the oocyte then establishes the anterior pole of the oocyte. One could therefore argue that eggs pass on their cytoplasmic organization by building adults that will recreate the same organization in their eggs. Here, the paragenetic information for the anteroposterior polarity of the organism again resides in a localized cytoplasmic determinant.

In order to prove that cytoplasmic localization is necessary for an animal to complete its life cycle, it must be shown that disturbances of localization processes interfere with survival or reproduction. This has been shown by both genetic and experimental means. For instance, there are maternal effect mutants of *Drosophila* that transfer normal amounts of functional bicoid

mRNA to the oocyte but fail to properly localize the mRNA at the entry point. Embryos developing from such eggs have abnormal body patterns and are not viable. Similarly, if anterior determinants in chironomid eggs are dislodged or inactivated by experimental means that do not affect nuclear DNA, the eggs nevertheless develop abnormal body patterns characterized by mirror-image duplications of the head or abdomen or by complete inversion of the anteroposterior polarity (see Fig. 8.25). Such embryos are inviable or infertile.

In summary, both genetic and paragenetic information are passed on throughout the life cycles of higher organisms, and both are indispensable for normal development. Some types of paragenetic information, such as cell shape and cytoskeletal organization, may be passed on directly during mitosis. Other types of paragenetic information, such as cytoplasmic localization in eggs, are propagated in a cyclical fashion by different, stage-specific means.

SUMMARY

Cells originate only from preexisting cells. During division, cells inherit two types of information: genetic and paragenetic. Genetic information, encoded in DNA, is used in daughter cells to synthesize molecular building blocks for maintenance and growth. Paragenetic information is present in structural aspects of cellular organization and is not encoded in DNA.

Many biomolecules undergo self-assembly, a spontaneous association for which all the necessary energy and information are contributed by those components that form the final product. During self-assembly, proteins may change in conformation as they release energy and assume a more stable state. Self-assembly generates oligomeric proteins, small organelles such as ribosomes, and simple viruses. More complex organelles and viruses, on the other hand, are formed by aided assembly, a process that requires auxiliary components—such as catalysts, energy sources, or scaffolds—that will not become part of the assembled product. Self-assembly and aided assembly do not strictly depend on paragenetic information. However, their initial phases are greatly accelerated by the presence of seed structures.

The most prevalent assembly process in cells is directed assembly. This term is used for assembly requiring seeds or larger organizing centers, or when a decision must be made because the same molecular building blocks can assemble into different products. Which product is formed then depends on paragenetic information present in the form of alternative seeds or organizing centers. In ciliated protozoa, basal bodies and other cortical structures are duplicated by directed assembly before cells undergo fission. Local irregularities in the cortex are passed on indefinitely in this process. The inheritance of such cortical abnormalities is not associated with any genetic changes, but is propagated locally by the directed assembly of cortical units.

In addition to replicating local cortical structures, ciliates also propagate their global body pattern during fission. Under certain experimental conditions, global patterning can be made to take place separately from the local assembly of ciliary structures. The mechanisms underlying global cell patterning appear to differ from those of locally directed assembly.

Metazoan cells pass on paragenetic information during cell division. Paragenetic patterns include the distribution of receptor proteins that target newly synthesized polypeptides to their destinations.

The eggs of most metazoa contain paragenetic information in the form of localized cytoplasmic determinants. The cytoplasmic organization of the egg is passed on through the embryonic and adult body pattern until the anatomy of the adult ovary regenerates the egg's cytoplasmic organization. The effects of genetically or experimentally induced changes in cytoplasmic localization show that both genetic and paragenetic information must be passed on properly throughout the life cycle.

SUGGESTED READINGS

Frankel, J. 1989. *Pattern Formation: Ciliate Studies and Models.* New York: Oxford University Press.

Grimes, G. W., and K. J. Aufderheide. 1991. *Cellular Aspects of Pattern Formation: The Problem of Assembly.* New York: Karger Press.

Prusiner, S. B. 1992. Chemistry and biology of prions. *Biochemistry* **31**:12277–12288.

Wood, W. B. 1980. Bacteriophage T4 morphogenesis as a model for assembly of subcellular structure. *Quart. Rev. Biol.* **55**:353–367.

Figure III.1 Mouse embryo at 11.5 days of development. This embryo harbors a transgene consisting of the transcribed region of the bacterial β-galactosidase gene and the regulatory region of the mouse *myogenin*$^+$ gene. The dark stain, caused by β-galactosidase, indicates that the transgene is expressed with the same pattern as the normal *myogenin*$^+$ gene, that is, in the myotomal regions of the somites and the limb buds. The *myogenin*$^+$ gene is necessary and sufficient for the differentiation of skeletal muscle tissue.

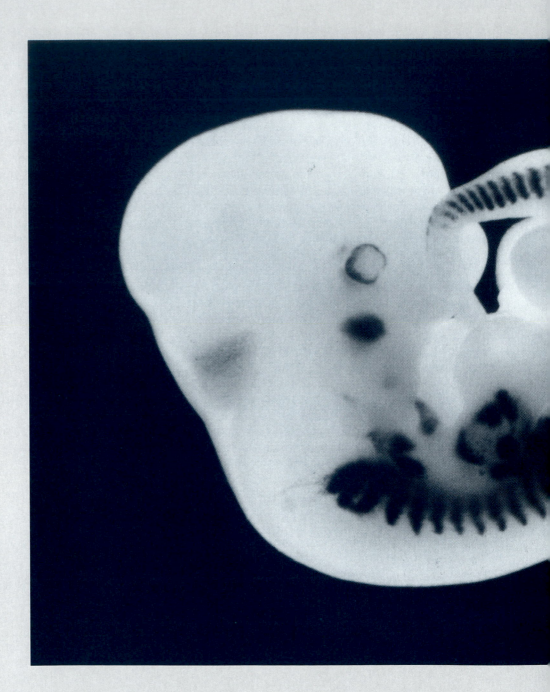

CURRENT TOPICS IN DEVELOPMENTAL BIOLOGY

In Part Three of this text, we will discuss a series of topics that are at once old and new. Although extensively investigated by early embryologists, these topics still provide new frontiers for researchers. Interest in these questions has been revitalized by the arrival of new research methods such as advanced microscopy, genetic analysis, and DNA cloning. Bringing these modern tools to bear on the long-standing problems of the discipline is what generates much of the current excitement in developmental biology.

Two of the topics at the core of developmental biology are cell differentiation and pattern formation. How can cells that originate from the same progenitor cell produce different tissues such as bone and muscle? And how are different pathways of cell differentiation orchestrated in time and space, so that a well-formed organ like the human hand results instead of a randomly aggregated mass of different tissues? These questions will be discussed in general terms and specifically with regard to five organisms that are amenable to genetic analysis: *Drosophila*, the house mouse, two flowering plants, and *Caenorhabditis elegans*.

Other questions center on morphogenesis, discussed in Part One of this text. For example, what are the roles of cell membranes and the extracellular matrix in morphogenesis? How are their properties controlled by the expression of specific genes? And how is gene expression in turn affected by the juxtaposition of cells as a result of morphogenetic movements? As we revisit these topics, our understanding will be enhanced by the genetic and molecular analysis covered in Part Two.

The topics of sex determination, hormonal control, and growth are presented in Part Three because their discussion relies heavily on genetic and molecular analysis. In particular, we will consider the following questions: How do genes or the environment determine sex? How do hormones regulate sexual development and metamorphosis? How are cell division and growth rates controlled in different tissues and in different body regions? What can we learn from tumors about the genes that control normal growth?

The genetic and molecular mechanisms underlying development are often complex. Developmental functions are controlled by autonomous programs within cells, by interactions among neighboring cells, and by hormones and other long-distance signals. Integrating all these controls requires networks of genetic interactions. Such networks were not designed; they evolved. In this respect, genetic networks resemble European cathedrals whose construction began in medieval times. Through the ages, these churches have been expanded, renovated, and redecorated according to the needs and tastes of each time. The cumulative results of these changes are still functional and often marvelous buildings. However, some of their design features are more intricate than their current functions require and can be understood only in a historical context.

CELL

DIFFERENTIATION

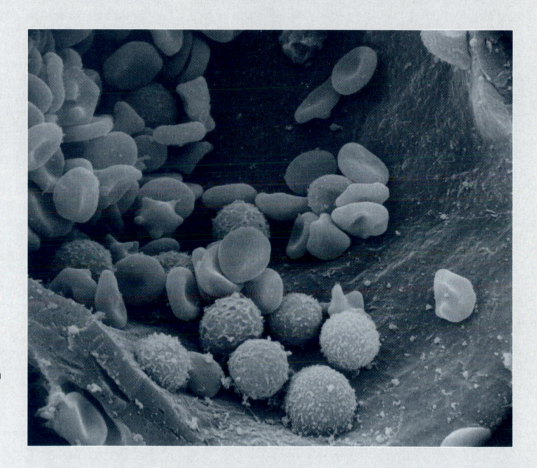

Figure 19.1 Scanning electron micrograph of mammalian blood cells in a small blood vessel. The doughnut-shaped cells are erythrocytes. The spherical cells in the foreground are leukocytes. All types of blood cells arise from one type of stem cell through asymmetric cell division. (From *Tissues and Organs: A Text Atlas of Scanning Electron Microscopy*, by Richard G. Kessel and Randy H. Kardon. W. H. Freeman & Co., 1979.)

In the course of its life cycle, an individual develops from a fertilized egg into an adult. The number of cells in adult multicellular organisms ranges from hundreds, as in *Caenorhabditis elegans,* to the tens of trillions found in large vertebrates. Regardless of the total number, each organism has a relatively small catalog of **cell types.** Each cell type is defined by its morphology and by characteristic molecules that can be detected with antibodies or cytochemical reactions. Most cell type designations, such as neuron or muscle fiber, refer to *mature* and *specialized* cells that have reached the final stage of development in their lineage. To emphasize the fact that organisms have different types of mature cells, this stage is called a cell's **differentiated state.** The process by which cells reach this state is called **cell differentiation.** It initially occurs during *histogenesis,* when tissues get ready for their special functions in the larva or adult (see Chapters 12 and 13), and it continues throughout the life of an organism as worn-out cells are replaced by new cells. Recent work has revealed some of the molecular mechanisms underlying cell differentiation. These mechanisms provide a basis for understanding two classical features of cell differentiation: the limited number of differentiated cell states, and their relative stability.

In this chapter, we will inspect three differentiating cell populations. First, we will examine *interstitial cells* and their derivatives in the small freshwater polyp *Hydra.* The main attractions of this organism are its relatively simple anatomy and the ease with which it can be experimentally manipulated. Next, we will consider the differentiation of mammalian blood cells (Fig. 19.1). Many of the proteins that regulate the development of these cells have recently been produced by cloning. Finally, we will explore the genetic control of muscle cell differentiation. This involves an interesting class of transcription factors that form heterodimers in which one factor enhances or inhibits the other.

The Principle of Cell Differentiation

Developmental biologists have long studied how various cell types reach their differentiated state. The general characteristics of this process are summarized in the *principle of cell differentiation:* multicellular organisms develop a limited catalog of distinct cell types that are usually stable once they have matured. This principle applies to all multicellular organisms and is basic to development.

Each Organism Has a Limited Number of Cell Types

The number of cell types in an organism is much smaller than its total number of cells. For instance, a hydrozoan like the freshwater polyp consists of tens of thousands of cells but has only seven cell types. Humans and other large vertebrates are made up of trillions of cells but

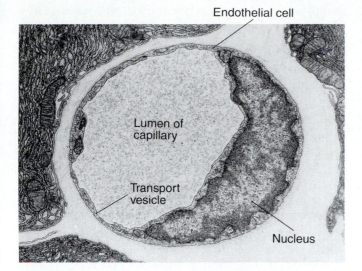

Figure 19.2 Endothelial cell forming a blood capillary. This electron micrograph shows a cross section of a single endothelial cell with joined margins to enclose the lumen of the capillary. The capillary is surrounded only by a basal lamina. The cell cytoplasm contains small vesicles that transport large molecules across the endothelium; the vesicles originate by endocytosis on one side and are given off by exocytosis on the other side.

Figure 19.3 The retina consists of several layers of cells. The slender elements at the top—known as rods and cones—are the outer segments of a layer of photoreceptors. Stimulation of the rods and cones by light is transmitted through subsequent layers of interneurons and ganglion cells to the brain. (From Gene Shih and Richard Kessel, *Living Images*. Science Books International, 1982.)

only about 200 cell types. Familiar examples of vertebrate cell types include epithelial cells, nerve cells, blood cells, and muscle cells. Each of these cell types is subdivided further. Muscle cells, for example, include skeletal muscle fibers, cardiac muscle cells, smooth muscle cells, and myoepithelial cells. Skeletal muscle fibers, in turn, are subdivided into red fibers, white fibers, intermediate fibers, muscle spindles, and satellite cells. Some cells take many different shapes. In particular, neurons have various processes of different lengths and branching patterns, which make classification difficult and somewhat arbitrary.

Each cell type has a distinctive set of structural and functional characteristics. The *endothelial cells* that line the inside of the cardiovascular system form a very thin cellular layer surrounded by a basal lamina (Fig. 19.2). These cells are adapted to the efficient transport of material between the lumen of the blood vessel and the surrounding tissue, and they regulate the transit of white blood cells in and out of the blood. In contrast, the *photoreceptors* in the retina of the eye are specialized for responding to light (Fig. 19.3). Their light-sensitive outer segments—known as rods and cones according to their shape—contain different complexes of proteins with visual pigments. There are many other distinct cell types, each with specific morphological and molecular features.

Mature cell types are *discrete*, with no intermediate forms. The names of some cell types, such as myoepithelial cells, suggest that the cells are intermediate between two distinct types. However, myoepithelial cells

are branched cells containing myofibrils, which typically surround gland cells like a basket and help to empty them by constricting (Fig. 19.4). Thus, myoepithelial cells are a discrete cell type and not intermediate between muscle cells and epithelial cells. Intermediates, when found, are usually in some stage of a develop-

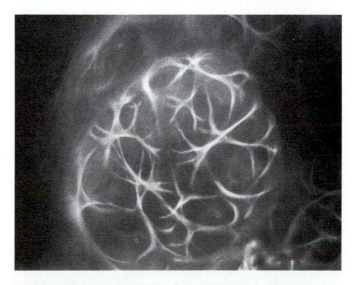

Figure 19.4 Myoepithelial cells in a mouse mammary gland, made visible by immunostaining (see Methods 4.1) of their abundant actin filaments. Myoepithelial cells are a specific type of muscle cell that is associated with ectodermal epithelia such as glands. The "arms" of each myoepithelial cell embrace a few gland cells and serve to expel milk and other secretions.

mental sequence. Gradual changes take place during the development of each cell type, but in their mature states cells are usually of discrete types.

The Differentiated State Is Generally Stable

A differentiated cell is generally stable. Most differentiated cells do not transform into other cell types as part of normal development. Bone cells do not normally become muscle cells, and muscle cells do not turn into epithelial cells. Under experimental conditions, one can observe dramatic exceptions to this rule. For instance, an isolated differentiated plant cell may *regenerate* entire new plants with *all* the differentiated cell types characteristic of the species (see Chapter 7). The iris of a salamander's eye can regenerate a new lens, whose cells have an entirely different morphology and synthesize a new set of proteins (T. Yamada, 1967). Typically, regeneration involves phases of *dedifferentiation,* during which cells lose their differentiated character and proliferate before they differentiate again in new ways. Finally, some cells, such as the follicle cells of ovarian egg chambers in insects, show *sequential polymorphism* (see Fig. 7.17). These cells assume different structures and functions at sequential stages of development. However, such cells represent exceptions rather than the rule.

Cell Differentiation Occurs in Steps and Depends on the Differentiation of Other Cells for Maintenance

Cell differentiation is a stepwise process, which may begin early in development. In many embryos, the animal blastomeres are smaller and contain less yolk than vegetal blastomeres. In mammalian embryos at the morula stage, the external cells are polar and form tight junctions, while the internal cells are apolar and connected by gap junctions. Some of these differences are directly visible under the microscope. Others, such as the synthesis of particular gene products, can be detected only with molecular probes. However, calling cells *differentiated* as soon as they become *distinguishable* would inflate the term "differentiated" and render it useless. We will therefore use this term only in reference to mature cells, which have reached the final stage of the differentiation process.

The concept of cell types pertains to the final *structure* of cells, regardless of their location in the body. For instance, the same types of skeletal muscle cells are found in the head, trunk, arms, and legs. Thus, two cells may be of the same cell type but have different developmental histories. Cells may even reach the same differentiated state through different mechanisms of cell determination: the same type of ascidian tail muscle cell may arise from primary muscle cells containing myoplasm or from secondary muscle cells without myoplasm (see Chapter 8).

Many differentiated cells are highly specialized and are dependent on other cells for their support and maintenance. For instance, the neurons in the central nervous system depend on *glial cells* for insulation. The entire nervous system is highly dependent on the respiratory and digestive systems for nourishment and gas exchange. Thus, maintaining the differentiated state of one cell type often depends on the differentiated state of other cells.

Cell Differentiation and Cell Division

The initial differentiation of many cell types occurs simultaneously during the embryonic period of *histogenesis.* However, cell differentiation is also an ongoing process that lasts throughout life. Since cells of most cell types have a shorter life span than the organism as a whole, cells in most tissues are continually dying and being replaced. The rate of this *turnover* differs from tissue to tissue. The epithelial cells of the small intestine are renewed every few days, whereas the turnover of cells in the pancreatic gland takes a year or more. Many tissues in which turnover is normally slow can be stimulated to replace cells faster when the need arises.

During adult life, new differentiated cells are produced in one of two ways. First, *differentiated* cells can divide, thus producing a pair of daughter cells of the same type. Second, new cells can differentiate from a pool of *undifferentiated* reserve cells called *stem cells.*

Some Cells Divide in the Differentiated State

In the epithelia of many glands, differentiated cells turn over slowly by cell division and cell death. The principal cells of the liver, called **hepatocytes,** renew in this fashion. Normally, hepatocytes are long-lived and divide very slowly. However, in response to food poisoning or injury, the surviving hepatocytes divide at a more rapid pace. For instance, if two-thirds of a rat's liver is removed surgically, the remaining one-third regenerates a nearly normal-sized liver within 2 weeks. (This capability of the liver was described in the Greek myth of Prometheus, long before the underlying mechanism was understood. When Prometheus stole the heavenly fire, Zeus chained him to a rock and sent an eagle to eat from his liver, which constantly renewed itself.)

Like other organs, the liver contains several cell types (Fig. 19.5). Besides the hepatocytes, it is composed of *endothelial cells* lining its blood vessels, *macrophages* that break down worn-out blood cells and other particulate matter in the bloodstream, and *fibroblasts*, which provide a framework of supporting connective tissue.

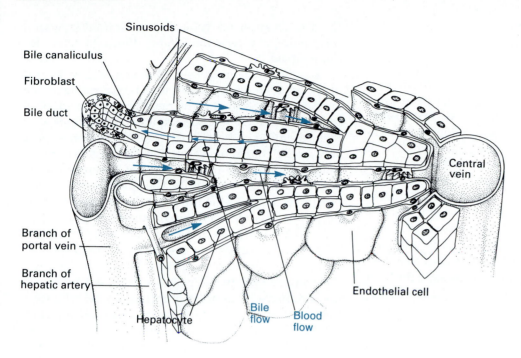

Figure 19.5 Structure of the liver. The hepatocytes are arranged in plates between blood-filled sinusoids connecting arteries and veins. The sinusoids are lined with endothelial cells. Bile formed by the hepatocytes drains into bile ducts. Portions of liver tissue are held together by connective tissue.

For balanced regeneration, each cell type must divide according to the losses sustained. Food poisoning affects hepatocytes more strongly than the other liver cells. If the liver is poisoned continually by alcohol or other drugs, the hepatocytes cannot divide fast enough, and the fibroblasts take over. As a result, the liver becomes irreversibly clogged with connective tissue—a condition known as cirrhosis.

The signals that stimulate and coordinate the division of liver cells are still under investigation. They are known to spread through the bloodstream, but their molecular nature is still unclear.

Cells of another type that divide in the differentiated state are the endothelial cells of the circulatory system. During growth, wound healing, and regeneration, blood vessels supply nearly all tissues according to their physiological needs. New blood vessels always originate from the endothelial cells of existing blood vessels. These cells first form a solid sprout, which then hollows out to form a tube (Fig. 19.6). This process is known as *angiogenesis,* and the substances stimulating it are called *angiogenic factors.* These factors are released by the surrounding cells in response to oxygen deprivation or a wound. Several angiogenic factors have been isolated, and the genes encoding them have been cloned (Folkman and Klagsbrun, 1987). Some of these factors, known as *fibroblast growth factors* and *transforming growth factors,* also stimulate the proliferation of fibroblasts and act as inductive signals during embryogenesis (see Chapter 9).

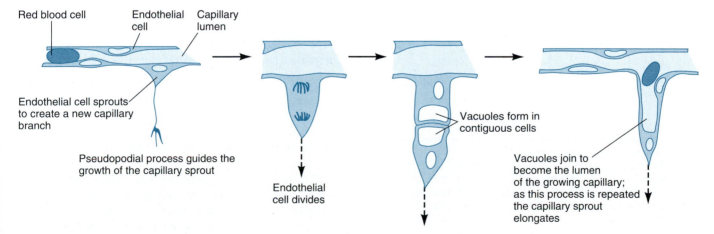

Figure 19.6 Schematic illustration of angiogenesis. A new blood capillary forms through the sprouting and mitosis of an endothelial cell in an existing capillary.

Other Cell Populations Are Renewed from Stem Cells

Tissues in which differentiated cells do not divide contain undifferentiated cells, known as *stem cells*, that are capable of dividing. In adults, stem cells have three main characteristics (P. A. Hall and F. M. Watt, 1989). First, they are undifferentiated. Second, they have a capacity for *unlimited self-renewal* throughout the life of the organism. Third, they have the potential for *asymmetric division*, in which one daughter cell remains a stem cell while the other becomes a **committed progenitor cell** (Fig. 19.7). The fraction of stem cell daughters that remain stem cells is known as the **self-renewal rate**. The commitment of the progenitor cells is unidirectional; their daughter cells do not normally become stem cells again. (Most investigators of stem cells use the term "commitment" as a synonym for *determination* as de-

fined in Chapter 6.) Typically, committed progenitor cells undergo several *amplifying divisions* before they differentiate.

Renewal by stem cells is observed in a wide range of cell populations, including blood cells, epithelial cells, and spermatogonia in vertebrates. Frequently, stem cells reside in relatively safe and sequestered positions, while their committed descendants move out to more exposed sites in the course of maturation. For instance, the epithelium lining the inner surface of the intestine consists of a single layer of cells. The epithelium covers the *villi* projecting into the lumen of the gut as well as the deep *crypts* that descend into the underlying connective tissue. The stem cells of the epithelium lie in the most protected positions near the bases of the crypts (Fig. 19.8). The committed progenitor cells, in the course of their amplifying divisions and subsequent differentiation, are pushed out of the crypt and up the villi, where they are most exposed to abrasion from food and chemical attack from digestive enzymes. After a few days, they reach the tip of the villus, from which they are discarded. Similarly, epidermal cells are "born" in the deepest layer of the epidermis, and blood cells are born in bone marrow.

Some authors use the term "stem cell" in the context of embryonic cell lineages. *Embryonic stem cells*, like their adult counterparts, undergo asymmetric mitoses that conserve the stem cell while producing a string of embryonic cells of a certain type. However, embryonic stem cells serve in this capacity only for a limited period, *generating* a series of descendants—which then live on as a group—rather than *renewing* a cell population in a steady turnover. For instance, the segmentally repeated mesoderm, nerve cells, and epidermal cells of leeches and other worms are given off sequentially by embryonic stem cells. Other authors refer to cultured cells from the inner cell mass of mammalian blastocysts as "embryonic stem cells" (see Chapter 14). The context will usually clarify which type of stem cell is under discussion.

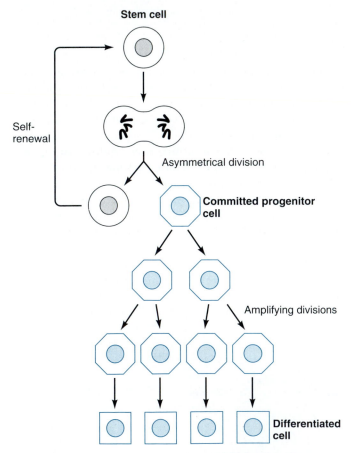

Figure 19.7 Characteristics of stem cells (black): undifferentiated state, unlimited capacity for self-renewal, and capacity for unequal division into another stem cell and a committed progenitor cell (color). The committed progenitor cell is not differentiated but is determined. After a few amplifying divisions, it will form a certain type of differentiated cell. The commitment is normally irreversible: committed progenitor cells do not return to the stem cell state. The type of stem cell drawn here is unipotent: it produces only one type of committed progenitor cell.

Stem Cells May Be Unipotent or Pluripotent

Some stem cells produce only one type of differentiated cell and are therefore called **unipotent**. Vertebrate stem cells forming the intestinal epithelium, the epidermis, or spermatogonia belong to this type. **Pluripotent** stem cells, on the other hand, give rise to several types of differentiated cells. When a pluripotent stem cell divides, the daughter cells can again be stem cells or any one of a small number of committed progenitor cell types; each progenitor cell then undergoes amplifying divisions and eventually produces a clone of differentiated cells. Most of the stem cells discussed in this chapter are pluripotent.

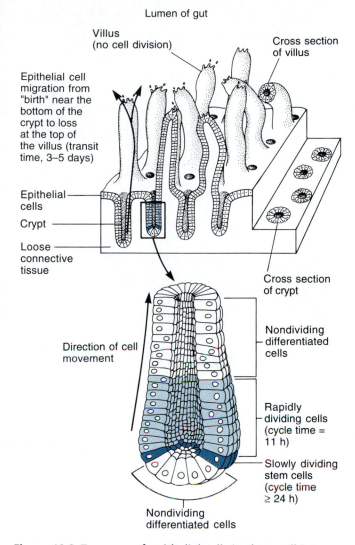

Lumen of gut

Villus
(no cell division)

Cross section
of villus

Epithelial cell
migration from
"birth" near the
bottom of the
crypt to loss
at the top of
the villus (transit
time, 3–5 days)

Epithelial
cells

Crypt

Loose
connective
tissue

Cross section
of crypt

Direction of cell
movement

Nondividing
differentiated
cells

Rapidly
dividing cells
(cycle time =
11 h)

Slowly dividing
stem cells
(cycle time
≥ 24 h)

Nondividing
differentiated cells

Figure 19.8 Turnover of epithelial cells in the small intestine. Stem cells located near the bases of crypts divide slowly and produce rapidly dividing committed progenitor cells. The cells differentiate as they move out of the crypt and onto the villi, where they are most exposed to digestive enzymes and soon die.

The frequency with which a certain type of differentiated cell is produced in a pluripotent stem cell system depends on two cell division rates and on a stochastic selection process. The two cell division rates—that is, numbers of cell divisions per unit time—are those of the pluripotent stem cell and of the committed progenitor cells. The other determining factor is the probability with which the division of a pluripotent stem cell produces the progenitor cell of interest. The signals underlying probabilities of this sort are unknown. The cell division rates, on the other hand, are often controlled by intercellular signals, some of which are being identified.

We will examine two pluripotent stem cell systems: the *interstitial* cells of freshwater polyps, and the *blood cells* of mammals.

Stem Cell Differentiation in *Hydra*

Small freshwater polyps of the genus *Hydra* have been the subject of experimental studies since 1740, when this creature caught the eye of Abraham Trembley, a Swiss naturalist, philosopher, and tutor of children from affluent families (Lenhoff and Lenhoff, 1988). With only magnifying glasses and boars' hairs as tools, Trembley discovered that *Hydra* could reproduce asexually by budding, and that cut parts would regenerate whole polyps.

The Organization of *Hydra* Is Relatively Simple

Since Trembley's time, many zoologists have noted that *Hydra* has several advantages for experimental work. First, *Hydra*'s powers of regeneration are indeed enormous. (The polyp is named after the mythical serpent with multiple heads, any of which, if cut off, would grow back as two.) Even in a small segment of a *Hydra* polyp, cells will reorganize to create a small but complete individual. Second, with only a handful of somatic cell types, the organization of *Hydra* is relatively simple. Third, when an *adult* polyp is dissociated into single cells, the cells will reaggregate and form a polyp again. This capability allows investigators to isolate *differentiated* cells and recombine them into entire organisms—an operation that in other animals is limited to embryonic cells.

The body of *Hydra* is essentially a tube with one open end, called the **hypostome**, which serves as both mouth and anus (Fig. 19.9). The hypostome is surrounded by a ring of tentacles, which are used for defense and for catching prey. The hypostome and the surrounding tentacles together are called the **head** of the hydra. At the opposite end of the tube there is a mucus-secreting **foot** with which the animal attaches to the substratum. Between the head and the foot extends the **body column.** Its central region, which is distended after ingestion of prey, is called the **gastric region.** *Hydra* normally reproduces asexually by budding, although under adverse circumstances individual organisms may reproduce sexually by means of eggs and sperm.

The cells of *Hydra* are arranged in two epithelia: an outer *epidermis*, and an inner **gastrodermis.** These epithelia are separated by the **mesoglea,** an extracellular membrane to which the two cellular layers adhere. Both cellular layers consist for the most part of **epitheliomuscular cells.** These cells have a broadened base with contractile filaments that allow the animal to contract and bend. Despite their muscular properties, the epitheliomuscular cells are usually called *epithelial cells* for short. In the gastrodermis, the epithelial cells can also absorb nutrients. In addition, both epidermis and gastrodermis contain nerve cells. Moreover, each ep-

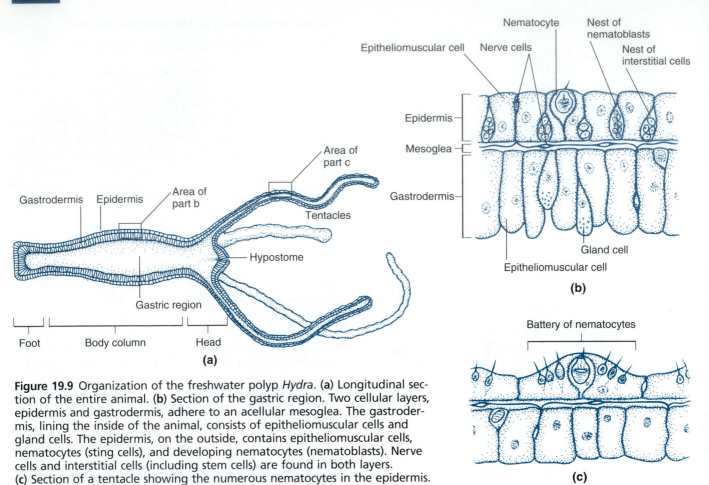

Figure 19.9 Organization of the freshwater polyp *Hydra*. (a) Longitudinal section of the entire animal. (b) Section of the gastric region. Two cellular layers, epidermis and gastrodermis, adhere to an acellular mesoglea. The gastrodermis, lining the inside of the animal, consists of epitheliomuscular cells and gland cells. The epidermis, on the outside, contains epitheliomuscular cells, nematocytes (sting cells), and developing nematocytes (nematoblasts). Nerve cells and interstitial cells (including stem cells) are found in both layers. (c) Section of a tentacle showing the numerous nematocytes in the epidermis.

ithelium contains cells to carry out its special functions. The gastrodermis of the hypostome contains gland cells, which secrete digestive enzymes; and the epidermis of the body column may hold gametes (eggs or sperm). The epidermis of the tentacles also contains numerous *nematocytes* (sting cells), which are equipped with elaborate devices to penetrate and hold prey.

Interspersed between the epithelial cells of both layers are *Hydra*'s most versatile cells, the *interstitial cells* (Fig. 19.10). There are large interstitial cells (about 10 μm in diameter) and small interstitial cells (about 5 μm in diameter). The large interstitial cells include *pluripotent stem cells* for nematocytes, gland cells, nerve cells, eggs, and sperm, as well as unipotent stem cells for eggs or sperm. The small interstitial cells are thought to be mostly precursor cells of nematocytes and nerve cells.

In summary, *Hydra* consists of three major cell lineages: the epithelial cells of the epidermis, the epithelial cells of the gastroderm, and the interstitial cells. The two epithelial lineages generate one cell type each, both of which divide as differentiated cells. The interstitial cells form a pluripotent stem cell system that generates five cell types: eggs, sperm, nerve cells, nematocytes, and gland cells.

Interstitial Cells Contain Pluripotent Stem Cells and Unipotent Stem Cells for Gametes

The interstitial cell lineage is more sensitive than the other two lineages to certain drugs including colchicine, hydroxyurea, and nitrogen mustard. Researchers have used these drugs to generate so-called epithelial animals that consist only of epidermal and gastrodermal epithelial cells with mesoglea in between (Bode et al., 1976; Marcum and Campbell, 1978). Such animals have swollen bodies because the absence of nerve cells leaves them unable to regulate their osmotic pressure (Fig. 19.11). They must be hand-fed because they cannot catch prey, but they reproduce by budding and can be maintained indefinitely as epithelial animals. Apparently, they form enough gland cells for food digestion from gastrodermis. By using marginal doses of those drugs that eliminate interstitial cells, researchers have also produced animals in which the interstitial cell lineage was reduced to unipotent stem cells that gave rise to eggs or sperm (depending on whether the animal was female or male). Alternatively, investigators have combined cells from epithelial animals with small numbers

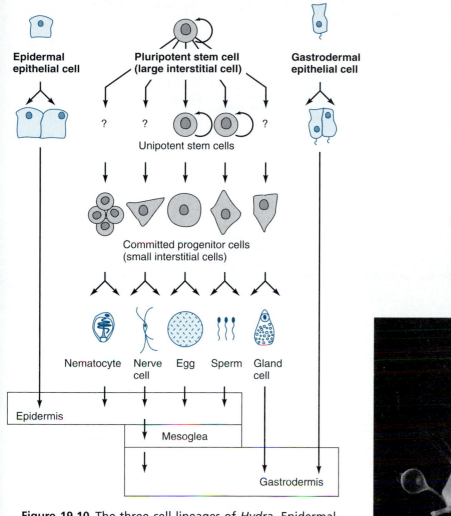

Figure 19.10 The three cell lineages of *Hydra*. Epidermal epithelial cells (top left) and gastrodermal epithelial cells (top right) divide as differentiated cells and form most of the epidermis and gastrodermis, with an acellular mesoglea in between (bottom). All other cells are formed by the interstitial cells, which include pluripotent and unipotent stem cells. The stem cells generate committed progenitor cells, which undergo amplifying divisions before they differentiate into nematocytes, nerve cells, gland cells, and, at least in some species, gametes.

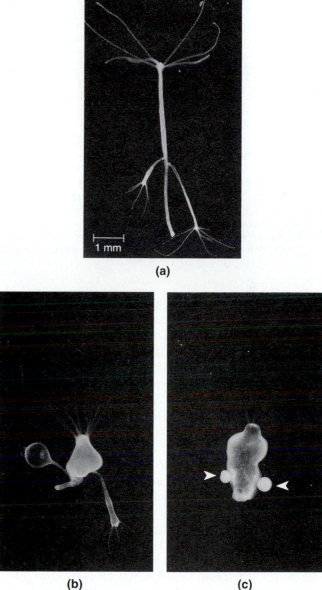

(b)   **(c)**

Figure 19.11 Normal and "epithelial" specimens of *Hydra oligactis*. (a) Normal individual with buds. (b) Animal from which all interstitial cells have been eliminated except for a few that are restricted to generating eggs. This specimen has the typical morphology of epithelial *Hydra:* a swollen body resulting from the inability to osmoregulate, and thin, translucent tentacles due to the absence of nematocytes. This individual has produced two buds. (c) Animal from the same clone as the specimen shown in part b. This animal has been kept at a low temperature to induce the formation of eggs (arrowheads).

of interstitial cells from normal donors to study the development of the transplanted interstitial cells in their hosts. These experiments have provided valuable insight into the development of the interstitial cell system of *Hydra.*

The interstitial cells of *Hydra* contain a population of pluripotent stem cells. Charles David and Susan Murphy (1977) showed this by analyzing clones of cells derived from single interstitial cells. They obtained such clones by using the procedure illustrated in Figure 19.12. Cells from normal *Hydra vulgaris* were mixed with an excess of cells from nitrogen mustard–treated individuals of the same species. Under these circumstances, *single* interstitial cells contained in the normal cell samples formed clones separated by large numbers of epithelial cells from the treated animals. Such clones of cells were identified by toluidine blue, which stains interstitial cells more intensely than it stains epithelial cells. To determine whether the clones were derived from pluripotent stem cells, nine clones were scanned for developing nematocytes and nerve cells. Each of the clones contained *both* cell types, indicating that they were derived from pluripotent stem cells.

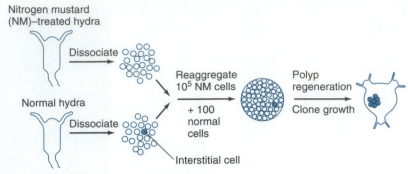

Figure 19.12 Procedure for obtaining clones derived from single interstitial cells in *Hydra*. Normal animals and animals treated with nitrogen mustard (NM) to eliminate interstitial cells were dissociated into single cells. Large numbers of cells (about 100,000) from NM-treated animals were mixed with small numbers of cells (20 to 400) from radiolabeled but otherwise normal animals. The cells were pelleted by centrifugation and allowed to regenerate polyps. Clones derived from interstitial cells in the regenerates were identified by their radiolabel and by staining with toluidine blue (shading). The cell types found in individual clones were identified morphologically.

In addition to pluripotent stem cells, at least some *Hydra* species also contain unipotent stem cells that are restricted to forming gametes. Such unipotent stem cells were first observed in *Hydra vulgaris* treated with colchicine or hydroxyurea. In most of the treated animals, the entire interstitial cell population was eliminated. However, a few animals were found that did not form nematocytes or nerve cells but occasionally produced either oocytes or sperm. In a similar experiment, C. Lynne Littlefield (1985, 1991) used *Hydra oligactis,* a species in which gamete formation is triggered by low temperature. From animals treated with marginal doses of hydroxyurea, she isolated lines of animals with low levels of interstitial cells (< 10% of normal). Such animals reproduced by budding and could be maintained for years without forming nerve cells or nematocytes. However, when exposed to low temperature, they reliably produced sperm or eggs (Fig. 19.11). Those interstitial cells that gave rise to sperm also stained specifically with a monoclonal antibody (Littlefield et al., 1985). These results demonstrate that the interstitial cells of *Hydra oligactis* contain a subpopulation of unipotent stem cells that are restricted to forming either eggs or sperm. Corresponding experiments with *Hydra magnipapillata* have revealed a similar subpopulation of interstitial cells that can differentiate into sperm but not into nerve cells or nematocytes (Nishimiya-Fujisawa and Sugiyama, 1993). Presumably, these unipotent stem cells are derived from pluripotent stem cells in the interstitial cell population.

Nerve Cell Differentiation Is Promoted by a Peptide Produced in the Head Region

One of the leading hypotheses about stem cell differentiation postulates that progenitor cell commitment occurs in response to microenvironments produced by other cells. *Hydra* is a convenient vehicle for testing this hypothesis.

To investigate the signals controlling *Hydra* nerve cell development, Thomas Holstein and colleagues (1986) used a bioassay based on regeneration. The researchers had observed that in polyps regenerated from segments of body columns the ratio of nerve cells to epithelial cells increased substantially between 12 and 24 h of regeneration. They concluded that most of the nerve cells they scored after 1 day must have formed from committed progenitor cells or amplification products of such cells that were already present when the regenerating tissue was isolated. Therefore, the ratio of nerve cells to epithelial cells after 1 day of regeneration could be used as a measure for the prevalence of committed nerve cell progenitors or their amplification products in the newly isolated piece of tissue.

To identify factors that stimulate nerve cell development in *Hydra,* the investigators pretreated donor polyps with various agents before they removed tissue from the body column for regeneration. Preincubation of the donor polyps in certain components extracted from *Hydra* enhanced nerve cell formation. The investigators purified the substance responsible for this activity and identified it as a peptide that had been defined previously by its ability to stimulate head regeneration and had therefore been termed the **head activator** (Schaller, 1981). Since the head activator consists of only 10 amino acids of known identity and sequence, it could be synthesized and tested for the ability to stimulate nerve cell commitment (Fig. 19.13). The synthetic peptide was fully effective at the extremely low concentration of 0.1 p*M*. Pretreatment of polyp tissue at this concentration nearly doubled the ratio of nerve cells to epithelial cells. Moreover, in normal *Hydra,* the rate of nerve cell differentiation is highest in the head region,

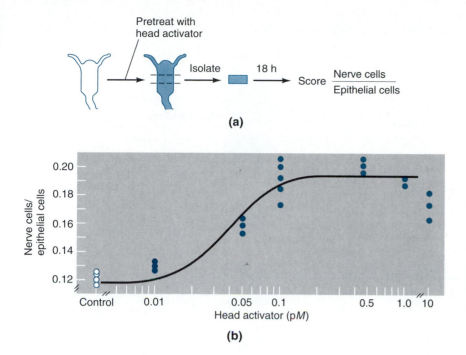

(a)

(b)

Figure 19.13 Analysis of nerve cell differentiation in *Hydra*. **(a)** Experimental strategy. Polyps are incubated with head activator, a synthetic peptide known to stimulate head regeneration. A piece of gastric region is removed from the pretreated animal and allowed to develop for 18 h before the ratio of nerve cells to epithelial cells is scored. **(b)** Nerve cell–epithelial cell ratio plotted versus head activator concentration used during pretreatment. A very low concentration (0.1 pM = 10⁻¹³ mole per liter) nearly doubles the density of nerve cells.

where the greatest concentration of head activator is also present. The researchers concluded that the peptide acting as head activator in *Hydra* also causes a commitment of stem cells to give rise to nerve cells (T. W. Holstein and C. N. David, 1990). Alternatively, head activator may enhance the proliferation of nerve cell precursors (Bode et al., 1990).

In summary, experiments with *Hydra* have shown that the interstitial cells include stem cells that give rise to all cell types of this organism except the epithelial cells of epidermis and gastrodermis. These stem cells are pluripotent, giving off different types of committed progenitors and unipotent stem cells for either eggs or sperm. A local chemical signal involved in head regeneration also enhances the commitment of nerve cell progenitors or their subsequent amplification. Similar characteristics are found in other stem cell lineages, including the mammalian blood cells, which will be discussed next.

Growth and Differentiation of Blood Cells

Although separated by nearly a billion years of evolution, the interstitial cells of hydrozoans and the blood cells of mammals are remarkably similar as stem cell systems. Analyzing blood cell development has been more difficult than studies of *Hydra*, but health concerns are fueling interest in it. Fatal illnesses such as leukemia can be understood as disorders in the pro-

gression from stem cells to committed progenitor cells and differentiated cells.

The blood of vertebrates contains many types of cells with different functions, ranging from the transport of oxygen to the destruction of foreign cells (Fig. 19.1). All mature blood cells are short-lived and must be replaced continuously in a process called **hematopoiesis** (Gk. *haimat-*, "blood"; *poiein*, "to make"). An adult human produces billions of blood cells each hour just to replace normal loss. All blood cells originate from pluripotent stem cells called **hematopoietic stem cells** (Fig. 19.14). Like other stem cell populations, hematopoietic cells migrate extensively during their development. In mammals, they originate in the embryonic yolk sac, from where they move via the liver to the spleen and the bone marrow (Metcalf and Moore, 1971). In most adult mammals, the principal hematopoietic tissue is the bone marrow.

A group of blood cells known as *lymphocytes* mature in the thymus gland, lymph nodes, and other so-called *lymphoid organs*. Accordingly, lymphocytes and their precursors are referred to as **lymphoid cells**, in contrast to all other blood cells, which are collectively called **myeloid cells**. More than 99% of the myeloid cells are red blood cells, or **erythrocytes**. They transport oxygen (O_2) and carbon dioxide (CO_2) and are confined to blood vessels. Other myeloid cells, as well as lymphoid cells, use blood vessels mainly for rapid transport to other tissues, where they combat infections and kill the body's own tumorous and aging cells. In addition, blood contains **platelets**, cell fragments without nuclei that pinch off from large bone marrow cells called *megakaryocytes*. Platelets are involved in blood clotting and blood vessel repair.

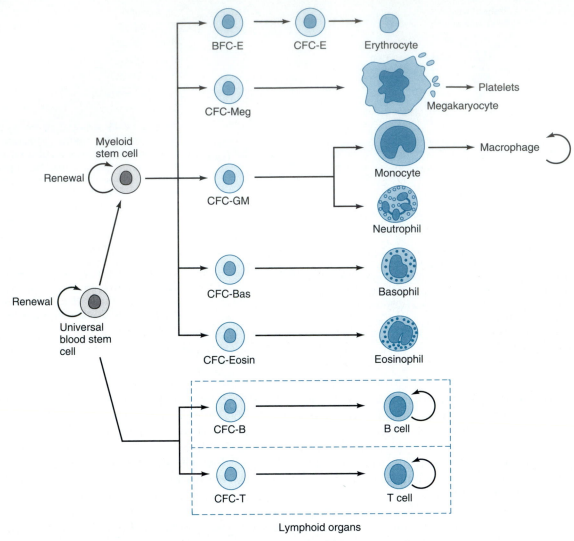

Figure 19.14 A tentative scheme of hematopoiesis. A universal blood stem cell divides infrequently to renew itself and to produce cells with more restricted potencies. The daughter cells are either myeloid stem cells or committed progenitor cells that give rise to lymphoid cells. The myeloid stem cells in turn renew themselves and produce different types of committed progenitors for myeloid cells. The committed progenitor cells in this system are referred to as CFCs (for colony-forming cells). The CFCs undergo amplifying divisions (not shown) under the control of specific regulatory proteins before they differentiate. Most CFCs produce only one type of differentiated cell. However, CFC-GM gives rise to both neutrophilic granulocytes and macrophages. The differentiated cells live for a few days or weeks and carry out specific functions.

The Same Universal Stem Cell Forms All Types of Blood Cells

Proving the existence of hematopoietic stem cells and describing some of their properties has been a difficult task (Dexter and Spooncer, 1987). All types of blood cells begin their development in the bone marrow and look very similar in their early stages. Thus, the existence of stem cells can only be inferred from clonal analysis, i.e., by tracing the progeny of single hematopoietic cells.

The *spleen colony assay* pioneered by J. E. Till and E. A. McCulloch (1961) has been used by many re-

searchers to assess the potency of bone marrow cells for forming different blood cell types. (The same procedure is also used by clinicians to reconstitute the hematopoietic systems of patients whose stem cells are deficient or have been killed as a side effect of chemotherapy or radiation therapy. See Golde, 1991.) For the spleen colony assay, an animal is exposed to a heavy dose of X-rays so that its hematopoietic cells can no longer form new blood cells. However, the animal is saved by a transfusion of bone marrow cells from an unirradiated and immunologically compatible donor (Fig. 19.15). The restoration of the hematopoietic system in the recipient can be observed in the blood and, in particular, in the spleen. Two

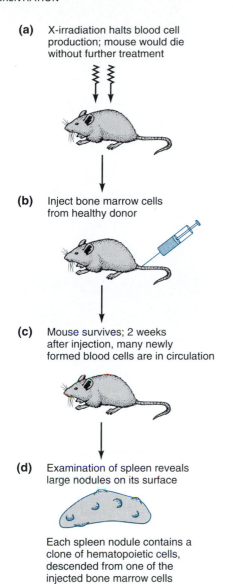

(a) X-irradiation halts blood cell production; mouse would die without further treatment

(b) Inject bone marrow cells from healthy donor

(c) Mouse survives; 2 weeks after injection, many newly formed blood cells are in circulation

(d) Examination of spleen reveals large nodules on its surface

Each spleen nodule contains a clone of hematopoietic cells, descended from one of the injected bone marrow cells

Figure 19.15 Spleen colony assay to assess the potency of injected bone marrow cells. **(a)** A mouse is heavily X-irradiated to prevent all resident hematopoietic stem cells from further division. **(b)** The mouse receives bone marrow cells from a healthy and immunologically compatible donor. **(c)** Two weeks after treatment, newly formed blood cells are in circulation. **(d)** The spleen of the reconstituted mouse has multiple nodules, each containing a clone of blood cells derived from a single injected bone marrow cell.

weeks after transfusion, the spleen shows several distinct nodules, each containing a colony of new blood cells. The spherical shape of the nodules and their isolation from each other suggest that each is derived from a single precursor cell. Microscopic examination of such nodules shows that they contain erythrocytes, granulocytes, and other myeloid cells. Some colonies contain only one type of blood cell, while others contain several myeloid cell types. These observations suggest that each nodule originates either from a pluripotent myeloid stem cell or from a committed progenitor cell.

To prove that the nodules that develop in a spleen colony assay are generated by single cells, Becker and coworkers (1963) individually marked the marrow cells used for transfusion. For this purpose, the experimenters induced chromosome abnormalities by X-irradiating the *donor* marrow cells after they had been taken from the donor and before they were transfused to the recipient. Spleen nodules founded by such cells still contained different types of myeloid cells, but *each* cell in a given nodule had the same chromosome anomaly. Therefore, all marked blood cells in a nodule had to be derived from a single founder cell, which the researchers termed a **colony-forming cell (CFC)**.

To learn whether CFCs are *stem cells,* researchers had to test their capacity for self-renewal. To this end, Jurášková and Tkadleček (1965) used cells from spleen nodules to rescue a second series of irradiated mice. Some of the secondary nodules contained only one type of myeloid cell, while others contained all types. The investigators concluded that the *primary* nodules contained multiple types of committed progenitor cells *and* self-renewing pluripotent cells. In other words, some of the primary CFCs had to be *pluripotent myeloid stem cells.*

An even greater potency of bone marrow cells was indicated by the observation that they could restore not only the myeloid but also the lymphoid cells of irradiated recipients. This raised the question whether there are universal blood stem cells that give rise to both myeloid and lymphoid cells.

▼

In search of a universal blood stem cell, Gordon Keller and coworkers (1985) used a marker that allowed them to trace both myeloid and lymphoid cell descendants in the same animal. These investigators infected the transfused marrow cells with a virus containing the *neo^R* gene as a selectable marker (Fig. 19.16). The virus was inserted randomly into the genomic DNA of bone marrow cells, conferring on them two valuable properties. First, the infected cell was resistant to neomycin and could therefore be selected from uninfected cells by culture in a neomycin-containing medium. Second, because the virus was replicated in the same genomic position whenever its host cell divided, each infected cell formed a clone carrying the insertion site of the virus as a unique marker. Cells containing the virus in the same genomic site therefore had to belong to the same clone. To characterize the virus insertion site in the blood cell DNA, the investigators used the *Southern blotting* technique (see Methods 14.3). When they tested DNA prepared from different tissues of mice with reconstituted hematopoietic systems, they found that bone marrow, spleen, lymph nodes, and thymus gland cells did indeed contain the marker virus in DNA fragments of the

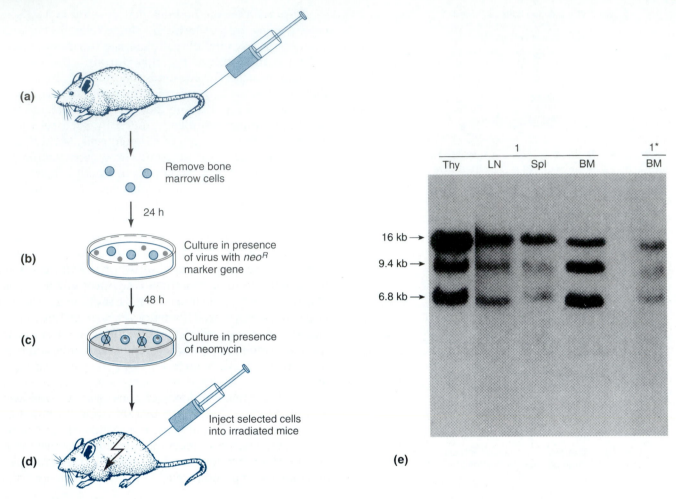

Figure 19.16 Tracing of blood cell clones founded by single bone marrow cells marked by the insertion of a virus. **(a)** Bone marrow cells were removed from a healthy mouse. **(b)** The cells were cultured in the presence of a virus carrying the *neo^R* gene as a selectable marker. The virus inserted itself into the genomic DNA of the bone marrow cells, conferring on them resistance to neomycin and marking each cell with the virus insertion site. **(c)** The cells were cultured in the presence of neomycin to select the virus-infected cells. **(d)** The selected cells were injected into X-irradiated mice as described in Fig. 19.15. **(e)** Genomic DNA was prepared from thymus (Thy), lymph node (LN), spleen (Spl), and bone marrow (BM) of the mouse with a reconstituted hematopoietic system (designated 1). Genomic DNA was also prepared from the bone marrow of another irradiated mouse (designated 1*), which had previously received bone marrow cells from mouse 1. DNA from each source was analyzed by Southern blotting (see Methods 14.3), using as a probe the *neo^R* sequence. In each case, three DNA restriction fragments of 6.8 kb, 9.4 kb, and 16 kb were labeled. The results indicate that a maximum of three transplanted bone marrow cells reconstituted the entire hematopoietic system and that each of these three cells contributed to all hematopoietic tissues.

same lengths. These results proved the existence of universal blood stem cells. Whether these cells produce the committed progenitors for different lymphoid cells directly, or whether there is a lymphoid stem cell in analogy to the myeloid stem cell, remains to be elucidated.

The reconstitution experiments just described show that a small number of cells can produce an entire generation of myeloid and lymphoid blood cells. To appreciate the significance of this result, one needs to keep in mind that the number of transplanted bone marrow cells in a reconstitution experiment is much smaller than the normal complement of bone marrow in a mouse. Moreover, an irradiated mouse saved by a transfusion of bone marrow cells can in turn be used as a donor of bone marrow cells to save another irradiated mouse, and the experiment can be repeated at least one more time. These results indicate an astounding capacity for self-renewal in those stem cells that come to be the founder cells of a reconstituted hematopoietic system.

Blood Cell Differentiation Can Be Studied in Vitro

The conclusions reached from reconstitution experiments in vivo have been confirmed by observations made in vitro (Metcalf, 1977). Since blood cells can be cultured in dishes with soft gels and defined media, it is possible to monitor the capacity of individual cells to divide and differentiate. Some cells behave in culture like pluripotent stem cells, forming successive colonies containing all myeloid cell types (Fig. 19.17). Other isolated cells reveal more limited potencies, forming only two differentiated cell types, such as macrophages and neutrophilic granulocytes. Still other cells form just one type of blood cell. As the potencies of blood cells become more limited, their capacity for self-renewal also decreases, at least normally.

Observations made in vitro also indicate that there is a stochastic element in the way an *individual* hematopoietic cell behaves. Cells isolated from apparently homogeneous populations produce colonies of remarkably different sizes and characters. Even sister cells cultured under identical conditions will often produce colonies with different numbers or types of cells. Thus, the regulation of division and commitment in each individual cell involves stochastic or very labile processes, the nature of which is still unknown. Only statistically do entire blood cell populations behave in predictable ways under the control of a class of growth factors, to be described shortly.

The results described so far can be summarized as follows (refer to Fig. 19.14). Universal hematopoietic stem cells give rise to myeloid stem cells and to lymphoid progenitor cells. The myeloid stem cells produce several classes of committed progenitor cells, which have lost the capacity for self-renewal, or at least have a very low probability of actually self-renewing. The committed progenitor cells undergo a potentially large but limited number of amplifying divisions before terminal differentiation. These divisions will be described further when erythrocyte formation is discussed.

Hematopoiesis Depends on Colony-Stimulating Factors (CSFs)

The development of hematopoietic cells depends on regulatory proteins, which have either inhibitory or enhancing effects. For instance, a hematopoietic stem cell inhibitor synthesized by bone marrow cells limits stem cell proliferation (G. J. Graham et al., 1990). The majority of the known regulatory proteins have stimulating effects and are known as *growth factors, cytokines,* or *colony-stimulating factors (CSFs).* Their presence is required for the survival, division, and differentiation of hematopoietic cells (Dexter and Spooncer, 1987; Metcalf, 1989). In the absence of CSFs, the cells die, regardless of their stage of development. A cell's ability to respond to a given CSF depends on the presence of matching receptors. Some progenitor cells have more than one type

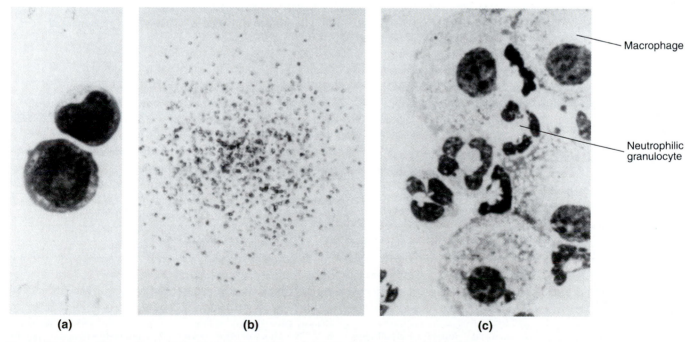

(a) (b) (c)

Figure 19.17 In vitro culture of hematopoietic cells. **(a)** Purified colony-forming cells (CFCs). **(b)** Stimulated by colony-stimulating factors, CFCs form large colonies of differentiated blood cells. **(c)** Neutrophilic granulocytes and macrophages form in one colony.

of receptor and develop according to which CSF is present in higher concentration. For example, certain progenitor cells can form two types of leukocytes, namely, macrophages and neutrophilic granulocytes (Figs. 19.14 and 19.17). Each of these cell types is enhanced by its own CSF. Thus, granulocyte CSF fosters granulocyte development while macrophage CSF favors macrophage development. There are also pluripotent CSFs that stimulate the formation of both cell types. Thus, the overall rate of progenitor cell production, and the degree to which each type is amplified, depend on the overall amount and mixture of CSFs acting on the progenitor cell.

Many CSFs have been cloned and characterized molecularly along with their receptors. Table 19.1 lists these factors along with their target cells. As can be seen from the table, most CSFs act on more than one target cell, and conversely, most target cells are stimulated by several CSFs. The overall result depends on the relative concentrations of different CSFs and their receptors. CSF concentrations may vary between local microenvironments created by surrounding tissues.

In living bone marrow, hematopoietic cells develop amid a supporting tissue known as **stroma.** When stroma and hematopoietic cells are cocultured in vitro, stroma cells show two interesting properties (Dexter, 1982). First, they support hematopoiesis even without

diffusible CSFs in the medium. Second, this support requires direct contact between the stroma and the hematopoietic cells. Only stroma cells from bone marrow—not those from other tissues—support the growth of hematopoietic cells. These observations suggest that bone marrow stroma cells may deploy CSFs in their membranes or in extracellular matrix. Indeed, two specific CSFs are adsorbed by heparan sulfate, a major component of the extracellular matrix in mouse bone marrow stroma (R. Roberts et al., 1988). This kind of CSF deployment would enable stroma cells to establish local microenvironments, or *niches,* each containing a unique mixture of immobilized CSFs in addition to the diffusible CSFs, which are presumably shared among the niches. Local niches in bone marrow should make it possible to expose hematopoietic cells to different CSF mixtures at successive stages of development. Some blood cells do require different CSFs in sequence.

The Development of Erythrocytes Depends on Two CSFs

Erythrocytes are by far the most common blood cells. Highly specialized, they are packed with hemoglobin for the transport of O_2 and CO_2. Erythrocytes mature in the bone marrow from precursor cells known as *erythroblasts.* In mammals, maturation of an erythroblast

---

TABLE 19.1

Hematopoietic Growth Factors and Their Target Cells

Factor	Symbol	Erythroid Cell BFC-E	CFC-E	Granulocyte Eosinophil	Neutrophil	Macrophage	Mast Cell	Mega-karyocyte	Lymphocyte B	T	Pluripotent Progenitor Cell	Stem Cell
Multipotential colony–stimulating factor	Multi-CSF, IL-3	X		X	X	X	X	X			X	X
Granulocyte-macrophage colony–stimulating factor	GM-CSF	X		X	X	X		X			X	
Granulocyte colony–stimulating factor	G-CSF				X							
Macrophage colony–stimulating factor	M-CSF					X						
Mast-cell growth factor	MGF	X			X		X	X				X
Erythropoietin	Epo		X									
Interleukin-1	IL-1											X
Interleukin-2	IL-2								X	X		
Interleukin-4	IL-4						X		X	X		
Interleukin-5	IL-5			X					X			
Interleukin-6	IL-6				X				X	X	X	
Leukemia inhibitory factor	LIF					X						

*For brevity, most target cell listings refer to the differentiated cell state, such as T lymphocyte. However, it must be understood that the growth factors act on stem cells and committed progenitor cells, such as T-lymphocyte colony–forming cells (CFC-Ts).
Sources: Dexter and Spooncer (1987), © 1987 Annual Reviews, Inc.; Metcalf (1989), © 1989 Macmillan Magazines Limited; and Copeland et al. (1990), © Cell Press. Used with permission.

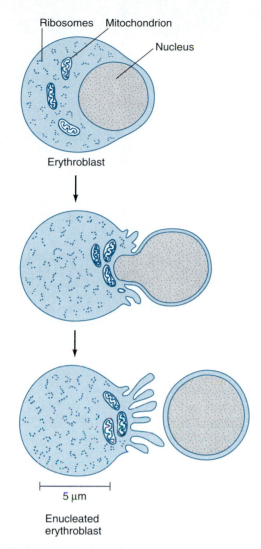

Ribosomes   Mitochondrion

Nucleus

Erythroblast

5 µm

Enucleated
erythroblast

Figure 19.18 Maturation of an erythroblast. While an erythroblast is still in the bone marrow, its nucleus is pinched off and digested by macrophages. The enucleate erythroblast is released into the circulation, where it also loses its mitochondria and ribosomes to become an erythrocyte.

the kidney; it binds to and activates a matching receptor on the cell surface of CFC-E's. The receptor in turn activates a *protein kinase*, which phosphorylates other proteins to eventually stimulate the mitotic division of CFC-E's (Witthuhn et al., 1993). Thus, in the presence of erythropoietin, a CFC-E divides five or six times, giving rise to about 30 to 60 erythroblasts, which then mature to erythrocytes.

The concentration of erythropoietin, which controls the number of final divisions, increases in response to any shortage of erythrocytes or lack of oxygen in the blood. In this fashion, the amount of erythrocytes is continuously adjusted to physiological conditions. For instance, with increasing height above sea level, the oxygen pressure falls and the red blood cell count increases in compensation. Similarly, local or systemic infections produce signals that lead to dramatic increases in the concentration of leukocytes.

In summary, studies on CSFs have shown that these factors cooperate, both simultaneously and in sequence, to regulate the proliferation and maturation of different types of blood cells in response to changing physiological conditions. While some CSFs are diffusible, others may be bound locally to stroma cell surfaces or extracellular matrix. The molecular characterization and cloning of CSFs has greatly enhanced our understanding of blood cell differentiation and the possibilities for the clinical correction of blood disorders. However, the molecular events underlying the commitment of stem cells to certain blood cell lineages are still unknown. Analysis of cell commitment in terms of specific gene activities has not yet been possible for blood cells, but an understanding of the genetic link is beginning to emerge from studies on muscle differentiation, as we will see next.

entails the loss of the nucleus, mitochondria, and ribosomes (Fig. 19.18). Once released into the bloodstream, erythrocytes have a life span of a few months before they are broken down in the liver and spleen.

The continuous formation of new erythrocytes occurs in two major steps that involve different CSFs (Fig. 19.19). Myeloid stem cells give off a type of committed progenitor cell called an *erythrocyte burst–forming cell (BFC-E).* The name derives from the fact that these cells can produce bursts of up to 5000 erythrocytes. The BFC-E's respond to a CSF called *interleukin 3 (IL-3),* which generally promotes the proliferation of hematopoietic stem cells and committed progenitor cells (Table 19.1). After about six divisions, the BFC-E descendants begin to respond to another CSF known as *erythropoietin.* These cells are then called *erythrocyte colony–forming cells (CFC-E).* Erythropoietin is a hormone formed in

Genetic Control of Muscle Cell Differentiation

In 1987, a research team led by Harold Weintraub reported that the overexpression of a single transgene in connective tissue cells caused these cells to differentiate as muscle cells (R. L. Davis et al., 1987). The investigators had transfected cultured mouse fibroblasts with cloned DNA containing the cDNA of *MyoD+*, a gene specifically expressed in *myoblasts.* The transformed cells proceeded to undergo *myogenesis;* that is, they developed into skeletal muscle fibers. This was an astounding result, because fibroblasts do not normally form skeletal muscle. As described in Chapter 13, fibroblasts can give rise to other connective tissue cells such as *osteoblasts, chondrocytes,* and *adipocytes.* Fibroblasts also form *smooth muscle cells* during normal development, but not cardiac or skeletal muscle. Thus, the activity of a single

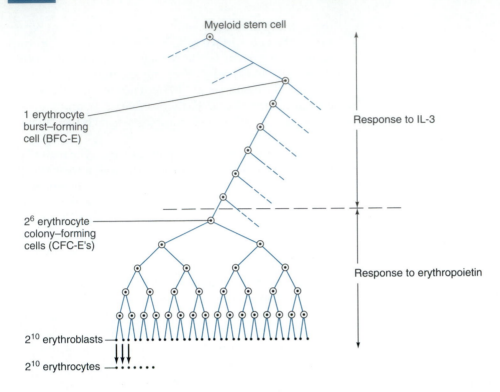

Myeloid stem cell

1 erythrocyte
burst–forming
cell (BFC-E)

Response to IL-3

2^6 erythrocyte
colony–forming
cells (CFC-E's)

Response to erythropoietin

2^{10} erythroblasts

2^{10} erythrocytes

Figure 19.19 Flow diagram of erythropoiesis. Pluripotent myeloid stem cells give rise to erythrocyte burst–forming cells (BFC-E's), which undergo up to six amplifying divisions. The formation of BFC-E's and their divisions are stimulated by a hormone called interleukin-3 (IL-3). Further development requires another hormone, erythropoietin. The cells that respond to erythropoietin are called erythrocyte colony–forming cells (CFC-E's). After further amplifying divisions, CFC-E's give rise to erythroblasts, which will differentiate into mature erythrocytes.

gene, $MyoD^+$, could completely redirect a cell and force it onto a new pathway of differentiation.

The deceptively simple picture of a single gene acting as a master switch has since become more complex. It now appears that a small family of genes cooperate in determining mesodermal cells to become myoblasts and to differentiate into muscle fibers.

MyoD Belongs to a Family of Myogenic bHLH Proteins

Cell transformations similar to the one just described have revealed that $MyoD^+$ belongs to a family of genes that also includes $myogenin^+$, $myf\text{-}5^+$, and $MRF\text{-}4^+$ (Olson, 1990; Olson and Klein, 1994; Weintraub et al., 1991; Weintraub, 1993). Each of these genes can initiate muscle development in nonmuscle cells. The proteins encoded by these genes are transcription factors that have about 80% sequence similarity in a domain of 70 amino acids known as the *basic helix-loop-helix (bHLH)* domain (Fig. 19.20). It contains a basic region (b), which acts as a weak base, and a helix-loop-helix (HLH) motif, where two α helices are connected by a peptide loop. The basic region is necessary for DNA binding and transcriptional activation, while the HLH motif is responsible for dimer formation. Because of the common feature of the bHLH domain, the MyoD, myogenin, myf-5, and MRF-4 proteins are collectively called the *myogenic bHLH proteins.* These proteins have been found in vertebrates, sea urchins, *Drosophila,* and *Caenorhabditis elegans.* In all animals investigated so far, myogenic bHLH proteins are synthesized specifically during muscle development.

The myogenic bHLH proteins and their genes form a network of regulatory interactions that seem to stabilize the expression of these genes, or subsets thereof, once their expression exceeds a certain threshold. In addition, the myogenic bHLH proteins control target genes that are expressed specifically in muscle, such as the gene for muscle creatine kinase.

The myogenic bHLH proteins share the bHLH motif with about a dozen other gene-regulatory proteins referred to as the *bHLH superfamily.* The superfamily includes so-called E proteins, which are expressed in a wide range of tissues and species and bind to specific DNA sequences known as E boxes (Murre et al., 1989; Ellenberger et al., 1994). Only a few amino acid sequences in the basic region distinguish the myogenic bHLH proteins from other bHLH proteins and account for their specific role in myogenesis (R. L. Davis et al., 1990; Brennan et al., 1991). Nevertheless, the myogenic effect of the myogenic bHLH proteins depends on their cooperation with more ubiquitous members of the bHLH superfamily.

Dimer Formation between Myogenic bHLH and Other Proteins Provides Gene-Regulatory Diversity

Like many other transcription factors, bHLH proteins bind to their target genes as dimers. In particular, MyoD and myogenin proteins form heterodimers with E proteins, and myogenin seems to trigger myogenesis *only* as a heterodimer (Lassar et al., 1991). Apparently, various dimers of bHLH proteins bind with different affinities to enhancer motifs and have different effects on the

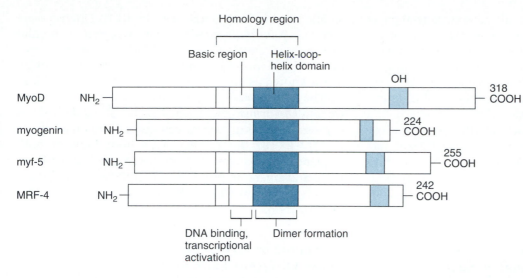

Figure 19.20 General structure of four mammalian myogenic bHLH proteins. The helix-loop-helix domain is required for dimer formation. The basic region is necessary for DNA binding and transcriptional activation. The number at the end of each box indicates the number of amino acid residues in the respective protein.

activity of the promoters they regulate. Thus, a small number of bHLH products can generate a wide variety of dimeric proteins with different regulatory properties.

Heterodimers of bHLH protein may be activating or inhibitory transcription factors. An example of the inhibitory effects of dimerization is seen in the heterodimer formed by MyoD protein and another protein called Id (for inhibitor of differentiation). The Id protein contains the HLH motif, which mediates dimer formation, but not the basic region, which binds to DNA. Most important, the Id protein dimerizes with MyoD and E proteins, but the heterodimers do not bind to DNA. A single basic region by itself is insufficient to bind the heterodimer to its recognition motif (Fig. 19.21). In vivo, Id prevents MyoD from activating a muscle-specific enzyme, creatine kinase (Benezra et al., 1990). Forced expression of Id in fibroblasts prevents their transformation to myoblasts and retards the formation of myotubes in muscle cell lines (Jen et al., 1992). The production of Id protein declines when *myoblasts* begin their final differentiation, the formation of muscle fibers. This explains nicely the apparent paradox that MyoD is present in muscle cell lineages long before myogenesis begins.

Examination of other vertebrate cell lines indicates that the Id^+ gene is expressed in a wide range of cell types. Generally, the Id protein seems to prevent terminal differentiation by tying up bHLH proteins in nonfunctional heterodimers. When cells mature, the levels of Id mRNA generally fall. The factors that modulate the expression of Id are not yet known.

Expression of Either $MyoD^+$ or myf-5^+ Is Necessary for Skeletal Muscle Development in Vivo

Although much has been learned about the function of myogenic bHLH proteins in cultured cells, the importance of these proteins in vivo would be demonstrated

most clearly by mutants homozygous for *null alleles* of myogenic bHLH genes. Such mutants were generated in mice, using the "knockout" technique of substituting an inactive transgene for the corresponding resident gene (see Fig. 14.12). Surprisingly, mice lacking MyoD protein were viable and fertile (Rudnicki et al., 1992). However, they had elevated levels of myf-5 mRNA, suggesting that the lack of MyoD protein was compensated by increased production of myf-5 protein. Transgenic mice without myf-5 protein were unable to breathe and died immediately after birth, but histological examination of skeletal muscle revealed no defects. Apparently, the inability to breathe was caused by the lack of distal parts of the ribs (T. Braun et al., 1992). In contrast, transgenic mice lacking *both* MyoD and myf-5 protein died after birth, and histological examination of

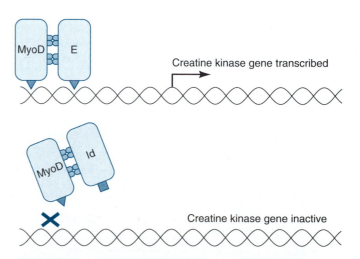

Figure 19.21 Gene regulation by heterodimers of basic helix-loop-helix proteins. Binding of the MyoD/E protein heterodimer enhances the transcription of the muscle creatine kinase gene and other muscle-specific target genes. Dimerization of the MyoD protein with the Id protein precludes binding to the creatine kinase enhancer, so that the gene is not activated.

these mice revealed a complete absence of skeletal muscle (Rudnicki et al., 1993). Thus, neither MyoD nor myf-5 protein is strictly required for skeletal muscle development, but at least one of the two must be present. Again illustrating the *principle of synergistic mechanisms,* each of the proteins can replace the other to a large extent with regard to myogenesis.

Both *MyoD+* and *myf-5+* activate their own expression, but each may be inhibiting the other (Olson and Klein, 1994). Either gene is thought to be activated when mesodermal cells are determined to become *myoblasts,* and the activity of either gene seems to stabilize this determined state.

Myogenin+ Expression Is Necessary for Histogenesis of Skeletal Muscle

The *myogenin+* gene appears to act later in the course of myogenesis, possibly during the formation of *myotubes.* Experiments knocking out the *myogenin+* gene had disastrous effects on skeletal muscle development (Hasty et al., 1993; Nabeshima et al., 1993). Mice without a functional copy of this gene did not breathe and died at birth, apparently because their diaphragms consisted of disorganized cells and no skeletal muscle fibers had developed. Similarly, other skeletal muscles were severely disorganized and reduced in mass. By comparison, the heart was formed and beat normally at birth, and other internal organs appeared normal as well. Apparently, the lack of myogenin interrupted the histogenesis of skeletal muscle, whereas cardiac and smooth muscle developed normally. Thus, myogenin is essential for skeletal muscle differentiation.

In order to identify genes that regulate myogenin, Tse-Chang Cheng and colleagues (1993) constructed *fusion genes* consisting of the 5' regulatory region of the mouse *myogenin+* gene and the *lacZ* reporter gene. Mouse embryos transgenic for this fusion gene showed lacZ staining exactly where and when myogenin is normally synthesized: in the *myotome* portions of the

somites and in limb buds (see Fig. III.1). *Deletion mapping* and *point mutagenesis* of the myogenin 5' regulatory region revealed separate binding sites for myogenic bHLH proteins and for another transcription factor known as *myocyte-specific enhancer factor-2 (MEF-2).* The mouse *MEF-2+* gene has 85% sequence similarity to a *Drosophila* gene termed *DMEF-2+. DMEF-2+* is expressed in the myogenic cell lineage of *Drosophila,* as is *MEF-2+* in the mouse. This homology in both sequence and expression domain suggests similar functions and opens the exciting possibility of combining the advantages of *Drosophila* and the mouse for further investigation.

Myogenic bHLH Genes Show Different Expression Patterns during Development

The expression of myogenic bHLH genes has been studied using *in situ hybridization* and *immunostaining* (see Methods 15.1 and 4.1). These studies have confirmed the central role of the *myogenin+* gene, for its products were found in all skeletal muscle lineages. Other results from these studies point out many unresolved questions.

Developing muscles vary in the timing and intensity with which different myogenic bHLH genes are expressed (Buckingham, 1992). For instance, in the developing mouse, the sequential activation patterns of *myf-5+, MyoD+,* and *myogenin+* differ between myotomes and limb buds (Fig. 19.22). In myotomes, myf-5 mRNA accumulates first at 8 days, whereas transcription of *myogenin+* in myotomes begins at 8.5 days and *MyoD+* is not transcribed until 10.5 days. By contrast, transcription of *myf-5+* in limb buds begins at 10.5 days, followed by the simultaneous transcription of *MyoD+* and *myogenin+* at 11 days. The transcription of *MRF-4* in

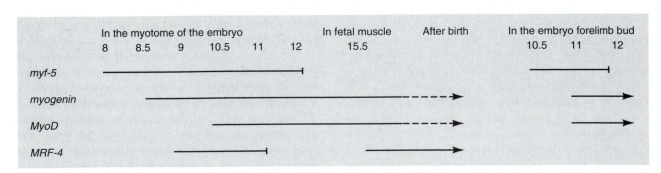

Figure 19.22 Expression of myogenic genes in mice at different stages of embryonic and fetal development. In situ hybridization (see Methods 15.1) reveals that mRNAs transcribed from four myogenic genes accumulate on different schedules.

myotomes extends from 9 to 11 days and then begins again at 15.5 days in fetal muscle. The nonsimultaneous transcription of myogenic bHLH genes indicates that these genes respond to different combinations of regulatory signals. This raises the question whether $MyoD^+$ and myf-5^+ are truly dispensable for skeletal muscle development, as suggested by the null phenotypes, or whether they have subtle functions that are not critical under laboratory conditions but may be adaptive in a natural environment.

The expression of $MyoD^+$ in early development is also reversible. In *Xenopus*, $MyoD^+$ is first transcribed at a low level throughout the blastula before RNA accumulation is restricted to the mesoderm, to somites, and finally to myotomes (Hopwood et al., 1989; Scales et al., 1990; Rupp and Weintraub, 1991). The accumulation of *Xenopus* MyoD protein closely follows that of MyoD mRNA (Hopwood et al., 1992). In the myotomes, MyoD protein accumulates in the cell nuclei, as expected of a transcription factor. In *Caenorhabditis elegans*, Michael Krause and colleagues (1990) have traced MyoD protein by immunostaining. The first staining appears at the 100-cell stage, and all stained cells later produce only body-wall muscle. Pharyngeal muscle and other cell lineages are not stained. However, when $MyoD^+$ transcription is traced using a *lacZ* gene under the MyoD promoter, the reporter gene is expressed as early as the 28-cell stage. This detection method marks only cells that have at least some muscle cells among their progeny. Those blastomeres that have *both* muscle and nonmuscle descendants are of particular interest. Evidently, they can transcribe the $MyoD^+$ gene and pass along this ability to some of their progeny while shutting off $MyoD^+$ in others. This shows that $MyoD^+$ transcription can be reversed as long as the concentration of MyoD protein is low.

The examples just discussed show that myogenic bHLH proteins are often produced early in development before muscle cell lineages begin differentiating or even before they are fully determined. This observation seems to contradict the view that the myogenic bHLH proteins activate genes expressed in differentiated muscle, such as myosin and creatine kinase genes. However, the paradox could be explained by the presence of myogenic bHLH proteins in inactive forms, such as dimers with Id proteins.

Finally, expression studies have shown that many structural genes expressed in skeletal muscle are also expressed in the heart, although the myogenic bHLH proteins are restricted to skeletal muscle. How these structural genes are regulated in the heart remains to be elucidated.

Although many questions about myogenic bHLH protein remain to be answered, the discovery of these proteins has been conceptually important. The finding that expression of individual myogenic bHLH genes can be necessary or sufficient for skeletal muscle development points out that small groups of switch genes may control cell differentiation. These switch genes seem to form networks that stabilize their expression and perhaps modulate their regulatory functions. If there were only a limited number of gene networks with these properties, then this would explain two classical features of cell differentiation: the limited catalog of mature cell types in each species and the general stability of differentiated cell states.

SUMMARY

Every species has a limited number of cell types, each with a distinct morphology and molecular makeup. The same cell type may appear in several body regions and may even develop from different embryonic cell lineages. Cells that have acquired the distinctive features of their type are called mature or differentiated cells. The differentiated state is generally stable: differentiated cells do not normally transform into other cell types. Many differentiated cells are highly specialized and depend on other kinds of cells for support and maintenance.

Cell differentiation occurs initially during the embryonic period of histogenesis. In some differentiated tissues, including nerve tissue and skeletal muscle, there is little turnover or addition of cells after the embryonic period. In other tissues, cell differentiation continues throughout adulthood and cells are turned over many times during the life of the organism. The replacement of differentiated cells occurs in one of two ways. In certain tissues, including liver and blood vessel endothelium, differentiated cells are still capable of dividing. In others, including epidermis and blood, worn-out cells are replaced from a pool of undifferentiated stem cells. Stem cells retain their capacity for self-renewal throughout the life of the organism. As stem cells divide, they produce both new stem cells and committed progenitor cells. Stem cells are either unipotent or pluripotent—that is, they give rise to either one or more types of committed progenitor cells.

The freshwater polyp *Hydra* consists of three cell lineages. Two of these lineages are epithelial; their cells divide in the differentiated state. The third cell lineage, the interstitial cells, forms nematocytes, nerve cells, gland cells, and gametes. The interstitial cells renew from pluripotent stem cells and from unipotent stem cells for either eggs or sperm. The commitment of nerve cell progenitors from stem cells and/or the differentiation of nerve cells from progenitors is enhanced by a peptide produced locally in the head region.

The blood cells of vertebrates are continuously renewed from a small pool of hematopoietic stem cells located mostly in bone marrow. A single pluripotent stem cell can form committed progenitors for all types of blood cells. Each progenitor undergoes several amplifying divisions before it gives rise to differentiated cells. The survival, division, and differentiation of blood cells depend on regulatory proteins, called colony-stimulating factors, which act on specific receptors in their target cells. Some colony-stimulating factors are diffusible, while others are bound to bone marrow stroma cells and extracellular matrix.

The development of skeletal muscle fibers is regulated by a small family of bHLH proteins, which act as transcription factors and have been highly conserved in evolution. The genes of the myogenic bHLH family regulate one another, giving the differentiated state of muscle fibers great stability. They also activate other muscle-specific genes such as the gene for muscle creatine kinase. The myogenic bHLH proteins need to form dimers in order to recognize their target genes. By forming heterodimers, these proteins enhance or inhibit one another's activity.

SUGGESTED READINGS

David, C. N., T. C. G. Bosch, B. Hobmayer, T. Holstein, and T. Schmidt. 1987. Interstitial stem cells in Hydra. In W. F. Loomis, ed., *Genetic Regulation of Development,* 389–408. New York: Alan R. Liss.

Dexter, T. M., and E. Spooncer. 1987. Growth and differentiation in the hemopoietic system. *Ann. Rev. Cell Biol.* **3**:423–441.

Hall, P. A., and F. M. Watt. 1989. Stem cells: The generation and maintenance of cellular diversity. *Development* **106**:619–633.

Weintraub, H. 1993. The MyoD family and myogenesis: Redundancy, networks, and thresholds. *Cell* **75**:1241–1244.

PATTERN FORMATION AND EMBRYONIC FIELDS

Figure 20.1 Argus pheasant displaying a dazzling array of eyespots on his tail feathers. The eyespots exemplify the ability of organisms to form patterns, harmonious arrays of different elements. In the case of an eyespot, the pattern arises through the synthesis of dark pigment in a circular territory of cells, and the synthesis of lighter pigments in drop-shaped territories of cells surrounding the dark spot. How are the signals that determine cells to make dark and light pigments coordinated in space so that a well-shaped eyespot results, rather than a pepper-and-salt mixture of dark and light cells?

By most accounts, *pattern formation* is the central topic in developmental biology. Patterns are harmonious arrays of different elements, such as the array of five fingers on a hand. Patterns can be seen at different levels of organization. For instance, the overall body pattern of birds features a pair of wings and a tail. Each of these organs shows a typical pattern of its own, including feathers of different sizes, shapes, and colors. Each feather, in turn, is composed of a shaft with barbs and barbules.

These consist of dead, pigmented epidermal cells, often arranged in brightly colored patterns (Fig. 20.1). Cells, too, have internal patterns, but in this chapter we will focus on patterns consisting of many cells.

Many patterns are formed when different types of cells mature in a coordinated fashion. For example, the eyespot pattern in a peacock feather results from the arrangement of cells containing different pigments in nested rings, with the darkest pigment occupying the center. The development of such patterns raises two questions. First, how do seemingly equal precursor cells give rise to different types of mature cells? This is the process of cell determination and cell differentiation discussed in previous chapters. Second, how are multiple pathways of cell determination coordinated so that an orderly array is formed rather than a random mixture of the same cell types? In other words, how does spatial coordination occur? The answer to one question may hold the key to the other: the same signals that bring about spatial coordination may also control, directly or indirectly, the determination of individual cell types.

Even though patterns become most obvious after cell differentiation, the process of pattern formation begins much earlier in embryogenesis. Consider the development of the vertebrate limb. An early limb bud consists of an ectodermal cover and a mesenchymal core (see Chapter 13). The core cells give rise to muscles and to the cartilage models that will later be replaced with bones. However, before the core cells become histologically different, they exhibit regular patterns of cell division and cell condensation. Cell division takes place more rapidly in a *progress zone* near the tip of the limb. Cell *condensations* are areas of higher cell density, which develop into either cartilage or muscle. The cartilage condensations proceed from proximal regions to distal and synthesize the extracellular matrix characteristic of cartilage (Fig. 20.2)

In addition to patterns based on cell division and condensation, there are patterns of *programmed cell death*, especially in appendages with long, separated digits, such as the human hand or the foot of a chicken. The cells between the developing digits of the limb bud synthesize proteolytic enzymes that kill these cells. As the dead cells are consumed by macrophages, the digits become separated. Thus, the processes of cell division, cell differentiation, and programmed cell death are all involved in generating spatial patterns. As each pattern is created and refined, the visible complexity of the organism increases.

The fate of a cell depends on its position within a territory of cells, such as a limb bud or another organ rudiment. This basic observation has led to the idea that each cell in a territory may have a *positional value* similar to the coordinates of a geometric point in a coordinate grid. This concept has become widely accepted, and the term "positional value" is often used as if it were a known physical

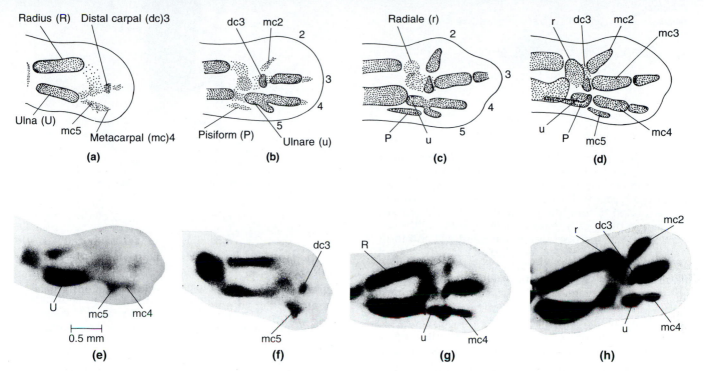

Figure 20.2 Pattern of precartilaginous cell condensations in the chick forelimb. **(a–d)** Drawings of precartilaginous cell condensations (stippled) and cartilaginous elements (outlined) in chicken wing buds at successive stages. The drawings were made from autoradiographs such as the ones shown in parts e through h. **(e–h)** Autoradiographs of sectioned chicken wing buds after labeling with $^{35}SO_4$. This label is incorporated into chondroitin sulfate, an extracellular matrix component characteristic of cartilage. Note that the intensity of the label progresses from proximal to distal with time.

entity, even if this is not the case. The term should be regarded as conceptual until the nature of the signals that control the behavior of cells in a particular territory has actually been revealed.

In many cases, the cellular territories in which patterns form have the properties of *fields.* A field is defined as a group of cells that can form a certain structure, such as a limb. The characteristic attribute of fields is known as *size invariance:* fields can be enlarged or reduced in size and still give rise to the same complete and normally proportioned structure, except that the structure forms on a larger or smaller scale. This implies that fields are capable of regulation, an attribute discussed in Chapter 6. Cells in fields can adjust their state of determination according to signals received from other cells.

In Chapters 20 through 24, we will examine concepts and experimental strategies researchers have used in attempting to understand pattern formation. In the present chapter, we will discuss some of the classic experiments in which cell isolation and transplantation have served as the principal methods. These experiments have provided the conceptual framework for the study of pattern formation.

In Chapters 21 through 24, we will explore how genetic and molecular tools have helped investigators to probe directly into the patterning mechanisms of fruit flies, vertebrates, flowers, and roundworms.

Regulation and the Field Concept

Most of the cells that engage in pattern formation are still capable of *regulation:* their *potency* is greater than their *fate.* As an example, let us consider the cells that form the forelimb in the salamander *Ambystoma maculatum.* A *fate map* for the future forelimb at an advanced embryonic stage is shown in Figure 20.3. A circular area of somatic lateral mesoderm and overlying epidermis is fated to form the free limb. The free limb area is surrounded by a ring of cells that normally form the shoulder girdle and the peribrachial flank from which the free limb extends. Together, these areas are called the *limb disc.*

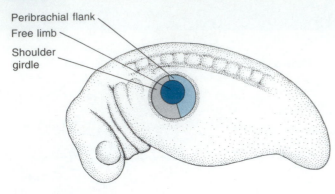

Peribrachial flank
Free limb
Shoulder
girdle

Figure 20.3 Forelimb field of the newt *Ambystoma maculatum.* The field is located in the somatic layer of the lateral mesoderm but also affects the development of the overlying ectoderm. The central area of the field contains the cells fated to form the free limb. These cells are surrounded by other cells that normally give rise to the shoulder girdle and peribrachial flank but will also form limb if the central cells are removed. A circle of cells outside this region does not normally contribute to the forelimb but may do so if all interior cells are removed.

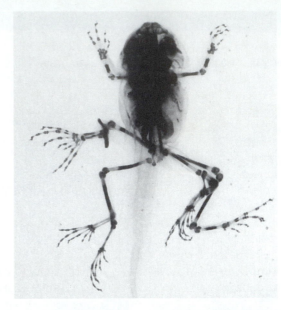

Figure 20.4 Adult frog (*Hyla regilla*) with multiple hindlimbs, from a pond inhabited by parasitic worms, which apparently partitioned the hindlimb buds of tadpoles. The adult frog was cleared and stained to make cartilage and bone visible.

A different picture arises when the developmental potencies of the limb disc and surrounding areas are tested by removal or transplantation. If the free limb area is removed, some of the cells that would normally contribute to shoulder girdle and peribrachial flank will deviate from their fate and form the free limb. If the entire limb disc is removed, an additional ring of cells surrounding the disc will form limb, shoulder girdle, and peribrachial flank, although with considerable delay. If the surrounding ring of embryonic cells is also removed, no limb will form. These results indicate that the potency of forming a complete limb is inherent in a larger group of cells than the area fated to form the limb. This larger area is called the *limb field,* and parts of the limb field can form a complete limb.

If a limb disc is transplanted to another location in the embryonic lateral mesoderm, it will give rise to a supernumerary limb. If the limb disc is removed and split before being reimplanted, *two* supernumerary limbs will develop. Such an outcome was observed in a natural experiment that took place in Santa Cruz, California. In a particular pond, both frogs and salamanders formed supernumerary limbs (Fig. 20.4). Investigation showed that parasitic flatworms (trematodes) were burrowing into developing limb buds (Sessions and Ruth, 1990). Presumably, the trematodes were effectively splitting the limb fields into two or more parts that gave rise to multiple limbs.

Similar regulation capabilities may be shown by entire embryos. As described in Chapter 6, many animal embryos are capable of *twinning.* For instance, when a sea urchin or frog embryo is split at the 2-cell stage, each blastomere will develop into a small but normally proportioned larva (see Figs. 6.8 and 8.2a). Conversely, when two or three mouse embryos are stuck together,

they will also give rise to one normally proportioned mouse (see Fig. 6.27).

The German embryologist Hans Driesch (1894), who studied twinning in sea urchins, described the embryo as a *harmonisches äquipotentielles System* (harmonious equipotential system), meaning that isolated blastomeres had the same ability as an entire embryo to form a whole larva. The American embryologist Ross G. Harrison (1918) adopted the English translation of this phrase to characterize the same properties of the amphibian limb disc. Later investigators coined the shorter term *field* to describe any group of cells that cooperate in forming a complete and well-proportioned organ or embryo. The ability of fields to form the same pattern after experimental size reduction or enlargement is now referred to as *size invariance.* This term is often misunderstood: it does *not* mean that the size of a field is invariant. Rather, it means the pattern is invariant regardless of field size.

Classical embryologists recognized the importance of fields to their understanding of development. Paul Weiss (1939), in his textbook *Principles of Development,* devoted 148 out of 573 pages to the topic of gradual determination and the field concept. Later on, fields fell out of fashion, presumably because they stood in the way of simplifying things for the purpose of analysis. However, the application of molecular probes may lead to a revival of the field concept: Eddy De Robertis and coworkers (1991) showed that an antibody raised against a particular *Xenopus* protein stains cells in the pectoral fin bud of the zebra fish embryo as well as cells in the fin disc before outgrowth (see Color Plate 6; for a description of *immunostaining*, see Methods 4.1). The antibody was directed against the protein's *homeodomain*, a specific DNA-binding domain that characterizes a family of transcription factors (see Chapter 15). This indicates that the cells in the fin field produce a protein that contains a very similar domain and presumably acts as a transcription factor. Such observations confirm the notion of a field as a group of cells programmed for similar gene activities.

Many territories in which pattern formation takes place exhibit the properties of fields. They have been found to do so in all embryos that are capable of regulation. For the embryo, fields are a marvelous way of compensating for growth irregularities and injuries. For the researcher, the field phenomenon is a formidable complication of the study of pattern formation. However, any serious model must take it into account.

Characteristics of Pattern Formation

In this section, we will examine a few key experiments that establish some major points regarding pattern formation: patterning depends on cellular interactions, on the genetic information of the interacting cells, and on their developmental history. The studies also indicate that at least some patterning signals are fairly general: they have been well conserved during evolution, and they are used in different parts of the body either simultaneously or successively.

Pattern Formation Depends upon Cellular Interactions

In embryos with regulative capabilities, cells are determined gradually and, to a large extent, by interactions with other cells. In such embryos, *embryonic induction* is a prevalent patterning mechanism (see Chapters 9 and 11). To investigate this process, Spemann (1938) performed numerous transplantation experiments with newt embryos. By grafting tissues between two closely related species, the light-colored *Triton cristatus* and the darkly pigmented *Triton taeniatus*, he could easily ascertain which of the developing structures were formed by the graft and which by the host. In a typical experiment of this kind, he exchanged a piece of prospective flank epidermis with a piece of prospective brain at the early gastrula stage (see Figs. 6.13a and 6.14). Each transplanted piece developed according to its *new* position in its host: prospective epidermis developed into brain, and vice versa. Because the structures formed by the transplants depended upon surrounding tissue, their development was classified as *dependent differentiation*.

Not all grafted pieces of early gastrula undergo dependent differentiation. In the famous organizer experiment of Spemann and Mangold (1924), dorsal blastopore lip of the early gastrula underwent *self-differentiation*: upon grafting, it did not adapt to its new location but stuck to its original fate of forming notochord. Moreover, the graft induced the surrounding host tissue to cooperate with the graft in forming an additional set of dorsal embryonic organs (see Fig. 11.15). In other words, the transplanted organizer set up an additional field for dorsal organs.

These experiments show with great clarity that the formation of the body pattern in amphibian embryos depends on cellular interactions. Cells in some embryonic regions are determined early, act as inducers on their surroundings, and undergo self-differentiation. Cells in other regions are determined later, respond to the inducing signals from other cells, and undergo dependent differentiation.

The Response to Patterning Signals Depends on Available Genes and Developmental History

As part of the reciprocal transplantation experiments performed in Spemann's laboratory, tissue grafts were exchanged between a salamander, *Triturus taeniatus*, and a frog, *Rana esculenta* (Spemann and Schotté, 1932). These species belong to two separate subclasses of amphibians, urodeles (tailed amphibians) and anurans (tailless amphibians). How would the great phylogenetic distance between these groups affect the results of tissue transplantations? For instance, if prospective belly ectoderm from a salamander gastrula were transplanted into the prospective mouth ectoderm of a frog gastrula, would the transplant still undergo dependent differentiation, that is, develop mouthparts? If so, would those mouthparts be teeth and balancers (characteristic of the salamander), or horn-covered jaws and a sucker (characteristic of the frog)?

The results of this experiment by Spemann and Schotté are illustrated in Figure 20.5 and summarized in Table 20.1. Indeed, prospective belly epidermis from the salamander gastrula, when transplanted into the prospective mouth region of the frog, gave rise to mouthparts. However, the structures formed by the graft were the balancers and teeth characteristic of the salamander. Conversely, prospective frog belly epidermis in the salamander mouth region gave rise to frog mouthparts. In control transplantations within the *same* species, prospective belly epidermis transplanted to the mouth region formed mouth structures as expected. In each combination of donor and host, the transplants underwent dependent differentiation, but the type of mouthparts formed was characteristic of the donor species.

The limitation of the transplants to the structures of their species could be ascribed, in principle, to either genetic or nongenetic factors. However, additional ex-

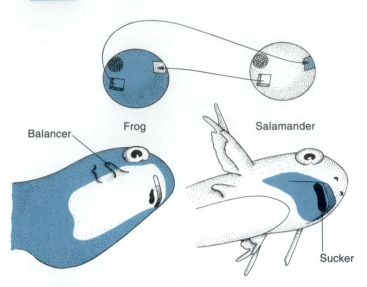

Figure 20.5 Dependence of inductive interactions on the genotype of the responding tissue. The diagram shows a reciprocal transplantation experiment between early gastrulae of a frog (*Rana esculenta*) and a salamander (*Triton taeniatus*). The two species differ in their mouth structures. The frog has horn-covered jaws and a sucker right behind the mouth. The salamander has teeth and paired "balancer" organs behind and below the eyes. Prospective belly epidermis from the salamander gastrula was transplanted into the prospective mouth region of the frog gastrula, and vice versa. Both transplants developed according to their new location, and both formed mouth structures that fit topographically with the mouth structures derived from host tissue. However, the mouth structures formed by the transplants had the morphological characteristics of the donor: the frog transplant in the salamander host formed horn-covered jaws and a sucker, and the salamander transplant in the frog host formed teeth and a balancer.

periments that will be described shortly indicate that the genome of the transplant is critical. Frog cells do not seem to have (or use) the genetic information for forming salamander structures, and vice versa. But the transplants from both species respond to similar patterning signals that prompt them to form mouthparts—whatever mouthparts are in their genetic repertoire.

TABLE 20.1		
Transplantation Experiment of Spemann and Schotté (1932) Involving Frog and Salamander Gastrulae		
	Structures Formed after Transplantation to Mouth Region of	
Donor Species of Prospective Belly Epidermis Graft	**Frog**	**Salamander**
Frog	Horned jaw, sucker	Horned jaw, sucker
Salamander	Teeth, balancers	Teeth, balancers

The response of a tissue to inductive signals is also limited by its developmental history. As we saw in previous chapters, induction is often a cumulative process involving multiple inducers. For instance, the lens of the eye is formed by head ectoderm in response to three inducing tissues. During gastrula stages, the head ectoderm that will eventually form the lens is underlain by the endoderm of the future pharynx. This endoderm is the first inducer of the lens. As gastrulation proceeds, mesoderm that will later form the heart comes to lie next to the future lens ectoderm. The heart mesoderm is the second inducer of the lens. Finally, as the optic vesicles bulge out from the embryonic brain, the future retina comes in contact with the overlying head ectoderm. The optic vesicle, then, is the third inducer of the lens, although it acts *indirectly* by displacing mesenchymal cells that would otherwise *inhibit* lens formation.

The response of the head ectoderm to the second and third lens inducers depends on its previous interactions with the first inducer. Antone Jacobson (1966) studied this phenomenon quantitatively in a salamander, *Taricha torosa*. In one experiment, he removed the earliest inducer, the pharyngeal endoderm (Fig. 20.6). The earlier the removal occurred, the lower was the percentage of successful lens inductions. Thus, the early history of inductive interactions clearly biased the response of head ectoderm to the later inducers.

In most inductive interactions, the *competence* of a tissue to respond to an inductive stimulus depends on the developmental history of the responding tissue. In normal development, competence typically arises some time before the responding tissue is exposed to the new stimulus and lasts for a while thereafter.

Patterning Signals Have a High Degree of Generality

The most enthusiastic students of pattern formation have hoped to find that patterning signals are universal, that is, the same in all organisms, as is the genetic code. Indeed, some patterning signals do show a remarkable degree of phylogenetic conservation. The transplantation experiment summarized in Figure 20.5 and Table 20.1 demonstrates that a salamander graft "understands" inductive signals provided by a frog host and vice versa. This is astounding, since frogs and salamanders have evolved independently for more than 200 million years.

The patterning signals exchanged between cells are also general in the sense that they are used in different regions of the same embryo—either simultaneously, or successively at different stages of development. This is illustrated by the phenotypes observed in certain *Antennapedia* alleles of *Drosophila*. The wild-type *Antennapedia*+ gene is most active in the thoracic segments and plays a key role in leg development. The

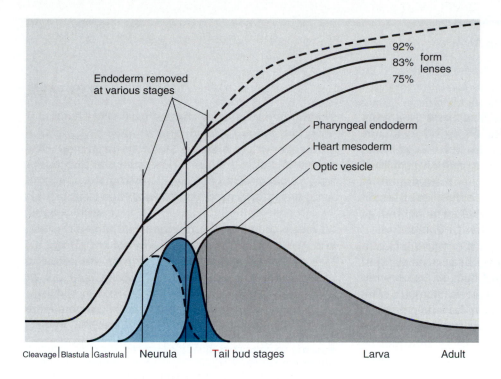

Endoderm removed at various stages

92%
83% form lenses
75%

Pharyngeal endoderm
Heart mesoderm
Optic vesicle

Cleavage | Blastula | Gastrula | Neurula | Tail bud stages | Larva | Adult

Figure 20.6 Dependence of inductive interactions on the developmental history of the responding tissue. The lens of the eye in the salamander *Taricha torosa* is induced in head ectoderm by successive interactions with pharyngeal endoderm (light color), heart mesoderm (dark color), and the optic vesicle (gray). The height of each curve (ordinate) represents the stage-dependent capacity to induce lens formation. Removal of the endoderm during the neurula and early tail bud stages (vertical lines) leaves some embryos unable to subsequently respond to inducers. The longer the endoderm is present, the greater the percentage of individuals that form lenses.

Antennapedia^R allele is characterized by a small chromosomal inversion that destabilizes the control of the *Antennapedia* gene. This mutation causes sporadic activation in the antenna, where the gene is normally silent. The ectopic gene activity causes the formation of leg structures instead of antennal parts. Rarely is the entire antenna transformed into an entire leg. More commonly, parts of the antenna are replaced with parts of the second leg. When John Postlethwait and Howard Schneiderman (1971) examined these replacements in detail, the found that the replacements are strictly position-specific (Fig. 20.7). The most distal part of the antenna, the *arista*, is always replaced with the last tarsal segment carrying the claws at the tip. The most proximal part of the antenna is always replaced by the most proximal leg structure, known as the *trochanter*. Intermediate parts of the antenna are replaced with intermediate leg parts.

The strict order of the homeotic transformations in the *Antennapedia*^{R} mutant suggests that the leg and the antenna respond to similar or identical patterning signals that specify the proximodistal order of their segments. The character of the structures (leg or antenna) formed in response to a signal depends on the local activity or inactivity of the *Antennapedia* gene. This interpretation is in accord with the conclusion drawn earlier that the response to patterning signals depends on gene activities in the reacting cells.

In summary, pattern formation depends on intercellular signals. At least some of these signals have been well conserved during evolution. The response of cells to patterning signals depends on the cells' genetic reper-

toire and developmental history. These characteristics of pattern formation have shaped the concepts used in its analysis.

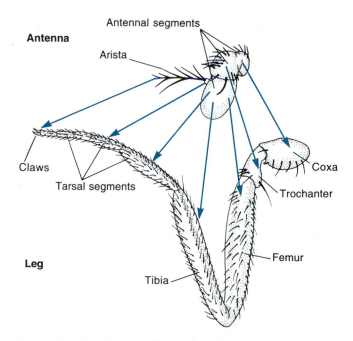

Antenna

Antennal segments

Arista

Claws

Tarsal segments

Coxa

Trochanter

Femur

Leg

Tibia

Figure 20.7 Position-specific replacement of antennal parts with corresponding parts of the second leg in the homeotic mutant *Antennapedia*^{R} of *Drosophila*. Sporadic expression of the deregulated gene in the antenna causes the transformation of antennal parts into leg parts. Each arrow points from the antennal area to the region of the leg that replaces it. Note that the replacements maintain the same proximodistal order.

The Concept of Positional Value

When Hans Driesch discovered what is now termed an embryonic field, he immediately realized that a cell's fate depends on its *position* in the field. Driesch made this a salient point in his treatise *Analytische Theorie der organischen Entwicklung* ("Analytical Theory of Organic Development," 1894). As a cautionary explanation, however, he added that it was not position itself that determined the fates of cells but rather the *different signals* received by the cells according to their positions. In discussions of this point, several biologists have used the analogy of a mountain, which supports a typical pattern of changing vegetation, from tropical or deciduous forests at the bottom, through grasslands and coniferous trees, to mosses and lichens at the top (Sander, 1990). A similar pattern of change in plant communities is observed as one moves at low altitude from the equator to the poles of the earth. At first glance, it may seem that position itself (altitude or latitude) favors the growth of some plants over others. However, further comparison and experimentation show that the factors of more direct importance are the temperature and the amount of precipitation prevailing in each position. Similarly, the analysis of embryonic pattern formation proceeds from the study of the effects of cell position to the study of molecular signals received in different positions.

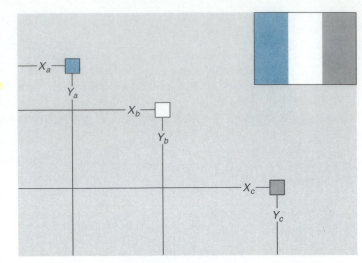

Figure 20.8 French flag model used by Wolpert (1978) to explain his concept of pattern formation. A two-dimensional territory of embryonic cells undergoes differentiation, forming cells with three kinds of pigment: blue (color), white, or red (gray shading). The process is thought to require two steps. First, each cell is assigned a positional value (X_a/Y_a, X_b/Y_b, X_c/Y_c) with regard to the two axes of the flag. The Y coordinate is unimportant in this case. In the second step of the patterning process, cells interpret their positional values. Cells with low X values produce blue pigment; cells with intermediate X values produce white pigment; cells with high X values produce red pigment. As a result, the array of differentiated cells resembles the French flag (shown as inset in upper right corner).

Pattern Formation Is Modeled as a Two-Step Process of Cells Receiving Positional Values and Interpreting Them

The apparent effect of a cell's position on its fate in the embryo was cast in modern terms by Lewis Wolpert (1969, 1978). According to Wolpert's concept, pattern formation is a two-step process: first the cells of a territory are assigned positional values, and then they interpret these values according to their genetic information and developmental history. The *positional value* of a cell specifies its position within a territory, in much the way geometric coordinates define a point in a grid. *Interpretation* is the general process whereby a positional value leads to a particular cell activity or differentiated state.

Using the French flag as a model, Wolpert presents the patterning process in a rectangular territory of cells as shown in Fig. 20.8. The positional value (coordinates; in this case, only the X coordinate is important) assigned to each cell specifies whether the cell is located in the left-hand third, the middle third, or the right-hand third of the territory. At first, the cells are still "embryonic," or unpigmented. Subsequently, the cells interpret their positional values by synthesizing blue, white, or red pigment. The actual interpretation, however, depends on the cells' genetic information, which

is "French." Cells with "American" genes would respond to the same positional values but would interpret them differently.

Wolpert's French flag analogy provides a model for the transplantation experiments discussed earlier. We can imagine a patch from the upper left-hand corner of an embryonic French flag being swapped with a patch from the lower margin of an embryonic American flag (Fig. 20.9). The transplants develop in accordance with their new positions in their hosts, indicating that the first step in pattern formation, the assignment of positional values, has been adjusted in the grafts. The second step, the interpretation of positional values, depends on the nationalities of the grafts, which have not been changed by the transplantation. The French graft develops vertical strips of white and red, while the American graft forms stars. Correspondingly, in the experiment of Spemann and Schotté, salamander belly ectoderm transplanted into the mouth region of a frog formed mouthparts, according to its new positional value (Fig. 20.5). However, the rules for interpreting the new positional values were set by the genetic information in the salamander graft. Likewise, when antennal parts of *Antennapedia^R* mutant fruit flies were replaced by leg structures of corresponding positional value, the change in interpretation resulted from ectopic activation of the *Antennapedia* gene (Fig. 20.7).

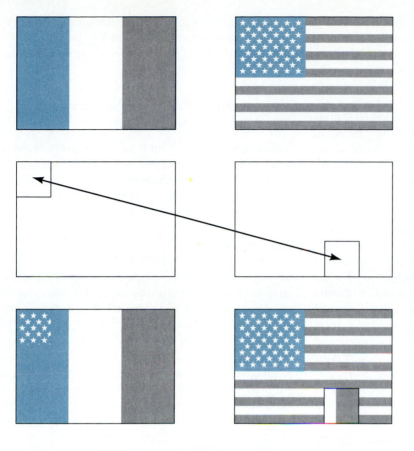

Figure 20.9 Interpretation of the experiments shown in Figures 20.5 and 20.7 in terms of the French flag model shown in Figure 20.8. A piece of embryonic French flag grafted onto an embryonic American flag develops in accordance with its new position but reveals its own genetic program. Near the bottom right, this calls for vertical white and red (gray) strips. Conversely, a patch of embryonic American flag grafted onto the upper left-hand corner of an embryonic French flag forms stars.

Positional Values Are Assigned Relative to Reference Points

Of critical importance in Wolpert's concept of pattern formation are the cells that provide the *reference points* for the assignment of positional values. A reference point corresponds to the zero point defining the position of any other point on a coordinate axis. Equally important is how cells compute their distance from the reference points of a territory. As we will see, some ways of using reference points impart field properties— that is, size invariance—to a territory, whereas other ways do not. We will explore this issue using a string of cells as a simplified version of the French flag model (Fig. 20.10). The question, then, is, What rules might a horizontal row of cells follow in order to generate three zones of equal width that form different pigments— blue, white, and red?

First, let us assume that cells acquire positional values simply according to their rank, taking the left-hand side of the string as a reference point. The nth cell from the left then has the positional value X_n. In terms of the French flag model, a simple set of interpretation rules would be to produce blue pigment for $n \leq 5$, white pigment for $6 \leq n \leq 10$, and red pigment for $11 \leq n$. This combination of a reference point and interpretative rules will produce a harmonious French flag only if the total number of cells in the string is exactly 15. Any decrease or increase in the total cell number will lead to the formation of either a truncated flag or a flag with an over-

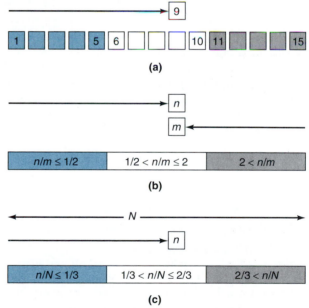

Figure 20.10 Different methods of specifying and interpreting positional value can generate the color pattern of the French flag. Since only the horizontal coordinate matters in this flag, it is assumed for simplicity that the flag is formed by a linear string of cells. **(a)** Positional value is specified simply by counting cell rank from the left. **(b)** Positional value is specified by counting cell rank from left and right and forming a ratio. **(c)** Positional value is specified by counting cell rank from the left and dividing by the total number of cells. With each method, positional values are interpreted by comparison with certain threshold values. The first method is not size-invariant, whereas the second and third methods are.

proportionate red region. Under these rules, then, the territory would lack the basic field property of size invariance.

In order to achieve size invariance in a territory, cells need a more sophisticated way of assigning themselves positional values and interpreting them. One way is for each cell to "count" its rank from both ends of the string. The positional value of each cell would then be $X_{n,m}$, with n indicating the rank from the left end and m indicating the rank from the right end of the string. A suitable set of interpretative rules would be to produce blue pigment for $n/m \leq 1/2$, white pigment for $1/2 < n/m \leq 2$, and red pigment for $2 < n/m$. This combination of two reference points and interpretative rules would produce a well-proportioned flag with any sufficiently large number of cells. The same purpose would be achieved if cells counted their rank only from the left but in addition could perceive the total number of cells (N) in the string. The positional value of each cell would be n/N; blue pigment would have to be made for $n/N \leq 1/3$, and so forth. Such rules of interpretation would be easy to program into a computer, which would then generate French flags of all sizes on its screen. How living cells could carry out such arithmetic operations is another matter. One possible mechanism—a morphogen gradient—will be discussed later in this chapter.

In the French flag model, the ends of a string of cells or the margins of a rectangle serve as natural reference points. In actual embryos, however, fields often overlap and lack anatomically conspicuous reference points. Consequently, before cells assign themselves positional values, they must establish their own reference points. It is important to distinguish between these two events, because the mechanisms used to establish reference points are easily mistaken for the mechanisms by which these points later confer positional value on other cells (see Chapter 22).

Limb Regeneration and the Polar Coordinate Model

The French flag model, with its rectangular coordinates, resembles a notebook or computer screen more than a real embryo. However, many embryonic fields, such as the limb field shown in Figure 20.3, have a well-defined focal area surrounded by marginal zones of decreasing potential that are not clearly demarcated. In such cases, a more suitable model of pattern formation is the *polar coordinate model,* in which positional values are assigned to cells according to their distance from a center and their angular orientation in an area subdivided like a clock face.

The polar coordinate model was conceived to explain the results of regeneration experiments with salamander limbs, roach legs, and *Drosophila* imaginal discs. We will first summarize some of the salamander limb experiments, to make the special features of the model more readily intelligible.

Regeneration Restores the Elements Distal to the Cut Surface

When a salamander limb or tail is amputated, epidermal cells spread over the wound surface and form an *apical epidermal cap.* Underneath the cap, all tissues undergo dedifferentiation and generate a cone of cells called a *regeneration blastema.* After a period of further proliferation, blastema cells redifferentiate and restore the missing limb.

When an appendage has been severed, only the stump—not the severed part—is supplied with blood, and thus only the stump can regenerate. For this reason, the parts that are considered missing—and are regenerated—are always the parts distal to the cut surface. What would the severed part do if it could be supplied with blood? Would an arm, if its life were sustained, regenerate a shoulder, or even an entire animal? To answer this question, Elmer G. Butler (1955) cut the arm of a salamander at the wrist and through the humerus and stuck the wrist end into an opening in the flank, where it connected with the nervous and circulatory systems. The humerous end formed a regeneration blastema and regenerated *all structures distal to the cut surface* (Fig. 20.11). Thus, most of the regenerate was a *mirror image* of the upper and lower arm parts already present. This result shows that the regenerated limb pattern is determined by the position of the cut and not by other properties of the stump. This rule, which has also been observed in other regenerating organisms, is known as the *rule of distal transformation:* regeneration restores those elements that are normally present distal to the cut surface.

Intercalary Regeneration Restores Missing Segments between Unlike Parts

Regeneration is also observed in experiments that juxtapose tissues having unlike fates. The confrontation of nonneighbor cells triggers a process called *intercalary regeneration,* in which cells proliferate and restore the intervening parts. Differences in pigmentation can be used to discern which of the juxtaposed tissues gives rise to the regenerate. In salamanders, intercalary regeneration is stimulated by grafting a blastema resulting from the amputation of a hand to the stump of an amputated arm (Fig. 20.12). The blastema heals onto the stump and then forms a hand as if it had remained in its original position at the wrist. In addition, cells from the stump divide and regenerate the intervening parts between the hand and the upper arm. This type of in-

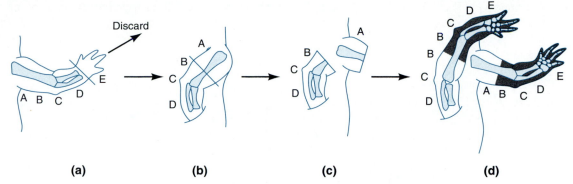

Figure 20.11 Distal transformation in a regenerating salamander limb. The capital letters mark different proximodistal levels of the forelimb. The forelimb is amputated at the wrist **(a)**, and the cut surface of the stump is inserted into the flank **(b)**. After healing, the inserted limb is cut again at the level of the humerus (upper arm) **(c)**. This cut generates two different stumps: one contains the proximal part of the upper arm; the other contains the lower arm and the distal part of the upper arm. However, both stumps end where the second cut was made through the upper arm. **(d)** Both stumps regenerate the same limb parts distal to the cut surface.

tercalary regeneration obeys the rule of distal transformation, since the intervening parts are distal to the cells from which they are derived. However, in similar experiments with insect legs, the rule of distal transformation is violated. Here, it is mostly the distal leg part that regenerates the intervening structures (Bohn, 1976).

Limb Regeneration Resembles Embryonic Limb Bud Development

A regenerating limb is similar to a normal embryonic limb bud in many respects (see Chapter 13). Both have a mesenchymal core covered by a layer of ectodermal cells. The apical epidermal cap of the regenerating limb corresponds to the apical ectodermal ridge of the limb bud. Removal of the ectodermal cover stops both

limb regeneration and normal development. The tip of the mesenchymal core, called a *progress zone* in the limb bud, is a zone of rapid mitotic divisions like the regeneration blastema of the regenerating limb. If mitoses are blocked by X-irradiation, neither normal limb development nor regeneration proceeds.

What is most important for pattern formation is that limb buds and regenerating limbs respond to similar if not identical patterning signals (Muneoka and Bryant, 1982). This is indicated by the results of transplantation experiments with axolotl larvae. Hindlimbs develop more slowly than forelimbs in axolotl, so that tips of developing hindlimb buds and blastemas of regenerating forelimbs can be swapped between animals at similar stages of development (Fig. 20.13). If the transplants were properly aligned, limb bud tips cooperated with

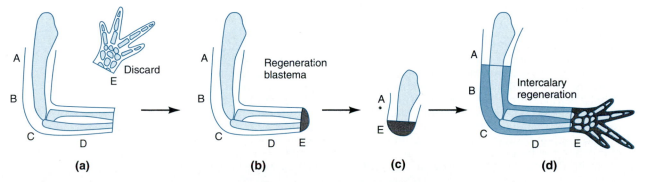

Figure 20.12 Intercalary regeneration in the salamander limb. The capital letters mark different proximodistal levels of the forelimb. **(a)** The limb is amputated at the wrist. **(b)** A regeneration blastema forms at the amputation site. **(c)** The regeneration blastema is grafted to the freshly severed stump of a limb amputated through the humerus (upper arm bone). This juxtaposes tissues from levels A and E. **(d)** The distal limb part is regenerated from the grafted blastema (level E), while the intervening levels are regenerated from stump tissue (level A).

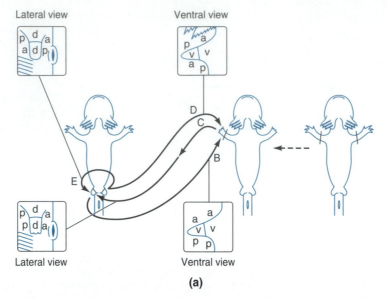

Lateral view   Ventral view

(a)

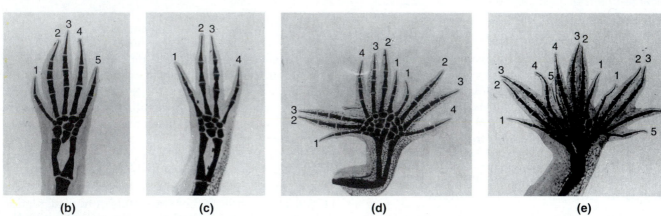

(b)　　　(c)　　　(d)　　　(e)

Figure 20.13 Combination of *regenerating* forelimb blastemas with *normal* hindlimb bud stumps and vice versa by grafting operations on axolotl larvae. The grafting operations are shown in ventral view except for insets marked as lateral views. Note that the forelimbs of axolotl larvae are fully differentiated while the hindlimbs are still growing out as buds. **(a)** The forelimbs were amputated through the humerus. After 10 to 12 days, the forelimb blastemas had regenerated to the point of early digit formation. These animals were matched with siblings that had hindlimb buds with sizes similar to the regenerating forelimbs. Some of the grafts (B and C) were made ipsilaterally (using grafts and stumps from the same side) so that graft and stump could be aligned with respect to both their anteroposterior and dorsoventral axes. Other grafts (D and E) were made contralaterally (using grafts and stumps from opposite sides) so that anterior graft was juxtaposed with posterior stump and vice versa. The insets show closer views of the grafts in their dorsal (d) or ventral (v) aspects to indicate the alignment or misalignment of anterior (a) and posterior (p) tissues. **(b–e)** Skeletal preparations made 4 to 6 weeks after grafting. **(b)** Normal right hindlimb formed after the ipsilateral graft (graft B) of a right hindlimb bud onto a regenerating right forelimb stump. **(c)** Normal forelimb formed after the ipsilateral graft (graft C) of a right regenerating forelimb blastema onto a right hindlimb bud stump. **(d)** Limb that resulted from the contralateral graft (graft D) of a left hindlimb bud to a right regenerating forelimb blastema stump. **(e)** Limb resulting from a contralateral graft (graft E) of a left hindlimb bud to a right hindlimb bud stump. Supernumerary limb structures developed when the anteroposterior axes of the graft and the stump were in opposite orientation.

stumps of regenerating limbs in forming new limbs. Likewise, regeneration blastemas cooperated with limb bud stumps. However, the most intriguing results were obtained when the grafts were attached with a reversed anteroposterior axis. Such misalignment caused characteristic pattern abnormalities that were the same in limb buds, regenerating limbs, and combinations of both. These results indicate that the patterning mechanisms in limb buds and regenerating limbs are similar. Presumably, cells in regeneration blastemas produce the same signals and activate the same genes that were involved in the original development of the limb. This conclusion is in line with the generality of patterning signals discussed previously.

The Polar Coordinate Model Is Based on Two Empirical Rules

Limb regeneration studies strongly support the concept that cells have positional values and that they follow certain rules in generating additional cells with other positional values. These rules were formulated by Vernon French, Peter Bryant, and Susan Bryant as part of their *polar coordinate model* (French et al., 1976; S. V. Bryant, 1977; S. V. Bryant et al., 1981). In this model, positional values are assigned to limb cells as if they were located in sections on the surface of a cone (Fig. 20.14). The base of the cone represents the proximal end of the limb, and the tip represents the distal end. The positional values along the proximodistal axis are designated A to E, with A at the proximal end (the base). A second set of positional values, designated by the numbers 1 to 12, are distributed around the circumference of the cone, with 6 marking the ventral midline and 12 the dorsal midline. The cone can be compressed into a disc, which

then looks like a clock face or a dartboard. This configuration is used to model the vertebrate limb disc before the limb bud grows out. The same clock face is applied to *imaginal discs* of insects that give rise to legs and other appendages.

Two simple, empirical rules govern the behavior of cells according to the polar coordinate model. The first is the *shortest intercalation rule.* It states that wherever cells with nonadjacent positional values confront one another, the incongruity triggers cell proliferation. The growth continues until cells with all intermediate positional values have been regenerated. In confrontations of cells with different circumferential values, it is always the shorter of the two possible sets of intermediate values that is intercalated (Fig. 20.15). For instance, if cells with positional values 1 and 4 are grafted next to each other, cells with values 2 and 3 will be intercalated rather than cells with values 5 through 12.

The second rule of the polar coordinate model is the *distalization rule.* It states that cells can regenerate only cells with more distal positional values. In a limb amputated at level A, blastema cells can give rise only to cells with B values, which in turn can produce only cells with C values, and so forth (Fig. 20.16). This rule forces the blastema cells to gradually adopt more distal positional values until the limb is completely regenerated.

The polar coordinate model makes no explicit assumptions about the molecular mechanisms by which cells acquire positional values. However, since the model is based on interactions between neighboring cells, the implication is that positional values are set by signals exchanged between nearest neighbors. One signal of this type would be cell adhesiveness, and indeed, cells that form the distal part of a limb adhere to one another more strongly than cells that form the proximal part (see Chapter 25).

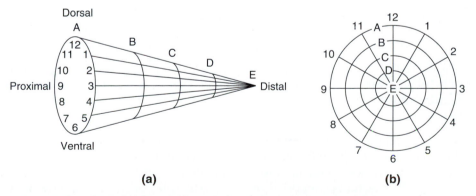

(a)   **(b)**

Figure 20.14 Positional values in a limb field according to the polar coordinate model. The limb field is modeled as a cone **(a)** or as a clock face **(b)**. Each cell is assumed to have positional values in the proximodistal direction (A to E) and in circumference (1 to 12). Positional value A belongs to the proximal part (the base) of the limb, and positional value E belongs to the distal part (the tip). Positional value 12 is on the dorsal midline of the limb, and positional value 6 is on the ventral midline.

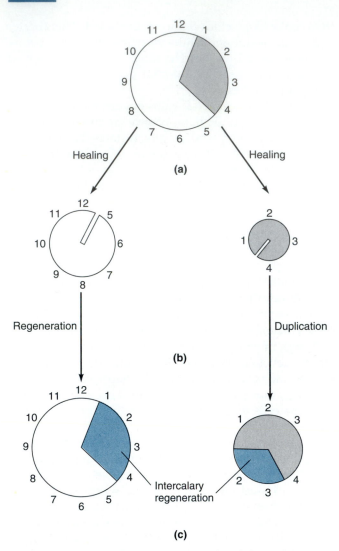

Figure 20.15 Shortest intercalation rule of the polar coordinate model. **(a)** A small segment, of circumferential values 1 to 4, is cut out. **(b)** Healing juxtaposes values 5 and 12 in the large piece as well as 1 and 4 in the small piece. **(c)** Intercalary regeneration restores the missing values (color) along the shortest circumferential route. This process regenerates the full set of values in the large piece but causes duplication in the small piece.

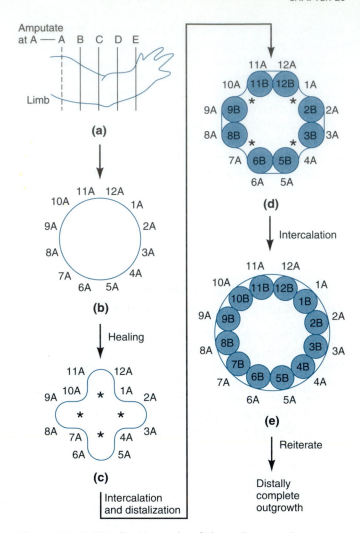

Figure 20.16 Distalization rule of the polar coordinate model. **(a)** Limb amputation removes levels B through E, leaving level A. **(b)** The circle represents the edge of the wound. **(c)** Healing involves the migration of dermal fibroblasts from the wound margin to the center. As a result, fibroblasts with different circumferential values confront one another (asterisks). **(d)** Intercalation restores the missing circumferential values according to the shortest intercalation rule. In addition, the distalization rule forces the new cells to adopt the next more distal value, B (shading). Among the new cells, several circumferential values are missing, creating more confrontations of nonneighboring cells (asterisks). **(e)** Intercalary regeneration completes a continuous set of circumferential values at the B level. The process is reiterated until the most distal level (E) is completed.

The Polar Coordinate Model Explains the Development of Supernumerary Limbs from Misaligned Regenerates

The polar coordinate model explains not only the progress of normal limb regeneration but also the formation of supernumerary limbs, which frequently grow out from misaligned grafts. One such case, reported by Susan Bryant and Laurie Iten (1976), is summarized in Figure 20.17. It shows that the model correctly predicts not only the number and location of the supernumerary limbs but also their orientation and handedness. This remarkable accuracy validates the polar coordinate model and the general concept of positional value.

However, it must be understood that positional value is a conceptual term and that its physical basis remains to be discovered.

The rules of shortest intercalation and distalization, on which the polar coordinate model is based, suggest that positional values of regenerated cells are determined by the positional values of their nearest neighbors. Such close-range interactions contrast with the long-range signals that are postulated by the gradient models discussed in the following section.

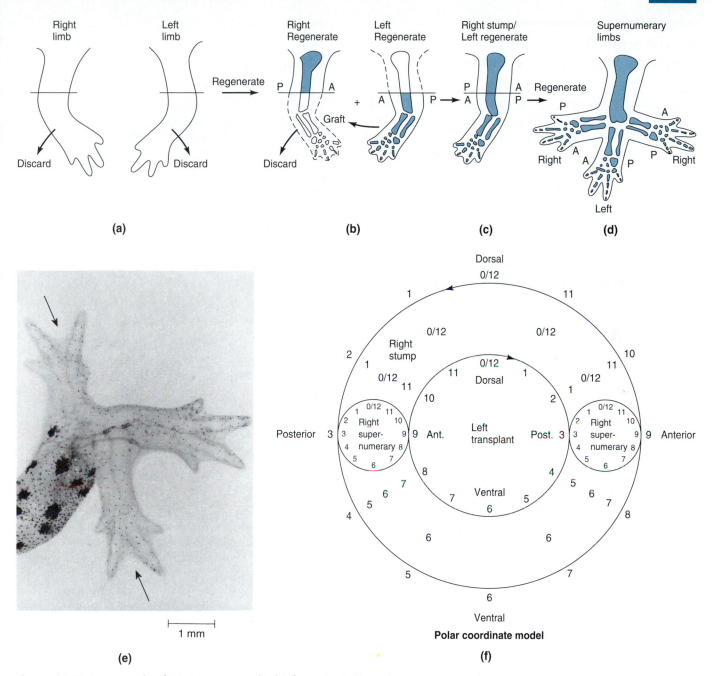

Figure 20.17 Outgrowth of supernumerary limbs from a misaligned regenerate in the salamander *Notophthalmus viridescens.* **(a)** Both hindlimbs are amputated through the femur and allowed to regenerate. **(b)** When the regenerates begin to form digits, the right regenerate is removed and the left regenerate is grafted to the right stump. **(c)** This transplantation juxtaposes anterior tissue of the stump and posterior tissue of the regenerate, and vice versa. **(d)** Two supernumerary limbs form at the sites of maximum disparity. **(e)** Photograph of the regenerated left hindlimb and two supernumerary right limbs (arrows). **(f)** Schematic cross section of the graft-host junction. The inner and outer circles represent the respective circumferences of the graft and the stump. The diameters of the graft and the stump are shown different for clarity only. The numbers are positional values according to the polar coordinate model. The values between the circles are generated according to the shortest intercalation rule. At the two sites of maximum incongruity, the shortest intercalation goes in either direction, depending on the neighboring intercalations. The resulting complete circles of positional values behave like regenerating limbs, causing the outgrowth of supernumeraries. The model predicts that the supernumeraries will have the same handedness and orientation as the stump. These predictions are confirmed by the results.

Morphogen Gradients as Mediators of Positional Value

Positional values may be imparted to the cells of a territory by various means. For instance, Wolpert and coworkers (1975) proposed that the values along the proximodistal axis of the chicken wing are specified by the number of mitotic cycles that a cell spends in the *progress zone* at the tip of the wing bud. Proximal cells leave the progress zone early and have low positional values. More distal cells leave the zone after more mitotic cycles and have correspondingly higher positional values.

The models proposed most frequently as a means of specifying positional values are known as *gradient models.* A gradient may be set up by any property that varies continuously depending on location. As an example, a heat source generates a gradient of decreasing temperature with increasing distance from the source. The gradients thought to guide pattern formation in embryos are usually associated with varying concentrations of diffusible signal molecules. Such molecules are called *morphogens,* because they are presumed to control pattern formation and morphogenesis.

A Morphogen Gradient Can Specify a Range of Positional Values and Polarity to a Field

Most gradient models propose that a distinct group of cells acts as the *source* of a morphogen (Fig. 20.18). The morphogen diffuses away from the source and is eventually destroyed. The destruction site may be another special group of cells acting as a *sink.* In this case, the morphogen concentration will decrease *linearly* between the source and the sink. Alternatively, *all* cells outside the source may destroy the morphogen, in which case its concentration will decrease exponentially with distance from the source. Models based on morphogen gradients imply that cells respond to long-range signals from distant reference points, rather than short-range signals from their nearest neighbors.

In addition to a range of positional values, a morphogen gradient also specifies the *polarity,* or the direction in which these values follow one another. The polarity of a one-dimensional gradient may be symbolized by an arrow indicating the direction in which the morphogen concentration decreases. This is the direction in which the positional values will follow one another in ascending order.

Positional Values Specified by Morphogen Gradients May Elicit Differential Gene Activities and Cell Behaviors

Cells can interpret positional values if some of their genes are activated or inhibited by the morphogen at

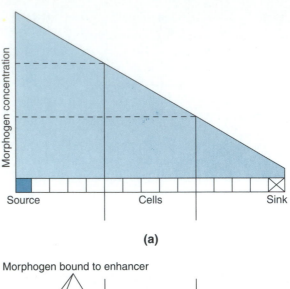

(a)

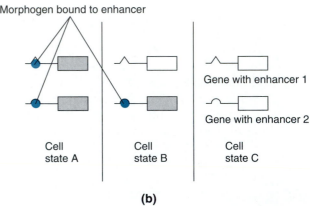

(b)

Figure 20.18 Model for specification of positional values by a morphogen gradient and interpretation of positional values by genes. **(a)** In a string of cells, the cell at one end is a morphogen source. The morphogen diffuses through the string until it is destroyed in the cell at the opposite end, which acts as a morphogen sink. At equilibrium, the morphogen concentration decreases linearly between the source and the sink. The local concentration then specifies a positional value to each cell in the string, with the source and sink serving as reference points. **(b)** Positional values can be interpreted by cells if different local morphogen concentrations cause different gene activities. It is assumed here that the morphogen is a transcription factor binding to two enhancers, 1 and 2, of certain target genes. Enhancer 1 has little affinity for the morphogen and is bound only at high concentration. Enhancer 2 has a higher affinity and is still bound at intermediate morphogen concentration. Neither enhancer is bound at low concentration. The threshold concentrations between these states define boundary lines at which the activity patterns of the target genes change. The resulting gene activity patterns define three cell states: state A has genes with enhancer 1 and with enhancer 2 active (gray shading), state B has only genes with enhancer 2 active, and state C has neither of these genes active.

concentrations within the range of the gradient. Each positional value then elicits a unique combination of active and inactive genes. If these genes were to control certain pathways of cell behavior and differentiation, each level of morphogen concentration would evoke a corresponding type of cell activity.

The interaction between a morphogen and its target genes is straightforward if the morphogen is a transcription factor. Such a factor will activate or inhibit different sets of target genes depending on its local concentration and its affinity for each target gene enhancer (Fig. 20.18). As an example, the bicoid protein forms an anteroposterior gradient in *Drosophila* eggs and activates the embryonic *hunchback* gene as well as other genes in a concentration-dependent way (see Chapters 15 and 21). However, transcription factors can act as morphogens only in embryos characterized by delayed cellularization or having cytoplasmic bridges that allow large molecules to pass between cells. In other systems, morphogens must be small molecules that can pass either through gap junctions or through cell membranes and the extracellular matrix. Such molecules may act as ligands that activate or inhibit transcription factors, as discussed previously in connection with steroid hormones (see Chapter 15). Alternatively, morphogen molecules may bind to cell membrane receptors that activate second messengers (see Chapter 2).

Although the gradient model is related to the older concept of induction, the gradient model is more specialized because it is limited to diffusible signal molecules whereas the induction concept also allows for nondiffusible signals between adjacent cells or between cells and extracellular matrix. On the other hand, the gradient model is broader than the induction concept because morphogen-sensitive cells can adopt one of at least three different states of determination: one default state and at least two more states triggered when the morphogen concentration exceeds certain thresholds (Fig. 20.18). This is at least one state more than is postulated in a simple induction model, where the responding tissue has only two responses: one default and one signal-dependent program. Both models seem to have many applications.

Morphogen Gradients Confer Size Invariance on Embryonic Fields

Morphogen gradients have held the interest of developmental biologists throughout the twentieth century for three reasons. First, they explain at least conceptually how *several qualitatively different cell states* (e.g., blue, white, and red cells, in the French flag model) are specified by *quantitative variations of a single parameter*, the morphogen concentration. Second, they offer a simple explanation for the polarized sequence with which the different elements of biological patterns arise. (Again in terms of the French flag, the blue strip is always next to the flagpole, followed by the white and then by the red, because the hypothetical morphogen source is located next to the pole and the colors form in order of decreasing morphogen concentration.) Third, morphogen gradients offer a plausible explanation for the size invariance that characterizes pattern formation in embry-

onic fields. This is especially important because size invariance is the hallmark of embryonic fields.

To understand how size invariance is conferred on patterns specified by a gradient, let us consider again a string of cells with a source at one end and a sink at the other. The gradient may specify a pattern of three cell states—A, B, and C—as explained in Figure 20.18. Removing or adding cells somewhere between the source and the sink will simply cause the formation of a gradient with a steeper or gentler slope, and thus a pattern that is compressed or spread out accordingly (Fig. 20.19a and b).

What happens if the cellular string is cut in half, leaving one half without a sink and the other half without a source (Fig. 20.19c)? If the missing reference points are not restored, the half without a sink will generate only cell state A (because maximal levels of morphogen will accumulate everywhere) and the half without a source will generate only state C (because the sink will drain the entire string of the morphogen). To restore the missing reference points, the cells need additional mech-

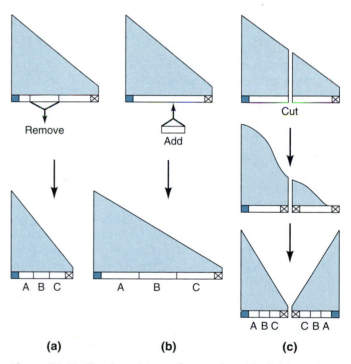

Figure 20.19 Size invariance of an embryonic field specified by a morphogen gradient. In a string of cells with a morphogen source at one end and a sink at the other end, a pattern of three cell states (A, B, and C) is specified, as explained in Figure 20.18. Note that the morphogen concentration at the sink is greater than zero. **(a, b)** Removal or addition of cells between source and sink makes the gradient steeper or gentler, with the resulting pattern of cell states compressed or spread out proportionally. **(c)** Cutting the cellular string in half is thought to cause the formation of new sinks at the cut end. In addition, the old sink is assumed to turn into a new source as the local morphogen concentration approaches zero. Under these conditions, both halves form two complete patterns in mirror-image orientation.

anisms. For instance, each cell might be programmed to become a sink if it finds itself at the end of the string, except if the morphogen concentration is near zero, in which case the end cell will become a source. Under these rules, each half will form a new sink at the cut end. In addition, the old sink will turn into a new source when the morphogen concentration has fallen close to zero. Other rules would be necessary to explain how a normally proportioned pattern is formed when two or more strings of cells are fused together.

In many cases, morphogen gradients offer elegant explanations not only for the specification of positional values, but also for the field properties that characterize many territories in which patterns are formed. Morphogen gradients have therefore been invoked to explain many cases of pattern formation in development. In the rest of this chapter, we will briefly consider two examples. For each one, we will ask what data are particularly well explained by a morphogen gradient, which problems are left unresolved by such a hypothesis or require additional assumptions, and what alternative models have been proposed to explain the same data. Additional examples will be discussed in Chapters 21 and 22.

Sea Urchin Development Has Been Modeled by Morphogen Gradients and by Sequential Induction

Sea urchin embryos have long been favorite objects of developmental studies, and data obtained from experiments with these embryos have been interpreted in terms of two different models. One is a gradient model; it postulates that the animal and vegetal poles of the embryo are sources of antagonistic gradients, and that blastomeres develop according to the local ratio of animal and vegetal morphogens. The other model relies on sequential induction. It assumes that the vegetalmost blastomeres are determined as mesodermal cells by cytoplasmic localization, that the prospective mesodermal blastomeres induce their nearest neighbors to develop as endodermal cells, and that the prospective endodermal blastomeres in turn induce the remaining blastomeres to become ectoderm. We will first examine the relevant data and then discuss how well they are explained by each model.

Many of the experiments in this area were carried out by the Swedish embryologist Sven Hörstadius (1973), who also established the *fate map* for the 60-cell stage (see Fig. 10.4). The cells of the animal hemisphere (an_1 and an_2) and the first vegetal layer (veg_1) give rise to the ectoderm. The cells of the second vegetal layer (veg_2) form the archenteron, or primitive gut. The most vegetal cells, called micromeres because of their small size, form the mesodermal organs of the pluteus larva.

In order to test the developmental *potencies* of sea urchin blastomeres, Hörstadius studied their develop-

ment in isolation and under various abnormal combinations. In one of his experiments, he isolated different layers of blastomeres at the 60-cell stage and combined them with increasing numbers of micromeres (Fig. 20.20). The addition of micromeres improved the development not only of the spicules, which are normally derived from micromeres, but also of the archenteron, which is normally formed from the veg_2 layer. Thus, the micromeres generally enhanced the formation of the structures derived from the vegetal half of the embryo. The most complete larvae were obtained when the number of micromeres added was gauged to the position of the layer they were combined with. Specifically, the most animal layer (an_1) formed a complete pluteus when combined with four micromeres. The an_2 layer needed only two micromeres for the same result, and the veg_1 layer needed just one.

Hörstadius interpreted his results in terms of a gradient model proposed by his mentor, John Runnström. According to this model, two cytoplasmic determinants are distributed as antagonistic gradients: one concentrated near the animal pole and decreasing toward the vegetal pole, and another concentrated near the vegetal pole and decreasing toward the animal pole. During cleavage, each blastomere incorporates both determinants in a ratio depending on its location. Development of a normal pluteus larva requires a balance of blastomeres with animal and vegetal determinants. The model explains particularly well the finding that blastomere layers of increasingly animal *position* are balanced by increasing *numbers* of micromeres.

If the concept of positional information is combined with the Runnström-Hörstadius model, each blastomere may be assigned a positional value according to the local ratio of animal to vegetal determinants. It is interesting to apply this double-gradient model to results from other experiments (Fig. 20.21). First, well-proportioned larvae can be obtained by combining the animal cell layer with the micromeres at the 16-cell stage. In terms of the double-gradient model, this amounts to combining the cells containing the most *extreme* ratios of animal and vegetal determinants. Conceivably, diffusion of the determinants or intercalary regeneration will restore the missing intermediate ratios and, hence, positional values. This interpretation implies that the animal and vegetal determinants, which are originally *localized* in the egg, later *diffuse* from cell to cell or produce diffusible products; only if this is so does the double-gradient model account for the formation of a complete larva.

A normal larva will also form from *intermediate* blastomeres after removal of the most extreme layers at the 32-cell stage. In this case, the missing positional values are the extreme ones; therefore, they cannot be restored by either diffusion or intercalary regeneration alone. To account for the development of a complete larva, one has to make an additional assumption. For instance,

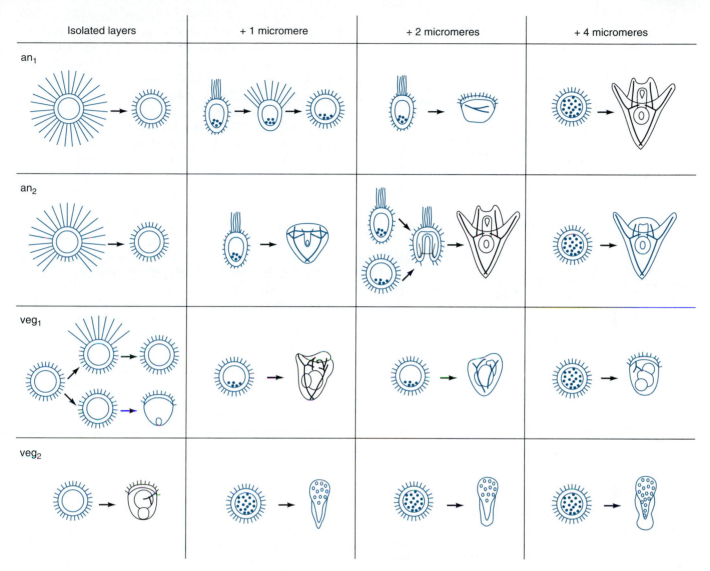

Figure 20.20 Complementation of animal and vegetal blastomeres in sea urchin embryos. The left-hand column shows the development of isolated tiers from 60-cell embryos. The other columns show the development of the same tiers after addition of one, two, or four micromeres. An an_1 tier by itself forms a ciliated blastula, which is considered an "animalized" larva. It takes the addition of four micromeres to form a complete pluteus. In contrast, only two micromeres must be added to an an_2 layer to produce a complete larva. A veg_1 layer by itself develops into either a ciliated blastula or a larva with a small archenteron. However, adding only one micromere causes development of a "vegetalized" larva with an overproportionate archenteron. A veg_2 layer by itself forms a disorganized larva with cilia, some spicules, and a small archenteron. Adding only one micromere produces an exogastrula with a large archenteron, considered a severely vegetalized larva.

the intermediate layers may sense their new position at the extreme ends of the field and become the new *reference points* by restoring the extreme ratios of animal to vegetal determinants or their products. After this step, the intermediate positional values might be regenerated, as before. The nature of the proposed animal and vegetal gradients or any diffusible products has not been defined.

An alternative model based on localization and a *sequence of inductive interactions* was proposed by Eric

Davidson (1989). According to this model, a small number of transcription factors are activated in a pattern corresponding to different cell lineages (Fig. 20.22). An active factor in the micromeres enhances the expression of certain genes in the primary mesenchyme lineage. The same factor is also present in all other blastomeres but is not normally active there. The gene activity stimulated by the micromere factor includes the synthesis of ligands that are incorporated into the plasma membrane of the micromeres. These ligands act on the neigh-

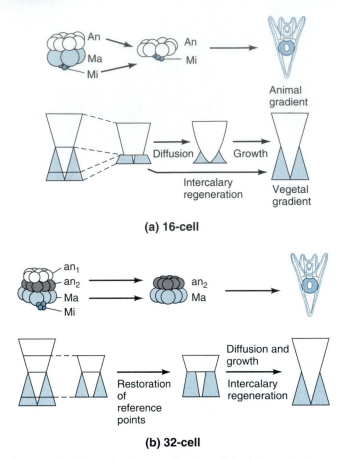

(a) 16-cell

(b) 32-cell

Figure 20.21 The double-gradient model of Runnström and Hörstadius postulates that the cytoplasm of sea urchin eggs contains antagonistic gradients of animal and vegetal determinants. **(a)** Extremely animal cells (An) combined with extremely vegetal cells (Mi) from 16-cell embryos form a complete larva. The double-gradient model explains the restoration of missing positional values either by diffusion and growth or by intercalary regeneration. **(b)** Intermediate layers of cells (an$_2$ and Ma, or macromeres) from 32-cell embryos form a complete larva. Additional assumptions about the establishment of new reference points are necessary to explain the restoration of the extreme positional values.

plain some of the quantitative data summarized in Figure 20.20. The result that an$_1$ cells need four micromeres to form a complete larva while an$_2$ cells need only two is not accounted for. Conceptually, the Davidson model differs from the Runnström-Hörstadius model by relying on one reference point, the micromeres, instead of two antagonistic poles. It is also based on a sequence of close-range inductions rather than on long-range control by diffusible morphogens. Last but not least, the Davidson model spells out the molecular function of its key elements as ligands and transcription factors. Since several lineage-specifically expressed genes of the sea urchin have been cloned, transcription factors acting on them can now be isolated and characterized. Whether these factors are distributed and activated in patterns according to the model can then be tested directly. By virtue of making testable predictions, this model should serve a useful purpose even if its molecular details need to be revised in the process.

Additional work has pointed up the role of inductive *inhibition* between animal blastomeres. Jonathan Henry and his colleagues (1989) compared the potential of entire layers of animal cells with that of isolated pairs of animal cells. Confirming the results of Hörstadius, they found that *intact layers* produced swollen ectodermal vesicles. In contrast, *isolated pairs* of animal cells formed mostly ectodermal but also endodermal and mesenchymal cells. Thus, pairs of animal cells showed a greater potential than intact layers of the same cells. The surplus potential was greater when the animal cells were isolated early in development. It must be concluded that the contact between animal cells helps to restrict their potential to ectodermal structures. In similar experiments, David Hurley and his colleagues (1989) found that blastomere isolation changed the accumulation of mRNAs transcribed from lineage-specifically expressed genes. Both sets of data indicate that inductive interactions occur within the same layer of blastomeres and between different layers. Evidently, these interactions can enhance or inhibit the lineage-specific expression of certain genes.

Insect Development Has Been Modeled by Morphogen Gradients and by Inductive Interactions

Insect embryos have also been favorite objects of embryologists interested in pattern formation (Sander, 1975, 1976). The data from insect embryos have been interpreted in terms of various models, including one postulating a pair of antagonistic morphogenetic gradients, as in the case of the sea urchin embryo. More recently, additional data have indicated that the actual patterning process in the embryo must be more complex than predicted by a pure gradient model.

Experiments demonstrating the action of a posterior cytoplasmic determinant were carried out by Klaus

boring veg$_2$ cells to inhibit the micromere factor and to activate another transcription factor that enhances the expression of certain genes in the endoderm lineage. The resulting gene activity in the endoderm lineage includes the synthesis of other ligands that act on their animal neighbors to inhibit the activity of the endoderm factor and to enhance the expression of genes characteristic of the ectodermal lineage.

The Davidson model accommodates particularly well the special role of the micromeres, which become determined early during cleavage and can induce any neighboring cells to invaginate and form an archenteron (Ransick and Davidson, 1993). The model also explains the ability of the veg$_2$ cells to functionally replace the micromeres by assuming that the micromere factor is activated in veg$_2$ cells when it is no longer repressed by the neighboring micromeres. The model does not ex-

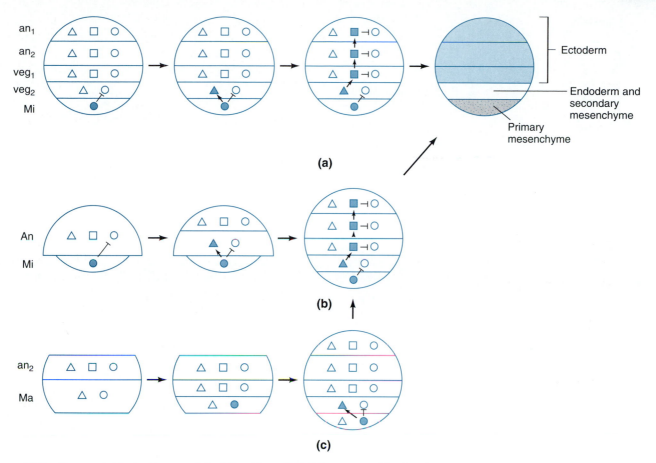

Figure 20.22 Sequential induction model of Davidson (1989) for sea urchin development. The blastomere fates are specified by localized transcription factors that are activated by a cascade of inductive events. **(a)** Normal development of the 60-cell stage. The micromeres (Mi) contain an active transcription factor (filled circle) for genes expressed in the primary mesenchyme and other micromere-derived cells. The same micromere factor in an inactive form (open circles) is also present in the other blastomeres but is inhibited (bar). The veg_2 layer contains an active transcription factor (filled triangle) for genes expressed in endodermal cells. The active state of this endoderm factor depends on induction by the micromeres, which expose an inducing ligand (arrow) on their surface. The endoderm factor is absent from the micromeres and is inactive (open triangles) in the more animal blastomeres. Similarly, the veg_1 layer and the animal layers contain an active transcription factor (filled squares) for genes expressed in the ectoderm. Its activity depends on induction by neighboring veg_2 cells or adjacent animal cells, which send out inducing ligands (arrows) on their surfaces. The localizations and inductive interactions that govern normal development also allow the restoration of the normal gene activity pattern in the abnormal blastomere combinations of parts b and c. **(b)** Extreme cell blastomere layers (An and Mi) combined at the 16-cell stage (see also Fig. 20.21a). Cell divisions and the hypothetical sequence of events outlined in part a restore the normal distribution of transcription factors. **(c)** Intermediate cell layers (an_2 and Ma) isolated at the 32-cell stage (see also Fig. 20.21b).

Sander (1960) with eggs of the leafhopper *Euscelis plebejus* (Fig. 20.23). The eggs, which the insect deposits in plant leaves, are limp and can be invaginated by poking with a blunt needle. Sander used this technique to move posterior cytoplasm anteriorly, using a cluster of symbiotic bacteria at the posterior pole as a convenient marker to monitor the position of the displaced material. Sander also separated anterior and posterior egg fragments by ligation with hair loops or by pinching the eggs between razor blades with blunted edges. The embryos that developed from such manipulated eggs

showed clearly that the posterior cytoplasm plays a key role in specifying the anteroposterior body pattern of *Euscelis*.

Large anterior egg fragments, generated by ligation during cleavage stages, by themselves produced only head structures (Fig. 20.23c). However, such fragments became capable of forming more segments or even complete embryos if posterior cytoplasm was shifted before ligation so that it was included in the anterior fragment (Fig. 20.23e and f). Was this dramatic increase in potency caused by the posterior cytoplasm itself, or by a

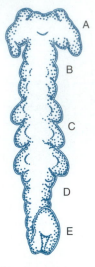

(a) Germ band stage

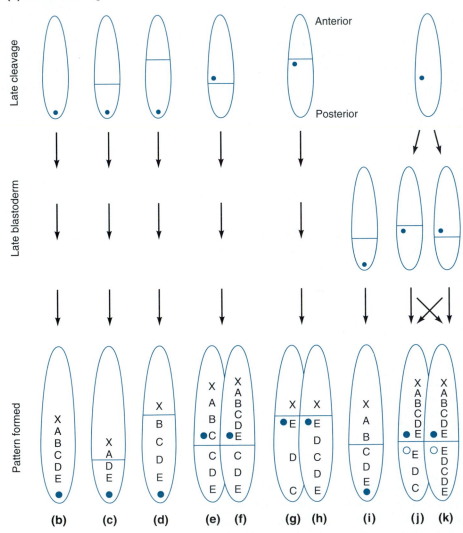

Figure 20.23 Ligation or combined ligation and translocation of posterior cytoplasm in leafhopper (*Euscelis plebejus*) eggs: anterior pole up, stages of operation indicated at left. **(a)** Embryo at the germ band stage, with capital letters assigned to body regions. A = procephalon; B = mouthparts; C = thoracic segments; D = abdominal segments; E = telson; X = extraembryonic membranes. The horizontal line within the egg outline represents ligation at the respective level. The black dot represents the ball of symbiotic bacteria used to monitor the position of posterior cytoplasm. The open circles in (j) and (k) represent alternative positions of the symbiotic bacteria observed in some experiments. **(b–k)** Different sets of experiments. The pairs **(e, f), (g, h)**, and **(j, k)** represent different results obtained in the same series. Capital letters represent the embryonic body patterns identified at the germ band stage shown in part a.

factor emanating from it? To answer this question, Sander shifted the posterior pole material anteriorly and left it there for several hours before placing a ligature in front of or behind the material (Fig. 20.23j and k). The potential of the anterior fragment increased all the same, indicating that a signal produced by the posterior cytoplasm played the key role. In posterior egg fragments, the combined transplantations and ligations sometimes caused the formation of abdomens with reversed polarity (Fig. 20.23g and j). In other cases, the same experiments produced double abdomens, consisting of two sets of thoracic and abdominal segments joined in mirror-image symmetry (Fig. 20.23h and k). Thus, the signal emanating from the posterior cytoplasm was spreading both anteriorly and posteriorly.

The signal spreading from the posterior cytoplasm of *Euscelis* eggs has two important characteristics. First, it does not always cause anterior fragments to form complete embryos. In cases when it does not, the additional segments formed are *not the terminal* abdominal segments normally formed next to the posterior pole. Instead, middle segments are added *continuously* to the head segments that develop in anterior fragments without posterior cytoplasm. Second, the length of a given set of segments formed may vary by a factor of more than 2. (Compare the arrays labeled CDE in Figure 20.23g, j, and k.) One can explain these characteristics by assuming that the signal emanating from the posterior cytoplasm has the form of a gradient with the highest concentration near the ball of symbiotic bacteria.

In addition to a posterior morphogen, Sander (1975) proposed that a second morphogen is produced in a source near the anterior pole. He suggested that the segments of the *Euscelis* embryo were specified by the ratio of anterior to posterior morphogen concentration—an idea he developed independently of the double-gradient model for sea urchins. He inferred the existence of the anterior morphogen partly from the effects of early ligation: posterior egg fragments form partial

embryos missing anterior parts. Specifically, these partial embryos have *fewer* segments than expected according to the fate map. Instead, the reduced sets of segments are stretched out farther than usual (Fig. 20.23b and d). Sander concluded that the posterior fragment failed to develop according to the fate map because it was cut off from the anterior morphogen source.

Sander's gradient model was challenged by Otto Vogel (1978, 1982), who pushed in both ends of the *Euscelis* egg with two blunt needles until the presumed anterior and posterior morphogen sources were next to each other (Fig. 20.24). Under these conditions, the ratio of anterior to posterior morphogen should have been nearly uniform throughout the egg. Nevertheless, about half of the badly deformed eggs formed complete embryos, which were often twisted in a spiral around the two needle tips. Vogel concluded that the body pattern of the embryo must be specified by signals other than the ratio of freely diffusible morphogens.

Vogel (1982) also ligated *Euscelis* eggs. Confirming Sander's results, he found that anterior fragments formed partial embryos. They began anteriorly with the procephalon containing brain and eyes, and terminated posteriorly in mouthparts or thoracic segments. The character of the posteriormost segment depended on the length of the anterior fragment. Short fragments contained only procephalon, while longer fragments

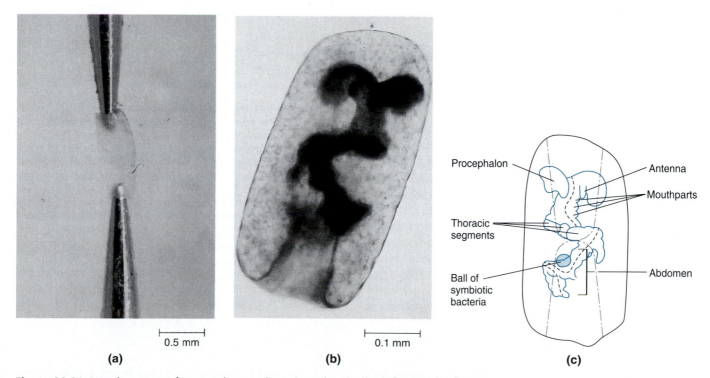

0.5 mm	0.1 mm
(a)	(b)

(c)

Figure 20.24 Development of a complete embryo in a drastically deformed leafhopper egg. **(a)** Photograph of egg being invaginated with blunt needle tips at both the anterior and the posterior pole. **(b)** Photograph of a fixed and stained embryo that developed in an egg that had been invaginated at both ends until the needle tips met. **(c)** Interpretative drawing of part b. The embryo (color) has formed in a spiral around the juxtaposed needle tips (broken lines).

terminated in more posterior segments. When he tabulated his results, Vogel found that the number of segments *did not increase steadily* with the length of the anterior fragment. The embryos tended to have either no mouthparts at all or all three of them (mandible, maxilla, and labium). This discontinuity in segment number was at odds with the smooth decrease of the anterior to posterior morphogen ratio implied in Sander's model. Vogel suggested that the body pattern might be specified in several steps, with broad anterior, posterior, and middle areas determined initially by localized cytoplasmic factors. Local interactions between these areas, instead of long-range signals from the poles, would then specify smaller groups of segments and eventually individual segment characters.

The results obtained with sea urchin and insect embryos indicate that the process of embryonic pattern formation is too complex to be described in terms of a single model. Experience has been the same with many other models of pattern formation (Held, 1992). It appears that patterning mechanisms have evolved into intricate networks that encompass morphogen gradients, sequential induction, and possibly other types of signaling. To unravel processes of this complexity, genetic and molecular methods have proven to be very powerful, particularly in studies of *Drosophila*. This work has considerably advanced our understanding of embryonic pattern formation and has brought us to a new era in which modern tools are applied to classical topics. Some of these exciting advances will be discussed in the following chapters.

SUMMARY

Pattern formation is the process by which harmonious arrays of different elements develop in living things. Many patterns in adult organisms are formed by the coordinated differentiation of various cell types. However, the formation of embryonic patterns often begins with distinct areas of mitotic division, cell movement, and cell death. Pattern formation often occurs in fields, groups of cells that cooperate in the formation of certain organs. The main characteristic of fields is their size invariance, that is, their ability to form the same pattern regardless of experimental size reduction or enlargement.

Pattern formation relies on interactions between cells, as seen most clearly during induction. The response of cells to patterning signals also depends on their genetic information and developmental history. Some patterning signals are used in similar ways in different body parts and during several stages of development. Transplantations between species show that many of these signals have been conserved during evolution.

The fate of a cell depends upon its position within a territory of cells. This observation has led to the concept of positional values, which are analogous to the coordinates of a geometric point. According to this concept, pattern formation is a two-step process. First, the cells within a territory receive positional values with regard to certain foci or boundaries that serve as reference points. Second, each cell interprets its positional value by engaging in activities that depend on its genetic makeup and developmental history.

The concept of positional value is implied in the polar coordinate model of limb development and regeneration. This model, based on the two rules of shortest intercalation and distalization, predicts the results of various transplantation experiments in great detail.

Although positional values may be specified by various means, they are most frequently associated with gradients of diffusible signal molecules called morphogens. Morphogens are thought to emanate from sources that serve as reference points for cellular territories. As a morphogen diffuses away from its source and is degraded, its concentration decreases. Different local concentration levels would then confer positional values on the cells within the territory. Interpretation of these values would be controlled by genes with enhancers that have different affinities for the morphogen. Gradient models propose that more than two cell states can be controlled by the quantitative variation of one signal.

Gradient models offer a simple rationale for the sequence and polarity with which the different elements of a pattern are formed. With some additional assumptions, they can also explain the size invariance of embryonic fields. Because of these properties, morphogen gradients have been invoked to explain pattern formation in sea urchin embryos, insect embryos, and many other developmental systems. However, more recent work has also led to consideration of alternative models that rely on sequential induction.

SUGGESTED READINGS

Bryant, S. V., V. French, and P. J. Bryant. 1981. Distal regeneration and symmetry. *Science* **212**:993–1002.

Hörstadius, S. 1973. *Experimental Embryology of Echinoderms.* Oxford: Clarendon Press.

Sander, K. 1975. Pattern specification in the insect embryo. In

Ciba Foundation Symposium 29 (new series): Cell Patterning, 241–263. Amsterdam: Elsevier Associated Scientific Publishers.

Wolpert, L. 1978. Pattern formation in biological development. *Scientific American* **239**, October, 154–164.

GENETIC AND MOLECULAR ANALYSIS OF PATTERN FORMATION IN THE *Drosophila* EMBRYO

Figure 21.1 Drosophila embryos stained to reveal the expression domains of an embryonic patterning gene, *paired⁺*, during different phases of development. An initial pattern of broad domains evolves into seven stripes as the embryo forms a superficial layer of cells. Later, a segmental pattern of 14 stripes appears. At mid-embryogenesis, *paired⁺* is expressed in specific regions of the head and central nervous system. The first critical function of the gene seems to occur during the period when seven stripes are visible: these stripes have a bisegmental periodicity, and every other segment is deleted in *paired*-deficient embryos.

The analysis of pattern formation has been revolutionized by the introduction of genetic and molecular techniques. These powerful tools have been brought to bear in studies on several species—the fruit fly *Drosophila melanogaster*, the house mouse, the roundworm *Caenorhabditis elegans,* and two flowering plants—that will be covered in this chapter and in those that follow. The zebra fish has also recently become the object of large mutagenesis screens; among vertebrates it may soon rival the mouse in usefulness for studies that involve genetics.

At present, genetic analysis of pattern formation has advanced furthest in the fruit fly *Drosophila melanogaster.* Two circumstances particularly favor study of this species: its genes have been extensively characterized (see Chapter 14), and its embryos are large enough to permit transplantation experiments. Indeed, *Drosophila* and other insect embryos have long been subjects of experimental analysis (Counce, 1973; Sander, 1976). Traditional and modern research methods in combination have made the *Drosophila* embryo the best-understood so far in the area of pattern formation (Lawrence, 1992).

Classical ligation, translocation, and irradiation experiments with insect embryos have generated strikingly abnormal body patterns such as mirror-image duplication of heads coupled with the absence of abdomens, and vice versa. These results have formed the basis for a general model of pattern formation involving cytoplasmic determinants localized near the anterior and posterior poles of the egg (see Chapter 20). However, traditional experiments affected so many egg components that it was difficult to ascertain which were critical to the results observed. The insight gained with these methods was therefore limited in detail and somewhat controversial.

In 1980, Christiane Nüsslein-Volhard and Eric Wieschaus reported on a new approach to the analysis of insect development. Using large mutagenesis screens, they identified and characterized specific *Drosophila* mutants and the abnormal forms of embryonic development they undergo. Their analysis was boosted tremendously by gene cloning methods, which came into wide use soon after. These techniques allowed researchers to view the expression patterns of embryonic genes directly (Fig. 21.1), to study the molecular nature of the proteins encoded by these genes, and to unravel the network of controls among these genes. The new capabilities generated tremendous enthusiasm, attracting dozens and then hundreds of laboratories to *Drosophila* research. About a decade later, most of the gene activities that direct the unfolding of the *Drosophila* egg into a segmented embryo were understood at a satisfying level. More recently, the gastrulation, organogenesis, and histogenesis of *Drosophila* have become the new frontiers of molecular embryology.

Review of *Drosophila* Oogenesis and Embryogenesis

In order to discuss the genetic control of development in *Drosophila,* we need to review some basic facts about oogenesis and embryogenesis. All the materials and information necessary to form a larva must be assembled in the egg. Three cell types within the *Drosophila* ovary are involved in this task: *oocytes, nurse cells,* and *follicle cells* (Fig. 21.2a–c). The oocyte is connected by cytoplasmic bridges to 15 sister cells, the nurse cells, which synthesize large amounts of RNA and cytoplasm for transfer into the oocyte. Both oocytes and nurse cells are germ line cells. The follicle cells originate independently of the germ line from embryonic mesoderm and sur-round the oocyte–nurse cell complex. In addition to playing a role in *vitellogenesis,* follicle cells synthesize the *vitelline envelope* and the *chorion,* which form the eggshell and help to establish the anteroposterior and dorsoventral polarity axes of the egg. For instance, the follicle cells on the dorsal anterior of the oocyte are especially tall and form the respiratory appendages that stick out like rabbit ears dorsally and anteriorly from the chorion.

Like other insect eggs, the *Drosophila* egg undergoes *superficial cleavage* (Fig. 21.2d and e). After nine rounds of mitosis, the first cells are formed at the posterior pole. These cells, known as *pole cells,* are the primordial germ cells. After 13 mitotic divisions, about 6000 nuclei are lined up beneath the egg surface but are not yet separated by cell membranes. This stage is called the *preblastoderm,* or syncytial blastoderm. Infoldings of plasma membrane then separate the nuclei from each other, thus generating a monolayer of cells called the *cellular blastoderm.* Up to the preblastoderm stage, the insect embryo is one large, multinucleate cell in which nuclei can communicate with one another directly using transcription factors and other large proteins as signal molecules. This feature may have allowed the evolution of genetic control mechanisms that differ from those of other embryos, in which blastomeres communicate through gap junctions and extracellular signals.

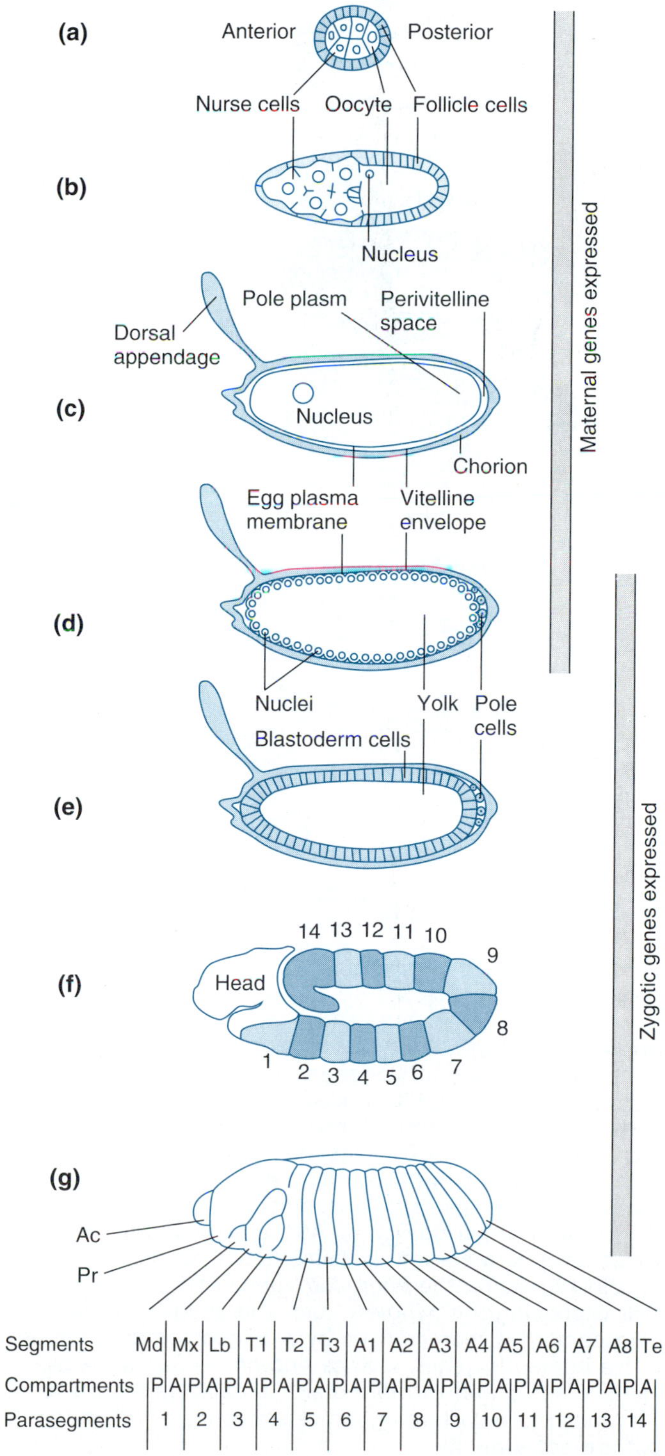

Figure 21.2 Schematic illustrations of oogenesis and embryogenesis in *Drosophila.* **(a)** Egg chamber at an early stage of oogenesis. The germ line–derived oocyte–nurse cell complex is surrounded by mesodermal follicle cells. **(b)** Egg chamber at an intermediate stage of oogenesis. The nurse cells are located anterior to the oocyte. The oocyte nucleus is displaced anteriorly and dorsally. **(c)** Newly laid egg. The egg cell is surrounded by two covers, the vitelline envelope and the chorion, both produced by the follicle cells. Near the anterior and posterior tips, the egg plasma membrane and the vitelline envelope are separated by a fluid-filled space, the perivitelline space. **(d)** Preblastoderm stage. Pole cells have budded off at the posterior pole. Most nuclei have moved to the egg periphery but are not yet separated by plasma membranes. Transcription of the embryonic genome has begun. **(e)** Blastoderm stage. Membrane infoldings between the nuclei have created a monolayer of about 5000 uniform cells. **(f)** Extended germ band stage. The rudiment of the embryo proper, called the germ band, has extended posteriorly and doubled up until the posterior tip touches the head rudiment dorsally. The germ band is subdivided into parasegments numbered from 1 to 14. **(g)** Advanced embryo after germ band shortening and dorsal closure. The embryo is now subdivided into segments corresponding to the segments of the larva and adult. Each segment consists of an anterior compartment (A), derived from one parasegment, and a posterior compartment (P), derived from the following parasegment. Egg covers have been deleted in parts f and g for clarity. Ac = acron; Pr = procephalon; Md = mandible; Mx = maxilla; Lb = labium; T1 = prothorax; T2 = mesothorax; T3 = metathorax; A1 to A8 = abdominal segments; Te = telson.

Blastoderm formation is followed by dramatic morphogenetic movements, including gastrulation, which generate an embryonic rudiment known as the *germ band.* It consists of the three germ layers and essentially represents the ventral part of the future larva. The germ band elongates until, doubling back on itself, it has become twice as long as the egg—a stage known as the *extended germ band* (Fig. 21.2f). Constrictions divide the germ band into units called *parasegments;* they have the length of segments but are out of register with the definitive segments of the advanced embryo and larva (Fig. 21.2g). The epidermis of each parasegment consists of two *compartments,* which are separated by clonal restriction lines (see Chapter 6). The germ band shortens again so that its posterior tip returns to the posterior egg pole. During germ band shortening, the parasegmental boundaries disappear, and segmental boundaries develop. The new boundaries are drawn so that the posterior compartment of one parasegment and the anterior compartment of the following parasegment together form one segment. Each definitive *segment* contains a pair of ganglia and paired portions of mesoderm. In a process called *dorsal closure,* the flanks of the embryo grow laterally and dorsally around the remaining yolk, generating a cylindrical embryo that resembles the future larva.

The body of *Drosophila* is subdivided into head, thorax, and abdomen. The head comprises a tip called the *acron,* two or more fused segments that together form the *procephalon* and contain most of the brain, and three segments that carry mouthparts: *mandible, maxilla,* and *labium.* The three thoracic segments are known as the *prothorax, mesothorax,* and *metathorax.* The abdomen, behind the thorax, has eight visible segments and a tip without overt segmentation called the *telson.*

All substances deposited in the egg have been synthesized under the control of the maternal genome. Maternal gene products support the *Drosophila* embryo until the preblastoderm stage, when transcription of the embryo's own genes begins. Mutations in either maternal or embryonic genes may cause abnormal pattern formation in the developing embryo.

Cascade of Developmental Gene Regulation

Epigenesis was originally put forth as a morphological principle: the generation of more complex forms from simpler precursors (see Chapter 1). In accord with this principle, the *Drosophila* embryo develops from a relatively uniform egg to a blastoderm stage in which there are two morphologically different cell types (pole cells and somatic cells), and then to an embryo with dozens of cell types. However, epigenesis also occurs at the level of gene activity. Here we see a cascade that begins

with a few localized gene products and develops into an ever more complex network of transcription factors and gene activities. To facilitate keeping track of the multitude of genes involved, Table 21.1 (pages 536–540) lists their names and main characteristics.

The epigenesis of gene regulation in *Drosophila* begins with a few localized products of *maternal effect genes* (Fig. 21.3). These products direct the segregation

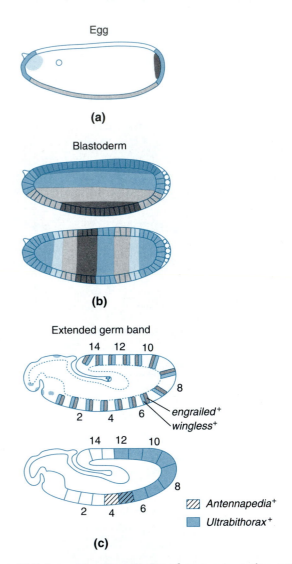

Figure 21.3 Increasing complexity of gene expression patterns during early embryogenesis in *Drosophila.* **(a)** The egg contains four localized products of maternal effect genes, which are deposited anteriorly (light color), posteriorly (dark gray), at both poles (dark color), or ventrally (light gray). **(b)** At the blastoderm stage, zygotic gene expression defines several longitudinal and transverse stripes without periodicity. **(c)** At the extended germ band stage, each parasegment is delineated anteriorly by a strip of cells expressing the *engrailed⁺* gene and posteriorly by cells expressing the *wingless⁺* gene. In addition, each parasegment is characterized by a unique combination of active homeotic genes, such as *Antennapedia⁺* in parasegments 4 and 5 and *Ultrabithorax⁺* in parasegments 5 to 12. Many more patterning genes are active at this stage but are not shown.

of pole cells and the activation of certain *zygotic genes* in specific domains of blastoderm cells. Some zygotic genes are expressed in longitudinal domains that are parallel to the anteroposterior axis of the embryo. These longitudinal domains include a ventral strip of invaginating endoderm and mesoderm, adjacent ventral neuroepidermis, and dorsal epidermis. Other zygotic genes are expressed in transverse bands of cells that are parallel to the dorsoventral axis (or circumference) of the embryo. These bands run parallel to the future segment boundaries on the blastoderm fate map but are much broader than one segment. By further genetic interactions, these broad transverse expression domains generate narrower expression domains, which eventually assume parasegmental periodicity.

The stepwise refinement of zygotic gene expression domains has been particularly well investigated for the anteroposterior pattern (Fig. 21.4). The first zygotic genes expressed in this class are called *gap genes*, because their loss-of-function phenotypes show gaps in the body pattern that are several segments wide. Combinations of gap gene activities control the expression of a second tier of zygotic genes, which are known as *pair-rule genes*, since their mutant alleles cause defects or abnormal development in every other segment. Combinations of pair-rule gene activities control the expression of a third tier of zygotic genes, called *segment polarity genes*, because their mutant alleles disturb the anteroposterior polarity of each segment. Collectively, gap genes, pair-rule genes, and segment polarity genes are called *segmentation genes*, because this entire class serves to subdivide the germ band into parasegments and later into segments. (Note that segment polarity genes and segmentation genes are *not synonymous*.)

Gap genes and pair-rule genes also control the expression of the *homeotic genes* already mentioned in previous chapters. Homeotic mutant phenotypes are characterized by striking transformations of certain body parts, such as the replacement of antennae with legs (see Color Plate 5). Finally, there must be a class of *realizator genes* that control morphological details such as the size of a segment and the position of its bristles. However, little is known about these genes.

The hierarchy depicted in Figure 21.4 is based mainly on three lines of evidence. First, comparing the effects of two mutations singly and in combination often reveals whether or not one of the mutated genes controls the expression of the other. Second, the expression pattern of any cloned gene can be assayed by *immunostaining* or *in situ hybridization* (see Methods 4.1 and 15.1). By comparing the expression patterns of the same gene (gene *A*) in wild-type embryos and in embryos mutant for another gene (gene *B*), one can see directly whether gene *B* affects the expression of gene *A*. For instance, the abnormal expression pattern of a pair-rule gene (*hairy*$^+$) in a mutant for a gap gene (*knirps*) shows that *knirps*$^+$ directly or indirectly controls *hairy*$^+$ (Fig. 21.5). Third, specific binding of a protein encoded by one gene to the enhancer region of another gene suggests that the first gene *directly* controls the expression of its target gene. Indeed, most patterning genes in *Drosophila* encode *transcription factors* controlling the expression of their subordinate genes.

The timing of the gene activities shown in Figure 21.4 offers some clues to the overall functions of the respective genes. The maternal gene products last from oogenesis to early embryogenesis. The gap genes and the pair-rule genes are expressed for short periods of time from the preblastoderm to the extended germ band stage, and the mRNAs transcribed from these genes have short half-lives. In contrast, segment polarity and homeotic genes are expressed continuously from the late blastoderm stage on. These observations suggest that maternal, gap, and pair-rule genes play an important but temporary role: they establish the expression patterns of both segment polarity and homeotic genes. Segment polarity and homeotic genes have the lasting function of defining and maintaining the anteroposterior body pattern of the organism. These critical genes begin their activity at the same time that the potency of blastoderm cells is restricted to individual segments (see Chapter 6). The coincidence suggests that establishing the expression patterns of segment polarity genes and homeotic genes is part of the cell determination process itself. This hypothesis is in accord with

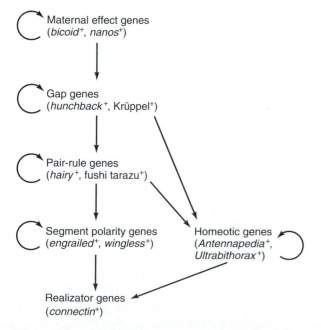

Figure 21.4 Cascade of gene regulation controlling the development of the anteroposterior body pattern in *Drosophila*. Under each group heading, two representative genes are named as examples. The straight arrows indicate the control of a gene of one group by genes of another group. The circular arrows symbolize interactions between genes of the same group.

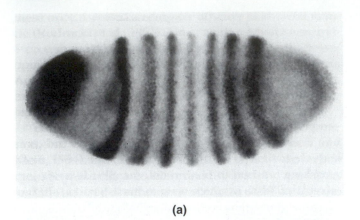

(a)

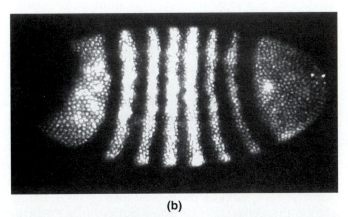

(b)

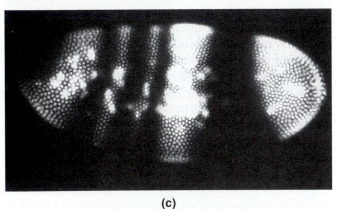

(c)

Figure 21.5 Establishing genetic hierarchies by comparing the expression pattern of the same gene in wild-type embryos and in embryos mutant for another gene. Here, expression of the pair-rule gene *hairy*⁺ is monitored in wild-type and in *knirps*⁻ mutant embryos at the blastoderm stage. Anterior pole is to the left. **(a)** Bright-field photograph of a wild-type embryo immunostained for hairy protein (see Methods 4.1). The dark stain reveals the presence of hairy protein in an anterior dorsal patch and in seven stripes around the embryo. **(b)** Same embryo, now under fluorescent illumination. All nuclei appear bright from a fluorescent DNA strain, except in the cells expressing the *hairy*⁺ gene, where the immunostain quenches the fluorescence. **(c)** Embryo homozygous for a *knirps* loss-of-function allele, also shown under fluorescent illumination. The *hairy*⁺ gene is expressed normally in the anterior patch and the first three stripes. The other stripes have run together into one broad stripe. Comparison of parts b and c shows that the normal function of the *knirps*⁺ gene directly or indirectly affects the expression of the *hairy*⁺ gene in the posterior four stripes.

models of cell determination based on bistable control circuits of switch genes (see Chapter 15).

In the remainder of this chapter, we will see how the cascade of embryonic patterning genes unfolds in time. First, we will examine the role of maternal effect genes, segmentation genes, and homeotic genes in forming the anteroposterior body pattern. This is the best-understood case of biological pattern formation to date, and it will therefore be presented in some detail. Next, we will look briefly at the genetic control of dorsoventral pattern formation. As we explore the control hierarchies revealed by genetic and molecular techniques, we will pause occasionally and ask how these new results compare with the older gradient model of embryonic pattern formation. Thereafter, we will discuss the compartment hypothesis, which links the activity of certain "selector genes" to *compartments* in the adult epidermis identified by clonal analysis. Toward the end of this chapter, we will describe the structure and function of the *homeodomain,* a functional domain that is shared by many of the proteins encoded by patterning genes in *Drosophila.* The homeodomain is characteristic of transcription factors and has been exceedingly well conserved during the evolution of all eukaryotes. Finally, we will consider which parts of the genetic hierarchy governing pattern formation in *Drosophila* have been found in other insects.

Maternal Genes Affecting the Anteroposterior Body Pattern

About 20 maternal effect genes involved in anteroposterior pattern formation in *Drosophila* have been isolated, and these are probably the majority of *all* maternal genes with this function (Nüsslein-Volhard, 1991). Females mutant in any of these genes produce offspring lacking particular body regions while other regions are sometimes enlarged or duplicated (Fig. 21.6). The number of mutant phenotypes is much smaller than the number of genes, suggesting that groups of maternal effect genes cooperate in generating certain parts of the embryo. Each group acts on specific gap genes, so that a group of maternal effect genes and their target gap genes form a *genetic control system.* There are three control systems, affecting respectively the anterior, the posterior, and the terminal parts of the body pattern. Each system acts independently of the others: even in double mutants, in which two systems are inactivated simultaneously, the third system still operates and forms its part of the body pattern (Nüsslein-Volhard et al., 1987).

Although the anterior, posterior, and terminal control systems employ different molecular mechanisms, they share three basic features (Fig. 21.7). First, the activity of each system starts with the *localization of a maternal signal* in the oocyte (Fig. 21.3a). In the anterior

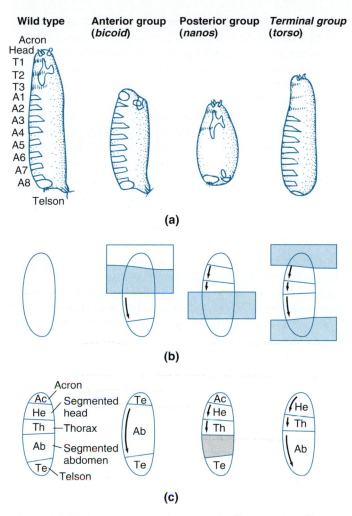

(a)

(b)

(c)

Figure 21.6 Three groups of maternal effect genes affect different regions of the embryonic body pattern. **(a)** For each group, a representative gene and its loss-of-function phenotype are indicated. **(b)** The rectangles indicate the embryonic parts on the blastoderm fate map that are not formed in mutant embryos. For instance, in *bicoid* embryos, head and thorax (shaded area) are missing and the acron area (open rectangle) forms a telson. **(c)** The remaining areas are usually enlarged or transformed, so that the entire fate map is changed in the mutant embryos. Arrows indicate anteroposterior polarity.

and posterior systems, this signal is carried by localized mRNAs. In the terminal system, the follicle cells release a spatially restricted protein into the perivitelline space near the egg poles. Second, in each system, the localized signal triggers a mechanism that causes the *uneven distribution of a transcription factor*. Third, the transcription factors set up *specific gap gene expression domains*.

The anterior and terminal transcription factors act as *morphogen gradients*, controlling two or more target genes in concentration-dependent ways. The posterior system, in contrast, acts by limiting the realm of an anterior gradient. For the most part, the phenotypes resulting from abnormal shapes of these gradients are *not* caused by cell death or cell movement. Rather, the cells receive abnormal determination signals, which alter the fate map of the entire embryo.

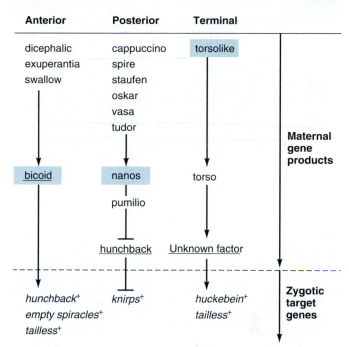

Figure 21.7 Three systems of genes and their products, each generating a part (anterior, posterior, or terminal) of the anteroposterior body pattern. Highlighted gene products are three of the four egg components also shown in Figure 21.3a. Underlined gene products are asymmetrically distributed transcription factors. Arrowheads symbolize activation; perpendiculars symbolize inhibition.

The Anterior Group Generates an Autonomous Signal

Because the nurse cells are adjacent to the future *anterior* pole of the egg, nurse cell cytoplasm enters the oocyte anteriorly; thus, the nurse cells' anterior location imposes an *anteroposterior polarity* on the developing egg. Messenger RNA transcribed from the maternal *bicoid⁺* gene in the nurse cells appears to be trapped upon entry into the oocyte (see Figs. 8.1 and 15.9). This localization mechanism depends on a large secondary structure formed by the trailer region of bicoid mRNA, as discussed in Chapter 8.

The localized bicoid mRNA sets up a gradient of bicoid protein, which acts as a *morphogen*, generating several cell states characterized by different gene activities (Color Plate 1 and Fig. 21.8). Above a certain threshold concentration, the protein activates transcription of *Krüppel⁺*, a gap gene. In normal embryos, the threshold for activating *Krüppel⁺* transcription is reached near 40% *egg length (EL)*, where 0% is defined as the posterior pole. At a higher concentration, normally reached around 50% egg length, bicoid protein also activates transcription of another gap gene, *hunchback⁺*, as described in detail in Chapter 15. At still higher concentrations, reached beyond 58% egg length, bicoid protein *inhibits* transcription of *Krüppel⁺*, thus limiting its expression to a band around the egg equator. In addition, the local concentration of bicoid protein controls the expression of *empty spiracles⁺*, a gene expressed in a band

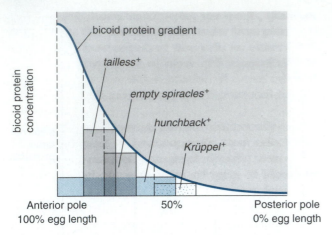

Figure 21.8 Activation of zygotic genes, including *tailless⁺*, *empty spiracles⁺*, *hunchback⁺*, and *Krüppel⁺*, by the bicoid protein. The bicoid protein concentration forms a gradient with a maximum near the anterior pole and an exponential decline toward the posterior. The bicoid protein acts as a morphogen, activating and repressing several target genes depending on its local concentration. The threshold concentrations (vertical lines) at which the target genes are activated or inhibited are shown here for normal embryos derived from mothers with two copies of the *bicoid⁺* gene. In embryos from mothers with one or four *bicoid⁺* copies, the bicoid protein gradient is generally lower or higher, so that the thresholds for target gene activation and inhibition are shifted anteriorly or posteriorly.

of blastoderm cells between 67 and 78% egg length (Walldorf and Gehring, 1992). Finally, bicoid concentrations normally reached between 75 and 88% egg length, in concert with a maternal gene product of the terminal system, activate the *tailless⁺* gene (Pignoni et al., 1992).

The outstanding feature of bicoid mRNA is its ability to establish an anterior pole at any location in the egg. This was shown by Wolfgang Driever and his colleagues (1990), who microinjected bicoid mRNA into different sites in eggs derived from wild-type and *bicoid* mutant mothers (Fig. 21.9). In each case the injected mRNA remained localized at the site of injection and gave rise to a gradient of bicoid protein. The developing embryos formed anterior structures near the injection sites. Eggs from wild-type mothers injected at the posterior pole gave rise to embryos with cephalic or thoracic structures on both ends. Eggs from mothers defective in both *bicoid* genes formed head structures wherever they were injected with functional bicoid mRNA. These results show that bicoid mRNA by itself establishes the anterior pole and organizes the embryonic body pattern so that the anteriormost structures are formed at the highest concentration of bicoid protein.

The localization of bicoid mRNA at the anterior pole of the egg occurs in four distinct phases during oogenesis and depends on the activity of several other genes within the anterior group (St. Johnston et al., 1989). For instance, loss of function in the maternal *exuperantia⁺* or *swallow⁺* genes leads to an almost uniform distribution of bicoid mRNA in the egg, and corresponding ab-

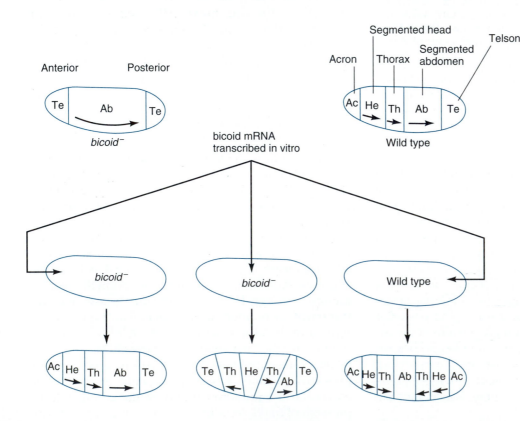

Figure 21.9 Microinjection of bicoid mRNA into *Drosophila* eggs at different sites. Embryos are oriented with anterior pole to the left, dorsal side up. The diagrams at the top are blastodermal fate maps of the recipient eggs obtained from wild-type or *bicoid⁻* mothers. The middle row illustrates the experimental setup. Active bicoid mRNA was transcribed in vitro from a cloned and modified bicoid cDNA. This RNA was injected into the anterior pole, the middle, or the posterior pole of the recipients. The bottom row shows how the fate maps of the recipients changed as a result. Arrows inside body regions indicate anteroposterior polarity.

normalities occur in the developing embryos (Frohnhöfer and Nüsslein-Volhard, 1987). The exuperantia product seems to interact with bicoid mRNA and with microtubules in the nurse cell cytoplasm during the transport of bicoid mRNA to the oocyte (Pokrywka and Stephenson, 1991; S. Wang and T. Hazelrigg, 1994). The swallow product appears to be involved in maintaining bicoid mRNA in the oocyte.

In addition to the bicoid gradient, the *Drosophila* embryo contains a gradient of maternally encoded hunchback protein. (The *hunchback*⁺ gene is exceptional in that it is expressed as a maternal effect gene *and* as a zygotic gap gene. The maternal and zygotic hunchback mRNAs are transcribed from different promoters but encode the same protein. See Figure 15.11.) The concentration of maternal hunchback protein is uniformly high in the anterior third of the egg and then declines to the posterior. This profile is shaped indirectly by the product of *nanos*⁺, a member of the posterior group described in the following section. When zygotic transcription begins at the preblastoderm stage, the gradient of maternal hunchback protein is replaced with a similar gradient of zygotic hunchback protein. Embryos lacking both maternal and zygotic hunchback protein show more severe anterior defects than embryos lacking only zygotic hunchback protein, indicating that maternal hunchback protein contributes to normal anterior development (Lehmann and Nüsslein-Volhard, 1987). Molecular studies indicate maternal hunchback and bicoid proteins act synergistically in controlling gap gene expression (Struhl et al., 1992; Simpson-Brose et al., 1994).

The Posterior Group Acts by Double Negative Control

The posterior group contains at least nine maternal effect genes. Females deficient in any of these genes lay eggs that develop into embryos without an abdomen. The key component of the posterior group is the *nanos*⁺ gene (Lehmann and Nüsslein-Volhard, 1991; C. Wang and R. Lehmann, 1991). As a counterpart to bicoid mRNA, nanos mRNA is localized at the posterior pole of the egg, where it acts as a source for a gradient of nanos protein (Fig. 21.10a).

The *nanos*⁺ function is required to activate the transcription of *knirps*⁺, a gap gene expressed in a broad posterior band of blastoderm cells. In analogy to the action of bicoid protein on the *hunchback*⁺ gene in the anterior system, it was first assumed that the nanos product itself would be a transcriptional activator of *knirps*⁺. Surprisingly, the nanos protein is not a transcription factor. Instead of directly activating *knirps*⁺, it acts by removing an inhibitor: the nanos protein reduces the concentration of maternal hunchback protein, which represses *knirps*⁺ at high concentration.

Initially, maternal hunchback mRNA is evenly distributed throughout the embryo. The presence of nanos protein blocks hunchback mRNA translation, and indirectly promotes the degradation of hunchback mRNA (R. P. Wharton and G. Struhl, 1991). Thus, the nanos protein prevents the accumulation of maternal hunchback protein in the posterior half of the egg, where it would otherwise inhibit the expression of *knirps*⁺ (Fig. 21.7). Experimenters demonstrated this effect by engineering females with germ lines deficient in both *hunchback* and *nanos* (Hülskamp et al., 1989; Irish et al., 1989; Struhl, 1989). Despite the absence of both maternal gene products, eggs from these females developed into normal, fertile flies.

The development of normal embryos from eggs without maternal nanos and hunchback products reveals an apparent redundancy. It seems that the subsequent *zygotic* expression of the *hunchback*⁺ gene in the anterior half of the embryo (see Fig. 15.10 and Color Plate 7) can substitute for the gradient of maternal hunchback protein. According to the *principle of synergistic mechanisms*, one could view the overlap as a safeguard that ensures normal development even if one control is not fully functional under certain environmental conditions or because of natural variation. Such overlaps may also have allowed the genetic control of anteroposterior pattern formation to evolve in different ways (see Patterning Genes in Other Insects at the end of this chapter).

The nanos mRNA localized at the posterior generates a gradient of nanos protein that spreads over the posterior half of the egg. The activity of the nanos protein requires the function of the *pumilio*⁺ gene (Barker et al., 1992). Loss of *pumilio*⁺ function mimics the *nanos*⁻ phenotype: the developing embryos have no abdomen. Conceivably, the pumilio protein acts as a *molecular chaperone* to stabilize and protect the nanos gene product (Macdonald, 1992).

Many Posterior-Group Genes Are Also Required for Pole Cell Formation

The localization and activity of nanos mRNA depend on at least six additional maternal effect genes. Eggs deficient in the products of the maternal *cappuccino*⁺, *oskar*⁺, *spire*⁺, *staufen*⁺, *vasa*⁺, or *tudor*⁺ genes all fail to form *polar granules*, the electron-dense ribonucleoprotein particles that normally assemble near the posterior pole during oogenesis (see Fig. 8.15). Embryos developing from eggs without polar granules show two conspicuous defects: they lack abdominal segments, and they do not form pole cells. Apparently, polar granules are repositories for both nanos mRNA and germ cell determinants (Lehmann and Rongo, 1993).

Some of the gene products assembled in polar granules are present as mRNAs, while others are localized there as proteins. The localization of these components requires their movement from the nurse cells into the oocyte and then across the length of the oocyte to the posterior pole. This elaborate process has been studied in detail for oskar mRNA, one of the components of polar granules. Jeongsil Kim-Ha and colleagues (1993) have identified several elements within the trailer region of oskar mRNA that are required for different legs of its journey to the posterior pole. The localization of nanos mRNA to the posterior pole also depends on its trailer region (Gavis and Lehmann, 1992).

Which of the polar granule components are merely for anchorage and which, if any, act as germ cell determinants is still unclear. However, some gene products have been ruled out as true germ cell determinants by an ingenious experiment in which oskar mRNA was *mislocalized* to the anterior pole (Fig. 21.10). Anne Ephrussi and Ruth Lehmann (1992) cloned a hybrid gene encoding an oskar mRNA with the 3′ untranslated region of bicoid mRNA. Females containing the cloned hybrid gene as a transgene produced eggs in which the modified oskar mRNA was localized to the anterior pole along with the normally supplied bicoid mRNA. The mislocalized oskar mRNA caused the formation of ectopic pole cells at the anterior pole and a mirror-image duplication of the posterior abdomen. By trans-

planting the ectopic pole cells to pole cell–deficient hosts, the researchers ascertained that the ectopic pole cells could produce functional gametes.

Having set up an experimental system for pole cell formation by mislocalizing oskar mRNA, the investigators could now test whether the genes required for normal pole cell formation were also required for ectopic pole cell formation. The idea was that the mislocalized oskar mRNA should function independently of all genes that are merely required for the normal localization of oskar mRNA at the posterior pole. For these tests, the *oskar/bicoid* hybrid gene was used to transform fly strains mutant for any of the other genes required for normal pole cell formation. The results showed that two of these genes, *vasa*$^+$ and *tudor*$^+$, were required for ectopic pole cell formation. These genes are presumably downstream of *oskar*$^+$ in the genetic hierarchy that controls pole cell development (Fig. 21.11). Three other genes, *cappuccino*$^+$, *spire*$^+$, and *staufen*$^+$, were dispensable for ectopic pole cell formation. These genes must therefore be involved in assembling the polar granules or other components that are required for the normal localization of germ cell determinants at the posterior pole.

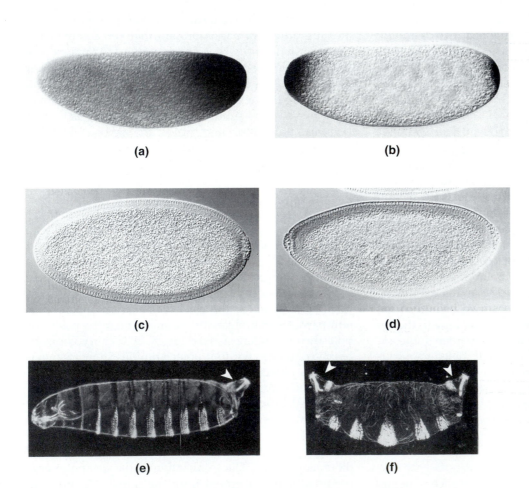

(a)

(b)

(c)

(d)

(e)

(f)

Figure 21.10 Effects of mislocalizing oskar mRNA to the anterior pole region of *Drosophila* embryos; anterior pole to left. **(a, c, e)** Wild-type embryos; **(b, d, f)** embryos derived from mothers containing a transgene that encodes a modified oskar mRNA with the 3′ untranslated region of bicoid mRNA. The bicoid segment causes mislocalization of the modified oskar RNA to the anterior pole. **(a)** Normal embryo during intravitelline cleavage immunostained for the nanos protein (see Methods 4.1). **(b)** Embryo from transgenic mother, with normal nanos mRNA at the posterior and mislocalized nanos mRNA at the anterior pole (labeled by in situ hybridization; see Methods 15.1). **(c)** Normal embryo at preblastoderm stage with pole cells at the posterior pole. **(d)** Embryo from transgenic mother, with pole cells at the anterior and posterior poles. **(e)** Normal larva, with head skeleton to the left and spiracles (arrowhead) at the posterior tip of the abdomen. **(f)** Embryo from transgenic mother, showing mirror-image duplication of the posterior abdominal segments.

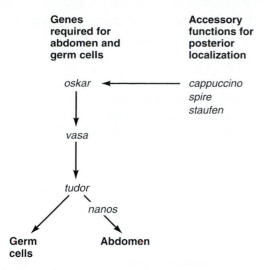

Figure 21.11 Hierarchy of genes involved in forming abdomen and germ cells, and in localizing abdomen and germ cell determinants to the posterior pole.

The Terminal Group Relies on the Local Activation of a Receptor

The anteriormost and posteriormost embryonic regions, including acron and telson, are affected by the terminal system of maternal effect genes. These genes set up a complex signal transduction pathway that begins in two particular patches of follicle cells located next to the anterior and posterior poles of the oocyte (Fig. 21.12a and b). Leslie Stevens and her colleagues (1990) showed

that the activity of the *torsolike*$^+$ gene is critical in polar follicle cells. Subsequent cloning of *torsolike*$^+$ revealed that this gene is expressed selectively in follicle cells at the anterior and posterior poles of the oocyte, and that the torsolike protein has the characteristics of a *ligand*,

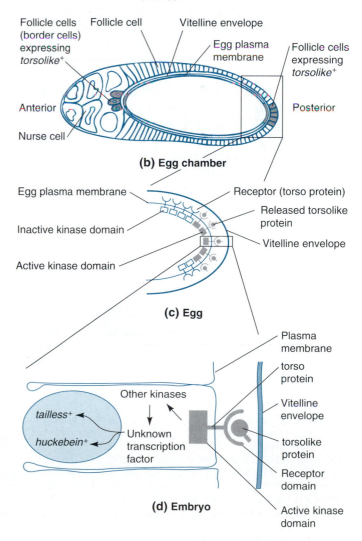

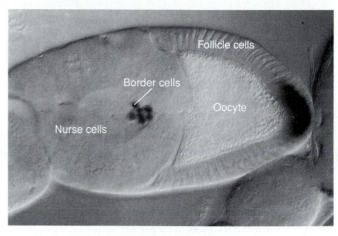

Figure 21.12 Model for the action of the terminal system of genes controlling the anteriormost and posteriormost parts of the body pattern in *Drosophila*. **(a)** Photograph of an egg chamber during mid-oogenesis. Nurse cells are to the left; the oocyte surrounded by follicle cells is to the right. The dark stain reveals the activity of a *lacZ* transgene inserted next to the promoter of a resident *torsolike*$^+$ gene. The transgene is expressed in the same region as the normal *torsolike*$^+$ gene: in posterior follicle cells and in the border cells, which will become anterior follicle cells. **(b)** Late during oogenesis, anterior and posterior patches of follicle cells express the *torsolike*$^+$ gene, which encodes a signal protein (gray shading). This signal protein, or part of it, appears to be released into the perivitelline space between the vitelline envelope and the egg plasma membrane. **(c)** In the egg, the maternally encoded torso protein is inserted into the plasma membrane. Its extracellular domain is a receptor for the torsolike protein. Binding of the torsolike protein activates the cytoplasmic domain of the torso protein, which has tyrosine kinase activity (gray). **(d)** When blastoderm cells are formed in the embryo, the receptor kinase initiates a sequence of protein kinase reactions, which are thought to activate a transcription factor encoded by an unknown gene. This factor is thought to be distributed as a gradient and to activate embryonic target genes (*tailless*$^+$ and *huckebein*$^+$) depending on its local concentration.

or secreted protein (Savant-Bhonsale and Montell, 1993). In the egg and early embryo, the torsolike protein forms a symmetrical gradient with maximum concentrations at the egg poles (Martin et al., 1994). Conceivably, the protein is linked to the vitelline envelope, and a fragment of the protein is later released into the perivitelline space.

The released portion of the torsolike protein binds to a receptor protein encoded by the *torso*[+] gene (Fig. 21.12c and d). The receptor is synthesized in the egg cytoplasm and is uniformly incorporated into the entire egg plasma membrane (Sprenger et al., 1989). However, the ligand (torsolike fragment) seems to be immediately trapped by an excess of torso receptor protein. Since the ligand is present mainly at the egg poles, the receptor is activated only in the pole regions (Sprenger and Nüsslein-Volhard, 1992; Casanova and Struhl, 1993).

The cytoplasmic domain of the torso receptor protein has the properties of a *tyrosine kinase,* that is, an enzyme phosphorylating certain tyrosine residues of other proteins (Lu et al., 1993). When the embryo reaches the cellular blastoderm stage, the kinase activity starts a sequence of protein phosphorylations, which is thought to eventually generate or activate an unknown transcription factor. This factor is expected to act as a morphogen gradient, controlling by its local concentration the activity of at least two embryonic gap genes, *tailless*[+] and *huckebein*[+], which are expressed in the terminal regions of the embryo (Casanova, 1990; Perkins et al., 1992). In addition, the signal transduction cascade started by torso protein inactivates bicoid protein, thus down-regulating the expression of zygotic *hunchback*[+] and other bicoid target genes (Ronchi et al., 1993).

In summary, the three groups of maternal effect genes control their zygotic target genes through different mechanisms. The anterior group generates localized bicoid mRNA, giving rise to a gradient of bicoid protein, which acts as a transcription factor for *hunchback*[+] and *Krüppel*[+]. The posterior group produces the localized nanos signal, which by double negative control allows the embryonic *knirps*[+] gene to become active. The terminal group causes the localized expression of the *torsolike*[+] gene in polar patches of follicle cells. This product acts as a ligand causing the local activation of an embryonic receptor encoded by the *torso*[+] gene. The torso receptor starts a signal transduction cascade that is thought to activate a transcription factor enhancing the expression of at least two target genes, *tailless*[+] and *huckebein*[+].

Segmentation Genes

The signals provided by the maternal gene products regulate a class of zygotic genes collectively called *segmentation genes* (Akam, 1987; Ingham, 1988). These genes have a dual function during embryonic development. First, they control the initial subdivision of the embryo into *parasegments.* Second, they play key roles in the organization of the nervous system later during development. We will focus on the early function of segmentation genes.

Three groups of segmentation genes are distinguished by the phenotypes of their null alleles (Fig. 21.13). A typical *gap gene* mutant is characterized by the loss of an entire series of segments. *Pair-rule* mutants have defects in or deletions of corresponding parts in every other segment, often leading to germ bands with half the normal number of segmental units. *Segment polarity* mutants show deletions and duplications in each

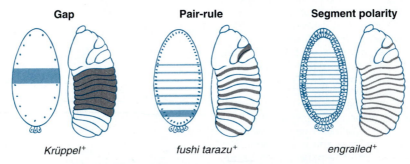

Gap **Pair-rule** **Segment polarity**

Krüppel[+] *fushi tarazu*[+] *engrailed*[+]

Figure 21.13 Three groups of segmentation genes are illustrated by their transcription patterns (color in left-hand diagram of each pair) and by those larval areas that are missing or transformed in loss-of-function phenotypes (gray in right-hand diagrams of each pair). Transcription patterns are revealed by in situ hybridization (see Methods 15.1). *Krüppel*[+], a gap gene, is first expressed after the 12th nuclear division in a belt near the middle of the embryo. Null mutants lack the thorax and anterior abdominal segments. *fushi tarazu*[+], a pair-rule gene, is transcribed in seven distinct stripes after the 13th nuclear division, before cellularization. In the null phenotype, alternate segment boundaries are deleted. *engrailed*[+], a segment polarity gene, is transcribed in 14 narrow stripes at the cellular blastoderm stage. Mutants have a pattern duplication at the posterior margin of each segment.

segment. The three groups form a genetic hierarchy. Maternal gene products control gap genes, gap genes control pair-rule genes, and these in turn control segment polarity genes (Fig. 21.4). At each level of the hierarchy, there are also interactions among certain genes of the same group.

The functions of some segmentation genes had been predicted while others came as a surprise. Since segments had been known as repetitive units of the insect body, it was not unexpected to find the segment polarity genes expressed in corresponding parts of each segment. However, the mutant phenotypes and deduced functions of gap and pair-rule genes were entirely unforeseen. Also unexpected, at least on the basis of the classical gradient models, was the multitude of genetic interactions necessary to build the basic body pattern of an insect. The analysis of these interactions is one of the greatest achievements in developmental biology.

Gap Genes Are Controlled by Maternal Products and by Interactions among Themselves

Gap genes are the critical links between localized maternal gene products and all other zygotic genes involved in specifying the anteroposterior body pattern (Hülskamp and Tautz, 1991). The name "gap genes" is derived from the observation of large gaps of several contiguous segments in mutant phenotypes of the first

gap genes to be described, which were those affecting the midbody region (Fig. 21.14). However, the classification of gap genes has changed somewhat with the growing understanding of their function. Genes are now considered to be gap genes if they are controlled by maternal gene products and affect other segmentation genes.

Among the known gap genes, we will discuss primarily four examples: *hunchback*⁺, *Krüppel*⁺, *knirps*⁺, and *tailless*⁺. Their mRNAs and proteins are synthesized in broad transverse domains that change considerably during development; the major expression domains are shown in Figure 21.15. The *hunchback*⁺ gene is expressed in a large anterior domain, and the null phenotype lacks the posteriormost head segment as well as all thoracic segments. The *Krüppel*⁺ gene is expressed in a central domain, and the phenotype of its null allele lacks the entire thorax and the first five abdominal segments. The *knirps*⁺ gene is expressed in a posterior domain, and its null phenotype lacks almost the entire abdomen. Finally, *tailless*⁺ is expressed in the two terminal domains near the egg poles, and its null phenotype lacks anterior head parts as well as the eighth abdominal segment plus telson.

Within each domain, the gap protein concentration peaks near the center and falls off gradually toward the margins (see Color Plate 7). If the expression domain includes an egg pole, the protein concentration is highest there. The gap domains overlap considerably at their

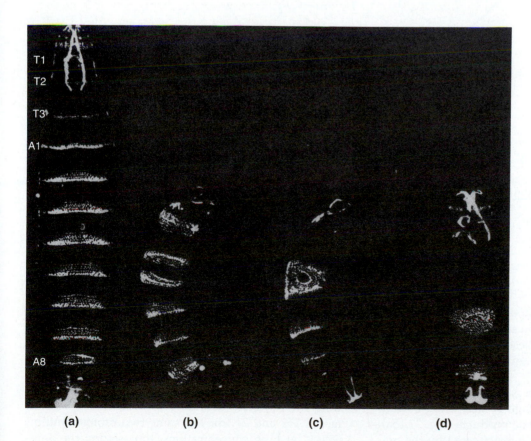

(a)   (b)   (c)   (d)

Figure 21.14 Phenotypes of gap mutants. The photographs show cuticle preparations under dark-field illumination. **(a)** Wild-type larva having eight abdominal segments (A1–A8), three thoracic segments (T1–T3), and an involuted head with mouth hooks (tucked inside T1 and T2). **(b)** *hunchback* embryo with thorax and head parts deleted. In this mutant allele, anterior abdominal segments are also replaced by mirror-image duplications of posterior abdominal segments. **(c)** *Krüppel* embryo missing thorax and A1 through A5 and having instead a mirror-image duplication of A6. **(d)** *knirps* embryo in which A2 through A6 are missing and A1 is fused to A7.

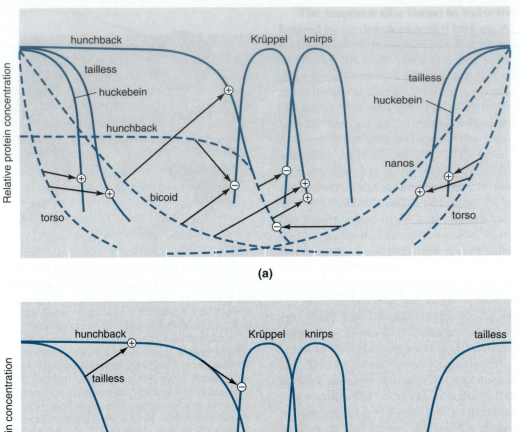

(a)

(b)

Figure 21.15 Simplified scheme of gap gene regulation. Some gap gene domains have been omitted. Several quantitative aspects have been estimated rather than measured. **(a)** Activating (+) and inhibitory (−) effects of maternal encoded proteins. Relative concentrations of maternal proteins are indicated by dashed lines; relative concentrations of gap gene proteins are shown as solid lines. The origins of the arrows indicate whether high, medium, or low levels of the respective protein are required for the regulatory effect. **(b)** Early zygotic interactions among gap genes and their proteins.

margins, and their profiles change as development proceeds. For instance, zygotic *hunchback*⁺ expression is observed in the anterior half of the embryo at the preblastoderm stage. At the blastoderm stage, a posterior domain is added, and the expression then recedes from both poles (see Fig. 15.10). These dynamic changes reflect a multiplicity of control signals that affect gap gene expression.

The earliest gap gene expression domains are delineated at the preblastoderm stage by activating and inhibitory effects of maternal gene products (Fig. 21.15a). Expression of the *tailless*⁺ gene is activated by the ma-

ternal *torso*⁺ function in both pole regions of the embryo. In the anterior pole region, *tailless*⁺ expression is also affected by the maternal bicoid protein (Pignoni et al., 1992). Zygotic *hunchback*⁺ gene expression is activated anteriorly by the bicoid protein and inhibited posteriorly by the nanos protein, as discussed earlier. The *Krüppel*⁺ gene is regulated by the maternally encoded bicoid and hunchback proteins. Either regulatory protein activates *Krüppel*⁺ at low concentrations, thus setting the posterior boundary of *Krüppel*⁺ expression (Figs. 21.15a and 21.16). The same two proteins inhibit *Krüppel*⁺ at high concentrations, thus setting the ante-

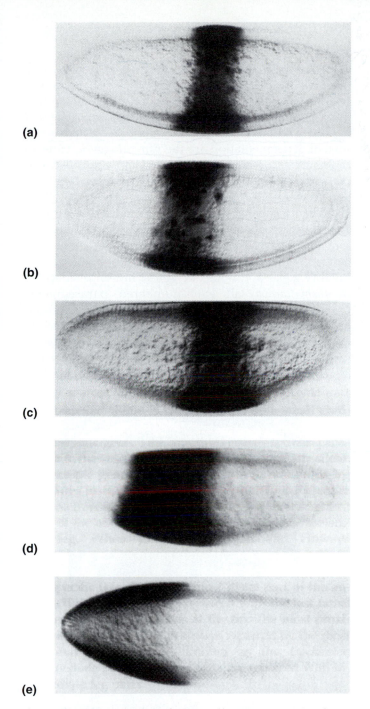

(a)

(b)

(c)

(d)

(e)

rior limit of its expression domain. The *knirps⁺* gene seems to become active spontaneously, since no activating maternal transcription factors have been found.

In addition to being controlled by maternal effect gene products, gap genes also interact among themselves (Fig. 21.15b). Zygotically expressed hunchback protein has the same activating and inhibitory effects on *Krüppel⁺* and *knirps⁺* expression as the maternally expressed hunchback protein. Thus, in *hunchback⁻* mutant embryos, the *Krüppel⁺* expression domain is broadened and shifted anteriorly (Fig. 21.16). Because the hunchback protein inhibits the *knirps⁺* gene more efficiently than the *Krüppel⁺* gene, *knirps⁺* is expressed more posteriorly. Likewise, in *tailless⁻* mutant embryos, the *Krüppel⁺* expression domain is broadened, indicating an inhibitory effect of the tailless protein on *Krüppel⁺* expression. Similarly, the expression domain of *knirps⁺* is limited by hunchback protein anteriorly and by tailless protein posteriorly. Both proteins seem to bind directly to enhancer regions of *knirps⁺* (Pankratz et al., 1992).

In the regulation of the *Krüppel⁺* gene, there is considerable overlap in both activating and inhibitory control functions (Fig. 21.17; see Tautz, 1992). Two regulatory proteins, bicoid and hunchback, are involved in setting both the anterior and the posterior boundaries of *Krüppel⁺* expression. In addition, the tailless, knirps, and giant proteins are involved as negative regulators of *Krüppel⁺*. The lack of any one of these five regulatory proteins causes only a small effect, as illustrated in Figure 21.16. However, if two or three regulatory proteins are missing, the expression of *Krüppel⁺* changes substantially. These observations illustrate the *principle of synergistic mechanisms* already discussed in Chapters 4 and 11. Each of the genes involved in regulating *Krüppel⁺* expression has additional functions; absence of function in any of them causes a distinct mutant phenotype. With respect to their common function—the regulation of *Krüppel⁺*—each of the five genes by it-

Figure 21.16 Regulation of *Krüppel⁺* gene expression by maternal and zygotic gene products (refer to Fig. 21.15). The expression of *Krüppel⁺* is assayed at the blastoderm stage by immunostaining (see Methods 4.1). Anterior pole to the left. **(a)** Wild-type embryo. Krüppel protein is present in a central band of cells. **(b)** *hunchback⁻* mutant embryo. The *Krüppel⁺* expression domain is slightly broadened and shifted anteriorly. **(c)** *tailless⁻* mutant embryo. The *Krüppel⁺* expression domain is slightly broadened in both directions. **(d)** Embryo derived from a *bicoid⁻* female. The *Krüppel⁺* domain is expanded far anteriorly, reflecting the lack of two regulators, bicoid protein and zygotic hunchback protein, which embryos from *bicoid⁻* mothers do not produce. **(e)** Embryo from a *bicoid⁻*, *torsolike⁻* double mutant female. The *Krüppel⁺* domain is shifted and extends to the anterior pole, reflecting the absence of bicoid, zygotic hunchback, and tailless proteins as regulators.

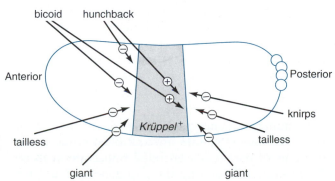

Figure 21.17 Synergistic regulation of the gap gene *Krüppel⁺* by multiple maternal and zygotic proteins. Note the inclusion of the protein encoded by the gap gene *giant⁺*, which does not appear in Figure 21.15.

self may be dispensable. However, their synergistic actions could make the control of *Krüppel*⁺ more reliable under a wide range of genetic and environmental circumstances.

The molecular basis of gap gene regulation has been investigated in considerable detail. All gap gene–regulating proteins have the characteristics of transcription factors, which bind directly to enhancer regions of their target genes. The binding of bicoid protein to *hunchback*⁺ enhancers has already been described in Chapter 15. Other *hunchback*⁺ enhancers are bound by the hunchback and Krüppel proteins, presumably mediating a positive feedback loop of the *hunchback*⁺ gene on its own expression and an inhibition of *hunchback*⁺ by *Krüppel*⁺ (Treisman and Desplan, 1989).

▼

In the regulatory region of the *Krüppel*⁺ gene, Michael Hoch and his colleagues (1991) delimited a 730-bp segment that is bound by at least four transcription factors: the bicoid, hunchback, knirps, and tailless proteins. Reporter genes containing this segment and the transcribed region of the bacterial *lacZ* gene were expressed correctly in the same domain as the normal *Krüppel*⁺ gene. Within the 730-bp segment, the investigators identified 6 binding sites for bicoid protein and 10 binding sites for hunchback protein. Through *DNase footprinting*, they determined that none of these binding sites were overlapping. The same 730-bp segment was also found to contain 1 binding site for knirps protein and 7 binding sites for tailless protein (Hoch et al., 1992). Most of these binding sites overlap with binding sites for bicoid protein. One of these regions, a 16-bp element bound by both bicoid and knirps proteins, was chosen for further analysis. Cultured cells were *transfected* with plasmids containing the chloramphenicol acetyltransferase (CAT) reporter gene under control of the 16-bp element. Addition of another plasmid encoding bicoid protein caused a dose-dependent increase in CAT expression. When knirps protein was introduced along with bicoid protein, the bicoid-dependent increase in CAT expression was reduced by an amount depending on the dosage of knirps protein. These results suggest a simple mechanism that can regulate genes by an activating and an inhibitory transcription factor: both factors may bind competitively and reversibly to identical or overlapping enhancer elements.

In summary, genetic and molecular analysis of gap genes has shown that they are controlled by morphogen gradients of maternally encoded transcription factors. The proteins encoded by the gap genes are also transcription factors, which mediate additional regulatory interactions among gap genes. These multiple regulatory proteins cooperate in limiting gap gene expression to broad transverse bands.

Pair-Rule Genes Are Controlled by Gap Genes and by Other Pair-Rule Genes

Gap gene expression domains divide the *Drosophila* embryo into broad transverse stripes. These stripes are irregular in width and do not correspond to the pattern of parasegments on the blastoderm fate map. The first zygotic genes expressed with a periodic pattern reminiscent of segmentation are the *pair-rule genes*. Three of them, *even-skipped*⁺, *hairy*⁺, and *runt*⁺, are called **primary pair-rule genes**, because they are expressed first and are controlled directly by the gap genes; the others are called **secondary pair-rule genes**, since they are controlled by the primary pair-rule genes. Of particular interest among the secondary pair-rule genes is *fushi tarazu*⁺, which is critical to the control of both segment polarity genes and homeotic genes. In null phenotypes of *fushi tarazu*, all even-numbered parasegments (parasegments 2, 4, and so on) are missing. Because each parasegment contains a future segment boundary, the mutant larvae have only half the normal number of segment boundaries (Fig. 21.18). Indeed, *fushi tarazu* is a Japanese phrase meaning "not enough segments."

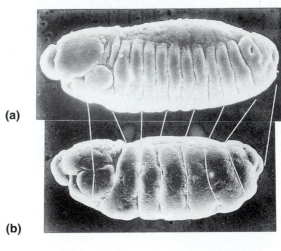

(a)

(b)

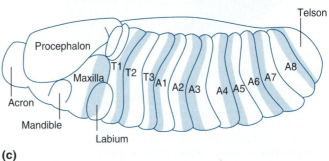

(c)

Figure 21.18 Phenotype of *fushi tarazu*. **(a)** Scanning electron micrograph of a wild-type embryo in lateral view. **(b)** *fushi tarazu*⁻ embryo at a corresponding stage of development. White lines between parts a and b connect homologous portions of the segmented germ band. **(c)** Diagram of a wild-type embryo. The colored areas are missing in the *fushi tarazu*⁻ embryo. These areas are the expression domains of *fushi tarazu*⁺; they correspond to even-numbered parasegments in early embryos (see Fig. 21.2f).

The expression of pair-rule genes begins in fuzzy, broad, and irregular stripes during the formation of the cellular blastoderm (Fig. 21.19). By the time cellularization is complete, pair-rule genes are expressed in seven or eight sharp, narrow, and evenly spaced stripes that encircle the entire embryo. A typical pair-rule stripe at this stage is as wide as four blastoderm cells or one prospective parasegment. The stripes are separated by interstripes—also initially four blastoderm cells wide— where the gene is not expressed. A stripe and an interstripe together are as wide as two segments, so the expression pattern of each pair-rule gene has a ***bisegmental periodicity.*** The expression patterns of *fushi tarazu*⁺ and *even-skipped*⁺ are in register with the parasegmental primordia, with *fushi tarazu*⁺ expressed in even-numbered parasegments and *even-skipped*⁺ in odd-numbered parasegments (Lawrence and Johnston, 1989). As development proceeds, the stripes become narrower while the interstripes become wider. Closer analysis of the *fushi tarazu*⁺ and *even-skipped*⁺ stripes shows that the expression of both genes decreases from the posterior margin while the anterior borders of these stripes, each marking the anterior border of a parasegment, remain sharp (see Color Plate 4). The expression patterns of *hairy*⁺ and *runt*⁺ are also complementary to each other but are out of register with both parasegmental and segmental primordia.

How are the periodic (bisegmental) patterns of primary pair-rule genes generated from the aperiodic (irregular) pattern of gap gene domains? The gap genes, themselves delimited in part by maternal morphogen gradients, in turn form gradients of zygotic morphogens. Being transcription factors, the gap gene proteins activate and inhibit primary pair-rule genes so that they are expressed in periodic patterns of stripes. This control is complicated because the gradient set up by any single gap gene is relatively short, so that each gap gene can control only part of the entire stripe pattern of a pair-rule gene (Fig. 21.5). Close investigation has shown that each pair-rule gene expression pattern is controlled by a combination of several gap genes. This piecemeal type of control is concealed by the reg-

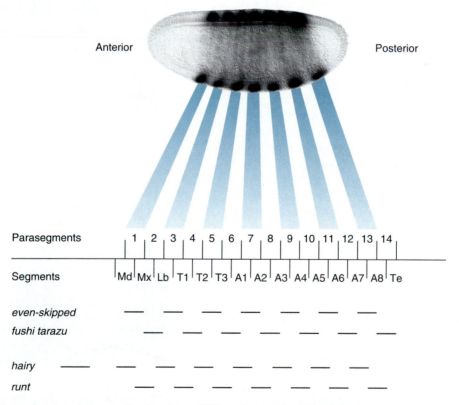

Figure 21.19 Expression of pair-rule genes in the *Drosophila* embryo. The photograph shows an embryo immunostained (see Methods 4.1) for the protein encoded by the *even-skipped*⁺ gene. The embryo is at an advanced stage of blastoderm formation; the plasma membrane infoldings have passed the nuclei (see Fig. 5.18). The even-skipped protein has accumulated in seven regular stripes, which will give rise to the odd-numbered parasegments (1, 3, 5, etc.). When cellularization is complete, expression remains strong in the anterior two cell rows of each parasegment but decreases toward the posterior margin (see Color Plate 4). The scales indicate the positions of segments and parasegments, with segment names abbreviated as in Figure 21.2. The lines at the bottom represent the initial expression domains of further pair-rule genes.

ularity of the striping patterns that it eventually brings forth. However, the complexity of the underlying regulatory mechanisms is betrayed by the large, irregular domains in which they are initially expressed and by the unusually large regulatory regions of the primary pair-rule genes, which span up to 20 kb.

▼

As an example of a primary pair-rule gene controlled by gap genes, let us consider the *hairy*[+] gene. Kenneth Howard and his colleagues (1988) analyzed the effects of various mutations in the regulatory region of *hairy*[+] on the expression pattern of the gene. Using in situ hybridization (see Methods 15.1) with a probe binding to hairy mRNA, they found that in each mutant a unique part of the normal pattern was missing (Fig. 21.20). They concluded that the expression of *hairy*[+] in different stripes is controlled by different *enhancers*. They further speculated that these enhancers may be bound by different combinations of gap gene products. In accord with this hypothesis, the eight *hairy*[+] stripes do not appear simultaneously, but develop in a peculiar sequence, which bears some relationship to the stepwise activation of the gap genes.

To test the notion of stripe-specific enhancers, Michael Pankratz and his colleagues (1990) isolated various *restriction fragments* from the upstream regulatory region of the *hairy*[+] gene and spliced each fragment in front of a *lacZ reporter gene* (Fig. 21.21). Each fusion gene was introduced into wild-type flies by *P-element transformation*. When embryos from transformant lines were assayed for *lacZ* expression, each fragment from the *hairy*[+] regulatory region was found to enhance a characteristic subset of stripes. One fragment located about 8 kb upstream of the *hairy*[+] promoter enhanced the formation of stripe 6; a neighboring fragment enhanced the formation of stripe 7. Some fragments from the *hairy*[+] enhancer region drove the expression of more than one stripe, but each fragment enhanced the expression of a characteristic subpattern.

To find out which gene products interact with the specific enhancer fragments of the *hairy*[+] gene, Pankratz and colleagues (1990) introduced their fusion gene constructs into mutant flies that were deficient in various gap genes (Fig. 21.22). They found that loss of *Krüppel*[+] activity caused stripe 5 of *hairy*[+] to disappear and stripe 6 to expand anteriorly. Loss of *knirps*[+] activity caused *hairy*[+] stripe 6 to disappear and *hairy*[+] stripe 7 to expand anteriorly. The researchers concluded that *Krüppel*[+] activity is required to express *hairy*[+] in stripe 5 and to set the anterior boundary of stripe 6 (Fig. 21.23). This conclusion implies that at a low concentration Krüppel protein interacts with one enhancer region to inhibit the expression of the *hairy*[+] gene in stripe 6 while at a higher concentration Krüppel protein interacts with another enhancer region to promote the expression of

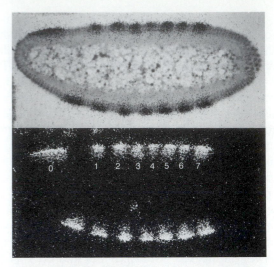

(a)

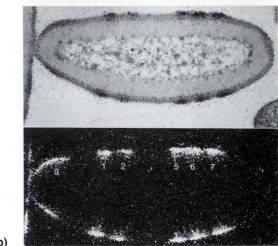

(b)

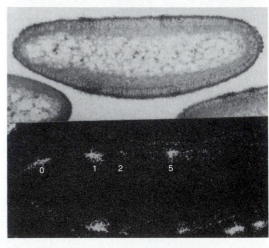

(c)

Figure 21.20 Effects of mutations in different regulatory regions of the *hairy*[+] gene. The panels show sections of blastoderm embryos hybridized in situ with a radiolabeled probe binding to hairy mRNA (see Methods 15.1). Each specimen is photographed in bright-field illumination (top) and in dark-field illumination (bottom). **(a)** In the wild-type pattern there are eight expression domains: an anterior dorsal patch (0) and seven stripes (1–7). **(b)** In the mutant *hairy*[m3] phenotype, stripes 3 and 4 are suppressed while stripes 5 through 7 are broadened. **(c)** In the *hairy*[m7] phenotype, stripes 3, 4, 6, and 7 are suppressed while stripe 2 is quite weak.

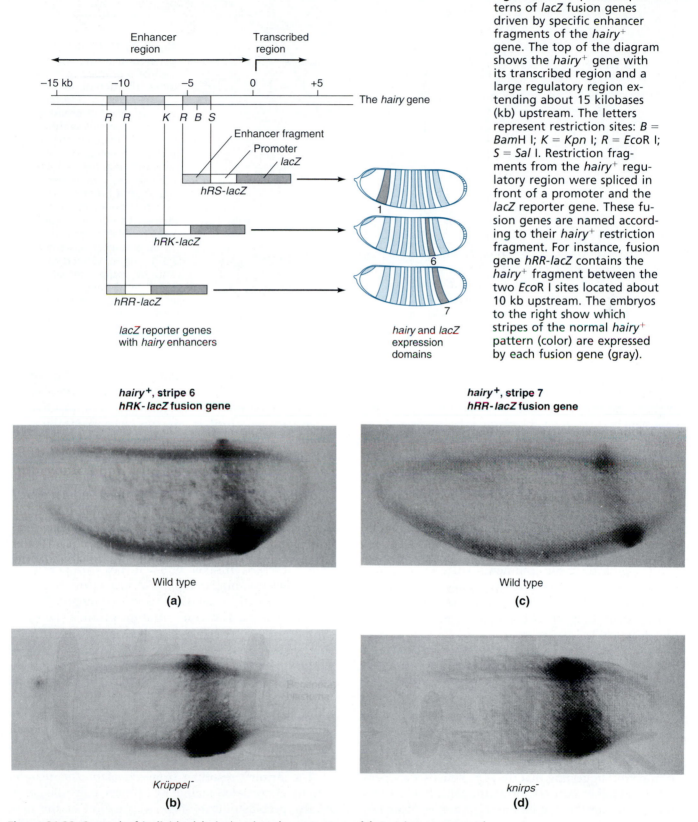

Figure 21.21 Expression patterns of *lacZ* fusion genes driven by specific enhancer fragments of the *hairy+* gene. The top of the diagram shows the *hairy+* gene with its transcribed region and a large regulatory region extending about 15 kilobases (kb) upstream. The letters represent restriction sites: *B* = *Bam*H I; *K* = *Kpn* I; *R* = *Eco*R I; *S* = *Sal* I. Restriction fragments from the *hairy+* regulatory region were spliced in front of a promoter and the *lacZ* reporter gene. These fusion genes are named according to their *hairy+* restriction fragment. For instance, fusion gene *hRR-lacZ* contains the *hairy+* fragment between the two *Eco*R I sites located about 10 kb upstream. The embryos to the right show which stripes of the normal *hairy+* pattern (color) are expressed by each fusion gene (gray).

Figure 21.22 Control of individual *hairy+* stripes by gap genes. **(a)** A stripe corresponding to stripe 6 of the normal *hairy+* phenotype is expressed by fusion gene *hRK-lacZ* in a wild-type host embryo (see Fig. 21.21). **(b)** In a *Krüppel⁻* mutant embryo transformed with *hRK-lacZ*, stripe 6 expands anteriorly. **(c)** A stripe corresponding to stripe 7 of the normal *hairy+* phenotype is expressed by fusion gene *hRR-lacZ* in a wild-type host embryo. **(d)** In a *knirps⁻* mutant embryo transformed with *hRR-lacZ*, stripe 7 expands anteriorly.

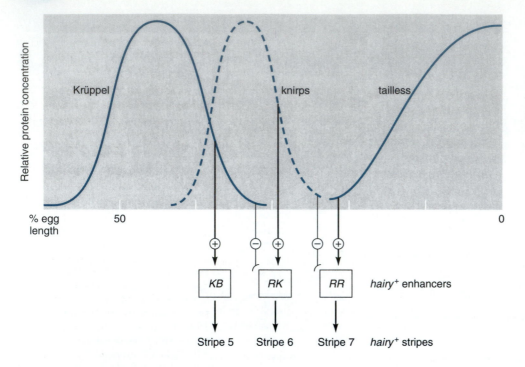

Figure 21.23 Model of gap gene control over the expression of a primary pair-rule gene. The gap gene proteins Krüppel, knirps, and tailless are expressed in broad, overlapping domains without periodicity. However, through unique combinations of positive and negative regulation, they control the expression of *hairy⁺* and other primary pair-rule genes in periodic stripes. The diagram summarizes regulatory effects on three enhancer regions of the *hairy⁺* gene, termed *KB, RK,* and *RR,* which control the expression of *hairy⁺* in stripes of 5, 6, and 7, respectively (see Fig. 21.21).

hairy⁺ in stripe 5. Likewise, a low concentration of knirps protein inhibits *hairy⁺* expression in stripe 7 while a higher concentration of knirps protein enhances *hairy⁺* expression in stripe 6. These results were confirmed and extended by other investigators, who identified gap gene products that control the expression of *hairy⁺* in the other stripes (Howard and Struhl, 1990; Riddihough and Ish-Horowicz, 1991), and who in addition observed that another primary pair-rule gene, *runt⁺*, contributes to the even spacing of the *hairy⁺* stripes.

To test whether the gap proteins act directly on the *hairy⁺* enhancer fragments, Pankratz and colleagues (1990) incubated these fragments with Krüppel and knirps proteins and then precipitated the proteins with appropriate antibodies. They observed that each gap protein bound to its specific target fragments of *hairy⁺* regulatory DNA.

In corresponding studies, Dusan Štanojević and his colleagues (1989) found that hunchback and Krüppel proteins bind to specific enhancer sequences of another primary pair-rule gene, *even-skipped⁺*. Expression of this gene in stripe 2 is activated by the maternal bicoid and hunchback proteins and inhibited by the gap gene proteins Krüppel and giant (Štanojević et al., 1991). The binding sites for the two activating proteins tend to occur in pairs, suggesting that these proteins act cooperatively (S. Small et al., 1991, 1992); the same holds for the binding sites for the two inhibitory proteins. Moreover, the binding sites for activating and inhibitory proteins overlap, indicating that the four transcription factors interact by competitive and reversible binding, as in the regulation of *Krüppel⁺* described previously.

Together, the available data indicate that the expression patterns of *primary pair-rule genes* are controlled by local combinations of different gap gene proteins. A given gap gene protein may enhance or inhibit transcription of the same pair-rule gene, depending on gap protein concentration and on what other transcription factors are bound simultaneously.

In addition to being controlled by gap genes, the primary pair-rule genes feed back on themselves and interact with one another to delineate their expression domains. Their interactions cause dynamic changes in the expression pattern. For instance, *even-skipped⁺* is initially expressed in seven broad stripes with poorly defined edges. Within 30 min, these stripes are sharpened and refined so that they are only two or three cells wide and show a marked concentration profile. The even-skipped protein concentration becomes highest in the anteriormost cells of each stripe and declines toward the posterior (see Color Plate 4). This modulation relies on an enhancer element of *even-skipped⁺* that is bound by the even-skipped protein itself and at least two other transcription factors (Jiang et al., 1991). Modulations of this kind, which are also observed for other pair-rule genes, are important for the control of segment polarity genes, as we will see.

The striping resulting from *secondary pair-rule gene* activity is regulated mainly by primary pair-rule genes. At the preblastoderm stage, expression of the *fushi tarazu⁺* gene seems to be controlled negatively by the *even-skipped⁺* gene, so that the two genes' striping patterns are complementary. At the blastoderm stage, *fushi tarazu⁺* expression is also inhibited by *hairy⁺*, so that the *fushi tarazu⁺* stripes are then narrowed to the first two or three rows of cells within each even-numbered parasegment (see Color Plate 4).

At the end of the blastoderm stage, the *Drosophila* embryo is encircled by the patterns of at least eight pair-rule genes, each expressed in seven or more stripes. These stripes are generally not in register, so each transverse row of blastoderm cells expresses a different combination of pair-rule genes (Fig. 21.24). The rudiment of each parasegment at the blastoderm stage is about four rows of cells wide. For instance, the anteriormost row of cells in a future odd-numbered parasegment expresses five pair-rule genes: *even-skipped⁺*, *odd-skipped⁺*, *paired⁺*, *hairy⁺*, and *sloppy-paired⁺*. The next row of cells to the posterior expresses three pair-rule genes, namely, *even-skipped⁺*, *paired⁺*, and *hairy⁺*.

Because of the bisegmental periodicity of the pair-rule patterns, corresponding rows in the rudiments of odd-numbered and even-numbered parasegments express different combinations of pair-rule genes. The anteriormost row of an even-numbered future parasegment expresses *fushi tarazu⁺*, *runt⁺*, and *odd-paired⁺*, a combination that is quite different from the genes expressed in the anteriormost cells of a future odd-numbered parasegment.

The fireworks of pair-rule gene activity last only for a few hours. During that time, the genes produce different combinations of regulatory proteins that entrain the activity of more permanently expressed genes. These are the segment polarity genes, discussed in the following section, and the homeotic genes, to be described thereafter.

Segment Polarity Genes Define the Boundaries and Polarity of Parasegments

Mutations in *segment polarity genes* cause deletions and duplications in corresponding portions of each segment. These deficiencies are often associated with a loss of anteroposterior polarity within the segment (Bejsovec and Wieschaus, 1993; Ingham and Martinez-Arias, 1992; Peifer and Bejsovec, 1992). This section will focus on three segment polarity genes, *engrailed⁺*, *wingless⁺*, and *hedgehog⁺*.

At the blastoderm stage, each future parasegment consists of approximately four transverse rows of cells. In each parasegment rudiment, the *engrailed⁺* gene is expressed in the anteriormost row and the *wingless⁺* gene in the posteriormost row (Fig. 21.24). Researchers have analyzed the regulation of these genes by examining their expression patterns in embryos mutant in various pair-rule genes. For instance, *wingless⁺* is expressed over the entire width of even-numbered parasegments in *fushi tarazu⁻* mutant embryos (Ingham et al., 1988). Thus, the normal activity of *fushi tarazu⁺* must inhibit the expression of *wingless⁺* in all but the posteriormost row in even-numbered parasegments. From this and many other observations arose the following model for the activation of *engrailed⁺* and *wingless⁺*.

The activation of *engrailed⁺* depends on the simultaneous expression of two pair-rule genes (Fig. 21.24). In odd-numbered parasegments, *engrailed⁺* is activated in those cells that contain both even-skipped and paired proteins. In even-numbered parasegments, the cells expressing *engrailed⁺* are those with both fushi tarazu and odd-paired proteins. Thus, *engrailed⁺* becomes active in the anteriormost row of cells in each parasegment. In contrast, the activation of the *wingless⁺* gene is under positive and negative control. The paired and odd-paired proteins activate *wingless⁺*; the even-skipped and fushi tarazu proteins inhibit *wingless⁺*. These controls allow the *wingless⁺* gene to be expressed only at the posterior margin of each parasegment, where the activities of *even-skipped⁺* and *fushi tarazu⁺* decrease as the cellularization of the blastoderm proceeds. Thus, *wingless⁺* becomes active in the anteriormost row of

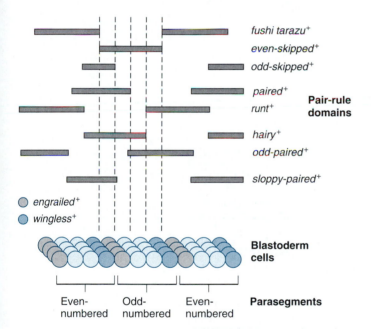

Figure 21.24 Expression domains of pair-rule genes (top) and segment polarity genes (bottom). The bars at the top represent the initial expression domains. (Some of these domains change as the blastoderm stage advances. For instance, the *fushi tarazu⁺* and *even-skipped⁺* domains shrink to bands two or three cells wide, as shown in Color Plate 4. The circles at the bottom represent individual blastoderm cells. The rudiment of each parasegment consists of about four transverse rows of blastoderm cells. Each row within a parasegment expresses a unique combination of pair-rule genes. The combinations of active pair-rule genes also differ between odd-numbered and even-numbered parasegments. As an example, only even-numbered parasegments express *fushi tarazu⁺*, while odd-numbered parasegments express *even-skipped⁺*. Nevertheless, the expression patterns of segment polarity genes are identical in all parasegments. In particular, the anteriormost row expresses the *engrailed⁺* gene, and the posteriormost row expresses the *wingless⁺* gene.

cells in each parasegment, right behind the row of cells expressing *engrailed*⁺ in the preceding parasegment.

Once activated, the segment polarity genes remain active through embryonic, larval, and pupal development. Throughout this time, *engrailed*⁺ remains involved in maintaining the anteroposterior body pattern, although its expression pattern undergoes subtle changes, as we will see. The *wingless*⁺ gene acquires a new function in the *imaginal discs* of the larva, epithelial bags that give rise to much of the adult epidermis (discussed later in this chapter; see also Chapter 6). In several imaginal discs, the *wingless*⁺ gene is expressed ventrally, and its ectopic expression in dorsal cells ventralizes the structures formed by these cells and

their neighbors (Couso et al., 1993; Struhl and Basler, 1993).

As the expression pattern and function of segment polarity genes change during development, so does their genetic control. The continued expression of *engrailed*⁺ is maintained by three different mechanisms operating at successive stages. These mechanisms include positive feedback, in which engrailed protein acts on its own gene, and a positive control by the product of the *wingless*⁺ gene (Heemskerk et al., 1991). The wingless product is a signal peptide that is secreted from cells and acts on matching receptors in the plasma membranes of neighboring cells (Fig. 21.25a). Released from the posteriormost cells in each parasegment, the wing-

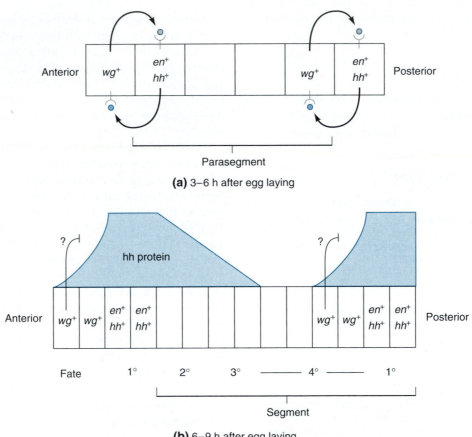

(a) 3–6 h after egg laying

(b) 6–9 h after egg laying

Figure 21.25 Interactions among the segment polarity genes *engrailed*⁺ (*en*⁺), *hedgehog*⁺ (*hh*⁺), and *wingless*⁺ (*wg*⁺) in the *Drosophila* embryo. **(a)** After formation of the cellular blastoderm, each parasegment consists of about four transverse rows of cells. The anteriormost cells within each parasegment express *engrailed*⁺ and *hedgehog*⁺ while the posteriormost cells express *wingless*⁺. During gastrulation and germ band extension (3 to 6 h after egg laying), the engrailed protein, a transcription factor, enhances the expression of *hedgehog*⁺ within the same cells. The hedgehog protein, a signal peptide, stabilizes the expression of *wingless*⁺ in neighboring cells that have been initiated to express *wingless*⁺. Conversely, the wingless protein, also a signal peptide, maintains the expression of *engrailed*⁺ in competent neighboring cells. **(b)** During the extended germ band stage and germ band retraction (6 to 9 h after egg laying), segments consist of about ten rows of cells, with the posteriormost rows expressing *engrailed*⁺ and *hedgehog*⁺. The hedgehog protein acts as a morphogen gradient. Depending on the local concentration of hedgehog protein, dorsal cells produce large denticles (1° fate), a smooth surface (2° fate), thick pigmented hair (3° fate), or fine hair (4° fate). The wingless protein may inhibit the diffusion of hedgehog protein to the anterior and the realization of 1°, 2°, and 3° fates.

less peptide maintains in particular the expression of *engrailed*[+] in the posteriorly adjacent cells across the parasegmental boundary (Noordermeer et al., 1994; Siegfried et al., 1994; Vincent and Lawrence, 1994).

Conversely, the cells expressing *engrailed*[+] maintain *wingless*[+] activity in their anterior neighbors through a feedback loop that involves the *hedgehog*[+] gene (Ingham and Hidalgo, 1993; Heemskerk and DiNardo, 1994). The engrailed protein is a transcription factor, which activates the *hedgehog*[+] gene within the same cells that express *engrailed*[+]. The hedgehog product, a signal peptide like the wingless product, stabilizes the activity of *wingless*[+] in the anteriorly adjacent cells. In addition, the hedgehog protein acts as a morphogen gradient as will be discussed in the next section.

The molecular mechanisms by which *wingless*[+] and *engrailed*[+] mutually stabilize their expression in adjacent rows of cells reflect the completion of the cellular blastoderm in the *Drosophila* embryo. During the preceding, syncytial blastoderm stage, transcription factors could freely diffuse and thus provide a very direct mechanism of control among nuclei and their surrounding jackets of cytoplasm. After cellularization, cells need to interact through signal peptides and their receptors (see Chapter 2).

Midway through Embryogenesis, Parasegmental Boundaries Are Replaced with Segmental Boundaries

The first morphologically identifiable boundaries in the *Drosophila* embryo are shallow grooves in the epidermis that appear after gastrulation and germ band elongation (Fig. 21.26). These grooves separate *parasegments*, which are significant as modular units of gene expression in the early embryo. For example, the initial expression domains of *fushi tarazu*[+] and *even-skipped*[+] are parasegmental, and the stripes of *engrailed*[+] and *wingless*[+] expression mark the anterior and posterior margins of parasegments (Fig. 21.24). Also, many of the homeotic genes to be discussed in the following section are initially expressed in parasegments.

A few hours after germ band elongation, the parasegmental boundaries are smoothed out and segmental boundaries are formed. The change in the register of segmentation can be observed conveniently in the position of the openings of the respiratory system, known as the *tracheal pits*. First located in the middle of each parasegment, the tracheal pits are later found in the grooves of the segmental boundaries. Thus, parasegmental and segmental boundaries have the same length but are out of register by about half a segment's length.

The transition from parasegments to segments is reflected in the changing expression pattern of *engrailed*[+] (Fig. 21.27). At the blastoderm stage, the *engrailed*[+] domain is only one or two cells wide and forms the sharp anterior border of each parasegment. After germ band elongation, which is associated with *convergent exten-*

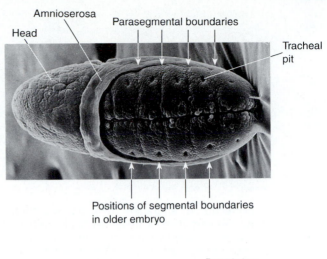

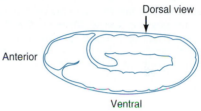

Figure 21.26 Parasegmental and segmental boundaries in the *Drosophila* embryo. The scanning electron micrograph shows a dorsal view of the elongated germ band; the diagram represents the same stage in lateral view. The abdomen is maximally extended and folded back on itself at this stage. The shallow grooves are parasegmental boundaries. In the middle of each parasegment are a pair of tracheal pits, openings of the respiratory system. The parasegmental boundaries will soon disappear and a new set of segmental boundaries will form where the tracheal pits are.

sion, the *engrailed*[+] stripes widen to two or three cells, still with a sharp anterior margin marking the beginning of a parasegment. When the parasegments disappear, the anterior margins of the *engrailed*[+] stripes become ragged and the posterior margins sharpen up, each now demarcating the posterior boundary of a segment. The new segmental boundaries establish the segments as the new repetitive units of the longitudinal body pattern.

The epidermal cells of a segment behave as *fields* in specifying the intrasegmental patterns (Ingham and Martinez-Arias, 1992). They cooperate in forming the pattern of pigments, bristles, and other structures that characterize the larval and adult segments. One *morphogen* that specifies the segmental pattern is the hedgehog peptide, which is still synthesized by the cells also expressing *engrailed*[+] at the posterior border of each segment (Fig. 21.25b). As hedgehog peptide crosses the segment boundary and enters the following segment, it forms an anteroposterior gradient that specifies the fates of cells in a concentration-dependent way (Heemskerk and DiNardo, 1994; Tabata and Kornberg, 1994). The mechanism that allows the hedgehog pep-

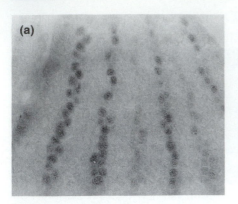
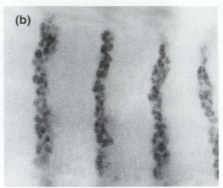
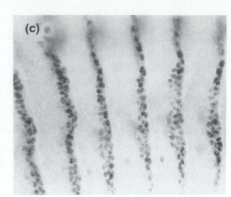

Figure 21.27 Expression of *engrailed*⁺ near parasegmental and segmental boundaries; anterior to left. The engrailed protein was made visible by immunostaining (see Methods 4.1). All photographs are at the same magnification. **(a)** At the cellular blastoderm stage and during gastrulation (3 h after egg laying), engrailed protein is synthesized in stripes that are one to two cells wide. **(b)** After germ band extension (5 h), the engrailed stripes are two to three cells wide, most likely because the germ band has undergone convergent extension. At this stage, the cells at the anterior margin of each stripe synthesize high levels of engrailed protein, which sharply delineate the anterior boundary of a parasegment. In contrast, cells at the posterior margins synthesize low and variable amounts of engrailed protein. **(c)** After germ band shortening (10 h), the cells at the posterior margin of each engrailed stripe form a sharp margin, marking the posterior boundary of a segment.

tide to spread more posteriorly than anteriorly, is still being investigated.

The segmental subdivision of the *Drosophila* larva carries over to the segmental subdivision of the adult (Fig. 21.28). The *imaginal discs*, which give rise to adult epidermis, originate after gastrulation when ectodermal cells segregate into those that will form larval epidermis and those that will form imaginal discs. Clonal analysis suggests that the earliest imaginal disc cells are already restricted in their fates to individual segments but not with regard to dorsal versus ventral organs, such as wings and legs (see Fig. 6.10). Although the imaginal discs form in register with segments, the older parasegmental boundaries can still be traced later as clonal restriction lines and boundaries of gene expression domains. For instance, the imaginal disc cells that express *engrailed*⁺ form the posterior compartment of a segment, as will be discussed later in this chapter.

In summary, the segmentation genes of *Drosophila* are arranged in a control hierarchy that proceeds from a set of aperiodic and simple expression domains to a periodic and detailed segmental pattern. In response to localized maternal gene products, the gap genes are activated in broad, overlapping domains. Various combinations of gap proteins generate individual stripes of primary pair-rule gene expression. Together, these stripes form regular patterns with bisegmental periodicity. The primary pair-rule genes control the secondary pair-rule genes, and together they regulate the segment polarity genes; these are expressed in narrow segmental stripes that initially delineate the anterior and posterior margins of parasegments. Midway through embryogenesis, the parasegmental boundaries are replaced with segmental boundaries that persist in the larva and adult. The segment polarity genes remain active in segments and imaginal discs, where they provide the basis for intrasegmental pattern formation.

Homeotic Genes

Segmentation genes may be sufficient to specify the body pattern of a centipede, in which nearly all segments are alike. In insects, however, most segments differ from one another in size, appendages, pigmentation, and other features. Such segment-specific traits are specified by *homeotic genes*. Like segmentation genes, the homeotic genes have two major functions. From the blastoderm stage on, they are involved in specifying the overall body pattern. Later, they also play specific roles in organizing the nervous system. We will focus here on their function in specifying the overall body pattern.

Homeotic mutants of *Drosophila* and other insects have long been known for their dramatic phenotypes. The hallmark of these mutants is the transformation of one body part into another that is normally formed elsewhere. Familiar examples of *Drosophila* homeotic mutants include *Ultrabithorax*, in which the metathorax is transformed into an extra mesothorax, thus producing a four-winged fly (see Fig. 1.21c), and *Antennapedia*, in which antennae are replaced with legs (see Color Plate 5). Such mutant phenotypes are caused by loss-of-function alleles, as in the case of *Ultrabithorax*, as well as by gain-of-function alleles, as in the case of *Antennapedia*. Thus, either loss or gain of homeotic gene activity may trigger a switch to another developmental pathway, in accord with the *principle of default programs* already discussed in Chapters 8, 15, and 16.

The first homeotic mutant was described by the German entomologist G. Kraatz (1876), who found a

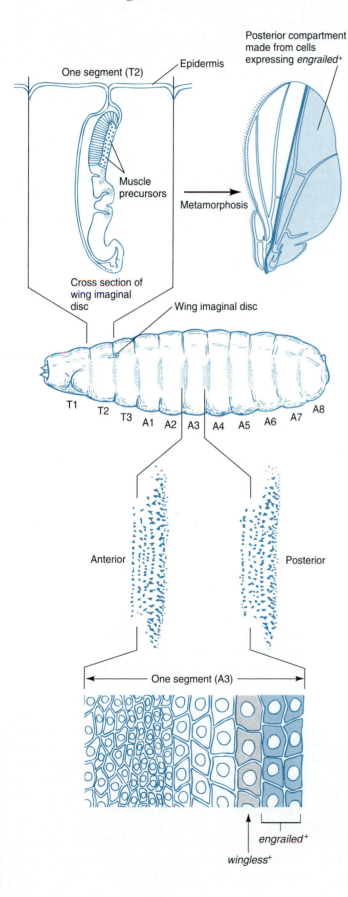

Figure 21.28 Development of epidermal structures and *engrailed*+ gene expression in *Drosophila*. At the center is a view of the ventral surface of the larva, anterior to the left. On the anterior portion of each abdominal segment are several rows of denticles with which the larva crawls. The cells expressing *wingless*+ (gray) and *engrailed*+ (color), which defined the parasegmental boundaries in the early embryo, are now in the posterior part of the segment; the posterior margin of the *engrailed*+ domain marks the posterior segment boundary. Attached to the larval epidermis are imaginal discs, epithelial bags that are everted during metamorphosis to form the adult epidermis. Only the disc forming the left wing is shown. (For other discs, see Figure 6.20.) The imaginal disc cells that express *engrailed*+ form the posterior compartment of the wing (color).

sawfly of the species *Cimbex axillaris,* in which the distal part of the left antenna was replaced by a foot (Fig. 21.29). The English zoologist William Bateson (1894) described Kraatz' discovery and many related observations in a book on morphological discontinuities and their role in the origin of species. He termed replacements of antennal parts with leg parts and similar transformations "homoeotic." Biologists at his time were very interested in **homologous organs,** that is, organs that have corresponding positions in different animals. For instance, the balancer organs (halteres) on the metathorax of a fly are homologous to the second pair of wings in other insects such as dragonflies or bees. Organs are said to be **serially homologous** if they occupy corresponding positions in different segments of the same animal. Thus, fly balancers are also serially homologous to fly wings, which occupy a corresponding position on the mesothorax. Since homeotic mutations cause transformations between homologous organs, changes in homeotic gene activities may have been key events in the evolution of insects and other arthropods (R. A. Raff and T. C. Kaufman, 1983).

The Homeotic Genes of *Drosophila* Are Clustered in Two Chromosomal Regions

The homeotic genes of *Drosophila* are clustered in two regions of the third chromosome (Fig. 21.30a). One region, the **Antennapedia complex,** contains four homeotic genes: *labial*+, *Deformed*+, *Sex combs reduced*+, and *Antennapedia*+. These genes specify the morphological characteristics of *anterior* parasegments up to parasegment 5. The *Antennapedia* complex also contains nonhomeotic patterning genes including *bicoid*+ and *fushi tarazu*+. The other chromosomal region, known as the **bithorax complex,** contains three homeotic genes: *Ultrabithorax*+, *abdominal A*+, and *Abdominal B*+. These genes specify the characteristics of *posterior* parasegments, beginning with parasegment 6 and extending into the abdomen.

Strikingly, the physical order of these genes on the chromosome is the same as the anteroposterior order of

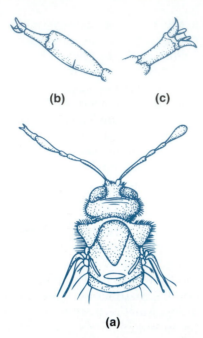

Figure 21.29 Homeotic transformation in a sawfly, *Cimbex axillaris.* **(a)** The anterior part of the fly is shown from a dorsal perspective. The distal portion of the left antenna developed as a foot. The right antenna is normal, ending in a club-shaped terminal joint. **(b)** The foot at the end of the left antenna seen from above. **(c)** The end of the left antenna seen from the front. The ectopic foot is smaller than a normal foot but is perfectly formed with claws and all other morphological characteristics.

their expression domains, especially with regard to the anterior boundary of each expression domain (Fig. 21.30b). The anterior region of each homeotic gene's expression domain usually gives rise to the body parts that are most strongly affected by loss-of-function alleles of the gene. The *labial*[+] gene, for example, is located at the 3′ end of the *Antennapedia* complex; the gene is expressed in the anterior head, and loss-of-function alleles affect the same area. In contrast, the *Antennapedia*[+] gene is located at the 5′ end of the *Antennapedia* complex, its expression domain begins in parasegment 4, and its loss-of-function allele most strongly affects the first two thoracic segments. The *Abdominal B*[+] gene is located farthest toward the 5′ end in the *bithorax* complex; the gene is expressed in parasegments 10 to 14, and loss-of-function alleles affect the same region. Thus, the more 3′ a gene is located on the chromosome, the farther anterior reaches its expression domain, and the more anterior are the body regions affected by loss-of-function alleles. This **colinearity rule** was first noted by Edward Lewis (1963, 1978); its significance is still uncertain, but it is observed in vertebrates and other animal phyla as well (see Chapter 22). (Lewis also proposed that all homeotic genes have arisen from a single ancestral gene by a process of duplication and modification. This hypothesis is borne out, at least in part, by the presence of a highly con-

served DNA sequence, the *homeobox*—discussed later in this chapter—in all of these genes.)

The molecular structure of homeotic genes suggests that they function in different contexts and that they are regulated by multiple signals. For instance, the *Antennapedia*[+] gene has two promoters and two polyadenylation sites (Kaufman et al., 1990). Although at least four different mRNAs are produced, they all encode the same protein. However, the two promoters and different *leader* and *trailer* regions of the mRNAs confer different regulatory properties. Another striking feature of the *Antennapedia*[+] gene is the enormous size of some of its introns. At 57 kb, the largest intron is more than 10 times longer than all exons together. It is estimated that the large introns add about an hour of lag time between the beginning and the end of the transcription of each pre-mRNA molecule, a feature that may be useful in regulation of the gene (Shermoen and O'Farrell, 1991). Similarly, the *Ultrabithorax*[+] gene is more than 300 kb long and produces several proteins by alternative splicing. Some of them are especially abundant during early embryogenesis, while others seem to function mostly in the organization of the central nervous system (Mann and Hogness, 1990).

Homeotic Genes Are Expressed Regionally and Specify Segmental Characteristics

The segment-specific defects observed in homeotic mutants suggest that homeotic genes have regional patterns of expression. For instance, a null allele of *Antennapedia* transforms the second leg into an antenna and causes abnormalities in the other parts of the thorax, while leaving the head and abdomen unaffected (Struhl, 1981). These observations show that the Antennapedia protein is necessary for normal thorax development. As expected, the gene is expressed especially in the thorax (Bermingham et al., 1990; see Fig. 15.21). Transcripts from both promoters are first detected during the early blastoderm stage in a ring of cells that will give rise to parasegments 4 through 6 (Fig. 21.31). At the extended germ band stage, transcripts are found in the epidermis and mesoderm of the thorax, mostly in parasegments 4 and 5, and in the central nervous system throughout the thorax and abdomen.

The *Antennapedia* alleles that gave the gene its name have the reverse effect of the null allele: they are gain-of-function alleles that transform antenna into leg (see Color Plate 5). Such alleles are dominant over the wild type and are associated with chromosomal inversions. These inversions bring the transcribed region of *Antennapedia* under the control of promoters of other genes, causing inappropriate synthesis of Antennapedia protein in the head, where the gene is normally silent. This was demonstrated by Stephan Schneuwly and his

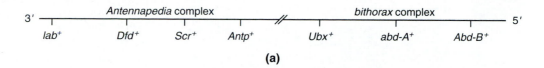

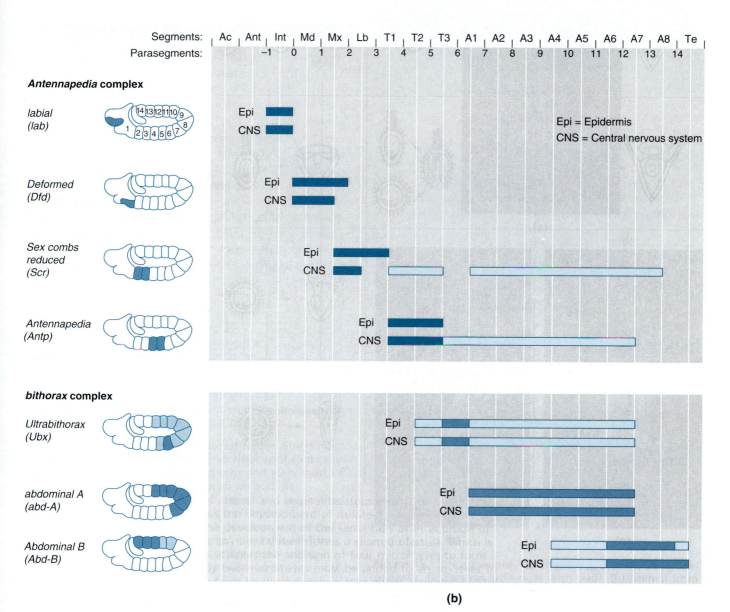

Figure 21.30 Chromosomal location and expression domains of homeotic genes in *Drosophila*. **(a)** The *Antennapedia* complex contains the *labial*[+], *Deformed*[+], *Sex combs reduced*[+], and *Antennapedia*[+] genes. The *bithorax* complex contains the *Ultrabithorax*[+], *abdominal A*[+], and *Abdominal B*[+] genes. **(b)** Domains of homeotic gene expression (mRNA and protein synthesis) in the embryonic epidermis and central nervous system. The darker color indicates the areas of strongest gene expression. Segments are labeled as in Figure 21.2, except that the procephalon is shown as an antennal (Ant) and an intercalary (Int) segment. In the diagrams to the left, epidermal expression is mapped relative to parasegmental boundaries. Note that the physical order of the genes from 3′ to 5′ on the chromosome correlates with the anterior boundary of the gene's expression domain in the embryo (colinearity rule).

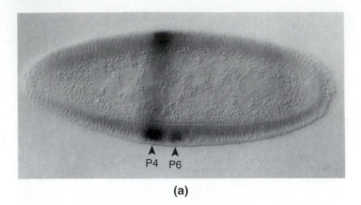

(a)

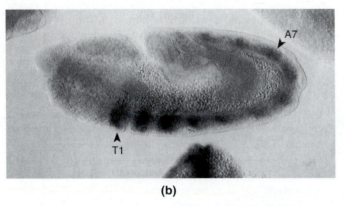

(b)

Figure 21.31 Transcription of the *Antennapedia*⁺ gene in the *Drosophila* embryo shown by in situ hybridization (see Methods 15.1). **(a)** Blastoderm stage. A suitable probe reveals transcription from promoter P₂ in the rudiments of parasegments 4 and 6. **(b)** Extended germ band stage: transcription from both promoters (P₁ and P₂) in the ectoderm and mesoderm of the thorax, and also in the nervous system of thorax and abdomen.

colleagues (1987), who spliced a heat shock promoter in front of an Antennapedia cDNA and introduced this construct into the *Drosophila* genome by P-element transformation. Heat activation of the transgene at certain larval stages caused the transformation of antennae into legs and the appearance of other thoracic structures in the head. This stunning redesign of the fly's body plan confirmed that the *Antennapedia*⁺ gene specifically promotes thorax development. Activation of this gene in the head region mimics a homeotic gene expression pattern that normally prevails in the thorax, and the morphological development changes accordingly.

Because of the staggered expression patterns of homeotic genes, more homeotic genes tend to be active in posterior parasegments than in anterior parasegments (Fig. 21.30b). This rule allows two general predictions about the phenotypes of homeotic mutants. First, a gain-of-function allele will mimic a homeotic gene activity pattern that normally prevails more posteriorly and will therefore transform structures into posterior serial homologues. The *Antennapedia* gain-of-function allele just discussed confirms this prediction. Second, a loss-of-function allele will generally mimic a homeotic gene activity pattern that normally prevails

more anteriorly and will therefore transform structures into serial homologues that normally occur farther anterior. We will examine this prediction as it applies to certain loss-of-function alleles of *Ultrabithorax*.

▼

The expression of *bithorax* complex genes begins in relatively simple patterns, but later on becomes quite complex. Ultrabithorax proteins are first synthesized at the blastoderm stage in parasegment 6 and, less strongly, in parasegments 8, 10, and 12 (Irvine et al., 1991). Later, these proteins form intricate mosaic patterns in parasegments 5 to 12 (Peifer et al., 1987). These expression domains are controlled by several enhancer regions located upstream and downstream of the promoters, and even in introns of the transcribed region (Fig. 21.32). Segments differ not only in the combination of active homeotic genes but also in the particular mosaic of cells expressing each gene.

Mutations that interfere with particular enhancers of *Ultrabithorax* change the mosaic of gene expression in one or more segments, often with striking effects on segmental morphology. For instance, mutant alleles inactivating two specific enhancers change the pattern of Ultrabithorax protein in the anterior metathorax (aT3) so that it resembles the distribution characteristic of the anterior mesothorax (aT2). In flies surviving to adulthood, aT3 is replaced with an extra copy of aT2. Mutations in another *Ultrabithorax* enhancer change the expression pattern in the posterior metathorax (pT3), so that it resembles the pattern characteristic of the posterior mesothorax (pT2). Correspondingly, pT2 develops in place of pT3. A combination of these mutations generates the "four-winged fly" phenotype mentioned previously (see Fig. 1.21).

The *Antennapedia* and *Ultrabithorax* mutants discussed here confirm that gain-of-function alleles of homeotic genes tend to mimic homeotic gene activity patterns that normally prevail more posteriorly and cause replacements of structures with posterior serial homologues. Conversely, most loss-of-function alleles mimic homeotic gene activity patterns that normally occur more anteriorly and cause replacements of structures with anterior serial homologues. There are homeotic genes in vertebrates that follow the same rules (see Chapter 22).

Homeotic Genes Are Controlled by Gap Genes, Pair-Rule Genes, the *Polycomb*⁺ Gene, and Interactions among Themselves

Transcription of homeotic genes begins just before the cellularization of the blastoderm, at about the same time as the pair-rule genes are being transcribed.

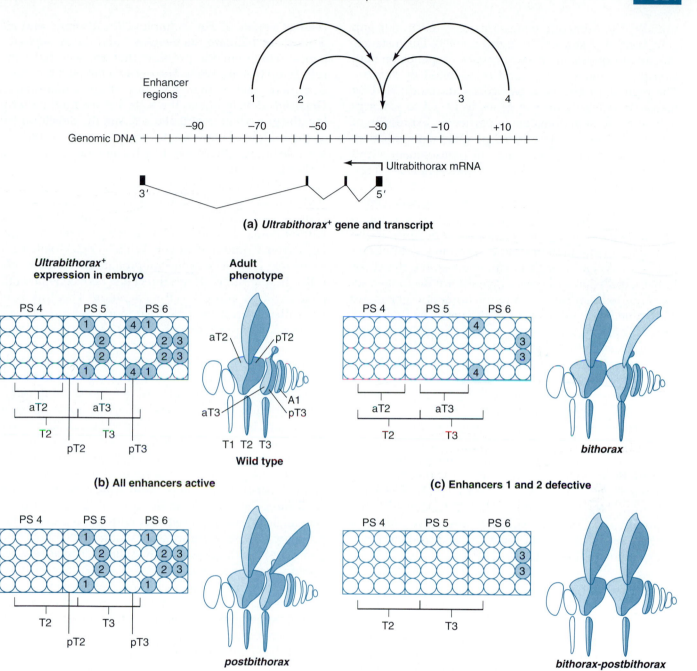

Figure 21.32 Gene expression patterns and phenotypes of mutants having loss of function in certain enhancers of the *Ultrabithorax*[+] gene. **(a)** Transcribed region of the gene and four enhancer regions (1 through 4). The numbers on the scale refer to kilobases (kb) of genomic DNA from an arbitrary reference point. The small blocks under the scale represent the exons; the angular brackets introns. The angular arrow indicates the beginning and direction of transcription; the curved arrows represent the actions of the enhancer regions. **(b–e)** The diagrams to the left represent simplified embryonic expression patterns of *Ultrabithorax*[+] in parasegments PS 4 to PS 6. Numerals in shaded areas indicate gene expression driven by the respective enhancers. Thoracic segments T2 and T3 are composed of anterior and posterior compartments (aT2, pT2, etc.), in keeping with their origin from two parasegments. The diagrams to the right represent homeotic transformations in the adult thoracic region. **(b)** Wild type. **(c)** Mutant alleles known as *bithorax* and *anterobithorax* show loss of function in enhancers 1 and 2. As a result, the aT3 compartment has the same lack of *Ultrabithorax*[+] expression and the same adult morphology as aT2. The loss of enhancer function also affects the abdominal segments, but the changes there are morphologically less dramatic. **(d)** Mutant alleles known as *postbithorax* have loss of function in enhancer 4. As a result, the pT3 compartment lacks *Ultrabithorax*[+] expression as does pT2 and has the same adult morphology. **(e)** Triple mutants (*anterobithorax, bithorax,* and *postbithorax*) are lacking function in enhancers 1, 2, and 4. As a result, the entire segment T3 has the same lack of *Ultrabithorax*[+] expression, and the same morphology, as T2.

Initially, the homeotic genes are activated under gap gene control (Akam, 1987; Ingham, 1988). For instance, the initial expression of *Antennapedia*⁺ depends on both *hunchback*⁺ and *Krüppel*⁺ and is inhibited by *knirps*⁺. The regulatory effects of gap genes are modulated by interactions with pair-rule genes to generate parasegmental expression domains. In particular, expression of *fushi tarazu*⁺ in even-numbered parasegments enhances the activation of *Antennapedia*⁺ in parasegment 4 and the activation of *Ultrabithorax*⁺ in parasegment 6 (Figs. 21.30b and 21.31a).

The homeotic genes, like the segment polarity genes, remain active throughout embryonic, larval, and pupal development. Both types of genes seem to be indispensable for maintaining the determined state of certain cells. When the segmentation genes that direct the initial expression of homeotic genes are no longer active, the expression domains of homeotic genes are maintained by a group of genes known as the *Polycomb*⁺ genes. The *Polycomb*⁺ genes produce negative regulators of transcription that bind to response elements associated with the homeotic genes (Lonie et al., 1994). These factors seem to provide a stable memory of the homeotic gene controls originally set by gap and pair-rule proteins (J. Simon et al., 1992, 1993).

In addition, homeotic genes maintain their expression domains by interactions among themselves. For instance, the Ultrabithorax proteins stimulate transcription of their own gene and inhibit transcription of *Antennapedia*⁺ (see Figs. 15.22 and 15.23). Similarly, abdominal A protein inhibits the expression of *Ultrabithorax*⁺. The *Deformed*⁺ gene remains active through a positive feedback loop in which its own protein binds to its enhancer (Regulski et al., 1991).

Homeotic Genes Control Different Types of Target Genes

The protein encoded by a single homeotic gene cannot by itself generate the size, shape, pigments, bristles, and all the other features that characterize a specific segment. It is evident that homeotic genes must control the expression of many other genes, called *realizator genes*. In accord with this notion, the proteins encoded by homeotic genes are transcription factors and bind to specific DNA sequences of their target genes. In some cases, the target genes also encode transcription factors, so that the homeotic genes are at the top of a regulatory cascade, several steps removed from the final realizator genes. For instance, the homeotic gene *Abdominal B*⁺ regulates the *empty spiracles*⁺ gene, and the empty spiracles protein in turn is a transcription factor regulating genes required for the formation of the filzkörper, a structural specialization of the eighth abdominal segment (B. Jones and W. McGinnis, 1993).

Other target genes of homeotic genes do not encode transcription factors but encode proteins for inter-

cellular signaling. For example, *Ultrabithorax*⁺ and *abdominal A*⁺ regulate the *wingless*⁺ and *decapentaplegic*⁺ genes, which encode proteins that are secreted from cells (Hursh et al., 1993). The decapentaplegic protein is a growth factor that has numerous functions in *Drosophila* development. It affects dorsoventral patterning, the growth of imaginal discs, and the development of the midgut. Likewise, the wingless protein plays a key role in the patterning of each segment, as discussed earlier in this chapter.

Finally, homeotic genes regulate genes for proteins that are embedded in the plasma membrane and control cell adhesion. Using an antibody against Ultrabithorax protein to identify target genes it regulates, Alex Gould and Robert White (1992) isolated the *connectin*⁺ gene. The connectin protein can mediate cell-cell adhesion, suggesting a direct link between homeotic gene function and cell-cell recognition. This link is of particular interest because it supports the *compartment hypothesis* discussed in a subsequent section of this chapter.

In summary, the homeotic genes of *Drosophila* are clustered in two chromosomal regions and have unusually large regulatory regions governing complex expression patterns. The overall effect of homeotic gene activities is to confer unique morphological characteristics on each body region along the anteroposterior axis. In homeotic mutants, certain body parts are replaced by homologous parts that are normally formed elsewhere. Gain-of-function alleles usually cause replacements with more posterior homologues, while loss-of-function alleles cause replacements with more anterior homologues. These transformations indicate that homeotic genes are switch genes used by cells to choose between alternative developmental pathways. In accord with this function, the proteins encoded by homeotic genes are transcription factors. They exert their effects by regulating target genes that encode other transcription factors, secreted proteins involved in intercellular signaling, and cell adhesion molecules.

The Dorsoventral Body Pattern

Like the anteroposterior pattern, the dorsoventral pattern in *Drosophila* develops through the activity of maternal effect genes that establish a morphogen gradient. The morphogen is a transcription factor encoded by the *dorsal*⁺ gene. In embryos derived from mothers with null alleles of *dorsal*, all ventral and lateral structures, including the mesoderm and central nervous system, are missing. Instead, the dorsalmost epidermal structure, a fleece of fine hair, develops around the entire dorsoventral circumference of these embryos (Fig. 21.33). As in the case of the anteroposterior pattern, the ventral and lateral pattern elements are missing not be-

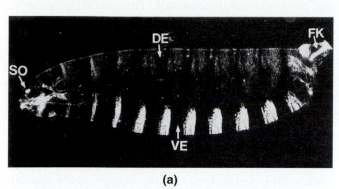

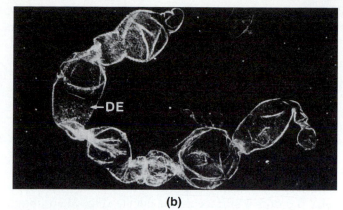

(a)

(b)

Figure 21.33 The *dorsal* phenotype. The photomicrographs show cuticle preparations under dark-field illumination. **(a)** Wild-type larva. Anterior end, marked by sense organ (SO), to the left; posterior end marked by Filzkörper (FK). Ventral side down. **(b)** Larva derived from a female homozygous for a *dorsal* null allele. The cuticle is a tube without dorsoventral polarity, having none of the ventral denticle belts that are prominent on the ventral epidermis (VE) of the wild type. The fine hair normally limited to the dorsal epidermis (DE) is seen around the circumference of the mutant larva.

cause their precursor cells died, but because blastoderm cells are determined abnormally. It follows that the normal function of the *dorsal*+ gene is required to specify the *ventral and lateral* elements of the body pattern. (As explained in Chapter 14, the names first given to genes usually epitomize a loss-of-function phenotype. Of course, these names are then contrary to the wild-type functions of the genes.)

The dorsal protein forms a morphogen gradient with highest concentration in the ventral nuclei of the blastoderm embryo (Fig. 21.34a). This gradient fails to form properly in *dorsalized* embryos, in which the dorsal protein is not synthesized, and in *ventralized* embryos, where synthesis or deployment is misregulated so that dorsal protein accumulates at high concentration in nuclei around the entire embryo (Fig. 21.34b and c). As a

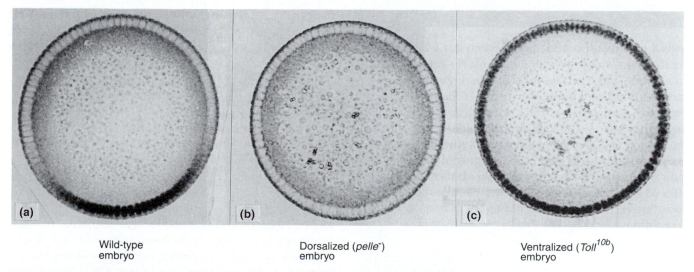

(a)

(b)

(c)

Wild-type
embryo

Dorsalized (*pelle*−)
embryo

Ventralized (*Toll*[10b])
embryo

Figure 21.34 Distribution of the dorsal protein in normal and mutant *Drosophila* embryos at the blastoderm stage. The photomicrographs show cross sections of embryos oriented with the ventral side down. The dorsal protein is detected by immunostaining (see Methods 4.1). **(a)** Wild-type embryo. There is a gradient in the nuclear concentration of dorsal protein with a maximum toward the ventral midline. **(b)** Dorsalized embryo. One of the genes (*pelle*) required for the synthesis of dorsal protein is deficient, so that the dorsal state of the wild type is expanded around the embryo. **(c)** Ventralized embryo. The *Toll* gene, required for the proper formation of the dorsal protein gradient, is active independently of its normal spatial clues, so that dorsal protein accumulates around the entire circumference of the embryo.

transcription factor, the dorsal protein activates or inhibits certain embryonic genes in dorsoventral order (K. V. Anderson, 1987; Nüsslein-Volhard, 1991; Govind and Steward, 1991). The earliest expression domains of these genes at the blastoderm stage correlate approximately with certain cell fates during later development.

The maternal gene products that eventually establish the gradient of dorsal protein interact in a complicated signal chain involving about 20 proteins and weaving from the oocyte to the follicle cells and back to the embryo (Fig. 21.35). The first sign of dorsoventral polarity is the eccentric position of the oocyte nucleus near the dorsal surface of the oocyte (Fig. 21.36). Eggs laid by females homozygous for loss-of-function alleles of *gurken* or *torpedo* are ventralized and have reduced dorsal structures (Schüpbach, 1987; Neuman-Silberberg and Schüpbach, 1994). These mutants are thought to be defective in transmitting a *dorsalizing signal* that originates in the oocyte nucleus and is received by the follicle cells. The follicle cells on the dorsal surface of the

oocyte would then receive a stronger signal because of the proximity of the nucleus. The gurken protein belongs to the *transforming growth factor* α family, proteins that bind as ligands to matching receptor proteins on the surfaces of other cells (Neuman-Silberberg and Schüpbach, 1993). The torpedo protein shows the molecular features of a growth factor receptor protein (Price et al., 1989). Most likely, the gurken protein is the dorsalizing signal and the torpedo protein is its receptor.

The activation of the torpedo receptor by its ligand has two major effects on the dorsal follicle cells. First, these cells generate the dorsal structures of the chorion, including the respiratory appendages. Second, the activity of the torpedo protein interferes with the expression of at least one of a group of genes including *nudel*+, *pipe*+, *windbeutel*+, and probably *gastrulation defective*+ (Fig. 21.36). Consequently, the combined activity of these genes, called *ventral polarizing activity,* is limited to ventral follicle cells (Stein et al., 1991; Stein and Nüsslein-Volhard, 1992).

The ventral polarizing activity provides the orienting clue later during embryonic cleavage, when precursor proteins for two proteases (encoded by the *snake*+ and *easter*+ genes) and for a signal peptide (encoded by the *spätzle*+ gene) are released into the *perivitelline space* (Fig. 21.37a and b). Under the influence of the ventral polarizing activity, the snake and easter proteases become locally active and cleave an active signal peptide from the spätzle precursor (C. L. Smith and R. DeLotto, 1994). The active spätzle signal peptide in turn binds as a ligand to a receptor protein encoded by the *Toll*+ gene (D. S. Schneider et al., 1991; Roth, 1993; Morisato and Anderson, 1994). Although the Toll protein is present around the entire egg plasma membrane, it is activated mostly in the ventral area by the bound spätzle peptide. (This activation mechanism is similar to the local activation of the torso receptor in the egg plasma membrane by the torsolike product of the follicle cells adjacent to the egg poles. See Figure 21.12.) *Toll*+ activity then occurs in a gradient around the egg periphery with a maximum at the ventral midline.

The *Toll*+ activity, through signal-transmitting proteins in the egg cytoplasm, determines the distribution of the maternally encoded dorsal protein (Figs. 21.34a and c and 21.37c). The dorsal protein is initially retained in the egg cytoplasm by its association with the protein encoded by *cactus*+, another maternal effect gene expressed in the early embryo (Kidd, 1992). The cactus protein binds to a domain of the dorsal protein that is necessary for transport into the nucleus (Whalen and Steward, 1993). The *Toll*+ activity in the ventral region of the embryo, along with products of *pelle*+ and other maternal effect genes, releases the dorsal protein from the cactus protein. The dorsal protein then moves into nuclei, directed by its own nuclear localization sequence

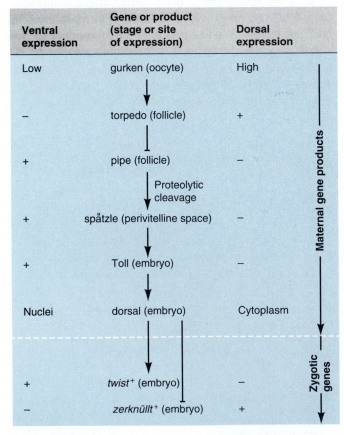

Ventral expression	Gene or product (stage or site of expression)	Dorsal expression	
Low	gurken (oocyte)	High	Maternal gene products
−	torpedo (follicle)	+	
+	pipe (follicle)	−	
	Proteolytic cleavage		
+	spätzle (perivitelline space)	−	
+	Toll (embryo)	−	
Nuclei	dorsal (embryo)	Cytoplasm	
+	*twist*+ (embryo)	−	Zygotic genes
−	*zerknüllt*+ (embryo)	+	

Figure 21.35 Cascade of maternal gene products and zygotic genes that establish dorsoventral polarity in the *Drosophila* embryo. The central column lists some of the key gene products and genes and indicates where they are expressed. The left and right columns indicate the level of expression of each gene in the ventral and dorsal areas.

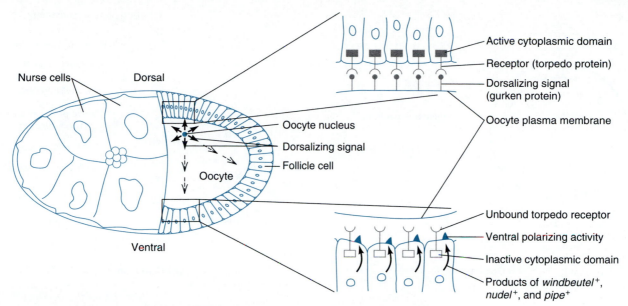

Figure 21.36 Origin of the dorsoventral polarity from the eccentric position of the oocyte nucleus in *Drosophila*. The diagram to the left shows a median section of an egg chamber; anterior to the left, dorsal side up. The *gurken*⁺ gene and perhaps others expressed in the oocyte nucleus generate a dorsalizing signal that is stronger toward the dorsal surface. The insets to the right show the dorsalizing signal as a ligand extending dorsally, but not ventrally, into the space between the oocyte and follicle cells. The ligand binds to a receptor protein encoded by the *torpedo*⁺ gene in the follicle cells. In the ventral follicle cells, where the torpedo protein remains inactive, the products of several genes, including *windbeutel*⁺, *nudel*⁺, *pipe*⁺, and presumably *gastrulation defective*⁺, assemble to generate a ventral polarizing activity. In the dorsal follicle cells, the activity of the torpedo protein prevents the generation of the ventral polarizing activity.

(Govind and Steward, 1991). Thus, the uneven distribution of *Toll*⁺ activity finally creates a gradient in the nuclear concentration of dorsal protein, with a maximum in the ventral nuclei.

The nuclear dorsal protein serves as a morphogen gradient, establishing a coarse dorsoventral pattern in the expression of several zygotic genes including *zerknüllt*⁺, *decapentaplegic*⁺, *twist*⁺, and *snail*⁺ (Fig. 21.37d). In wild-type embryos, *zerknüllt*⁺ is initially expressed in the dorsal half of the embryo but later restricted to a dorsal band five to six cells wide. Expression of *zerknüllt*⁺ is inhibited even at low levels of dorsal protein, which binds with high affinity to the regulatory region of the *zerknüllt*⁺ gene (Ip et al., 1991). Similarly, the dorsal protein restricts expression of *decapentaplegic*⁺ to the dorsal 40% of the embryo's circumference (Fig. 21.38a). Conversely, dorsal protein activates the expression of *twist*⁺ ventrally (Thisse et al., 1991). However, the binding sites for dorsal protein in the *twist*⁺ promoter have only low affinity, so that sufficient binding occurs only at high levels of dorsal protein in the ventral 20% of the egg circumference (Fig. 21.38b). Both twist and dorsal proteins cooperate in activating the *snail*⁺ gene in an expression domain that almost coincides with the *twist*⁺ domain except that it

has sharper edges: cells with low amounts of dorsal or twist protein produce no snail protein (Ip et al., 1992b).

The transcription domains of these genes at the blastoderm stage correspond approximately to five major areas on the fate map (Figs. 21.37 and 21.39). The ventralmost cells, transcribing *twist*⁺ and *snail*⁺, invaginate and form the embryonic endoderm and mesoderm. The small bands of cells expressing low levels of twist but no snail protein are known as *mesectodermal cells*; they give rise to the neurons positioned along the ventral midline. The adjacent ventrolateral cells, expressing none of the zygotic genes discussed here, respond to a combination of low levels of dorsal protein, the presence of other transcription factors, and the absence of snail protein (Ip et al., 1992a; Jiang and Levine, 1993). These cells develop as the neuroectoderm, which forms the remainder of the central nervous system and ventral epidermis. The dorsolateral cells express *decapentaplegic*⁺, which encodes a secreted protein that accumulates as an extracellular morphogen gradient (Ferguson and Anderson, 1992; K. A. Wharton et al., 1993). Low levels of decapentaplegic protein promote dorsal epidermis development; high levels of the same protein, along with expression of *zerknüllt*⁺, direct the formation of the *amnioserosa*, an extraembryonic membrane.

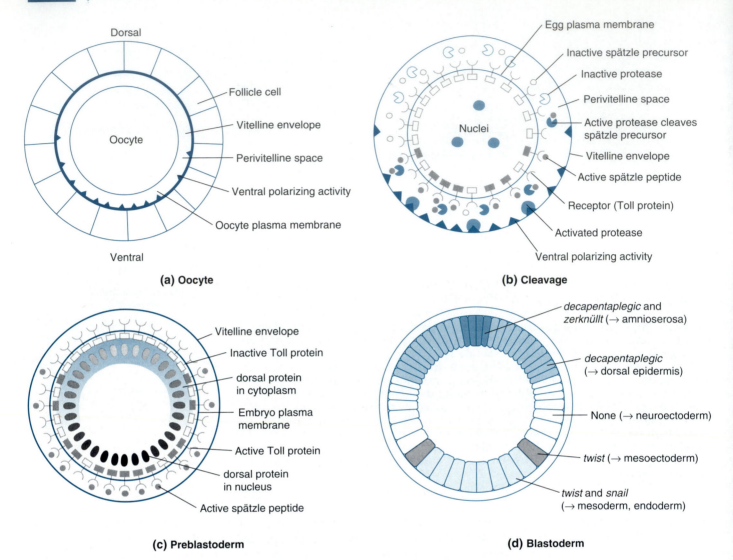

(a) Oocyte

(b) Cleavage

(c) Preblastoderm

(d) Blastoderm

Figure 21.37 Development of the dorsoventral polarity in the *Drosophila* embryo. The schematic cross sections are oriented with the ventral side down. **(a)** Follicle cells synthesize the vitelline envelope surrounding the oocyte. The polarizing activity generated by the ventral follicle cells is present ventrally, but not dorsally, on the inside of the vitelline envelope (see Fig. 21.36). **(b)** After fertilization, precursor proteins for two proteases encoded by *snake*[+] and *easter*[+], and for a signal peptide encoded by *spätzle*[+], are all released into the perivitelline space between the vitelline envelope and the egg plasma membrane. The ventral polarizing activity activates the proteases, which then cleave an active spätzle peptide from its precursor. The spätzle peptide binds as a ligand to a receptor protein on the egg plasma membrane. The receptor protein, which is encoded by the *Toll*[+] gene, is deployed around the entire plasma membrane of the embryo. However, because active spätzle ligand is produced mostly ventrally, Toll receptor is activated ventrally, to a lesser extent laterally, but not dorsally. **(c)** Dependent on the graded Toll activity, the maternally encoded dorsal protein is transported from the cytoplasm into the nuclei of the preblastoderm embryo. The Toll-dependent transport results in a gradient of nuclear dorsal protein with a maximum concentration ventrally. **(d)** Acting as a transcription factor, the dorsal protein inhibits the zygotic *zerknüllt*[+] and *decapentaplegic*[+] genes, limiting their expression to dorsal and dorsolateral cells. In ventral nuclei, the high dorsal protein concentration activates the zygotic *twist*[+] and *snail*[+] genes. The expression domains of the zygotic genes coincide approximately with the major fates of these cells to become amnioserosa, dorsal epidermis, neuroectoderm, mesectoderm, and mesoderm or endoderm.

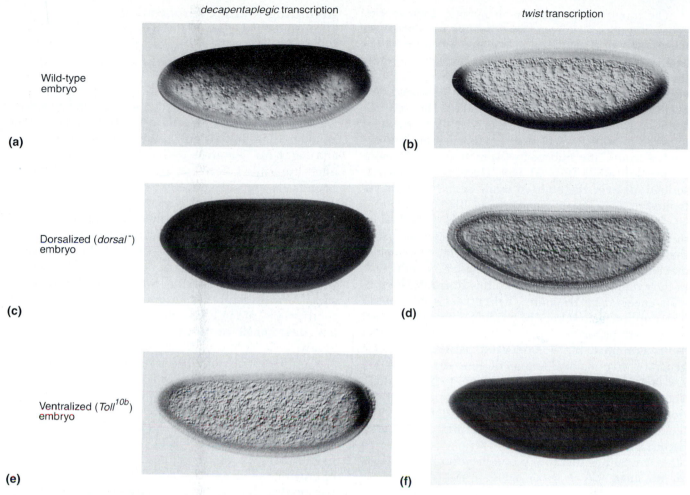

decapentaplegic transcription

twist transcription

Wild-type embryo

(a) **(b)**

Dorsalized (*dorsal⁻*) embryo

(c) **(d)**

Ventralized (*Toll¹⁰ᵇ*) embryo

(e) **(f)**

Figure 21.38 Expression patterns of the zygotic *decapentaplegic⁺* and *twist⁺* genes revealed by in situ hybridization (dark stain; see Methods 15.1); dorsal is up and anterior pole to left. In wild-type embryos **(a, b)**, *decapentaplegic⁺* is transcribed in the dorsal 40% of the embryo's circumference and *twist⁺* in the ventral 20%. These expression patterns are changed dramatically in embryos dorsalized by a maternal *dorsal⁻* allele **(c, d)**, and in embryos ventralized by a maternal *Toll¹⁰ᵇ* gain-of-function allele **(e, f)**.

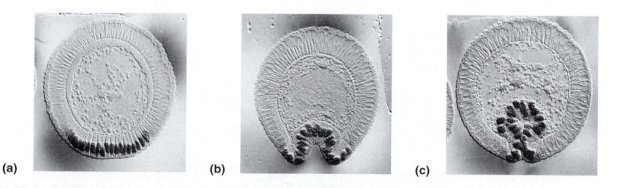

(a) **(b)** **(c)**

Figure 21.39 Expression of the *twist⁺* gene in the *Drosophila* embryo. The photomicrographs show transverse sections, ventral side down, immunostained with an antibody to twist protein (see Methods 4.1). The twist protein, a transcription factor, accumulates in the nuclei of the producing cells. The nuclei of unstained cells appear as depressed ovals. **(a)** Beginning of gastrulation. The cells on the ventral side form a plate. Note the decrease of twist protein concentration toward the sides. **(b)** Invagination of the ventral plate forms the embryonic mesoderm. **(c)** Mesoderm invagination almost complete. The outermost cells stained for twist protein are mesectodermal cells, which will give rise to neurons.

The transcription patterns of zygotic dorsoventral genes change as expected in dorsalized embryos, in which the low dorsal level of nuclear dorsal protein is extended around the embryo by a maternal loss-of-function allele (Fig. 21.34b). In such embryos, *twist*⁺ is not expressed at all, while *decapentaplegic*⁺ is transcribed over the entire embryo (Fig. 21.38c and d). The converse changes are observed in ventralized embryos, where the high ventral level of nuclear dorsal protein is extended around the embryo by the maternal *Toll*^{10b} gain-of-function allele (Fig. 21.34c). Here, *twist*⁺ is transcribed all over the embryo while *decapentaplegic*⁺ is limited to the terminal regions (Figs. 21.38e and f).

The embryonic genes that are directly controlled by the dorsal morphogen in turn control realizator genes that regulate cell division patterns, morphogenetic movements, and the eventual differentiation of the cells in their domains. In accord with this function, the proteins encoded by *zerknüllt*⁺, *twist*⁺, and *snail*⁺ are all transcription factors, and the product of *decapentaplegic*⁺ is a growth factor.

In summary, the dorsoventral patterning system generates five major domains of gene expression and cell fates on both sides of the embryo (right and left). These domains are superimposed on the anteroposterior pattern of different parasegments determined by homeotic genes. The anteroposterior and dorsoventral patterns are specified almost independently of each other; mutations disrupting one have no major effects on the other.

The Compartment Hypothesis

The dorsoventral and anteroposterior patterning genes subdivide the *Drosophila* blastoderm into about 100 patches, each filled with a few dozen to several hundred cells. Each patch is characterized by a specific combination of active patterning genes, including segmentation genes such as *engrailed*⁺, homeotic genes such as *Antennapedia*⁺, and zygotic dorsoventral genes such as *twist*⁺. How does this patchwork turn into an embryo, and later on into an adult? Does each patch develop independently, or do cells from different patches mix with one another? The relationship between patches of embryonic blastoderm cells and areas of adult tissue is complicated by the process of imaginal disc formation. Only subpopulations of embryonic cells contribute to imaginal discs, which in turn form most of the adult epidermis (Fig. 21.28; see also Chapter 6). Nevertheless, the adult cells that stem from the same embryonic patch stay together, and their fate is related to the embryonic patterning genes that remain active throughout development.

According to the Compartment Hypothesis, Areas of Prospective Epidermis Are Characterized by the Activity of Specific Selector Genes

In the context of fate mapping and cell determination, we saw that labeled clones do not transgress certain boundary lines in the adult epidermis (see Methods 6.1 and Fig. 6.6). The areas defined by clonal restriction lines are called *compartments*. Some compartment boundaries are anatomically prominent, such as segmental boundaries or the boundary between the upper and lower faces of a wing. Other compartment boundaries are anatomically inconspicuous. For instance, each body segment of *Drosophila* is subdivided into an anterior and a posterior compartment (Fig. 21.2g). In the wing, this anteroposterior compartment boundary strikes across featureless terrain between the third and fourth veins, along a line that does not seem to have any anatomical or physiological significance (Fig. 21.40a).

Compartments may become further subdivided in the course of development: a boundary that is crossed by clones originating at an early stage may be respected by clones originating later (Crick and Lawrence, 1975). Each time, though, a compartment is filled by several clones, not just one. Compartments in insect epidermis have striking properties. First, they are the basic units of growth and shape regulation. If one clone in a compartment is allowed by genetic trickery to grow larger than usual, the remaining clones of the same compartment grow correspondingly smaller, and the compartment as a whole maintains its size and shape (see Fig. 6.6b). Second, adjacent compartments differ in their cell surface properties, and cells from different compartments do not intermingle. The result is that compartment boundaries are smooth, like the interface between oil and water.

Although compartments are originally defined by clonal restriction lines, Antonio Garcia-Bellido (1975) has linked compartments to patterning genes by the *compartment hypothesis*. He proposed that each compartment is characterized by a specific combination of active patterning genes, which he called *selector genes*. The selector genes are thought to control many *realizator genes*, which confer the different surface properties and other morphogenetic features that give each compartment its unique character. According to this hypothesis, selector genes have the following properties. First, the expression domains of selector genes coincide with certain compartments. Second, mutations in selector genes cause major changes in the morphology of compartments. Third, selector genes—directly or indirectly—affect cell surfaces. Fourth, the products of selector genes are gene-regulatory factors.

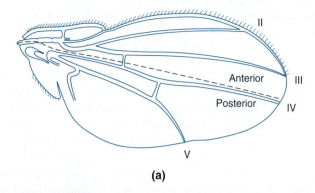

(a)

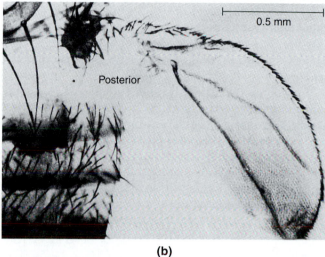

(b)

Figure 21.40 Coincidence of a compartment boundary, as defined by clonal restriction, with a functional domain of the *Ultrabithorax*⁺ gene. **(a)** Schematic drawing of a *Drosophila* wing. Roman numerals indicate major wing veins. The compartment boundary (dashed line) between veins III and IV is a line of clonal restriction between the anterior and posterior compartments. **(b)** Composite appendage of an adult fly mutant in an enhancer of the *Ultrabithorax*⁺ gene. (Expression of this allele, a *bithorax* allele, was illustrated in Figure 21.32c.) The photomicrograph shows the anterior half of a wing attached to the posterior half of a haltere. Note that the posterior margin of the wing half coincides with the compartment boundary shown in the drawing.

Ultrabithorax⁺, *engrailed*⁺, and *apterous*⁺ Fulfill the Criteria of Selector Genes

Among the patterning genes that have the characteristics of selector genes is the homeotic gene *Ultrabithorax*⁺, which is required for generating the morphology of the metathorax instead of the mesothorax morphology. Flies carrying the *bithorax*³ mutant allele, which interferes with an enhancer of *Ultrabithorax*⁺, have an anterior mesothorax in place of the anterior metathorax (Fig. 21.32). The borderline of this transformation follows the anteroposterior compartment boundary very closely

(Fig. 21.40b). Thus, the function of the *Ultrabithorax*⁺ gene that is driven by the mutated enhancer is necessary in the anterior compartment of the metathorax but not in the posterior compartment. Loss of this function causes a major morphological transformation. It also changes the surface properties of cells as assayed by cell-mixing experiments (Garcia-Bellido and Lewis, 1976) and by the direct regulation of the *connectin*⁺ gene by Ultrabithorax protein as discussed previously. Finally, the Ultrabithorax protein is a transcription factor that, directly or indirectly, regulates the activity of a large number of realizator genes. Similar experiments indicate that other homeotic genes function as selector genes as well.

Another patterning gene with selector gene properties is the segment polarity gene *engrailed*⁺. Flies homozygous for the mutant allele *engrailed*¹ have wings with posterior margins that look like anterior wing margins. Other *engrailed* alleles are lethal in homozygous individuals, but heterozygous flies with homozygous mutant clones are viable. When such clones are generated in the anterior compartment of any segment, development is completely normal. However, similar clones in the posterior compartments cause severe morphological distortions (Kornberg, 1981). These results indicate that the *engrailed*⁺ function is required for normal morphogenesis of the posterior—but not the anterior—compartment of each segment.

Using *clonal analysis*, Gines Morata and Peter Lawrence (1975) provided evidence that the anteroposterior compartment boundary within each segment is maintained by *engrailed*⁺ expression on one side of the boundary and nonexpression on the other side. By X-ray–induced somatic crossover (see Methods 6.1), the researchers generated cell clones marked with a simple cuticular marker (Fig. 21.41a–c). When generated in flies with at least one wild-type allele of *engrailed*⁺, such clones respected the anteroposterior compartment boundary in the wing. However, in flies homozygous for the *engrailed*¹ allele, marked clones straddled the boundary between the anterior and the posterior half of the wing. These results showed that the *engrailed*⁺ function is required for maintaining the compartment boundary.

Another experiment provided evidence that the compartment boundary is maintained when the *engrailed*⁺ gene is active in the posterior compartment and inactive in the anterior compartment. In this experiment, the labeled clones were homozygous for both the mutant *engrailed*¹ allele and a cuticular marker gene. Such clones crossed from the posterior into the anterior compartment but not in the opposite direction (Fig. 21.41d and e). On the basis of these results, the re-

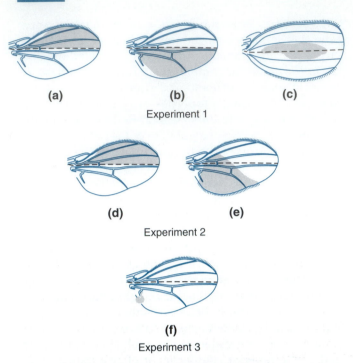

(a)   (b)   (c)

Experiment 1

(d)   (e)

Experiment 2

(f)

Experiment 3

Figure 21.41 Dependence of the anteroposterior compartment boundary in the *Drosophila* wing on the function of the *engrailed*⁺ gene. The diagrams summarize the results from three experiments. In experiment 1, the labeled clones (gray shading) were homozygous for a mutated cuticular marker gene, *multiple wing hair.* In *engrailed¹/+* flies, carrying one wild-type and one mutant allele of *engrailed,* such clones respected the anteroposterior compartment boundary (dashed lines), whether they were generated in the anterior compartment **(a)** or in the posterior compartment **(b).** When similar labeled clones were generated in homozygous *engrailed¹/engrailed¹* mutants, the clones crossed the boundary between the anterior half and the posterior half of the wing **(c).** In experiment 2, the labeled clones were homozygous for both *engrailed¹* and a mutated cuticular marker gene, *pawn.* (These two genes are located next to each other on the chromosome and are therefore rarely separated during somatic crossover.) In flies that were *pawn/+* and *engrailed¹/+* , labeled clones generated in the anterior compartment respected the compartment boundary and did not affect wing morphology **(d).** Similar clones in the posterior compartment crossed into the anterior compartment and locally transformed the posterior wing margin into an anterior margin with sensory bristles **(e).** In experiment 3, clones of cells with two different loss-of-function alleles (*engrailed¹/engrailed^{C2}*) were generated. These clones showed a more severe loss of function than *engrailed¹/engrailed¹* clones. If such clones originated in the anterior compartment, they respected the compartment boundary. When originating in the posterior compartment close to the boundary, the clones crossed into the anterior compartment as in part e. When originating deep in the posterior compartment, the mutant clone segregated as a vesicle from the wing blade **(f).**

searchers concluded that the *engrailed*⁺ gene is normally expressed in the posterior but not in the anterior compartment. They proposed further that the *engrailed¹* al-

lele causes a partial loss of function. (This assumption is supported by the *engrailed¹* phenotype, which shows *incomplete* transformation of a posterior wing half into an anterior one.) Consistent with these proposals, Morata and Lawrence argued that an *engrailed¹/engrailed¹* clone originating in the posterior compartment (which expresses *engrailed¹*) has cell surface properties halfway between anterior cells (which do not express any *engrailed* allele) and posterior cells (which express both alleles). Such a clone therefore mixes with *engrailed¹/+* cells in both compartments and crosses the boundary. In contrast, an *engrailed¹/engrailed¹* clone originating in the anterior compartment expresses zero *engrailed*⁺ activity because the gene is not active anteriorly. Its cells therefore do not mix with posterior cells expressing the *engrailed*⁺ allele and do not cross the boundary.

Subsequent work with other *engrailed* alleles causing a more severe loss of function confirmed the critical effect of *engrailed*⁺ on cell affinity. When clones with two different loss-of-function alleles (*engrailed¹/engrailed^{C2}*) originate near the compartment border, they cross into the anterior compartment, as seen in Figure 21.41e. However, when such a clone originates away from the compartment border, the margin of the clone constricts, and the narrowing ring cleanly pinches out the clone so that it separates from the wing blade as a vesicle (Fig. 21.41f).

The experiments just described indicate that the *engrailed*⁺ gene affects the surface properties of the cells expressing it. When two groups of cells, one expressing *engrailed*⁺ and the other lacking it, contact each other, they remain separate and form a smooth boundary. Because of this property, the morphological distortions caused by the gene's loss-of-function alleles, and the gene's transcription factor characteristics, the *engrailed*⁺ gene qualifies as a selector gene.

The compartment hypothesis places stringent requirements on the expression of selector genes. If selector gene activities are to specify the character of each compartment including its cell surface properties, the expression domains of selector genes should follow compartment boundaries *exactly.* Because compartments are defined by clonal restriction lines established as early as the blastoderm stage, it follows that cell clones labeled at later stages should fall entirely inside or outside a selector gene domain. In other words, such clones should not straddle the margins of selector gene expression domains. This requirement seems at odds with the observation that homeotic genes as well as *engrailed*⁺ are still subject to regulation by other patterning genes past the blastoderm stage, as described in earlier sections of this chapter.

To resolve the apparent discrepancy, Jean-Paul Vincent and Patrick O'Farrell (1992) labeled single cells at the blastoderm stage and later evaluated the resulting clones with regard to the expression domains of *engrailed+*. Most of the labeled clones were entirely inside an *engrailed+* stripe or entirely outside. In particular, no clones straddled the anterior margins of *engrailed+* stripes. However, some clones straddled the posterior boundaries of *engrailed+* expression domains at the extended germ band stage. Such straddling clones became rare later during the shortened germ band stage. The investigators concluded that the straddling clones were in a phase of transition, during which entire clones lost the expression of *engrailed+*. These clones presumably arose from cells that received insufficient amounts of the wingless product from across the compartment boundary to stabilize *engrailed+* expression. This explanation reconciles the continued regulation of *engrailed+* with the clonal inheritance of *engrailed+* expression as required by the compartment hypothesis.

In a related study, Seth Blair (1992) used X-ray–induced somatic crossover (see Methods 6.1) to map the border between the anterior and posterior compartments in wing *imaginal discs* of larvae and pupae. Comparison with the expression domain of *engrailed+* showed that engrailed protein was generally limited to the posterior compartment of the wing disc, as expected. In older larvae and pupae, however, low levels of en-grailed protein also extended into the anterior compartment. In contrast, other genes were expressed strictly in the anterior compartment. Blair's study also revealed a boundary of morphologically aligned cells in wing imaginal discs that coincided with the border between the anterior and posterior compartments (Fig. 21.42). While most of the cells in the disc epithelium are polygonal in shape, cells at the boundary are more rectangular and form an irregular but still distinctly smoother line of apposition with adjacent cells. This boundary of cell alignment is very suggestive of a difference in cell affinity, as postulated by the compartment hypothesis. Blair concluded that the late expression of *engrailed+* in the anterior compartment was probably too late to alter cell affinity or fate, or that any such effect was counteracted by genes expressed strictly in the anterior compartment.

The *apterous+* gene of *Drosophila* also has all the properties of a selector gene (Blair, 1993; Diaz-Benjumea and Cohen, 1993). Flies lacking *apterous+* activity fail to form wings and halteres, although the remainder of the thorax is almost normal. The *apterous+* gene is expressed dorsally, but not ventrally, in the imaginal discs for wings and halteres (see Color Plate 9). Indeed, the boundary of *apterous+* expression coincides exactly with the clonal restriction line that separates the dorsal from the ventral compartment of the wing imaginal disc. Cells that are deficient in *apterous* as a result of somatic crossover (see Methods 6.1) and are located in the dorsal wing compartment become ventral in character.

If such cells arise near the dorsoventral compartment boundary, they are included in the ventral compartment, apparently because they adhere more strongly to cells without apterous protein than to cells with apterous protein. If the *apterous−* cells originate deep within dorsal territory, they form a clone of cells that are ventral in morphology. Moreover, such clones may extend out of the wing surface like an extra winglet. It appears that wing tissue grows out from the thorax wherever cells expressing *apterous+* are juxtaposed with cells that do not express this gene. Finally, the apterous protein has the molecular characteristics of a transcription factor.

In summary, the *Ultrabithorax+*, *engrailed+*, and *apterous+* genes of *Drosophila* function as selector genes. These genes are thought to represent a small group of genes that act in concert to control batteries of realizator genes. Selector genes are linked to epidermal compartments, which are defined by clonal restriction, through the compartment hypothesis. According to this hypothesis, compartments differ from one another in their unique combinations of selector genes that control specific patterns of gene expression and morphogenesis while preventing cell mixing between compartments.

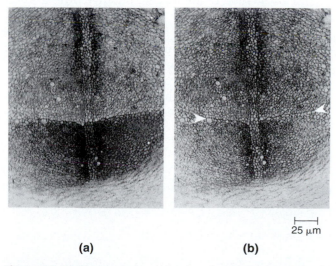

(a)   (b)

⊢——⊣
25 μm

Figure 21.42 Photomicrographs showing the boundary of morphologically aligned cells (arrowheads) on the surface of a *Drosophila* wing imaginal disc. **(a)** Microfilaments have been stained to make cell boundaries visible. **(b)** Same specimen also showing the expression of an *engrailed-lacZ* fusion gene in the posterior wing compartment (dark stain below the line of cell alignment). Note that the line of cell alignment coincides with the boundary of *engrailed-lacZ* expression. (See also Color Plate 8.)

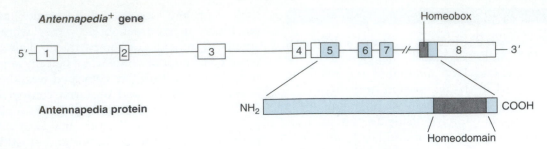

Figure 21.43 Structural organization of the *Antennapedia*⁺ gene and the Antennapedia protein. Exons A through H are separated by introns, which are not drawn to scale. The homeobox, located in exon 8, is 180 nucleotides long. Within the Antennapedia protein, the homeodomain is located near the carboxyl terminus (COOH). The homeodomain encompasses 60 amino acids in a sequence that has been highly conserved during evolution.

The Homeodomain

Molecular analysis of *Drosophila* patterning genes revealed that many of them share a consensus sequence of 180 nucleotides (W. McGinnis et al., 1984; Scott and Weiner, 1984). This sequence was dubbed the *homeobox* because it was discovered in two homeotic genes, *Antennapedia*⁺ and *Ultrabithorax*⁺. Since then, homeoboxes have been found in virtually all homeotic genes of *Drosophila* and also in many other patterning genes. (The gap genes, most of which are characterized by a similarly conserved motif known as *zinc fingers* are a major exception.) A typical homeobox-containing gene, the *Antennapedia*⁺ gene of *Drosophila*, is shown in Figure 21.43. The homeobox is located near the 3′ end of the gene, which encodes the carboxyl end of the protein product.

The protein domain encoded by the homeobox is termed the *homeodomain*. Sequence comparisons have revealed two important features of homeodomain-containing proteins. First, these proteins occur in all eu-

karyotes, from yeasts to plants and in all animal phyla. Second, each species produces several homeodomain-containing proteins, and in all analyzed cases they function as transcription factors (Scott et al., 1989; Gehring et al., 1994).

The similarity of homeodomains from different species, and among the homeodomain-containing proteins of a given species, is astounding (Fig. 21.44). The conservation suggests that these proteins act cooperatively in ways that have strongly constrained their evolution. Even the nucleotide sequences of different homeoboxes are so similar that a homeobox can serve as a probe to screen *recombinant DNA libraries* for more homeobox-containing genes or cDNAs (see Methods 14.2). This method of identifying putative transcription factors has opened the way to the molecular analysis of gene-regulatory networks in all higher organisms including mammals and plants (see Chapters 22 and 23).

The conservation of the homeodomain and its occurrence in known transcription factors suggests that it has a central function in this type of protein, such as binding to specific DNA sequences. This hypothesis

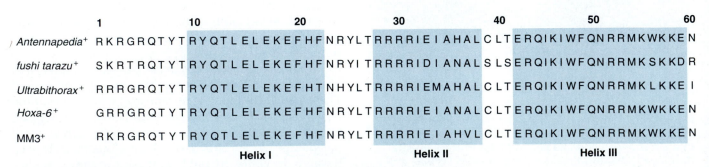

Figure 21.44 Amino acid sequences of homeodomains in proteins encoded by the *Drosophila* genes *Antennapedia*⁺, *fushi tarazu*⁺, and *Ultrabithorax*⁺; the mouse gene *Hoxa-6*⁺; and the *Xenopus* gene *MM3*⁺. The shaded areas indicate the three α helices formed within the homeodomain. Note that the three homeodomains from *Drosophila* differ in several amino acids, even in helix III, which generally shows the highest degree of conservation.

was confirmed by physicochemical analyses revealing the three-dimensional structure of the homeodomain and they way it binds to DNA (Gehring et al., 1994; Wolberger et al., 1991). The homeodomain is characterized by three α helices (Figs. 21.44 and 21.45). Helix III is called the *recognition helix:* it is aligned with the major DNA groove, where it interacts with specific nucleotide sequences. This helix is also the best-conserved segment of all homeodomains. Helices I and II are at right angles with helix III and farther away from the DNA. However, the loop between helices I and II, as well as amino acids at the start of helix II, contacts the DNA backbone. In addition, the N terminus of the homeodomain, which precedes domain I, is in close contact with the DNA minor groove.

The great similarity between the homeodomains of different transcription factors within the same organism raises the question, How can these factors nevertheless have very specific regulatory effects? Does all their specificity reside in the homeodomains, or do other protein domains contribute? If the homeodomain ac-counts for all or most of the specificity, are certain portions of the homeodomain more critical than others? To answer these questions, investigators first transformed flies with engineered transgenes that encoded the homeodomain of one protein and other domains from a different protein. These experiments showed that the regulatory specificity of a homeodomain-containing protein resides for the most part in the homeodomain itself. For instance, the *bicoid*[+] homeodomain attached to a transcription-activating sequence from yeast can substitute for the entire normal bicoid protein, completely restoring the normal development of offspring from *bicoid*[−] mutant mothers (Driever et al., 1989b).

In a related series of experiments, homeodomains were swapped between the Antennapedia and Ultrabithorax proteins. The hybrid proteins, with normal proteins as controls, were expressed from inducible promoters in transgenic flies (Mann and Hogness, 1990). Because these proteins were expressed *everywhere* in transgenic flies, they caused dramatic transformations, such as the replacement of antennae with legs as described earlier in this chapter. A hybrid protein that is mostly Ultrabithorax, but has the Antennapedia homeo-odomain and carboxyl terminal, causes almost the same transformations as a complete Antennapedia protein. A different set of transformations is caused by either a complete Ultrabithorax protein or a hybrid protein that is mostly Antennapedia except for a homeodomain and a carboxyl terminal from Ultrabithorax. Thus, the homeo-domains make almost all the difference in transformations, while the remaining parts of the proteins have only small effects. Similar results were obtained in homeodomain-swapping experiments between the Antennapedia and Sex combs reduced proteins (Gibson et al., 1990).

In extensions of these experiments, hybrid proteins with swapped *parts* of homeodomains were tested for their biological activity. Such experiments with Antennapedia and Sex combs reduced proteins showed that the N-terminal portions of the two homeodomains, which differ by four amino acids, determine the regulatory specificity of these two proteins (Zeng et al., 1993; Furukubo-Tokunaga et al., 1993). The N-terminal portion precedes helix I and is in contact with the minor DNA groove. Corresponding experiments with Ultra-bithorax and Antennapedia proteins also revealed that part of their specificity resides in the N termini of the homeodomains. Additional amino acids within the homeodomain and beyond the C terminus of the homeo-domain contribute to the specificity, although none of them is likely to contact DNA (Chan and Mann, 1993). Thus, the differences between homeotic gene products may be determined by a difference in DNA binding and/or by selective association with other transcription factors.

Even subtle differences among homeodomains can strongly affect their affinity for DNA binding sites. For

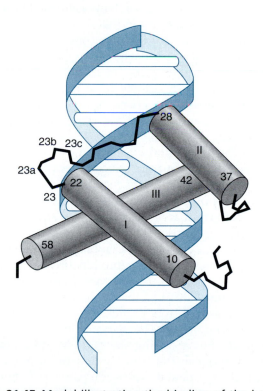

Figure 21.45 Model illustrating the binding of the homeodomain to its target DNA. This model is based on an analysis of the yeast MATα2 protein, but it is thought to apply to other homeodomains as well. The three α helices of the homeodomain are represented by cylinders (gray shading) with Roman numerals. Arabic numerals indicate the first and last amino acid of each helix. The ribbons depict the sugar-phosphate backbone of the DNA. Helix III is docked in the major groove of the DNA, where it interacts with specific nucleotides. The N-terminal region, which precedes domain I, is in close contact with the minor DNA groove. The loop between helices I and II contacts the DNA backbone.

instance, a single amino acid substitution in the recognition helix of the fushi tarazu homeodomain changes its preferred DNA-binding motif from CCATTA to GGATTA (Schier and Gehring, 1992). And indeed, almost all homeodomains within a given species differ by at least one amino acid within the recognition helix and show additional substitutions elsewhere (Fig. 21.44; Treisman et al., 1992). Thus, the small differences between homeodomains can account, at least in part, for their binding to different target genes.

Another set of observations indicates that cooperation and competition among homeodomain proteins contribute significantly to their specific effects on different target genes (Gehring et al., 1990; Hayashi and Scott, 1990). This hypothesis is in accord with data on similar interactions between other transcription factors, as described earlier in this chapter for the regulation of primary pair-rule genes. Cooperativeness between homeodomain proteins and other gene regulators is indicated by both genetic and molecular data. Mutations in the *extradenticle* gene, for example, cause transformations in thoracic and abdominal segments without altering the expression patterns of homeotic genes (Peifer and Wieschaus, 1990). In particular, *Antennapedia*[+] and *Ultrabithorax*[+] are correctly expressed in *extradenticle*[−] embryos, in which thoracic segments are transformed anteriorly and abdominal segments posteriorly.

Protein-DNA binding studies show that extradenticle protein binds alongside Ultrabithorax protein to an enhancer of a gene regulated by *Ultrabithorax*[+], and that the cooperative binding selectively increases the affinity of Ultrabithorax protein, but not Antennapedia protein, for the target gene of *Ultrabithorax*[+] (Chan et al., 1994). Similar studies indicate that a region N-terminal of the extradenticle homeodomain is required for cooperation with Ultrabithorax and abdominal A proteins, whereas a region C-terminal of the extradenticle homeodomain is necessary for cooperation with the engrailed protein (van Dijk and Murre, 1994). Thus, homeotic gene products achieve some of their biological specificity by cooperative DNA binding with other transcription factors.

Patterning Genes in Other Insects

Do other insects have the same hierarchies of patterning genes as *Drosophila*? As discussed in the previous section, the homeobox is a highly conserved DNA motif. Moreover, an entire cluster of homeotic genes, corresponding to the combined *Antennapedia* complex and *bithorax* complex of *Drosophila*, is found in a single complex in the flour beetle *Tribolium castaneum* (Beeman et

al., 1993). Comparisons of both DNA sequences and mutant phenotypes indicate that the biological functions of these genes have been largely conserved between the two species, which have evolved separately for more than 300 million years. For instance, *Drosophila* embryos homozygous for loss-of-function alleles of *abdominal A* show transformations of parasegments in the anterior abdomen causing them to resemble parasegment 6 (the posterior of T3 and the anterior of A1). The *Tribolium* gene with the greatest sequence similarity to this gene produces similar loss-of-function phenotypes. However, the transformations to parasegment 6 morphology extend to the end of the abdomen in *Tribolium*, whereas in *Drosophila* the posterior abdomen is unaffected.

A high degree of conservation is also observed for the *engrailed*[+] gene. When engrailed protein is traced by *immunostaining*, its expression patterns in a grasshopper, *Schistocerca*, and other arthropods are very similar to the *Drosophila* pattern (Patel, 1993). The engrailed protein accumulates in the posterior portion of each segment. The engrailed-positive stripes appear sequentially as the grasshopper germ band is formed head and thorax first, and as the abdominal segments are formed sequentially from a proliferation zone near the posterior end (Fig. 21.46). Similarly, the expression patterns of *even-skipped*[+] and *wingless*[+] in *Tribolium* resemble the patterns observed in *Drosophila* in most respects, although the segments in *Tribolium* are formed sequentially as in the grasshopper (Patel et al., 1994; Nagy and Carrol, 1994).

Given the segmental expression of the *engrailed*[+] gene in *Schistocerca* and the critical importance of *even-skipped*[+] for the regulation of *engrailed*[+] in *Drosophila*, it came as a surprise when it was discovered that the expression pattern of *even-skipped*[+] in *Schistocerca* differs greatly from the pattern observed in *Drosophila* (Patel et al., 1992). In the grasshopper embryo, *even-skipped*[+] is expressed but never resolves into repetitive stripes, and even-skipped protein disappears before engrailed mRNA becomes detectable. Evidently, *even-skipped*[+] does not regulate the expression of *engrailed*[+] in *Schistocerca* as it does in *Drosophila*. Only the function of *even-skipped*[+] in generating the nervous system seems conserved between the fruit fly and the grasshopper. Likewise, the segmental expression pattern of *engrailed*[+] in the honeybee seems to be controlled without a *fushi tarazu*[+] homologue, since a search for this homologue has been unsuccessful (Walldorf et al., 1989). It appears that the overlapping control mechanisms of segmentation gene expression have allowed the regulatory hierarchy to evolve in different ways.

Some of the maternal genes that are expressed during oogenesis and early embryogenesis also seem to be less well conserved. In close relatives of the fruit fly, such as the housefly *Musca domestica*, conservation is

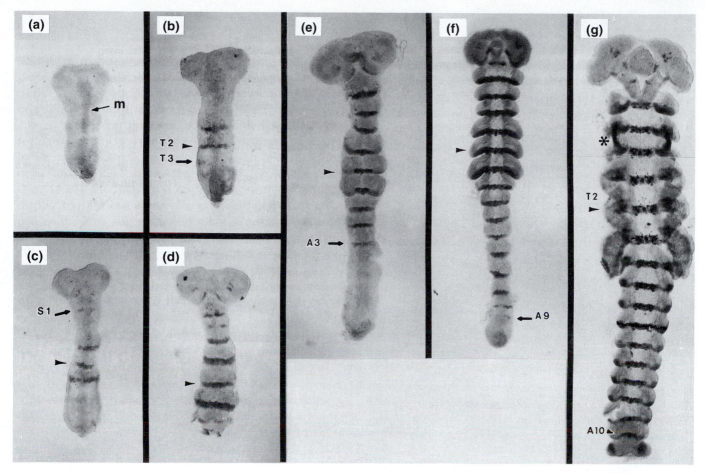

Figure 21.46 Expression of the *engrailed⁺* gene in the embryo of the grasshopper *Schistocerca gregaria*. Gene expression is shown by immunostaining using a monoclonal antibody to engrailed protein. The protein accumulates near the posterior margin of each segment. The photos are all at the same magnification, and anterior is always up. The position of the stripe belonging to the second thoracic segment is indicated by an arrowhead. **(a)** Early germ band primordium during mesoderm (m) formation. **(b)** Thoracic stripes are forming. **(c–f)** Additional stripes appear in head and abdomen as more segments are formed sequentially. **(g)** The final engrailed stripe has appeared in the 10th abdominal segment (A10).

still very good (Sommer and Tautz, 1991). However, even within the order of the diptera (two-winged insects), some of these genes appear to have diverged. The anterior determinants in eggs of the midge *Chironomus samoensis* differ functionally and molecularly from bicoid mRNA. Inactivation of the anterior determinants in *Chironomus* results in the *double-abdomen* phenotype, which features a mirror-image duplication of the abdomen, while the *bicoid⁻* phenotype is highly asymmetrical. Mimicking the double-abdomen phenotype in *Drosophila* requires transplanting nanos activity to the anterior pole (Ephrussi and Lehmann, 1992). Moreover, rescue experiments indicate that ante-

rior determinant activity in *Chironomus* eggs is associated with a localized maternal mRNA that is much smaller than bicoid mRNA (Elbetieha and Kalthoff, 1988).

It is difficult to generalize before more genes have been investigated in a comparative way, but so far, it appears that the most-conserved patterning genes belong to the homeotic and segment polarity classes. These genes function not only in establishing the segmental body pattern, but also in maintaining it throughout development. The patterning genes that are expressed only temporarily during the establishment of the body pattern seem more prone to evolutionary change.

TABLE 21.1

Patterning Genes of *Drosophila*

This table lists those patterning genes of *Drosophila* that are mentioned in this chapter. Expression patterns are usually given for transcription. If a protein is mentioned explicitly, then its pattern differs from that of the mRNA. Otherwise, the protein pattern can be assumed to follow the RNA pattern after a short time interval. Mutant phenotypes are usually those of homozygous loss-of-function alleles. Partial loss-of-function or gain-of-function alleles are mentioned explicitly. The molecular nature of a gene product is usually inferred from sequence comparisons with databases and in some cases supported by direct experiments. Developmental functions are mainly inferred from looking at one expression pattern in embryos mutant for another gene. The directness of the observed effects is therefore uncertain.

Gene	Expression Pattern	Mutant Phenotype	Molecular Nature	Developmental Function
Maternal anteroposterior axis				
Anterior system				
bicoid (*bcd*)	Transcription in nurse cells. mRNA localized in anterior of oocyte and egg. Translation in cleavage and blastoderm stages. Protein forms anteroposterior gradient.	Deletion of head and thorax; substitution of second telson for head	Genetic regulator (homeobox and paired repeat)	Activates *hb* transcription; activates *Kr* at low level and represses *Kr* at higher level; acts as morphogen on anterior genes
exuperantia (*exu*)	Maternal	Reduction of head; replacement by telson	Protein binding bcd mRNA?	Needed to transport bcd mRNA from nurse cells
hunchback (*hb*)	See Zygotic anteroposterior gap class			
swallow (*swa*)	Maternal	Reduction of head; cellularization defects	Cytoskeletal component?	Needed to maintain localization of bcd mRNA
Posterior System				
cappuccino (*capu*)	Maternal	A2–A7 missing; no pole cells		Required for polar granule assembly
nanos (*nos*)	Transcription in oocyte and nurse cells. Localized to posterior in late oocyte and egg, then to pole cells.	Absence of abdomen; pole cells normal		Inhibits translation of maternal hb mRNA, hence allows activation of *kni* in posterior; regulates *h, eve, ftz*
oskar (*osk*)	Maternal	Similar to *capu*		Required for pole cell formation
pumilio (*pum*)	Transcription maternal. Protein probably localized to posterior, 20–0% egg length (EL).	Most of abdomen missing; pole cells normal	Molecular chaperone?	Needed for transport of nos protein from pole plasm to prospective abdomen
spire (*spir*)	Maternal	Similar to *capu*		Required for polar granule assembly
staufen (*stau*)	Maternal	Similar to *capu* but with head defect		Required for polar granule assembly
tudor (*tud*)	Maternal	Similar to *capu*		Required for pole cell formation
vasa (*vas*)	Transcription in nurse cells and oocyte; mRNA uniform in early embryo. Protein localized in posterior of oocyte, egg, pole cells.	Similar to *capu*	Translation factor?	Required for pole cell formation
Terminal System				
torso (*tor*)	Transcription in nurse cells. Transport to oocyte, where distributed uniformly in plasma membrane.	Absence of labrum, reduction of head skeleton, absence of A8 and telson; gain of function: suppression of segmentation in thorax and abdomen	Cell surface receptor (Tyr kinase)	Activates *tll, hkb*

(continued on next page)

Gene	Expression Pattern	Mutant Phenotype	Molecular Nature	Developmental Function
torsolike (tsl)	Maternal; required in anterior and posterior follicle cells	Similar to *tor* (loss of function only)	Secreted protein	Ligand for tor

Zygotic anteroposterior gap class

Gene	Expression Pattern	Mutant Phenotype	Molecular Nature	Developmental Function
empty spiracles (ems)	Transcription during blastoderm 78–67% EL; later, in cells forming tracheal pits and spiracles	Defects in procephalon, tracheal system	Genetic regulator (homeobox)	
giant (gt)	Initial transcription at preblastoderm stage 82–60% and 33–0% EL; changes later	Defects in or absence of labrum, labium, A5–A7	Genetic regulator (Leu zipper)	Activates *Scr*; represses *Kr*
huckebein (hkb)		Defects in termini; *tor* effects not attributable to *tll*		
hunchback (hb)	Maternal mRNA initially uniform; protein forms antero-posterior gradient. Zygotic transcription 100–55% EL and 20–10% EL. Posterior domain long-lasting.	Zygotic: deletion of labium, thorax, aA7/pA8 (PS 13). No pure maternal effect, but if mother and sperm are *hb* $^-$, then deletion of all mouth-parts and thoracic segments and A1–A3, with mirror duplication of abdomen.	Genetic regulator (finger)	Represses *kni*; represses *Kr* at high level and activates at low level; regulates *eve*; regulates *h, run, ftz*; activates *Antp* P^2 (with *ftz*); represses *Ubx*
knirps (kni)	Transcription starts at preblastoderm stage forming zone 45–30% EL. Ventral patch and narrow ring form in anterior by cellular blastoderm stage.	Replacement of A1–A7 by single A-type segment	Genetic regulator (finger)	Represses *Kr*; regulates *h, run, eve, ftz*; represses *Abd-B* and *Antp*
Krüppel (Kr)	Transcription starts at preblastoderm stage in central zone, eventually expanding to 60–30% EL. Later also posterior patch and patch in head.	Purely zygotic. Head normal, deletion of thorax and A1–A5. Duplicated inverted A6.	Genetic regulator (finger)	Activates *kni*; represses *gt*; regulates *eve*; regulates *h, run, ftz*; represses *Scr*; activates *Antp* P^1; represses *Abd-B*
tailless (tll)	Transcription in termini of preblastoderm embryo	Head skeleton reduced but labrum present. Hindgut and Malpighian tubules absent.	Steroid receptor family	Activates *hb*; represses *Kr*; represses *kni*; activates *h*

Zygotic pair-rule class

Gene	Expression Pattern	Mutant Phenotype	Molecular Nature	Developmental Function
even-skipped (eve)	Transcription starts at preblastoderm stage; 7 stripes corresponding to parasegments 1, 3, 5, etc., by cellularization stage. Fades during gastrulation.	Abolishes segmentation, giving lawn of denticles. Weak alleles delete odd parasegments.	Genetic regulator (homeobox)	Represses *ftz, odd, run*; activates itself; activates *en*; represses *wg*; activates *Dfd*
fushi tarazu (ftz)	Transcription starts at preblastoderm stage; 7 stripes corresponding to parasegments 2, 4, 6, etc., by cellularization. Fades during germ band extension.	Deletes even parasegments	Genetic regulator (homeobox)	Activates itself; activates *en*; activates *Antp* P^2; activates *Ubx*; represses *wg*
hairy (h)	Transcription starts at preblastoderm stage; 7 stripes + dorsal head patch by time of cellularization. Fades during gastrulation.	Range from deletion of odd parasegments to formation of lawn of denticles. Also loss of labral tooth. Other alleles give extra hairs in adult.	Genetic regulator (helix-loop-helix)	Represses *run, ftz*; activates *eve*

(continued on next page)

Gene	Expression Pattern	Mutant Phenotype	Molecular Nature	Developmental Function
odd-paired (*opa*)		Opposite to *prd*		Activates *wg*, *en*
odd-skipped (*odd*)	Transcription starts at syncytial blastoderm stage; stripes 2 cells wide.	Deletions of less than 1 segment width including posterior part of denticle bands in odd abdominal segments, with mirror duplications of anterior part	Genetic regulator (finger)	Represses *en*
paired (*prd*)	Transcription starts at preblastoderm stage; 8 stripes with dorsal head patch by cellularization. During germ band extension stripes 2–7 split and then fade.	Deletions about 1 segment wide starting in middle of denticle band affecting mainly odd abdominal segments	Genetic regulator (homeobox)	Activates *wg*, *en*
runt (*run*)	Transcription starts at preblastoderm stage; 7 stripes out of phase with *h*. After cellularization, pattern becomes single-segment stripes, which persist through germ band extension.	Deletions more than 1 segment wide centered on T2, A1, A3, A5, A7, with mirror duplication of what is left	Nuclear protein	Represses *h*, *eve*; activates *ftz*
sloppy-paired (*slp*)		Like *prd* but weaker		

Zygotic segment polarity class

Gene	Expression Pattern	Mutant Phenotype	Molecular Nature	Developmental Function
engrailed (*en*)	Transcription starts at blastoderm stage and produces 14 narrow stripes by extended germ band stage, marking the anterior of each parasegment. Expression in posterior compartments of larval imaginal discs.	Ventral cuticle a continuous lawn of denticles. Less severe alleles delete even parasegments.	Genetic regulator (homeobox)	Maintains *hh* and indirectly *wg*; selector gene defining posterior compartment
hedgehog (*hh*)	Posterior compartments of larval segments and imaginal discs	Posterior compartment replaced with mirror image of anterior compartment	Secreted signaling molecule	Maintains *wg* and indirectly *en*; segmental morphogen
wingless (*wg*)	Transcription starts at blastoderm stage and produces 14 narrow stripes by extended germ band stage, marking the posterior of each parasegment.	Ventral cuticle a continuous lawn of denticles, indicating deletion of $\frac{3}{4}$ of each segmental repeating unit	Signaling molecule (member of large family of Wnt gene products)	Maintains *en*

Homeotic genes and related genes

Gene	Expression Pattern	Mutant Phenotype	Molecular Nature	Developmental Function
abdominal A (*abd-A*)	Transcribed in PS 7–12 by extended germ band stage	Transforms PS 7–9 to PS 6	Genetic regulator (homeobox)	Represses *Ubx*; activates *wg*, *dpp*; selector gene
Abdominal B (*Abd-B*)	Transcribed in PS 10–13 and PS 14 (separate promoters) by extended germ band stage	Transforms PS 10–14 to PS 9	Genetic regulator (homeobox)	Represses *Ubx*; selector gene
Antennapedia (*Antp*)	Transcription starts in blastoderm; mainly PS 4 and 5 in extended germ band (separate promoters).Thoracic imaginal discs.	Transforms PS 4, 5 to PS 3	Genetic regulator (homeobox)	Selector gene

(continued on next page)

Gene	Expression Pattern	Mutant Phenotype	Molecular Nature	Developmental Function
apterous (ap)	Dorsal compartments of wing and haltere discs	Wings and halteres missing	Genetic regulator	Required for wing (haltere) outgrowth; selector gene
Deformed (Dfd)	Transcription in preblastoderm stage in PS 1; later also PS 0. Eye-antennal disc.	Deletion of mandibular and maxillary segments	Genetic regulator (homeobox)	Activates itself; selector gene
labial (lab)	Transcription in extended germ band anterior to cephalic furrow and in posterior midgut	Deletes labial derivatives	Genetic regulator (homeobox)	Selector gene
Polycomb (Pc)		Multiple homeotic transformations		Maintains expression domain of homeotic genes
Sex combs reduced (Scr)	Transcription in PS 2 in blastoderm; later PS 3 epidermis and mesoderm, and abdominal ganglia	Transforms PS 3 to PS 4 and PS 2 to PS 1	Genetic regulator (homeobox)	Selector gene
Ultrabithorax (Ubx)	Transcription starts in cellular blastoderm, mainly in PS 6, but at lower levels in PS 5–13; more in even-numbered parasegments. Neuromeres; metathoracic discs.	Transforms PS 5, 6 to PS 4	Genetic regulator (homeobox)	Represses Scr; represses Antp; activates itself; activates wg, dpp; selector gene

Maternal dorsoventral system

Gene	Expression Pattern	Mutant Phenotype	Molecular Nature	Developmental Function
cactus (cac)	Maternal; action in embryo	Ventralizing	Genetic regulator	Inhibits entry of dl protein to nuclei
dorsal (dl)	Transcription in nurse cells; uniform mRNA in oocyte and egg; ventral-dorsal gradient of protein in preblastoderm nuclei	Embryos become tubes of dorsal epidermis. Some haploinsufficient alleles.	Genetic regulator	Activates twi, sna; represses zen, dpp; morphogen for dorsoventral pattern
easter (ea)	Maternal, needed in embryo	Similar to dl. Also weak ventralizing gain-of-function alleles.	Serine protease	Activation of Tl ligand (spätzle protein)
gurken (grk)	Maternal; needed in oocyte	Ventralizing; affects eggshell and embryo	Growth factor-like	Ligand for top protein; dorsalizing signal from oocyte
nudel (ndl)	Maternal; needed in follicle cells	Similar to dl		Ventral polarizing activity
pelle (pll)	Maternal, acts in embryo	Similar to dl	Protein kinase	Contributes to activation of dl protein
pipe (pip)	Maternal; needed in follicle cells	Similar to dl		Ventral polarizing activity
snake (snk)	Maternal, needed in embryo	Similar to dl	Serine protease	Activation of Tl ligand (spätzle protein)
spätzle (spz)	Maternal, needed in embryo	Similar to dl		Ligand for Tl protein
Toll (Tl)	Maternal; uniform mRNA in oocyte and egg	Similar to dl. Gain-of-function alleles ventralize, producing denticle belts all around.	Cell surface protein	Releases dl protein from cac protein, permits dl protein to enter nucleus
torpedo (top)	Maternal; needed in follicle cells	Similar to grk	Growth factor receptor	Follicle cell receptor for dorsalizing signal (gurken peptide)

(continued on next page)

Gene	Expression Pattern	Mutant Phenotype	Molecular Nature	Developmental Function
windbeutel (*wbl*)	Maternal; needed in follicle cells	Similar to *dl*		Ventral polarizing activity
Zygotic dorsoventral system				
decapentaplegic (*dpp*)	Transcription in preblastoderm on dorsal side, curling around to ventral at poles. Later, visceral mesoderm, fore- and hindgut. Imaginal discs.	Loss of amnioserosa and reduction of dorsal epidermis. Viable alleles produce multiple defects in imaginal disc derivatives.	Signaling molecule (TGF β homologue)	
snail (*sna*)	Transcription in midventral strip of prospective mesoderm	Loss of mesoderm		
twist (*twi*)	Transcription in midventral strip of prospective mesoderm and mesectoderm	Loss of mesoderm	Genetic regulator (helix-loop-helix)	
zerknüllt (*zen*)	Transcription in preblastoderm stage on dorsal side, concentrated in prospective amnioserosa by the beginning of gastrulation	Loss of amnioserosa and optic lobe	Genetic regulator (homeobox)	

Source: After Slack (1991). Reprinted with permission of Cambridge University Press.

SUMMARY

Large mutagenesis screens have uncovered about 200 genes governing embryonic pattern formation in *Drosophila*. These genes form a regulatory cascade, which begins anew in each generation. During oogenesis, the products of a few maternally expressed genes are deposited in specific areas of the egg. These localized mRNAs and proteins generate uneven distributions of transcription factors, which in turn act on a small number of zygotic target genes. These first zygotic patterning genes are expressed in broad transverse stripes, which are the beginning of an anteroposterior body pattern, and in longitudinal stripes that form the rudiment of a dorsoventral body pattern.

The first zygotic genes involved in anteroposterior patterning are known as gap genes. Loss of function in a typical gap gene causes deletion of several contiguous segments. Gap gene expression domains differ in width and overlap to varying degrees, but none of them repeats periodically. Nevertheless, gap gene products direct the expression of a second tier of zygotic genes, the pair-rule genes, in periodic stripes. Each stripe and an adjacent interstripe together are as wide as two segments but are not necessarily congruent with prospective segments. Several pair-rule genes jointly control a third tier of zy-

gotic genes, called segment polarity genes. Their expression domains delineate parasegments, metameric units of the early embryo that each correspond to the posterior compartment of a definitive segment plus the anterior compartment of the following segment.

Whereas the boundaries of parasegments are defined by the functions of segment polarity genes, the individual character of each segment is determined by a unique combination of homeotic genes. Homeotic genes are characterized by dramatic mutant phenotypes in which certain body parts are transformed into other parts normally formed elsewhere. The expression of segment polarity and homeotic genes begins at the blastoderm stage, when the fate of embryonic cells becomes restricted to their future segments. These genes are expressed throughout development and are the genes that maintain segmental boundaries and the individual characteristics of each segment.

The dorsoventral body pattern is specified in a similar manner. Maternal gene products set up an uneven distribution of a transcription factor that acts on a small number of zygotic target genes. The dorsoventral pattern is generated nearly independently of the anteroposterior pattern.

Some of the patterning genes identified in *Drosophila* have been linked by the compartment hypothesis to identifiable areas of epidermis. Compartments are delineated by clonal restriction lines and appear to be the basic units of growth control and morphogenesis. According to the compartment hypothesis, each compartment is characterized by a specific combination of active selector genes that maintain its boundaries and generate its specific morphology by controlling batteries of realizator genes.

Sequence analysis of many patterning genes in *Drosophila* has revealed a highly conserved DNA sequence called the homeobox. It encodes a domain of 60 amino acids, known as the homeodomain. Homeodomain-containing proteins act as transcription factors, with the homeodomain itself binding to specific DNA sequences. The most striking feature of the homeodomain is its extreme evolutionary conservation from yeasts to plants and in all groups of animals.

The genetic and molecular analysis of pattern formation in *Drosophila* embryos has confirmed, to an extent, the classical gradient model of morphogens. At least three of the maternally encoded transcription factors—and several segmentation gene products as well—control their target genes in a concentration-dependent way. However, most gradients in *Drosophila* govern only part of a body axis, and the number of target genes in each case is rather small. Local interactions between nuclei and the epigenetic generation of complex patterns from simpler precursors have been found to play a larger role than expected.

Some of the genetic hierarchies that govern pattern formation in *Drosophila* have synergistic regulatory mechanisms. Such mechanisms seem to be common among insects, and somewhat different systems of control have evolved from them in different insect groups.

SUGGESTED READINGS

Gehring W. J., Y. Q. Qian, M. Billeter, K. Furukubo-Tokunaga, A. F. Schier, D. Resendez-Perez, M. Affolter, G. Otting, and K. Wüthrich. 1994. Homeodomain-DNA recognition. *Cell* **78**:211–223.

Hülskamp, M. and D. Tautz. 1991. Gap genes and gradients—The logic behind the gaps. *BioEssays* **13**:261–268.

Ingham, P. W., and A. Martinez-Arias. 1992. Boundaries and fields in early embryos. *Cell* **68**:221–235.

Lawrence, P. A. 1992. *The Making of a Fly: The Genetics of Animal Design.* Cambridge, Mass.: Blackwell Scientific Publications.

Nüsslein-Volhard, C. 1991. Determination of the embryonic axes of *Drosophila. Development* **Supplement 1**:1–10.

GENETIC

AND MOLECULAR

ANALYSIS

OF VERTEBRATE

DEVELOPMENT

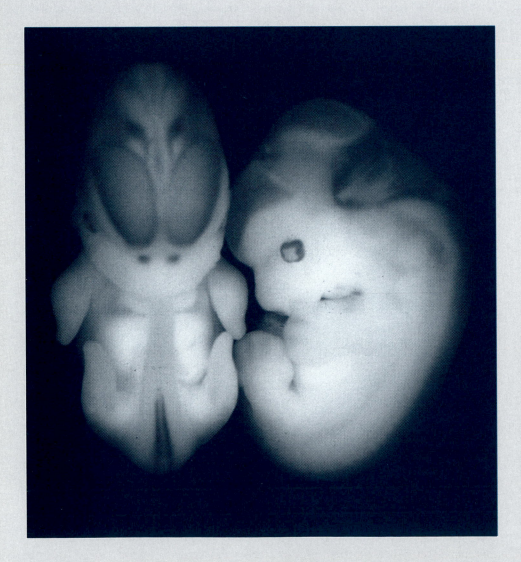

Figure 22.1 A method for identifying mouse genes involved in embryonic pattern formation relies on a transgene that produces a blue stain in fixed host tissues where the transgene has been active. Because the transgene has no promoter of its own, it reveals the activity pattern of a resident promoter located nearby in the host genome. The photograph shows transgenic fetuses with blue stain in the brain and spinal cord, indicating that the transgene is driven by a resident promoter that is active specifically in the central nervous system.

The genetic and molecular analysis of development in *Drosophila* has been a spectacular success. The application of these methods has yielded concrete and detailed answers to long-standing questions in development. But how general are the lessons learned from the fruit fly? As humans, we have a particular interest in the development of vertebrates, especially mammals. Naturally, we are eager to know how much of the insight gathered from flies applies to our closer phylogenetic relatives.

Past work with *Drosophila* has shown that mutants are the most valuable tools for studying the functions of genes.

Unfortunately, the most tractable vertebrates for experimentation—amphibians and birds—are unsuitable for mutational analysis because of their large genomes and long generation times. The zebra fish has recently been recognized as a vertebrate on which large-scale mutagenesis screens can be carried out (Rossant and Hopkins, 1992). If such screens are successful, zebra fish research may pay rich dividends.

Genetically, the best-known vertebrate so far is the mouse. More than 2000 mouse genes have been mapped, and a growing number of these genes have been assigned to cloned DNA segments (M. C. Green, 1989; Wagner, 1990). This number is believed to be less than 5% of all mouse genes, but more mutants are constantly being generated by *insertional mutagenesis* and *gene targeting* (Fig. 22.1).

In this chapter, we will explore the use of genetic and molecular methods for the analysis of pattern formation in vertebrate embryos. Featured prominently in this analysis is a set of homeobox-containing genes. After describing the evolution and expression patterns of these genes in mice, we will examine whether they act as homeotic selector genes like their counterparts in *Drosophila*. Finally, we will consider the role of retinoic acid, a small signal molecule related to vitamin A, in the regulation of homeobox gene activity.

Strategies for Identifying Patterning Genes in Vertebrates

A variety of strategies have been used to identify patterning genes in mice (M. Kessel and P. Gruss, 1990). One strategy starts with mutant alleles that identify a genetic *locus,* which is mapped on the basis of frequency of recombination with known marker genes. With luck, a DNA segment close to the new locus of interest has already been cloned. In this case, one can "walk" from the previously cloned DNA segment to the new locus by cloning the intervening DNA in a series of large overlapping segments. In an alternative strategy, known as *insertional mutagenesis,* researchers insert a bacterial *reporter gene,* such as the *lacZ* gene, randomly into the mouse genome (see Chapters 14 and 15). If by chance the reporter gene "lands" next to a promoter or enhancer that drives a patterning gene, then the reporter gene will be expressed in the transgenic mouse with the same tissue- and stage-specificity as the endogenous patterning gene would be in a normal mouse (Fig. 22.1). Because it is not a natural part of the mouse genome, the reporter gene also "tags" its insertion site, which then can easily be cloned.

While traditional strategies proceed from analyzing mutant alleles to cloning the affected segments of genomic DNA, alternative strategies known as *reverse genetics* begin with the cloning of a gene and then employ molecular techniques to generate mutant alleles or their equivalents. Researchers taking the reverse genetics approach typically use known patterning genes from other organisms, primarily *Drosophila,* as probes to identify similar genes in *libraries* of mouse DNA. The ultimate goal of both traditional and reverse genetics is to infer the biological role of a gene from its molecular features and from the phenotypes generated by mutant alleles.

Genetic Loci Are Matched Up with Cloned Genes

Several of the known mouse mutants affect embryonic patterning. A well-known gene on chromosome 17 is the *T* locus, which was first detected as a spontaneous mutation named *Brachyury* (Dobrovolskaia-Zavovskaia, 1927). Mutant alleles cause short, kinky tails in heterozygotes (*T/+*), while homozygotes (*T/T*) die before birth. Gastrulating *T/T* embryos produce insufficient mesoderm and fail to form a notochord. The *T* locus interacts with other genes in a segment of chromosome 17 known as the *t complex,* and the interactions have revealed several other genes in the *t* complex (Bennett, 1975). Chromosomal deletions and duplications covering the *T* locus have allowed researchers to clone a gene, *me75+,* that appears to be identical with the *T* locus (Herrmann et al., 1990). *In situ hybridization* (see Methods 15.1) showed that *me75+* is expressed in epiblast cells, early mesoderm, and notochord (Wilkinson et al., 1990). This expression pattern is in accord with earlier histological observations indicating that *T* homozygotes are defective in the formation of the notochord and other mesodermal structures. Definitive proof that *me75+* and *T* are allelic came from a phenotypic rescue of *T/+* mutants by microinjection of the cloned *me75+* gene into the male pronuclei of fertilized eggs (Stott et al., 1993; see Chapter 14 for the experimental techniques).

Mouse DNA clones that match up with known mutants have been identified by screening mouse DNA libraries with labeled probes from known *Drosophila* patterning genes (see Methods 14.2). For example, the murine *Pax 1+* gene was obtained by hybridization with the *paired box,* a repetitive element contained in the *paired+* gene of *Drosophila* (M. Kessel and P. Gruss, 1990). In situ hybridization revealed that *Pax 1+* is expressed in intervertebral discs, the sternum (breastbone), and the thymus gland. Mice mutant in the *undulated* gene suffer from defective intervertebral discs. Analysis of the wild type and three mutant alleles of *undulated* revealed that each mutant allele of *undulated* was also mutated in its *Pax 1* DNA sequence and expressed *Pax*

1 in an abnormal pattern. These observations suggest very strongly that the *undulated+* gene and the cloned *Pax 1+* DNA are identical.

Thus, by matching up genetic loci with DNA clones, researchers can correlate the molecular features and expression patterns of genes with their mutant phenotypes. From such correlations they can often infer the biological function of a gene. However, this approach has been limited by the relatively small number of available mutants in mice and the virtual lack of mutants in other vertebrates.

Promoter Trapping Identifies Patterning Genes

In order to generate additional mouse mutants, investigators use *embryonic stem (ES) cells,* totipotent cells that are derived from the inner cell mass of *blastocysts.* ES cells can be genetically transformed with transgenes—a procedure that in a fraction of cases causes *insertional mutagenesis* (see Chapter 14). ES cells with stably inserted transgenes are injected into host blastocysts carrying a set of marker genes that differ from those of the ES cells. In some of the resulting *chimeras,* the transformed ES cells contribute to the germ line, as revealed by transmission of the ES cell marker genes to offspring. These offspring also carry one copy of the transgene, which can then be bred to homozygosity.

The following method is particularly useful for identifying patterning genes because it makes their expression domains visible even before mutants have been bred to homozygosity (Friedrich and Soriano, 1991). This method, called *promoter trapping,* identifies promoters in the mouse genome by their ability to drive the expression of a reporter transgene such as *NeoR-lacZ* (Fig. 22.1). The transgene has no promoter of its own, so it is expressed only when inserted near the promoter of a host gene. Depending upon the site of insertion, the pre-mRNA transcribed from the transgene is processed as part of the host gene's transcript. The transgene encodes a protein with a dual function: it confers resistance to a toxic agent (neomycin, or G418), and it generates a blue stain from a colorless chromogenic substrate. Any ES cell that synthesizes the transgenic protein acquires three properties, which are passed on to its descendants (Fig. 22.2). First, the cell will survive in the presence of G418, which kills nontransformed stem cells. Second, the cell will turn blue after appropriate fixation and addition of the chromogenic substrate. Third, because the transgene interrupts the host gene at the point of insertion, the host gene will frequently lose part or all of its normal function.

Transformed ES cells are injected into wild-type blastocysts, where they join the inner cell mass to generate chimeric mice (see Fig. 14.11). Chimeric males in which the ES cells have contributed to the germ line are mated with wild-type females. Half of their offspring are het-

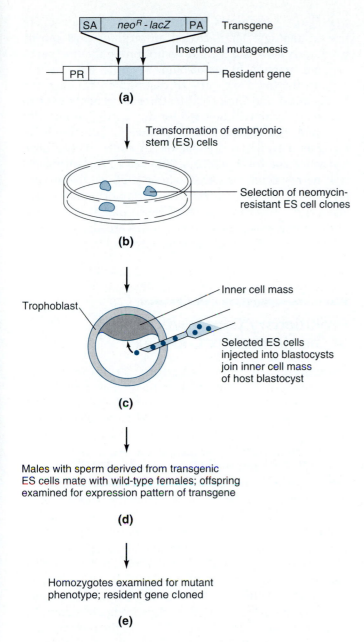

Figure 22.2 Promoter trapping of mouse genes by insertional mutagenesis. **(a)** An engineered transgene encodes a hybrid protein conferring resistance (*NeoR*) to a toxic drug (neomycin, or G418) and the enzymatic activity (*lacZ*) to produce blue stain from a chromogenic substrate. The coding region is flanked by splice acceptor (SA) and polyadenylation (PA) sites. If inserted into a resident gene, the transgene is transcribed under control of the resident gene's promoter (PR). **(b)** Embryonic stem (ES) cells are transformed with the transgene and selected by culture in the presence of G418, which kills all nontransformed cells. **(c)** Selected ES cells are injected into mouse blastocysts to generate chimeras. **(d)** Chimeric males containing the transgene in their sperm are mated with wild-type females to generate offspring heterozygous for the transgene. The expression pattern of the transgene during embryogenesis is detected with chromogenic substrate. **(e)** Heterozygotes with specific expression patterns are bred to homozygosity to analyze mutant phenotypes resulting from the interruption of the resident gene. The mutated gene is cloned using the transgene as a tag.

erozygous for both the transgene and the interrupted host gene. Fetuses are removed and fixed at different stages of development to be stained. The blue stain reveals the activity pattern of the promoter that drives the expression of the transgene. Presumably, the transgene is expressed with the same pattern as the interrupted host gene would have been. Under this assumption, the blue staining pattern already provides a clue to the function of the host gene. For instance, blue stain in the neural tube suggests that the host gene may play a specific role in the development of the central nervous system. Siblings of heterozygotes showing specific staining patterns are bred to homozygosity so that the phenotypic effects of interrupting the host gene can be studied. If the homozygotes display specific abnormalities, then the defects provide further clues to the interrupted gene's role in development. Finally, the transgene is used as a tag to clone the interrupted host gene from a genomic DNA library (see Methods 14.2). The function of the cloned gene can then be inferred from the amino acid sequence of its protein product, its expression pattern at different stages, and the defects found in the mutant phenotype.

Using this technique, Friedrich and Soriano (1991) generated 24 mouse lines having various patterns of transgene expression. In some lines, all cells of an embryo expressed the transgene, while in others transgene expression was restricted to specific regions or tissues. In only 9 of the 24 lines did the homozygous transgene cause embryonic lethality. The lack of overt phenotypes in the remaining cases suggests that the transgene insertion did not interfere with resident gene function or that loss of function in the interrupted resident gene can be compensated by other genes.

The latter explanation is in accord with the *principle of synergistic mechanisms* already discussed in Chapters 4, 11, and 21. In mice, this principle seems to preclude the discovery of many genes, because their loss of function does not generate a discernible phenotype.

Vertebrate Genes Can Be Isolated with Molecular Probes from Other Organisms

Another method for identifying previously unknown genes relies on the fact that complementary nucleic acids hybridize even if their nucleotide sequences do not match exactly. Therefore, a well-characterized gene from a species like *Drosophila* can be used as a hybridization probe to screen a DNA library of another organism such as the mouse. Under proper reaction conditions, the *Drosophila* probe will hybridize with any cloned segment of mouse DNA that contains a nearly complementary sequence. Such segments of mouse DNA can be subcloned and sequenced. The predicted amino acid sequence of the encoded protein can be compared with the known sequence of the protein encoded by the *Drosophila* probe. In many cases, the proteins from the two species are strikingly similar in certain functional domains. The more extensive their similarities are, the more compelling is the argument that the two proteins have similar *biochemical* functions in the fly and the mouse. This strategy has allowed molecular biologists to "fish" for vertebrate genes that might be related to well-characterized genes from *Drosophila*, *Caenorhabditis elegans*, or yeasts.

This method has been used with astounding success for genes containing sequences that have been well conserved during evolution. Genes with extensive sequence similarity are often referred to as **homologous genes** or **homologues.** (We used this term in Chapters 3 and 14 to refer to the corresponding pairs of genes and chromosomes inherited from both parents. In strict usage, calling two genes or other structures homologous implies that both have evolved from the same ancestral structure. However, many molecular biologists use the term simply to express sequence similarity.) Strong sequence similarities between genes or between their protein products tend to be found mostly in certain functional domains such as the *homeodomain*, which has been exceedingly well conserved during evolution.

As we will see, the homeotic genes of *Drosophila*, such as *Antennapedia⁺*, have clearly identifiable homologues in virtually every eukaryotic organism that has been screened. The *Antennapedia⁺* homologue of the mouse, for example, is designated *Hoxb-6⁺*. The proteins encoded by the two genes are nearly identical within their homeodomains, although they differ greatly in other parts. Nevertheless, the Hoxb-6 protein can substitute functionally for the Antennapedia protein. This was shown when the *Hoxb-6⁺* gene was overexpressed under the control of a heat-inducible promoter in transgenic flies (Malicki et al., 1990). In the flies' heads, where the resident *Antennapedia⁺* gene is silent, expression of the *Hoxb-6⁺* transgene caused the same antenna-to-leg transformations also observed after overexpression of an *Antennapedia⁺* transgene (see Color Plate 5). Similarly, the mouse *Hoxa-5⁺* gene is functionally equivalent to the *Sex combs reduced⁺* gene in *Drosophila* (Zhao et al., 1993). Other experiments have shown that the human *HOXD4⁺* gene can substitute for the regulatory function of its *Drosophila* homologue, *Deformed⁺* (N. McGinnis et al., 1990). These results indicate that the homeotic genes of *Drosophila* and their mammalian homologues encode proteins that share both the conserved homeodomain and certain biochemical functions. In addition to homeotic genes, *Drosophila* genes containing the *paired box, zinc fingers,* or other conserved sequences have been used to isolate similar vertebrate genes.

Proteins with similar *biochemical* functions in different species do not necessarily have corresponding *biological* functions. Not even the fact that the mouse *Hoxb-6⁺* gene promotes leg formation in *Drosophila* proves that *Hoxb-6⁺* promotes leg formation in the mouse. (To use an analogy: The fact that an airplane part can be used to fix a lawn mower does not prove that the part's original function in the airplane was the same as it is now in the lawn mower.) To distinguish between the biochemical and biological features of genes, the term **homeobox gene** will be used for any gene that encodes a protein with a homeodomain, which presumably functions as a transcription factor. In contrast, the term *homeotic gene* (or *homeotic selector gene*, for emphasis) will be reserved for genes whose mutant phenotypes replace certain body parts with parts normally formed elsewhere. Experiments have shown that at least some of the homeobox genes in vertebrates are also homeotic selector genes like their counterparts in *Drosophila*.

Evolutionary Conservation of the Homeobox Complex

Throughout the animal kingdom, certain homeobox genes are found in clusters that have been exceedingly well conserved during evolution (Krumlauf, 1994; W. McGinnis and R. Krumlauf, 1992). As discussed in the previous chapter, the homeotic genes of *Drosophila* are clustered in two complexes, the *Antennapedia complex (ANTP-C)* and the *bithorax complex (BX-C)*, both located on the right arm of the third chromosome. In other insects, a single **homeobox complex (HOM-C)** contains genes homologous to the *Antennapedia* complex and *bithorax* complex together. *Amphioxus*, representing the cephalochordates, a primitive sister group of the vertebrates, has a similar homeobox complex (Garcia-Fernàndez and Holland, 1994). Mammals, including the mouse and the human, have four clusters of homeobox genes, each equivalent to the single homeobox complex of insects and *Amphioxus* (Fig. 22.3). In the mouse, these complexes are called **Hox complexes,** designated **Hox A, Hox B, Hox C,** and **Hox D.** Each *Hox* complex is located on a different chromosome and contains about 10 homeobox genes, called **Hox genes.**

In addition to the genes located in the *bithorax* and *Antennapedia* complexes, many other *Drosophila* genes contain homeoboxes and are involved in embryonic pattern formation (see Chapter 21). Likewise, several mouse genes have homeoboxes and are expressed in embryos but are located outside the *Hox* complexes. There are striking parallels between these genes, with respect to their nucleotide sequences and their expression domains (Simeone et al., 1992). This chapter will focus on genes within the *Hox* complexes.

Most *Hox* genes have been cloned, and from their nucleotide sequences one can predict the amino acid sequences of the encoded proteins, including their homeo-

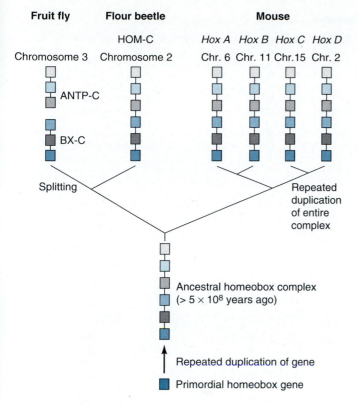

Figure 22.3 Simplified model of the evolution of the homeobox complex. A primordial homeobox gene was duplicated and diversified several times to generate a complex consisting of several homeobox-containing genes. This ancestral homeobox complex must have existed in organisms that lived before the evolutionary separation of vertebrates and arthropods. During the evolution of the mouse, the entire complex was duplicated repeatedly, and four different chromosomes (2, 6, 11, and 15) retained one copy each. During the evolution of *Drosophila*, the homeobox complex was split into the *Antennapedia* complex (ANTP-C) and the *bithorax* complex (BX-C), both located on chromosome 3. The flour beetle (*Tribolium*) and several other animal species still have one homeobox complex (HOM-C).

domains. Sequence comparisons of homeodomains led to the exciting conclusion that all homeobox complexes found in different animals today presumably originated from one primordial homeobox gene (A. Graham et al., 1989; Kappen et al., 1989; Murtha et al., 1991). This primordial gene, through a series of duplication and divergence steps, must have generated the **ancestral homeobox complex** (Fig. 22.3). This ancestral complex must have been present in animals that lived before vertebrates and arthropods (including insects) began to evolve separately more than 500 million years ago. The ancestral homeobox complex was duplicated repeatedly during the evolution of vertebrates, and four copies of the complex persist on different chromosomes in the mouse. During the evolution of the fruit fly, a chromosome break must have split the homeobox complex into the *Antennapedia* and *bithorax* com-

plexes, a step that probably occurred late during insect evolution, since in other insects a unified homeobox complex is still present.

Not only have individual homeotic genes and *Hox* genes been conserved during evolution; the physical order in which they are arranged in their complex has also been retained. On the basis of detailed sequence comparisons, these genes and their encoded proteins can be grouped into 13 *subfamilies* (Figs. 22.4 and 22.5). Each subfamily is represented in different *Hox* complexes, and each *Hox* complex contains genes from several subfamilies. Thus, each *Hox* gene has been given a dual designation: a letter from A through D to indicate the *Hox* complex, and a number between 1 and 13 to indicate the subfamily. The gene names are given as $Hoxa\text{-}1^+$, $Hoxc\text{-}4^+$, and so on, in mice, and $HOXA1^+$, $HOXC4^+$, and so on, in humans (Scott, 1992).

The most striking observation is that genes from different subfamilies are arranged in the same physical order within all *Hox* complexes, with gene 1 next to the 3' end and gene 13 next to the 5' end of each complex (Fig. 22.5). Moreover, the *Antennapedia* and *bithorax* complexes of *Drosophila* can be aligned with the *Hox* complexes of mammals in such a way that each *Drosophila* homeotic gene is assigned to the *Hox* subfamily with which it has the greatest sequence similarity. Strong cases for these assignments can be made on the basis of sequence similarities between the *Drosophila* homeotic genes $labial^+$, $proboscipedia^+$, $Deformed^+$, and $Abdominal$ B^+ and *Hox* subfamilies 1, 2, 4, and 9. The conservation of this order argues strongly for the evolution of the homeobox complexes in flies and mice from a common ancestor. The assignment of the other *Drosophila* homeotic genes to *Hox* subfamilies on the basis of sequence similarities is less compelling. These genes may also have evolved by additional duplication and divergence steps *after* the evolutionary separation of arthropods and vertebrates.

Expression Patterns of *Hox* Genes

Not only do *Drosophila* homeotic genes and mouse *Hox* genes share their chromosomal order and sequence information in their critical domains; mouse *Hox* genes also follow the same *colinearity rule* already discussed in the previous chapter for the homeotic genes of *Drosophila*: the physical order of *Hox* genes within each complex is related to the order of their expression domains along the anteroposterior axis of the embryo. Moreover, some boundaries of *Hox* gene expression are also *clonal restriction lines,* just as the expression domains of some homeotic genes delineate *compartment* boundaries in *Drosophila*.

Amino acid sequences

Antp	RKRGRQTYTRYQTLELEKEFHFNRYLTRRRRIEIAHALCLTERQIKIWFQNRRMKWKKENK
Hoxb-9	SRKK-CP--K---------L--M----D--H-V-RL-N-S---V---------M--M--
Hoxa-9	TRKK-CP--KH--------L--M----D--Y-V-RL-N----V---------M--I--
Hoxc-9	TRKK-CP--K---------L--M----D--Y-V-RV-N----V---------M--M--
Hoxd-9	TRKK-CP--K---------L--M----D--Y-V-RI-N----V---------M--MS-
Abd-B	VRKK-KP-SKF--------L--A-VSKQK-W-L-RN-Q----V---------N--NSQ
Hoxb-8	-R------S----------L--P----K----VS---G----V-------------N
Hoxc-8	-RS-----S----------L--P----K----VS---G----V-------------N
Hoxb-7	--------------------------Y----------T------------------
Hoxa-7	---H-
Hoxb-6	GR------------------------Y-----------------------------S-
Hoxa-6	GR--------------------------------N---------------------
Hoxc-6	-R----I-S-------------------------N---------------------SN
abd-A	-R-------F-----------H-------------------------L---LR
Ubx	-R-----------------T-H--------M--------------------L---IQ
Hoxb-5	G--A-TA--------------------------------S----------------D--
Hoxa-5	G--A-TA--------------------------------S----------------D--
Scr	T--Q-TS--L---H-
Hoxb-4	P--S-TA---Q-V-------Y--------V--------S----------------DH-
Hoxa-4	P--S-TA---Q-V-----------------T---S---V-------------DH-
Hoxd-4	P--S-TA---Q-V-----------------T---P----------------DH-
Dfd	P--Q-TA---H-I-------Y---------T-V-S--------------------D--
Hoxb-3	S--A-TA--SA-LV--------------C-P--V-M-NL-N-S-----------Y--DQ-
Hoxa-3	S----TA--P-LV-------------M-P--V-M-NL-N--------------Y--DQ-
Hoxd-3	S--A-TA--SA-LV-----------FV-P--VQM-NL-N-S-----------Y--DQ-
zenZ1	L--S-TAF-SV-LV---N--KS-M--Y-T------QR-S-C---V--------F--DIQ
zenZ2	S--S-TAFSSL-LI---R---L-K--A-T----SQR-A----V----------L--STN
Hoxa-1	PNAV-TNS-TK-LT--------K----AA-V---AS-Q-N-T-V-----------Q--RE-
lab	NNS--TNF-NK-LT--------------A-----NT-Q-N-T-V-----------Q--RV-

Figure 22.4 Amino acid sequence comparisons of mouse and *Drosophila* homeodomains, showing the existence of subfamilies. The *Antennapedia*[+] sequence of *Drosophila* shown at the top is used as the basis for all comparisons. Each of the letters in this sequence represents one amino acid. Dashes represent amino acids identical to the *Antennapedia*[+] standard. The homeodomains are grouped in subfamilies based on the common deviations from the *Antennapedia*[+] standard.

The Order of Mouse *Hox* Genes on the Chromosome Is Colinear with Their Expression in the Embryo

The colinearity rule was already noted for the homeotic genes of *Drosophila*: as one progresses from 3′ to 5′ through the *bithorax* complex and the *Antennapedia* complex, the anterior boundary of each gene's expression domain moves farther to the posterior (see Fig. 21.30). The same colinearity is observed for the expression of *Hox* genes in the mouse. The *Hox B* genes provide an example.

All genes of the mouse *Hox B* complex are transcribed in the central nervous system from neurulation until birth and beyond. The expression domain of each *Hox B* gene begins at a well-defined anterior boundary and extends posteriorly into the spinal cord, where the expression boundary is usually not well defined. For genes near the 3′ end of the complex, such as *Hoxb-3*[+],

the anterior boundary maps in the hindbrain (Fig. 22.6). For the genes near the 5′ end of the complex, such as *Hoxb-9*[+], the anterior boundary maps in the spinal cord (A. Graham et al., 1989; Wilkinson et al., 1989). An exception is *Hoxb-1*[+], a gene with a chromosomal location closer to 3′ than *Hoxb-2*[+] but an anterior boundary behind that of *Hoxb-2*[+].

What is the significance of this intriguing colinearity between the physical order of genes within a homeobox complex and the anterior borders of their expression domains? Why has this correlation been conserved during hundreds of millions of years? Its adaptive value, if any, appears to be subtle, because the chromosomal position of homeobox genes is not an absolute requirement for their function. For instance, the *bithorax* complex of *Drosophila* can be split without disturbing normal development (Struhl, 1984). However, in accord with the *principle of synergistic mechanisms*, the regulation of each homeobox gene might still be *facilitated* by

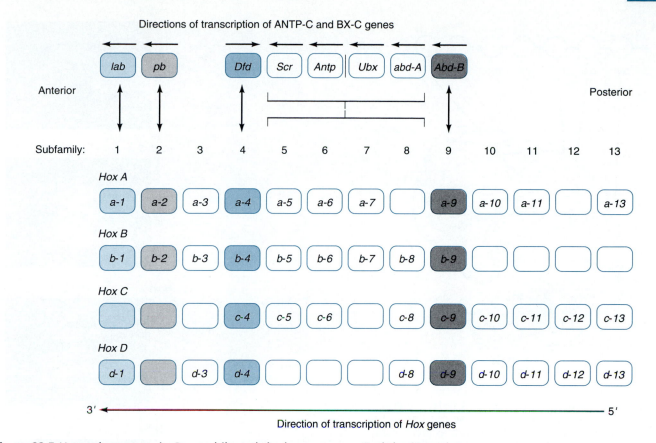

Figure 22.5 Homeobox genes in *Drosophila* and the house mouse. Each horizontal row represents a homeobox complex, with genes shown in their true physical order on the chromosome. Remarkably, the genes can be aligned so that each vertical column represents a subfamily based on maximum sequence similarity among the homeodomains encoded (see Fig. 22.4). Thus, subfamily members are arranged in the same order in all homeobox complexes. The abbreviations for *Drosophila* genes refer to the homeotic genes illustrated in Figure 21.30. In the abbreviations for mouse genes, the letter refers to the *Hox* complex in which the gene is located, and the number indicates the subfamily. The arrows indicate the 5'-3' polarity of the respective genes, that is, the direction of transcription. The long arrow also indicates the direction in which the anterior boundary of the embryonic expression domain moves as one proceeds farther in the 3' direction to the next gene. The data show that homeobox complexes have evolved from a common ancestral cluster that already contained many of the homeobox genes present today. Only *Sex combs reduced*[+] *(Scr)*, *Antennapedia*[+] *(Antp)*, *Ultrabithorax*[+] *(Ubx)*, *abdominal A*[+] *(abd-A)*, and the genes located closer to the 5' end than the *Abdominal-B*[+] *(Abd-B)* subfamily, as well as the corresponding *Hox* genes, may have diverged later.

its normal position within its complex. As discussed in Chapter 15, the gene-regulatory regions known as *enhancers* may act over large distances. Thus, enhancers may act on more than one promoter. Conceivably, neighboring genes may share enhancer regions, which then link their target genes into functional groups that facilitate the development of the correct expression pattern. Indeed, there are indications for shared regulatory elements among genes of the *Hox B* complex (Whiting et al., 1991; Sham et al., 1992).

The transcription factors that control the expression of *Hox* genes are still under investigation (Wilkinson, 1993). As an example, the expression of *Hoxb-2*[+] in the

hindbrain is promoted by the *Krox-20*[+] gene, and the Krox-20 protein, a transcription factor of the zinc finger type, binds to a specific enhancer region upstream of *Hoxb-2*[+] (Sham et al., 1993). Other candidates for transcription factors involved in *Hox* gene regulation are a family of receptors that function in a ligand-dependent way similar to steroid receptors (see Chapter 15). These potential regulators of *Hox* genes will be discussed in a subsequent section on the functional analysis of the *Hox* genes.

The boundaries of *Hox B* gene expression domains in the hindbrain (*rhombencephalon*) were mapped with respect to a repeating pattern of bulges called **rhom-**

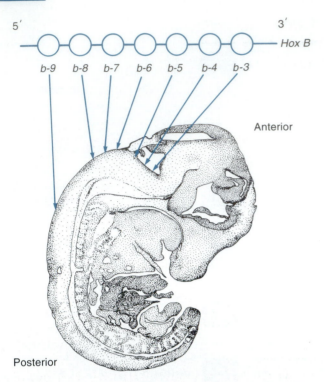

5′   3′

Hox B

b-9   b-8   b-7   b-6   b-5   b-4   b-3

Anterior

Posterior

Figure 22.6 Anterior boundaries for the expression of *Hox B* genes in the mouse central nervous system, and the correlation of these boundaries with gene position in the *Hox B* complex. The upper diagram represents several genes from the *Hox B* complex. The lower diagram shows a 12.5-day embryo in median section, head to upper right. The dots indicate the respective anterior limits of transcription of the genes, as revealed by in situ hybridization (see Methods 15.1).

bomeres (Fig. 22.7). In histological sections cut parallel to the axis of the hindbrain, the anterior boundaries of transcription were detected by in situ hybridization

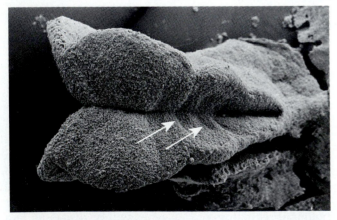

Figure 22.7 Rhombomeres in the developing mammalian brain. This scanning electron micrograph shows a mouse embryo during closure of the neural tube. The periodic ridges (arrows) in the bottom of the hindbrain are called rhombomeres.

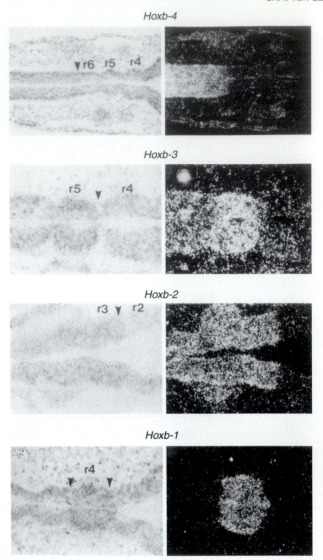

Hoxb-4

r6 r5 r4

Hoxb-3

r5 ▼ r4

Hoxb-2

r3 ▼ r2

Hoxb-1

r4

Figure 22.8 Transcription patterns of *Hox B* genes in the rhombencephalon (hindbrain) of a 9.5-day mouse embryo. Accumulated mRNA is made visible by in situ hybridization (see Methods 15.1) using gene-specific probes. In each pair of photomicrographs, the left was taken with bright-field illumination to show the sectioned tissue, and the right was taken with dark-field illumination to show the labeled probes as white grains. The serial bulges (rhombomeres) of the hindbrain are numbered, from anterior to posterior, r1, r2, etc. Cranial ganglia VII and VIII, adjacent to r4, were used as landmarks. Arrowheads mark the anterior boundaries of the expression domains of four *Hox B* genes.

(Fig. 22.8). Clearly, the investigated genes have expression domains that terminate near the boundaries between rhombomeres. Corresponding results were obtained with genes from the other mouse *Hox* complexes. *Hox* genes belonging to the same subfamily tend to have similar but not always identical anterior boundaries in their expression domains.

In addition to rhombomeres, *Hox* genes are also expressed in embryonic mesoderm and in the *neural crest cells* that contribute to *cranial nerves* and *pharyngeal arches*. Remarkably, *Hox* genes have anterior limits of expression in these structures similar to those in rhombomeres (Hunt et al., 1991b). Thus, the central nervous system, cranial nerves, and pharyngeal arches formed in the hindbrain region all tend to have the same combination of active *Hox* genes depending on their anteroposterior position (Fig. 22.9). This situation is reminiscent of the parasegments in *Drosophila*, which are also characterized by similar combinations of active homeotic and segment polarity genes in ectoderm, mesoderm, and endoderm.

Boundaries of *Hox* Gene Expression Are Also Clonal Restriction Lines

The coincidence of some gene expression boundaries with rhombomere and segmental boundaries in mice parallels the properties of epidermal compartments in insects. However, insect compartments are defined by clonal restriction lines, that is, lines not transgressed by cell clones generated after a certain stage of development (see Chapters 6 and 21). Are there similar clonal restriction lines between the rhombomeres of the vertebrate brain, or between groups of neural crest cells?

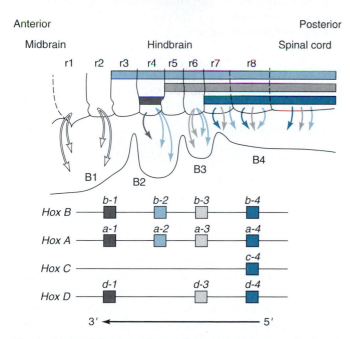

Figure 22.9 Summary of *Hox* gene expression in the head region of the mouse embryo. The top diagram shows a side view of the head region. The bottom diagram shows four subfamilies of homeobox genes (vertical rows). All subfamilies are expressed posteriorly; progressively fewer, anteriorly. Expression domains of these subfamilies include the spinal cord, hindbrain rhombomeres r3 to r8, and neural crest cells (arrows), which contribute to cranial ganglia and pharyngeal arches (B1 to B4).

This question is difficult to answer in regard to mice, because mammalian embryos past the implantation stage are difficult to manipulate and maintain. However, such experiments have been carried out in chickens, which are similar to mice in their modes of gastrulation and organogenesis. Chickens also have homeobox genes with expression domains similar to those of mice. For instance, a chicken homologue of *Hoxb-1*[+] is expressed in rhombomere 4, exactly like its mouse counterpart (Sundin and Eichele, 1990).

In order to test chicken rhombomeres for clonal restriction lines, Scott Fraser and his colleagues (1990) injected a fluorescent dye into individual cells of living chicken embryos. After a period of development, clones formed by the descendants of the labeled cells were inspected and mapped to the hindbrain. The critical question was whether a marked clone would respect the boundaries between rhombomeres or transgress them. The answer depended on the stage when the clone founder cell was labeled, just as in insects. When a founder cell was labeled before the appearance of morphological rhombomere boundaries, some of the resulting clones were found to overlap neighboring rhombomeres (Fig. 22.10). However, when founder cells were labeled after boundary appearance, most of the clones were restricted at the rhombomere boundaries (Birgbauer and Fraser, 1994). Thus, rhombomeres in the chicken brain seem to undergo a stage-dependent clonal restriction similar to that of epidermal compartments in insects.

To explore whether the clonal restriction lines in chicken rhombomeres were based on cell affinities, Sarah Guthrie and Andrew Lumsden (1991) transplanted rhombomere pieces. When pieces from different rhombomeres were juxtaposed, they reestablished a boundary between them, especially if a piece from an odd-numbered rhombomere was juxtaposed with a piece from an even-numbered rhombomere. In contrast, pieces from the same rhombomere combined without forming a boundary and showed extensive cell mixing (Guthrie et al., 1993). Thus, cells from different chicken rhombomeres minimized contact, whereas cells from the same rhombomere mixed freely, again in parallel with cells from insect epidermal compartments.

The transplanted tissue pieces also behaved autonomously—that is, were unaffected by their new environment—with regard to gene expression and cell differentiation. Rhombomere 4 tissue transplanted anteriorly into rhombomere 2 continued to express *Hoxb-1*[+] and proceeded to form the nerve nuclei and cranial nerve roots characteristic of the rhombomere 4 level (Guthrie et al., 1992; Kuratani and Eichele, 1993). These observations show that certain brain regions are determined at the time they become clonally restricted units. In addition, the results suggest that the determined state of brain regions may be linked to the expression of their region-specific *Hox* genes.

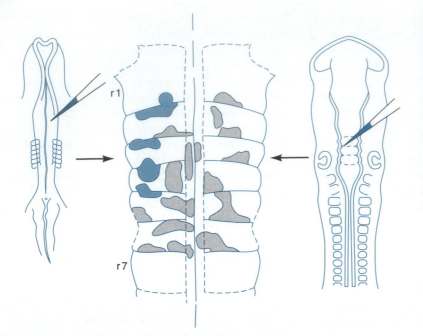

Figure 22.10 Clonal analysis of the rhombencephalon (hindbrain) development in the chick embryo. The center diagram shows the ventral aspect of a chicken hindbrain as if it had been cut open dorsally and spread out flat. The diagram on the left illustrates the marking of clone founder cells at the 5-somite stage before the appearance of rhombomeres. Some of the resulting clones (gray) respect the rhombomere boundaries, while other clones (color) overlap adjacent rhombomeres. The drawing on the right shows the result of the same procedure performed after the appearance of rhombomeres. In this case, most clones respect the rhombomere boundaries.

The results described so far show that the homeotic genes of insects and the *Hox* genes of mice share several properties. First, their protein products contain homeodomains, which are characteristic of transcription factors. Second, these genes are clustered in chromosomal complexes that have been conserved exceedingly well during evolution. Third, these genes are expressed in staggered domains, and the anterior boundaries of their expression domains are colinear with the physical order of the genes within their complex. Fourth, some of the anterior expression boundaries coincide with compartment lines that seem to be based on different cell affinities.

In addition, homeotic genes in insects act as *selector genes* directing the *morphological* development of each compartment (see Chapter 21). This is proven by the striking phenotypes of homeotic mutants, which show characteristic transformations of one compartment into a duplication of another compartment, such as the replacement of halteres with wings or antennae with legs. The following section describes attempts to generate equivalents of homeotic mutants in vertebrates by reverse genetics.

Functional Analysis of *Hox* Genes in Vertebrates

To determine whether the *Hox* genes of mice and the corresponding genes of other vertebrates control the morphology of certain body regions, one needs to analyze mutants or their equivalents. Unfortunately, no spontaneous mutants for these genes have yet been found in vertebrates. Therefore, several groups of investigators have used reverse genetics to obtain the equivalents of mutant phenotypes (Krumlauf, 1994). The following examples represent four experimental strategies for analyzing the function of a previously cloned vertebrate homeobox gene.

Inactivation of a Homeodomain Protein Causes a Transformation toward a More Anterior Fate

One way of generating the equivalent of a loss-of-function mutant is to *inactivate the product* of the gene of interest. This can be achieved by injecting an antibody to the protein encoded by the gene. The antibody binds to its target protein in vivo and interferes with its function. This approach was used by Eddy De Robertis and his coworkers (1990) to study *Xenopus* embryos. Soon after the discovery of the homeobox in *Drosophila*, they used a cloned *Antennapedia*[+] probe to screen a library of *Xenopus* DNA. They isolated a *Xenopus* gene now called *XlHbox 1*[+], a homologue of the murine *Hoxb-6*[+]. Having confirmed that *XlHbox 1*[+] encodes a homeodomain-containing protein, they raised antibodies directed against specific domains of this protein.

The purified antibodies were used to localize the expression domain of the *XlHbox 1*[+] gene by immunostaining (see Methods 4.1). The gene is expressed in a transverse band including the forelimb field in the neck region of the tadpole (Fig. 22.11). The role of other homeobox genes in limb development will be discussed in a later section of this chapter.

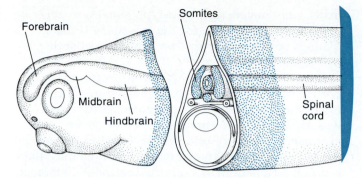

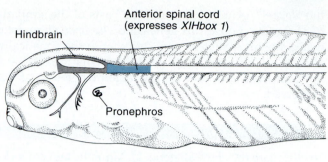

(a) Control tadpole

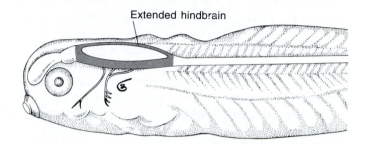

(b) Antibody-injected tadpole

Figure 22.11 Expression of the *XlHbox 1*[+] gene in the *Xenopus* tadpole. Immunostaining (see Methods 4.1) reveals the accumulation of XlHbox 1 protein in a discrete band (color) in the anterior trunk of the tadpole. The protein, which is present in both mesodermal and ectodermal tissues, accumulates in cell nuclei, in accord with its presumed function as a transcription factor.

Figure 22.12 Transformation of anterior spinal cord into hindbrain in *Xenopus* tadpoles after injection with antibodies against XlHbox 1 protein. **(a)** Normal embryo. The hindbrain has a large cavity closed dorsally by a thin roof, while the spinal cord has only a narrow canal (not shown) surrounded on all sides by thick neural tissue. The anterior part of the spinal cord adjacent to the hindbrain naturally produces XlHbox 1 protein. **(b)** Embryo developed from a zygote injected with antibody to XlHbox 1 protein. The anterior part of the spinal cord is transformed into an extension of the hindbrain. The closed roof of the hindbrain extension shows that the transformation does not result from a failure of the neural tube to close. The anteroposterior level of the transformed area was identified by morphological markers including the pronephros.

In order to interfere with the function of XlHbox 1 protein, the investigators used antibodies directed specifically against the protein sequence upstream of the homeodomain (C. V. E. Wright et al., 1989). These antibodies were injected into zygotes before first cleavage. The antibody remained intact in the embryo and was distributed to all cells during cleavage. In some of the developing tadpoles, there was a specific transformation in the same part of the central nervous system where *XlHbox 1*[+] is normally expressed. In normal embryos, this area contains the anterior part of the spinal cord, which is adjacent to the hindbrain. In antibody-injected embryos, the anterior spinal cord was replaced with an extension of the hindbrain (Fig. 22.12). The replacement could be readily identified because the hindbrain has a large cavity, the *fourth ventricle*, with a thin dorsal roof (see Chapter 12). In contrast, the spinal cord has only a small central canal surrounded on all sides by a thick layer of neural tissue. This transformation from spinal cord to hindbrain was found in 26 of 92 tadpoles (28%) developing from zygotes injected with anti-XlHbox 1 protein. Among 106 control tadpoles raised from zygotes injected with other antibodies, none experienced this type of transformation.

The experiment just described shows that XlHbox 1 protein is necessary for forming the anterior part of the spinal cord. If the protein is absent or reduced in amount, this part of the central nervous system develops as hindbrain. In other words, the lack of XlHbox 1 protein causes the transformation toward a more anterior fate. The experiments described in the following section will show that similar results can be obtained in mice by inactivating certain *Hox* genes.

Elimination of *Hox* Genes May Cause Transformations toward More Anterior Fates

In mice, experimenters can analyze the biological function of cloned genes by using a strategy known as *gene targeting*, or more colloquially, as *gene knockout*. First, they replace the gene of interest in an *embryonic stem (ES) cell* with an inactive allele of the same gene. Then, they use descendants of the transformed ES cell to create chimeric mice, in which the ES cells often contribute to the germ line. Finally, they breed chimeric males so that the altered gene will be passed on to all cells in some of their progeny. As illustrated in Figure 14.12, mice in which both copies of the *Hoxb-4*[+] gene were knocked out developed an abnormal second cervical vertebra that was morphologically similar to the first cervical vertebra. In a corresponding experiment, Hervé Le Mouellic and his colleagues (1990, 1992) eliminated the *Hoxc-8+* gene in mice. On the basis of the similarities between mouse *Hox* genes and *Drosophila* homeotic genes, the investigators expected that in the genetically altered mice there would be transformations toward more anterior fates in the body region where *Hoxc-8+* is expressed.

▼

To eliminate the *Hoxc-8+* gene, Le Mouellic and coworkers transformed ES cells with a DNA construct containing the bacterial *lacZ* and *NeoR* genes flanked by sequences from the transcribed region of the *Hoxc-8+* gene (Fig. 22.13). The flanking sequences were selected so that homologous recombination would replace a transcribed portion of the resident *Hoxc-8+* gene with the *lacz* and *NeoR* genes. This replacement would eliminate the function of the resident gene and bring the *lacZ* gene under the control of the resident *Hoxc-8+* regulatory region.

Transformed ES cells were identified by their ability to grow in a medium containing the poisonous drug G418. Such cells were grown into clones, and their DNA was tested by *Southern blotting* (see Methods 14.3) to determine whether the transgene had become inserted into one of the two resident *Hoxc-8+* genes. (This insertion site is favored over other sites by the presence of *Hoxc-8+* sequences in the transgene, which facilitate homologous recombination between identical DNA sequences.) Stem cell clones testing positive in Southern blotting were used to generate chimeric males producing sperm derived from the transformed ES cells. These

males were mated with wild-type females to generate offspring heterozygous for the transgene (*Hoxc-8+/−*). Matings of heterozygotes then produced offspring homozygous for the transgene (*Hoxc-8−/−*).

In both homozygotes and heterozygotes, the *lacZ* portion of the transgene was expressed under the control of the resident *Hoxc-8+* promoter. As expected, *lacZ* was expressed more strongly in homozygotes, and the expression pattern of *lacZ* corresponded to the normal pattern of *Hoxc-8+* expression. In wild-type mice there is generally a pattern of 7 cervical vertebrae, 13 thoracic vertebrae with ribs, and 6 lumbar vertebrae before the hipbone (called *sacrum*). For easy reference, these vertebrae are numbered consecutively from 1 to 26 in Figure 22.14. In 12.5-day embryos, *Hoxc-8+* expression begins with a sharp boundary behind the rudiment of vertebra 12, continues strongly until the future vertebra 22, and then decreases toward the posterior. Within this expression domain the researchers expected to find transformations toward more anterior fates.

The most obvious transformation in the homozygous mutant (*Hoxc-8−/−*) mice affected vertebra 21. This is normally the first lumbar vertebra; in homozygotes, however, this vertebra had the morphological characteristics of a thoracic vertebra, with a pair of ribs attached to it (Fig. 22.15a). This vertebra was followed by five vertebrae without ribs, so that the total number of vertebrae before the sacrum was 26 as in wild-type mice. The development of a 14th vertebra with ribs must therefore be viewed as a homeotic transformation of the 21st vertebra from lumbar to thoracic and not as the development of an additional thoracic vertebra.

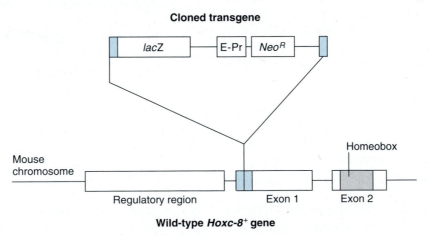

Cloned transgene

Wild-type *Hoxc-8+* gene

Figure 22.13 Inactivation of the mouse *Hoxc-8+* gene in embryonic stem cells by homologous recombination. The bottom diagram shows the wild-type mouse *Hoxc-8+* gene. The transcribed region has two exons, with the second exon containing the homeobox. The upper diagram shows the cloned transgene containing the *lacZ* gene (transcribed region only) and the *NeoR* gene (including the promoter, Pr, and enhancer, E) flanked by sequences from the first exon of *Hoxc-8+*. The flanking sequences, which are present in both the transgene and the resident *Hoxc-8+* gene, facilitate transgene insertion by homologous recombination. The *lacZ* gene of a properly inserted transgene is expressed under the control of the *Hoxc-8+* regulatory region.

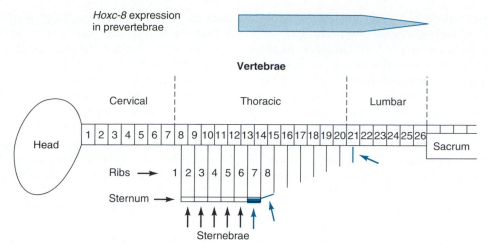

Figure 22.14 Homeotic transformations in mice homozygous for a null allele of *Hoxc-8*. In wild-type 12.5-day embryos, *Hoxc-8*⁺ expression begins with a sharp boundary behind prevertebra 12, continues strongly until prevertebra 22, and then decreases toward the posterior. This region develops abnormally in mutants homozygous for a null allele of this gene (*Hoxc-8*⁻/⁻). Wild-type mice have seven cervical vertebrae (vertebrae 1–7), 13 thoracic vertebrae (8–20) with ribs, and 6 lumbar vertebrae (21–26) without ribs. Homozygous mutants have the abnormalities drawn in color and marked with colored arrows. Vertebra 21, normally the first lumbar vertebra, is transformed into a thoracic vertebra. The sternum (breastbone) has an extra element (sternebra) between the attachment sites of ribs 6 and 7. Rib 8, normally unattached, is attached to rib 7 or to the sternum. These abnormalities can be interpreted as transformations of each affected segment toward the morphological character of its anterior neighbor.

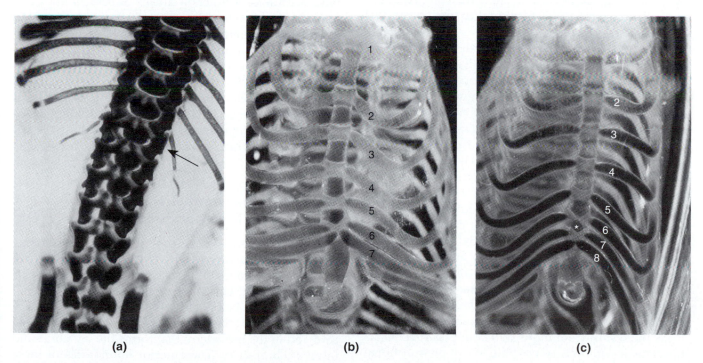

Figure 22.15 Homeotic transformations in mice homozygous for a null allele of a *Hox* gene (see Fig. 22.14). **(a)** Homozygous mutant (*Hoxc-8*⁻/⁻) having a pair of ribs attached to vertebra 21 (arrow). **(b)** Sternum region of the wild type in ventral view. The first seven ribs (numbered 1 through 7) are attached to the sternum, with the attachment sites of the first six ribs separated by sternebrae. The eighth rib is floating. **(c)** Sternum region of a homozygous mutant. Eight ribs are attached to the sternum, and an extra sternebra (white asterisk) has developed between ribs 6 and 7.

Further transformations of a homeotic type were observed in the area of the breastbone (sternum). In wild-type animals, the first seven ribs* are attached to the sternum (Fig. 22.15b). The attachment sites of the first six ribs are separated by bony elements (sternebrae), while the sixth and seventh ribs are attached to the sternum at the same point. In homozygous mutants, an additional sternebra developed between the sixth and seventh ribs, and the eighth rib was fused with the sternum at the same point as the seventh rib (Fig. 22.15c). In addition, the 12th rib of homozygous mutants was transformed into a likeness of the 11th rib by the criteria of overall length and bone:cartilage ratio. Moreover, other ribs and the transverse processes of vertebrae seemed anteriorized. All of the transformations described here were observed in most of the homozygous mutants and occasionally in heterozygotes.

In their experiments, Le Mouellic and his coworkers found that elimination of $Hoxc-8^+$ function in mice causes morphological transformations toward anterior fates in the body region where the gene is normally expressed. Similar transformations toward more anterior fates were observed in the neck region after knocking out the $Hoxb-4^+$ gene, as described in Chapter 14. Corresponding results were obtained after targeted disruption of $Hoxd-3^+$ (Condie and Capecchi, 1993). Likewise, knocking out the $Hoxa-2^+$ gene was found to cause homeotic transformations of second to first pharyngeal arch elements (Rijli et al., 1993).

Not all knockout experiments with Hox genes have led to anterior transformations. Inactivation of $Hoxa-5^+$ or $Hoxa-11^+$ caused posterior transformations, indicating that loss of function in either gene generates or mimics selector gene activities that normally prevail more posteriorly (Jeannottee et al., 1993; K. S. Small and S. Potter, 1993). Mice homozygous for null alleles of $Hoxa-3^+$ (Chisaka and Capecchi, 1991) or $Hoxa-1^+$ (Lufkin et al., 1991) had specific deficiencies and malformations but underwent no clear transformations toward anterior fates. These results can be interpreted in different ways. First, some Hox genes might not act as homeotic selector genes. Second, partial overlap of functions of Hox genes within a subfamily might obscure the consequences of losing the activity of one subfamily member. In accord with this partial backup hypothesis, the areas affected by knocking out $Hoxa-3^+$ or $Hoxa-1^+$ were smaller than the expression domains of those genes.

As the results from more knockout experiments accumulate, it will become possible to explore functional relationships between Hox genes. Since Hox genes are most similar to other genes within their own subfamily with regard to amino acid sequence and expression pattern, it had been thought that genes in the same subfamily would also carry out similar functions. It was therefore surprising that targeted disruption of $Hoxa-3$ and $Hoxd-3$ led to defects in different tissues. Mice deficient in the former gene had disorders in structures derived from neural crest cells, whereas mice lacking the latter gene function were defective in derivatives of paraxial and lateral mesoderm (Chisaka and Capecchi, 1991; Condie and Capecchi, 1993). Nevertheless, a synergistic interaction between the two genes was revealed by double mutants. In mice homozygous for both $Hoxa-3^-$ and $Hoxd-3^-$, the entire atlas is deleted, whereas in mice homozygous for $Hoxd-3^-$ alone the atlas is fused with the base of the skull (Condie and Capecchi, 1993, 1994). However, it has become apparent that the effects of gene knockouts may depend on the exact nature of the disruption in the targeted gene and on the genetic background provided by the mouse strain used in the experiment (Ramírez-Solis et al., 1993).

Although much more needs to be learned about the functions of Hox genes, the experiments noted so far confirm and extend the parallelism between mouse Hox genes and $Drosophila$ homeotic genes. The elimination of at least some homeobox genes or their products in vertebrates causes transformations of body elements into anterior serial homologues. These transformations are similar to the phenotypes of most $Drosophila$ mutants with loss-of-function alleles of certain homeotic genes, such as a transformation from metathorax to mesothorax (see Fig. 21.32). Conversely, $gain$-of-$function$ alleles of certain homeotic genes in $Drosophila$ cause mimicking of an activity pattern normally found more posteriorly. Such mutants, including the dominant $Antennapedia$ alleles, result in replacements of structures with more posterior serial homologues, such as a transformation from antenna to leg. As we will see in the following, similar transformations occur in mice as well.

Overexpression of Hox Genes May Cause Transformations toward More Posterior Fates

Many experiments in the functional analysis of Hox genes rely on *overexpression*, that is, the expression of a transgene at stages or in areas where the resident pair of genes is silent. This strategy generates the equivalent of a gain-of-function allele.

The normal $Hoxa-7^+$ gene is expressed in the spinal cord and in the mesoderm that forms the vertebrae and intervertebral discs. The anterior boundaries of these expression domains are in the lower neck, for the spinal cord, and in the thorax, for the mesoderm (Püschel et al., 1991). To cause overexpression of the $Hoxa-7^+$ gene, Michael Kessel and his colleagues (1990) engineered a transgene containing the transcribed region of $Hoxa-7^+$ under the control of an actin regulatory region. In mice transgenic for this construct, $Hoxa-7$ was expected to be

---

* In anatomical terminology, the world "rib" often refers to a *pair* of rib structures. Thus, "rib 7" in this context refers to the seventh pair of ribs.

expressed in the entire embryo, because actin promoters are active in all cells. The transgene's ectopic activity was expected to be critical in the head and the parts of the neck where the resident *Hoxa-7⁺* genes are silent. If the *Hoxa-7⁺* gene acted as a homeotic selector gene, then its overexpression should transform head and neck structures into structures normally found farther posterior. Such transformations were indeed observed.

The overt segmentation of vertebrates begins with the formation of *somites*, epithelial balls of paraxial mesoderm (see Chapter 13). Generally, the *sclerotome* portions of the somite form vertebral rudiments called *prevertebrae* (Fig. 22.16b). However, somite development is modified in the head and neck region (Jenkins, 1969; Romer, 1976; Verbout, 1985). The first somites, called *occipital somites*, do not generate vertebrae but form the *occipital bones* (Lat. *occiput,* "back of the head"). These bones surround the opening in the base of the skull, through which the spinal cord exits into the vertebral column. The fourth somite forms the rudiment of the *proatlas,* which develops into a separate vertebra in certain reptiles. In mammals, however, the proatlas rudiment contributes to the basioccipital bone and to the second vertebra.

The first seven vertebrae in mammals do not carry ribs. They are called *cervical vertebrae* (Lat. *cervix,* "neck"). The first two cervical vertebrae are specially adapted to allow the head to move relative to the vertebral column. The first vertebra is called the *atlas,* after the Greek mythological figure who supported the sky with his shoulders. The atlas lacks the vertebral body, that is, the weight-bearing, drum-shaped part of

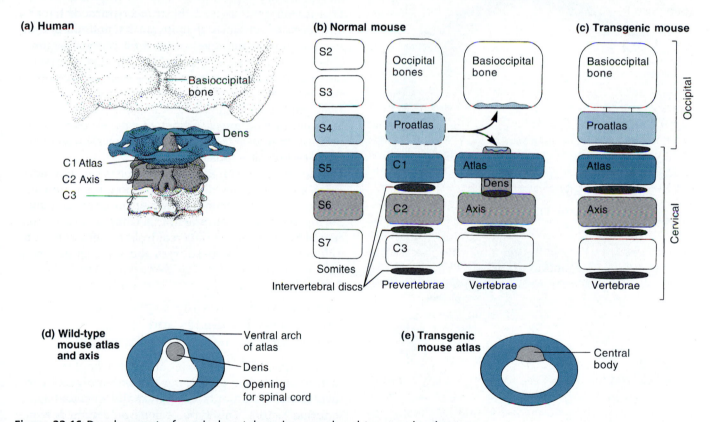

Figure 22.16 Development of cervical vertebrae in normal and transgenic mice overexpressing a *Hoxa-7* transgene. **(a)** Anatomical drawings of part of the skull and anteriormost vertebrae of a normal human in dorsal view. **(b)** Schematic diagrams showing development of six somites (S2 to S7, left column) in a normal mouse. The somites develop into occipital bones and two cervical prevertebrae (C1 and C2), with a transient rudiment (proatlas) contributing to both occipital bones and cervical vertebrae. C1 and C2 form the anteriormost cervical vertebrae, the atlas and the axis. Part of C1, instead of forming the body of the atlas, gives rise to the dens of the axis. **(c)** Cervical vertebrae in postnatal transgenic mice overexpressing a *Hoxa-7* transgene. The proatlas persists as a vertebra, separated from the atlas by an extra intervertebral disc. The atlas has a body, and the axis has no dens. **(d)** Anterior (cranial) view of the normal atlas and axis. The atlas has no central body. It rotates around the dens of the axis. Note the slim ventral arch of the atlas. **(e)** Anterior view of the atlas in transgenic mice expressing a *Hoxa-7* transgene. The atlas has a central body like a more posterior cervical vertebra. The underlying axis (not shown) has no dens.

a standard vertebra (Fig. 22.16a and d). Instead, the atlas is held together ventrally by a slim *ventral arch*. What would have been the body of the atlas has become part of the second cervical vertebra, the **axis**. The axis has a central process called the *dens* (Lat., "tooth"), which develops from cells that would otherwise have formed the body of the atlas. This modification allows the atlas, along with the skull, to rotate around the dens.

In the experiment of M. Kessel and colleagues (1990), the atlas of transgenic mice overexpressing the *Hoxa-7* transgene showed a striking abnormality: there was a vertebral body inside the ventral arch (Figs. 22.16c and

e, and 22.17). Correspondingly, the axis of transgenic mice had no dens. Moreover, the small patch of intervertebral disc tissue that is normally embedded between the dens and the body of the axis was broadened into a full-sized intervertebral disc separating the atlas from the axis. In addition, transgenic mice had another intervertebral disc anterior to the atlas; and in front of this disc, a supernumerary vertebra corresponding to the proatlas was attached to the occipital bones (Fig. 22.16c). In sum, the characteristic adaptations of the atlas and axis had altogether disappeared, and the anterior cervical vertebrae had assumed the structure normally seen in *more posterior* cervical vertebrae.

The experiment just described shows that overexpression of the *Hoxa-7* gene in mouse embryos changes the development of vertebrae in the head-neck zone so that they resemble more posterior vertebrae. Similar transformations were observed in mice with a transgene combining the transcribed region of *Hoxd-4* with the regulatory region of *Hoxa-1* (Lufkin et al., 1992). This fusion gene caused the transcription of Hoxd-4 mRNA more anteriorly than its normal anterior boundary, which is at the level of the first cervical somites. This ectopic expression resulted in the expected posterior transformation: occipital bones were replaced with structures resembling cervical vertebrae. This result again confirms the general parallelism between mouse *Hox* genes and *Drosophila* homeotic genes: overexpressions of both tend to cause morphological transformations toward more posterior fates.

The dramatic effects of misexpressing homeotic genes and *Hox* genes raise the question, How are these genes activated in exactly their appropriate domains during normal development? In Chapter 21, we discussed the relevant genetic controls in *Drosophila*. The following section will address the same question in mice.

Retinoic Acid Shifts the Expression Domains of *Hox* Genes and Causes Morphological Transformations

In the *Drosophila* embryo, as noted in Chapter 21, the earliest embryonic patterning genes are regulated by *concentration* gradients of maternally encoded transcription factors. This type of gene regulation is facilitated by the superficial cleavage of insect eggs, which leaves large molecules free to diffuse even after 13 rounds of mitotic division.

This mechanism is not found in mice; the holoblastic cleavage of mammals precludes it. However, gradients of transcription factor *activity* might be generated by a ligand-dependent transcription factor produced evenly in all cells and a ligand that is produced locally and diffuses across cell membranes. Such a system may be provided by retinoic acid and various retinoic acid

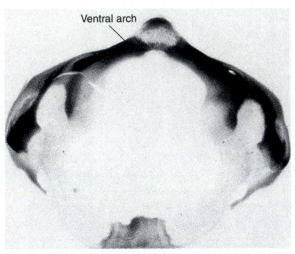

(a) Wild type

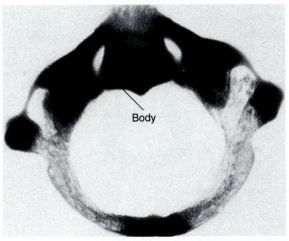

(b) Transgenic

Figure 22.17 The atlas bone of a normal mouse **(a)** and a mouse overexpressing a *Hoxa-7* transgene **(b)**; refer to Figure 22.16d and e, respectively. The photographs show the vertebrae in anterior view. Note the slim ventral arch of the normal atlas and the solid body of the transgenic atlas.

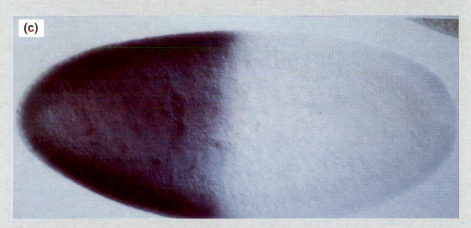

Color Plate 1   Regional control of embryonic gene expression by maternally encoded transcription factors in *Drosophila melanogaster*. Anterior pole to the left. **(a)** Maternal bicoid mRNA, stained dark here by in situ hybridization, becomes localized near the anterior pole while the oocyte develops in the ovary. **(b)** Bicoid protein, made visible by immunostaining, is synthesized after egg deposition. The protein decays as it diffuses away from its mRNA, thus forming a concentration gradient. **(c)** When the embryo begins to express its own genes, bicoid protein acts as a transcription factor. Above a threshold value that is normally surpassed in the anterior egg half, bicoid protein activates the embryonic *hunchback*⁺ gene, whose transcripts are stained dark here by in situ hybridization.

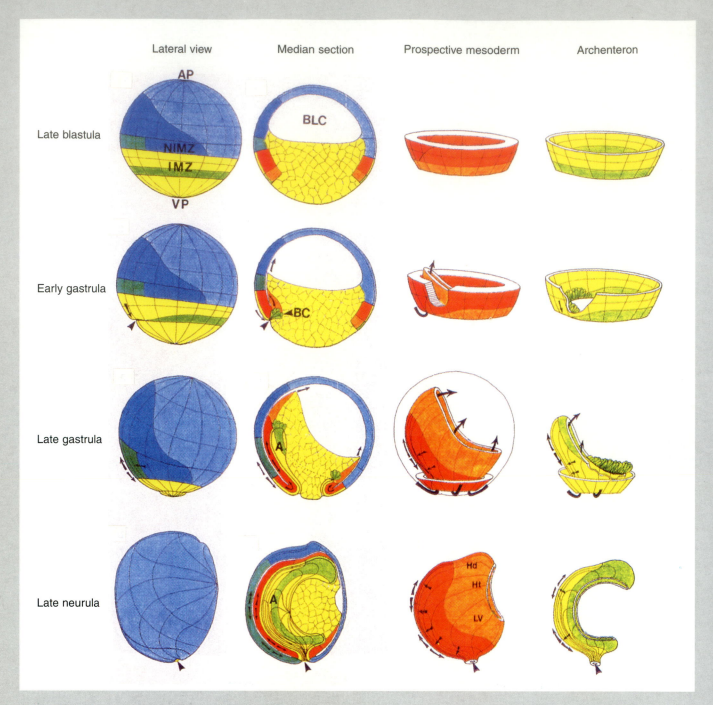

Color Plate 2 Gastrulation in *Xenopus laevis*. Dorsal side to the left, animal pole at the top. Prospective tissues are color-coded: epidermis (light blue), neural tissue (dark blue) including notoplate (blue-green), mesoderm (red and orange), archenteron roof and floor (yellow), and archenteron tip and sides (light green). Arrowheads point to the blastopore; arrows indicate the direction of movement. A = archenteron; AP = animal pole; BC = bottle cells; BLC = blastocoel; Hd = head mesoderm; Ht = heart mesoderm; IMZ = involuting marginal zone, composed of a deep layer (red) and a superficial layer (yellow); LV = lateral and ventral mesoderm; NIMZ = noninvoluting marginal zone; VP = vegetal pole.

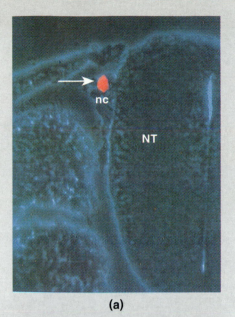

(a)

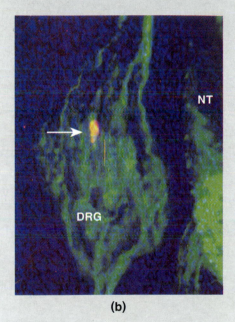

(b)

Color Plate 3 Pluripotency of migrating neural crest cells in chicken embryos. **(a)** Individual neural crest cells (nc) migrating away from the neural tube (NT) were microinjected with a red fluorescent dye; the descendants of the injected cell formed a clone of labeled cells. **(b)** Two days later embryos were fixed, sectioned, and immunostained with a primary antibody against neurofilaments and a secondary antibody conjugated with a green fluorescent dye. A cell (arrow) stained with both dyes was found in a dorsal root ganglion (DRG). Other neural crest cells from the same clone were located in a sympathetic ganglion and in a ventral root. Thus, the founder cell of the clone was not restricted to a single fate.

(a)

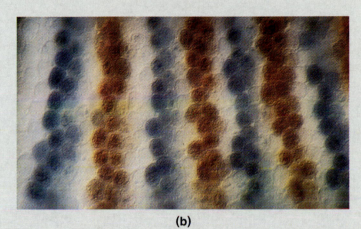

(b)

Color Plate 4 Expression of pair-rule genes in the *Drosophila* embryo. Anterior pole to the left, ventral side down. Embryos were immunostained for fushi tarazu protein (brown) and even-skipped protein (blue). Cells having these proteins will form the anterior margins of even-numbered and odd-numbered parasegments, respectively. Both proteins are transcription factors and accumulate in cell nuclei, where they bind to the regulatory regions of their target genes. Each parasegment at this stage is about four cells wide, and each pair-rule gene is expressed with a bisegmental periodicity. **(a)** Embryo at the cellular blastoderm stage. **(b)** Similar embryo at higher magnification.

Color Plate 5 *Antennapedia* phenotype of *Drosophila*. The *Antennapedia+* gene normally is expressed in the thorax. Certain gain-of-function alleles are also expressed in the head, causing the antennae to be replaced by legs.

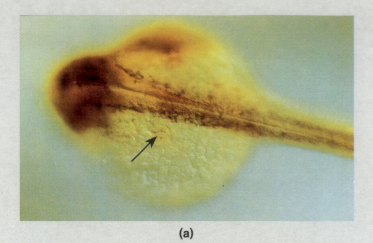

(a)

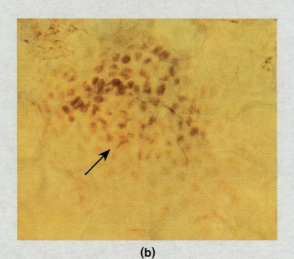

(b)

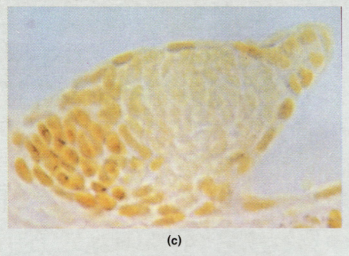

(c)

Color Plate 6 Fin field in the zebra fish embryo. **(a)** Zebra fish embryo immunostained with a primary antibody against the homeodomain from *Xenopus* XlHbox 1 protein. The arrow points to a circular area containing the cells that will form the pectoral fin (fin field). **(b)** High-power magnification shows that the fin field cells express XlHbox 1 antigen. **(c)** Histological section indicating that the antigen is expressed in anterior and proximal mesoderm as well as in overlying epidermal ectoderm.

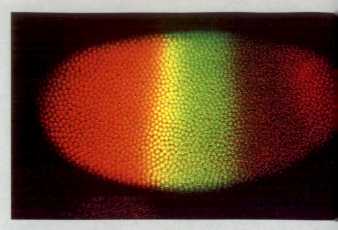

Color Plate 7 Expression of two gap genes, *hunchback*+ (orange) and *Krüppel*+ (green), in a *Drosophila* embryo at the cellular blastoderm stage. Anterior to the left, ventral side at the bottom. The embryo was immunostained for each gap gene protein, and the two immunofluorescent images were superimposed by computer. Note the overlap between two expression domains (yellow band) and the gradual decrease of expression toward the margins.

Color Plate 8 Cellular compartments in *Drosophila* imaginal discs. The photo shows the apical surface of a wing imaginal disc. The disc has been stained with a green fluorescent dye that outlines cell boundaries and a red fluorescent dye that marks cells expressing the *engrailed*+ gene. The anterior limit of *engrailed*+ expression coincides with a relatively smooth and straight line of cell boundaries, indicating that cells expressing *engrailed*+ minimize their contact with cells that do not express this gene.

Color Plate 9 The imaginal discs for wing (left) and haltere (right) in *Drosophila* are subdivided into compartments defined by clonal restriction lines. Each compartment is characterized by the expression of certain selector genes. One gene (*vestigial*+, green stain) is expressed in the appendages, as distinct from the thorax. Another gene (*cubitus interruptus dominus*+, blue stain) marks the anterior compartments. A third gene (*apterous*+, red stain) is expressed only in the dorsal compartments. Overlap between red and green gives rise to yellow-orange; red and blue produce purple.

(a) (b) (c)

(d) (e)

Color Plate 10 Wild-type and homeotic mutants of the wall cress *Arabidopsis thaliana*. **(a)** Wild type with sepals (hidden), four petals, six stamens, and a pistil fused from two carpels. **(b)** *agamous* mutant showing extra petals and sepals in the absence of stamens and carpels. **(c)** *apetala3-1* mutant, with the outer two whorls (both sepals) removed to show the third whorl organs, which develop as carpels rather than stamens. **(d)** *apetala2-2* mutant showing first whorl carpels and absence of second whorl organs. **(e)** *apetala2-1* mutant with first whorl leaves and second whorl stamen-petal intermediates.

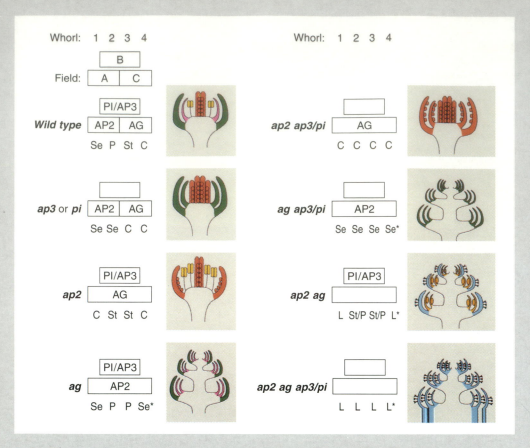

Color Plate 11 Genetic model of whorl determination in *Arabidopsis* flowers. The four whorls of floral organs are determined by three sets of homeotic genes, expressed in three ring-shaped or circular fields (A, B, and C), which are represented as boxes in a schematic cross section of half a floral primordium (top left). In wild-type flowers, the *apetala2⁺* (*ap2⁺*) gene is expressed in field A, and its product (AP2) determines the development of sepals (Se). The *pistillata⁺* (*pi⁺*) and *apetala3⁺* (*ap3⁺*) genes are expressed in field B, and their products (AP3 and PI), together with AP2, determine the development of petals (P). The *agamous⁺* (*ag⁺*) gene is expressed in field C, and its product (AG), together with PI/AP3, determines the formation of stamens (St). By itself, AG determines the formation of carpels (C). The activities of *apetala2⁺* and *agamous⁺* inhibit each other, and *agamous⁺* also terminates the formation of additional flower organs after carpel formation. For single, double, and triple loss-of-function mutants (identified in boldface at left), the model predicts the distribution of gene products and correctly explains the phenotypes (schematic drawings at right). The asterisk is a reminder that loss of the *agamous⁺* function causes the formation of additional whorls inside whorl 4. L = leaves or carpel-like leaves.

(a)

(b)

(c)

(d)

Color Plate 12 Phenotypes of double and triple homeotic mutants of *Arabidopsis*. **(a)** Flowers homozygous for both *apetala2-1* and *apetala3-1*. All organs are carpels or carpel-like leaves. **(b)** Flowers homozygous for both *apetala3-1* and *agamous*. All floral organs are sepals, and supernumerary whorls form. **(c)** Flowers homozygous for both *apetala2-1* and *agamous*. The organs of whorls 1 and 4 are leaves, the organs of whorls 2 and 3 are intermediate between petals and stamens, and additional whorls are formed. **(d)** Flowers homozygous for the three mutations *apetala2-1*, *pistillata*, and *agamous*. All floral organs develop as leaves, and additional whorls form.

Color Plate 13 Phenotype of the *superman* mutant of *Arabidopsis*. This flower shows 4 sepals, 4 petals, 10 stamens, and a reduced pistil. See also Figure 23.16.

(a)

(e)

(b)

(f)

(c)

(g)

(d)

Color Plate 14 Parallel series of homeotic mutants of *Antirrhinum* (left) and *Arabidopsis* (right). **(a, e)** Wild type. **(b, f)** Class A mutants *ovulata* and *apetala2*, both with carpels in place of sepals and no petals. **(c, g)** Class B mutants *deficiens* and *pistillata*, both without petals and stamens. **(d, h)** Class C mutants *plena* and *agamous*, both with a reiterated sequence of sepal-petal-petal.

(h)

Color Plate 15 Flowers of *Antirrhinum* shown in face view. The wild type (left) is bilaterally symmetrical; the *cycloidea* mutant (right) is radially symmetrical.

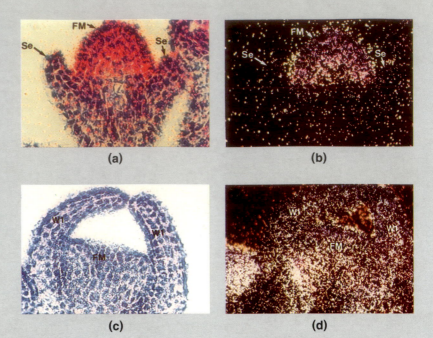

(a)

(b)

(c)

(d)

Color Plate 16 Distribution of agamous mRNA in *Arabidopsis* flower primordia. All panels show sections of flower primordia hybridized in situ with a radiolabeled agamous antisense RNA probe (see Methods 15.1). The photographs in parts a and c were taken with bright-field illumination; radioactivity signals appear as red dots. Parts b and d show the same sections in dark-field illumination; radioactivity signals appear as white grains. **(a, b)** Wild-type flower primordium. The sepal primordia (Se) are unlabeled; the central floral meristem cells (FM) contain agamous mRNA. These are the cells from which the stamen and carpel primordia will arise. **(c, d)** *apetala2* mutant flower primordium at a corresponding stage. The organ primordia in whorl 1 (W1), which form carpels instead of sepals, contain agamous mRNA.

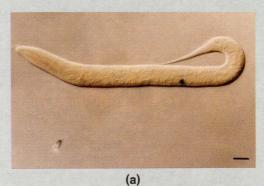

(a)

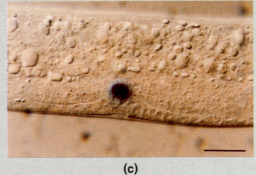

(c)

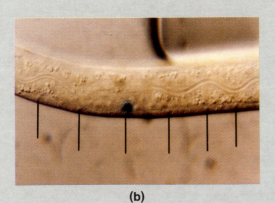

(b)

Color Plate 17 Expression of a *lin-3-lacZ* fusion gene in *Caenorhabditis elegans*. **(a)** Molting L2 larva showing *lacZ* expression (blue stain) in the anchor cell of the gonad rudiment. **(b)** Molting L2 larva showing the six vulval precursor cells (marked by vertical lines) and *lacZ* expression in the anchor cell. **(c)** L4 larva with the anchor cell on top of the induced vulva. Scale bars 20 μm.

receptors, which occur naturally in embryos and function as ligand-dependent transcription factors similar to steroid receptors (see Chapter 15).

Retinoic acid and other retinoids such as vitamin A have long been known as teratogens (Gk. *teras*, "monster"; thus, a severely malformed embryo). For instance, amputation of *tails* from frog tadpoles and subsequent application of retinyl palmitate to the wound surface may lead to the formation of supernumerary *hindlimbs* (Maden, 1993). This property, along with the presence of retinoic acid and its receptors in vertebrates, has made retinoic acid an important tool in the study of *Hox* gene expression.

To analyze the effects of retinoids on *Hox* gene expression, Michael Kessel and Peter Gruss (1991) fed retinoic acid to pregnant mice and monitored their fetuses for resulting malformations and for changes in *Hox* gene expression patterns. When given during day 7 of pregnancy, retinoic acid caused transformations of cervical vertebrae similar to those observed after overexpression of *Hoxa-7* (Figs. 22.16 and 22.17). In addition,

vertebra 7, which normally develops as a cervical vertebra without ribs, developed as a thoracic vertebra with ribs. These transformations of vertebrae toward more *posterior* fates were correlated with an *anterior* shift in expression of the (resident) *Hoxa-7*$^+$ genes during the development of prevertebrae. In fetuses exposed to retinoic acid, Hoxa-7 mRNA was detectable in prevertebra 9, and strong accumulation was found in prevertebra 10 and more posterior prevertebrae (Fig. 22.18a and b). In untreated control fetuses, a faint *Hoxa-7*$^+$ signal was seen in prevertebra 10, and a strong signal in prevertebra 11 and in more posterior prevertebrae (Fig. 22.18c and d). Thus, exposure to retinoic acid shifted the expression domain of *Hoxa-7*$^+$ by one segment toward the anterior, generating in each segment a situation that normally occurred one segment farther posterior. Correspondingly, each segment developed morphological characteristics of its posterior neighbor.

Similar effects of retinoic acid applied at 7.5 days were observed in the development of the hindbrain (H. Marshall et al., 1992; Conlon and Rossant, 1992). Again

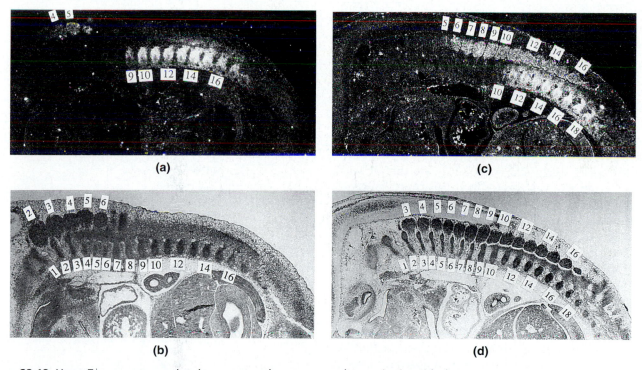

(a)　　　　　　　　　　　　(c)

(b)　　　　　　　　　　　　(d)

Figure 22.18 *Hoxa-7*$^+$ gene expression in mouse embryos exposed to retinoic acid; dorsal surface up, anterior to left. Females were fed with retinoic acid on day 7 of pregnancy, and their embryos were prepared for in situ hybridization at 12.5 days. The labeled probe hybridized specifically with Hoxa-7 mRNA, which accumulated in prevertebrae (lower row of numbers, 1 being the atlas) and spinal ganglia (upper row). The photographs on top were taken with dark-field illumination to show the signal generated by the radiolabeled probe as white grains. The photographs below show the same specimens in transmitted light to make organ rudiments more visible. **(a, b)** Embryo exposed to retinoic acid during day 7 of pregnancy. **(c, d)** Normal control embryo. Exposure to retinoic acid shifts *Hoxa-7*$^+$ expression in vertebrae as well as spinal ganglia anteriorly by about one segment to the anterior.

the expression domains of several *Hox* genes shifted anteriorly, and rhombomeres 2 and 3 developed with the morphological characteristics of rhombomeres 4 and 5. This transformation also involved the associated cranial nerve, which developed not as a trigeminal nerve but as a facial nerve.

It is apparent that retinoic acid *can* interfere with the process that activates *Hox* genes in their specific expression domains. Whether retinoids are *natural* activators of *Hox* genes—and, if so, how they act in the normal activation process—is still being investigated (S. V. Bryant and D. M. Gardiner, 1992). In *Xenopus* embryos at the neurula stage, retinoids are present in an anteroposterior gradient, with the highest level at the posterior end (Chen et al., 1994). It is tempting to speculate that retinoids act as *morphogens* activating specific combinations of *Hox* genes along the anteroposterior axis. However, in the experiments of Kessel and Gruss described previously, retinoic acid did not always cause posterior transformations. When administered at 8.5 days, retinoic acid had effects that were opposite to those observed after application on day 7. Retinoic acid now caused anterior transformations in another region of the vertebral column, similar to those observed after elimination of the *Hoxc-8*+ gene. For instance, the 8th pair of ribs reached the sternum, whereas normally these ribs are the first without a sternal junction (Fig. 22.15b and c).

The perplexing and sometimes paradoxical effects of retinoids are consistent with the general observation that a given transcription factor, such as a retinoic acid receptor, may have different effects on different target genes, effects that depend on cooperative or inhibitory actions of other transcription factors.

The results discussed so far indicate that at least some *Hox* genes in vertebrates have the same properties that characterize the homeotic genes of *Drosophila*. Such *Hox* genes are expressed in domains defined by clonal restriction lines, and their expression affects cell adhesion. Most important, the morphological development of a given embryonic area depends on its specific combination of active *Hox* genes. The *Hox* genes in turn encode transcription factors acting on subordinate genes. Thus, the common function of homeotic genes in flies, mice, and probably most other animals is to define anteroposterior spatial patterns of selector gene activities, which in turn produce their morphogenetic effects through various types of target genes.

Pattern Formation and *Hox* Gene Expression in Limb Buds

In their pursuit of answers to questions of pattern formation, morphogen gradients, and—more recently—

homeobox gene activation, many scientists have examined the limb buds of vertebrates, especially chickens and mice (Tabin, 1991; Izpisúa-Belmonte and Duboule, 1992). The chicken wing bud has the advantage of being relatively large and easily accessible for transplantation experiments.

The wing buds of vertebrates consist of a mesenchymal core and an ectodermal epithelium. The ectoderm is necessary for bud growth, but the mesenchyme determines the character of the limb to be formed (see Chapter 13). The mesenchymal core is derived from two mesodermal sources: *somite* cells that give rise to muscle, and *lateral plate* cells that contribute cartilage and other connective tissues. The lateral plate cells determine the limb pattern: in the absence of somites, the lateral mesoderm will form a muscleless limb with a normally structured skeleton and the rudiments of tendons (Kieny and Chevallier, 1979).

As the limb bud grows it develops three polarity axes: anteroposterior, dorsoventral, and proximodistal (Fig. 22.19). These axes are set up at different stages of

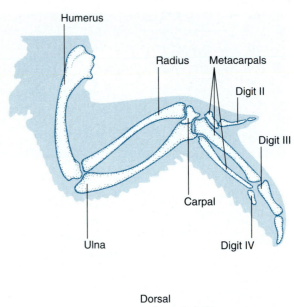
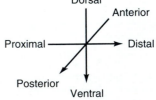

Figure 22.19 Cartilage elements in the chicken wing. The diagram shows a dorsal view of the right wing. The digits are numbered II, III, and IV, according to their homology with the digits of reptiles and mammals. Digit II is homologous to the index finger of the human hand; digit IV is homologous to the human ring finger. The wing has three polarity axes: anteroposterior, dorsoventral, and proximodistal.

development and apparently by different mechanisms. We will focus on the anteroposterior axis (from thumb to little finger, in the human hand). This axis is already specified in the limb field of the lateral plate mesoderm before the limb bud grows out (Zwilling, 1956). The most conspicuous feature of the anteroposterior limb pattern is a series of digits, each of which has a unique size and shape.

A Zone of Polarizing Activity Determines the Anteroposterior Pattern of the Chicken Wing

Transplantation experiments on chicken wing buds have revealed a key element in the establishment of anteroposterior polarity. John Saunders and Mary Gasseling (1968) removed a bit of tissue from the *posterior rim* of a donor wing bud and inserted it into the *anterior rim* of a host wing bud (Fig. 22.20). As a result, the normal sequence of digits in a chicken wing (II-III-IV) was changed to a mirror-image duplication (IV-III-II-III-IV). When the posterior tissue was implanted at the *tip* of the donor wing bud, the resulting pattern of digits was II-III-IV-III-IV (from anterior to posterior). It appeared that posterior wing bud tissue had the ability to establish the posterior pole of the anteroposterior axis. This ability was later termed *polarizing activity.*

Other researchers repeated and extended these experiments, mapping the polarizing activities of differ-

ent limb bud areas at different stages of development. In early limb buds, the activity is strongest at the posterior edge of the bud near its junction with the body wall (Fig. 22.21). In more advanced limb buds, the activity is greatest in a distal portion of the posterior edge. At each stage, the most active area is called the *zone of polarizing activity (ZPA).* The polarizing activity, as measured in transplantation experiments, is correlated with the expression of *Sonic hedgehog*[+], a chicken gene with sequence similarity to the *hedgehog*[+] gene of *Drosophila* (Riddle et al., 1993).

Polarizing activity was also found in the corresponding zones of reptilian and mammalian limb buds, indicating that the polarizing signal has been well conserved during evolution. Transplantations between different types of vertebrates also showed that the extra digits were formed by cells of the host limb bud rather than by the transplant itself. Additional experiments revealed that polarizing activity is present in embryonic tissues other than limb buds, such as *Hensen's node* in chicken gastrulae, the *notochord*, the *floor plate* of the developing spinal cord, and the *genital tubercle.*

Lewis Wolpert (1978) interpreted the polarizing activity in terms of *positional values.* He proposed that the ZPA produces a morphogen that diffuses away from its source, thus forming a morphogen gradient. According to Wolpert's model, digit IV of the wing is formed posteriorly, where the morphogen concentration is highest; digit III, at an intermediate level; and digit II, anteriorly, where the morphogen level is lowest (Fig. 22.22). Thus, adding extra ZPA material anteriorly or at the wing bud tip changes the morphogen gradient, and the pattern of digits formed changes accordingly. This model provided an elegant explanation for the results described and triggered an active search for the morphogen.

Retinoic Acid Mimics the Effect of a Transplanted Zone of Polarizing Activity

A candidate for the putative morphogen was discovered when synthetic retinoic acid applied to the anterior wing bud margin mimicked the effect of a ZPA: it generated mirror-image duplicate wing patterns (Tickle et al., 1982; Summerbell, 1983). Thus, retinoic acid could be naturally involved in establishing the ZPA, or it could be the putative morphogen generated by the ZPA, or it could just be an exogenous drug that mimics either of these activities. To distinguish among these possibilities, Christina Thaller and Gregor Eichele (1987, 1988) measured the concentrations of natural retinoids in anterior and posterior halves of chicken wing buds. Their results were compatible with the hypothesis that the zone of polarizing activity at the posterior margin of the wing bud acts as a natural source of retinoic acid, that retinoic acid is present as a morphogen gradient decreasing from posterior to anterior, and that the dif-

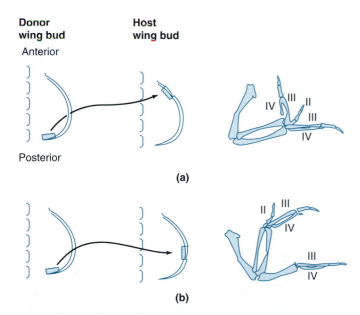

Figure 22.20 Effects of grafting tissue from the posterior margin of a chicken wing bud. The digits are numbered as in Figure. 22.19. **(a)** Implantation of posterior tissue into the anterior margin of a host wing bud results in a mirror-image duplication of the digital pattern (IV-III-II-III-IV). **(b)** Implantation into the tip of a host wing bud results in a supernumerary ulna with additional digits III and IV.

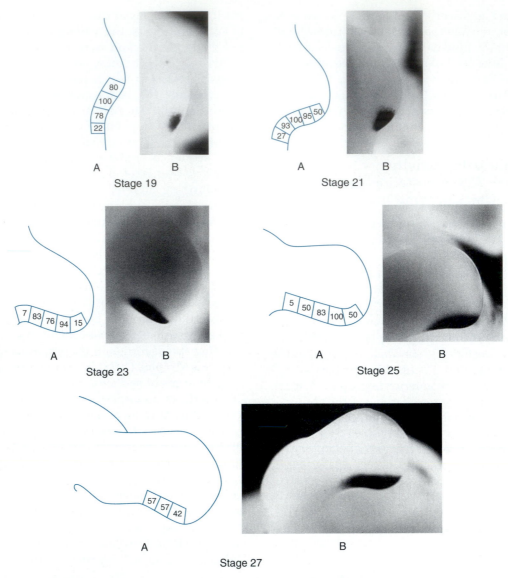

Figure 22.21 Polarizing activity and *Sonic hedgehog*⁺ gene expression in chicken wing buds. The numbered stages of development are as defined by Hamburger and Hamilton (1951). For each stage, a diagram (A) and a photograph (B) of a limb bud are shown. In the diagrams, the number in each square indicates the relative strength (%) of the polarizing activity determined in transplantation experiments by Honig and Summerbell (1985). The photos show expression of the *Sonic hedgehog*⁺ gene as revealed by in situ hybridization (Riddle et al., 1993).

ferent digits of the limb are specified by different local concentrations of retinoic acid. However, the results were equally compatible with the view that relatively high concentrations of retinoic acid *establish* a zone of polarizing activity, which then specifies the anteroposterior pattern by other means.

Retinoic Acid Does Not Act as a Morphogen in Limb Buds

Subsequent experiments cast serious doubts on the proposition that retinoic acid is the morphogen respon-

sible for the patterning effects generated by the zone of polarizing activity in limb buds. First, the endogenous concentration of retinoic acid decreases after the early limb bud stage and remains low during subsequent development (Thaller and Eichele, 1990). In contrast, the activity of the polarizing zone remains at relatively high levels past the decline of retinoic acid concentration (Fig. 22.21). Second, wing bud cells respond differently to retinoic acid and implanted zones of polarizing activity (Noji et al., 1991). When a resin bead soaked in retinoic acid is implanted into a chick limb bud, the synthesis of retinoic acid receptor protein in neighbor-

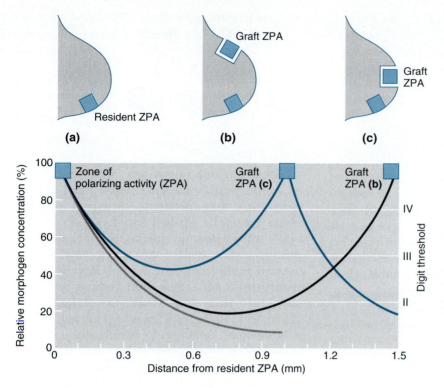

Figure 22.22 Gradient model to explain the results of the transplantation experiments shown in Fig. 22.20. **(a)** In a normal chicken wing bud, the zone of polarizing activity (ZPA) at the posterior margin is thought to set up a morphogen gradient that declines exponentially toward the anterior margin (gray line in bottom diagram). At certain threshold concentrations (horizontal lines), cells respond by forming digits IV, III, and II. **(b)** If an additional ZPA is implanted anteriorly, the two morphogen sources generate a symmetrical gradient (black line) specifying the digit pattern IV-III-II-II-III-IV. **(c)** Grafting a ZPA to the tip of the wing bud generates an asymmetrical morphogen profile (colored line), specifying the digit pattern IV-III-III-IV-III-II.

ing cells strongly increases even at the minimum concentration of retinoic acid required for digit duplications. In contrast, the polarizing zone does not provoke such an increase in retinoic acid receptor deployment in its normal position or after transplantation to the anterior wing bud margin.

The foregoing results suggest that retinoic acid may actually *establish* the posterior location of polarizing activity. This hypothesis was tested by a serial transplantation experiment in which a retinoic acid carrier was first implanted into the anterior of a limb bud. The tissue *next to* the implant was then transplanted to a second limb bud to test its polarizing activity (Fig. 22.23). If the transplanted tissue was in contact with the retinoic acid source for more than 15 h, it acquired polarizing activity itself and caused digit pattern duplications in the secondary host limbs (Wanek et al., 1991). The extra digits in the secondary limb were shown to be derived from the host tissue, and not from the transplanted tissue. The acquired polarizing activity could not be ascribed to the transfer of absorbed retinoic acid, because it appeared after a time delay and because the donors were washed for 90 min before the transplantation to remove any exogenous retinoic acid.

Taken together, the available data indicate that exogenous retinoic acid may act by inducing the formation of an ectopic ZPA. In the general terminology used in Chapter 20, exogenous retinoic acid establishes a boundary of the field rather than acting as a morphogen. The natural distribution of retinoic acid in limb buds suggests that retinoic acid may also initiate ZPA forma-

tion in normal development. In accord with this hypothesis, the highest level of endogenous retinoic acid is found during early limb bud formation (Thaller and Eichele, 1990). If the ZPA produces a morphogen, its nature has to be different from retinoic acid. New investigations into the molecular basis of the polarizing activity again involve the use of cloned homeobox genes.

Hox Gene Expression Controls Limb Patterning

The identification of homeobox genes has provided a new basis for studying pattern formation in vertebrates. Because some of these genes are expressed in limb buds and encode transcription factors, they are likely to play key roles in the genetic control of limb development (Izpisúa-Belmonte and Duboule, 1992). The homeobox genes with the most intriguing expression patterns described so far in mouse limb buds are *Hoxd-13*[+], *Hoxd-12*[+], *Hoxd-11*[+], *Hoxd-10*[+], and *Hoxd-9*[+] (Fig. 22.24). Within each limb bud, they are expressed in a colinear pattern. The gene at the 3' end of the cluster, *Hoxd-9*[+], is expressed first at the posterior and distal margin, an area coinciding approximately with the zone of polarizing activity. The *Hoxd-9*[+] expression domain expands anteriorly and proximally as the limb bud grows. Next, expression of *Hoxd-10*[+] begins at the same posterior distal spot and spreads anteriorly and proximally, but not as far as that of *Hoxd-9*[+]. Genes are sequentially activated but each is successively more restricted in its spreading, so that the final expression domains form a

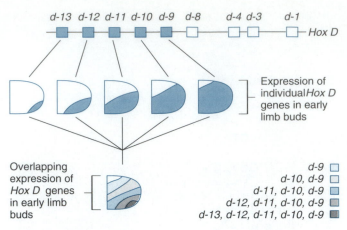

Figure 22.24 Expression of *Hox D* genes in the mouse limb bud. Genes *Hoxd-9*[+] through *Hoxd-13*[+] are expressed as a nested set, all domains beginning at the posterior distal margin and extending more or less toward the anterior and proximal.

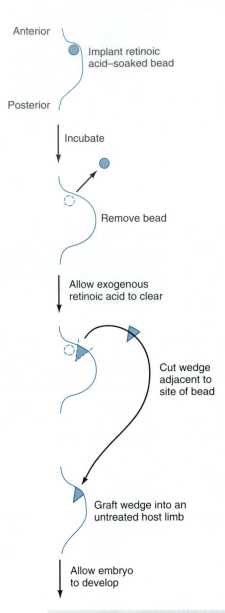

nested set, with all genes expressed in the zone of polarizing activity.

In the chicken, there is a homologous cluster of genes, chicken *Hox D*, which is expressed correspondingly in the wing bud (Izpisúa-Belmonte et al., 1991; Nohno et al., 1991). The expression patterns of these genes were monitored not only in normal wing buds but also after transplantation of ZPAs or retinoic acid–soaked resin beads to the anterior wing bud margin (Fig. 22.25). During the first 12 h after transplantation, no ectopic gene expression was observed. At 16 h, the chicken *Hoxd-9*[+] gene (which is also the first to be expressed in normal development) acquired a small extra expression domain at the anterior margin. At 20 h, an ectopic anterior chicken *Hoxd-11*[+] domain appeared. Finally, at 24 h, chicken *Hoxd-13*[+] was also expressed ectopically at the anterior margin. The ectopic expression domains were generally smaller than their normal

Figure 22.23 Conversion by retinoic acid of anterior wing bud cells into zones of polarizing activity. Resin beads were soaked in retinoic acid (RA) so that anterior implantation of a bead would mimic the effect of a transplanted zone of polarizing activity (see Fig. 22.20a). The soaked beads were implanted into the anterior margin of early chicken wing buds and left there for 12 to 24 h. After bead removal, embryos were incubated for another 90 min to clear any surplus retinoic acid remaining in the tissue. A wedge of tissue adjacent to the former site of the bead was transplanted to the anterior margin of an untreated host. Hosts were allowed to develop for 5 days before they were fixed and stained to analyze their digit patterns. The transplants changed the digit pattern of the developing wings in a manner similar to a transplanted zone of polarizing activity taken from the posterior wing bud margin of an untreated donor.

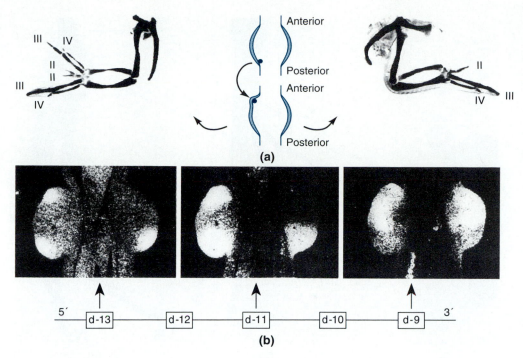

Figure 22.25 Chicken *Hox D* expression domains in wing buds after grafting of a polarizing region. **(a)** The center diagram shows the transplantation of a polarizing zone (colored dot) from the posterior margin of a donor wing bud (top) to the anterior margin of a host wing bud (bottom). The photographs show typical wing patterns that developed within a week from manipulated wing buds (left) and control wing buds (right). **(b)** Transcription domains of chicken *Hoxd-13*[+], *Hoxd-11*[+], and *Hoxd-9*[+] in manipulated wing buds (to the left) and controls (to the right) 24 h after grafting. Note the ectopic anterior expression domains of chicken *Hoxd-13*[+] and *Hoxd-11*[+] in the manipulated wing buds.

counterparts but fell into a similarly nested arrangement. Thus, the experimental treatments that caused the eventual duplication of the digit pattern also duplicated the temporal and spatial appearance of the *Hox D* expression domains.

These experiments just described indicate a strict correlation between the expression pattern of the *Hox D* genes and the digit pattern that is subsequently formed. However, the correlation does not prove that the chicken *Hox D* gene activity pattern determines the morphological pattern of digits that becomes visible later. Alternatively, chicken *Hox D* gene activities might be gratuitous signs of the ongoing determination process. To distinguish between these possibilities, chicken *Hox D* gene expression patterns need to be altered directly, and subsequent morphological developments need to be monitored. Such an experiment was made possible by a new technique for *local* genetic transformation. Bruce Morgan and his colleagues (1992) used a *retrovirus* vector to introduce a cloned mouse *Hoxd-11*[+] gene into the genome of embryonic chicken cells. After injection into an early limb bud, the virus spread locally and infected all cells of the bud as the embryo developed. Using in situ hybridization (see Methods 15.1), the researchers ascertained that the transgene was expressed over the entire bud but did not affect the activity of resident chicken *Hox* genes.

Overexpression of mouse *Hoxd-11*[+] in chicken limb buds changed the expression of homeotic genes as diagrammed in Fig. 22.26. In a normal chicken leg, the "big toe" (toe I) pointing backward on the foot develops from bud cells expressing chicken *Hoxd-9*[+] and *Hoxd-10*[+]. In an infected leg bud, these cells express the same resident genes plus the *Hoxd-11*[+] transgene. Such a combination of chicken *Hox* genes in a normal leg bud is associated with the formation of the "index" toe (toe II). It differs from toe I by being longer, having three bone elements instead of two, and by pointing forward. As expected, toe I from an infected leg bud showed the same morphological characteristics as toe II (Fig. 22.27). A similar transformation, this one associated with the development of an extra digit, was observed in infected wing buds. These results show that *Hox* gene expression is directly involved in forming the pattern of digits in chicken limbs. Changes in the combination of *Hox* genes expressed are *sufficient* to cause corresponding changes in the digit pattern.

Given the functional role of *Hox* genes in specifying digit characters, it may be significant that the expression domains of the *Hox D* genes in the limb bud define exactly five stripes with different combinations of *Hox D* gene activity (Fig. 22.24). Since *Hox* genes and their functions have been exceedingly well conserved during evolution, the availability of five *Hox D* gene expression domains may have acted as a constraint that has prevented the evolution of land-dwelling vertebrates with more than five *different* fingers or toes (Tabin, 1992). Mutant phenotypes with more than five fingers arise frequently, but an extra finger tends to have the same morphological features as one of its neighbors. In other words, the *number* of fingers is increased, but not the number of finger *characters*. A mere increase in number,

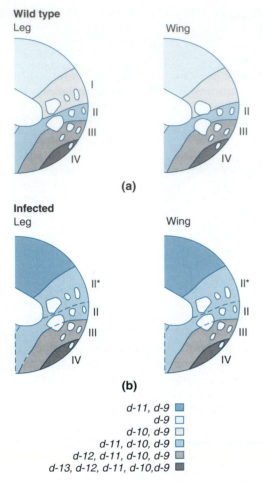

Wild type

Leg — Wing

(a)

Infected

Leg — Wing

(b)

d-11, d-9
d-9
d-10, d-9
d-11, d-10, d-9
d-12, d-11, d-10, d-9
d-13, d-12, d-11, d-10, d-9

Figure 22.26 Ectopic expression of a mouse *Hoxd-11*+ transgene in chicken limb buds. **(a)** Normal expression patterns of chicken *Hox D* genes in the leg bud and the wing bud, superimposed on a fate map of the cartilage elements. The Roman numerals indicate prospective digits. **(b)** Altered expression patterns produced after limb bud infection with a virus containing the mouse *Hoxd-11*+ gene. (Mouse *Hoxd-11*+ and chicken *Hoxd-11*+ encode similar proteins and are not distinguished here.) Note that the leg bud region forming digit I expresses chicken *Hoxd-9*+ and chicken *Hoxd-10*+ in wild-type buds, whereas the same region in infected leg buds expresses *Hoxd-11*+ in addition. This is the same gene combination normally expressed in the region forming digit II, and correspondingly, digit I is transformed into an extra digit II (designated II*) in legs developing from infected buds. A corresponding change in the homeobox gene expression pattern of wing buds is followed by the formation of an extra digit II.

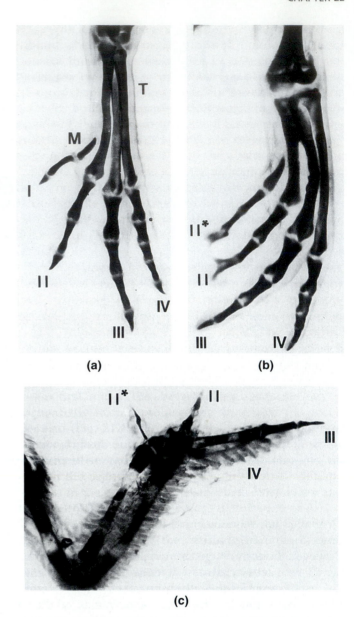

(a) **(b)**

(c)

Figure 22.27 Homeotic transformations caused by ectopic expression of a mouse *Hoxd-11*+ transgene in chicken limb buds as diagrammed in Figure 22.26. **(a)** Skeleton of a wild-type leg. T = tarsometarsals; M = anterior metatarsal. Toe I, pointing backward, consists of two digital elements on a metatarsal (M). Toe II has three digital elements. **(b)** Leg derived from an infected leg bud expressing the transgene. Toe I is replaced with another toe (toe II*) resembling toe II. **(c)** Right wing derived from an infected wing bud expressing the mouse transgene. The wing has an extra digit (II*) anterior to digit II.

however, without the potential for gene-regulatory and thus anatomical diversification, has apparently not been adaptive.

The *Sonic hedgehog*+ Gene Mediates the Polarizing Activity

In search of a mechanism by which the zone of polarizing activity may activate *Hox* genes in a nested set,

researchers again took advantage of the fact that many elements of pattern formation have been conserved between invertebrates and vertebrates. The *Drosophila* segment polarity gene *hedgehog*+ is expressed in the posterior cells of each segment, along with *engrailed*+ (see Chapter 21). The hedgehog protein has the molecular properties of a secreted protein, and it appears to form a morphogen gradient, specifying different cell states in the subsequent segment of the fly embryo (Heemskerk and DiNardo, 1994).

Screening of genomic and cDNA libraries (see Methods 14.2) from mouse, chicken, and zebra fish with a *Drosophila* hedgehog probe has revealed vertebrate genes with similar sequences, including one termed *Sonic hedgehog*$^+$ (J. C. Smith, 1994). In vertebrate embryos, *Sonic hedgehog*$^+$ is expressed in Hensen's node, the notochord, the floor plate of the spinal cord and hindbrain, and the posterior margins of limb buds, all tissues that can act as inducers. In particular, the expression domain of *Sonic hedgehog*$^+$ in chicken limb buds, as revealed by in situ hybridization, coincides closely with the zone of polarizing activity, as assayed by transplantation experiments (Fig. 22.21).

To test the involvement of Sonic hedgehog protein in the process of wing pattern specification, Robert Riddle and his colleagues (1993) implanted retinoic acid–soaked beads in the anteriors of limb buds and assayed for the expression of *Sonic hedgehog*$^+$ by in situ hybridization (see Methods 15.1). Beginning a day after bead implantation, the anterior margin of the wing bud began to mirror the posterior margin in expressing *Sonic hedgehog*$^+$. Thus, the polarizing activity induced by retinoic acid is accompanied by *Sonic hedgehog*$^+$ expression.

To test whether *Sonic hedgehog*$^+$ actually establishes polarizing activity, the researchers transfected cultured cells with the *Sonic hedgehog*$^+$ gene and implanted a pellet of transfected cells anteriorly into a limb bud. This operation was sufficient to cause duplications of the bone pattern in the wing similar to those obtained after transplanting posterior wing bud tissue or beads soaked in retinoic acid.

Finally, the investigators tested the effect of ectopic *Sonic hedgehog*$^+$ expression on *Hox* gene activation. As expected, *Hox D* genes were activated in mirror-image duplications of their normal expression patterns in control limbs.

The experiments on limb buds show that *Hox* genes and their counterparts in other vertebrates play key roles in patterning the limbs. A polarizing signal associated with the expression of the *Sonic hedgehog*$^+$ gene is established at the posterior margin of the limb bud. This signal activates *Hox* genes in a nested set of expression domains, and the activities of these genes control the limb pattern to be formed. Since *Hox* genes encode transcription factors, they are expected to exert their morphogenetic effects by controlling various types of target genes.

SUMMARY

Various strategies are being used to identify patterning genes in vertebrates. In the mouse, a small number of such genes have been found by mutant analysis and have subsequently been cloned. Additional mutants are generated by promoter or enhancer trapping. Alternative strategies—known as reverse genetics—begin with the cloning of genes that are likely to play key roles in pattern formation. One way of identifying such genes in vertebrates is by their sequence similarity to well-characterized patterning genes from other organisms such as *Drosophila*. This strategy works well even if the sequence similarity is limited to certain motifs, such as the homeobox. Once the cloned genes are in hand, investigators proceed with molecular strategies to generate the equivalents of mutants. The goal of both genetics and reverse genetics is to characterize the biochemical and morphological functions of certain control genes in embryonic pattern formation.

In all animals investigated so far, homeobox-containing genes play key roles in development. Subgroups of these genes are clustered in homeobox complexes. Most invertebrates seem to have one homeobox complex, while vertebrates have four complexes located on four different chromosomes. Each homeobox complex contains about 10 genes, each encoding a protein with a homeodomain characteristic of transcription factors. Sequence comparisons indicate that all homeobox complexes have evolved from the same ancestral homeobox complex, which must have originated more than 500 million years ago. The greatest sequence similarities are found between genes that are in corresponding positions within different homeobox complexes.

Each gene of a vertebrate homeobox complex is expressed during embryogenesis in a characteristic area with a sharp anterior boundary. The location of this boundary in the embryo and the chromosomal position of the gene are colinear: as one progresses from 3' to 5' through the homeobox complex, the anterior boundary of the expression domain shifts posteriorly. Within the hindbrain of the chicken, the expression boundaries coincide with the limits of morphological units called rhombomeres. The rhombomeres are separated by clonal restriction lines similar to epidermal compartments in insects.

At least some mouse homeobox genes act as homeotic selector genes, similar to the homeotic genes of

Drosophila. In the clawed frog *Xenopus,* functional inactivation of the *XlHbox 1*[+] gene transforms a segment of spinal cord into an extension of the hindbrain. Similarly, elimination of the *Hoxc-8*[+] gene in mice causes transformations toward more anterior fates. Conversely, overexpression of the *Hoxa-7*[+] gene transforms anterior neck vertebrae so that they resemble more posterior vertebrae.

Retinoic acid and related compounds have dramatic effects on both the expression of homeobox genes and the morphological development of the anteroposterior body axis. Experimental application of retinoic acid also affects the zone of polarizing activity, which is normally established at the posterior margin of limb buds. The polarizing activity is associated with the expression of the *Sonic hedgehog*[+] gene, which in turn activates certain homeobox genes in a distinct temporal and spatial order. Thus, homeobox genes play key roles in establishing the anteroposterior body pattern and the limb patterns of vertebrates.

SUGGESTED READINGS

Kessel, M., and P. Gruss. 1990. Murine development control genes. *Science* **249**:374–379.

Krumlauf, R. 1994. *Hox* genes in vertebrate development. *Cell* **78**:191–201.

McGinnis, W., and R. Krumlauf. 1992. Homeobox genes and axial patterning. *Cell* **68**:283–302.

Tabin, C. 1991. Retinoids, homeoboxes, and growth factors: Toward molecular models for limb development. *Cell* **66**: 199–217.

GENETIC AND MOLECULAR ANALYSIS OF PATTERN FORMATION IN PLANTS

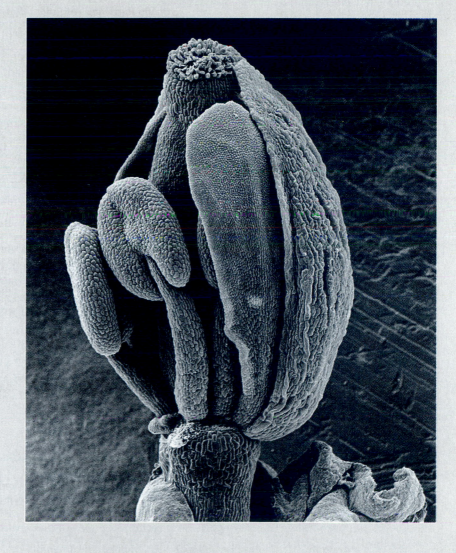

Figure 23.1 This scanning electron micrograph shows a nearly mature flower of the wall cress *Arabidopsis thaliana*. Parts of the flower have been removed to show the internal organs. The central flower organ, the pistil, is surrounded by six stamens, in which the pollen grains will be produced. The stamens are enveloped by four petals (one visible in front), which become large and white in the adult flower. The leaflike sepals (one visible to the right) on the outside protect the budding flower but are later outgrown by the other flower organs. *Arabidopsis thaliana* is used as a model plant because of its favorable traits for genetic studies.

Reproduction and Growth of Flowering Plants

Spore-Forming Generations Alternate with Gamete-Forming Generations

Plant Embryo Develop inside the Flower

Meristems at the Tips of Root and Shoot Allow Plants to Grow Continuously

Genetic Analysis of Pattern Formation in Plant Embryos

Similar Methods Are Used for the Genetic Analysis of Plant and Animal Development

Pattern Formation in Plant Embryos Involves 25 to 50 Specific Gene Functions

Groups of Genes Control Distinct Patterning Events

Genetic Studies of Flower Development

Flower Development Occurs in Four Steps

Arabidopsis Has Homeotic Genes

Homeotic Phenotypes Can Be Explained by a Genetic Model

Double Homeotic Phenotypes Confirm the Genetic Model

Homeotic Genes Are Controlled by Regulator Genes

Other Plants Have Patterning Genes Similar to Those of *Arabidopsis*

Antirrhinum Has Genes Controlling Organ Variations within the Same Whorl

Molecular Analysis of Homeotic Plant Genes

The *agamous*⁺ Gene Encodes a Transcription Factor

The *agamous*⁺ Gene Is Expressed in Stamen and Carpel Primordia

Similar Mutant Phenotypes, Spatial Expression Patterns, and Protein Products Reveal Homologous Genes between *Arabidopsis* and *Antirrhinum*

The Known Plant Homeotic Genes Have No Homeobox

Many Questions about Pattern Formation in Flowers Are Still Unanswered

The great success of genetic and molecular studies on fly and mouse development has enticed several researchers to apply the same methods to entirely different systems—plants. Plant genetics has a long history, not only in agriculture but also in science: the Mendelian laws of genetics are based on Gregor Mendel's cross-pollinations of peas. The snapdragon *Antirrhinum majus* and the corn

plant, *Zea mays*, have been favorite subjects of genetic research in the twentieth century. During recent years, the wall cress *Arabidopsis thaliana* has emerged as a model plant with very favorable traits for genetic studies (Fig. 23.1). It belongs to the flowering plants, or **angiosperms**, and among these it represents the **dicotyledons** (or dicots, for short), that is, the plants with two **cotyledons**, or seed leaves. *Arabidopsis* seems to be fairly representative dicot, so the knowledge it yields is likely to apply to other dicots as well.

The application of genetic and molecular tools to plant development has proved very rewarding. In particular, investigators have isolated mutants with specific abnormalities in embryonic formation. Some of these mutants show striking similarities to patterning mutants of *Drosophila*. For example, in certain mutants of *Arabidopsis*, entire segments of the apical-basal body pattern are deleted, much as anteroposterior segments are in gap mutants of *Drosophilia*. Another set of mutants has revealed the existence of *homeotic genes* in plants. Like their animal counterparts, plant homeotic mutants are characterized by transformations of certain organs into other organs that are normally formed elsewhere. For instance, one of these mutants transforms *stamens* (the organs that produce pollen grains) into *petals* (the colored leaflike organs of flowers). Also like animal homeotic genes, plant homeotic genes have restricted expression domains and encode transcription factors.

In this chapter, we will first introduce some basic aspects of plant development. Then we will make a quick survey of plant mutagenesis and the general use of mutants in studying plant development. Next we will discuss the mutagenesis screen that yielded the embryonic patterning mutants mentioned above. Another major topic will be a genetic model based on the phenotypes of a set of homeotic genes in *Arabidopsis* and *Antirrhinum*. Predictions derived from this model have been tested with genetic and molecular techniques, and examples of these studies will be presented. At the end of this chapter we will consider the relationship between homeotic genes and homeobox-containing genes.

Reproduction and Growth of Flowering Plants

To better appreciate the genetics of pattern formation in plants, it will be useful to review some basic aspects of plant reproduction and growth. More information on these subjects may be found in the reviews of Steeves and Sussex (1989) and Lyndon (1990).

Spore-Forming Generations Alternate with Gamete-Forming Generations

In the life cycles of plants, a spore-forming generation alternates with a gamete-forming generation (Fig. 23.2). The spore-forming generation is diploid and is called the *sporophyte* (Gk: *phyton*, "plant"). In the grown sporophyte, haploid *spores* are produced by meiosis. The spores develop into haploid *gametophytes*, which in turn form *gametes*. Male and female gametes unite in fertilization and form a diploid *zygote*, which develops into the next sporophyte generation. In mosses and ferns, these alternating generations are quite distinct. In angiosperms, the sporophyte is all we see, but the male and female gametophytes are still present as tiny organisms within the flower.

The parts of the flower in which the spores and gametophytes develop are shown in Figure 23.3. Flowers produce two types of spores, microspores and megaspores. The microspores, better known as *pollen grains*, arise in flower organs called *stamens*. The megaspores develop inside another set of flower organs called *carpels*, which fuse to form the central part of the flower, the *pistil*. Each carpel has one or more spore-forming tissues known as *ovules*. Each ovule contains a megaspore mother cell, which undergoes meiosis to produce one haploid megaspore. The megaspore divides mitotically and develops into a female gametophyte called the *embryo sac*. The embryo sac is a microscopic organism containing one egg cell and several other haploid cells.

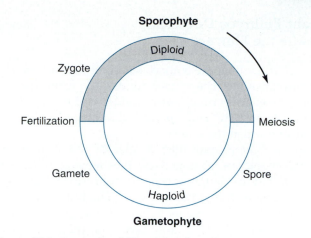

Figure 23.2 Generalized life cycle of a plant. In angiosperms, the gametophytes are reduced to microscopic organisms inside the flower.

In the process of *pollination*, pollen grains are transferred by wind or by animals to the tip of the pistil. Here each pollen grain divides mitotically, producing a male gametophyte that consists only of two sperm cells and a tube cell. The tube cell extends a long tip into an embryo sac, where one of the sperm cells fertilizes the egg to form the *zygote*. The other sperm cell unites with two of the other embryo sac cells to form a triploid *endosperm cell*. This kind of double fertilization is unique to angiosperms.

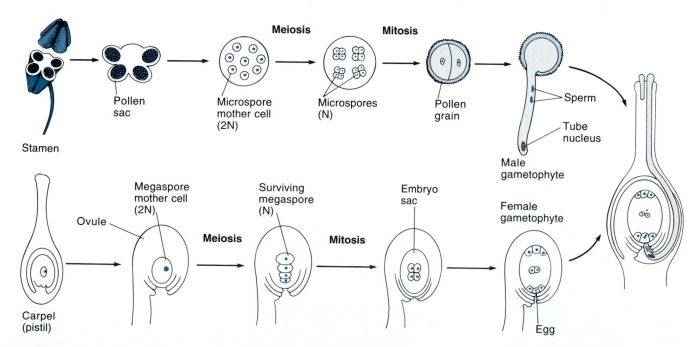

Figure 23.3 Spore and gametophyte formation in an angiosperm.

Plant Embryos Develop inside the Flower

After fertilization, the embryo sac continues to develop within the ovule. The zygote divides and forms the *embryo*. The endosperm cell also divides and forms a temporary tissue, the **endosperm**, which is subsequently used to nourish the embryo.

In *Arabidopsis* and other dicots, the initial divisions of the zygote are analogous to the cleavage of animal eggs in that the daughter cells do not restore their original volume before dividing again. The first division of the *Arabidopsis* zygote as asymmetrical, generating a small apical cell and a large basal cell (Fig. 23.4). The apical cell gives rise to the embryo proper except for the root, which derives from the basal cell. The basal cell also forms the extraembryonic *suspensor*, which pushes the growing embryo deep into the endosperm. At the 16-cell stage, the embryo consists of 8 outer cells, which give rise to epidermis, and 8 inner cells. This stage also marks the beginning of the **globular stage**, during which the embryo proper is spherical. With the transition to the **heart stage**, the embryo establishes a bilateral symmetry by initiating cotyledon formation. When the embryo consists of about 100 cells, the primordia of root, shoot, vascular tissue, and ground tissue are discernible. Thus, many features of the basic plant pattern are already established at the heart stage.

As the embryo develops further, the ovule in which it resides is transformed into a *seed* (Fig. 23.5). The

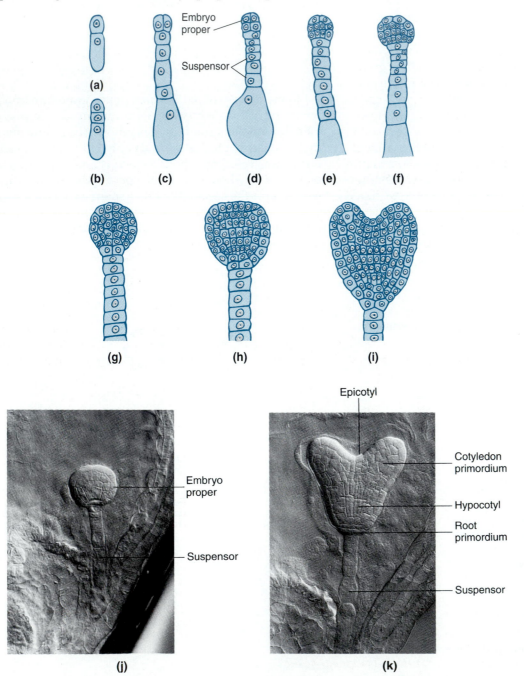

(a)

(b)

Embryo proper

Suspensor

(c)

(d)

(e)

(f)

(g)

(h)

(i)

Embryo proper

Suspensor

(j)

Epicotyl

Cotyledon primordium

Hypocotyl

Root primordium

Suspensor

(k)

Figure 23.4 Early stages of embryogenesis in the wall cress *Arabidopsis*. **(a–d)** Early development of the embryo and the suspensor. **(e)** At the 16-cell stage, the embryo proper consists of 8 outer cells and 8 inner cells. **(f–h)** Globular stages. **(i)** Heart stage. **(j, k)** Photographs of an early globular stage and the heart stage.

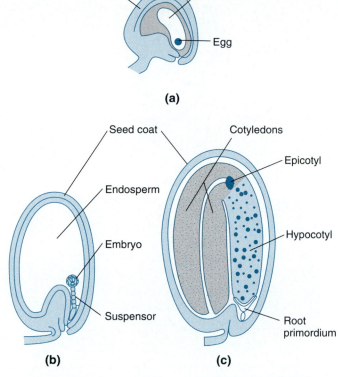

Figure 23.5 Seed development in *Arabidopsis*. **(a)** Ovule with embryo sac containing egg cell. **(b)** Immature seed with embryo at the globular stage, suspensor, and endosperm tissue. **(c)** Mature seed (0.5 mm long) with embryo. Visible are the two cotyledons, epicotyl, hypocotyl, and root primordium. The endosperm and suspensor have been used up and are degenerating. The seed is surrounded by a tough coat.

cells, vascular cells, and storage cells. The shoot primordium between the two cotyledons and the root primordium at the basal tip of the hypocotyl remain small and quiescent until *germination*. The covering layers of the ovule develop into the tough coat of the seed, and the surrounding pistil gives rise to the *fruit*.

Plants that have only one cotyledon, such as grasses, cereals, and maize, develop similarly except that the seed retains much of its endosperm and the embryo is comparatively smaller. In all plants, the seed is a quiescent embryonic stage adapted for dispersal and survival outside the mother plant.

The embryo resumes its growth at *germination*, when the embryo sprouts from the seed. Germination depends on both internal and environmental factors. The food reserves for the early phase of germination are stored in the cotyledons or in the endosperm. The root usually emerges first to anchor the seedling and to take up water and dissolved minerals. The hypocotyl stretches before the shoot primordium begins to grow and form the shoot. The young seedling shows the same pattern that was already visible at the heart stage of the embryo (Fig. 23.6). From top to bottom we can distinguish four main elements: the shoot primordium, now called *epicotyl*; the *cotyledons*; the *hypocotyl*; and the *root primordium*. These elements constitute the *apical-basal body pattern* of the plant seedling. The seedling also has a *radial pattern*, which is most clearly seen in the hypocotyl and involves three major tissues: the outer epidermis, the inner mass of ground tissue, and the centrally located vascular strands.

Meristems at the Tips and Root and Shoot Allow Plants to Grow Continuously

While most cells of a seedling proceed to differentiate for functions in photosynthesis, transport, and storage, certain nests of cells remain forever embryonic and ac-

cotyledons, and the *hypocotyl* beneath the cotyledons, grow substantially using the endosperm and the suspensor as sources of nutrients. At the same time, embryonic cells continue to differentiate into epidermal

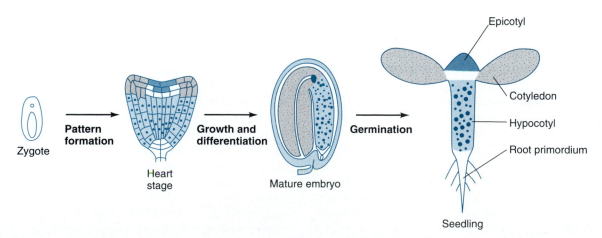

Figure 23.6 Pattern formation in the plant embryo. Four major elements (epicotyl, two cotyledons, hypocotyl, and root primordium) are laid down at the heart stage.

tive in cell division. These nests of cells, called *meristems,* are located at the tips of the root and shoot primordia. Plant meristem cells are like animal *stem cells:* as they divide, they form various progenitor cells of other tissues and regenerate new meristem cells. Postembryonic plant development, which generates entire organ systems including the entire flowering part of a plant, depends largely on meristems. After germination, the root meristem forms the root of the mature plant while the shoot meristem gives rise to the stem and leaves. The meristems keep adding on new sections of root, stem, and branches. Thus, unlike many animals, plants can grow throughout their lives.

The apical meristem of a growing plant pushes the tip of the plant upward as the meristem cells divide and grow. At regular intervals, the meristem also produces lateral swellings that develop into leaves, branches, or flowers. The points where leaf primordia arise from the stem are called *nodes,* and the length of stem between two nodes is called an *internode.* Internodes are short near the meristem and elongate later on. Because of its short internodes, the meristem is typically covered with leaf primordia (Fig. 23.7). The resulting compound structure, consisting of the meristem, leaf primordia, and short internodes, is called a *bud.* Roots grow similarly except that the root meristem is covered by a protective cap of nondividing cells that produce a lubricating substance.

The pattern in which an apical shoot meristem gives off the primordia of leaves, branches, and flowers generates much of the overall appearance of a plant. After germination of an *Arabidopsis* seed, the shoot meristem gives off a set of leaf primordia in a spiral arrangement with short internodes (Fig. 23.8). Thus, the basal portion of the plant develops as a rosette. Subsequently, the pattern of cell division in the apical meristem changes to produce a few small leaves separated by long internodes. After this transient phase, the apical meristem produces floral primordia, still in a spiral pattern. Each floral primordium forms a flower, which later develops into a fruit.

In summary, plant development differs from animal development in several respects. Plant zygotes are very small and develop inside maternal tissue, a situation paralleled only by mammals and a few other groups of animals. The apical meristems allow continuous growth of plants throughout their lives. No *germ line* is set aside early during plant development: the flower organs that give rise to megaspores and microspores develop from the same meristem cells that have also produced the entire shoot. Correspondingly, many nonreproductive plant cells remain *totipotent,* retaining the capacity of forming an entire new plant (see Chapter 7). Unlike animal embryos, plant embryos do not undergo major morphogenetic movements, and they produce a smaller number of histologically distinct cell types.

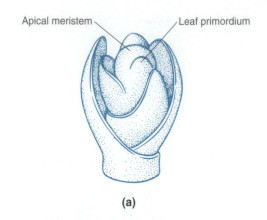

(a)

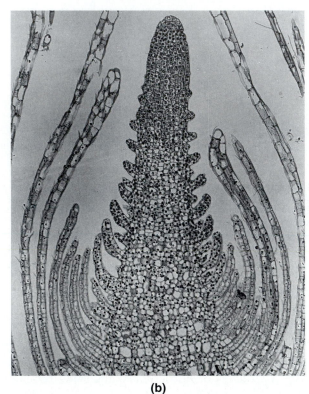

(b)

Figure 23.7 Growing bud of a marsh plant, *Elodea* sp. **(a)** Schematic diagram. **(b)** Photograph of a histological section. The apical meristem forms nodes with leaf primordia and short internodes. The growing leaf primordia curve upward and enclose the meristem in a bud.

Some of the differences between plant development and animal development are significant form a genetic point of view. The development of a male gametophyte from the haploid microspore requires gene expression and should therefore cause the elimination of many mutant alleles before fertilization can occur. The small size and maternal nourishment of the plant embryo make it less dependent than most animal embryos on maternally provided RNA. Instead, one might expect that comparatively more zygotic genes are activated early during plant embryogenesis.

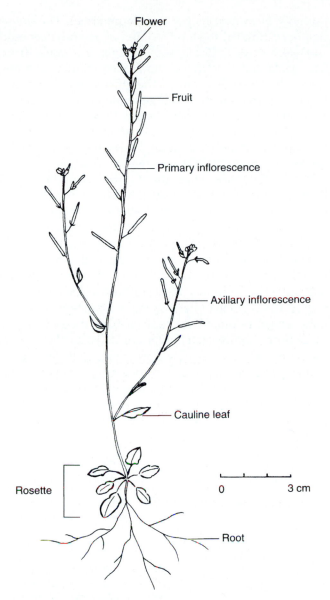

Figure 23.8 Drawing of the wall cress *Arabidopsis thaliana*, showing the overall appearance of the plant. Initially, the apical meristem generates a basal rosette of leaves that are arranged in a spiral pattern with short internodes. Later-formed leaves (cauline leaves) are separated by longer internodes. Upon floral induction, the apical meristem changes into the primary inflorescence meristem; it forms flowers instead of leaves, still in a spiral pattern. Each flower matures into an elongated fruit. Each of the cauline leaves contains an axillary inflorescence meristem that reiterates the development of the primary inflorescence.

Genetic Analysis of Pattern Formation in Plant Embryos

Most of the plants that lend themselves to genetic analysis have long generation times and large genomes, making studies based on both classical and molecular genetics very time-consuming. An exception is *Arabidopsis,* which has several advantages as a model plant for genetic analysis. It is small enough to be grown in large numbers in the laboratory. Its generation time is about 2 months, and each plant produces many seeds. It self-fertilizes, so that mutant alleles are easily made homozygous. Most important, its genome size (7×10^7 nucleotide pairs) is about as small as those of *Caenorhabditis elegans* and *Drosophila.* These features combine to make saturation screens for certain classes of genes feasible and allow investigators to clone any gene on the basis of its phenotype and map position. *Arabidopsis* is therefore well on its way to becoming the *Drosophila* of plant developmental biology.

Similar Methods Are Used for the Genetic Analysis of Plant and Animal Development

The methods of plant mutagenesis are essentially the same as those used for animals. Geneticists have employed X-rays as well as chemical mutagens. Researchers working with maize have taken advantage of transposons similar to the P elements of *Drosophila* (Chandlee, 1991; see also Chapter 14). Many plants are also susceptible to genetic transformation using a naturally occurring bacterial plasmid as a vector. The bacterium *Agrobacterium tumefaciens* transfers a plasmid, called Ti (for tumor-inducing), to cells of its natural host plants. This step is followed by the integration of part of the plasmid, called **T-DNA** (for transferred DNA), into the plant genome. The T-DNA encodes products that induce crown gall tumors, and also make the plant produce certain amino acids for the bacterium.

The ability of T-DNA to stably integrate into its host's genome has made it a useful vector for transforming plant cells with engineered genes. For this purpose, the T-DNA has been disarmed so that it is still transferred but no longer induces tumors. Instead, it may contain genes that confer resistance to antibiotics so that transformed cells can be selected in culture. Most important, any cloned genes included into the T-DNA will be introduced into the genome of host plant cells. The same system is also suitable for insertional mutagenesis, in which the inserted DNA is used to introduce a selectable marker and a tag to pull out the mutated gene from a genomic library (see Methods 14.2).

Mutants have been used in several ways to study plant development (Meinke, 1991a, b). For example, genetic marking of single cells has been employes for clonal analysis just as in insects (see Methods 6.1). Studies employing this technique indicate that shoot meristem cells do not always divide in the same manner or contribute to precisely the same portion of the adult plant (McDaniel and Poethig, 1988; Jegla and Sussex, 1989). Thus, cellular interactions must be involved in pattern formation, while cell lineage seems to play only a minor role in plant development. In addition, mutagenesis of entire plants has revealed a wide range of genes involved in embryogenesis as well as later development. The following sections describe genetic and molecular studies of embryonic pattern formation and flower development in *Arabidopsis*.

Pattern Formation in Plant Embryos Involves 25 to 50 Specific Gene Functions

Encouraged by the successful genetic analysis of embryonic development in *Drosophila*, Gerd Jürgens and his coworkers (1991) set out on a similar analysis of plant embryogenesis. At the start of their work, they asked two basic questions. First, how many genes are involved in determining the body pattern of the plant embryo? Second, can the process of pattern formation be broken down into distinct events that are controlled by particular subsets of genes? To answer these questions, the researchers carried out a saturation mutagenesis screen with *Arabidopsis* seeds.

▼

To saturate the genome of an organism with mutations, each gene has to be mutated at least once. The number of mutagenized lines that must be screened to achieve saturation depends on the number of genes per genome and on the efficiency of the mutagenizing treatment

(number of mutations per treated genome). The investigators estimated the total number of essential genes in *Arabidopsis* at 5000, based on genetic data from *Drosophila* and *C. elegans*, which have similar genome sizes. They also determined the efficiency of their mutagenizing treatment, which generated about 0.7 mutations per treated genome. Finally, they used a mathematical formula (the Poisson distribution) indicating that an *average of five* mutant alleles per gene was required to obtain *at least one mutant allele for nearly all* (99%) of the genes. Thus, the number of mutagenized lines to be screened was 5000 × 5 ÷ 0.7 = 36,000. To be on the safe side, they screened about 44,000 lines.

The screening procedure is outlined in Figure 23.9. Wild-type seeds are mutagenized with ethylmethylsulfonate (EMS) and grown into adult plants. The embryo contained in each seed has only two precursor cells that later give rise to spores. If one of these cells is mutated, it produces a clone of heterozygous meristem cells that, on average, populate the reproductive tissue in every other flower. Since each flower normally fertilizes itself, the seeds produced by one flower represent one mutagenized line. Seeds from each line are germinated and screened for abnormal development. Lines producing seedlings with distinct pattern abnormalities are inspected and categorized, and their normal-looking siblings are used to breed the mutant lines.

Among 44,000 mutagenized lines screened, Jürgens and coworkers (1991) found 25,000 embryonic lethal mutants, which they did not analyze in detail. About

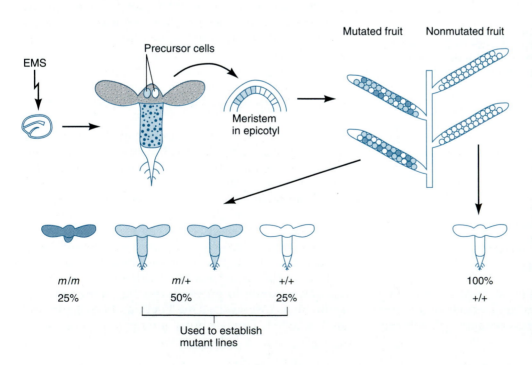

Figure 23.9 Mutagenesis screen to identify patterning genes in *Arabidopsis*. Seeds are exposed to the chemical mutagen ethylmethylsulfonate (EMS) and raised to adult plants. The embryo contained in each seed has only two precursor cells that later give rise to spores. One of these cells, on average, is mutagenized (light color) and produces a clone of heterozygous meristem cells that statistically populate the reproductive tissue in every flower. Since each flower normally fertilizes itself, about one-quarter of the seeds (bold color) from a mutated fruit will be homozygous for the mutant allele and give rise to an abnormal seedling (labeled *m/m*). Their normal-looking siblings (labeled *m/+* and *+/+*) are then used to establish a mutant line.

5000 lines produced seedlings with mutant phenotypes having abnormal pigmentation or abnormal morphology or both. The majority of the morphologically abnormal seedlings had reduced or misshapen cotyledons, and their roots tended to be short or less differentiated than normal. These phenotypes did not seem to provide any clues to patterning events that might be affected specifically, although this possibility could not be completely ruled out.

About 250 lines showed specific morphological abnormalities. These phenotypes were most likely the result of mutations in genes critical for distinct events during embryonic pattern formation. The investigators therefore classified these lines as putative patterning mutants. Lines with similar phenotypes were crossed in *complementation tests* to determine whether they were *allelic*, that is, representing different mutations of the same gene. For each of the first genes to be identified, nine mutant alleles were obtained on average (Table 23.1). Thus, the total number of patterning genes identified in this screen is expected to be around 30 when the analysis is complete.

Groups of Genes Control Distinct Patterning Events

In the first sample of mutant phenotypes they studied, Jürgens and coworkers found mutations that affect three aspects of the body organization: the *apical-basal pattern*, the *radial pattern*, and the *shape* (Table 23.1; Mayer et al., 1991). The mutations affecting the apical-basal body pattern generate large deletions, which can be classified as *apical, basal, central,* and *terminal* (Fig. 23.10). Four mutant phenotypes illustrating such deletions are shown in Figure 23.11, along with several phenotypes having radial-pattern and shape-change defects.

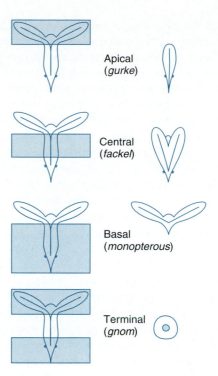

Figure 23.10 Apical-basal patterning mutants of *Arabidopsis*. Four types of pattern deletion, and representative mutant alleles causing them, are indicated. The wild-type seedling patterns with the deleted parts shaded are shown on the left, and the resulting mutant phenotypes are shown on the right. Note the complementarity between apical and basal deletions, as well as between central and terminal.

Two pairs of deletions—apical and basal, as well as central and terminal—add up to the entire apical-basal body pattern, suggesting that the corresponding gene activities may define boundaries along the axis. Each of these deletions can readily be detected in heart-stage embryos, indicating that the apical-basal pattern is formed before this stage.

TABLE 23.1

Mutations Affecting Body Organization in the *Arabidopsis Embryo*

Mutant Phenotype	Gene	Number of Alleles
Apical-basal–pattern defect		
Apical	*gurke*	9
Central	*fackel*	5
Basal	*monopteros*	11
Terminal	*gnom*	15
Radial-pattern defect	*keule*	9
	knolle	2
Shape-change defect	*fass*	12
	knopf	6
	mickey	8

Source: From Mayer et al. (1991). Used with permission, © 1991 Macmillan Magazines Limited.

▼

In the *apical deletion* class, mutant alleles of the *gurke* gene (G. Gurke, "cucumber") affect the shoot meristem and the cotyledons (Fig. 23.11b). The internal phenotype matches the external appearance: the vascular strands terminate apically without branching. The *central deletion* class is represented by *fackel* mutants (G. Fackel, "torch"); they lack the hypocotyl. The cotyledons appear directly attached to the root, and the vascular strands separate at the upper end of the root and diverge into the cotyledons (Fig. 23.11c). In the class of *basal deletions*, the phenotype of *monopterous* mutant seedlings is lacking both the root and the hypocotyl (Fig. 23.11d). This phenotype has been traced back to the early embryonic stage, when cells of the embryo proper and the uppermost suspensor cell fail to establish the division patterns that would normally generate the basal body structures (Berleth and Jürgens, 1993).

The *terminal class* of pattern deletions affects structures located at the two opposite ends of the seedling. Mutations in the *gnom* gene (G. *Gnom*, "dwarf") ¨cause deletion of both the root and the cotyledons, leaving cone- or ball-shaped seedlings with no apparent axial organization (Fig. 23.11e). Although vascular cells are present and well differentiated, they are not connected to form strands. The *gnom* phenotype has been traced back to the first division of the zygote, which generates two cells of nearly the same size instead of a small apical cell and a large basal cell (Mayer et al., 1993).

Radial pattern defects are caused by mutations in two genes, *knolle* and *keule* (G. *Knolle*, "bulb," and *Keule*, "club"; Fig. 23.11f and g). In *knolle* mutants, one cannot morphologically distinguish an outer epidermal cell layer from the internal layers. Mutations in *keule* lead to a bloated and irregular appearance of epidermal cells. Both mutant phenotypes can already be recognized at the globular stage, when in the wild-type embryo an outer group of epidermal cells would surround an inner group of cells.

Shape mutants, in contrast to apical-basal and radial mutants, do not affect specific pattern elements but rather cause grossly abnormal overall shapes. Seedlings with mutant alleles of the *fass* gene (G. *Fass*, "barrel") contain the normal pattern elements but are compressed in the apical-basal axis (Fig. 23.11h). Mutant *knopf* seedlings (G. *Knopf*, "button") look small, round, and pale (Fig. 23.11). Mutations in the *mickey* gene cause thickened, disc-shaped cotyledons, which appear disproportionately large relative to the short hypocotyl and root of the seedling (Fig. 23.11j).

Compared to the work already done with *Drosophila*, the genetic analysis of pattern formation in plant embryos has just begun. However, the mutant lines of *Arabidopsis* that have been collected are raw material for further research, which will probably unfold rapidly. Researchers will next want to generate double mutants, because their phenotypes will provide clues to the genetic hierarchy that controls pattern formation in *Arabidopsis*. When some of the catalogued genes have been cloned, their expression patterns can be studied by *in situ hybridization* and *immunostaining*. The application of these techniques to mutant and wild-type embryos will provide a direct test for hypotheses about control interactions between patterning genes. At the same time, sequencing data will help to characterize the molecular nature of the proteins encoded by these genes. Such results should allow investigators to piece together a molecular model of pattern formation in plant embryos. Indeed, work is already proceeding along these lines in another area of plant molecular genetics, which will be described next.

Genetic Studies of Flower Development

The seasonal cycle of angiosperms culminates with the development of flowers, which contain the reproductive organs of the plant. Flowers consist of four rings of organs called **whorls.** From outside to inside, the organs of the four whorls are known as sepals, petals, stamens, and carpels (Fig. 23.12). The **sepals,** forming the outermost whorl, are most similar to leaves. The **petals,** which form the next whorl, are often brightly colored and attract insects for pollination. Inside the petals is a whorl of *stamens,* in which pollen grains develop as described earlier. In the central whorl, formed by the *carpels,* megaspores develop inside ovules and form embryo sacs. The carpels fuse into the central organ of the flower, the *pistil,* which later gives rise to the fruit.

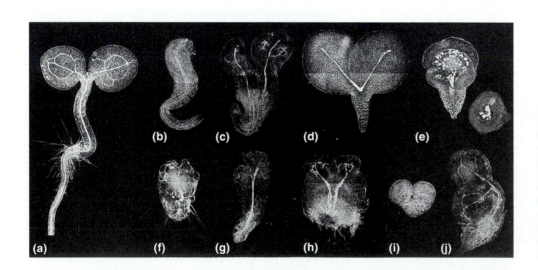

Figure 23.11 Phenotypes of mutant *Arabidopsis* seedlings: **(a)** wild-type; **(b)** apical deletion (*gurke*); **(c)** central deletion (*fackel*); **(d)** basal deletion (*monopterous*); **(e)** terminal deletion (*gnom*); **(f, g)** radial-pattern defects caused by mutations in *knolle* and *keule*; **(h, i, j)** shape-change mutants *fass, knopf,* and *mickey*.

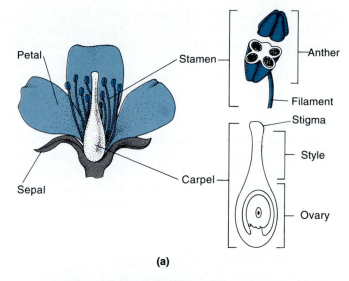

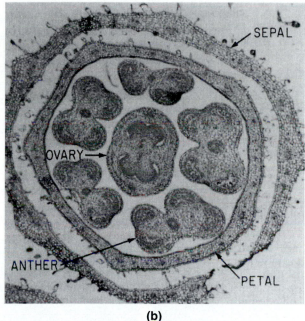

Figure 23.12 Flower structure. **(a)** Schematic lateral view of a generalized flower. **(b)** Photomicrograph of a histological section across a tobacco flower. A complete flower contains four whorls of organs: sepals, petals, stamens, and carpels. The leaflike sepals and the colored petals form the outer two whorls. These organs are often fused at their bases. Inside is a whorl of stamens, each consisting of a thin filament and a thicker organ called the anther, in which the pollen grains develop. The central organ of the flower, the pistil, is formed from a whorl of fused carpels. The pistil consists of a basal ovary containing the ovules; a top part called the stigma, to which the pollen grains attach; and a connecting tube, the style.

For centuries, naturalists and morphologists have recognized intuitively that all organs of the flower are modified leaves. As we will see, it takes only the activity of a few genes to transform leaves into flower organs.

Flower Development Occurs in Four Steps

After a period of vegetative growth, plants become capable of forming flowers. The process of flower development can be subdivided into four steps, each of which is affected by certain genes (Fig. 23.13; see also Weigel and Meyerowitz, 1993; H. Ma, 1994). The first step is *floral induction.* It is triggered by a combination of environmental and internal signals including day length, temperature, the age of the plant, and its nutritional state. Floral induction results in the reorganization of the apical shoot meristem from a *vegetative meristem* producing leaves into an *inflorescence meristem* giving rise to flower primordia. It takes a few days for this florally induced state to be achieved, but thereafter it remains stable even if environmental conditions revert to those that promote vegetative growth (Steeves and Sussex, 1989). Floral induction is affected by more than 10 *flowering genes;* mutations in these genes accelerate or delay flowering (Koorneef et al., 1991). For example, the *embryonic flower+* gene is required for the vegetative state of the apical shoot meristem; mutants with loss-of-function alleles of this gene form a single flower right after germination (Sung et al., 1992).

The second step in flower development is the actual formation of flower primordia, which arise through distinct patterns of cell division in the inflorescence meristem. The pattern in which the flower primordia are generated varies among different groups of plants. In *Arabidopsis,* the flower primordia arise in a spiral pattern on the flanks of the inflorescence meristem (Fig.

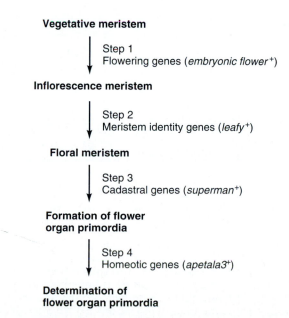

Figure 23.13 Flower development can be broken down into four major steps, each depending on the functions of certain genes.

23.14a). Correspondingly, the flowers and later the fruits of *Arabidopsis* are attached to the stem in a spiral (Fig 23.8). Because young flower primordia show some meristem properties, they are also called *floral meristems*. The genes that control the formation of floral meristems from the inflorescence meristem are called *meristem identity genes*. A representative of this group in *Arabidopsis* is the *leafy*[+] gene, which is strongly expressed in floral meristems (Weigel et al., 1992). In *leafy* mutants, flowers are transformed into secondary inflorescence shoots. Thus, the normal *leafy*[+] activity promotes the transition from an inflorescence meristem to a floral meristem. Correspondingly, leafy RNA accumulates in floral primordia but not in the floral meristem proper. However, *leafy*[+] acts in concert with *apetala1*[+] and other meristem identity genes (Mandel et al., 1992; Bowman et al., 1993; Schultz and Haughn, 1993).

The third step in flower development is the formation of flower *organ primordia* from the floral meristem. In contrast to the unlimited number and spiral arrangement of leaves and floral meristems, the flower organ primordia of *Arabidopsis* are produced in four *whorls*. In the wild type, the floral meristem forms four sepal primordia, then four petal primordia, and finally six stamen primordia and the primordium of the pistil, which is composed of two carpels. Genes related to the formation of flower organ primordia are called *cadastral*

genes. Mutations in one cadastral gene, the *clavata1* gene, affect the number of carpels and other floral organs without changing the morphological features of these organs (Leyser and Furner, 1992; Clark et al., 1993). Other cadastral genes are known to limit the expression of certain target genes to certain domains. A representative of this group is *superman*[+], which will be described later.

The final stage of flower development is the *determination* of the organ primordia and their subsequent *differentiation* into organs appropriate for their positions. The determination of flower organ primordia is controlled by *homeotic genes*, which will be the focus of the following discussion.

Arabidopsis Has Homeotic Genes

A particularly rewarding set of studies has focused on homeotic mutations, which change the morphological character of the organs formed in certain whorls (Coen and Meyerowitz, 1991; Meyerowitz et al., 1991; Weigel and Meyerowitz, 1994). In such mutants, as a rule, the first three steps of flower development are completed normally, so that all organ primordia form in the numbers and positions as they would in the wild types. However, the subsequent determination and differentiation steps are abnormal. In one mutant, for instance, the outermost whorl forms carpels instead of sepals,

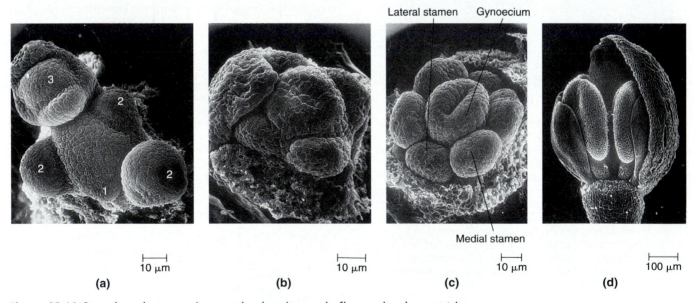

Figure 23.14 Scanning electron micrographs showing early flower development in *Arabidopsis*. **(a)** Inflorescence meristem of a plant. The apical part of the meristem is surrounded by flower primordia (also called floral meristems) numbered 1, 2, and 3 with increasing age. Sepal formation is beginning in the most advanced flower primordium (3). **(b)** Individual flower primordium at a later stage. Three sepals have been removed to show the underlying whorls. **(c)** Another developing flower with sepals and petals removed to reveal the primordia of two medial and four lateral stamens and two carpels, which have fused into one pistil or gynoecium. **(d)** Developing flower with two sepals removed to show the underlying petals, stamens, and pistil.

and the following whorl forms stamens instead of petals. In all mutants of this class, certain organs are replaced by other organs that are normally formed elsewhere. This is the main characteristic of *homeotic mutants*, which we have discussed in the context of *Drosophila* and mouse development (see Chapters 21 and 22). The genes defined by the homeotic mutants of *Arabidopsis* are expressed in restricted domains, and they control in a combinatorial way the morphological development of these domains (Fig. 23.15 and Color Plate 10). We therefore refer to these plant genes as *homeotic genes*.

All known homeotic mutations in *Arabidopsis* affect sets of two neighboring whorls: whorls 1 and 2 (normally forming sepals and petals), whorls 2 and 3 (normally petals and stamens), or whorls 3 and 4 (normally stamens and carpels). For each pair of whorls, one or two pertinent mutants will be described briefly here. More detailed descriptions and additional illustrations may be found in the accounts of Bowman et al. (1991)

and Meyerowitz et al. (1991). Nearly all homeotic mutants are *recessive*, so that an abnormal phenotype develops only if both copies of the gene are mutated. Presumably these phenotypes result from a partial or complete loss of gene function.

Whorls 1 and 2 develop abnormally in mutants of the *apetala2* gene. In most alleles, including *apetala2-2*, the organs of whorl 1 (normally sepals) are carpels, leaves, or absent (Fig. 23.15 and Color Plate 10d). The organs of whorl 2 (normally petals) are usually absent. In the *apetala2-1* allele, whorl 1 consists of leaves and whorl 2 consists of four stamens (or stamen-petal intermediates) in place of petals (Fig. 23.15 and Color Plate 10e).

Whorls 2 and 3 are affected by mutations in two different genes, *apetala3* and *pistillata*, which cause similar phenotypes. (Despite their similar names, *apetala2* and *apetala3* are different genes with different functions.) Mutants homozygous for *apetala3* or *pistillata* have a normal first whorl of sepals, but the organs of whorl 2 develop as additional sepals rather than petals. In ad-

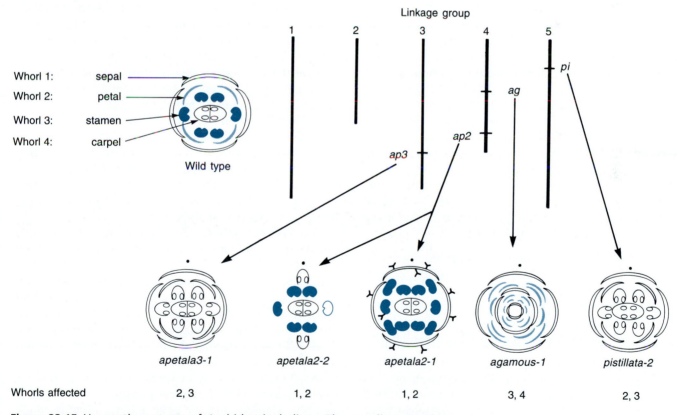

Figure 23.15 Homeotic mutants of *Arabidopsis thaliana*. The top diagrams show a schematic representation of the wild-type flower and the five linkage groups (chromosomes) of *Arabidopsis* with the map positions of four homeotic genes. The bottom diagrams represent the phenotypes of certain mutant alleles of those genes. In *apetala3–1*, petals are replaced with sepals, and stamens with carpels. In one *apetala2–2* phenotype, two sepals are replaced with carpels while the other sepals and all petals are missing. In *apetala2–1*, the organs of whorl 1 develop as leaves, and petals are replaced with stamens. In *agamous–1*, all six stamens are replaced with additional petals, and the carpels are replaced with another flower of the same mutant phenotype. The *pistillata–2* phenotype is very similar to that of *apetala3–1*.

dition, the six organs of whorl 3 develop as carpels rather than stamens (Fig. 23.15 and Color Plate 10c).

Whorls 3 and 4 are affected by mutations in the *agamous* gene. The six positions of whorl 3 are occupied by petals instead of stamens. Moreover, the cells that normally form whorl 4 develop into another flower consisting of one whorl of sepals and two whorls of petals (Fig. 23.15 and Color Plate 10b). This process of forming flowers within flowers is reiterated several times.

In summary, each of the four homeotic genes described here affects two adjacent whorls of flower primordia while leaving the organ identity in the other two whorls nearly unaffected. The mutant phenotypes are characterized by transformations of flower organs into organs that normally form in different whorls. Most of these mutations affect only the morphological character of the plant organs and not their number or positioning within the whorl.

Homeotic Phenotypes Can Be Explained by a Genetic Model

On the basis of the homeotic phenotype just described, John Bowman and his colleagues (1989, 1991) have proposed the following model for the specification of the four organ characters in flowers. The homeotic genes affecting flower formation fall into three classes and are active in three overlapping rings (labeled A, B, and C in Color Plate 11). The *apetala2*[+] gene represents class A, which affects whorls 1 and 2. The *apetala3*[+] and *pistillata*[+] genes represent class B, affecting whorls 2 and 3. The *agamous*[+] gene, which represents class C, affects whorls 3 and 4.

The critical postulate of the model is that the activity of these homeotic genes *determines* the character of the flower organs formed, regardless of where these genes are expressed. Thus, activity of

- *apetala2*[+] alone determines sepal development.
- *apetala2*[+], *apetala3*[+], and *pistillata*[+] together determine petal development.
- *agamous*[+], *apetala3*[+], and *pistillata*[+] together determine stamen development.
- *agamous*[+] alone determines carpel development.

In addition, the model postulates that *apetala2*[+] represses *agamous*[+] and vice versa. Moreover, *agamous*[+] is thought to *terminate* the development of additional flower organ primordia once the carpel primordia have been formed.

The determination of each flower organ character by its specific combination of gene activities is thought to occur even if the pattern of homeotic gene expression is changed by mutation. For instance, in a mutant without the *apetala2*[+] function, the *agamous*[+] gene is no longer inhibited in whorls 1 and 2. Consequently, whorl 1 now expresses *agamous*[+] and forms carpels, and whorl

2 expresses *agamous*[+], *apetala3*[+], and *pistillata*[+], causing stamen formation. In this fashion, the model explains the homeotic phenotypes described in this section.

The homeotic plant genes considered here, like their counterparts in animals, are thought to direct the morphological development of flower organs by controlling batteries of realizator genes. Accordingly, the homeotic genes are expected to become active before organ differentiation, and their products are expected to show the molecular characteristics of gene-regulatory proteins. Both these expectations are met in animals, as discussed in Chapters 21 and 22, and in plants as well. Very little is known about the hypothetical realizator genes, presumably because the phenotypes generated by mutants in each single realizator gene are so subtle that they tend to go undetected.

Double Homeotic Phenotypes Confirm the Genetic Model

To further test the validity of their genetic model, Bowman and colleagues (1989, 1991) generated double and triple homeotic mutants of *Arabidopsis* and compared their phenotypes with the predictions of the model as shown in Color Plate 11. In mutants homozygous for loss-of-function alleles of *apetala2* and *apetala3* (or *apetala2* and *pistillata*), all whorls should express only *agamous*[+], and all organ primordia should therefore develop into carpels. This is indeed the phenotype of the double mutant (Color Plate 12a).

A double mutant for *apetala3* (or *pistillata*) and *agamous* retains only the *apetala2*[+] function, which should be expressed in all four whorls. Accordingly, all organ primordia should develop into sepals. Also, the loss of *agamous*[+] function should cause the formation of additional whorls. This is exactly what takes place in the actual phenotype (Color Plate 12b).

The *apetala2, agamous* double mutant is the most interesting double mutant. According to the model, whorls 1 and 4 will not express any of the homeotic genes discussed here. Therefore, if at least one of these genes is necessary for developing a flower organ, then whorl-1 and whorl-4 primordia should develop unlike any of the organs normally seen in flowers. Indeed, these primordia develop as leaves, by many morphological criteria (Color Plate 12c). The whorl-2 and whorl–3 organ primordia express the *apetala3*[+] and *pistillata*[+] genes. They have the potential to develop as petals or stamens, but the gene activities that normally decide between these two options are missing. Actually, the organs of whorls 2 and 3 develop as intermediates between petals and stamens. In addition, the loss of the *agamous*[+] function in whorl 4 again causes the formation of additional whorls.

Finally, the triple mutant combinations *apetala2, pistillata, agamous,* and *apetala2, apetala3, agamous* should reveal the ground state in every whorl and should not contain any floral organs. Indeed, all organs in the triple mutant flowers develop as leaves, or slightly carpelloid leaves (Color Plate 12d).

In conclusion, the model of Bowman and colleagues (1989, 1991) has proved very effective in explaining the morphological character of the flower organs in double and triple mutants for homeotic genes.

Homeotic Genes Are Controlled by Regulator Genes

While homeotic genes are thought to control batteries of realizator genes, they must themselves be under some sort of regulation to establish their expression domains. Some of this regulation is between the homeotic genes, as in the case of *apetala2*[+] and *agamous*[+], which inhibit each other. However, since *apetala2*[+] is expressed prior to *agamous*[+], the original limitation of *apetala2*[+] to whorls 1 and 2 still needs to be explained. Similarly, the restriction of *apetala3*[+] and *pistillata*[+] to whorls 2 and 3 calls for an explanation. Thus, plant homeotic genes, like their animal counterparts, are seen as pat of a genetic hierarchy but not as its top tier. Most likely, the regulators of homeotic genes are to be found among the genes that control earlier steps in flower development, including the *cadastral genes* and the *meristem identity genes*.

A cadastral gene that defines the boundary of *apetala3*[+] and *pistillata*[+] expression between whorls 3 and 4 has been revealed by the *superman* mutant (Bowman et al., 1992). The phenotype of the recessive *superman-1* allele is an expansion of whorl 3 at the expense of whorl 4 (Color Plate 13 and Fig. 23.16). Whorls 1, 2, and 3 contain the normal numbers of sepals, petals, and stamens, while the center whorl forms additional stamens and a reduced pistil. The extra stamens and the reduced pistil are often joined to each other.

The *superman* phenotype suggests that the normal function of the *superman*[+] gene includes inhibiting the expression of *apetala3*[+] and *pistillata*[+] in whorl 4. If the model shown in Color Plate 11 is extended to incorporate this function of the *superman*[+] gene, additional predictions can be made about the phenotypes of double mutants for *superman* and any of the homeotic mutants (Fig. 23.17). These predictions are confirmed by the actual phenotypes of such double mutants. In situ hybridizations also show that accumulation of apetala3 mRNA, which is restricted to whorls 2 and 3 in the wild type, expands in *superman* flowers to include most of whorl 4. In summary, both genetic and molecular data indicate that *superman*[+] is a regulator gene of *apetala3*[+].

A meristem identity gene, *leafy*[+], is also involved in regulating homeotic genes (Weigel et al., 1992). Strong *leafy*, *apetala3*, and *pistillata* mutant alleles share similar phenotypes: their flowers consist mostly of sepals and carpels. Transcription of the *apetala3*[+] gene is also reduced in *leafy* mutants, indicating that the normal *leafy*[+] function activates the expression of *apetala3*[+].

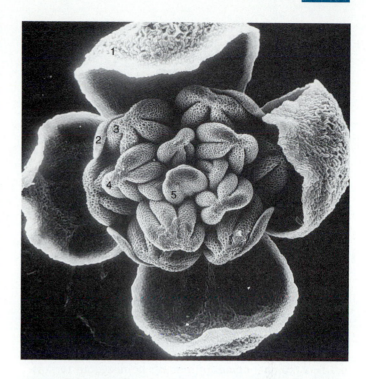

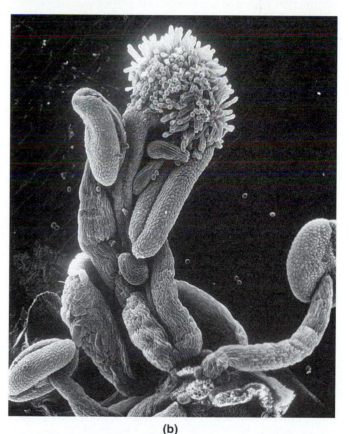

(b)

Figure 23.16 Scanning electron micrographs of developing *superman* flowers. **(a)** This specimen has four sepals, four petals, and six stamens in their normal positions in the first three whorls (1–3). Eight additional stamens occupy two additional rings (4 and 5) in the center, from which carpels are missing. **(b)** Mosaic organ of fused carpels and stamens at the center of a *superman* flower. All other organs were removed.

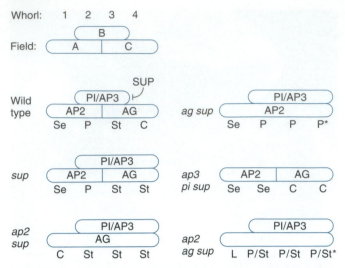

Figure 23.17 Extended model of whorl determination in *Arabidopsis* flowers including the restrictive action of the superman gene product (SUP) on the expression of the *pistillata*[+] and *apetala3*[+] genes (see also Color Plate 11). The extended model does not reflect the development of a reduced pistil in whorl 4. However, the model correctly predicts the phenotypes of mutants homozygous for loss-of-function alleles of *superman* and one or two homeotic genes. A = expression domain of *apetala2*[+], B = expression domain of *apetala3*[+] and *pistillata*[+], C = expression domain of *agamous*[+]; SUP, AP2, PI, AP3, AG = gene products of *superman*[+], *apetala*[+], *pistillata*[+], *apetala3*[+], and *agamous*[+]; *sup, ap2, ag, ap3, pi* = mutants of *superman, apetala2, agamous, apetala3,* and *pistillata;* Se = sepals; P = petals; St = stamens; C = carpels; L = leaves. An asterisk indicates that loss of the *agamous*[+] function causes the formation of additional whorls inside whorl 4.

As more double mutants in floral patterning genes are constructed, and as more of these genes are cloned, additional regulatory interactions are rapidly emerging (Ma, 1994). Some of the controls appear to be not only transcriptional but also posttranscriptional (Jack et al., 1994).

Other Plants Have Patterning Genes Similar to Those of *Arabidopsis*

The insight derived from the genetic analysis of flower development in *Arabidopsis* probably applies to a wide range of plants. This is indicated by the isolation of very similar mutants in the snapdragon *Antirrhinum majus.* Within the dicotyledonous plants, *Arabidopsis* and *Antirrhinum* are only distantly related, so characteristics shared by these species are likely to be common to the entire group.

Both *Arabidopsis* and *Antirrhinum* are well suited to genetic research (Coen and Meyerowitz, 1991). The advantages of *Arabidopsis* have already been explained. *Antirrhinum* is a much bigger plant, large enough to allow collection of floral tissues in quantity for biochemical analysis. *Antirrhinum* also has naturally occurring

transposons, which facilitate insertion mutagenesis but also make isolated strains genetically unstable. Recent molecular studies indicate a high degree of DNA sequence conservation between the two species. It is therefore possible to combine the advantages of both species by using cloned genes from one species as probes to screen DNA libraries of the other species.

Morphologically, *Antirrhinum* differs from *Arabidopsis* in the number and shape of floral organs (Fig. 23.18). The snapdragon has five sepals and five petals; one of the five stamens, however, is aborted during development. The petals are fused at the base, forming a

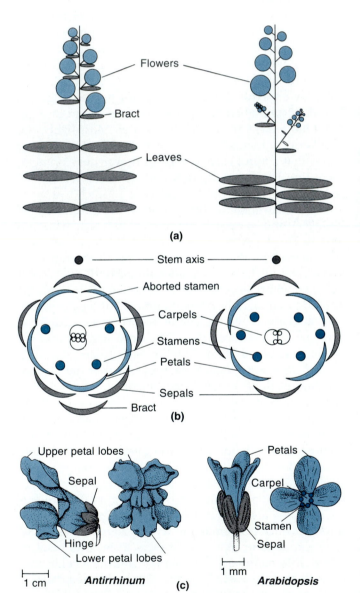

Figure 23.18 Comparison of the snapdragon *Antirrhinum majus* and the wall cress *Arabidopsis thaliana.* **(a)** Schematic diagrams of the entire plants. **(b)** Floral diagrams indicating the numbers and positions of floral organs. **(c)** Drawings of flowers. Note that *Antirrhinum* flowers are about 10-fold larger in linear dimension and 1000-fold larger in mass than *Arabidopsis* flowers.

TABLE 23.2

Phenotypes of Floral Homeotic Mutants in *Arabidopsis* and *Antirrhinum*

Mutant Class	Organ Type in Whorl				Mutants Obtained in:	
	1	2	3	4	*Antirrhinum*	*Arabidopsis*
—	Sepals	Petals	Stamens	Carpels	Wild type	Wild type
A	Carpels	Stamens	Stamens	Carpels	*ovulata* *macho*	*apetala2*
B	Sepals	Sepals	Carpels	Carpels	*deficiens* *globosa*	*apetala3* *pistillata*
C	Sepals	Petals	Petals	Sepals*	*pleniflora* *plena*	*agamous*

*Another *agamous* flower with sepals in the outer whorl (*Arabidopsis*) or extra whorls of petaloid structures (*Antirrhinum*).
Source: After Drews et al. (1991b). Used with permission.

tube that terminates in five lobes. The lowest lobe is larger and folds back to the outside, giving the flower a bilateral symmetry. The pistil, like the one in *Arabidopsis*, consists of two united carpels.

The existence of transposons in the genome of *Antirrhinum* facilitates the isolation of mutants and cloning of the corresponding genes, including a series of homeotic genes that produce phenotypes very similar to those of *Arabidopsis*. The similarity is especially noteworthy in view of the differences between the two species with respect to number and position of floral organs. This finding underlines the specific role of the homeotic genes in determining the *morphological character of entire whorls* of flower organs, not the number or specific variations of organs within a whorl. The parallel features of corresponding mutants are compiled in Table 23.2 and illustrated in Color Plate 14.

The similarity between known genes of *Arabidopsis* and *Antirrhinum* extends to nonhomeotic patterning genes and is also observed at the molecular level, as we will see shortly. Thus, it appears that patterning genes have been conserved very well during the evolution of dicotyledonous plants.

Antirrhinum Has Genes Controlling Organ Variations within the Same Whorl

The bilateral symmetry of *Antirrhinum* depends on genes that control organ variations within the same whorl. Several mutations of the *cycloidea* complex are known to have flowers with a more radial symmetry. Extreme *cycloidea* mutants produce flowers in which all petals resemble the lowest petal of a wild type (Color Plate 15). The stamen that is aborted in the wild-type flower also develops in *cycloidea* mutants so that whorls 2 and 3 both have a fivefold radial symmetry.

The radializing effect of *cycloidea* mutations on whorls 2 and 3 seems independent of the organs formed

in these whorls. This can be seen in double mutants combining *cycloidea* with homeotic mutations affecting whorl 2 or 3. For instance, the *ovulata* mutation of *Antirrhinum*, like *apetala2-1* in *Arabidopsis*, transforms petals into stamens. In *cycloidea*, *ovulata* double mutants, whorls 2 and 3 both form five stamens in radial symmetry.

These observations show that in addition to the homeotic genes specifying the morphological characters of entire whorls, there are homeotic genes that specify the characters of different organs within a whorl. These two classes of homeotic genes seem to act both independently of each other and in combination.

Molecular Analysis of Homeotic Plant Genes

Although the genetic model of flower development is very effective in explaining mutant phenotypes, it makes no predictions about the molecular mechanisms involved in the actions of the homeotic genes. It does not explain the nature of the signals to which the homeotic genes respond, nor does it predict how the homeotic genes act on their target genes. To study these mechanisms, researchers had to clone those genes.

The *agamous*[+] Gene Encodes a Transcription Factor

The first homeotic gene of *Arabidopsis* to be cloned was the *agamous*[+] gene. The cloning strategy took advantage of a mutant allele, *agamous-2*, generated by insertional mutagenesis with *T-DNA* (Feldman et al., 1989). Using the T-DNA as a tag, Martin Yanofsky and his colleagues (1990) cloned the plant genomic DNA next to the insert and used this DNA as a probe to isolate a corresponding clone from a wild-type *Arabidopsis* genomic DNA library (see Methods 14.2). This clone, when used to transform *agamous-2* mutant plants, restored the wild phenotype to the transformants. This proved that the cloned DNA did contain the *agamous*+ gene. Nucleotide sequences from the gene and cDNAs showed that the gene consists of at least nine exons and eight introns. The agamous protein shares a domain of sequence similarity with two previously identified transcription factors, the human serum response factor (SRF) and the product of the *MCM1*+ gene in yeast. The domain comprises 58 amino acids and mediates the DNA binding in these transcription factors (Fig. 23.19). The putative function of the agamous protein as a transcription factor is consistent with the role of the *agamous*+ gene as a selector gene that controls the expression of many other genes.

The *agamous*+ Gene Is Expressed in Stamen and Carpel Primordia

The genetic model of flower development also predicts the *minimum* domains and stages for the expression of the *agamous*+ gene. (A gene can be expressed beyond the area and time of its biological activity if the latter depends on the presence of other factors. For instance, a transcription factor may be synthesized but not active because it needs to cooperate with another transcription factor to be active.) Northern blotting and in situ hybridization experiments (see Methods 14.3 and 15.1) indicate that agamous mRNA is present only in flowers. Transcription begins with the formation of the first organ primordia and continues throughout flower development. In developing wild-type flowers, the *agamous*+ gene is transcribed only in stamen and carpel primordia, not in sepal and petal primordia (Color Plate 16). Thus, expression of the *agamous*+ gene is restricted to those primordia that are affected in loss-of-function mutant alleles, in accord with the genetic model.

The availability of cloned agamous cDNA was used to test predictions derived from the genetic model. The complementary phenotypes generated by loss-of-function mutations in *agamous* and *apetala2* suggest that these two genes inhibit each other's expression. To test this hypothesis, Yukiko Mizukami and Hong Ma (1992) transformed *Arabidopsis* plants with a fusion gene consisting of a viral promoter and agamous cDNA. The fusion gene was expressed in all parts of the transgenic plants, which formed flowers resembling those of *apetala2* loss-of-function mutants. This result indicates that *agamous*+ does indeed inhibit *apetala2*+ function.

As another test of the genetic model, Gary Drews and his colleagues (1991a) examined whether *apetala2*+ gene activity inhibits expression of the *agamous*+ gene as postulated. If so, absence of *apetala2*+ function should allow the *agamous*+ gene to expand its expression domain to whorls 1 and 2. This expansion was indeed observed when a labeled probe for agamous mRNA was hybridized in situ with sections of *apetala2* flower primordia. Here the agamous mRNA had accumulated in all four whorls, instead of being restricted to whorls 3 and 4 (see Color Plate 16). This result confirmed the prediction derived from the genetic model that *apetala2*+ (directly or indirectly) controls *agamous*+ expression. (Note that the *Arabidopsis* investigators used the same method employed by *Drosophila* researchers to establish a regulatory hierarchy among embryonic patterning genes. See Figure 21.4.)

Similar Mutant Phenotypes, Spatial Expression Patterns, and Protein Products Reveal Homologous Genes between *Arabidopsis* and *Antirrhinum*

When Szuszanna Schwarz-Sommer and her colleagues (1990) cloned and sequenced the *deficiens*+ gene of *Antirhinnum*, they found the same DNA-binding domain in the deficiens protein that had also been found in the agamous protein of *Arabidopsis* and in other transcription factors (Fig. 23.19). They dubbed this domain the **MADS box** because it had been identified in *MCM1*, *agamous*, *deficiens*, and *SRF* transcription factors. The researchers observed that the *deficiens*+ gene was tran-

AP3	RGKIQIKRIENQTNRQVTYSKRRNGLFKKAHELTVLCDARVSIIMFSSSNKLHEYISP	
DEF	– S – – – – – K – – – – – i – – TQ – – – – – – – –	53/58
AG	– – – – E – – – – – – – t – – – – – – Fc – – – – – – – I – – – y – – S – – – – – e – AL – V – – – rgR – y – – snn	40/58
MCM1	– r – – E – – f – – – k – r – h – – F – – – Kh – Im – – – f – – S – – tgTq – ILLVv – eTgIVytFsT –	26/58
SRF	– v – – kMef – D – kir – yt – F – – – Kt – Im – – – y – – St – tgTq – ILLVa – eTgHVytFaTr	19/58

Figure 23.19 Amino acid sequence comparison for the putative DNA-binding region of several proteins: apetala3 (AP3) of *Arabidopsis*, deficiens (DEF) of *Antirrhinum*, agamous (AG) of *Arabidopsis*, yeast MCM1 gene product, and human serum response factor (SRF). The top line is the AP3 sequence in single-letter code. In the sequences below, only amino acids diverging from the AP3 standard are shown, while identical amino acids are represented by a dash. Capitals represent amino acids that are similar to their counterparts in AP3; lowercase letters represent amino acids unlike their AP3 counterparts. The numbers to the right indicate the fraction of amino acids a protein has in common with AP3.

scribed most actively in those wild-type organ primordia that were transformed in the mutant phenotype: petals and stamens (Fig. 23.20). These results confirmed the expectation that homeotic plant genes in general encode regionally expressed transcription factors.

Mutations in the *deficiens* gene of *Antirrhinum* and the *apetala3* gene of *Arabidopsis* generate similar phenotypes: they transform petals into sepals and stamens into carpels (Color Plate 14c and g, and Table 23.2). Using *deficiens* cDNA of *Antirrhinum* as a molecular probe to screen *Arabidopsis* DNA libraries, Thomas Jack and his colleagues (1992) identified a clone containing the *apetala3*[+] gene of *Arabidopsis*. The deduced sequence of the apetala3 protein included the MADS box and further regions of sequence similarity (Fig. 23.19). As expected, the investigators found the strongest accumulation of apetala3 mRNA in the petal and stamen primordia of developing wild-type flowers. Thus, the two genes turned out to produce equivalent mutant phenotypes, to have corresponding expression patterns, and to encode similar proteins. Most likely, therefore, these genes are homologous; that is, they have evolved from a gene in a common ancestor of *Antirrhinum* and *Arabidopsis*.

Another homologous pair of genes is the *leafy*[+] gene of *Arabidopsis* and the *floricaula*[+] gene of *Antirrhinum* (Coen et al., 1991; Weigel et al., 1992). Mutations in either gene cause the transformation of flowers into entire inflorescence shoots, although the phenotype is more severe in *Antirrhinum*. Both *leafy*[+] and *floricaula*[+] are expressed in floral meristems, and the proteins encoded by the two genes show 70% sequence identity. These observations indicate that not only the homeotic genes but also other floral patterning genes are conserved among dicotyledonous plants.

The Known Plant Homeotic Genes Have No Homeobox

The known homeotic plant genes share the main characteristics of homeotic genes in *Drosophila* and the mouse (see Chapters 21 and 22): mutations in these genes cause homeotic transformations, their expression is restricted to certain organ primordia, and they encode putative transcription factors. On the basis of these similarities, one might expect plant homeotic genes to share two other features with their counterparts in fruit flies and mice: the clustering in complexes and the *homeobox*. However, the plant homeotic genes cloned so far are scattered over several chromosomes (Fig. 23.15). The MADS box shared by many plant homeotic proteins has about the same length as the *homeodomain*, and it is thought to form similar α helices that bind to specific DNA recognition motifs, but there is no sequence similarity to the homeodomain (compare Figs. 22.4 and 23.19).

This is not to say that plants do not produce proteins with homeodomains. A gene encoding such a protein, *Knotted-1*, was isolated from maize (Vollbrecht et al., 1991). The gene was identified by gain-of-function alleles that alter leaf development. Nests of cells in the mutant leaves fail to differentiate properly and continue to divide, throwing the leaf blades into folds and distorting their vein patterns. The sequence similarity between the homeodomain encoded by the *Knotted-1*[+] gene and homeodomains of other organisms is rather low, ranging from 22/64 identity with the human Prl (pre-B-cell leukemia) protein to 12/64 identity with the Antennapedia protein of *Drosophila*. When a maize genomic DNA library is screened with the *Knotted-1* homeobox as a labeled probe, a family of homeobox-containing genes is identified. The functions of these genes remain to be elucidated.

On the basis of the few homeotic genes and homeobox genes so far identified in plants, it appears that they may form two distinct groups. The main characteristic of *homeotic genes* is their *biological function*: they specify, by their activity or inactivity, a pattern of determined states along an embryonic axis. Apparently, this function can be carried out by transcription factors with a homeodomain and by other transcription factors with a different DNA-binding domain, such as the MADS box. The main characteristic of *homeobox genes* is a *biochemical feature*: they encode transcription factors with a homeodomain. In animals, most homeotic genes

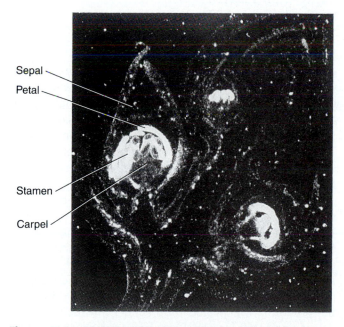

Figure 23.20 Spatial expression pattern of the *deficiens*[+] gene in *Antirrhinum* shown by in situ hybridization (see Methods 15.1). The tip of an inflorescence is shown in longitudinal section and with dark-field illumination. Most of the label (white grains) accumulates over the primordia of petals and stamens. The other parts of the inflorescence, including the primordia of sepals and carpels, are labeled less strongly.

Sepal
Petal
Stamen
Carpel

are homeobox genes that are arranged in highly conserved *Hox*-like clusters as discussed in Chapters 21 and 22. Indeed, Jonathan Slack and his colleagues (1993) consider the *Hox* complex and its expression pattern in the embryo as the defining character of the animal kingdom and propose that it be called the *zootype*. Plants clearly do not show this character. Here, homeotic genes and homeobox genes have evolved independently. The disjunction indicates that *each* of the two functions, generating patterns of determined states and producing transcription factors with homeodomains, was *by itself* sufficiently adaptive and/or interactive to be conserved through a billion years of evolution.

Many Questions about Pattern Formation in Flowers Are Still Unanswered

Molecular analysis of the first homeotic plant genes has confirmed the genetic model, but many fundamental questions are still unanswered. The genes controlling the homeotic plant genes are largely unknown, as are the realizator genes presumed to be controlled by the homeotic genes. It is also not clear whether the homeotic genes act in all cell types of a developing flower organ or only in specific cells, such as the epidermal cells.

Perhaps the most disturbing observations—because they are unexplained by the model—are organ deficiencies found in some of the homeotic phenotypes. For instance, certain *apetala2* alleles affect the numbers and positions of stamens and carpels in whorls 3 and 4, although according to the model these whorls should develop normally. Also, the organs of whorl 2 are missing in *apetala2-2* mutants, and their absence results from a failure of the organ primordia ever to appear, not from a failure of such primordia to develop (Bowman et al., 1991). Conceivably, these organ losses may be related to the expanded *agamous*$^+$ activity, which in *apetala2* mutants is not restricted by the normal *apetala2*$^+$ function. The reiterations of flower organs in the *agamous* mutant phenotype suggest that the normal *agamous*$^+$ function includes the termination of further organ formation once whorl 4 has been established. This termination function of the *agamous*$^+$ gene, if expanded beyond its normal realm, might account for the organ losses in whorls 2 and 3 of *apetala* mutants.

In any event, the apparent termination function of *agamous*$^+$ shows that this homeotic gene is involved not only in the determination of flower organ primordia but also in limiting the number of whorls. In this sense, *agamous*$^+$ is also a meristem identity gene. It will be interesting to see how many patterning genes are involved in more than one step of flower development.

SUMMARY

Plant development differs from animal development in several respects. The fruit that develops from the central part of a flower contains many seeds, and each seed contains one embryo. The embryo already shows the basic pattern of the seedling that will emerge when the seed germinate: it contains an epicotyl, one or two cotyledons, a hypocotyl, and a root primordium. The epicotyl and the root primordium contain meristem cells, which retain their ability to divide throughout the life of the plant. Meristem cells are similar to the stem cells of animals, in that they give rise to the primordia of different plant organs while at the same time generating new meristem cells. Depending on internal and environmental factors, the apical shoot meristem produces the primordia of leaves, branches, or flowers. Compared with animal development, plant development relies less on maternal RNA and more on early embryonic gene activity. Also, the overall appearance of a plant is generated primarily by the timing and pattern of meristem cell divisions; morphogenetic movements of cells or epithelia play a minor role.

The wall cress *Arabidopsis thaliana* has emerged as a model for genetic and molecular studies of the development of angiosperms. A saturation mutagenesis screen for abnormal embryogenesis has yielded 250 mutant lines, which probably represent about 30 genes involved in embryonic pattern formation. In the mutant phenotypes, there are specific deletions along the apical-basal axis, defects in the radial stem pattern, or major shape changes. These observations suggest that it will be possible to identify genes controlling specific steps of the patterning process.

Other studies have revealed a set of homeotic genes that control the determination of flower organ primordia. These genes bear many similarities to their counterparts in fruit flies and mice: they encode transcription factors, their expression is restricted to certain domains, and their mutant phenotypes are characterized by homeotic transformations. For instance, one type of homeotic mutant, the primordia that would form stamens in the wild type develop as petals instead. Plant homeotic genes are regulated by other patterning genes that control earlier steps in floral development. In addition, there are regulatory interactions among homeotic plant genes.

A simple genetic model explains most phenotypic traits of the known homeotic mutants and their combinations. The regulatory interactions implied in the model are confirmed by the available molecular data. As expected, the proteins encoded by the plant homeotic genes have the molecular characteristics of transcription factors. However, the homeotic plant genes cloned so far do not have a homeobox. This observation suggests that the specification of developmental pathways by homeotic gene activities has evolved independently of the biochemical feature of the homeodomain, which is common to a family of transcription factors.

The snapdragon *Antirrhinum majus* has floral patterning genes corresponding to those of *Arabidopsis* with respect to mutant phenotypes, expression domains, and DNA sequence. Thus, these genes have been well conserved during the evolution of dicotyledonous plants. In addition to those homeotic genes that control the organ character of entire whorls, *Antirrhinum* also has genes that control organ variations within the same whorl.

SUGGESTED READINGS

Coen, E. S., and R. M. Meyerowitz. 1991. The war of the whorls: Genetic interactions controlling flower development. *Nature* **353**:31–37.

Coen, E. S., S. Doyle, J. M. Romero, R. Elliott, R. Magrath, and R. Carpenter. 1991. Homeotic genes controlling flower development in *Antirrhinum. Development* supplement 1:149–155.

Ma, H. 1994. The unfolding drama of flower development: Recent results from genetic and molecular analyses. *Genes & Development* **8**:745–756.

Mayer, U., R. A. Torres Ruiz, T. Berleth, S. Miséra, and G. Jürgens 1991. Mutations affecting body organization in the *Arabidopsis* embryo. *Nature* **353**:402–407.

Weigel, D., and E. M. Meyerowitz. 1993. Genetic hierarchy controlling flower development. In M. Bernfield, ed., *Molecular Basis of Morphogenesis,* 93–107. New York: Liss.

EXPERIMENTAL AND GENETIC ANALYSIS OF *Caenorhabditis elegans* DEVELOPMENT

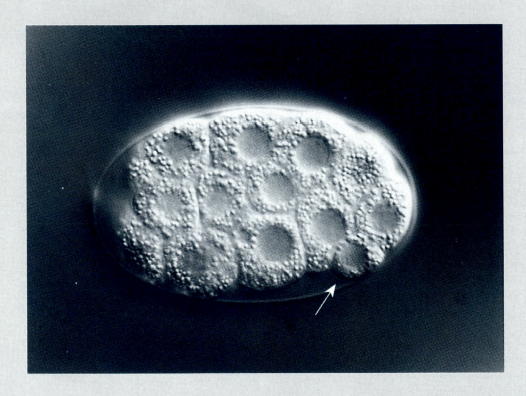

Figure 24.1 Living embryo of the roundworm *Caenorhabditis elegans* at the 28-cell stage; anterior pole to left. The arrow marks the primordial germ cell. The other cells are somatic cells. The entire cell division pattern from zygote to adult is invariant between individuals. The cell nuclei appear as depressions in this photograph because of the contrast enhancement used.

Normal Development

After *Drosophila* and the house mouse, the soil nematode *Caenorhabditis elegans* is the third most popular animal subject for the genetic analysis of development. Living on agar plates or in liquid medium and feeding on bacteria, this tiny worm is easy to keep in the laboratory. Because *C. elegans* is transparent at all stages of its life, cell nuclei are readily observed in live specimens under the microscope (Fig. 24.1). And because an adult consists of only about 1000 somatic cells, the anatomy of *C. elegans* is comparatively simple.

From a research standpoint, the most important attribute of *C. elegans*—which it shares with other nematodes—is a striking invariance of development that makes individuals from the same species *exactly* alike (Sulston et al., 1983). The number of cell divisions, as well as their timing and orientation, is strictly controlled, so that the number of cells per individual is constant. Most cells do not move much during development, and those that do will wind up at specific destinations. Thus, the neighbors with which any given cell interacts during its development are also the same from one individual to another.

Advantages of *C. elegans* for the genetic analysis of development include a short life cycle, a small genome, and relatively little repetitive DNA (Brenner, 1974; see also Table 14.1). Moreover, the fact that most adults are self-fertilizing makes it particularly easy to breed alleles to homozygosity.

The roundworm's unique combination of developmental and genetic advantages has allowed researchers to study some basic questions in development with unparalleled precision. Typically, they can carry out such studies on *C. elegans* at the level of single cells, since the life history of each cell in the animal is known. We will illustrate this type of analysis by returning to two familiar topics, cytoplasmic localization and embryonic induction; and by looking at two new topics, the timing of developmental events and programmed cell death. In each case, mutant analysis and molecular techniques have confirmed and extended the conclusions reached from earlier studies.

Normal Development

Hermaphrodites and Males

Adult *C. elegans* worms are either hermaphroditic or male (Fig. 24.2). *Hermaphrodites* are defined as individuals that have both male and female reproductive organs. In the case of *C. elegans*, hermaphrodites can fertilize their eggs either with their own sperm or with sperm obtained by copulating with males. The two-armed gonad of the hermaphrodite produces sperm

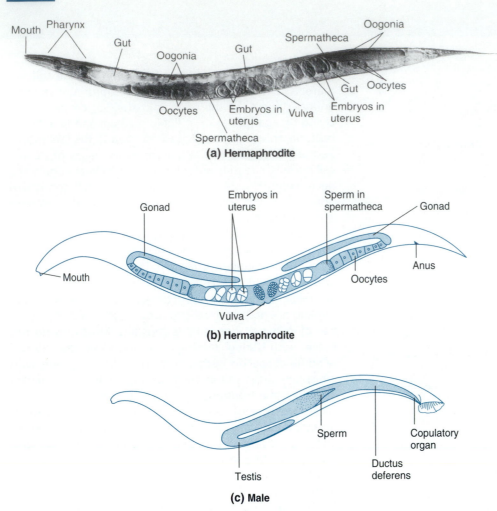

Mouth Pharynx
Gut
Oogonia
Gut
Spermatheca
Oogonia
Oocytes
Gut Oocytes
Oocytes
Embryos in Vulva
uterus
Embryos in
uterus
Spermatheca

(a) Hermaphrodite

Gonad
Embryos in
uterus
Sperm in
spermatheca
Gonad
Mouth
Anus
Oocytes
Vulva

(b) Hermaphrodite

Sperm
Copulatory
organ
Ductus
deferens
Testis

(c) Male

Figure 24.2 Adult morphology and reproductive organs of the roundworm *C. elegans.* **(a)** Light micrograph of a gravid hermaphrodite. The mouth, pharynx, and anterior gut are seen to the left of the vulva, a ventral opening for copulation and egg laying. The two arms of the gonad (see panel b) contain oogonia, which develop into mature oocytes and are fertilized with sperm stored in the spermatheca. The embryos undergo cleavage while still in the uterus. **(b)** Schematic diagram of the adult hermaphrodite's reproductive organs. Each gonadal arm, now functioning as an ovary, is connected to a spermatheca, where eggs are fertilized by stored sperm derived either from the hermaphroditic gonad during the larval stage or obtained as an adult from copulating males. In the uterus, the fertilized eggs undergo cleavage before they are laid through the vulva. **(c)** Male reproductive organs. A single testis releases sperm via a ductus deferens to a copulatory organ at the posterior end.

during the last larval instar (growth stage). These sperm are stored in **spermathecae** (sing., *spermatheca*), receptacles located between each gonad and the uterus. In the adult hermaphrodite, the gonads function as ovaries and produce oocytes. The mature eggs are fertilized as they pass the spermathecae on their way to the uterus. Fertilized eggs develop in the uterus until around the 32-cell stage, when they are born through a ventral genital opening, the *vulva*. An adult *C. elegans* hermaphrodite consists of 959 somatic cells and produces about 2000 gametes (Wood, 1988a).

The male of *C. elegans* has a one-armed gonad that produces only sperm, which are delivered to the vulva of the hermaphrodite via the copulatory organ at the male's posterior end. The adult male consists of 1031 cells and produces about 1000 sperm. Since hermaphrodites have two sex chromosomes and males only one, the self-fertilization of hermaphrodites produces almost exclusively more hermaphrodites; occasional males that develop after self-fertilization result from chromosomal nondisjunction during meiosis. However, sperm from a male outcompete sperm from a hermaphrodite at fertilization, so that hermaphrodites that have mated produce predominantly male-fertilized offspring, of which 50% are male and 50% hermaphroditic.

Fertilization, Cleavage, and Axis Formation

From fertilization to hatching, the embryogenesis of *C. elegans* takes only 14 h (Fig. 24.3; Sulston et al., 1983). The anteroposterior polarity of the egg becomes visible after fertilization. The male pronucleus is located near the posterior pole, but it is not known whether sperm entry establishes the posterior pole, or whether an anteroposterior polarity is already present before fertilization. Once fertilized, the egg forms a tough and nearly impermeable outer shell. The female pronucleus resumes meiosis and gives off polar bodies at the anterior pole. Next, the egg undergoes *pseudocleavage,* a set of constrictions near the anterior pole and equator that subsequently regress. At the same time, the female pronucleus moves posteriorly toward the male pronucleus until the two meet in the posterior egg half.

The early cleavage in *C. elegans* are asymmetrical and asynchronous. The first mitotic spindle forms and

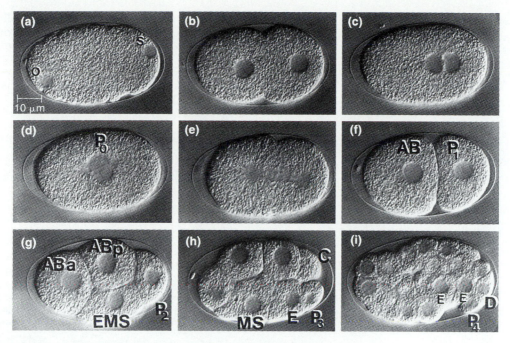

Figure 24.3 Cleavage of *C. elegans*. In each photograph, the anterior pole is oriented to the left. **(a)** About 30 min after fertilization, the female pronucleus (o) is visible anteriorly and the male pronucleus (s) posteriorly. **(b, c)** The pronuclei migrate until they meet in the posterior half of the egg, while the egg constricts temporarily in pseudocleavage. **(d)** The pronuclei rotate and fuse near the egg center. **(e)** The mitotic spindle orients parallel to the anteroposterior axis. **(f)** The first cleavage separates a large somatic cell (AB) from a smaller germ line cell (P_1). **(g)** After division of AB into ABa (anterior) and ABp (posterior), P_1 cleaves into a large somatic cell, EMS, and a smaller new germ line cell, P_2. **(h)** The division of both AB cells and EMS is followed by the unequal cleavage of P_2 into a somatic cell, C, and a smaller new germ line cell, P_3. **(i)** The somatic cells divide again. In the germ line, the last unequal cleavage generates the somatic founder cell D and the primordial germ cell P_4. Soon after, gastrulation begins with the immigration of the two E cells.

positions itself parallel to the anteroposterior axis and slightly posterior to the egg center. The pronuclei fuse, and the first mitosis and cleavage follow immediately, about 35 min after fertilization. Cytokinesis produces a larger anterior blastomere called AB and a smaller posterior blastomere called P_1. About 10 min later, the AB cell divides, producing an anterior daughter cell (ABa) and a posterior daughter cell (ABp). Shortly thereafter, P_1 divides into P_2 at the posterior pole and EMS opposite ABp. EMS defines the future ventral side of the embryo and ABp the future dorsal side. Shortly after the cleavage of AB, EMS divides into the E cell and the MS cell.

Founder Cells

The first cleavages can be viewed as an embryonic stem cell pattern. The zygote undergoes a series of asymmetric divisions, each producing an anterior somatic cell (AB, EMS, C, and D) and a posterior germ line cell labeled P_1 after the first division, P_2 after the second division, and so forth until P_4 (Fig. 24.4). The resulting five cells (six after the division of EMS into E and MS) are called *founder cells*, because each of them develops in a characteristic way. The founder cells and their respective progeny continue to cleave at a rate that is roughly proportional to the founder cells' size and follows their order of origin. AB cleaves fastest, D slowest, and P_4—the primordial germ cell—cleaves only once more during embryogenesis. The timing and orientation of these cleavages are nearly invariant from one individual to another.

Given the importance of cell lineage in *C. elegans* development, one might expect each founder cell to give rise only to certain cell types and, conversely, all cells of a certain type to be derived from a single founder cell. However, lineage analysis has revealed that this is not generally true (Fig. 24.4). While intestinal cells and germ cells are indeed derived from single founder cells, this is not the case for hypodermis cells, neurons, and muscle. Thus, cells of similar type are not necessarily closely related, and closely related cells may turn into different types. In these cases, then, the founder cells do *not* produce separate lineages of specific cell types.

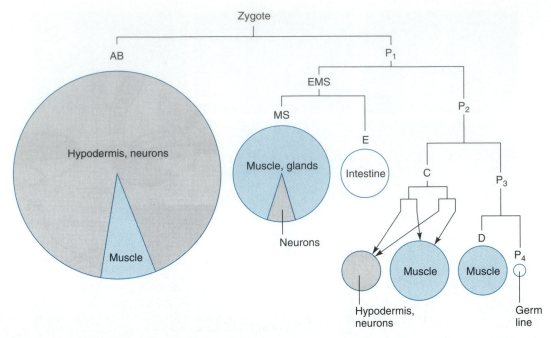

Figure 24.4 Generation of founder cells in *C. elegans*, and a summary of cell types derived from them. Areas of circles and sectors are proportional to numbers of cells. Gray shading represents ectodermal tissues, and color represents mesodermal tissue. Founder cells E, D, and P₄ give rise to one cell type each, while all other founder cells have mixed progeny. Conversely, a given cell type, such as muscle, may originate from two or more founder cells.

Gastrulation, Organogenesis, and Histogenesis

Gastrulation begins at the 28-cell stage, when the sixteen AB descendants lie anteriorly and laterally, the four MS derivatives lie ventrally, and the four C descendants lie posteriorly and dorsally. The two E derivatives (Ea and Ep) as well as D and P₄ lie ventrally and posteriorly (Fig. 24.3). At the start of gastrulation, the two E cells sink inward, followed by P₄ and the daughter cells of MS. As the entry zone widens, prospective pharynx cells and body muscle cells derived from the AB, C, and D lineages follow into the interior. Further divisions of the E cells and pharynx precursor cells generate the cells for the digestive tube. The body muscle cells are positioned between the digestive tube and the outer cell layer, which forms the epidermis (in roundworms, caller the hypodermis) and the nervous system. Except during gastrulation, cell movements play a minor role in *C. elegans* development; most cells are formed near their final locations.

Cell division and organogenesis continue until about 6 h after fertilization, when the hermaphrodite embryo consists of 558 cells and the male embryo of 560 cells. During the next 6 h, the embryo changes from a spheroid to an elongated shape as microfilaments and microtubules squeeze the embryo around its circumference (Priess and Hirsh, 1986). Subsequently, the worm's elongated shape is maintained by a collagenous cuticle that is secreted by the hypodermis. During the final 2 h of embryogenesis, the pharynx begins to pump and the eggshell is softened enzymatically so that the worm can hatch.

Larval Development

After hatching, the larva takes 3 days to grow into an adult. During this period, the worm increases in length from 0.24 to 1.2 mm while molting and replacing its cuticle four times to accommodate its increasing size. The first molt is from the first larval stage (designated L1) to the second larval stage (L2), and the fourth molt is from L4 to the adult. From the time of hatching, the average life span is 18 days. However, under conditions of starvation or crowding, worms can enter an alternative, longer-lived larval stage termed the *dauer larva* (G. *dauern*, "to last" or "persist"). Dauer larvae are resistant to desiccation and other harsh treatments and can survive for months. When favorable conditions return, dauer larvae molt into L4's and then into adults.

During larval development, reproductive organs and other adult structures arise from a class of cells known as **postembryonic blast cells**. The postembryonic blast cells of *C. elegans*, like the *imaginal discs* of *Drosophila*, develop during the larval stages and give rise to descendants that do not function until the adult stage is reached. Figure 24.5 shows the positions of postembryonic blast cells in a young L1 larva. The fur-

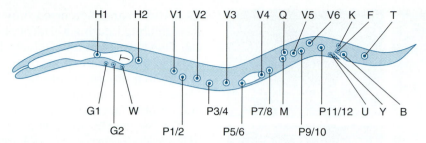

Figure 24.5 Stage L1 larva of *C. elegans* with postembryonic blast cells labeled. Most blast cells are paired, but only the left cell of each pair is shown. G1, G2, B, U, Y, and F are located medially, P1/2 through P11/12 originate bilaterally and then migrate ventrally, at which time they are designated in sequence P1, P2, . . . , P12. (These cells are unrelated to the embryonic cells P_1 through P_4.) B, U, Y, and F are found in both sexes but divide only in males.

ther divisions of these blast cells are as precisely timed and oriented as the embryonic cell divisions. During postembryonic development, about 200 cells fuse into *syncytia* of various sizes, including the hypodermis, certain muscles, and the gut.

Localization and Induction during Early Cleavage

The invariance of nematode development allows investigators to analyze how *single cells* are determined. As in other animals, the two major mechanisms of cell determination are cytoplasmic localization and embryonic induction. Both phenomena have been studied in *C. elegans* by experimental and genetic means.

The First Cleavage Generates Blastomeres with Different Potentials

The first cleavage in *C. elegans* generates two blastomeres AB and P_1. These two cells differ in size (AB is larger), timing of further cleavage (AB divides faster), and *fate* (Fig. 24.4). AB and P_1 also differ in *potency*, as shown by isolation experiments in which one of the two blastomeres is destroyed inside the eggshell (Laufer et al., 1980) or removed through a hole made in the eggshell (Priess and Thompson, 1987). The surviving blastomere, which remains inside the eggshell, is then monitored for the types of daughter cells it generates. The cell types produced are identified by their division patterns, by natural markers such as the birefringent *rhabditin* granules that are characteristic of gut cells, and/or by *immunostaining* with tissue-specific antibodies (see Methods 4.1). Under these experimental conditions, an isolated P_1 blastomere gives rise to pharynx, body-wall muscle, and but cells, whereas the descendants of isolated AB blastomeres form hypodermis cells and neurons.

What is the origin of these differences in cellular size, behavior, fate, and potency? Clearly, the fertilized egg has an anteroposterior polarity, whose origin is not yet clear but which is either preformed in the oocyte or oriented by sperm entry. Most likely this polarity entails an uneven distribution of cytoplasmic components that are partitioned asymmetrically during first cleavage and then act as *cytoplasmic determinants*. While the nature of these determinants is still unknown, attention has focused on granules resembling the *polar granules* that are associated with germ cell determinants in *Drosophila* and other animals (see Chapter 8). Although the determinative function of these granules has not been proven in *C. elegans*, their easy visibility makes them an interesting model for studying the localization process itself (Wood, 1988b; Strome, 1989).

P Granules Are Segregated into Germ Line Cells

In attempts to identify unevenly distributed cytoplasmic components, Susan Strome and William Wood (1983) raised monoclonal antibodies from mice injected with homogenized *C. elegans* embryos and screened them using an immunostaining procedure. Several independently obtained antibodies reacted with cytoplasmic granules termed **P granules** because they are segregated into the P cells during cleavage (Fig. 24.6). Later during development, P granules are present only in germ line cells but not in mature sperm.

Possible mechanisms of P-granule segregation include cytoplasmic transport as well as selective stabilization at the posterior pole. To further investigate this process, Strome and Wood (1983) treated *C. elegans* embryos immediately after fertilization with microtubule inhibitors. These treatments blocked the migration of pronuclei, but the P granules nevertheless became localized near the posterior pole. By contrast, in embryos treated with microfilamemt inhibitors, the P granules coalesced near the center of the embryo instead of ac-

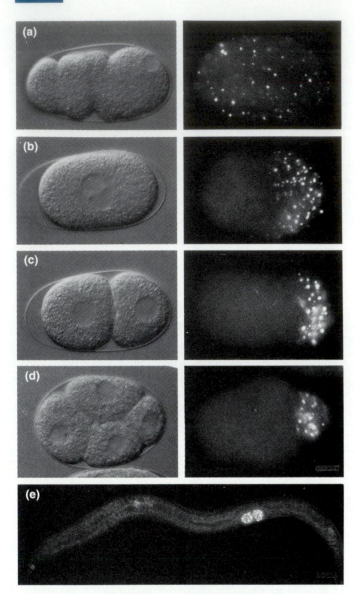

Figure 24.6 Segregation of P granules during early cleavage of *C. elegans*. The left panels show living embryos photographed with contrast enhancement; the right panels and bottom panel show fixed specimens with immunofluorescent P granules (see Method 4.1). Anterior to left, posterior to right. **(a)** In the fertilized egg during pseudocleavage, P granules are dispersed throughout the cytoplasm. **(b)** After pronuclear migration, P granules are localized to the posterior cortex. **(c, d)** P granules are distributed to the P_1 cell of the 2-cell embryo and later to the P_2 cell of the 4-cell embryo. **(e)** P granules are densely packed in the two promordial germ cells of a newly hatched larva.

these disturbances occurred only when microfilaments were inhibited during a critical phase of the first cell cycle. These observations suggest that fertilization activates two cytoskeletal systems: a microtubule-based system that moves the pronuclei, and a microfilament-based system for transporting the P granules and generating other asymmetries in the zygote.

For the genetic analysis of cytoplasmic localization in *C. elegans* embryos, researchers have carried out *mutagenesis screens* for *maternal effect* mutations that interfere with embryonic development. One team isolated four genes, *par-1⁺*, *par-2⁺*, *par-3⁺*, and *par-4⁺*, which are involved in the partitioning of cytoplasmic components during early cleavage (Kemphues et al., 1988; Kirby et al., 1990). Offspring from *par* mutant hermaphrodites show abnormalities in pseudocleavage, positioning of the first mitotic spindle, cytoplasmic streaming, cell division timing, and the segregation of P granules and possibly of other components related to gut cell development. Because mutations in any of the four *par⁺* genes generate similar phenotypes, it appears that the products of these genes exert different types of control over the same cellular system, such as the microfilaments. Cloning and sequencing the *par⁺* genes will provide clues to the biochemical function of their protein products, and possibly their biological roles.

The Maternal Gene *skn-1⁺* Encodes a Protein That Acts as a Localized Determinant for the EMS Blastomere

The pharynx of *C. elegans* is formed by descendants of two cell lineages. At the 28-cell stage, there are five blastomeres that will produce all of the cells in the pharynx (Fig. 24.7). Three of these cells, derived via ABa from AB, give rise to the anterior pharynx. This contribution depends on inductive interactions with P_1 decendants, as will be discussed shortly. The other two blastomeres, which form the posterior pharynx, are derived via MS and EMS from P_1. These cells can give rise to pharyngeal cells in isolation, that is, without inductive interaction. P_1 passes on the ability to form pharynx autonomously to one of its daughter cells, EMS, which in turn passes it on to MS. Thus, some factor or property required for autonomous pharyngeal cell development appears to be partitioned asymmetrically during the first three cleavage divisions.

A candidate for conferring this ability is the product of the *skn-1⁺* gene, which was discovered in a screen for maternal effect mutants that lack pharyngeal cells (Bowerman et al., 1992, 1993). In offspring from *skn-1* mutant hermaphrodites, the EMS descendants form hypodermis and body-wall muscle instead of producing pharyngeal and intestinal cells as they do in normal embryos. The skn-1 protein has the molecular characteristics of a transcription factor and in the wild type accumulates mainly in the nuclei of P_1 and its descendants

cumulating at the posterior pole. Also missing in these embryos were other signs of anteroposterior polarity, such as anterior contractions during pseudocleavage. Moreover, the pronuclei moved together but met in the center instead of posteriorly. Extending these experiments, David Hill and Susan Strome (1990) found that microfilament inhibitors also abolished the differences in the subsequent cleavage patterns of AB and P_1. All

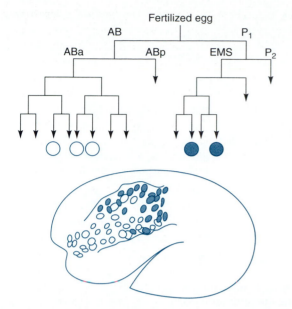

Figure 24.7 Lineage and position of pharynx cells in *C. elegans*. The upper diagram represents the lineage of the five blastomeres at the 28-cell stage that give rise to the pharynx. Three of these blastomeres (open circles) are descendants of AB, and the other three blastomeres (filled circles originate from P_1. The lower diagram shows an embryo, about midway through its development, with the descendants of the AB cell and the P_1 cell that contribute to the anterior and posterior pharynx, respectively.

through the 8-cell stage (Fig. 24.8). Because of its strictly maternal effect, skn-1 protein must be translated from maternal mRNA in the oocyte or egg. Conceivably, the skn-1 mRNA could be localized posteriorly, although this has not been directly demonstrated.

The restriction of skn-1 protein to the nuclei of P_1 and its descendants depends on the function of two other maternal effect genes, namely, the $par-1^+$ gene mentioned previously and the $mex-1^+$ gene. In embryos from *par-1* or *mex-1* mothers, all cell nuclei contain levels of skn-1 protein that are intermediate between the skn-1 levels of the P_1 and AB lineages in wild-type embryos (Fig. 24.8). This observation suggests that the par-1 and mex-1 products are required for the localization of skn-1 mRNA or protein, so that failure of the localization process would lead to intermediate skn-1 levels throughout the egg.

Not only is the skn-1 protein *necessary* for the ability of the EMS cells to form pharynx and intestine, but it also seems *sufficient* to convert other cells to form pharyngeal and muscle cells. This is suggested by the phenotypes of embryos from hermaphrodites deficient in the maternal effect gene $mex-1^+$ (Mello et al., 1992). Such

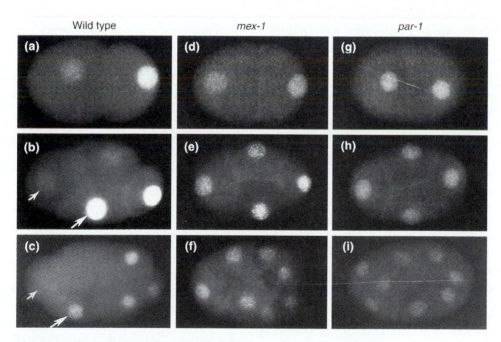

Figure 24.8 Accumulation of skn-1 protein in wild-type and *mex-1* or *par-1* mutant embryos of *C. elegans*. The top, middle, and bottom rows show 2-cell, 4-cell, and 8-cell embryos, respectively. Anterior is to the left. All embryos were immunostained for skn-1 protein (see Methods 4.1). **(a-c)** Wild-type embryos. The skn-1 protein accumulates in the nuclei of P_1 and its descendants. At the 4-cell stage, there is a marked difference between skn-1 levels in P_2 and EMS (long arrow) as compared with ABa (short arrow) and ABp (top). **(d–f)** Embryos from *mex-1* mutant mothers and **(g–i)** embryos from *par-1* mutant mothers. In both mutants, all nuclei of an embryo contain about the same level of skn-1 protein, intermediate between the levels of the P_1 lineage and the AB lineage in the wild type.

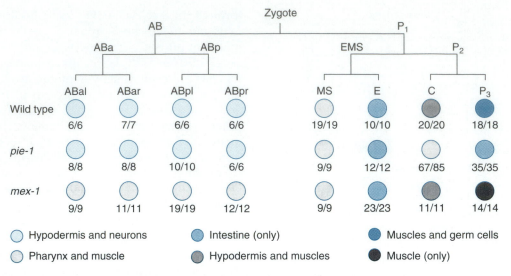

Figure 24.9 Altered blastomere fates in offspring of hermaphrodites with mutations in the maternal effect gene *pie-1* or *mex-1*. The diagram shows the lineage and names of all *C. elegans* blastomeres up to the 8-cell stage. For instance, ABpr is the right daughter of the posterior daughter of AB. The coded circles indicate the types of cells produced by each blastomere after all other blastomeres were killed with a microbeam. In *pie-1* offspring, P_2 is transformed into another EMS, so that the E and MS lineages are duplicated in the posterior. In *mex-1* offspring, all four AB granddaughters are transformed into extra MS blastomeres, so that an excess of pharyngeal and muscle cells is formed in the anterior. In addition, P_3 forms only muscle cells. The numbers indicate how many times a particular result was obtained: thus, 67 out of 85 isolated C blastomeres from *pie-1* offspring gave rise to pharynx and muscle cells.

embryos produce too many pharyngeal and muscle cells (*mex* stands for *muscle excess*). Lineage analysis shows that the extra cells are formed because all AB descendants at the 8-cell stage deviate from their fates and develop like MS blastomeres (Fig. 24.9). The ability of transformed blastomeres to develop like MS cells depends on the presence of the skn-1 gene product. (Offspring from hermaphrodites with both *mex-1* and *skn-1* loss of function do not develop extra pharyngeal muscle.) Because skn-1 is mislocalized to the AB descendants in offspring of *mex-1* hermaphrodites, it appears that the ectopic presence of skn-1 protein in these cells may be sufficient to convert them into MS-like cells.

For normal development of the P_1 cell lineage, it seems necessary to limit the activity of skn-1 protein to the EMS cell. This restriction requires the function of yet another maternal effect gene, *pie-1⁺* (*pie* stands for *pharyngeal and intestinal excess*). In offspring from *pie-1* loss-of-function hermaphrodites, the C blastomere produces pharyngeal and muscle cells instead of hypodermis and muscle as in normal embryos (Fig. 24.9). This conversion of the C blastomere depends on the maternal skn-1 function, as in the case of the converted AB descendants. However, the distribution of skn-1 protein in embryos from *pie-1* hermaphrodites is normal. Thus, *pie-1⁺* seems to restrict the activity of skn-1 protein by means other than localization.

Taken together, the data suggest that the maternally supplied skn-1 protein is necessary and sufficient to specify the fate of the EMS blastomere, and that the maternal mex-1 and pie-1 products localize and regulate the skn-1 product so that its activity is restricted to EMS.

Determination of Anterior Pharyngeal Muscle Cells Requires Inductive Interactions

The discovery of maternal effect genes that provide localized cytoplasmic markers or limit particular gene activities to certain blastomeres has confirmed the traditional view of the *C. elegans* embryo as a "mosaic" in which the determination of early blastomeres relies on cytoplasmic localization rather than on induction. This view has been based to a large extent on blastomere separation experiments, in which isolated founder cells of *C. elegans* developed according to their fates (Laufer et al., 1980; L. G. Edgar and J. D. McGhee, 1986); the results seem to exclude any role for cellular interactions. However, the method used in most of these blastomere isolation experiments was to kill all blastomeres except one, and this approach has a serious limitation: the remnants of the killed blastomeres might still interact with the surviving cell and restrict its potency. (This problem is illustrated dramatically by the different results one obtains after separating frog blastomeres at the 2-cell stage and after killing one blastomere; see Chapter 6.) Therefore, if the development of a blastomere *is* changed when its neighbors are killed, then cellular interaction probably occurs. On the other hand, an indication that the development of a blastomere *is not* affected when its neighbors are killed must be interpreted with great caution.

The need for cellular interaction in determining the ABa blastomere of *C. elegans* was revealed in a set of experiments by James Priess and J. Nichol Thompson (1987). The researchers focused on the formation of pharyngeal muscle cells, which they identified by immunostaining. From lineage analysis it was known that ABa generates anterior pharyngeal muscle while EMS gives rise to posterior pharyngeal muscle (Fig. 24.7). To learn whether each subset of pharyngeal muscle cells arises independently of the other, the investigators removed specific blastomeres at the 2-cell and 4-cell stages and monitored the development of pharyngeal muscle in the remainder of the embryo. Their results are summarized in Table 24.1.

▼

Individual blastomeres were removed by puncturing the egg with a fine glass needle and applying pressure to extrude the unwanted blastomeres through the opening in the eggshell. After removal of AB at the 2-cell stage, the remaining P_1 blastomere generated pharyngeal muscle, as it normally does. Conversely, if the P_1 blastomere was extruded, the remaining AB blastomere did not produce pharyngeal muscle. This result showed that P_1 or some of its descendants were necessary for cells in the AB progeny to form anterior pharyngeal muscle. To determine which daughter cell of P_1 fulfilled this role, the researchers removed either P_2 or EMS at the 4-cell stage. Without EMS, the remaining embryo (ABa, ABP, and P_2) did not form pharyngeal muscle. In contrast, after removal of P_2, the remaining embryo (ABa, ABP, and EMS) formed anterior as well as posterior pharyngeal muscle cells. These results showed that EMS gives rise to posterior pharyngeal cells itself and induces part of the AB progeny to form anterior pharyngeal cells.

To see whether only ABa—which normally produces anterior pharyngeal muscle cells—is competent to be induced by EMS the investigators removed ABa. In this case, the anterior pharyngeal cells were formed by ABp. Moreover, if ABa and ABp were exchanged inside the eggshell with a blunt needle, the embryos proceeded to develop normally, with the original ABp blastomere (now in the ABa position) giving rise to anterior pharyngeal muscle.

The experiments of Priess and Thompson (1987) indicate that descendants of AB depend on inductive in-

TABLE 24.1

Development of Pharyngeal Muscle Cells in *C. elegans* Embryos after Removal of Individual Blastomeres at the 2-Cell Stage or the 4-Cell Stage

Removed Cell	Cells Remaining in Eggshell	Development of Pharyngeal Muscle Cells	
		Anterior	Posterior
—	AB, P_1	+	+
P_1	AB	−	−
AB	P_1	−	+
—	ABa, ABp, EMS, P_2	+	+
EMS	ABa, ABp, P_2	−	−
P_2	ABa, ABp, EMs	+	+
ABp	ABa, EMS, P_2	+	+
ABa	ABp, EMS, P_2	+	+

Source: From data of Priess and Thompson (1987). Used with permission. © *Cell Press.*

teractions with WMS or some of its descendants to form anterior pharyngeal muscle. The two daughter cells of AB—ABa and ABp—are equivalent through the 4-cell stage, and progeny of either daughter can be induced to form pharyngeal muscle cells. However, only certain descendants of ABa are normally induced. Which cells among the EMS descendants act as inducers of which AV descendants remains to be found.

The induction of ABa descendants to form anterior pharyngeal muscle depends on the function for the *glp-1*[+] gene, which was originally discovered in a screen for mutations affecting germ line development (*glp* stands for *germ line proliferation*). In *glp-1* mutant embryos, cells that normally give rise to anterior pharynx seem to form hypodermis and/or extra neurons, the same cell types also formed by isolated AB cells (Priess et al., 1987). The use of *temperature-sensitive* alleles in temperature-shift experiments indicates that *glp-1*[+] function is required between the 4-cell and 28-cell stage. This is also the period during which AB descendants must be induced to form anterior pharynx cells, because after this period these cells develop autonomously. Molecular data indicate that *glp-1*[+] encodes a receptor protein, suggesting that this protein acts in the AB descendants that receive the inducing signal from EMS or its descendants (J. Austin et al., 1989).

P_2 Induces EMS to Form the Gut Cell Lineage

Another example of induction during the determination of founder cells in *C. elegans* involves the E blastomere, which gives rise to the entire gut. Certain experiments have been taken as evidence that E is determined autonomously, without the need for inductive interactions with its neighbors. However, these experiments were based on blastomere isolation that involved killing neighbor cells, a problematic method, as discussed earlier. The development by Lois Edgar of a new technique for culturing *C. elegans* blastomeres outside the eggshell made it possible to reinvestigate E cell determination without the limitations of the older method.

▼

Using Edgar's technique, Bob Goldstein (1992) removed the eggshells from cleavage-stage embryos, separated the blastomeres by forcing embryos through a narrow pipette, and monitored the development of individual blastomeres in culture. In particular, he assayed the descendants of isolated EMS blastomeres for the ability to form *rhabditin* granules, which are characteristic of gut cells. He found that this ability depended on the time of blastomere separation: EMS blastomeres isolated during the first half of the 4-cell stage never gave rise to gut cells, whereas EMS blastomeres isolated later during the 4-cell stage did (Fig. 24.10a). In cases in which no gut cells formed, both EMS daughters proceeded to di-

vide in synchrony and with the rhythm characteristic of MS blastomeres, whereas normally the E blastomere divides more slowly than its MS sister.

These results could be interpreted two ways. First, the determination of E as the gut founder cell may depend on a previous interaction of EMS with one of its neighbors. Second, the failure of early isolated EMS blastomeres to produce gut cells may reflect a greater sensitivity of younger embryos to damage from handling. To decide which interpretation was correct, Goldstein (1992) recombined early isolated EMS blastomeres with one or two AB daughters or with P_2 (Fig. 24.10b and c). Recombination with P_2, but not with AB daughters, restored the ability of early isolated EMS blastomeres to produce gut cells. This result rules out the sensitivity hypothesis and confirms that the formation of gut cells by isolated EMS cells requires direct contact between EMS and P_2.

In the normal embryo, the P_2 blastomere sits at the posterior end (the "E" end) of EMS. The experiments of Goldstein (1992) suggest that P_2 induces EMS to divide

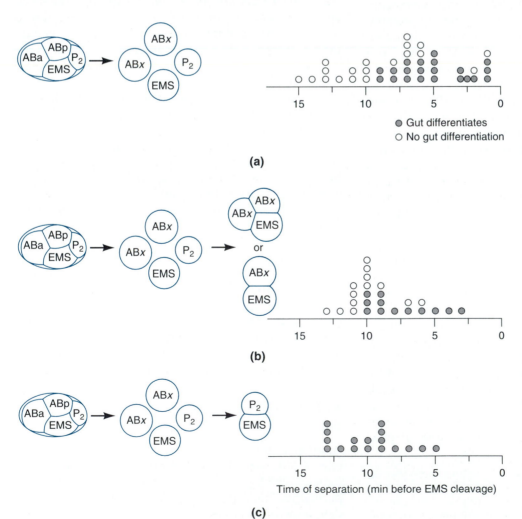

Figure 24.10 Development of gut cells from isolated and recombined EMS blastomeres of *C. elegans*. The experiments are diagrammed to the left, and the results are tallied to the right. Each filled circle represents an experiment in which an isolated EMS cell gave rise to gut cells, as indicated by the presence of rhabditin granules. Each open circle represents an experiment in which an isolated EMS cell survived but did not form gut cells. The circles are stacked over a time axis that indicates how much time elapsed between isolation of an EMS cell and its cleavage into E and MS daughter cells. **(a)** Cell isolation only. **(b)** Cell isolation and subsequent recombination (within 1 to 2 min) of the EMS cell with ABa or ABp or both. **(c)** Cell isolation and subsequent recombination (within 1 to 2 min) of the EMS cell with P_2. Note that EMS cells isolated early did not form gut cells unless they were recombined with P_2.

asymmetrically so that the daughter blastomere next to P_2 becomes the gut founder cell E. To test this hypothesis, Goldstein (1993) placed P_2 in random positions on an isolated EMS. Gut cells consistently differentiated from the EMS daughter that was contacting P_2, indicating that any half of EMS can respond to gut induction. In addition, moving P_2 around to the opposite side of EMS in an otherwise intact embryo caused the two EMS daughters to switch the timing of their subsequent divisions—and gut cells to originate from what would normally be the MS blastomere.

To determine whether EMS is the only blastomere that can be induced by P_2 to form a gut founder cell, Goldstein (1993), separated blastomeres as described earlier and recombined P_2 with ABa, ABP, or both. Since none of these combinations gave rise to gut cells, it appears that only EMS is competent to respond to the inductive signal from P_2 that generates the E blastomere. To test whether this particular competence of EMS requires a previous inductive interaction between AB and P_1, Goldstein (1993) separated AB from P_1 immediately upon the first cleavage. Isolated P_1 blastomeres gave rise to gut cells in most cases, while isolated AB blastomeres never did. Thus, it seems that EMS acquires its competence for induction of E by P_2 independently of AB, presumably by the localized action of the skn-1 protein, as discussed earlier.

Heterochronic Genes

Our previous discussions of pattern formation centered on the *spatial* aspect of embryonic development and its genetic control. For example, one of the major questions about segmentation genes and homeotic genes in *Drosophila* was how their expression is limited to certain spatial domains. Here we shall focus on the *timing* of development. Timing is critical in morphogenetic movements such as gastrulation as well as in life cycle events like larval molts or pupation. Changes in the timing of such events are thought to underlie many steps in evolution (Gould, 1977).

In *C. elegans,* each cell division occurs at a particular time in development, and every cell lineage is characterized by the timing as well as the orientation and symmetry of cell divisions. Like other aspects of development, the temporal sequence of events is under genetic control. This is revealed by the existence of **heterochronic mutations,** which cause cells to undergo patterns of cell division and differentiation that are normally observed at a different stage (Ambros and Horvitz, 1984). Heterochronic mutations are remarkable similar to homeotic mutations. In our definition of heterochronic mutations, one only needs to replace "heterochronic" with "homeotic" and "stage" with "place" to obtain the correct definition of a homeotic mutation.

We can extend this parallel by saying that *morphogen gradients* may exist in time as well as in space. A *temporal gradient* of morphogenetic signal would activate different target genes according to its *current* concentration rather than according to its *local* concentration as a spatial gradient does (Fig. 24.11; see Fig. 20.18 for comparison). Thus, the temporal and spatial dimensions of development may be controlled by similar genetic mechanisms. We will examine in detail one gene that generates a temporal gradient.

Mutations in the *lin-14*$^+$ Gene Are Heterochronic

The *lin-14*$^+$ gene of *C. elegans* plays a central role in the development of hypodermal and intestinal cells as well as neurons (Ruvkun et al., 1991). We focus here on a hypodermal cell lineage that originates from the *postembryonic blast cell* T (Fig. 24.5). *lin-14*$^+$ is a heterochronic gene, mutations in which change both the synthesis of lin-14 protein and the division pattern of T (Fig. 24.12). In the wild type, lin-14 protein is synthesized early during the first larval stage (L1), and cell divisions in the T lineage begin during L1 and end during L2. In animals with loss-of-function alleles of *lin-14*, the span of lin-14 protein activity is abbreviated. The same alleles cause precocious execution of cell divisions normally observed in descendant cells at a later stage. Conversely, gain-of-function alleles of *lin-14* affect the T cell lineage in the opposite way: cell divisions are normal during L1, but later divisions repeat the same division pattern over and over. These alleles also cause an extended span of lin-14 protein synthesis. Thus, abbreviated activity of lin-14 protein causes a precocious "aging" of the cell division pattern, whereas an extended synthesis of lin-14 protein keeps the lineage permanently "young."

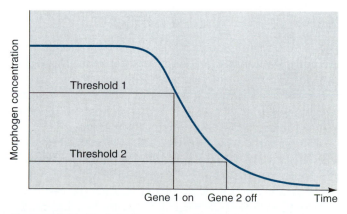

Figure 24.11 Model of a temporal morphogen gradient. The concentration of a morphogenetic signal produced in a given cell or embryonic region varies over time. At certain times during development, the concentration of the signal reaches threshold values at which genes regulated by the morphogen are switched on or off.

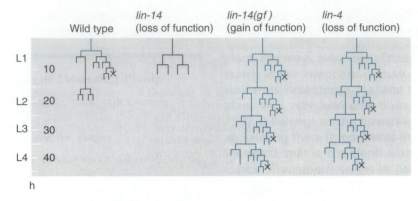

Figure 24.12 The effect of heterochronic mutations on the cell lineage formed by the postembryonic blast cell T, which contributes to the hypodermis of *C. elegans*. The colored portions of the lineage diagrams represent those cells that express lin-14 protein. The time axis to the left indicates hours after hatching and the four larval stages (L1 to L4). In the wild type, T expresses lin-14 protein early during L1 and divides during L1 and L2. A loss-of-function mutant, *lin-14*, lacks any detectable lin-14 protein, and its blast cell T develops a precocious L2-specific lineage at stage L1. A gain-of-function mutant, *lin-14(gf)*, expresses lin-14 protein at all larval stages and reiterates the cell lineage of T characteristic of L1. The same phenotype is produced by a loss-of-function mutation in another gene, *lin-4*.

Other genes act in the same pathway as *lin-14*⁺ to control the temporal patterns of postembryonic blast cell divisions. A loss-of-function mutation in *lin-4* causes the same repetitive lineage as *lin-14* gain-of-function mutations (Fig. 24.12). However, the *lin-4* loss-of-function phenotype depends on a functioning *lin-14*⁺ gene: mutants defective in both genes show the same phenotype as *lin-14* loss-of-function mutants. These observations indicate that *lin-4*⁺ normally down-regulates the *lin-14*⁺ gene after the early L1 stage. In contrast, the gene *lin-28*⁺ up-regulates lin-14 protein levels during L1. Loss-of-function alleles of *lin-28* cause a precocious decrease in lin-14 protein levels. Thus, the *lin-4*⁺ and *lin-28*⁺ genes act antagonistically in regulating *lin-14*⁺.

The *lin-14*⁺ Gene Encodes a Nuclear Protein That Forms a Temporal Concentration Gradient

The observed relationship between lin-14 protein and the cell division pattern suggests that lin-14 protein may form a *temporal gradient*, with a high concentration early during the L1 stage and a declining concentration thereafter. The gradient may orchestrate the normal sequence of cell divisions in the T lineage.

To test for a temporal gradient, Gary Ruvkun and John Giusto (1989) cloned the *lin-14*⁺ gene and raised antibodies against the lin-14 protein (Fig. 24.13). The antibodies stained specific nuclei in early larvae from wild-type strains but not from strains bearing *lin-14* null alleles, showing that the antibody was directed specifically against lin-14 protein. As expected, lin-14 protein was present in the T cell lineage and in all other postembryonic blast cell lineages that are affected by *lin-14* mutations.

In wild-type animals, the staining was most intense in late embryos just before hatching, and in newly hatched L1 larvae. The blast cells containing lin-14 protein underwent their first L1-specific division, but then the lin-14 concentration fell before the next division (Fig. 24.12). By stage L2, the concentration had decreased by a factor of more than 25. In contrast, L3 larvae of a gain-of-function mutant designated *lin-14(gf)* had continuously high concentrations of lin-14 protein in their blast cell nuclei.

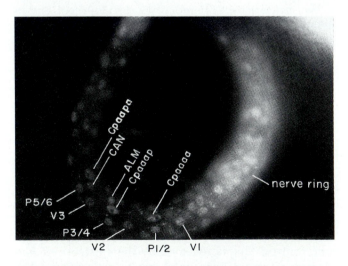

Figure 24.13 Expression of the *lin-14*⁺ gene in *C. elegans* larvae. This photograph shows a wild-type L1 larva immunostained for lin-14 protein (see Methods 4.1). The lin-14 protein accumulates in the nuclei of postembryonic blast cells, such as P5/6, and descendants of embryonic blastomeres, such as Cpaaap. (The latter designation refers to the posterior daughter of the anterior daughter of the anterior daughter of the anterior daughter of the posterior daughter of blastomere C.)

The nuclear localization of the lin-14 protein suggests that it regulates the transcription of certain target genes or the processing of their transcripts. Most likely, the lin-14 protein cooperates with other regulatory factors in eliciting different responses in different cell lineages.

The *lin-4*+ Gene Encodes Small Regulatory RNAs with Antisense Complementarity to lin-14 mRNA

Cloning and sequencing the *lin-14*+ gene showed that three lin-14 mRNAs are formed by differential splicing. Comparisons with other sequences in databases did not turn up any similarities between the L1 proteins and other known proteins. However, it was possible to determine the nature of two gain-of-function mutations that keep expressing lin-14 protein beyond the time of its normal decline (Wightman et al., 1991). These mutations interfere with the exon that encodes the *trailer* region of the mRNAs. One of these mutations is a large insertion, and the other is a deletion in the same region.

Since the trailer regions of other mRNAs have been shown to interact with proteins that regulate their nucleocytoplasmic transport, half-life, and translation (see Chapters 16 and 17), it might have been expected that lin-4+ would encode such a regulatory protein. It therefore came as a surprise when it was discovered that *lin-4*+ does not encode a protein (R. C. Lee et al., 1993; Wightman et al., 1993). Instead, *lin-4*+ encodes two small transcripts, about 22 and 61 nucleotides long, with sequences complementary to repeated sequence elements in the trailer region of lin-14 mRNA. These data suggest that the temporal gradient of lin-14 protein is formed by hybridization of the lin-4 regulatory RNAs with the repeated sequence elements in lin-14 mRNA. How the duplex formation in the trailer region of lin-14 mRNA down-regulates protein synthesis remains to be elucidated.

Down-Regulation of *lin-14*+ Expression Is Initiated by a Developmental Cue

In an experiment to investigate what triggers the down-regulation of *lin-14*+ gene expression in normal development, newly hatched larvae were starved. Under these circumstances, lin-14 protein concentration remained high beyond the normal period of 12 h of postembryonic development after which the lin-14 protein disappears from normal larvae. Thus, *lin-14*+ down-regulation is not a function of time but instead depends on some cue that is disturbed by starvation. Furthermore, the continued high level of lin-14 proteins in starved larvae depends on protein synthesis and is as-

sociated with a high concentration of lin-14 mRNAs. These results suggest that lin-14 protein synthesis continues in starved animals, possibly because the *lin-4*+ gene, which is required for down-regulating lin-14 protein, is activated by a stimulus that depends on feeding.

In summary, the role of the heterochronic gene *lin-14*+ in the temporal regulation of postembryonic cell lineages can be envisioned as follows. A feeding-dependent stimulus activates the *lin-14*+ gene, which encodes small regulatory RNAs that hybridize with the trailer regions of lin-14 mRNAs with the effect of down-regulating their translation. As the concentration of lin-14 proteins falls, it generates a decreasing temporal gradient, which in turn controls a set of target genes that regulate the stage-specific cell division patterns.

If the expression of a heterochronic gene is curtailed or extended, cell lineages or other developmental processes are terminated precociously or are sustained longer than normal. The effects of heterochronic mutations on developmental timing are analogous to the effects of homeotic mutations on spatial organization: otherwise-normal (or nearly normal) processes occur either at the wrong time or in the wrong place. Both types of mutation can cause dramatic changes in development that are nevertheless restricted in time or space and not downright lethal. In particular, the effects of gain-of-function mutations in these genes are sometimes rather limited, presumably because the products of these genes act in conjunction with other gene products that may be temporally or locally restricted.

Gain-of-function mutations tend to be genetically dominant and are therefore of particular importance in evolution. Unlike recessive mutations, they change the phenotype of heterozygous carriers, so that selection begins before a mutant allele is bred to homozygosity. Consequently, evolution based on gain-of-function mutations can occur rapidly even in large outbred populations. Conceivably, then, both homeotic and heterochronic mutations have played major roles in evolution (Gould, 1977).

Programmed Cell Death

During most animals' development, many cells arise that do not develop further but instead die. Since cell death often occurs in a distinct pattern, it seems to have a morphogenetic role (Saunders, 1982). For instance, the developing digits of the limbs of higher vertebrates are molded by the programmed death of cells between them. During the development of the vertebrate nervous system, up to 50% of many types of neurons normally die after they have already formed synaptic connections with their target cells (Raff et al., 1993). This massive cell death is ascribed to the failure of the dying neurons to obtain the amounts of trophic factors from their target cells that they need to survive. Likewise, hematopoietic cells die rapidly in the absence of appropriate growth factors (see Chapter 19). Similar strategies are observed in other cell types that require signals from neighboring cells to survive (Williams and Smith, 1993). The survival signals seem to act by suppressing an intrinsic cell suicide program, the protein components of which appear to be expressed by default in all cell types that are so programmed.

The phenomenon of controlled cell death is called *programmed cell death* or *apoptosis* (a Greek word that applies to a flower losing its petals or a tree its leaves). Cells undergoing apoptosis show characteristic features, including shrinkage, chromatin condensation, and blebbing, that distinguish them from cells undergoing random, nonprogrammed cell death. Single genes that are involved in the control of programmed cell death have been found in *Drosophila* and in humans, but only in *C. elegans* has an entire network of control genes been identified.

Of the 1090 somatic cells generated in the hermaphrodite of *C. elegans*, 131 die (Sulston et al., 1983). Similarly, 148 of 1179 somatic cells die in the male. In every individual the same cells die, each in its own time and place. Because of this feature and its favorable genetic properties, *C. elegans* is a particularly useful organism for analyzing apoptosis. We will explore the genetic controls that specify which cells live and which cells die, and discuss some of the cellular mechanisms that cause cell death.

Programmed Cell Death in *C. elegans* Is Controlled by a Genetic Pathway

More than a dozen genes in *C. elegans* have specific functions connected with programmed cell death. Two key genes are *ced-3*$^+$ and *ced-4*$^+$, which are required for almost all programmed cell deaths in both embryos and larvae (*ced* stands for *cell death*). The available data suggest that these genes control the production of *cytotoxins*, substances that kill the cells in which they are made. Loss-of-function mutants in these genes are nearly normal in morphology and behavior, showing that the prevention of programmed cell death is not detrimental to *C. elegans* (H. M. Ellis and H. R. Horvitz, 1986). The extra surviving cells in such mutants do not divide, but at least some of them differentiate and form additional neurons. A larger group of genes, including *ced-1*$^+$, cause the dead cells to be engulfed and degraded by their neighbors. Loss-of-function mutations in these genes allow dead cells to remain present (Fig. 24.14). Yet another gene, *nuc-1*$^+$ (for *nuclease-deficient*), is necessary for the breakdown of DNA in the engulfed cells.

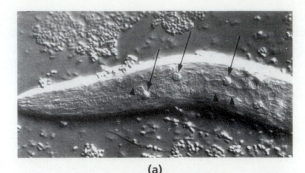

(a)

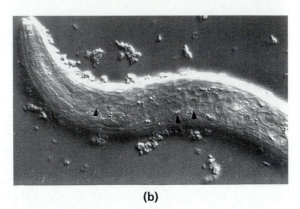

(b)

Figure 24.14 Absence of cell death in *ced-3* mutants of *C. elegans*. **(a)** Photograph of a newly hatched larva that is mutant in the *ced-1* gene. Arrows indicate dying cells, which remain visible because the *ced-1* mutation interferes with the removal of dying cells by other cells that would occur in the wild type. **(b)** Similar larva mutant in the *ced-1* gene and the *ced-3* gene. No dying cells are detectable, because the *ced-3* mutation interferes with programmed cell death. Arrowheads indicate the same nuclei in both photographs, demonstrating that both have the same focal plane.

The *ced-3*$^+$ and *ced-4*$^+$ genes, the group represented by *ced-1*$^+$, and the *nuc-1*$^+$ gene define a pathway of cell destruction and removal (Fig. 24.15). However, these genes do not control which cells survive and which die. This selection is made by control genes acting on *ced-3*$^+$ and *ced-4*$^+$. One control gene is *ced-9*$^+$, which protects *all* cells that normally escape programmed cell death. Other genes, including *ces-1*$^+$ and *ces-2*$^+$ (*ces* is for *cell death specification*), prevent the programmed death of *specific* cells without affecting the death of other cells.

In the following, we will first explore how *ced-3*$^+$ and *ced-4*$^+$ work and then see how they are regulated.

The Genes *ced-3*$^+$ and *ced-4*$^+$ Act Cell-Autonomously to Initiate Programmed Cell Death

One fundamental question is whether *ced-3*$^+$ and *ced-4*$^+$ act cell-autonomously—that is, whether these genes act within the cells that die or in other cells that control cell

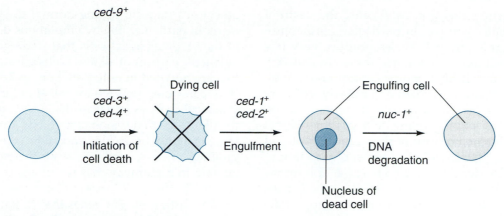

Figure 24.15 Model of a general genetic pathway controlling programmed cell death in *C. elegans*. In each cell, the normal function of the genes *ced-3*[+] and *ced-4*[+] is to initiate cell death. In most cells, however, the function of these genes is inhibited by the activity of another gene, *ced-9*[+]. Dying cells are engulfed by neighboring cells, an activity that requires the function of a group of genes represented by *ced-1*[+] and *ced-2*[+]. Degradation of the DNA of the engulfed cell depends on the activity of yet another gene, *nuc-1*[+].

death from outside. To answer this question, Junying Yuan and H. Robert Horvitz (1990) analyzed *genetic mosaics*, that is, animals in which some cells were wild-type and others were mutant for either *ced-3* or *ced-4*.

▼

Such mosaic individuals of *C. elegans* can be generated by a technique using chromosomal fragments that are not well integrated into mitotic spindles and are therefore randomly lost during cleavage (Herman, 1984). The fragments carry marker genes with known cell-specific functions, from which investigators can infer which cell lineages have lost the chromosome fragment and which have kept it.

A suitable fragment carrying the wild-type allele of *ced-3*[+] was introduced into individuals homozygous for a loss-of-function allele of *ced-3*. Thus, cell lineages that had lost the fragment were genotypically mutant for *ced-3* while those retaining the fragment were heterozygous. The results showed that programmed cell death requires a wild-type allele of *ced-3*[+] in the dying cell itself. This was particularly clear in a pair of neurons, CEMDR and CEMDL, which are physically adjacent in the larva although they originate from two different founder cells, ABa and ABp. One of these cells often survived (if its lineage had lost the chromosome fragment) while the adjacent neuron died (if its lineage had kept the chromosome fragment). A corresponding analysis of the *ced-4*[+] gene yielded the same results.

The observations of Yuan and Horvitz (1990) argue strongly that programmed cell death in *C. elegans* is ini-

tiated by the activity of *ced-3*[+] and *ced-4*[+] in the dying cells themselves and not in neighboring cells. This conclusion implies that the proteins encoded by these genes are themselves cytotoxic or promote the synthesis of cytotoxic substances.

To begin the molecular analysis of the ced-3 and ced-4 proteins, Yuan and Horvitz (1992) cloned the *ced-4*[+] gene. Transcripts from this gene accumulate at high levels during embryogenesis, when most programmed cell death occurs. The predicted sequence of the ced-4 protein includes two domains that occur in calcium-activated proteins, suggesting that the activity of the ced-4 protein may also be regulated by calcium. Subsequent cloning of the *ced-3*[+] gene indicated that the encoded protein is a proteolytic enzyme (Yuan et al., 1993).

The *ced-9*[+] Gene Protects Cells from Programmed Death

How are potentially lethal activities like those of *ced-3*[+] and *ced-4*[+] controlled so that only the appropriate cells are killed? Genetic analysis shows that a regulatory gene, *ced-9*[+], inhibits the activities of *ced-3*[+] and *ced-4*[+] (Hengartner et al., 1992).

The cells that normally die as a result of *ced-3*[+] and *ced-4*[+] action survive in mutants carrying a gain-of-function allele of *ced-9*[+], designated *ced-9(gf)*. This lack of cell death is most clearly indicated by the absence of undergraded cell corpses in *ced-9(gf)* larvae that are also mutant in *ced-1*. At least some of the extra cells that survive in *ced-9(gf)* individuals are functional, as they are in individuals carrying loss-of-function alleles of *ced-3* and *ced-4*. These observations suggest that *ced-9*[+] inhibits the activities of *ced-3*[+] and *ced-4*[+].

Loss-of-function alleles in *ced-9* cause the death of cells that normally survive. In particular, hermaphrodites homozygous for such alleles produce very few eggs, all of which die. Furthermore, these mutants lack many neurons involved in motor control and exhibit uncoordinated body movements. Similarly, cells are missing from the tails of homozygous mutant males. Direct observation of developing embryos reveals that the missing cells are born by normal cell division but later die. These abnormalities do not show up in mutants that are also deficient in *ced-3* or *ced-4*. This means that *ced-9*[+] acts above *ced-3*[+] and *ced-4*[+] in the genetic pathway for cell death. Together, all results support the hypothesis that *ced-9*[+], directly or indirectly, protects cells from programmed cell death by inhibiting the activities of *ced-3*[+] and *ced-4*[+].

The Genes *ces-1*[+] and *ces-2*[+] Control the Programmed Death of Specific Cells in the Pharynx

While the *ced-9*[+] gene prevents death in a wide spectrum of cells, other genes affect the programmed death of specific cells. For instance, *ces-1*[+] and *ces-2*[+] control the death of specific neurons (R. E. Ellis and H. R. Horvitz, 1991). Both loss-of-function alleles of *ces-2* and a gain-of-function allele of *ces-1*, designed *ces-1(gf)*, affect in particular the sister cells of neurons designated NSM and I2 (Fig. 24.16). Normally, these sister cells die, and their corpses are clearly visible in larvae that are defective in the genes required for dead cell engulfment. However, in *ces-1 (gf)* mutants, the sister cells of NSM and I2 survive and differentiate into neurons, just as the NSM and I2 cells do.

The *ces-1* and *ces-2* mutations prevent the programmed death of specific cells but do not affect the other cells that die during normal *C. elegans* development. In particular, the two mutations do not affect any of the 18 pharyngeal cells that undergo programmed death: the corpses of all these cells are visible in animals that are mutated in either *ces-1* or *ces-2* and are also defective in genes required for cell engulfment. Moreover, statistical analysis indicates that *ces-1*[+] and *ces-2*[+] affect the life-or-death decision of each NSM and I2 sister cell independently (R. E. Ellis and H. R. Horvitz, 1991). The products of *ces-1*[+] and *ces-2*[+] are therefore expected to operate in a pathway that is cell-autonomous. In each cell, *ces-2*[+] and *ces-1*[+] seem to form a series of inhibitory controls acting on the genes that initiate programmed cell death, *ced-3*[+] and *ced-4*[+] (Fig. 24.17).

The Product of a Human Gene, *bcl-2*[+], Prevents Programmed Cell Death in *C. elegans*

At least two of the *C. elegans* genes that control programmed cell death, *ced-3*[+] and *ced-9*[+], have close functional and molecular counterparts in mammals including humans. If these similarities were extended to more of the control hierarchy known from *C. elegans*, this would be another striking example of evolutionary conservation of molecular mechanisms in development.

The *C. elegans* gene *ced-9*[+] supports a similar function as the human gene *bcl-2*[+]. In mammals, *bcl-2*[+] prevents apoptosis in various cell types including embryonic neurons (Allsopp et al., 1993) and hematopoietic cells (Fairbairn et al., 1993). As a functional test of the similarity between ced-9 and bcl-2 proteins, David Vaux and his colleagues (1992) transformed *C. elegans* with a cloned human *bcl-2*[+] gene driven by a *heat shock promoter*. To make any changes in programmed cell death clearly visible, the researchers transformed the *ced-1*

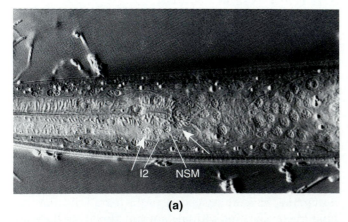

(a)

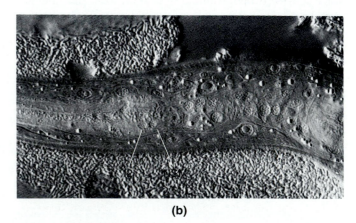

(b)

Figure 24.16 Programmed cell death of specific neurons in third-instar larvae of *C. elegans*. The photographs show a *ces-1(gf)* mutant (a) and a wild-type animal (b). The mutant has two extra neurons (arrows) that are not present in the wild type. The extra neurons, which undergo programmed cell death in the wild type, are sister cells of the neurons I2 and NSM.

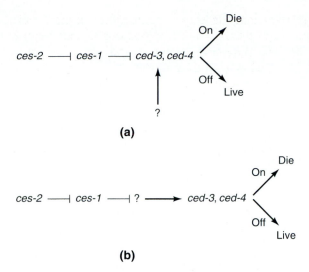

(a)

(b)

Figure 24.17 Beyond the general genetic control of programmed cell death as shown in Figure 24.15, additional genes regulate the death of specific cells in *C. elegans*. This diagram shows two similar models for the control of cell death in the sister cells of NSM neurons shown in Figure 24.16. **(a)** *ces-2*⁺ and *ces-1*⁺ form a series of inhibitions that act on the genes that initiate programmed cell death, *ced-3*⁺ and *ced-4*⁺, which are activated by an unknown process. **(b)** Alternatively, *ces-2*⁺ and *ces-1*⁺ inhibit an unknown gene that in turn activates *ced-3*⁺ and *ced-4*⁺.

mutant of *C. elegans*, in which dead cells are not engulfed.

The heat shock to start the expression of the *bcl*⁺ transgene was applied during early gastrulation, well before the onset of any cell deaths. The transgene activity significantly reduced the number of cell corpses visible in late embryos. An average of 38 cell corpses turned up in nontransformed heat-shocked *ced-1* embryos, whereas in transformed embryos there were only 16 cell corpses on average. Continuous observation of several transformed embryos revealed that their lower number of dead cells did not result from a delay in the onset of cell deaths or from more effective engulfment of dead cells. Thus, the human bcl-2 protein clearly interferes with programmed cell death in *C. elegans*.

The results of Vaux and colleagues suggests that the human *bcl-2*⁺ gene may be homologous to the *ced-9*⁺ gene of *C. elegans*. Subsequent cloning of *ced-9*⁺ indeed revealed considerable sequence similarity between the proteins encoded by the two genes (Hengartner and Horvitz, 1994). The bcl-2 protein has been detected in mitochondria and also in the nuclear envelope and the endoplasmic reticulum, but the molecular mechanism of its action remains to be elucidated (Hockenbery et al., 1993).

Because *ced-9*⁺ inhibits *ced-3*⁺ and *ced-4*⁺ in *C. elegans*, *bcl-2*⁺ may antagonize corresponding genes in mammals. The mammalian gene *Nedd2*⁺ (for *n*eural *d*evelopmental *d*own-regulation) is strongly expressed in

tissues undergoing programmed cell death and encodes a protease that causes apoptosis, just as *ced-3*⁺ does. In order to test whether *bcl-2*⁺ expression inhibits *Nedd2*⁺ activity, Sharad Kumar and colleagues (1994) transfected cultured mammalian cells with cDNAs representing these genes. While cells transfected with *Nedd2*⁺ alone underwent apoptosis, cells transfected with both *Nedd2*⁺ and *bcl-2*⁺ survived to a large extent. Thus, at least part of the genetic hierarchy that controls apoptosis in *C. elegans* has been well conserved in evolution.

Vulva Development

The vulva of *C. elegans* is a ventral opening that is used for copulation and egg laying. It develops in the hermaphrodite during the larval stages. The formation of the vulva epitomizes many basic processes in development, especially embryonic induction (Horvitz and Sternberg, 1991; R. J. Hill and P. W. Sternberg, 1993).

The Vulval Precursor Cells Form an Equivalence Group

The cells giving rise to the vulva belong to the *postembryonic blast cells* that develop during the larval stages to form those adult structures that do not function during the larval stages (Fig. 24.5). Six blast cells, all derived from the Abp blastomere, are lined up along the ventral midline of the larva and have the potency to form the vulva of the adult (Fig. 24.18a). These cells are designated P3p, P4p, P5p, P6p, P7p, and P8p, because they are the posterior daughter cells from a ventral series of blast cells labeled P1 through P12. (These cells are unrelated to the embryonic cells P_1 through P_4.) All six blast cells, P3p through P8p, are called ***vulval precursor cells (VPCs)***, although only three VPCs are normally fated to form the vulva. The other three VPCs contribute to the hypodermis. Which VPCs actually form the vulva depends on their proximity to a particular cell in the gonad rudiment, the ***anchor cell***, which will link the vulva to the uterus. The VPC next to the anchor cell, normally P6p, forms the internal part of the vulva, and the two neighboring VPCs, normally P5p and P7p, form the external vulva.

The VPC closest to the anchor cell undergoes three divisions and forms eight daughter cells in a circular arrangement (Fig. 24.18b–e). This fate is called the *primary (1°) fate*. Each of the two neighboring VPCs produces seven descendants in a semicircular arrangement. This fate is referred to as the *secondary (2°) fate*. The remaining VPCs undergo only one division, producing two descendants that fuse with other cells forming the hypodermal syncytium. This fate is called the *tertiary (3°) fate*. The 22 descendants of the VPCs with primary and secondary fates form a doughnut-shaped syn-

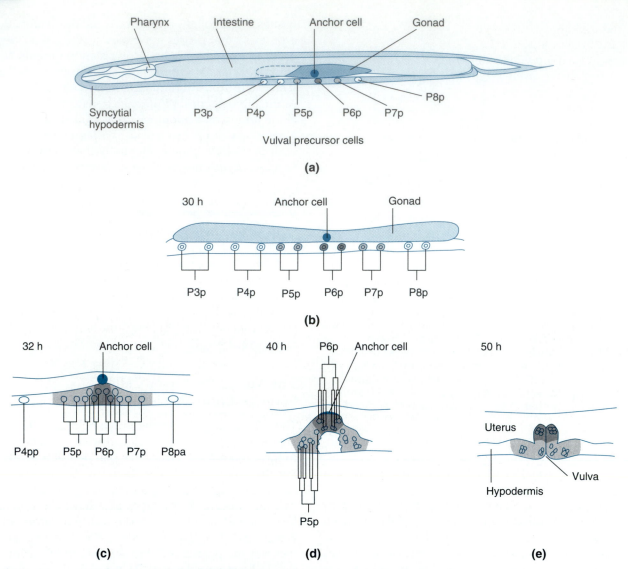

Figure 24.18 Formation of the vulva in *C. elegans*. Six vulval precursor cells (VPCs) have the potency of forming a vulva, but only three normally do. The fate of the VPCs depends on their proximity to the gonadal anchor cell, which will link the vulva to the uterus. The VPC next to the anchor cell forms the inner part of the vulva, and the two neighboring VPCs form the outer vulva. The remaining three VPCs contribute to the hypodermis. **(a)** Early third-stage larva, with the six VPCs, designated P3p to P8p, aligned along the ventral surface of the developing gonad, which contains the anchor cell. **(b–d)** VPC divisions. P6p is normally fated to have eight descendants forming the interior vulva (primary fate). P5p and P7p are normally fated to have seven descendants each, together forming the exterior vulva (secondary fate). The vulva becomes syncytial like the hypodermis. **(e)** Vulval development is complete in the adult.

cytium, the center of which forms the vulval opening. The vulva interacts with specific nerve, muscle, and uterus cells during copulation and egg laying.

Although the six VPCs in the *C. elegans* larva express different fates, they have the same potencies, since each of them may assume the primary, secondary, or tertiary fate. For such a group of cells that have the same potencies but different fates, scientists studying *C. elegans* have coined the term ***equivalence group***. The properties of an equivalence group are similar to those

of an *embryonic field*. Thus, destroyed cells may be functionally replaced by alternative cells from the same equivalence group. Typically, the cells of an equivalence group are related by lineage but are not clones.

Vulva Development Requires Three Intercellular Signals

The six VPCs in *C. elegans* were revealed as an equivalence group in the course of cell removal experiments and through study of certain mutant phenotypes. The available data indicate that normal vulval development requires at least three intercellular signals originating in the gonadal anchor cell, in the VPCs themselves, and in the adjacent hypodermis (Fig. 24.19a).

Vulva development clearly depends on an inductive signal from the gonadal anchor cell. If this cell is destroyed with a laser microbeam, no vulva is formed (Fig. 24.19b). All VPCs then express the tertiary fate, and the resulting animal cannot copulate or lay eggs (Kimble, 1981). Conversely, if all gonadal cells except the anchor cell are destroyed, a vulva still develops. Therefore, the anchor cell is both necessary and sufficient for vulva formation—at least in wild-type *C. elegans*.

Further proof of an inductive signal emanating from the anchor cell comes from certain mutant phenotypes with displaced gonads (J. H. Thomas et al., 1990). If the gonad is displaced anteriorly, so that the anchor cell is next to P5p, this cell instead of P6p adopts the primary fate. Thus, according to the criterion of *heterotopic trans-*

plantation, the gonadal anchor cell *induces* the adjacent VPC to assume the primary fate and the next two VPCs to assume the secondary fate. If the gonad is displaced dorsally, so that the anchor cell is not in direct contact with any VPC, the closest three VPCs nevertheless form a vulva, indicating that the inductive signal is diffusible or can be transmitted through other cells.

A signal from the hypodermis seems to inhibit vulva formation from noninduced VPCs. This is indicated by the phenotype of mutants that are deficient in the gene *lin-15* (for cell *lin*eage abnormal). In these mutants, all six VPCs adopt the primary or secondary fate, and multiple vulvalike structures develop (Ferguson et al., 1987). Destroying the entire gonad in *lin-15* mutants does not prevent multiple vulva formation. Therefore the mutation does not act by causing overproduction of the inducing signal in the gonad.

To analyze where the *lin-15*+ gene is normally required, Herman and Hedgecock (1990) used the genetic mosaic analysis described earlier in the context of programmed cell death.

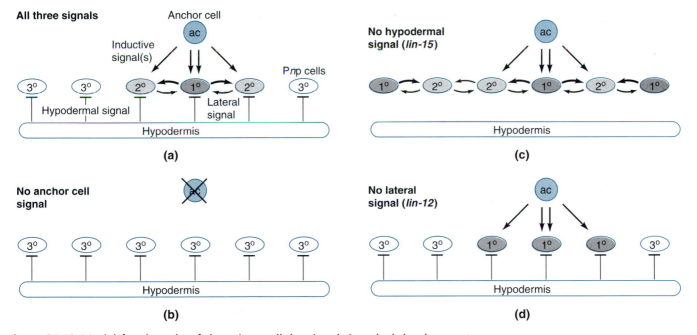

Figure 24.19 Model for the role of three intercellular signals in vulval development. **(a)** General model. First, the anchor cell induces vulva formation in three vulval precursor cells (VPCs). Second, the induced VPCs exchange lateral signals so that a cell adopting the primary fate prevents its neighbors from doing the same and makes them adopt the secondary fate instead. The inductive and lateral signals cooperate so that the primary fate is always expressed next to the gonadal anchor cell. Third, the hypodermis sends an inhibitory signal that prevents the VPCs from assuming primary and secondary fates except when induced by the anchor cell. **(b)** In the absence of an anchor cell, all VPCs assume tertiary fates, because there is no inductive signal to override the inhibitory signal from the hypodermis. **(c)** The mutant allele *lin-15* acting in the hypodermis allows the formation of multiple vulvae, apparently because the inhibitory signal from the hypodermis is missing. In this case, the pattern of primary and secondary fates among the VPCs varies, but the primary fate is never expressed by two neighboring VPCs. **(d)** In individuals carrying the mutant allele *lin-12*, the three VPCs closest to the anchor cell adopt the primary fate, presumably because the mutant is deficient in lateral signaling.

The investigators engineered hermaphrodites with two X chromosomes, each carrying a recessive loss-of-function allele of *lin-15*[+,] plus an unstable X chromosome fragment carrying the wild-type allele *lin-15*[+]. Cell-specific marker genes associated with the different *lin-15* alleles allowed the investigators to deduce from which lineages the X fragment had been lost. The fates adopted by VPCs in various mosaic individuals showed that normal vulva formation requires expression of the *lin-15*[+] gene outside the VPCs and the anchor cell, most likely in the hypodermis.

The researchers concluded that a signal dependent on *lin-15*[+] expression in the hypodermis inhibits an intrinsic tendency of *all* VPCs to adopt the primary or the secondary fate and instead promotes their assumption of the tertiary fate. This inhibition appears to be overridden selectively by the inductive signal(s) from the anchor cell, which restore the intrinsic tendency of the VPCs to realize their potential for the primary or the secondary fate.

Lateral signals between VPCs also influence their fates, as the following observations demonstrate. In *lin-15* mutants with or without an anchor cell, the six VPCs generally express the primary and secondary fates in an alternating pattern—for example, 2°-1°-2°-1°-2°-1°. Exceptions to this rule are adjacent pairs of secondary fates; adjacent pairs of primary fates are not observed (Fig. 24.19c). These observations suggest that a VPC expressing the primary fate prevents neighboring VPCs from doing the same, thus causing them to express the secondary fate.

This lateral inhibition hypothesis receives further support from cell removal experiments with *lin-15* mutants (Sternberg, 1988). If the anchor cell and all VPCs except P7p are killed, then P7p invariably expresses the primary fate. By contrast, if P8p is also left alive, P8p sometimes assumes the primary fate, in which case P7p expresses the secondary fate. These results indicate that the VPCs communicate with one another so that only one of a neighboring pair expresses the primary fate.

The lateral signal depends on the activity of the *lin-12*[+] gene: in animals lacking the *lin-12* function, no VPC expresses the secondary fate; in animals with increased *lin-12*[+] function, all six VPCs can express the secondary fate (Greenwald et al., 1983). Genetic mosaic analysis and molecular data indicate that the lin-12 protein is a receptor located in the membrane of the responding cells (Seydoux and Greenwald, 1989; Yochem et al., 1988). The simplest interpretation of these data is that the lin-12 protein acts as a receptor for the lateral signal sent by VPCs expressing the primary fate.

The lateral signal cooperates with the inductive sig-

nal from the anchor cell to determine one VPC for the primary fate and its two neighbors for the secondary fate. In *lin-12* mutants, the anchor cell signal induces three VPCs to express the primary fate (Fig. 24.19d). This result shows that the signal(s) from the anchor cell by themselves are not sufficient to specify the normal pattern of 2°-1°-2° fates. On the other hand, the fact that in wild-type animals the primary fate is always expressed next to the anchor cell indicates that the anchor cell has a special impact on its nearest neighbor. This effect might be caused by a single inductive signal or by two such signals. A single inductive signal might be spatially or temporally graded so that a high concentration (or early reception) induces the primary fate while lower concentration (or later reception) induces the secondary fate. Alternatively, the anchor cell might produce one short-range signal to induce the primary fate and a longer-range signal to induce the secondary fate. In either case, the inductive signal(s) from the anchor cell seem to generate only a bias in the closest VPC to express the primary fate. This bias needs to be reinforced by the lateral signal so that the nearest neighbor of the anchor cell is unequivocally determined to express the primary fate and to prevent its neighbors from doing the same.

The *lin-3*[+] Gene Encodes an Inductive Signal for Vulva Development

The model of vulval induction shown in Figure 24.19 has been confirmed in a grand way by the molecular characterization of some of the key gene products involved (Fig. 24.20). In particular, Russell Hill and Paul Sternberg (1992) have cloned the gene that encodes an inductive signal for vulva development. The gene they chose for analysis was *lin-3*[+], since loss of function in this gene causes the *vulvaless* phenotype, that is, hermaphrodites lacking a vulva.

The cloned *lin-3*[+] gene, when introduced as a transgene into *lin-3* mutants, restored these to normal development. Also, overexpression of *lin-3*[+] from multiple transgenes produced a dominant *multivulva* phenotype, in which all VPCs had primary or secondary fates (Fig. 24.21). To test whether the transgenes' ability to cause the multivulva phenotype depended upon the inductive signal, the researchers removed the anchor cell from animals with multiple transgenes. Most of these animals did not form a vulva. All these results were compatible with the hypothesis that *lin-3*[+] encodes a signal for vulval induction.

The lin-3 protein predicted from the DNA sequence is a membrane-spanning protein, that is, a protein that has a *cytoplasmic domain,* a *transmembrane domain* embedded in the phospholipid bilayer of the plasma membrane, and an *extracellular domain* extending outside the cell. The extracellular domain of the lin-3 protein contains a consensus sequence, called the *EGF repeat,* which

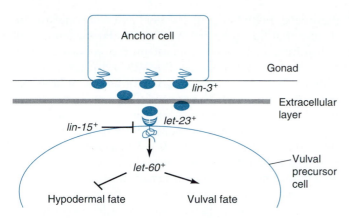

Figure 24.20 Molecular model of vulval induction. The *lin-3*⁺ gene is expressed in the anchor cell, and the lin-3 protein induces neighboring vulval precursor cells (VPCs) to assume vulval fates. The lin-3 protein may remain attached to the anchor cell membrane or may be released into the body cavity between the embryonic gonad and the VPCs. The *let-23*⁺ gene is expressed in VPCs and encodes the receptor for the lin-3 protein. The binding of the lin-3 protein activates its receptor, which is otherwise inhibited by the product of the *lin-15*⁺ gene. The activated let-23 protein in turn activates a signal protein encoded by the *lin-60*⁺ gene. The let-60 protein, through other signal proteins, acts as a molecular switch. When active, it makes VPCs realize vulval fates. If inactive, it allows VPCs to assume the hypodermal fate.

is also present in epidermal growth factor (EGF) and related growth factors known primarily from vertebrates (see Chapter 29). This sequence similarity, and the apparently similar functions of vertebrate growth factors and the *C. elegans* lin-3 protein as inductive signals, again underline the evolutionary conservation of many molecular mechanisms in development.

The lin-3 protein occurs in two forms that are encoded by alternatively spliced transcripts and differ by a stretch of 15 amino acids in the extracellular domain between the EGF repeat and the transmembrane domain. Possibly, the protein forms differ in their ability to be cleaved from the inducing cell by proteases. In this case, one form of lin-3 protein would stay attached to the producing cell and could therefore act only on adjacent cells. In contrast, the other form of lin-3 protein could be clipped to produce a diffusible signal that could act on nonadjacent cells as well.

To find out where the *lin-3*⁺ gene is expressed, the investigators constructed a *reporter gene* by inserting the bacterial *lacZ* gene into the first exon of the cloned *lin-3*⁺ gene (Color Plate 17). The fusion gene was expected to encode a protein consisting of the extracellular and transmembrane domains of lin-3 followed by the galactosidase domain encoded by *lacZ*. Indeed, in transgenic animals this fusion protein was still able to induce multivulva development and produced the galactosidase activity that turns a suitable substrate into a blue stain. The stain was precisely where it should be if *lin-3*⁺ en-

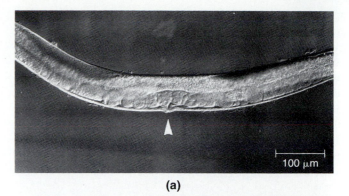

(a)

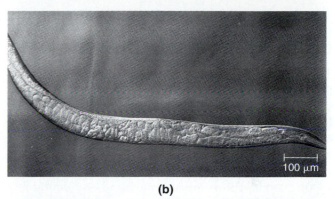

(b)

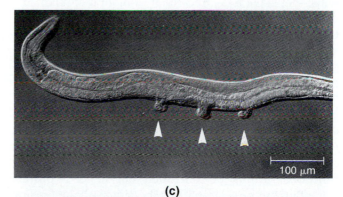

(c)

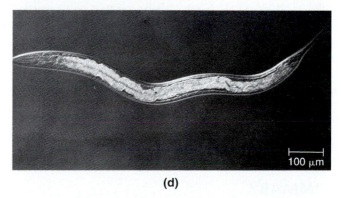

(d)

Figure 24.21 Photomicrographs of *C. elegans* lin-3 phenotypes. **(a)** Adult wild-type hermaphrodite. The arrowhead marks the position of the vulva. **(b)** Vulvaless hermaphrodite in which the *lin-3* gene has been interrupted by a transposon. **(c)** Hermaphrodite with multiple copies of a *lin-3*⁺ transgene. The animal has three vulvalike structures (arrowheads). **(d)** Hermaphrodite with multiple copies of a *lin-3*⁺ transgene but without a gonad. Most of these animals do not form a vulva.

codes a vulva-inducing signal: in the gonadal anchor cell. Moreover, the fusion protein began to accumulate when the vulva was induced: as the L2 larva molted into L3. No expression of the fusion gene was detectable outside the anchor cell. Thus, the lin-3 protein is produced exactly where and when we would expect if it were a vulva-inducing signal.

In summary, the biological effects of the lin-3 protein as well as its expression pattern and molecular characteristics are compatible with its possible role as a vulva-inducing signal.

The *let-23*[+] and *let-60*[+] Genes Encode Receptor and Signaling Proteins for Vulval Induction

The molecular characteristics of the lin-3 protein correspond very well to those of two other gene products involved in vulval induction: the let-23 protein and the let-60 protein ("let" for the *let*hal phenotypes of some alleles). The let-23 protein bears the hallmarks of a receptor protein that is bound to and activated by a ligand, such as EGP or the lin-3 protein (Aroian et al., 1990). Such receptors are also membrane-spanning proteins: they have an extracellular domain that binds to a matching ligand, a transmembrane domain, and an intracellular domain that has tyrosine kinase activity (Fig. 24.20). When a suitable ligand binds to the extracellular domain, it activates the cytoplasmic domain to phosphorylate tyrosine residues of target proteins.

Loss-of-function alleles of *let-23*, like those of *lin-3*, generate the *vulvaless* phenotype. This phenotype persists even in *let-23* mutants that have been transformed with multiple *lin-3*[+] transgenes. Thus, *lin-3*[+] transgenes can determine VPCs for vulval fates only if *let-23*[+] is functional. It must be concluded that the let-23 protein acts after the lin-3 protein in the signal chain causing vulval induction. This conclusion, drawn from mutant analysis, fits the molecular data indicating that *lin-3*[+] encodes an EGF-like molecule and *let-23*[+] a matching receptor.

The identification of the let-23 protein as a tyrosine kinase fits, in turn, with the molecular characteristics of the protein encoded by the *let-60*[+] gene (Han and

Sternberg, 1990; Beitel et al., 1990). Genetic analysis indicates that this gene acts as a molecular switch in vulval induction. Loss-of-function mutations cause a *vulvaless* phenotype, because the VPCs do not generate vulval cells, even in the presence of an active lin-3 signal and an intact let-23 receptor. Conversely, mutations that increase *let-60*[+] activity cause a *multivulva* phenotype, because all VPCs generate vulval cells, even under a weak inductive signal (*lin-3* loss-of-function mutants) or an inefficient receptor (*let-23* loss-of-function mutants). Together, these observations indicate that vulval induction activates the let-60 protein, presumably through the tyrosine kinase activity of let-23 protein.

Several gene products interact with the let-23–let-60 pathway of vulval induction (Fig. 24.20). One, of course, is the lin-3 protein, which binds as a ligand to the receptor domain of let-23. The lin-15 product, which prevents vulva development without an inductive signal, also seems to feed into this pathway. Because the phenotypes caused by *let-23* and *let-60* mutations override those caused by *lin-15* mutations, it appears that lin-15 protein dampens the activity of let-23 protein. Downstream of *let-60*[+], the *lin-31*[+] gene acts near the end of the vulval signaling pathway (L. M. Miller et al., 1993). In loss-of-function alleles of *lin-31*, each vulval precursor cell may adopt any of the three possible vulval cell fates. This rather intriguing phenotype suggests that *each* vulval precursor cell requires *lin-31*[+] activity to adopt its appropriate fate. In other words, the absence of *lin-31*[+] activity is a state that does not normally occur in any wild-type vulval precursor cell. Thus, unlike many other regulatory genes, *lin-31*[+] does not act as a switch that specifies one cell fate when it is active (on) and another state when it is inactive (off). How else the gene acts remains to be elucidated.

Taken together, the molecular characterizations of the gene products that bring about vulva formation provide striking confirmation of the three-signal model that was originally based on mutant analysis and cell removal experiments. Further analysis of the intracellular signaling processes within the VPCs that integrate these three signals should greatly enhance our understanding of embryonic induction in molecular terms.

SUMMARY

The soil nematode C. elegans is very suitable for developmental analysis because of its transparency, simple anatomy, and genetic properties. Most important, the development of the worm is invariant: the number of cell divisions, their timing and orientation, and the final number of somatic cells are the same for each hermaphrodite and for each male of the species.

During early cleavage, blastomeres are determined by cytoplasmic localization as well as by induction. The segregation of P granules into the germ line cells depends on a microfilament-based cellular transport system during a critical phase of the first cell cycle. The nuclei of the P_1 descendants accumulate the maternally expressed skn-1 protein, which is required for determining the EMS blastomere. The EMS blastomere or some of its descendants induce descendants of the AB blastomere to form anterior pharyngeal muscle. Similarly, the P₂ blastomere induces the EMS blastomere to form the gut precursor cell E.

Heterochronic genes in *C. elegans* control the timing of developmental events. Mutations in heterochronic genes cause cells to undergo patterns of cell division and differentiation that are normally observed at a different stage. For instance, the lin-14 protein forms a temporal gradient, with a high concentration during the first larval stage and a declining concentration thereafter. Mutations that shorten the period of lin-14 protein synthesis cause the cell division pattern to age prematurely. Conversely, extended synthesis of lin-14 protein keeps the cell lineage permanently young. The normal down-regulation of lin-14 protein depends on the activity of another gene that is activated by a feeding-dependent stimulus.

Of the 1090 cells generated in the hermaphrodite of *C. elegans,* 131 do not divide or differentiate; instead, these cells die. This process of programmed cell death is controlled by a genetic pathway. The pathway is initiated by two genes, *ced-3⁺* and *ced-4⁺*, which seem to encode cytotoxic proteins or regulators of such proteins. Whether a cell lives or dies is determined autonomously in each cell and is controlled by genes that act on *ced-3⁺* and *ced-4⁺*. One of these regulatory genes acts in all cells undergoing programmed cell death, while other genes affect specific cells individually.

The formation of the vulva in *C. elegans* has been used to study the role of cell interactions during larval development. Six vulval precursor cells (VPCs) located along the ventral midline have the potency to form the vulva. Only three VPCs, those closest to the anchor cell in the gonad rudiment, actually contribute to the vulva. Vulva development requires at least three signals from the anchor cell, the VPCs, and the hypodermis. The anchor cell produces an inductive signal that activate a matching receptor on the three nearest VPCs. These three VPCs interact via another signal to ensure proper vulva formation. In the remaining VPCs, the receptor for the inductive signal is dampened by a third signal from the hypodermis.

The invariant cell lineage and other features of *C. elegans* have allowed scientists to analyze the development of single cells with unparalleled precision. Studies of mutants of *C. elegans* have helped to uncover some of the molecular mechanisms involved in such basic developmental processes as cytoplasmic localization, induction, timing, and programmed cell death. At least some of the molecules involved—such as growth factors, their receptors, and regulatory signals for cell death—are shared with vertebrates and presumably other organisms. Thus, many of the data gathered from the tiny roundworm will promote a better understanding of basic molecular mechanisms that underlie the development of many organisms.

SUGGESTED READINGS

Ellis, R. E., and H. R. Horvitz. 1991. Two *C. elegans* genes control the programmed deaths of specific cells in the pharynx. *Development* **112**:591–603.

Hill, R. J., and P. W. Sternberg. 1993. Cell fate patterning during *C. elegans* vulval development. *Development* 1993 supplement, 9–18.

Ruvkun, G., B. Wightman, T. Bürglin, and P. Arasu. 1991. Dominant gain-of-function mutations that lead to misregulation of the *C. elegans* heterochronic gene *lin-14,* and the evolutionary implications of dominant mutations in pattern-formation genes. *Development* supplement 1, 47–54.

Wood, W. B., 1988. *The Nematode Caenorhabditis elegans.* Cold Spring Harbor, N.Y.: Cold Spring Harbor Laboratory.

CELL ADHESION
AND
MORPHOGENESIS

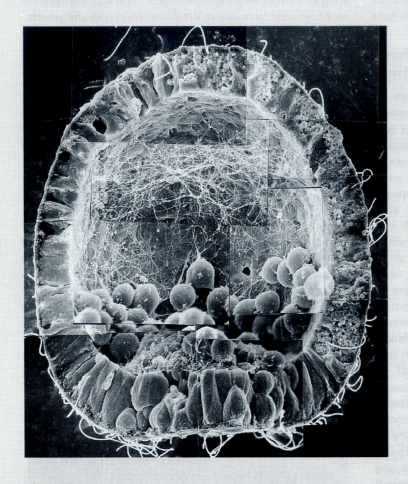

Figure 25.1 Scanning electron micrograph showing the inside of an early sea urchin gastrula. The round cells near the bottom of the embryonic cavity (blastocoel) are the primary mesenchyme cells. They were part of the outer epithelium before they lost adhesiveness to their neighbors and moved into the blastocoel. Changes in cell adhesiveness are a driving force in morphogenesis.

Cell adhesion plays a critical role in the basic phenomenon of morphogenesis. Differences in adhesion can account, at least in part, for the spatial order in which embryonic cells become arranged. In classic studies, researchers dissociated embryos into cells, mixed and reaggregated them. The cells then separated into different tissues, which arose in spatial configurations that were often strikingly similar to those of normal embryos. Measurements of cell adhesiveness showed that the cells occupying the central position in mixed aggregates were those that adhered to one another most strongly. In recent experiments, genetically transformed cells that differed only in the amount of cell adhesion molecules on their surfaces behaved the same way.

The adhesiveness between cells changes in the course of development. Embryos consist of closely knit epithelia and loose mesenchyme, but these conformations wax and wane as development proceeds. Epithelial cells become mesenchymal in the sea urchin embryo at the beginning of gastrulation, when cells from the vegetal plate ingress

and form the primary mesenchyme cells (Fig. 25.1). Conversely, mesenchymal cells may band together and form epithelia as, for example, in chicken embryos, where individual cells ingress through the primitive streak and later form the somites and lateral plates of the mesoderm.

Even within an intact epithelium, cells may change neighbors. Morphogenetic movements such as convergent extension and epiboly are associated with cell intercalations, in which cells detach from their neighbors and insert themselves between other cells. Thus, the organized making and breaking of cell contacts is an integral part of morphogenesis.

The role of cell associations in morphogenesis was first revealed by experiments in which sponges were broken down into single cells and small groups of cells and were then allowed to reaggregate (H. V. Wilson, 1907). The dissociated cells actively migrated until they adhered to one another, forming clusters that grew by addition of more cells or by fusion with other clusters. Eventually, the aggregated cells reconstituted new sponges, complete with the flagellated chambers and branched canals that characterize the sponge's internal structure. These experiments were a striking demonstration that dissociated cells can reassemble into an organism. Moreover, when sponge cells from different species were mixed, they sorted out into separate aggregates, showing that species differ in their cell affinities.

More recently, the use of antibodies and DNA cloning have allowed investigators to isolate several types of *cell adhesion molecules (CAMs)* that mediate cell adhesion. Cloning and sequencing of the genes that encode CAMs have provided detailed information about the nature of these molecules and have allowed researchers to design experiments that reveal the role of individual CAMs in specific developmental processes.

In the course of this chapter, we will first summarize the classic experiments that have pointed out the central role of cell adhesion in embryonic development. Next we will examine the molecular characteristics of three families of CAMs. Then we will explore the functions of CAMs in major developmental processes including cell determination, the formation of boundaries between tissues, and cell migration. We will conclude with some preliminary evidence of the role of CAMs in the regulatory circuits by which gene expression controls morphogenetic events and vice versa.

Cell Aggregation Studies in Vitro

To analyze cell adhesion, investigators often disaggregate embryonic tissues into single cells. Depending on the species, disaggregation may be achieved mechanically by forcing the tissue through a fine-meshed sieve, by changing the pH of the medium, by removing calcium ions (Ca^{2+}) and magnesium ions (Mg^{2+}) from the medium, or by adding proteases to partially digest the tissue. After any of these treatments, cells are allowed to recover in a normal medium before their behavior is studied.

Cells from Different Tissues Display Selective Affinities

The most dramatic morphogenetic movements in embryonic development occur during gastrulation and organogenesis. At those stages, do cells have different affinities that may guide their movements? This question was addressed in a classic study by Philip Townes and Johannes Holtfreter (1955) using amphibian embryos.

Amphibian embryonic cells are especially suited to reaggregation experiments. They are relatively large, and each cell contains its own nutrient reservoir in the form of yolk. Cells from different germ layers or regions typically differ in size, shape, or pigmentation, so that they can be distinguished with relative ease. Moreover, cells readily dissociate if the pH of the culture medium is raised briefly.

In one experiment, Townes and Holtfreter dissociated neural plate cells and prospective epidermis cells from early neurulae and mixed them together (Fig. 25.2). The cells initially formed aggregates in which the two cell types were randomly mixed. However, the neural plate cells subsequently vanished from the surface and moved inside while the prospective epidermal cells moved to the outside of the aggregate. When axial mesoderm cells were added, they were sandwiched between the neural cells on the inside and the future epidermal cells on the outside. In similar experiments, a mixture of prospective epidermal cells and mesodermal cells sorted out so that mesoderm was formed inside and epidermis outside. Likewise, mesodermal and endodermal cells segregated with mesoderm inside. However, when epidermal cells were added to the mixture, the reaggregated endoderm was surrounded by mesoderm, and the mesoderm in turn was enveloped by epidermis. Generally, the epithelia reconstituted by the cells showed the appropriate polarity, with an apical surface on the outside and a basal surface on the inside.

Two conclusions from these experiments are important here. First, cells from different germ layers sort out from one another, thus revealing distinct surface properties. Second, the positions occupied by the cells in

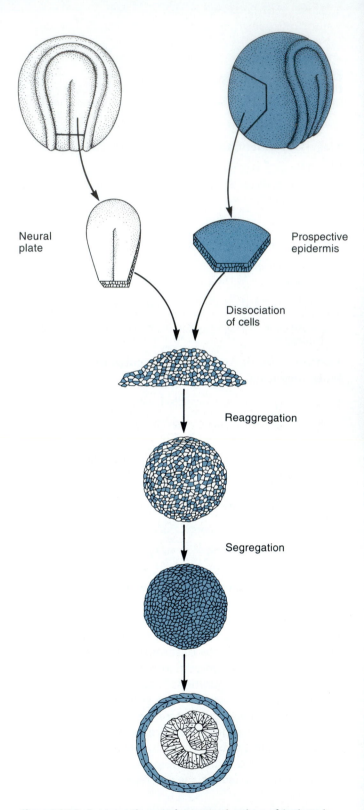

Neural plate

Prospective epidermis

Dissociation of cells

Reaggregation

Segregation

Figure 25.2 Segregation and reorganization of isolated cells from an amphibian neurula. Neural plate from an unpigmented embryo and prospective epidermis from a pigmented embryo were excised and dissociated into single cells. When mixed together, the cells first formed a mixed aggregate but eventually sorted out so that the neural cells occupied the central position.

reaggregates reflect their relative positions in normal embryos. (In interpreting the case where endoderm surrounds mesoderm, remember that the intestinal epithelium is part of the body surface while the mesoderm and its derivatives are internal.) The behavior of reaggregating cells in vitro suggests that selective cell affinities help establish and maintain the spatial order of different tissues in the embryo.

Tissues Form Hierarchies of Adhesiveness

Mixtures of different cell types sort out not only during gastrulation and organogenesis but also during later stages. When cells from embryonic heart, liver, retina, cartilage, and other tissues are mixed, they sort out and arrange themselves in concentric layers. The aggregates mimic the normal relative positions of cells; for instance, muscle cells surround chondrocytes (cartilage cells). Even when the participating cells would not normally occur next to each other, they reaggregate in a reproducible order. For example, when pigmented retina cells and heart muscle cells from chicken embryos are mixed, the heart cells first form islets surrounded by retina cells until eventually all heart cells are together on the inside; never do the heart cells envelope the retina cells.

On the basis of their behavior in mixed aggregates, cells from different tissues can be ranked in a hierarchy reflecting their ability to occupy the center position. Chondrocytes, heart cells, and liver are ranked in this order because in mixed aggregates with either heart or liver cells, chondrocytes move to the inside, and heart cells in turn are always enveloped by liver cells (Fig. 25.3). This hierarchy was interpreted by Malcolm Steinberg (1963, 1970) in terms of quantitative differences in *cell adhesion*, that is, the energy that must be expended to separate two adhering cells. Steinberg proposed that cells in a mixed aggregate move randomly until the sum of all cell adhesions has reached a maximum. This is also the state of minimum free energy, at which further cell rearrangements require the input of energy instead of giving off energy. This state is reached when the cell type with maximum adhesiveness has formed a central sphere surrounded by concentric spheres of cells with progressively lower adhesiveness. This *differential adhesion hypothesis* is in accord with the general thermodynamic principle that all closed systems naturally move toward a state of minimum free energy.

As a direct test of the differential adhesion hypothesis, Herbert Phillips and Malcolm Steinberg (1969) centrifuged round cell aggregates and took the resistance of the aggregates to deformation as a measure of the

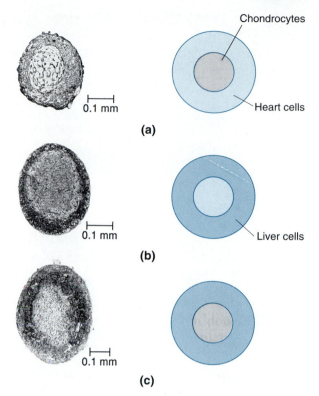

Figure 25.3 Sorting out and spatial ordering in mixed aggregates of different chicken cell types. Each photograph shows the segregation and reorganization of the two types of cells indicated in the diagram to the right. **(a)** Chondrocytes move to the center after mixing with heart cells. **(b)** Heart cells move to the center when mixed with liver cells. **(c)** Chondrocytes move to the center when mixed with liver cells. The three cell types form a hierarchical order with regard to their ability to occupy the center position.

cells' adhesiveness. They chose centrifugation conditions that compressed originally round aggregates while allowing flat aggregates ("pancakes") of the same cells to assume a somewhat rounder shape (that of a hamburger bun). They reasoned that aggregates starting from either a round or a flat shape should approach the same intermediate shape at which the forces of cell adhesion, which work toward imposing a round shape, would just balance the centrifugal force flattening the aggregate. Such centrifugation conditions were indeed found, and as expected, the shapes of aggregates of different cell types at equilibrium reflected their behavior in aggregation experiments. Cells that tended to occupy the center of mixed aggregates—chondrocytes, in Steinberg's experiments—formed the roundest bun shape, indicating that they had the strongest adhesion (Fig. 25.4). Conversely, cells that tended to be displaced to the outside in mixed aggregates—the liver cells—formed the flattest buns, indicating the weakest adhesion. These results confirmed the hypothesis that the behavior of different cell types in mixed aggregates reflects quantitative differences in adhesiveness.

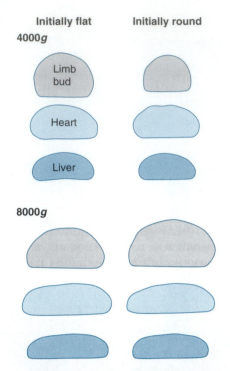

Initially flat   **Initially round**
4000*g*

Limb
bud

Heart

Liver

8000*g*

Figure 25.4 Profiles of centrifuged aggregates of different chicken cells. Starting from either round or flat (pancake-like) shapes, the aggregates assumed a shape somewhat like a hamburger bun. This shape represents a state of equilibrium at which the flattening centrifugal force just balances the forces of cell adhesion, which work toward imposing a spherical shape. The exact shape of the aggregates at equilibrium depended on the strength of the centrifugal force (4000*g* or 8000*g*) and on the cell type used. Limb bud cells (mostly chondrocytes) formed the most-rounded buns, indicating greatest adhesiveness, while liver cells were least adhesive and heart cells in between. This order of adhesiveness correlates with the ability of these cell types to occupy the central position in mixed aggregates (see Fig. 25.3).

The differential adhesion hypothesis does not predict what or how many molecular mechanisms cell adhesion is based on. It just states that the *overall* strength of adhesion dictates the position of a cell type in a mixed aggregate and, where applicable, in a normal embryo. Molecular studies described later in this chapter show clearly that *several types of cell surface molecules* contribute to the phenomenon of cell adhesion.

Stage-Dependent and Region-Specific Cell Adhesion in Vivo

If cell adhesion plays a significant role in morphogenesis, then changes in cell adhesion as measured in vitro should correlate with the morphogenetic movements observed in vivo. To look for such correlations, researchers isolated embryonic cells from specific regions

and at defined stages, measured their adhesiveness in culture, and tested whether regional or temporal changes in adhesiveness correlated with changing cellular behaviors in the intact embryo.

Cell Adhesiveness Changes during Gastrulation

Some of the most dramatic morphogenetic movements occur during gastrulation. Changes in cell adhesiveness during this period have been observed in several organisms, including fishes and sea urchins.

In fish and in other groups that undergo discoidal cleavage, the pregastrula embryo forms a disc-shaped *blastoderm* consisting of several cell layers on top of the uncleaved yolk. The blastoderm cells are nearly round and do not seem to adhere to one another very strongly (see Fig. 5.13). During gastrulation and organogenesis, these cells spread over the uncleaved yolk reservoir, in a movement known as *epiboly*. The cells undergoing epiboly become extremely flat. This shape change might be a passive response to stretching, but it could also reflect an increase in the cells' adhesiveness to the underlying yolk cytoplasm.

To decide between these two hypotheses, John Trinkaus (1963) examined whether isolated blastoderm cells flattened in culture. He dissociated blastoderm cells of the killifish *Fundulus heteroclitus* and observed their behavior in a culture medium under the microscope. Pregastrula cells remained spheroid for 6 h after culturing. In contrast, many gastrula cells flattened extensively within an hour or two on the glass bottom of the culture chamber. Similar flattening occurred when cells from pregastrula embryos were kept in prolonged culture until control intact embryos underwent epiboly. Thus, the flattening of isolated cells in vitro was correlated with the onset of epiboly in vivo. Trinkaus concluded that the gastrula cells were not flattened passively by stretching. Instead, he ascribed the flattening to an increase in the cells' adhesiveness to glass and to their natural substratum in the intact embryo. Furthermore, he proposed that the flattening of individual cells may actively contribute to the morphogenetic movement of epiboly.

Another study correlating morphogenetic movements with apparent changes in cell adhesiveness was carried out by David McClay and Charles Ettensohn (1987) on the *primary mesenchyme cells* of sea urchin embryos. These cells undergo *ingression* at the beginning of gastrulation, when they leave the *vegetal plate* epithelium and move into the blastocoel (see Chapter 10). In the course of this movement, the mesenchyme cells lose contact with the cells that remain in the vegetal plate. Just before the primary mesenchyme cells ingress into the blastocoel, their apical parts become very elongated as if the cells were still adhering to the *hyaline layer* surrounding the embryo before letting go (Fig. 25.5). If this

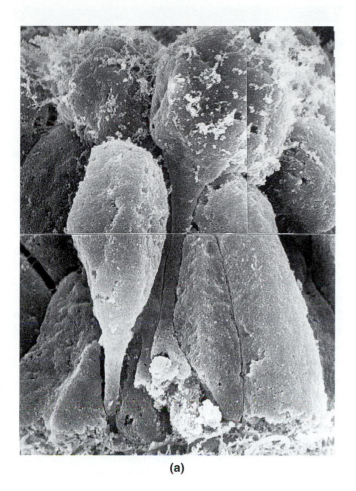

(a)

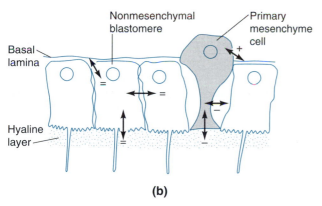

(b)

Figure 25.5 Ingression of primary mesenchyme cells during sea urchin gastrulation. **(a)** Scanning electron micrograph shows the vegetal plate viewed from inside the blastocoel. An ingressing mesenchyme cell is still connected with a stalk to the hyaline layer on the outside of the embryo. **(b)** Diagram showing a strip of vegetal plate cells in lateral view. The blastocoel, into which the primary mesenchyme cells move, is oriented to the top. The hyaline layer surrounding the embryo is at the bottom. The symbols reflect the changes in cell adhesiveness compiled in Table 25.1. As primary mesenchyme cells ingress into the blastocoel, they lose adhesiveness to the hyaline layer and to neighboring blastomeres. At the same time, the mesenchymal cells gain adhesiveness to the basal lamina, which is formed on the basal cell surfaces and appears in the micrograph as flocculent material.

ingression movement were based on changes in cell adhesion, the primary mesenchyme cells should lose adhesiveness to vegetal plate cells and to the hyaline layer while gaining adhesiveness to the *basal lamina* underneath the vegetal plate and to the extracellular material in the blastocoel. These predictions were tested in the following experiment.

▼

To assess the adhesiveness of primary mesenchyme cells to neighboring cells and to extracellular materials, Rachel Fink and David McClay (1985) used a quantitative assay that was designed to measure the force necessary to separate radiolabeled test cells from a layer to which they adhered. The test cells were allowed to settle briefly in small dishes whose bottoms were covered with a layer of cultured cells to which the test cells adhered, or with a coat of isolated extracellular material. After the test cells attached, the dishes were centrifuged with the test cells facing outward so that the centrifugal force dislodged them from the layer on which they had settled. By running the centrifuge at different speeds, the researchers determined the force necessary to dislodge half of the test cells.

The assay was applied to migrating primary mesenchyme cells isolated from late sea urchin blastulae and to nonmigratory precursors of mesenchyme cells. The precursor cells were grown in culture from micromeres isolated at the 16-cell stage. For comparison, the adhesion assay was also performed with ectoderm and endoderm cells isolated from gastrulae. Each type of test cell was allowed to settle on a layer of cultured gastrula cells, on a layer of isolated basal lamina, or on a layer of hyalin (a protein from the hyaline layer). The measurement results are compiled in Table 25.1.

TABLE 25.1

Binding Strength of Sea Urchin Micromeres, Primary Mesenchyme Cells, and Control Cells to Other Gastrula Cells, Basal Lamina, and Hyaline Layer

Test Cells	Force (μdyn) Required to Dislodge Test Cells from:		
	Hyalin	Gastrula Cells	Basal Lamina
16-cell–stage micromeres	58.0	68.0	0.5
Migratory-stage mesenchyme cells	0.1	0.1	15.0
Gastrula ectoderm and endoderm cells	50.0	50.0	0.5

Binding strength is calculated from the centrifugal force that removes 50% of the test cells from the substrate.
Source: From Fink and McClay (1985). Used with permission.

Micromeres adhered to hyalin, but in their descendants, the migratory mesenchyme cells, adhesiveness decreased to less than 1% of the original force. In contrast, gastrula ectoderm and endoderm cells showed no such decrease during the same time interval. Micromeres also adhered to gastrula cell layers, but their adhesiveness was lost around the time when the primary mesenchyme cells ingressed in vivo. Conversely, micromeres adhered very weakly to basal lamina, but this adhesiveness increased about 30-fold in migratory mesenchyme cells. When tested against isolated components of basal lamina, the wandering mesenchyme cells were found to have acquired a particular adhesiveness to fibronectin (see Chapter 26).

The experiments of Fink and McClay (1985) show that the ingression of primary mesenchyme cells in vivo is accompanied by dramatic changes in cell adhesiveness (Fig. 25.5b). As the cells ingress, they lose adhesiveness to other cells and to hyalin while gaining adhesiveness to basal lamina and presumably to other extracellular materials in the blastocoel. These data confirm the hypothesis that changes in cell adhesion contribute to the ingression process.

It must be remembered, however, that cell adhesion is only one aspect of morphogenetic movements. For a cell to migrate, it must not only adhere to a new substrate (cell or extracellular material), but it must also couple cytoskeletal elements to the adhesion sites, it must pull toward the new adhesion sites, and it must let go of previous adhesion sites.

Cell Adhesiveness Changes along the Proximodistal Axis of a Limb

The regenerating salamander limb is another system in which cell adhesiveness in vitro has been correlated with morphogenesis in vivo. As discussed in Chapter 20, regeneration begins with the formation of a *regeneration blastema*, a mound of dedifferentiated cells that forms at the end of an amputated limb stump. The blastema cells proliferate and undergo a process of pattern formation and redifferentiation that eventually restores the lost limb parts. Thus, when a limb is amputated through the upper arm, the regeneration blastema gives rise to the missing part of the upper arm plus the entire lower arm, wrist, and hand. However, if the hand is amputated at the wrist, the regeneration blastema restores only the hand.

How does the regeneration blastema "know" its position along the proximodistal axis? The following experiments indicate that cell adhesion is at least a partial answer.

Cells adhere more strongly to one another in regeneration blastemas formed *distally* than in those formed *proximally*. This was shown by James Nardi and David Stocum (1983), who cultured regeneration blastemas that had originated at different proximodistal levels. For instance, the researchers juxtaposed a blastema from the wrist of a right forelimb with a blastema from the elbow of a left forelimb. The regeneration blastemas were obtained from axolotls or newts, and only the mesenchymal cores of the blastemas were used in the experiments. To monitor some of the interactions between the paired blastemas, the researchers labeled one of each pair with tritiated thymidine; as another method of labeling, matched axolotl blastemas were also taken from animals with different pigmentation. The paired blastemas healed together within 2 to 4 hours and assumed a more or less spherical shape during the following 4 days of culture. At the end of the culture period, the blastemas were fixed, sectioned, and prepared for *autoradiography* (see Methods 3.1).

Microscopic inspection of the sections showed that blastemas from the same proximodistal level tended to fuse along a straight boundary (Fig. 25.6). In contrast, blastemas from different levels engulfed each other in a regular way: the more proximal blastema tended to surround the more distal blastema. In terms of the differential adhesion hypothesis, the researchers concluded that cells in regeneration blastemas are more adhesive when the blastema is formed in a more distal position. Extending this interpretation, they suggested that *intercalary regeneration* (see Chapter 20) might be triggered by a disparity in cell adhesiveness between proximal and distal tissues.

In subsequent experiments, Karen Crawford and David Stocum (1988) showed that the proximodistal gradient in cell adhesiveness observed in vitro also controlled the behavior of regeneration blastemas in vivo. In these experiments, an axolotl forelimb was amputated through the upper arm, elbow, or wrist, and a hindlimb was amputated through the thigh (Fig. 25.7). When both limbs had formed regeneration blastemas, the forelimb blastema was excised and grafted to the dorsal surface of the hindlimb at the blastema-stump junction. All grafts healed within 24 h and established blood circulation during the following day.

All developing forelimb grafts formed recognizable forelimbs. The upper arm blastemas formed arms beginning at the upper arm level and branched off from the thigh level, where they had been implanted. However, the grafted elbow blastemas were *displaced distally from the implantation site* until they reached the knee level of the regenerating hindlimb, at which point they formed arms beginning at the elbow. The grafted wrist blastemas were displaced even further, to the ankle level of the regenerating hindlimb, where they formed hands. Thus, each grafted forelimb blastema

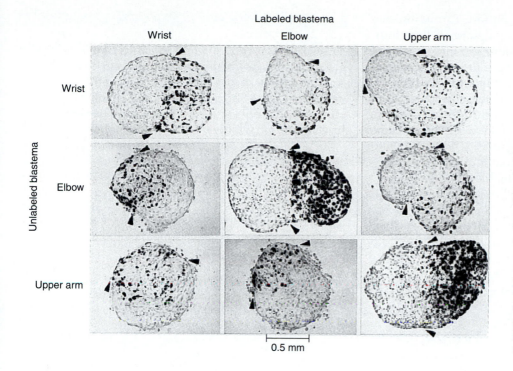

Labeled blastema

Wrist   Elbow   Upper arm

Unlabeled blastema

Wrist

Elbow

Upper arm

0.5 mm

Figure 25.6 Cellular rearrangements in pairs of regeneration blastemas cultured together. Blastemas were obtained from newts or axolotls amputated at the level of the wrist, elbow, or upper arm. One of each pair was labeled with tritiated thymidine, which generates dark grains in autoradiographs (see Methods 3.1). Blastemas from the same proximodistal level fuse along a straight line, whereas in pairs from different levels the proximal blastema tends to engulf the distal blastema.

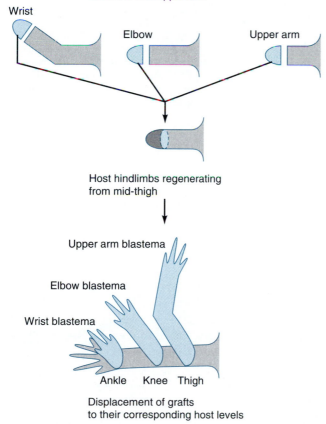

Donor forelimbs regenerating from wrist, elbow, or mid-upper arm

Wrist

Elbow   Upper arm

Host hindlimbs regenerating from mid-thigh

Upper arm blastema

Elbow blastema

Wrist blastema

Ankle   Knee   Thigh

Displacement of grafts to their corresponding host levels

Figure 25.7 Limb bud transplantations designed to detect a proximodistal gradient of cell adhesiveness in regeneration blastemas of the axolotl. Forelimbs were amputated through the upper arm, elbow, or wrist, and hindlimbs were amputated through the thigh. After the limbs had formed regeneration blastemas, a forelimb blastema (color) was excised and grafted to the dorsal surface of the hindlimb at the blastema-stump junction. As the hindlimb grew out in its continuing regeneration, the graft was displaced distally until it reached the corresponding hindlimb level in the host. Thus, a forelimb starting at the upper arm was regenerated at the level of the thigh, a forelimb starting at the elbow was regenerated from the knee, and a hand was regenerated from the ankle.

was displaced distally to the position in the hindlimb that corresponded to the forelimb position from which the graft had been obtained. These results indicate that there are similar differences in cell affinity along the proximodistal axes of forelimbs and hindlimbs. Because of these differences, each grafted forelimb blastema moves with the outgrowing hindlimb blastema until the cell surface properties of the graft match those of its host environment.

In the conceptual terms of pattern formation discussed in Chapter 20, the results described here indicate that the positional value along the proximodistal axis may be encoded in cell surface properties, including cell adhesiveness. To further test this possibility, Crawford and Stocum (1988) used retinoic acid, an agent known to proximalize regeneration patterns, presumably by changing the activity of *selector genes* such as the murine *Hox* genes (see Chapter 22). When forelimb blastemas were pretreated with retinoic acid before they were transplanted to regenerating hindlimbs, the grafts formed arms beginning at the shoulder girdle, regardless of the forelimb level at which the grafted blastema had originated. Correspondingly, none of the grafts was displaced distally with the outgrowing hindlimb blastema. These results confirmed that, whatever the physical basis of positional value along the proximodistal axis may be, it is linked to cell adhesiveness.

In summary, cell adhesion studies in vitro and in vivo support the concept that cells adhere selectively in ways that are region-specific and stage-dependent. Thus, it appears that cell adhesiveness is an integral part of the overall process of cell determination and cell differentiation. In terms of differential gene expression, this means that cells constantly adjust and readjust the assortment of gene products that mediate cell adhesion. Conversely, the cell contacts, shape changes, and mechanical stresses that result from cell adhesion may feed back into a cell's program of gene expression. In view of the critical functions of cell adhesion, researchers are searching intensively for the molecules by which cells adhere to one another and to extracellular materials.

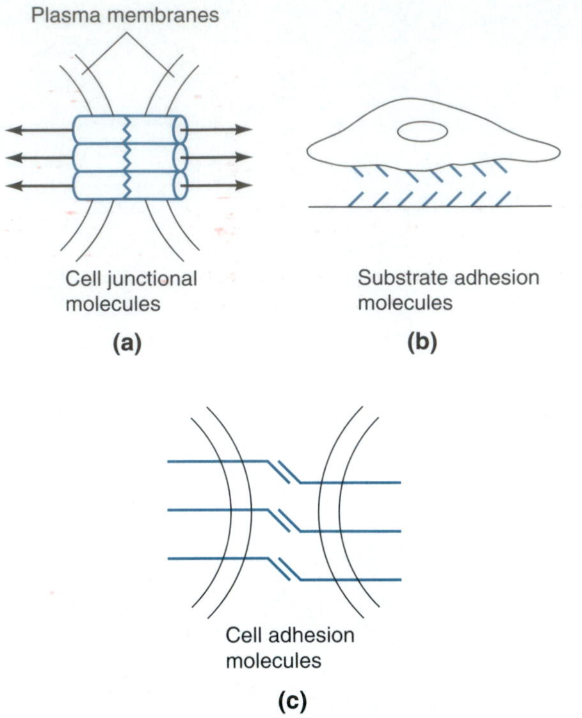

Figure 25.8 Three types of molecules (color) involved in connections between cells. **(a)** Cell junctional molecules form communication channels (gap junctions), seals between the intercellular space and the external environment (tight junctions), or stable adhesion sites (desmosomes). **(b)** Substrate adhesion molecules (slanted lines) include extracellular molecules and the corresponding cellular receptors whose interaction causes cells to adhere to extracellular substrates. **(c)** Cell adhesion molecules (bent lines) are embedded in the cell membranes and link cells to each other quickly but weakly.

Cell Adhesion Molecules

Cells adhere to other cells and to extracellular materials by means of *cell junctions, substrate adhesion molecules,* and *cell adhesion molecules* (Fig. 25.8). Cell junctions are large molecular structures that form slowly and are generally strong and durable (see Chapter 2). Substrate adhesion molecules, which will be discussed in Chapter 26, are a group made up of components of the extracellular matrix and their matching receptors in cell membranes. *Cell adhesion molecules (CAMs)* are single molecules embedded in cell plasma membranes, by which cells adhere to one another quickly, selectively, and relatively weakly.

A typical CAM is a glycoprotein with three large domains: a large *extracellular domain,* a *transmembrane domain,* and a *cytoplasmic domain* (Fig. 25.9). By means of their extracellular domains, CAMs bind to other CAMs. *Homotypic binding* is the binding that occurs between CAMs of the same type, whereas *heterotypic binding* is that between different types of CAMs. The binding

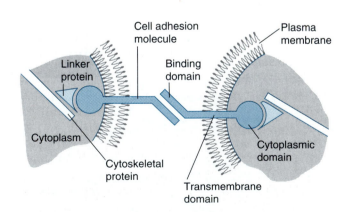

Figure 25.9 Schematic representation of two cell adhesion molecules (CAMs) binding to each other. A typical CAM consists of an extracellular binding domain, a transmembrane domain, and a cytoplasmic domain, which is connected, via one or more linker proteins, to a cytoskeletal protein.

Isolating Cell Adhesion Molecules and Their Genes with Antibodies

Antibodies, also called *immunoglobulins,* are vertebrate blood serum proteins that are produced by B lymphocytes. Each clone of lymphocytes makes one type of antibody, directed against a specific *epitope* (binding site) on an *antigen* (a foreign molecule or cell). A given antigen may have several different epitopes and may therefore react with different antibodies. *Polyclonal antibodies* are products of two or more lymphocyte clones; they are prepared from blood serum of immunized mammals, usually rabbits, into which an antigenic preparation has been injected. Typically, many clones of lymphocytes in the immunized animal produce antibodies directed against the various epitopes present in the antigenic preparation. *Monoclonal antibodies* are the products of single lymphocyte clones, which are obtained by fusing lymphocytes from immunized mice with tumor cells. The fusion cells, called *hybridoma cells,* are capable of unlimited mitosis, like tumor cells, but also retain the characteristic function of a normal lymphocyte: to generate a clone of cells that produce the same type of antibody against a specific epitope (Köhler and Milstein, 1975). From a single immunized mouse, one can obtain many hybridoma cell clones that can be screened on the basis of the antibodies they produce. Even if a mouse is immunized with a mixture of many antigenic molecules, each hybridoma cell clone produces only one type of antibody.

To obtain polyclonal antibodies against CAMs, the cell type of interest or a preparation of plasma membranes from such cells is injected into a rabbit. The antibodies obtained are tested for the ability to inhibit aggregation of the cells. (For this test, the antibodies must be cleaved into fragments—known as *Fab fragments* or *univalent antibodies*—having only one antigen-binding site. Normal antibodies are bivalent, having two binding sites that agglutinate the cells to which they bind.) The inhibition of cell aggregation by Fab fragments is then used as a screening procedure to identify one or more CAMs. Cell surface proteins from the cells of interest are separated by electrophoresis and tested individually for the ability to *overcome* the inhibition of cell aggregation by the antibody (Fig. 25.10). A protein that has this effect must compete with a CAM on the cells for the binding sites of antibodies and must therefore be the same CAM.

Monoclonal antibodies against cell surface glycoproteins can also be tested for the ability to inhibit cell aggregation. An antibody that does so is again most likely directed against a CAM. Such a monoclonal an-

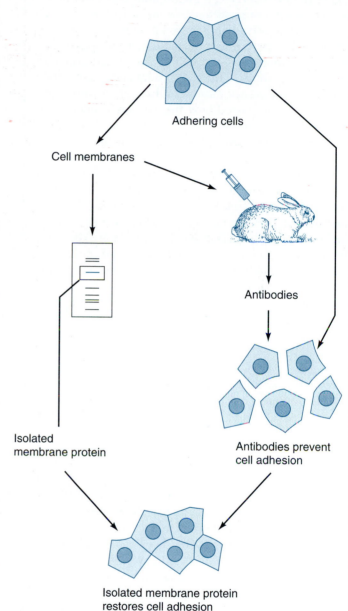

Figure 25.10 Use of polyclonal antibodies to identify cell adhesion molecules (CAMs). Membranes from cells that adhere in culture are injected into a rabbit. Univalent antibodies prepared from the rabbit's blood serum prevent cell adhesion, indicating that the rabbit is producing antibodies against one or more CAMs. Cell membrane proteins isolated by electrophoresis are then tested individually for the ability to overcome the rabbit antibodies' inhibition of cell adhesion. A protein that has this ability must be competing with a CAM on the cell surfaces for the antigen-binding sites of antibodies and most likely will be the CAM itself.

Continued on next page

tibody can be used to isolate the CAM itself from the mixed proteins of the cell membranes that were used as an antigenic preparation. For this purpose, the antibody is immobilized on a chromatography column. The antigenic preparation is passed over the column under conditions that allow the immobilized antibody to bind to its antigen, the CAM, while all other components of the preparation flow through. A different medium is then passed over the column to release the CAM from the antibody.

Either of these strategies yields a purified CAM and an antibody directed against it. These preparations can be used in turn to clone the gene and cDNAs encoding the CAM. One strategy is to determine part of the CAM's amino acid sequence and then use the corresponding nucleotide sequences for screening DNA libraries (see Methods 14.2). Alternatively, a labeled monoclonal antibody itself can be used as a probe to screen *expression libraries,* which are prepared similarly to genomic and cDNA libraries except that the host bacteria synthesize the protein encoded by the cloned DNA.

specificity depends on both the protein and the carbohydrate moieties of CAMs. The transmembrane domain links a CAM to the plasma membrane by hydrophobic interactions. The cytoplasmic domain is often connected, via one or more *linker proteins,* to microfilaments or other cytoskeletal structures. This connection is important because the phospholipid bilayer that constitutes most of the plasma membrane is fluid and offers no resistance to lateral floating. Without linkage to the cytoskeleton, a cell adhesion molecule behaves like an unanchored buoy. Such is the case for some CAMs that lack both the transmembrane and cytoplasmic domains. Their extracellular domain is bound covalently to special membrane phospholipids.

The molecular characterization of CAMs is based to a large extent on the nucleotide sequences of their genes and cDNAs. To clone the genes and cDNAs, it was necessary to obtain purified CAMs first. For this task, researchers have made ingenious use of antibodies against CAMs (Methods 25.1).

Using polyclonal and monoclonal antibodies, investigators have isolated several CAMs and characterized them biochemically and by immunostaining (see Methods 4.1). The structure and function of those CAMs for which the corresponding genes have been cloned and sequenced have been analyzed further. Cloning these genes has also set the stage for exciting new research on how their expression is regulated as part of the overall processes of cell determination and morphogenesis.

On the basis of their molecular characteristics, CAMs can be subdivided into three families: immunoglobulin-like CAMs, cadherins, and lectins or enzymes that bind to the carbohydrate moieties of CAMs. Each family will be discussed in turn.

Immunoglobulinlike CAMs May Allow or Hinder Cell Adhesion

The first-known cell adhesion molecule was characterized in the laboratory of Gerald Edelman and named the *neural cell adhesion molecule (N-CAM),* because it was isolated from the neural retina of chickens (Brackenbury et al., 1977; Thiery et al., 1977; Edelman, 1983). N-CAM function is *necessary* for neural development: antibodies against N-CAM interfere with the development of chicken retina in culture (Buskirk et al., 1980), with the establishment of proper connections between the optic nerve and the brain in *Xenopus* (S. E. Fraser et al., 1988), and with many other neural processes. N-CAM is also *sufficient* for cell adhesion: mouse L cells, tumor cells that do not express any known cell adhesion molecule and do not adhere to one another, are transformed into adhesive cells by transfection with cDNA encoding N-CAM (Mege et al., 1988).

Molecular analysis of N-CAM has revealed several intriguing features (Cunningham et al., 1987, Rutishauser et al., 1988). One of them is the complexity of differential splicing and posttranslational modification events that generate more than 100 different N-CAM proteins from a single gene. Proteins derived from the same gene are called *isoforms.* In the brain, the three most abundant N-CAM isoforms have molecular weights of approximately 180, 140, and 120 kd. The two larger isoforms are transmembrane proteins that differ in the length of their cytoplasmic domain, which includes the C terminus of the molecule (Fig. 25.11). The third isoform is linked covalently to a glycosyl-phosphatidylinositol (GPI) molecule in the outer phospholipid leaf of the plasma membrane (He et al., 1986).

The large extracellular domain of all N-CAM proteins contains several carbohydrate moieties and five smaller repetitive domains that are also found in immunoglobulins. These domains are thought to fold into complementary tertiary structures that allow *homotypic binding.* The extensive similarities between N-CAM and immunoglobulins indicate that both evolved from a common precursor. Because insects do not have immunoglobulins but do have CAMs similar to vertebrate N-CAMs, it is likely that the vertebrate immunoglobulins evolved from an N-CAM-like cell adhesion molecule (Grenningloh et al., 1990). In addition to the immunoglobulinlike domains, the region located on the C-terminal side of the fifth immunoglobulin domain seems especially important for the orientation of growing axons (Pollerberg and Beck-Sickinger, 1993).

A unique characteristic of N-CAM is the molecule's high content of *polysialic acid (PSA),* a negatively charged

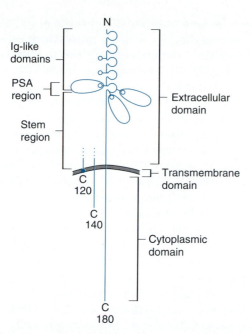

Figure 25.11 Schematic representation of N-CAM. The major isoforms of N-CAM (with molecules measuring 180, 140, and 120 kd) differ only in their C-terminal part; the N-terminal parts are identical. The two long isoforms have transmembrane and cytoplasmic domains; the shortest isoform is linked to the outer phospholipid layer of the plasma membrane. The large extracellular domain, beginning at the N terminus, has several carbohydrate chains (small circles) and five small immunoglobulinlike domains (loops). Three polysialic acid (PSA) chains (depicted as clubs) are attached to carbohydrates in the area of the fifth immunoglobulin domain.

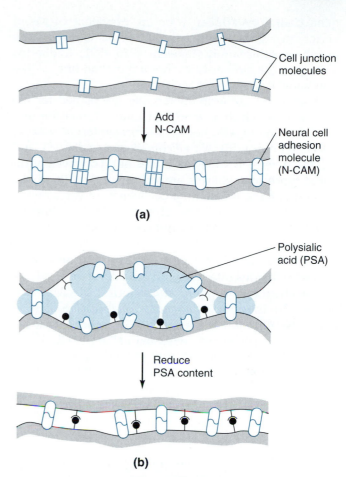

Figure 25.12 Two mechanisms for the regulation of cell-cell interactions by N-CAM. **(a)** N-CAM forms with low polysialic acid (PSA) content allow cells to adhere by homotypic binding of the N-CAM molecules themselves and also facilitate cellular interactions between other cell adhesion molecules or between cell junction molecules (rectangles). **(b)** N-CAM forms with high PSA content (color shading) prevent cells from adhering to each other. Subsequent removal of PSA permits homotypic binding of N-CAM molecules and interactions between other cell adhesion molecules (balls and sockets).

polysaccharide present primarily, if not exclusively, in N-CAM (Rutishauser et al., 1988). Three long PSA chains are attached to carbohydrate moieties in the region of the fifth immunoglobulin domain (Fig. 25.11). These chains vary in length during development and account for 10 to 30{pct} of the total weight of the N-CAM. High PSA content *interferes* with cell adhesion mediated by N-CAM itself and also by other types of cell adhesion molecules (Rutishauser, 1992). In this sense, N-CAM functions as a cell adhesion molecule or as an anti-cell adhesion molecule, depending on PSA content (Fig. 25.12).

Generally, embryos have high-PSA forms of N-CAM, which facilitate cellular movements, whereas adult tissues express low-PSA forms, which promote more stable cell adhesion (Rieger et al., 1985). In the brains of normal mice, the transition from high-PSA to low-PSA N-CAMs is completed 21 days after birth. In contrast, homozygous *staggerer* mutants do not complete this transition on time and also experience failures in synapse formation and severe neurological disorders (Edelman and Chuong, 1982). These observations suggest that the neurological disorders in this mutant may result from failures in critical cell adhesion events due to the extended presence of high-PSA N-CAMs.

Since N-CAM was discovered, researchers have characterized several other CAMs that have multiple immunoglobulinlike and fibronectinlike domains. Members of this family include *Ng-CAM* and *L1* in vertebrates and *neuroglian* and *fasciclins* in arthropods (Grenningloh et al., 1990).

Cadherins Mediate Ca²⁺-Dependent Cell Adhesion

The *cadherins* are a family of CAMs defined by their dependence on calcium ions (Takeichi, 1977). In the absence of Ca^{2+}, they undergo a conformational change that renders them susceptible to breakdown by proteases. The biological significance of this Ca^{2+} dependence is not known; other CAMs are not Ca^{2+}-dependent. Cadherins appear to be the prevalent CAMs in vertebrates: cells in virtually all tissues express some members of this family (Takeichi, 1988, 1991). Cadherins seem to work primarily by homotypic binding, as the immunoglobulinlike CAMs do.

Four main cadherin subfamilies are named for the tissues in which they were first found: *E-cadherin* (epithelial cadherin, also known as *uvomorulin*), *P-cadherin* (placental cadherin), *N-cadherin* (neural cadherin), and *L-CAM* (liver cell adhesion molecule). However, additional types of cadherins are still being identified. All cadherins have the same basic structure (Fig. 25.13). They are transmembrane glycoproteins of approximately 125 kd, with the cytoplasmic domain ending in the carboxyl terminus.

Cadherins are *necessary* for cell adhesion: antibodies against cadherins interfere with cell adhesion. Similarly, the inactivation of maternal cadherin mRNA in frog oocytes by injection of *antisense* oligonucleotides reduces the adhesion of blastomeres in embryos that develop from such oocytes (Heasman et al., 1994).

Cadherins are also *sufficient* to confer adhesiveness on cells that would otherwise not adhere. This has been demonstrated in genetically transformed mouse L cells, which normally do not express cell adhesion molecules

or adhere to one another (Nagafuchi et al., 1987). A transgene encoding E-cadherin causes such cells to aggregate in a Ca^{2+}-dependent way and to form compact colonies (Fig. 25.14). The presence or absence of E-cadherin also changes the affinity of a given cell type for other cells. In reconstituted embryonic lung tissue, L cells expressing an E-cadherin transgene associate with epithelial cells that naturally express E-cadherin, whereas untransfected L cells associate with mesenchymal cells (Nose et al., 1988). Moreover, when transfected cells expressing P-cadherin in substantially differing amounts are mixed, the cells expressing more cadherin are enveloped by the cells expressing less (Steinberg and Takeichi, 1994). Thus, differences in the expression levels of cadherin can mediate—at least in part—the ability of different cell types to organize themselves in concentric spheres that was demonstrated so strikingly by classic cell reaggregation experiments.

To determine which domains in the cadherin molecule are responsible for its binding specificity, Akinao Nose and his colleagues (1990) constructed chimeric

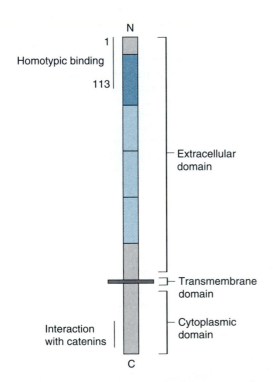

Figure 25.13 The basic structure of cadherins. In the extracellular domain, the first 113 amino acids from the N terminus are responsible for the binding specificity of the molecule. The segments shaded with light color represent three repeated regions in the molecule. In the cytoplasmic domain, a region close to the carboxyl terminus interacts with proteins called catenins, which link cadherins to the cytoskeleton.

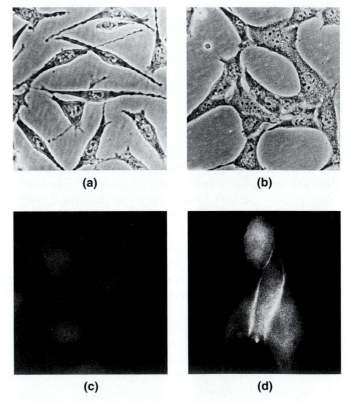

Figure 25.14 Genetic transformation of mouse L cells with a cDNA for E-cadherin. The top two photographs were taken in transmitted light. The bottom photographs show fluorescence after immunostaining for E-cadherin. **(a, c)** Normal L cells, which produce little cadherin and do not adhere. **(b, d)** L cells transformed with E-cadherin cDNA. Note the formation of cell clusters and the strong staining for E-cadherin where cells touch.

molecules consisting of E-cadherin and P-cadherin domains. They found that the binding specificity resides within the first 113 amino acids of the extracellular domain (Fig. 25.13). Most of these amino acids are conserved between different cadherins, and changing just a few of the variable ones alters the specificity of cadherin binding. Also, antibodies capable of blocking cadherin activity recognize epitopes close to the amino terminus.

The cytoplasmic domain of cadherins is the most conserved part of the molecule. This fact suggests that the cytoplasmic domain interacts with other molecules and that this interaction is critical to cadherin function. Indeed, *deletion mapping* by transfecting cells with truncated cadherin cDNAs showed that a region close to the C terminus is necessary for cell adhesion (Fig. 25.13). Cadherins missing this region do insert into the plasma membrane but do not accumulate near cell junctions like normal cadherins (Nagafuchi and Takeichi, 1988). Such truncated cadherins are also easier to extract from cells than complete cadherins. Furthermore, truncated cadherins are not associated with *catenins,* cytoplasmic linker proteins that join cadherins to cytoskeletal components (McCrea and Gumbiner, 1991; S. Schneider et al., 1993). Taken together, these observations indicate that catenins and possibly other proteins link cadherins to the cytoskeleton as shown in Figure 25.9.

The role of cytoplasmic linker proteins in cadherin-mediated cell adhesion raises the possibility that these linker proteins may be present in limiting amounts and so have a regulatory function. To test this possibility, Chris Kintner (1992) injected mRNA encoding an N-cadherin without a functional extracellular domain into one blastomere of *Xenopus* embryos at the 4-cell stage. The truncated N-cadherin transcribed from this mRNA was expected to bind to cytoplasmic linker proteins without contributing to cell adhesion. If the linker proteins were in limiting supply, then the nonfunctional cadherin should displace its functional counterpart from linker proteins, resulting in reduced cell adhesion. Indeed, neural tissue derived from the injected blastomere was severely disturbed in its morphology (Fig. 25.15). Subsequent deletion mapping of the cytoplasmic domain of the nonfunctional cadherin attributed the antiadhesion effect to regions involved in catenin binding. These results suggest that cell adhesion can be regulated by interactions between cadherins and cytoplasmic components, such as catenins, and that this regulation may affect tissue integrity and morphogenesis.

Lectins and Glycosyltransferases Act as Cell Adhesion Molecules

A third family of CAMs is made up of two types of proteins: *lectins* and *glycosyltransferases*. Both bind *heterotypically* to *oligosaccharides,* short chains of sugar molecules extending from other cell surface glycoproteins.

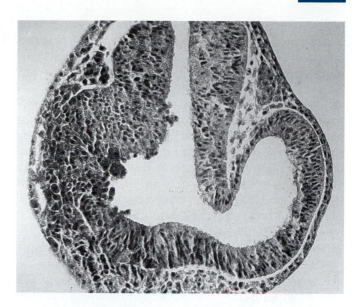

Figure 25.15 The role of cytoplasmic linker proteins in cadherin-mediated cell adhesion. This photograph shows a transverse section of the brain of a *Xenopus* embryo in which one blastomere at the 4-cell stage was injected with mRNA encoding a truncated N-cadherin that had no functional extracellular domain. In the right half, brain development is normal; the section passes through the forming eye vesicle. A tracer coinjected with the mRNA reveals that the left half is derived from the injected blastomere. Here the neural tissue is severely disturbed, with detaching cells entering the brain ventricle. No such disturbances were observed in embryos injected with mRNAs encoding similar truncated cadherins with additional deletions in the cytoplasmic domain. Thus, the truncated N-cadherin must compete with normal cadherin for a limiting cytoplasmic component, presumably linker proteins that anchor cadherin to the cytoskeleton.

Lectins bind to their oligosaccharide counterparts without modifying them. Lectin binding plays an important role when eukaryotes are infected by viruses, bacteria, or protozoa, and when lymphocytes fight infections (Sharon and Lis, 1993). Lectins are also present on certain neurons, although their function there is unknown. *Glycosyltransferases* are enzymes known primarily from their role in the endoplasmic reticulum and Golgi apparatus, where they transfer monosaccharides from activated donor molecules to the oligosaccharide moieties of glycoproteins. When such enzymes were also found on cell surfaces, Saul Roseman (1970) proposed that they function as cell adhesion molecules. Initially a controversial idea, this proposal has since been supported by much evidence (Shur, 1991).

Most attention has focused on the membrane localization and function of *galactosyltransferases (GalTases),* a subfamily of glycosyltransferases that transfer galactose from a nucleotide donor (such as uridinediphosphate galactose, or UDPGal) to an oligosaccharide. A simple model proposed for the function of GalTase as a CAM is shown in Figure 25.16. Presumably, GalTase adheres to an oligosaccharide moiety of a glycoprotein in the plasma membrane of an adjacent cell. (Alternatively, the oligosaccharide may be part of an extracellular matrix, as in the mouse zona pellucida.) The adhesion persists as long as no UDPGal or other galactose donor is present. If such a donor is added to adhering cells in vitro, the GalTase performs its enzymatic function of transferring galactose to its substrate, and then the cells detach. Whether this reaction normally occurs in vivo is not known; it would seem to be an elegant way of making cell adhesion reversible, but the presence of nucleotide sugars in the extracellular space is not well documented.

Like most other CAMs, GalTase is a transmembrane protein, although in contrast to CAMs and cadherins, the cytoplasmic domain of GalTase begins with the amino terminus and its extracellular domain ends with the carboxyl terminus (Shaper et al., 1988). The mouse GalTase gene encodes two mRNAs, which are translated into two different proteins. One isoform, 399 amino acids long, seems to be targeted to the plasma membrane, where it functions as a CAM; while the other isoform, which is 13 amino acids shorter, is targeted to the Golgi apparatus, where it functions in protein glycosylation (Lopez et al., 1991).

The function of GalTase as a CAM is particularly well established in mouse sperm, where it interacts with the ZP3 protein of the egg (D. J. Miller et al., 1992; see also Chapter 4). Also, GalTase has been localized by immunostaining to cultured mouse F9 cells, preimplantation embryos, and uterine epithelial cells (Shur, 1991). In all these instances, antibodies and Fab fragments directed against GalTase inhibit cell adhesion and reverse adhesions already formed.

As expected of a CAM, GalTase appears to be linked to the cytoskeleton. This is indicated by the results of experiments in which cells were extracted with detergents that remove phospholipids from all cell membranes but leave the cytoskeleton intact (Eckstein and Shur, 1992). In such preparations, most of the GalTase remained associated with the cytoskeleton instead of being washed out with the phospholipids. A linkage between GalTase and the cytoskeleton is also suggested by cell transformations with cDNAs encoding truncated GalTase without extracellular domains (S. C. Evans et al., 1993). The result was the same as in the analogous experiments with truncated cadherin described earlier: the transformed cells lost their adhesiveness to one another and to extracellular substrates.

Most of the CAMs described in this section act as single transmembrane molecules, are linked to the cytoskeleton, and mediate a selective adhesiveness of relatively low strength. The weak binding allows cells to easily join and dissociate again in morphogenetic movements. In addition to the three families of CAMs described here, there are other molecules that have similar properties. Some *growth factors* and their receptors are also membrane-bound (see Chapter 29). Expression of these molecules from transgenes can mediate selective cell adhesion, at least in vitro. Thus, some intercellular signal molecules also function as cell adhesion molecules. Conversely, the CAMs discussed here do more than just hold cells together: they are part of those signal chains that link gene expression with morphogenesis, as will be discussed in the following section.

The Role of Cell Adhesion Molecules in Morphogenesis

As we explore the role of CAMs in developing embryos, we must remember that morphogenesis is a complex phenomenon involving cell movements, local control of cell proliferation, cell shape changes, changes in CAMs, and changes in the composition of extracellular material (Bard, 1990; see also Chapters 2, 10, 26, and 29 of this text). We will focus here on the involvement of the CAMs described in the previous section. Are their expression patterns consistent with the view that they are the critical agents of cell adhesion? Do perturbations of CAMs have the expected effects on development? What are the roles of CAMs in cell movement and cell growth? How is the expression of CAMs regulated, and do CAMs have regulatory functions themselves?

CAM Expression Is Correlated with Cell Fates

The accumulation of CAMs is developmentally regulated, as shown by *in situ hybridization* or *immunostaining* (see Methods 4.1 and 15.1). CAMs are present in certain embryonic areas, and at specific stages, and their expression patterns are generally dynamic. These pat-

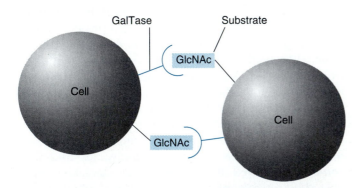

Figure 25.16 Model for cell adhesion mediated by a galactosyltransferase (GalTase) located on the cell surface. The GalTase located in the plasma membrane of one cell adheres to one of its substrates, N-acetylglucosamine (GlcNAc), which is part of a glycoprotein in the plasma membrane of another cell. Structural limitations prevent GalTase from binding to substrates on the same cell.

terns change in particular during critical steps that cells take toward their eventual fates. Thus, cells with different fates often express different CAMs. For instance, in a chicken embryo before gastrulation, the epiblast cells show an intricate pattern of N-CAM and E-cadherin expression (Edelman, 1984). The highest concentration of N-CAM is found in prospective neural plate, while lower concentrations are found in future notochord, somites, and lateral mesoderm (Fig. 25.17). Prospective urinary tract cells express both N-CAM and E-cadherin. All these areas are surrounded by endoderm and prospective epidermis, which express E-cadherin. In each area, CAM is expressed in a dynamic pattern. For instance, the notochord at first does not stain for N-CAM, then it stains intensely, and finally the stain disappears.

Many CAM expression patterns change during *induction* events, indicating that the accumulation of specific CAMs is an early step in the determination of embryonic tissues. In *Xenopus* embryos, neural induction triggers N-cadherin synthesis in prospective neural epithelium prior to neurulation (Detrick et al., 1990). Similarly, the expression of N-CAM is an early response to neural induction (Kintner and Melton, 1987). N-CAM mRNA is first detected at the early neurula stage. In situ hybridization shows that N-CAM expression is restricted to the neural plate (Fig. 25.18) and later to the neural tube. N-CAM is also synthesized when animal cap and vegetal base from *Xenopus* blastulae are cultured together, a situation that leads first to mesoderm induction and then to neural induction. In control experiments in which animal cap and vegetal base are cultured separately, no neural tissue forms and no N-CAM is synthesized. Therefore, N-CAM synthesis is a reliable marker of neural induction.

Expression of Distinct CAMs Generates Boundaries between Tissues

Embryonic tissues often express distinct CAMs: the patterns of N-CAM and E-cadherin accumulation in the

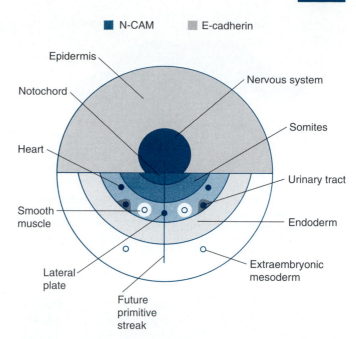

Figure 25.17 Fate map of a chicken embryo superimposed on an expression map of two CAMs: N-CAM (color) and E-cadherin (gray). The map shows a dorsal view of the epiblast before primitive streak formation. The vertical line indicates the position of the future primitive streak. The labels indicate the fates of different epiblast areas. Most areas express either E-cadherin or N-CAM. Prospective urinary tract expresses both CAMs, while prospective smooth muscle and extraembryonic mesoderm express neither CAM.

chicken epiblast are one striking example (Fig. 25.17). Later, during neurulation in the chicken, the neural tube expresses N-CAM and N-cadherin while the surrounding epidermis expresses E-cadherin. The neural crest cells stop expressing all cadherins when they separate from the ectodermal epithelium and become migratory (Takeichi, 1988). Thus, the three embryonic tissues that originate from the ectoderm exhibit different patterns of CAM expression as they separate from one another

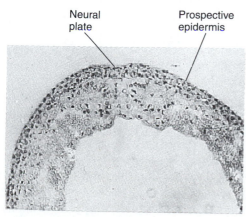

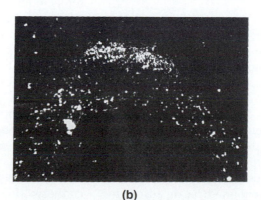

(a) (b)

Figure 25.18 Photomicrographs showing restriction of N-CAM expression to the neural plate: **(a)** transmitted light; **(b)** dark-field illumination. Transverse sections of *Xenopus* embryos are shown, with dorsal side up. In situ hybridization (see Methods 15.1) with a radiolabeled probe for N-CAM mRNA reveals that this mRNA accumulates in the neural plate but is absent from prospective epidermis.

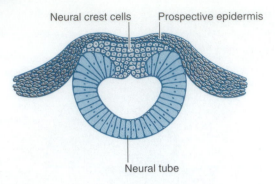

Figure 25.19 Differential expression of cadherin subfamilies in ectoderm during neurulation in chickens. The prospective epidermis expresses E-cadherin, the neural tube expresses N-cadherin, and the neural crest cells express neither of these CAMs.

(Fig. 25.19). A similar dichotomy in CAM expression exists in the follicles of developing feathers (Chuong and Edelman, 1985a, b). Such follicles, like hair follicles (see Fig. 12.42), consist of an ectodermal epithelium that expresses E-cadherin and a mesodermally derived mesenchyme that expresses N-CAM. Throughout the complex process of feather formation, these two tissues express clearly distinct types of CAMs.

These observations suggest that the expression of specific CAMs may actually *generate and maintain* boundaries between different embryonic tissues. Such boundaries may serve several important functions. First, they may prevent mixing between cells of different fates, as observed for *compartment* boundaries in insects and mammals (see Chapters 21 and 22). Second, boundaries may reduce the transfer of signals between tissues with different fates and allow them to be governed by separate genetic control circuits, as postulated by the compartment hypothesis. Third, boundaries may be lines of reduced cell adhesion that facilitate the separation of tissues in morphogenetic movements. To test some of these possibilities, investigators have forced the expression of certain CAMs in areas where they are not normally produced.

▼

Toshihiko Fujimori and his colleagues (1990) injected N-cadherin mRNA into one blastomere of *Xenopus* embryos at the 2-cell stage. The exogenous mRNA was later translated in various regions of the treated embryos, generally in a mosaic distribution pattern. In epidermis, overexpression of N-cadherin converted the cell shape from squamous to cuboidal or even to the columnar shape normally seen in neural plate. When exogenous N-cadherin was expressed in both the neural tube and the adjacent dorsal epidermis, the two tissues did not separate from each other, as they normally do for neural tube closure, but remained partly fused. In

normal development, the epidermis that separates from the neural tube does not express N-cadherin (Fig. 25.19). Therefore the failure of the two tissues to separate in the treated embryos must be ascribed to the faulty expression of N-cadherin in the cells that would normally form epidermis. This means that the normal lack of N-cadherin expression in prospective epidermis is necessary for development of the normal squamous shape of epidermal cell and for the separation of epidermis from the neural tube.

Further observations showed that even quantitative differences in N-cadherin expression generated demarcation lines between cells. When high ectopic expression of N-cadherin occurred in parts of the neural tube, the cells in these regions condensed into clumps and formed sharp boundaries with neighboring cells that expressed only endogenous cadherin. Thus, not only the presence of N-cadherin but also its concentration is important for normal neurulation.

The results of Fujimori and coworkers (1990) indicate that, indeed, qualitative and quantitative differences in CAM expression generate boundaries between embryonic tissues that prevent cell mixing and facilitate the separation of these tissues in morphogenetic movements. Similar results were obtained by R. Jennifer Detrick and her colleagues (1990). They found that in *Xenopus* embryos into which N-cadherin mRNA had been injected, the blastoderm cells expressing N-cadherin stayed together as coherent patches and did not mix with the other blastoderm cells to the extent normal for untreated embryos.

CAMs Allow Cells to Separate and Reattach

In several morphogenetic events, cells that begin as part of an epithelium break away from their neighbors and migrate individually. In sea urchins, mesenchyme cells separate from the vegetal plate and ingress during gastrulation; in vertebrates, neural crest cells separate from the ectodermal epithelium, and somites release mesenchymal cells that form *sclerotomes*. In all these cases, the cells that break away lose adhesion to their former neighbors. The changes in adhesiveness in the primary mesenchyme cells in sea urchins have actually been measured (Table 25.1).

If CAMs account for loss of cell adhesion, they should disappear or be modified when cells break away from epithelia. The disappearance of CAMs has actually been observed when somitic cells and neural crest cells become migratory (Takeichi, 1988). As long as they are tight epithelial spheres, somites strongly express N-cadherin (Fig. 25.20). However, when parts of each

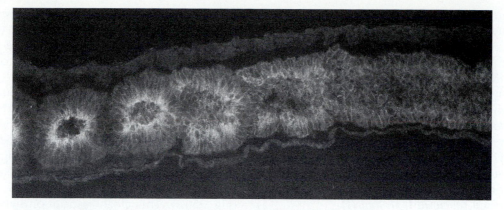

Figure 25.20 N-cadherin expression during somite formation in the chicken embryo. The photograph shows a paramedian section at the somite level, with anterior to the left. The somites are formed as Hensen's node regresses from anterior to posterior (see Chapter 10). Somite formation is therefore more advanced in the anterior of the embryo. The formation of somites as epithelial spheres correlates with the accumulation of N-cadherin, revealed by immunostaining (see Methods 4.1) with a fluorescent secondary antibody.

somite are converted into mesenchymal cells and form sclerotomes, they cease production of N-cadherin.

During the development of neural crest cells, the expression of N-CAM also changes as expected. In chicken embryos, neural crest cells express N-CAM until they detach from the epithelium (Thiery et al., 1982). While migratory, they fail to bind to antibodies against N-CAM, indicating that N-CAM is no longer present or has been modified to the point of losing its antigenicity. After the migration has ceased, the neural crest cell derivatives again express N-CAM and N-cadherin. Conceivably, the microenvironment at their destination elicits the reexpression of these CAMs so that the cells will become trapped. The neural crest cells of *Xenopus* show the same pattern of CAM expression before and after migration and nonexpression during migration (Balak et al., 1987).

CAMs Guide the Migration of Nerve Cells and Axons

In the developing nervous system, neuroblasts migrate to specific destinations, and neurons form highly specific synaptic connections. CAMs are critical in the guided cell migration and axonal growth that generate most of the neuronal connections (Hynes and Lander, 1992). For instance, migrating neural crest cells, which do not express cadherins or N-CAM, do utilize GalTase (Hathaway and Shur, 1992). Immunostaining reveals GalTase on migrating neural crest cells and on the basal lamina that surrounds the neural tube. Moreover, antibodies recognizing GalTase inhibit neural crest cell migration in chickens, whereas immunoglobulins from preimmune serum do not have this effect. Similarly, a monoclonal antibody against a specific extracellular re-

gion of N-CAM interferes with the development of chicken retina in culture (Pollerberg and Beck-Sickinger, 1993); the antibody perturbs the orientation of axons within the retina and away from the retina.

A promising system for studies on neuronal recognition is the central nervous system (CNS) of insects. Large insects such as the grasshopper *Schistocerca americana* offer relatively large, individually identifiable neurons for experimental manipulation. *Drosophila* is much smaller, but the knowledge of its genetics is a powerful tool. Fortunately, the anatomy of the nervous systems of the two insects, and the CAMs guiding their development, are very similar, so that researchers can benefit from the advantages of both systems.

The insect CNS develops from the neuroepithelium, which is located along the ventral surface of the embryo and forms a pair of ganglia in each segment. Axons growing out from the neurons connect the ganglia, so that the CNS comes to look like a rope ladder. The first neurons to send out their axons are called *pioneer neurons*. The *growth cones* (see Fig. 2.19) of their axons are guided anteriorly, posteriorly, or laterally by arrays of special glial cells. As these pioneer axons contact each other, they establish a ladder-shaped scaffold along which all other axons will grow. As new axons are added to existing ones, they form bundles, or *fascicles*, of closely associated axons. Each fascicle is formed by the same axons in all individuals of a species. This means that all nonpioneer axons select one specific fascicle from the many available fascicles to guide their growth, an ability known as *selective fasciculation*. Indeed, if an axon cannot find its appropriate fascicle after experimental perturbation, it stops growing instead of joining one of the nearby fascicles (Bastiani et al., 1986). These observations led to the formulation of the *labeled pathways hypothesis,* which states that axon pathways are differentially labeled by CAMs in such a way that axonal growth cones can distinguish among them.

The search for CAMs that might function as the guiding labels for growing axons has focused on proteins that are synthesized specifically in certain fascicles. Three such immunoglobulinlike CAMs were isolated from *Schistocerca* and *Drosophila,* and the genes encoding them were cloned. Immunostaining showed that these proteins were made in different but overlapping sets of fascicles, especially on fasciculating axons and growth cones. Because of these properties, the CAMs were termed *fasciclin I, II, and III.*

In *Drosophila* mutants that are deficient in the *fasciclin II* gene, the pioneer neurons for certain fascicles stall and the fascicle is not formed properly (Grenningloh et al., 1991). Mutations in the other fasciclin genes cause misrouting of axons from sensory neurons in the wing (Whitlock, 1993). In addition, ectopic sensory neurons develop in the wing. These results indicate that the fasciclins are required for neuron development and axon guidance. Whether fasciclins are sufficient to guide certain neurons in specific directions remains to be established.

CAMs Facilitate the Formation of Cell Junctions

The forces holding CAMs together are weak. Like the hooks and loops of Velcro, they are designed for quick attachment and detachment, and their strength is based on large numbers. Thus, the main function of CAMs seems to be to provide reversible cell adhesions that allow for cell migration and intercalation.

More stable connections between cells are provided by gap junctions, desmosomes, and tight junctions (see Chapter 2). These junctions are large, complex structures, designed not only to hold cells together but also to form channels of communication and tightly seal off the internal intracellular spaces of an organism from the external environment. Before such stable junctions are established, cells typically adhere to each other by means of cell adhesion molecules. Two examples will serve to illustrate this point.

A dramatic event in early mammalian development is *compaction,* when blastomeres become polarized and form tight junctions near their apical surfaces (see Chapter 5). Before compaction, mouse embryos express E-cadherin (also known as uvomorulin, in this context). In embryos immunostained for E-cadherin, staining is especially strong where the blastomeres touch one another (Johnson et al., 1986; Vestweber et al., 1987). When raised in media with antibodies to E-cadherin, embryos fail to undergo compaction; no tight junctions form, and their blastomeres do not become polarized. Similar effects are noticeable after incubation with antibodies against GalTase (Bayna et al., 1988). It appears that both CAMs are required for the formation of tight junctions

and that tight junctions in turn are necessary for blastomeres to become polarized, so that they acquire an apical surface with certain membrane proteins facing the outside and different surfaces with other membrane proteins facing the inside.

Polarized cells connected by tight junctions are also the hallmark of adult epithelia. Parallel to the tight junctions, there is often a *zonula adherens,* characterized by rings of microfilaments that circle beneath the plasma membrane of each cell like a purse string (see Fig. 2.21a). Across the cell membrane from the microfilament rings, the intercellular space stains positive for E-cadherin (Fig. 25.21). This close association between microfilament rings on the cytoplasmic side and E-cadherin on the extracellular side is resistant to *cytochalasin D* and to cell extraction with detergents (Hirano et al., 1987). These observations indicate that the cytoplasmic domains of the E-cadherins are linked to the microfilaments, presumably by linker proteins such as the catenins described earlier (Fig. 25.13). Thus, the zonula adherens appears to be a specialized case of general cell adhesion based on E-cadherin. Indeed, a zonula adherens forms rapidly, and dependent on E-cadherin, when isolated epithelial cells are allowed to make contact again (Gumbiner, 1990).

CAMs Act as Morphoregulatory Molecules

According to the compartment hypothesis, specific combinations of *selector genes* control *realizator genes,* which in turn regulate morphogenesis and cell differentiation in each compartment (see Chapter 21). Since cells adhere more strongly to cells from the same compartment than to cells from other compartments, CAM genes should be among the realizator genes. This concept was extended by Gerald Edelman (1988), who described CAMs as *morphoregulatory molecules* mediating between gene activities and morphogenetic events.

A simplified four-step model of the morphoregulatory role of CAMs is shown in Figure 25.22. First, selector genes regulate the expression of CAM genes, as well as other realizator genes, by transcriptional control. Second, CAM gene expression results in synthesis of CAMs, which control cell adhesiveness. Third, cell adhesion affects cell shape and cell movements, two of the basic cellular properties that influence morphogenesis. The fourth and most intriguing step in the morphoregulatory cycle is feedback from cell adhesion, or from morphogenetic events, to the activity of selector genes. This step may involve changes in *inter*cellular signals as morphogenetic movements generate new contacts between cells that were originally separated. Alternatively, new cell shapes or movements may change *intra*cellular signaling by triggering a response in receptors or ion channels in the plasma membrane that are sensitive to mechanical stress.

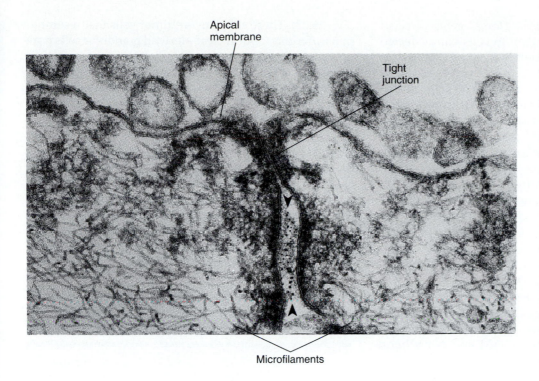

Apical membrane

Tight junction

Microfilaments

Figure 25.21 Presence of E-cadherin in the zonula adherens of mouse intestinal epithelium. In this transmission electron micrograph, two abutting epithelial cells are linked by a tight junction. Beneath the tight junction is the zonula adherens, characterized by rings of microfilaments (mi) circling the inside of each cell like a purse string (refer to Fig. 2.21a). The extracellular space across the cell membranes from the microfilament rings is filled with E-cadherin (between arrowheads). This is shown here by immunostaining using a primary antibody against E-cadherin and a secondary antibody that is conjugated with fine gold particles. These particles absorb electrons and thus generate black dots in electron micrographs.

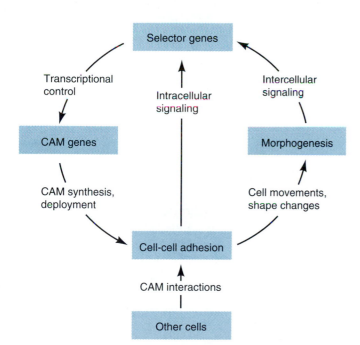

Figure 25.22 Model showing the possible role of CAMs in a morphoregulatory cycle. Selector genes control the expression of CAM genes. The synthesis and deployment of CAMs modulates cell-cell adhesion, which is also affected by the CAMs of other cells. Altered cell-cell adhesion causes changes in cell shape and cell movements, two major cell attributes that affect morphogenesis. Morphogenetic movements may in turn generate new cell contacts, which set the stage for the exchange of new intercellular signals. In addition, altered cell shapes and movement may affect intracellular signaling through membrane receptors or ion channels that respond to mechanical stimuli. Both intercellular and intracellular signals feed back into the expression of selector genes, thus starting a new morphoregulatory cycle.

What evidence is there that CAMs act as morphoregulatory molecules? Two steps in the proposed morphoregulatory cycle—the synthesis of CAMs from their genes and the effects of cell adhesion on morphogenetic events—have already been discussed. The remainder of this chapter will deal with the other two steps in the cycle: the transcriptional control of CAM genes by selector genes, and the feedback of cell adhesion and morphogenetic events to selector gene activity.

Direct transcriptional control of CAM genes by selector genes implies that transcription factors encoded by selector genes bind to CAM gene enhancers. Because at least some selector genes contain a homeobox, it is a reasonable hypothesis that certain enhancer regions in CAM genes may be bound by homeodomain proteins. This hypothesis is in accord with the observation that retinoic acid, a teratogenic agent acting on homeobox genes, changes cell affinity, as shown by the experiments of Crawford and Stocum (1988) described earlier.

Further evidence for CAM gene control by homeodomain proteins was obtained by Frederick Jones and his colleagues (1992a). They cotransfected cultured human 3T3 cells with two plasmids in an experiment similar to the one illustrated in Figure 15.22. The *reporter plasmid*, containing the first exon and 59 flanking sequences of the mouse N-CAM gene and the bacterial chloramphenicol acetyltransferase (CAT) gene, directed the synthesis of CAT under control of N-CAM promoter and enhancer sequences. The *effector plasmid*, containing a viral promoter and the *Xenopus Hoxb-9* gene, directed the synthesis of Hoxb-9 protein. In cells cotransfected with both plasmids, CAT activity was much higher than in cells transfected only with reporter plasmid. However, when an additional effector plasmid directing the synthesis of *Xenopus* Hoxb-8 protein was introduced, the CAT activity went back down. These results show that the mouse N-CAM gene contains enhancer regions modulating this gene's expression positively or negatively under the influence of two different homeodomain-containing proteins from *Xenopus*, at least in the context of other transcription factors present in the human host cells. Whether such regulation takes place normally in the mouse or *Xenopus* remains to be seen.

The effects of cell adhesion on selector gene activity via intracellular signaling were demonstrated in an experiment by Patrick Doherty and his colleagues (1991). These investigators observed a given type of test cell as it interacted with other cells expressing N-CAM, or N-cadherin, or neither of the two. The test cells were PC 12 cells, which express both N-CAM and N-cadherin

(Fig. 25.23). These cells are rat tumor cells that resemble *chromaffin cells* (neural crest–derived glandular cells normally found in the adrenal medulla). With nerve growth factor and other stimuli, these cells can be converted to a neuronal phenotype with extended *neurites*. PC 12 cells were cultured on a layer of human 3T3 cells, which do not by themselves produce any stimuli that would convert PC 12 cells from the chromaffin to the neuronal phenotype. However, PC 12 cells cultured on 3T3 cells that had been transfected with N-CAM or N-cadherin cDNA converted to the neuronal phenotype. The conversion was nearly independent of transcription but could be blocked by antibodies directed against N-CAM (or N-cadherin) and by drugs that are known to interfere with Ca^{2+} channels. The investigators concluded that the PC 12 cells interact homotypically with the transformed 3T3 cells, using whatever CAM the 3T3 cells are expressing. This interaction triggers a Ca^{2+}-dependent change in intracellular signaling that promotes the conversion to the neuronal phenotype.

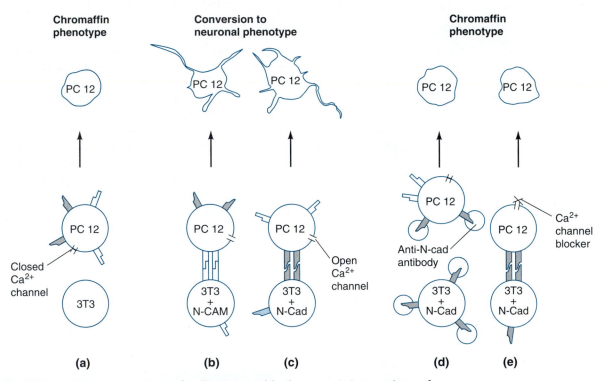

Figure 25.23 Phenotypical conversion of cells triggered by homotypic interactions of CAMs. A rat tumor cell line, PC 12, is grown on a layer of human cells, 3T3. The PC 12 cells express both N-CAM and N-cadherin, whereas the 3T3 cells by themselves do not express either of these CAMs. **(a)** PC 12 cells grown on nontransformed 3T3 cells resemble chromaffin cells, which are normally found in the adrenal medulla. **(b, c)** If PC 12 cells are grown on 3T3 cells transfected with cDNA encoding N-CAM or N-cadherin, the PC 12 cells are converted to a neuronal phenotype characterized by neurites. **(d, e)** The conversion is inhibited by antibodies to the CAM expressed by the 3T3 cells and by drugs that block the opening of Ca^{2+} channels. Apparently, the homotypic interaction of PC 12 cells with N-CAM or N-cadherin on the transfected 3T3 cells causes a Ca^{2+}-dependent change in intracellular signaling that promotes the phenotypical conversion.

The results described here afford us only a first glimpse at the morphoregulatory functions of CAMs. Clearly, these molecules are central to the process of generating a three-dimensional organism from a linear string of genetic information encoded in DNA and from the spatial clues provided by the egg architecture. Without the reliable function of CAMs, morphological features like the shapes of ears and noses would not be heritable. Obtaining a more complete picture of CAMs as morphoregulatory molecules will take much more research. This research will also include another class of morphoregulatory molecules, the substrate adhesion molecules, which will be discussed in the following chapter.

SUMMARY

When cells from different embryonic tissues are mixed, they segregate and form concentric arrangements in which the most adhesive cells occupy the central position. Morphogenetic movements such as epiboly and ingression are associated with measurable changes in cell adhesiveness. Cell adhesiveness also changes along the proximodistal axis of a regenerating limb.

Cell adhesion molecules are single molecules in the plasma membrane by means of which cells adhere quickly, selectively, and relatively weakly. The weak binding allows cells to dissociate again easily. A typical cell adhesion molecule (CAM) has three domains: a large extracellular binding domain that adheres to another CAM, a transmembrane domain, and a cytoplasmic domain. The cytoplasmic domain is connected to the cytoskeleton by linker proteins. Some CAMs, however, have only an extracellular domain that is attached to the plasma membrane by a phosphatidylinositol anchor.

There are three major families of CAMs. One family, which includes N-CAM, is immunoglobulinlike; its polysialic acid chains, which are attached to the extracellular domain, prevent or permit homotypic binding, depending on their quantity. The second family, called cadherins, includes the most prevalent CAMs. They are Ca^{2+}-dependent and also bind homotypically. A third family of CAMs, the lectins and glycosyltransferases, bind heterotypically to oligosaccharide moieties of glycoproteins.

The expression patterns of CAMs are correlated with cell fates. They change dynamically, especially as cells change from epithelial to mesenchymal or vice versa. CAM expression also changes during inductive events, indicating that CAM expression is closely associated with cell determination. Qualitative or quantitative differences in CAM expression generate boundaries between embryonic tissues. Regulating cell attachment and detachment, CAMs are involved in cell migration and growth cone extension. Axons and growth cones appear to be labeled by CAMs that allow them to fasciculate selectively. Adhesion by means of CAMs seems to precede the formation of more elaborate, stable junctions between cells.

CAMs, particularly the ones expressed in developing nervous systems, are very well conserved between invertebrates and vertebrates. As morphoregulatory molecules, they play key roles in the regulatory circuits by which gene expression controls morphogenetic events and vice versa. CAM expression may be controlled by homeotic genes, and CAM functions may in turn affect certain gene activities through intracellular and intercellular signals.

SUGGESTED READINGS

Cunningham, B. A. 1991. Cell adhesion molecules and the regulation of development. *American Journal of Obstetrics and Gynecology* 164:939–948.

Hynes, R. O., and A. D. Lander. 1992. Contact and adhesive specificities in the associations, migrations, and targeting of cells and axons. *Cell* 68:303–322.

Shur, B. D. 1991. Cell surface β-1,4-galactosyltransferase: Twenty years later. *Glycobiology* 1:563–575.

Takeichi, M. 1991. Cadherin cell adhesion receptors as a morphogenetic regulator. *Science* 251:1451–1455.

EXTRACELLULAR MATRIX AND MORPHOGENESIS

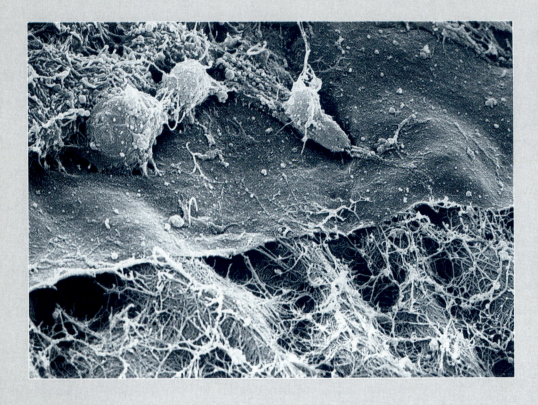

Figure 26.1 Scanning electron micrograph of the cornea of a chicken embryo. The specimen was torn to expose the interface between the outer embryonic epithelium (top left) and the underlying mesenchyme (embedded in fibrous material, bottom). The small epithelial cells have deposited a dense sheet of extracellular matrix, called a basal lamina (center), on their basal surface. The interstices between the mesenchymal cells and, to a lesser extent, between the epithelial cells are filled with a different kind of extracellular matrix. Its fibrous components, mostly collagen, are well preserved in the fixation process that precedes electron microscopy. The amorphous components of the matrix—mostly large polysaccharides—tend to be washed out or to collapse around the fibers during fixation.

ECM Molecules and Their Cellular Receptors

Glycosaminoglycans and Proteoglycans Form an Amorphous, Hydrophilic Ground Substance

Fibrous Glycoproteins Make Up the Dynamic Meshwork of the ECM

Cell Surface Proteoglycans and Integrins Mediate Cell Adhesion to ECM Molecules

Interactions between Cells and ECM

Cells Create and Re-create Their ECM Environment

Cells May Orient ECM Fibers through Integrins or by Mechanical Force

Cells Adhere Preferentially to ECM Components

Fibrous ECM Components Provide Contact Guidance to Cells

The Role of Fibronectin during Amphibian Gastrulation

The Role of ECM in Neural Crest Cell Migration and Differentiation

Neural Crest Cells of the Trunk Avoid the Notochord Area and the Posterior Halves of Sclerotomes

Local ECMs Determine the Region-Specific Fates of Neural Crest Cells

The *white* Axolotl Mutant Interferes with the Local Maturation of an ECM Component

Several ECM Components Are Critical for Cranial Neural Crest Cell Development

Cranial and Trunk Neural Crest Cells Attach to ECM Components through Different Mechanisms

Position-Specific Integrins in *Drosophila melanogaster*

Morphoregulatory Roles of the ECM

ECM Components Trigger Cellular Reorganizations That Affect Gene Expression

Expression of the Tenascin Gene Is Controlled by a Homeodomain-Containing Transcription Factor

---

Organisms do not consist entirely of packed cells. The interstices between cells are filled with materials that are collectively called the *extracellular matrix (ECM)*. The ECM consists essentially of two components: an amorphous ground substance and a meshwork of fibers. The ground substance attracts water by osmotic pressure and forms a kind of gel, whereas the fibers reinforce the ground substance while resisting its tendency to expand. Because the ECM components are linked together by multiple binding domains they form a matrix, which serves a wide variety of functions.

The ECM is most plentiful in connective tissues. When calcified, it forms hard tissues, as in bones and teeth. In tendons, it accumulates tough fibers that maximize tensile strength. The ECM also forms transparent layers, as in the cornea. Because of such functions, the ECM has traditionally been thought of as an inert scaffolding that determines primarily the physical properties of a tissue. However, it is now clear that the ECM is very interactive (Hay, 1991a; Hynes, 1992). The ECM influences cell division, shape, movement, and differentiation, through receptor proteins that bind specifically to some of its components. In addition, some ECM components act as repositories for inductive signaling molecules and growth factors that bind to matching receptors in cell plasma membranes. The occupancy of cell receptors by ECM components or ECM-bound signals feeds back on the behavior and gene activity of a cell. The interactive nature of the ECM is of particular interest in embryos, where the ECM helps to direct cellular and epithelial movements and to localize inductive signals.

The principal embryonic tissues, mesenchyme and epithelia, have characteristic types of ECM (Fig. 26.1; see also Fig. 2.20). Mesenchymal cells are surrounded on all sides by interstitial spaces filled with ECM. Beneath epithelia, the ECM takes the form of a dense sheet called a *basal lamina,* which separates the basal surface of the epithelium from the underlying mesenchyme.

The ECM is composed of a variety of molecules, which are released by *exocytosis* from the cells that produce them into the extracellular space, where they *self-assemble*. Thus, cells create their own extracellular environment by synthesizing the ECM building blocks and by maintaining appropriate conditions for their self-assembly. Many molecules found in the ECM have been very well conserved during evolution. For instance, collagen fibers are found not only in vertebrates but also in sea urchins, roundworms, and many other invertebrates.

In this chapter, we will summarize the molecular properties of the most common ECM components and then examine how cells interact with the ECM in morphogenesis. In particular, we will focus on the function of fibronectin in amphibian gastrulation, on the roles of ECM in neural crest cell migration and differentiation, and on the genetic analysis of integrins in *Drosophila*. To conclude, we will consider ECM molecules and their receptors together as a larger class of morphoregulatory molecules.

ECM Molecules and Their Cellular Receptors

The ECM components interact with one another and with specific receptor molecules in the plasma membranes of cells. This section describes the major players in these interactions: the *glycosaminoglycans* and *proteoglycans* that form the gel-like component of the ECM, the fibrous *glycoproteins* that make up the meshwork of the ECM, and the receptors in the cell surface that mediate cell adhesion to ECM components.

Glycosaminoglycans and Proteoglycans Form an Amorphous, Hydrophilic Ground Substance

Glycosaminoglycans are long, unbranched polysaccharide chains composed of repeating disaccharide units (Wight et al., 1991). One sugar in each disaccharide is an amino sugar (*N*-acetylglucosamine or *N*-acetylgalactosamine), and the other sugar is a uronic acid (Fig. 26.2). Different glycosaminoglycans are distinguished by the nature of their sugar monomers, the type of linkage between them, and the number and location of added sulfate groups. The most abundant glycosaminoglycans are *hyaluronic acid, chondroitin sulfate, dermatan sulfate, heparin, heparan sulfate,* and *keratan sulfate.* Most glycosaminoglycans are covalently linked to core proteins, which in turn form the side chains of a long glycosaminoglycan (Fig. 26.3). The resulting large aggregates, called ***proteoglycans***, may have molecular weights of 10^8 or more. An exception is *hyaluronic acid,* which is just one large glycosaminoglycan.

Glycosaminoglycans have many negative charges, so that they do not fold tightly but form extended random coils. They attract cations such as the sodium ion (Na^+), which in turn attract large clouds of water. As a result, glycosaminoglycans occupy much space and resist compression. Their most common function seems to be filling space while allowing cell migration and the diffusion of water-soluble molecules. This is the case particularly for hyaluronic acid, which is most abundant in embryos and in regenerating and healing tissues.

Fibrous Glycoproteins Make Up the Dynamic Meshwork of the ECM

The fibrous components of ECM consist mostly of *glycoproteins,* that is, proteins with attached oligosaccharides. In contrast to *proteoglycans,* the sugar moieties of glycoproteins are more diverse and account for less than half of the total molecular mass.

Collagens   A family of stiff, ropelike glycoproteins that are found in all metazoa are known as *collagens* (Linsenmayer, 1991). Their critical importance is underlined by the lethality of several mutations in collagen genes of mammals and other organisms. In mammals, collagens are the most abundant proteins, accounting for 25% of the animal's total protein. Each collagen protein is composed of three polypeptides that are extremely rich in glycine, proline, and hydroxyproline. The polypeptides wind around one another in a triple helix and are cross-linked by covalent bonds (see Fig. 17.19).

There are several collagen genes, each encoding a particular polypeptide. Moreover, many collagen pre-mRNAs are spliced alternatively, so that even more varieties of collagen peptides are generated. The polypeptides combine to form a wide range of different collagen proteins, which assemble into fibrils or meshworks that differ greatly in their physical properties. Correspondingly, collagens as well as other ECM components are distributed in a tissue-specific way. Type I collagens occur in most adult vertebrate tissues. They are particularly abundant in skin, bone, tendon, and cornea; these tissues are most strongly affected in mutants that produce abnormal type I collagens. Type II collagens are characteristic of cartilage and intervertebral discs, and mutations affecting type II collagen cause cartilage disorders. Type III is present in skin, blood vessels, and internal organs. Type IV is characteristic of

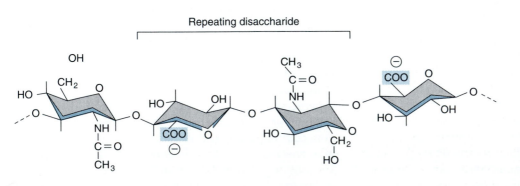

Repeating disaccharide

D-Glucuronic acid   *N*-Acetyglucosamine

Figure 26.2 Portion of hyaluronic acid, a glycosaminoglycan. Hyaluronic acid consists of thousands of repeating disaccharides, each composed of D-glucuronic acid and *N*-acetylglucosamine. Note the negatively charged carboxyl groups. In glycosaminoglycans with sulfate groups, the density of negative charges is even higher.

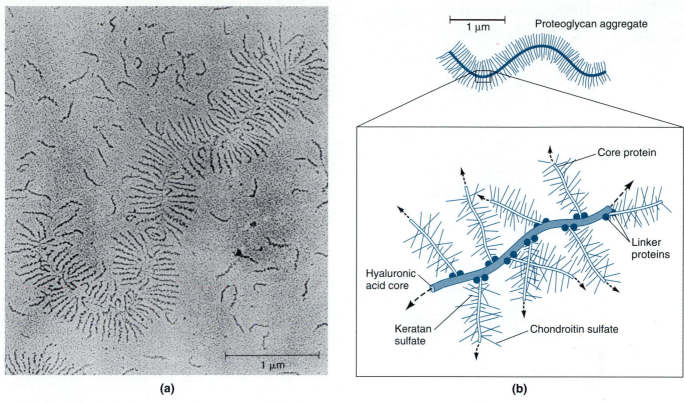

Figure 26.3 Large proteoglycan aggregate from fetal bovine cartilage. **(a)** Electron micrograph of molecules contrasted with platinum. **(b)** Interpretative drawing of the large aggregate shown in the micrograph. Two types of glycosaminoglycans, keratan sulfate and chondroitin sulfate, are linked to core proteins, which in turn are linked to a hyaluronic acid core.

basal laminae, and mutations affecting a particular chain of type IV collagen cause a kidney deficiency from ultrastructural defects in the basal lamina that separates the *glomerulus* of each nephron from the lumen of *Bowman's capsule* (see Chapter 13).

All collagen molecules are secreted from cells as *procollagens*, that is, as triple helices with globular ends. However, the procollagen types differ in their processing and self-assembly in the extracellular space. In the processing of most procollagens, including types I through III, the globular ends are cleaved off to generate collagen molecules, which then self-assemble into long, cablelike fibrils (see Fig. 17.20). Such fibrils have great tensile strength and are abundant in bones, tendons, the dermal layer of the skin, and other connective tissues. In contrast, type IV procollagen retains its nonhelical ends and self-assembles into multilayered networks, which form the core of basal laminae (Fig. 26.4).

Fibronectins Several ECM glycoproteins have specific binding domains for cells and other ECM components (K. M. Yamada, 1991). Their multiple binding sites enable these glycoproteins to cross-link other ECM elements and cells into three-dimensional meshworks.

Best characterized among such interactive glycoproteins is *fibronectin,* which plays a major role in cell migration. The fibronectin molecule is a dimer of two long polypeptides linked by disulfide bonds near their carboxyl termini (Fig. 26.5). Each polypeptide is folded into several globular domains connected by flexible chains. The globular domains have binding sites for cells and for collagen, heparin, and other ECM components. Several different variants of fibronectin originate through differential splicing of pre-mRNA. These variants differ in solubility. Soluble fibronectins are found in blood serum, where they play a role in blood clotting. Insoluble fibronectins are abundant in the ECM, where they form extended fibrils that facilitate the migration of embryonic cells as discussed later in this chapter.

The cell-binding activity of fibronectin depends primarily on a tripeptide sequence known as the *RGD sequence* (arginine-glycine-asparagine), which is recognized by a family of cell surface receptors called *integrins.* In solution, synthetic peptides that contain the RGD sequence compete with fibronectin for cell receptors and so inhibit cell adhesion to fibronectin. However, cell adhesion to the RGD sequence in fibronectin is only

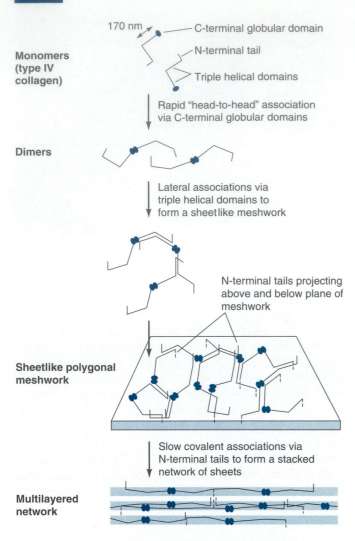

Monomers (type IV collagen)

170 nm — C-terminal globular domain

N-terminal tail

Triple helical domains

Rapid "head-to-head" association via C-terminal globular domains

Dimers

Lateral associations via triple helical domains to form a sheetlike meshwork

N-terminal tails projecting above and below plane of meshwork

Sheetlike polygonal meshwork

Slow covalent associations via N-terminal tails to form a stacked network of sheets

Multilayered network

Figure 26.4 Model of type IV collagen self-assembly into a multilayered network as found in basal laminae. The model is based on the electron-microscopic analysis of supramolecular structures that self-assemble in vitro.

part of a more complex set of interactions between cells and the ECM. The RGD sequence occurs not only in fibronectin but also in other ECM components, and domains of the fibronectin molecule other than the RGD sequence contribute to cell binding.

Laminin Another large interactive glycoprotein, called *laminin,* is especially plentiful in basal laminae. Laminin consists of three polypeptides arranged in the shape of a cross and held together by disulfide bridges (Fig. 26.6). As in the case of collagen, the component polypeptides of laminin are encoded by small families of different genes, and resulting combinations are tissue-specific. Like fibronectin, laminin has several domains that bind to other ECM components or cells. Some of these domains bind to type IV collagen; another, to heparin. There are at least two cell-binding regions in the center of the molecule and several additional sites in the long arm. Laminin promotes adhesion of many cell types and in particular the extension of *neurites* from nerve cells. In basal laminae, laminin interacts with collagen and epithelial cells on one side while fibronectin interacts with collagen and mesenchymal cells on the other.

Tenascin Yet another large interactive glycoprotein, *tenascin,* is shaped like a six-armed star (K. M. Yamada, 1991). Each arm contains repetitive domains that allow tenascin to interact with cells as well as with other ECM molecules. As a cell interaction protein, tenascin is unusual in that it can mediate both adhesive and repulsive interactions. Its synthesis is often correlated with morphogenetic events, including cell migration, gastrulation, organogenesis, wound healing, and tumor development.

In addition to its structural function, the ECM also confers signals that affect cell proliferation and differ-

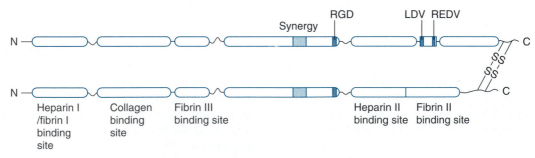

Synergy   RGD   LDV REDV

N   C

N   C

Heparin I /fibrin I binding site

Collagen binding site

Fibrin III binding site

Heparin II binding site

Fibrin II binding site

Figure 26.5 Schematic diagram of a fibronectin molecule. The large glycoprotein is a dimer of two similar polypeptides, linked by a pair of disulfide bonds near their carboxyl terminals. Each polypeptide is divided into several domains that bind to other ECM components or to cells. The cell-binding domains (color bands, labeled only at the top) include the arginine-glycine-asparagine (RGD) sequence, the synergy region that acts together with the RGD sequence, and an alternatively spliced region (rectangle) with a leucine-asparagine-valine (LDV) and an arginine-glutamine-asparagine-valine (REDV) sequence. The major ECM components bound by each domain are labeled only at the bottom.

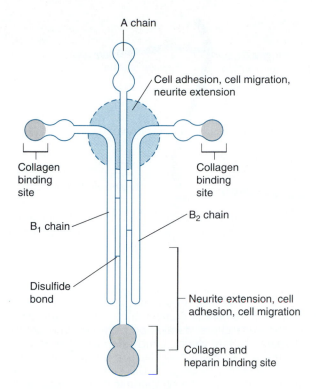

A chain

Cell adhesion, cell migration, neurite extension

Collagen binding site

Collagen binding site

B_1 chain

B_2 chain

Disulfide bond

Neurite extension, cell adhesion, cell migration

Collagen and heparin binding site

Figure 26.6 Schematic representation of a typical laminin molecule. It consists of three polypeptides arranged in the form of a cross and held together by disulfide bonds. The longest polypeptide, called the A chain, extends the entire length of the molecule and terminates in two globular binding domains for heparin and collagen at the base of the cross. The two other chains, B_1 and B_2, are linked with the A chain and extend as the side arms of the cross. Putative binding regions are labeled.

entiation. ECM molecules carry out this signal function in different ways (Hynes, 1992; J. C. Adams and F. M. Watt, 1993). First, ECM molecules can serve as repositories for *growth factors,* that is, signal peptides that promote the survival and proliferation of certain target cells (see Chapter 29). Thus, the ECM prevents loss of growth factors through diffusion and keeps them in a relatively persistent local supply. For instance, certain hematopoietic growth factors bound to heparan sulfate are thought to contribute to bone marrow microenvironments that promote specific steps in blood cell development (see Chapter 19). Second, some ECM molecules function as growth factors themselves. Third, ECM molecules interact with cell membrane receptors that have both structural and signal-transducing functions. These receptors will be discussed next.

Cell Surface Proteoglycans and Integrins Mediate Cell Adhesion to ECM Molecules

The complexity of the ECM is mirrored by a heterogeneous array of cell surface receptors. Cells adhere to ECM components via two major classes of receptors

that act synergistically and thereby confer specificity on cell-ECM adhesions. First, a family of transmembrane glycoproteins, called the *integrins,* bind to RGD and other recognition sequences. Integrins appear to be the major connectors by which cells adhere to ECM molecules (Ruoslahti, 1991; Hynes, 1992). The second mechanism relies on *cell surface proteoglycans,* which are also components of the plasma membrane; their core proteins are either inserted across the phospholipid bilayer or linked to it (Wight et al., 1991). These integral membrane proteoglycans in turn bind to the interactive glycoproteins of the ECM.

Integrins are a family of transmembrane receptor glycoproteins, each consisting of two peptides, or chains, one α chain and one β chain (Fig. 26.7). The α chain contains three or four binding sites for divalent cations; the affinity and binding specificity between integrins and their ligands depends on the presence of magnesium or calcium ions. The basic integrin structure has been highly conserved in evolution and is very similar in vertebrates and invertebrates. In mammals, at least fourteen α chains and eight β chains allow for the formation of a wide variety of dimeric integrins. Each chain is encoded by a different gene, and alternative splicing of pre-mRNAs introduces a further level of complexity.

The extracellular domains of the α and β chains combine to form binding sites for RGD or other specific amino acid sequences. Because these sequences occur not only in fibronectin but also in other ECM molecules, most integrins bind to multiple ligands. Conversely, most ECM molecules have several binding sites for different integrins. These overlapping binding specificities allow cells to respond to various combinations of molecules in their extracellular environment.

Under most circumstances, the cytoplasmic domains of both integrin chains seem to be connected, via linker proteins such as *talin* and *α-actinin,* to microfilament bundles inside the cell (Fig. 26.7). Ning Wang and colleagues (1993) demonstrated the connection between integrins and cytoskeletal elements directly by applying a magnetic twisting device to cell surfaces. Twisting of integrins caused a force-dependent stiffening response of the cells. This response was abolished by cytoskeletal inhibitors such as *cytochalasin D,* and no response was observed when the twisting device was applied to nonadhesion receptors. Thus, integrins seem to link a cell's cytoskeleton to the ECM just as cell adhesion molecules connect the cytoskeletons of two cells.

Cell surface proteoglycans are proteoglycans that are linked to cell plasma membranes. Typically, the core proteins of these proteoglycans have a cytoplasmic domain, a transmembrane domain, and an extracellular domain, just as most cell adhesion molecules have (see Chapter 25). One of the best-investigated cell surface proteoglycans, **syndecan,** is diagrammed in Figure 26.8. Its extracellular domain comprises different glycos-

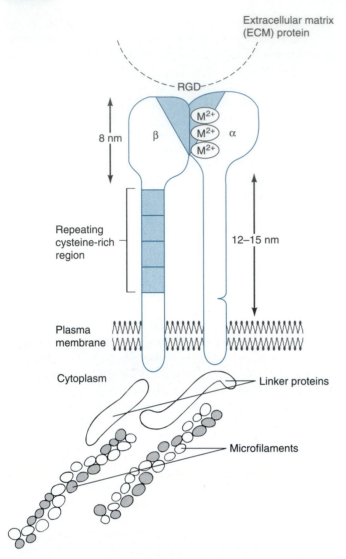

Figure 26.7 Structural features of integrins. The integrins are receptor molecules embedded in cell plasma membranes. Each integrin consists of two glycosylated polypeptides, the α chain and the β chain, which are held together noncovalently. The α chain has been processed posttranslationally into two disulfide-linked fragments. The α chain has several binding sites for divalent cations (labeled M^{2+}), which stabilize the connection to the β chain. The β chain has repetitive cysteine-rich sequences. Both chains cross the plasma membrane. Together, their extracellular domains form a binding domain for ligands such as the arginine-glycine-asparagine (RGD) sequence (see Fig. 26.5). The cytoplasmic domains of the two chains are connected to microfilaments by linker proteins, including talin and α-actinin.

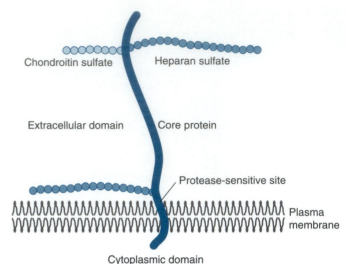

Figure 26.8 Schematic drawing of syndecan, a cell surface proteoglycan. The core protein has an extracellular domain, a transmembrane domain, and a cytoplasmic domain. The extracellular domain bears chondroitin sulfate and heparan sulfate chains, which bind to other ECM components. A protease-sensitive site is right outside the plasma membrane; cleavage at this site releases the entire extracellular domain. The cytoplasmic domain associates with microfilaments when the extracellular domain is cross-linked with other ECM molecules.

aminoglycans that interact with collagens, fibronectin, and possibly other ECM components (Bernfield and Sanderson, 1990). Close to the plasma membrane, the core protein has a protease-sensitive site that, when cleaved, releases the extracellular domain. The cytoplasmic domains of syndecan and several other cell surface proteoglycans may be connected with microfilaments, especially when their extracellular domains are cross-linked with other ECM components.

In addition to their mechanical function of linking the cytoskeleton to extracellular matrix molecules, integrins and cell surface proteoglycans also act as signal-transducing molecules (Damsky and Werb, 1992; Hynes, 1992). Signal transfer from the ECM into cells is indicated by the dependence of the differentiation of certain cells on the functions of specific integrins. For example, osteocyte differentiation can be prevented by antibodies to the integrin β_1 subunit. Signal transfer from the cells to the ECM occurs through modulation of integrin affinities and specificities. For instance, circulating blood platelets do not normally bind to the serum protein fibrinogen, which is a good thing because such binding would lead to blood clotting. In order to start fibrinogen binding, platelets must first be activated by thrombin or other agents; this activation step includes a conformational change of an integrin that binds to fibrinogen.

The nature of the signal transmission from integrins or cell surface proteoglycans to the cell cytoplasm is still being investigated. The cytoplasmic domains of these receptors are small and have no known enzymatic activity. Instead, they seem to interact with other cytoplasmic factors to generate *kinase* activity and second messenger activities similar to those elicited by membrane receptors for hormones and growth factors.

Interactions between Cells and ECM

Cells and ECM affect each other in many ways. In this section, we will explore how cells synthesize and replace ECM components as development progresses, and how cells may orient ECM fibers by exerting mechanical force. Conversely, we will see that the composition of the ECM, and even the orientation of its fibers, influence the growth and migration of cells.

Cells Create and Re-create Their ECM Environment

Cells create an ECM environment for themselves that suits their particular needs. The cell divisions, rearrangements, and migrations that characterize embryonic development occur in an ECM that is loose and facilitates cell movement; an ECM with the same characteristics is produced during regeneration and wound healing. The ECM that is present under any of these conditions contains abundant hyaluronic acid. Subsequently, as tissues differentiate or as other functional requirements arise, hyaluronic acid is invariably replaced with other ECM components.

The replacement of hyaluronic acid in regenerating limbs was studied by Bryan Toole and Jerome Gross (1971). Newt limbs were amputated and allowed to regenerate. At intervals during the regeneration period, the newts were injected with a radioactive precursor that is incorporated into glycosaminoglycans. One day after the injection, the regenerating limbs were removed, and the radioactivity in different types of glycosaminoglycans was measured. The radioactivity of each glycosaminoglycan sample is a combined measure of the amount synthesized and the amount degraded. As summarized in Figure 26.9, the amount of radiolabeled hyaluronic acid increased steeply during the first 10 days of regeneration and dropped sharply thereafter. Most of it accumulated in the distal portion of the regenerating limb, which contains the regeneration blastema. In contrast, the amount of radiolabeled chondroitin sulfate rose more slowly over a longer period of time and reached its maximum during cartilage formation in the regenerating limb. The sharp decrease in hyaluronic acid after 10 days coincided with a significant increase in the activity of *hyaluronidase,* an enzyme that degrades hyaluronic acid. Thus, the decrease in hyaluronic acid must result, at least in part, from regulated enzymatic degradation.

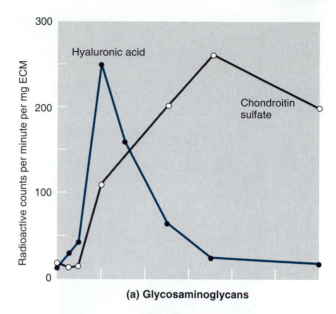

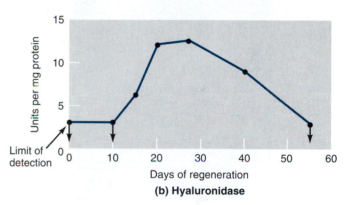

Figure 26.9 Exchange of glycosaminoglycans in regenerating newt limbs. **(a)** Incorporation of injected [³H]acetate precursor into hyaluronic acid and chondroitin sulfate. The ordinate represents the radioactivity per sample contained in each glycosaminoglycan after a 24-h incubation period with the labeled precursor. **(b)** Accumulation of hyaluronidase, an enzyme that degrades hyaluronic acid, measured in units of activity per milligram of protein. Note that the amount of hyaluronic acid decreases after the activity of hyaluronidase becomes measurable. The amount of labeled chondroitin sulfate increases during the same time.

The results just described indicate that the composition of the ECM in a regenerating limb changes dramatically over time. During the formation of the regeneration blastema, the prevalent glycosaminoglycan is hyaluronic acid. Apparently, this ECM component facilitates cell proliferation and migration in the regeneration blastema. Upon completion of this phase, most of the hyaluronic acid is replaced with chondroitin sulfate, which is characteristic of cartilage. This example illustrates how the ECM environment of cells changes in stage-dependent, region-specific ways and in accord with the prevailing cell activities.

Cells May Orient ECM Fibers through Integrins or by Mechanical Force

Cells not only synthesize different mixtures of ECM molecules in a stage-dependent way, but they also orient fibrous ECM components along preferred axes. Cultured fibroblasts, for example, produce extracellular fibronectin fibrils that align with their intracellular microfilaments (Fig. 26.10; Hynes and Destree, 1978). If

such cells are treated with *cytochalasin*, a drug that disrupts microfilaments, the fibronectin fibrils separate from the cells (Ali and Hynes, 1977). The most likely explanation for this outcome is that the alignment between microfilaments on the inside and fibronectin fibrils on the outside is mediated by integrins. Thus, integrins may align fibronectin fibrils according to the preexisting arrangement of cellular microfilaments. Conversely, a preexisting orientation of fibronectin fibrils could be imposed through integrins on the microfilaments of adjacent cells.

There are also indications that cells align fibrous components in the ECM by exerting traction. This possibility was tested by David Stopak and Albert Harris (1982), who cultured tissue pieces from chicken embryos in gels containing randomly oriented collagen fibrils. When two or more tissue explants were close to each other in the same culture dish, collagen fibrils between them aligned parallel to the axis connecting the explants (Fig. 26.11). Collagen alignment occurred between tissues containing migratory fibroblasts and between aggregates of fibroblasts, but not between aggregates of transformed fibroblasts that migrate very little. The researchers concluded that migratory fibroblasts attach to collagen fibrils and orient them by mechanical forces (traction) as they migrate. Again, the molecules by which cells attach to collagen fibrils and orient them are likely to be integrins.

Cells Adhere Preferentially to ECM Components

The ECM environment feeds back to the cells that created it and to any other cells that move into it. The following observations show how ECM components can affect the outgrowth of neurites from cultured neurons.

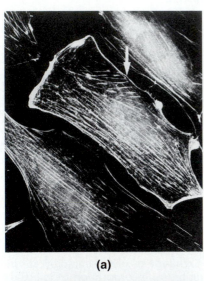

(a)

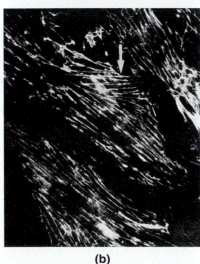

(b)

Figure 26.10 Alignment of intracellular microfilament bundles in cultured rat fibroblasts with extracellular fibronectin fibrils. Microfilaments **(a)** and fibronectin fibrils **(b)** of the same cells are made visible selectively by immunostaining (see Methods 4.1) using secondary antibodies with different fluorescent dyes. Note the parallel alignment of the intracellular and extracellular bundles (arrows).

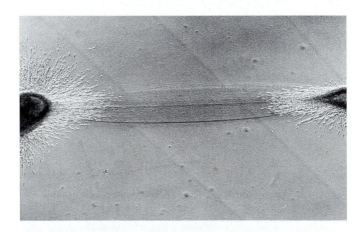

Figure 26.11 Alignment of collagen fibrils between two pieces of embryonic chicken heart cultured in a gel with randomly oriented collagen. This photograph was taken with oblique illumination after 90 h of culture, when tracts of aligned collagen fibrils became visible between neighboring tissue pieces.

The ECM of the central and peripheral nervous systems contains type IV collagen, fibronectin, and laminin. To determine whether the neurites that extend from developing neurons grow preferentially on any of these three ECM components, Ross Gundersen (1987) cultured neurons on surfaces coated with these components. Glass coverslips were first covered with type IV collagen as a background substrate before dots or stripes of laminin were added. When neurons were cultured on such coverslips, their neurites grew out in revealing patterns. Once a neurite had found a laminin surface, it would not leave it. On round laminin dots, the neurites grew in endless circles, exactly outlining the margin of the dot but not venturing outside it (Fig. 26.12). Only after 1 or 2 days, when neurites had fasciculated into substantial bundles, did they cross back onto the collagen background. This behavior could be ascribed to the chemical difference between laminin and collagen or to some mechanical barrier that might have been created as the laminin dried onto the collagen layer.

To decide whether the neurites were confined by chemical or mechanical means to the laminin dot, Gundersen (1987) mixed polylysine into the collagen background on some coverslips and applied dots of fibronectin, instead of laminin, to others. In both cases, the neurites crossed easily out of the dots and onto the background material. Apparently, the cells were not contained by a mechanical barrier but responded to the alternative ECM molecules that were available. Given a choice between laminin and collagen, they definitely preferred laminin, and they also grew at the fastest rate on laminin. These observations suggest that the spatial and temporal distribution of laminin might be a guiding factor for neurite growth in vivo.

Fibrous ECM Components Provide Contact Guidance to Cells

Many morphogenetic processes in embryos involve oriented cell migration. During amphibian gastrulation, for example, mesodermal cells of the *deep zone* and the *involuting marginal zone* migrate on the basal surface of the ectoderm (blastocoel roof) toward the animal pole (see Chapter 10). ECM fibrils found on this surface are scarce at blastula stage, but become more numerous in early gastrulae and continue to accumulate during gastrulation (Fig. 26.13). These fibrils have a tendency to become oriented along meridional lines between the dorsal blastopore and the animal pole (K. E. Johnson et al., 1990). Thus, the preferred orientation of the fibrils is along the route taken by mesodermal cells during gastrulation. A possible cause of the fibril alignment is the attachment of the fibrils to integrins or other anchors in cell plasma membranes. As the dorsal ectodermal cells undergo *convergent extension*, the integrins may be rearranged so that the attached fibrils become oriented along the blastopore–animal pole axis.

The *filopodia* of the mesoderm cells that migrate on the inside of the blastocoel roof are attached to the extracellular fibrils between the dorsal blastopore and the animal pole (Nakatsuji et al., 1982). This behavior suggests that

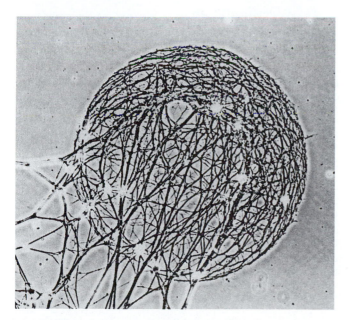

Figure 26.12 Preferential growth of neurites on laminin in cell culture. Neurons were cultured on glass coverslips coated with type IV collagen, and with a dot of laminin added on top. When a neurite grew onto the laminin dot, it did not leave the dot for 1 or 2 days. It migrated in circles, exactly demarcating the perimeter of the dot and not venturing out of it until the neurites had formed fascicles.

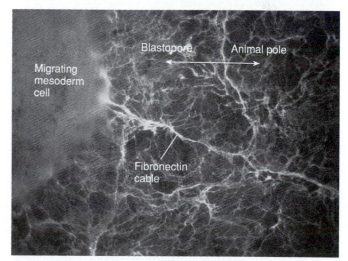

Figure 26.13 Migrating mesoderm cell (large body at lower left) inside an early gastrula of a salamander (*Ambystoma maculatum*). The scanning electron micrograph shows the basal (inner) surface of cells forming the blastocoel roof. The fibrous ECM cables produced by these cells are immunostained (see Methods 4.1) for fibronectin. The preferred orientation of the fibers is along the blastopore–animal pole axis (double-headed arrow).

the fibrils may orient the cell migration. The phenomenon of cells' being guided by fibril orientation or similar mechanical clues, which has been reported by several authors, is known as *contact guidance* (Weiss, 1958).

To test the role of contact guidance in amphibian gastrulation, Norio Nakatsuji and Kurt Johnson (1983) developed culture conditions under which gastrula mesodermal cells could be observed in isolation but still migrated at approximately the same rate and assumed much the same shape as they would in whole embryos. The researchers transferred ECM from the inside of a blastocoel roof to a glass coverslip by isolating an ectoderm piece and placing it basal surface down on the glass surface. After 5 h of culture, they marked the margins of the ectoderm piece and the blastopore–animal pole axis on the coverslip before they flushed off the tissue. As a result of this procedure, the glass surface was now coated, or *conditioned*, with ECM residue. When mesodermal cells from gastrulae of a salamander, *Ambystoma maculatum*, or a frog, *Xenopus laevis*, were placed on the conditioned area, the cells quickly attached and formed lamellipodia. The researchers recorded the subsequent cell movements by videomicroscopy and then fixed the coverslips for examination with a scanning electron microscope. They compared the recorded cell movements with the orientations of ECM fibrils seen under the electron microscope. Indeed, the cell migrations tended to follow the local fibril orientation: in areas where the cell trails were unoriented, the fibrils were not significantly aligned; and where the cell trails were oriented, so were the fibrils.

In an extension of this experiment, De-Li Shi and colleagues (1989) conditioned the surfaces of plastic culture dishes with ECM deposits from large rectangular explants of blastocoel roofs from early newt gastrulae (*Pleurodeles waltlii*). The perimeter of each explant used for conditioning, and its animal pole–blastopore axis, were marked on the dish (Fig. 26.14). A smaller explant comprising marginal zone and adjacent ectoderm was then placed in the middle of the conditioned area with its animal pole–blastopore axis perpendicular to that of the ECM deposit. During the following culture period, mesoderm cells from the second explant migrated on the ECM deposited by the first explant. In 53 out of 78 such trials, the new mesoderm cells migrated toward the animal pole of the ECM deposit. In 2 cases, they migrated toward the blastopore pole of the ECM, and in 15 cases the cells moved in both directions. No migration was observed in the remaining 8 cases.

Similar experiments were also carried out with mesoderm explants and isolated cells from *Xenopus* gastrulae. In this and other frog species, migration of the involuted mesoderm on the inside of the blastocoel roof

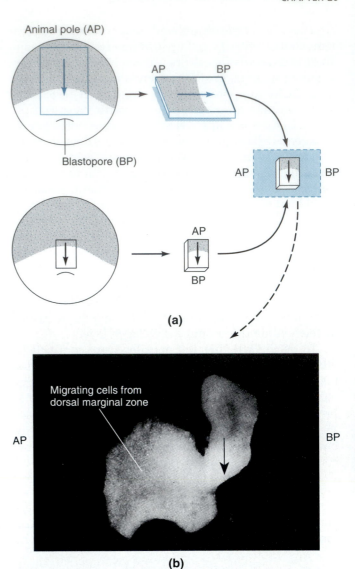

(a)

(b)

Figure 26.14 ECM components deposited by the blastocoel roof orient migrating mesoderm cells in the newt *Pleurodeles waltlii*. **(a)** The bottom of a plastic culture dish was conditioned with a large explanted section of blastocoel roof left basal surface down in the dish for 2 h. The perimeter of the explant and its animal pole (AP)–blastopore (BP) axis (colored arrow) were marked on the dish. When the explant was flushed off, it left behind a deposit of ECM material. A smaller explant of dorsal marginal zone was placed in the center of the conditioned area with its animal pole–blastopore axis (black arrow) perpendicular to that of the ECM deposit. **(b)** Photograph of the area represented by the colored rectangle in the diagram. During the 24-h period in which the explant was cultured, mesodermal cells from the dorsal marginal zone migrated out. In most cases, these cells moved toward the animal pole of the ECM deposit.

seems to be dispensable for gastrulation, because gastrulation proceeds nearly normally even if the blastocoel roof has been removed (see Chapter 10). However, in vitro experiments indicate that head mesoderm and axial mesoderm cells migrate independently of each other on the inside of isolated blastocoel roof and on plastic surfaces coated with ECM from the blastocoel roof (Winklbauer and Nagel, 1991). The cells migrate toward the animal pole side, provided that the fibronectin fibrils in the ECM coat are left intact. If fibronectin fibril formation is disturbed by addition of cytochalasin B or GRGDSP peptide *while the ECM coat is being laid down by the ectoderm cells,* then the mesoderm cells will still adhere to the coated surface and migrate, but the direction of their migration will be random.

The experiments just described demonstrate that the direction in which mesoderm cells from amphibian gastrulae migrate is biased by contact with ectodermal ECM components. The orienting clue seems to come from a preferred orientation of fibronectin fibrils along the blastopore–animal pole axis. How the mesodermal cells migrating along these fibrils distinguish the opposing directions toward the animal pole and toward the blastopore is not understood. Perhaps the fibro-nectin fibrils have a molecular polarity, with the C termini of all peptides oriented in one direction. In vivo, contact guidance by ECM fibrils also acts in concert with additional signals. For instance, amphibian mesoderm cells and many other cells demonstrate **contact inhibition:** when two cells collide, they both stop and then move in different directions. Statistically, contact inhibition should cause mesoderm cells to move away from the crowded blastopore region. Thus, contact inhibition by other cells and contact guidance by ECM fibrils together may direct the cells toward the animal pole.

Together, the observations discussed in this section indicate that cells create and re-create their ECM environments in accord with stage-specific requirements. Moreover, cells can orient the fibrous components of their ECM using integrins coupled to their microfilaments and/or traction. Conversely, cells respond to chemical differences between available ECM components and are guided by contact with oriented ECM fibrils.

The Role of Fibronectin during Amphibian Gastrulation

The extracellular matrix plays a major role in the morphogenetic processes of gastrulation, neural crest cell migration, and organogenesis. ECM fibrils seem to guide the migration of mesoderm cells on isolated gastrula ectoderm and conditioned surfaces in vitro, as described in the previous section. To probe the role of ECM materials on the inside of the blastocoel roof during gastrulation *in vivo*, Jean-Claude Boucaut and his

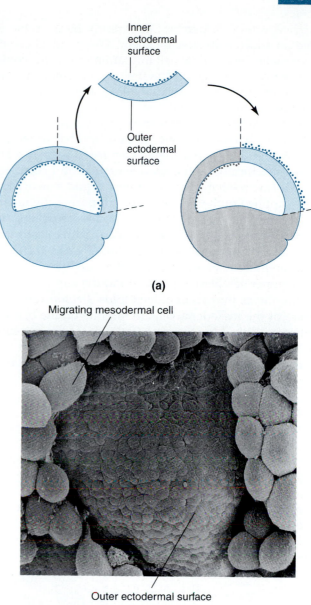

(a)

Migrating mesodermal cell

Outer ectodermal surface

(b)

Figure 26.15 Role of the inner surface of the blastocoel roof in mesodermal cell migration during gastrulation of the newt *Pleurodeles waltlii.* **(a)** Grafting procedure. A segment of dorsal ectoderm was excised and reinserted with the inner ectodermal surface turned outside. **(b)** Scanning electron micrograph showing the inside of a gastrula 3 h after grafting. The migrating mesodermal cells adhered to the inside of the ectoderm except for the graft. Apparently, the outer ectodermal surface did not provide suitable attachment sites.

coworkers (1984) carried out grafting experiments with gastrulae of the newt *Pleurodeles waltlii.* The investigators cut out patches of dorsal ectoderm from early gastrulae and reinserted them inside out (Fig. 26.15). Consequently, during gastrulation, migrating mesoderm cells faced the outer ectodermal surface instead of the inner ectodermal surface, which is their normal substratum. The mesodermal cells did not migrate on the inverted ectoderm patches, apparently because there were no attachment sites for them. This result showed that proper mesodermal cell migration requires a component that is present on the inner ectodermal surface but not on the outer one. Most likely, the critical components were ECM fibers.

The ECM fibers and integrins on the inner ectodermal surface include fibronectin and fibronectin receptors, as revealed by *immunostaining* (see Methods 4.1). To test specifically for the involvement of fibronectin in mesodermal cell migration, Boucaut and coworkers (1984) applied probes that disrupt cell-fibronectin interaction. First, they injected *univalent antibodies* against fibronectin into the blastocoels of early newt gastrulae (Fig. 26.16). This treatment completely upset the gastrulation process. Few if any mesodermal cells migrated on the inner ectodermal surface, and the surface of the ectoderm was thrown into deep folds. Presumably, the failure of the mesoderm cells to migrate on the blastocoel roof blocked the normal involution of the involuting marginal zone, which therefore made no room for the epiboly of the noninvoluting marginal zone and the animal cap. Control gastrulae injected with immunoglobulins from *preimmune* serum developed normally. Thus, gastrulation was not inhibited by some unspecific immunoglobulin activity but by the specific binding of antibodies to fibronectin.

Antifibronectin antibody injected into the blastocoel might inhibit gastrulation by blocking the interaction of the fibronectin RGD sequence with its receptors on the migrating mesoderm cells. To test this hypothesis, Boucaut and coworkers (1985) injected synthetic peptides containing the RGD sequence into the blastocoels of early gastrulae. The effects were very similar to those obtained with antifibronectin antibodies. Control embryos injected with a different peptide representing the collagen-binding domain of fibronectin developed normally. The researchers concluded that the injected RGD sequences blocked the fibronectin receptors of the migrating mesoderm cells, thus preventing them from interacting with fibronectin. Confirming this conclusion, the researchers found that univalent antibodies to fibronectin receptor inhibited gastrulation in the same way (Darribère et al., 1988).

Together, the observations described in this section indicate that fibronectin fibrils provide contact guidance to migrating cells during amphibian gastrulation. This guidance, along with other clues, such as contact inhibition, seems to herd migrating mesoderm cells toward the animal pole.

The critical importance of fibronectin in embryonic cell migration is not limited to the amphibians. In a related study on mouse embryos, the fibronectin gene was eliminated by homologous recombination (for method, see Fig. 14.12). Homozygous mutant embryos implanted normally and initiated gastrulation, but subsequently displayed severe defects in mesodermally derived tissues (George et al., 1993). Notochord and somites were absent, the heart and embryonic blood vessels were deformed, and extraembryonic mesodermal structures were also defective. These abnormalities suggest that the lack of fibronectin severely disturbs mesodermal cell migration, adhesion, proliferation, and/or differentiation.

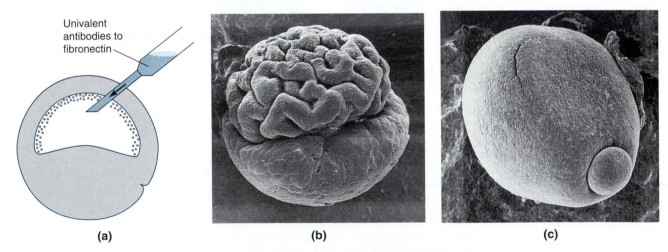

(a)   (b)   (c)

Figure 26.16 Dependence of gastrulation in the newt *Pleurodeles waltlii* on cell-fibronectin interaction. **(a)** The blastocoels of early gastrulae were injected with univalent antibodies to fibronectin. **(b)** The normal gastrulation movement of involution did not occur, the blastocoel roof was thrown into convoluted folds, and the entire vegetal hemisphere remained at the surface of the embryo. Similar effects were observed after injection of peptides containing the RGD sequence and after injection of univalent antibodies to fibronectin receptor. **(c)** Control embryo (blastopore to the lower right), injected with univalent immunoglobulins from preimmune serum, developed normally.

The Role of ECM in Neural Crest Cell Migration and Differentiation

Neural crest cells are a particularly intriguing system for studying the role of ECM in cell migration and cell determination (Bronner-Fraser, 1989, 1993). Neural crest cells arise from the dorsal aspect of the neural tube during neurulation (see Chapter 11). They migrate along several pathways and give rise to a wide spectrum of cell types, ranging from pigment to cartilage cells. Because the fate of a neural crest cell is correlated with its migration route, it is possible that both are determined by the same signals. The nature of these signals is largely unknown, but studies clearly show that ECM components and integrins are involved. These studies also confirm a conclusion already drawn from transplantation studies that the neural crest cells of the head and the trunk differ in many respects.

Neural Crest Cells of the Trunk Avoid the Notochord Area and the Posterior Halves of Sclerotomes

The migration routes and fates of head neural crest cells differ from those of the trunk. The cranial neural crest cells migrate underneath the ectoderm. Some of them enter the *pharyngeal arches,* where they form cartilaginous elements of the head skeleton. Others contribute to the *ciliary ganglion* of the eye, to cranial sensory ganglia, and to the *parasympathetic ganglia,* which innervate internal organs. In contrast, trunk neural crest cells migrate along two major pathways: dorsolaterally under the ectoderm, and ventrally through the anterior sclerotome (Fig. 26.17). The cells that migrate dorsolaterally give rise to pigment cells in the skin. Those that take the ventral pathway produce *dorsal root ganglia, sympathetic ganglia,* and *adrenal medulla* cells.

Are there specific signals that cause trunk neural crest cells to follow either the dorsolateral or the ventral pathway? Normally, neural crest cells migrate first along the ventral path and then enter the dorsolateral path a day later. After surgical removal of the *dermamyotome,* crest cells enter the dorsolateral pathway precociously as if an inhibition had been removed (C. A. Erickson et al., 1992). The temporary inhibition is correlated with the presence of chondroitin-6-sulfate and another moiety—presumably also a glycosaminoglycan—that is bound by a specific *lectin* known as peanut agglutinin lectin (Oakley et al., 1994). These markers disappear normally at the time when neural crest cells enter the dorsolateral pathway, and they disappear precociously after dermamyotome removal. Thus, the glycosaminoglycans themselves or coregulated molecules seem to inhibit neural crest cell migration.

Trunk neural crest cells migrating ventrally pass through anterior sclerotomes but avoid posterior scle-

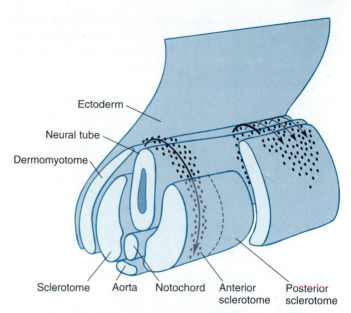

Figure 26.17 Schematic diagram illustrating the early pathways of trunk neural crest cell migration in the chicken. From atop the neural tube, the trunk neural crest cells migrate either dorsolaterally (small curved arrows) between the ectoderm and the dermomyotome or ventrally (large curved arrows) through the somitic sclerotome. The ventrally migrating cells move only through the anterior half of each sclerotome and not through the posterior half. Neural crest cells also avoid the region around the notochord.

rotomes. A similar restriction to the anterior half of each segment is seen in the motor axons growing out of the neural tube. To determine which tissue imposes this pattern, researchers reversed the anteroposterior axis of either the neural tube, to which the neural crest cells are originally attached, or one of the *segmental plates,* the paraxial mesoderm bands from which the somites arise (Keynes and Stern, 1984; Bronner-Fraser and Stern, 1991). They found that neural tube reversal did not change the restriction of axons and neural crest cells to the anterior half of each segment. In contrast, reversing the anteroposterior axis of the segmental plate on one side of a chicken embryo did have a marked effect. The neural crest cells on the operated side migrated through the posterior halves of the sclerotomes, that is, the halves that would have been anterior in their original orientation. These results show that the restriction of neural crest cell migration to the anterior sclerotome is imposed by the somitic mesoderm.

Attempts to identify an ECM component that accumulates only in anterior or only in posterior somitic halves have led to the discovery of several such components. Thus, no single molecule has yet been found that would be sufficient to facilitate neural crest cell migration through the anterior half or to prohibit neural crest cells from passing through the posterior half.

Another conspicuous behavior of migrating neural

crest cells is their avoidance of a space around the notochord that is about 85 μm wide. To test whether the neural crest cell–free space is established by an inhibitory signal from the notochord, Zoé Pettway and her colleagues (1990) implanted an additional notochord parallel to the resident one and along the ventral pathway of neural crest cells. Two days later, the area around the graft was free of neural crest cells. This inhibitory effect was lost when the extra notochord was pretreated with trypsin or chondroitinase. These results indicate that the notochord releases a chondroitin sulfate–containing glycoprotein that inhibits neural crest migration.

The foregoing results indicate that neural crest migration in the trunk region is guided by interactions with other tissues or ECM components that either inhibit or facilitate the passage of neural crest cells.

Local ECMs Determine the Region-Specific Fates of Neural Crest Cells

ECM components determine not only the migration routes but also the region-specific *fates* of neural crest cells. This effect was demonstrated by Roberto Perris and his colleagues (1988), who removed regional ECM samples from living salamander embryos and tested the effects of these samples on premigratory neural crest cells in vitro.

▼

The researchers used miniature sheets of nitrocellulose as microcarriers for ECM materials. The microcarriers were implanted into axolotl embryos, where they were left to adsorb ECM materials. The implantation sites were along the main migration routes of trunk neural crest cells: along the dorsolateral pathway, where pigment cells are formed; and along the ventral pathway, where the dorsal root ganglia arise (Fig. 26.18). After they had adsorbed ECM materials for several hours, the microcarriers were removed from the embryos and transferred to small tissue culture dishes, where they were to serve as a substrate for culturing neural crest cells. Premigratory (naive) neural crest cells were detached from the dorsal side of spinal cords and deposited on the ECM-covered microcarriers.

After several days in culture, the neural crest cell derivatives were classified as pigment cells or neurons according to the following criteria. Cells were scored as *xanthophores* (yellow pigment cells) if they contained pteridines, pigments having a characteristic fluorescence in ultraviolet light. Cells containing tyrosinase, an enzyme involved in synthesizing black pigment, were identified as *melanocytes*. Both substances were found among the derivatives of neural crest cells cultured on

microcarriers coated with subepidermal ECM; cells in these cultures showed no signs of developing as nerve cells. Some of the pigment cells had moved away from their microcarriers during the culture period, indicating that temporary contact with subepidermal ECM was sufficient for subsequent differentiation as pigment cells.

Cells were classified as neurons, on the other hand, if they showed projections (neurites) terminating in

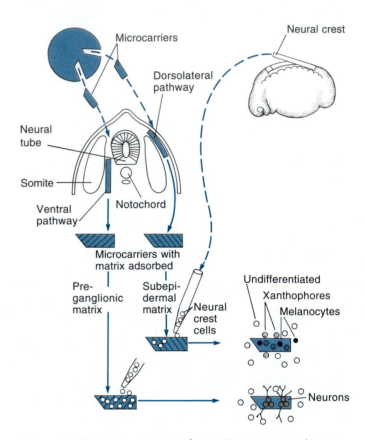

Figure 26.18 Determination of premigratory neural crest cells by region-specific ECM components. Microcarriers (approximate size 400 × 150 × 4 μm) made from sterile nitrocellulose were implanted into the trunk region of axolotl embryos. The implantation sites were along the two main migration routes of neural crest cells: along the dorsolateral pathway in the subepidermal space, and along the ventral pathway at the site where the dorsal root ganglia form. The carriers were left in the embryos to adsorb ECM materials for 10 to 12 h. Just before the onset of neural crest cell migration, the carriers were recovered from the embryos and transferred to small cell culture dishes. Premigratory neural crest cells removed from neural tubes were deposited on the ECM-covered microcarriers and incubated for up to 5 days. Neural crest cells cultured on microcarriers from the subepidermal space (dorsolateral pathway) developed into melanocytes (black circles) or xanthophores (dotted circles). In contrast, premigratory neural crest cells cultured on microcarriers from the dorsal root ganglion area (ventral pathway) developed into neurons (black circles with projections). Neural crest cells cultured without microcarrier contact remained undifferentiated (white circles).

growth cones and if they immunostained for neural cell adhesion molecules or neural intermediate filaments. These traits were observed in derivatives of neural crest cells that had been cultured on microcarriers coated with ECM from the ventral pathway. Such cells never showed the pteridine fluorescence or tyrosinase activity characteristic of pigment cells.

Control neural crest cells grown in the same culture dishes but having no contact with the ECM-covered microcarriers did not form aggregates or express any of the traits indicative of either pigment cell or neuron development.

The results obtained by the experimenters indicate that groups of premigratory neural crest cells cultured in vitro can give rise to either pigment cells or nerve cells. The type of cell that develops depends on the source of the ECM with which the developing neural crest cells are in contact. Even temporary contact with subepidermal ECM from the dorsolateral pathway promotes pigment cell development. Apparently, region-specific ECMs are involved in determining or selecting neural crest cells during their migration along different pathways (see Fig. 12.28). In addition to ECM, other embryonic signals might be involved according to the *principle of synergistic mechanisms.*

The *white* Axolotl Mutant Interferes with the Local Maturation of an ECM Component

The ECM environments that affect the migration routes and fates of neural crest cells change and mature in intricate local patterns. A glimpse at the complexity of this process was already provided by the delayed entry of neural crest cells into the dorsolateral pathway as discussed earlier in this section. Another example is the *white* mutant of the axolotl *Ambystoma mexicanum:* the pigmentation of the mutant larva is restricted to a strip of epidermis at the level of the spine (Fig. 26.19). The normal function of the mutated gene seems to be required for the wild-type distribution of the pigment cell precursors, which normally migrate along the dorsolateral pathway and spread under most of the epidermis. In homozygous mutants, the pigment cell precursors stay near their site of origin at the neural crest. Jan Löfberg and his colleagues (1989) analyzed this mutant (using the microcarrier technique described in Figure

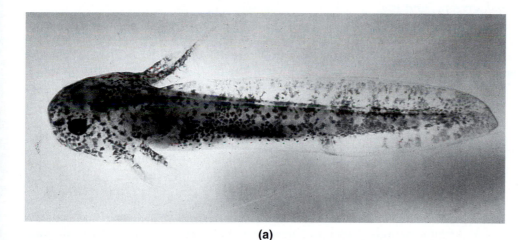

(a)

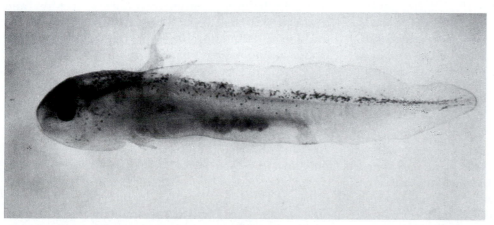

(b)

Figure 26.19 Swimming tadpoles of the axolotl *Ambystoma mexicanum.* **(a)** Wild phenotype. Pigment cells are widespread over the entire body. The black pigment cells (melanocytes) are most conspicuous, while the yellow pigment cells (xanthophores) are in the lighter areas in between. **(b)** Homozygote for a mutant allele of the *white* gene, which is required for the proper spreading of pigment cell precursors. Note that the pigment cells of the larva are restricted to an irregular stripe at the level of the spine. The flanks as well as the dorsal and ventral fins are unpigmented.

26.18) by transferring ECM samples between wild-type and *white* mutant embryos, and between embryos at different stages of neural crest cell migration. They found that *subepidermal* ECM obtained from *white* mutants *at the time of normal pigment cell migration* failed to support the migration of pigment cells in *white* or wild-type hosts. In contrast, dorsolateral ECM from *white* donors *aged beyond the normal pigment cell migration period* did stimulate pigment cell migration.

The results indicate that the wild-type function of the mutated gene is required locally in the subepidermal space during the period of normal pigment cell migration. In the *white* mutant, this function seems to mature late so that the pigment cells no longer respond. The temporal specificity of this genetic function suggests that many such functions may be involved in the entire process of neural crest cell migration and differentiation. *Heterochronic mutants* like the one described here, which shift the relative timing of developmental events, are thought to underlie many evolutionary processes (see Chapter 24).

Several ECM Components Are Critical for Cranial Neural Crest Cell Development

The migratory pathways of neural crest cells are lined with numerous glycoproteins, proteoglycans, and glycosaminoglycans. To explore whether these components are necessary for neural crest cell migration, researchers perturbed the interactions between neural crest cells and specific ECM components with *univalent antibodies* (Bronner-Fraser, 1989, 1993).

In several studies, antibodies were injected into the space lateral to the mesencephalon of chicken embryos. The antibodies diffused freely on the injected side, but were barely detectable on the uninjected side, as if they did not readily cross the midline. Antibodies against the β_1 subunit of integrin caused major defects, including neural tube anomalies, reduction in the number of neural crest cells on the injected side, neural crest cells within the lumen of the neural tube, and ectopic neural crest cells outside the neural tube (Fig. 26.20). Similar defects observed after injection of antibodies against fibronectin and synthetic peptides containing the RGD sequence indicate that cranial neural crest cells also need to interact with fibronectin for normal development.

Antibodies against several other ECM components have also been shown to interfere with cranial neural crest cell development. For instance, antibodies against a complex of laminin and heparan sulfate cause defects similar but not identical to those previously mentioned. In addition, antibodies against tenascin inhibit cranial neural crest cell migration. Still other disturbances were observed after injection of antibodies against cell adhesion molecules such as *N-cadherin* or *N-CAM* (Bronner-

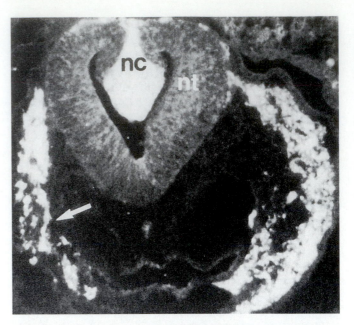

Figure 26.20 Effects of a univalent antibody against the β chain of an integrin on neural crest cell development in the chicken embryo. This fluorescence micrograph shows a transverse section of the head of an embryo fixed 18 h after antibody injection. The brightly stained cells are neural crest cells. On the injected side (arrow), neural crest cells are less numerous than on the uninjected side and are distributed differently. A large mass of neural crest cells (nc) is also found abnormally within the neural tube (nt).

Fraser et al., 1992). Taken together, these observations corroborate the conclusion drawn from other results that neural crest cell migration is a complex process involving several ECM components as well as cell adhesion molecules.

Cranial and Trunk Neural Crest Cells Attach to ECM Components through Different Mechanisms

Interestingly, the same antibodies that interfere with the development of *cranial* neural crest cells in vivo have no comparable effects on the development of *trunk* neural crest cells. In vitro experiments confirm that cranial and trunk neural crest cells attach to ECM components through different mechanisms (Lallier et al., 1992). Under certain culture conditions, trunk neural crest cells attach to type I and type IV collagens, but cranial neural crest cells do not. For adhesion to fibronectin substrate, trunk neural crest cells require divalent cations such as the calcium ion (Ca^{2+}), whereas cranial neural crest cells bind with or without such ions. For laminin substrate, trunk neural crest cells require two integrins while cranial neural crest cells use only one.

In vivo perturbation experiments indicate that fibronectin, integrins, laminin–heparan sulfate complex, and tenascin play a critical role in the head but seem to have only secondary or overlapping functions in the trunk. Conversely, tissue-derived clues such as differences between anterior and posterior sclerotomes and inhibitory signals from the notochord appear to function mainly in the trunk region. Moreover, transplantation experiments described in Chapter 12 show that cranial neural crest cells have a wider range of potencies than trunk neural crest cells. Thus, the two populations of neural crest cells differ markedly with respect to their control mechanisms for both migration and determination.

Position-Specific Integrins in *Drosophila melanogaster*

Drosophila integrins, like their counterparts in vertebrates, play important roles in cell differentiation and tissue cohesion (Bunch and Brower, 1993; N. H. Brown, 1993). Like their vertebrate homologues, the *Drosophila* integrins consist of two chains, an α and a β chain, both of which are large transmembrane glycoproteins (Fig. 26.7). The two known *Drosophila* integrins are called **PS1** and **PS2**, because their expression in imaginal discs is *position-specific*. PS1 and PS2 have the same β chain, β_{PS}, but different α chains, α_{PS1} and α_{PS2}. As in vertebrates, integrins in *Drosophila* have alternatively spliced regions and typically seem to be anchored by their cytoplasmic domains to microfilaments.

All available data indicate that the PS integrins function in ECM binding. As a functional test of this notion, Thomas Bunch and Danny Brower (1992) transfected a line of cultured *Drosophila* cells that express little if any endogenous PS integrin. After transfection with cDNAs for β_{PS} and α_{PS2}, the cells expressed PS2 on their surfaces, as shown by immunostaining. Unlike nontransfected control cells, the transfected cells spread on coats of purified vertebrate fibronectin and a similar interactive glycoprotein, *vitronectin*. The spreading effect was abolished when RGD peptide was added to the culture medium. An ECM component that functions as a ligand for *Drosophila* PS2 integrin is **tiggrin**, a large polypeptide that contains the RGD sequence and is present at muscle attachments and other sites where cells express integrins (Fogerty et al., 1994).

The subcellular distribution of PS integrins is also consistent with a role in cell-ECM binding, since the strongest immunostaining is observed at the basal cell surfaces (Brower et al., 1984, 1985). Moreover, the distribution of PS integrins in the wing *imaginal disc* reflects the subdivision of the wing into dorsal and ventral *compartments*: PS1 is expressed mainly in dorsal cells and PS2 primarily in ventral cells (Fig. 26.21). Indeed, the region-specific expression of PS1 and PS2 is controlled by the *apterous+* gene, which is expressed only in the dorsal wing compartment and maintains the dorsal-ventral compartment boundary in the wing

(Blair et al., 1994). Ectopic ventral expression of *apterous+* causes ectopic ventral synthesis of α_{PS1} mRNA and PS1 integrin, while loss of apterous function can cause the ectopic dorsal synthesis of PS2 integrin. Thus, the function of *apterous+* as a *selector gene* includes the region-specific synthesis of PS1 and PS2.

The analysis of PS integrin function in *Drosophila* is greatly facilitated by the availability of mutants. The β_{PS} chain is encoded by the *myospheroid+* gene, and the α_{PS2} chain is encoded by the *inflated+* gene. Both genes were known from their mutant phenotypes before it was recognized that they encode integrin subunits. Loss of function in either gene is lethal in the embryonic stage and associated with similar phenotypic defects. The gene encoding the α_{PS1} chain is also under active investigation.

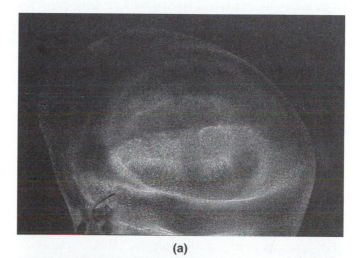

(a)

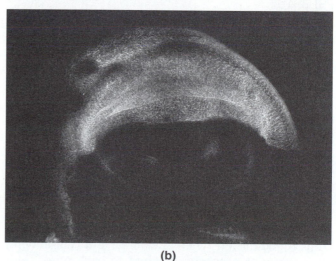

(b)

Figure 26.21 Expression of PS1 and PS2 integrins in *Drosophila* imaginal discs. **(a)** Late third-instar disc immunostained with a monoclonal antibody recognizing PS1 integrin, which is confined mostly to the dorsal (lower) half of the disc. **(b)** Similar disc stained with a monoclonal antibody recognizing PS2 integrin, which is restricted to the ventral (upper) half of the disc.

The *myospheroid*[+] gene was so named because in mutant embryos carrying loss-of-function alleles of the gene, the skeletal muscles pull away from the body wall and form round bodies of muscle tissue (T. R. F. Wright, 1960; Newman and Wright, 1981). The skeletal muscles appear to develop normally and even attach to the correct places on the epidermis, but when the muscles begin to contract, they detach and become rounded. The interactions between the visceral muscles and the gut are similarly abnormal. In accord with these phenotypic defects, wild-type embryos express PS integrins at the interface between embryonic mesoderm and ectoderm, and later at the attachment sites of muscles to epidermis and gut (Leptin et al., 1989). Another prominent defect in *myospheroid* embryos is a large hole in the epidermis on the dorsal surface, where epidermal sheets from the right and left sides would normally grow together and fuse in *dorsal closure* (see Chapter 21). In the mutant embryos, the two epidermal sheets grow together but are pulled apart again when muscular contractions begin. Together, wild-type expression patterns and mutant phenotypes indicate that PS integrins in *Drosophila* are required for the attachment of various tissues to one another.

Loss of function in the *inflated* gene, which encodes the α_{PS2} chain, generates mutant phenotypes that are similar to though not quite as drastic as those caused by loss of function in *myospheroid*, which encodes the β_{PS} chain (N. H. Brown, 1994). This difference reflects the synthetic role of the two gene products: loss of α_{PS2} still leaves the embryo with the ability to produce PS1 integrin, whereas embryos without β_{PS} cannot synthesize PS1 or PS2.

Although complete loss of either *myospheroid* or *inflated* function is lethal during embryogenesis, mutants with alleles that retain part of the normal gene function may survive to adulthood (Brower and Jaffe, 1989). From such individuals it has been learned that these two genes have critical functions as well during postembryonic development of wing, eye, and muscle (Zusman et al., 1993). We will focus on wing development.

When the *Drosophila* adult emerges from the pupal case, the wing is converted from an epithelial bag into a flat, bilayered blade. This process juxtaposes the dorsal wing layer, which expresses PS1, and the ventral wing layer, which expresses PS2 (Fristrom et al., 1993). Mutants with partial loss of function in *inflated* have large, round blisters where the dorsal and ventral wing layers are not properly attached to each other (Fig. 26.22). Similar blisters form when cell clones homozygous for *myospheroid* or *inflated* loss-of-function alleles are generated by *somatic crossover* (see Methods 6.1). Such experiments are especially instructive if cell clones are made homozygous for both a loss-of-function allele of a PS integrin gene and a closely linked cuticular

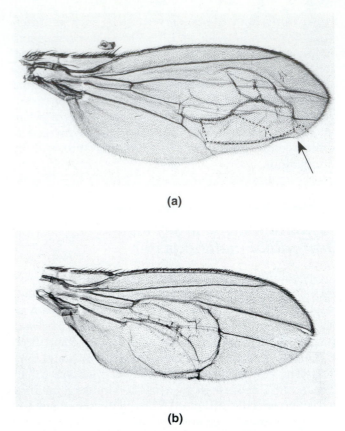

(a)

(b)

Figure 26.22 Blistering in *Drosophila* wings caused by mutations in PS integrin genes. **(a)** Blistering caused by X-ray–induced clones homozygous for a null mutation in the *myospheroid* gene. The blister is larger than the clone (marked by dotted lines). **(b)** Wing from a fly homozygous for a partial loss-of-function allele of the *inflated* gene.

marker allele. The results show that *inflated* clones on the ventral wing surface are always associated with blisters whereas corresponding clones on the dorsal wing surface are never associated with blisters (Brabant and Brower, 1993). It follows that the function of *inflated*[+], which is necessary for synthesis of PS2 but not PS1, is required on the ventral wing layer but not on the dorsal.

The functional significance of the alternative splicing of myospheroid pre-mRNA was tested by P-element–mediated transformation to introduce transgenes producing only one of the two spliced forms of β_{PS} (Zusman et al., 1993). Either isoform is sufficient to rescue the *myospheroid*[−] mutant phenotype in the wing, but both spliced forms are necessary to rescue the embryonic defects.

Taken together, the available data indicate that the *myospheroid*[+] and *inflated*[+] genes have critical functions during embryonic and postembryonic development. At each step, the products of these genes are required to hold different tissues together. In particular, the dorsal and ventral layers of the *Drosophila* wing are attached to each other by the formation of PS1 in the upper wing layer and PS2 in the lower wing layer, and presumably by the binding of both integrins to as yet unidentified ECM components between the two layers.

The presence of PS integrins in *Drosophila* is significant in several respects. First, it demonstrates that integrins have been conserved in evolution for more than 600 million years, while insects and vertebrates have evolved independently. Second, mutant alleles and transgenes in *Drosophila* are powerful research tools that are not available in most other organisms. Finally, it appears that the number of integrins and their substrates is smaller in *Drosophila* than in vertebrates. With fewer components, this system should require less effort for a satisfactory analysis.

Morphoregulatory Roles of the ECM

As discussed in Chapter 25, cell adhesion molecules (CAMs) function as *morphoregulatory molecules* (see Fig. 25.22). Gerald Edelman (1988) proposed similar morphoregulatory functions for **substrate adhesion molecules (SAMs),** a collective term he coined for ECM molecules and their cellular receptors. The morphoregulatory role of SAMs can be summarized in a simple model (Fig. 26.23). A few selector genes are thought to control the stage-dependent and region-specific activity of SAM genes. SAM gene expression generates SAMs, which mediate cell-ECM adhesion. Cell-ECM interactions affect cell shape and movements, two determinants of morphogenesis. Closing the cycle, cell-ECM adhesion and the resulting morphogenetic events feed back into selector gene expression. The feedback may occur through *inter*cellular signals between new cell neighbors and/or through *intra*cellular signals that depend on cell shape or integrin-ECM binding. The altered selector gene activity then starts a new morphoregulatory cycle.

A possible shortcut in the morphoregulatory cycle is a direct feedback from an altered ECM composition to selector gene activity. Such a shortcut would be a consequence of the ECMs' ability to bind *growth factors* that regulate gene activity without the need for physical alterations in cell shape or behavior. Gene regulation by ECM-bound growth factors was mentioned previously in the context of hematopoiesis (see Chapter 19) and will be discussed further in Chapter 29.

Two steps in the morphoregulatory cycle—the synthesis and deployment of SAMs and the effects of SAM activities on cellular behavior—have already been illustrated in this chapter. In the remainder of the chapter,

we will consider some experiments that demonstrate the other two steps in the morphoregulatory cycle.

ECM Components Trigger Cellular Reorganizations That Affect Gene Expression

Changes in the ECM often cause rearrangements in cells' surfaces and cytoskeletons, particularly in the cortex of microfilaments beneath the plasma membrane. It appears that the organization of these microfilaments, which are linked to the cytoplasmic domains of integrins, depends on the binding of the extracellular integrin domains to ECM components. Changes in the cytoskeletal organization, in turn, can have dramatic effects on gene expression. The following examples will illustrate these connections.

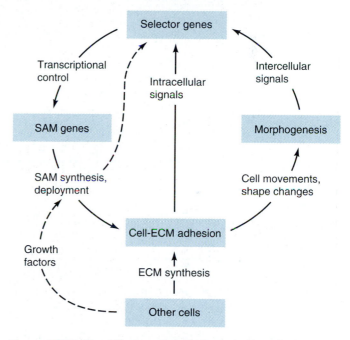

Figure 26.23 Simplified model of the function of substrate adhesion molecules (SAMs) in a morphoregulatory cycle. Selector genes control the activity of SAM genes. SAM gene expression modulates cell-ECM adhesion, which is also affected by ECM synthesized by other cells. Cell-ECM adhesion directs cellular behaviors such as cell shape changes and cell migration, two activities that contribute to morphogenesis. Morphogenetic events in turn may generate new cell contacts, which provide for the exchange of new intercellular signals. In addition, altered cell shapes and movements may affect intracellular signaling through membrane receptors or ion channels that respond to mechanical stimuli. Both intercellular and intracellular signals feed back into the expression of selector genes, thus starting a new morphoregulatory cycle. A possible shortcut from SAM gene activities back to selector genes (represented by dashed arrows) may be based on the ability of ECM molecules to bind gene-regulatory signal substances such as growth factors. See also Figure 25.22.

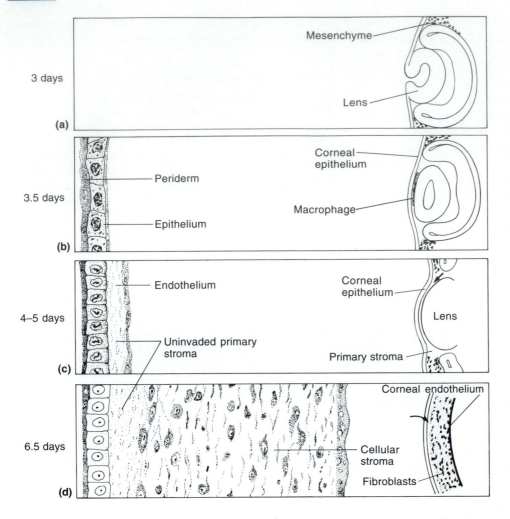

Figure 26.24 Cornea development in the chick embryo. The drawings to the left show corneas at successive stages; the diagrams to the right outline the overall development of the eye at the same stages. **(a)** 3 days of incubation. The lens vesicle pinches off, and the overlying epithelium begins to form the cornea. **(b)** 3.5 days. Macrophages clean up debris from lens vesicle formation. Prospective cornea consists of epithelium and overlying periderm. **(c)** 4–5 days. The corneal epithelium has secreted the primary corneal stroma, and surrounding mesenchymal cells invade the area (straight arrow) to form the corneal endothelium. **(d)** 6.5 days. Mesenchymal cells have invaded the entire stroma except for a shallow zone next to the epithelium (curved arrow). See also Figure 26.1.

The *cornea* of the eye is a multilayered structure consisting of an outer epithelium, an inner endothelium, and an intervening tissue called *stroma.* The stroma contains fibroblasts in a collagen-rich ECM. Cornea formation involves several steps. When the lens pinches off as a vesicle, the overlying epithelium closes up, and any debris left from this process is removed by macrophages (Fig. 26.24a and b). At this stage, the epithelium overlying the lens vesicle differs little from the surrounding ectoderm, which will form epidermis. However, under the influence of the lens vesicle, the prospective cornea cells secrete at their basal surface the *primary stroma,* a collagenous layer of ECM (Fig. 26.24c). Next, mesenchymal cells invade the corneal area and form the corneal endothelium, which discharges abundant glycosaminoglycans into the stroma. Subsequently, mesenchymal cells derived from neural crest cells invade the stroma as fibroblasts (Fig. 26.24c and d).

emanating from the *lens capsule,* a thick basal lamina surrounding the lens. Isolated corneal epithelium placed on a lens capsule in vitro produces primary stroma, while corneal epithelium placed on plastic does not. When the lens capsule is replaced by collagen-containing substrata, the corneal epithelium still forms stroma, but noncollagenous substrata do not induce stroma formation (Fig. 26.25).

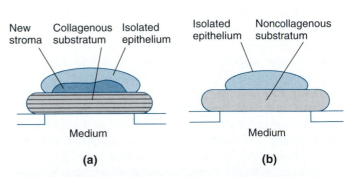

Figure 26.25 Stroma formation by corneal epithelium in vitro. **(a)** Isolated corneal epithelium placed on substrata containing collagen is induced to form stroma as it does in vivo (see Fig. 26.24). **(b)** In contrast, corneal epithelium on noncollagenous substrata does not form stroma.

A series of experiments by Elizabeth Hay (1981) and her coworkers has shown that cornea formation is induced by the underlying lens and depends on signals

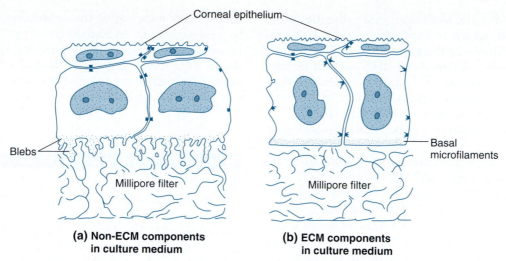

Figure 26.26 Effect of soluble ECM components on the behavior of corneal epithelial cells in culture. Corneal epithelium is isolated and placed on a millipore filter that is soaked from below with culture medium. After the adhering stroma has been removed enzymatically, the basal surfaces of the epithelial cells form blebs. **(a)** Bleb formation continues if the culture medium contains non-ECM components such as albumin or immunoglobulin. **(b)** If the culture medium contains ECM components such as soluble collagen (types I to IV), fibronectin, or laminin, the blebbing subsides, the basal cell surfaces become smooth, and the basal microfilaments in the cell cytoplasm re-form.

In an extension of these experiments, corneal epithelium was isolated and placed on millipore filters that were soaked from below with culture medium (Hay, 1991b). Before being placed on the filter, the epithelium was immersed in enzymes to remove any adhering stroma. In response to this treatment, the basal surfaces of the epithelial cells formed blebs. Bleb formation continued during subsequent culture unless the culture medium contained certain ECM components (Fig. 26.26). When soluble collagen, laminin, or fibronectin was present in the medium, it bound to the basal surface of the corneal epithelium, and the blebs disappeared. The corneal epithelium then formed a smooth basal surface, reorganized its basal microfilaments, and synthesized large amounts of collagen as it would do in vivo as part of stroma formation. These changes were correlated. Their extent depended on the concentration of the ECM component in the medium, and the same maximum effect was obtained with collagen, laminin, or fibronectin (Sugrue and Hay, 1986).

The results of Hay and coworkers confirm that the lens capsule of the eye induces the corneal epithelium to synthesize the primary stroma. The first detectable step in the induction process is a reorganization of the basal surface and the underlying microfilaments in the epithelial cells. This event seems to stem from the extracellular binding of collagen, laminin, or fibronectin to matching integrins, which are also linked to micro-

filaments in the basal cytoplasm of the epithelial cells. The reorganization of the basal cell surface and microfilaments in turn appears to activate gene expression in the corneal epithelium—in particular, the synthesis of collagen as an ingredient of the primary stroma.

Similar observations have been made with other organisms and cells. In sea urchin blastulae, β-aminopropionitrile (BAPN) arrests gastrulation by interfering with the normal assembly of collagen fibrils. In the arrested embryos, most genes are transcribed normally, but a few genes are inactivated selectively, presumably as a result of the suppressed gastrulation (Wessel et al., 1989). In cultured vertebrate cells, both cell shape and the types of protein synthesized may depend on whether the cells are suspended in gels or kept on flat surfaces. For example, rabbit chondrocytes (cartilage-forming cells) are round and synthesize large amounts of type II collagen and chondroitin sulfate in vivo. They retain the same shape and synthesize these same proteins in large quantities when suspended in an agarose gel (Benya and Shaffer, 1982). However, when kept on flat surfaces in monolayer cultures, the cells are spindle-shaped like fibroblasts and synthesize mostly type I collagen and low levels of proteoglycans. Thus, these cells can be driven back and forth between two states of cell shape and gene expression depending on the substrate used for cell culture.

These observations lend themselves to three different interpretations, each of which may be valid under certain conditions in vivo. First, ECM binding may affect cell shape and gene expression independently of each other. Second, ECM binding may directly affect gene expression, which in turn may affect cell shape. Third, ECM binding—or the substrate used for cell culture—may directly influence cell shape, which in turn may affect gene expression. The third possibility is supported by the fact that integrins link the ECM with microfilaments and by the following experiments, which show that the microfilament system can regulate gene expression.

In a study on the connection between the state of the cytoskeleton and gene activity, Nina Zanetti and Michael Solursh (1984) monitored the effects of cytoskeletal inhibitors on *chondrogenesis* (cartilage formation) in cultured chicken limb bud cells. To detect chondrogenesis, the researchers used immunostaining of type II collagen, and alcian blue staining of acidic glycosaminoglycans. Without cytoskeletal inhibitors, the cells spread out on plastic culture dishes, developed numerous microtubules and actin bundles, and did not initiate chondrogenesis. When cytochalasin D, a microfilament inhibitor, was added to the medium, the cells became round, lost their actin cables, and began chondrogenesis. Agents that disrupt microtubules had no apparent effect. These results indicate that the state of the microfilament system affects gene activities involved in the synthesis of type II collagen and glycosaminoglycans.

The molecular mechanisms by which the state of the microfilament system might influence gene expression are not clear. The mechanical tension generated by microfilaments between different ECM adhesion sites may change the permeability of stretch-activated membrane channels, which in turn affects gene expression. More directly, the binding of ECM to its cellular receptors may yield intracellular signals similar to those generated when other membrane receptors are occupied by their chemical or mechanical ligands (see Chapters 2 and 4). In accord with this speculation, integrins were found to regulate the phosphatidylinositol cycle, sodium/hydrogen ion exchange pumps, and specific protein phosphorylations (Hynes, 1992).

Expression of the Tenascin Gene Is Controlled by a Homeodomain-Containing Transcription Factor

The model of a morphoregulatory cycle for substrate adhesion molecules presented in Figure 26.23 implies that selector genes drive the expression of SAM genes. Several selector genes encode transcription factors with homeodomains that bind to matching recognition motifs in their target genes. To identify SAM genes containing such enhancer motifs, researchers have used the same type of cotransfection experiment described in Chapter 25.

Frederick Jones and his colleagues (1992b) cotransfected cultured human 3T3 cells with a *reporter plasmid* and an *effector plasmid*. The reporter plasmid contained the 5′ end and flanking sequences of the chicken tenascin gene spliced to the transcribed portion of the bacterial chlorramphenicol acetyltransferase (CAT) gene. The effector plasmid, containing a viral promoter and the mouse *Evx-1*⁺ gene, caused the host cells to produce Evx-1 protein, a homeodomain-containing transcription factor, similar to the even-skipped protein of *Drosophila*. A control effector plasmid was constructed similarly, except that it encoded another homeodomain-containing transcription factor, the Hoxa-5 protein. Cells cotransfected with the reporter plasmid and the *Evx-1*⁺ effector plasmid showed high levels of CAT activity, whereas control cells transfected with the same reporter plasmid and the *Hoxa-5*⁺ effector had no detectable CAT activity. Evidently, Evx-1 protein activated the chicken tenascin gene. *Deletion mapping* indicated that the activation occurred indirectly through transcription factors encoded by the host cell genome. Despite this complication, the results support the hypothesis that homeobox-containing selector genes may control the local activation of SAM genes such as tenascin.

Together, the experiments discussed in this section are compatible with the concept that ECM molecules and integrins act as morphoregulatory molecules. These molecules affect the shape and behavior of cells so that they exchange new signals with one another or modify their own intracellular signaling. The altered signals then bring about new combinations of active selector genes. Selector genes, in turn, control the expression of multiple realizator genes, including the genes encoding ECM components and integrins.

SUMMARY

Cells synthesize an extracellular matrix (ECM). In epithelia, the ECM is formed as a basal lamina, whereas in other tissues the ECM surrounds cells on all sides. The ECM consists of a meshwork of fibers embedded in an amorphous ground substance. The ground substance consists of glycosaminoglycans and proteoglycans, which form extended random coils that hold water by osmotic pressure. The fibrous ECM components are glycoproteins that reinforce the ground substance and resist its expansive forces. Because the ECM components are linked to one another by multiple binding regions, they form a stable, multifunctional matrix.

Cells are linked to ECM components by surface proteoglycans that are inserted into the plasma membrane and by heterodimeric receptor proteins called integrins. The α and β subunits of integrins have extracellular domains, which bind to certain ECM sites; transmembrane domains; and cytoplasmic domains linked to microfilaments. ECM molecules and integrins have been highly conserved during evolution.

Cells and ECM interact in many ways. Cells synthesize stage-dependent and region-specific ECM components in accord with prevailing cell activities. Cells also orient ECM fibers by means of their integrins or by exerting tension. Conversely, ECM components guide cell movements by adhesion and contact guidance.

ECM components, cell surface proteoglycans, and integrins are collectively called substrate adhesion molecules (SAMs). Like cell adhesion molecules, SAMs are thought to play key roles in morphoregulatory cycles. Their stage-dependent and region-specific distributions cause cell shape changes and migrations that generate new intracellular and intercellular signals. These signals may alter the activity of certain selector genes, which in turn regulate SAM genes and other realizator genes.

SUGGESTED READINGS

Bronner-Fraser, M. 1993. Environmental influences on neural crest cell migration. Journal of Neurobiology **24**:233–247.

Brown, N. H. 1993. Integrins hold Drosophila together. BioEssays **15**:383–390.

Hay, E. D. 1991. Collagen and other matrix glycoproteins in embryogenesis. In E. D. Hay, ed., Cell Biology of Extracellular Matrix, 419–462. New York: Plenum.

Hynes, R. O. 1992. Integrins: Versatility, modulation, and signaling in cell adhesion. Cell **69**:11–25.

Johnson, K. E., N. Nakatsuji, and J.-C. Boucaut. 1990. Extra-cellular matrix control of cell migration during amphibian gastrulation. In G. M. Malacinski, ed., Cytoplasmic Organization Systems, 349–374. New York: McGraw-Hill.

SEX

DETERMINATION

Figure 27.1 Sexual dimorphism. In mammals and many other groups of animals, males and females differ not only in their reproductive organs but also in other physical and behavioral traits. In deer, males are considerably larger and have conspicuous antlers, which they use for fighting. The development of sex-specific traits in mammals is controlled by the SRY^+ gene on the Y chromosome, which is present in males but absent in females.

Most animal species are **dioecious** (Gk. *di-*, "double"; *oikos*, "house"), meaning that individuals belong to two sexes with different reproductive organs. Typically these are females producing eggs and males producing sperm. In many dioecious species, the differences between the sexes are not limited to reproductive organs and behavior but extend to other characteristics, such as size, color, ornaments, and weaponry. This situation is called **sexual dimorphism** (Gk. *di-*; *morphe*, "form"). The degree of sexual dimorphism varies: among vertebrates, mammals (Fig. 27.1) and birds tend to be more sexually dimorphic than fishes and reptiles.

A large number of invertebrates are **monoecious**, meaning that all individuals of a species look alike and produce both eggs and sperm. Such individuals, which have both testes and ovaries, are called *hermaphrodites*. However, some dioecious species include hermaphrodites, too. For instance, roundworms of the species *Caenorhabditis elegans* are either hermaphrodites or males, as described in Chapter 24. Males and hermaphrodites of this species are as distinct as males and females of other types of animals.

The natural mechanism by which an individual of a dioecious species becomes either male or female (or hermaphroditic) is called **sex determination**. An organism's sex has profound consequences for the course of its development. It determines whether the germ cells will develop into eggs or sperm, and—especially in sexually dimorphic species—it is responsible for many somatic traits.

Biologists have uncovered many mechanisms of sex determination, which fall into two categories (Table 27.1). In some species, sex is determined *after* fertilization by environmental factors such as temperature or the presence of sex partners. This type of sex determination is known as **environmental sex determination.** In other species, sex is determined *at* fertilization by the combination of genes that the zygote inherits. This type of sex determination is called **genotypic sex determination.**

In this chapter, we will first describe some examples of environmental sex determination and then focus on genotypic sex determination. The latter has been investigated especially well in three types of animals: the fruit fly *Drosophila*, the roundworm *C. elegans*, and placental mammals, represented by the human and the mouse. A comparison of sex determination in these organisms reveals that certain biological processes are common to all species with genotypic sex determination. However, the genetic and molecular mechanisms that control these processes differ widely and seem to have evolved separately in many taxonomic groups.

Environmental Sex Determination

A well-known case of environmental sex determination is *Bonellia viridis*, a marine echiuroid worm (Leutert, 1975). Adult females of this species attach to rocks in the ocean. The body of the female measures more than 10 cm, and a long proboscis extends from it for feeding. The males, which at 1 to 3 mm are tiny by comparison, live as parasites inside the females. The larvae of *Bonellia* live as part of the *plankton,* that is, the small organisms that passively drift in the water. Sex determination occurs when the larvae settle on a substrate and metamorphose. Those that settle in isolation develop into fe-

TABLE 27.1

Examples of Different Sex Determination Systems

Species	Mechanism	Sexes	
Mouse, human	GSD: dominant Y	XX: female	XY: male
Birds	GSD: dominant W?	ZW: female	ZZ: male
Turtles	ESD: temperature	Warm: female	Cool: male
Alligators	ESD: temperature	Cool: female	Warm: male
Insects:			
Drosophila melanogaster	GSD: X:A ratio	XX: female	XY: male
Musca domestica	GSD: dominant *M* locus	*m/m*: female	*M/m*: male
Apis mellifera	GSD: haplodiploidy	Diploid: female	Haploid: male
Nematodes:			
Caenorhabditis elegans	GSD: X:A ratio	XX: hermaphrodite	XO: male
Meloidogyne incognita	ESD: population density	Sparse: female	Crowded: male
Echiuroid worm:			
Bonellia viridis	ESD: female presence	Yes: male	No: female

GSD = genotypic sex determination; ESD = environmental sex determination.
Source: Modified from Hodgkin (1992). Used with permission from Bio Essays. © *FCSU Press.*

males, whereas those that fall on an adult female become males.

Two hypotheses could explain this single observation: environmental sex determination, or genotypic sex determination coupled with differential mortality. According to the latter hypothesis, genotypically male larvae would survive if they settled on females but die if they settled in isolation, and genotypically female larvae would survive in isolation but die if they landed on existing females. To test this hypothesis, Leutert (1975) kept *Bonellia* in laboratory culture. He observed that *more than half* of a known number of larvae developed as males under the influence of females, whereas *more than half* of the initial number of larvae developed as females in isolation. This result supported the environmental hypothesis, because if the hypothesis of genotypic sex determination and differential mortality were valid, then each of these numbers should have been *half or fewer*. Substances causing male development have been extracted from the proboscis and the gut of females but remain to be characterized.

Dependence of sex determination on social context is also observed in several species of fish (Crews, 1994). Orange and white anemone fish are born male and later develop into females. The opposite course is observed in several coral reef fish, which start out female and later become male. The timing of the sex change depends on a social trigger, such as the disappearance of a dominant male or female. Still other fish species, in-

cluding the butter hamlet, have gonads that produce eggs and sperm simultaneously; individuals may alternate between male and female behavior during successive matings. All types of sex determination that depend on social context seem to maximize the chance of reproduction in small groups of individuals.

An environmental sex determination strategy that maximizes the number of offspring that can be produced under favorable conditions is observed in a nematode, *Meloidogyne incognita*, which is a plant parasite. If nutrients are plentiful and the population density is low, the worms become female; under adverse conditions, they become male.

A puzzling type of environmental sex determination is observed in many reptiles. In all crocodilians, many turtles, and some lizards, sex is determined during embryogenesis by the incubation temperature (J. J. Bull, 1980, 1983; Crews, 1994). These reptiles are egg layers that place their eggs in the ground or in constructed mounds. The temperature of the external environment is therefore the sex-determining factor, and a small temperature difference may cause dramatic changes in the *sex ratio* (proportion of males among all individuals of a population). For instance, in certain map turtles, the sex ratio drops from 1.0 (all males) to 0.0 (all females) as the ambient temperature increases from 28 to 30°C (Fig. 27.2a). This pattern of cool temperatures producing males and warm temperatures yielding females is common among turtles. The pattern is reversed in

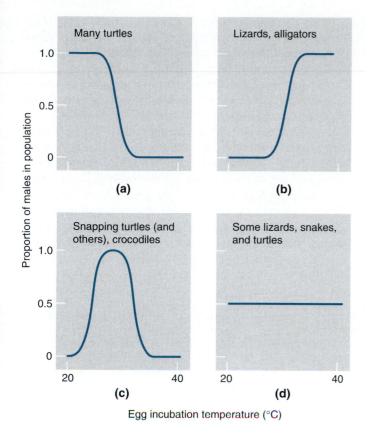

Figure 27.2 Sex ratio (proportion of males in total population) as a function of egg incubation temperature in different reptiles. Four general patterns are recognized. **(a)** In many turtles, males develop at low temperatures and females at high temperatures. **(b)** The reverse is seen in many lizards and alligators: females develop at low temperatures, males at high temperatures. **(c)** A pattern of females at low and high temperatures and males at intermediate temperatures occurs in the snapping turtle, the leopard gecko, and crocodiles. **(d)** In other reptiles, the sex ratio is not significantly influenced by the incubation temperature.

as ovaries or as testes. After this event, the sex of the animal is fixed for life. This pattern has many similarities with the sexual development of mammals, which also have a sexually indifferent period, which lasts through the sixth week of pregnancy in humans (see Chapter 28). However, the critical event that directs the indifferent gonads to become either testes or ovaries is triggered by temperature in reptiles as opposed to gene activities in mammals.

Genotypic Sex Determination

Environmental sex determination was the dominant model until the beginning of the twentieth century (Mittwoch, 1985). In particular, human sex determination was ascribed to environmental factors ranging from nutrition to the heat of passion during coitus. However, in humans as well as most other organisms, the sex ratio is close to 0.5 and is remarkably resistant to environmental manipulation. After Mendel's laws gained widespread recognition and chromosomes were recognized as the carriers of genetic information, geneticists began to speculate that a sex ratio of 0.5 could be achieved simply if one sex was homozygous and the other sex heterozygous for some sex-determining gene. Soon thereafter, sexually dimorphic chromosomes (called *sex chromosomes*, for short) were found in several species, including insects (McClung, 1902; E. B. Wilson, 1905) and humans (Painter, 1923).

The well-known mechanism of sex determination in humans and other mammals is diagrammed in Figure 27.3. Here, males have two different sex chromosomes, designated X and Y, whereas females have two X chromosomes. All other chromosomes are indistinguishable

lizards and alligators (Fig. 27.2b). A third pattern of temperature-dependent sex determination, in which females develop at low and high temperature extremes and males develop at intermediate temperatures, is observed in snapping turtles, the leopard gecko, and crocodiles (Fig. 27.2c). In still other reptiles, the sex ratio is not significantly affected by incubation temperature (Fig. 27.2d). The adaptive values of these differing patterns are not obvious (J. J. Bull, 1983). Presumably, these patterns allow males to develop at temperatures that benefit males more than females, and vice versa. It will be interesting to find out whether females adaptively modify the sex ratio of their offspring by choosing appropriate nest sites.

In reptiles with temperature-dependent sex determination, the sex is determined by the temperature that prevails midway through embryonic development, when the originally indifferent gonads develop either

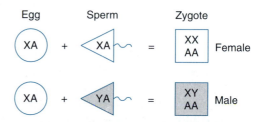

Figure 27.3 Sex determination in mammals. In addition to two sets of autosomes (AA), males have two different sex chromosomes, designated X and Y, whereas females have two X chromosomes. Consequently, males produce two types of sperm with different sets of chromosomes, XA and YA, while females produce only one chromosomal type of egg, XA. Fertilization by an XA sperm results in an XXAA zygote, which develops into a female. Fertilization by a YA sperm results in an XYAA zygote, which develops into a male. This system generates a sex ratio of 0.5, provided that XA and YA sperm have equal chances at fertilization, and that XXAA and XYAA zygotes are equally viable.

between the sexes and are called *autosomes*. A complete haploid set of autosomes is designated A. Consequently, males produce two types of sperm with different sets of chromosomes, XA and YA, and are therefore called *heterogametic*. Females produce only one chromosomal type of egg, XA, and are called *homogametic*. Fertilization by an XA sperm results in an XXAA zygote, which develops into a female. Fertilization by a YA sperm results in an XYAA zygote, which develops into a male.

The heterogametic sex is not necessarily the male. In birds, for instance, females are heterogametic, and their sex chromosomes are traditionally designated ZW. Male birds are homogametic, and their sex chromosomes are designated ZZ.

Instead of occurring in two *forms*, sex chromosomes may also occur in two *numbers*—typically, two chromosomes in one sex and one chromosome in the other sex. In the most common system of this type, exemplified by *C. elegans*, females have XXAA diploid cells and produce XA eggs, whereas males have XOAA diploid cells and produce sperm that are either XA or OA, with O representing *no* chromosome.

In yet another system of genotypic sex determination, females arise from fertilized eggs whereas males arise from unfertilized eggs. This mechanism, known as **arrhenotoky** (Gk. *arrhen*, "male"; *tokos*, "childbirth"), is common among ants, bees, and wasps. Finally, there are species whose chromosomes are not visibly dimorphic but whose sex determination is nevertheless genotypic. In the simplest case, sex is determined by two alleles of the same gene. One sex is homozygous (*m/m*) for one allele and the other sex is heterozygous (*M/m*). This system is found in the housefly, *Musca domestica*, which has *m/m* females and *M/m* males. In more complex cases, sex is determined by several genes, each having a small effect.

As this survey suggests, genotypic sex determination systems have changed frequently during evolution. A survey of sex-determining systems in amphibians revealed widespread variation in the heterogametic sex and in the extent of sex chromosome dimorphism (Hillis and Green, 1990). Among 63 species of frogs and salamanders, researchers found ZW female heterogamety, OW female heterogamety, and XY male heterogamety. Phylogenetic analysis suggests that the ancestral state for amphibians was female heterogamety, and that XY male heterogamety has evolved independently at least seven times. In one case, the data suggest that male heterogamety has reversed to female heterogamety.

The apparent instability of genotypic sex-determining mechanisms can be explained by the accumulation of deleterious mutant alleles on the sex chromosome that occurs only in the heterogametic sex (Hodgkin, 1992; Rice, 1994). The following model assumes that genotypic sex determination starts simply with an allelic

difference in one sex gene in a homologous pair of chromosomes (*M/m* versus *m/m*, as shown in Figure 27.4). Under conditions of sexual reproduction, the chromosomal region around the *M* gene would never be homozygous and would therefore accumulate recessive loss-of-function alleles because they are not selected against. In addition, there would be a positive selection for alleles near *M* that are beneficial to the heterogametic sex but harmful to the homogametic sex. As a result of both these trends, the originally homologous chromosomes would become increasingly different, thus generating an X/Y chromosomal dimorphism and making *crossing over* events between the two chromosomes less and less likely. Eventually, the Y chromosome would have no functional genes on it except for those that have a use in sex determination or a specific role in the het-

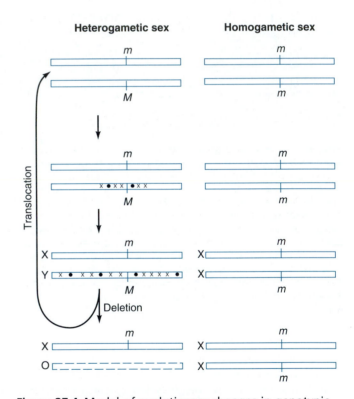

Figure 27.4 Model of evolutionary changes in genotypic sex determination. As a starting point, the model assumes an allelic difference in one sex gene (*M/m* versus *m/m*) located on a homologous pair of chromosomes. The chromosomal region around the *M* allele never becomes homozygous and therefore accumulates recessive loss-of-function alleles (crosses) and alleles that are beneficial to *M/m* but harmful to *m/m* individuals (dots). This process generates an X/Y chromosomal dimorphism, and the Y chromosome eventually has no genes left except for a few that are involved in sex determination or functions specific to the heterogametic sex. Finally, the *M* allele itself may become unreliable in its expression, so that the species could lose its heterogametic sex. This consequence would be avoided if the *M*-carrying chromosome region translocated to an autosome, thus restarting the cycle just described, or if the sex-determining mechanism changed to an XX/XO system.

erogametic sex. Being located on an almost nonfunctional chromosome, the *M* allele itself might become unreliable in its expression, so that the species would lose its heterogametic sex. Two events could avert this consequence. First, the *M*-carrying chromosome region might translocate to an autosome, an event that would restart the cycle discussed so far. Second, the sex-determining mechanism could become an XX/XO system, in which sex is determined by the number (dosage) of *m* or other genes located on the X chromosome.

The model just described is supported by several lines of evidence. First, the genotypic sex-determining mechanisms have changed frequently during evolution, as we have seen. Second, the Y and W chromosomes of many species contain few genes other than the sex-determining gene(s). Third, transpositions of a male-determining gene seem to occur naturally in the housefly (Franco et al., 1982). Fourth, XX/XO systems are widespread and have probably evolved independently in many species.

Sex Determination in Mammals

Sex determination in mammals is genotypic, and the mechanism seems to be fairly well conserved between mice and humans. This conservation has facilitated the genetic and molecular analysis of mammalian sex determination, because data gathered from humans and mice complement each other. On the one hand, abnormal sex chromosomes that cause sterility are readily discovered in humans when the afflicted individuals seek medical attention. On the other hand, mice are suitable for experiments that would be unethical with humans.

Maleness in Mammals Depends on the Y Chromosome

The XX/XY system of sex determination is found in mammals and many other taxonomic groups. Yet the molecular mechanisms underlying this system may differ. This is illustrated by the sexual phenotypes of individuals with abnormal combinations of sex chromosomes (Table 27.2). For example, individuals with only one X chromosome (designated XO) are infertile *males* in *Drosophila* but are *females* in humans and mice. Human XO females show *Turner's syndrome*, which includes short stature, ovarian degeneration, lack of secondary sex characteristics, and mental retardation. The chromosome status XXY generates fertile females in *Drosophila* but sterile males in humans. Such human males show *Klinefelter's syndrome*, which is characterized by long legs, wide hips, narrow shoulders, and lack of spermatogenesis.

From these and other abnormal chromosome configurations, it is clear that development of a male human or mouse depends on the presence of a Y chromosome (Ford et al., 1959; Jacobs and Strong, 1959; Welshons and Russell, 1959). Conversely, the mere absence of a Y chromosome is sufficient for development of a female, which is therefore considered the *default program* of mammalian sex determination. In contrast, the Y chromosome is not required for male sex determination in *Drosophila* (but it is necessary for spermatogenesis). The sex of *Drosophila* is determined by the $X:A$ ratio (discussed later in this chapter; see also Chapter 16).

Dosage Compensation in Mammals Occurs by X Chromosome Inactivation

The X chromosome of mammals contains a large number of genes that have nothing to do with sex determi-

---

TABLE 27.2			
Chromosome Status and Sexual Phenotype in Three Organisms with XX-XY Sex Determination			
Sex Chromosome Status	Human Phenotype	Mouse Phenotype	*Drosophila* Phenotype
XO	Sterile female*	Fertile female	Sterile male
XX	Normal female	Normal female	Normal female
XXX	Fertile female	Unknown	Sterile female
XY	Normal male	Normal male	Normal male
XXY	Sterile male†	Sterile male	Fertile female
XYY	Fertile male	Semisterile male	Fertile male

*Turner's syndrome.
†Klinefelter's syndrome.
Source: After Sutton (1988). Used with permission. © 1988 Harcourt Brace & Company.

nation but nevertheless are not present on the Y chromosome. Consequently, these genes are present in two doses in females and one dose in males, generating a serious imbalance. Studies on losses and duplications of chromosome fragments in various diploid animals show that haploidy or triploidy for more than a few percent of all genes is deleterious or lethal. Apparently, many genes are involved in regulatory networks that are finely tuned and cannot tolerate a major reduction or increase in gene activity. It appears that each group of animals has found some mechanism of *dosage compensation* to equalize the expression of non-sex-related genes located on sex chromosomes.

Mammals achieve dosage compensation by inactivating one of the two X chromosomes in nearly each cell of a female during the blastocyst stage (Fig. 27.5). In the inner cell mass, which gives rise to the embryo proper, the choice of which of the two X chromosomes in any given cell is inactivated seems to be random

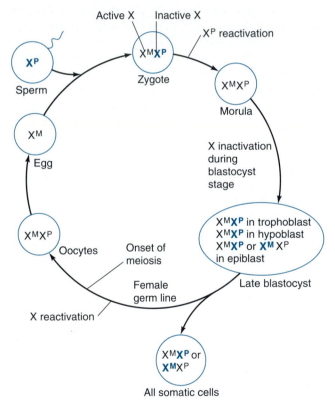

Figure 27.5 Cycle of X chromosome inactivation in the mouse. Inactive chromosomes are shown in color. Both X chromosomes are active in the morula. Inactivation occurs during the blastocyst stage. The paternally inherited X chromosome (X^P) is inactivated selectively in the trophoblast and the hypoblast. In the epiblast, which gives rise to the embryo proper, inactivation occurs randomly in either X^P or the maternally inherited X chromosome (X^M). These inactivations are passed on clonally and are permanent in somatic cells. In the female germ line, the inactivated chromosome is reactivated shortly before meiosis. In the male germ line, the X chromosome is inactivated between meiosis and fertilization.

(Gartler and Riggs, 1983). In female somatic cells, the inactivated X chromosome of a cell will replicate during mitosis but will remain inactive in all the cell's progeny. In the female germ line, inactivated X chromosomes are reactivated as oogonia enter meiotic prophase (Kratzer and Chapman, 1981). Even in the male germ line, the X chromosome is temporarily inactivated between meiosis and fertilization. The inactivated X chromosomes are visible in interphase nuclei as deeply staining bodies of *heterochromatin*, known as *Barr bodies* or *sex chromatin*. Nearly all genes in sex chromatin are shut down, but a few genes escape inactivation. These genes are located primarily in the *pseudoautosomal region*, which is similar in X and Y chromosomes and therefore behaves like an autosomal region in genetic tests. Presumably, the extra dosage of pseudoautosomal genes in XXY males accounts for Klinefelter's syndrome, whereas the reduced dosage of these genes in XO females causes Turner's syndrome.

Random inactivation of one mammalian X chromosome as a mechanism of dosage compensation was first proposed by Mary Lyon (1961). She based her hypothesis primarily on the observation of sex chromatin and on her analysis of coat color patterns in mice. If a female mouse is heterozygous for two alleles (*Pm* and *Pp*) of an X-linked pigmentation gene, then patches expressing *Pm* will alternate in her coat with patches expressing *Pp*. Subsequent studies on cultured cell clones confirmed Lyon's conclusions. Ronald Davidson and coworkers (1963) showed that human females heterozygous for two variants of the enzyme glucose-6-phosphate dehydrogenase (G6PD-A and G6PD-B) have two types of skin cells: one cell type expresses *only* G6PD-A, and the other type expresses *only* G6PD-B (Fig. 27.6). Corresponding results were obtained for another human X-linked enzyme, hypoxanthine phosphoribosyltransferase (Migeon, 1971).

The molecular mechanism of X chromosome inactivation is still under investigation. In all cell types with an inactivated X chromosome, there is a specific transcript, known as Xist (for *X-inactive-specific transcript*) in the mouse and XIST in humans (McCarrey and Dilworth, 1992). The synthesis of Xist RNA, which is encoded by a gene on the X chromosome, precedes X inactivation (Kay et al., 1993). It appears, therefore, that *Xist^+* expression is the cause rather than an effect of X inactivation. The expression of *XIST^+* is somehow triggered by the number of X chromosomes present, independent of the Y chromosome, because in both normal XX females and XXY males one X chromosome is inactivated. Indeed, the inactivation mechanism affects all X chromosomes except one: in XXX females, two of the three X chromosomes are inactivated. Inactive X chromosomes are also distinguished by the lack of acetylation in histone H4 (Jeppesen and Turner, 1993). Apparently, acetylated H4 isoforms promote a chromatin configuration that is conducive to gene expression.

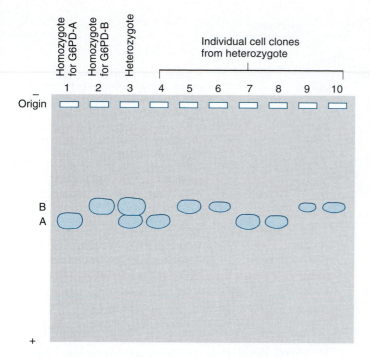

Figure 27.6 Occurrence of two genetic variants of the enzyme glucose-6-phosphate dehydrogenase (G6PD-A and G6PD-B) in human females as revealed by electrophoresis. Enzyme from females homozygous for either variant A or variant B appears in only one band after electrophoresis and specific staining (lanes 1 and 2). Both variants are seen in cultures of mixed skin cells from heterozygous females (lane 3). In contrast, each clone raised from a single skin cell of a heterozygous female (lanes 4 to 10) contains only one variant.

The Earliest Sexual Difference Appears in the Gonad

Mammalian development begins with a *sexually indifferent stage,* which in the human lasts through the first six weeks of gestation. The first difference in the development of males versus females appears when the indifferent gonad rudiments begin to form either testes or ovaries. The emergence of this difference is referred to as **primary sex differentiation;** it occurs independently of sex hormones but includes the differentiation of those gonadal cells that will produce sex hormones during the following phase, known as **secondary sex differentiation** (see Chapter 28).

The observation that maleness in mammals depends on the presence of a Y chromosome indicates that some signal encoded by the Y chromosome directs the formation of the testes. The part of the Y chromosome that promotes testis development has been termed the *testis-determining factor,* abbreviated as **TDF** in humans and *Tdy* in mice.

When and where in the developing gonad does TDF act? One way of investigating this problem is to identify the gonadal cell type that shows the first visible sexual difference. The gonad develops from two cell

lineages: somatic cells and germ line cells. The somatic cells, originating from the *intermediate mesoderm,* form a *genital ridge* on the *mesonephros,* whereas the primordial germ cells arise from the posterior yolk sac (see Fig. 3.3). The primordial germ cells migrate around the hindgut and through the dorsal mesentery into the genital ridge, where they arrive during the 6th week of development in the human and during the 11th or 12th day after mating in the mouse. Shortly before and during the arrival of the primordial germ cells, the coelomic epithelium of the genital ridge proliferates, and columns of epithelial cells penetrate the underlying mesenchyme. Here they surround the germ cells and form irregularly shaped cords called *primitive sex cords.* Through this stage, there is no morphological difference between the gonads of XX and XY individuals, so the gonad at this stage is known as the *indifferent gonad.*

The first cells to show a sexual difference are the somatic cells derived from the primitive sex cords. These cells, called the *supporting cells,* eventually give rise either to *Sertoli cells* in the testis or to *follicle cells* in the ovary (Fig. 27.7). These cell lineages develop in close interaction with the germ cells (McLaren, 1991).

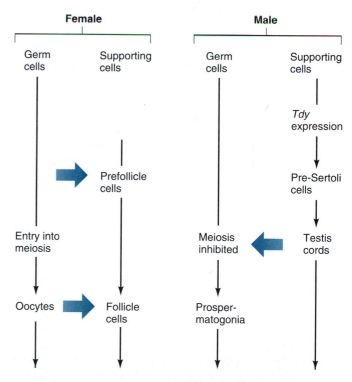

Figure 27.7 Development of supporting cells and germ cells in the fetal mouse gonad. Arrows indicate interactions between the two cell lineages. In females, prefollicle cells are formed in the absence of germ cells but depend on germ cells for further development. Even after entry into meiosis and follicle formation, the maintenance of follicle cells depends on the continued presence of oocytes. In males, expression of the testis-determining factor (Tdy) in the supporting cells causes the development of pre-Sertoli cells, which inhibit meiosis and induce the germ line cells to form prospermatogonia.

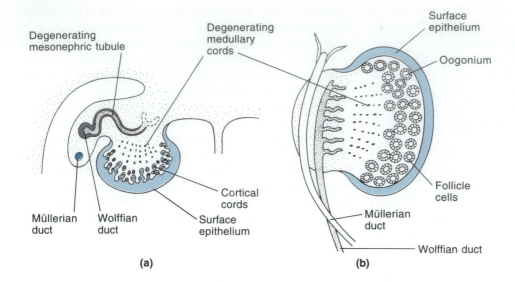

Degenerating mesonephric tubule

Degenerating medullary cords

Surface epithelium

Oogonium

Müllerian duct

Wolffian duct

Surface epithelium

Cortical cords

Follicle cells

Müllerian duct

Wolffian duct

(a)

(b)

Figure 27.8 Ovary development in the human. **(a)** Transverse section of the developing ovary during the seventh week of development, showing the persistence of primary sex cords in the cortex and their degeneration in the medulla (core). **(b)** Frontal section of the ovary and genital ducts in the fifth month of development. The cortical sex cords have broken up into individual follicles, each containing an oogonium surrounded by follicle cells.

In XX embryos, the primitive sex cords persist only in the cortex (periphery) of the gonad, and the supporting cells form *prefollicle* cells (Fig. 27.8). Thereafter, the germ cells enter meiosis, an event that begins after 12 weeks of development in the human female but not until the onset of puberty in males (see Fig. 3.6). Later, the cortical cords derived from the primitive sex cords break up into primordial follicles.

In XY embryos, the primitive sex cords persist in the medulla (core) of the gonad, where they form horseshoe-shaped *testis cords* (Fig. 27.9). The supporting cells form *pre-Sertoli cells*, which are the first male-specific cell type to differentiate. The germ cells associated with pre-Sertoli cells do not enter meiosis—a clear sign that they have departed from female development.

The testis cords continue to develop into *seminiferous tubules* of the testis.

Supporting cells and germ cells develop according to the *principle of reciprocal interactions* (double-stemmed arrows in Figure 27.7; see also Chapter 13). If germ cells are absent, as in certain mouse mutants or after experimental interference with germ cell migration, prefollicular cells are formed, but they degenerate later. If germ cells are lost from developing ovaries after meiosis, the follicle cells transdifferentiate into cells that resemble Sertoli cells. Conversely, the formation of pre-Sertoli cells and testis cords inhibits the germ cells from entering meiosis.

The observations described in this section are compatible with the hypothesis that TDF may determine

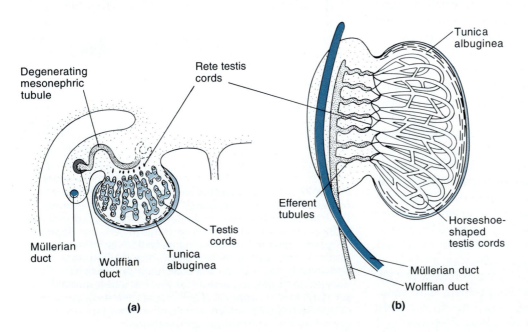

Degenerating mesonephric tubule

Rete testis cords

Tunica albuginea

Müllerian duct

Wolffian duct

Tunica albuginea

Testis cords

Efferent tubules

Horseshoe-shaped testis cords

Müllerian duct

Wolffian duct

(a)

(b)

Figure 27.9 Testis development in the human. **(a)** Transverse section of the testis during the eighth week of development, showing the persistence of the primary sex cords in the medulla while the cortex develops into a white, fibrous coat, the tunica albuginea. **(b)** Frontal section of the testis and genital ducts in the fourth month of development. The medullary sex cords have formed horseshoe-shaped testis cords. They are connected by a network of tubules, the rete testis, and via efferent tubules, to the former mesonephros, which develops into the epididymis.

the supporting cell lineage of the gonad to develop as pre-Sertoli cells. However, morphological data cannot exclude the possibility that TDF acts first in a different type of cell without causing a morphological change, and that a signal coming from this type of cell triggers the morphological changes in the supporting cells.

The Testis-Determining Factor Acts Primarily in the Prospective Sertoli Cells

Taking a genetic approach to locating the primary site of Tdy action, researchers have analyzed mouse *chimeras*, that is, individuals consisting of two types of cells that are genotypically different (say, genotype A and genotype B). Chimeras are routinely obtained by fusing two embryos from different mouse strains before *compaction* or by injecting *inner cell mass* cells from a donor blastocyst into a host blastocyst from a different strain. The chimeras are allowed to develop in foster mothers until they reach the stage of interest for analysis. Then each cell or tissue can be scored as genotype A or B by means of suitable genetic markers, such as enzyme variants or transgenes. Chance dictates that 50% of all chimeras will be **XX↔XY chimeras**, that is, will be composed of XX and XY cells; these chimeras can be identified by microscopic inspection of their sex chromosomes. If mouse strains with morphologically different Y chromosomes are used, one can also tell which strain contributed the XY cells to a chimera.

Early experiments with mouse chimeras indicated that Tdy does *not* act in germ line cells: an oocyte in a chimeric mouse was found to have an XY karyotype, indicating that Tdy is not *sufficient* for male germ line development (E. P. Evans et al., 1977). Tdy is also not *necessary* for male germ line initiation, because germ cells without Y chromosomes were seen to develop into spermatogonia, even though they did not proceed with meiosis and spermiogenesis (Burgoyne et al., 1988). Therefore, the *initiation* of male gametogenesis must be an indirect effect of the Y chromosome brought about by creation of a testicular environment, rather than a direct effect of Y chromosome expression in the germ line. Only meiosis and the following stages of spermatogenesis require expression of Y-linked genes in germ line cells, but these genes are different from Tdy (K. Ma et al., 1993).

Once germ cells could be excluded as primary sites of Tdy action, the interest focused on somatic gonadal cells—in particular, the *Sertoli cells* already described and the *interstitial cells* (Leydig cells), which later produce testosterone.

▼

To determine whether prospective Sertoli cells or Leydig cells were the primary sites of Tdy action, Paul

Burgoyne and his colleagues (1988) created XX↔XY chimeras from two mouse strains that had two electrophoretic variants of the enzyme glucose phosphate isomerase, designated GPI-A and GPI-B. Chimeras were killed after 12 to 15 days of development, a bone marrow sample was removed for chromosome inspection, and the testes were removed for tissue separation. The external layer of connective tissue (tunica albuginea) was removed manually. The remainder of the testes was dissociated into cells by mild digestion with protease and collagenase, and the isolated cells were separated by sedimentation into one fraction containing Sertoli cells and another fraction containing Leydig cells. Proteins from each fraction were separated by gel electrophoresis and stained for GPI activity (Fig. 27.10). The results showed that XX cells contributed randomly to all testicular tissues except the Sertoli cells, which were almost exclusively derived from XY cells. These data confirmed the hypothesis that Tdy is necessary and sufficient for Sertoli cell differentiation, but not for the development of other testicular cells.

In a follow-up study, Stephen Palmer and Paul Burgoyne (1991) used XX↔XY chimeras in which one component was marked by a globin transgene, which

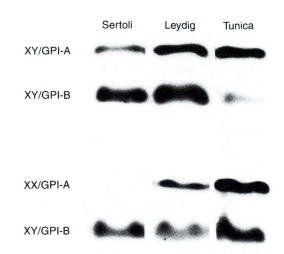

Figure 27.10 Analysis of testicular cell types from chimeric mice for the proportion of cells with Y chromosomes. Proteins from three cell types (top) obtained from individual chimeras were separated by gel electrophoresis and stained for glucose phosphate isomerase (GPI) activity. The two mouse strains used for generating each chimera had different GPI variants, designated GPI-A and GPI-B, and morphologically different sex chromosomes. Upper rows: An XY ↔ XY chimera composed of two males had both GPI variants in each testicular tissue. Lower rows: An XX ↔ XY chimera of one male and one female had both GPI variants in the tunica albuginea and the Leydig cell fraction. In contrast, the Sertoli cell fraction showed a strong GPI-B band, and a very weak GPI-A band that was detectable only by scanning the gel with a densitometer. Because the GPI-A variant in this chimera was derived from XX cells, it was apparent that very few of these cells contributed to the Sertoli cell fraction.

could be detected by *in situ hybridization* (see Methods 15.1). This marker allowed the investigators to determine the contributions of XX and XY cells to any testicular cell lineage in microscopic sections. They found that most cell types in fetal XX↔XY testes were random mixtures of XX and XY cells. Only the Sertoli cells were derived mostly from the XY component of each chimera. This bias became even stronger at later stages of testicular development. However, small proportions of XX Sertoli cells were consistently observed.

Apparently, the supporting cell lineage—which gives rise to the Sertoli cells in males—is the initial site of Tdy expression, and Tdy activity in this lineage generates an environment in which other gonadal cell lineages will develop along testicular pathways, regardless of their own sex chromosome status. The small contributions of XX cells to the Sertoli cell lineage in chimeric testes indicate that Sertoli cell determination is subject to the *neighborhood effect* discussed in Chapter 6. If there are enough XY cells to get the Tdy-initiated process of Sertoli cell determination under way, a few XX cells can be recruited to join in. This indicates that in the development of a normal male, with all cells having a Y chromosome, the supporting cells in the indifferent gonad promote and stabilize one another's expression of Tdy.

Mapping and Cloning of the Testis-Determining Factor

The discovery in 1959 that the Y chromosome is necessary for testis development in humans and mice was only the beginning of a long quest to identify the testis-determining factor (Affara, 1991; Bogan and Page, 1994). Exactly where on the Y chromosome is TDF located? Is it one gene or more? What is the nature of the signal encoded by the gene(s)?

Translocated Y Chromosome Fragments Causing Sex Reversal Are Used to Map TDF

The critical data for mapping TDF on the Y chromosome were provided by human patients whose phenotypic sex did not seem to match their chromosomal constitution. These *sex-reversed individuals* were sterile men carrying what appeared to be two X chromosomes and no Y, and women with Turner's syndrome carrying one X and one apparent Y. Closer analysis showed, however, that most of these cases involved the translocation of a Y chromosome fragment. Data from such patients were used to map TDF by delineating the smallest Y chromosome fragment that caused sex reversal.

The observed sex reversals can be explained as the results of rare crossover events between the X and the Y chromosome during male meiosis (Fig. 27.11). However, the postulated exchanges of chromosome fragments are

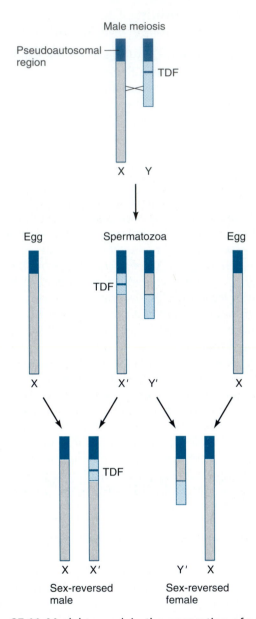

Figure 27.11 Model to explain the generation of sex-reversed humans by rare X-Y recombination during male meiosis. The top diagram shows one chromatid each from the X and Y chromosomes during meiotic prophase I. The pseudoautosomal regions are nearly identical in the two chromosomes. Only this region pairs during meiosis, and recombination events outside this region are correspondingly rare. If they occur, however, and if the exchanged fragment of the Y chromosome contains the testis-determining factor (TDF), then the crossover generates an X-like chromosome (X') containing TDF and a Y-like chromosome (Y') lacking TDF. If sperm with such chromosomes fertilize an egg, the resulting zygotes will develop into sex-reversed males and females, respectively.

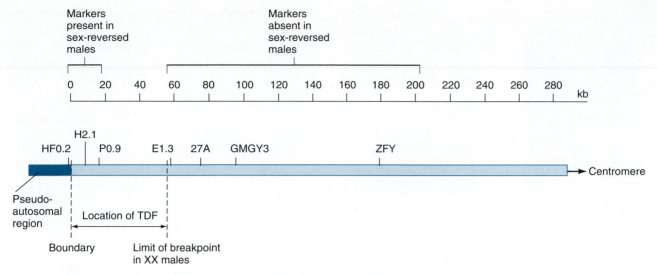

Figure 27.12 Map of the human Y chromosome (colored bar) near the pseudoautosomal region. The scale at the top indicates the distance from the pseudoautosomal boundary in kilobases (kb). The acronyms above the chromosome indicate the positions of cloned DNA segments used as labeled probes on Southern blots of genomic DNA from sex-reversed XX males (see Table 27.3 and Fig. 27.13). The results from those blots indicate that TDF maps to within 60 kb of the pseudoautosomal boundary.

cytologically invisible and genetically difficult to detect because of the paucity of genes on the Y chromosome. Fortunately, a large number of cloned DNA probes from the human Y chromosome became available in the 1980s. By applying such probes to *Southern blots* (see Methods 14.3) of total genomic DNA from sex-reversed human patients, one can directly demonstrate the postulated exchange of chromosome fragments. The cross-

over points can then be mapped, and the area containing TDF can be narrowed down.

Using this method, Mark S. Palmer and his colleagues (1989) mapped the crossover points of four sex-reversed males who had sought medical attention for sterility and irregularities in their genitalia (Fig. 27.12; Table 27.3). Each of the four men seemed to have two X chromosomes, but Southern blots of their genomic DNA

TABLE 27.3

Mapping of Genomic DNA Breakpoints in XX Sex-Reversed Patients

The data indicate that in each case the breakpoint is located between the DNA segments bound by probes P0.9 and E1.3.

Patient	Genitalia	Testis Biopsy	Genomic Markers Present[a]					
			HFO.2	H2.1	P0.9	E1.3	27A	*ZFY*
ZM	Vaginal pouch; hypospadia[b]	Leydig cells; Sertoli cells; no spermatogonia	+	+	+	−	−	−
MB	Cryptorchidism[c]	Leydig cells; Sertoli cells; no spermatogonia	+	+	+	−	−	−
DL[d]	Hypospadia[b]	Bilateral ovotestis	+	+	+	−	−	−
TL[d]	Uterus; hypospadia[b]	Sertoli cells; no spermatogonia	+	+	+	−	−	−

[a] See map in Figure 27.12.
[b] Urethra opens on lower surface of penis instead of at the tip.
[c] Testis fails to descend into scrotum.
[d] Familial case.
Source: After M. S. Palmer et al. (1989). Used with permission. © 1989 Macmillan Magazines Limited.

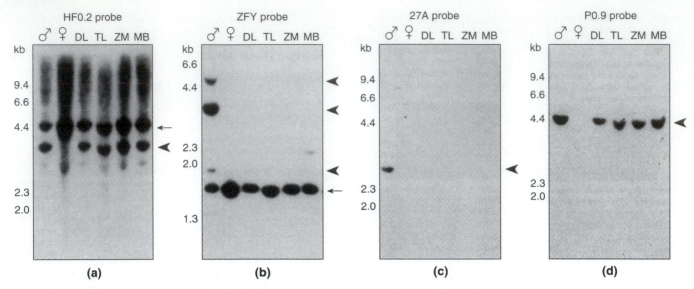

Figure 27.13 Hybridization of sex-specific DNA probes with Southern blots of human genomic DNA (see Methods 14.3). The lanes in this experiment contained restriction digests of genomic DNA from a normal human male (♂), a normal human female (♀), and the four sex-reversed XX males (DL, TL, ZM, and MB) also listed in Table 27.3. **(a)** DNA probed with Hf0.2, a pseudoautosomal probe. The probe bound to a 4.5-kb fragment (arrow) of X-specific DNA and a 3.2-kb fragment (arrowhead) of Y-specific DNA reaching across the pseudoautosomal boundary. The latter fragment was present in all sex-reversed patients but missing from normal female genomic DNA. (The label at the 3.2-kb level in the female DNA lane is a smear caused by DNA overdigestion, as indicated by the lack of a distinct bulge in the labeling pattern. **(b)** DNA probed with a cDNA clone representing part of the *ZFY*⁺ gene. The probe bound to three Y-specific bands (arrowheads) appearing only in the normal male, indicating that these DNA fragments were not exchanged in the sex-reversed patients. (The weak 2.3-kb band in lane MB was not reproducible and was ascribed to incomplete digestion of the genomic DNA.) There was also a common 1.6-kb band (arrow), confirming that *ZFY* has a close homologue on the X chromosome. **(c)** DNA probed with clone 27A. Only the normal male was positive, indicating that the corresponding portion of genomic DNA is Y-specific but not exchanged in the sex-reversed patients. **(d)** DNA probed with probe P0.9. The probe bound to DNA from the normal male and the sex-reversed patients, indicating that this DNA is Y-specific and exchanged in the sex-reversed patients.

revealed Y-specific DNA fragments that were not present in genomic DNA of normal females. Clearly, at least one of the chromosomes in each sex-reversal patient contained DNA segments that are normally found on the Y chromosome. In each case, the exchanged DNA included a segment from the pseudoautosomal region, which is homologous in X and Y chromosomes (Figs. 27.12 and 27.13a). None of the exchanged DNA fragments included the *ZFY*⁺ gene, which researchers had previously thought might be the TDF (Fig. 27.13b). Further probing showed that DNA segments located 10 kb and 20 kb away from the pseudoautosomal region were included in the exchanged DNA whereas segments located 60 kb or farther away were not part of the exchanged DNA (Fig. 27.12c and d).

Based on the DNA samples from four sex-reversed XX males, M. S. Palmer and colleagues (1989) mapped the gene(s) acting as TDF within 60 kb of DNA directly adjacent to the pseudoautosomal region. In a follow-up study using DNA from the same XX males but additional markers, Andrew Sinclair and colleagues (1990) further delimited the location of TDF to 35 kb of DNA next to the pseudoautosomal region. This result set the stage for isolating genes from this region and testing them for the properties expected of TDF.

The SRY/Sry Function Is Necessary for Testis Determination

Within the 35 kb of human DNA delimited as the location of TDF, Andrew Sinclair and his colleagues (1990) cloned and characterized a gene that encodes a DNA-binding protein and is highly conserved among mammals. They termed this gene *SRY*⁺ (for sex-determining region Y). John Gubbay and his colleagues (1990) found that a homologous mouse gene, referred to as *Sry*⁺, was absent in a sex-reversed XY female mouse. These researchers then proceeded to test various adult and fetal mouse tissues for transcripts from the Sry^+ gene. In particular, they analyzed genital ridges from 11.5-day-old mouse fetuses, because the testis-determining factor (Tdy) is expected to be active at this time and in this tissue in males. Indeed, the researchers found Sry transcripts in RNA from the genital ridges of male fetuses but not female fetuses. The same transcripts were found in adult testis but not in liver.

In an extension of this study, Peter Koopman and his colleagues (1990) found that Sry transcripts were present in male genital ridges from 10.5- through 12.5-day fetuses but none earlier or later. Thus, transcription of the Sry^+ gene was not a general aspect of gonad development but correlated specifically with testis determination. Moreover, Sry transcripts were found in fetuses homozygous for the *white spotting* allele, which interferes with primordial germ cell migration so that the testes develop without germ cells. Thus, the Sry^+ gene was transcribed in the somatic tissue of the gonad rather than in the germ line. In summary, these observations showed that Sry^+ was expressed exactly when and where Tdy was expected to be active.

The claim that SRY^+ (Sry^+) was the elusive TDF (Tdy) received further support when a sex-reversed XY human female patient was found to have a Y chromosome in which SRY was not deleted but was *newly mutated*. Philippe Berta and his colleagues (1990) discovered this patient's point mutation when they were screening sex-reversed patients for alterations in the SRY gene. This mutation explained how the patient could have a Y chromosome and still be phenotypically female. Most interestingly, her father—the donor of the mutated Y chromosome—did not carry this mutation and was a normal male, and the same held for her brother; paternity was confirmed with several unrelated DNA probes. Evidently, then, a Y chromosome had undergone a point mutation in the SRY gene as it was passed on from father to daughter and had in the

process lost its testis-determining capacity. In a similar case involving another XY female, Ralf Jäger and his colleagues (1990) found that a new frameshift mutation had taken place in the SRY gene. Corresponding observations were made on female XY mice (Lovell-Badge and Robertson, 1990; Gubbay et al., 1992). These observations show that the SRY/Sry function is *necessary* for testis determination.

The Mouse Sry^+ Gene Can Be Sufficient for Testis Determination

To demonstrate that the SRY^+ (or Sry^+) gene by itself is *sufficient* for the testis-determining function, it has to be shown that adding SRY^+ or Sry^+ to a normal XX genome causes the development of a male. This test was carried out by Peter Koopman and his colleagues (1991) using the transgenic mouse technique described in Chapter 14. The investigators injected mouse eggs with a 14-kb fragment of mouse genomic DNA that contained the Sry^+ gene and, on the basis of sequence analysis, no other genes. The transgenic individuals were tested for their sex chromosome status and for their sexual development.

Fertilized eggs were injected with the Sry^+ transgene and allowed to develop in foster mothers. Some of the embryos were removed after 14 days and assayed for their chromosomal status, the integration of the transgene, and their phenotypic sex (Table 27.4). The chromosomal status (karyotype) was assayed by staining for *sex chromatin*, which indicates the presence of more than one X chromosome and therefore most likely an XX karyotype. In addition, the researchers used Southern blot analysis (see Methods 14.3) to test for the

TABLE 27.4

Analysis of Mouse Embryos Injected with an Sry^+ Transgene

The embryos were assayed for the presence of sex chromatin (Barr body), the Sry^+ transgene, the Zfy^+ gene (located on the normal Y chromosome), and the phenotypic sex (testis cords). Among eight chromosomal females (XX karyotype) transgenic for Sry^+, two were phenotypic males while six were phenotypic females. ND = not determined.

Number of Embryos	Sex Chromatin	Sry^+	Zfy^+	Deduced Karyotype	Transgenic	Phenotypic Sex
63	+	−	−/ND	XX	−	♀
27	−	+	+	XY	ND	♂
58	−	ND	ND	XY	ND	♂
2	−	−	−	XO	−	♀
6	+	+	−	XX	+*	♀
2	+	+	−	XX	+	♂

* In four of these cases, the transgene may not have been present in all cells.

Source: After Koopman et al. (1991). Used with permission. © 1991 Macmillan Magazines Limited.

Zfy gene, which is present on normal Y chromosomes but was not part of the transgene. Thus, sex chromatin–positive and *Zfy*⁺-negative individuals were classified as XX karyotypes while sex chromatin–negative and *Zfy*⁺-positive individuals were classified as XY karyotypes.

Among seventy-one XX individuals, eight were identified as transgenic by the presence of the *Sry*⁺ gene in Southern blots. Among these eight XX transgenics, two were phenotypically male, as indicated by the formation of testis cords (Fig. 27.14). Histological examination of the transgenics showed that their testis cords were normal and that their testes were indistinguishable from testes of their normal XY siblings. The other six XX transgenics were phenotypically female.

To test the adult phenotype of *Sry*⁺ transgenic mice, some of the injected eggs were allowed to develop to term. Among 93 mice born, three were XX transgenics. Of these, two were XX females, and one was an XX male (Fig. 27.15). The XX transgenic male was similar in size to his normal XY littermates. His copulatory behavior was normal, but none of the females with which he mated became pregnant. However, this was no surprise, since males with two X chromosomes are always sterile. Internal examination revealed a normal male reproductive tract with no sign of hermaphroditism. The

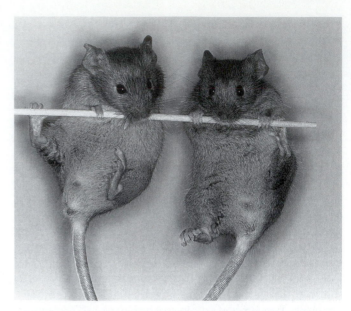

Figure 27.15 Normal and transgenic males. The mouse to the left is a normal male. The mouse to the right is chromosomally female (XX) but made transgenic for the *Sry*⁺ gene, which is normally located on the Y chromosome. The transgenic mouse has the external and internal genitalia of a normal male. He also copulates normally but is sterile. The *Sry*⁺ gene can be sufficient for testis formation and normal male sexual development in a chromosomal female, except that spermatogenesis requires one or more additional genes located on the Y chromosome.

Normal males   XX transgenic embryos   Normal females

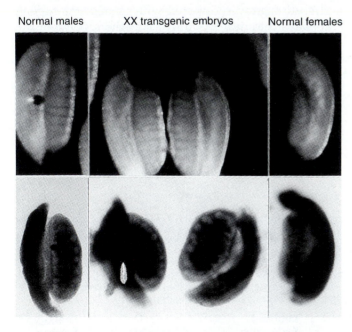

Figure 27.14 Testis development of two mouse embryos with XX karyotype and multiple copies of an *Sry*⁺ transgene (see Table 27.4). The two center panels show pairs of embryonic gonads and adjacent kidneys, excised from two transgenic XX embryos. The two left panels show single gonads and attached kidneys from normal males. The two panels at the right show single kidneys and gonads from normal females. The gonads of the two XX transgenic embryos have the characteristic stripes associated with testis cord formation.

testes were smaller than normal, but again this is normal for sex-reversed XX males. Histologically, the testes of the XX transgenic male contained seminiferous tubules, clearly defined Sertoli cells, Leydig cells, and other testicular cells, but no germ cells undergoing spermatogenesis.

Finally, the investigators made mice transgenic for *SRY*⁺, the human homologue of mouse *Sry*⁺, in order to determine whether *SRY*⁺ can also be sufficient to cause testis formation and male development in mice. The two transgenic males that were obtained passed on the transgene to about half of their offspring, as expected for a transgene residing in only one chromosome of a homologous pair. Most but not all of the transgenic progeny expressed the human transgene in the embryonic gonads at levels that were greater than those of the endogenous *Sry*⁺ gene. However, no testis cord development was observed in XX transgenic fetuses.

The experiments of Koopman and coworkers (1991) show that a 14-kb segment of DNA containing *Sry*⁺ can be sufficient to determine testis formation and male development. No other genes were detected in the 14-kb segment, so it appears that *Sry*⁺ alone can act as the testis-determining factor in mice. This means that any other genes *required* for the male sex determination pathway must be X-linked or autosomal. However, the fact that only a fraction of XX individuals identified as transgenic for Sry^+ actually developed into males is puzzling. One possible explanation is that the 14-kb fragment used in the experiment did not contain all regulatory regions of the Sry^+ gene. Also, the Sry^+ transgene may have been present only in some tissues, but not in the testes, of some of the transgenic mice. Finally, the expression of transgenes in mice depends on their chromosomal location (see Chapter 14). Future experiments in which the level of transgene expression is monitored may clarify these points.

The fact that none of the XX mice identified as transgenic for human SRY^+ developed as a male can be interpreted several ways. The human transgene may be even more deficient in regulatory regions than the 14-kb mouse fragment. Also, the human SRY transcript may not be properly processed and translated, or the human SRY protein may not match the gene-regulatory components and target genes of the mouse.

Even in a human XX individual, normal SRY protein by itself may be insufficient for complete male sex differentiation. This possibility is suggested by the phenotypes of the four sex-reversed males who possessed the SRY^+ gene in a small segment of Y-chromosomal DNA (Fig. 27.12; Table 27.3). All of these males had formed testicular tissue but none of them developed fully normal male genitalia. These observations suggest that in the human, complete male sex differentiation requires additional Y-linked genes that act in concert with, or are regulated by, the SRY^+ gene.

The Regulators and Target Genes of SRY^+ Still Need to Be Identified

By analogy with the genetic control of other developmental processes, SRY^+ and Sry^+ are thought to be parts of genetic pathways that directly determine the sex of the supporting cells in the gonad, and indirectly affect sex determination of the germ line cells. The existence of such pathways is indicated by several lines of evidence, including the nature of the SRY/Sry proteins, the expression pattern of Sry^+, unusual cases of sex-reversal in humans, and the existence of a sex hormone that is produced in the Sertoli cells (Bogan and Page, 1994).

The SRY^+ gene encodes a protein of the high mobility group (HMG), a family of DNA-binding proteins. Their DNA-binding domain, the HMG domain, binds to DNA in a sequence-specific manner. The HMG domain may be the only functional domain of the protein, since no activation domain reminiscent of other transcription factors has been found. Also, whereas the HMG domain is highly conserved among marsupials and placental mammals, other parts of the SRY protein

are quite divergent even among related species (Whitfield et al., 1993; P. K. Tucker and B. L. Lundrigan, 1993). The HMG domain is thought to considerably distort the bound DNA, perhaps activating its target genes by altering chromatin structure or by bringing together distant DNA sequences (Harley et al., 1992).

The restriction of Sry^+ expression to the gonad during two days of embryonic development implies that there are signals regulating Sry^+.

The existence of target genes that are regulated by SRY^+ is suggested by rare cases of sex-reversed XX males who do *not* have the SRY^+ gene. These cases can be ascribed to gain-of-function mutations in autosomal or X-linked genes that normally require activation by SRY^+. Conversely, cases of sex-reversed XY females who have an apparently intact SRY^+ could be explained as having loss-of-function mutations in one or more target genes of SRY^+.

A prime candidate for being a target gene of SRY^+ is the gene encoding the *anti-Müllerian duct hormone (AMH)*, which inhibits the development of female genital ducts in males (see Chapter 28). AMH is produced in the same Sertoli cells in which Sry^+ becomes first active. However, since AMH mRNA appears with a delay of two days after Sry mRNA, and because AMH mRNA continues to be produced after Sry mRNA disappears, the regulation of AMH^+ by SRY^+ may be indirect (Münsterberg and Lovell-Badge, 1991).

Aside from the leads just mentioned, the genetic pathways that link SRY^+/Sry^+ to their regulators and targets are unknown, and major discoveries are still to be made. For example, McElreavy and colleagues (1993) have predicted, on the basis of a pedigree analysis of three human families, a gene termed Z^+, which is envisioned as a female master regulator that would be active in wild-type females and repressed by SRY^+ in normal males.

Sex Determination and Sex Differentiation in *Drosophila, Caenorhabditis elegans,* and Mammals

In the last section of this chapter, we will compare sex determination and sex differentiation in the fruit fly *Drosophila melanogaster*, the roundworm *Caenorhabditis elegans*, and placental mammals, represented by the house mouse and the human. We will see that genotypic sex determination controls the same biological functions in all species but that the genetic and molecular control mechanisms differ widely (Hodgkin, 1990, 1992; Parkhurst and Meneely, 1994).

For the remainder of this chapter and the following chapter, we will use the notation SRY^+ for the human

TABLE 27.5

Sex Determination Features in *Drosophila*, in *C. elegans*, and in Humans and Mice

Feature	*Drosophila*	*C. elegans*	Humans and Mice
Sexual dimorphism	Male/female	Male/hermaphrodite	Male/female
Sex chromosomes	XY/XX	XO/XX	XY/XX
Sex-determining signal	X : A ratio	X : A ratio	SRY^+ activity
Master control gene	$Sexlethal^+$	$xol\text{-}1^+$, $sdc\text{-}1^+$, $sdc\text{-}2^+$	
Dosage compensation	Male hypertranscription	Female hypotranscription	X inactivation
Somatic sex coordination	RNA splicing cascade, cell autonomous	Gene inactivation cascade, induction	Sex hormones
Germ line sex differentiation	$Sexlethal^+$ activity, induction by somatic cells	$tra\text{-}1^+$ activity (fem^+ activity)	Induction by somatic cells

SRY^+ gene as well as the mouse Sry^+ gene and its equivalents in other mammals.

Genotypic Sex Determination Provides a Primary Regulatory Signal

The three types of organisms we will consider here are sexually dimorphic: *Drosophila* and mammals have males and females, and *C. elegans* has males and hermaphrodites. In each case, sex determination is genetic, with XX/XY chromosomes in *Drosophila* and mammals and with an XX/XO system in *C. elegans*. At the most general level, genetic sex determination involves a primary signal that affects a set of regulatory genes, which in turn control sex differentiation (Table 27.5). The primary signal is the $X : A$ ratio in *Drosophila* and *C. elegans*, and in mammals it is the SRY^+ gene product. The regulatory genes that respond to the primary signal form a hierarchy in *Drosophila* and *C. elegans*; the corresponding genes in mammals are still under investigation. At the top of the hierarchy are the $Sexlethal^+$ gene of *Drosophila* and the $xol\text{-}1^+$, $sdc\text{-}1^+$, and $sdc\text{-}2^+$ genes in *C. elegans*. These hierarchies diverge into three pathways controlling the three basic aspects of sexual development: dosage compensation, somatic sex differentiation, and germ line sex differentiation. Typically, a pathway contains a set of *regulatory genes*, which in turn control a large number of *realizator genes*. The realizator genes are directly involved in sexually dimorphic functions such as pigmentation, gametogenesis, and mating behavior.

The X : A Ratio Is Measured by Numerator and Denominator Elements

The primary sex-determining signal in mammals is carried by the SRY^+ gene product, as discussed earlier. In both *Drosophila* and *C. elegans*, the sex-determining signal is the X : A ratio. This was established by a classical

study, in which Calvin Bridges (1921) carried out crossing experiments to obtain fruit flies carrying different numbers of X chromosomes over two or three sets of autosomes (Fig. 27.16). Corresponding experiments were carried out by James Madl and Robert Herman (1979) with *C. elegans*. The results showed that the sex-determining signal in both organisms is the X : A ratio, rather than the absolute number of X chromosomes. X:A ratios of 0.33 and 0.5 cause male development, ratios near 0.67 give rise to intersexes, and ratios between 0.75 and 1.5 support the development of females in *Drosophila* and hermaphrodites in *C. elegans*. Individuals with X : A ratios smaller than 0.33 or greater than 1.5 are not viable.

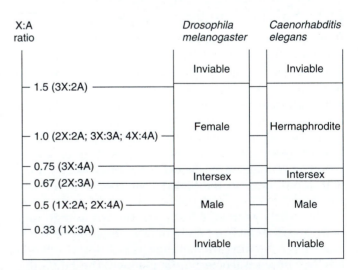

Figure 27.16 Control of the sexual phenotype by the X:A ratio in *Drosophila* and *Caenorhabditis elegans*. Individuals carrying different numbers of X chromosomes and sets of autosomes were obtained by suitable genetic crosses. The columns indicate the ranges of X:A ratios that support the development of males and females (*Drosophila*) or hermaphrodites (*C. elegans*).

The data shown in Figure 27.16 reveal a rather sharp threshold between X:A ratios of less than 0.67 supporting male development and ratios of 0.75 or more supporting female development. How can embryos respond to small differences in the X:A ratio so consistently? What both *Drosophila* and *C. elegans* measure as the numerator of the X:A ratio is the activity of certain X-linked genes. Obviously, the amounts of products encoded by these so-called **numerator genes** increase in proportion to the number of X chromosomes present in a cell. This implies that the numerator genes must escape *dosage compensation,* which equalizes the cellular amount of most X-linked gene products in the two sexes. Apparently, either the dosage compensation mechanism does not act on numerator genes, or the X:A ratio is measured and stored before dosage compensation sets in.

Some of the genes involved in computing the X:A ratio in *Drosophila* have been characterized, including the X-linked numerator genes *sisterless-a*$^+$ and *sisterless-b*$^+$ as well as the autosomal gene *deadpan*$^+$, which may act as a denominator gene. All three genes encode transcription factors of the *helix-loop-helix (HLH)* superfamily, which form homodimers and heterodimers that may enhance or inhibit transcription (see Chapters 15 and 19). Depending on their relative amounts, HLH factors activate, or fail to activate, the *Sexlethal*$^+$ gene, which is the master regulator of sexual development in *Drosophila*. Activation of *Sexlethal*$^+$ initiates development of a female, whereas inactivity of *Sexlethal*$^+$ allows males to develop by default (see Chapter 16).

The two genes *sisterless-a*$^+$ and *sisterless-b*$^+$ have been so named because loss-of-function alleles are lethal in females but not in males. In females with loss-of-function alleles of these genes, the *Sexlethal*$^+$ gene is not activated and the female-specific dosage compensation does not occur. In these XX individuals, as a result, X-linked genes are transcribed at the high rate that is appropriate only for XY individuals. Such XX individuals die.

The critical role of the *sisterless-b*$^+$ locus was demonstrated by James Erickson and Thomas Cline (1991), who rescued females carrying a temperature-sensitive lethal allele of *sisterless-b* by genetic transformation with a cloned genomic DNA segment containing the *sisterless-b*$^+$ allele. The *sisterless-b*$^+$ transgene also rescued *sisterless-a* mutant females, although less efficiently. The sisterless-b protein, also known as T4, turned out to be a transcription factor of the HLH superfamily. Because of the known cooperative properties of HLH factors, the researchers suggested that T4 may act as a numerator in computing the X:A ratio and that T4 may interact with other HLH factors in activating *Sexlethal*$^+$.

A candidate for the denominator function in the X:A ratio was found by Susan Younger-Shepherd and her colleagues (1992). They investigated the role of the *deadpan*$^+$ gene, which is located on an autosome and encodes an HLH protein. The researchers found that XY individuals with reduced dosages of *deadpan*$^+$ or increased dosages of *sisterless-b*$^+$ died during various stages of development, except when there was a null mutation in the *Sexlethal*$^+$ gene. They concluded that the lethality resulted from inappropriate activation of *Sexlethal*$^+$ and consequent inappropriate dosage compensation. Conversely, XX individuals with increased dosages of *deadpan*$^+$ or reduced dosages of *sisterless-b*$^+$ were deficient in *Sexlethal*$^+$ expression and died, presumably from insufficient dosage compensation.

Further studies revealed that additional HLH proteins encoded by maternally expressed genes are required for proper communication of the X:A ratio.

Drosophila and *C. elegans* Have Master Regulatory Genes for Sex Determination

The three basic aspects of sexual development—dosage compensation, somatic sex differentiation, and gametic sex differentiation—do not have to be controlled by the same mechanism. Indeed, we have already seen that in mammals the testis-determining factor controls the somatic sex and, indirectly, the gametic sex, whereas dosage compensation is controlled by X-linked factors. In *Drosophila*, however, all aspects of sexual development are uniformly controlled by the *Sexlethal*$^+$ gene. A similar master switch in *C. elegans* is formed by three genes known as *xol-1*$^+$, *sdc-1*$^+$, and *sdc-2*$^+$.

The *Sexlethal*$^+$ gene of *Drosophila* responds to the relative concentrations of HLH transcription factors that measure the X:A ratio, as discussed previously. A high X:A ratio activates an *establishment promoter* causing the transcription of a pre-mRNA that is spliced and translated into functional Sexlethal protein (Keyes et al., 1992). The function of the establishment promoter is quickly taken over by the *maintenance promoter* of the *Sexlethal*$^+$ gene. The pre-mRNA transcribed from the maintenance promoter can be spliced in two alternative ways, yielding male-specific or female-specific mRNAs (see Chapter 16). The male-specific mRNA contains an exon with a stop codon and is translated into a truncated, nonfunctional protein. The female-specific mRNA does not contain this exon and is translated into a full-length, functional protein. The male-specific splicing pattern is the default program, whereas the female-specific splicing pattern requires the presence of Sexlethal protein. In females, the establishment promoter provides a start-up supply of Sexlethal protein, which then perpetuates its own synthesis by regulating the splicing of the pre-mRNA transcribed later from the maintenance promoter (Bell et al., 1991). In males, no start-up Sexlethal protein is produced, and all pre-mRNA transcribed from the maintenance promoter is spliced unproductively.

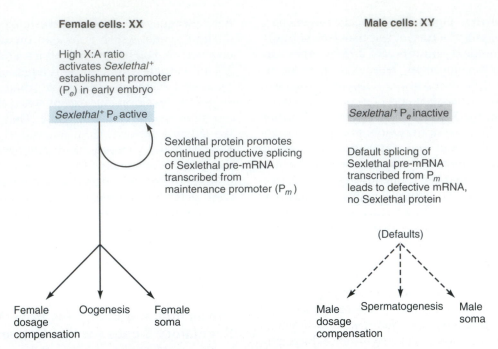

Figure 27.17 The *Drosophila* gene *Sexlethal*⁺ as a master regulator of sex determination. Pre-mRNA transcribed from the maintenance promoter of *Sexlethal*⁺ is spliced in the default pattern in males and in a regulated pattern in females. The male-specific pattern results in the synthesis of a truncated Sexlethal protein. The female-specific pattern leads to the synthesis of full-length Sexlethal protein, which continues to direct the productive splicing of Sexlethal pre-mRNA. The resulting positive feedback loop perpetuates the synthesis of functional Sexlethal protein, which directs female somatic development, oogenesis, and female dosage compensation. Lack of functional Sexlethal protein allows the default programs for male somatic development, spermatogenesis, and male dosage compensation to occur.

The Sexlethal protein controls not only the splicing of its own pre-mRNA, but also the expression of subordinate genes that direct female somatic development, oogenesis, and female dosage compensation (Fig. 27.17). Lack of Sexlethal protein results in the default expression patterns, which lead to male somatic development, spermatogenesis, and male dosage compensation. Some features of these regulatory pathways will be described farther on.

Master regulators of somatic sex differentiation and dosage compensation have also been identified in *C. elegans*. These genes differ from *Sexlethal*⁺ in their mode of action and in the type of protein they encode. Instead of a single gene with positive feedback control, three genes without apparent feedback loops make up the master regulatory system (Fig. 27.18). The gene *xol-1*⁺ (for XO-lethal) is required in XO animals for male somatic development and proper dosage compensation.

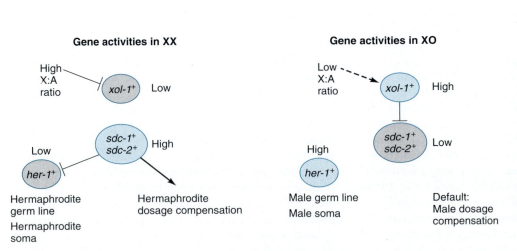

Figure 27.18 Control of somatic sex and dosage compensation in *C. elegans* by three master regulatory genes: *xol-1*⁺, *sdc-1*⁺, and *sdc-2*⁺. In XX (hermaphrodite) embryos, the high X:A ratio leads to low activity of the *xol-1*⁺ gene, which permits the *sdc*⁺ genes to be active. The sdc products regulate dosage compensation genes and inhibit the gene *her-1*⁺ to allow female somatic development. In XO (male) embryos, the *xol-1*⁺ gene is active and down-regulates the *sdc*⁺ genes, thus permitting male somatic development and dosage compensation.

Mutations in *xol-1* cause XO animals to develop as hermaphrodites and their transcription of X-linked genes to be improperly reduced, with death resulting (L. M. Miller et al., 1988). Two other genes, *sdc-1*[+] and *sdc-2*[+] (for *sex* and *dosage* compensation), have the converse effects: loss-of-function mutations in these genes cause male somatic sex differentiation and hyperexpression of X-linked genes in XX animals. Because loss of function in the *sdc* genes counteracts the lethality of *xol-1* mutations in males, the *sdc*[+] genes must be downstream of *xol-1*[+]. It appears that *xol-1*[+] responds to the X:A ratio, and that the *sdc*[+] genes regulate separate pathways for somatic sex differentiation and dosage compensation. The *sdc-1*[+] gene encodes a protein with seven *zinc fingers*, indicating that it acts as a transcription factor (Nonet and Meyer, 1991).

In summary, the master regulatory genes controlling sexual development in *Drosophila* and *C. elegans* serve the same biological functions but differ in their molecular and regulatory characteristics. These observations indicate that these master regulatory genes are not homologous but have evolved in parallel.

Dosage Compensation Mechanisms Vary Widely

As discussed earlier, the difference in X-linked gene dosages between the two sexes must be compensated for so that both sexes receive equal amounts of gene products. In mammals, as we have seen, all X chromosomes except one are inactivated, so that both sexes are left with one active X chromosome.

A different mechanism, known as **male hypertranscription**, achieves the same goal in *Drosophila*. Transgenes inserted into the X chromosome of a male are transcribed at nearly twice the rate of the same genes inserted elsewhere (Spradling and Rubin, 1983; Lucchesi and Manning, 1987). Male hypertranscription relies on two types of regulatory elements. First, there are relatively short DNA sequences on the X chromosome that act as twofold enhancers, apparently capable of placing any nearby gene under dosage compensation control. Second, the products of at least four autosomal genes, known as the *MSL* set (for *male-specific lethal*), are required to *allow* X chromosome hypertranscription in males. Loss-of-function mutations in any of these genes have no effect on XX individuals but cause the death of XY individuals by insufficient X chromosome transcription. Conversely, the Sexlethal function is necessary to *prevent* X chromosome hypertranscription in females; loss of Sexlethal function in XX individuals causes death by hypertranscription of X-linked genes.

One of the *MSL* genes, the *maleless*[+] gene, was cloned, and antibodies were raised against the maleless protein (Kuroda et al., 1991). *Immunostaining* of polytene chromosomes with these antibodies (see Methods 4.1) revealed that the maleless protein is present at hundreds of different sites on the X chromosome in males but not in females (Fig. 27.19). This result, together with the phenotype of the *maleless*[−] mutants, indicates that the maleless protein facilitates male hypertranscription by interacting with the X-linked enhancer sequences. The binding of maleless protein to the X chromosome in males depends on the function of three other *MSL* genes, designated *msl-1*[+], *msl-2*[+], and *msl-3*[+] (for *male sex lethal*). Loss of function in any of these genes abolishes the specific immunostaining of the male X chromosome with antibodies against maleless protein (Gorman et al., 1993). Moreover, the *Sexlethal* function, which prevents

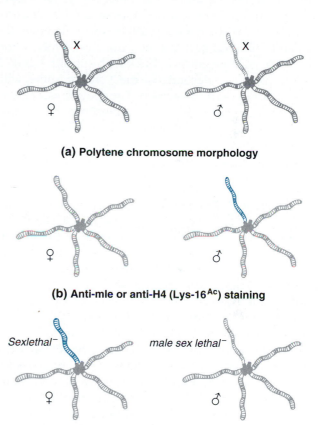

(a) Polytene chromosome morphology

(b) Anti-mle or anti-H4 (Lys-16^Ac) staining

(c) Anti-mle staining of mutants

Figure 27.19 Sex-specific association of the X chromosome in *Drosophila* males with maleless (mle) protein and a histone H4 isoform with acetylated lysine-16 (H4(Lys-16Ac)). **(a)** All diagrams show a complete set of polytene chromosomes from salivary gland cells. The chromosomes adhere to one another with their heterochromatic centromeres. The X chromosome (chromosome I) and the right and left arms of chromosomes II and III are long. The Y chromosome and chromosome IV are small. The autosomes and the female X represent pairs of homologous chromosomes whereas the male X represents only one chromosome. **(b)** In wild-type larvae, male but not female X chromosomes immunostain (color) for mle protein and H4(Lys-16Ac). **(c)** The inhibition of anti-mle staining in female X chromosomes depends on the function of the *Sexlethal*[+] gene. The anti-mle staining of male X chromosomes depends on the function of three autosomal genes termed *male sex lethal*[+] (*msl-1*[+], *msl-2*[+], and *msl-3*[+]).

hypertranscription of female X chromosomes, prevents the binding of maleless protein to the male X chromosome. In mutant females with unstable *Sexlethal* expression, maleless protein binds to the X chromosome in cells lacking Sexlethal protein.

The maleless protein has the molecular characteristics of a *helicase*, that is, an enzyme that unwinds DNA and RNA helices (Gorman et al., 1993). The maleless protein may therefore act to decondense chromosomal DNA. This hypothesis is supported by further observations. First, in polytene chromosomes, the single male X chromosome is more diffuse in appearance than the diploid female X chromosome and the autosomes. Second, a histone H4 isoform acetylated at lysine-16 is found at numerous sites along a male polytene X chromosome but not in male autosomes or any female chromosomes (B. M. Turner et al., 1992). Taken together, the available data suggest that the maleless protein and other *MSL* gene products interact with specific X-linked DNA sequences to decondense the X chromosome, with the result of hypertranscription. The specific association of maleless protein with the X chromosome is prevented in females under the influence of the Sexlethal protein through an as yet unknown mechanism.

In *C. elegans*, dosage compensation involves at least four genes, collectively known as the *DCD* set (for *dosage compensation dumpies*, referring to the dumpy body shape of sublethal alleles). Loss of function in these genes is deleterious or lethal to hermaphrodites but not to males. Apparently these genes act by reducing the transcription of X-linked genes in XX individuals but not in XO individuals. Thus, the *DCD* set in *C. elegans* and the *MSL* set in *Drosophila* seem to work in opposite directions: the *MSL* set causes hypertranscription of the single X chromosome in males, whereas the *DCD* set reduces transcription of both X chromosomes in hermaphrodites. Both mechanisms differ from X inactivation in mammals, demonstrating again that various mechanisms have evolved independently to satisfy the general need for dosage compensation.

The Somatic Sexual Differentiation Occurs with Different Degrees of Cell Autonomy

The most apparent effect of sex determination is the somatic sexual differentiation. In mammals, the second phase of this process is dominated by sex hormones, as will be described further in Chapter 28. The male fetus produces two sex hormones: *testosterone*, which is synthesized in the *interstitial cells* (or *Leydig cells*), promotes male secondary sex differentiation; and *anti-Müllerian duct hormone (AMH)*, produced in the Sertoli cells, inhibits the development of female genital rudiments.

The primary sex-determining signal in mammals, the SRY protein produced in the Sertoli cells, presumably activates the synthesis of AMH and testosterone, but the regulatory mechanisms are not yet understood.

In the fruit fly, somatic sex differentiation is controlled by a hierarchy of genes involving, in order, *Sexlethal*[+], *transformer*[+] plus *transformer-2*[+], and *doublesex*[+] (Fig. 27.20). The regulatory steps in this cascade are all based on differential splicing of pre-mRNAs, as detailed in Chapter 16. The default program at each step results in the development of somatic male characteristics. For instance, XX individuals with loss-of-function alleles of *transformer* develop as somatic males. Conversely, XY individuals carrying a transgenic cDNA for the female-specific transformer mRNA develop as sterile females (McKeown et al., 1988). Differential splic-

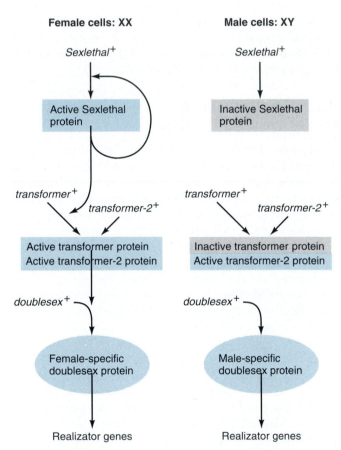

Figure 27.20 Somatic sex determination in *Drosophila* by a cascade of differential splicing (see also Fig. 27.17). In females, the presence of Sexlethal protein regulates the splicing of transformer pre-mRNA so that an active protein is synthesized. The transformer (tra) protein and the (non-sex-specific) transformer-2 (tra-2) protein together regulate the splicing of doublesex pre-mRNA so that the female-specific doublesex (dsx) protein is produced. In males, the absence of functional Sexlethal and transformer proteins allows the default splicing of doublesex pre-mRNA to occur. The resulting male-specific mRNA encodes a functional male-specific doublesex protein. Both doublesex proteins regulate sex-specific realizator genes—at least some of them directly.

ing of doublesex pre-mRNA, in contrast to the other differential splicing steps in this cascade, yields two functional proteins: one male-specific and one female-specific. These proteins are thought to control realizator genes so that the sexually dimorphic characters develop. At least some of the realizator genes are directly controlled by doublesex proteins. Both male and female doublesex proteins, for example, bind directly to the enhancer regions of two genes encoding yolk proteins, with the female-specific protein causing activation and the male-specific protein causing inhibition (Burtis et al., 1991; Coschigano and Wensink, 1993). These genes are expressed in the fat body of females, but not of males, indicating that the regulatory cascade for somatic sex differentiation is continuously active.

In *C. elegans*, the regulatory cascade for somatic sex differentiation involves more genes and more steps. This greater complexity may reflect the dual roles of these genes in somatic and germ line development. Beginning with the *her-1*⁺ gene, which is controlled by the master regulatory system of *C. elegans* sex determination, there are at least four groups of autosomal genes in the cascade, with negative regulation at each step (Fig. 27.21). Some of these genes are named *tra-1*⁺, and so forth (for transformer), like their namesakes in the corresponding regulatory cascade in *Drosophila*. And as in *Drosophila*, loss-of-function alleles in any of the *tra* genes make XX individuals develop like males, whereas transgenic expression of these genes in XO individuals causes them to develop as hermaphrodites.

Beyond these similarities in the names and biological functions of some genes, there are no further similarities between the regulatory cascades of *Drosophila* and *C. elegans*. DNA sequences and molecular mechanisms are different, and in particular, no alternative splicing seems to be involved in *C. elegans*.

The various control mechanisms for somatic sex differentiation confer different degrees of *cell autonomy*. That is, cells of the three types of animals discussed here have different abilities to become male or female independently of their neighbors. In mammals, cell autonomy is limited to the expression of the testis-determining factor in Sertoli cells, and even there a neighborhood effect is observed, as described earlier. However, once the sex hormones are produced, they are carried throughout the body in the blood, and all target cells in a normal individual respond to them. Depending on testosterone level, the target organs of a fetus will become male or female. Even in cases of marginal testosterone level, all target organs develop a similar degree of intersexuality. This occurs, for instance, when female fetuses of some mammals are connected by placental blood circulation to male siblings.

In *C. elegans* and *Drosophila*, no sex hormones are carried all over the body as a systemic signal. Instead, the genetic control mechanisms act independently in each cell. The question then arises just how independent each cell is in its sex determination. Could a single cell develop as a female when all its neighbors are male, or vice versa?

In *C. elegans*, this question has been tested with the genetic mosaic technique described in Chapter 24. One of the genes tested for cell-autonomous expression was *her-1*⁺, the first gene in the regulatory cascade for somatic sexual development (Fig. 27.21). Expression of *her-1*⁺ is necessary for development of a male but not of a hermaphrodite. An extra chromosome fragment carrying both *her-1*⁺ and a cell-autonomous marker gene (*ncl-1*⁺), if spontaneously lost from an individual chromosomally mutant for these two genes, generates a genetic mosaic for *her-1* and *ncl-1*. Using such a mosaic, Craig Hunter and William Wood (1992) found that *her-1*⁺ expression in a sexually dimorphic cell is not necessary for that cell to adopt a male fate. Cells lacking the *her-1*⁺ gene can be masculinized by neighboring cells expressing the gene. In accord with this result, the her-1 protein has characteristics of a signal peptide exchanged between cells (Perry et al., 1993). Conversely, cells expressing the *her-1*⁺ gene can be feminized by neighboring cells lacking this gene activity. This interaction appears to involve the product of the *tra-2*⁺ gene, which has the molecular characteristics of a membrane receptor protein and seems to be *inactivated* by the her-1 protein (Kuwabara and Kimble, 1992). Conceivably, then, a cell that expresses *her-1*⁺ itself but is surrounded by cells failing to express this gene retains enough active tra-2 receptor to undergo female development.

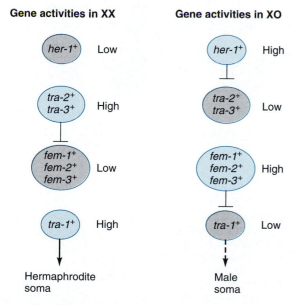

Gene activities in XX

her-1⁺ — Low
tra-2⁺ tra-3⁺ — High
fem-1⁺ fem-2⁺ fem-3⁺ — Low
tra-1⁺ — High
Hermaphrodite soma

Gene activities in XO

her-1⁺ — High
tra-2⁺ tra-3⁺ — Low
fem-1⁺ fem-2⁺ fem-3⁺ — High
tra-1⁺ — Low
Male soma

Figure 27.21 Somatic sex determination in *C. elegans*. Activity of the *her-1*⁺ gene is transduced through three further regulatory steps to control the activity of the terminal gene *tra-1*⁺. High activity of this gene promotes female somatic development, whereas low activity fosters male development. The same genes control germ line development.

The neighborhood effects observed for the expression of *her-1*[+], and presumably *tra-2*[+], suggest a possible adaptive value for the long cascade of inhibitory gene interactions that control somatic sex development in *C. elegans*. The genes active in male cells allow these cells to recruit neighbors to male development, and the genes active in female cells promote the converse recruitment. Presumably, these interactions help prevent the formation of sexual chimeras, which are detrimental to a population.

The highest degree of cell autonomy is observed in the somatic sex differentiation of *Drosophila* and other insects. This is shown dramatically by **gynandromorphs** (Gk. *gyne*, "woman"; *andros*, "man"; *morphe*, "form"). Such individuals are sexual chimeras, combining body regions that are fully male with others that are completely female (Fig. 27.22). Gynandromorphs arise from XX embryos when one X chromosome is not integrated properly into a mitotic spindle and consequently is lost from a nucleus (A. J. Sturtevant, 1929). The progeny of this nucleus are then XO and form male cells while the other nuclei in the same individual are still XX and form female cells. Recessive mutant alleles for cuticular markers on the persisting X chromosome in male cells delineate the boundary between XO tissues, where the mutant alleles are *hemizygous* and therefore expressed, and XX tissues, where the mutant alleles are covered by wild-type alleles on the other X chromosome. If one X chromosome is lost during the first mitosis, about half of the gynandromorph is of the mutant phenotype and the other half, the wild phenotype. If the chromosome loss occurs later, the mutant portion of the gynandromorph is correspondingly smaller. In any case, wherever a body area is sexually dimorphic, the mutant portion has clearly male characteristics, such as abdominal pigment or *sex combs* on the foreleg. These observations indicate that sex determination occurs autonomously in each somatic cell according to its X : A ratio.

Germ Line Sex Differentiation Involves Interactions with Somatic Gonadal Cells

The third aspect of sex determination, in addition to dosage compensation and somatic sexual development, is germ line sex differentiation. Whether germ cells enter oogenesis or spermatogenesis is determined by interactions between germ cells and somatic gonadal cells. The interactions between the supporting cell lineage and the germ line in the mammalian gonad have been discussed earlier in this chapter.

In *Drosophila*, the genes regulating germ line sex differentiation differ considerably from those controlling somatic sexual development. Of the latter, only *Sex-*

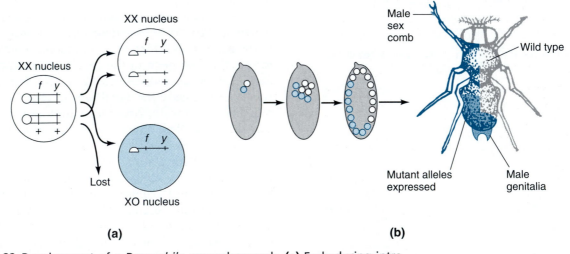

(a)   (b)

Figure 27.22 Development of a *Drosophila* gynandromorph. **(a)** Early during intravitelline cleavage of an XX individual, a nucleus loses one X chromosome during mitosis, thus generating one XX and one XO daughter nucleus. If the individual was heterozygous for recessive X-linked alleles, such as *forked (f)* and *yellow (y)*, these alleles may now be hemizygous in the XO nucleus and its progeny, where they will later be expressed as forked bristles and yellow cuticle. The same alleles are still masked by wild-type alleles in the XX nucleus and its progeny, so that they will not be expressed. **(b)** As the gynandromorph develops into an adult, the phenotypic expression of the X-linked markers clearly delineates XO tissues (color) from XX tissues. Wherever a body area is sexually dimorphic, the XX tissues are fully female whereas the XO tissues have distinctly male characteristics, such as abdominal pigmentation or sex combs at the forelegs. These observations show that the sex of each somatic cell is determined according to its X:A ratio.

lethal⁺ is required *within* the germ line (Steinmann-Zwicky et al., 1989). In the germ line, as in the soma, the presence of Sexlethal protein determines female development, whereas absence of Sexlethal protein determines male development. Moreover, splicing of the Sexlethal pre-mRNA occurs at the same alternative sites in the germ line as in the soma (Oliver et al., 1993). However, the position of *Sexlethal*⁺ is at the bottom of the control hierarchy in the germ line, whereas it is at the top in the soma (Fig. 27.23). In the germ line, *Sexlethal*⁺ is controlled by several other genes, which in turn are regulated by an *inductive signal* from the somatic cells of the gonad. This inductive signal depends on the somatic hierarchy discussed earlier, with *Sexlethal*⁺ at the top and *doublesex*⁺ at the bottom.

▼

The role of an inductive signal from somatic to germ line cells in *Drosophila* first became apparent after transplantation of *pole cells,* that is, the primordial germ cells. Because the sex of living *Drosophila* embryos is not apparent to experimenters, pole cell transplantations randomly generate individuals with XX soma and XY pole cells, or vice versa, in addition to correctly matched males and females (Steinmann-Zwicky et al., 1989). In XX somatic gonads, germ cells develop according to their X:A ratio; that is, XX germ cells undergo oogenesis, and XY germ cells take the spermatogenic pathway, although they are arrested as spermatocytes. The results are different in XY gonads. Here, both XX and XY germ cells enter the spermatogenic pathway, although the XX germ cells do not complete it. These results show that the completion of spermatogenesis as well as oogenesis depends on inductive signals from gonadal to somatic cells, and that these signals function appropriately only between XX soma and XX germ line and between XY soma and XY germ line.

In addition, XX germ cells depend on an inductive signal from somatic gonadal cells to even begin oogenesis. This signal requires the activity of the *transformer*⁺ and *doublesex*⁺ genes, as indicated by transplantations of XX germ cells into normal and transgenic XY hosts (Steinmann-Zwicky, 1994). In an independent study, the final step in the genetic cascade of the germ line—the splicing of the Sexlethal pre-mRNA—was tested in individuals carrying mutant alleles for genes known or suspected to be involved in germ line sexual development (Oliver et al., 1993). The results from both sets of experiments are compatible with the working model shown in Figure 27.23.

The complexities of germ line sex differentiation in *C. elegans* are somewhat different. On the one hand, the

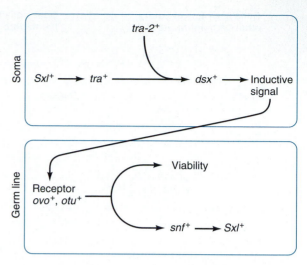

Figure 27.23 A working model of female germ line sex determination in *Drosophila*. Germ line sex is determined by regulatory hierarchies in both the soma and the germ line. The key gene, *Sexlethal*⁺ (*Sxl*⁺), is at the top of the regulatory hierarchy in the soma and at the bottom of a different hierarchy in the germ line. Female-specific splicing of Sexlethal pre-mRNA in the soma starts the cascade of regulation shown in Figure 27.20. This cascade leads to the synthesis of female-specific doublesex (dsx) protein, which in turn promotes the synthesis of an inductive signal. The signal binds to a receptor in the germ line cells, which through genes including *ovo*⁺, *ovarian tumor*⁺ (*otu*⁺), and *sans fille*⁺ (*snf*⁺) ensures the viability of XX germ line cells and the female-specific expression of *Sexlethal*⁺ in the germ line.

regulatory cascade that controls somatic sex also controls germ line sex. On the other hand, in the young adult hermaphrodite, sperm develop in an animal that is otherwise female. The simplest model that accommodates the available data is shown in Figure 27.24. It involves the *fog-2*⁺ gene, so called because its loss of function causes *feminization* of the germ line. Its wild-type function is to down-regulate the activities of *tra-2*⁺ and *tra-3*⁺ in the germ line of the young adult. This allows the temporary activation of the *fem*⁺ genes, which inhibit *tra-1*⁺ so that spermatogenesis can occur. Later in adult life, *fog-2*⁺ activity ceases, *tra-2*⁺ and *tra-3*⁺ inhibit the *fem*⁺ genes, *tra-1*⁺ becomes active, and oogenesis ensues.

The influence of *tra-1*⁺ on both germ line and somatic sex is strong although not exclusive (Hodgkin, 1983). Gain-of-function alleles, symbolized as *tra-1(gf)*, have a feminizing effect: both XX and XO individuals heterozygous for *tra-1(gf)* develop, for the most part, as females; a few are hermaphrodites. Homozygotes for *tra-1(gf)* are invariably females. Null alleles, symbolized as *tra-1(0)*, have a masculinizing effect: XO individuals develop as normal males, while XX individuals develop as somatic males with irregular gonads. Only about a third of these XX individuals are fertile males, indicat-

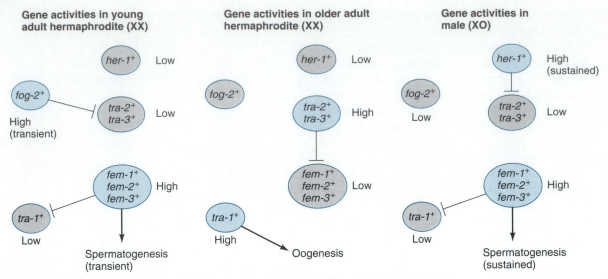

Figure 27.24 Simplified model for the determination of germ line sex in *C. elegans*. Germ line sex determination involves many of the genes in the regulatory cascade that directs somatic sex determination (see Fig. 27.21). An additional gene, *fog-2⁺*, inhibits *tra-2⁺* and *tra-3⁺* in the germ line of the young adult hermaphrodite, so that the *fem⁺* genes become temporarily active and promote spermatogenesis, as they do in the male. When *fog-2⁺* activity ceases, the *fem⁺* genes are repressed and oogenesis can take place.

ing that other genes, such as the *fem⁺* group, have some effect on gametogenesis.

Even though the *tra-1⁺* gene of *C. elegans* does not determine germ line sex exclusively, mutations in this gene can radically change the entire sex determination system in *C. elegans*, as shown in Figure 27.25. Crosses between *tra-1(0)/tra-1(gf)* females and *tra-1(0)/tra-1(0)* males result in progeny with the same sexual genotypes and phenotypes as their parents. The two mutant alleles of the *tra-1* locus, located on the third autosomal chromosome, have in effect generated an *Mm/mm* system as diagrammed in Figure 27.4. The *tra-1(gf)* allele acts as *M* and its homologous *tra-1(0)* allele acts as *m*. The X chromosome has now become irrelevant to sex determination and just takes care of its own dosage compensation as shown in Figure 27.18. The genetically engineered male/female *mm/Mm* strain grows poorly compared with the wild-type male/hermaphrodite XO/XX strain, presumably because the engineered strain depends on mating for propagation and the fertility of the *mm* males is low. However, the strain is viable and can be maintained for many generations, demonstrating how easily one system of genetic sex determination can be converted into another. This experiment reinforces the notion that sex determination systems are relatively unstable and may evolve rapidly. This is in stark contrast to other developmental processes, including embryonic pattern formation and programmed cell death, which rely on exceedingly well-conserved genetic and molecular mechanisms.

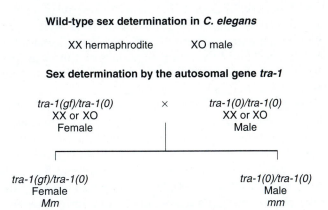

Figure 27.25 Artificial sex determination system for *C. elegans*. The normal mode of sex determination, dictated by the X:A ratio, can be overridden by mutations in the autosomal *tra-1⁺* gene. Gain-of-function mutations, designated *tra-1(gf)*, are dominant and feminizing. Null mutations, designated *tra-1(0)*, are recessive and masculinizing. Consequently, a stable strain can be constructed in which the autosomal *tra-1(gf)* behaves as a feminizing *M* allele while its *tra-1(0)* homologue acts like an *m* allele in an *Mm/mm* system.

SUMMARY

Most animals are dioecious and more or less sexually dimorphic: they exist as males or females, or in some species as males or hermaphrodites. Sex determination is the natural mechanism by which an animal is determined to become either male or female (or hermaphroditic).

In some species, sex determination is environmentally controlled; sex is determined after fertilization by factors such as temperature or the proximity of sex partners. For most species, however, sex determination is a function of genotype; sex is determined at fertilization by the combination of genes that the zygote inherits. In the simplest case, one sex is heterozygous for two alleles of the same gene and the other sex is homozygous (the *Mm/mm* system). This situation easily evolves into a system with entire chromosomes that are sexually dimorphic, such as the X and Y chromosomes of mammals. Sexually dimorphic chromosomes are called sex chromosomes to distinguish them from all other chromosomes, known as autosomes. Instead of having XX females and XY males, other taxonomic groups have XX females (or hermaphrodites) and XO males, or ZW females and ZZ males. The common feature of these different systems is that one sex is heterogametic (producing two types of gametes that differ in their sex-linked genes) while the other sex is homogametic (producing gametes that are identical in their sex-linked genes).

Genotypic sex determination controls three aspects of sexual development: somatic sex differentiation, germ line sex differentiation, and dosage compensation. Somatic sex differentiation brings forth the somatic dichotomy between males and females (or hermaphrodites). Germ line sex differentiation generates either sperm or eggs. Dosage compensation equalizes the expression of non-sex-related genes located on the sex chromosome that is present in two copies in one sex and one copy in the other sex. These three aspects of sexual development are common to mammals, insects, and roundworms. However, the genes and molecular mechanisms that control these pathways of development vary widely and appear to have evolved independently.

The critical element of sex determination in mammals is a testis-determining factor (TDF) located on the Y chromosome. In mice, this factor can be generated through the activity of a single gene, known as Sry^+. TDF activity causes the supporting cells of the indifferent gonad to develop into pre-Sertoli cells, which in turn induce the germ line cells to become spermatogonia. TDF also causes somatic gonadal cells to produce two hormones, testosterone and anti-Müllerian duct hormone, which will direct further male sexual development. Lack of TDF activity allows the germ line cells of the indifferent gonad to develop as oocytes; they direct the supporting cells to become ovarian follicle cells, which will contribute to the production of female sex hormones. Dosage compensation in mammals is initiated by one or more X-linked loci that cause the virtual inactivation of all X chromosomes in a cell except one.

In the fruit fly *Drosophila melanogaster,* the primary sex-determining signal is the X:A ratio, that is, the number of X chromosomes divided by the number of autosomal sets. An X:A ratio greater than 0.75 activates a master control gene, $Sexlethal^+$. This gene feeds back positively into its own expression and directs all three genetic pathways of sex determination toward female development. An X:A ratio of 0.67 or smaller fails to activate $Sexlethal^+$ and causes male development by default. Dosage compensation occurs through an increase in the transcription rate of X-linked genes in males.

In the roundworm *Caenorhabditis elegans,* the X:A ratio acts on a small group of genes, $xol\text{-}1^+$, $sdc\text{-}1^+$, and $sdc\text{-}2^+$, which act together as a master control switch. This switch acts on a set of genes regulating dosage compensation by reducing the transcription rate of X-linked genes in XX individuals. Independently, the master switch genes act on a genetic cascade of inhibitory interactions that eventually regulate both somatic and germ line sex differentiation. The unified control of somatic and germ line sex makes it possible to change the entire sex determination mechanism in *C. elegans* by engineering mutations in autosomal control genes.

SUGGESTED READINGS

Bogan, J. S., and D. C. Page. 1994. Ovary? Testis? A mammalian dilemma. *Cell* **76:**603–607.

Bull, J. J. 1983. *Evolution of Sex Determining Mechanisms.* Menlo Park, Calif.: Benjamin-Cummings.

Hodgkin, J. 1992. Genetic sex determination mechanisms and evolution. *BioEssays* **14:**253–261.

Keyes, L. N., T. W. Cline, and P. Schedl. 1992. The primary sex determination signal of *Drosophila* acts as the level of transcription. *Cell* **68:**933–943.

HORMONAL

CONTROL

OF DEVELOPMENT

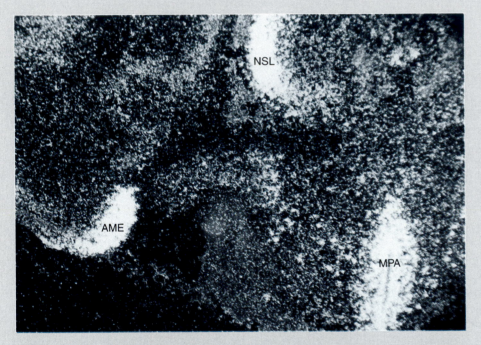

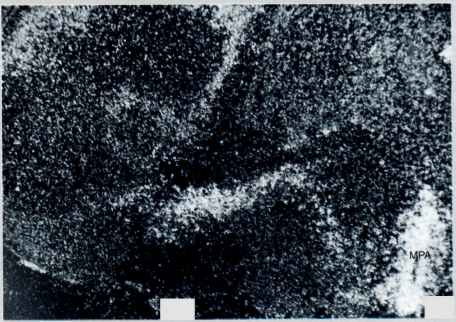

Figure 28.1 In vertebrate brains, some groups of neurons are activated by androgens, hormones circulating primarily in males, whereas other brain areas are affected by the female sex hormones estrogen and progesterone. This pair of dark-field photographs shows frontal sections of the brain of the whiptail lizard *Cnemidophorus inornatus*. The sections were hybridized in situ with radiolabeled probes that bind either to mRNA for androgen receptor (top) or to mRNA for progesterone receptor (bottom). The white grains reveal the distribution of the mRNA bound by either probe. Some brain areas, including the medial periventricular area (MPA) of the diencephalon, contain the mRNAs for both hormone receptors. The mRNA for androgen receptor is concentrated in two telencephalon areas: the nucleus septalis lateralis (NSL), and the external nucleus of the amygdala (AME). The latter is known to be involved in aggression and male copulatory behavior. The mRNAs for estrogen and progesterone are selectively present in other brain areas (not shown) that are associated with the regulation of ovulation and with female sexual receptivity.

Cells in a developing organism exchange signals that coordinate their actions and allow the organism to adapt to external events. Three principal agents of cell communication have evolved in metazoa: nerve cells, short-range chemical signals such as inducers and growth factors, and long-range chemical signals called *hormones*. Nerve cells send electrical impulses through long extensions, which selectively reach their targets. Inducers and growth factors act locally because they bind rapidly to cell membranes and extracellular materials. In contrast, **hormones** are chemical signals that are produced by specialized gland cells and then carried throughout the organism in the blood or equivalent body fluids.

Like broadcasts, hormonal signals are propagated everywhere but act only on receivers that are appropriately tuned. In other words, hormones elicit a response only from *target cells*, which have matching hormone receptors and other cofactors located on their plasma membranes or in their cytoplasm. Because a hormone accumulates at the receptors to which it binds, the target cells of a hormone can be identified by injection of radiolabeled hormone and subsequent autoradiography. Alternatively, target cells can be revealed by immunostaining of the hormone receptor, or by in situ hybridization with a labeled probe that binds to the mRNA encoding the receptor (Fig. 28.1).

Typically, a hormone coordinates the activities of multiple target tissues or organs. For instance, in the human female, progesterone prepares the inner lining of the uterus (the endometrium) during each menstrual cycle for implantation of an egg; if pregnancy occurs, progesterone also maintains the endometrium, stimulates the growth of the uterine wall, and promotes the formation of mammary ducts. The coordinating action of a hormone often depends on concentration changes over time. Such changes are typically controlled by the rate of synthesis in the hormone-producing glands, which is controlled by feedback loops, other hormones, or nervous system input. The synthesis of progesterone in the ovary is regulated by luteinizing hormone from the pituitary gland.

In this chapter, we will review some general aspects of hormone function before turning to the hormonal control of sex differentiation in mammals, including humans. These discussions will cover the development of the reproductive organs as well as the development of the brain and behavior. Next we will explore the hormonal control of insect metamorphosis, including exciting new work on the molecular mechanisms of hormone action. Finally, we will examine the role of hormones in amphibian metamorphosis and highlight some of the astounding parallels to hormonal control of insect metamorphosis.

General Aspects of Hormone Action

In order to act on a target cell, hormones first bind to specific proteins that act as receptors. A bound hormone activates its receptor by conferring on it a specific ability to interact with other molecules. Animal hormones can be divided into two groups: water-soluble and lipid-soluble. The *water-soluble hormones* are mostly peptides, such as the growth hormone released by the anterior pituitary gland. These hormones are unable to cross cell plasma membranes; instead they bind to the *extracellular domains of receptor proteins*. The loaded receptors then act as ion channels or protein kinases, or they interact with other membrane proteins to increase the levels of second messengers such as cAMP or calcium ions (Ca^{2+}); see Chapter 2. The *lipid-soluble hormones* comprise steroids such as testosterone, terpenoids such as retinoic acid, and the thyroxines released by the thyroid gland. These hormones diffuse readily through cell plasma membranes and bind to *receptors in the cytoplasm*. The loaded receptors accumulate in the cell nuclei, where they act as transcription factors (see Chapter 15).

The dependence of hormone action on receptors, and the specific interaction of the receptors with regulatory gene elements or intracellular signaling molecules, confer a remarkable degree of specificity on the action of hormones. This specificity is enhanced when receptors consist of two different polypeptides, or when the ultimate effect of a hormone depends on the presence of additional signaling components or on enzymes to synthesize a particular product. In such cases, the hormone action is limited to those cells in which *all* of these cofactors are available. The receptors and cofactors are deployed in an epigenetic process during development: as an organism becomes more complex, more types of target cells with specific combinations of hormone receptors and assorted cofactors originate.

Hormonal Control of Sex Differentiation in Mammals

As discussed in the preceding chapter, mammalian sex determination is controlled by the SRY^+ gene on the Y chromosome. SRY^+ expression in the supporting-cell lineage causes the indifferent gonad to develop as a testis, whereas lack of SRY^+ expression allows the gonad to develop as an ovary. We have already discussed this first phase of sexual development, known as *primary sex differentiation*, in Chapter 27. In the present chapter, we will focus on the subsequent phase, called *secondary sex differentiation*, which is governed by hormonal control.

The Synthetic Pathways for Male and Female Sex Hormones Are Interconnected

Many vertebrate hormones, including most sex hormones, are steroids, consisting of four connected car-

bon rings with different substitutions. All steroid hormones are derived from cholesterol, as shown in Figure 28.2. Four synthetic steps convert cholesterol into *progesterone*, a female sex hormone. Three additional steps transform progesterone into **testosterone**, the key male sex hormone. Testosterone in turn is converted by an enzyme, *5α-reductase*, into *5α-dihydrotestosterone (DHT)*, another male sex hormone. Collectively, testosterone and DHT are referred to as **androgens** (Gk. *andro-*, "man"; *gennan*, "to produce"). Another enzyme, **aromatase**, converts testosterone into *17β-estradiol*, a female sex hormone commonly called **estrogen**. Proges-

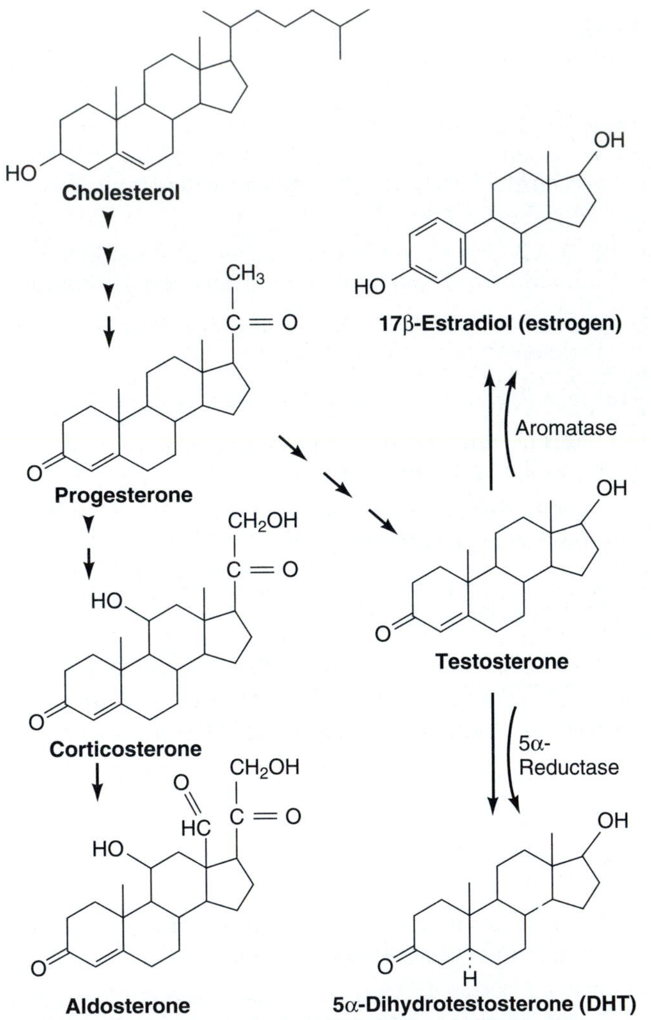

Figure 28.2 Synthetic pathways for vertebrate steroid hormones. All steroids are synthesized from cholesterol. The number of arrows indicates the number of synthetic steps in each pathway. Note that testosterone, the key male sex hormone, is a derivative of the female sex hormone progesterone and the precursor of another female sex hormone, 17β-estradiol (estrogen). The conversion of testosterone into estrogen is catalyzed by the enzyme aromatase. An alternative enzyme, 5α-reductase, catalyzes the conversion of testosterone into 5α-dihydrotestosterone (DHT). These two enzymatic conversions are practically irreversible, because the reverse reactions require much energy.

terone is also a precursor of corticosterone and aldosterone, steroid hormones that are involved in the metabolism of minerals and glucose, respectively.

The principal organs of steroid hormone synthesis are the gonads and the cortex of the adrenal glands, but certain synthetic steps also occur in other tissues. Testosterone is synthesized mainly in the *interstitial cells*, or *Leydig cells*, of the testes. In male mammals, androgens are produced at significant levels during fetal development under the influence of gonadotropic hormones from the placenta. Estrogens are synthesized, via testosterone, in the *thecal cells* and *granulosa cells* of the ovarian follicles. In many mammals, including humans, however, the level of estrogen synthesis in female fetuses is low, and female sexual development during this phase occurs mostly by default, that is, in the absence of androgens. With the beginning of adulthood (puberty, in mammals), gonadotropins from the anterior pituitary gland stimulate testes and ovaries to produce testosterone and estrogen, respectively. These hormones control the development and maintenance of *secondary sex characteristics*, such as breasts in human females and larger muscles in males. In addition to the gonads, the adrenal glands continue to synthesize significant levels of sex hormones.

The pathways of steroid biosynthesis have important biological implications. First, because the testes have to synthesize progesterone in order to produce testosterone and the ovaries have to synthesize testosterone as a precursor of estrogen, the hormonal differences between males and females are more complex than they appear. Second, some critical steps of sex hormone synthesis occur outside the gonads. The pathways that convert testosterone into either DHT or estrogen are especially important. Each conversion depends on a specific enzyme, 5α-reductase or aromatase, respectively. These enzymes are present not only in the gonads but also in target tissues of DHT and estrogen. For instance, significant concentrations of 5α-reductase are present in the indifferent external genitalia of the early embryo. Here testosterone is converted into DHT, which is required for development of the male external genitalia. Later during development, certain areas of the brain in both sexes contain significant amounts of aromatase, the enzyme that converts testosterone into estrogen.

Figure 28.3 provides an overview of sex differentiation and of the role of hormones in the process.

The Genital Ducts Develop from Parallel Precursors, the Wolffian and Müllerian Ducts

As discussed in the previous chapter, the mammalian testis and ovary develop from *complementary* portions of the indifferent gonad: the testis develops from the medulla, whereas the ovary arises from the cortex. As we have seen, this decision is controlled by the *SRY+*

gene. Once the embryonic gonads develop as either testes or ovaries, the following steps of sexual differentiation are controlled entirely by hormones.

The genital ducts that transport and nurture sperm, eggs, and embryos develop from two pairs of embryonic precursors, the *Wolffian duct* and the *Müllerian duct* (see Figs. 27.8 and 27.9). Both ducts are present in parallel at the end of the sexually indifferent stage, which lasts through the sixth week of human development. Their future development is controlled by the presence or absence of two male sex hormones, *testosterone* and **anti-Müllerian duct hormone (AMH)**, the latter a glycoprotein produced by the *Sertoli cells* of the testis.

The Wolffian duct is originally part of the *mesonephros* (embryonic kidney), and it persists in this capacity in lower vertebrates of both sexes (see Chapter 13). In mammals and birds, the kidney function of the mesonephros is superseded by the development of the definitive kidney (metanephros), and the Wolffian duct degenerates in females. In males, however, the Wolffian duct becomes part of the reproductive system (Fig. 28.4). Here, the duct gives rise to the *epididymis*, where sperm mature; the *ductus deferens*, which transports the sperm; and the *seminal vesicle*, where sperm are stored. Beyond the seminal vesicle, the ductus deferens is called the *ejaculatory duct*. The Müllerian duct in the male degenerates except for small vestiges, and the degeneration process depends on AMH and its receptors.

The Müllerian duct persists in females, since females produce no AMH. It gives rise to the oviduct, the uterus, and, in mammals, the upper portion of the vagina (Fig. 28.5). The Wolffian duct in the female degenerates except for minor vestiges.

The hormonal control of secondary sex differentiation was first established by observations on cows and rabbits. Lillie (1917) found that genetically female calves were masculinized in their external genitalia, their genital ducts, and even in their gonadal development if they shared the uterine circulatory system with a male twin. Such masculinized female calves were known as **freemartins**; the conditions of their development suggested that some male factor(s) can pass through the circulatory system and redirect genetically female sex organs toward male development.

In later studies, Jost (reviewed 1965) removed testes or ovaries from rabbit fetuses and monitored the animals' sexual development. In *both* sexes, removal of the gonads was followed by development of the Müllerian ducts into oviducts and a uterus, and by the formation of female external genitalia. These results indicated that the female organs arise by default, whereas the development of male organs requires some testicular factor(s). By additional experiments, Jost showed that testes produce at least two such factors. If he grafted a testis next to the ovary of a female rabbit embryo, *two effects* were observed on the operated side: the Wolffian duct developed, and the Müllerian duct degenerated. If instead

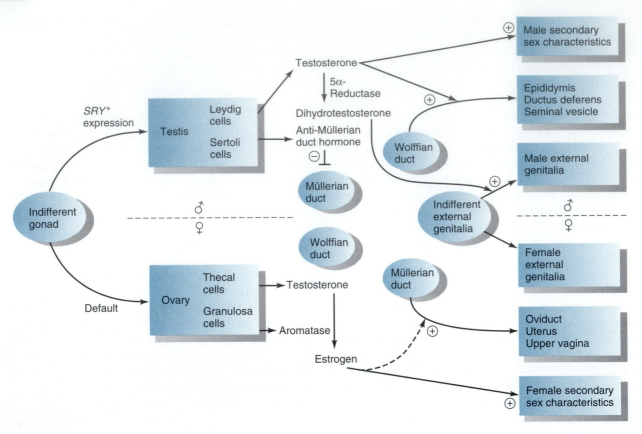

Figure 28.3 Somatic sexual development in mammals. Primary sex differentiation is under genetic control: expression of the SRY⁺ gene in the indifferent gonad causes testis development, whereas lack of SRY⁺ expression allows ovary development. The following phase, known as secondary sex differentiation, is controlled by hormones. The developing testis produces testosterone in the interstitial, or Leydig, cells and anti-Müllerian duct hormone (AMH) in the Sertoli cells. During fetal development, testosterone causes the Wolffian duct to develop into the epididymis, ductus deferens, and seminal vesicle. Some of the testosterone is converted by 5α-reductase to dihydrotestosterone (DHT). DHT brings about male development of the originally indifferent external genitalia. AMH causes degeneration of the Müllerian duct. During puberty, testosterone promotes the development of secondary male sex characteristics. In the developing ovary, thecal cells surrounding the follicle cells produce testosterone, which is converted to estrogen by aromatase in the granulosa cells surrounding the oocyte. However, estrogen levels are low during fetal development, in most species, so that the development of the external female genitalia occurs by default. Likewise, absence of AMH is sufficient for development of the Müllerian duct into the oviduct, uterus, and upper vagina. After puberty, high levels of estrogen promote the development of female secondary sex characteristics.

of a testis he transplanted a crystal of testosterone, only *one effect* was observed: the Wolffian duct developed, but the Müllerian duct did not degenerate. Subsequent investigations identified the testicular factor that is not replaced by testosterone as AMH (Josso et al., 1976).

More detailed knowledge was obtained through the analysis of human congenital abnormalities such as *androgen insensitivity*. This syndrome is caused by a mutation in the gene for the androgen receptor, that is, the receptor for testosterone and DHT (Meyer et al., 1975). The mutation is X-linked and is transmitted by female carriers who are heterozygous for it. XY individuals who inherit the mutated allele develop testes that

produce testosterone and AMH as in normal males. Nevertheless, these individuals develop as phenotypic females (Fig. 28.6). Because of the lack of androgen receptors, there is no response to testosterone in the target organs. As a result, the Wolffian duct degenerates, and female external genitalia develop. In the absence of a testosterone response, the level of estrogen produced in the adrenal glands is sufficient to promote full development of secondary female sex characteristics. This is an important observation, because it indicates that estrogen receptors are formed in both males and females. However, the production of AMH in the testes causes the elimination of the Müllerian duct, so that

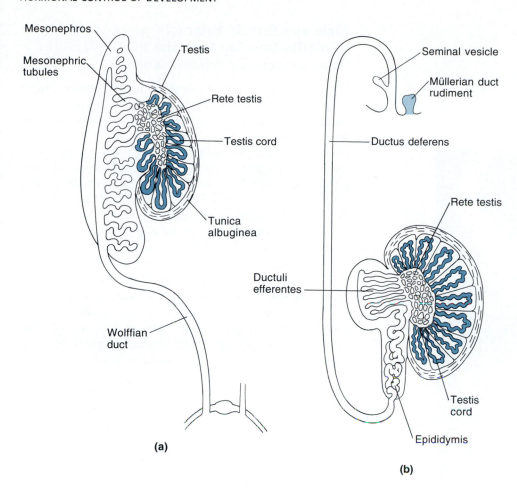

(a)

- Mesonephros
- Mesonephric tubules
- Testis
- Rete testis
- Testis cord
- Tunica albuginea
- Wolffian duct

(b)

- Seminal vesicle
- Müllerian duct rudiment
- Ductus deferens
- Rete testis
- Ductuli efferentes
- Testis cord
- Epididymis

Figure 28.4 Development of the genital ducts in the human male. **(a)** Fourth month of development. The distal portion of the Wolffian duct is still embedded in the mesonephros as the mesonephric tubules connect via the rete testis to the testis cords. The Müllerian duct has already degenerated. **(b)** After the descent of the testis, the mesonephric tubules have become ductuli efferentes, which will conduct sperm after puberty. Much of the proximal Wolffian duct has been transformed into the convoluted tube of the epididymis, from which the ductus deferens will carry the sperm to the seminal vesicle. The most distal section of the Wolffian duct will form the ejaculatory duct that connects the seminal vesicle with the penile urethra.

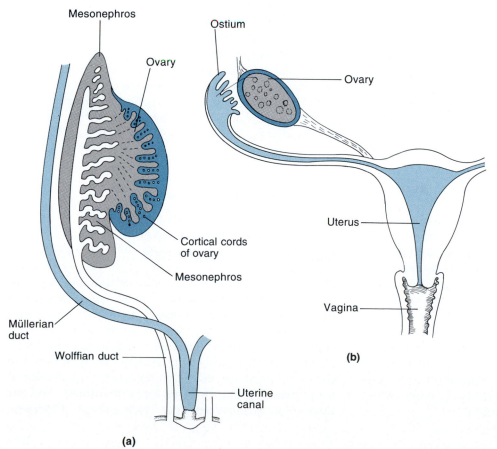

(a)

- Mesonephros
- Ovary
- Cortical cords of ovary
- Mesonephros
- Müllerian duct
- Wolffian duct
- Uterine canal

(b)

- Ostium
- Ovary
- Uterus
- Vagina

Figure 28.5 Development of the genital ducts in the human female. **(a)** Third month of development. Both the Wolffian and the Müllerian ducts are still in place. **(b)** In the newborn female, the Wolffian duct has disappeared except for a few vestiges. The proximal part of each Müllerian duct has formed one oviduct with an opening (ostium) into the abdominal cavity, and the distal portions of the two Müllerian ducts have joined to form the uterus and the upper vagina.

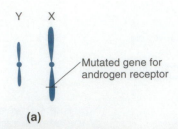

(a)

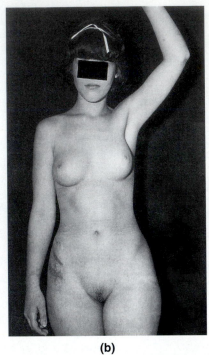

(b)

Figure 28.6 Androgen insensitivity in genetically male individuals. Such individuals have normal male sex chromosomes (XY) except for a loss-of-function mutation in the X-linked gene that encodes the androgen receptor **(a).** The individuals' cells are unable to bind androgens but respond to estrogens produced in the adrenal glands. Correspondingly, they develop a distinctly female appearance **(b).** Internally, however, they lack the derivatives of the Müllerian duct, because they produce and respond to anti-Müllerian duct hormone.

the vagina remains shallow and no menstruations set in at puberty.

The action of AMH in humans is revealed by the *persistent Müllerian duct syndrome (PMDS)*, which is characterized by the persistence of oviducts and a uterus in otherwise nearly normal men. The condition is usually discovered during surgery for inguinal hernia or cryptorchidism (failure of the testes to descend into the scrotum). In some PMDS cases, the gene encoding AMH is mutated so that no functional AMH is made (Knebelmann et al., 1991). Other PMDS patients produce apparently normal AMH, which is still made in small quantities in adult testes. Delineation of the molecular defect in these cases will probably require prior characterization of the AMH receptor.

Male and Female External Genitalia Develop from the Same Embryonic Primordia

Unlike the male and female gonads and genital ducts, which develop from *complementary* embryonic rudiments, the male and female external genitalia develop from *the same* embryonic rudiments.

At the end of the indifferent stage—after 6 weeks, in the human embryo—the *cloaca* has divided into the *urogenital sinus* and the *anus*, and both openings are closed off by temporary membranes (Fig. 28.7). *Urethral folds* on both sides of the urogenital sinus unite anteriorly to form the *genital tubercle*. In the male, the urethral folds close ventrally and, together with the genital tubercle, form the *penis*. In the female, the genital tubercle forms the *clitoris*, while the urethral folds develop into the *labia minora*. Thus, the urogenital sinus becomes either the *penile urethra* in the male or the *vestibule* of the vagina in the female. On both sides of the urethral folds at the indifferent stage, a pair of *genital swellings* arise. They form either the *labia majora* of the female or the *scrotum* surrounding the testes in the male.

Hormonal control of external genital differentiation in the human is illustrated by a heritable form of male pseudohermaphroditism that is frequently observed in an area of the Dominican Republic (Imperato-McGinley et al., 1974). The anomaly is caused by a *5α-reductase deficiency*, which renders XY fetuses unable to convert testosterone to DHT. Because testosterone does not activate androgen receptors to the same extent as DHT, such individuals are born with ambiguous external genitalia including a blind vaginal pouch, labia with testes inside, and an enlarged clitoris. These babies are usually raised as girls until puberty, when the greatly increased level of androgen causes the penis and scrotum to grow to a near normal state of development. As the secondary male sex characteristics also develop, most of these individuals begin to live like males and may be fertile.

Hormonal Control of Brain Development and Behavior in Vertebrates

Having seen how sex hormones control the development of the reproductive organs, we will now explore the effects of the same hormones on brain development and on behavior. First, we will see how hormone receptors—that is, those molecules that account for much of the specificity in hormone action—are distributed in the brain. Next we will examine courtship behavior of songbirds, copulatory behavior in mammals, and the association of these activities with sexually dimorphic

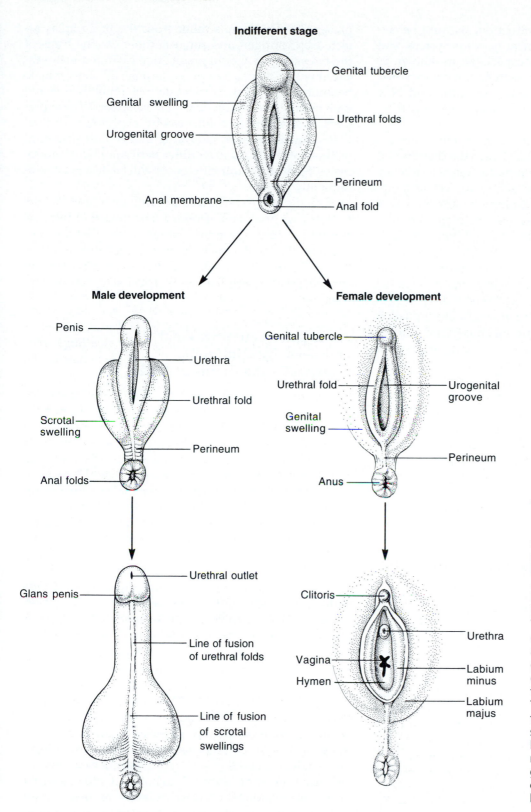

Indifferent stage

Genital tubercle

Genital swelling

Urethral folds

Urogenital groove

Perineum

Anal membrane

Anal fold

Male development

Penis

Urethra

Urethral fold

Scrotal swelling

Perineum

Anal folds

Urethral outlet

Glans penis

Line of fusion of urethral folds

Line of fusion of scrotal swellings

Female development

Genital tubercle

Urethral fold

Urogenital groove

Genital swelling

Perineum

Anus

Clitoris

Urethra

Vagina

Hymen

Labium minus

Labium majus

Figure 28.7 Development of the external genitalia in the human male (left) and female (right) from the indifferent stage (top). At the indifferent stage, the urogenital groove becomes separated from the anus by the perineum, and both openings are closed off by temporary membranes. The genital tubercle forms either most of the penis or all of the clitoris. The urogenital groove becomes either the penile urethra or the vestibule of the vagina. The urethral folds form either the part of the penis that surrounds the urethra or the labia minora (sing., labium minus). The genital swellings give rise to either the scrotum or the labia majora (sing., labium majus).

brain areas. Finally, we will compare the performance of human males and females in certain standardized tests, and discuss how performance depends on hormone levels present during the tests and during fetal development.

Androgen and Estrogen Receptors Are Distributed Differently in the Brain

Males and females differ in mating and parenting behavior and, more subtly, in some behaviors that are less directly associated with reproduction. Because behaviors are controlled by the central nervous system, and since reproductive behaviors especially are known to depend on hormone levels, researchers have explored how sex hormones interact with the brain, and how these interactions might differ between males and females.

As discussed earlier in this chapter, hormones act through specific receptor proteins, and the distribution of such receptors largely accounts for the selective action of hormones on specific target organs. It was therefore logical for researchers to explore the distribution of estrogen and androgen receptors in the brain. Since localizing the receptors themselves turned out to be difficult, they injected radiolabeled estrogen, testosterone, and other hormones into various vertebrates and monitored the accumulation pattern of each hormone in the brain by *autoradiography* (see Methods 3.1). Each hormone showed a characteristic pattern of accumulation in certain brain regions, including those known to be involved in reproductive behavior and aggression. Comparing accumulation patterns revealed an interesting result: while there were major differences in accumulation patterns of *different* hormones for both sexes, the accumulation patterns of a *given* hormone differed in only subtle ways between the sexes. Because steroid hormones normally accumulate in cells by binding to their receptors, the observed hormone accumulation patterns are likely to reflect hormone receptor distributions. This conclusion was confirmed by investigations that revealed directly the distribution of mRNAs for sex steroid *receptors*.

▼

To compare the localizations of androgen receptor, estrogen receptor, and progesterone receptor in the brains of reptiles, Larry Young and his colleagues (1994) used *in situ hybridization* with radiolabeled probes that each bound specifically to one of the corresponding mRNAs. Largely in agreement with earlier results on hormone accumulation, the three mRNAs were strikingly different in their distribution (Fig. 28.1). The mRNA for androgen receptor was especially concentrated in brain areas involved in aggression, mounting, and copulation. The mRNAs for estrogen and progesterone were more concentrated in brain areas associated with the regulation of ovulation and with female sexual receptivity. Also in accord with earlier results on hormone accumulation, the mRNA distributions showed only subtle differences between the sexes.

The available data on sex hormone receptors in the brain indicate that the major differences are not between the sexes but between the receptors for different hormones. Therefore, the brain areas stimulated in the presence of androgens differ from the brain areas affected by estrogen and progesterone. On the basis of these results, one would predict that castrated males injected with estrogen and progesterone will show female behaviors and that ovariectomized females injected with androgens will display male behaviors. These predictions are largely confirmed by experimental data, with a few exceptions indicating that certain structural features of the brain itself differ between males and females. Such structural differences will be discussed subsequently.

Exactly how sex steroids activate or inhibit the activity of neurons is still under investigation. Because steroid receptors are transcription factors that are activated by their matching hormones, target genes of these hormones may well encode proteins affecting the ion channels that determine the electrical activities of neurons.

Androgens Cause Seasonal and Sexually Dimorphic Changes in the Brains of Birds

Songbirds are a large and diverse group of birds that produce complex but stereotypical sounds. Generally, the males sing to establish and defend territories and to attract females. Those songbirds that live in temperate and subtropical latitudes breed seasonally and sing most actively early in the breeding season. Male songbirds learn their songs from older males. During this early phase of *song acquisition*, the songs of the novice are initially crude and variable, but with continued tutoring and practice they become well defined and invariant in structure. This latter phase of song development is referred to as *song crystallization*.

Like other courtship behaviors, birdsong is strongly influenced by circulating hormone levels (Bottjer, 1991; Kelley and Brenowitz, 1992). In male zebra finches, there is a linear correlation between singing and the concentration of testosterone in the blood serum (Pröve, 1978). Castration of a male reduces the rate of singing from about 45 songs to 8 songs per 15-min observation session (Fig. 28.8; Arnold, 1975). If the same male then receives an implant that releases testosterone propionate, his rate of singing increases again to about 27 songs in 15 min. In other songbirds, testosterone injections induce females to sing and cause males to sing out of season (Thorpe, 1958; Nottebohm, 1980). Together, these observations show clearly that the singing of songbirds depends on the level of testosterone or one of its metabolites.

To test which specific hormones promote singing in zebra finches, researchers injected different testosterone metabolites into castrated male zebra finches. Combinations of estrogen and DHT restored singing to its precastration level, whereas either hormone by itself was less effective (Harding, 1983). Similar investiga-

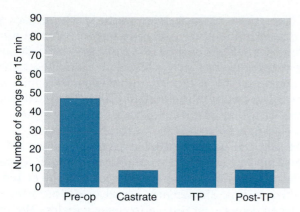

Figure 28.8 Hormonal control of singing in male zebra finches. The rate of song production (number of songs during a 15-min observation session) is shown for normal males (Pre-op), after castration (Castrate), after implantation in castrates of pellets releasing testosterone propionate (TP), and after removal of the implants (Post-TP). The song rate is reduced dramatically after castration but is increased significantly by the hormone implant.

tions with sparrows indicate that estrogen facilitates song acquisition whereas testosterone is required for song crystallization (Marler et al., 1988). Thus, it appears that different testosterone metabolites promote different phases of song development.

There are also anatomical differences between the brains of male and female zebra finches and canaries, especially during the breeding season. Certain brain nuclei that are involved in singing are up to 6 times larger in males than in females (Fig. 28.9; Arnold, 1980). In autoradiographic studies, radiolabeled testosterone accumulated selectively in these song control nuclei (Arnold et al., 1976). Three of the nuclei are known as the *higher vocal center (HVC)*, the *robustus archistriatalis (RA)*, and the *magnocellular nucleus of the anterior neostriatum (MAN)*. The RA is innervated by both the HVC and the MAN and depends on these innervations for growth and neuronal survival (F. Johnson and S. W. Bottjer, 1994). In males, these nuclei are much larger in the spring than in other seasons (Nottebohm, 1981); very similar changes in volume occur in females treated with testosterone. Moreover, the proportion of cells in the male HVC and MAN that accumulate labeled androgens increases during the time of song crystallization (Bottjer, 1987). These data indicate that testosterone or its metabolites cause a seasonal and sexually dimorphic increase in the volume of the song control nuclei of songbirds. The larger nuclei have more neurons, neurons of larger size, and more synapses.

The larger number of neurons in the male song control nuclei may result from the action of testosterone on the neuron generation and migration process. Adult

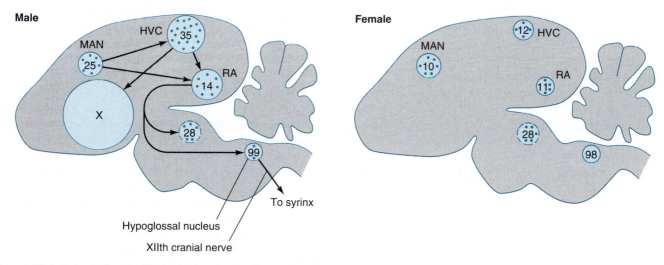

Figure 28.9 Sexual dimorphism of song control nuclei in the songbird brain. The schematic drawings show paramedian sections of male and female songbird brains; anterior to left, dorsal surface up. Circles and arrows represent brain areas and projections thought to be involved in controlling song. The final projection is from the hypoglossal nucleus through the XIIth cranial nerve to the syrinx, the tracheal organ where the birdsong is actually produced. The size of each circle is proportional to the volume of the corresponding brain region in the zebra finch. Circles with dashed lines indicate brain regions whose volumes have not been estimated. Dots mark regions that become labeled after injection of radioactive testosterone. The numbers within each circle indicate the approximate percentage of labeled cells within each region. Note the differences in size and labeling index of the nuclei called the higher vocal center (HVC), the robustus archistriatalis (RA), and the magnocellular nucleus of the anterior neostriatum (MAN). Area X cannot be seen in stained sections of brains from female zebra finches.

birds—in contrast to mammals—continue to generate new neurons from the neuroepithelial layer that lines the brain ventricles (see Chapter 12). DNA labeling with injected radioactive thymidine indicates that neuroblasts born in the neuroepithelial layer move to the higher vocal center, where they become functional neurons before most of them disappear within less than a year. Thus, there seems to be a constant turnover of neurons in the higher vocal center. But such a turnover of neurons presumably takes place in other areas of the brain as well, both in sexually dimorphic areas and in parts of the brain that do not differ between the sexes (Nottebohm, 1989).

The exact function of the neurons of the larger nuclei in the sexually dimorphic areas also needs to be clarified. Since steroid receptors are transcription factors activated by their cognate hormones, it would not be surprising if sex hormones affected not only the number but also the functions of neurons in a sexually dimorphic way.

Prenatal Exposure to Sex Hormones Affects Adult Reproductive Behavior and Brain Anatomy in Mammals

The hormonal control of reproductive behavior and brain development has also been studied in mammals (Breedlove, 1992; Crews, 1994). Some of the most stereotypical and easily quantifiable behaviors occur during copulation of rodents. A female rodent arches her back to elevate her rump and head and moves her tail to one side. This posture is called *lordosis,* and it is taken as a measure of the female's receptivity to mating. Normally, female rodents exhibit lordosis in response to a mounting male only during a particular phase of the ovulatory cycle. When the ovaries are removed, a female ceases to exhibit lordosis. However, her lordotic behavior can be reactivated by injections of estrogen and progesterone.

When the same hormonal regimen that reactivates lordotic behavior in an ovariectomized female rat is given to a castrated male rat, the male shows almost no lordosis in response to another male's mounting. The fact that castrated male rats and ovariectomized female rats behave differently after the same hormonal treatment indicates that their previous conditioning must be different. Because the central nervous system mediates behavior, the difference presumably lies therein.

One difference between the brains of males and females is the prenatal exposure of the male brain to androgens. If this difference were critical, then female rodents exposed to testosterone prenatally should not show lordosis as adults when given estrogen and progesterone treatment and mounted by a male. This prediction was confirmed in tests with guinea pigs exposed prenatally to testosterone propionate (Phoenix et al., 1959).

The effects of fetal exposure to sex hormones are also seen in mammals that produce litters of multiple young during each pregnancy (vom Saal, 1989). Because the growing fetuses exchange steroid hormones through the placental blood circulation, females developing between two males in the uterus (called 2M females) are exposed to higher levels of testosterone than females developing next to one male (1M females) or no males (0M females). As compared with 0M females, 2M females have masculinized genitalia, take longer to reach puberty, and as adults have shorter and fewer reproductive cycles. 2M females are also less attractive to males and more aggressive to other females. The converse effects are observed for males developing next to females in utero in species where female fetuses have higher concentrations of estrogen than males, a condition not found in all mammals. In mice, 2F males have smaller seminal vesicles and are less aggressive and sexually more active than 0F males.

These observations inspired the *organizational hypothesis*, according to which prenatal exposure to androgens permanently alters the developing brain, causing it to function in male-specific ways later in life. This hypothesis has since been tested for other behaviors and with different mammals and has been confirmed with few exceptions. Interestingly, the sensitive period during which a postnatal behavior is masculinized by prenatal exposure to androgens depends on the behavior and the species studied. Thus, it appears that specific parts of brains have different phases during which they are most sensitive to hormonal conditioning.

Having confirmed the hormonal control of reproductive *behavior,* researchers looked for sexual dimorphism in those areas of the rodent *brain* that are involved in reproductive behaviors. They identified one such area in rats as the *sexually dimorphic nucleus in the preoptic area (SDN-POA)* of the hypothalamus (Raisman and Field, 1973; Gorski et al., 1978). This nucleus is discernible from the 20th day of gestation and becomes sexually dimorphic during the first 10 days after birth. During this period, the volume of the SDN-POA becomes 5 to 6 times larger in normal males than in females (Fig. 28.10). In accord with the organizational hypothesis, the SDN-POA of male rats castrated after birth remains small, and the SDN-POA of females treated with androgens shortly before or after birth grows larger. After day 10, hormonal manipulations no longer change the volume of the SDN-POA.

Treatment of newborn rat females with a synthetic estrogen also masculinizes the SDN-POA, indicating that testosterone exerts its effect through its estrogen derivative. This conclusion is confirmed by the following observation. Rats with genetic androgen insensitivity, which lack androgen receptors but do have estrogen receptors and aromatase, develop a male-sized SDN-POA that contrasts with their otherwise female appear-

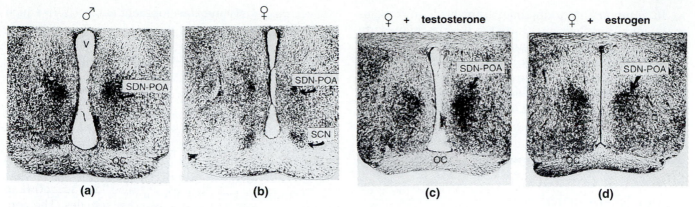

Figure 28.10 The sexually dimorphic nucleus of the preoptic area (SDN-POA) of the rat hypothalamus. The SDN-POA is located on both sides of the third ventricle (V) above the suprachiasmatic nucleus (SCN) on top of the optic chiasm (OC). The volume of the SDN-POA is about 5 times larger in male rats **(a)** than in female rats **(b)**. Perinatal treatment of female rats with either testosterone **(c)** or a synthetic estrogen **(d)** enlarges (masculinizes) the SDN-POA.

ance (Rosenfeld et al., 1977). Female rats escape the masculinizing effect of estrogen on the SDN-POA through the action of α-fetoprotein (AFP), a fetal serum protein that binds to estrogen and prevents its passage through the *blood-brain barrier* into the brain. Testosterone, which does not bind to AFP, circumvents this barrier and is converted to estrogen within the brain.

Certain Nonsexual Human Behaviors Depend on Prenatal Exposure to Androgens

Has the organizational hypothesis been confirmed for humans, and does it apply to behaviors and brain regions that are not directly related to reproduction? In other words, does prenatal exposure to androgens permanently alter the human brain, and does the alteration affect male and female behaviors in nonsexual contexts?

The human hypothalamus does contain a sexually dimorphic nucleus, known as the INAH3, which stands for third interstitial nucleus of the anterior hypothalamus (LeVay, 1991). The homology between the human INAH3 and the rat SDN-POA has not been established. The male INAH3 is only about twice as large, on average, as the female INAH3, and the difference probably would have gone undetected if the rat SDN-POA had not been studied first. Thus, it is not clear how many more subtle sexual dimorphisms there are in other parts of the human brain.

Subtle differences between nonreproductive behaviors of men and women are suggested by the different distributions of scores in standardized tests (Hampson and Kimura, 1992). Men tend to score higher on manipulating three-dimensional images, guiding or intercepting projectiles, finding particular shapes in complex figures, and mathematical reasoning. In contrast,

women tend to perform better on manual precision tasks, perceptual speed, verbal fluency, and arithmetic tests. These test scores may be biased by different role models and social expectations for boys and girls. Nevertheless, a biological component to these abilities is indicated by cyclic changes in the test scores of individual men with the annual fluctuations of their androgen levels, and by changes in the performance of individual women with their reproductive cycle. However, these correlations are with *current* steroid levels rather than with prenatal exposure.

Evidence for organizational effects of steroids on the *prenatal* brain are derived from human females affected by a genetic defect called **congenital adrenal hyperplasia (CAH)**, an adrenal enzyme deficiency that causes excessive androgen levels. Girls with CAH are born with masculinized external genitalia. This condition can be corrected surgically, and the genetically caused overproduction of androgens can be treated with cortisone. However, the prenatal exposure to androgens has a lasting effect on the behavior of girls with CAH. They grow up more "tomboyish" than their unaffected sisters, according to the girls' own assessments, teachers' accounts, and observations of the girls' behavior. Because parents are likely to encourage traditional feminine behavior in their CAH daughters at least as much as in their unaffected daughters, the unusual preferences of the CAH daughters cannot be ascribed to parental expectations. In standardized tests, the CAH daughters are also superior to their unaffected sisters in spatial manipulations and in discovering hidden figures, both tasks on which males tend to do better. Very similar observations were made on daughters born to mothers who had been given progestin medications during pregnancy to prevent miscarriages. Together, the observations on these girls strongly suggest that their brains were conditioned by their fetal hormonal environment.

In summary, data obtained from birds and mammals, including humans, indicate that hormones are powerful regulators of sexual development. Their effects are seen most clearly in physical sexual dimorphisms and adult reproductive behaviors, but they also seem to reach into the development of brain functions that are not directly related to reproduction.

Hormonal Control of Insect Metamorphosis

While the hormonal control of vertebrate development is of great interest for understanding our own sexuality, the molecular mechanisms of hormone action have been analyzed most successfully for the insect molting hormone *ecdysone*. Studying this hormone has several advantages. First, it is a steroid hormone, so its action on target genes is straightforward, with the hormone-receptor complex acting directly as a transcription factor. Second, the effects of ecdysone on many of its target genes can be seen directly in the puffing patterns of the *polytene chromosomes* of dipteran larvae. Third, the availability of *Drosophila* mutants has facilitated the functional analysis of ecdysone target genes. Before tackling the molecular aspects of ecdysone action, we will briefly survey postembryonic development in insects, in which ecdysone plays a central role.

Insect Molting, Pupation, and Metamorphosis Are Controlled by Hormones

The development of insects and other arthropods is punctuated by a series of *molts*. These are necessary for growing, because arthropods are covered by a *cuticle*, a hardened extracellular shell secreted by the epidermis. Since the cuticle does not expand, it needs to be shed periodically to allow growth. At each molt, the epidermis withdraws from the old cuticle, expands, and then forms a new, larger cuticle that is initially soft. The newly hatched individual moves and exerts pressure with its body fluid to shed the old cuticle and stretch the new one.

During molts, insects may also change in body form, a process called *metamorphosis*. Primitive insects, such as grasshoppers, have a gradual or *incomplete metamorphosis*: with each molt, the insect comes to look more like an adult (see Fig. 1.13). More highly evolved insects such as butterflies, flies, ants, and beetles undergo a dramatic or *complete metamorphosis* (see Fig. 1.15). The juvenile stages of these insects are generically termed *larvae* (sing., *larva*); the larvae of individual insect orders have more specific common names, such as caterpillars, maggots, and grubs. The larva and adult of a

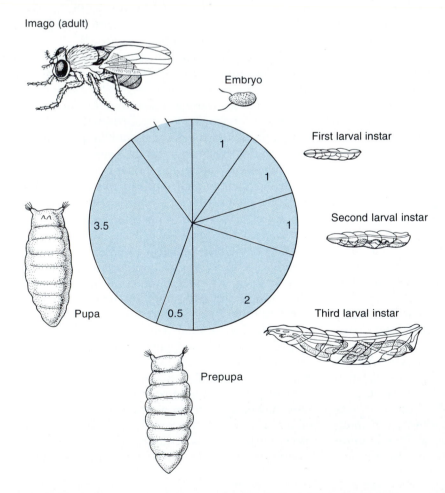

Figure 28.11 Life cycle of *Drosophila*, a holometabolous insect. Seven development stages of the life cycle are shown: the embryo, three larval instars, prepupa, pupa, and adult. The numbers represent the duration of each stage in days.

species differ widely in their morphology and lifestyle, and they fulfill different functions in the life cycle of the species. Insect larvae are primarily eating machines: the caterpillars of large moths increase their body weight several hundredfold. Adult insects typically fly, thus achieving dispersal in addition to reproduction.

Insects with complete metamorphosis are known as **holometabolous insects;** they go through a number of larval stages, called **instars**, which are separated by molts (Fig. 28.11). After the last larval instar, they molt into an immobile, nonfeeding stage known as the *pupa*, during which metamorphosis occurs. During the pupal stage, most larval organs are broken down. Only some larval organs, including the gut and the nervous system, survive at least partially in the adult. While most of the larval tissues are dying, the adult structures are being formed. Most of the adult epidermis is made from *imaginal discs*, folded epithelial bags that are set aside during larval development and evert during metamorphosis (see Fig. 6.19). At the end of metamorphosis, the insect undergoes a final or adult molt and hatches from the pupal cuticle. The adult insect is also called the

imago (hence the term *imaginal disc*). The group of flies to which *Drosophila* belongs has a special feature called the **puparium.** This is the hardened cuticle of the last larval instar. Instead of being shed during the pupal molt, it forms a barrel-shaped case around the pupa.

The salient feature of metamorphosis in insects (and many other animals) is its tight hormonal control (Etkin and Gilbert, 1968). The timing of insect molts as well as their character (larval, pupal, or adult) is regulated by hormone-secreting glands that are controlled in turn by the brain. Figure 28.12 shows a simplified model of this process in the silk moth, but the model seems applicable to a wide range of insects (Schneiderman and Gilbert, 1964).

In response to environmental and endogenous stimuli, neurosecretory cells in the brain release a peptide hormone called **prothoracicotropic hormone (PTTH)**. Through a complex system of *second messengers*, PTTH stimulates the **prothoracic gland** to synthesize a steroid hormone, *ecdysone*. (Ecdysone is a prohormone that is converted in other tissues to its active form, 20-hydroxyecdysone. For simplicity, we will use the term

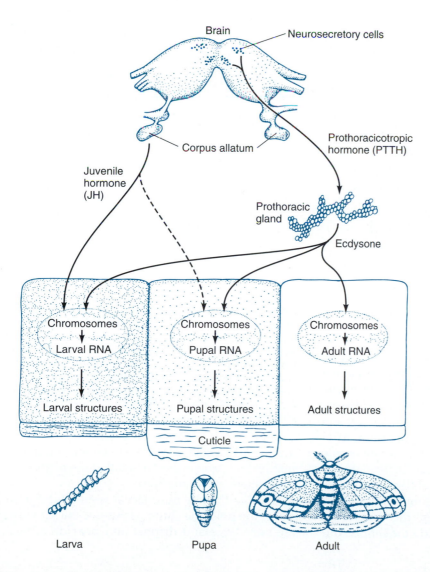

Figure 28.12 Hormonal control of metamorphosis in the silk moth *Cecropia*. All molts seem to be initiated by prothoracicotropic hormone (PTTH) from neurosecretory cells in the brain. PTTH stimulates the prothoracic gland to release the molting hormone ecdysone. The character of the molt is determined by the concentration of juvenile hormone (JH). JH is produced in the corpora allata, a pair of glands attached to and controlled by the brain. At high JH levels, molts generate additional larval instars. In contrast, a reduced JH level causes a pupal molt, and complete absence of JH leads to an adult molt.

"ecdysone" for both forms.) Each molt is triggered by a surge of ecdysone in the hemolymph (blood equivalent) of the insect. The character of the molt is determined by the concentration of *juvenile hormone (JH)*, a lipid-soluble hormone produced by a pair of glands called the *corpora allata* (sing., *corpus allatum*). The corpora allata are attached to and controlled by the brain. JH essentially inhibits metamorphosis. At high JH concentration, a molt results in another larval instar. A reduced JH level leads to a pupal molt, and complete absence of JH to an adult molt.

▼

The hormonal control of insect metamorphosis was first analyzed in the pioneering experiments of Vincent Wigglesworth (reviewed 1972). He worked with a bloodsucking bug, *Rhodnius prolixus*, a species with incomplete metamorphosis. The bug goes through five juvenile instars without wings before molting into a winged adult. Wigglesworth used the technique of *parabiosis*, that is, a connection between two living organisms that allows the exchange of blood or an equivalent body fluid (Fig. 28.13). When a molting fifth-instar juvenile and a first-instar juvenile were joined in parabiosis, they both underwent metamorphosis. Regardless of its minute size, the first instar molted into a precocious adult. In another experiment, Wigglesworth removed the corpora allata from a third-instar juvenile. As a result, the next molt turned this juvenile into a precocious adult. Conversely, when the corpora allata from a fourth-instar juvenile were implanted into a fifth-instar juvenile, the recipient molted into an oversized "sixth-instar" juvenile instead of becoming an adult.

The mechanisms of JH action are still largely unresolved. However, new work on the molecular basis of ecdysone action, which we will discuss next, has generated much excitement.

Puffs in Polytene Chromosomes Reveal Genes Responsive to Ecdysone

Each molt in the life cycle of *Drosophila* is preceded by a pulse of ecdysone; the pulse that is best investigated occurs at the end of the third larval instar, right before puparium formation (Fig. 28.14). A few hours before this pulse, the larva stops feeding and wanders around in search of a place in which to form a puparium. After finding a suitable spot, the larva cements its cuticle onto the substratum with a glue produced in its salivary glands. The larva then assumes a barrel-like shape while its cuticle hardens to form the puparium. Twelve hours later, another ecdysone pulse triggers the eversion of the head, which has been tucked inside the thorax dur-

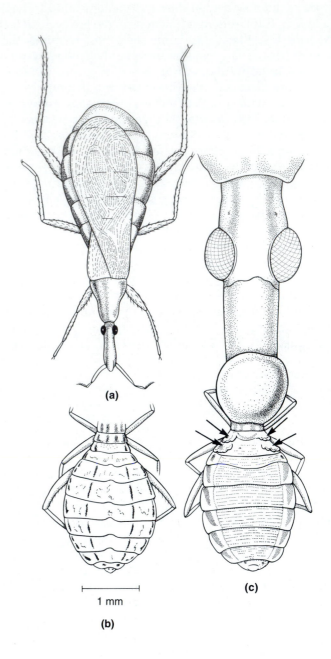

(a)

(b)

1 mm

(c)

Figure 28.13 Control of metamorphosis by blood-borne signals in the bug *Rhodnius prolixus*. **(a)** Normal adult. **(b)** Thorax and abdomen of a normal second-instar larva at higher magnification. **(c)** Precocious "adult" produced from a first-instar juvenile that was joined with paraffin wax to the head of a molting fifth-instar juvenile. The precocious adult has rudimentary wings (arrows) and the pigmentation pattern of the adult rather than that of the second-instar juvenile.

ing the larval stages. The stage between puparium formation and head eversion is called the *prepupa*, and the subsequent stage, the *pupa*.

During the late third larval instar and the prepupal stage, the polytene chromosomes of the salivary glands are very large and distinct and provide a remarkable

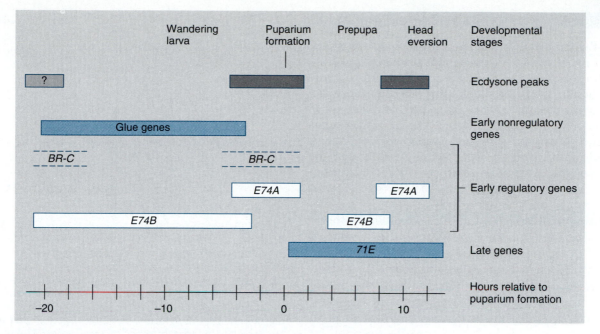

Figure 28.14 Ecdysone peaks and gene activity at the onset of metamorphosis in *Drosophila*. Developmental time is indicated in hours before and after puparium formation. Relative to this time axis, the diagram shows developmental stages, ecdysone peaks, and the activity periods of a few of the genes that respond to ecdysone.

picture of the effects of ecdysone on gene expression. The chromosomal puffs, which signal gene activity, can easily be observed at successive stages, because each puff can be assigned a unique position on the cytogenetic map of *Drosophila* (see Figs. 14.5 and 15.18). The changing pattern of puffs thus directly reveals which genes are transcribed at successive stages.

Repeated observations have shown that each stage of metamorphosis is characterized by a specific combination of puffs (Figs. 28.14 and 28.15; Andres and Thummel, 1992). In third-instar larvae only a few *intermolt puffs* are visible prior to the wandering stage. As the ecdysone level increases, signaling puparium formation, a small set of fewer than 10 *early puffs* appears. These early puffs disappear several hours later while more than 100 *late puffs* appear in a reproducible sequence as ecdysone concentration recedes.

To test whether the complex puffing patterns at the beginning of metamorphosis are all triggered by ecdysone, researchers injected the hormone prematurely. Indeed, the injected ecdysone induced puffs that would normally be seen only later in development (Clever and Karlson, 1960). Extending this work, Michael Ashburner (1972) devised a method for maintaining salivary glands from third-instar larvae in vitro, thus obtaining a degree of control not possible in the intact organism. Again, the controlled addition of ecdysone stimulated the changes in puffing activity normally associated with early metamorphosis (Fig. 28.15).

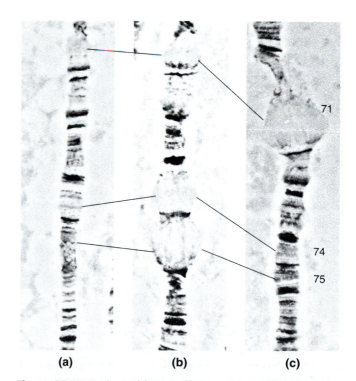

Figure 28.15 Early and late puffs in polytene chromosomes of *Drosophila melanogaster* salivary glands cultured in vitro. The photographs show a section of the right arm of chromosome III. **(a)** No puffs are visible before ecdysone is added. **(b)** About 5 h after ecdysone addition, two early puffs (74 and 75) are very large, and a late puff (71) is beginning to form. **(c)** About 10 h after ecdysone addition, the early puffs have regressed, and the late puff has grown to full size.

Using this culture system, Ashburner and coworkers studied the effects of ecdysone addition at various concentrations, and of ecdysone removal. They also examined whether the resulting puffing patterns depended on the ability of the salivary glands to synthesize new proteins (Ashburner, 1990). They confirmed that a small number of "early genes" became puffed within minutes after addition of ecdysone, and they found that inhibitors of protein synthesis did not affect the appearance of these puffs. They concluded that the *activation* of the early genes was independent of newly synthesized proteins. However, inhibiting protein synthesis did have an effect on the *regression* of at least some of the early puffs. In the continuous presence of ecdysone, these puffs reached a maximum size after 4 h and then disappeared. However, if protein synthesis was inhibited, the puffs persisted. This observation suggested that inactivation of the early genes might depend on the proteins they encode or on proteins encoded by their target genes.

In contrast to the small set of early genes, the large group of late genes failed to be activated by ecdysone in the absence of protein synthesis. This result suggested that early-gene proteins might not only limit the activity period of the early genes but also activate the late genes. However, this hypothesis by itself did not explain another remarkable feature of the puffing pattern: the late puffs appeared and regressed in a very strict temporal sequence over a period of about 12 h. A clue to this puzzle was obtained from another experiment: if salivary glands were incubated in ecdysone for a few hours and then washed free of the hormone, many of the late puffs were induced prematurely. It appeared that late puffs were initially repressed by ecdysone until increasing amounts of early-gene proteins offset the hormone's repressive effects.

From their in vitro studies, Ashburner and coworkers (1974) derived a simple model for the control of chromosome puffing by ecdysone. This model has since been extended on the basis of additional data, and an expanded version of the model is shown in Figure 28.16. According to this model, ecdysone binds to a cytoplasmic receptor as other steroid hormones do (see Chapter 15). The hormone-receptor complex then acts as a transcription factor on different groups of target genes. Some of these target genes are *early nonregulatory genes*. For instance, the genes for the glue proteins used by the third-instar larva to cement its cuticle to a substrate are shut off by the ecdysone-receptor complex. Other nonregulatory genes are presumably activated in tissue-specific ways. In addition, the ecdysone-receptor complex activates a small number of *early regulatory genes*.

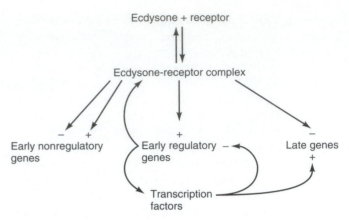

Figure 28.16 Model for gene regulation by ecdysone in *Drosophila*. Ecdysone forms a complex with a receptor protein, and the ecdysone-receptor complex has multiple effects. First, it acts directly on early nonregulatory genes that have tissue-specific functions. Second, the ecdysone-receptor complex activates a small number of early regulatory genes. After a few hours, the transcription factors encoded by these genes inhibit their own genes and activate a large number of late genes. The early regulatory genes also encode ecdysone receptor peptides. Finally, the ecdysone-receptor complex inactivates the late genes until this inhibition is overcome by the early regulatory gene proteins.

These genes were originally identified as early puffs in salivary gland chromosomes and were later shown to be activated in virtually all other tissues. Some of these early regulatory genes encode proteins that cause the regression of the early puffs and the appearance of the late puffs described previously. Another early regulatory gene encodes ecdysone receptors. Finally, the ecdysone-receptor complex inhibits the *late genes* until the early regulatory proteins overcome the inhibition and trigger transcription of the late genes.

Multiple Ecdysone Receptors Have Tissue-Specific Distributions

Molecular studies to test the Ashburner model began with the genetic and molecular characterization of genes located in early puffs. Of particular interest was the cloning of an early regulatory gene termed EcR^+ (Koelle et al., 1991). The EcR peptide acts as an ecdysone receptor, or as part of such a receptor, according to the following criteria. First, it binds active ecdysone and is immunologically indistinguishable from a known ecdysone-binding protein. Second, EcR binds to DNA with high specificity at known enhancer motifs that are ecdysone-responsive. Third, EcR is present in ecdysone-responsive cultured cells and absent in derivatives of such cells that have become nonresponsive to ecdysone. Such nonresponsive cells are restored to ecdysone responsiveness by transfection with cDNA encoding EcR. Fourth, EcR is synthesized at all stages and in all tissues that are ecdysone-responsive. In particular, EcR is already present when the ecdysone peak occurs during the third larval instar.

In accord with the Ashburner model, EcR binds to early puffs and many late puffs. In particular, the ecdysone-receptor complex induces the synthesis of more ecdysone receptors. This *autoinduction* of the ecdysone receptor is indicated by the following data. First, the synthesis of EcR mRNA increases during each ecdysone peak throughout the life cycle of *Drosophila*. Second, EcR mRNA accumulates rapidly after an ecdysone pulse. When cultured larval organs are incubated with ecdysone, EcR mRNA concentration increases after 15 min and reaches a maximum about 1 h after ecdysone addition (Karim and Thummel, 1992). Given a pre-mRNA size of 36 kb and an average transcription rate of about 1.2 kb per min, the transcriptional response certainly appears to be direct.

Further analysis of the EcR^+ gene showed that it includes two overlapping transcription units encoding three peptides, termed EcR-A, EcR-B1, and EcR-B2. All three peptides have identical DNA-binding, and ligand-binding domains but different transactivational domains (Fig. 28.17; Talbot et al., 1993). Antibodies to EcR-A and EcR-B1 reveal characteristic expression patterns of these isoforms at the beginning of metamorphosis. For example, the imaginal discs, which give rise to the adult epidermis, show a uniformly high A:B1 ratio of receptor peptides. In contrast, most of the larval organs that are broken down during metamorphosis exhibit a low A:B1 ratio of receptor peptides. Such correlations suggest that different combinations of ecdysone receptor peptides may activate specific target genes, which in turn initiate different metamorphic responses, such as disc eversion and organ degradation.

A similar correlation between metamorphic response and the type of ecdysone receptor synthesized is observed at the level of individual neurons in the central nervous system of *Drosophila* and a moth, *Manduca sexta* (Truman et al., 1994). Generally, most larval neurons produce high levels of EcR-B1 at the start of metamorphosis while they lose larval features in response to ecdysone. Later, during the pupal-adult transformation, they switch to EcR-A expression and respond by maturing to their adult form. A particular correlation is observed in a set of about 300 neurons that produce a 10-fold higher level of EcR-A than other neurons in *Drosophila* pupae (Fig. 28.18). Although these neurons are heterogeneous in their location and function, they all share the same fate of undergoing rapid degeneration as the adult emerges from the pupal case (Robinow et al., 1993). One prerequisite for this death is the decline of ecdysone at the end of metamorphosis: treatment of flies with ecdysone at least 3 h before the normal time of cell degeneration prolongs the life of the cells.

The synthesis of a receptor variant in a tissue is suggestive of its function but does not prove it. This diffi-

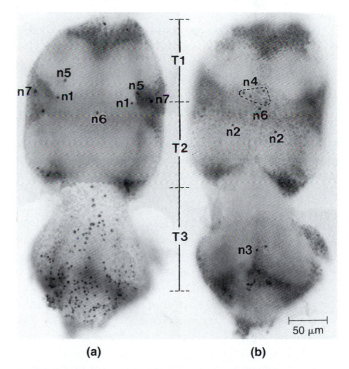

(a)   **(b)**

Figure 28.18 Selective synthesis of the ecdysone receptor isoform EcR-A by specific neurons in the central nervous system of a newly hatched *Drosophila* adult. The neurons producing the EcR-A polypeptide have been immunostained (see Methods 4.1) and appear dark in this photograph. These neurons share the fate of rapid cell death within a few hours after eclosion, thus illustrating a correlation between EcR isoform synthesis and cell fate. T1, T2, and T3 designate the approximate extents of the CNS portions (neuromeres) that innervate the respective thoracic segments. **(a)** Dorsal view; **(b)** ventral view. The labels over the neuromeres designate individual neurons that can be identified reproducibly on the basis of their position.

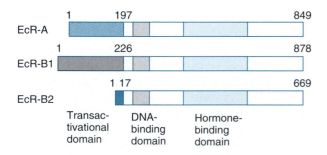

Figure 28.17 Schematic representation of the three peptides encoded by the EcR^+ gene of *Drosophila*. All three peptides have identical DNA- and hormone-binding domains. The peptides differ in their transactivational domains, which control their activating or inhibitory activities as transcription factors. Numbers indicate amino acid residues, counted from the amino terminus.

culty has been overcome with the isolation and characterization of *EcR* mutants that are deficient in the EcR-B1 receptor peptide while synthesizing the other two peptides normally. In these mutants, the larval tissues (which synthesize mostly the B1 peptide) fail to form the puparium, whereas imaginal discs (which produce mostly the A peptide) begin their normal eversion (Bender et al., 1993). Thus, the EcR-B1 receptor peptide appears to function selectively in larval tissues as expected.

The data discussed so far indicate that the synthesis of different combinations of ecdysone receptor isoforms is one way in which the response to ecdysone is made tissue-specific. Another possible mechanism for modulating the response to ecdysone is based on the fact that all steroid receptors bind to their DNA response elements as dimers of two peptides (see Fig. 15.14). While some receptor peptides bind as homodimers, others combine with different peptides to form heterodimers (Mangelsdorf and Evans, 1992). The evidence cited previously, that the EcR peptide is part of an ecdysone receptor, does not show whether it acts as a homodimer, or as a heterodimer in conjunction with another peptide, or both.

Additional studies indicate that EcR does form a heterodimer with the peptide encoded by another *Drosophila* gene, *ultraspiracle*[+] (Yao et al., 1992). Sequence comparisons show that the ultraspiracle peptide shares much sequence similarity with the mammalian retinoid X receptor peptide (RXR), which forms heterodimers with thyroid hormone receptor and other receptor peptides (Oro et al., 1990). Indeed, EcR-RXR heterodimers are as effective as EcR-ultraspiracle heterodimers in making cells ecdysone-responsive and in binding to DNA response elements (H. E. Thomas et al., 1993).

The sequence homology and shared function of the RXR and ultraspiracle peptides show that these versatile receptor peptides have been conserved in evolution since vertebrates and arthropods diverged more than 500 million years ago. Presumably, such peptides play a very important role in diversifying the regulatory effect of one signal—the hormone or equivalent ligand—on a wide range of target cells. For instance, the RXR peptide is known for its ability to form heterodimers that activate some target genes in mammalian cells while inhibiting others. Presumably, the homologous ultraspiracle peptide of *Drosophila* confers the same ability on the ecdysone receptor.

In summary, the molecular data on ecdysone receptors indicate that they act as a competitive and versatile group of transcription factors with a wide range of target gene specificities and activational properties. It will be interesting to see how the target genes of ecdysone receptor peptides interact with earlier-acting patterning genes such as homeotic genes—which decide, for instance, whether an imaginal disc will form a leg or an antenna.

Early Regulatory Genes Coordinate the Timing and Intensity of the Response to Ecdysone

The cloning of several other genes activated directly by ecdysone has revealed more about the molecular mechanisms underlying the Ashburner model (Andres and Thummel, 1992). Some of the early response genes are activated by ecdysone only in specific tissues, placing them in the group of presumed *early nonregulatory genes*. Other early genes are activated in virtually all tissues including larval tissues doomed to be lysed during metamorphosis and imaginal discs programmed to build the new epidermis. This feature and the puffing patterns of these genes in salivary glands place them in the *early regulatory group*.

Early regulatory genes located in three puffs, designated E74, E75, and BR-C, have been characterized genetically and molecularly. While E74 and E75 seem to contain single genes, BR-C contains several genes. Mutations in these genes are lethal during late larval, prepupal, or pupal stages, in accord with the model. Analysis of the phenotypes caused by mutations in BR-C genes revealed very specific defects (Restifo and White, 1991, 1992). Some mutations interfere with the dramatic morphogenetic movements that occur in the *Drosophila* brain during metamorphosis, including the fusion of the right and left brain hemispheres. In addition, BR-C mutant individuals show abnormalities in the epidermis, flight muscles, and gut, and their larval salivary glands fail to break down. These observations demonstrate that early regulatory genes, which are the first to respond to ecdysone, are in turn necessary for a wide range of specific metamorphic processes.

Molecular analysis of early regulatory genes has revealed several common features, including their unusually long (60- to 100-kb) pre-mRNAs, their differential splicing, and the nature of their products: they are all transcription factors of the zinc finger type (Andres and Thummel, 1992). The significant delay inherent in the synthesis of long pre-mRNAs suggests that their lengths may contribute to the timing of gene expression in the regulatory hierarchy. The differentially spliced mRNAs give rise to isoforms of transcription factors, suggesting that each early regulatory gene controls multiple target genes with different specificities. It was shown directly that the function of one gene located in the BR-C puff is required for the transcription of six other genes located in a late puff, E71 (Guay and Guild, 1991). This was a direct demonstration that early genes in the ecdysone regulatory hierarchy stimulate late genes, in accord with the Ashburner model.

Further analysis showed that the early regulatory genes produce different transcripts in response to different ecdysone concentrations, thereby setting up a sequential activation of late genes in response to the surges and declines of ecdysone.

One of the early regulatory genes, the $E74^+$ gene, is transcribed from two promoters, both of which are activated directly by the ecdysone-receptor complex (Thummel et al., 1990). This arrangement leads to the synthesis of two mRNAs, E74A and E74B, which encode two related proteins that have unique amino-terminal regions joined to a common carboxyl-terminal region. The synthesis of E74A and E74B mRNAs is coordinately regulated by each ecdysone pulse. For instance, the pulses that mark the ends of the larval and prepupal stages coincide with high rates of E74 mRNA synthesis (Fig. 28.14). In each case, however, E74A mRNA synthesis is preceded by a period of E74B mRNA synthesis. While this coordination is explained to a small extent by the greater length of the E74A pre-mRNA, the more important observation is that the $E74B^+$ promoter responds to a much lower ecdysone concentration.

▼

To determine the effects of ecdysone concentration on the synthesis of E74A mRNA and E74B mRNA, Felix Karim and Carl Thummel (1991) maintained late third-instar larval organs of *Drosophila* in culture, where they could be exposed to controlled concentrations of ecdysone for controlled periods. *Northern blots* (see Methods 14.3) of total RNA extracted from these organs showed that the two types of mRNA accumulate in response to different ecdysone concentration ranges (Fig. 28.19). Photometric processing of the data indicated that E74B mRNA reaches 50% of its maximum prevalence at an ecdysone concentration of 8×10^{-9} M whereas E74A mRNA reaches 50% maximum prevalence at 2×10^{-7} M ecdysone. The researchers concluded that the E74A promoter requires a 25-fold higher ecdysone concentration to become activated than the E74B promoter.

Using the same set of techniques, Karim and Thummel (1991) studied the time course of E74A and E74B mRNA prevalence at different ecdysone concentrations. Provided that the ecdysone concentration was high enough, E74A mRNA was always synthesized with a delay of 1 h after ecdysone addition and rose to a constant level. In contrast, the profile of E74B mRNA accumulation changed dramatically with ecdysone concentration. At 5×10^{-9} M ecdysone, E74B mRNA was present at a uniformly low level above background for more than 8 h. At ecdysone concentrations above 10^{-7} M, E74B mRNA disappeared after a 2-h burst as E74A mRNA rose to its constant high level. The simplest interpretation of these profiles would be that E74A protein inhibits the E74B promoter, but further tests showed that the regulation must be more complex.

Finally, the investigators showed that the relative levels and ecdysone induction profiles of E74A and E74B mRNA vary among different tissues. For instance,

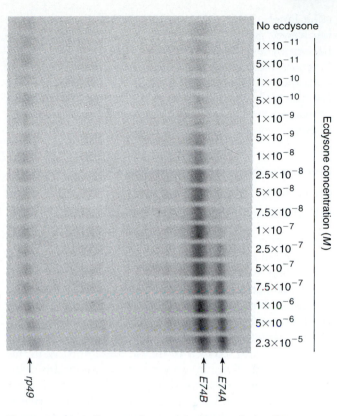

Ecdysone concentration (M):
No ecdysone
1×10^{-11}
5×10^{-11}
1×10^{-10}
5×10^{-10}
1×10^{-9}
5×10^{-9}
1×10^{-8}
2.5×10^{-8}
5×10^{-8}
7.5×10^{-8}
1×10^{-7}
2.5×10^{-7}
5×10^{-7}
7.5×10^{-7}
1×10^{-6}
5×10^{-6}
2.3×10^{-5}

← rp49   ← E74B   ← E74A

Figure 28.19 Ecdysone dose-response analysis of E74A and E74B mRNA prevalence. Late third-instar larval organs were cultured and treated for 1.5 h with different ecdysone concentrations. Total RNA was extracted and analyzed by Northern blotting (see Methods 14.3) using a labeled probe representing the common region of both mRNAs. Nevertheless, the two RNAs could be distinguished by their different sizes. Another labeled probe hybridizing with the mRNA for a ribosomal protein (rp49) was added as a control for equal RNA loading in each lane. The results show that two E74 mRNAs accumulate in response to different ecdysone concentrations. The prevalence of E74B mRNA increases above a constant background level as ecdysone concentrations rise between 1×10^{-9} and 7.5×10^{-7} M. In contrast, E74A mRNA has no detectable background level and increases between ecdysone concentrations of 5×10^{-8} and 7.5×10^{-7} M.

larval brains contain only small amounts of E74A mRNA in the absence of ecdysone. Anterior gut has low levels of E74A mRNA, while E74B mRNA is still present 6 h after ecdysone addition, as if it were not repressed in this tissue.

It appears that the $E74^+$ gene, and by extension other early regulatory genes, act as diversifiers for the ecdysone signal. Each ecdysone pulse generates a *temporal morphogen gradient* (see Chapter 24), to which the early regulatory genes respond by generating different combinations of regulatory proteins at different ecdysone levels. Time delays and interactions among the regulatory genes generate waxing and waning combinations of several gene products. Because they are transcription factors, these products presumably cooperate with the ecdysone receptor isoforms and other tissue-specific transcription factors to generate the complex choreography of gene regulation that must underlie a phenomenon as dramatic as insect metamorphosis.

Hormonal Control of Amphibian Metamorphosis

Amphibians were the first class of vertebrates to conquer the land, but most present-day amphibians still return to the water to reproduce. Consequently, the water is still the habitat of most amphibian larvae. As the larvae make the transition to land-dwelling adults, they undergo metamorphosis.

Anurans (tailless amphibians, including frogs and toads), typically have a dramatic metamorphosis that affects nearly all organ systems and major bodily functions. Locomotion changes from swimming with a tail fin to jumping with legs (Fig. 28.20). Gills are replaced with lungs, and blood cells produce a different hemoglobin that binds oxygen less avidly. Mouth, tongue, and gut are greatly modified as the diet changes from herbivorous to carnivorous. The liver and kidney are retooled so that instead of excreting mostly ammonia, the adult excretes mostly urea. The senses of sight and hearing adapt from the physical conditions of water to those of air. Remarkably, all these changes occur while the animal is moving and feeding. There is no quiescent stage, like the insect pupa, devoted exclusively to reconstruction. In other respects, the metamorphoses of flies and frogs share many similarities (Tata, 1993).

In urodeles (tailed amphibians, including newts and salamanders), the metamorphic changes are generally less dramatic and vary among species.

The typical frog metamorphosis is divided into three stages (Fig. 28.20). *Premetamorphosis* is the period before metamorphosis when the tadpole has no limbs. *Prometamorphosis* is the stage when tadpoles have small hindlimbs but no forelimbs. The *climax of metamorphosis* begins with the emergence of the forelimbs and extends through the most dramatic changes, including full limb growth and tail resorption. These morphologically defined stages are also characterized by different hormonal and genetic activities.

The Hormonal Response in Amphibian Metamorphosis Is Organ-Specific

Early investigators of amphibian metamorphosis took advantage of the feasibility of transplantation experi-

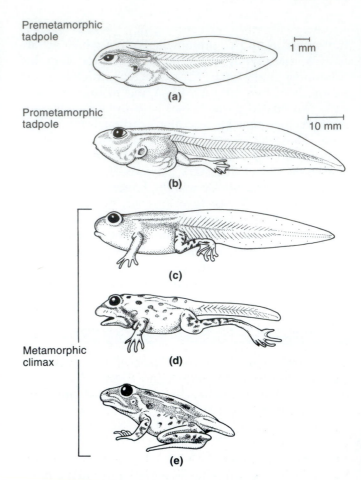

Figure 28.20 Metamorphosis in a typical frog (*Rana pipiens*). **(a)** Premetamorphic tadpole; **(b)** prometamorphic tadpole showing growth of hindlimbs; **(c)** onset of metamorphic climax with eruption of forelimbs; **(d, e)** later climax stages showing the appearance of the froglike organization.

ments with these animals. One of these experiments is of particular interest because it illustrates in a graphic way that the hormonal response varies among target organs.

A given type of tissue may fare quite differently during metamorphosis depending on its location. For instance, bone and skeletal muscle will grow if located in the legs. However, the same tissues will be degraded in the tail. To test whether the response depends on the location of the responding tissue *at the time of metamorphosis*, researchers transplanted a tail from a donor tadpole into the flank of a host tadpole (Fig. 28.21). The new location of the graft was one where there is normally no degradation of bone or muscle. Nevertheless, the grafted tail was degraded and resorbed at about the same pace as the host's own tail (Geigy, 1941). Conversely, when a tadpole eye was transplanted to the tail of a tadpole host, the eye escaped the destruction of the tail and remained attached to the host's rear end at the conclusion of metamorphosis (Schwind, 1933). Thus,

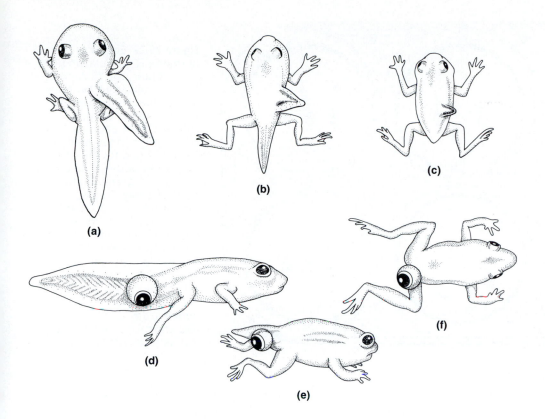

(a)

(b)

(c)

(d)

(e)

(f)

Figure 28.21 Organ specificity of the metamorphic response in frogs. **(a–c)** A tail grafted to the flank is degraded and resorbed along with the host tail. **(d–f)** An eye transplanted to the tail remains unaffected by the degradation of the tail.

the metamorphic response of a tissue is independent of its location during metamorphosis. Rather, the response depends on the *organ* in which the tissue resides, which, of course, is correlated with the location in which the tissue originally developed. In molecular terms, these results indicate that the distribution of hormone receptors and other cofactors that control the metamorphic events is organ-specific and depends on developmental history.

Amphibian Metamorphosis Is Controlled by Two Hormones: Thyroid Hormone and Prolactin

As in insects, the metamorphosis of amphibians is controlled by two hormone-producing glands that are regulated in turn by the brain. The principal metamorphic hormone in amphibians is *thyroid hormone (TH)*, which is comparable to insect ecdysone in its actions. It is produced in the thyroid gland, located ventrally of the trachea in the neck region. The active components of TH are *thyroxine (T_4)* and *triiodothyronine (T_3)*, both iodinated derivatives of the amino acid tyrosine. T_3, which generally seems to be the more active component, is also synthesized from T_4 in tissues other than the thyroid gland. When the thyroid gland is removed from young tadpoles, they grow into giant tadpoles that never metamorphose. Conversely, when TH is applied to young tadpoles with food or by injection, they meta-

morphose prematurely. TH acts on its target cells through *thyroid receptors (TRs)*, cytoplasmic receptors that belong to the same superfamily as the ecdysone receptors.

The second metamorphic hormone in amphibians is *prolactin*, a peptide produced in the anterior pituitary gland that is comparable in its effects to insect juvenile hormone. Amphibian prolactin is the same hormone that in mammals promotes mammary duct development during pregnancy and milk production during nursing. In amphibians, however, prolactin acts as a larval growth hormone and as an antagonist to TH.

The identification of TH and prolactin as amphibian metamorphic hormones has a long history, beginning with the observation that tadpoles metamorphosed prematurely when their food contained sheep thyroid gland (Gudernatsch, 1912). The direct action of both hormones on tail regression was demonstrated in vitro: isolated tails kept in organ culture regressed at the same pace seen in vivo when T_3 was added to the culture medium. However, this effect was inhibited when prolactin was present simultaneously (Derby and Etkin, 1968).

▼

These observations were confirmed and extended by Jamshed Tata and his coworkers (1991), who studied the modulating effects of prolactin in T_3-induced regression

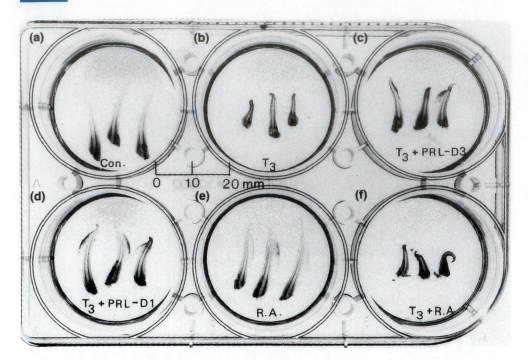

Figure 28.22 Effects of thyroid hormone (T₃), prolactin (PRL), and retinoic acid (R.A.) on isolated tails from *Xenopus* larvae. Tails from premetamorphic tadpoles were cultured for 1 day in a chemically defined medium before hormones or morphogens were added. The tails were photographed after a total of 8 days in culture. **(a)** Control (no additions). **(b)** T₃ added. **(c)** PRL added 3 days after T₃. **(d)** PRL added 1 day after T₃. **(e)** R.A. **(f)** T₃ and R.A. added simultaneously. Note that the regression caused by T₃ was counteracted effectively by PRL added 1 day after T₃, and less effectively by PRL added 3 days after T₃.

of isolated tails and on the outgrowth of isolated limbs from *Xenopus* tadpoles. Tails removed from premetamorphic tadpoles retained their size and morphology during 8 days of culture in a serum-free, chemically defined medium (Figs. 28.22 and 28.23). However, the addition of $2 \times 10^{-9}\,M$ T₃ caused complete loss of the dorsal and ventral tail fins as well as considerable loss of skin, connective tissue, and musculature. Prolactin added on the same day as T₃ completely prevented the disappearance of the fins and, to a large extent, the loss of muscle and connective tissue. Even if prolactin was added after tail regression had already begun, part of the dorsal fin was retained and the loss in connective tissue and muscle was less severe. Additional experiments showed that the tail regression was not caused simply by dehydration but was a true destruction of tissue reflected by loss of wet weight and DNA.

In control experiments, a biologically inactive analogue of T₃, 3,5-diiodothyronine, showed no effect. Retinoic acid, known as a potent morphogen from other experiments, had little effect by itself and did not modulate the effect of T₃.

Another major event in frog metamorphosis, the outgrowth of limbs, is also promoted by T₃ and inhibited by prolactin in vitro (Fig. 28.24). Isolated hindlimb buds just kept in medium grew little, and their ends remained paddle-shaped. However, the addition of T₃ to the medium had dramatic results: the limbs grew by 4 to 5 mm and underwent apparently normal morphogenesis complete with digit formation and histological differentiation of cartilage, muscle, and skin. Again, all these effects of T₃ were inhibited by prolactin, and prolactin was most effective if applied together with T₃ or

soon thereafter. As in the experiments with isolated tails, retinoic acid did not mimic T₃, nor did it significantly modify the effect of T₃.

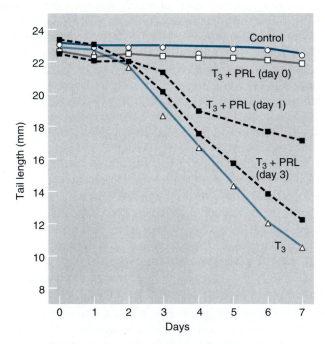

Figure 28.23 Tail regression induced by thyroid hormone (T₃) and its inhibition by prolactin (PRL), in the experiments pictured in Figure 28.22. The tail lengths were measured daily. Relative to the controls kept in medium without hormones, addition of T₃ caused steady tail regression. PRL added simultaneously with T₃ inhibited the regression completely. The inhibitory effect of PRL was diminished when PRL was added 1 day or 3 days after T₃.

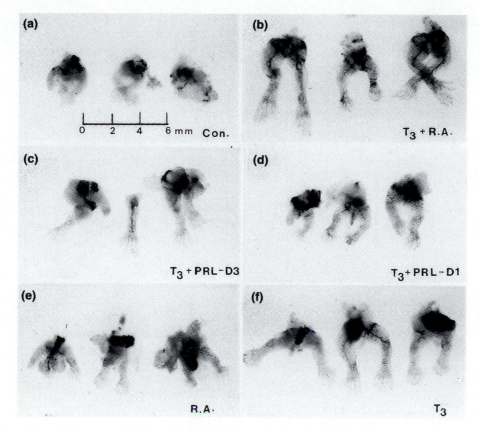

Figure 28.24 Effects of thyroid hormone (T_3), prolactin (PRL), and retinoic acid (R.A.) on isolated hindlimb buds from *Xenopus* larvae. Hindlimb buds and the intervening tissue were removed from premetamorphic tadpoles and cultured for 1 day in a chemically defined medium before hormones or morphogens were added. The duration of hormone or morphogen treatment shown in panels **(a)** through **(f)** was the same as described for the corresponding panels of Figure 28.22. The limbs were photographed at the end of the culture period. Note that the limbs treated with T_3 alone formed fingers. The growth and differentiation caused by T_3 was inhibited effectively by PRL added 1 day after T_3: after 8 days, the ends of the resulting limbs were still paddle-shaped. The inhibition by PRL added 3 days after T_3 was less effective.

The experiments of Tata and his group (1991) confirmed that T_3 can promote key events of frog metamorphosis in vitro, and they further showed that these effects are hormone-specific: they are not mimicked by analogues of T_3 or another potent morphogen. In addition, the experiments confirm that prolactin inhibits the action of T_3. The inhibitory action of prolactin in vitro is complete if it is administered *together with* T_3, and prolactin is still partially effective when T_3 is already present but its results have just set in.

The Production of the Amphibian Metamorphic Hormones Is Controlled by the Brain

As is the case with insect metamorphosis, the glands that produce the amphibian metamorphic hormones are in turn controlled by the brain (Fig. 28.25). The syn-

thesis of TH in the thyroid gland is initiated by *thyroid-stimulating hormone (TSH)*, a peptide produced in the anterior pituitary gland. The production of TSH in the pituitary is stimulated by the hypothalamic portion of the amphibian brain through *thyroid-stimulating hormone releasing factor (TSH-RF)*. At the same time, the hypothalamus suppresses prolactin synthesis in the pituitary gland through a prolactin inhibitor, possibly dopamine. These regulatory mechanisms mature during metamorphosis, setting in motion a complex system of interactions that unfold as follows (Etkin, 1968).

During premetamorphosis, the signals from the hypothalamus to the pituitary are weak, and the thyroid gland is still in the process of development, so that the TH level is very low while prolactin is produced spontaneously in the pituitary. This leads to a low TH:prolactin ratio, which characterizes the premetamorphic growth phase. Prometamorphosis begins when the hypothalamus becomes sensitive to a positive TH feedback loop. In response to TH, the hypothalamus re-

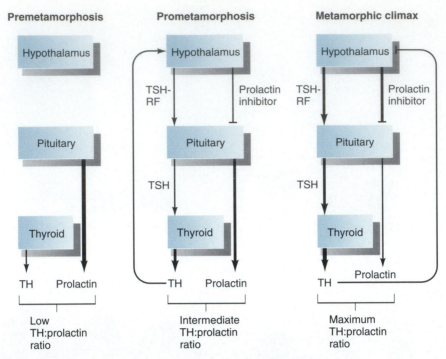

Premetamorphosis Prometamorphosis Metamorphic climax

Figure 28.25 Hormonal control of frog metamorphosis. The diagrams show the hypothalamus-pituitary-thyroid axis at consecutive stages. During premetamorphosis, there is no regulatory input from the hypothalamus to the anterior pituitary gland, which then produces much prolactin spontaneously. The thyroid gland produces minimal amounts of thyroid hormone (TH). During prometamorphosis, the hypothalamus becomes sensitive to a positive feedback loop from TH. The resulting release of thyroid-stimulating hormone releasing factor (TSH-RF) and prolactin inhibitor increases the production of thyroid-stimulating hormone (TSH) and TH while diminishing prolactin. The escalating production of TH triggers the metamorphic climax before a reduction in the size of the thyroid gland and a negative feedback loop to the hypothalamus lead to a new balance of TH and prolactin.

leases both TSH-RF and prolactin inhibitor, so the production of TSH—and consequently TH—increases while the output of prolactin diminishes. The resulting increase in the TH:prolactin ratio characterizes prometamorphosis. The escalating activity of the hypothalamus-pituitary-thyroid axis leads to the climax of metamorphosis, during which the TH:prolactin ratio reaches its maximum and the metamorphic changes proceed most rapidly. After metamorphosis, the TH:prolactin ratio decreases again as a new hormonal balance is reached. The synthesis of TH then decreases through a negative feedback loop from high TH concentrations on the hypothalamus and through a partial degeneration of the thyroid gland as part of the metamorphic changes. At the same time, the prolactin concentration increases again.

Various Defects in Metamorphosis Cause Paedomorphic Development in Salamanders

The hormonal control of metamorphosis in amphibians has provided the basis for an interesting evolutionary

phenomenon called **paedomorphosis** (Gk. *paido-*, "child"; *morphe*, "form"). Animals of paedomorphic species reach sexual maturity while retaining a juvenile appearance in nonreproductive organ systems. Juvenile characteristics are retained to varying degrees in different salamander species. For example, the axolotl *Ambystoma mexicanum* becomes sexually mature while still living in water and retaining the external gills and large tail fins that are typical of a urodele larva (Fig. 28.26). It also develops some adult characteristics, such as legs, adult hemoglobin, and excretion of urea, but normally it does not metamorphose completely. Similar examples of paedomorphosis based on partial metamorphosis are observed in other salamander species.

Experimental analysis has shown that in the course of evolution each paedomorphic species has become defective at some point along the signal chain that leads from the hypothalamus via the pituitary and thyroid glands to the target organs of TH (Frieden, 1981). The tiger salamander, *Ambystoma tigrinum*, depends on a certain environmental input to its brain for metamorphosis. In the high and cold parts of the species' range, in the Rocky Mountains, the adults are water-dwelling

Figure 28.26 Artificial metamorphosis in the axolotl *Ambystoma mexicanum.* The large individual has been raised normally, is aquatic and has external gills and a tail fin. The smaller specimen has been treated with thyroxine to induce metamorphosis; individuals like this are usually smaller, lack gills, and are terrestrial.

and paedomorphic; in the lower and warmer parts of its range, the same species undergoes metamorphosis, and the adults become land-living before they mate. Larvae from the high part of the range will metamorphose if they are simply placed in warmer water. It seems that below a certain temperature the hypothalamus of this species does not produce enough TSH-RF to induce metamorphosis. The axolotl *Ambystoma mexicanum* does not normally metamorphose, as noted earlier. However, experimental application of TH or TSH will generate an adult of this species that does not occur in nature (Fig. 28.26). Thus, *A. mexicanum* must be defective in the release of TSH or at an earlier step in the control hierarchy. Finally, some other salamanders, including *Necturus* and *Siren* species, cannot be induced to metamorphose with TH. Presumably, they lack functional TH receptors.

The Receptors for Ecdysone and Thyroid Hormone Share Many Properties

Like insect ecdysone, amphibian TH acts through receptor proteins belonging to the same superfamily of ligand-dependent transcription factors as the receptors of other steroid hormones and of retinoic acid. Besides their general molecular structure, the ecdysone receptors (EcR) and the TH receptors (TR) share several properties, including autoinduction, heterodimer formation, and tissue-specific distribution.

In *Xenopus laevis*, there are two families of TH receptors, TRα and TRβ (Yaoita and Brown, 1990). The expression of both receptor types begins when the tadpole hatches, but TRα mRNAs are generally more abundant. The prevalence of TRα mRNAs increases during premetamorphosis, is maximal by prometamorphosis, and falls to a lower level after the climax of metamorphosis (Fig. 28.27). In contrast, TRβ mRNAs are barely detectable during premetamorphosis but rise in synchrony with the level of TH in the blood serum. Both TH and TRβ reach a maximum during the metamorphic climax and drop to about 10% of the peak value thereafter. The correlation suggests that TH may induce the synthesis of TRβ mRNAs. Indeed, the expression of TRβ genes can be *autoinduced* precociously in premetamorphic tadpoles if they are kept in water with exogenous T_3 for 2 days. Further experiments show that the autoinduction of TRβ by T_3 is totally abolished, and the autoinduction of TRα diminished, if tadpoles are kept in water with prolactin before they are exposed to T_3 (Baker and Tata, 1992). However, little is known about the molecular basis of prolactin action.

In an intriguing parallel with the ecdysone receptor discussed earlier, the thyroid hormone receptor peptide is greatly enhanced in its activity by heterodimer formation with retinoid X receptor (RXR) peptide (Kliewer et al., 1992; Zhang et al., 1992). The TR-RXR heterodimer binds more effectively to its target DNA recognition sequences (thyroid hormone response elements) and enhances transcription more strongly than homodimers of either peptide. This interaction is also reminiscent of the bHLH family of transcription factors, which includes MyoD and E protein (see Chapter 19). In each case, heterodimer formation allows small families of structurally related peptides to form a wide range of transcription factors with distinct functional properties.

The regional distribution of TR mRNAs was analyzed by in situ hybridization (see Methods 15.1). In premetamorphic tadpoles, the strongest hybridization signals are seen in the brain and spinal cord (Kawahara et al., 1991). High TR mRNA concentrations are also present in the intestinal epithelium and the tail tip, organs marked for regression. From prometamorphosis on, strong hybridization signals are exhibited in limb buds, organs programmed for growth. Thus, TR mRNAs accumulate selectively in the target tissues of TH, as expected from previous observations on other hormone receptors.

In summary, the hormonal controls of metamorphosis in insects and amphibians share several common features. Are these control mechanisms *homologous*, that

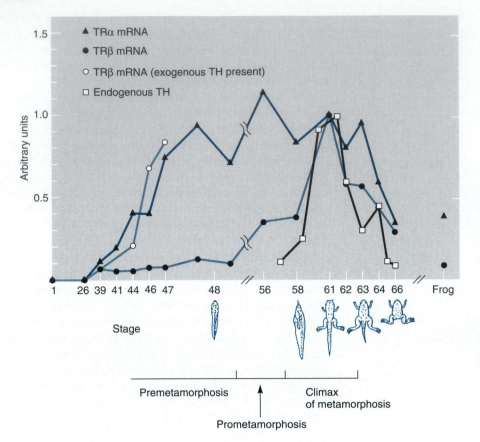

Figure 28.27 Summary of thyroid hormone receptor (TRα and TRβ) gene expression during *Xenopus* development. Developmental stages (abscissa) are represented by diagrams and by metamorphic phases. The prevalence of TRα mRNA, TRβ mRNA, and thyroid hormone is indicated in relative units, with the amount at the peak of metamorphic climax taken as 1.0. Note that the prevalence of TRβ mRNA closely follows the TH concentration. The levels of TRβ mRNA in premetamorphic tadpoles kept in water with exogenous T₃ for 2 days are also shown.

is, based on a common ancestral mechanism? On the one hand, the overall regulation by two antagonistic hormones and the control of the glands producing these hormones by input from the brain are striking in their parallelism (compare Figs. 28.12 and 28.25). There are also many shared details in the action of ecdysone receptors and thyroid receptors, as we have seen. On the other hand, features like autoinduction and heterodimer formation are general characteristics of hormone action and transcriptional regulation and are not unique to

metamorphosis control. Also, the chemical nature of the metamorphosis hormones is quite different in insects and amphibians and does not support homology. On balance, it seems more likely that the similarities observed are the result of *convergence*; that is, although amphibian and insect metamorphosis evolved separately, they nevertheless acquired analogous features because they were selected to fulfill similar functional requirements.

SUMMARY

Hormones are chemical signals that are produced by specialized glands and transmitted through blood or equivalent body fluids. However, a hormone elicits a response only from those cells that have the appropriate receptors and other cofactors; such cells are called the target cells of the hormone. Water-soluble hormones bind to transmembrane receptors and act through second messengers. Lipid-soluble hormones bind to cytoplasmic receptors, which then accumulate in the nucleus and act as transcription factors. Typically, a hormone coordinates the activities of several target tissues or organs.

Secondary sex differentiation in mammals is directed by hormonal control mechanisms. Testicular cells synthesize the principal male sex hormones, testosterone and anti-Müllerian duct hormone. Testosterone and its metabolites direct the male development of originally indifferent external genitalia. In the presence of testosterone, the mesonephric (Wolffian) duct develops into the epididymis, ductus deferens, and seminal vesicle, while anti-Müllerian duct hormone makes the Müllerian duct degenerate. In the absence of both hormones, the Wolffian duct disappears while the Müllerian duct gives rise to the oviduct, uterus, and upper vagina. Ovaries produce estrogen, a female sex hormone. Estrogen levels are low during fetal development but rise after puberty under the influence of gonadotropic hormones from the pituitary gland.

Hormones also control the sexually dimorphic development of the brain and behavior. The receptors for each sex hormone in the brain have a distinct distribution, so that the brain regions stimulated by androgens differ from the brain regions affected by estrogen or progesterone. In songbirds, testosterone or its metabolites promote both male singing and an increase in the number of neurons in certain brain nuclei. According to the organizational hypothesis, prenatal exposure of mammals to testosterone or its metabolites permanently alters the developing brain, causing it to function in male-specific ways later in life. This hypothesis receives its strongest support from the development of reproductive behavior and sexually dimorphic brain areas in rodents. However, the hypothesis seems to hold as well for some human behaviors.

Molecular aspects of hormone action have been studied most extensively in animals that undergo indirect development. Insect metamorphosis is controlled by two antagonistic hormones, ecdysone and juvenile hormone. The glands producing these hormones are regulated by the brain. The ecdysone-receptor complex activates a small number of early response genes, which encode transcription factors that regulate a larger number of late response genes. The early genes include, in particular, the gene encoding ecdysone receptor peptides. These peptides form heterodimers that respond to different ecdysone concentrations, have a wide range of target gene specificities, and occur selectively in certain tissues.

Amphibian metamorphosis is also controlled by two antagonistic hormones, thyroid hormone and prolactin, the production of which is likewise controlled by the brain. Defects in this control hierarchy have led to the paedomorphic development of several salamander species. Amphibian thyroid hormone receptors share many features with the insect ecdysone receptors, but these similarities seem to be convergent rather than based on homology.

SUGGESTED READINGS

Andres, A. J., and C. S. Thummel. 1992. Hormones, puffs and flies: The molecular control of metamorphosis by ecdysone. *Trends in Genetics* **8**:132–138.

Breedlove, S. M. 1992. Sexual differentiation of the brain and behavior. In J. B. Becker, S. M. Breedlove, and D. Crews, eds., *Behavioral Endocrinology*, 39–70. Cambridge, Mass.: MIT Press.

Kimura, D. 1992. Sex differences in the brain. *Scientific American*, September, 119–125.

Tata, J. R. 1993. Gene expression during metamorphosis: An ideal model for post-embryonic development. *BioEssays* **15**:239–248.

ORGANISMIC GROWTH AND ONCOGENES

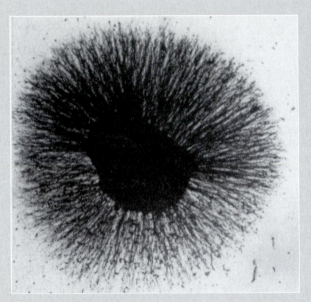
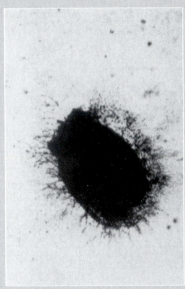

Figure 29.1 Effect of nerve growth factor (NGF) on the growth and differentiation of neurons. The photographs show sensory ganglia removed from chicken embryos and cultured for 24 h in a semisolid medium. The ganglion to the left was cultured in the presence of purified NGF, a peptide that promotes the survival, growth, and differentiation of nerve cells. The halo around the ganglion consists of fibers that have grown out of the neurons in the ganglion. The ganglion to the right was cultured in the same way but without NGF. NGF was discovered by Rita Levi-Montalcini and Stanley Cohen, who in 1986 received the Nobel Prize in medicine for their discovery.

Growth is a basic component of development. This is obvious if one compares the size of a human egg, a sphere about 0.1 mm in diameter, with the bulk of a human adult. *Growth* of an organism is defined as a change in its mass. Growth is usually positive—that is, the mass of living tissue increases—but we also speak of negative growth when the mass decreases.

A tissue can grow by several means. Most common is cell division followed by enlargement of the daughter cells until each has reached the previous size of the mother cell; this type of growth is called *hyperplasia* (Gk. *hyper*, "above"; *plassein*, "to form"). Another way for growth to occur is by an increase in mass without cell division, a process called *hypertrophy* (Gk. *trophe*, "nourishment"). A third means of growth is the deposition of *extracellular matrix*. The following examples illustrate these mechanisms.

Embryos typically do not grow during cleavage; they undergo cell division, but the daughter cells do not enlarge. When an embryo begins obtaining nutrients from elsewhere, by feeding or by tapping into maternal blood, it starts undergoing hyperplasia. This is the standard mechanism of growth, which may greatly increase the number of cells in an organism. Consequently, this chapter will deal for the most part with hyperplasia. The negative form of hyperplasia is *cell death,* an important mechanism of morphogenesis. For example, fingers and toes become separated as a result of programmed cell death in the intervening tissue of the limb buds; also, vertebrates routinely produce more neurons than eventually persist; pruning away excess neurons is a way of making nerve connections more selective.

Hypertrophy is observed in cells that no longer divide. For instance, *adipocytes* (fat cells) store lipids in large cytoplasmic droplets (see Fig. 13.17). If an organism consumes a rich diet, these lipid droplets may enlarge enormously without division of the adipocyte. Seasonal positive and negative hypertrophy of adipocytes is common among hibernating animals. A different type of hypertrophy occurs in skeletal muscle fibers. Under conditions of regular exercise and adequate nutrition, skeletal muscle fibers increase in girth through addition of more actin and myosin fibrils. The negative form of this type of growth is known as *dystrophy.*

Growth by addition or subtraction of extracellular matrix is observed mainly in connective tissues such as cartilage and bones. As described in Chapter 13, the extracellular matrix of bones is produced by *osteoblasts* and destroyed by cells of a different type called *osteoclasts.* The dynamic equilibrium between adding and subtracting extracellular matrix, which is under hormonal control, leads to a net positive growth early in life but tends to create a negative balance later, leading to a condition known as *osteoporosis* (Gk. *osteon*, "bone"; *poros*, "passage," "pore").

Superimposed on the growth mechanisms described so far are processes of tissue renewal, which essentially replace worn-out cells with new cells. The new cells are generated by cell divisions, and if a slight imbalance in the turnover generates more new cells than dying cells, then the process amounts to hyperplasia. It is therefore appropriate that some of the chemical signals that regulate these cellular turnovers have been termed *growth factors.*

In a growing organ, such as a human leg during puberty, all of the processes discussed here are going on at the same time. Many cells, including the endothelial cells of the blood vessels and the cartilage cells in the *epiphyseal discs* of the long bones, undergo hyperplasia. Elsewhere in the bones and in other connective tissues, growth is achieved by the addition of extracellular matrix. Muscle fibers increase in length by recruiting additional myoblasts into the muscle fibers, and they grow in girth by hypertrophy. Blood cells and epidermal cells increase in number by a boost in their renewal divisions.

Unraveling the mechanisms that control growth is an ongoing challenge. The purpose of this chapter is to discuss the major lines of investigation in this area and to gain an appreciation of the unresolved problems. After introducing the methods used to measure growth, we will discuss transplantation experiments revealing an intrinsic growth potential that limits growth during normal development. Then we will turn to extrinsic regulators such as hormones and growth factors. In particular, we will recount the discovery of the first growth factor, known as *nerve growth factor*, which promotes the survival and differentiation of certain neurons (Fig. 29.1). Thereafter, we will discuss the regulation of cell division and the role in this process of genes that have the potential to cause tumors. The characterization of these genes has greatly enhanced our understanding of both tumor development and normal growth control.

The Measurement of Growth

By weighing a developing organism at intervals and plotting the weight versus time in a coordinate grid, one obtains a *growth curve*. The shape of a growth curve is often *sigmoid*; that is, it resembles the letter S (Fig. 29.2a). The earlier part of the growth curve is easily understood, whereas the latter part is difficult to explain.

Initially, the mass increase per unit time is proportional to the mass already present. This relation can be expressed by the equation

$$\frac{dM}{dt} = g \cdot M$$

which is equivalent to

$$\frac{1}{M} \cdot \frac{dM}{dt} = g$$

where M represents mass, t is time, and g is a coefficient called the *relative growth rate*. As long as the relative growth rate is constant, the growth curve rises exponentially; if the mass is plotted on a logarithmic scale, the exponential growth appears as a straight line (Fig. 29.2b). This type of curve is obtained not only for organisms, but also for freshly seeded cell cultures, and for animal populations that are not limited by predation or by the carrying capacity of the environment. Thus, the exponential part of a growth curve reflects a growth phase during which all cells in an organism divide at constant intervals.

The exponential phase is followed by a phase of slower growth, which reflects inhibitory interactions or competition over resources in cell cultures and animal populations. Similar forces appear to limit the growth of individual organisms, but what exactly these forces are remains to be found. In species with limited growth,

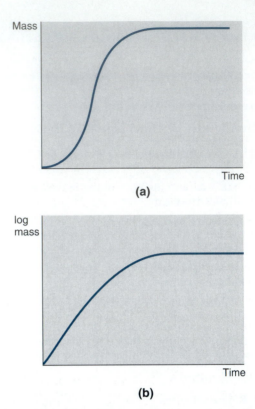

Figure 29.2 Growth curves show the mass of an organism or organ plotted versus time. If mass is plotted on a linear scale **(a)**, the curve is typically sigmoid. If mass is plotted on a logarithmic scale **(b)**, the initial portion of the curve is often linear.

or *determinate growth*, a group that includes humans and most other mammals, the growth curve approaches a horizontal line. In forms with unlimited growth, or *indeterminate growth*, which include plants and the majority of animals, the growth curve is still sigmoid but ends with a slow incline.

Different organs or regions of an animal may grow at the same relative growth rate, a situation described as **isometric growth**. More common is **allometric growth**, or a process in which some organs grow faster than others. Julian Huxley (1932), who studied allometric growth extensively, used the following equation to describe allometric growth:

$$y = b \cdot x^k$$

which is equivalent to

$$\log y = b + k \cdot \log x$$

where y is the mass of one part at a given time, x is the mass of another part (or the entire organism) at the same time, b is a coefficient that gives the mass of y when $x = 1$, and—most important—k is the **growth ratio**, that is, the relative growth rate of y divided by the relative growth rate of x. If $k = 1$, then x and y grow isometrically; and if $k > 1$, then y is *positively allometric* relative to *x*. The equation is based on the assumption that the growth ratio itself is constant over extended periods of time, and Huxley found empirically that this is true in many cases. In these cases, *k* can be determined by plotting *y* versus *x* on a double logarithmic coordinate grid and finding the straight line of best fit.

An example of allometric growth is seen in the human fetus, where the head is disproportionately large while the legs are proportionally shorter than in the adult (see Fig. 1.12). Thus, the legs grow faster than the head; in other words, their growth ratio relative to the head is greater than 1. Within the head, the facial parts are positively allometric relative to the braincase. This allometry, which is quite obvious in the human, is even greater in other mammals and tends to be more pronounced in males (Fig. 29.3).

An extreme example of allometric growth, also sexually dimorphic, is seen in the fiddler crab *Uca pugnax* (Fig. 29.4). The adult male of this species uses one of its chelae (claws) for fighting and display. In females and in young males, both chelae are the same size and relatively small, each representing about 8% of the crab's entire mass. However, as a male approaches adulthood, one chela grows with a growth ratio of 1.62 relative to the entire body, so that eventually this chela accounts for about 38% of the male's total body weight.

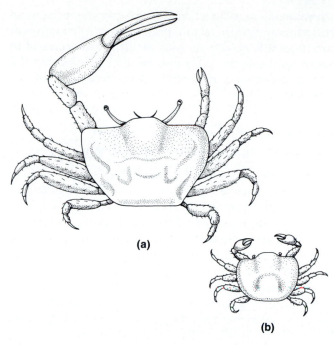

(a)

(b)

Figure 29.4 Allometric growth of one chela (claw) in the fiddler crab *Uca pugnax*. **(a)** Adult male with disproportionately large left chela; **(b)** juvenile male with chelae of equal size.

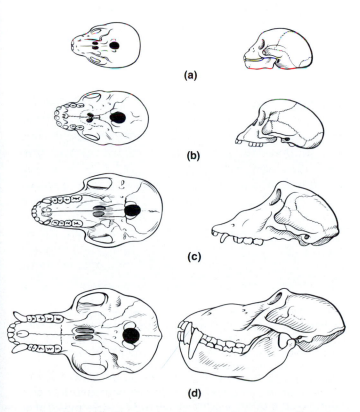

(a)

(b)

(c)

(d)

Figure 29.3 Allometric growth of the face relative to the braincase in the baboon: **(a)** newborn; **(b)** juvenile; **(c)** adult female; **(d)** adult male. Note that the face is initially smaller than the braincase but later grows larger, especially in the male.

These observations show that growth ratios and relative growth rates may change in stage-dependent and region-specific ways. Some authors have ascribed these changes, at least in part, to physical constraints such as the surface-to-volume ratio and the distance molecules can effectively travel through tissue by diffusion. However, the fact that some allometries are sexually dimorphic and bilaterally asymmetric also indicates that genetic and/or hormonal control mechanisms are probably involved.

Growth Analysis by Heterospecific Transplantation

The relative growth rate of an organ, and its final size, are influenced by a variety of factors, which are partly *intrinsic* to the organ and partly *extrinsic*. Intrinsic factors include the genotype of the organ and its previous developmental history. Intrinsic factors define a *growth potential*, that is, the maximum rate at which the organ can grow and, in species with determinate growth, its final size under favorable circumstances. Extrinsic factors include those provided by the organismic environment, such as nutrition and hormones, as well as factors in the external environment, such as temperature and social stress. To assess the relative importance of intrinsic and extrinsic factors, researchers have replaced organs from one species with the corresponding organs from another species, a procedure known as *heterospecific* (or *heteroplastic) transplantation.* These experiments have revealed that intrinsic factors play a major role in growth control, but they also showed that different parts of an organ may interact with one another in defining the growth potential of the entire organ (Weiss, 1939; R. G. Harrison, 1935).

The Growth Potential of a Limb Is a Property of the Limb's Mesoderm

Among salamanders of the genus *Ambystoma*, the forelimbs of certain species differ in their final sizes and in their growth rates at various stages. Two species used in several studies on growth regulation are *Ambystoma punctatum* and *Ambystoma tigrinum*. *A. tigrinum* is more active and voracious, grows more rapidly, and has an adult length twice that of *A. punctatum*. In *A. punctatum*, the forelimbs appear early in development: they are fully formed, complete with digits, by the time the larva begins to swim and feed (Fig. 29.5). In contrast, the forelimbs of *A. tigrinum* at the corresponding stage are small nodules without visible differentiation. Once they have started to grow, however, they proceed at a much faster rate and eventually grow to a much greater size than *A. punctatum* forelimbs.

To test whether these specific time characteristics and final sizes are intrinsic to the limbs or to the entire organism, Ross G. Harrison (1924) carried out reciprocal heterospecific transplantations. He transplanted forelimb buds before their outgrowth began, during the embryonic period, from the right side of an *A. punctatum* to the right side of an *A. tigrinum* individual and vice versa (Fig. 29.6). The left limb buds remained in place as controls.

The *A. punctatum* limb bud graft on the *A. tigrinum* host developed rapidly during the embryonic period and forged ahead of the normal *A. tigrinum* limb on the opposite side. Its movements were somewhat delayed, but it soon became functional and remained so throughout its life. During the larval period, however, the grafted limb fell behind the normal limb, and at metamorphosis, the grafted limb was about half as long as the control limb on the other side of the host (Fig. 29.6b).

The reciprocal experiment gave corresponding results. The *A. tigrinum* limb bud graft on the *A. punctatum* host embryo grew slowly at first, but once it started, it grew rapidly and eventually became about twice as long as its contralateral control limb (Fig. 29.6e). The grafted limb was also fully functional despite its oversize. Thus the growth of the transplanted limb buds corresponded unmistakably to the intrinsic time schedule and final size of the donor species.

To further analyze whether the intrinsic growth characteristics resided in a particular part of the grafted limb bud, R. G. Harrison (1935) extended these experiments and grafted only the ectodermal cap or the mesenchymal core of a bud. The effects of transplanting *ectoderm* were only subtle (Fig. 29.6a and d). However, when limb bud *mesoderm* was grafted alone, the outcome was very similar to that obtained by grafting the entire limb bud (Fig. 29.6c and f). Together, these results

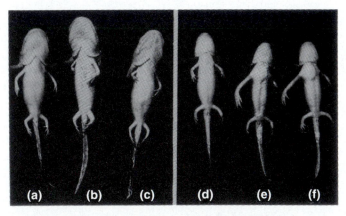

Figure 29.6 Growth of limbs exchanged between two urodele species with different growth characteristics. Limb buds were exchanged reciprocally on the right sides of the embryos (the left sides in the photographs, which are ventral views taken about 3 months later). **(a–c)** *Ambystoma tigrinum;* **(d–f)** *Ambystoma punctatum.* **(a, d)** Ectoderm of the limb bud was replaced with limb bud ectoderm of the other species. All limbs developed according to host size. **(b, e)** Whole limb buds were exchanged. The grafts developed according to the growth characteristics of the donor. **(c, f)** Mesoderm of the limb buds was exchanged. The grafts developed according to the growth characteristics of the donor.

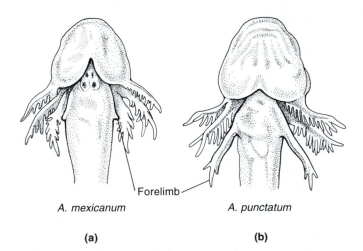

Figure 29.5 Larvae of *Ambystoma mexicanum* **(a)** and *A. punctatum* **(b)** at the beginning of the swimming and feeding stage. Ventral views. Note that *A. punctatum* has well-differentiated forelimbs with digits whereas *A. mexicanum* has only small nodules at this stage. In this respect, *A. mexicanum* is similar to *A. tigrinum*, which was used in the experiment shown in Figure 29.6.

showed that the intrinsic growth properties of a limb bud reside mostly in its mesenchymal core.

The growth of the limb grafts was analyzed, for the most part, through comparison with the control limb on the opposite side of the *host*. This control limb belonged to a different species and therefore had a different set of intrinsic growth factors, whereas the extrinsic factors were exactly the same for both the transplant and the host control. However, comparisons were also made between the grafted limb and its nontransplanted counterpart in the *donor*. This control limb belonged to the same individual, so it shared the same intrinsic factors, but the extrinsic factors, such as hormones and nutritional state, were different. These extrinsic factors proved to be difficult to analyze, because they entail a large number of variables including neuromuscular and digestive mechanisms. For instance, a diet of beef and liver favored the growth of *A. tigrinum* over *A. punctatum*, whereas small crustacea and worms had the reverse effect. Under conditions of maximal feeding, however, the limb transplants and their donor controls grew at more or less the same rate.

Harrison's experiments showed that salamander limb buds grow into functional limbs after heterospecific transplantation. The growth potential of the grafted limb is intrinsic to the donor species, and it is fully realized in the host under conditions of maximal feeding. This result is of particular interest when the grafted limb grows to a larger size than its host control limb even though both limbs are supplied with the same nutrients, hormones, and other systemic factors of the host. It must be concluded that under these circumstances the grafted limb, because of its intrinsic properties, utilizes the host environment more efficiently than the host control limb. For the most part, these intrinsic growth properties of the limb reside in the mesenchymal core of the limb bud.

The Optic Cup and the Lens of the Eye Adjust Their Growth Rates to Each Other

Heterospecific transplantations with other organs, including gills, heart, ear, spinal cord, and eye, yielded results similar to those obtained with limb buds. Eye transplantations are of particular interest because they illustrate that different organ parts may contribute to the intrinsic growth characteristics of an organ.

There are two main constituents of the developing eye: the *optic cup*, which develops from the *optic vesicle*; and the *lens*, which is formed by the overlying *lens epidermis* (see Chapter 12). It is possible to transplant the entire eye rudiment or only the lens epidermis or only the optic vesicle.

In separate experiments, R. G. Harrison (1929) and Twitty and Elliott (1934) transplanted entire eye rudiments reciprocally between embryos of *Ambystoma punctatum* and *A. tigrinum*. In both studies, the transplants grew to a final size that was characteristic of the donor species. Thus, the *A. punctatum* host grew to its modest overall size with a large *A. tigrinum* eye, whereas the *A. tigrinum* host grew to its much larger overall size with a small *A. punctatum* eye (Fig. 29.7). The grafted eyes were normally shaped, moved normally, and in many cases seemed to have connected properly to the brain.

As in the limb bud transplantations, the most revealing result was the growth of the *A. tigrinum* transplant on the *A. punctatum* host. Both the grafted eye rudiment and its host control were supplied with the same nutrients, hormones, and other factors from the host, but the graft made the more efficient use of these supplies and grew much larger than the host's own eye. This outcome shows that the growth of the host eye was limited by intrinsic factors rather than by extrinsic ones.

To analyze the separate contributions of the eye lens and the optic cup to the overall size of the eye, R. G. Harrison (1929, 1935) carried out similar transplantations except that he grafted only lens epidermis or optic vesicle instead of the entire eye rudiment. As the recipients developed, he periodically measured the sizes

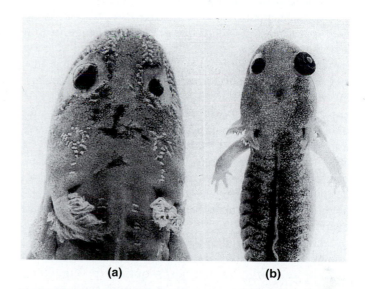

(a)　　　　　　　　　　　(b)

Figure 29.7 Reciprocal transplantation of embryonic eyes between two salamander species of different sizes. **(a)** *Ambystoma tigrinum* with right eye from *A. punctatum*; **(b)** *A. punctatum* with right eye from *A. tigrinum*. During the months after the operation, *A. tigrinum* grew much faster than *A. punctatum*, but the grafted eyes acquired the sizes characteristic of their donors. The photographs show preserved specimens, with skin removed around the eyes to reveal their sizes.

TABLE 29.1

Growth Ratios of Normal, Transplanted, and Chimeric Salamander Eyes

Material Transplanted	Combination	Number of		Growth Ratio k	k (P/T) × k (T/P)	Eye Index i			i (P/T) × i (T/P)
		Cases	Survivors			Max*	Min	Average	
Whole eye	P/T	11	5	0.73	0.99	0.62	0.67	0.64	1.15
	T/P	24	12	1.36		2.00	1.63	1.80	
Optic vesicle	P/T	14	2	0.87	0.97	0.77	0.77	0.77	1.01
	T/P	13	2	1.11		1.33	1.28	1.31	
Lens epidermis	P/T	16	7	0.87	1.06	0.79	0.83	0.81	1.03
	T/P	11	5	1.22		1.31	1.21	1.27	

*The terms maximum and minimum are used with respect to the greatest effect, that is, the greatest deviation from unity (1.0).
P/T = *Ambystoma punctatum* graft on *Ambystoma tigrinum.*
T/P = *Ambystoma tigrinum* graft on *Ambystoma punctatum.*
k = relative growth rate of grafted eye (diameter) and normal control on opposite side of the host.
i = ratio of diameter of grafted and host control eyes at close of experiment.
Source: After R. G. Harrison (1929, 1935). Used with permission.

of their lenses and optic cups using a microscope with a scale built into one eyepiece. Some specimens were sacrificed for histological sectioning.

From his measurements, Harrison computed the *growth ratio k* of the grafted eye diameter relative to the control eye on the opposite side of the host. He also calculated the *eye index i*, that is, the ratio of the diameters of the grafted and the host control eye at the close of the experiment. The results are compiled in Table 29.1. The number of analyzable cases was small, especially for the optic vesicle transplantations, which were survived by only two individuals in each series. However, the range in the measurements for the eye index in these cases was narrow, suggesting that the average from a larger number of measurements would probably be close to the average from the cases at hand. For both *k* and *i*, the greatest departures from unity (1.0)—meaning the greatest differences between the transplanted eye and the host control—were observed after transplantation of entire eye rudiments. Smaller but still significant effects were observed after transplantation of either optic vesicle or lens epidermis. As expected, the products of *k* values for reciprocal experiments (*A. punctatum* graft on *A. tigrinum* and vice versa) and the products of *i* values for reciprocal experiments were close to unity.

When *A. tigrinum* lens epidermis was transplanted to an *A. punctatum* embryo, the newly formed lens was initially too large for the host's optic cup (Fig. 29.8). However, the cup nearly caught up with the lens once the host larva began to feed. After 3 to 4 months, the lens had grown less than it would have in a transplant of an entire eye rudiment, and the optic cup had grown more than its host control. The resulting eye was intermediate in its overall size between normal *A. tigrinum* and *A. punctatum* eyes, and the lens was nearly in nor-

mal proportion to the eye. There were seven analyzable cases, and these observations were made in all seven.

Five cases of the reciprocal experiment were useful. In each, when *A. punctatum* lens epidermis was transplanted to an *A. tigrinum* embryo, the newly formed lens was initially too small for the host optic cup. At the close of the experiment, the lens had grown some more than it would have in a transplant of an entire eye rudiment, and the optic cup had grown considerably less

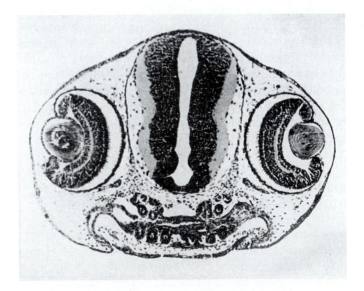

Figure 29.8 Transverse section through the head of an *Ambystoma punctatum* embryo, 8 days after it received a graft of *A. tigrinum* lens epidermis over the right optic vesicle. In the chimeric right eye (to the left in the photograph), the lens was derived from the graft, and the optic cup from the host. In the other eye, both components are derived from the host. Note that the lens is disproportionately large in the chimeric eye.

than its host control. Again, the final size of the chimeric eye was intermediate between the normal eye sizes of the two species, and the lens was nearly in normal proportion to the eye.

When an *A. tigrinum* optic vesicle was transplanted to an *A. punctatum* embryo, the grafted optic cup initially grew too fast for the lens, which was of host origin. In one case analyzed 19 days after transplantation, the lenses in both eyes were still the same size, indicating that the graft had not yet affected the growth of the lens. After a few months, however, the lenses in the two chimeric eyes had grown by an average of 13% in each dimension. At the same time, the host lenses retarded the growth of the graft optic cups, which grew only to an average eye index of 1.31 instead of the 1.80 observed after transplantation of an entire eye (see Table 29.1). In the reciprocal experiment, the grafted *A. punctatum* optic cup was too small for the lens formed by the *A. tigrinum* host. At the close of the experiment, however, the graft optic cups accelerated in growth, reaching an average eye index of 0.77 instead of the 0.68 value observed after transplantation of an entire eye. At the same time, the *A. tigrinum* lenses in the chimeric eyes were retarded in their growth, so that they reached an average of only 84% of their host control diameter.

The heterospecific transplantation experiments just cited reveal that the optic cup and the lens in chimeric eyes affect each other's growth rate. This growth regulation of *eye components* is in stark contrast to the lack of growth regulation of the *entire eye*, which, upon transplantation, stubbornly grows according to its intrinsic potential.

When heterospecific transplantations generate a size disparity, the oversized component down-regulates its growth while the undersized component up-regulates its growth. Of particular interest is the up-regulation in the growth rate of the undersized component, because this component exceeds what seemed to be its intrinsic growth potential. It must be concluded that the *A. punctatum* eye components keep each other's growth potential low when they are together. Conversely, the *A. tigrinum* eye components together seem to keep each other's growth potential at a high level.

Taken together, the transplantation experiments discussed in this section reveal the existence of intrinsic factors that impose species-specific limits on both the growth rate and the final size of organs. However, the establishment and maintenance of these limits may depend on the completeness of the organ and on interactions between its components. The same experiments show that the actual growth of organs also depends on extrinsic factors such as nutrients and other blood-borne components. Such extrinsic factors will be discussed in the following sections.

Growth Hormones

Exogenous growth factors in vertebrates include several hormones. Among these, the one with the strongest and most widespread effects is known as *somatotropin* or *growth hormone (GH)*. It is produced by the anterior pituitary gland under the control of *growth hormone releasing factor (GHRF)* from the *hypothalamus*. GH deficiencies, which may result from mutations or from diseases affecting the pituitary, cause dwarfism, whereas GH excess causes gigantism (Fig. 29.9).

The action of GH was demonstrated most clearly by the genetic transformation of mice with a transgene encoding rat growth hormone. This experiment showed that an elevated level of growth hormone alone was sufficient for mice to grow to twice the size of their littermates (see Figs. 14.1 and 14.27). Thus, the intrinsic growth potential that limits the augmenting effects of

Figure 29.9 Gigantism and dwarfism in humans. The photograph shows a normal-sized man (left) walking next to a "giant" (center), whose pituitary gland was overactive, and a "dwarf," whose pituitary was underactive.

nutrition can be expanded by excess amounts of growth hormone.

In mammals, GH has little effect on fetal growth; transgenic mice with enhanced GH levels before birth grow like their normal littermates (Palmiter et al., 1983). However, GH is the single most important stimulus for postnatal growth. It acts on a number of target tissues by stimulating mitosis. In particular, GH causes the longitudinal growth of bones by acting on the *epiphyseal discs*. Here, GH prompts the division and maturation of *chondrocytes*, thereby adding more cartilaginous material for bone formation. Muscle, bone marrow, liver, and kidney tissue also respond to GH.

The GH gene is expressed in the *somatotropic cells*, a specific subset of anterior pituitary gland cells; no other site of GH synthesis has so far been detected (Karin et al., 1990). The specificity of this control contrasts with the expression of other hormones, which is less restricted. The regulatory region of the GH gene has a cAMP response element, which mediates the transcriptional control by GHRF through a mechanism that involves cAMP (Fig. 29.10). The regulatory region of the GH gene also includes enhancer sequences that bind to other regulatory proteins, including the glucocorticoid receptor and—at least in rats—the thyroid receptor. However, these regulatory proteins are also present in other cells and cannot by themselves account for the selective synthesis of GH in somatotropic cells.

Only one regulatory protein, designated GHF-1, has been found exclusively in these cells. GHF-1 seems to unlock the GH promoter so that it responds to the other, more ubiquitous regulatory proteins. The amino acid

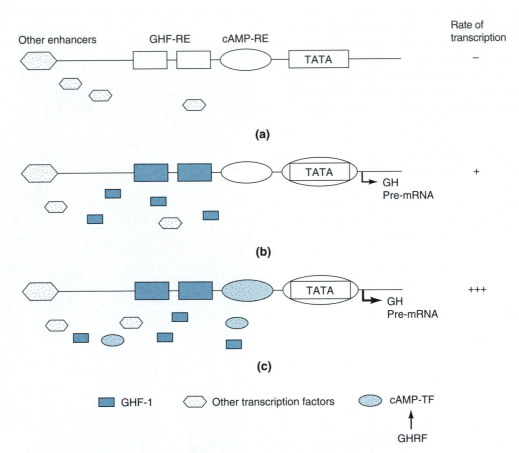

Figure 29.10 Model illustrating the role of a regulatory protein, designated GHF-1, in the cell-specific activation of the growth hormone (GH) gene in the mouse. The diagram shows the regulatory region of the GH gene, with the promoter (represented by TATA box), a cAMP response element (cAMP-RE), two GHF-1–specific response elements (GHF-RE), and other, nonspecific enhancer sequences. **(a)** Before embryonic day 16, the GH-producing cells contain some of the ubiquitous transcription factors that bind to the nonspecific enhancer sequences. However, in the absence of the specific transcription factor GHF-1, none of the regulatory proteins can activate the GH promoter. **(b)** During embryonic day 16, GHF-1 appears and binds to GHF-RE, thus unlocking the promoter. **(c)** The activity of the transcription complex (oval around TATA box) can be further enhanced by growth hormone releasing factor (GHRF) from the hypothalamus, which acts via a cAMP-dependent protein kinase (see Fig. 2.26) that activates a transcription factor (cAMP-TF).

sequence of GHF-1 includes a *homeodomain* and a *POU domain,* both of which are also known from other transcription factors. However, the *transcriptional activation domain* of GHF-1 is distinct from other activation domains, suggesting that GHF-1 may interact with different parts of the transcriptional complex than those acted on by other transcription factors.

Growth hormone seems to act both directly and indirectly. The indirect effects are mediated by *somatomedins,* insulinlike peptides that are synthesized by the liver in response to GH. The blood plasma levels of *somatomedin C (insulinlike growth factor I, or IGF-I)* and *somatomedin A (insulinlike growth factor II, or IGF-II)* are generally correlated with GH level. Moreover, IGF-I can substitute for GH in causing bone growth and weight gain; this was shown when IGF-I was injected into hypophysectomized rats, that is, rats whose pituitary glands had been removed so that their blood contained no endogenous GH (Schoenle et al., 1982).

In addition to GH and somatomedins, thyroid hormone is also important in regulating the postnatal growth of mammals. Thyroid hormone deficiency causes reduced body weight, whereas hyperthyroidism results in accelerated growth, although these may be indirect effects of thyroid receptor action on GH gene expression. Insulin may also play a role in controlling both prenatal and postnatal growth. Untreated young diabetics suffer retarded growth, and infants born to diabetic mothers are larger than normal, presumably because the fetus produces excess insulin in response to the elevated glucose level of the maternal blood. Other hormones specifically control the growth of certain target organs. For instance, sex hormones stimulate or inhibit the growth of reproductive organs, as described in Chapter 28.

Growth Factors

In addition to hormones, which by definition are produced in endocrine glands and transported by the bloodstream, there are a wide range of so-called *growth factors,* which are produced by many tissues and act locally as *paracrine* or *autocrine* signals (see Fig. 2.22).

Nerve Growth Factor Promotes the Growth and Differentiation of Certain Neurons

The first growth factor to be discovered was called *nerve growth factor (NGF)* because it is necessary for the survival and differentiation of neurons. NGF was discovered by Rita Levi-Montalcini and Stanley Cohen in an exciting set of experiments they carried out between 1950 and 1958. The importance of their discoveries was acknowledged in 1986 when both scientists were awarded the Nobel Prize in medicine.

Levi-Montalcini's contribution to the discovery of NGF began with classical embryological experiments (reviewed by Levi-Montalcini, 1964, 1988). She and her mentor, Viktor Hamburger (a student of Spemann's), were interested in the general question how neurons make selective contact with their appropriate target organs. To examine the role of the target organs, the researchers transplanted extra limb buds to chicken embryos and observed the resulting perturbations in the developing nervous system. The transplantations affected in particular the sensory ganglia in the dorsal roots of the spinal nerves (see Fig. 12.19). In the vicinity of the extra limbs, the dorsal root ganglia came to be larger than they normally would have become.

In normal development, the neurons of the sensory ganglia send fibers outward to the periphery of the body and inward to the spinal cord. The rudiments of the segmental ganglia are initially similar in size, but the ganglia that innervate limbs retain a larger volume than those ganglia that innervate segments without limbs. This disparity is brought about mainly by cell death; the more target tissue there is for a ganglion to innervate, the more of its neurons survive (Hamburger and Levi-Montalcini, 1949). Thus, part of the strategy by which ganglia innervate their targets is to produce an excess of neurons and allow only those that have made the most functional connections to survive.

What kind of signal may emanate from a developing limb that promotes the survival and differentiation of neurons? Following up on an earlier observation on the nerve growth-promoting effects of certain tumors made in Hamburger's laboratory, he and Levi-Montalcini implanted pieces of assorted mouse tumors into chicken embryos. One particular tumor, a sarcoma (connective tissue tumor) designated S.180, provoked a stunning reaction in its host: nerve fibers from nearby ganglia grew into the tumor before they innervated the adjacent limb bud. Moreover, the ganglia that sent axons into the tumor grew to a larger-than-normal size. The tumor especially promoted the development of the *sympathetic ganglia*—that is, the visceral ganglia forming chains on both sides of the vertebral column—and the *sensory ganglia* located in the dorsal roots of the spinal nerves. Both types of ganglia are derived from neural crest cells as discussed in Chapter 12. The tumor exerted this effect even when transplanted on the *chorioallantoic membrane* of the chicken embryo so that the tumor and the ganglia were not in direct contact. The investigators concluded that the tumor released a humoral (Lat. *humor,* "body fluid") factor that traveled through blood vessels to the sympathetic neurons. They termed this novel factor *nerve growth factor.*

Fascinated by the effects of NGF, Levi-Montalcini learned to maintain sympathetic and sensory embryonic ganglia in a semisolid culture medium, a rare technique at the time. This allowed her to test the effect of the S.180 factor on cultured ganglia, an assay that proved to be much simpler and quicker than the use of whole embryos. When cultured in the vicinity of S.180 tumor fragments, the ganglia produced a stunning halo of nerve cell processes within 10 h (Fig. 29.1). Control ganglia cultured in the same way but without S.180 cells displayed only a sparse and irregular outgrowth of nerve fibers. Now the investigators had a *bioassay* that allowed them to screen within hours large numbers of tissues, organic fluids, and chemicals for nerve growth factor activity.

At this point, Hamburger's research group was joined by Stanley Cohen, a biochemist. Using the bioassay developed by Levi-Montalcini, he tested fractions prepared from various tissues for their NGF activity, that is, for their ability to promote the formation of nerve fibers by cultured ganglia. He determined that NGF had the properties of a nucleoprotein, that is, a complex consisting of a protein and a nucleic acid moiety. In order to test which moiety was responsible for the NGF activity, Cohen and Levi-Montalcini pretreated the active nucleoprotein fraction with snake venom, a known source of phosphodiesterase enzymes, which degrade nucleic acids. They expected either that the added venom would not affect the activity of the nucleoprotein fraction, indicating that the nucleic acid moiety was dispensable, or that the venom would somehow poison the effect. Much to their surprise, the investigators found that the added snake venom *enhanced* the growth-promoting effect of the nucleoprotein fraction.

Indeed, small amounts of snake venom alone, injected into embryos or tested in the bioassay, promoted the growth and differentiation of sympathetic and sensory ganglia. The results showed that the snake venom itself was a rich source of NGF.

Next, Cohen found that mouse salivary glands, which are homologous to snake venom glands, are an even better source from which to isolate pure NGF. Cohen also found that a variety of normal and tumor cells secreted NGF.

When injected into newborn mice, purified NGF caused sympathetic ganglia to grow 4 to 6 times larger than those in untreated control mice (Levi-Montalcini, 1964). Cell measurements and cell counts showed that NGF promoted an increase in cell number as well as an increase in the size of individual neurons (Fig. 29.11). The increase in the number of neurons per ganglion relative to control ganglia does not necessarily mean that NGF stimulates the division of neurons in the ganglia; more likely, NGF prevents or reduces the degree of programmed cell death that normally occurs in the ganglia. Injection of NGF did not seem to produce similar changes in *parasympathetic* ganglia or in the central nervous system.

In addition to increasing the number and size of neurons, NGF also has an *orienting* effect on the outgrowth of neurites (nerve cell extensions) from the sympathetic ganglia. When purified nerve growth factor was injected into the brains of newborn rats, nerve fibers sprouted from the sympathetic ganglia and invaded the brain and spinal cord (Levi-Montalcini and Calissano, 1979). Apparently, the NGF injected into the brain dif-

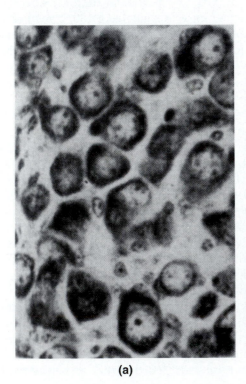 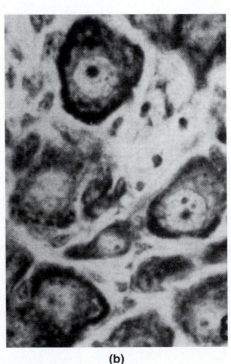

(a) (b)

Figure 29.11 Increased size of neurons in the sympathetic ganglia of mice injected with purified nerve growth factor (NGF). The photographs show histological sections of sympathetic ganglia from two mouse littermates, one injected with saline solution only **(a)**, and the other with the same solution containing NGF **(b)**. Note that the neurons from the NGF-treated mouse are much larger.

fused through the motor and sensory roots of the spinal nerves and reached the sympathetic chain ganglia flanking the vertebral column, where the NGF induced the outgrowth of nerve fibers that are not seen in normal rats (Fig. 29.12). This finding implies that the tip of a sprouting sympathetic nerve fiber elongates along a pathway that is determined, at least in part, by a gradient of NGF. The tips then grow in the direction of increasing NGF concentration; this phenomenon is called the **neurotropic effect** of NGF (Gk. *tropikos,* "turning"). This effect was confirmed in vitro: in a setup where neurites growing out from cultured neurons could grow into a chamber containing NGF and another chamber without NGF, neurites grew only into the chamber with NGF.

The NGF fractions isolated from mouse tumor cells and salivary glands had identical biochemical and bio-physical characteristics and were both secreted from the producing cells. Over the following years, Cohen and others characterized NGF as a dimer of two peptides, each 118 amino acids long. NGF binds to specific receptors, located in particular on the *growth cones* of axons. This binding appears to be critical to the extension of the growth cone. In addition, NGF is taken up by *receptor-mediated endocytosis* (see Fig. 2.14). The endocytotic vesicles containing NGF move by retrograde transport from the tip of the axon to the cell body, where they seem to influence neuron survival.

The experiments described so far showed that NGF is *sufficient* to promote the growth and differentiation of certain neurons. To test whether NGF is also *necessary* for this process, Cohen and Levi-Montalcini prepared rabbit anti-NGF antibodies and injected them into newborn mice. A month later, the sympathetic and sensory ganglia in the treated mice were smaller than in the controls. In particular, the sympathetic ganglia were barely visible and contained only a few neurons amounting to 3 to 5% of the normal cell population (Fig. 29.13). Similarly, anti-NGF antibody added to cultured sympathetic ganglion cells caused a massive degeneration of neurons while the glial and connective tissue cells remained unaffected. The degenerative effect of anti-NGF antibody on neurons was overcome by adding more NGF.

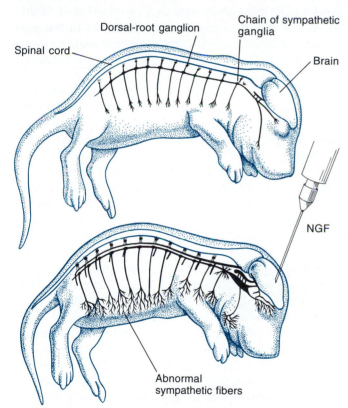

Figure 29.12 Neurotropic effect of nerve growth factor (NGF). NGF purified from mouse salivary glands was injected into the brains of newborn rats; control animals were injected with saline only. The injected fluid diffused into the spinal cord and from there into the spinal nerve roots and the sympathetic ganglia. In the NGF-treated animals (bottom), the sympathetic ganglia sent out abnormally long and numerous fibers that grew out into the periphery and back through the spinal nerve roots into the spinal cord and as far as the brain. In the saline-injected controls, the sympathetic ganglia sent out their normal complement of fibers into peripheral organs. The results indicate that NGF enhances the growth of sympathetic fibers and causes their orientation toward the highest NGF concentration.

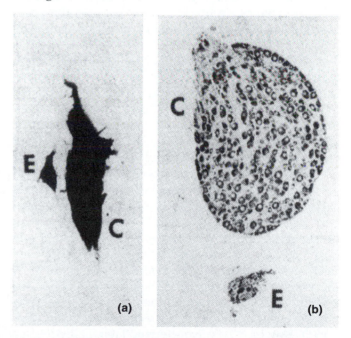

Figure 29.13 Inhibitory effect of antibodies against nerve growth factor on the growth and maintenance of sympathetic ganglia. Experimental (E) mice were injected with antibodies for 3 days after birth, whereas control (C) mice were injected with saline only. Wholemounts **(a)** and histological sections **(b)** of superior cervical ganglia were prepared after 3 months. The ganglia of the experimental mice were much smaller than those of control mice, and their neurons were fewer in number.

Joining their embryological and biochemical expertise in a brilliant series of experiments, Levi-Montalcini and Cohen showed that various tissues, both tumorous and normal, secrete a nerve growth factor protein that promotes the growth and differentiation of neurons. Nerve growth factor has two general functions. First, it is critical for the survival, growth, and differentiation of neurons in sympathetic and dorsal root ganglia. Second, it has a *tropic* effect that directs the fibers of these neurons to any target that releases it.

Other Growth Factors May Affect Multiple Target Cells in a Variety of Ways

Since the discovery of NGF, several related peptides have been characterized. Like NGF, they are released from cells and typically act on membrane-bound receptors in the producing cells and/or other target cells. Most of these peptides diffuse over short distances, but some are bound to extracellular matrix components, where they can be stored in a localized form for extended periods of time. Some of these signal peptides act on specific target cells, often causing cell proliferation. The *interleukins (ILs)* and other *hematopoietic factors* discussed in Chapter 19 belong in this category.

Other cellular signal peptides have a wide range of target cells and multiple functions. Several of these peptides are termed "growth factors" because they promoted cell division in the studies in which they were first identified, even though further investigations showed that under different conditions the same factors may inhibit cell proliferation. In fact, some growth factors have additional effects that are entirely unrelated to cell division and growth. Many growth factors are named after the cell types and assays that led to their initial isolation, despite the fact that their activity was found in other cell types as well. The following examples illustrate this somewhat bewildering situation.

While he was still characterizing NGF, Stanley Cohen discovered another growth factor. When newborn mice—whose eyes would normally be closed until 10 days after birth—were injected with incompletely purified NGF from mouse salivary glands, they opened their eyes a week ahead of schedule. Cohen found that the precocious eye opening was due to accelerated growth and keratinization of the epidermis of the eyelids. He ascribed this effect to an *epidermal growth factor (EGF),* which was characterized later. EGF acts on a variety of cell types and may promote or inhibit growth. For instance, EGF promotes the growth of mammary gland ducts, which are derived from epidermis, but also stimulates the division of fibroblasts, which are completely unrelated to epidermal cells. In addition, EGF accelerates the maturation of the pulmonary epithelium and is necessary for the formation of the hard palate. On the other hand, EGF inhibits the proliferation of hair follicle cells and certain cancer cells (Carpenter and Cohen, 1978).

Fibroblast growth factor (FGF) stimulates the proliferation of many cell types, including fibroblasts and endothelial cells. Along with EGF, FGF also promotes the division of *myoblasts* before they fuse to form multinuclear *myotubes* (Olwin and Hauschka, 1988). On the other hand, FGF has an antiproliferative effect on some tumor cells.

Yet another growth factor is known as *platelet-derived growth factor (PDGF),* because platelets release it during blood clotting. In culture, PDGF stimulates the division of two related cell types, fibroblasts and smooth muscle cells. In vivo, this process seems to facilitate wound closure. PDGF also stimulates the proliferation of glial cells and prevents their premature differentiation. Moreover, EGF and PDGF cooperate with *somatomedins* to stimulate the proliferation of fat cells and connective tissue cells.

A large superfamily of more than 20 growth factors is known as the *transforming growth factor-β (TGF-β)* superfamily, so named after its first member, TGF-β1, which was found to transform cultured cells from normal to tumorous growth. It appears now that the mitogenic effect of TGF-β1 and other TGF-β factors is caused indirectly through the stimulation of other growth factors. On the other hand, TGF-β inhibits the growth of various tumors and nontumorous epithelia, and again, it appears that TGF-β exerts this effect by acting upon other molecules that control the cell cycle (Moses et al., 1990).

Some of the most intriguing members of the TGF-β superfamily, including the decapentaplegic, activin, Vg1, nodal, and dorsalin-1 proteins, are involved in pattern formation (Kingsley, 1994). In *Drosophila,* decapentaplegic controls the formation of dorsal structures in the embryo, the patterning of the gut, and the growth and patterning of many imaginal discs in the larva (see Chapter 21). In *Xenopus,* Vg1 and activin appear to be key signals in the induction of dorsal mesoderm (see Chapter 9). In the mouse, nodal seems to be involved in mesoderm formation and axis organization (Zhou et al., 1993). In the chicken, dorsalin-1 is synthesized in the dorsal portion of the spinal cord, where it promotes the formation of neural crest cells and inhibits the induction of motor neurons by the floor plate (see Chapter 12).

These examples highlight three striking features of members of the TGF-β superfamily. First, many TGF-β members are involved in setting up polarity axes and specifying embryonic pattern elements. Second, a given family member may play multiple roles in a given organism. Third, TGF-β members show up in these functions throughout the animal kingdom. It appears that these proteins are extremely versatile and effective, and that the genes encoding them have been co-opted—through the addition of new control elements—for a wide spectrum of functions in development. We will continue the discussion of one of these features, the versatility of growth factors, in the following section, and return to another aspect, the linkage between growth control and pattern formation, at the end of this chapter.

The Effects of a Growth Factor May Depend on the Presence of Other Growth Factors

Even in a single cell type, the action of a growth factor may depend on the context set by other growth factors. For example, TGF-β stimulates growth of certain cultured fibroblasts in the presence of PDGF but inhibits their growth if EGF is present (A. B. Roberts et al., 1985). TGF-β can also stimulate replication of *osteoblasts,* but this proliferative effect can be reversed by EGF and FGF (Centrella et al., 1987). Similarly, IL-4 may either enhance or antagonize the growth of an entire group of hematopoietic progenitor cells, depending on the presence of other growth factors, notably IL-3 (Rennick et al., 1987). The action of a growth factor may also depend on the differentiated state of a target cell. For instance, TGF-β stimulates cartilage formation early in the development of embryonic mesenchyme and chondroblasts but inhibits it later. In accord with the multiple and context-dependent actions of growth factors, most cells have receptors for several growth factors (Cross and Dexter, 1991).

On the basis of these observations, Sporn and Roberts (1988) have suggested that growth factors are part of a complex cellular signaling language, in which individual growth factors are the equivalents of the letters that compose words. According to this analogy, informational content lies not in an individual growth factor, but in the entire set of growth factors and other signals to which a cell is exposed. The adaptive value of such a combinatorial system would be that a small number of signals can transmit a wide range of different instructions.

Learning how different growth factor combinations are translated into distinct patterns of gene expression remains a major challenge. Several interactions can occur between signaling pathways that lead from the binding of a receptor by its ligand to the action of transcription factors or other regulators of gene expression. One possible interaction, at the level of growth factor receptors, is named *transmodulation* or *receptor cross talk.* In particular, it has been observed that the binding of one growth factor to its receptor changes the distribution of another receptor or the binding affinity of another receptor for its factor; for example, PDGF can decrease the affinity of the EGF receptor for its ligand (Bowen-Pope et al., 1983). In addition, there may be interaction between the second messenger pathways through which growth factors exert their effects on cells. Finally, the transcription factors that are activated by growth factors have combinatorial effects, as discussed in Chapters 15 and 21. Analysis of these interactions has been boosted by the cloning of genes that encode

growth control molecules. These genes will be discussed later in this chapter.

Cell Cycle Control

How do growth hormones and growth factors control growth? Studies on this subject have focused mostly on *hyperplasia,* and in particular on the control of the cell cycle. Such investigations are typically carried out on cultured cell lines that proliferate indefinitely. Although such cell lines have been selected for survival in vitro, they are widely regarded as models for the behavior of cells in vivo. A commonly used line of cells, known as NIH 3T3 cells, is derived from mouse fibroblasts (Fig. 29.14a). These cells proliferate most rapidly when set-

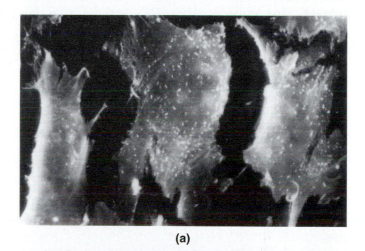

(a)

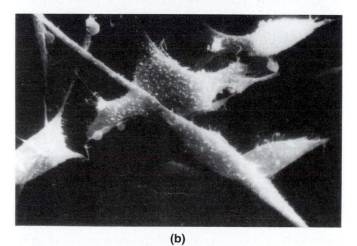

(b)

Figure 29.14 Normal and transformed mouse fibroblasts viewed under the scanning electron microscope. **(a)** Normal cells lie flat on the bottom of the culture dish and stop dividing when they cover most of the bottom as a single cell layer (monolayer). **(b)** After transformation with genes that interfere with normal growth control, the same cells assume a tumorlike phenotype. Transformed cells become rounded and climb on top of each other, and their further division is less dependent on the presence of growth factors in the medium.

tled on the bottom of a culture dish and covered with a medium containing blood *serum* (the liquid that rises to the top when clotted blood is centrifuged). Serum contains not only nutrients for the cells but also growth factors. The growth factors are present in serum in concentrations of about 10^{-10} M; when they are used up, normal cells stop dividing whereas tumorous cells keep growing.

As described in Chapter 2, the cell cycle is divided into four major phases known as M (mitosis), G_1 (RNA synthesis), S (DNA synthesis), and G_2 (RNA synthesis). The longest phase is G_1, during which the cell synthesizes most of the components necessary for DNA replication and mitosis. When certain conditions, such as nutrients or growth factors, are lacking, cells pause at a so-called *restriction point (R)* located between G_1 and S. Once a cell has moved beyond R, the cell is committed to entering S, G_2, and M. Other restriction points between G_2 and M are important for embryonic cells, but less so in postembryonic cells.

Nondividing cells, such as cultured cells deprived of serum, enter a modified G_1 state known as *quiescent state* or G_0 *state.* In this state, a cell does not synthesize the proteins needed for DNA replication and mitosis. If cultured cells are supplied with fresh serum, they can pass the restriction point in G_0 and shift back into G_1. In multicellular organisms, many cells enter the G_0 state as part of their terminal differentiation, and most of these cells never divide again. The overall growth of an adult tissue by hyperplasia depends on how many of its cells are "parked" in G_0, how many are cycling, and which proportion of them die per unit time.

Whether a cycling cell passes the R point before S phase, and whether a cell in G_0 moves back into G_1, depends on the combination of growth factors available (Pardee, 1989). The molecular mechanisms underlying this control are only partially understood, but a tentative scheme is shown in Figure 29.15. The first step in the regulatory cascade is the binding of one or more growth factors to their matching receptors, which are transmembrane proteins. Binding of a ligand to the extracellular receptor domain activates the cytoplasmic receptor domain or other membrane-associated proteins. Some of these proteins are *kinases,* enzymes that phosphorylate specific amino acid residues in cytoplasmic proteins including *SIF-A* and *P91*. When phosphorylated, these proteins forge a direct pathway to the nucleus, where they interact with other proteins in gene regulation. Another receptor-activated protein, known as the *ras* protein, activates a cascade of mitogen-activated protein kinases (MAPKs), each of which phosphorylates and thereby activates the kinase downstream. The MAPK cascade activates transcription factors, which in turn regulate genes encoding proteins thought to be necessary for a cell to overcome certain R points in the cell cycle. The nature of these cell cycle

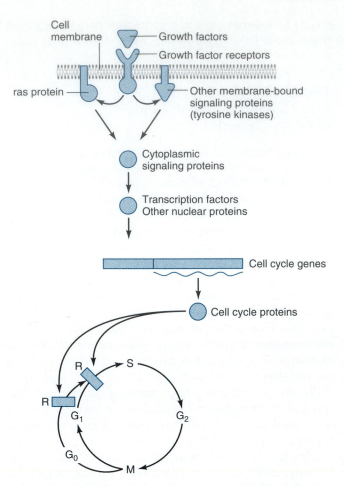

Figure 29.15 Cell cycle control by growth factors. In this simplified scheme, the colored components are those that transfer the signal for the cell to divide (the mitogenic signal). Growth factors bind to receptor proteins located in the cell membrane. The bound receptors activate membrane-associated signal proteins, such as the ras protein or tyrosine kinases, which in turn activate signaling proteins located in the cytoplasm. The cytoplasmic proteins then transmit the mitogenic signal to transcription factors and other nuclear proteins, which regulate the expression of cell cycle genes. The products of these genes control the progress of the cell past restriction points (R) in the cell cycle—in particular, from G_0 to G_1 phase and from G_1 to S phase.

proteins and the R points that they help to overcome is poorly understood. One possible scenario is that an R-point barrier is set up by an inactive protein that must be phosphorylated before the cell can pass the R point. The activating cell cycle protein then would have to be another protein kinase.

Fibroblasts and most other cultured cells cycle most actively when they are attached to a solid surface, such as the bottom of a culture dish or some extracellular matrix in vivo. Cells cultured in suspension become rounded and hardly ever divide, revealing a poorly understood aspect of cell cycle control known as *anchorage dependence.* This phenomenon lends further support to the concept of morphoregulatory cycles (see Figs. 25.22

and 26.23), which implies that cell adhesion may affect gene expression.

Following the general strategy of studying the abnormal to understand the normal, cell cycle researchers are making more and more use of tumors that arise from genetic defects. Analysis of the affected genes has greatly improved our understanding of both tumorigenesis and normal growth control.

Tumor-Related Genes

Cells that escape from their normal growth controls develop into **tumors** (Lat. *tumor*, "a swelling"), or **neoplasms** (Gk. *neo-*, "new," "recent"; *plasma*, "a thing formed"). Tumors may be *benign* (Lat., "mild") or *malignant* (Lat., "injurious"). Malignant tumors are also called **cancers**. Besides growing unrestrained, cancers also invade other tissues and give off cells that spread via the circulatory system and start additional cancers elsewhere. This chapter will focus on the uncontrolled growth aspect of tumors.

Tumors can originate because of *changes in gene expression* or through alterations in the genomic DNA itself, that is, by *mutation*. This discussion will be limited to tumors originating by mutation, because they have lent themselves better to analysis.

There are two classes of normal eukaryotic genes that can mutate or become translocated so that they promote tumor formation. Genes of the two classes are distinguished by the nature of their tumor-causing mutant alleles (Bishop, 1991; T. Hunter, 1991). Genes of the first class, known as **proto-oncogenes,** give rise to dominant *gain-of-function* alleles that cause deregulated growth. Genes of the second class, called **tumor suppressor genes,** have tumor-causing *loss-of-function* alleles, which are typically recessive. Thus, both copies of a tumor suppressor gene in a diploid cell must be mutated for the tumorous phenotype to appear. The examination of tumors often reveals the loss or inactivation of both copies of a tumor suppressor gene and changes in one or more proto-oncogenes.

The term "proto-oncogene" is derived from the term **oncogene** (Gk. *oncos*, "mass") and reflects the circuitous way in which proto-oncogenes were discovered. Oncogenes are defined operationally as genes that can transform cultured cells to anchorage independence and uncontrolled growth (Fig. 29.16). The first known oncogenes were transmitted by viruses, and it took an ingenious analysis to show that these oncogenes are not intrinsic to the genomes of their viral carriers. Instead, they are eukaryotic DNA segments that hitched a ride in viral DNAs and became mutated or deregulated in the process. Other oncogenes that were identified later are not virus-borne but are simply mutant alleles of normal cellular genes involved in growth control. To

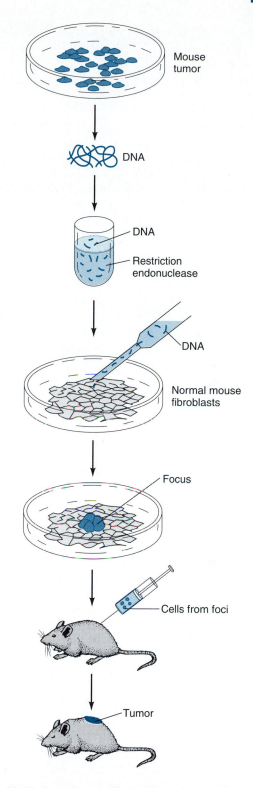

Figure 29.16 Genetic transformation of normal mouse fibroblasts into tumor cells. DNA from mouse tumor cells is extracted and cut into large fragments with a restriction endonuclease. These DNA fragments are added to nontumorous mouse fibroblasts, and the cells are subjected to a treatment that promotes the uptake of exogenous DNA. As a result, some of the recipient cells grow into small tumors called foci (sing., focus). Cells from such foci are shown in Figure 29.14b. When transformed cells are transplanted to mice, they continue to grow and may kill the host.

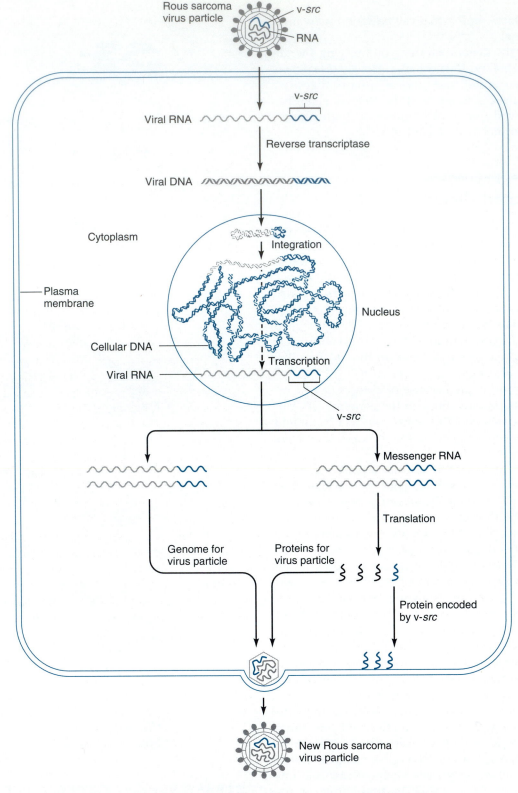

Figure 29.17 Life cycle of the Rous sarcoma virus, which belongs to the class of retroviruses. When a retrovirus infects a eukaryotic host cell, an enzyme known as reverse transcriptase is translated from the single-stranded RNA that serves as the viral genome. The enzyme reverse-transcribes the viral genome into DNA. The DNA is integrated into the host's genomic DNA, so that the viral genes are transcribed along with the active host genes. Some of the viral RNA transcripts provide the genomes for a new generation of viruses, while others are translated into viral proteins in the host cell cytoplasm. The viral RNA and proteins are then assembled into new virus particles that are eventually released from the host cell. The Rous sarcoma virus contains, in addition to its own genes, the v-*src* gene, which is derived from a eukaryotic gene that encodes a protein kinase, that is, an enzyme phosphorylating other proteins. Deregulated activity of the v-*src* gene transforms the host cell into a cancer cell.

emphasize the allelic relationship between oncogenes (virus-borne or not) and their unaltered cellular counterparts, the latter are called proto-oncogenes.

Oncogenes Are Deregulated or Mutated Proto-oncogenes

The existence of tumor viruses was first suspected at the start of the twentieth century. A critical discovery was made in 1910 by Francis Peyton Rous, who showed that a cell-free extract from a chicken sarcoma (tumor of connective tissue), when injected into healthy chickens, caused them to develop sarcomas. However, his work was not well received, and because of his peers' disapproval, Rous abandoned this line of work and took up another project. Decades later, when other researchers established the existence of tumor viruses beyond doubt, Rous was vindicated. In 1966, at the age of 85, he was awarded the Nobel Prize. The virus he discovered is now known as the *Rous sarcoma virus (Rous SV)*.

The Rous SV belongs to the class of **retroviruses,** which are the only tumor viruses with an RNA genome. Their name derives from a unique feature of their life cycle: their genomic RNA must be reverse-transcribed into DNA in order for these viruses to propagate (Varmus, 1982). This unusual step is accomplished by an enzyme, *reverse transcriptase*, which synthesizes single-stranded DNA on a single-stranded RNA template. Thereafter, the RNA template is replaced with a second DNA strand, and the double-stranded DNA is inserted into the genomic DNA of the host cell. The life cycle of Rous SV, which is a typical retrovirus, is diagrammed in Figure 29.17. If the virus carries only its own genes, the host cell may survive the virus infection, provided that the insertion of the viral DNA into the host genome does not interrupt a critical host gene. However, the retroviral DNA may also contain a wayward segment of the host's genomic DNA or a cDNA that has been reverse-transcribed from an mRNA in the host cytoplasm. In either case, the host gene that has become part of the virus is now transcribed at the same level as the viral genes, and the resulting overexpression may be deleterious to the host. In particular, overexpression may derange the host cell's cycle control, in which case the viral copy of the host gene would be called an oncogene.

The oncogene of the Rous SV was the first oncogene to yield to experimental analysis (Bishop, 1982, 1989). Investigators cut the Rous sarcoma virus DNA into restriction fragments and tested each fragment for the ability to transform normal fibroblasts into cancer cells. This method revealed a gene at the end of the Rous sarcoma virus DNA, which was named the viral *src* gene, or v-*src*. The gene encodes a protein with a molecular weight of 60,000 daltons, designated pp60v-*src*. Further analysis showed that pp60v-*src* is a *tyrosine kinase*, an enzyme that adds phosphate groups specifically to tyrosine residues of proteins. Additional studies have shown that many oncogenes encode tyrosine kinases.

The greatest reward from the work on the v-*src* gene came with the identification of its eukaryotic homologue, the *src*$^+$ proto-oncogene, or cellular *src*$^+$ gene, abbreviated c-*src*$^+$.

According to a hypothesis originally proposed by Robert Huebner and George Todaro (1969), oncogenes are part of the normal genome of eukaryotes, and they are harmless unless they are overexpressed or mutated by a carcinogenic agent such as a virus. To test this hypothesis, a research team consisting of Dominique Stehelin, Harold Varmus, J. Michael Bishop, and Peter Vogt (1976) set out to explore whether the genome of healthy chickens contained a gene that was similar to v-*src*. They reasoned that if normal chicken DNA contained a homologue of v-*src*, then single-stranded v-*src* DNA should hybridize to single-stranded genomic DNA from a normal chicken. To test this prediction, they reverse-transcribed Rous sarcoma virus RNA in the presence of radiolabeled deoxyribonucleotides and isolated the DNA segment transcribed from the v-*src* gene (Fig. 29.18). When they mixed this labeled v-*src* segment with single-stranded DNA from healthy chickens, they found that the v-*src* segment hybridized to a complementary segment of unlabeled chicken DNA. Additional experiments showed that DNA from other birds, fishes, and mammals—including humans—all contained c-*src*$^+$ homologues that hybridize to the v-*src* probe.

Closer inspection, however, revealed a major difference between the v-*src* gene and its c-*src*$^+$ homologues: like most eukaryotic genes, the c-*src*$^+$ genes contain several *introns* (Fig. 29.19). Apart from their introns, the c-*src*$^+$ variants found in fishes, birds, and mammals are all closely similar to v-*src* and to one another.

The seminal study of Stehelin and her colleagues (1976) showed that the v-*src* gene isolated from the Rous SV has close homologues in the normal genomes of all vertebrates. It appears that the Rous SV at some point integrated a c-*src*$^+$ gene copy from a host cell. This could have been an actual gene from which all introns were subsequently lost. More likely, the eukaryotic DNA that hitched a ride in the virus was a cDNA transcript made from host mRNA by the viral reverse transcriptase. Since this research was done, similar investigations have revealed cellular proto-oncogenes for almost all retroviral oncogenes.

Establishing the basic relationship between oncogenes and proto-oncogenes was a significant step to a better understanding of both cancer and normal growth control. The transformation of cells from normal to tumorous growth provides a straightforward bioassay for

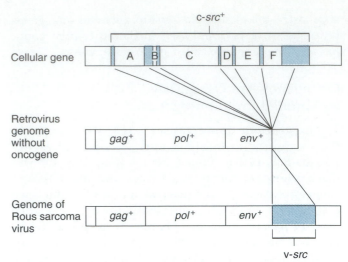

Figure 29.19 Comparison of the cellular *src*⁺ proto-oncogene (*c-src*⁺, shown at the top) with the viral *src* oncogene (*v-src*, shown at the bottom). The *c-src*⁺ gene is split by several introns (A to F), whereas its *v-src* homologue is not. The complete Rous sarcoma virus contains *v-src* and three genes encoding proteins for the viral capsid (*gag*⁺), reverse transcriptase (*pol*⁺), and the spikes of the viral envelope (*env*⁺). The mutant Rous sarcoma virus strain shown in the middle does not contain the *v-src* gene.

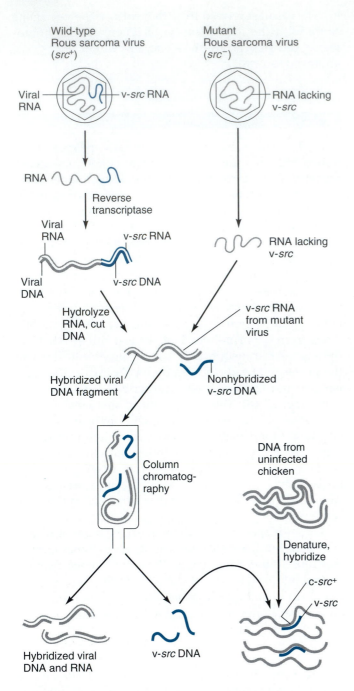

Figure 29.18 Demonstration that the *v-src* gene from the Rous sarcoma virus (Rous SV) has a homologue in the genomic DNA of healthy chickens. RNA from Rous SV carrying the *v-src* gene was reverse-transcribed into single-stranded, labeled DNA. After hydrolysis of the RNA template and fragmentation of the DNA transcript, the labeled DNA fragments were hybridized with RNA extracted from a mutant Rous SV that lacked the *v-src* gene. Most of the DNA hybridized with the deletion-mutant RNA, but the *v-src* DNA fragments found no RNA partners. The hybrids were separated by column chromatography from the single-stranded DNA, which now represented only the *v-src* gene. This probe was hybridized with denatured genomic DNA from a normal chicken. The probe hybridized, revealing that DNA from normal chicken cells contained an *src*⁺ proto-oncogene, or cellular *src*⁺ gene (*c-src*⁺).

the isolation of oncogenes, which in turn can be used as probes to isolate the corresponding proto-oncogenes. This strategy has revealed many factors in the network of growth control that were previously unknown. For their pioneering work in this area, Michael Bishop and Harold Varmus received the Nobel Prize in physiology or medicine in 1989.

The great similarity among the *c-src*⁺ genes from different classes of vertebrates indicates that the gene carries out important functions. The proteins encoded by *c-src*⁺ and *v-src* are very similar, but the exact functions of pp60c-*src* are not yet understood. The target proteins that are phosphorylated by pp60c-*src* share large domains that are bound by pp60c-*src* (Waksman et al., 1992). These recognition domains help to identify target proteins of *c-src*⁺, but the functions of these target proteins remain to be elucidated.

Oncogenes Arise from Proto-oncogenes through Various Genetic Events

Viruses may cause tumors through the overexpression of an oncogene that has become part of the viral genome but encodes a normal or nearly normal gene product. In fact, some viruses turn proto-oncogenes into oncogenes by simply inserting a strong viral promoter next to the proto-oncogene, thus enhancing the proto-oncogene's activity. However, viruses are not the only tumorigenic agents. It has been known for some time that certain types of chemicals and radiation are both mutagenic and carcinogenic. Thus, it was not surprising to find that oncogenes can be derived from the homologous proto-oncogenes through a variety of genetic events.

In some cases, a single point mutation is enough to turn a useful proto-oncogene into a deadly oncogene. For example, the oncogene that causes human bladder carcinoma is derived from its normal proto-oncogene by one base substitution that changes a GGC codon, which codes for glycine, into a GTC codon, which codes for valine (Reddy et al., 1982; Weinberg, 1983).

Larger chromosomal rearrangements can also generate oncogenes from proto-oncogenes. For instance, *Burkitt's lymphoma*, a tumor of human B lymphocytes, has been found in association with three independent chromosomal translocations (Fig. 29.20). In each case, the c-*myc*$^+$ proto-oncogene, which encodes a nuclear cell cycle regulator protein, is translocated next to an immunoglobulin gene (Croce et al., 1983; Croce, 1987; Dalla-Favera et al., 1982; Erikson et al., 1983). The immunoglobulin genes are transcribed most actively in B lymphocytes, as are the inserted c-*myc*$^+$ genes. The forced expression of the cell cycle protein then makes the cells proliferate at a higher-than-normal rate.

Finally, proto-oncogenes can become oncogenes by selective amplification. As a rule, genomic DNA is replicated evenly, without selective omissions or extra copies. However, since this rule is easily broken under selection pressure for multiple gene copies, the molecular mechanisms for selective replication of certain DNA segments must be present in eukaryotic cells but not normally used (see Chapter 7). Some proto-oncogenes have a propensity for selective amplification. For instance, the c-*myc*$^+$ gene is selectively amplified in several independent tumors.

Proto-oncogenes Encode Growth Factors, Growth Factor Receptors, Signal Proteins, and Transcription Factors

The isolation and analysis of proto-oncogenes has confirmed and extended the general scheme of cell cycle control that had previously been based on the behavior of cultured fibroblasts (Fig. 29.15). The proteins encoded by proto-oncogenes generate the *mitogenic signals* prompting cells to divide. Mitogenic proteins are classified as growth factors, growth factor receptors, membrane-bound signaling proteins, cytoplasmic signaling proteins, and nuclear proteins, including transcription factors (Table 29.2).

Growth Factors Several proto-oncogenes encode secreted proteins that are known growth factors or are likely to be growth factors (Cross and Dexter, 1991). They include the *sis*$^+$ gene encoding the B chain of platelet-derived growth factor (PDGF) and the *FGF-5*$^+$ gene encoding a protein related to fibroblast growth

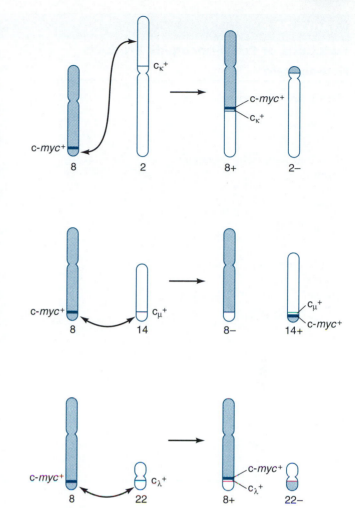

Figure 29.20 Chromosomal translocations giving rise to Burkitt's lymphoma. The diagram depicts human chromosomes; only one of each homologous pair is shown. The c-*myc*$^+$ gene, normally located on human chromosome 8, promotes DNA replication and mitosis. The other chromosomes each carry an immunoglobulin gene, designated c_κ^+, c_λ^+, or c_μ^+; these genes are strongly expressed in lymphocytes. If c-*myc*$^+$ is translocated near the promoter of an immunoglobulin gene, c-*myc*$^+$ is expressed more strongly than from its own promoter, and tumorous growth results. Three different translocations of this type have been found between chromosome 8 and chromosome 2, 14, or 22.

factor (FGF). Proto-oncogene products that are likely growth factors include int-2, which is similar to fibroblast growth factor. The *int-2*$^+$ gene causes mouse mammary tumors when it is hyperactivated, and thus turned into an oncogene, by the nearby insertion of a virus.

Growth Factor Receptors Other proto-oncogenes encode growth factor receptors (T. Hunter, 1991). Their oncogene homologues encode defective receptors that constantly give off the mitogenic signal produced by a normal receptor when bound by a ligand. For example,

TABLE 29.2

Functions of Proto-oncogene Products

Proto-oncogene	Product
Class 1: growth factors	
sis^+	PDGF B-chain growth factor
$int\text{-}2^+$	FGF-related growth factor
hst^+ (KS3)	FGF-related growth factor
Class 2: growth factor receptors	
$erbB^+$	EGF receptor protein—tyrosine kinase
neu^+	Receptorlike protein—tyrosine kinase
fms^+	CSF-1 receptor protein—tyrosine kinase
met^+	HGF receptor protein—tyrosine kinase
trk^+	NGF receptor
mas^+	Angiotensin receptor, not a protein kinase
Class 3: membrane-bound signal proteins	
$H\text{-}ras^+$	Membrane-associated GTP-binding/GTPase
$K\text{-}ras^+$	Membrane-associated GTP-binding/GTPase
src^+	Nonreceptor protein—tyrosine kinase
yes^+	Nonreceptor protein—tyrosine kinase
fgr^+	Nonreceptor protein—tyrosine kinase
lck^+	Nonreceptor protein—tyrosine kinase
Class 4: cytoplasmic signal proteins	
MEK^+	Mitogen-activated protein kinase
$pim\text{-}1^+$	Cytoplasmic protein—serine kinase
mos^+	Cytoplasmic protein—serine kinase (cytostatic factor)
cot^+	Cytoplasmic protein—serine kinase
Class 5: transcription factors and other nuclear proteins	
myc^+	Sequence-specific DNA-binding protein
myb^+	Sequence-specific DNA-binding protein
fos^+	Combines with c-jun product to form AP-1 transcription factor
jun^+	Sequence-specific DNA-binding protein; part of AP-1
$erbA^+$	Thyroxine (T_3) receptor
rel^+	NF-κB-related protein
ets^+	Sequence-specific DNA-binding protein

Note: The table is incomplete. Some of these proto-oncogenes were originally detected as retroviral oncogenes or tumor oncogenes. Others were identified at the boundaries of chromosomal translocations and at sites of retroviral insertions in tumors, or they were found in the form of amplified genes in tumors. HGF = hepatocyte growth factor.
Source: After T. Hunter (1991). Used with permission. © Cell Press.

the normal receptor for epidermal growth factor (EGF) has low tyrosine kinase activity when no ligand is bound, and much more tyrosine kinase activity when EGF is bound. The viral oncogene v-*erbB* encodes a truncated version of the EGF receptor that seems to have a constitutively high tyrosine kinase activity, which stimulates the cell to divide continuously. Another proto-oncogene that encodes a tyrosine kinase receptor is trk^+. The *trk* oncogene was discovered in a human colon cancer, but expression studies in mice showed that transcription of the trk^+ proto-oncogene is limited to cranial sensory ganglia and dorsal root ganglia (Martin-Zanca et al., 1990). These ganglia respond to nerve growth factor (NGF), and indeed, binding studies with labeled NGF revealed that the trk protein acts as an NGF receptor (Kaplan et al., 1991).

Membrane-Bound Signal Proteins A large number of proto-oncogenes encode proteins that transmit mitogenic signals from growth factor receptors to the nucleus (Table 29.2; T. Hunter, 1991; Cantley et al., 1991). Many of these signal proteins are membrane-bound tyrosine kinases, including the pp60c-*src* protein already discussed. Another membrane-associated signal protein is encoded by the c-*ras*⁺ proto-oncogene, the homologue of the v-*ras* oncogene of two sarcoma viruses. The c-ras protein receives mitogenic signals from activated growth factor receptors through other proteins (Egan et al., 1993). In its active, that is, signal-transmitting form, c-ras protein is associated with guanosine triphosphate (GTP). This active form of c-ras protein is normally short-lived, because the c-ras protein has a GTPase activity that hydrolyzes GTP to GDP (McCormick, 1989). However, the protein encoded by the v-*ras* oncogene is defective in its GTPase activity, so that the v-ras protein passes on a mitogenic signal continuously.

Cytoplasmic Signal Proteins The membrane-bound signaling proteins pass the mitogenic signal on to cytoplasmic signaling proteins. These include a cascade of mitogen-activated protein kinases (MAPKs), each phosphorylating and thereby activating the protein downstream. Activated ras protein starts this cascade through another kinase, known as the Raf-1 protein. Oncogenic forms of Raf-1 lack certain N-terminal sequences and are permanently activated. The last signal protein in the MAP kinase series activates a transcription factor or another nuclear protein.

Transcription Factors Several proto-oncogene products are components of transcription factors (Lewin, 1991). For instance, the peptides encoded by the c-*fos*[+] and c-*jun*[+] genes both have *leucine zippers* and form dimers that act as transcription factors. The peptides can form both c-jun homodimers and c-fos/c-jun heterodimers. The heterodimer, also known as AP-1 transcription factor, is more stable. Both genes were discovered as viral oncogenes. The v-jun protein differs from c-jun by a deletion of 27 amino acids near the transcriptional activation domain and three single substitutions. Similarly, v-fos is derived from c-fos by the loss of 50 amino acids near the C terminus and four substitutions.

The AP-1 transcription factor affects the expression of a wide range of target genes, many of which are not directly involved in cell cycle control. Nevertheless, the amount of c-fos protein seems to be especially critical for the regulation of bone development: overproduction of v-fos protein leads to bone cancer. The selective effect of oncogene expression on a tissue may result from a specific combination of the affected transcription factor with other factors present in that tissue; other tissues with different transcription factor combinations may be unaffected (see Chapters 15 and 21). The selective effect of mutations in proto-oncogenes on specific tissues is also observed in the tumor suppressor genes, which will be discussed next.

Tumor Suppressor Genes Limit the Frequency of Cell Divisions

Whereas oncogenes behave as dominant gain-of-function alleles, other tumor-related genes have been discovered through loss-of-function alleles, which are usually recessive. These genes are referred to as *tumor suppressor genes*. Their normal function is to limit the frequency of cell divisions (Weinberg, 1988; C. J. Marshall, 1991). In terms of the scheme drawn in Figure 29.15, tumor suppressor gene products may inhibit the tran-

scription of cell cycle genes or may be involved in setting up the restriction points that cells must pass to go from G_0 to G_1 phase or from G_1 to S phase. The existence of tumor suppressor genes is indicated by several lines of evidence, including cell hybrids and familial cancers.

Hybrid cells generated by fusing normal cells with tumor cells do not give rise to tumors unless certain chromosomes are subsequently lost from the hybrids. Thus, the loss of a chromosome or chromosome fragment can turn a nontumor cell into a tumor cell. This observation indicates that certain genes suppress the tumorigenic potential of cell hybrids and, by extension, of normal cells. By analyzing the effects of chromosome losses from hybrid cells, and through the use of many transfer techniques for single chromosomes, it has been possible to identify the chromosomal locations of some of these suppressor genes.

It has long been known that predispositions to certain types of cancers run in families. One of these cancers is *retinoblastoma*, a rare childhood cancer in which immature retina cells proliferate. There are two forms of this cancer: hereditary and nonhereditary. The hereditary form afflicts both eyes and is often associated with tumors in other organs, whereas the nonhereditary form affects only one eye. Some patients with retinoblastoma have a visibly abnormal karyotype, with a specific deletion on one copy of chromosome 13 (Fig. 29.21). These

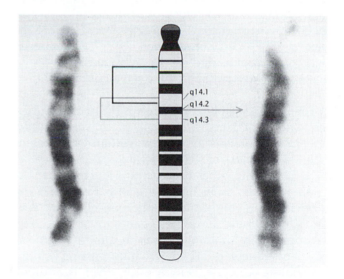

Figure 29.21 Identification of a tumor suppressor gene by means of chromosomal deletions. The photograph to the left shows a normal human chromosome 13, and the drawing in the center is an interpretative diagram. The photograph to the right shows a chromosome 13 from a human retinoblastoma; it lacks the segment that is marked by the narrow bracket in the center diagram. The arrow points to where the segment should be. A different deletion, observed in a chromosome 13 from a different retinoblastoma, is represented by the wide bracket in the diagram. The two deletions overlap in a small segment within band 13.q14.1. The retinoblastoma gene has been cloned from this segment.

observations suggest that patients with hereditary retinoblastoma have a predisposition to this and other cancers because they lack one copy of a tumor suppressor gene, designated $Rb1^+$, in all of their cells. A single mutation in the remaining intact copy of the gene in a developing retina cell should then be sufficient to start a cancer. Indeed, the normal allele, on the chromosome 13 inherited from the unaffected parent, has been shown to be missing or mutated in retinoblastomas and several other cancers.

Mice in which both copies of the $Rb1$ gene have been interrupted by *insertional mutagenesis* (see Fig. 14.12) die as embryos, presumably because of irregular mitoses of neurons or blood cells (E. Y.-H. P. Lee et al., 1992). With only one copy of the normal $Rb1^+$ gene, mice are phenotypically normal at birth but die from pituitary tumors within 10 months. Conversely, overexpression of the human $Rb1^+$ gene in transgenic mice results in dwarfism (Bignon et al., 1993). These results indicate that the normal function of tumor suppressor genes is to limit cell growth and that the loss of this function leads to tumor formation.

How does the Rb1 gene product restrict cell growth? The protein encoded by $Rb1^+$ accumulates in the nucleus. The phosphorylation of Rb1 protein changes with the cell cycle: the protein is underphosphorylated in G_0 phase and during most of G_1 phase; it becomes phosphorylated during late G_1 phase and remains in this state through S phase, G_2, and M. The late G_1 stage, when Rb1 protein becomes phosphorylated, is a restriction point in the cell cycle (Fig. 29.15). Before this point, progress of a cell through the mitotic cycle depends on mitogens and can be inhibited by antimitogens. Past the restriction point, a cell is committed to divide.

The underphosphorylated form of Rb1 is the active form, and it interacts with a transcription factor called E2F. The interaction is blocked by certain viral proteins that mimic the loss of $Rb1^+$ function (Chellappan et al., 1991). These and other observations suggest that Rb1 protein normally limits the activity of E2F, which would otherwise promote the transcription of cell cycle genes. Loss of Rb1 would allow these genes to be transcribed at inappropriate times and result in tumorous growth. Rb1 also sequesters cyclin D1, thus putting another brake on the cell cycle (Dowdy et al., 1993). In addition, Rb1 acts in combination with other transcription factors to promote cell differentiation. For instance, Rb1 bound to MyoD promotes myogenesis and inhibits cell proliferation (Gu et al., 1993). Conversely, Id-2, a helix-loop-helix protein that heterodimerizes with MyoD, inhibits myogenesis, and promotes cell proliferation, is bound or sequestered by Rb1 (Iavarone et al., 1994). These interactions provide another mechanism by which cell proliferation and cell differentiation are regulated antagonistically.

Another tumor suppressor gene, designated $p53^+$ for the molecular weight (53 kd) of the encoded protein, is similarly involved in many regulatory pathways that control cell division. The $p53^+$ gene is located on the human chromosome 17 in a region that is lost or mutated in a wide range of human cancers, including the common colon, lung, and breast cancers (A. J. Levine et al., 1991). In most of these cancers there is a loss of both $p53^+$ alleles. Introduction of the normal $p53^+$ gene into tumor cell lines with little or no endogenous p53 protein suppresses growth. Conversely, like mice without Rb1, mice deficient for p53 are born phenotypically normal but are susceptible to spontaneous tumors (Donehower et al., 1992).

In addition to preventing tumors, p53 is involved in programmed cell death, or *apoptosis* (Lowe et al., 1993).

Some of the functions of p53 protein in normal cells, and of its malfunctions in tumor cells, have been revealed. Again in parallel with Rb1, p53 exerts multiple controls on cell division by interacting with other proteins. For instance, p53 binds to and sequesters a DNA-binding protein complex, designated RPA, which is necessary for DNA replication (Dutta et al., 1993). Also, p53 binds specifically to certain DNA sequences, thus inhibiting or enhancing the transcription of genes. Thus, p53 cooperates with the TATA box–binding protein, TBP, a subunit of the general transcription factor TFIID, in binding to TATA box–containing promoters (X. Chen et al., 1993). In addition, p53 binds to an enhancer of the $WAF1^+$ gene (for *w*ild-type p53 *a*ctivated DNA *f*ragment 1) and activates its transcription (El-Deiry et al., 1993). Transfection of cultured cells from human colon, lung, and brain cancers with $WAF1$ cDNA suppresses the cells' tumorous growth. The WAF1 protein, isolated independently and named Cip1 (for *c*dk-*i*nteracting *p*rotein 1) by Harper and colleagues (1993), binds and inactivates cdk2, a cyclin-dependent kinase that is required for cell cycle progression to S phase. In a final parallel with Rb1, p53 is bound and sidetracked by certain proteins encoded by tumorigenic viruses, such as the human papilloma virus.

Tumor suppressor proteins, the downstream proteins synthesized under their influence, and any drug that mimics the actions of these proteins are of great medical interest. These agents are therefore being investigated very actively, and studies of them should greatly enhance our understanding of both tumor development and normal growth control.

Unresolved Questions about Growth

Despite great advances in our understanding of organismic growth control, major issues in this area are still unresolved. One of these issues is the relationship between cell division and cell differentiation. For many cell types studied, growth and differentiation seem to be mutually exclusive, especially in tissues that renew from *stem cells*. Tumor cells are generally undifferentiated, suggesting that oncogene activity may prevent cells from achieving or maintaining a differentiated state. Growth factors are known to inhibit *myogenesis* of cultured cells (Olson, 1992). Similarly, the phenotype of leukemic cells is usually that of an immature blood precursor cell (Sawyers et al., 1991). Some of the molecular mechanisms by which cell proliferation and cell differentiation exclude each other are known. For instance, *fibroblast growth factor*, a mitogen, inhibits *myogenin*, a transcription factor activating many muscle-specific genes, by phosphorylating its DNA-binding site (Li et al., 1992). Conversely, *MyoD*, another transcription factor promoting myogenesis, inhibits cell division by blocking a serum-response element in the enhancer region of the c-*fos*$^+$ gene (Trouche et al., 1993). Also, Rb1 and p53 suppress cell division and promote certain cell differentiations, as mentioned earlier.

In other cases, the relationship between signaling pathways that promote cell proliferation and those promoting cell differentiation seems more complex. Many growth factors are also involved in cell determination and cell differentiation. NGF promotes not only the survival of neurons but also neurite formation, which is part of the differentiation of neurons. Similarly, TGF-β in adult cells promotes cell division or cell differentiation, depending on the developmental state of the cells and the presence of other growth factors; whereas other members of the TGF-β family act as morphogens and inductive signals.

There may also be a connection between growth control and regeneration. According to models of pattern formation discussed in Chapter 20, the cells of a developing organ have different *positional values*, analogous to geometric points in a coordinate grid. If wounds or transplantations confront cells that are not normally next neighbors, the juxtaposition of disparate positional values is thought to cause *intercalary regeneration*. This process continues until enough cells with intervening positional values have been generated. Applying this model to normal growth, Vernon French, Peter Bryant, and Susan Bryant (1976) suggested that when an organ primordium is established in an embryo, neighboring cells may receive disparate positional values so that the confrontations stimulate growth (Fig. 29.22). The growth stimulus would disappear and growth would stop when enough cells with intervening positional values

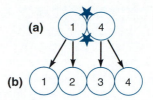

Figure 29.22 Intercalation and growth control according to a hypothesis proposed by French and colleagues (1976). **(a)** Cells in an organ primordium receive discontinuous positional values (1 and 4). The confrontation between the disparate values stimulates intercalary regeneration. **(b)** When enough cells with intervening values (2 and 3) have been generated, growth stops.

were intercalated. Observations on *Drosophila* imaginal discs suggest that their growth in larvae may indeed be controlled by intrinsic factors that are based on cell interactions (P. J. Bryant and P. Simpson, 1984).

Finally, growth control may be linked to, or may at least use the same signals as, pattern formation. This linkage has already been pointed out in our previous discussion of the TGF-β superfamily. A similar dual role is seen in a pair of homologous proteins encoded by the mouse proto-oncogene *int-1*$^+$ and the *Drosophila* patterning gene *wingless*$^+$. The *int-1*$^+$ gene was discovered through its oncogenic effect: if the mouse mammary tumor virus inserts itself near *int-1*$^+$, the proto-oncogene is overexpressed and causes tumors (Nusse et al., 1984). When investigators used *int-1*$^+$ to search for a *Drosophila* homologue, they found an astounding degree of similarity with *wingless*$^+$ (Rijsewijk et al., 1987). The *wingless*$^+$ gene is best known for its action as a *segment polarity gene* in establishing the parasegmental boundary (see Chapter 21). Genetic and expression studies indicate that the wingless protein acts as a paracrine signal on cells on the other side of the parasegmental boundary, which express *engrailed*$^+$. Because of the molecular similarity between *wingless*$^+$ and *int-1*$^+$, it is possible that the proteins encoded by these genes may have similar biological functions. Thus, int-1 may play a role in both cell cycle control and pattern formation.

The examples mentioned here suggest that growth control is more closely linked to cell differentiation, regeneration, and pattern formation than previously thought. Therefore, future research into growth control holds the promise not only of working out some unknown molecular mechanisms but also of leading to new concepts.

SUMMARY

Growth in an organism is defined as a change in mass. The most common form of growth, called hyperplasia, is cell division followed by enlargement of the daughter cells. Other forms of growth are an increase in cell mass without division, and the deposition of extracellular matrix. The change in mass per unit time relative to the mass already present is called the relative growth rate. Isometric growth occurs when all parts of an organism grow at the same relative rate. More common is allometric growth, or the growth of different regions at different rates.

Heterospecific transplantations indicate that most organisms and their organs have an intrinsic growth potential, that is, a maximum rate at or size to which they can grow under favorable circumstances in normal development. The growth potential of an organ may depend on the completeness of the organ and on interactions between its components.

Exogenous growth factors include nutrients, growth hormones, and growth factors. The single most important growth hormone in vertebrates, called somatotropin or simply growth hormone, is produced in the anterior pituitary gland under the influence of the hypothalamus. Other hormones, including somatomedins, thyroid hormone, insulin, testosterone, and estrogen, also affect overall or local growth. In addition to hormones, which by definition are produced in endocrine glands and transported widely to target cells by blood or other body fluids, there are a wide range of growth factors that are produced by many tissues and act locally as paracrine or autocrine signals. These factors have growth-related effects and other effects on multiple target cells. The action of a given growth factor may depend on the mix of other growth factors present.

Growth hormones and growth factors act on their target cells through specific receptor proteins in the plasma membrane. The binding of a growth factor receptor to its ligand starts a chain of signals that eventually promote mitosis. The cytoplasmic domains of the receptors often act as tyrosine kinases, or they activate other membrane-bound tyrosine kinases. From here, mitogenic signals are propagated through second messengers and cytoplasmic signal proteins to nuclear proteins, including transcription factors. These factors control the expression of genes encoding proteins that are required for DNA synthesis and for passing certain restriction points in the cell cycle.

The study of cell cycle control has been greatly enhanced by the isolation and characterization of tumor-related genes. Two classes of these genes are distinguished by their mutant phenotypes. One class, known as proto-oncogenes, gives rise to dominant gain-of-function alleles that cause deregulated growth. The mutant alleles, called oncogenes, are defined by their ability to transform normal cells into tumor cells. Oncogenes are derived from proto-oncogenes by mutation or other genetic mechanisms that cause either overexpression of the gene or the synthesis of an altered product. Proto-oncogenes encode all the proteins that transmit mitogenic signals. The second class of tumor-related genes, called tumor suppressor genes, is defined by loss-of-function alleles. The normal function of these genes is to limit the frequency of cell division. Deletions and point mutations in tumor suppressor genes are common in human cancers. The proteins encoded by tumor suppressor genes may also contribute to a mutual exclusion between cell division and cell differentiation.

SUGGESTED READINGS

Bishop, J. M. 1991. Molecular themes in oncogenesis. *Cell* **64**:235–248.

Kingsley, D. M. 1994. The TGF-β superfamily: New members, new receptors, and new genetic tests of function in different organisms. *Genes Dev.* **8**:133–146.

Levi-Montalcini, R. 1964. Growth control of nerve cells by a protein factor and its antiserum. *Science* **143**:105–110.

Marshall, C. J. 1991. Tumor suppressor genes. *Cell* **64**:313–326.

Abel T., Bhatt R. and Maniatis T. (1992) A *Drosophila* CREB/ATF transcriptional activator binds to both fat body- and liver-specific regulatory elements. *Genes and Dev.* **6:** 466–480.

Adams C.C. and Workman J.L. (1993) Nucleosome displacement in transcription. *Cell* **72:** 305–308.

Adams J.C. and Watt F.M. (1993) Regulation of development and differentiation by the extracellular matrix. *Development* **117:** 1183–1198.

Affara N.A. (1991) Sex and the single Y. *BioEssays* **13:** 475–478.

Akam M. (1987) The molecular basis for metameric pattern in the *Drosophila* embryo. *Development* **101:** 1–22.

Alberts B., Bray D., Lewis J., Raff M., Roberts K. and Watson J.D. (1983) *Molecular Biology of the Cell.* New York: Garland Publishing, Inc.

Alberts B., Bray D., Lewis J., Raff M., Roberts K. and Watson J.D. (1989) *Molecular Biology of the Cell.* 2nd ed. New York: Garland Publishing, Inc.

Alberts B., Bray D., Lewis J., Raff M., Roberts K. and Watson J.D. (1994) *Molecular Biology of the Cell.* 3rd ed. New York: Garland Publishing, Inc.

Albrecht-Buehler G. (1977) Daughter 3T3 cells: Are they mirror images of each other? *J. Cell Biol.* **72:** 595–603.

Ali I.U. and Hynes R.O. (1977) Effects of cytochalasin B and colchicine on attachment of a major surface protein of fibroblasts. *Biochim. Biophys. Acta* **471:** 16–24.

Allen R.D. (1987) The microtubule as an intracellular engine. *Scientific American* **256** (February): 42–49.

Allsopp T.E., Wyatt S., Paterson H.F. and Davies A.M. (1993) The proto-oncogene *bcl-2* can selectively rescue neurotrophic factor-dependent neurons from apoptosis. *Cell* **73:** 295–307.

Amaya E., Musci T.J. and Kirschner M.W. (1991) Expression of a dominant negative mutant of the FGF receptor disrupts mesoderm formation in *Xenopus* embryos. *Cell* **66:** 257–270.

Amaya E., Stein P.A., Musci T.J. and Kirschner M.W. (1993) FGF signalling in the early specification of mesoderm in *Xenopus*. *Development* **118:** 477–487.

Ambros V. and Horvitz H.R. (1984) Heterochronic mutants of the nematode *Caenorhabditis elegans*. *Science* **226:** 409–416.

Anderson C.L. and Meier S. (1981) The influence of the metameric pattern in the mesoderm on migration of cranial neural crest cells in the chick embryo. *Dev. Biol.* **85:** 383–402.

Anderson K.V. (1987) Dorsal-ventral embryonic pattern genes of *Drosophila*. *Trends in Genetics* **3:** 91–97.

Andres A.J. and Thummel C.S. (1992) Hormones, puffs and flies: The molecular control of metamorphosis by ecdysone. *Trends in Genetics* **8:** 132–138.

Anstrom J.A., Chin J.E., Leaf D.S., Parks A.L. and Raff R.A. (1987) Localization and expression of msp130, a primary mesenchyme lineage-specific cell surface protein of the sea urchin embryo. *Development* **101:** 255–265.

Arnold A.P. (1975) The effects of castration and androgen replacement on song, courtship, and aggression in zebra finches. *J. Exp. Zool.* **191:** 309–326.

Arnold A.P. (1980) Sexual differences in the brain. *Am. Sci.* **68:** 165–173.

Arnold A.P., Nottebohm F. and Pfaff D. (1976) Hormone concentrating cells in vocal control and other areas of the brain of the zebra finch (*Poephila guttata*). *J. Comp. Neurol.* **165:** 487–512.

Aroian R.V., Koga M., Mendel J.E., Ohshima Y. and Sternberg P.W. (1990) The *let-23* gene necessary for *Caenorhabditis elegans* vulval induction encodes a tyrosine kinase of the EGF receptor subfamily. *Nature* **348:** 693–699.

Asakura S., Eguchi G. and Iino T. (1966) *Salmonella* flagella: *In vitro* reconstruction and over-all shapes of flagellar filaments. *J. Mol. Biol.* **16:** 302–316.

Asashima M., Nakano H., Uchiyama H., Sugino H., Nakamura T., Eto Y., Ejima D., Nishimatsu S.-I., Ueno N. and Konoshita K. (1991) Presence of activin (erythroid differentiation factor) in unfertilized eggs and blastulae of *Xenopus laevis*. *Proc. Nat. Acad. Sci. USA* **88:** 6511–6514.

Ashburner M. (1972) Patterns of puffing activity in the salivary gland chromosomes of *Drosophila*. *Chromosoma* **38:** 255–281.

Ashburner M. (1990) Puffs, genes, and hormones revisited. *Cell* **61:** 1–3.

Ashburner M. and Gelbart W. (1991) *Drosophila* genetic maps. *Drosophila Information Service* **69** (May).

Ashburner M., Chihara C., Meltzer P. and Richards G. (1974) Temporal control of puffing activity in polytene chromosomes. *Cold Spring Harbor Symp. Quant. Biol.* **38:** 655–662.

Ault K.T., Durmowicz G., Galione A., Harger P.L. and Busa W.B. (1994) Modulation of mesoderm induction and body axis determination by the polyphosphoinositide cycle signal transduction pathway. (Submitted)

Austin C.R. (1952) The "capacitation" of the mammalian spermatozoa. *Nature* **170:** 326.

Austin C.R. (1965) *Fertilization.* Englewood Cliffs, NJ: Prentice Hall.

Austin J., Maine E.M. and Kimble J. (1989) Genetics of intercellular signalling in *C. elegans*. *Development* **107** (Supplement): 53–57.

Avery O.T., MacLeod C.M. and McCarty M. (1944) Studies on the chemical nature of the substance inducing transformation of pneumococcal types: Induction of transformation by a desoxyribonucleic acid fraction isolated from pneumococcus type III. *J. Exp. Med.* **79**: 137–158.

Baker B.S. (1989) Sex in flies: The splice of life. *Nature* **340**: 521–524.

Baker B.S. and Tata J.R. (1992) Prolactin prevents the autoinduction of thyroid hormone receptor mRNAs during amphibian metamorphosis. *Dev. Biol.* **149**: 463–467.

Balak K., Jacobson M., Sunshine J. and Rutishauser U. (1987) Neural cell adhesion molecule expression in *Xenopus* embryos. *Dev. Biol.* **119**: 540–550.

Balinsky B.I. (1975) *An Introduction to Embryology.* Philadelphia: Saunders.

Bandziulis R.J., Swanson M.S. and Dreyfuss G. (1989) RNA-binding proteins as developmental regulators. *Genes and Dev.* **3**: 431–437.

Bard J. (1990) *Morphogenesis: The Cellular and Molecular Processes of Developmental Anatomy.* Cambridge: Cambridge University Press.

Barker D.D., Wang C., Moore J., Dickinson L.K. and Lehmann R. (1992) Pumilio is essential for function but not for distribution of the *Drosophila* abdominal determinant nanos. *Genes and Dev.* **6**: 2312–2326.

Baroffio A., Dupin E. and Le Douarin N.M. (1991) Common precursors for neural and mesectodermal derivatives in the cephalic neural crest. *Development* **112**: 301–305.

Basler K., Edlund T., Jessell T.M. and Yamada T. (1993) Control of the cell pattern in the neural tube: Regulation of cell differentiation by *dorsalin-1*, a novel TGFβ family member. *Cell* **73**: 687–702.

Bastiani M.J., du Lac S. and Goodman C.S. (1986) Guidance of neuronal growth cones in the grasshopper embryo. I. Recognition of a specific axonal pathway by the pCC neuron. *J. Neurosci.* **6**: 3518–3531.

Bateson W. (1894) *Materials for the Study of Variation.* London: Macmillan.

Bautzmann H., Holtfreter J., Spemann H. and Mangold O. (1932) Versuche zur Analyse der Induktionsmittel in der Embryonalentwicklung. *Naturwissenschaften* **20**: 971–974.

Bayna E.M., Shaper J.H. and Shur B.D. (1988) Temporally specific involvement of cell surface β-1,4 galactosyltransferase during mouse embryo morula compaction. *Cell* **53**: 145–157.

Beachy P.A., Krasnow M.A., Gavis E.R. and Hogness D.S. (1988) An *Ultrabithorax* protein binds sequences near its own and the *Antennapedia* P1 promoters. *Cell* **55**: 1069–1081.

Beadle G.W. and Tatum E.L. (1941) Genetic control of biochemical reactions in *Neurospora*. *Proc. Nat. Acad. Sci. USA* **27**: 499–506.

Beams H.W. and Kessel R.G. (1974) The problem of germ cell determinants. *Int. Rev. Cytol.* **39**: 413–479.

Beams H.W. and Kessel R.G. (1976) Cytokinesis: A comparative study of cytoplasmic division in animal cells. *Am. Sci.* **64**: 279–290.

Beatty R.A. (1967) Parthenogenesis in vertebrates. In: *Fertilization*, C.B. Metz and A. Monroy (eds.), vol. 1, 413–440. New York: Academic Press.

Becker A.J., McCulloch E.A. and Till J.E. (1963) Cytological demonstration of the clonal nature of spleen cells derived from transplanted mouse marrow cells. *Nature* **197**: 452–454.

Beddington R.S.P. (1994) Induction of a second neural axis by the mouse node. *Development* **120**: 613–620.

Beeman R.W., Stuart J.J., Brown S.J. and Denell R.E. (1993) Structure and function of the homeotic gene complex (HOM-C) in the beetle, *Tribolium castaneum*. *BioEssays* **15**: 439–444.

Beermann W. (1952a) Chromosomes and genes. In: *Developmental Studies on Giant Chromosomes*, W. Beermann (ed.), 1–33. New York: Springer-Verlag.

Beermann W. (1952b) Chromomerenkonstanz und spezifische Modifikationen der Chromosomenstruktur in der Entwicklung und Organdifferenzierung von *Chironomus tentans*. *Chromosoma* **5**: 139–198.

Beisson J. and Sonneborn T.M. (1965) Cytoplasmic inheritance of the organization of the cell cortex in *Paramecium aurelia*. *Proc. Nat. Acad. Sci. USA* **53**: 275–282.

Beitel G.J., Clark S.G. and Horvitz H.R. (1990) *Caenorhabditis elegans ras* gene *let-60* acts as a switch in the pathway of vulval induction. *Nature* **348**: 503–509.

Bejsovec A. and Wieschaus E. (1993) Segment polarity gene interactions modulate epidermal patterning in *Drosophila* embryos. *Development* **119**: 501–517.

Bell L.R., Main E.M., Schedl P. and Cline T.W. (1988) *Sex-lethal*, a Drosophila sex determination switch gene, exhibits sex-specific RNA splicing and sequence similarity to RNA binding proteins. *Cell* **55**: 1037–1046.

Bell L.R., Horabin J.I., Schedl P. and Cline T.W. (1991) Positive autoregulation of *Sex-lethal* by alternative splicing maintains the female determined state in *Drosophila*. *Cell* **65**: 229–239.

Bender M., Talbot W. and Hogness D. (1993) A family of ecdysone receptor proteins controls distinct tissue-specific responses to ecdysone. Abstract, 34th Annual Drosophila Research Conference, San Diego, CA.

Benezra R., Davis R.L., Lockshon D., Turner D.L. and Weintraub H. (1990) The protein Id: A negative regulator of helix-loop-helix DNA binding proteins. *Cell* **61**: 49–59.

Bennett D. (1975) The T-locus of the mouse. *Cell* **6**: 441–454.

Bentley J.K., Shimomura H. and Garbers D.L. (1986) Retention of a functional resact receptor in isolated sperm plasma membranes. *Cell* **45**: 281–288.

Benya P.D. and Shaffer J.D. (1982) Dedifferentiated chondrocytes reexpress the differentiated collagen phenotype when cultured in agarose gels. *Cell* **30**: 215–224.

Berleth T. and Jürgens G. (1993) The role of the *monopteros* gene in organising the basal body region of the *Arabidopsis* embryo. *Development* **118**: 575–587.

Berleth T., Burri M., Thoma G., Bopp D., Richstein S., Frigerio G., Noll M. and Nüsslein-Volhard C. (1988) The role of localization of *bicoid* RNA in organizing the anterior pattern of the *Drosophila* embryo. *EMBO J.* **7**: 1749–1756.

Bermingham J.R., Jr., Martinez-Arias A., Petitt M.G. and Scott M.P. (1990) Different patterns of transcription from the two *Antennapedia* promoters during *Drosophila* embryogenesis. *Development* **109**: 553–566.

Bernfield M. and Sanderson R.D. (1990) Syndecan, a developmentally regulated cell surface proteoglycan that binds extracellular matrix and growth factors. *Phil. Trans. R. Soc. London B* **327**: 171–186.

Berridge M.J., Downes C.P. and Hanley M.R. (1989) Neural and developmental actions of lithium: A unifying hypothesis. *Cell* **59**: 411–419.

Berta P., Hawkins J.R., Sinclair A.H., Taylor A., Griffiths B.L., Goodfellow P.N. and Fellous M. (1990) Genetic evidence equating *SRY* and the testis-determining factor. *Nature* **348**: 448–449.

Beyer A.L. and Osheim Y.N. (1990) Ultrastructural analysis of the ribonucleoprotein substrate for pre-mRNA processing. In: *The Eukaryotic Nucleus: Molecular Biochemistry and Macromolecular Assemblies,* vol. 2, P.R. Strauss and S.H. Wilson (eds.), 431–445. Caldwell, NJ: Telford Press.

Bienz M. and Tremml G. (1988) Domain of *Ultrabithorax* expression in *Drosophila* visceral mesoderm from autoregulation and exclusion. *Nature* **333**: 576–578.

Bier K.H. (1963) Autoradiographische Untersuchungen über die Leistungen des Follikelepithels und der Nährzellen bei der Dotterbildung und Eiweissynthese im Fliegenovar. *Wilhelm Roux's Arch.* **154**: 552–575.

Bier K.H. (1964) Die Kern-Plasma-Relation und das Riesenwachstum der Eizellen. *Zool. Anz.* (Supplement 27): 84–91.

Bignon Y.-J., Chen Y., Chang C.-Y., Riley D.J., Windle J.J., Mellon P.L. and Lee W.-H. (1993) Expression of a retinoblastoma transgene results in dwarf mice. *Genes and Dev.* **7**: 1654–1662.

Birgbauer E. and Fraser S.E. (1994) Violation of cell lineage restriction compartments in the chick hindbrain. *Development* **120**: 1347–1356.

Bishop J.M. (1982) Oncogenes. *Scientific American* **246** (March): 81–92.

Bishop J.M. (1989) Viruses, genes and cancer. *Am. Zool.* **29**: 653–666.

Bishop J.M. (1991) Molecular themes in oncogenesis. *Cell* **64**: 235–248.

Blair S.S. (1992) Engrailed expression in the anterior lineage compartment of the developing wing blade of *Drosophila*. *Development* **115**: 21–33.

Blair S.S. (1993) Mechanisms of compartment formation: Evidence that non-proliferating cells do not play a critical role in defining the D/V lineage restriction in the developing wing of *Drosophila*. *Development* **119**: 339–351.

Blair S.S., Brower D.L., Thomas J.B. and Zavortink M. (1994) The role of *apterous* in the control of dorsoventral compartmentalization in PS integrin gene expression in the developing wing of *Drosophila*. *Development* **120**: 1805–1815.

Bleil J.D. and Wassarman P.M. (1980) Mammalian sperm and egg interaction: Identification of a glycoprotein in mouse-egg zonae pellucidae possessing receptor activity for sperm. *Cell* **20**: 873–882.

Bleil J.D. and Wassarman P.M. (1986) Autoradiographic visualization of the mouse egg's sperm receptor bound to sperm. *J. Cell Biol.* **102**: 1363–1371.

Bleil J.D. and Wassarman P.M. (1988) Galactose at the nonreducing terminus of O-linked oligosaccharides of mouse egg zona pellucida glycoprotein ZP3 is essential for the glycoprotein's sperm receptor activity. *Proc. Nat. Acad. Sci. USA* **85**: 6778–6782.

Bleil J.D., Greve J.M. and Wassarman P.M. (1988) Identification of a secondary sperm receptor in the mouse egg zona pellucida: Role in maintenance of binding of acrosome-reacted sperm to eggs. *Dev. Biol.* **128**: 376–385.

Blobel C.P., Wolfsberg T.G., Turck C.W., Myles D.G., Primakoff P. and White J.M. (1992) A potential fusion peptide and an integrin ligand domain in a protein active in sperm-egg fusion. *Nature* **356**: 248–252.

Blondeau J.-P. and Baulieu E.E. (1984) Progesterone receptor characterized by photoaffinity labeling in the plasma membrane of *Xenopus laevis* oocytes. *Biochem. J.* **219**: 785–792.

Bloom W. and Fawcett D.W. (1975) *A Textbook of Histology.* 10th ed. Philadelphia: Saunders.

Blum M., Gaunt S.J., Cho K.W.Y., Steinbeisser H., Blumberg B., Bittner D. and De Robertis E.M. (1992) Gastrulation in the mouse: The role of the homeobox gene *goosecoid*. *Cell* **69**: 1097–1106.

Blumberg B., Wright C.V.E., De Robertis E.M. and Cho K.W.Y. (1991) Organizer-specific homeobox genes in *Xenopus laevis* embryos. *Science* **253**: 194–196.

Bode H.R., Flick K.M. and Smith G.S. (1976) Regulation of interstitial cell differentiation in *Hydra attenuata*. I. Homeostatic control of interstitial cell population size. *J. Cell Sci.* **20**: 29–46.

Bode H.R., Gee L.W. and Chow M.A. (1990) Neuron differentiation in Hydra involves dividing intermediates. *Dev. Biol.* **139**: 231–243.

Bodemer C.W. (1968) *Modern Embryology.* New York: Holt, Rinehart and Winston.

Bogan J.S. and Page D.C. (1994) Ovary? Testis? A mammalian dilemma. *Cell* **76**: 603–607.

Boggs R.T., Gregor P., Idriss S., Belote J.M. and McKeown M. (1987) Regulation of sexual differentiation in *D. melanogaster* via alternative splicing of RNA from the *transformer* gene. *Cell* **50**: 739–747.

Bohn H. (1976) Regeneration of proximal tissues from a more distal amputation level in the insect leg (*Blaberus craniifer, Blattaria*). *Dev. Biol.* **53**: 285–293.

Bottjer S.W. (1987) Ontogenetic changes in the pattern of androgen accumulation in song-control nuclei of male zebra finches. *J. Neurobiol.* **18**: 125–139.

Bottjer S.W. (1991) Neural and hormonal substrates for song learning in zebra finches. *The Neurosciences* **3**: 481–488.

Boucaut J.-C., Darribère T., Boulekbache H. and Thiery J.P. (1984) Prevention of gastrulation but not neurulation by antibodies to fibronectin in amphibian embryos. *Nature* (London) **307**: 364–367.

Boucaut J.-C., Darribère T., Shi D.L., Boulekbache H., Yamada K.M. and Thiery J.P. (1985) Evidence for the role of fibronectin in amphibian gastrulation. *J. Embryol. exp. Morphol.* **89** (Supplement): 211–227.

Boveri T. (1887) Ueber die Differenzierung der Zellkerne während der Furchung des Eies von *Ascaris megalocephala. Anat. Anz.* **2**: 688–693.

Boveri T. (1902) On multipolar mitosis as a means of analysis of the cell nucleus. In: *Foundations of Experimental Embryology*, B.H. Willier and J.M. Oppenheimer (eds.), 74–97. New York: Hafner.

Boveri T. (1907) Zellenstudien VI. Die Entwicklung dispermer Seeigeleier. Ein Beiträg zur Befruchtungslehre und zur Theorie des Kernes. *Jenaer Zeitsch. Naturwiss.* **43**: 1–292.

Bowen-Pope D.F., DiCorleto P.E. and Ross R.J. (1983) Interactions between the receptors for platelet-derived growth factor and epidermal growth factor. *J. Cell Biol.* **96**: 679–683.

Bowerman B., Eaton B.A. and Priess J.R. (1992) *skn-1*, a maternally expressed gene required to specify the fate of ventral blastomeres in the early *C. elegans* embryo. *Cell* **68**: 1061–1075.

Bowerman B., Draper B.W., Mello C.C. and Priess J.R. (1993) The maternal gene *skn-1* encodes a protein that is distributed unequally in early *C. elegans* embryos. *Cell* **74**: 443–452.

Bowman J.L., Smyth D.R. and Meyerowitz E.M. (1989) Genes directing flower development in *Arabidopsis. Plant Cell* **1**: 37–52.

Bowman J.L., Sakai H., Jack T., Weigel D., Mayer U. and Meyerowitz E.M. (1992) *SUPERMAN*, a regulator of floral homeotic genes in *Arabidopsis. Development* **114**: 599–615.

Bowman J.L., Alvarez J., Weigel D., Meyerowitz E.M. and Smyth D.R. (1993) Control of flower development in *Arabidopsis thaliana* by *APETALA1* and interacting genes. *Development* **119**: 721–743.

Bowman J.L., Smyth D.R. and Meyerowitz E.M. (1991) Genetic interactions among floral homeotic genes of *Arabidopsis. Development* **112**: 1–20.

Brabant M.C. and Brower D.L. (1993) PS2 integrin requirements in *Drosophila* embryo and wing morphogenesis. *Dev. Biol.* **157**: 49–59.

Brackenbury R., Thiery J.-P., Rutishauser U. and Edelman G.M. (1977) Adhesion among neural cells of the chick embryo. I. Immunological assay for molecules involved in cell-cell binding. *J. Biol. Chem.* **252**: 6835–6840.

Bradley A. (1987) Production and analysis of chimeric mice. In: *Teratocarcinomas and Embryonic Stem Cells: A Practical Approach*, E.J. Robertson (ed.), 113–151. Oxford: IRL Press.

Brandeis M., Ariel M. and Cedar H. (1993) Dynamics of DNA methylation during development. *Bioassays* **15**: 709–713.

Branden C. and Tooze J. (1991) *Introduction to Protein Structure*. New York: Garland Publishing, Inc.

Braun R.E., Peschon J.J., Behringer R.R., Brinster R.L. and Palmiter R.D. (1989) Protamine 3′-untranslated sequences regulate temporal translational control and subcellular localization of growth hormone in spermatids of transgenic mice. *Genes and Dev.* **3**: 793–802.

Braun T., Rudnicki M.A., Arnold H.-H. and Jaenisch R. (1992) Targeted inactivation of the muscle regulatory gene *Myf-5* results in abnormal rib development and perinatal death. *Cell* **71**: 369–382.

Bray D. and White J.G. (1988) Cortical flow in animal cells. *Science* **239**: 883–888.

Breedlove S.M. (1992) Sexual differentiation of the brain and behavior. In: *Behavioral Endocrinology*, J.B. Becker, S.M. Breedlove, and D. Crews (eds.), 39–70. Cambridge, MA: MIT Press.

Brennan T.J., Chakraborty T. and Olson E.N. (1991) Mutagenesis of the myogenin basic region identifies an ancient protein motif critical for activation of myogenesis. *Proc. Nat. Acad. Sci. USA* **88**: 5675–5679.

Brenner S. (1974) The genetics of *Caenorhabditis elegans. Genetics* **77**: 71–94.

Bridges C.B. (1921) Triploid intersexes in *Drosophila melanogaster. Science* **54**: 252–254.

Briggs R. (1977) Genetics of cell type determination. In: *Cell Interactions in Differentiation*. M. Karkinen-Jääskeläinen, L. Saxen and L. Weiss (eds.), *6th Sigrid Jusélius Foundation Symp.*, 23–43. New York: Academic Press.

Briggs R. and King T.J. (1952) Transplantation of living nuclei from blastula cells into enucleated frogs' eggs. *Proc. Nat. Acad. Sci. USA* **38**: 455–463.

Brinster R.L. (1976) Participation of teratocarcinoma cells in mouse embryo development. *Cancer Res.* **36**: 3412–3414.

Bronner-Fraser M. (1989) Role of the extracellular environment in neural crest migration. In: *Cytoplasmic Organization Systems*, G.M. Malacinski (ed.), 375–402. New York: McGraw–Hill.

Bronner-Fraser M. (1993) Environmental influences on neural crest cell migration. *J. Neurobiol.* **24**: 233–247.

Bronner-Fraser M. and Fraser S.E. (1991) Cell lineage analysis of the avian neural crest. *Development* (Supplement 2): 17–22.

Bronner-Fraser M. and Stern C. (1991) Effects of mesodermal tissues on avian neural crest cell migration. *Dev. Biol.* **143**: 213–217.

Bronner-Fraser M., Wolf J.J. and Murray B.A. (1992) Effects of antibodies against N-cadherin and N-CAM on the cranial neural crest and neural tube. *Dev. Biol.* **153**: 291–301.

Browder L.W. (1984) *Developmental Biology*. 2nd ed. Philadelphia: Saunders.

Browder L.W., ed. (1985) *Developmental Biology*, vol. 1, *Oogenesis*. New York: Plenum.

Browder L.W., Erickson C.A. and Jeffery W.A. (1991) *Developmental Biology*. Philadelphia: Saunders.

Brower D.L. and Jaffe S.M. (1989) Requirement for integrins during *Drosophila* wing development. *Nature* **324**: 285–287.

Brower D.L., Wilcox M., Piovant M., Smith R.J. and Reger L.A. (1984) Related cell-surface antigens expressed with positional specificity in *Drosophila* imaginal discs. *Proc. Nat. Acad. Sci. USA* **81**: 7485–7489.

Brower D.L., Piovant M. and Reger L.A. (1985) Developmental analysis of *Drosophila* position-specific antigens. *Dev. Biol.* **108**: 120–130.

Brown D.D. and Dawid I.B. (1968) Specific gene amplification in oocytes. *Science* **160**: 272–280.

Brown J.E. and Weiss M.C. (1975) Activation of production of mouse liver enzymes in rat hepatoma-mouse lymphoid hybrids. *Cell* **6**: 481–494.

Brown M. and Goldstein J.L. (1984) How LDL receptors influence cholesterol and atherosclerosis. *Scientific American* **251** (November): 58–66.

Brown N.A. and Wolpert L. (1990) The development of handedness in left/right asymmetry. *Development* **109**: 1–9.

Brown N.H. (1993) Integrins hold *Drosophila* together. *BioEssays* **15**: 383–390.

Brown N.H. (1994) Null mutations in the α_{PS2} and β_{PS} integrin subunit genes have distinct phenotypes. *Development* **120**: 1221–1231.

Brueckner M., D'Eustacio P. and Horwich A.L. (1989) Linkage mapping of a mouse gene, *iv*, that controls left-right asymmetry of the heart and viscera. *Proc. Nat. Acad. Sci. USA* **86**: 5035–5038.

Brun R.B. (1978) Developmental capacities of *Xenopus* eggs, provided with erythrocyte or erythroblast nuclei from adults. *Dev. Biol.* **65**: 271–284.

Bryant P.J. and Simpson P. (1984) Intrinsic and extrinsic control of growth in developing organs. *Quart. Rev. Biol.* **59**: 387–415.

Bryant S.V. (1977) Pattern regulation in amphibian limbs. In: *Vertebrate Limb and Somite Morphogenesis*, D.A. Ede, J.R. Hinchliffe and M. Balls (eds.), *3rd Symp. Brit. Soc. Dev. Biol.*, 319–327. Cambridge: Cambridge University Press.

Bryant S.V. and Gardiner D.M. (1992) Retinoic acid, local cell-cell interactions, and pattern formation in vertebrate limbs. *Dev. Biol.* **152**: 1–25.

Bryant S.V. and Iten L.E. (1976) Supernumerary limbs in amphibians: Experimental production in *Notophthalmus viridescens* and a new interpretation of their formation. *Dev. Biol.* **50**: 212–234.

Bryant S.V., French V. and Bryant P.J. (1981) Distal regeneration and symmetry. *Science* **212**: 993–1002.

Buckingham M. (1992) Making muscle in mammals. *Trends in Genetics* **8**: 144–149.

Büeler H., Fischer M., Lang Y., Bluethmann H., Lipp H.-P., DeArmond S.J., Prusiner S.B., Aguet M. and Weissmann C. (1992) Normal development and behaviour of mice lacking the neuronal cell-surface PrP protein. *Nature* **356**: 577–582.

Büeler H., Aguzzi A., Sailer A., Greiner R.-A., Autenried P., Aguet M. and Weissmann C. (1993) Mice devoid of PrP are resistant to scrapie. *Cell* **73**: 1339–1347.

Bull A. (1966) *Bicaudal*, a genetic factor which affects the polarity of the embryo in *Drosophila melanogaster*. *J. Exp. Zool.* **161**: 221–241.

Bull J.J. (1980) Sex determination in reptiles. *Quart. Rev. Biol.* **55**: 3–21.

Bull J.J. (1983) *Evolution of Sex Determining Mechanisms*. Menlo Park, CA: Benjamin/Cummings.

Bunch T.A. and Brower D.L. (1992) *Drosophila* PS2 integrin mediates RGD-dependent cell-matrix interactions. *Development* **116**: 239–247.

Bunch T.A. and Brower D.L. (1993) Drosophila cell adhesion molecules. *Current Topics Dev. Biol.* **28**: 81–123.

Buratowski S. (1994) The basics of basal transcription by RNA polymerase II. *Cell* **77**: 1–3.

Burgoyne P.S., Buehr M., Koopman P., Rossant J. and McLaren A. (1988) Cell-autonomous action of the testis-determining gene: Sertoli cells are exclusively XY in XX↔XY chimaeric mouse testes. *Development* **102**: 443–450.

Burnside B. (1971) Microtubules and microfilaments in newt neurulation. *Dev. Biol.* **26**: 416–441.

Burnside B. (1973) Microtubules and microfilaments in amphibian neurulation. *Am. Zool.* **88**: 353–372.

Burnside B. and Jacobson A.G. (1968) Analysis of morphogenetic movements in the neural plate of the newt *Taricha torosa*. *Dev. Biol.* **18**: 537–552.

Burtis K.C. and Baker B.S. (1989) Drosophila *doublesex* gene controls somatic sexual differentiation by producing alternatively spliced mRNAs encoding related sex-specific polypeptides. *Cell* **56**: 997–1010.

Burtis K.C., Coschigano K.T., Baker B.S. and Wensink P.C. (1991) The Doublesex proteins of *Drosophila melanogaster* bind directly to a sex-specific yolk protein gene enhancer. *EMBO J.* **10**: 2577–2582.

Busa W.B. and Gimlich R.L. (1989) Lithium-induced teratogenesis in frog embryos prevented by a polyphosphoinositide cycle intermediate or a diacylglycerol analog. *Dev. Biol.* **132**: 315–324.

Buskirk D.R., Thiery J.-P., Rutishauser U. and Edelman G.M. (1980) Antibodies to a neural cell adhesion molecule disrupt histogenesis in cultured chick retinae. *Nature* (London) **285**: 488–489.

Butler E.G. (1955) Regeneration of the urodele forelimb after reversal of its proximo-distal axis. *J. Morphol.* **96**: 265–281.

Butler P.J. and Klug A. (1978) The assembly of a virus. *Scientific American* **239** (November): 62–69.

Cantley L.C., Auger K.R., Carpenter C., Duckworth B., Graziani A., Kapeller R. and Soltoff S. (1991) Oncogenes and signal transduction. *Cell* **64**: 281–302.

Carlson B.M. (1988) *Patten's Foundations of Embryology*. 5th ed. New York: McGraw-Hill.

Carlson G.A., Hsiao K., Oesch B., Westaway D. and Prusiner S.B. (1991) Genetics of prion infections. *Trends in Genetics* **7**: 61–65.

Carola R., Harley J.P. and Noback C.R. (1990) *Human Anatomy and Physiology*. New York: McGraw-Hill.

Carpenter G. and Cohen S. (1978) Epidermal growth factors. In: *Biochemical Actions of Hormones*, G. Litwack (ed.), vol. V, 203–247. New York: Academic Press.

Carroll S.B. and Scott M.P. (1986) Zygotically active genes that affect the spatial expression of the *fushi tarazu* segmentation gene during early Drosophila embryogenesis. *Cell* **45:** 113–126.

Carter K.C., Bowman D., Carrington W., Fogarty K., McNeil J.A., Fay F.S. and Lawrence J.B. (1993) A three-dimensional view of precursor messenger RNA metabolism within the mammalian nucleus. *Science* **259:** 1330–1335.

Casanova J. (1990) Pattern formation under the control of the terminal system in the *Drosophila* embryo. *Development* **110:** 621–628.

Casanova J. and Struhl G. (1993) The *torso* receptor localizes as well as transduces the spatial signal specifying terminal body pattern in *Drosophila*. *Nature* **362:** 152–155.

Casey J.L., Koeller D.M., Ramin V.C., Klausner R.D. and Harford J.B. (1989) Iron regulation of transferrin receptor messenger RNA levels requires iron-responsive elements and a rapid turnover determinant in the 3' untranslated region of the messenger RNA. *EMBO J.* **8:** 3693–3700.

Cather J.N., Render J.A. and Freeman G. (1986) The relation of time to direction and equality of cleavage in *Ilyanassa* embryos. *Int. J. Invertebrate Reprod. and Dev.* **9:** 179–194.

Centrella M., McCarthy T.L. and Canalis E. (1987) Mitogenesis in fetal rat bone cells simultaneously exposed to type β transforming growth factor and other growth regulators. *FASEB J.* **1:** 312–317.

Chambon P. (1981) Split genes. *Scientific American* **244** (May): 60–77.

Chan S.-K. and Mann R.S. (1993) The segment identity functions of Ultrabithorax are contained within its homeo domain and carboxy-terminal sequences. *Genes and Dev.* **7:** 796–811.

Chan S.-K., Jaffe L., Capovilla M., Botas J. and Mann R.S. (1994) The DNA binding specificity of Ultrabithorax is modulated by cooperative interactions with extradentile, another homeoprotein. *Cell* **78:** 603–615.

Chandlee J.M. (1991) Analysis of developmentally interesting genes cloned from higher plants by insertional mutagenesis. *Dev. Genet.* **12:** 261–271.

Chellappan S.P., Hiebert S., Mudryj M., Horowitz J.M. and Nevins J.R. (1991) The E2F transcription factor is a cellular target for the RB protein. *Cell* **65:** 1053–1061.

Chen X., Farmer G., Zhu H., Prywes R. and Prives C. (1993) Cooperative DNA binding of p53 with RFIID (TBP): A possible mechanism for transcriptional activation. *Genes and Dev.* **7:** 1837–1849.

Chen Y.P., Huang L. and Solursh M. (1994) A concentration gradient of retinoids in the early *Xenopus laevis* embryo. *Dev. Biol.* **161:** 70–76.

Cheng T.-C., Wallace M.C., Merlie J.P. and Olson E.N. (1993) Separable regulatory elements governing *myogenin* transcription in mouse embryogenesis. *Science* **261:** 215–218.

Chisaka O. and Capecchi M.R. (1991) Regionally restricted developmental defects resulting from targeted disruption of the mouse homeobox gene *Hox-1.5*. *Nature* **350:** 473–479.

Cho K.W.Y., Blumberg B., Steinbeisser H. and De Robertis E.M. (1991) Molecular nature of Spemann's organizer: The role of the *Xenopus* homeobox gene *goosecoid*. *Cell* **67:** 1111–1120.

Christ B., Jacob H.J. and Jacob M. (1977) Experimental analysis of the origin of the wing musculature in avian embryos. *Anat. Embryol.* **150:** 171–186.

Christian J.L., Olson D.J. and Moon R.T. (1992) Xwnt-8 modifies the character of mesoderm induced by bFGF in isolated *Xenopus* ectoderm. *EMBO J.* **11:** 33–41.

Christy B. and Scangos G. (1986) Control of gene expression by DNA methylation. In: *Molecular Genetics of Mammalian Cells: A Primer in Developmental Biology*, G.M. Malacinski (ed.), 3–28. New York: Macmillan.

Chuong C.-M. and Edelman G.M. (1985a) Expression of cell adhesion molecules in embryonic induction. I. Morphogenesis of nestling feathers. *J. Cell Biol.* **101:** 1009–1026.

Chuong C.-M. and Edelman G.M. (1985b) Expression of cell adhesion molecules in embryonic induction. II. Morpho-genesis of adult feathers. *J. Cell Biol.* **101:** 1027–1043.

Clack J.A. (1989) Discovery of the earliest known tetrapod stapes. *Nature* **342:** 425–427.

Clark S.E., Running M.P. and Meyerowitz E.M. (1993) *CLAVATA1*, a regulator of meristem and flower development in *Arabidopsis*. *Development* **119:** 397–418.

Clement A.C. (1952) Experimental studies on germinal localization in *Ilyanassa*. I. The role of the polar lobe in determination of the cleavage pattern of its influence in later development. *J. Exp. Zool.* **121:** 593–626.

Clement A.C. (1962) Development of *Ilyanassa* embryo following removal of the D-macromere at successive cleavage stages. *J. Exp. Zool.* **149:** 193–216.

Clever U. and Karlson P. (1960) Induktion von Puff-Veränderungen in den Speicheldrüsenchromosomen von *Chironomus tentans* durch Ecdyson. *Exp. Cell Res.* **20:** 623–626.

Coen E.S. and Meyerowitz E.M. (1991) The war of the whorls: Genetic interactions controlling flower development. *Nature* **353:** 31–37.

Coen E.S., Doyle S., Romero J.M., Elliott R., Magrath R. and Carpenter R. (1991) Homeotic genes controlling flower development in *Antirrhinum*. *Development* (Supplement 1): 149–155.

Cohen A.M. and Konigsberg I.R. (1975) A clonal approach to the problem of neural crest determination. *Dev. Biol.* **46:** 262–282.

Colbert E.H. (1980) *Evolution of the Vertebrates: A History of the Backboned Animals through Time.* 3rd ed. New York: Wiley.

Colgan J., Wampler S. and Manley J.L. (1993) Interaction between a transcriptional activator and transcription factor IIB *in vivo*. *Nature* **362:** 549–553.

Collinge J., Whittington M.A., Sidle K.C.L., Smith C.J., Palmer M.S., Clarke A.R. and Jeffreys J.G.T. (1994) Prion protein is necessary for normal synaptic function. *Nature* **370:** 295–297.

Colwin L.H. and Colwin A.L. (1960) Formation of sperm entry holes in the vitelline membrane of *Hydroides hexagonis* (Annelida) and evidence of their lytic origin. *J. Biophys. Biochem. Cytol.* **7:** 315–320.

Condie B.G. and Capecchi M.R. (1993) Mice homozygous for a targeted disruption of *Hoxd-3 (Hox-4.1)* exhibit anterior transformations of the first and second cervical vertebrae, the atlas and the axis. *Development* **119:** 579–595.

Condie B.G. and Capecchi M.R. (1994) Mice with targeted disruptions in the paralogous genes *hoxa-3* and *hoxd-3* reveal synergistic interactions. *Nature* **370:** 304–307.

Conklin E.G. (1905) The organization and cell lineage of the ascidian egg. *J. Acad. Nat. Sci. Philadelphia* **13:** 1–119.

Conlon R.A. and Rossant J. (1992) Exogenous retinoic acid rapidly induces anterior ectopic expression of murine *Hox-2* genes in vivo. *Development* **116:** 357–368.

Constantini F., Chada K. and Magram J. (1986) Correction of murine ß-thalassemia by gene transfer into the germ line. *Science* **233:** 1192–1194.

Cooley L., Kelley R. and Spradling A. (1988) Insertional mutagenesis of the *Drosophila* genome with single P elements. *Science* **239:** 1121–1128.

Copeland N.G., Gilbert D.J., Chao B.C., Donovan P.J., Jenkins N.A., Cosman D., Anderson D., Lyman S.D. and Williams D.E. (1990) Mast cell growth factor maps near the steel locus on mouse chromosome 10 and is deleted in a number of steel alleles. *Cell* **63:** 175–183.

Coschigano K.T. and Wensink P.C. (1993) Sex-specific transcriptional regulation by the male and female doublesex proteins of *Drosophila. Genes and Dev.* **7:** 42–54.

Counce S.J. (1973) The causal analysis of insect embryogenesis. In: *Developmental Systems: Insects*, S.J. Counce and C.H. Waddington (eds.), vol. 2, 1–156. London/New York: Academic Press.

Couso J.P., Bate M. and Martínez-Arias A. (1993) A *wingless*-dependent polar coordinate system in *Drosophila* imaginal discs. *Science* **259:** 484–489.

Cowan A.E. and Myles D.G. (1993) Biogenesis of surface domains during spermiogenesis in the guinea pig. *Dev. Biol.* **155:** 124–133.

Craig E.A. (1993) Chaperones: Helpers along the pathways to protein folding. *Science* **260:** 1902–1903.

Crampton H.E. (1894) Reversal of cleavage in a sinistral gastropod. *Ann. New York Acad. Sci.* **8:** 167–170.

Crawford K. and Stocum D.L. (1988) Retinoic acid coordinately proximalizes regenerate pattern and blastema differential affinity in axolotl limbs. *Development* **102:** 687–698.

Crews D. (1994) Animal sexuality. *Scientific American* **270** (January): 108–114.

Crick F.H.C. and Lawrence P.A. (1975) Compartments and polyclones in insect development. *Science* **189:** 340–347.

Croce C.M. (1987) Role of chromosome translocations in human neoplasia. *Cell* **49:** 155–156.

Croce C.M., Thierfelder W., Erickson J., Nishikura K., Finan J., Lenoir G.M. and Nowell P.C. (1983) Transcriptional activation of an unrearranged and untranslocated c-*myc* oncogene by translocation of a C$_\lambda$ locus in Burkitt lymphoma cells. *Proc. Nat. Acad. Sci. USA* **80:** 6922–6926.

Cross M. and Dexter T.M. (1991) Growth factors in development, transformation, and tumorigenesis. *Cell* **64:** 271–280.

Crossley I., Whalley T. and Whitaker M.J. (1991) Guanosine 5-thiotriphosphate may stimulate phosphoinositide messenger production in sea urchin eggs by a different route than the fertilizing sperm. *Cell Regulation* **2:** 121–133.

Cunningham B.A. (1991) Cell adhesion molecules and the regulation of development. *Am. J. Ob. Gyn.* **164:** 939–948.

Cunningham B.A., Hemperly J.J., Murray B.A., Prediger E.A., Brackenbury R. and Edelman G.M. (1987) Neural cell adhesion molecule: Structure, immunoglobulin-like domains, cell surface modulation, and alternative RNA splicing. *Science* **236:** 799–806.

Czihak G., Langer H. and Ziegler H. (1976) *Biologie.* Heidelberg: Springer.

Dale L. and Slack J.M.W. (1987a) Fate map for the 32-cell stage of *Xenopus laevis. Development* **99:** 527–551.

Dale L. and Slack J.M.W. (1987b) Regional specification within the mesoderm of early embryos of *Xenopus laevis. Development* **100:** 279–295.

Dale L., Smith J.C. and Slack J.M.W. (1985) Mesoderm induction in *Xenopus laevis*: A quantitative study using a cell lineage label and tissue-specific antibodies. *J. Embryol. exp. Morphol.* **89:** 289–312.

Dale L., Matthews G., Tabe L. and Colman A. (1989) Developmental expression of the protein product of Vg1, a localized maternal mRNA in the frog *Xenopus laevis. EMBO J.* **8:** 1057–1065.

Dale L., Matthews G. and Colman A. (1993) Secretion and mesoderm-inducing activity of the TGF-β-related domain of *Xenopus* Vg1. *EMBO J.* **12:** 4471–4480.

Dalla-Favera R., Bregni M., Erikson J., Patterson D., Gallo R.C. and Croce C.M. (1982) Human c-*myc* onc gene is located on the region of chromosome 8 that is translocated in Burkitt lymphoma cells. *Proc. Nat. Acad. Sci. USA* **79:** 7824–7827.

Damsky C.H. and Werb Z. (1992) Signal transduction by integrin receptors for extracellular matrix: Cooperative processing of extracellular information. *Current Biol.* **4:** 772–781.

Dan J.C. (1967) Acrosome reaction and lysins. In: *Fertilization*, C.B. Metz and A. Monroy (eds.), vol. 1, 237–367. New York: Academic Press.

Dan K. (1979) Studies on unequal cleavage in sea urchins. I. Migration of the nuclei to the vegetal pole. *Dev. Growth Differ.* **21:** 527–535.

Danilchik M.V. and Denegre J.M. (1991) Deep cytoplasmic rearrangements during early development in *Xenopus laevis. Development* **111:** 845–856.

Darnell J., Lodish H. and Baltimore D. (1990) *Molecular Cell Biology*. 2nd ed. New York: Scientific American Books.

Darribère T., Yamada K.M., Johnson K.E. and Boucaut J.-C. (1988) The 140 kD fibronectin receptor complex is required for mesodermal cell adhesion during gastrulation in the amphibian *Pleurodeles waltlii*. *Dev. Biol.* **126**: 182–194.

Darwin C. (1868) *The Variation of Animals and Plants under Domestication*. New York: Orange Judd.

Davenport R.W., Dou P., Rehder V. and Kater S.B. (1993) A sensory role for neuronal growth cone filopodia. *Nature* **361**: 721–724.

David C.N. and Murphy S. (1977) Characterization of interstitial stem cells in Hydra by cloning. *Dev. Biol.* **58**: 372–383.

David C.N., Bosch T.C.G., Hobmayer B., Holstein T. and Schmidt T. (1987) Interstitial stem cells in Hydra. In: *Genetic Regulation of Development*, W.F. Loomis (ed.), 289–408. New York: Alan R. Liss.

Davidson E.H. (1976) *Gene Activity in Early Development*. 2nd ed. New York: Academic Press.

Davidson E.H. (1986) *Gene Activity in Early Development*. 3rd ed. New York: Academic Press.

Davidson E.H. (1989) Lineage-specific gene expression and the regulative capacities of the sea urchin embryo: A proposed mechanism. *Development* **105**: 421–445.

Davidson E.H. (1990) How embryos work: A comparative view of diverse models of cell fate specification. *Development* **108**: 365–389.

Davidson R.G., Nitowsky H.M. and Childs B. (1963) Demonstration of two populations of cells in the human female heterozygous for glucose-6-phosphate dehydrogenase variants. *Proc. Nat. Acad. Sci. USA* **50**: 481–485.

Davis B.K. (1981) Timing of fertilization in mammals: Sperm cholesterol/phospholipid ratio as determinant of capacitation interval. *Proc. Nat. Acad. Sci. USA* **78**: 7560–7564.

Davis R.L., Weintraub H. and Lassar A.B. (1987) Expression of a single transfected cDNA converts fibroblasts to myoblasts. *Cell* **51**: 987–1000.

Davis R.L., Cheng P-F., Lassar A.B. and Weintraub H. (1990) The MyoD DNA binding domain contains a recognition code for muscle-specific gene activation. *Cell* **60**: 733–746.

Deininger P.L. and Daniels G.R. (1986) The recent evolution of mammalian repetitive DNA elements. *Trends in Genetics* **2**: 76–80.

DeLeon D.V., Cox K.H., Angerer L.M. and Angerer R.C. (1983) Most early-variant histone mRNA is contained in the pronucleus of sea urchin eggs. *Dev. Biol.* **100**: 197–206.

Denny P.C. and Tyler A. (1964) Activation of protein synthesis in non-nucleate fragments of sea urchin eggs. *Biochem. Biophys. Res. Commun.* **145**: 245–249.

Derby A. and Etkin W. (1968) Thyroxine induced tail resorption *in vitro* as affected by anterior pituitary hormones. *J. Exp. Zool.* **169**: 1–8.

Derman E., Krauter K., Walling L., Weinberger C., Ray M. and Darnell J.E., Jr. (1981) Transcriptional control in the production of liver-specific mRNAs. *Cell* **23**: 731–739.

De Robertis E.M. and Gurdon J.B. (1977) Gene activation in somatic nuclei after injection into amphibian oocytes. *Proc. Nat. Acad. Sci. USA* **74**: 2470–2474.

De Robertis E.M. and Gurdon J.B. (1979) Gene transplantation and the analysis of development. *Scientific American* **241** (December): 74–82.

De Robertis E.M., Oliver G. and Wright C.V.E. (1990) Homeobox genes and the vertebrate body plan. *Scientific American* **263** (July): 46–52.

De Robertis E.M., Morita E.A. and Cho K.W.Y. (1991) Gradient fields and homeobox genes. *Development* **112**: 669–678.

Detrick R.J., Dickey D. and Kintner C.R. (1990) The effects of N-cadherin misexpression on morphogenesis in *Xenopus* embryos. *Neuron* **4**: 493–506.

Dexter T.M. (1982) Stromal cell associated haemopoiesis. *J. Cell Physiol.* (Supplement 1): 87–94.

Dexter T.M. and Spooncer E. (1987) Growth and differentiation in the hemopoietic system. *Ann. Rev. Cell Biol.* **3**: 423–441.

Diamond M.I., Miner J.N., Yoshinaga S.K. and Yamamoto K.R. (1990) Transcription factor interactions: Selectors of positive or negative regulation from a single DNA element. *Science* **249**: 1266–1272.

Diaz-Benjumea F.J. and Cohen S.M. (1993) Interaction between dorsal and ventral cells in the imaginal disc directs wing development in *Drosophila*. *Cell* **75**: 741–752.

DiBerardino M.A. and Hoffner N. (1971) Development and chromosomal constitution of nuclear transplants derived from male germ cells. *J. Exp. Zool.* **176**: 61–72.

DiBerardino M.A. and Hoffner N.J. (1983) Gene reactivation in erythrocytes: Nuclear transplantation in oocytes and eggs of *Rana*. *Science* **219**: 862–864.

DiBerardino M.A., Hoffner N.J. and Etkin L.D. (1984) Activation of dormant genes in specialized cells. *Science* **224**: 946–952.

Dickerson R.E. and Geis I. (1983) *Hemoglobin*. Menlo Park, CA: Benjamin/Cummings.

Ding D. and Lipshitz H.D. (1993) Localized RNAs and their functions. *BioEssays* **15**: 651–658.

Ding D., Parkhurst S.M. and Lipshitz H.D. (1993) Different genetic requirements for anterior RNA localization revealed by the distribution of Adducin-like transcripts during *Drosophila* oogenesis. *Proc. Nat. Acad. Sci. USA* **90**: 2512–2516.

Dingwall C. and Laskey R. (1992) The nuclear membrane. *Science* **258**: 942–947.

Dirksen M.L. and Jamrich M. (1992) A novel, activin-inducible, blastopore lip-specific gene of *Xenopus laevis* contains a *fork head* DNA-binding domain. *Genes and Dev.* **6**: 599–608.

Dobrovolskaia-Zavadskaia N. (1927) Sur la mortification spontanée de la queue chez la souris nouveau-née et sur l'existence d'un caractère (facteur) héréditaire "non-viable." *Comp. Rend. Soc. Biol.* **97**: 114–116.

Doherty P., Ashton S.V., Moore S.E. and Walsh F.S. (1991) Morphoregulatory activities of NCAM and N-cadherin can be accounted for by G protein-dependent activation of L- and N-type neuronal Ca^{2+} channels. *Cell* **67**: 21–33.

Donehower L.A., Harvey M., Slagle B.L., McArthur M.J., Montgomery C.A., Jr., Butel J.S. and Bradley A. (1992) Mice deficient for p53 are developmentally normal but susceptible to spontaneous tumours. *Nature* **356**: 215–221.

Doniach T., Phillips C.R. and Gerhart J.C. (1992) Planar induction of anteroposterior pattern in the developing central nervous system of *Xenopus laevis. Science* **257**: 542–545.

Dowdy S.F., Hinds P.W., Louie K., Reed S.I., Arnold A. and Weinberg R.A. (1993) Physical interaction of the retinoblastoma protein with human D cyclins. *Cell* **73**: 499–511.

Draetta G., Brizuela L., Potashkin J. and Beach D. (1987) Identification of p34 and p13, human homologs of the cell cycle regulators of fission yeast encoded by *cdc2*[+] and *suc1*[+]. *Cell* **50**: 319–325.

Draetta G., Luca F., Wetendorf J., Brizuela L., Ruderman J. and Beach D. (1989) cdc2 protein kinase is complexed with both cyclin A and B: Evidence for proteolytic inactivation of MPF. *Cell* **56**: 829–838.

Drews G.N. and Goldberg R.B. (1989) Genetic control of flower development. *Trends in Genetics* **5**: 256–261.

Drews G.N., Bowman J.L. and Meyerowitz E.M. (1991a) Negative regulation of the Arabidopsis homeotic gene *AGAMOUS* by the *APETALA2* product. *Cell* **65**: 991–1002.

Drews G.N., Weigel D. and Meyerowitz E.M. (1991b) Floral patterning. *Current Opinion in Genetics and Development* **1**: 174–179.

Dreyer C., Scholz E. and Hausen P. (1982) The fate of oocyte nuclear proteins during early development of *Xenopus laevis. Wilhelm Roux's Arch.* **191**: 228–233.

Driesch H. (1892) The potency of the first two cleavage cells in echinoderm development. Experimental production of partial and double formations. In: *Foundations of Experimental Embryology*, B.H. Willier and J.M. Oppenheimer (eds.), 38–50. New York: Hafner.

Driesch H. (1894) *Analytische Theorie der organischen Entwicklung*. Leipzig: Engelmann.

Driesch H. (1898) Über rein-mütterliche Charaktere und Bastardlarven von Echiniden. *Wilhelm Roux's Arch.* **7**: 65–102.

Driesch H. (1909) *Philosophie des Organischen*. 2 vols. Leipzig: Engelmann.

Driever W. and Nüsslein-Volhard C. (1988) A gradient of *bicoid* protein in *Drosophila* embryos. *Cell* **54**: 83–93.

Driever W. and Nüsslein-Volhard C. (1989) The bicoid protein is a positive regulator of *hunchback* transcription in the early *Drosophila* embryo. *Nature* **337**: 138–143.

Driever W., Thoma G. and Nüsslein-Volhard C. (1989a) Determination of spatial domains of zygotic gene expression in the *Drosophila* embryo by the affinity of binding sites for the bicoid morphogen. *Nature* **340**: 363–367.

Driever W., Ma J., Nüsslein-Volhard C. and Ptashne M. (1989b) Rescue of *bicoid* mutant *Drosophila* embryos by bicoid fusion proteins containing heterologous activating sequences. *Nature* **342**: 149–154.

Driever W., Siegel V. and Nüsslein-Volhard C. (1990) Autonomous determination of anterior structures in the early *Drosophila* embryo by the maternal *bicoid* morphogen. *Development* **109**: 811–820.

Dunphy W.G. and Kumagai A. (1991) The cdc25 protein contains an intrinsic phosphatase activity. *Cell* **67**: 189–196.

Dutta A., Ruppert J.M., Aster J.C. and Winchester E. (1993) Inhibition of DNA replication factor RPA by p53. *Nature* **365**: 79–82.

Eckstein D.J. and Shur B.D. (1992) Cell surface β-1,4-galactosyltransferase is associated with the detergent-insoluble cytoskeleton on migrating mesenchymal cells. *Exp. Cell Res.* **201**: 83–90.

Eddy E.M. (1988) The spermatozoon. In: *The Physiology of Reproduction*, E. Knobil and J.D. Neill (eds.), 27–68. New York: Raven Press.

Eddy E.M., Clark J.M., Gong D. and Fenderson B.A. (1981) Origin and migration of primordial germ cells in mammals. *Gamete* **4**: 333–362.

Edelman G.M. (1983) Cell adhesion molecules. *Science* **219**: 450–457.

Edelman G.M. (1984) Cell adhesion molecules: A molecular basis for animal form. *Scientific American* **250** (April): 118–129.

Edelman G.M. (1988) *Topobiology: An Introduction to Molecular Embryology*. New York: Basic Books.

Edelman G.M. and Chuong C.-M. (1982) Embryonic to adult conversion of neural cell adhesion molecules in normal and staggerer mice. *Proc. Nat. Acad. Sci. USA* **79**: 7036–7040.

Edgar B.A. and O'Farrell P.H. (1989) Genetic control of cell division patterns in the Drosophila embryo. *Cell* **57**: 177–187.

Edgar B.A. and O'Farrell P.H. (1990) The three postblastoderm cell cycles of Drosophila embryogenesis are regulated in G2 by *string. Cell* **62**: 469–480.

Edgar B.A. and Schubiger G. (1986) Parameters controlling transcriptional activation during early Drosophila development. *Cell* **44**: 871–877.

Edgar B.A., Kiehle C.P. and Schubiger G. (1986) Cell cycle control by the nucleo-cytoplasmic ratio in early Drosophila development. *Cell* **44**: 365–372.

Edgar B.A., Sprenger F., Duronio R.J., Leopold P. and O'Farrell P.H. (1994) Distinct molecular mechanisms regulate cell cycle timing at successive stages of *Drosophila* embryogenesis. *Genes and Dev.* **8**: 440–452.

Edgar L.G. and McGhee J.D. (1986) Embryonic expression of a gut-specific esterase in *Caenorhabditis elegans. Dev. Biol.* **114**: 109–118.

Egan S.E., Giddings B.W., Brooks M.W., Buday L., Sizeland A.M. and Weinberg R.A. (1993) Association of Sos Ras exchange protein with

Grb2 is implicated in tyrosine kinase signal transduction and transformation. *Nature* **363**: 45–51.

Eisenbach M. and Ralt D. (1992) Precontact mammalian sperm-egg communication and role in fertilization. *Am. J. Physiol.* **262** (*Cell Physiol.* **31**): C1095–C1101.

Elbetieha A. and Kalthoff K. (1988) Anterior determinants in embryos of *Chironomus samoensis*: Characterization by rescue bioassay. *Development* **104**: 61–75.

El-Deiry W.S., Tokino T., Velculescu V.E., Levy D.B., Parsons R., Trent J.M., Lin D., Mercer W.E., Kinzler K.W. and Vogelstine B. (1993) WAF1, a potential mediator of p53 tumor suppression. *Cell* **75**: 817–825.

Elinson R.P. (1986) Fertilization in amphibians: The ancestry of the block to polyspermy. *Int. Rev. Cytol.* **101**: 59–100.

Elinson R.P. and Pasceri P. (1989) Two UV-sensitive targets in dorsoanterior specification of frog embryos. *Development* **106**: 511–518.

Elinson R.P. and Rowning B. (1988) A transient array of parallel microtubules in frog eggs: Potential tracks for a cytoplasmic rotation that specifies the dorso-ventral axis. *Dev. Biol.* **128**: 185–197.

Ellenberger T., Fass D., Arnaud M. and Harrison S.C. (1994) Crystal structure of transcription factor E 47: E-box recognition by basic region helix-loop-helix dimer. *Genes and Dev.* **8**: 970–980.

Ellis H.M. and Horvitz H.R. (1986) Genetic control of programmed cell death in the nematode C. elegans. *Cell* **44**: 817–829.

Ellis R.E. and Horvitz H.R. (1991) Two *C. elegans* genes control the programmed deaths of specific cells in the pharynx. *Development* **112**: 591–603.

Emeson R.B., Hedjran F., Yeakley J.M., Guise J.W. and Rosenfeld M.G. (1990) Alternative production of calcitonin and CGRP mRNA is regulated at the calcitonin-specific splice acceptor. *Nature* **341**: 76–80.

Epel D. (1977) The program of fertilization. *Scientific American* **237** (November): 128–138.

Ephrussi A. and Lehmann R. (1992) Induction of germ cell formation by *oskar*. *Nature* **358**: 387–392.

Epperlein H.H. and Löfberg J. (1984) Xanthophores in chromatophore groups of the premigratory neural crest initiate the pigment pattern of the axolotl larva. *Wilhelm Roux's Arch.* **193**: 357–369.

Erickson C.A., Duong T.D. and Tosney K.W. (1992) Descriptive and experimental analysis of the dispersion of neural crest cells along the dorsolateral path and their entry into ectoderm in the chick embryo. *Dev. Biol.* **151**: 251–272.

Erickson J.W. and Cline T.W. (1991) Molecular nature of the *Drosophila* sex determination signal and its link to neurogenesis. *Science* **251**: 1071–1074.

Erickson J.W. and Cline T.W. (1993) A bZIP protein, sisterless-a, collaborates with bHLH transcription factors early in *Drosophila* development to determine sex. *Genes and Dev.* **7**: 1688–1702.

Erikson J., Nishikura K., Ar-Rushdi A., Finan J., Emanuel B., Lenoir G., Nowell P.C. and Croce C.M. (1983) Translocation of an immunoglobulin κ locus to a region 3' of an unrearranged c-*myc* oncogene enhances c-*myc* transcription. *Proc. Nat. Acad. Sci. USA* **80**: 7581–7585.

Etkin W. (1968) Hormonal control of amphibian metamorphosis. In: *Metamorphosis*, W. Etkin and L.I. Gilbert (eds.), 313–348. New York: Appleton-Century-Crofts.

Etkin W. and Gilbert L.I. (1968) *Metamorphosis*. New York: Appleton-Century-Crofts.

Ettensohn C.A. (1985a) Gastrulation in the sea urchin embryo is accompanied by the rearrangement of invaginating epithelial cells. *Dev. Biol.* **112**: 383–390.

Ettensohn C.A. (1985b) Mechanisms of epithelial invagination. *Quart. Rev. Biol.* **60**: 289–307.

Ettensohn C.A. (1992) Cell interactions and mesodermal cell fates in the sea urchin embryo. *Development* (Supplement): 43–51.

Ettensohn C.A. and Ingersoll E.P. (1992) Morphogenesis of the sea urchin embryo. In: *Morphogenesis*, E.F. Rossomando and S. Alexander (eds.), 189–262. New York: Marcel Dekker.

Ettensohn C.A. and McClay D.R. (1988) Cell lineage conversion in the sea urchin embryo. *Dev. Biol.* **125**: 396–409.

Evans E.P., Ford C.E. and Lyon M.F. (1977) Direct evidence of the capacity of the XY germ cell in the mouse to become an oocyte. *Nature* **267**: 430–431.

Evans R.M. (1988) The steroid and thyroid hormone receptor superfamily. *Science* **240**: 889–895.

Evans S.C., Lopez L.C. and Shur B.D. (1993) Dominant negative mutation in cell surface β-1,4 galactosyltransferase inhibits cell-cell and cell-matrix interactions. *J. Cell Biol.* **120**: 1045–1057.

Eyal-Giladi H. (1984) The gradual establishment of cell commitments during the early stages of chick development. *Cell Differentiation* **14**: 245–255.

Fagard R. and London I.M. (1981) Relationship between phosphorylation and activity of heme-regulated eukaryotic initiation factor 2 kinase. *Proc. Nat. Acad. Sci. USA* **78**: 866–870.

Fairbairn L.J., Cowling G.J., Reipert B.M. and Dexter T.M. (1993) Suppression of apoptosis allows differentiation and development of a multipotent hemopoietic cell line in the absence of added growth factors. *Cell* **74**: 823–832.

Falb D. and Maniatis T. (1992) A conserved regulatory unit implicated in tissue-specific gene expression in *Drosophila* and man. *Genes and Dev.* **6**: 454–465.

Fawcett D.W. (1975) The mammalian spermatozoon. *Dev. Biol.* **44**: 394–436.

Feldherr C.M. (1992) *Nuclear Trafficking*. San Diego: Academic Press.

Feldman K.A., Marks M.D., Christianson M.L. and Quatrano R.S. (1989) A dwarf mutant of *Arabidopsis* generated by T-DNA insertion mutagenesis. *Science* **243**: 1351–1354.

Felsenfeld G. (1992) Chromatin as an essential part of the transcriptional mechanism. *Nature* **355**: 219–224.

Ferguson E.L. and Anderson K.V. (1992) *decapentaplegic* acts as a morphogen to organize dorsal-ventral pattern in the *Drosophila* embryo. *Cell* **71**: 451–461.

Ferguson E.L., Sternberg P.W. and Horvitz H.R. (1987) A genetic pathway for the specification of the vulval cell lineages of *Caenorhabditis elegans*. *Nature* **326**: 259–267.

Fink R.D. and McClay D.R. (1985) Three cell recognition changes accompany the ingression of sea urchin primary mesenchyme cells. *Dev. Biol.* **107**: 66–74.

Fisher R.A. (1930) *The Genetical Theory of Natural Selection*. Oxford: Clarendon Press.

Fleischmann G. (1975) Origin and embryonic development of fertile gonads with and without pole cells of *Pimpla turionellae* L. (Hymenoptera, Ichneumonidae). *Zool. Jahrb. Anat.* **94**: 375–412.

Fleming T.P. (1987) A quantitative analysis of cell allocation to trophectoderm and inner cell mass in the mouse blastocyst. *Dev. Biol.* **119**: 520–531.

Fleming T.P., Pickering S.J., Qasim F. and Maro B. (1986) The generation of cell surface polarity in mouse 8-cell blastomeres: The role of cortical microfilaments analyzed using cytochalasin D. *J. Embryol. exp. Morphol.* **95**: 169–191.

Florman H.M. and Wassarman P.M. (1985) O-linked oligosaccharides of mouse egg ZP3 account for its sperm receptor activity. *Cell* **41**: 313–324.

Foe V.E. (1989) Mitotic domains reveal early commitment of cells in *Drosophila* embryos. *Development* **107**: 1–22.

Foe V.E. and Alberts B.M. (1983) Studies of nuclear and cytoplasmic behavior during the five mitotic cycles that precede gastrulation in *Drosophila* embryogenesis. *J. Cell Sci.* **61**: 31–70.

Fogerty F.J., Fessler L.I., Bunch T.A., Yaron Y., Parker C.G., Nelson R.E., Brower D.L., Gullberg D. and Fessler J.H. (1994) Tiggrin, a novel *Drosophila* extracellular matrix protein that functions as a ligand for *Drosophila* $\alpha_{PS2}\beta_{PS}$ integrins. *Development* **120**: 1747–1758.

Folkman J. and Klagsbrun M. (1987) Angiogenic factors. *Science* **235**: 442–447.

Foltz K.R. and Lennarz W.J. (1993) The molecular basis of sea urchin gamete interactions at the egg plasma membrane. *Dev. Biol.* **158**: 46–61.

Foltz K.R., Partin J.S. and Lennarz W.J. (1993) Sea urchin egg receptor for sperm: Sequence similarity of binding domain and hsp70. *Science* **259**: 1421–1425.

Ford C.E., Jones K.W., Polani P.E., deAlmeida J.C. and Briggs J.H. (1959) A sex-chromosome anomaly in a case of gonadal dysgenesis (Turner's syndrome). *Lancet* **1**: 711–713.

Fox R.O., Evans P.A. and Dobson C.M. (1986) Multiple conformations of a protein demonstrated by magnetization transfer NMR spectroscopy. *Nature* **320**: 192–194.

Franco M.G., Rubini P.G. and Vecchi M.L. (1982) Sex determinants and their distribution in various populations of *Musca domestica* L. of Western Europe. *Genet. Res.* **40**: 279–293.

Frank E. and Sanes J.R. (1991) Lineage of neurons and glia in chick dorsal root ganglia: Analysis *in vivo* with a recombinant retrovirus. *Development* **111**: 895–908.

Frankel J. (1989) *Pattern Formation: Ciliate Studies and Models*. New York: Oxford University Press.

Fraser S., Keynes R. and Lumsden A. (1990) Segmentation in the chick embryo hindbrain is defined by cell lineage restrictions. *Nature* **344**: 431–435.

Fraser S.E., Carhart M.S., Murray B.A., Chuong C.-M. and Edelman G.M. (1988) Alterations in the *Xenopus* retinotectal projection by antibodies to *Xenopus* N-CAM. *Dev. Biol.* **129**: 217–230.

Freeman G. (1976) The role of cleavage in the localization of developmental potential in the ctenophore *Mnemiopsis leidyi*. *Dev. Biol.* **49**: 143–177.

Freeman G. (1979) The multiple roles which cell division can play in the localization of developmental potential. In: *Determinants of Spatial Organization*, S. Subtelny and I.R. Konigsberg (ed.), *37th Symp. Soc. Dev. Biol.*, 53–76. New York: Academic Press.

Freeman G. (1983) The role of egg organization in the generation of cleavage patterns. In *Time, Space, and Pattern in Embryonic Development*, W.R. Jeffery and R.A. Raff (eds.), 171–196. New York: Alan R. Liss.

Freeman G. (1988) The factors that promote the development of symmetry properties in aggregates from dissociated echinoid embryos. *Wilhelm Roux's Arch.* **197**: 394–405.

Freeman G. and Lundelius J.W. (1982) The developmental genetics of dextrality and sinistrality in the gastropod *Lymnaea peregra*. *Wilhelm Roux's Arch.* **191**: 69–83.

Freeman G. and Miller R.L. (1982) Hydrozoan eggs can only be fertilized at the site of polar body formation. *Dev. Biol.* **94**: 142–152.

French V., Bryant P.J. and Bryant S.V. (1976) Pattern regulation in epimorphic fields. *Science* **193**: 969–981.

Frieden E. (1981) The dual role of thyroid hormones in vertebrate development and calorigenesis. In: *Metamorphosis: A Problem in Developmental Biology*, 2nd ed., L.I. Gilbert and E. Frieden (eds.), 545–564. New York: Plenum.

Friedrich G. and Soriano P. (1991) Promoter traps in embryonic stem cells: A genetic screen to identify and mutate developmental genes in mice. *Genes and Dev.* **5**: 1513–1523.

Frigerio G., Burri M., Bopp D., Baumgartner S. and Noll M. (1986) Structure of the segmentation gene *paired* and the Drosophila PRD gene set as part of a gene network. *Cell* **47**: 735–746.

Fristrom D. (1976) The mechanism of evagination of imaginal discs of *Drosophila melanogaster*. III. Evidence for cell rearrangement. *Dev. Biol.* **54**: 163–171.

Fristrom D., Wilcox M. and Fristrom J. (1993) The distribution of PS integrins, laminin A and F-actin during key stages in *Drosophila* wing development. *Development* **117**: 509–523.

Frohnhöfer H.G. and Nüsslein-Volhard C. (1986) Organization of anterior pattern in the *Drosophila* embryo by the maternal gene *bicoid*. *Nature* **324**: 120–125.

Frohnhöfer H.G. and Nüsslein-Volhard C. (1987) Maternal genes required for the anterior localization of *bicoid* activity in the embryo of *Drosophila*. *Genes and Dev.* **1**: 880–890.

Frydman J., Nimmesgern E., Ohtsuka K. and Hartl F.U. (1994) Folding of nascent polypeptide chains in a high molecular mass assembly with molecular chaperones. *Nature* **370**: 111–117.

Fujimori T., Miyatani S. and Takeichi M. (1990) Ectopic expression of N-cadherin perturbs histogenesis in *Xenopus* embryos. *Development* **110**: 97–104.

Fujisue M., Kobayakawa Y. and Yamana K. (1993) Occurrence of dorsal axis-inducing activity around the vegetal pole of an uncleaved *Xenopus* egg and displacement to the equatorial region by cortical rotation. *Development* **118**: 163–170.

Fuller M.T. and Wilson P.G. (1992) Force and counterforce in the mitotic spindle. *Cell* **71**: 547–550.

Fullilove S.L. and Jacobson A.G. (1971) Nuclear elongation and cytokinesis in *Drosophila montana*. *Dev. Biol.* **26**: 560–577.

Furukubo-Tokunaga K., Flister S. and Gehring W.J. (1993) Functional specificity of the Antennapedia homeodomain. *Proc. Nat. Acad. Sci. USA* **90**: 6360–6364.

Gall J.G. (1991) Spliceosomes and snurposomes. *Science* **252**: 1499–1500.

Gallicano G.I., Schwarz S.M., McGaughey R.W. and Capco D.G. (1993) Protein kinase C, a pivotal regulator of hamster egg activation, functions after elevation of intracellular free calcium. *Dev. Biol.* **156**: 94–106.

Garcia-Bellido A. (1975) Genetic control of wing disc development in *Drosophila*. In: *Cell Patterning, Ciba Foundation Symposia* **29**: 161–182. Amsterdam: Associated Scientific Publishers.

Garcia-Bellido A. and Lewis E.B. (1976) Autonomous cellular differentiation of homeotic bithorax mutants of *Drosophila melanogaster*. *Dev. Biol.* **48**: 400–410.

Garcia-Bellido A., Lawrence P.A. and Morata G. (1979) Compartments in animal development. *Scientific American* **241** (July): 102–110.

Garcia-Fernàndez J. and Holland P.W.H. (1994) Archetypal organization of the amphioxus *Hox* gene cluster. *Nature* **370**: 563–566.

Garrod A.E. (1909) *Inborn Errors of Metabolism.* Reprinted 1963 with a supplement by Harry Harris. London: Oxford University Press.

Gartler S.M. and Riggs A.D. (1983) Mammalian X-chromosome inactivation. *Ann. Rev. Genet.* **17**: 155–190.

Gasser C.S. and Fraley R.T. (1992) Transgenic crops. *Scientific American* **266** (June): 62–69.

Gasser R.F. (1975) *Atlas of Human Embryos.* New York: Harper and Row.

Gaul U. and Jäckle H. (1990) Role of gap genes in early *Drosophila* development. *Advances in Genetics* **27**: 239–275.

Gautier J., Solomon M.J., Booher R.N., Bazan J.F. and Kirschner M.W. (1991) cdc25 is a specific tyrosine phosphatase that directly activates p34^{cdc2}. *Cell* **87**: 197–211.

Gavis E.R. and Lehmann R. (1992) Localization of *nanos* RNA controls embryonic polarity. *Cell* **71**: 301–313.

Gehring W. (1968) The stability of the determined state in cultures of imaginal discs in Drosophila. In: *The Stability of the Differentiated State*, H. Ursprung (ed.), 136–154. New York: Springer-Verlag.

Gehring W.J. (1985) Homeotic genes, the homeobox, and the genetic control of development. *Cold Spring Harbor Symp. Quant. Biol.* **50**: 243–251.

Gehring W.J. (1992) The homeobox in perspective. *Trends in Biochemical Sciences* **17**: 277–280.

Gehring W.J., Müller M., Affolter M., Percival-Smith A., Billeter M., Qian Y.Q., Otting G. and Wüthrich K. (1990) The structure of the homeodomain and its functional implications. *Trends in Genetics* **6**: 323–329.

Gehring W.J., Qian Y.Q., Billeter M., Furukubo-Tokunaga K., Schier A.F., Resendez-Perez D., Affolter M., Otting G. and Wüthrich K. (1994) Homeodomain-DNA recognition. *Cell* **78**: 211–223.

Geigy R. (1941) Die Metamorphose als Folge gewebsspezifischer Determination. *Revue Suisse de Zoologie* **48**: 483–494.

Geisler R., Bergmann A., Hiromi Y. and Nüsslein-Volhard C. (1992) *Cactus*, a gene involved in dorsoventral pattern formation of *Drosophila*, is related to IκB gene family of vertebrates. *Cell* **71**: 613–621.

George E.L., Georges–Labouesse E.N., Patel-King R.S., Rayburn H. and Hynes R.O. (1993) Defects in mesoderm, neural tube and vascular development in mouse embryos lacking fibronectin. *Development* **119**: 1079–1091.

Gerace L. and Burke B. (1988) Functional organization of the nuclear envelope. *Ann. Rev. Cell Biol.* **4**: 335–374.

Gerhart J. and Keller R. (1986) Region-specific cell activities in amphibian gastrulation. *Ann. Rev. Cell Biol.* **2**: 201–229.

Gerhart J., Ubbels G., Black S., Hara K. and Kirschner M. (1981) A reinvestigation of the role of the grey crescent in axis formation in *Xenopus laevis*. *Nature* (London) **292**: 511–516.

Gerhart J., Danilchik M., Doniach T., Roberts S., Rowning B. and Stewart R. (1989) Cortical rotation of the *Xenopus* egg: Consequences for the antero-posterior pattern of embryonic dorsal development. *Development* **107** (Supplement): 37–51.

Gibson G., Schier A., LeMotte P. and Gehring W.J. (1990) The specificities of Sex combs reduced and Antennapedia are defined by a distinct portion of each protein that includes the homeodomain. *Cell* **62**: 1087–1103.

Gierer A. (1974) Hydra as a model for the development of biological form. *Scientific American* **231** (December): 44–54.

Gilbert S.F. (1988) *Developmental Biology.* 2nd ed. Sunderland, MA: Sinauer Associates.

Gilbert S.F. (1991) *Developmental Biology.* 3rd ed. Sunderland, MA: Sinauer Associates.

Giles R.E., Blanc H., Cann H.M. and Wallace D.C. (1980) Maternal inheritance of human mitochondrial DNA. *Proc. Nat. Acad. Sci. USA* **77**: 6715–6719.

Gillespie L.L., Paterno G.D. and Slack J.M.W. (1989) Analysis of competence: Receptors for fibroblast growth factor in early *Xenopus* embryos. *Development* **106**: 203–208.

Gimlich R.L. (1986) Acquisition of developmental autonomy in the equatorial region of the *Xenopus* embryo. *Dev. Biol.* **115**: 340–352.

Gimlich R.L. and Cook J. (1983) Cell lineage and the induction of nervous systems in amphibian development. *Nature* **306**: 471–473.

Gimlich R.L. and Gerhart J.C. (1984) Early cellular interactions promote embryonic axis formation in *Xenopus laevis. Dev. Biol.* **104**: 117–130.

Glabe C.G. (1985) Interaction of the sperm adhesive protein, bindin, with phospholipid vesicles. II. Bindin induces the fusion of mixed-phase vesicles that contain phosphatidylcholine and phosphatidylserine in vitro. *J. Cell Biol.* **100**: 800–806.

Glabe C.G. and Clark D. (1991) The sequence of the *Arbacia punctulata* bindin cDNA and implications for the structural basis of species-specific sperm adhesion and fertilization. *Dev. Biol.* **143**: 282–288.

Glabe C.G. and Lennarz W.J. (1979) Species-specific sperm adhesion in sea urchins: A quantitative investigation of bindin-mediated egg agglutination. *J. Cell Biol.* **83**: 595–604.

Glover D.M. and Hames B.D. (1989) *Genes and Embryos.* Oxford: IRL Press.

Glover D.M., Gonzalez C. and Raff J.W. (1993) The centrosome. *Scientific American* **268** (June): 62–68.

Golde D.W. (1991) The stem cell. *Scientific American* **265** (December): 86–93.

Goldschmidt R. (1938) *Physiological Genetics.* New York: McGraw-Hill.

Goldstein B. (1992) Induction of gut in *Caenorhabditis elegans* embryos. *Nature* **357**: 255–257.

Goldstein B. (1993) Establishment of gut fate in the E lineage of *C. elegans*: The role of lineage-dependent mechanisms and cell interactions. *Development* **118**: 1267–1277.

Goossen B., Caughman S.W., Harford J.B., Klausner R.D. and Hentze M.W. (1990) Translational repression by a complex between the iron-responsive element of ferritin messenger RNA and its specific cytoplasmic binding protein is position-dependent *in vivo. EMBO J.* **9**: 4127–4133.

Gordon J.W. (1988) Transgenic mice. In: *Developmental Genetics of Higher Organisms: A Primer in Developmental Biology,* G.M. Malacinski (ed.), 477–498. New York: Macmillan.

Gordon J.W., Scangos G.A., Plotkin D.J., Barbosa J.A. and Ruddle F.H. (1980) Genetic transformation of mouse embryos by microinjection of purified DNA. *Proc. Nat. Acad. Sci. USA* **77**: 7380–7384.

Gordon R. (1985) A review of the theories of vertebrate neurulation and their relationship to the mechanics of neural tube birth defects. *J. Embryol. Exp. Morphol.* **89** (Supplement): 229–255.

Gordon R. and Brodland G.W. (1987) The cytoskeletal mechanics of brain morphogenesis: Cell state splitters cause primary neural induction. *Cell Biophysics* **11**: 177–238.

Gordon R. and Jacobson A.G. (1978) The shaping of tissues in embryos. *Scientific American* **238** (June): 106–113.

Gorman M., Kuroda M.I. and Baker B.S. (1993) Regulation of the sex-specific binding of the maleless dosage compensation protein to the male X chromosome in *Drosophila. Cell* **72**: 39–49.

Gorski R.A., Gordon J.H., Shryne J.E. and Southam A.M. (1978) Evidence for a morphological sex difference within the medial preoptic area of the rat brain. *Brain Res.* **143**: 333–346.

Gould A.P. and White R.A.H. (1992) *Connectin,* a target of homeotic gene control in *Drosophila. Development* **116**: 1163–1174.

Gould S.J. (1977) *Ontogeny and Phylogeny.* Cambridge, MA: Belknap/Harvard.

Gould S.J. (1990) An earful of jaw. *Natural History* **3**: 12–23.

Goulding M.D., Lumsden A. and Gruss P. (1993) Signals from the notochord and floor plate regulate the region-specific expression of two Pax genes in the developing spinal cord. *Development* **117**: 1001–1016.

Govind S. and Steward R. (1991) Dorsoventral pattern formation in *Drosophila*: Signal transduction and nuclear targeting. *Trends in Genetics* **7**: 119–125.

Grace M., Bagchi M., Ahmad F., Yeager T., Olson C., Chakravarty I., Nasron N., Banerjee A. and Gupta N.K. (1984) Protein synthesis in rabbit reticulocytes: Characteristics of the protein factor RF that reverses inhibition of protein synthesis in heme-deficient reticulocyte lysates. *Proc. Nat. Acad. Sci. USA* **79**: 6517–6521.

Graham A., Papalopulu N. and Krumlauf R. (1989) The murine and Drosophila homeobox gene complexes have common features of organization and expression. *Cell* **57**: 367–378.

Graham C.F. and Wareing P.F. (1976) *The Developmental Biology of Plants and Animals.* Philadelphia: Saunders.

Graham G.J., Wright E.G., Hewick R., Wolpe S.D., Wilkie N.M., Donaldson D., Lorimore S. and Pragnell I.B. (1990) Identification and characterization of an inhibitor of haemopoietic stem cell proliferation. *Nature* **344**: 442–444.

Grainger J.L. and Winkler M. (1987) Fertilization triggers unmasking of maternal mRNAs in sea urchin eggs. *Mol. Cell. Biol.* **7**: 3947–3954.

Grainger R.M., Henry J.J. and Henderson R.A. (1988) Reinvestigation of the role of the optic vesicle in embryonic lens induction. *Development* **102:** 517–526.

Grant P. (1978) *Biology of Developing Systems.* New York: Holt, Rinehart and Winston.

Green G.R. and Poccia E.L. (1985) Phosphorylation of sea urchin sperm H1 and H2B histones precedes chromatin decondensation and H1 exchange during pronuclear formation. *Dev. Biol.* **108:** 235–245.

Green M.C. (1989) Catalog of mutant genes and polymorphic loci. In: *Genetic Variants and Strains of the Laboratory Mouse,* 2nd ed., M.F. Lyon and A.G. Searle (eds.), 12–403. New York: Oxford University Press.

Greenwald I.S., Sternberg P.W. and Horvitz H.R. (1983) The *lin-12* locus specifies cell fates in *Caenorhabditis elegans. Cell* **34:** 435–444.

Grenningloh G., Bieber A.J., Rehm E.J., Snow P.M., Traquina Z.R., Hortsch M., Patel N.H. and Goodman C.S. (1990) Molecular genetics of neuronal recognition in *Drosophila*: Evolution and function of immunoglobulin superfamily cell adhesion molecules. *Cold Spring Harbor Symp. Quant. Biol.* **55:** 327–340.

Grenningloh G., Rehm E.J. and Goodman C.S. (1991) Genetic analysis of growth cone guidance in *Drosophila*: Fasciclin II functions as a neuronal recognition molecule. *Cell* **67:** 45–57.

Griffith F. (1928) The significance of pneumococcal types. *J. Hyg.* (London) **27:** 113–159.

Grimes G.W. (1990) Inheritance of cortical patterns in ciliated protozoa. In: *Cytoplasmic Organization Systems,* G.M. Malacinski (ed.), 23–43. New York: McGraw-Hill.

Grimes G.W. and Aufderheide K.J. (1991) *Cellular Aspects of Pattern Formation: The Problem of Assembly.* Basel/New York: Karger.

Grimes G.W. and Hammersmith R.L. (1980) Analysis of the effects of encystment and excystment on incomplete doublets of *Oxytricha fallax. J. Embryol. exp. Morphol.* **59:** 19–26.

Grimes G.W., McKenna M.E., Goldsmith-Spoegler C.M. and Knaupp E.A. (1980) Patterning and assembly of ciliature are independent processes in hypotrich ciliates. *Science* **209:** 281–283.

Gross P.R. and Cousineau G.H. (1964) Macromolecular synthesis and the influence of actinomycin D on early development. *Exp. Cell Res.* **33:** 368–395.

Groudine M. and Weintraub H. (1981) Activation of globin genes during chicken development. *Cell* **24:** 393–401.

Groudine M., Eisenman R. and Weintraub H. (1981) Chromatin structure of endogenous retroviral genes and activation by an inhibitor of DNA methylation. *Nature* (London) **292:** 311–317.

Grunstein M. (1992) Histones as regulators of genes. *Scientific American* **267** (October): 68–74B.

Grunz H. and Tacke L. (1986) The inducing capacity of the presumptive endoderm of *Xenopus laevis* studied by transfilter experiments. *Wilhelm Roux's Arch.* **195:** 467–473.

Gu W., Schneider J.W., Condorelli G., Kaushal S., Mahdavi V. and Nadal-Ginard B. (1993) Interaction of myogenic factors and the retinoblastoma protein mediates muscle cell commitment and differentiation. *Cell* **72:** 309–324.

Guay P.S. and Guild G.M. (1991) The ecdysone-induced puffing cascade in *Drosophila* salivary glands: A broad-complex early gene regulates intermolt and late gene transcription. *Genetics* **129:** 169–175.

Gubbay J., Collignon J., Koopman, P., Capel B., Economou A., Münsterberg A., Vivian N., Goodfellow P. and Lovell-Badge R. (1990) A gene mapping to the sex-determining region of the mouse Y chromosome is a member of a novel family of embryonically expressed genes. *Nature* **346:** 245–250.

Gubbay J., Vivian N., Economou A., Jackson D., Goodfellow P. and Lovell-Badge R. (1992) Inverted repeat structure of the *Sry* locus in mice. *Proc. Nat. Acad. Sci. USA* **89:** 7953–7957.

Gudernatsch J.F. (1912) Feeding experiments on tadpoles. I. The influence of specific organs given as food on growth and differentiation. A contribution to the knowledge of organs with internal secretion. *Wilhelm Roux's Arch.* **35:** 457–483.

Gulyas B.J. (1975) A reexamination of the cleavage patterns in eutherian mammalian eggs: Rotation of the blastomere pairs during second cleavage in the rabbit. *J. Exp. Zool.* **193:** 235–248.

Gumbiner B. (1990) Generation and maintenance of epithelial cell polarity. *Current Opinion in Cell Biology* **2:** 881–887.

Gundersen R.W. (1987) Response of sensory neurites and growth cones to patterned substrata of laminin and fibronectin *in vitro. Dev. Biol.* **121:** 423–431.

Günther J. (1971) Entwicklungsfähigkeit, Geschlechtsverhältnis und Fertilität von *Pimpla turionellae* L. (Hymenoptera, Ichneumonidae) nach Röntgenbestrahlung oder Abschnürung des Eihinterpols. *Zool. Jahrb. Anat. Bd.* **88:** 1–46.

Gurdon J.B. (1962) The developmental capacity of nuclei taken from intestinal epithelial cells of feeding tadpoles. *J. Embryol. exp. Morphol.* **10:** 622–640.

Gurdon J.B. (1968) Transplanted nuclei and cell differentiation. *Scientific American* **219** (December): 24–35.

Gurdon J.B. (1988) A community effect in animal development. *Nature* **336:** 772–774.

Gurdon J.B., Fairman S., Mohun T.J. and Brennan S. (1985) Activation of muscle-specific actin genes in Xenopus development by an induction between animal and vegetal cells of a blastula. *Cell* **41:** 913–922.

Guthrie S. and Lumsden A. (1991) Formation and regeneration of rhombomere boundaries in the developing chick hindbrain. *Development* **112:** 221–229.

Guthrie S., Muchamore I., Kuroiwa A., Marshall H., Krumlauf R. and Lumsden A. (1992) Neuroectodermal autonomy of *Hox-2.9* expression revealed by rhombomere transpositions. *Nature* **356:** 157–159.

Guthrie S., Prince V. and Lumsden A. (1993) Selective dispersal of avian rhombomere cells in orthotopic and heterotopic grafts. *Development* **118**: 527–538.

Guyette W.A., Matusik R.J. and Rosen J.M. (1979) Prolactin-mediated transcriptional and post-transcriptional control of casein gene expression. *Cell* **17**: 1013–1023.

Gwatkin R.B.L. (1977) *Fertilization Mechanisms in Man and Mammals.* New York: Plenum.

Hadek R. (1969) *Mammalian Fertilization: An Atlas of Ultrastructure.* New York: Academic Press.

Hadorn E. (1948) Genetische und entwicklungsphysiologische Probleme der Insektenontogenese. *Folia Biotheoretica* **3**: 109–126.

Hadorn E. (1963) Differenzierungsleistungen wiederholt fragmentierter Teilstücke männlicher Genitalscheiben von *Drosophila melanogaster* nach Kultur *in vivo. Dev. Biol.* **7(1)**: 617–629.

Hadorn E. (1968) Transdetermination in cells. *Scientific American* **219** (November): 110–120.

Haecker V. (1912) Untersuchungen über Elementareigenschaften. *Zeitschr. indukt. Abst. Vererb.* **8**: 36–47.

Hafen E., Levine M. and Gehring W. (1984) Regulation of *Antennapedia* transcript distribution by the *bithorax* complex in *Drosophila. Nature* **307**: 287–289.

Hagedorn H.H. and Kunkel J.G. (1979) Vitellogenin and vitellin in insects. *Ann. Rev. Entomol.* **24**: 474–505.

Hall B.K. and Hörstadius S. (1988) *The Neural Crest.* London: Oxford University Press.

Hall P.A. and Watt F.M. (1989) Stem cells: The generation and maintenance of cellular diversity. *Development* **106**: 619–633.

Halprin K.M. (1972) Epidermal "turnover time"—A reexamination. *J. Invest. Dermatol.* **86**: 14–19.

Hamaguchi M.S. and Hiramoto Y. (1980) Fertilization process in the heart urchin, *Clypeaster japonicus,* observed with a differential interference microscope. *Dev. Growth Differ.* **22**: 517–530.

Hamburger V. (1960) *A Manual of Experimental Embryology.* Rev. ed. Chicago/London: University of Chicago Press.

Hamburger V. (1988) *The Heritage of Experimental Embryology: Hans Spemann and the Organizer.* New York: Oxford University Press.

Hamburger V. and Hamilton H.L. (1951) A series of normal stages in the development of the chick embryo. *J. Morph.* **88**: 49–92.

Hamburger V. and Levi-Montalcini R. (1949) Proliferation, differentiation and degeneration in the spinal ganglia of the chick embryo under normal and experimental conditions. *J. Exp. Zool.* **111**: 457–501.

Hampson E. and Kimura D. (1992) Sex differences and hormonal influences on cognitive function in humans. In: *Behavioral Endocrinology,* J.B. Becker, S.M. Breedlove and D. Crews (eds.), 347–400. Cambridge, MA: MIT Press.

Han M. and Sternberg P.W. (1990) *let-60,* a gene that specifies cell fates during C. elegans vulval induction, encodes a *ras* protein. *Cell* **63**: 921–931.

Hanahan D. (1989) Transgenic mice as probes into complex systems. *Science* **246**: 1265–1274.

Handyside A.H. (1978) Time of commitment of inside cells isolated from preimplantation mouse embryos. *J. Embryol. exp. Morphol.* **45**: 37–53.

Hara K. (1977) The cleavage pattern of the axolotl egg studied by cinematography and cell counting. *Wilhelm Roux's Arch.* **181**: 73–87.

Hardin J. (1988) The role of secondary mesenchyme cells during sea urchin gastrulation studied by laser ablation. *Development* **103**: 317–324.

Hardin J. (1989) Local shifts in position and polarized motility drive cell arrangement during sea urchin gastrulation. *Dev. Biol.* **136**: 430–445.

Hardin J. and Keller R. (1988) The behavior and function of bottle cells during gastrulation of *Xenopus laevis. Development* **103**: 211–230.

Hardin J. and McClay D.R. (1990) Target recognition by the archenteron during sea urchin gastrulation. *Dev. Biol.* **142**: 86–102.

Hardin J.D. and Cheng L.Y. (1986) The mechanisms and mechanics of archenteron elongation during sea urchin gastrulation. *Dev. Biol.* **115**: 490–501.

Harding C.F. (1983) Hormonal specificity and activation of social behavior in the male zebra finch. In: *Hormones and Behaviour in Higher Vertebrates,* J. Balthazart, E. Prove and R. Gilles (eds.), 275–289. Berlin: Springer-Verlag.

Harley V.R., Jackson D.I., Hextall P.J., Hawkins J.R., Berkovitz G.D., Sockanathan S., Lovell-Badge R. and Goodfellow P.N. (1992) DNA binding activity of recombinant *SRY* from normal males and XY females. *Science* **255**: 453–456.

Harper J.W., Adami G.R., Wei N., Keyomarsi K. and Elledge S.J. (1993) The p21 Cdk-interacting protein Cip1 is a potent inhibitor of G1 cyclin-dependent kinases. *Cell* **75**: 805–816.

Harris H. (1974) *Nucleus and Cytoplasm.* 3rd ed. Oxford: Clarendon Press.

Harrison P.R., Birnie G.D., Hell A., Humphries S., Young R.D. and Paul J. (1974) Kinetic studies of gene frequency. I. Use of a DNA copy of reticulocyte 9 S RNA to estimate globin gene dosage in mouse tissues. *J. Mol. Biol.* **84**: 539–554.

Harrison R.G. (1918) Experiments on the development of the forelimb of *Ambystoma,* a self-differentiating equipotential system. *J. Exp. Zool.* **25**: 413–461.

Harrison R.G. (1924) Some unexpected results of the heteroplastic transplantation of limbs. *Proc. Nat. Acad. Sci. USA* **10**: 69–74.

Harrison R.G. (1929) Correlation in the development and growth of the eye studied by means of heteroplastic transplantation. *Wilhelm Roux's Arch.* **120**: 1–55.

Harrison R.G. (1935) Heteroplastic grafting in embryology. *The Harvey Lectures, 1933–1934:* 116–157.

Harvey E.B. (1940) A comparison of the development of nucleate and non-nucleate eggs of *Arbacia punctulata. Biol. Bull.* **79**: 166–187.

Harvey E.B. (1956) *The American Arbacia and Other Sea Urchins.* Princeton, NJ: Princeton University Press.

Hasty P., Bradley A., Morris J.H., Edmondson D.G., Venuti J.M., Olson E.N. and Klein W.H. (1993) Muscle deficiency and neonatal death in mice with a targeted mutation in the *myogenin* gene. *Nature* **364**: 501–506.

Hathaway H.J. and Shur B.D. (1992) Cell surface β1,4-galactosyltransferase mediates neural crest cell migration and neurulation *in vivo. J. Cell Biol.* **117**: 369–382.

Hatta K., Kimmel C.B., Ho R.K. and Walker C. (1991) The cyclops mutation blocks specification of the floor plate of the zebrafish central nervous system. *Nature* **350**: 339–441.

Hay E.D. (1981) Collagen and embryonic development. In: *Cell Biology of Extracellular Matrix*. E.D. Hay (ed.), 379–409. New York: Plenum.

Hay E.D., ed. (1991a) *Cell Biology of Extracellular Matrix* 2nd ed. New York: Plenum.

Hay E.D. (1991b) Collagen and other matrix glycoproteins in embryogenesis. In: *Cell Biology of Extracellular Matrix*, 2nd ed., E.D. Hay (ed.), 419–462. New York: Plenum.

Hayashi S. and Scott M.P. (1990) What determines the specificity of action of Drosophila homeodomain proteins? *Cell* **63**: 883–894.

He H.T., Barbet J., Chaix J.-C. and Goridis C. (1986) Phosphatidyl-inositol is involved in the membrane attachment of N-CAM 120, the smallest component of the neural cell adhesion molecule. *EMBO J.* **5**: 2489–2494.

Heasman J., Ginsberg D., Geiger B., Goldstone K., Pratt T., Yoshida-Noro C. and Wylie C. (1994) A functional test for maternally inherited cadherin in *Xenopus* shows its importance in cell adhesion at the blastula stage. *Development* **120**: 49–57.

Heemskerk J. and DiNardo S. (1994) *Drosophila hedgehog* acts as a morphogen in cellular patterning. *Cell* **76**: 449–460.

Heemskerk J., DiNardo S., Kostriken R. and O'Farrell P.H. (1991) Multiple modes of *engrailed* regulation in the progression towards cell fate determination. *Nature* **352**: 404–410.

Hegner R.W. (1927) *College Zoology*. New York: Macmillan.

Held L.I., Jr. (1992) *Models for Embryonic Periodicity*. Basel: Karger.

Hemmati-Brivanlou A. and Melton D.A. (1992) A truncated activin receptor inhibits mesoderm induction and formation of axial structures in *Xenopus* embryos. *Nature* **359**: 609–614.

Hemmingsen S.M., Woodford C., van der Vies S.M., Tilly K., Dennis D.T., Georgopoulos C.P., Hendrix R.W. and Elliz R.J. (1988) Homologous plant and bacterial proteins chaperone oligomeric protein assembly. *Nature* **333**: 330–334.

Hengartner M.O. and Horvitz H.R. (1994) *C. elegans* survival gene *ced-9* encodes a functional homolog of the mammalian proto-oncogene *bcl-2. Cell* **8**: 1613–1626.

Hengartner M.O., Ellis R.E. and Horvitz H.R. (1992) *Caenorhabditis elegans* gene *ced-9* protects cells from programmed cell death. *Nature* **356**: 494–499.

Henikoff S. and Meneely P.M. (1993) Unwinding dosage compensation. *Cell* **72**: 1–2.

Henry J.J., Amemiya S., Wray G.A. and Raff R.A. (1989) Early inductive interactions are involved in restricting cell fates of mesomeres in sea urchin embryos. *Dev. Biol.* **136**: 140–153.

Henry J.J., Wray G.A. and Raff R.A. (1990) The dorsoventral axis is specified prior to first cleavage in the direct developing sea urchin *Heliocidaris erythrogramma. Development* **110**: 875–884.

Herman R.K. (1984) Analysis of genetic mosaics of the nematode *Caenorhabditis elegans. Genetics* **108**: 165–180.

Herman R.K. and Hedgecock E.M. (1990) Limitation of the size of the vulval primordium of *Caenorhabditis elegans* by *lin-15* expression in surrounding hypodermis. *Nature* **348**: 169–171.

Herrmann B.G., Labeit S., Poustka A., King T.R. and Lehrach H. (1990) Cloning of the T gene required in mesoderm formation in the mouse. *Nature* **343**: 617–622.

Hill D.P. and Strome S. (1990) Brief cytochalasin-induced disruption of microfilaments during a critical interval in 1-cell *C. elegans* embryos alters the partitioning of developmental instructions to the 2-cell embryo. *Development* **108**: 159–172.

Hill R.J. and Sternberg P.W. (1992) The gene *lin-3* encodes an inductive signal for vulval development in *C. elegans. Nature* **358**: 470–476.

Hill R.J. and Sternberg P.W. (1993) Cell fate patterning during *C. elegans* vulval development. *Development* (Supplement): 9–18.

Hillis D.M. and Green D.M. (1990) Evolutionary changes of heterogametic sex in the phylogenetic history of amphibians. *J. Evol. Biol.* **3**: 49–64.

Hinchliffe J.R. and Johnson D.R. (1980) *The Development of the Vertebrate Limb: An Approach through Experiment, Genetics, and Evolution*. Oxford: Clarendon Press.

Hinton H.E. (1970) Insect egg shells. *Scientific American* **223** (August): 84–91.

Hirano S., Nose A., Hatta K., Kawakami A. and Takeichi M. (1987) Calcium-dependent cell-cell adhesion molecules (cadherins): Subclass-specificities and possible involvement of actin bundles. *J. Cell Biol.* **105**: 2501–2510.

Hiromi Y. and Gehring W.J. (1987) Regulation and function of the Drosophila segmentation gene *fushi tarazu. Cell* **50**: 963–974.

His W. (1874) *Unsere Körperform und das physiologische Problem ihrer Enstehung: Briefe an einen befreundeten Naturforscher*. Leipzig: F.C.W. Vogel.

Hisatake K., Roeder R.G. and Horikoshi M. (1993) Functional dissection of TFIIB domains required for TFIIB-TFIID-promoter complex formation and basal transcription activity. *Nature* **363**: 744–747.

Hoch M., Seifert E. and Jäckle H. (1991) Gene expression mediated by *cis*-acting sequences of the *Krüppel* gene in response to the *Drosophila* morphogens *bicoid* and *hunchback. EMBO J.* **10**: 2267–2278.

Hoch M., Gerwin N., Taubert H. and Jäckle H. (1992) Competition for overlapping sites in the regulatory region of the *Drosophila* gene *Krüppel*. *Science* **256**: 94–97.

Hockenbery D.M., Oltvai Z.N., Yin X.–M., Milliman C.L. and Korsmeyer S.J. (1993) *Bcl-2* functions in an antioxidant pathway to prevent apoptosis. *Cell* **75**: 241–251.

Hodgkin J. (1983) Two types of sex determination in a nematode. *Nature* **304**: 267–268.

Hodgkin J. (1985) Males, hermaphrodites, and females: Sex determination in *Caenorhabditis elegans*. *Trends in Genetics* **1**: 85–88.

Hodgkin J. (1990) Sex determination compared in *Drosophila* and *Caenorhabditis*. *Nature* **344**: 721–728.

Hodgkin J. (1992) Genetic sex determination mechanisms and evolution. *BioEssays* **14**: 253–261.

Hoffman A., Sinn E., Yamamoto T., Wang J., Roy A., Horikoshi M. and Roeder R.G. (1990) Highly conserved core domain and unique N terminus with presumptive regulatory motifs in a human TATA factor (TFIID). *Nature* **346**: 387–390.

Holowacz T. and Elinson R.P. (1993) Cortical cytoplasm, which induces dorsal axis formation in *Xenopus*, is inactivated by UV irradiation of the oocyte. *Development* **119**: 277–285.

Holstein T., Schaller H.C. and David C.N. (1986) Nerve cell differentiation in Hydra requires two signals. *Dev. Biol.* **115**: 9–17.

Holstein T.W. and David C.N. (1990) Putative intermediates in the nerve cell differentiation pathway in Hydra have properties of multipotent stem cells. *Dev. Biol.* **142**: 401–405.

Holtfreter J. (1933) Die totale Exogastrulation, eine Selbstablösung des Ektoderms vom Entomesoderm. *Wilhelm Roux's Arch.* **129**: 669–793.

Holtfreter J. (1936) Regionale Induktionen in xenoplastisch zusammengesetzten Explantaten. *Wilhelm Roux's Arch.* **134**: 446–550.

Holtfreter J. (1938) Differenzierungspotenzen isolierter Teile der Anurengastrula. *Wilhelm Roux's Arch.* **138**: 657–738.

Holtfreter J. (1943) A study of the mechanics of gastrulation: Part I. *J. Exp. Zool.* **94**: 261–318.

Holtfreter J. (1944) A study of the mechanics of gastrulation: Part II. *J. Exp. Zool.* **95**: 171–212.

Holtfreter J. (1945) Neuralization and epidermization of gastrula ectoderm. *J. Exp. Zool.* **98**: 161–208.

Holtfreter J. (1946) Structure, motility and locomotion in isolated embryonic amphibian cells. *J. Morph.* **79**: 27–62.

Holwill S., Heaseman J., Crawley C.R. and Wylie C.C. (1987) Axis and germ line deficiencies caused by u.v. irradiation of *Xenopus* oocytes cultured *in vitro*. *Development* **100**: 735–743.

Honig L.S. and Summerbell D. (1985) Maps of strength of positional signalling activity in the developing chick wing bud. *J. Embryol. exp. Morphol.* **87**: 163–174.

Hopper A.F. and Hart N.H. (1985) *Foundations of Animal Development.* 2nd ed. New York: Oxford University Press.

Hopwood N.D., Pluck A. and Gurdon J.B. (1989) MyoD expression in the forming somites is an early response to mesoderm induction in *Xenopus* embryos. *EMBO J.* **8**: 3409–3417.

Hopwood N.D., Pluck A., Gurdon J.B. and Dilworth S.M. (1992) Expression of XMyoD protein in early *Xenopus laevis* embryos. *Development* **114**: 31–38.

Horder T.J., Witkowski J.A. and Wylie C.C., eds. (1985) *A History of Embryology.* Cambridge: Cambridge University Press.

Hörstadius S. (1973) *Experimental Embryology of Echinoderms.* Oxford: Clarendon Press.

Horvitz H.R. and Sternberg P.W. (1991) Multiple intercellular signalling systems control the development of the *Caenorhabditis elegans* vulva. *Nature* **351**: 535–541.

Hoshijima K., Inoue K., Higuchi I., Sakamoto H. and Shimura Y. (1991) Control of *doublesex* alternative splicing by *transformer* and *transformer-2* in *Drosophila*. *Science* **252**: 833–836.

Houliston E. (1994) Microtubule translocation and polymerization during cortical rotation in *Xenopus* eggs. *Development* **120**: 1213–1220.

Houliston E. and Elinson R.P. (1991) Evidence for the involvement of microtubules, ER, and kinesin in the cortical rotation of fertilized frog eggs. *J. Cell Biol.* **114**: 1017–1028.

Howard K., Ingham P. and Rushlow C. (1988) Region-specific alleles of the *Drosophila* segmentation gene *hairy*. *Genes and Dev.* **2**: 1037–1046.

Howard K.R. and Struhl G. (1990) Decoding positional information: Regulation of the pair-rule gene *hairy*. *Development* **110**: 1223–1231.

Hozumi N. and Tonegawa S. (1976) Evidence for somatic rearrangement of immunoglobulin genes coding for variable and constant regions. *Proc. Nat. Acad. Sci. USA* **73**: 3628–3632.

Huebner R.J. and Todaro G.J. (1969) Oncogenes of RNA tumor viruses as determinants of cancer. *Proc. Nat. Acad. Sci. USA* **64**: 1087–1094.

Hülskamp M. and Tautz D. (1991) Gap genes and gradients—The logic behind the gaps. *BioEssays* **13**: 261–268.

Hülskamp M., Schröder C., Pfeifle C., Jäckle H. and Tautz D. (1989) Posterior segmentation of the *Drosophila* embryo in the absence of a maternal posterior organizer gene. *Nature* **338**: 629–632.

Hunt P., Guilsano M., Cook M., Sham M., Faiella A., Wilkinson D., Boncinelli E. and Krumlauf R. (1991a) A distinct *Hox* code for the branchial region of the vertebrate head. *Nature* **353**: 861–864.

Hunt P., Whiting J., Muchamore I., Marshall H. and Krumlauf R. (1991b) Homeobox genes and models for patterning the hindbrain and branchial arches. *Development* (Supplement 1): 187–196.

Hunter C.P. and Wood W.B. (1992) Evidence from mosaic analysis of the masculinizing gene *her-1* for cell interactions in *C. elegans* sex determination. *Nature* **355**: 551–555.

Hunter T. (1991) Cooperation between oncogenes. *Cell* **64**: 249–270.

Hunter T. and Karin M. (1992) The regulation of transcription by phosphorylation. *Cell* **70**: 375–387.

Hurley D.L., Angerer L.M. and Angerer R.C. (1989) Altered expression of spatially regulated embryonic genes in the progeny of separated sea urchin blastomeres. *Development* **106**: 567–579.

Hursh D.A., Padgett R.W. and Gelbart W.M. (1993) Cross regulation of *decapentaplegic* and *Ultrabithorax* transcription in the embryonic visceral mesoderm of *Drosophila*. *Development* **117**: 1211–1222.

Huxley J.S. (1932) *Problems of Relative Growth*. New York: Dial.

Hynes R.O. (1992) Integrins: Versatility, modulation, and signaling in cell adhesion. *Cell* **69**: 11–25.

Hynes R.O. and Destree A.T. (1978) Relationships between fibronectin (LETS protein) and actin. *Cell* **15**: 875–886.

Hynes R.O. and Lander A.D. (1992) Contact and adhesive specificities in the associations, migrations, and targeting of cells and axons. *Cell* **68**: 303–322.

Iatrou K., Spira A.W. and Dixon G.H. (1978) Protamine messenger RNA: Evidence for early synthesis and accumulation during spermatogenesis in rainbow trout. *Dev. Biol.* **64**: 82–98.

Iavarone A., Garg P., Lasorella A., Hsu J. and Israel M.A. (1994) The helix-loop-helix protein Id-2 enhances cell proliferation and binds to the retinoblastoma protein. *Genes and Dev.* **8**: 1270–1284.

Illmensee K. (1973) The potentialities of transplanted early gastrula nuclei of *Drosophila melanogaster*. Production of their imago descendants by germline transplantation. *Wilhelm Roux's Arch.* **171**: 331–343.

Illmensee K. and Mahowald A.P. (1974) Transplantation of posterior polar plasm in *Drosophila*. Induction of germ cells at the anterior pole of the egg. *Proc. Nat. Acad. Sci. USA* **71**: 1016–1020.

Illmensee K., Mahowald A.P. and Loomis M.R. (1976) The ontogeny of germ plasm during oogenesis in *Drosophila*. *Dev. Biol.* **49**: 40–65.

Imperato-McGinley J., Guerrero L., Gautier T. and Peterson R.E. (1974) Steroid 5-alpha-reductase deficiency in man: An inherited form of male pseudohermaphroditism. *Science* **186**: 1213–1215.

Ingham P.W. (1988) The molecular genetics of embryonic pattern formation in *Drosophila*. *Nature* **335**: 25–34.

Ingham P.W. (1990) Genetic control of segmental patterning in the *Drosophila* embryo. In: *Genetics of Pattern Formation and Growth Control*, A.P. Mahowald (ed.), *48th Symp. Soc. Dev. Biol.*, 181–196. New York: Wiley/Alan R. Liss.

Ingham P.W. and Hidalgo A. (1993) Regulation of *wingless* transcription in the *Drosophila* embryo. *Development* **117**: 283–291.

Ingham P.W. and Martinez-Arias A. (1992) Boundaries and fields in early embryos. *Cell* **68**: 221–235.

Ingham P.W., Baker N.E. and Martinez-Arias A. (1988) Regulation of segment polarity genes in the *Drosophila* blastoderm by *fushi tarazu* and *even skipped*. *Nature* **331**: 73–75.

Ip Y.T., Kraut R., Levine M. and Rushlow C.A. (1991) The *dorsal* morphogen is a sequence-specific DNA-binding protein that interacts with a long-range repression element in *Drosophila*. *Cell* **64**: 439–446.

Ip Y.T., Park R.E., Kosman D., Yazdanbakhsh K. and Levine M. (1992a) *dorsal-twist* interactions establish *snail* expression in the presumptive mesoderm of the *Drosophila* embryo. *Genes and Dev.* **6**: 1518–1530.

Ip Y.T., Park R.E., Kosman D., Bier E. and Levine M. (1992b) The *dorsal* gradient morphogen regulates stipes of *rhomboid* expression in the presumptive neuroectoderm of the *Drosophila* embryo. *Genes and Dev.* **6**: 1728–1739.

Irish V., Lehmann R. and Akam M. (1989) The *Drosophila* posterior-group gene *nanos* functions by repressing *hunchback* activity. *Nature* **338**: 646–648.

Irvine K.D., Helfand S.L. and Hogness D.S. (1991) The large upstream control region of the *Drosophila* homeotic gene *Ultrabithorax*. *Development* **111**: 407–424.

Isaacs H.V., Tannahill D. and Slack J.M.W. (1992) Expression of a novel FGF in the *Xenopus* embryo. A new candidate inducing factor for mesoderm formation and anteroposterior specification. *Development* **114**: 711–720.

Izpisúa-Belmonte J.-C. and Duboule D. (1992) Homeobox genes and pattern formation in the vertebrate limb. *Dev. Biol.* **152**: 26–36.

Izpisúa-Belmonte J.C., Tickle C., Dolle P., Wolpert L. and Duboule D. (1991) Expression of the homeobox *Hox-4* genes and the specification of position in chick wing development. *Nature* **350**: 585–589.

Jack T., Brockman L.L. and Meyerowitz E.M. (1992) The homeotic gene *APETALA3* of Arabidopsis thaliana encodes a MADS-box and is expressed in petals and stamens. *Cell* **68**: 683–697.

Jack T., Fox G.L. and Meyerowitz E.M. (1994) Arabidopsis homeotic gene *APETALA3* ectopic expression: Transcriptional and posttranscriptional regulation determine floral organ identity. *Cell* **76**: 703–716.

Jäckle H. and Eagleson G.W. (1980) Spatial distribution of abundant proteins in oocytes and fertilized eggs of the Mexican axolotl (*Ambystoma mexicanum*). *Dev. Biol.* **75**: 492–499.

Jäckle H., Tautz D., Schuh R., Seifert E. and Lehmann R. (1986) Cross-regulatory interactions among the gap genes of *Drosophila*. *Nature* **324**: 668–670.

Jackson I.J. (1989) The mouse. In: *Genes and Embryos*, D.M. Glover and B.D. Hames (eds.), 165–221. Oxford: IRL Press.

Jackson R.J. (1993) Cytoplasmic regulation of mRNA function: The importance of the 3' untranslated region. *Cell* **74**: 9–14.

Jackson R.J. and Standart N. (1990) Do the poly(A) tail and 3' untranslated region control mRNA translation? *Cell* **62**: 15–24.

Jacob F. and Monod J. (1961) Genetic regulatory mechanisms in the synthesis of proteins. *J. Mol. Biol.* **3**: 318–356.

Jacobs P.S. and Strong J.A. (1959) A case of human intersexuality having a possible XXY sex determining mechanism. *Nature* **183:** 302–303.

Jacobson A.G. (1966) Inductive processes in embryonic development. *Science* **152:** 25–34.

Jacobson A.G. (1978) Some forces that shape the nervous system. *Zoon* **6:** 13–21.

Jacobson A.G. (1981) Morphogenesis of the neural plate and tube. In: *Morphogenesis and Pattern Formation*, T.G. Connelly, L.L. Brinkley and B.M. Carlson (eds.), 233–263. New York: Raven Press.

Jacobson A.G. (1985) Adhesion and movement of cells may be coupled to produce neurulation. In: *The Cell in Contact: Adhesions and Junctions as Morphogenetic Determinants*, G.M. Edelman and J.P. Thiery (eds.), 49–65. New York: Wiley.

Jacobson A.G. (1988) Somitomeres: Mesodermal segments of vertebrate embryos. *Development* **104** (Supplement): 208–220.

Jacobson A.G. (1994) Normal neurulation in amphibia. In: *Neural Tube Defects*, G. Bock and J. Marsh (eds.), 6–24. New York: Wiley.

Jacobson A.G. and Gordon R. (1976) Changes in the shape of the developing vertebrate nervous system analyzed experimentally, mathematically and by computer simulation. *J. Exp. Zool.* **197:** 191–246.

Jacobson A.G. and Sater A.K. (1988) Features of embryonic induction. *Development* **104:** 341–359.

Jacobson A.G., Oster G.F., Odell G.M. and Cheng L.Y. (1986) Neurulation and the cortical tractor model for epithelial folding. *J. Embryol. exp. Morphol.* **96:** 19–49.

Jaffe L.A. (1976) Fast block to polyspermy in sea urchins is electrically mediated. *Nature* **261:** 68–71.

Jaffe L.A. and Gould M. (1985) Polyspermy-preventing mechanisms. In: *Biology of Fertilization*, C.B. Metz and A. Monroy (eds.), vol. 3, 223–249. New York: Academic Press.

Jaffe L.F. (1966) Electrical currents through the developing *Fucus* egg. *Proc. Nat. Acad. Sci. USA* **56:** 1102–1109.

Jäger R.J., Anvret M., Hall K. and Scherer G. (1990) A human XY female with a frame shift mutation in the candidate testis-determining gene *SRY. Nature* **348:** 452–454.

Jeannottee L., Lemieux M., Charron J., Poirier F. and Robertson E. (1993) Specification of axial identity in the mouse: Role of the *Hoxa-5* (*Hox1.3*) gene. *Genes and Dev.* **7:** 2085–2096.

Jeffery W.R. (1984) Pattern formation by ooplasmic segregation in ascidian eggs. *Biol. Bull.* **166:** 277–298.

Jeffery W.R. and Meier S. (1983) A yellow crescent cytoskeletal domain in ascidian eggs and its role in early development. *Dev. Biol.* **96:** 125–143.

Jegla D.E. and Sussex I.M. (1989) Cell lineage patterns in the shoot meristem of the sunflower embryo in the dry seed. *Dev. Biol.* **131:** 215–225.

Jen Y., Weintraub H. and Benezra R. (1992) Overexpression of Id protein inhibits the muscle differentiation program: In vivo association of Id with E2A proteins. *Genes and Dev.* **6:** 1466–1479.

Jenkins F.A. (1969) The evolution and development of the dens of the mammalian axis. *Anat. Rec.* **164:** 173–184.

Jeppesen P. and Turner B.M. (1993) The inactive X chromosome in female mammals is distinguished by a lack of histone H4 acetylation, a cytogenetic marker for gene expression. *Cell* **74:** 281–289.

Jiang J. and Levine M. (1993) Binding affinities and cooperative interactions with bHLH activators delimit threshold responses to the dorsal gradient morphogen. *Cell* **72:** 741–752.

Jiang J., Hoey T. and Levine M. (1991) Autoregulation of a segmentation gene in *Drosophila*: Combinatorial interaction of the *even-skipped* homeo box protein with a distal enhancer element. *Genes and Dev.* **5:** 265–277.

Johnson F. and Bottjer S.W. (1994) Afferent influences on cell death and birth during development of a cortical nucleus necessary for learned vocal behavior in zebra finches. *Development* **120:** 13–24.

Johnson K.E., Nakatsuji N. and Boucaut J.-C. (1990) Extracellular matrix control of cell migration during amphibian gastrulation. In: *Cytoplasmic Organization Systems*, G.M. Malacinski (ed.), 349–374. New York: McGraw-Hill.

Johnson M.H., Maro B. and Takeichi M. (1986) The role of cell adhesion in the synchronization and orientation of polarization in 8-cell mouse blastomeres. *J. Embryol. exp. Morphol.* **93:** 239–255.

Johnston M.C. (1966) A radioautographic study of the migration and fate of cranial neural crest cells in the chick embryo. *Anat. Rec.* **156:** 143–156.

Jones B. and McGinnis W. (1993) The regulation of empty spiracles by abdominal-B mediates an abdominal segment identity function. *Genes and Dev.* **7:** 229–240.

Jones F.S., Prediger E.A., Bittner D.A., De Robertis E.M. and Edelman G.M. (1992a) Cell adhesion molecules as targets for *Hox* genes: Neural cell adhesion molecule promoter activity is modulated by cotransfection with *Hox-2.5* and *-2.4. Proc. Nat. Acad. Sci. USA* **89:** 2086–2090.

Jones F.S., Chalepakis G., Gruss P. and Edelman G.M. (1992b) Activation of the cytotactin promoter by the homeobox-containing gene *Evx-1. Proc. Nat. Acad. Sci. USA* **89:** 2091–2095.

Josso N., Picard M.-Y. and Tran D. (1976) The antimüllerian hormone. *Rec. Prog. Horm. Res.* **33:** 117–167.

Jost A. (1965) Gonadal hormones in the sex determination of the mammalian foetus. In: *Organogenesis*, R.L. DeHann and H. Ursprung (eds.), 611–628. New York: Holt, Rinehart and Winston.

Judd B.H. and Young M.W. (1973) An examination of the one cistron-one chromomere concept. *Cold Spring Harbor Symp. Quant. Biol.* **38:** 573–579.

Jurášková V. and Tkadleček L. (1965) Character of primary and secondary colonies of hematopoiesis in the spleen of irradiated mice. *Nature* **206:** 951–952.

Jürgens G., Mayer U., Ruiz R.A.T., Berleth T. and Miséra S. (1991) Genetic analysis of pattern formation in the *Arabidopsis* embryo. *Development* (Supplement 1): 27–38.

Kafatos F.C. (1976) Sequential cell polymorphism: A fundamental concept in developmental biology. *Adv. Insect Physiol.* **12**: 1–15.

Kafatos F.C., Regier J.C., Mazur G.D., Nadel M.R., Blau H.M., Petri W.H., Wyman A.R., Gelinas R.E., Moore P.B., Paul M., Efstratiadis A., Vournakis J.N., Goldsmith M.R., Hunsley J.R., Baker B., Nardi J. and Koehler M. (1977) The egg shell of insects: Differentiation-specific proteins and the control of their synthesis and accumulation during development. In: *Results and Problems in Cell Differentiation*, W. Beerman (ed.), vol. 8, 45–143. Berlin/New York: Springer-Verlag.

Kahn C.R., Coyle J.T. and Cohen A.M. (1980) Head and trunk neural crest *in vitro*: Autonomic neuron differentiation. *Dev. Biol.* **77**: 340–348.

Kalthoff K. and Elbetieha A. (1986) Transplantation of localized anterior determinants in *Chironomus* eggs by microinjection. *J. Embryol. Exp. Morphol.* **97** (Supplement): 181–196.

Kalthoff K., Rau K.-G. and Edmond J.C. (1982) Modifying effects of ultraviolet irradiation on the development of abnormal body patterns in centrifuged insect embryos (*Smittia* sp., Chironomidae, Diptera). *Dev. Biol.* **91**: 413–422.

Kandel E.R., Schwartz J.H. and Jessell T.M. (1991) *Principles of Neuroscience.* 3rd ed. New York: Elsevier.

Kao K.R., Masui Y. and Elinson R.P. (1986) Lithium-induced respecification of pattern in *Xenopus laevis* embryos. *Nature* **322**: 371–373.

Kaplan D.R., Hempstead B.L., Martin-Zanca D., Chao M.V. and Parada L.F. (1991) The *trk* proto-oncogene product: A signal transducing receptor for nerve growth factor. *Science* **252**: 554–558.

Kappen C., Schughart K. and Ruddle F.H. (1989) Two steps in the evolution of Antennapedia-class vertebrate homeobox genes. *Proc. Nat. Acad. Sci. USA* **86**: 5459–5463.

Karim F.D. and Thummel C.S. (1991) Ecdysone coordinates the timing and amounts of E74A and E74B transcription in *Drosophila*. *Genes and Dev.* **5**: 1067–1079.

Karim F.D. and Thummel C.S. (1992) Temporal coordination of regulatory gene expression by the steroid hormone ecdysone. *EMBO J.* **11**: 4083–4093.

Karin M., Castrillo J.-L. and Theill L.E. (1990) Growth hormone gene regulation: A paradigm for cell-type-specific gene activation. *Trends in Genetics* **6**: 92–96.

Karp G. and Berrill N.J. (1981) *Development.* 2nd ed. New York: McGraw-Hill.

Kauffman S.A. (1973) Control circuits for determination and transdetermination. *Science* **181**: 310–318.

Kauffman S.A. (1980) Heterotopic transplantation in the syncytial blastoderm of *Drosophila*: Evidence for anterior and posterior nuclear commitments. *Wilhelm Roux's Arch.* **189**: 135–145.

Kaufman T.C., Seeger M.A. and Olsen G. (1990) Molecular and genetic organization of the Antennapedia gene complex of *Drosophila melanogaster*. *Advances in Genetics* **27**: 309–362.

Kawahara A., Baker B.S. and Tata J.R. (1991) Developmental and regional expression of thyroid hormone receptor genes during *Xenopus* metamorphosis. *Development* **112**: 933–943.

Kay G.F., Penny G.D., Patel D., Ashworth A., Brockdorff N. and Rastan S. (1993) Expression of *Xist* during mouse development suggests a role in the initiation of X chromosome inactivation. *Cell* **72**: 171–182.

Keeton W.T. and Gould J.L. (1986) *Biological Science.* 4th ed. New York: Norton.

Keller G., Paige C., Gilboa E. and Wagner E.R. (1985) Expression of a foreign gene in myeloid and lymphoid cells derived from multipotent hematopoietic precursors. *Nature* **318**: 149–154.

Keller R. (1991) Early embryonic development of *Xenopus laevis*. In: *Methods in Cell Biology*, B.K. Kay and H.B. Peng (eds.), vol. 36, 61–113. New York: Academic Press.

Keller R. and Danilchik M. (1988) Regional expression, pattern and timing of convergence and extension during gastrulation of *Xenopus laevis*. *Development* **103**: 193–209.

Keller R., Shih J. and Domingo C. (1992a) The patterning and functioning of protrusive activity during convergence and extension of the Xenopus organiser. *Development* (Supplement): 81–91.

Keller R., Shih J. and Sater A. (1992b) The cellular basis of the convergence and extension of the *Xenopus* neural plate. *Developmental Dynamics* **193**: 199–217.

Keller R., Shih J., Sater A. and Moreno C. (1992c) Planar induction of convergence and extension of the neural plate by the organizer of *Xenopus*. *Developmental Dynamics* **193**: 218–234.

Keller R.E. (1978) Time-lapse cinemicrographic analysis of superficial cell behavior during and prior to gastrulation in *Xenopus laevis*. *J. Morphol.* **157**: 223–248.

Keller R.E. (1980) The cellular basis of epiboly: An SEM study of deep cell rearrangement during gastrulation of *Xenopus laevis*. *J. Embryol. exp. Morphol.* **60**: 201–243.

Keller R.E. (1981) An experimental analysis of the role of bottle cells and the deep marginal zone in gastrulation of *Xenopus laevis*. *J. Exp. Zool.* **216**: 81–101.

Keller R.E. (1986) The cellular basis of amphibian gastrulation. In: *Developmental Biology: A Comprehensive Synthesis*, vol. 2; *The Cellular Basis of Morphogenesis*, L.W. Browder (ed.), 241–327. New York: Plenum.

Keller R.E. and Jansa S. (1992) Xenopus gastrulation without a blastocoel roof. *Developmental Dynamics* **195**: 162–176.

Keller R.E. and Tibbetts P. (1989) Mediolateral cell intercalation in the dorsal, axial mesoderm of *Xenopus laevis*. *Dev. Biol.* **131**: 539–549.

Keller R.E., Danilchik M., Gimlich R. and Shih J. (1985) The function and mechanism of convergent extension during gastrulation of *Xenopus laevis*. *J. Embryol. exp. Morphol.* **89** (Supplement): 185–209.

Keller S.H. and Vacquier V.D. (1994) The isolation of acrosome-reaction-inducing glycoproteins from sea urchin egg jelly. *Dev. Biol.* **162:** 304–312.

Kelley D.B. and Brenowitz E. (1992) Hormonal influences on courtship behaviors. In: *Behavioral Endocrinology*, J.B. Becker, S.M. Breedlove and D. Crews (eds.), 187–218. Cambridge, MA: MIT Press.

Kelly R.O. (1981) The developing limb. In: *Morphogenesis and Pattern Formation*, T.G. Connelly, L.L. Brinkley and B.M. Carlson (eds.), 49–85. New York: Raven Press.

Kelly S.J. (1977) Studies of the developmental potential of 4- and 8-cell stage mouse blastomeres. *J. Exp. Zool.* **200:** 365–376.

Kemphues K.J. (1989) *Caenorhabditis*. In: *Genes and Embryos*, D.M. Glover and B.D. Hames (eds.), 95–126. Oxford: IRL Press.

Kemphues K.J., Priess J.R., Morton D.G. and Cheng N. (1988) Identification of genes required for cytoplasmic localization in early C. elegans embryos. *Cell* **52:** 311–320.

Kessel M. and Gruss P. (1990) Murine development control genes. *Science* **249:** 374–379.

Kessel M. and Gruss P. (1991) Homeotic transformations of murine vertebrae and concomitant alteration of *Hox* codes induced by retinoic acid. *Cell* **67:** 89–104.

Kessel M., Balling R. and Gruss P. (1990) Variations of cervical vertebrae after expression of a *Hox 1.1* transgene in mice. *Cell* **61:** 301–308.

Keyes L.N., Cline T.W. and Schedl P. (1992) The primary sex determination signal of Drosophila acts at the level of transcription. *Cell* **68:** 933–943.

Keynes R.J. and Stern C.D. (1984) Segmentation in the vertebrate nervous system. *Nature* **310:** 786–789.

Kidd S. (1992) Characterization of the *Drosophila cactus* locus and analysis of interactions between cactus and dorsal proteins. *Cell* **71:** 623–635.

Kieny M. and Chevallier A. (1979) Autonomy of tendon development in the embryonic chick wing. *J. Embryol. exp. Morphol.* **49:** 153–165.

Kim Y., Geiger J.H., Hahn S. and Sigler P.B. (1993) Crystal structure of a yeast TBP/TATA-box complex. *Nature* **365:** 512–520.

Kimble J. (1981) Alterations in cell lineage following laser ablation of cells in the somatic gonad of C. elegans. *Dev. Biol.* **87:** 286–300.

Kimelman D., Kirschner M. and Scherson T. (1987) The events of the midblastula transition in *Xenopus* are regulated by changes in the cell cycle. *Cell* **48:** 399–407.

Kimelman D., Abraham J.A., Haaparanta T., Palsi T.M. and Kirschner M.W. (1988) The presence of fibroblast growth factor in the frog egg: Its role as a natural mesoderm inducer. *Science* **242:** 1053–1056.

Kimelman D., Christian J.L. and Moon R.T. (1992) Synergistic principles of development: Overlapping patterning systems in *Xenopus* mesoderm induction. *Development* **116:** 1–9.

Kim-Ha J., Webster P.J., Smith J.L. and Macdonald P.M. (1993) Multiple RNA regulatory elements mediate distinct steps in localization of *oskar* mRNA. *Development* **119:** 169–178.

Kimura D. (1992) Sex differences in the brain. *Scientific American* **267** (September): 119–125.

King R.C. (1970) *Ovarian Development in Drosophila melanogaster*. New York: Academic Press.

King T.J. (1966) Nuclear transplantation in amphibia. *Methods Cell Physiol.* **2:** 1–36.

Kingsley D.M. (1994) The TGF-ß superfamily: New members, new receptors, and new genetic tests of function in different organisms. *Genes and Dev.* **8:** 133–146.

Kintner C.R. (1992) Regulation of embryonic cell adhesion by the cadherin cytoplasmic domain. *Cell* **69:** 225–236.

Kintner C.R. and Melton D.A. (1987) Expression of *Xenopus* N-CAM RNA in ectoderm is an early response to neural induction. *Development* **99:** 311–325.

Kirby C., Kusch M. and Kemphues K. (1990) Mutations in the *par* genes of *Caenorhabditis elegans* affect cytoplasmic reorganization during the first cell cycle. *Dev. Biol.* **142:** 203–215.

Kirkpatrick M. and Jenkins C.D. (1989) Genetic segregation and the maintenance of sexual reproduction. *Nature* **339:** 300–302.

Kirschner M.W., Gerhart J.C., Hara K. and Ubbels G.A. (1980) Initiation of the cell cycle and establishment of bilateral symmetry in *Xenopus* eggs. *38th Symp. Soc. Dev. Biol.,* 187–216.

Kliewer S.A., Umesono K., Mangelsdorf D.J. and Evans R.M. (1992) Retinoid X receptor interacts with nuclear receptors in retinoic acid, thyroid hormone and vitamin D_3 signalling. *Nature* **355:** 446–449.

Kline D., Simoncini L., Mandel G., Maue R.A., Kado R.T. and Jaffe L.A. (1988) Fertilization events induced by neurotransmitters after injection of mRNA in *Xenopus* eggs. *Science* **241:** 464–467.

Knebelmann B., Boussin L., Guerrier D., Legeai L., Kahn A., Josso N. and Picard J.-Y. (1991) Anti-Müllerian hormone Bruxelles: A nonsense mutation associated with the persistent Müllerian duct syndrome. *Proc. Nat. Acad. Sci. USA* **88:** 3767–3771.

Kocisko D.A., Come J.H., Priola S.A., Chesebro B., Raymond G.J., Lansbury P.T. and Caughey G. (1994) Cell-free formation of protease-resistant prion protein. *Nature* **370:** 471–474.

Koelle M.R., Talbot W.S., Segraves W.A., Bender M.T., Cherbas P. and Hogness D.S. (1991) The Drosophila EcR gene encodes an ecdysone receptor, a new member of the steroid receptor superfamily. *Cell* **67:** 59–77.

Köhler G. and Milstein C. (1975) Continuous cultures of fused cells secreting antibody of predefined specificity. *Nature* **256:** 495–497.

Kondrashov A.S. (1988) Deleterious mutations and the evolution of sexual reproduction. *Nature* **336:** 435–440.

Koopman P., Münsterberg A., Capel B., Vivian N. and Lovell-Badge R. (1990) Expression of a candidate sex-determining gene during mouse testis differentiation. *Nature* **348**: 450–452.

Koopman P., Gubbay J., Vivian N., Goodfellow P. and Lovell-Badge R. (1991) Male development of chromosomally female mice transgenic for *Sry*. *Nature* **351**: 117–121.

Koorneef M., Hanhart C.J. and van der Veen J.H. (1991) A genetic and physiological analysis of late flowering mutants in *Arabidopsis thaliana*. *Mol. Gen. Genet.* **229**: 57–66.

Kornberg T. (1981) *engrailed*: A gene controlling compartment and segment formation in *Drosophila*. *Proc. Nat. Acad. Sci. USA* **78**: 1095–1099.

Kraatz G. (1876) *Deutsche Entomol. Zeitschr.* **20**: 377.

Krasnow M.A., Saffman E.E., Kornfeld K. and Hogness D.S. (1989) Transcriptional activation and repression by *Ultrabithorax* proteins in cultured Drosophila cells. *Cell* **57**: 1031–1043.

Kratzer P.G. and Chapman V.M. (1981) X-chromosome reactivation in oocytes of *Mus caroli*. *Proc. Nat. Acad. Sci. USA* **78**: 3093–3097.

Krause M., Fire A., Harrison S.W., Priess J. and Weintraub H. (1990) CeMyoD accumulation defines the body wall muscle cell fate during *C. elegans* embryogenesis. *Cell* **63**: 907–919.

Kropf D.L. (1992) Establishment and expression of cellular polarity in fucoid zygotes. *Microbiol. Rev.* **56**: 316–336.

Kropf D.L., Kloareg B. and Quatrano R.S. (1988) Cell wall is required for fixation of the embryonic axis in *Fucus* zygotes. *Science* **239**: 187–190.

Kropf D.L., Berge S.K. and Quatrano R.S. (1989) Actin localization during *Fucus* embryogenesis. *Plant Cell* **1**: 191–200.

Krumlauf R. (1992) Evolution of the vertebrate *Hox* homeobox genes. *BioEssays* **14**: 245–252.

Krumlauf R. (1994) *Hox* genes in vertebrate development. *Cell* **78**: 191–201.

Kühn A. (1961) *Grundriss der allgemeinen Zoologie*. 14th ed. Stuttgart: Georg Thieme.

Kühn A. (1971) *Lectures on Developmental Physiology*. Translated by R. Milkman. Berlin/Heidelberg/New York: Springer.

Kuhn R., Schäfer U. and Schäfer M. (1988) *Cis*-acting regions sufficient for spermatocyte-specific transcriptional and spermatid-specific translational control of the *Drosophila melanogaster* gene *mst(3)gl-9*. *EMBO J.* **7**: 447–454.

Kumar S., Kinoshita M., Noda M., Copeland N.G. and Jenkins N.A. (1994) Induction of apoptosis by the mouse *Nedd2* gene, which encodes a protein similar to the *C. elegans* cell death gene *ced-3* and the mammalian IL-1ß-converting enzyme. *Genes and Dev.* **8**: 1613–1626.

Kuratani S.C. and Eichele G. (1993) Rhombomere transplantation repatterns the segmental organization of cranial nerves and reveals cell-autonomous expression of a homeodomain protein. *Development* **117**: 105–117.

Kuroda M.I., Kernan M.J., Kreber R., Ganetzky B. and Baker B.S. (1991) The *maleless* protein associates with the X chromosome to regulate dosage compensation in *Drosophila*. *Cell* **66**: 935–947.

Kuwabara P.E. and Kimble J. (1992) Molecular genetics of sex determination in *C. elegans*. *Trends in Genetics* **8**: 164–168.

Lallier T., Leblanc G., Artinger K.B. and Bronner-Fraser M. (1992) Cranial and trunk neural crest cells use different mechanisms for attachment to extracellular matrices. *Development* **116**: 531–541.

Lamb T.M., Knecht A.K., Smith W.C., Stachel S.E., Economides A.N., Stahl N., Yancopolous G.D. and Harland R.M. (1993) Neural induction by the secreted polypeptide noggin. *Science* **262**: 713–718.

Lane M.C., Koehl M.A.R., Wilt F. and Keller R.E. (1993) A role for regulated secretion of apical extracellular matrix during epithelial invagination in the sea urchin. *Development* **117**: 1049–1060.

Langman J. (1981) *Medical Embryology*. 4th ed. Baltimore: Williams and Wilkins.

Larabell C. and Nuccitelli R. (1992) Inositol lipid hydrolysis contributes to the Ca^{2+} wave in the activating egg of *Xenopus laevis*. *Dev. Biol.* **153**: 347–355.

Larsen W.J. (1993) *Human Embryology*. New York: Churchill Livingstone.

Laskey R.A. (1974) Biochemical processes in early development. In: *Companion to Biochemistry*, A.T. Bull, J.R. Lagnado, J.O. Thomas and K.F. Tipton (eds.), vol. 2, 137–160. London: Longman.

Laskey R.A. and Gurdon J.B. (1970) Genetic content of adult somatic cells tested by nuclear transplantation from cultured cells. *Nature* **228**: 1332–1334.

Laskey R.A., Mills A.D., Gurdon J.B. and Partington G.A. (1977) Protein synthesis in oocytes of *Xenopus laevis* is not regulated by the supply of messenger RNA. *Cell* **11**: 345–351.

Lassar A.B., Davis R.L., Wright W.E., Kadesch T., Murre C., Voronova A., Baltimore D. and Weintraub H. (1991) Functional activity of myogenic HLH proteins requires hetero-oligomerization with E12/E47-like proteins in vivo. *Cell* **66**: 305–315.

Laufer J.S., Bazzicalupo P. and Wood W.B. (1980) Segregation of developmental potential in early embryos of *Caenorhabditis elegans*. *Cell* **19**: 569–577.

Lawrence P.A. (1992) *The Making of a Fly: The Genetics of Animal Design*. Cambridge, MA: Blackwell.

Lawrence P.A. and Johnston P. (1989) Pattern formation in the *Drosophila* embryo: Allocation of cells to parasegments by *even-skipped* and *fushi tarazu*. *Development* **105**: 761–767.

Lawrence P.A. and Morata G. (1993) A no-wing situation. *Nature* **366**: 305–306.

Layton W.M., Jr. (1976) Random determination of a developmental process: Reversal of normal visceral asymmetry in the mouse. *J. Heredity* **67**: 336–338.

Lazaris-Karatzas A., Montine K.S. and Sonenberg N. (1990) Malignant transformation by a eukaryotic initiation factor subunit that binds to mRNA 5′ cap. *Nature* **345**: 544–547.

Le Douarin N.M. (1969) Particularités du noyau interphasique chez la caille japonaise (*Coturnix coturnix japonica*). Utilisation des ces particularités comme "marquage biologique" dans les recherches sur les interactions tissulaires et les migrations cellulaires au course de l'ontogenèse. *Bull. Biol. Fr. Belg.* **103**: 435–452.

Le Douarin N.M. (1986) Cell line segregation during peripheral nervous system ontogeny. *Science* **231**: 1515–1522.

Le Douarin N.M., Ziller C. and Couly G.F. (1993) Patterning of neural crest derivatives in the avian embryo: *In vivo* and *in vitro* studies. *Dev. Biol.* **159**: 24–49.

Lee E.Y.-H.P., Chang C.-Y., Hu N., Wang Y.-C.J., Lai C.-C., Herrup K., Lee W.-H. and Bradley A. (1992) Mice deficient for Rb are nonviable and show defects in neurogenesis and haematopoiesis. *Nature* **359**: 288–294.

Lee R.C., Feinbaum R.L. and Ambros V. (1993) The *C. elegans* heterochronic gene *lin-4* encodes small RNAs with antisense complementarity to *lin-14*. *Cell* **75**: 843–854.

Lee S., Gilula N.B. and Warner A.E. (1987) Gap junctional communication and compaction during preimplantation stages of mouse development. *Cell* **51**: 851–860.

Lehmann R. and Nüsslein-Volhard C. (1986) Abdominal segmentation, pole cell formation, and embryonic polarity require the localized activity of *oskar*, a maternal gene in *Drosophila*. *Cell* **47**: 141–152.

Lehmann R. and Nüsslein-Volhard C. (1987) *hunchback*, a gene required for segmentation of an anterior and posterior region of the *Drosophila* embryo. *Dev. Biol.* **119**: 402–417.

Lehmann R. and Nüsslein-Volhard C. (1991) The maternal gene *nanos* has a central role in posterior pattern formation in the *Drosophila* embryo. *Development* **112**: 679–691.

Lehmann R. and Rongo C. (1993) Germ plasm formation and germ cell determination. *Seminars Dev. Biol.* **4**: 149–159.

Le Mouellic H., Lallemand Y. and Brulet P. (1990) Targeted replacement of the homeobox gene *Hox-3.1* by the *Escherichia coli* lacZ in mouse chimeric embryos. *Proc. Nat. Acad. Sci. USA* **87**: 4712–4716.

Le Mouellic H., Lallemand Y. and Brulet P. (1992) Homeosis in the mouse induced by a null mutation in the *Hox-3.1* gene. *Cell* **69**: 251–264.

Lenhoff H.M. and Lenhoff S.G. (1988) Trembley's polyps. *Scientific American* **258** (April): 108–113.

Leptin M. and Grunewald B. (1990) Cell shape changes during gastrulation in *Drosophila*. *Development* **110**: 73–84.

Leptin M., Bogaert T., Lehmann R. and Wilcox M. (1989) The function of PS integrins during *Drosophila* embryogenesis. *Cell* **56**: 401–408.

Leutert R. (1975) Sex determination in *Bonellia*. In: *Intersexuality in the Animal Kingdom*, R. Reinboth (ed.), 84–90. Berlin/New York: Springer-Verlag.

LeVay S. (1991) A difference in hypothalamic structure between heterosexual and homosexual men. *Science* **253**: 1034–1037.

Levi-Montalcini R. (1964) Growth control of nerve cells by a protein factor and its antiserum. *Science* **143**: 105–110.

Levi-Montalcini R. (1988) *In Praise of Imperfection*. New York: Basic Books.

Levi-Montalcini R. and Calissano P. (1979) The nerve-growth factor. *Scientific American* **240** (June): 68–77.

Levine A., Bashan-Ahrend A., Budai-Hadrian O., Gartenberg D., Menasherow S. and Wides R. (1994) odd Oz: A novel *Drosophila* pair rule gene. *Cell* **77**: 587–598.

Levine A.J., Momand J. and Finlay C.A. (1991) The p53 tumour suppressor gene. *Nature* **351**: 453–456.

Levine M., Garen A., Lepesant J.-A. and Lepesant-Kejzlarova J. (1981) Constancy of somatic DNA organization in developmentally regulated regions of the *Drosophila* genome. *Proc. Nat. Acad. Sci. USA* **78**: 2417–2421.

Lewin B. (1980) *Gene Expression*. 2nd ed. Vol. 2. New York: Wiley/Interscience.

Lewin B. (1990) *Genes IV*. Oxford: Oxford University Press.

Lewin B. (1991) Oncogenic conversion by regulatory changes in transcription factors. *Cell* **64**: 303–312.

Lewis E.B. (1963) Genes and developmental pathways. *Am. Zool.* **3**: 33–56.

Lewis E.B. (1978) A gene complex controlling segmentation in *Drosophila*. *Nature* **276**: 565–570.

Leyser H.M.O. and Furner I.J. (1992) Characterisation of three shoot apical meristem mutants of *Arabidopsis thaliana*. *Development* **116**: 397–403.

Leyton L. and Saling P. (1989) Evidence that aggregation of mouse sperm receptors by ZP3 triggers the acrosome reaction. *J. Cell Biol.* **108**: 2163–2168.

Li L., Zhou J., James G., Heller-Harrison R., Czech M.P. and Olson E.N. (1992) FGF inactivates myogenic helix-loop-helix proteins through phosphorylation of a conserved protein kinase C site in their DNA-binding domains. *Cell* **71**: 1181–1194.

Lillie F.R. (1917) The freemartin: A study of the action of sex hormones in foetal life of cattle. *J. Exp. Zool.* **23**: 371–452.

Linsenmayer T.F. (1991) Collagen. In: *Cell Biology of Extracellular Matrix*, 2nd ed., E.D. Hay (ed.), 7–44. New York: Plenum.

Littlefield C.L. (1985) Germ cells in *Hydra oligactis* males. I. Isolation of a subpopulation of interstitial cells that is developmentally restricted to sperm production. *Dev. Biol.* **112**: 185–193.

Littlefield C.L. (1991) Cell lineages in *Hydra*: Isolation and characterization of an interstitial stem cell restricted to egg production in *Hydra oligactis*. *Dev. Biol.* **143**: 378–388.

Littlefield C.L., Dunne J.F. and Bode H.R. (1985) Spermatogenesis in *Hydra oligactis*. I. Morphological description and characterization using a monoclonal antibody specific for cells of the spermatogenic pathway. *Dev. Biol.* **110**: 308–320.

Lodish H.F. and Small B. (1976) Different lifetimes of reticulocyte messenger RNA. *Cell* **7**: 59–65.

Löfberg J., Perris R. and Epperlein H. (1989) Timing in the regulation of neural crest cell migration: Retarded "maturation" of regional extracellular matrix inhibits pigment cell migration in embryos of the white axolotl mutant. *Dev. Biol.* **131**: 168–181.

Lohs-Schardin M. (1982) *Dicephalic*—A *Drosophila* mutant affecting polarity in follicle organization and embryonic patterning. *Wilhelm Roux's Arch.* **191**: 28–36.

Lohs-Schardin M., Cremer C. and Nüsslein-Volhard C. (1979) A fate map for the larval epidermis of *Drosophila melanogaster*: Localized cuticle defects following irradiation of the blastoderm with an ultraviolet laser microbeam. *Dev. Biol.* **73**: 239–255.

London C., Akers R. and Phillips C. (1988) Expression of Epi 1, an epidermis-specific marker in *Xenopus laevis* embryos, is specified prior to gastrulation. *Dev. Biol.* **129**: 380–389.

Longo F.J. (1989) Egg cortical architecture. In: *The Cell Biology of Fertilization*, H. Schatten and G. Schatten (eds.), 105–138. San Diego: Academic Press.

Lonie A., D'Andrea R., Paro R. and Saint R. (1994) Molecular characterisation of the *Polycomblike* gene of *Drosophila melanogaster*, in *trans*-acting negative regulator of homeotic gene expression. *Development* **120**: 2629–2636.

Lopez L.C., Bayna E.M., Litoff D., Shaper N.L., Shaper J.H. and Shur B.D. (1985) Receptor function of mouse sperm surface galactosyltransferase during fertilization. *J. Cell Biol.* **101**: 1501–1510.

Lopez L.C., Youakim A., Evans S.C. and Shur B.D. (1991) Evidence for a molecular distinction between golgi and cell surface forms of β1,4-galactosyltransferase. *J. Biol. Chem.* **266**: 15984–15991.

Lopez-Antunez L. (1971) *Atlas of Human Anatomy*. Philadelphia: Saunders.

Lorca T., Cruzalegui F.H., Fesquet D., Cavadore J.-C., Méry J., Means A. and Dorée M. (1993) Calmodulin-dependent protein kinase II mediates inactivation of MPF and CSF upon fertilization in *Xenopus* eggs. *Nature* **366**: 270–273.

Lovell-Badge R. and Robertson E. (1990) XY female mice resulting from a heritable mutation in the primary testis-determining gene, *Tdy*. *Development* **109**: 635–646.

Lowe S.W., Schmitt E.M., Smith S.W., Osborne B.A. and Jacks T. (1993) p53 is required for radiation-induced apoptosis in mouse thymocytes. *Nature* **362**: 847–849.

Lu X., Perkins L.A. and Perrimon N. (1993) The torso pathway in *Drosophila*: A model system to study receptor tyrosine kinase signal transduction. *Development* (Supplement): 47–56.

Lucchesi J. and Manning J.E. (1987) Gene dosage compensation in *Drosophila melanogaster*. *Advances in Genetics* **24**: 371–429.

Luckett W.P. (1978) Origin and differentiation of the yolk sac and extraembryonic mesoderm in presomite human and rhesus monkey embryos. *Am. J. Anat.* **152**: 59–58.

Lufkin T., Dierich A., LeMeur M., Mark M. and Chambon P. (1991) Disruption of the *Hox-1.6* homeobox gene results in defects in a region corresponding to its rostral domain of expression. *Cell* **66**: 1105–1119.

Lufkin T., Mark M., Hart C.P., Dollé P., LeMeur M. and Chambon P. (1992) Homeotic transformation of the occipital bones of the skull by ectopic expression of a homeobox gene. *Nature* **359**: 835–841.

Lumsden A. (1991) Motorizing the spinal cord. *Cell* **64**: 471–473.

Luna E.J. and Hitt A.L. (1992) Cytoskeleton-plasma membrane interactions. *Science* **258**: 955–964.

Lund E.J. (1922) Electrical control of polarity in an egg. *Proc. Soc. Exp. Biol.* **20**: 113.

Lutz D.A. and Inoué S. (1982) Colcemid but not cytochalasin inhibits asymmetric nuclear positioning prior to unequal cell division. *Biol. Bull.* **163**: 373–374.

Lutz D.A., Hamaguchi Y. and Inoué S. (1988) Micromanipulation studies of the asymmetric positioning of the maturation spindle in *Chaetopterus* sp. oocytes. I. Anchorage of the spindle to the cortex and migration of a displaced spindle. *Cell Motility and the Cytoskeleton* **11**: 83–96.

Lyndon R.F. (1990) *Plant Development*. London: Unwin Hyman.

Lynn J.W., McCulloh D.H. and Chambers E.L. (1988) Voltage clamp studies on fertilization in sea urchin eggs. II. Current patterns in relation to sperm entry, nonentry, and activation. *Dev. Biol.* **128**: 305–323.

Lyon M.F. (1961) Gene action in the X-chromosome of the mouse (*Mus musculus L.*). *Nature* **190**: 372–373.

Ma H. (1994) The unfolding drama of flower development: Recent results from genetic and molecular analyses. *Genes and Dev.* **8**: 745–756.

Ma K., Inglis J.D., Sharkey A., Bickmore W.A., Hill R.E., Prosser E.J., Speed R.M., Thomson E.J., Jobling M., Taylor K., Wolfe J., Cooke H.J., Hargreave T.B. and Chandley A.C. (1993) A Y chromosome gene family with RNA-binding protein homology: Candidates for the azoospermia factor AZF controlling human spermatogenesis. *Cell* **75**: 1287–1295.

Macdonald P.M. (1990) *bicoid* mRNA localization signal: Phylogenetic conservation of function and RNA secondary structure. *Development* **110**: 161–171.

Macdonald P.M. (1992) The *Drosophila pumilio* gene: An unusually long transcription unit and an unusual protein. *Development* **114**: 221–232.

Macdonald P.M. and Struhl G. (1988) *Cis*-acting sequences responsible for anterior localization of *bicoid* mRNA in *Drosophila* embryos. *Nature* **336**: 595–598.

Maden M. (1993) The homeotic transformation of tails into limbs in *Rana temporaria* by retinoids. *Dev. Biol.* **159**: 379–391.

Madl J.E. and Herman R.K. (1979) Polyploids and sex determination in *Caenorhabditis elegans*. *Genetics* **93:** 393–402.

Mahowald A.P. (1971) Polar granules of *Drosophila*. IV. Cytochemical studies showing loss of RNA from polar granules during early stages of embryogenesis. *J. Exp. Zool.* **176:** 345–352.

Mahowald A.P. (1972) Oogenesis. In: *Developmental Systems: Insects*, S.J. Counce (ed.), 1–47. New York: Academic Press.

Mahowald A.P., Illmensee K. and Turner F.R. (1976) Interspecific transplantation of polar plasm between *Drosophila* embryos. *J. Cell Biol.* **70:** 358–373.

Mahowald A.P., Allis C.D., Karrer K.M., Underwood E.M. and Waring G.L. (1979) Germ plasm and pole cells of *Drosophila*. In: *Determinants of Spatial Organization*, S. Subtelny and I.R. Konigsberg (eds.), *37th Symp. Soc. Dev. Biol.*, 127–146. New York: Academic Press.

Malacinski G.M., Brothers A.J. and Chung H.-M. (1977) Destruction of components of the neural induction system of the amphibian egg with ultraviolet irradiation. *Dev. Biol.* **56:** 24–39.

Malicki J., Schughart K. and McGinnis W. (1990) Mouse *Hox-2.2* specifies thoracic segmental identity in *Drosophila* embryos and larvae. *Cell* **63:** 961–967.

Mandel M.A., Gustafson-Brown C., Savidge B. and Yanofsky M.F. (1992) Molecular characterization of the *Arabidopsis* floral homeotic gene *APETALA1*. *Nature* **360:** 273–277.

Mangelsdorf D.J. and Evans R.M. (1992) Retinoid receptors as transcription factors. In: *Transcriptional Regulation*, K. Yamamoto and S.L. McKnight (eds.), chapt. 42. Cold Spring Harbor, NY: Cold Spring Harbor Laboratory Press.

Mangiarotti G., Zuker C., Chisholm R.L. and Lodish H.F. (1983) Different mRNAs have different nuclear transit times in *Dictyostelium discoideum* aggregates. *Mol. Cell. Biol.* **3:** 1511–1517.

Mangold O. (1933) Über die Induktionsfähigkeit der verschiedenen Bezirke der Neurula von Urodelen. *Naturwissenschaften* **21:** 761–766.

Maniatis T. (1991) Mechanisms of alternative pre-mRNA splicing. *Science* **251:** 33–34.

Mann R.S. and Hogness D.S. (1990) Functional dissection of *Ultrabithorax* proteins in *D. melanogaster*. *Cell* **60:** 597–610.

Marcum B.A. and Campbell R.D. (1978) Development of Hydra lacking nerve and interstitial cells. *J. Cell Sci.* **29:** 17–33.

Markert C.L. and Petters R.M. (1978) Manufactured hexaparental mice show that adults are derived from three embryonic cells. *Science* **202:** 56–58.

Marler P., Peters S., Ball G.F., Dufty A.M., Jr. and Wingfield J.C. (1988) The role of sex steroids in the acquisition and production of birdsong. *Nature* **336:** 770–772.

Marshall C.J. (1991) Tumor suppressor genes. *Cell* **64:** 313–326.

Marshall H., Nonchev S., Sham M.H., Muchamore I., Lumsden A. and Krumlauf R. (1992) Retinoic acid alters hindbrain *Hox* code and induces transformation of rhombomeres 2/3 into a 4/5 identity. *Nature* **360:** 737–741.

Martin J.-R., Raibaud A. and Ollo R. (1994) Terminal pattern elements in *Drosophila* embryo induced by the torso-like protein. *Nature* **367:** 741–745.

Martindale M.Q., Doe C.Q. and Morrill J.B. (1985) The role of animal-vegetal interaction with respect to the determination of dorsoventral polarity in the equal-cleaving spiralian, *Lymnaea palustris*. *Wilhelm Roux's Arch.* **194:** 281–295.

Martinez-Arias A. and Lawrence P.A. (1985) Parasegments and compartments in the *Drosophila* embryo. *Nature* **313:** 639–642.

Martin-Zanca D., Barbacid M. and Parada L.F. (1990) Expression of the *trk* proto-oncogene is restricted to the sensory cranial and spinal ganglia of neural crest origin in mouse development. *Genes and Dev.* **4:** 683–694.

Maslanski J.A., Leshko L. and Busa W.B. (1992) Lithium-sensitive production of inositol phosphates during amphibian embryonic mesoderm induction. *Science* **256:** 243–245.

Mason A.J., Pitts S.L., Nikolics K., Szonyi E., Wilcox J.N., Seeburg P.H. and Stewart T.A. (1986) The hypogonadal mouse: Reproductive functions restored by gene therapy. *Science* **234:** 1372–1378.

Masuda M. and Sato H. (1984) Asynchronization of cell division is concurrently related with ciliogenesis in sea urchin blastulae. *Dev. Growth Differ.* **26:** 281–294.

Masui Y. and Markert C.L. (1971) Cytoplasmic control of nuclear behavior during meiotic maturation of frog oocytes. *J. Exp. Zool.* **19:** 129–146.

Mayer U., Torres Ruiz R.A., Berleth T., Miséra S. and Jürgens G. (1991) Mutations affecting body organization in the *Arabidopsis* embryo. *Nature* **353:** 402–407.

Mayer U., Büttner G. and Jürgens G. (1993) Apical-basal pattern formation in the *Arabidopsis* embryo: Studies on the role of the *gnom* gene. *Development* **117:** 149–162.

Maynard Smith J. (1989) *Evolutionary Genetics*. New York: Oxford University Press.

McCain E.R. and McClay D.R. (1994) The establishment of bilateral asymmetry in sea urchin embryos. *Development* **120:** 395–404.

McCarrey J.R. and Dilworth D.D. (1992) Expression of *Xist* in mouse germ cells correlates with X-chromosome inactivation. *Nature Genetics* **2:** 200–203.

McClay D.R. and Ettensohn C.A. (1987) Cell recognition during sea urchin gastrulation. In: *Genetic Regulation of Development*, W.F. Loomis (ed.), 111–128. New York: Alan R. Liss.

McClay D.R., Armstrong N.A. and Hardin J. (1992) Pattern formation during gastrulation in the sea urchin embryo. *Development* (Supplement): 33–41.

McClintock B. (1952) Chromosome organization and genic expression. *Cold Spring Harbor Symp. Quant. Biol.* **16:** 13–47.

McClung C.E. (1902) The accessory chromosome—Sex determinant? *Biol. Bull.* **3:** 43–84.

McCormick F. (1989) *ras* GTPase activating protein: Signal transmitter and signal terminator. *Cell* **56:** 5–8.

McCrea P.D. and Gumbiner B.M. (1991) Purification of a 92-kDa cytoplasmic protein tightly associated with the cell-cell adhesion molecule E-cadherin (uvomorulin). *J. Biol. Chem.* **266:** 4514–4520.

McDaniel C.N. and Poethig R.S. (1988) Cell-lineage patterns in the shoot apical meristem of the germinating maize embryo. *Planta* **175:** 13–22.

McElreavey K., Vilain E., Abbas N., Herskowitz I. and Fellous M. (1993) A regulatory cascade hypothesis for mammalian sex determination: SRY represses a negative regulator of male development. *Proc. Nat. Acad. Sci. USA* **90:** 3368–3372.

McGinnis N., Kuziora M.A. and McGinnis W. (1990) Human *Hox-4.2* and *Drosophila deformed* encode similar regulatory specificities in *Drosophila* embryos and larvae. *Cell* **63:** 969–976.

McGinnis W. and Krumlauf R. (1992) Homeobox genes and axial patterning. *Cell* **68:** 283–302.

McGinnis W., Levine M.S., Hafen E., Kuroiwa A. and Gehring W.J. (1984) A conserved DNA sequence in homoeotic genes of the *Drosophila* Antennapedia and bithorax complexes. *Nature* **308:** 428–433.

McGrath J. and Solter D. (1984) Completion of mouse embryogenesis requires both the maternal and the paternal genome. *Cell* **37:** 179–183.

McGrew L.L., Dworkin-Rastl E., Dworkin M.B. and Richter J.D. (1989) Poly(A) elongation during *Xenopus* oocyte maturation is required for translational recruitment and is mediated by a short sequence element. *Genes and Dev.* **3:** 803–815.

McIntosh J.R. and McDonald K.L. (1989) The mitotic spindle. *Scientific American* (October) **261:** 48–56.

McKeown M., Belote J.M. and Boggs R.T. (1988) Ectopic expression of the female *transformer* gene product leads to female differentiation of chromosomally male Drosophila. *Cell* **53:** 887–895.

McKinnell R.G. (1978) *Cloning: Nuclear Transplantation in Amphibia.* Minneapolis: University of Minnesota Press.

McKinney M.L. and McNamara K.J. (1991) *Heterochrony: The Evolution of Ontogeny.* New York: Plenum.

McKnight G.S., Pennequin P. and Schimke R.T. (1975) Induction of ovalbumin mRNA sequences by estrogen and progesterone in chick oviduct as measured by hybridization to complementary DNA. *J. Biol. Chem.* **250:** 8105–8110.

McKnight S.L. (1991) Molecular zippers in gene regulation. *Scientific American* (April) **264:** 54–64.

McLaren A. (1976) *Mammalian Chimeras.* Cambridge: Cambridge University Press.

McLaren A. (1991) Development of the mammalian gonad: The fate of the supporting cell lineage. *BioEssays* **13:** 151–156.

Meedel T.H., Crowthier R.J. and Whittaker J.R. (1987) Determinative properties of muscle cell lineages in ascidian embryos. *Development* **100:** 245–260.

Mege R.-M., Matsuzaki F., Gallin W.J., Goldberg J.I., Cunningham B.A. and Edelman G.M. (1988) Construction of epithelioid sheets by transfection of mouse sarcoma cells with cDNAs for chicken cell adhesion molecules. *Proc. Nat. Acad. Sci. USA* **85:** 7274–7278.

Meier S. (1979) Development of the chick embryo mesoblast. *Dev. Biol.* **73:** 25–45.

Meinke D.W. (1991a) Genetic analysis of plant development. In: *Plant Physiology: A Treatise*, vol. 10, *Growth and Development*, F.C. Steward and R.G.S. Bidwell (eds.), 437–490. New York: Academic Press.

Meinke D.W. (1991b) Perspectives on genetic analysis of plant embryogenesis. *Plant Cell* **3:** 857–866.

Mello C.C., Draper B.W., Krause M., Weintraub H. and Priess J.R. (1992) The *pie-1* and *mex-1* genes and maternal control of blastomere identity in early C. elegans embryos. *Cell* **70:** 163–176.

Melton D.A. (1987) Translocation of a localized maternal mRNA to the vegetal pole of *Xenopus* oocytes. *Nature* **328:** 80–82.

Mermod J.J., Schatz G. and Crippa M. (1980) Specific control of messenger translation in *Drosophila* oocytes and embryos. *Dev. Biol.* **75:** 177–186.

Meshcheryakov V. (1978) Orientation of cleavage spindles in pulmonate molluscs. I. Role of blastomere shape in orientation of second cleavage spindles. *Ontogenez* **9:** 559–566.

Metcalf D. (1977) *Hemopoietic Colonies.* Berlin/Heidelberg: Springer-Verlag.

Metcalf D. (1989) The molecular control of cell division, differentiation commitment and maturation in hemopoietic cells. *Nature* **339:** 27–30.

Metcalf D. and Moore M.A.S. (1971) *Hemopoietic Cells.* Amsterdam: North-Holland Publishing Company.

Meyer W.J., III, Migeon B.R. and Migeon C.J. (1975) Locus on human X chromosome for dihydrotestosterone receptor and androgen insensitivity. *Proc. Nat. Acad. Sci. USA* **72:** 1469–1472.

Meyerowitz E.M., Bowman J.L., Brockman L.L., Drews G.N., Jack T., Sieburth L.E. and Weigel D. (1991) A genetic and molecular model for flower development in *Arabidopsis thaliana. Development* (Supplement 1): 157–167.

Michod R.E. and Levin B.R., eds. (1987) *The Evolution of Sex: An Examination of Current Ideas.* Sunderland, MA: Sinauer Associates.

Migeon B.R. (1971) Studies of skin fibroblasts from ten families with HGPRT deficiency, with reference to X-chromosomal inactivation. *Am. J. Hum. Genet.* **23:** 199–209.

Millar S.E., Chamow S.M., Baur A.W., Oliver C., Robey F. and Dean J. (1989) Vaccination with a synthetic zona pellucida peptide produces long-term contraception in female mice. *Science* **246:** 935–938.

Miller A.D. (1992) Human gene therapy comes of age. *Nature* **357:** 455–460.

Miller D.J., Macek M.B. and Shur B.D. (1992) Complementarity between sperm surface β-1,4-galactosyltransferase and egg-coat ZP3 mediates sperm-egg binding. *Nature* **357:** 589–593.

Miller L.M., Plenefisch J.D., Casson L.P. and Meyer B.J. (1988) *xol-1*: A gene that controls the male modes of both sex determination and dosage compensation in *C. elegans*. *Cell* **55**: 167–183.

Miller L.M., Gallegos M.E., Morisseau B.A. and Kim S.K. (1993) *lin-31*, a *Caenorhabditis elegans* HNF-3/ *fork head* transcription factor homolog, specifies three alternative cell fates in vulval development. *Genes and Dev.* **7**: 933–947.

Miller M., Park M.K. and Hanover J.A. (1991) Nuclear pore complex: Structure, function, and regulation. *Physiol. Rev.* **71**: 909–949.

Miller R. (1966) Chemotaxis during fertilization in the hydroid *Campanularia*. *J. Exp. Zool.* **161**: 23–44.

Miller R. (1978) Site-specific agglutination and the timed release of a sperm chemoattractant by the egg of the leptomedusan, Orthopyxis caliculata. *J. Exp. Zool.* **205**: 385–392.

Miller R. (1985) Sperm chemoattraction in the metazoa. In: *Biology of Fertilization*, C.B. Metz, Jr. and A. Monroy (eds.), vol. 2, 275–337. Orlando, FL: Academic Press.

Minshull J. (1993) Cyclin synthesis: Who needs it? *BioEssays* **15**: 149–155.

Mintz B. (1965) Experimental genetic mosaicism in the mouse. In: *Preimplantation Stages of Pregnancy*, G.E. Wolstenholme and M. O'Connor (eds.), *Ciba Foundation Symp.*, 194–207. Boston: Little, Brown.

Mintz B. and Illmensee K. (1975) Normal genetically mosaic mice produced from malignant teratocarcinoma cells. *Proc. Nat. Acad. Sci. USA* **72**: 3585–3589.

Mitchison T.J. (1989) Mitosis—From molecules to machine. *Am. Zool.* **29**: 523–535.

Mittenthal J.E. and Jacobson A.G. (1990) The mechanics of morphogenesis in multicellular embryos. In: *Biomechanics of Active Movement and Deformation of Cells*, NATO ASI Series, vol. H42, N. Akkas (ed.), 320–366. Berlin: Springer-Verlag.

Mittwoch U. (1985) Erroneous theories of sex determination. *J. Med. Genet.* **22**: 164–170.

Miyazaki S.-I. (1988) Inositol 1,4,5-trisphosphate-induced calcium release and guanine nucleotide-binding protein-mediated periodic calcium rises in golden hamster eggs. *J. Cell Biol.* **106**: 345–353.

Miyazaki S.-I., Yuzaki M., Nakada K., Shirakawa H., Nakanishi S., Nakade S. and Mikoshiba K. (1992) Block of Ca^{2+} wave and Ca^{2+} oscillation by antibody to the inositol 1,4,5-trisphosphate receptor in fertilized hamster eggs. *Science* **257**: 251–255.

Mizukami Y. and Ma H. (1992) Ectopic expression of the floral homeotic gene *AGAMOUS* in transgenic *Arabidopsis* plants alters floral organ identity. *Cell* **71**: 119–131.

Moore A.R. and Burt A.S. (1939) On the locus and nature of the forces causing gastrulation in the embryos of *Dendraster excentricus*. *J. Exp. Zool.* **82**: 159–171.

Moore G.D., Kopf G.S. and Schultz R.M. (1993) Complete mouse egg activation in the absence of sperm by stimulation of an exogenous G protein-coupled receptor. *Dev. Biol.* **159**: 669–678.

Moore K.L. (1982) *The Developing Human: Clinically Oriented Embryology*. 3rd ed. Philadelphia: Saunders.

Morata G. and Lawrence P.A. (1975) Control of compartment development by the *engrailed* gene in *Drosophila*. *Nature* **255**: 614–617.

Morata G. and Lawrence P.A. (1978) Cell lineage and homeotic mutants in the development of imaginal discs of *Drosophila*. In: *The Clonal Basis of Development*, S. Subtelny and I.M. Sussex (eds.), 45–60. New York: Academic Press.

Morgan B.A., Izpisúa-Belmonte J.-C., Duboule D. and Tabin C.J. (1992) Targeted misexpression of *Hox-4.6* in the avian limb bud causes apparent homeotic transformations. *Nature* **358**: 236–239.

Morgan T.H. (1903) The relation between normal and abnormal development of the embryo of the frog, as determined by the effects of lithium chloride in solution. *Wilhelm Roux's Arch.* **16**: 691–712.

Morgan T.H. (1927) *Experimental Embryology*. New York: Columbia University Press.

Morisato D. and Anderson K.V. (1994) The *spätzle* gene encodes a component of the extracellular signal pathway establishing the dorsal-ventral pattern of the Drosophila embryo. *Cell* **76**: 677–688.

Morrill J.B. (1986) Scanning electron microscopy of embryos. In: *Methods in Cell Biology*, T. Schroeder (ed.), 263–292. New York: Academic Press.

Morrill J.B. and Santos L.L. (1985) A scanning electron micrographical overview of cellular and extracellular patterns during blastulation and gastrulation in the sea urchin, *Lytechinus variegatus*. In: *The Cellular and Molecular Biology of Invertebrate Development*, R.H. Sawyer and R.M. Showman (eds.), 3–33. Columbia: University of South Carolina Press.

Moses H.M., Yang E.Y. and Pietenpol J.A. (1990) TGF–ß stimulation and inhibition of cell proliferation: New mechanistic insights. *Cell* **63**: 245–247.

Mouches C., Pauplin Y., Agarwal M., Lemieux L., Herzog M., Abadon M., Beyssat-Arnaouty V., Hyrien O., de Saint Vincent B.R., Georghiou G.P. and Pasteur N. (1990) Characterization of amplification core and esterase B1 gene responsible for insecticide resistance in *Culex*. *Proc. Nat. Acad. Sci. USA* **87**: 2574–2578.

Moury J.D. and Jacobson A.G. (1989) Neural fold formation at newly created boundaries between neural plate and epidermis in the axolotl. *Dev. Biol.* **133**: 44–57.

Moury J.D. and Jacobson A.G. (1990) The origins of neural crest cells in the axolotl. *Dev. Biol.* **141**: 243–253.

Mowry K.L. and Melton D.A. (1992) Vegetal messenger RNA localization directed by a 340-nt RNA sequence element in *Xenopus* oocytes. *Science* **255**: 991–994.

Moy G.W. and Vacquier V.D. (1979) Immuno-peroxidase localization of bindin during the adhesion of sperm to sea urchin eggs. *Current Topics Dev. Biol.* **13**: 31–44.

Mullins M.C., Hammerschmidt M., Haffter P. and Nüsslein-Volhard C. (1994) Large-scale mutagenesis in the zebrafish: In search of genes controlling development in a vertebrate. *Current Biol.* **4**: 189–202.

Muneoka K. and Bryant S.V. (1982) Evidence that patterning mechanisms in developing and regenerating limbs are the same. *Nature* **298**: 369–371.

Münsterberg A. and Lovell-Badge R. (1991) Expression of the mouse anti-Müllerian hormone gene suggests a role in both male and female sexual differentiation. *Development* **113**: 613–624.

Murray A.W. (1992) Creative blocks: Cell-cycle checkpoints and feedback controls. *Nature* **359**: 599–604.

Murray A.W. and Kirschner M.W. (1989) Cyclin synthesis drives the early embryonic cell cycle. *Nature* **339**: 275–280.

Murre C., McCaw P.S. and Baltimore D. (1989) A new DNA binding and dimerization motif in immunoglobulin enhancer binding, *daughterless*, *MyoD*, and *myc* proteins. *Cell* **56**: 777–783.

Murtha M.T., Leckman J.F. and Ruddle F.H. (1991) Detection of homeobox genes in development and evolution. *Proc. Nat. Acad. Sci. USA* **88**: 10711–10715.

Myles D.G. (1993) Molecular mechanisms of sperm-egg membrane binding and fusion in mammals. *Dev. Biol.* **158**: 35–45.

Nabeshima Y., Hanaoka K., Hayasaka M., Esumi E., Li S., Nonaka I. and Nabeshima Y.-I. (1993) *Myogenin* gene disruption results in perinatal lethality because of severe muscle defect. *Nature* **364**: 532–535.

Nagafuchi A. and Takeichi M. (1988) Cell binding function of E-cadherin is regulated by the cytoplasmic domain. *EMBO J.* **7**: 3679–3684.

Nagafuchi A., Shirayoshi Y., Okazaki K., Yasuda K. and Takeichi M. (1987) Transformation of cell adhesion properties of exogenously introduced E-cadherin cDNA. *Nature* **329**: 341–343.

Nagoshi R.N., McKeown M., Burtis K.C., Belote J.M. and Baker B.S. (1988) The control of alternative splicing at genes regulating sexual differentiation in *D. melanogaster*. *Cell* **53**: 229–236.

Nagy L.M. and Carroll S. (1994) Conservation of *wingless* patterning functions in the short-germ embryos of *Tribolium castaneum*. *Nature* **367**: 460–463.

Nakatsuji N. and Johnson K.E. (1983) Conditioning of a culture substratum by the ectodermal layer promotes attachment and oriented locomotion by amphibian gastrula mesodermal cells. *J. Cell Sci.* **59**: 43–60.

Nakatsuji N., Gould A. and Johnson K.E. (1982) Movement and guidance of migrating mesodermal cells in *Ambystoma maculatum* gastrulae. *J. Cell Sci.* **56**: 207–222.

Namenwirth M. (1974) The inheritance of cell differentiation during limb regeneration in the axolotl. *Dev. Biol.* **41**: 42–56.

Nameroff M. and Munar E. (1976) Inhibition of cellular differentiation by phospholipase C. II. Separation of fusion and recognition among myogenic cells. *Dev. Biol.* **49**: 288–293.

Nardi J.B. and Stocum D.L. (1983) Surface properties of regenerating limb cells: Evidence for gradation along the proximodistal axis. *Differentiation* **25**: 27–31.

Needham J. (1959) *A History of Embryology*. London/New York: Abelard-Schuman.

Neuman-Silberberg F.S. and Schüpbach T. (1993) The *Drosophila* dorsoventral patterning gene *gurken* produces a dorsally localized RNA and encodes a TGFα-like protein. *Cell* **75**: 165–174.

Neuman-Silberberg F.S. and Schupbach T. (1994) Dorsoventral axis formation in *Drosophila* depends on the correct dosage of gene *gurken*. *Development* **120**: 2457–2463.

Newman S.M., Jr. and Wright T.R.F. (1981) A histological and ultrastructural analysis of developmental defects produced by the mutation, *lethal(1)myospheroid*, in *Drosophila melanogaster*. *Dev. Biol.* **86**: 393–402.

Newport J.W. and Kirschner M.W. (1982a) A major developmental transition in early *Xenopus* embryos. I. Characterization and timing of cellular changes at midblastula stage. *Cell* **30**: 675–686.

Newport J.W. and Kirschner M.W. (1982b) A major developmental transition in early *Xenopus* embryos. II. Control of the onset of transcription. *Cell* **30**: 687–696.

Nicolet G. (1971) Avian gastrulation. *Adv. Morphog.* **9**: 231–262.

Niehrs C., Keller R., Cho K.W.Y. and De Robertis E.M. (1993) The homeobox gene *goosecoid* controls cell migration in Xenopus embryos. *Cell* **72**: 491–503.

Nieuwkoop P.D. (1969a) The formation of the mesoderm in urodelean amphibians. I. Induction by the endoderm. *Wilhelm Roux's Arch.* **162**: 341–373.

Nieuwkoop P.D. (1969b) The formation of the mesoderm in urodelean amphibians. II. The origin of the dorsoventral polarity of the mesoderm. *Wilhelm Roux's Arch.* **163**: 298–315.

Nieuwkoop P.D. and Faber J. (1967) *Normal Table of Xenopus laevis (Daudin)*. 2nd ed. Amsterdam: North-Holland Publishing Company.

Nieuwkoop P.D. and Sutasurya L.A. (1979) *Primordial Germ Cells in the Chordates: Embryognesis and Phylogenesis*, vol. 7. Cambridge: Cambridge University Press.

Nieuwkoop P.D. and Sutasurya L.A. (1981) *Primordial Germ Cells in the Invertebrates: From Epigenesis to Preformation*, vol. 10. Cambridge: Cambridge University Press.

Nijhout H.F. (1980) Pattern formation on lepidopteran wings: Determination of an eyespot. *Dev. Biol.* **80**: 267–274.

Nishida H. (1987) Cell lineage analysis in ascidian embryos by intracellular injection of a tracer enzyme. III. Up to the tissue restricted stage. *Dev. Biol.* **121**: 526–541.

Nishimiya-Fujisawa C. and Sugiyama T. (1993) Genetic analysis of developmental mechanisms in Hydra. *Dev. Biol.* **157**: 1–9.

Noden D.M. (1983) The role of the neural crest in patterning of avian cranial, skeletal, connective, and muscle tissues. *Dev. Biol.* **96**: 144–165.

Nohno T., Noji S., Koyama E., Ohyama K., Myokai F., Kuroiwa A., Saito T. and Taniguchi S. (1991) Involvement of the *Chox-4* chicken homeobox genes in determination of anteroposterior axial polarity during limb development. *Cell* **64:** 1197–1205.

Noji S., Nohno T., Koyama E., Muto K., Ohyama K., Aoki Y., Tamura K., Ohsugi K., Ide H., Taniguchi S. and Saito T. (1991) Retinoic acid induces polarizing activity but is unlikely to be a morphogen in the chick limb bud. *Nature* **350:** 83–86.

Nomura M. (1973) Assembly of bacterial ribosomes. *Science* **179:** 864–873.

Nonet M.L. and Meyer B.J. (1991) Early aspects of *Caenorhabditis elegans* sex determination and dosage compensation are regulated by a zinc-finger protein. *Nature* **351:** 65–68.

Noordermeer J., Klingensmith J. and Nusse R. (1994) *dishevelled* and *armadillo* act as the wingless signalling pathway in *Drosophila*. *Nature* **367:** 80–83.

Nose A., Nagafuchi A. and Takeichi M. (1988) Expressed recombinant cadherins mediate cell sorting in model systems. *Cell* **54:** 993–1001.

Nose A., Tsuji K. and Takeichi M. (1990) Localization of specificity determining sites in cadherin cell adhesion molecules. *Cell* **61:** 147–155.

Nöthiger R. (1972) Larval development of imaginal disks. In: *Results and Problems in Cell Differentiation,* vol. 5, H. Ursprung and R. Nöthiger (eds.), 1–34. Berlin: Springer-Verlag.

Nottebohm F. (1980) Testosterone triggers growth of brain vocal control nuclei in adult female canaries. *Brain Res.* **189:** 429–436.

Nottebohm F. (1981) A brain for all seasons: Cyclical anatomical changes in song control nuclei of the canary brain. *Science* **214:** 1368–1370.

Nottebohm F. (1989) From bird song to neurogenesis. *Scientific American* **260** (February): 74–79.

Nuccitelli R. (1978) Ooplasmic segregation and secretion in the *Pelvetia* eggs is accompanied by a membrane-generated electrical current. *Dev. Biol.* **62:** 13–33.

Nuccitelli R. and Grey R.D. (1984) Controversy over the fast, partial, temporary block to polyspermy in sea urchins: A reevaluation. *Dev. Biol.* **103:** 1–17.

Nuccitelli R. and Jaffe L. (1974) Spontaneous current pulses through developing fucoid eggs. *Proc. Nat. Acad. Sci. USA* **71:** 4855–4859.

Nurse P. (1990) Universal control mechanism regulating onset of M-phase. *Nature* **344:** 503–508.

Nusse R. and Varmus H.E. (1992) Wnt genes. *Cell* **69:** 1073–1087.

Nusse R., van Ooyen A., Cox D., Fung Y.K.T. and Varmus H. (1984) Mode of proviral activation of a putative mammary oncogene (*int-1*) on mouse chromosome 15. *Nature* **307:** 131–136.

Nüsslein-Volhard C. (1991) Determination of the embryonic axes of *Drosophila*. *Development* (Supplement 1): 1–10.

Nüsslein-Volhard C. and Wieschaus E. (1980) Mutations affecting segment number and polarity in *Drosophila*. *Nature* **287:** 795–801.

Nüsslein-Volhard C., Wieschaus E. and Kluding H. (1984) Mutations affecting the pattern of the larval cuticle in *Drosophila melanogaster*. *Wilhelm Roux's Arch.* **193:** 267–282.

Nüsslein-Volhard C., Frohnhöfer H.G. and Lehmann R. (1987) Determination of anteroposterior polarity in *Drosophila*. *Science* **238:** 1675–1681.

Oakley R.A., Lasky C.J., Erickson C.A. and Tosney K.W. (1994) Glycoconjugates mark a transient barrier to neural crest migration in the chicken embryo. *Development* **120:** 103–114.

Okada M., Kleinman I.A. and Schneiderman H.A. (1974) Restoration of fertility in sterilized *Drosophila* eggs by transplantation of polar cytoplasm. *Dev. Biol.* **37:** 43–54.

Oliver B., Kim Y.-J. and Baker B.S. (1993) Sex-lethal, master and slave: A hierarchy of germ-line sex determination in *Drosophila*. *Development* **119:** 897–908.

Olson E.N. (1990) MyoD family: A paradigm for development? *Genes and Dev.* **4:** 1454–1461.

Olson E.N. (1992) Interplay between proliferation and differentiation within the myogenic lineage. *Dev. Biol.* **154:** 261–272.

Olson E.N. and Klein W.H. (1994) bHLH factors in muscle development: Deadlines and commitments, what to leave in and what to leave out. *Genes and Dev.* **8:** 1–8.

Olwin B.B. and Hauschka S.D. (1988) Cell surface fibroblast growth factor and epidermal growth factor receptors are permanently lost during skeletal muscle terminal differentiation in culture. *J. Cell Biol.* **107:** 761–769.

Oppenheimer J.M. (1974) Asymmetry revisited. *Am. Zool.* **14:** 867–879.

Oro A.E., McKeown M. and Evans R.M. (1990) Relationship between the product of the *Drosophila ultraspiracle* locus and vertebrate retinoid X receptor. *Nature* **347:** 298–301.

Oster G.F. (1984) On the crawling of cells. *J. Embryol. exp. Morphol.* **83** (Supplement): 329–364.

Otte A.P., Koster C.H., Snoek G.T. and Durston A.J. (1988) Protein kinase C mediates neural induction in *Xenopus laevis*. *Nature* **334:** 618–620.

Otte A.P., van Run P., Heideveld M., van Driel R. and Durston A.J. (1989) Neural induction is mediated by cross-talk between the protein kinase C and cyclic AMP pathways. *Cell* **58:** 641–648.

Packard D.S., Jr. and Meier S. (1983) An experimental study of the somitomeric organization of the avian segmental plate. *Dev. Biol.* **97:** 191–202.

Painter T.S. (1923) Studies in mammalian spermatogenesis. II. The spermatogenesis of man. *J. Exp. Zool.* **37:** 291–335.

Palmer M.S., Sinclair A.H., Berta P., Ellis N.A., Goodfellow P.N., Abbas N.E. and Fellous M. (1989) Genetic evidence that *ZFY* is not the testis-determining factor. *Nature* **342:** 937–942.

Palmer M.S., Dryden A.J., Hughes J.T. and Collinge J. (1991) Homozygous prion protein genotype predisposes to sporadic Creutzfeldt-Jakob disease. *Nature* **352:** 340–342.

Palmer S.J. and Burgoyne P.S. (1991) *In situ* analysis of fetal, prepuberal and adult XX↔XY chimaeric mouse testes: Sertoli cells are predominantly, but not exclusively, XY. *Development* **112:** 265–268.

Palmiter R.D., Brinster R.L., Hammer R.E., Trumbauer M.E., Rosenfeld M.G., Birnberg N.C. and Evans R.M. (1982) Dramatic growth of mice that develop from eggs microinjected with metallothionein-growth hormone fusion genes. *Nature* **300:** 611–615.

Pan K.-M., Baldwin M., Nguyen J., Gasset M., Serban A., Groth D., Mehlhorn I., Huang Z., Fletterick R.J., Cohen F.E. and Prusiner S.B. (1993) Conversion of α-helices into β-sheets features in the formation of scrapie prion proteins. *Proc. Nat. Acad. Sci. USA* **90:** 10962–10966.

Pang P.P. and Meyerowitz E.M. (1987) *Arabidopsis thaliana*: A model system for plant molecular biology. *Bio/Technology* **5:** 1177–1181.

Pankratz M.J., Seifert E., Gerwin N., Billi B., Nauber U. and Jäckle H. (1990) Gradients of *Krüppel* and *knirps* gene products direct pair-rule gene stripe patterning in the posterior region of the *Drosophila* embryo. *Cell* **61:** 309–317.

Pankratz M.J., Busch M., Hoch M., Seifert E. and Jäckle H. (1992) Spatial control of the gap gene *knirps* in the *Drosophila* embryo by posterior morphogen system. *Science* **255:** 986–989.

Pardee A.B. (1989) G_1 events and regulation of cell proliferation. *Science* **246:** 603–608.

Parkhurst S.M. and Meneely P.M. (1994) Sex determination and dosage compensation: Lessons from flies and worms. *Science* **264:** 924–932.

Pasteels J. (1964) The morphogenetic role of the cortex of the amphibian egg. *Adv. Morphog.* **3:** 363–389.

Patel N.H. (1993) Evolution of insect pattern formation: A molecular analysis of short germband segmentation. In: *Evolutionary Conservation of Developmental Mechanisms*, A.D. Spradling (ed.), 85–110. New York: Wiley/Alan R. Liss.

Patel N.H., Ball E.E. and Goodman C.S. (1992) Changing role of *even-skipped* during the evolution of insect pattern formation. *Nature* **357:** 339–342.

Patel N.H., Condron B.G. and Zinn K. (1994) Pair-rule expression patterns of *even-skipped* are found in both short- and long-germ beetles. *Nature* **367:** 429–434.

Patten B.M. (1964) *Foundations in Embryology*. New York: McGraw-Hill.

Patten B.M. (1971) *Early Embryology of the Chick*. New York: McGraw-Hill.

Peifer M. and Bejsovec A. (1992) Knowing your neighbors: Cell interactions determine intrasegmental patterning in *Drosophila*. *Trends in Genetics* **8:** 243–249.

Peifer M. and Wieschaus E. (1990) Mutations in the *Drosophila* gene *extradenticle* affect the way specific homeodomain proteins regulate segmental identity. *Genes and Dev.* **4:** 1209–1223.

Peifer M., Karch F. and Bender W. (1987) The bithorax complex: Control of segmental identity. *Genes and Dev.* **1:** 891–898.

Pelling C. (1959) Chromosomal synthesis of ribonucleic acid as shown by incorporation of uridine labeled with tritium. *Nature* **184:** 655–656.

Percy J., Kuhn K.L. and Kalthoff K. (1986) Scanning electron microscopic analysis of spontaneous and UV-induced abnormal segment patterns in *Chironomus samoensis* (Diptera, Chironomidae). *Wilhelm Roux's Arch.* **195:** 92–102.

Perkins L.A., Larsen I. and Perrimon N. (1992) *corkscrew* encodes a putative protein tyrosine phosphatase that functions to transduce the terminal signal from the receptor tyrosine kinase torso. *Cell* **70:** 225–236.

Perona R.M. and Wassarman P.M. (1986) Mouse blastocysts hatch *in vitro* by using a trypsin-like proteinase associated with cells of mural trophectoderm. *Dev. Biol.* **114:** 42–52.

Perris R. and Bronner-Fraser M. (1989) Recent advances in defining the role of the extracellular matrix in neural crest development. *Comm. Dev. Neurobiol.* **1:** 61–83.

Perris R., von Boxberg Y. and Löfberg J. (1988) Local embryonic matrices determine region-specific phenotypes in neural crest cells. *Science* **241:** 86–89.

Perry M.D., Li W., Trent C., Robertson B., Fire A., Hageman J.M. and Wood W.B. (1993) Molecular characterization of the *her-1* gene suggests a direct role in cell signaling during *Caenorhabditis elegans* sex determination. *Genes and Dev.* **7:** 216–228.

Pettway Z., Guillory G. and Bronner-Fraser M. (1990) Molecular mechanisms of avian neural crest cell migration on fibronectin and laminin. *Dev. Biol.* **136:** 335–345.

Phelps B.M., Koppel D.E., Primakoff P. and Myles D.G. (1990) Evidence that proteolysis of the surface is an initial step in the mechanism of formation of sperm cell surface domains. *J. Cell Biol.* **111:** 1839–1847.

Phillips H.M. and Steinberg M.S. (1969) Equilibrium measurements of embryonic chick cell adhesiveness. I. Shape equilibrium in centrifugal fields. *Proc. Nat. Acad. Sci. USA* **64:** 121–127.

Phoenix C.H., Goy R.W., Gerall A.A. and Young W.C. (1959) Organizing action of prenatally administered testosterone propionate on the tissues mediating mating behavior in the female guinea pig. *Endocrinology* **65:** 369–382.

Piatigorsky J. (1981) Lens differentiation in vertebrates: A review of cellular and molecular features. *Differentiation* **19:** 134–153.

Piechaczyk M., Yang J.-Q., Blanchard J.M., Jeanteur P. and Marcu K.B. (1985) Posttranscriptional mechanisms are responsible for accumulation of truncated c–myc RNAs in murine plasma cell tumors. *Cell* **42:** 589–597.

Pignoni F., Steingrimsson E. and Lengyel J.A. (1992) *bicoid* and the terminal system activate *tailless* expression in the early *Drosophila* embryo. *Development* **115:** 239–251.

Placzek M., Jessell T.M. and Dodd J. (1993) Induction of floor plate differentiation by contact-dependent, homeogenetic signals. *Development* **117**: 205–218.

Placzek M., Tessier-Lavigne M., Yamada T., Jessell T. and Dodd J. (1990) Mesodermal control of neural cell identity: Floor plate induction by the notochord. *Science* **250**: 985–988.

Pokrywka N.J. and Stephenson E.C. (1991) Microtubules mediate the localization of *bicoid* RNA during *Drosophila* oogenesis. *Development* **113**: 55–66.

Pollerberg G.E. and Beck-Sickinger A. (1993) A functional role for the middle extracellular region of the neural cell adhesion molecule (NCAM) in axonal fasciculation and orientation. *Dev. Biol.* **156**: 324–340.

Postlethwait J.H. and Schneiderman H.A. (1971) Pattern formation and determination in the antenna of the homeotic mutant *Antennapedia* of *Drosophila melanogaster*. *Dev. Biol.* **25**: 606–640.

Postlethwait J.H., Johnson S.L., Midson C.N., Talbot W.S., Gates M., Ballinger E.W., Africa D., Andrews R., Carl T., Eisen J.S., Horne S., Kimmel C.B., Hutchinson M., Johnson M. and Rodriguez A. (1994) A genetic linkage map for the zebrafish. *Science* **264**: 699–703.

Prescott D.M. (1988) *Cells*. Boston: Jones and Bartlett.

Price J.V., Clifford R.J. and Schüpbach T. (1989) The maternal ventralizing locus *torpedo* is allelic to *faint little ball*, an embryonic lethal, and encodes the *Drosophila* EGF receptor homolog. *Cell* **56**: 1085–1092.

Priess J.R. and Hirsh E.I. (1986) *Caenorhabditis elegans* morphogenesis: The role of the cytoskeleton in elongation of the embryo. *Dev. Biol.* **117**: 156–173.

Priess J.R. and Thompson J.N. (1987) Cellular interactions in early *C. elegans* embryos. *Cell* **48**: 241–250.

Priess J.R., Schnabel H. and Schnabel R. (1987) The *glp-1* locus and cellular interactions in early *C. elegans* embryos. *Cell* **51**: 601–611.

Primakoff P., Lathrop W., Woolman L., Cowan A. and Myles D. (1988) Fully effective contraception in male and female guinea pigs immunized with the sperm protein PH-20. *Nature* **335**: 543–547.

Pröve E. (1978) Courtship and testosterone in male zebra finches. *Zeitschrift für Tierpsychologie* **48**: 47–67.

Prusiner S.B. (1982) Novel proteinaceous infectious particles cause scrapie. *Science* **216**: 136–144.

Prusiner S.B. (1992) Chemistry and biology of prions. *Biochemistry* **31**: 12277–12288.

Prusiner S.B., Groth D., Serban A., Koehler R., Foster D., Torchia M., Burton D., Yang S.-L. and DeArmond S.J. (1993) Ablation of the prion protein (PrP) gene in mice prevents scrapie and facilitates production of anti-PrP antibodies. *Proc. Nat. Acad. Sci. USA* **90**: 10608–10612.

Pursel V.G., Pinkert C.A., Miller K.F., Bolt D.J., Campbell R.G., Palmiter R.D., Brinster R.L. and Hammer R.E. (1989) Genetic engineering of livestock. *Science* **244**: 1281–1288.

Purves D. and Lichtman J.W. (1985) *Principles of Neural Development*. Sunderland, MA: Sinauer Associates.

Püschel A.W., Balling R. and Gruss P. (1991) Separate elements cause lineage restriction and specify boundaries of *Hox-1.1* expression. *Development* **112**: 279–287.

Quatrano R.S. (1990) Polar axis fixation and cytoplasmic localization in *Fucus*. In: *Genetics of Pattern Formation and Growth Control*, A.P. Mahowald (ed.), 31–46. New York: Wiley/Alan R. Liss.

Raff M.C., Barres B.A., Burne J.F., Coles H.S., Ishizaki Y. and Jacobson M.D. (1993) Programmed cell death and the control of cell survival: Lessons from the nervous system. *Science* **262**: 695–700.

Raff R.A. and Kaufman T.C. (1983) *Embryos, Genes, and Evolution: The Developmental-Genetic Basis of Evolutionary Change*. New York: Macmillan.

Raisman G. and Field P.M. (1973) Sexual dimorphism in the neuropil of the preoptic area of the rat and its dependence on neonatal androgen. *Brain Res.* **54**: 1–20.

Ralt D., Goldenberg M., Fetterolf P., Thompson D., Dor J., Mashiach S., Garbers D.L. and Eisenbach M. (1991) Sperm attraction to a follicular factor(s) correlates with human egg fertilizability. *Proc. Nat. Acad. Sci. USA* **88**: 2840–2844.

Ramírez-Solis R., Zheng H., Whiting J., Krumlauf R. and Bradley A. (1993) *Hoxb-4 (Hox-2.6)* mutant mice show homeotic transformation of a cervical vertebra and defects in the closure of the sternal rudiments. *Cell* **73**: 279–294.

Ransick A. and Davidson E.H. (1993) A complete second gut induced by transplanted micromeres in the sea urchin embryo. *Science* **259**: 1134–1138.

Rappaport R. (1967) Cell division: Direct measurement of maximum tension exerted by furrow of echinoderm eggs. *Science* **156**: 1241–1243.

Rappaport R. (1974) Cleavage. In: *Concepts of Development*, J. Lash and J. Whittaker (eds.), 76–98. Stamford, CT: Sinauer Associates.

Rau K.-G. and Kalthoff K. (1980) Complete reversal of antero-posterior polarity in a centrifuged insect embryo. *Nature* **287**: 635–637.

Ray R.P., Arora K., Nüsslein-Volhard C. and Gelbart W.M. (1991) The control of cell fate along the dorsal-ventral axis of the *Drosophila* embryo. *Development* **113**: 35–54.

Readhead C., Popko B., Takahashi N., Shine H.D., Saavedra R.A., Sidman R.L. and Hood L. (1987) Expression of a myelin basic protein gene in transgenic shiverer mice: Correction of the dysmyelinating phenotype. *Cell* **48**: 703–712.

Rebagliati M.R., Weeks D.L., Harvey R.P. and Melton D.A. (1985) Identification and cloning of localized maternal RNAs from Xenopus eggs. *Cell* **42**: 769–777.

Reddy E.P., Reynolds R.K., Santos E. and Barbacid M. (1982) A point mutation is responsible for the acquisition of transforming properties by the *T24* human bladder cancer oncogene. *Nature* **300**: 149–152.

Reeves O.R. and Laskey R.A. (1975) *In vitro* differentiation of a homogeneous cell population—the epidermis of *Xenopus laevis. J. Embryol. exp. Morphol.* **34:** 75–92.

Regulski M., Dessain S., McGinnis N. and McGinnis W. (1991) High-affinity binding sites for the *Deformed* protein are required for the function of an autoregulatory enhancer of the *Deformed* gene. *Genes and Dev.* **5:** 278–286.

Render J. (1983) The second polar lobe of the *Sabellaria cementarium* embryo plays an inhibitory role in apical tuft formation. *Wilhelm Roux's Arch.* **192:** 120–129.

Rennick D., Yang G., Muller-Sieburg C., Smith C., Arai N., Takabe Y. and Gennell L. (1987) Interleukin 4 (B-cell stimulatory factor 1) can enhance or antagonize the factor-dependent growth of hemopoietic progenitor cells. *Proc. Nat. Acad. Sci. USA* **84:** 6889–6893.

Restifo L.L. and White K. (1991) Mutations in a steroid hormone-regulated gene disrupt the metamorphosis of the central nervous system in *Drosophila. Dev. Biol.* **148:** 174–194.

Restifo L.L. and White K. (1992) Mutations in a steroid hormone-regulated gene disrupt the metamorphosis of internal tissues in *Drosophila*: Salivary glands, muscle, and gut. *Wilhelm Roux's Arch.* **201:** 221–234.

Reuss C. and Saunders J.W., Jr. (1965) Inductive and axial properties of prospective limb mesoderm in the early chick embryo. *Am. Zool.* **5:** 214.

Reyer R.W. (1954) Regeneration of the lens in the amphibian eye. *Quart. Rev. Biol.* **29:** 1–46.

Ribbert D. (1979) Chromomeres and puffing in experimentally induced polytene chromosomes of *Calliphora erythrocephala. Chromosoma* **74:** 269–298.

Rice W.R. (1994) Degeneration of a nonrecombining chromosome. *Science* **263:** 230–232.

Richter J.D. and Smith L.D. (1984) Reversible inhibition of translation by *Xenopus* oocyte-specific proteins. *Nature* **309:** 378–380.

Riddihough G. and Ish-Horowicz D. (1991) Individual stripe regulatory elements in the *Drosophila hairy* promoter respond to maternal, gap, and pair-rule genes. *Genes and Dev.* **5:** 840–854.

Riddle R.D., Johnson R.L., Laufer E. and Tabin C. (1993) *Sonic hedgehog* mediates the polarizing activity of the ZPA. *Cell* **75:** 1401–1416.

Rieger F., Grumet M. and Edelman G.M. (1985) N-CAM at the vertebrate neuromuscular junction. *J. Cell Biol.* **101:** 285–293.

Riggs A.D. (1990) DNA methylation and late replication probably aid cell memory, and type I DNA reeling could aid chromosome folding and enhancer function. *Phil. Trans. R. Soc. London B* **326:** 285–297.

Rijli F.M., Mark M., Lakkaraju S., Dierich A., Dollé P. and Chambon P. (1993) A homeotic transformation is generated in the rostral branchial region of the head by disruption of *Hoxa-2*, which acts as a selector gene. *Cell* **75:** 1333–1349.

Rijsewijk R., Schuermann M., Wagenaar E., Parren P., Weigel D. and Nusse R. (1987) The Drosophila homolog of the mouse mammary oncogene *int*-1 is identical to the segment polarity gene *wingless. Cell* **50:** 649–657.

Ritter W. (1976) Fragmentierungs und Bestrahlungsversuche am Ei von *Smittia* spec. (Diptera, Chironomidae). Master's thesis, Albert Ludwigs Universität, Freiburg.

Roberts A.B., Anzano M.A., Wakefield L.M., Roche N.S., Stern D.F. and Sporn M.B. (1985) Type β transforming growth factor: A bifunctional regulator of cellular growth. *Proc. Nat. Acad. Sci. USA* **82:** 119–123.

Roberts R., Gallagher J., Spooncer E., Allen T.D., Bloomfield F. and Dexter T.M. (1988) Heparan sulphate bound growth factors: A mechanism for stromal cell mediated haemopoiesis. *Nature* **332:** 376–378.

Roberts S.G.E., Ha I., Maldonado E., Reinberg D. and Green M.R. (1993) Interaction between an acidic activator and transcription factor TFIIB is required for transcriptional activation. *Nature* **363:** 741–744.

Robinow S., Talbot W.S., Hogness D.S. and Truman J.W. (1993) Programmed cell death in the *Drosophila* DNS is ecdysone-regulated and coupled with a specific ecdysone receptor isoform. *Development* **119:** 1251–1259.

Robinson K.R. and Jaffe L.F. (1975) Polarizing fucoid eggs drive a calcium current through themselves. *Science* **187:** 70–72.

Roelink H., Augsburger A., Heemskerk J., Korzh V., Norlin S., Ruiz i Altaba A., Tanabe Y., Placzek M., Edlund T., Jessell T.M. and Dodd J. (1994) Floor plate and motor neuron induction by *vhh-1*, a vertebrate homolog of *hedgehog* expressed by the notochord. *Cell* **76:** 761–775.

Rogers S., Wells R. and Rechsteiner M. (1986) Amino acid sequences common to rapidly degraded proteins: The PEST hypothesis. *Science* **234:** 364–368.

Romanes G.J. (1901) *Darwin and after Darwin.* London: Open Court Publishing.

Romer A.S. (1976) *The Vertebrate Body.* 5th ed. Philadelphia: Saunders.

Ronchi E., Treisman J., Dostatni N., Struhl G. and Desplan C. (1993) Down-regulation of the *Drosophila* morphogen bicoid by the torso receptor-mediated signal transduction cascade. *Cell* **74:** 347–355.

Rong P.M., Teillet M.-A., Ziller C. and Le Douarin N.M. (1992) The neural tube/notochord complex is necessary for vertebral but not limb and body wall striated muscle differentiation. *Development* **115:** 657–672.

Rorvik D.M. (1978) *In His Image: The Cloning of a Man.* Philadelphia: Lippincott.

Roseman S. (1970) The synthesis of complex carbohydrates by multi-glycosyltransferase systems and their potential function in intra-cellular adhesion. *Chem. Phys. Lipids* **5:** 270–297.

Rosenfeld J.M., Daley J.D., Ohno S. and YoungLai E.V. (1977) Central aromatization of testosterone in testicular feminized mice. *Experientia* **33:** 1392–1393.

Rosenquist G.C. (1966) An autoradiographic study of labelled grafts in the chick blastoderm. Development from primitive-streak stages to stage 12. *Contr. Embryol.* **38**: 71–110.

Rosenthal E.T., Hunt T. and Ruderman J.V. (1980) Selective translation of mRNA controls the pattern of protein synthesis during early development of the surf clam, *Spisula solidissima*. *Cell* **20**: 487–494.

Rosenthal E.T., Tansey T.R. and Ruderman J.V. (1983) Sequence-specific adenylations and deadenylations accompany changes in the translation of maternal messenger RNA after fertilization of *Spisula* oocytes. *J. Mol. Biol.* **166**: 309–327.

Ross J. (1989) The turnover of messenger RNA. *Scientific American* **260** (April): 48–55.

Rossant J. and Hopkins N. (1992) Of fin and fur: Mutational analysis of vertebrate embryonic development. *Genes and Dev.* **6**: 1–13.

Rossant J. and Lis W.T. (1979) Potential of isolated mouse inner cell masses to form trophectoderm derivatives *in vivo*. *Dev. Biol.* **70**: 255–261.

Roth S. (1993) Mechanisms of dorsal-ventral axis determination in *Drosophila* embryos revealed by cytoplasmic transplantations. *Development* **117**: 1385–1396.

Roux W. (1885) Beiträge zur Entwicklungsmechanik des Embryo. *Zeitschrift für Biologie* **21**: 411–524.

Rubin G.M. and Spradling A.C. (1982) Genetic transformation of *Drosophila* with transposable element vectors. *Science* **218**: 348–353.

Rudnicki M.A., Braun T., Hinuma S. and Jaenisch R. (1992) Inactivation of *MyoD* in mice leads to up-regulation of the myogenic HLH gene *Myf-5* and results in apparently normal muscle development. *Cell* **71**: 383–390.

Rudnicki M.A., Schnegelsberg P.N.J., Stead R.H., Braun T., Arnold H.-H. and Jaenisch R. (1993) MyoD or Myf-5 is required for the formation of skeletal muscle. *Cell* **75**: 1351–1359.

Ruiz i Altaba A. (1992) Planar and vertical signals in the induction and patterning of the *Xenopus* nervous system. *Development* **116**: 67–80.

Ruoslahti E. (1991) Integrins as receptors for extracellular matrix. In: *Cell Biology of Extracellular Matrix*, 2nd ed., E.D. Hay (ed.), 343–363. New York: Plenum.

Rupp R.A. and Weintraub H. (1991) Ubiquitous MyoD transcription at the midblastula transition precedes induction-dependent MyoD expression in presumptive mesoderm of X. laevis. *Cell* **65**: 927–937.

Rutishauser U. (1992) NCAM and its polysialic acid moiety: A mechanism for pull/push regulation of cell interactions during development? *Development* (Supplement): 99–104.

Rutishauser U., Acheson A., Hall A.K., Mann D.M. and Sunshine J. (1988) The neural cell adhesion molecule (NCAM) as a regulator of cell-cell interactions. *Science* **240**: 53–57.

Ruvkun G. and Giusto J. (1989) The *Caenorhabditis elegans* heterochronic gene *lin-14* encodes a nuclear protein that forms a temporal developmental switch. *Nature* **338**: 313–319.

Ruvkun G., Wightman B., Bürglin T. and Arasu P. (1991) Dominant gain-of-function mutations that lead to misregulation of the *C. elegans* heterochronic gene *lin-14*, and the evolutionary implications of dominant mutations in pattern-formation genes. *Development* (Supplement 1): 47–54.

Sachs A.B. (1993) Messenger RNA degradation in eukaryotes. *Cell* **74**: 413–421.

Sadhu K., Reed S.I., Richardson H. and Russell P. (1990) Human homolog of fission yeast cdc25 mitotic inducer is predominantly expressed in G2. *Proc. Nat. Acad. Sci. USA* **87**: 5139–5143.

St. Johnston D., Driever W., Berleth T., Richstein S. and Nüsslein-Volhard C. (1989) Multiple steps in the localization of *bicoid* RNA to the anterior pole of the *Drosophila* oocyte. *Development* **107** (Supplement): 13–19.

Sagata N., Oskarsson M., Copeland T., Brumbaugh J. and Vande Woude G.F. (1988) Function of c-*mos* proto-oncogene product in meiotic maturation in *Xenopus* oocytes. *Nature* **335**: 519–525.

Sagata N., Watanabe N., Vande Woude G.F. and Ikawa Y. (1989) The c-*mos* proto-oncogene product is a cytostatic factor responsible for meiotic arrest in vertebrate eggs. *Nature* **342**: 512–518.

Sander K. (1960) Analyse des ooplasmatischen Reaktionssystems von *Euscelis plebejus* Fall. (Cicadina) durch Isolieren und Kombinieren von Keimteilen. II. Mitteilung: Die Differen-zierungsleistungen nach Verlagern von Hinterpolmaterial. *Wilhelm Roux's Arch.* **151**: 660–707.

Sander K. (1975) Pattern specification in the insect embryo. In: *Cell Patterning, Ciba Foundation Symp.* **29** (new series): 241–263. Amsterdam: Elsevier, Excerpta Medica, North-Holland Publishing Company, Associated Scientific Publishers.

Sander K. (1976) Specification of the basic body pattern in insect embryogenesis. *Adv. Insect Physiol.* **12**: 125–238.

Sander K. (1983) The evolution of patterning mechanisms: Gleanings from insect embryogenesis and spermatogenesis. In: *Development and Evolution*, B.C. Goodwin, N. Holder and C.C. Wylie (eds.), 123–159. New York: Cambridge University Press.

Sander K. (1990) Von der Keimplasmatheorie zur synergetischen Musterbildung-Einhundert Jahre entwicklungsbiologischer Ideengeschichte. *Verh. Dtsch. Zool. Ges.* **83**: 133–177.

Sanford J.P., Clark H.J., Chapman V.M. and Rossant J. (1987) Differences in DNA methylation during oogenesis and spermatogenesis and their persistence during early embryogenesis in the mouse. *Genes and Dev.* **1**: 1039–1046.

Sardet C., Speksnijder J., Terasaki M. and Chang P. (1992) Polarity of the ascidian egg cortex before fertilization. *Development* **115**: 221–237.

Sater A.K., Alderton J.M. and Steinhardt R.A. (1994) An increase in intracellular pH during neural induction in *Xenopus*. *Development* **120**: 433–442.

Saunders J.W., Jr. (1970) *Patterns and Principles of Animal Development*. New York: Macmillan.

Saunders J.W., Jr. (1982) *Developmental Biology: Patterns, Problems, and Principles.* New York: Macmillan.

Saunders J.W., Jr. and Gasseling M.T. (1968) Ectoderm-mesenchymal interactions in the origin of wing symmetry. In: *Epithelial-Mesenchymal Interactions,* R. Fleischmajer and R.E. Billingham (eds.), 78–97. Baltimore: Williams and Wilkins.

Savage R. and Phillips C.R. (1989) Signals from the dorsal blastopore lip region during gastrulation bias the ectoderm toward a nonepidermal pathway of differentiation in *Xenopus laevis. Dev. Biol.* **133:** 157–168.

Savant-Bhonsale S. and Montell D.J. (1993) *torso-like* encodes the localized determinant of *Drosophila* terminal pattern formation. *Genes and Dev.* **7:** 2548–2555.

Sawada T. and Schatten G. (1988) Microtubules in ascidian eggs during meiosis, fertilization, and mitosis. *Cell Motility and the Cytoskeleton* **9:** 219–230.

Sawyers C.L., Denny C.T. and Witte O.N. (1991) Leukemia and the disruption of normal hematopoiesis. *Cell* **64:** 337–350.

Saxén L. and Toivonen S. (1962) *Primary Embryonic Induction.* London: Logos Press and Academic Press.

Scales J.B., Olson E.N. and Perry M. (1990) Two distinct *Xenopus* genes with homology to MyoD1 are expressed before somite formation in early embryogenesis. *Mol. Cell Biol.* **10:** 1516–1524.

Schaller H.C. (1981) Morphogenetic substances in Hydra. In: *Fortschritte der Zoologie,* vol. 26, H.W. Sauer (ed.), 153–162. Stuttgart/New York: Gustav Fischer Verlag.

Scharf S.R. and Gerhart J.C. (1983) Axis determination in eggs of *Xenopus laevis:* A critical period before first cleavage, identified by the common effects of cold, pressure, and ultraviolet irradiation. *Dev. Biol.* **99:** 75–87.

Schatten H. and Schatten G., eds. (1989a) *The Cell Biology of Fertilization.* San Diego: Academic Press.

Schatten H. and Schatten G., eds. (1989b) *The Molecular Biology of Fertilization.* San Diego: Academic Press.

Scheer U., Dabauvalle M.-C., Merkert H. and Denavente R. (1988) The nuclear envelope and the organization of the pore complexes. *Cell Biol. Int. Rep.* **12:** 669–689.

Scherson T., Serbedzija G., Fraser S. and Bronner-Fraser M. (1993) Regulative capacity of the cranial neural tube to form neural crest. *Development* **118:** 1049–1061.

Schier A.F. and Gehring W.J. (1992) Direct homeo-domain-DNA interaction in the autoregulation of the *fushi tarazu* gene. *Nature* **356:** 804–807.

Schierenberg E. (1986) Developmental strategies during early embryogenesis of *Caenorhabditis elegans. J. Embryol. exp. Morphol.* **97** (Supplement): 31–44.

Schneider D.S., Hudson K.L., Lin T.-Y. and Anderson K.V. (1991) Dominant and recessive mutations define functional domains of *Toll,* a transmembrane protein required for dorsal-ventral polarity in the *Drosophila* embryo. *Genes and Dev.* **5:** 797–807.

Schneider S., Herrenknecht K., Butz S., Kemler R. and Hausen P. (1993) Catenins in *Xenopus* embryogenesis and their relation to the cadherin-mediated cell-cell adhesion system. *Development* **118:** 629–640.

Schneiderman H.A. and Gilbert L.I. (1964) Control of growth and development in insects. *Science* **143:** 325–333.

Schneuwly S., Klemenz R. and Gehring W.J. (1987) Redesigning the body plan of *Drosophila* by ectopic expression of the homoeotic gene *Antennapedia. Nature* **325:** 816–818.

Schoenle E., Zapf J., Humbel R.E. and Froesch E.R. (1982) Insulin-like growth factor I stimulates growth in hypophysectomized rats. *Nature* **296:** 252–253.

Schoenwolf G.C. (1991) Cell movements driving neurulation in avian embryos. *Development* (Supplement 2): 157–168.

Schoenwolf G.C. and Smith J.L. (1990) Mechanisms of neurulation: Traditional viewpoint and recent advances. *Development* **109:** 243–270.

Schröder C., Tautz D., Seifert E. and Jäckle H. (1988) Differential regulation of the two transcripts from the *Drosophila* gap segmentation gene *hunchback. EMBO J.* **7:** 2881–2887.

Schroeder T.E. (1972) The contractile ring. II. Determining its brief existence, volumetric changes and vital role in cleaving *Arbacia* eggs. *J. Cell Biol.* **53:** 419–434.

Schroeder T.E. (1981) Development of a "primitive" sea urchin (*Eucidaris tribuloides*): Irregularities in the hyaline layer, micromeres, and primary mesenchyme. *Biol. Bull.* **161:** 141–151.

Schroeder T.E. (1987) Fourth cleavage of sea urchin blastomeres: Microtubule patterns and myosin localization in equal and unequal cell divisions. *Dev. Biol.* **124:** 9–22.

Schüle R., Umesono K., Mangelsdorf D.J., Bolado J., Pike J.W. and Evans R.M. (1990) Jun-Fos and receptors for vitamins A and D recognize a common response element in the human osteocalcin gene. *Cell* **61:** 497–504.

Schultz E.A. and Haughn G.W. (1993) Genetic analysis of the floral initiation process (FLIP) in *Arabidopsis. Development* **119:** 745–765.

Schüpbach T. (1987) Germ line and soma cooperate during oogenesis to establish the dorsoventral pattern of egg shell and embryo in *Drosophila melanogaster. Cell* **49:** 699–707.

Schwarz-Sommer Z., Huijser P., Nacken W., Saedler H. and Sommer H. (1990) Genetic control of flower development by homeotic genes in *Antirrhinum majus. Science* **250:** 931–936.

Schweizer G., Ayer-Le Lievre C. and Le Douarin N.M. (1983) Restrictions in developmental capacities in the dorsal root ganglia during the course of development. *Cell Differ.* **13:** 191–200.

Schwind J.L. (1933) Tissue specificity at the time of metamorphosis in frog larvae. *J. Exp. Zool.* **66:** 1–14.

Scott M.P. (1992) Vertebrate homeobox gene nomenclature. *Cell* **71:** 551–553.

Scott M.P. and O'Farrell P.H. (1986) Spatial programming of gene expression in early *Drosophila* embryogenesis. *Ann. Rev. Cell Biol.* **2**: 49–80.

Scott M.P. and Weiner A.J. (1984) Structural relationships among genes that control development: Sequence homology between the *Antennapedia, Ultrabithorax,* and *fushi tarazu* loci of *Drosophila. Proc. Nat. Acad. Sci. USA* **81**: 4115–4119.

Scott M.P., Tamkun J.W. and Hartzell G.W., III (1989) The structure and function of the homeodomain. *Biochim. Biophys. Acta* **989**: 25–48.

Selleck M.A.J. and Stern C.D. (1991) Fate mapping and cell lineage analysis of Hensen's node in the chick embryo. *Development* **112**: 615–626.

Selleck M.A.J., Scherson T.Y. and Bronner-Fraser M. (1993) Origins of neural crest cell diversity. *Dev. Biol.* **159**: 1–11.

Sessions S.K. and Ruth S.B. (1990) Explanation for naturally occurring supernumerary limbs in amphibians. *J. Exp. Zool.* **254**: 38–47.

Seydoux G. and Greenwald I. (1989) Cell autonomy of *lin-12* function in a cell fate decision in *C. elegans. Cell* **57**: 1237–1245.

Sham M.-H., Hunt P., Nonchev S., Papalopulu N., Graham A., Boncinelli R. and Krumlauf R. (1992) Analysis of the murine *Hox-2.7* gene: Conserved alternative transcripts with differential distributions in the nervous system and the potential for shared regulatory regions. *EMBO J.* **22**: 1825–1837.

Sham M.H., Vesque C., Nonchev S., Marshall H., Frain M., Das Gupta R., Whiting J., Wilkinson D., Charnay P. and Krumlauf R. (1993) The zinc finger gene *Krox20* regulates *HoxB2* (*Hox2.8*) during hindbrain segmentation. *Cell* **72**: 183–196.

Shaper N.L., Hollis G.F., Douglas J.G., Kirsch I.R. and Shaper J.H. (1988) Characterization of the full length cDNA for murine β-1,4-galactosyltransferase. Novel features at the 5′-end predict two translational start sites at two in-frame AUGs. *J. Biol. Chem.* **263**: 10420–10428.

Shapiro D.J., Blume J.E. and Nielsen D.A. (1987) Regulation of mRNA stability in eukaryotic cells. *BioEssays* **6**: 221–226.

Sharon N. and Lis H. (1993) Carbohydrates in cell recognition: Telltale surface sugars enable cells to identify and interact with one another. New drugs aimed at those carbohydrates could stop infection and inflammation. *Scientific American* **268** (January): 82–89.

Sharpe C.R., Fritz A., De Robertis E.M. and Gurdon J.B. (1987) A homeobox-containing marker of posterior neural differentiation shows the importance of predetermination in neural induction. *Cell* **50**: 749–758.

Shaw G. and Kamen R. (1986) A conserved AU sequence from the 3′ untranslated region of GM-CSF mRNA mediates selective mRNA degradation. *Cell* **46**: 659–667.

Sheets M.D., Fox C.A., Hunt T., Vande Woude G. and Wickens M. (1994) The 3′-untranslated regions of c-*mos* and cyclin mRNAs stimulate translation by regulating cytoplasmic polyadenylation. *Genes and Dev.* **8**: 926–938.

Shepard J.F. (1982) The regeneration of potato plants from leaf-cell protoplasts. *Scientific American* **246** (May): 154–165.

Shepherd G.M. (1988) *Neurobiology.* 2nd ed. New York: Oxford University Press.

Shermoen A.W. and O'Farrell P.H. (1991) Progression of the cell cycle through mitosis leads to abortion of nascent transcripts. *Cell* **67**: 303–310.

Sherr C.J. (1993) Mammalian G_1 cyclins. *Cell* **73**: 1059–1065.

Shi D.-L., Darribère T., Johnson K.E. and Boucaut J.-C. (1989) Initiation of mesodermal cell migration and spreading relative to gastrulation in the urodele amphibian *Pleurodeles waltl. Development* **105**: 351–363.

Shi X., Lu L., Qiu Z., He W. and Frankel J. (1991) Microsurgically generated discontinuities provoke heritable changes in cellular handedness of a ciliate, *Stylonychia mytilus. Development* **111**: 337–356.

Shulman K. (1974) Defects of the closure of the neural plate. In: *Neurology of Infancy and Childhood,* S. Carter and A.P. Gold (eds.), 20–31. New York: Appleton-Century-Crofts.

Shur B.D. (1989) Galactosyltransferase as a recognition molecule during fertilization and development. In: *The Molecular Biology of Fertilization,* H. Schatten and G. Schatten (eds.), 37–71. San Diego: Academic Press.

Shur B.D. (1991) Cell surface β1,4 galactosyltransferase: Twenty years later. *Glycobiology* **1**: 563–575.

Sieber-Blum M. and Cohen A.M. (1980) Clonal analysis of quail neural crest cells: They are pluripotent and differentiate *in vitro* in the absence of noncrest cells. *Dev. Biol.* **80**: 96–106.

Siegfried E., Wilder E.L. and Perrimon N. (1994) Components of *wingless* signalling in *Drosophila. Nature* **367**: 76–80.

Silver P.A. (1991) How proteins enter the nucleus. *Cell* **64**: 489–497.

Simcox A.A. and Sang J.H. (1983) When does determination occur in *Drosophila* embryos? *Dev. Biol.* **97**: 212–221.

Simeone A., Acampora D., Gulisano M., Stornaiuolo A. and Boncinelli E. (1992) Nested expression domains of four homeobox genes in developing rostral brain. *Nature* **358**: 687–690.

Simon J., Chiang A. and Bender W. (1992) Ten different *Polycomb* group genes are required for spatial control of the *abdA* and *AbdB* homeotic products. *Development* **114**: 493–505.

Simon J., Chiang A., Bender W., Shimell M.J. and O'Connor M. (1993) Elements of *Drosophila* bithorax complex that mediate repression by *Polycomb* group products. *Dev. Biol.* **158**: 131–144.

Simon M.I., Strathmann M.P. and Gautam N. (1991) Diversity of G proteins in signal transduction. *Science* **252**: 802–808.

Simpson-Brose M., Treisman J. and Desplan C. (1994) Synergy between the hunchback and bicoid morphogens is required for anterior patterning in *Drosophila. Cell* **78**: 855–865.

Sinclair A.H., Berta P., Palmer M.S., Hawkins J.R., Griffiths B.L., Smith M.J., Foster J.W., Frischauf A.-M., Lovell-Badge R. and Goodfellow P.N. (1990) A gene from the human sex-determining region encodes a protein with homology to a conserved DNA-binding motif. *Nature* **346:** 240–244.

Sippel A.E., Borgmeyer A.W., Püschel A.W., Rupp R.A.W., Stief A., Strech-Jurk U. and Theisen M. (1987) Multiple nonhistone protein-dNA complexes in chromatin regulate the cell- and stage-specific activity of an eukaryotic gene. In: *Results and Problems in Cell Differentiation: Structure and Function of Eukaryotic Chromosomes,* vol. 14, W. Hennig (ed.), 255–269. New York: Springer-Verlag.

Sive H.L. (1993) The frog princess: A molecular formula for dorsoventral patterning in *Xenopus. Genes and Dev.* **7:** 1–12.

Slack J.M.W. (1991) *From Egg to Embryo: Determinative Events in Early Development.* 2nd ed. Cambridge: Cambridge University Press.

Slack J.M.W. and Tannahill D. (1992) Mechanism of anteroposterior axis specification in vertebrates. *Development* **114:** 285–302.

Slack J.M.W., Holland P.W.H. and Graham C.F. (1993) The zootype and the phylotypic stage. *Nature* **361:** 490–492.

Small K.S. and Potter S. (1993) Homeotic transformations and limb defects in *Hoxa-11* mutant mice. *Genes and Dev.* **7:** 2318–2328.

Small S., Kraut R., Hoey T., Warrior R. and Levine M. (1991) Transcriptional regulation of a pair-rule stripe in *Drosophila. Genes and Dev.* **5:** 827–839.

Small S., Blair A. and Levine M. (1992) Regulation of *even-skipped* stripe 2 in the *Drosophila* embryo. *EMBO J.* **11:** 4047–4057.

Smith C.L. and DeLotto R. (1994) Ventralizing signal determined by protease activation in *Drosophila* embryogenesis. *Nature* **368:** 548–551.

Smith J.C. (1987) A mesoderm-inducing factor is produced by a *Xenopus* cell line. *Development* **99:** 3–14.

Smith J.C. (1993) Mesoderm-inducing factors in early vertebrate development. *EMBO J.* **12:** 4463–4470.

Smith J.C. (1994) *hedgehog,* the floor plate, and the zone of polarizing activity. *Cell* **76:** 193–196.

Smith J.C., Price B.M.J., Van Nimmen K. and Huylebroeck D. (1990) Identification of a potent *Xenopus* mesoderm-inducing factor as a homologue of activin A. *Nature* **345:** 729–731.

Smith L.D. (1989) The induction of oocyte maturation: Transmembrane signaling events and regulation of the cell cycle. *Development* **107:** 685–699.

Smith W.C. and Harland R.M. (1992) Expression cloning of noggin, a new dorsalizing factor localized to the Spemann organizer in Xenopus embryos. *Cell* **70:** 829–840.

Smith W.C., Knecht A.K., Wu M. and Harland R.M. (1993) Secreted *noggin* protein mimics the Spemann organizer in dorsalizing *Xenopus* mesoderm. *Nature* **361:** 547–549.

Sokol S. and Melton D.A. (1991) Pre-existent pattern in *Xenopus* animal pole cells revealed by induction with activin. *Nature* **351:** 409–411.

Sokol S.Y. and Melton D.A. (1992) Interaction of Wnt and activin in dorsal mesoderm induction in *Xenopus. Dev. Biol.* **154:** 348–355.

Solnica-Krezel L., Schier A.F. and Driever W. (1994) Efficient recovery of ENU-induced mutations from the zebrafish germline. *Genetics* **136:** 1401–1420.

Solomon F. (1981) Specification of cell morphology by endogenous determinants. *Cell Biol.* **90:** 547–553.

Sommer R. and Tautz D. (1991) Segmentation gene expression in the housefly *Musca domestica. Development* **113:** 419–430.

Sonneborn T.M. (1963) Does preformed cell structure play an essential role in cell heredity? In: *The Nature of Biological Diversity,* J.M. Allen (ed.), 165–221. New York: McGraw-Hill.

Sonneborn T.M. (1970) Gene action in development. *Proc. R. Soc. London B* **176:** 347–366.

Southern E.M. (1975) Detection of specific sequences among DNA fragments separated by gel electrophoresis. *J. Mol. Biol.* **98:** 503–517.

Spemann H. (1927) Neue Arbeiten über Organisatoren in der tierischen Entwicklung. *Naturwissenschaften* **15:** 946–951.

Spemann H. (1928) Die Entwicklung seitlicher und dorso-ventraler Keimhälften bei verzögerter Kernversorgung. *Zeitschr. f. wiss. Zool.* **132:** 105–134.

Spemann H. (1931) Über den Anteil von Implantat und Wirtskeim an der Orientierung und Beschaffenheit der induzierten Embryonalanlage. *Wilhelm Roux's Arch.* **123:** 389–417.

Spemann H. (1938) *Embryonic Development and Induction.* New Haven: Yale University Press. Reprinted 1967. New York: Hafner.

Spemann H. and Mangold H. (1924a) Induction of embryonic primordia by implantation of organizers from a different species. In: *Foundations of Experimental Embryology,* B.H. Willier and J.M. Oppenheimer (eds.), 144–184. New York: Hafner.

Spemann H. and Mangold H. (1924b) Über Induktion von Embryonalanlagen durch Implantation artfremder Organisatoren. *Wilhelm Roux's Arch.* **100:** 599–638.

Spemann H. and Schotté O. (1932) Über xenoplastiche Transplantation als Mittel zur Analyse der embryonalen Induktion. *Naturwissenschaften* **20:** 463–467.

Sporn M.B. and Roberts A.B. (1988) Peptide growth factors are multifunctional. *Nature* **332:** 217–219.

Spradling A.C. (1981) The organization and amplification of two chromosomal domains containing *Drosophila* chorion genes. *Cell* **27:** 193–201.

Spradling A.C. (1993) Germ line cysts: Communes at work. *Cell* **72:** 649–651.

Spradling A.C. and Mahowald A.P. (1980) Amplification of genes for chorion proteins during oogenesis in *Drosophila melanogaster. Proc. Nat. Acad. Sci. USA* **77:** 1096–1100.

Spradling A.C. and Rubin G.M. (1983) The effect of chromosomal position on the expression of the Drosophila xanthine dehydrogenase gene. *Cell* **34:** 47–57.

Sprenger F. and Nüsslein-Volhard C. (1992) Torso receptor activity is regulated by a diffusible ligand produced at the extracellular terminal regions of the *Drosophila* egg. *Cell* **71**: 987–1001.

Sprenger F., Stevens L.M. and Nüsslein-Volhard C. (1989) The *Drosophila* gene *torso* encodes a putative receptor tyrosine kinase. *Nature* **338**: 478–483.

Stalder J., Groudine M., Dodgson J.B., Engel J.D. and Weintraub H. (1980) Hb switching in chickens. *Cell* **19**: 973–980.

Štanojević D., Hoey T. and Levine M. (1989) Sequence-specific DNA-binding activities of the gap proteins encoded by *hunchback* and *Krüppel* in *Drosophila*. *Nature* **341**: 331–335.

Štanojević D., Small S. and Levine M. (1991) Regulation of a segmentation stripe by overlapping activators and repressors in the *Drosophila* embryo. *Science* **254**: 1385–1387.

Starck D. (1965) *Embryologie: Ein Lehrbuch auf allgemein biologischer Grundlage*. Stuttgart: Georg Thieme.

Stearns T. and Kirschner M. (1994) In vitro reconstitution of centrosome assembly and function: The central role of τ-tubulin. *Cell* **76**: 623–637.

Steeves T.A. and Sussex I.M. (1989) *Patterns in Plant Development*. 2nd ed. Cambridge: Cambridge University Press.

Stehelin D., Varmus H.E., Bishop J.M. and Vogt P. (1976) DNA related to the transforming gene(s) of avian sarcoma viruses is present in normal avian DNA. *Nature* **260**: 170–173.

Stein D. and Nüsslein-Volhard C. (1992) Multiple extracellular activities in *Drosophila* egg perivitelline fluid are required for establishment of embryonic dorsal-ventral polarity. *Cell* **68**: 429–440.

Stein D., Roth S., Vogelsang E. and Nüsslein-Volhard C. (1991) The polarity of the dorsoventral axis in the Drosophila embryo is defined by an extracellular signal. *Cell* **65**: 725–735.

Steinberg M.S. (1963) Reconstruction of tissues by dissociated cells. *Science* **141**: 401–408.

Steinberg M.S. (1970) Does differential adhesion govern self-assembly processes in histogenesis? Equilibrium configurations and the emergence of a hierarchy among populations of embryonic cells. *J. Exp. Zool.* **173**: 395–434.

Steinberg M.S. and Takeichi M. (1994) Experimental specification of cell sorting, tissue spreading, and specific spatial patterning by quantitative differences in cadherin expression. *Proc. Nat. Acad. Sci. USA* **91**: 206–209.

Steinmann-Zwicky M. (1994) Sex determination of the *Drosophila* germ line: *tra* and *dsx* control somatic inductive signals. *Development* **120**: 707–716.

Steinmann-Zwicky M., Schmid H. and Nöthiger R. (1989) Cell-autonomous and inductive signals can determine the sex of the germ line of Drosophila by regulating the gene *Sxl*. *Cell* **57**: 157–166.

Steller H. and Pirrotta V. (1985) A transposable P vector that confers selectable G418 resistance to *Drosophila* larvae. *EMBO J.* **4**: 167–171.

Sternberg P.W. (1988) Lateral inhibition during vulval induction in *Caenorhabditis elegans*. *Nature* **335**: 551–554.

Stevens L.M., Frohnhöfer H.G., Klingler M. and Nüsslein-Volhard C. (1990) Localized requirement for *torsolike* expression in follicle cells for the development of terminal anlagen of the *Drosophila* embryo. *Nature* **346**: 660–663.

Steward F.C. (1970) From cultured cells to whole plants: The induction and control of their growth and morphogenesis. *Proc. R. Soc. London B* **175**: 1–30.

Stocum D.L. and Fallon J.F. (1982) Control of pattern formation in urodele limb ontogeny: A review and a hypothesis. *J. Embryol. exp. Morphol.* **69**: 7–36.

Stopak D. and Harris A.K. (1982) Connective tissue morphogenesis by fibroblast traction. I. Tissue culture observations. *Dev. Biol.* **90**: 383–398.

Stott D., Kispert A. and Herrmann B.G. (1993) Rescue of the tail defect of *Brachyury* mice. *Genes and Dev.* **7**: 197–203.

Strehlow D. and Gilbert W. (1993) A fate map for the first cleavages of the zebra fish. *Nature* **361**: 451–453.

Strome S. (1989) Generation of cell diversity during early embryogenesis in the nematode *Caenorhabditis elegans*. *Int. Rev. Cytol.* **114**: 81–123.

Strome S. (1993) Determination of cleavage planes. *Cell* **72**: 3–6.

Strome S. and Wood W.B. (1983) Generation of asymmetry and segregation of germ-line granules in early C. elegans embryos. *Cell* **35**: 15–25.

Struhl G. (1981) A homoeotic mutation transforming leg to antenna in *Drosophila*. *Nature* **292**: 635–638.

Struhl G. (1984) Splitting the bithorax complex of *Drosophila*. *Nature* **308**: 454–457.

Struhl G. (1989) Differing strategies for organizing anterior and posterior body pattern in *Drosophila* embryos. *Nature* **338**: 741–744.

Struhl G. and Basler K. (1993) Organizing activity of wingless protein in *Drosophila*. *Cell* **72**: 527–540.

Struhl G., Johnston P. and Lawrence P.A. (1992) Control of *Drosophila* body pattern by the hunchback morphogen gradient. *Cell* **69**: 237–249.

Sturtevant A.H. (1923) Inheritance of direction of coiling in *Limnaea*. *Science* **58**: 269–270.

Sturtevant A.H. (1929) The claret mutant type of *Drosophila simulans*: A study of chromosome elimination and of cell lineage. *Zeitschr. f. wiss. Zool.* **135**: 323–356.

Sudarwati S. and Nieuwkoop P.D. (1971) Mesoderm formation in the anuran *Xenopus laevis* (Daudin). *Wilhelm Roux's Arch.* **166**: 189–204.

Sugrue S.P. and Hay E.D. (1986) The identification of extracellular matrix (ECM) binding sites on the basal surface of embryonic corneal epithelium and the effect of ECM binding on epithelial collagen production. *J. Cell Biol.* **102**: 1907–1916.

Sullivan D., Palacios R., Stavnezer J., Taylor J.M., Faras A.J., Diely M.L., Summers N.M., Bishop J.M. and Schimke R.T. (1973) Synthesis of a deoxyribonucleic acid sequence complementary to ovalbumin messenger ribonucleic acid and quantification of ovalbumin genes. *J. Biol. Chem.* **248**: 7530–7539.

Sulston J.E., Schierenberg E., White J.G. and Thomson J.N. (1983) The embryonic cell lineage of the nematode *Caenorhabditis elegans*. *Dev. Biol.* **100**: 64–119.

Summerbell D. (1983) The effects of local application of retinoic acid to the anterior margin of the developing chick limb. *J. Embryol. exp. Morphol.* **78**: 269–289.

Summerbell D. and Lewis J.H. (1975) Time, place and positional value in the chick limb bud. *J. Embryol. exp. Morphol.* **33**: 621–643.

Summers R.G., Morrill J.B., Leith A., Marko M., Piston D.W. and Stonebraker A.T. (1993) A stereometric analysis of karyokinesis, cytokinesis and cell arrangements during and following fourth cleavage period in the sea urchin, *Lytechinus variegatus*. *Dev. Growth Differ.* **35**: 41–57.

Sundin O. and Eichele G. (1990) A homeo domain protein reveals the metameric nature of the developing chick hindbrain. *Genes and Dev.* **4**: 1267–1276.

Sung Z.R., Belachew A., Shunong B. and Bertrand-Garcia R. (1992) *EMF*, an *Arabidopsis* gene required for vegetative shoot development. *Science* **258**: 1645–1647.

Surani M.A.H., Barton S.C. and Norris M.L. (1986) Nuclear transplantation in the mouse: Heritable differences between the parental genomes after activation of the embryonic genome. *Cell* **45**: 127–136.

Sutton H.E. (1988) *An Introduction to Human Genetics.* 4th ed. New York: Harcourt Brace Jovanovich.

Swain J.L., Stewart T.A. and Leder P. (1987) Parental legacy determines methylation and expression of an autosomal transgene: A molecular mechanism for parental imprinting. *Cell* **50**: 719–727.

Swanson L.W., Simmons D.M., Arriza J., Hammer R., Brinster R.L., Rosenfeld M.G. and Evans R.M. (1985) Novel developmental specificity in the nervous system of transgenic animals expressing growth hormone fusion genes. *Nature* **317**: 363–366.

Tabata T. and Kornberg T.B. (1994) Hedgehog is a signaling protein with a key role in patterning *Drosophila* imaginal discs. *Cell* **76**: 89–102.

Tabin C.J. (1991) Retinoids, homeoboxes, and growth factors: Toward molecular models for limb development. *Cell* **66**: 199–217.

Tabin C.J. (1992) Why we have (only) five fingers per hand: Hox genes and the evolution of paired limbs. *Development* **116**: 289–296.

Taira M., Jamrich M., Good P.J. and Dawid I.B. (1992) The *LIM* domain-containing homeo box gene *Xlim-1* is expressed specifically in the organizer region of *Xenopus* gastrula embryos. *Genes and Dev.* **6**: 356–366.

Takeichi M. (1977) Functional correlation between cell adhesive properties and some cell surface proteins. *J. Cell Biol.* **75**: 464–474.

Takeichi M. (1988) The cadherins: Cell-cell adhesion molecules controlling animal morphogenesis. *Development* **102**: 639–655.

Takeichi M. (1991) Cadherin cell adhesion receptors as a morphogenetic regulator. *Science* **251**: 1451–1455.

Talbot W.S., Swyryd E.A. and Hogness D.S. (1993) Drosophila tissues with different metamorphic responses to ecdysone express different edcysone receptor isoforms. *Cell* **73**: 1323–1337.

Tanaka Y. (1976) Effects of surfactants on the cleavage and further development of the sea urchin embryos. I. The inhibition of micromere formation at the fourth cleavage. *Dev. Growth Differ.* **18**: 113–122.

Tannahill D. and Melton D.A. (1989) Localized synthesis of the Vg1 protein during early *Xenopus* development. *Development* **106**: 775–785.

Tata J.R. (1993) Gene expression during metamorphosis: An ideal model for post-embryonic development. *BioEssays* **15**: 239–248.

Tata J.R., Kawahara A. and Baker B.S. (1991) Prolactin inhibits both thyroid hormone-induced morphogenesis and cell death in cultured amphibian larval tissues. *Dev. Biol.* **146**: 72–80.

Tautz D. (1988) Regulation of the *Drosophila* segmentation gene *hunchback* by two maternal morphogenetic centres. *Nature* **332**: 281–284.

Tautz D. (1992) Redundancies, development and the flow of information. *BioEssays* **14**: 263–266.

Taylor M.A. and Smith L.D. (1987) Induction of maturation in small *Xenopus laevis* oocytes. *Dev. Biol.* **121**: 111–118.

Telfer W.H. (1965) The mechanism and control of yolk formation. *Ann. Rev. Entomol.* **10**: 161–184.

Telfer W.H. (1975) Development and physiology of the oocyte-nurse cell syncytium. *Adv. Insect Physiol.* **11**: 223–319.

Telfer W.H., Huebner E. and Smith D.S. (1982) The cell biology of vitellogenic follicles in *Hyalophora* and *Rhodnius*. In: *Insect Ultrastructure*, R.C. King and H. Akai (eds.), vol. 1, 118–149. New York: Plenum.

Thaller C. and Eichele G. (1987) Identification and spatial distribution of retinoids in the developing chick limb bud. *Nature* **327**: 625–628.

Thaller C. and Eichele G. (1988) Characterization of retinoid metabolism in the developing chick limb bud. *Development* **103**: 473–483.

Thaller C. and Eichele G. (1990) Isolation of 3,4-didehydroretinoic acid, a novel morphogenetic signal in the chick wing bud. *Nature* **345**: 815–819.

Theriot J.A. and Mitchison T.J. (1991) Actin microfilament dynamics in locomoting cells. *Nature* **352**: 126–132.

Thiery J.-P., Brackenbury R., Rutishauser U. and Edelman G.M. (1977) Adhesion among neural cells of the chick embryo. II. Purification and characterization of a cell adhesion molecule from neural retina. *J. Biol. Chem.* **252**: 6841–6845.

Thiery J.-P., Duband J.L. and Delouvee A. (1982) Pathways and mechanisms of avian trunk neural crest cell migration and localization. *Dev. Biol.* **93**: 324–343.

Thisse C., Perrin-Schmitt F., Stoetzel C. and Thisse B. (1991) Sequence-specific transactivation of the Drosophila *twist* gene by the *dorsal* gene product. *Cell* **65**: 1191–1201.

Thomas H.E., Stunnenberg H.G. and Stewart A.F. (1993) Heterodimerization of the *Drosophila* ecdysone receptor with retinoid X receptor and *ultraspiracle*. *Nature* **362**: 471–475.

Thomas J.H., Stern M.J. and Horvitz H.R. (1990) Cell interactions coordinate the development of the C. elegans egg-laying system. *Cell* **62**: 1041–1052.

Thomsen G., Woolf T., Whitman M., Sokol S., Vaughan J., Vale W. and Melton D.A. (1990) Activins are expressed early in *Xenopus* embryogenesis and can induce axial mesoderm and anterior structures. *Cell* **63**: 485–493.

Thomsen G.H. and Melton D.A. (1993) Processed Vg1 protein is an axial mesoderm inducer in *Xenopus*. *Cell* **74**: 433–441.

Thorpe W.H. (1958) The learning of song patterns by birds, with especial reference to the song of the chaffinch, *Fringilla coelebs*. *Ibis* **100**: 535–570.

Thummel C.S., Burtis K.C. and Hogness D.S. (1990) Spatial and temporal patterns of E74 transcription during *Drosophila* development. *Cell* **61**: 101–111.

Tian M. and Maniatis T. (1992) Positive control of pre-mRNA splicing in vitro. *Science* **256**: 237–240.

Tickle C., Alberts B.M., Wolpert L. and Lee J. (1982) Local application of retinoic acid to the limb bud mimics the action of the polarizing region. *Nature* **296**: 564–565.

Tijo J.H. and Puck T.T. (1958) The somatic chromosomes of man. *Proc. Nat. Acad. Sci. USA* **44**: 1229–1237.

Till J.E. and McCulloch E.A. (1961) A direct measurement of the radiation sensitivity of normal mouse bone marrow cells. *Radiat. Res.* **14**: 215–222.

Tindall D.J., Rowley D.R., Murthy L., Lipshultz L.I. and Chang C.H. (1985) Structure and biochemistry of the Sertoli cell. *Ann. Rev. Cytol.* **94**: 127–149.

Tjian R. and Maniatis T. (1994) Transcriptional activation: A complex puzzle with few easy pieces. *Cell* **77**: 5–8.

Toole B.P. and Gross J. (1971) The extracellular matrix of the regenerating newt limb: Synthesis and removal of hyaluronate prior to differentiation. *Dev. Biol.* **25**: 57–77.

Torrey T.W. and Feduccia A. (1991) *Morphogenesis of the Vertebrates*. 5th ed. New York: Wiley.

Townes P.L. and Holtfreter J. (1955) Directed movements and selective adhesion of embryonic amphibian cells. *J. Exp. Zool.* **128**: 53–120.

Treisman J. and Desplan C. (1989) The products of the *Drosophila* gap genes *hunchback* and *Krüppel* bind to the *hunchback* promoters. *Nature* **341**: 335–337.

Treisman J., Harris E., Wilson D. and Desplan C. (1992) The homeodomain: A new face for the helix-turn-helix? *BioEssays* **14**: 145–150.

Trinkaus J.P. (1963) The cellular basis of *Fundulus* epiboly. Adhesivity of blastula and gastrula cells in culture. *Dev. Biol.* **7**: 513–532.

Trinkaus J.P. (1984) *Cells into Organs: The Forces That Shape the Embryo*. 2nd ed. Englewood Cliffs, NJ: Prentice Hall.

Trouche D., Grigoriev M., Lenormand J.-L., Robin P., Leibovitch S.A., Sassone-Corsi P. and Harel-Bellan A. (1993) Repression of c-*fos* promoter by MyoD on muscle cell differentiation. *Nature* **363**: 79–82.

Truman J.W., Talbot W.S., Fahrbach S.E. and Hogness D.S. (1994) Ecdysone receptor expression in the CNS correlates with stage-specific responses to ecdysteroids during *Drosophila* and *Manduca* development. *Development* **120**: 219–234.

Tsukiyama T., Becker P.B. and Wu C. (1994) ATP-dependent nucleosome disruption at a heat-shock promoter mediated by binding of GAGA transcription factor. *Nature* **367**: 525–532.

Tuchmann-Duplessis H., David G. and Haegel P. (1972) *Illustrated Human Embryology*, vol. 1, *Embryogenesis*. New York: Springer-Verlag.

Tucker G.C., Ayoyama H., Lipinski M., Tursz T. and Thiery J.-P. (1984) Identical reactivity of monoclonal antibodies N-1 and NC-1: Conservation in vertebrates on cells derived from neural primordium and on some leukocytes. *Cell Differ.* **14**: 223–230.

Tucker P.K. and Lundrigan B.L. (1993) Rapid evolution of the sex determining locus in Old World mice and rats. *Nature* **364**: 715–717.

Turner B.M., Birley A.J. and Lavender J. (1992) Histone H4 isoforms acetylated at specific lysine residues define individual chromosomes and chromatin domains in *Drosophila* polytene nuclei. *Cell* **69**: 375–384.

Turner F.R. and Mahowald A.P. (1976) Scanning electron microscopy of *Drosophila* embryogenesis. I. The structure of the egg envelopes and the formation of the cellular blastoderm. *Dev. Biol.* **50**: 95–108.

Turner P.R. and Jaffe L.A. (1989) G-proteins and the regulation of oocyte maturation and fertilization. In: *The Cell Biology of Fertilization*, H. Schatten and G. Schatten (eds.), 297–318. San Diego: Academic Press.

Twitty V.C. and Elliott H.A. (1934) The relative growth of the amphibian eye, studied by means of transplantation. *J. Exp. Zool.* **68**: 247–291.

Uzzell T.M. (1964) Relations of the diploid and triploid species of the *Ambystoma jeffersonianum* complex. *Copeia* (1964): 257–300.

Vacquier V.D. (1980) The adhesion of sperm to sea urchin eggs. In: *The Cell Surface: Mediator of Developmental Processes*, S. Subtelny and N.K. Wessels (eds.), 151–168. New York: Academic Press.

Vacquier V.D. and Moy G.W. (1977) Isolation of bindin: The protein responsible for the adhesion of sperm to sea urchin eggs. *Proc. Nat. Acad. Sci. USA* **74**: 2456–2460.

Vakaet L. (1984) Early development of birds. In: *Chimeras in Developmental Biology*, N.M. Le Douarin and A. McLaren (eds.), 71–88. London: Academic Press.

Valcárcel J., Singh R., Zamore P.D. and Green M.R. (1993) The protein Sex-lethal antagonizes the splicing factor U2AF to regulate alternative splicing of transformer pre-mRNA. Nature 362: 171–175.

van den Biggelaar J.A.M. and Guerrier P. (1983) Origin of spatial organization. In: The Mollusca, K.M. Wilbur (ed.), vol 3, 179–213. New York: Academic Press.

van Dijk M.A. and C. Murre (1994) extradenticle raises the DNA binding specificity of homeotic selector gene products. Cell 78: 617–624.

Varmus H.E. (1982) Form and function of retroviral proviruses. Science 216: 812–820.

Vasil V. and Hildebrandt A.C. (1965) Differentiation of tobacco plants from single isolated cells in microcultures. Science 150: 889–892.

Vaux D.L., Weissman I.L. and Kim S.K. (1992) Prevention of programmed cell death in Caenorhabditis elegans by human bcl-2. Science 258: 1955–1957.

Verbout A.J. (1985) The development of the vertebral column. Adv. Anat. Embryol. Cell. Biol. 90: 1–120.

Verdonk N.H. and Cather J.N. (1983) Morphogenetic determination and differentiation. In: The Mollusca, K.M. Wilbur (ed.), vol 3, 215–252. New York: Academic Press.

Vestweber D., Gossler A., Boller K. and Kemler R. (1987) Expression and distribution of cell adhesion molecule uvomorulin in mouse preimplantation embryos. Dev. Biol. 124: 451–456.

Villee C.A., Solomon E.P., Martin C.E., Martin D.W., Berg L.R. and Davis P.W. (1989) Biology. 2nd ed. Philadelphia: Saunders.

Vincent J.-P. and Lawrence P.A. (1994) Drosophila wingless sustains engrailed expression only in adjoining cells: Evidence from mosaic embryos. Cell 77: 909–915.

Vincent J.-P. and O'Farrell P.H. (1992) The state of engrailed expression is not clonally transmitted during early Drosophila development. Cell 68: 923–931.

Vincent J.-P., Oster G.F. and Gerhart J.C. (1986) Kinematics of gray crescent formation in Xenopus eggs: The displacement of subcortical cytoplasm relative to the egg surface. Dev. Biol. 113: 484–500.

Vogel O. (1978) Pattern formation in the egg of the leafhopper Euscelis plebejus fall. (Homoptera): Developmental capacities of fragments isolated from the polar egg regions. Dev. Biol. 67: 357–370.

Vogel O. (1982) Development of complete embryos in drastically deformed leafhopper eggs. Wilhelm Roux's Arch. 191: 134–136.

Vogt W. (1929) Gestaltungsanalyse am Amphibienkeim mit örtlicher Vitalfärbung. II. Teil. Gastrulation und Mesodermbildung bei Urodelen und Anuren. Wilhelm Roux's Arch. 120: 384–706.

Vollbrecht E., Veit B., Sinha N. and Hake S. (1991) The developmental gene Knotted-1 is a member of a maize homeobox gene family. Nature 350: 241–243.

vom Saal F.S. (1989) Sexual differentiation in litter-bearing mammals: Influence of sex of adjacent fetuses in utero. J. Animal Sci. 67: 1824–1840.

von Baer K.E. (1828) Entwicklungsgeschichte der Thiere: Beobachtung und Reflexion. Königsberg: Bornträger.

Von Brunn A. and Kalthoff K. (1983) Photoreversible inhibition by ultraviolet light of germ line development in Smittia sp. (Chironomidae, Diptera). Dev. Biol. 100: 426–439.

Wabl M.R., Brun R.B. and DuPasquier L. (1975) Lymphocytes of the toad Xenopus laevis have the gene set for promoting tadpole development. Science 190: 1310–1312.

Wagner E.F. (1990) Mouse genetics meets molecular biology at Cold Spring Harbor. New Biologist 2: 1071–1074.

Wakahara M. (1990) Cytoplasmic localization and organization of germ cell determinants. In: Cytoplasmic Organization in Development, G.M. Malacinski (ed.), vol. 4, 219–242. New York: McGraw-Hill.

Wakimoto T. and Karpen G.H. (1988) Transposable elements and germ-line transformation in Drosophila. In: Developmental Genetics of Higher Organisms: A Primer in Developmental Biology, G.M. Malacinski (ed.), 275–303. New York: Macmillan.

Waksman G., Kominos D., Robertson S.C., Pant N., Baltimore D., Birge R.B., Cowburn D., Hanafusa H., Mayer B.J., Overduin M., Resh M.D., Rios C.B., Silverman L. and Kuriyan J. (1992) Crystal structure of the phosphotyrosine recognition domain SH2 of v-src complexed with tyrosine-phosphorylated peptides. Nature 358: 646–653.

Walbot V. and Holder N. (1987) Developmental Biology. New York: Random House.

Wallace H., Maden M. and Wallace B.M. (1974) Participation of cartilage grafts in amphibian limb regeneration. J. Embryol. exp. Morphol. 32: 391–404.

Walldorf U. and Gehring W.J. (1992) Empty spiracles, a gap gene containing a homeobox involved in Drosophila head development. EMBO J. 11: 2247–2259.

Walldorf U., Fleig R. and Gehring W.J. (1989) Comparison of homeobox-containing genes of the honeybee and Drosophila. Proc. Nat. Acad. Sci. USA 86: 9971–9975.

Wanek N., Gardiner D.M., Muneoka K. and Bryant S.V. (1991) Conversion by retinoic acid of anterior cells into ZPA cells in the chick wing bud. Nature 350: 81–83.

Wang C. and Lehmann R. (1991) Nanos is the localized posterior determinant in Drosophila. Cell 66: 637–647.

Wang J. and Bell L.R. (1994) The Sex-lethal amino terminus mediates cooperative interactions in RNA binding and is essential for splicing regulation. Genes and Dev. 8: 2072–2085.

Wang N., Butler J.P. and Ingber D.E. (1993) Mechanotransduction across the cell surface and through the cytoskeleton. Science 260: 1124–1127.

Wang S. and Hazelrigg T. (1994) Implications for bcd mRNA localization from spatial distribution of exu protein in Drosophila oogenesis. Nature 369: 400–403.

Ward C.R. and Kopf G.S. (1993) Molecular events mediating sperm activation. Dev. Biol. 158: 9–34.

Ward G.E., Brokaw C.J., Garbers D.L. and Vacquier V.D. (1985) Chemotaxis of *Arbacia punctulata* spermatozoa to resact, a peptide from the egg jelly layer. *J. Cell Biol.* **101**: 2324–2329.

Warner A.E., Guthrie S.C. and Gilula N.B. (1984) Antibodies to gap junctional protein selectively disrupt junctional communication in the early amphibian embryo. *Nature* **311**: 127–131.

Wassarman P.M. (1987) The biology and chemistry of fertilization. *Science* **235**: 553–560.

Wassarman P.M. (1990) Profile of a mammalian sperm receptor. *Development* **108**: 1–17.

Wasserman W.J. and Smith L.D. (1978) Oocyte maturation in nonmammalian vertebrates. In: *The Vertebrate Ovary*, R.E. Jones (ed.), 443–468. New York: Plenum.

Watanabe N., Vande Woude G.F., Ikawa Y. and Sagata N. (1989) Specific proteolysis of the c-*mos* proto-oncogene product by calpain on fertilization of *Xenopus* eggs. *Nature* **342**: 505–511.

Watson J.D. and Crick F.H.C. (1953) Molecular structure of nucleic acids: A structure for deoxyribose nucleic acid. *Nature* (London) **171**: 737–738.

Weeks D.L., Rebagliati M.R., Harvey R.P. and Melton D.A. (1985) Localized maternal mRNAs in *Xenopus laevis* eggs. *Cold Spring Harbor Symp. Quant. Biol.* **L**: 21–30.

Weigel D. and Meyerowitz E.M. (1993) Genetic hierarchy controlling flower development. In: *Molecular Basis of Morphogenesis*, M. Bernfield (ed.), 93–107. New York: Alan R. Liss.

Weigel D. and Meyerowitz E.M. (1994) The ABCs of floral homeotic genes. *Cell* **78**: 203–209.

Weigel D., Alvarez J., Smyth D.R., Yanofsky M.F. and Meyerowitz E.M. (1992) *LEAFY* controls floral meristem identity in Arabidopsis. *Cell* **69**: 843–859.

Weinberg R.A. (1983) A molecular basis of cancer. *Scientific American* **249** (November): 126–142.

Weinberg R.A. (1988) Finding the anti-oncogene. *Scientific American* **259** (September): 44–51.

Weintraub H. (1993) The MyoD family and myogenesis: Redundancy, networks, and thresholds. *Cell* **75**: 1241–1244.

Weintraub H., Davis R., Tapscott S., Thayer M., Krause M., Benezra R., Blackwell T.K., Turner D., Rupp R., Hollenbert S., Zhuang Y. and Lassar A. (1991) The *myoD* gene family: Nodal point during specification of the muscle cell lineage. *Science* **251**: 761–766.

Weisblat D.A., Zackson S.L., Blair S.S. and Young J.D. (1980) Cell lineage analysis by intracellular injection of fluorescent tracers. *Science* **209**: 1538–1541.

Weiss P. (1939) *Principles of Development: A Text in Experimental Embryology.* New York: Henry Holt and Company.

Weiss P. (1958) Cell contact. *Int. Rev. Cytol.* **7**: 391–423.

Weissmann C. (1994) Molecular biology of prion diseases. *Trends in Cell Biology* **4**: 10–14.

Weliky M., Minsuk S., Keller R. and Oster G. (1991) Notochord morphogenesis in *Xenopus laevis*: Simulation of cell behavior underlying tissue convergence and extension. *Development* **113**: 1231–1244.

Wells D.E., Showman R.M., Klein W.H. and Raff R.A. (1981) Delayed recruitment of maternal histone H3 mRNA in sea urchin embryos. *Nature* **292**: 477–478.

Welshons W.J. and Russell L.B. (1959) The Y chromosome as the bearer of male determining factors in the mouse. *Proc. Nat. Acad. Sci. USA* **45**: 560–566.

Went D.F. and Krause G. (1973) Normal development of mechanically activated, unlaid eggs of an endoparasitic hymenopteran. *Nature* **244**: 454–455.

Wessel G.M., Zhang W., Tomlinson C.R., Lennarz W.J. and Klein W.H. (1989) Transcription of the Spec 1-like gene of *Lytechinus* is selectively inhibited in response to disruption of the extracellular matrix. *Development* **106**: 355–365.

Westaway D., Zuliani V., Cooper C.M., Da Costa M., Neuman S., Jenny A.L., Detwiler L. and Prusiner S.B. (1994) Homozygosity for prion protein alleles encoding glutamine-171 renders sheep susceptible to natural scrapie. *Genes and Dev.* **8**: 959–969.

Weston J.A. (1963) A radioautographic analysis of the migration and localization of trunk neural crest cells in the chick. *Dev. Biol.* **6**: 279–310.

Whalen A.M. and Steward R. (1993) Dissociation of the *dorsal-cactus* complex and phosphorylation of the dorsal protein correlate with the nuclear localization of *dorsal*. *J. Cell Biol.* **123**: 523–534.

Wharton K.A., Ray R.P. and Gelbart W.M. (1993) An activity gradient of *decapentaplegic* is necessary for the specification of dorsal pattern elements in the *Drosophila* embryo. *Development* **117**: 807–822.

Wharton R.P. and Struhl G. (1991) RNA regulatory elements mediate control of Drosophila body pattern by the posterior morphogen *nanos*. *Cell* **67**: 955–967.

Whitaker M. and Irvine R.G. (1984) Inositol 1,4,5-trisphosphate microinjection activates sea urchin eggs. *Nature* **312**: 636–639.

Whitaker M. and Patel R. (1990) Calcium and cell cycle control. *Development* **108**: 525–542.

Whitaker M. and Steinhardt R. (1985) Ion signalling in the sea urchin egg at fertilization. In: *Biology of Fertilization*, C.B. Metz and A. Monroy (eds.), vol. 3, 167–221. Orlando, FL: Academic Press.

Whitaker M. and Swann K. (1993) Lighting the fuse at fertilization. *Development* **117**: 1–12.

Whitfield L.S., Lovell-Badge R. and Goodfellow P.N. (1993) Rapid sequence evolution of the mammalian sex-determining gene *SRY*. *Nature* **364**: 713–715.

Whiting J., Marshall H., Cook M., Krumlauf R., Rigby P., Stott D. and Allemann R. (1991) Multiple spatially-specific enhancers are required to reconstruct the pattern of *Hox-2.6* gene expression. *Genes and Dev.* **5**: 2048–2059.

Whitlock K.E. (1993) Development of *Drosophila* wing sensory neurons in mutants with missing or modified cell surface molecules. *Development* **117**: 1251–1260.

Whittaker J.R. (1973) Segregation during ascidian embryogenesis of egg cytoplasmic information for tissue-specific enzyme development. *Proc. Nat. Acad. Sci. USA* **70**: 2096–2100.

Whittaker J.R. (1979) Cytoplasmic determinants of tissue differentiation in the ascidian egg. In: *Determinants of Spatial Organization*, S. Subtelny and I.R. Konigsberg (eds.), *37th Symp. Soc. Dev. Biol.*, 29–51. New York: Academic Press.

Whittaker J.R. (1980) Acetylcholinesterase development in extra cells caused by changing the distribution of myoplasm in ascidian embryos. *J. Embryol. exp. Morphol.* **55:** 343–354.

Wickens M. (1990) In the beginning is the end: Regulation of poly(A) addition and removal during early development. *Trends in Biochemical Science* **15:** 320–324.

Wieschaus E. and Gehring W. (1976) Clonal analysis of primordial disc cells in the early embryo of *Drosophila melanogaster*. *Dev. Biol.* **50:** 249–263.

Wigglesworth V.B. (1943) The fate of hemoglobin in *Rhodnius prolixus* and other blood-sucking arthropods. *Proc. R. Soc. London B* **131:** 313–339.

Wigglesworth V.B. (1972) *The Principles of Insect Physiology*. 7th ed. London: Chapman and Hall.

Wight T.N., Heinegård D.K. and Hascall V.C. (1991) Proteoglycans: Structure and function. In: *Cell Biology of Extracellular Matrix*, 2nd ed., E.D. Hay (ed.), 45–78. New York: Plenum.

Wightman B., Bürglin T.R., Gatto J., Arasu P. and Ruvkun G. (1991) Negative regulatory sequences in the *lin-14* 3′-untranslated region are necessary to generate a temporal switch during *Caenorhabditis elegans* development. *Genes and Dev.* **5:** 1813–1824.

Wightman B., Ha I. and Ruvkun G. (1993) Posttranscriptional regulation of the heterochronic gene *lin-14* by *lin-4* mediates temporal pattern formation in *C. elegans*. *Cell* **75:** 855–862.

Wikramanayake A.H., Uhlinger K., Griffin F.J. and Clark W.H., Jr. (1992) Sperm of the shrimp *Sicyonia ingentis* undergo a bi-phasic capacitation that is accompanied by morphological changes. *Dev. Growth Differ.* **34:** 347–355.

Wilkins A.S. (1986) *Genetic Analysis of Animal Development*. New York: Wiley.

Wilkins A.S. (1993) *Genetic Analysis of Animal Development*. 2nd ed. New York: Wiley.

Wilkinson D.G. (1993) Molecular mechanisms of segmental patterning in the vertebrate hindbrain and neural crest. *BioEssays* **15:** 499–505.

Wilkinson D.G., Bhatt S., Cook M., Boncinelli E. and Krumlauf R. (1989) Segmental expression of *Hox-2* homeobox-containing genes in the developing mouse hindbrain. *Nature* **341:** 405–409.

Wilkinson D.G., Bhatt S. and Hermann B.G. (1990) Expression pattern of the mouse T gene and its role in mesoderm formation. *Nature* **343:** 657–659.

Willadsen S.M. (1986) Nuclear transplantation in sheep embryos. *Nature* **320:** 63–65.

Williams G.T. and Smith C.A. (1993) Molecular regulation of apoptosis: Genetic controls on cell death. *Cell* **74:** 777–779.

Willier B.H. and Oppenheimer J.M. (1964) *Foundations of Experimental Embryology*. Englewood Cliffs, NJ: Prentice Hall.

Wilson E.B. (1904) Experimental studies on germinal localization. I. The germ regions in the egg of *Dentalium*. *J. Exp. Zool.* **1:** 1–72.

Wilson E.B. (1905) The chromosomes in relation to the determination of sex in insects. *Science* **22:** 500–502.

Wilson E.B. (1925) *The Cell in Development and Heredity*. 3rd ed. New York: Macmillan.

Wilson H.V. (1907) On some phenomena of coalescence and regeneration in sponges. *J. Exp. Zool.* **5:** 245–258.

Winklbauer R. and Nagel M. (1991) Directional mesoderm cell migration in the *Xenopus* gastrula. *Dev. Biol.* **148:** 573–589.

Winkler M. (1988) Translational regulation in sea urchin eggs: A complex interaction of biochemical and physiological regulatory mechanisms. *BioEssays* **8:** 157–161.

Winkler M., Nelson E.M., Lashbrook C. and Hershey J.W.B. (1985) Multiple levels of regulation of protein synthesis at fertilization in sea urchin eggs. *Dev. Biol.* **107:** 290–300.

Witschi E. (1956) *Development of Vertebrates*. Philadelphia: Saunders.

Witthuhn B.A., Quelle F.W., Silvennoinen O., Yi T., Tang B., Miura O. and Ihle J.N. (1993) JAK2 associates with the erythropoietin receptor and is tyrosine phosphorylated and activated following stimulation with erythropoietin. *Cell* **74:** 227–236.

Wolberger C., Vershon A.K., Liu B., Johnson A.D. and Pabo C.O. (1991) Crystal structure of a MAT α2 homeodomain-operator complex suggests a general model for homeodomain-DNA interactions. *Cell* **67:** 517–528.

Wolpert L. (1969) Positional information and the spatial pattern of cellular differentiation. *J. Theoret. Biol.* **25:** 1–47.

Wolpert L. (1978) Pattern formation in biological development. *Scientific American* **239** (October): 154–164.

Wolpert L., Lewis J. and Summerbell D. (1975) Morphogenesis of the vertebrate limb. In: *Cell Patterning, Ciba Foundation Symp.* **29** (new series): 95–130. Amsterdam: Elsevier, Excerpta Medica, North-Holland Publishing Company, Associated Scientific Publishers.

Wood W.B. (1980) Bacteriophage T4 morphogenesis as a model for assembly of subcellular structure. *Quart. Rev. Biol.* **55:** 353–367.

Wood W.B., ed. (1988a) *The Nematode Caenorhabditis elegans*. Cold Spring Harbor, NY: Cold Spring Harbor Laboratory Press.

Wood W.B. (1988b) Determination of pattern and fate in early embryos of *Caenorhabditis elegans*. In: *Developmental Biology: A Comprehensive Synthesis*, L.W. Browder (ed.), vol. 5, 57–78. New York: Plenum.

Wood W.B. (1991) Evidence from reversal of handedness in *C. elegans* embryos for early cell interactions determining cell fates. *Nature* **349:** 536–538.

Wright C.V.E., Cho K.W.Y., Hardwicke J., Collins R.H. and De Robertis E.M. (1989) Interference with function of a homeobox gene in Xenopus embryos produces malformations of the anterior spinal cord. *Cell* **59**: 81–93.

Wright T.R.F. (1960) The phenogenetics of the embryonic mutant, lethal myospheroid, in *Drosophila melanogaster. J. Exp. Zool.* **143**: 77–99.

Wylie C.C., Heasman J., Snape A., O'Driscoll M. and Hollwill S. (1985) Primordial germ cells of *Xenopus laevis* are not irreversibly determined early in development. *Dev. Biol.* **112**: 66–72.

Xing Y., Johnson C.V., Dobner P.R. and Lawrence J.B. (1993) Higher level organization of individual gene transcription and RNA splicing. *Science* **259**: 1326–1330.

Yajima H. (1964) Studies on embryonic determination of the harlequin-fly, *Chironomus dorsalis*. II. Effects of partial irradiation of the egg by ultra-violet light. *J. Embryol. exp. Morphol.* **12**: 59–100.

Yajima H. (1983) Induction of longitudinal double malformations by centrifugation or by partial irradiation of eggs in the chironomid species, *Chironomus samoensis* (Diptera: Chironomidae). *Entomol. Gen.* **8**: 171–191.

Yamada K.M. (1991) Fibronectin and other cell interactive glycoproteins. In: *Cell Biology of Extracellular Matrix*, 2nd ed., E.D. Hay (ed.), 111–146. New York: Plenum.

Yamada T. (1967) Cellular and subcellular events in Wolffian lens regeneration. *Current Topics Dev. Biol.* **2**: 247–283.

Yamada T., Placzek M., Tanaka H., Dodd J. and Jessell T.M. (1991) Control of cell pattern in the developing nervous system: Polarizing activity of the floor plate and notochord. *Cell* **64**: 635–647.

Yamada T., Pfaff S.L., Edlund T. and Jessell T.M. (1993) Control of cell pattern in the neural tube: Motor neuron induction by diffusible factors from notochord and floor plate. *Cell* **73**: 673–686.

Yamamoto K.R. (1985) Steroid receptor regulated transcription of specific genes and gene networks. *Ann. Rev. Genet.* **19**: 209–252.

Yamamoto K.R. (1989) A conceptual view of transcriptional regulation. *Am. Zool.* **29**: 537–547.

Yamamoto K.R., Pearce D., Thomas J. and Miner J.N. (1992) Combinatorial regulation at a mammalian composite response element. In: *Transcriptional Regulation*, S.L. McKnight and K.R. Yamamoto (eds.), 1169–1192. Plainview, NY: Cold Spring Harbor Laboratory Press.

Yanagimachi R. and Noda Y.D. (1970) Electron microscope studies of sperm incorporation into the golden hamster egg. *Am. J. Anat.* **128**: 429–462.

Yanofsky M.F., Ma H., Bowman J.L., Drews G.N., Feldmann K.A. and Meyerowitz E.M. (1990) The protein encoded by the *Arabidopsis* homeotic gene *agamous* resembles transcription factors. *Nature* **346**: 35–39.

Yao T.-P., Segraves W.A., Oro A.E., McKeown M. and Evans R.M. (1992) Drosophila ultraspiracle modulates ecdysone receptor function via heterodimer formation. *Cell* **71**: 63–72.

Yaoita Y. and Brown D.D. (1990) A correlation of thyroid hormone receptor gene expression with amphibian metamorphosis. *Genes and Dev.* **4**: 1917–1924.

Yasuda G.K., Baker J. and Schubiger G. (1991) Temporal regulation of gene expression in the blastoderm *Drosophila* embryo. *Genes and Dev.* **5**: 1800–1812.

Yew N., Mellini M.L. and Vande Woude G.F. (1992) Meiotic initiation by the *mos* protein in *Xenopus*. *Nature* **355**: 649–652.

Yisraeli J.K., Sokol S. and Melton D.A. (1990) A two-step model for the localization of maternal mRNA in *Xenopus* oocytes: Involvement of microtubules and microfilaments in the translocation and anchoring of Vg1 mRNA. *Development* **108**: 289–298.

Yochem J., Weston K. and Greenwald I. (1988) The *Caenorhabditis elegans lin-12* gene encodes a transmembrane protein with overall similarity to *Drosophila Notch*. *Nature* **335**: 547–551.

Yokoyama T., Copeland N.G., Jenkins N.A., Montgomery C.A., Elder F.F.B. and Overbeek P.A. (1993) Reversal of left-right asymmetry: A situs inversus mutation. *Science* **260**: 679–682.

Yost H.J. (1992) Regulation of vertebrate left-right asymmetries by extracellular matrix. *Nature* **357**: 158–161.

Young L.J., Lopreato G., Horan K. and Crews D. (1994) Cloning and *in situ* hybridization analysis of estrogen receptor, progesterone receptor and androgen receptor expression in the brain of whiptail lizards (*Cnemidophorus uniparens* and *C. inornatus*). *J. Comp. Neurol.* **347**: 288–300.

Younger-Shepherd S., Vaessin H., Bier E., Jan L.Y. and Jan Y.N. (1992) *deadpan*, an essential pan-neural gene encoding an HLH protein, acts as a denominator in Drosophila sex determination. *Cell* **70**: 911–922.

Yuan J. and Horvitz H.R. (1990) The *Caenorhabditis elegans* genes *ced-3* and *ced-4* act cell autonomously to cause programmed cell death. *Dev. Biol.* **138**: 33–41.

Yuan J. and Horvitz H.R. (1992) The *Caenorhabditis elegans* cell death gene *ced-4* encodes a novel protein and is expressed during the period of extensive programmed cell death. *Development* **116**: 309–320.

Yuan J., Shasham S., Ledoux S., Ellis H. and Horvitz H.R. (1993) The *C. elegans* death gene *ced-3* encodes a protein similar to mammalian interleukin-1ß-converting enzyme. *Cell* **75**: 641–652.

Yuge M., Kobayakawa Y., Fujisue M. and Yamana K. (1990) A cytoplasmic determinant for dorsal axis formation in an early embryo of *Xenopus laevis*. *Development* **110**: 1051–1056.

Zalokar M. and Erk I. (1976) Division and migration of nuclei during early embryogenesis of *Drosophila melanogaster. J. Microsc. Biol. Cell.* **25**: 97–106.

Zanetti N.C. and Solursh M. (1984) Induction of chondrogenesis in limb mesenchymal cultures by disruption of the actin cytoskeleton. *J. Cell Biol.* **99:** 115–123.

Zeng W., Andrew D.J., Mathies L.D., Horner M.A. and Scott M.P. (1993) Ectopic expression and function of the *Antp* and *Scr* homeotic genes: The N terminus of the homeodomain is critical to functional specificity. *Development* **118:** 339–352.

Zhang X.-K., Hoffmann B., Tran P.B.-V., Graupner G. and Pfahl M. (1992) Retinoid X receptor is an auxiliary protein for thyroid hormone and retinoic acid receptors. *Nature* **355:** 441–446.

Zhao J.J., Lazzarini R.A. and Pick L. (1993) The mouse *Hox-1.3* gene is functionally equivalent to the *Drosophila Sex combs reduced* gene. *Genes and Dev.* **7:** 343–354.

Zhou X., Sasaki H., Lowe L., Hogan B.L.M. and Kuehn M.R. (1993) *Nodal* is a novel TGF-ß-like gene expressed in the mouse node during gastrulation. *Nature* **361:** 543–547.

Zimmerman K., Shih J., Bars J., Collazo A. and Anderson D.J. (1993) *XASH-3*, a novel *Xenopus achaete-scute* homolog, provides an early marker of planar neural induction and position along the mediolateral axis of the neural plate. *Development* **119:** 221–232.

Ziomek C.A. and Johnson M.H. (1982) The roles of phenotype and position in guiding the fate of 16-cell mouse blastomeres. *Dev. Biol.* **91:** 440–447.

Zouros E., Freeman K.R., Ball A.O. and Pogson G.H. (1992) Direct evidence for extensive paternal mitochondrial DNA inheritance in the marine mussel *Mytilus. Nature* **359:** 412–414.

Zusman S., Grinblat Y., Yee G., Kafatos F.C. and Hynes R.O. (1993) Analyses of PS integrin functions during *Drosophila* development. *Development* **118:** 737–750.

Zwilling E. (1955) Ectoderm-mesoderm relationship in the development of the chick embryo limb bud. *J. Exp. Zool.* **128:** 423–441.

Zwilling E. (1956) Interaction between limb bud ectoderm and mesoderm in the chick embryo. I. Axis establishment. *J. Exp. Zool.* **132:** 157–172.

PHOTOGRAPH CREDITS

COLOR PLATES **1:** From Nüsslein-Volhard C. (1991) *Development* (Supplement 1): 1–10. Photographs courtesy of Christiane Nüsslein-Volhard. **2:** From Keller R. (1991) In: B.K. Kay and H.B. Peng (eds.), *Methods in Cell Biology*, vol. 36, 61–113. New York: Academic Press. Color print courtesy of Ray Keller. **3:** From Bronner-Fraser M. and Fraser S.E. (1991) *Development* (Supplement 2): 17–22. Print courtesy of Marianne Bronner-Fraser. **4:** From Lawrence P.A. and Johnston P. (1989) *Development* 105: 761–767. Photographs provided by Peter Lawrence. **5:** Courtesy of Walter Gehring. **6:** From De Robertis E.M., Morita E.A. and Cho K.W.Y. (1991) *Development* 112: 669–678. Prints courtesy of Eddy De Robertis. **7:** Print courtesy of Stephen Paddock, James Langeland and Sean Carroll. **8:** From Blair S.S. (1992) *Development* 115: 21–33. Print courtesy of Seth Blair. **9:** From Lawrence P.A. and Morata G. (1993) *Nature* 366: 305–306. Print provided by Jim Williams, Stephen Paddock and Sean Carroll. **10:** From Meyerowitz E.M., Bowman J.L., Brockman L.L., Drews G.N., Jack T., Sieburth L.E. and Weigel D. (1991) *Development* (Supplement 1): 157–167. Photographs courtesy of Elliot Meyerowitz. **11:** From Bowman J.L., Smyth D.R. and Meyerowitz E.M. (1991) *Development* 112: 1–20. Print provided by Elliot Meyerowitz. **12:** From Meyerowitz E.M., Bowman J.L., Brockman L.L., Drews G.N., Jack T., Sieburth L.E. and Weigel D. (1991) *Development* (Supplement 1): 157–167. Photographs courtesy of Elliot Meyerowitz. **13:** From Meyerowitz E.M., Bowman J.L., Brockman L.L., Drews G.N., Jack T., Sieburth L.E. and Weigel D. (1991) *Development* (Supplement 1): 157–167. Photograph courtesy of Elliot Meyerowitz. **14:** From Coen E.S. and Meyerowitz E.M. (1991) *Nature* 353: 31–37. Photographs courtesy of Elliot Meyerowitz. **15:** From Coen E.S. and Meyerowitz E.M. (1991) *Nature* 353: 31–37. Photograph courtesy of Enrico Coen. **16:** From Drews G.N., Bowman J.L. and Meyerowitz E.M. (1991a) *Cell* 65: 991–1002. Photographs courtesy of Gary Drews. **17:** From Hill R.J. and Sternberg P.W. (1992) *Nature* 358: 470–476. Photographs courtesy of Russell Hill.

PART OPENERS **I.1:** From Craig M.M. (1992) *Cell* 68(2): cover. Courtesy of Michel M. Craig. **II.1:** From Brinster R.L. and Palmiter R.D. (1986) Introduction of genes into the germ line of animals. *The Harvey Lectures*, **Series 80:** 1–38. New York: Alan R. Liss, Inc. Photograph courtesy of R.L. Brinster. **III.1:** From Cheng T.-C., Wallace M.C., Merlie J.P. and Olson E.N. (1993) *Science* 261(July 9): cover. Photo courtesy of T.-C. Cheng and E.N. Olson.

CHAPTER 1 **1.1:** From Wassarman P.M. (1990) *Development* 108: 1–17. Courtesy of Paul M. Wassarman. **1.2:** From Maro B., Gueth-Hallonet C., Aghion J. and Antony C. (1991) *Development* (Supplement 1): 17–25. Photograph courtesy of Bernard Maro. **1.3, 1.4:** Reproduced from Blechschmidt E. (1961) *The Stages of Human Development before Birth*. Philadelphia: W.B. Saunders. **1.5:** From Moore J.A. (1972) *Heredity and Development*, 2nd ed., 257. New York: Oxford University Press. Nicolaus Hartsoeker (1694) redrawn by J.A. Moore. **1.7: (a–i)** From Johnson K.E., Nakatsuji N. and Boucaut J.-C. (1990) Extracellular matrix control of cell migration during amphibian gastrulation. In: *Cytoplasmic Organization Systems*, G.M. Malacinski (ed.), 349–374. New York: McGraw-Hill. Photograph provided by De-Li Shi. **1.9:** From Ettensohn C.A. and Ingersoll E.P. (1992) Morphogenesis of the sea urchin embryo. In: *Morphogenesis*, E.F. Rossomando and S. Alexander (eds.), 189–262. New York: Marcel Dekker. Photograph courtesy of Charles A. Ettensohn. **1.13: (a–f)** Reproduced from Saunders J.W., Jr. (1970) *Patterns and Principles of Animal Development*. New York: Macmillan. **1.14:** Courtesy of Muriel Voter Williams. **1.15: (a–f)** Reproduced from Daly H.V., Doyen J.T. and Ehrlich P.R. (1978) *Introduction to Insect Biology and Diversity*. New York: McGraw-Hill. **1.16:** Photograph provided by Phil DeVries. **1.20: (a–b)** Reproduced from Nüsslein-Volhard C. (1977) *Wilhelm Roux's Arch. Dev. Biol.* 183: 249–268. **1.21 (b–c)** Courtesy of E.B. Lewis.

CHAPTER 2 **2.1:** From Rhodin J.A.G. (1975) *An Atlas of Histology*. New York: Oxford University Press. Photograph provided by

Johannes Rhodin. **2.4:** From Beams H.W. and Kessel R.G. (1976) *Am. Sci.* 64: 279–290. Courtesy of R.G. Kessel. **2.9: (a–d)** Courtesy of Shinya Inoué. **2.10: (a–e)** From Seegmiller R., Ferguson R. and Sheldon H. (1972) *J. Ultrastr. Res.* 2: 288. Provided by Robert E. Seegmiller and Huntington Sheldon. **2.15: (a–d)** From Perry M.M. and Gilbert A.B. (1979) *J. Cell Sci.* 39: 257–272. Courtesy of M.M. Perry and A.B. Gilbert. **2.17:** David M. Phillips/Visuals Unlimited. **2.19: (b–c)** From Smith S.J. (1988) *Science* 242: 709. Courtesy of Stephen J. Smith.

CHAPTER 3 **3.1:** Reproduced from Needham J. (1959) *A History of Embryology*. London/New York: Abelard-Schuman. **3.7: (b)** Courtesy of D.W. Fawcett. **3.10:** David M. Phillips/Visuals Unlimited. **3.11: (a–e)** From Dettlaff T.A. and Vassetzky S.G. (eds.) (1988) *Oocyte Growth and Maturation*. Translation edited by Frank Billett. New York: Consultants Bureau. Photograph courtesy of S.G. Vassetzky. **3.12:** Courtesy of Joseph G. Gall, Carnegie Institution. **3.13: (a)** From Brown D.D. and Dawid I.B. (1968) *Science* 160: 272–280. Courtesy of D.D. Brown and I.B. Dawid. **(b)** Courtesy of Ulrich Scheer. **3.18: (a–b)** From K. Bier; courtesy of D. Ribbert. **3.19:** From Telfer W.H., Huebner E. and Smith D.S. (1982) The cell biology of vitellogenic follicles in *Hyalophora* and *Rhodnius*. In: *Insect Ultrastructure*, vol. I, 118–149. New York: Plenum. Courtesy of W.H. Telfer. **3.24: (d)** D.W. Fawcett/Visuals Unlimited. **3.25: (a)** From Bloom W. and Fawcett D.W. (1975) *A Textbook of Histology*. 10th ed. Philadelphia: Saunders. **(b)** Courtesy of E.A. Anderson and David Albertini. **(c)** From Hinton H.E. (1970) *Scientific American* 223 (August): 84–91.

CHAPTER 4 **4.1:** David M. Phillips/Visuals Unlimited. **4.5: (a)** Reproduced from Hadek R. (1969) *Mammalian Fertilization: An Atlas of Ultrastructure*. New York: Academic Press. **4.7: (a–b)** From Vacquier V.D. (1979) *Current Topics Dev. Biol.* 13: 31–44. Courtesy of V.D. Vacquier, University of California, San Diego. **4.11:** From Perona R.M. and Wassarman P.M. (1986) *Dev. Biol.* 114: 42–52. Courtesy of Paul Wassarman. **4.12: (a)** From Lopez L.C., Bayna E.M., Litoff D., Shaper N.L., Shaper J.H. and Shur B.D. (1985) *J. Cell Biol.* 101: 1501–1510. Courtesy of Barry D. Shur. **4.13:** From Epel D. (1977) *Scientific American* 237 (November): 128–138. Provided by Mia Tegner and David Epel. **4.14: (a–b)** From Epel D. (1977) *Scientific American* 237 (November): 128–138. Photographs courtesy of Gerald Schatten. **(c)** Transmission electron micrograph of sperm uteralization through the fertilization core, courtesy of F.J. Longo. **4.15:** From Hamaguchi M.S. and Hiramoto Y. (1980) *Dev. Growth Differ.* 22: 517–530. Courtesy of M.S. Hamaguchi and Y. Hiramoto, Tokyo Institute of Technology. **4.19: (a–m)** From Gilkey J.C., Jaffe L.F., Ridgway E.G. and Reynolds G.T. (1978) *J. Cell. Biol.* 76: 448–466. Provided by Lionel Jaffe. **4.22: (b–c)** From Schuel H. (1985) Functions of egg cortical granules. In: *Biology of Fertilization*, C.B. Metz and A. Monroy (eds.), vol. 3. New York: Academic Press. Photograph courtesy of Herbert Schuel. **(d)** From Vacquier V.D. (1975) *Dev. Biol.* 43: 64–65. Photograph courtesy of Victor Vacquier. **4.23: (a–c)** From Tegner M.J. and Epel D. (1973) *Science* 179: 685–688. Provided by Mia Tegner and David Epel.

CHAPTER 5 **5.1:** From Beams H.W. and Kessel R.G. (1976) *Am. Sci.* 64: 279–290. Courtesy of R.G. Kessel. **5.10: (a–b)** From Fleming T.P., Pickering J., Qasim F. and Maro B. (1986) *J. Embryol. exp. Morphol.* 95: 169–191. Courtesy of Tom Fleming. **5.12: (a–f)** From Bavister B.D. (ed.) (1987) *The Mammalian Preimplantation Embryo: Regulation of Growth and Differentiation in Vitro*. New York: Plenum. Photograph courtesy of D.E. Boatman and B.D. Bavister. **5.13:** From Beams H.W. and Kessel R.G. (1976) *Am. Sci.* 64: 279–290. Courtesy of R.G. Kessel. **5.14:** Reproduced from Patten B.M. (1971) *Early Embryology of the Chick*. New York: McGraw-Hill. **5.17: (a–g)** From Turner F.R. and Mahowald A.P. (1976) *Dev. Biol.* 50: 95–108. Courtesy of F.R. Turner, Indiana University. **5.21: (a–d)** From Rappaport R. (1961) *J. Exp. Zool.* 148: 81–89. Photographs courtesy of R. Rappaport. **5.22: (a–c)** From Schroeder T.E. (1987) *Dev. Biol.* 124: 9–22. Courtesy of T.E. Schroeder. **5.23: (a–d)** From Lutz D.A., Hamaguchi Y. and Inoué S. (1988) *Cell Motility and the Cytoskeleton* 11: 83–96. Courtesy of Shinya Inoué. **5.25: (a–d)** From Harvey E.B.

(1956) *The American Arbacia and Other Sea Urchins*. New Jersey: Princeton Univ. Press. **5.29:** (a, c) From Edgar B.A. and O'Farrell P.H. (1989) *Cell* **57:** 177–187. Provided by Bruce Edgar.

CHAPTER 6 6.1: From Spemann H. (1928) *Zeitschrift für wiss. Zool.* **132:** 105–134. Print courtesy of Klaus Sander. **6.3:** From Strehlow D. and Gilbert W. (1993) *Nature* **361:** 451–453. Photograph courtesy of David Strehlow. **6.5:** (a–b) Reproduced from Garcia-Bellido A., Lawrence P.A. and Morata G. (1979) *Scientific American* **241** (July): 102–110. **6.8:** (a–b) Reproduced from Spemann H. (1938) *Embryonic Development and Induction*. New Haven: Yale University Press. Reprinted 1967. New York: Hafner. **6.14:** Reproduced from Spemann H. (1938) *Embryonic Development and Induction*. New Haven: Yale University Press. Reprinted 1967. New York: Hafner. **6.19:** (b–c) Photographs courtesy of Dianne Fristrom. **6.27:** (a–e) From Le Douarin N.L. and McLaren A. (eds.) (1984) *Chimeras in Developmental Biology*. London/Orlando: Academic Press. Photographs courtesy of V.E. Papaioannou.

CHAPTER 7 7.1: Reproduced from Steward F.C. (1963) *Scientific American* **209** (October): 104–115. **7.4:** (a) Courtesy of B.P. Kaufman. (b) Courtesy of Veikko Sorsa. **7.5:** Reproduced from Beermann W. (1952a) Chromosomes and genes. In: *Developmental Studies on Giant Chromosomes*, W. Beermann (ed.), 1–33. New York: Springer-Verlag. **7.6:** Preparation and photograph by A. von Brunn. Print supplied by K. Kalthoff. **7.7:** (a–i) Reproduced from Shepard J.F. (1982) *Scientific American* **246** (May): 154–165. **7.8:** (a–d) Reproduced from Spemann H. (1938) *Embryonic Development and Induction*. New Haven: Yale University Press. Reprinted 1967. New York: Hafner. **7.13:** Photographs courtesy of John Gurdon. **7.14:** (b) From Laskey R.A. and Gurdon J.B. (1970) *Nature* **228:** 1332–1334. Courtesy of J.B. Gurdon and R.A. Laskey. By permission of *Nature*, Macmillan Magazines Limited. **7.15:** (a–g) From Reyer R.W. (1954) *Quart. Rev. Biol.* **29:** 1–46. Courtesy of R.W. Reyer. **7.19:** (a–b) From De Robertis E.M. and Gurdon J.B. (1979) *Scientific American* **241** (December): 74–82. Courtesy of E.M. De Robertis and J.B. Gurdon. **7.22:** (a–c) From Harris H. (1974) *Nucleus and Cytoplasm*. 3rd ed. Oxford: Clarendon Press. Photograph courtesy of Henry Harris and the Company of Biologists.

CHAPTER 8 8.1: (a–c) From Frigerio G., Burri M., Bopp D., Baumgartner S. and Noll M. (1986) *Cell* **47:** 735–746. Photographs courtesy of Markus Noll. **8.4:** Courtesy of M.R. Dohmen. **8.8:** From Melton D.A. (1987) *Nature* **328:** 80–82. Photograph courtesy of Douglas Melton. **8.9:** (c) From Lohs-Schardin M. (1982) *Wilhelm Roux's Arch. Dev. Biol.* **191:** 28–36. Courtesy of Klaus Sander and Springer-Verlag Heidelberg. **8.11:** (a–b) From Jeffery W.R. (1988) *Dev. Biol.* **5:** 3–56. Courtesy of William R. Jeffery. **8.15:** (d) From Mahowald A.P. (1971) *J. Exp. Zool.* **176:** 329–344. Courtesy of Anthony P. Mahowald. **8.16:** (b–d) From Okada M., Kleinman I.A. and Schneiderman H.A. (1974) *Dev. Biol.* **37:** 43–54. Photograph courtesy of M. Okada and S. Kobayashi. **8.18:** (g–h) From Whittaker J.R. (1980) *J. Embryol. Exp. Morphol.* **55:** 343–354. Courtesy of J. Richard Whittaker. **8.19:** (a–b) Courtesy of Klaus Kalthoff. **8.23:** From Render J. (1983) *Wilhelm Roux's Arch. Dev. Biol.* **192:** 120–129. Photograph courtesy of Jo Ann Render. **8.26:** (a–d) From Rau K.-G. and Kalthoff K. (1980). *Nature* **287:** 635–637. Courtesy of Klaus Kalthoff.

CHAPTER 9 9.1: (a–b) Photograph courtesy of Dan Kessler and Doug Melton. **9.6:** (a–c) From Kropf D.L., Berge S.K. and Quatrano R.S. (1989) *Plant Cell* **1:** 191–200. Photograph provided by Darryl Kropf. **9.9:** (a–c) From Vincent J.-P., Oster G.F. and Gerhart J.C. (1986) *Dev. Biol.* **113:** 484–500. Courtesy of J.C. Gerhart. **9.13:** (b–e) From Yost, H.J. (1992) *Nature* **357:** 158–161. Courtesy of H. Joseph Yost. **9.14:** (a–c) From Danilchik M.V. and Denegre J.M. (1991) *Development* **111:** 845–856. Courtesy of Michael V. Danilchik. **9.15:** (a–b) From Fujisue M., Kobayakawa Y. and Yamana K. (1993) *Development* **118:** 163–170. Photograph courtesy of Megumi Fujisue. **9.18:** (a–b) From Gimlich R.L. and Gerhart J.C. (1984) *Dev. Biol.* **104:** 117–130. **9.19:** (c) From Sudarwati S. and Nieuwkoop P.D. (1971) *Wilhelm Roux's Arch. Dev. Biol.* **166:** 189–204. Photograph courtesy of Pieter D. Nieuwkoop.

CHAPTER 10 10.1: From Morrill J.B. and Santos L.L. (1985) A scanning electron micrographical overview of cellular and extracellular patterns during blastulation and gastrulation in the sea urchin, *Zytechinus variegatus*. In: *The Cellular and Molecular Biology of Invertebrate Development*, R.H. Sawyer and R.M. Showman (eds.), 3–33. Columbia: University of South Carolina Press. Photograph courtesy of John Morrill. **10.5:** (a–b) From Solursh M. (1986)

Migration of sea urchin primary mesenchyme cells. In: *Developmental Biology—A Comprehensive Synthesis*, L.W. Browder (ed.), vol. 2, 391–400. New York: Plenum. Photograph courtesy of Michael Solursh. **10.6:** Courtesy of John B. Morrill. **10.9:** (a–b) From Ettensohn C.A. and Ingersoll E.P. (1992) Morphogenesis of the sea urchin embryo. In: *Morphogenesis*, E.F. Rossomando and S. Alexander (eds.), 189–262. New York: Marcel Dekker. Photograph courtesy of Charles Ettensohn. **10.11:** (a) From Hardin J. (1989) *Dev. Biol.* **136:** 430–445. Photograph courtesy of Jeff Hardin. **10.21:** (b–c) From Keller R.E. and Tibbetts P. (1989) *Dev. Biol.* **131:** 539–549. Photograph courtesy of Ray Keller. **10.24:** (a–f) From Keller R.E. (1980) *J. Embryol. Exp. Morphol.* **60:** 201–243. Photograph courtesy of Ray Keller. **10.28:** From Solursh M. and Revel J.P. (1978) *Differentiation* **11:** 185–190. Courtesy of Michael Solursh. **10.31:** (b) From Carlson B.M. (1988) *Patten's Foundations of Embryology*. 5th ed. New York: McGraw-Hill. Courtesy of Bruce M. Carlson.

CHAPTER 11 11.1: From Schoenwolf G.C. (1991) *Development* (Supplement 2): 160. Photograph courtesy of Gary Schoenwolf. **11.9:** Reproduced from Gordon R. and Jacobson A.G. (1978) *Scientific American* **238** (June): 106–113. **11.17:** (a–b) Photographs courtesy of Lauri Saxén. **11.18:** Reproduced from Bodemer C.W. (1968) *Modern Embryology*. New York: Holt, Rinehart and Winston. **11.22:** (c) From Keller R. and Danilchik M. (1988) *Development* **103:** 193–209. Photograph courtesy of Ray Keller. **11.23:** (a–b) From Doniach et al. (1992) *Dev. Biol.* **104:** 117–130. Courtesy of J.C. Gerhart. **11.26:** (a–c) From Cho K.W.Y., Blumberg B., Steinbeisser H. and De Robertis E.M. (1991) *Cell* **67:** 1111–1120. Courtesy of E.M. De Robertis.

CHAPTER 12 12.1: Reproduced from Palay S.L. and Chan-Palay V. (1974) *Cerebellar Cortex: Cytology and Organization*, Fig. 7, p. 12. Berlin and Heidelberg: Springer-Verlag. **12.6:** H. Webster, D. Fawcett/Visuals Unlimited. **12.9:** From Goulding M.D., Lumsden A. and Gruss P. (1993) *Development* **117:** 1001–1016. Photograph courtesy of Andrew Lumsden. **12.21:** From Tosney K.W. (1978) *Dev. Biol.* **62:** 327. Photograph provided by Kathryn W. Tosney. **12.22:** (b–c) From Moury J.D. and Jacobson A.G. (1989) *Dev. Biol.* **133:** 44–57. Courtesy of J.D. Moury and A.G. Jacobson. **12.23:** (b) From Weston J.A. (1963) *Dev. Biol.* **6:** 279–310. With permission of Academic Press. **12.24:** (a–c) From Le Douarin N.M. (1986) *Science* **231:** 1515–1522. Courtesy of N. Le Douarin. **12.26:** Reproduced from Carlson B.M. (1988) *Patten's Foundations of Embryology*. 5th ed. New York: McGraw-Hill. **12.29:** From Alberts B., Bray D., Lewis J., Raff M., Roberts K. and Watson D. (1989) *Molecular Biology of the Cell*. 2nd ed. New York: Garland Publishing, Inc. Photograph provided by Jan Löfberg. **12.32:** Reproduced from Balinsky B.I. (1975) *An Introduction to Embryology*. Philadelphia: Saunders.

CHAPTER 13 13.1: Courtesy of Chester F. Reather, RPB, FBPA, Carnegie Institution of Washington, Davis, CA. **13.14:** (b) From Meier S. (1979) *Dev. Biol.* **73:** 25–45. Photograph by Stephen Meier, courtesy of Antone Jacobson. **13.18:** From Fawcett D.W. (1986) *Textbook of Histology*. 11th ed. Philadelphia: Saunders. **13.20:** From Rhodin J.A.G. (1974) *Histology: A Text and Atlas*. New York: Oxford University Press. Photograph provided by Johannes A.G. Rhodin. **13.21:** (a–b) From Shimada Y., Fischman D.A. and Moscona A.A. (1967) *J. Cell Biol.* **35:** 445–453. Photograph courtesy of Yutaka Shimada. **13.24:** From Poole T. and Steinberg M. (1981) *J. Embryol. Exp. Morphol.* **63:** 3. Photograph courtesy of Malcolm S. Steinberg. **13.30:** From Hurle J.M., Icardo J.M. and Ojeda J.L. (1980) *J. Embryol. Exp. Morphol.* **56:** 211–223. Photograph courtesy of J.M. Hurle. **13.34:** (b) D.W. Fawcett/Photo Researchers. **13.35:** From Kelly R.O. (1981) The developing limb. In: *Morphogenesis and Pattern Formation*, T.G. Connelly, L.L. Brinkley and B.M. Carlson (eds.), 49–85. New York: Raven Press. Photograph courtesy of Robert O. Kelly.

CHAPTER 14 14.1: Courtesy of Ralph Brinster. **14.2:** (a–b) From Frohnhöfer H.G. and Nüsslein-Volhard C. (1986) *Nature* **324:** 120–125. Photograph courtesy of C. Nüsslein-Volhard. **14.5:** From Bridges C.B. (1935) *J. Heredity* **26:** 60–64. Reproduced with permission from the American Genetic Association. **14.10:** From Chisholm A.D. and Hodgkin J. (1989) *Genes and Dev.* **3:** 1414. Photograph courtesy of Andrew Chisholm. **14.22:** From Rubin G.M. and Spradling A.C. (1982) *Science* **218:** 348–353. Courtesy of Gerald M. Rubin. **14.23:** From Rubin G.M. and Spradling A.C. (1982) *Science* **218:** 348–353. Courtesy of Gerald M. Rubin. **14.28:** From Readhead C., Popko B., Takahashi N., Shine H.D., Saavedra R.A., Sidman R.L. and Hood L. (1987) *Cell* **48:** 703–712. Courtesy of Carol Readhead.

CHAPTER 15 **15.1:** From Old R.W., Callan H.G. and Gross K.W. (1977) *J. Cell Sci.* **27:** 57–79. Photograph courtesy of R.W. Old. **15.2: (a–b)** From Hume C.R. and Dodd J. (1993) *Development* **119:** 1147–1160. Photograph provided by Jane Dodd. **15.3: (a–b)** From Tautz D. (1988) *Nature* **332:** 281–284. Courtesy of D. Tautz. **15.6: (b–c)** From Gehring W. (1987) *Cell* **50:** 963–974. Courtesy of Walter Gehring. **15.9: (a–b)** From Driever W. and Nüsslein-Volhard C. (1988) *Cell* **54:** 83–93. Photographs courtesy of C. Nüsslein-Volhard. **15.10: (a–c)** From Tautz D. (1988) *Nature* **332:** 281–284. Courtesy of D. Tautz. **15.12: (a–d)** From Driever W., Thoma G. and Nüsslein-Volhard C. (1989a) *Nature* **340:** 363–367. Photographs courtesy of C. Nüsslein-Volhard. **15.16:** Reproduced from Littau V.C., Allfrey V.G., Frenster J.H. and Mirsky A.E. (1964) *Proc. Nat. Acad. Sci. USA* **52:** 93–100. **15.17: (a–c)** Reproduced from Moore K.L. (1982) *The Developing Human: Clinically Oriented Embryology.* 3rd ed. Philadelphia: Saunders. **15.18: (a–b)** From Lezzi M. and Gilbert L.I. (1969) *Proc. Nat. Acad. Sci. USA* **64:** 498–503. Photograph courtesy of Lawrence I. Gilbert. **15.19: (a)** Photograph courtesy of Victoria Foe. **(b)** From Alberts B., Bray D. Lewis J., Raff M., Roberts K. and Watson J.D. (1983) *Molecular Biology of the Cell,* 388. New York/London: Garland Publishing, Inc. Photograph provided by Barbara Hamkalo. **15.20:** Reproduced from Groudine M. and Weintraub H. (1981) *Cell* **24:** 393–401. **15.21: (a–d)** From Hafen E., Levine M. and Gehring W. (1984) *Nature* **307:** 287–289. Courtesy of E. Hafen.

CHAPTER 16 **16.1:** From Akey C.W. (1992) The nuclear pore complex: A macromolecular transporter. In: *Nuclear Trafficking,* C.M. Feldherr (ed.), San Diego: Academic Press. Photograph courtesy of Christopher W. Akey. **16.2: (a)** From Chambon P. (1982) *Scientific American* **244** (May): 60–77. Courtesy of Pierre Chambon. **16.5: (c)** From Reed R., Griffith J. and Maniatis T. (1988) *Cell* **53:** 955. Photograph provided by Jack Griffith. **16.9: (a–b)** From Wilkins A.S. (1986) *Genetic Analysis of Animal Development.* New York: Wiley. Photographs courtesy of Bruce Baker. **(c)** From Wilkins A.S. (1986) *Genetic Analysis of Animal Development.* New York: Wiley. Photograph courtesy of D. Gubb. **16.11:** From Nagoshi R.N., McKeown M., Burtis K.C., Belote J.M. and Baker B.S. (1988) *Cell* **53:** 229–236. Photograph courtesy of Michael McKeown. **16.15: (b)** Adapted from Emeson R.B., Hedjran F., Yeakley J.M., Guise J.W. and Rosenfeld M.G. (1990) *Nature* **341:** 76–80. Courtesy of Ronald Emeson. **16.17:** From Scheer U., Dabauvalle M.-C., Merkert H. and Denavente R. (1988) *Cell Biol. Int. Rep.* **12:** 669–689. Photograph courtesy of Ulrich Scheer. **16.19: (a–d)** From Angerer R. and Angerer L. (1983) *Dev. Biol.* **100:** 197–206. Courtesy of Lynne M. Angerer. **16.20: (b–g)** Photographs courtesy of J.T. Bonner.

CHAPTER 17 **17.1: (a–b)** From Rhodin J.A.G. (1974) *Histology: A Text and Atlas.* New York: Oxford University Press. Photograph provided by Johannes A.G. Rhodin. **17.11:** From Rosenthal E.T., Hunt T. and Ruderman J.V. (1980) *Cell* **20:** 487–494. Courtesy of Joan Ruderman. **17.12:** From Rosenthal E.T., Hunt T. and Ruderman J.V. (1980) *Cell* **20:** 487–494. Courtesy of Joan Ruderman. **17.20:** From Marchini M., Morocutti M., Ruggeri A., Koch M.H.J., Bigi A. and Roveri N. (1986) *Connective Tissue Res.* **15:** 269–281. Photograph courtesy of Maurizio Marchini.

CHAPTER 18 **18.1:** From Tamm S.L. (1972) *J. Cell Biol.* **55:** 250–255. Photograph provided by Sidney L. Tamm. **18.3: (a–b)** S.L. Flegler/Visuals Unlimited. **(c)** Reproduced from Dykes G., Crepeau R.H. and Edelstein S.J. (1978) *Nature* **272:** 509. **18.4: (a)** From Butler P.J.G. and Klug A. (1978) *Scientific American* **239** (November): 62–69. Provided by P.J.G. Butler. **18.11: (a–b)** Reproduced from Asakura S., Eguchi G. and Iino T. (1966) *J. Mol. Biol.* **16:** 302–316. **18.15:** Photograph courtesy of Mary Osborn and Klaus Weber. **18.16: (b)** Reproduced from Goodenough U.W. and Heuser J.E. (1982) *J. Cell Biol.* **95:** 798–815. **18.17: (b)** From Alberts B., Bray D., Lewis J., Raff M., Roberts K. and Watson J.D. (1989) *Molecular Biology of the Cell.* 2nd ed. New York: Garland Publishing, Inc. Photograph courtesy of I.R. Gibbons. **18.18: (b)** Reproduced from Macleod A.G. (1973) *Cytology: The Cell and Its Nucleus.* Kalamazoo, MI: The Upjohn Company. **18.19: (b)** Reproduced from Ehret C.F. and McArdle E.W. (1974) The structure of *Paramecium* as viewed from its constituent levels of organization. In: *Paramecium—A Current Survey,* W.J. van Wagtendonk (ed.), 263–338. Amsterdam: Elsevier Scientific Publishing. **18.22: (a–c)** From Grimes G.W., McKenna M.E., Goldsmith-Spoegler C.M. and Knaupp-Walvogel E.A. (1980) *Science* **209:** 281–283. Provided by Gary W. Grimes. **18.25: (a–d)** From Albrecht-Buehler G. (1977) *J. Cell Biol.* **72:** 595–603. Courtesy of Guenter Albrecht-Buehler.

CHAPTER 19 **19.1:** From Kessel R.G. and Kardon R.H. (1979) *Tissues and Organs: A Text-Atlas of Scanning Electron Microscopy,* 37. San Francisco: W.H. Freeman. Courtesy of Richard G. Kessel. **19.2:** From Bolender R.P. (1974) *J. Cell Biol.* **61:** 272. Courtesy of Robert P. Bolender. **19.3:** From Shih G. and Kessel R. (1982) *Living Images.* Boston, MA: Science Books International. **19.4:** From Bissell M.J. and Hall H.G. (1987) Form and function in the mammary gland: The role of the extracellular matrix. In: *The Mammary Gland: Development, Regulation, and Function,* M.C. Neville and C.W. Daniel (eds.), 128. New York: Plenum. Photograph courtesy of Joanne Emerman. **19.11:** From Littlefield C.L. (1991) *Dev. Biol.* **143:** 378–388. Photograph courtesy of C. Lynne Littlefield. **19.16: (e)** Reproduced from Keller G., Paige C., Gilboa E. and Wagner E.R. (1985) *Nature* **318:** 149–154. **19.17: (a–c)** Reproduced from Metcalf D. (1989) *Nature* **339:** 27–30.

CHAPTER 20 **20.1:** Arthur Glauberman/Photo Researchers. **20.2: (e–h)** From Ede D.A., Hinchliffe J.R. and Balls M., eds. (1977b) *Vertebrate Limb and Somite Morphogenesis. 3rd Symp. Brit. Soc. Dev. Biol.* Cambridge: Cambridge University Press. Photographs courtesy of J.R. Hinchliffe. **20.4:** Photograph courtesy of Stanley K. Sessions. **20.13: (b–e)** From Muneoka K. and Bryant S.V. (1982) *Nature* **298:** 369–371. Photograph courtesy of Susan Bryant. **20.17: (e)** From Bryant S. (1977) *3rd Symp. Brit. Soc. Dev. Biol.* Courtesy of Susan Bryant. **20.24: (a–d)** From Vogel O. (1982) *Wilhelm Roux's Arch. Dev. Biol.* (1982) **191:** 134–136. Photograph courtesy of Otto Vogel.

CHAPTER 21 **21.1:** From Gutjahr T., Frei E. and Noll M. (1993) *Development* **117:** 609–623. Photograph courtesy of T. Gutjahr and M. Noll. **21.5: (a–c)** From Carroll S.B., Laughon L., and Thalley B.S. (1988) *Genes and Dev.* **2:** 883–890. Photograph provided by Stephen Paddock, James Langeland and Sean Carroll. **21.10: (a–f)** From Ephrussi A. and Lehmann R. (1992) *Nature* **358:** 387–392. Photographs courtesy of Ruth Lehmann. **21.12: (a)** From Savant-Bhonsale S. and Montell D.J. (1993) *Genes and Dev.* **7:** 2548–2555. Photograph courtesy of Denise Montell. **21.14: (a–d)** Reproduced from Jäckle H., Tautz D., Schuh R., Seifert E. and Lehmann R. (1986) *Nature* **324:** 668–670. **21.16: (a–e)** Reproduced from Gaul U. and Jäckle H. (1990) *Advances in Genetics* **27:** 239–275. **21.18: (a–b)** From Kaufman T.C., Seeger M.A. and Olsen G. (1990) *Advances in Genetics* **27:** 309–362. Photographs courtesy of Thomas C. Kaufman. **21.19: (a)** From Lawrence P.A. (1992) *The Making of a Fly: The Genetics of Animal Design.* Cambridge, MA: Blackwell. Photograph provided by Peter A. Lawrence. **21.20: (a–b)** From Howard K., Ingham P. and Rushlow C. (1988) *Genes and Dev.* **2:** 1037–1046. Photographs courtesy of P.W. Ingham. **(c)** Reproduced from Howard K., Ingham P. and Rushlow C. (1988) *Genes and Dev.* **2:** 1037–1046. **21.22: (a–d)** Reproduced from Pankratz M.J., Seifert E., Gerwin N., Billi B., Nauber U. and Jäckle H. (1990) *Cell* **61:** 309–317. **21.26:** From Ingham P.W. and Martinez-Arias A. (1992) *Cell* **68:** 221–235. Photograph courtesy of P.W. Ingham. **21.27: (a–c)** From Ingham P.W. and Martinez-Arias A. (1992) *Cell* **68:** 221–235. Photograph courtesy of A. Martinez-Arias. **21.29:** Reproduced from Bateson W. (1894) *Materials for the Study of Variation.* London: Macmillan. **21.31: (a–b)** From Bermingham J.R., Jr., Martinez-Arias A., Petitt M.G. and Scott M.P. (1990) *Development* **109:** 553–566. Photograph courtesy of M.G. Petitt and M.P. Scott. **21.33: (a–b)** From Nüsslein-Volhard C. (1991) *Development* (Supplement 1): 1–10. Photographs courtesy of C. Nüsslein-Volhard. **21.34: (a–c)** From Nüsslein-Volhard C. (1991) *Development* (Supplement 1): 1–10. Photographs courtesy of C. Nüsslein-Volhard. **21.38: (a–c)** From Leptin M. and Grunewald B. (1990) *Development* **110:** 73–84. Photographs courtesy of B. Grunewald and Maria Leptin. **21.39: (a–f)** From Ray R.P., Arora K., Nüsslein-Volhard C. and Gelbart W.M. (1991) *Development* **113:** 35–54. Photographs courtesy of Robert P. Ray. **21.40: (b)** Reproduced from Crick F.H.C. and Lawrence P.A. (1975) *Science* **189:** 340–347. **21.42: (a–b)** From Blair S.S. (1992) *Development* **115:** 21–33. Photograph courtesy of Seth Blair. **21.46: (a–g)** From Patel N.H. (1993) Evolution of insect pattern formation: A molecular analysis of short germband segmentation. In: *Evolutionary Conservation of Developmental Mechanisms,* Spradling A.D. (ed.), 85–110. New York: Wiley/Alan R. Liss. Photographs courtesy of Nipam Patel.

CHAPTER 22 **22.1:** From Friedrich G. and Soriano P. (1991) *Genes and Dev.* **5:** 1518. Photograph courtesy of G. Friedrich and P. Soriano. **22.7:** From Hunt P. and Krumlauf R. (1991) *Cell* **66:** 1075–1078. Photograph courtesy of Robb Krumlauf. **22.8: (a–h)** From Wilkinson D.G., Bhatt S., Cook M., Boncinelli E. and Krumlauf R. (1989) *Nature* **341:** 405–409. Photographs courtesy of Robb

Krumlauf. **22.15: (a–c)** From Le Mouellic H., Lallemand Y. and Brulet P. (1992) *Cell* **69**: 251–264. Photographs courtesy of Hervé Le Mouellic and Yvan Lallemand. **22.17: (a–b)** From Kessel M., Balling R. and Gruss P. (1990) *Cell* **61**: cover. Photographs courtesy of Michael Kessel. **22.18: (a–d)** From Kessel M. and Gruss P. (1991) *Cell* **67**: 89–104. Photographs courtesy of Michael Kessel. **22.21:** From Riddle R.D., Johnson R.L., Laufer E. and Tabin C. (1993) *Cell* **75**: 1401–1416. Photographs courtesy of R.L. Johnson. **22.23:** From Wanek N., Gardiner D.M., Muneoka K. and Bryant S.V. (1991) *Nature* **350**: 81–83. Photograph courtesy of N. Wanek. **22.25:** From Izpisúa-Belmonte J.C., Tickle C., Dolle P., Wolpert L. and Duboule D. (1991) *Nature* **350**: 585–589. Photograph courtesy of J.C. Izpisúa-Belmonte. **22.27: (a–c)** From Morgan B.A., Izpisúa-Belmonte J.-C., Duboule D. and Tabin C.J. (1992) *Nature* **358**: 236–239. Photographs courtesy of Bruce A. Morgan.

CHAPTER 23 **23.1:** From Bowman J.L., Alvarez J., Weigel D., Meyerowitz E.M. and Smyth D.R. (1993) *Development* **119**: 721–743. Photograph courtesy of John L. Bowman. **23.4: (j–k)** Photographs courtesy of Gerd Jürgens. **23.7: (b)** Photograph provided by Thomas Eisner, Cornell University. **23.8:** From Pang P.P. and Meyerowitz E.M. (1987) *Bio/Technology* **5**: 1177–1181. Drawing courtesy of Elliot Meyerowitz. **23.11:** From Mayer U., Torres Ruiz R.A., Berleth T., Miséra S. and Jürgens G. (1991) *Nature* **353**: 402–407. Photograph courtesy of Gerd Jürgens. **23.12: (b)** Reproduced from Drews G.N. and Goldberg R.B. (1989) *Trends in Genetics* **5**: 256–261. **23.14: (a–d)** From Meyerowitz E.M., Bowman J.L., Brockman L.L., Drews G.N., Jack T., Sieburth L.E. and Weigel D. (1991) *Development* (Supplement 1): 157–167. Photographs courtesy of Elliot Meyerowitz. **23.16: (a–b)** From Bowman J.L., Sakai H., Jack T., Weigel D., Mayer U. and Meyerowitz E.M. (1992) *Development* **114**: 599–615. Photographs courtesy of J.L. Bowman. **23.20:** From Schwarz-Sommer Z., Huijser P., Nacken W., Saedler H. and Sommer H. (1990) *Science* **250**: 931–936. Photograph courtesy of Zsuzsanna Schwarz-Sommer.

CHAPTER 24 **24.1:** From Sulston J.E., Schierenberg E., White J.G. and Thomson J.N. (1983) *Dev. Biol.* **100**: 64–119. Photograph courtesy of J.E. Sulston. **24.2: (a)** Photograph courtesy of Einhard Schierenberg. **24.3: (a–e)** From Schierenberg E. (1986) *J. Embryol. Exp. Morphol.* **97** (Supplement): 31–44. Photograph courtesy of Einhard Schierenberg. **24.6: (a–i)** From Strome S. (1989) *Int. Rev. Cytol.* **114**: 81–123. Academic Press: Florida. Photographs courtesy of Susan Strome. **24.8: (a–i)** From Bowerman B., Draper B.W., Mello C.C. and Priess J.R. (1993) *Cell* **74**: 443–452. Photographs courtesy of Bruce Bowerman. **24.13:** From Ruvkun G. and Giusto J. (1989) *Nature* **338**: 313–319. Photograph courtesy of Gary Ruvkun. **24.14: (a–b)** From Ellis H.M. and Horvitz H.R. (1986) *Cell* **44**: 817–829. Photographs courtesy of Hilary Ellis. **24.16: (a–b)** From Ellis R.E. and Horvitz H.R. (1991) *Development* **112**: 591–603. Photographs courtesy of Ronald Ellis. **24.21: (a–d)** From Hill R.J. and Sternberg P.W. (1992) *Nature* **358**: 470–476. Photographs courtesy of Russell J. Hill.

CHAPTER 25 **25.1:** From Morrill J.B. (1986) Scanning electron microscopy of embryos. In: *Methods in Cell Biology*, T. Schroeder (ed.), 263–292. New York: Academic Press. Photograph courtesy of John B. Morrill. **25.3: (a–c)** From Steinberg M.S. (1963) *Science* **141**: 401–408. Photographs courtesy of Malcolm S. Steinberg. **25.5: (a)** Photograph courtesy of J.B. Morrill. **25.6:** From Nardi J.B. and Stocum D.L. (1983) *Differentiation* **25**: 27–31. Photographs courtesy of David Stocum. **25.14: (a–d)** From Nagafuchi A., Shirayoshi Y., Okazaki K., Yasuda K. and Takeichi M. (1987) *Nature* **329**: 341–343. Photographs courtesy of Masatoshi Takeichi. **25.15:** From Kintner C.R. (1992) *Cell* **69**: 225–236. Photograph courtesy of Chris Kintner. **25.18: (a–b)** From Kintner C.R. and Melton D.A. (1987) *Development* **99**: 311–325. Photograph courtesy of Chris Kintner. **25.20:** From Takeichi M. (1988) *Development* **102**: 639–655. Photograph courtesy of Masatoshi Takeichi. **25.21:** From Takeichi M. (1988) *Development* **102**: 639–655. Photograph courtesy of Masatoshi Takeichi.

CHAPTER 26 **26.1:** Photograph courtesy of Kimiko Hayashi and Robert Trelstad. **26.3: (a)** Reproduced from Alberts B., Bray D., Lewis J., Raff M., Roberts K. and Watson J.D. (1989) *Molecular Biology of the Cell*. 2nd ed. New York: Garland Publishing, Inc. **26.10: (a–b)** From Hynes R.O. and Destree A.T. (1978) *Cell* **15**: 875–886. Photograph courtesy of Richard O. Hynes. **26.11:** From Stopak D. and Harris A.K. (1982) *Dev. Biol.* **90**: 383–398. Photograph courtesy of David Stopak and Albert Harris. **26.12:** Reproduced

from Gundersen R.W. (1987) *Dev. Biol.* **121**: 423–431. **26.13:** Photograph provided by Kurt E. Johnson. **26.14: (b)** From Shi D.-L., Darribère T., Johnson K.E. and Boucaut J.-C. (1989) *Development* **105**: 351–363. Photograph provided by De-Li Shi. **26.15: (b)** From Boucaut J.-C., Darribère T., Boulekbache H. and Thiery J.P. (1984) *Nature* **307**: 364–367. Photograph provided by J.-C. Boucaut. **26.16: (b–c)** From Boucaut J.-C., Darribère T., Boulekbache H. and Thiery J.P. (1984) *Nature* (London) **307**: 364–367. Photograph provided by J.-C. Boucaut. **26.19: (a–b)** From Löfberg J., Perris R. and Epperlein H. (1989) *Dev. Biol.* **131**: 168–181. Photographs provided by Jan Löfberg. **26.20:** Reproduced from Bronner-Fraser M. (1986) *Dev. Biol.* **117**: 528–536. **26.21: (a–b)** From Brower D.L., Wilcox M., Piovant M., Smith R.J. and Reger L.A. (1984) *Proc. Nat. Acad. Sci. USA* **81**: 7485–7489. Photographs courtesy of Danny Brower. **26.22: (a–b)** From Brower D.L. and Jaffe S.M. (1989) *Nature* **324**: 285–287. Photographs courtesy of Danny Brower.

CHAPTER 27 **27.1:** Toni Angermayer/Photo Researchers. **27.10:** From Burgoyne P.S., Buehr M., Koopman P., Rossant J. and McLaren A. (1988) *Development* **102**: 443–450. Photograph courtesy of Paul S. Burgoyne. **27.13: (a–d)** Reproduced from Palmer M.S., Sinclair A.H., Berta P., Ellis N.A., Goodfellow P.N., Abbas N.E. and Fellous M. (1989) *Nature* **342**: 937–942. Reprinted. Copyright 1989 Macmillan Magazines Limited. **27.14:** From Koopman P., Gubbay J., Vivian N., Goodfellow P. and Lovell-Badge R. (1991) *Nature* **351**: 117–121. Photograph courtesy of R. Lovell-Badge. **27.15:** From Koopman P., Gubbay J., Vivian N., Goodfellow P. and Lovell-Badge R. (1991) *Nature* **351**: 117–121. Photograph courtesy of R. Lovell-Badge.

CHAPTER 28 **28.1:** From Young L.J., Lopreato G., Horan K. and Crews D. (1994) *J. Comp. Neurol.* **347**: 288–300. Photographs courtesy of Larry Young and David Crews. **28.6:** Photograph courtesy of Leonard Pinsky, McGill University. **28.10:** From Breedlove S.M. (1992) Sexual differentiation of the brain and behavior. In: *Behavioral Endocrinology*, J.B. Becker, S.M. Breedlove and D. Crews (eds.), 39–70. Cambridge, MA: MIT Press. Photograph courtesy of Roger Gorski. **28.15:** Photographs courtesy of Michael Ashburner. **28.18: (a–b)** From Robinow S., Talbot W.S., Hogness D.S. and Truman J.W. (1993) *Development* **119**: 1251–1259. Photographs courtesy of Steven Robinow. **28.19:** From Karim F.D. and Thummel C.S. (1991) *Genes and Dev.* **5**: 1067–1079. Photograph courtesy of Felix D. Karim and Carl S. Thummel. **28.22:** From Tata J.R., Kawahara A. and Baker B.S. (1991) *Dev. Biol.* **146**: 72–80. Photograph courtesy of Jamshed R. Tata. **28.24:** From Tata J.R., Kawahara A. and Baker B.S. (1991) *Dev. Biol.* **146**: 72–80. Photograph courtesy of Jamshed R. Tata. **28.26:** Photograph courtesy of S. Duhon and G. Malacinski.

CHAPTER 29 **29.1:** Reproduced from Levi-Montalcini R. (1964) *Science* **143**: 105–110. **29.6: (a–f)** Reproduced from Harrison R.G. (1935) *Harvey Lectures, 1933–1934*, 116–157. **29.7: (a–b)** Reproduced from Twitty V.C. and Elliot H.A. (1934) *J. Exp. Zool.* **68**: 247. **29.8:** Reproduced from Harrison R.G. (1935). **29.9:** Globe Photos. **29.11:** Reproduced from Levi-Montalcini R. (1964) *Science* **143**: 105–110. **29.13: (a–b)** Reproduced from Levi-Montalcini R. (1964) *Science* **143**: 105–110. **29.14: (a–b)** Reproduced from Weinberg R.A. (1983) *Scientific American* **249** (November): 127. **29.21:** Reproduced from Weinberg R.A. (1988) *Scientific American* **259** (September): 46.

LINE ART CREDITS

The figures listed below are based on numerical data and/or diagrams of the authors quoted. For each entry, the bibliographical data may be found in the reference list. Credits for interpretative drawings of photos are listed among the photo credits.

COVER De Robertis et al. (1990).

CHAPTER 1 **1.5:** Moore (1982). **1.6:** Gilbert (1991). **1.11:** Carlson (1988). **1.12:** Huxley (1932).

CHAPTER 2 **2.3:** McIntosh and McDonald (1989). **2.5:** Draetta et al. (1989). **2.6:** Burnside (1971). **2.11:** Alberts et al. (1989). **2.12:** Luna and Hitt (1992). **2.13:** Alberts et al. (1989). **2.14:** Brown and Goldstein (1984). **2.16:** Alberts et al. (1989). **2.18:** Alberts et al. (1989). **2.20:** Hay (1991b). **2.21:** Keeton and Gould (1986). **2.22:** Darnell et al. (1990). **2.23:** Darnell et al. (1990). **2.24:** Darnell et al.

(1990). **2.25:** Darnell et al. (1990). **2.26:** Darnell et al. (1990). **2.27:** Darnell et al. (1990).

CHAPTER 3 **3.2:** Wilson (1925). **3.3:** Langman (1981). **3.6:** Tuchmann-Duplessis et al. (1972). **3.7:** (c) Carlson (1988). **3.8:** Bloom and Fawcett (1975). **3.9:** Bloom and Fawcett (1975). **3.14:** King (1970), Mahowald (1972). **3.15:** Alberts et al. (1983). **3.21:** Austin (1965). **3.23:** Minshull (1993). **3.26:** Carlson (1988). **3.27:** (a–b) Hinton (1970).

CHAPTER 4 **4.2:** Epel (1977). **4.3:** Miller (1966). **4.5:** Yanagimachi and Noda (1970). **4.8:** Glabe and Lennarz (1979). **4.10:** Bleil and Wassarman (1980), Florman and Wassarman (1985). **4.12:** Miller et al. (1992). **4.16:** Langman (1981). **4.17:** Turner and Jaffe (1989). **4.22:** (a) Gilbert (1988).

CHAPTER 5 **5.3:** Czihak et al. (1976). **5.4:** Hörstadius (1973), Summers et al. (1993). **5.5:** Alberts et al. (1989). **5.6:** Balinsky (1975). **5.7:** Grant (1978). **5.8:** Gasser (1975). **5.9:** Gulyas (1975). **5.11:** Carlson (1988). **5.15:** Saunders (1970). **5.16:** Foe and Alberts (1983). **5.18:** Fullilove and Jacobson (1971). **5.21:** Rappaport (1974). **5.24:** Morgan (1927). **5.26:** Freeman (1983). **5.28:** Foe and Alberts (1983), Edgar and Schubiger (1986). **5.29:** Edgar and O'Farrell (1989). **5.32:** Kühn (1971).

CHAPTER 6 **6.2:** Dale and Slack (1987a). **6.6:** Garcia-Bellido et al. (1979). **6.7:** Willier and Oppenheimer (1964). **6.9:** Spemann (1938), Gilbert (1988). **6.10:** Lawrence (1992). **6.12:** Balinsky (1975), Saunders (1970). **6.13:** Saxén and Toivonen (1962). **6.17:** Slack (1991). **6.18:** Slack (1991). **6.20:** Nöthiger (1972). **6.21:** Gehring (1968). **6.22:** Hadorn (1968). **6.23:** Gehring (1968). **6.24:** Gehring (1968). **6.25:** Kauffman (1973). **6.26:** Langman (1981).

CHAPTER 7 **7.3:** Boveri (1887). **7.4:** (c) Kühn (1961). **7.5:** Beermann (1952b). **7.7:** Shepard (1982). **7.9:** King (1966). **7.10:** McKinnell (1978). **7.11:** Gurdon (1968). **7.12:** Gurdon (1968). **7.14:** (a) Laskey and Gurdon (1970). **7.17:** King (1970). **7.18:** Gurdon (1968). **7.20:** De Robertis and Gurdon (1979).

CHAPTER 8 **8.2:** Bodemer (1968). **8.3:** Saunders (1982). **8.6:** Verdonk and Cather (1983). **8.7:** Martindale et al. (1985). **8.10:** Browder et al. (1991). **8.12:** Jeffery and Meier (1983). **8.13:** Freeman (1979). **8.14:** Freeman (1979). **8.17:** Illmensee and Mahowald (1974), Mahowald et al. (1979). **8.18:** Whittaker (1980). **8.23:** Render (1983).

CHAPTER 9 **9.3:** Walbot and Holder (1987). **9.4:** Jaffe (1966), Quatrano (1990). **9.5:** Jaffe (1966). **9.7:** Kimelman et al. (1992). **9.9:** Vincent et al. (1986). **9.11:** Kirschner et al. (1980). **9.12:** McCain and McClay (1994). **9.13:** (a) Yost (1992). **9.16:** Gimlich and Gerhart (1984). **9.20:** Dale and Slack (1987b), Gilbert (1991). **9.22:** Dale and Slack (1987b). **9.23:** Dale and Slack (1987b). **9.24:** Kao et al. (1986). **9.25:** Berridge et al. (1989).

CHAPTER 10 **10.4:** Hörstadius (1973), Ettensohn (1992). **10.7:** Moore and Burt (1939). **10.8:** Lane et al. (1993). **10.10:** Fristrom (1976). **10.11:** Hardin (1989). **10.12:** Gerhart and Keller (1986). **10.13:** Holtfreter (1943). **10.14:** Balinsky (1975). **10.15:** Keller (1981). **10.16:** Vogt (1929). **10.17:** Balinsky (1975). **10.18:** R.E. Keller (1981). **10.19:** Holtfreter (1933). **10.20:** R.E. Keller et al. (1985). **10.21:** (a) R.E. Keller et al. (1985). **10.22:** R.E. Keller et al. (1985, 1992a). **10.23:** R.E. Keller et al. (1992a). **10.24:** R.E. Keller (1980). **10.25:** Carlson (1988). **10.26:** Patten (1971), Balinsky (1975). **10.27:** Patten (1971), Balinsky (1975). **10.29:** Mittenthal and Jacobson (1990). **10.30:** Patten (1971). **10.32:** Luckett (1978). **10.33:** Langman (1981). **10.34:** Langman (1981).

CHAPTER 11 **11.2:** Langman (1981). **11.3:** Hamburger (1960). **11.4:** Lopez-Antunez (1971), Moore (1982). **11.5:** Burnside (1971), Carlson (1988). **11.6:** Balinsky (1975). **11.7:** Gordon and Jacobson (1978). **11.8:** Gordon and Jacobson (1978). **11.9:** Gordon and Jacobson (1978). **11.10:** Jacobson and Gordon (1976), Gordon and Jacobson (1978). **11.12:** Jacobson (1981). **11.13:** Jacobson et al. (1986). **11.15:** Spemann (1938), Hamburger (1988). **11.16:** Carlson (1988). **11.18:** Bodemer (1968). **11.21:** Holtfreter (1943). **11.22:** Keller and Danilchik (1988). **11.24:** Sharpe et al. (1987).

CHAPTER 12 **12.2:** Carlson (1988). **12.3:** Langman (1981). **12.4:** Carlson (1988). **12.5:** Shepherd (1988). **12.7:** Langman (1981). **12.8:** Lumsden (1991). **12.10:** Langman (1981). **12.11:** Carlson (1988). **12.12:** Langman (1981). **12.13:** Langman (1981). **12.14:** Langman (1981). **12.15:** Langman (1981). **12.16:** Langman (1981). **12.17:** Langman (1981). **12.18:** Langman (1981). **12.19:** Hopper and Hart (1985). **12.20:** Patten (1964). **12.22:** Moury and Jacobson (1989). **12.23:** Weston (1963). **12.25:** Carlson (1988). **12.26:** Carlson (1988). **12.27:** Le Douarin (1986). **12.28:** Perris and Bronner-Fraser (1989). **12.30:** Carlson (1988). **12.31:** Langman (1981). **12.32:** Balinsky (1975). **12.33:** Langman (1981). **12.34:** Carlson (1988). **12.35:** Carlson (1988). **12.36:** Langman (1981). **12.37:** Langman (1981). **12.39:** Langman (1981). **12.40:** Langman (1981). **12.41:** Bloom and Fawcett (1975). **12.42:** Langman (1981). **12.43:** Langman (1981), Carlson (1988).

CHAPTER 13 **13.2:** Langman (1981). **13.3:** Langman (1981). **13.4:** Langman (1981). **13.5:** Torrey and Feduccia (1991). **13.6:** Kühn (1961). **13.7:** (a) Starck (1965). **13.7:** (b) Langman (1981). **13.8:** Langman (1981). **13.9:** Romanes (1901). **13.10:** Carlson (1988). **13.11:** Langman (1981). **13.12:** Patten (1971). **13.13:** Mittenthal and Jacobson (1990). **13.14:** (a) Meier (1979). **13.15:** Langman (1981). **13.16:** Jacobson (1988). **13.17:** Alberts et al. (1989). **13.19:** Langman (1981). **13.23:** Carlson (1988). **13.25:** Langman (1981). **13.26:** Langman (1981). **13.27:** Langman (1981). **13.28:** Carlson (1988). **13.29:** Langman (1981). **13.30:** (a–d) Langman (1981). **13.31:** Romer (1976). **13.32:** (a) Romer (1976). **13.32:** (b) Langman (1981). **13.33:** Langman (1981). **13.34:** (a) Bloom and Fawcett (1975). **13.36:** Hinchliffe and Johnson (1980). **13.37:** Summerbell and Lewis (1975). **13.38:** Carlson (1988). **13.39:** Moore (1982). **13.40:** Carlson (1988). **13.41:** Langman (1981).

CHAPTER 14 **14.4:** (c–d) Wilson (1925). **14.8:** Scott and O'Farrell (1986). **14.9:** Kemphues (1989). **14.11:** McLaren (1976). **14.13:** Villee et al. (1989). **14.15:** Browder et al. (1991). **14.27:** Palmiter et al. (1982).

CHAPTER 15 **15.5:** Lewin (1990). **15.6:** (a) Hiromi and Gehring (1987). **15.11:** Driever and Nüsslein-Volhard (1989). **15.12:** Driever et al. (1989a). **15.13:** McKnight et al. (1975). **15.14:** Darnell et al. (1990). **15.23:** Krasnow et al. (1989). **15.24:** Alberts et al. (1989).

CHAPTER 16 **16.5:** (a–b) Alberts et al. (1989). **16.10:** Baker (1989). **16.13:** Darnell et al. (1990). **16.18:** Wells et al. (1981). **16.20:** (a) Alberts et al. (1989). **16.21:** Mangiarotti et al. (1983). **16.22:** Guyette et al. (1979). **16.23:** Alberts et al. (1989).

CHAPTER 17 **17.2:** Darnell et al. (1990). **17.3:** Alberts et al. (1989). **17.7:** Richter and Smith (1984). **17.9:** Harvey (1940). **17.10:** Gross and Cousineau (1964). **17.13:** Iatrou et al. (1978). **17.15:** Darnell et al. (1990). **17.16:** Alberts et al. (1989). **17.17:** Branden and Tooze (1991). **17.18:** Darnell et al. (1990). **17.19:** Darnell et al. (1990).

CHAPTER 18 **18.2:** Grant (1978). **18.4:** Butler and Klug (1978). **18.7:** Butler and Klug (1978). **18.8:** Butler and Klug (1978). **18.9:** Grant (1978). **18.10:** Wood (1980). **18.12:** Weissmann (1994). **18.13:** Prusiner et al. (1993). **18.14:** Alberts et al. (1989). **18.16:** (a) Darnell et al. (1990). **18.18:** (a) Alberts et al. (1989). **18.21:** (b) Sonneborn (1970). **18.23:** Frankel (1989). **18.24:** Alberts et al. (1989).

CHAPTER 19 **19.5:** Bloom and Fawcett (1975), Alberts et al. (1989). **19.6:** Alberts et al. (1989). **19.8:** Alberts et al. (1989). **19.9:** Gierer (1974). **19.12:** David and Murphy (1977). **19.13:** David et al. (1987). **19.14:** Golde (1991). **19.15:** Alberts et al. (1989). **19.16:** G. Keller et al. (1985). **19.18:** Alberts et al. (1989). **19.19:** Alberts et al. (1989). **19.20:** Olson (1990). **19.22:** Buckingham (1992).

CHAPTER 20 **20.3:** Stocum and Fallon (1982). **20.5:** Graham and Wareing (1976). **20.6:** Jacobson (1966). **20.7:** Postlethwait and Schneiderman (1971). **20.9:** Wolpert (1978). **20.11:** Walbot and Holder (1987). **20.12:** Saunders (1982). **20.14:** Bryant et al. (1981). **20.16:** Bryant et al. (1981). **20.17:** Bryant (1977). **20.20:** Hörstadius (1973). **20.21:** Kühn (1971). **20.22:** Davidson (1989). **20.23:** Sander (1975).

CHAPTER 21 **21.2:** (a–e) Nüsslein-Volhard (1991). **21.2:** (f) Akam (1987). **21.2:** (g) Martinez-Arias and Lawrence (1985). **21.3:** (a–b) Nüsslein-Volhard (1991). **21.3:** (c) Akam (1987). **21.6:** Nüsslein-Volhard et al. (1987). **21.7:** Nüsslein-Volhard (1991). **21.8:** Driever et al. (1989a), Pignoni et al. (1992), Walldorf and Gehring (1992). **21.9:** Driever et al. (1990). **21.11:** Ephrussi and Lehmann (1992).

21.13: Scott and O'Farrell (1986). **21.15:** Hülskamp and Tautz (1991). **21.17:** Tautz (1992). **21.21:** Pankratz et al. (1990), Lawrence (1992). **21.23:** Pankratz et al. (1990), Riddihough and Ish-Horowicz (1991). **21.24:** Ingham (1990). **21.28:** Lawrence (1992), Peifer and Bejsovec (1992). **21.30:** Kaufman et al. (1990), Gilbert (1991), Slack (1991). **21.32:** Lewis (1963), Peifer et al. (1987). **21.40: (a)** Morata and Lawrence (1978). **21.41:** Morata and Lawrence (1975), Lawrence (1992). **21.43:** Gehring et al. (1990). **21.44:** Gehring (1985), Graham et al. (1989). **21.45:** Wolberger et al. (1991).

CHAPTER 22 **22.4:** Graham et al. (1989). **22.5:** Scott (1992). **22.9:** Hunt et al. (1991a). **22.10:** Fraser et al. (1990). **22.11:** De Robertis et al. (1990). **22.12:** De Robertis et al. (1990). **22.13:** Le Mouellic et al. (1990). **22.14:** Le Mouellic et al. (1992). **22.16:** Kessel et al. (1990). **22.18:** Kessel and Gruss (1991). **22.19:** Saunders (1982). **22.20:** Saunders (1982). **22.21:** Honig and Summerbell (1985). **22.22:** Wolpert (1978). **22.23:** Noji et al. (1991). **22.24:** Tabin (1991). **22.26:** Morgan et al. (1992).

CHAPTER 23 **23.3:** Drews and Goldberg (1989). **23.4: (a–i)** Meinke (1991b). **23.5:** Meinke (1991b). **23.6:** Jürgens et al. (1991). **23.8:** Pang and Meyerowitz (1987). **23.9:** Jürgens et al. (1991). **23.10:** Mayer et al. (1991). **23.12:** Drews and Goldberg (1989). **23.15:** Meyerowitz et al. (1991). **23.17:** Bowman et al. (1992). **23.18:** Coen and Meyerowitz (1991). **23.19:** Jack et al. (1992).

CHAPTER 24 **24.2: (b–c)** Hodgkin (1985). **24.4:** Sulston et al. (1983). **24.5:** Wilkins (1986). **24.7:** Priess et al. (1987). **24.9:** Mello et al. (1992). **24.10:** Goldstein (1992). **24.12:** Ruvkun et al. (1991). **24.15:** Ellis and Horvitz (1986). **24.17:** Ellis and Horvitz (1991). **24.18:** Horvitz and Sternberg (1991). **24.19:** Horvitz and Sternberg (1991). **24.20:** Hill and Sternberg (1992).

CHAPTER 25 **25.2:** Townes and Holtfreter (1955). **25.4:** Phillips and Steinberg (1969). **25.5: (b)** Fink and McClay (1985). **25.7:** Crawford and Stocum (1988). **25.8:** Cunningham (1991). **25.11:** Pollerberg and Beck-Sickinger (1993). **25.12:** Rutishauser et al. (1988). **25.13:** Takeichi (1991). **25.16:** Shur (1991). **25.17:** Edelman (1984). **25.19:** Takeichi (1988).

CHAPTER 26 **26.2:** Alberts et al. (1989). **26.3:** Alberts et al. (1989). **26.4:** Alberts et al. (1989). **26.5:** Yamada (1991). **26.6:** Yamada (1991). **26.7:** Hynes (1992). **26.8:** Bernfield and Sanderson (1990). **26.9:** Toole and Gross (1971). **26.14: (a)** Shi et al. (1989). **26.15: (a)** Boucaut et al. (1984). **26.16: (a)** Boucaut et al. (1984). **26.17:** Bronner-Fraser (1993). **26.18:** Perris et al. (1988). **26.24:** Carlson (1988), Hay (1991b). **26.25:** Hay (1991b). **26.26:** Hay (1991b).

CHAPTER 27 **27.2:** Bull (1983). **27.4:** Hodgkin (1992). **27.5:** Gartler and Riggs (1983), McCarrey and Dilworth (1992). **27.6:** Davidson et al. (1963). **27.7:** McLaren (1991). **27.8:** Langman (1981). **27.9:** Langman (1981). **27.11:** Affara (1991). **27.12:** Palmer et al. (1989). **27.16:** Hodgkin (1990). **27.17:** Hodgkin (1992). **27.18:** Hodgkin (1992). **27.19:** Henikoff and Meneely (1993). **27.20:** Hodgkin (1992). **27.21:** Hodgkin (1992). **27.22:** Walbot and Holder (1987). **27.23:** Oliver et al. (1993). **27.24:** Hodgkin (1992). **27.25:** Hodgkin (1992).

CHAPTER 28 **28.2:** Breedlove (1992). **28.4:** Langman (1981). **28.5:** Langman (1981). **28.7:** Langman (1981). **28.8:** Kelley and Brenowitz (1992). **28.9:** Arnold (1980). **28.11:** Andres and Thummel (1992). **28.12:** Schneiderman and Gilbert (1964). **28.13:** Wigglesworth (1972). **28.14:** Andres and Thummel (1992). **28.16:** Ashburner et al. (1974), Andres and Thummel (1992). **28.17:** Talbot et al. (1993). **28.20:** Witschi (1956). **28.21:** Geigy (1941), Schwind (1933). **28.23:** Tata et al. (1991). **28.27:** Yaoita and Brown (1990).

CHAPTER 29 **29.3:** Weiss (1939). **29.4:** Morgan (1927). **29.5:** Harrison (1935). **29.10:** Karin et al. (1990). **29.12:** Levi-Montalcini and Calissano (1979). **29.16:** Weinberg (1988). **29.17:** Bishop (1982). **29.18:** Bishop (1982). **29.19:** Bishop (1982). **29.20:** Croce et al. (1983), Dalla-Favera et al. (1982), Erikson et al. (1983). **29.22:** Bryant and Simpson (1984).